Lecture Notes in Computer Science 16448

Founding Editors

Gerhard Goos
Juris Hartmanis

Editorial Board Members

Elisa Bertino, USA
Wen Gao, China
Bernhard Steffen, Germany
Moti Yung, USA

Advanced Research in Computing and Software Science
Subline of Lecture Notes in Computer Science

Subline Series Editors

Giorgio Ausiello, *University of Rome 'La Sapienza', Italy*
Vladimiro Sassone, *University of Southampton, UK*

Subline Advisory Board

Susanne Albers, *TU Munich, Germany*
Benjamin C. Pierce, *University of Pennsylvania, USA*
Bernhard Steffen, *University of Dortmund, Germany*
Deng Xiaotie, *Peking University, Beijing, China*
Jeannette M. Wing, *Microsoft Research, Redmond, WA, USA*

The series Lecture Notes in Computer Science (LNCS), including its subseries Lecture Notes in Artificial Intelligence (LNAI) and Lecture Notes in Bioinformatics (LNBI), has established itself as a medium for the publication of new developments in computer science and information technology research, teaching, and education.

LNCS enjoys close cooperation with the computer science R & D community, the series counts many renowned academics among its volume editors and paper authors, and collaborates with prestigious societies. Its mission is to serve this international community by providing an invaluable service, mainly focused on the publication of conference and workshop proceedings and postproceedings. LNCS commenced publication in 1973.

Jakub Kozik · Alexander Wolff

Editors

SOFSEM 2026: Theory and Practice of Computer Science

51st International Conference on Current Trends
in Theory and Practice of Computer Science, SOFSEM 2026
Kraków, Poland, February 9–13, 2026
Proceedings

Editors
Jakub Kozik
Jagiellonian University in Kraków
Krakow, Poland

Alexander Wolff
Universität Würzburg
Würzburg, Germany

ISSN 0302-9743   ISSN 1611-3349 (electronic)
Lecture Notes in Computer Science
ISBN 978-3-032-17800-8   ISBN 978-3-032-17801-5 (eBook)
https://doi.org/10.1007/978-3-032-17801-5

© The Editor(s) (if applicable) and The Author(s), under exclusive license
to Springer Nature Switzerland AG 2026

This work is subject to copyright. All rights are solely and exclusively licensed by the Publisher, whether the whole or part of the material is concerned, specifically the rights of translation, reprinting, reuse of illustrations, recitation, broadcasting, reproduction on microfilms or in any other physical way, and transmission or information storage and retrieval, electronic adaptation, computer software, or by similar or dissimilar methodology now known or hereafter developed.
The use of general descriptive names, registered names, trademarks, service marks, etc. in this publication does not imply, even in the absence of a specific statement, that such names are exempt from the relevant protective laws and regulations and therefore free for general use.
The publisher, the authors and the editors are safe to assume that the advice and information in this book are believed to be true and accurate at the date of publication. Neither the publisher nor the authors or the editors give a warranty, expressed or implied, with respect to the material contained herein or for any errors or omissions that may have been made. The publisher remains neutral with regard to jurisdictional claims in published maps and institutional affiliations.

This Springer imprint is published by the registered company Springer Nature Switzerland AG
The registered company address is: Gewerbestrasse 11, 6330 Cham, Switzerland

If disposing of this product, please recycle the paper.

Preface

These proceedings contain the papers that were presented at the 51st International Conference on Current Trends in Theory and Practice of Computer Science (SOFSEM 2026), held on February 9–13, 2026 in Kraków. The conference started with a one-day Ph.D. school on *Algorithms with Predictions* organized by Christian Coester and Adam Polak.

SOFSEM is an annual winter conference devoted to the theory and practice of computer science. The conference traditionally focuses on the latest results and developments of fundamental research in computer science (informatics), inspired by the algorithmic challenges of our time. SOFSEM has a long tradition as a high-quality research conference, and a venue where researchers from academia and industry in all stages of their career can share their insights. The series of SOFSEM conferences began in 1974, and was only interrupted in 2022 due to the COVID-19 pandemic. The Call for Papers of SOFSEM 2026 invited submissions from a broad range of topics, from algorithms and data structures via combinatorics of words to the theory of artificial intelligence and data science.

This year, there were 133 full submissions. The large number continues the positive trend of the last few years. We desk-rejected nine submissions that were clearly out of scope or of insufficient scientific level. In a single-blind review process, every other submission was reviewed by at least three people – program committee members or external reviewers (except for one submission that was reviewed only twice). The committee decided to accept 48 papers. The contributed talks were grouped into ten sessions and two parallel tracks. The program also included a poster session (with seven posters) and the following six invited talks:

- Jarosław Błasiok (ETH Zürich): *Semirandom Planted Clique*
- Jarosław Byrka (University of Wrocław): *On the Bidirected Cut Relaxation for Steiner Tree and Steiner Forest*
- Sandra Kiefer (University of Oxford): *Long-Refinement Graphs*
- Linda Kleist (University of Hamburg): *Snapshots of Reconfiguration*
- Jukka Suomela (Aalto University): *Distributed Quantum Advantage*
- Philip Wadler (University of Edinburgh): *Propositions as Types*

The program committee decided to give the best paper award to Therese Biedl for her contribution *Face-Hitting Dominating Sets in Plane Graphs: Alternative Proof and Linear-Time Algorithm* and the best student paper award to Jan Jedelský for his paper *k-Planar and Fan-Crossing Drawings and Transductions of Planar Graphs* (with Petr Hliněný). Congratulations!

We thank all members of the program committee and the external reviewers for their joint effort in selecting the best submissions. Our special thanks go to the local organizing committee – Grzegorz Gutowski (local chair), Bartosz Podkanowicz, and Grzegorz Ryn – for their excellent organization and for enriching the conference schedule with many unique activities, such as the morning run around and the morning swim in Zakrzówek, and excursions to the University Museum and the Kraków Arcade Game

Museum. A highlight of the social program was the workshop "Retroprogramming with fig-FORTH on an 8-bit Atari computer", organized by the Polish Society for Preserving the Technical Heritage (PTODT).

The conference took place on the new campus of the Jagiellonian University in Kraków. We gratefully acknowledge financial support from the Faculty of Mathematics and Computer Science and the Theoretical Computer Science Department of the Jagiellonian University. We also thank Springer for sponsoring the best paper awards, partially covering the costs of invited speakers, and publishing these proceedings in the Advanced Research in Computing and Software Science (ARCoSS) subseries of Lecture Notes in Computer Science (LNCS).

Finally, we thank all authors who submitted their work to SOFSEM 2026 and hope that future readers of these proceedings will appreciate the quality of the selected papers as much as we do.

December 2025   Jakub Kozik
                                                    Alexander Wolff

Organization

Program Committee

Marthe Bonamy	University of Bordeaux, France
Pierre Fraigniaud	CNRS and Université Paris Cité, France
Jakub Gajarský	University of Warsaw, Poland
Serge Gaspers	University of New South Wales, Australia
Bernhard Gittenberger	TU Wien, Austria
Pascal Gollin	University of Primorska, Slovenia
Gwenaël Joret	Université libre de Bruxelles, Belgium
Marcin Jurdzinski	University of Warwick, UK
Jarkko Kari	University of Turku, Finland
Eun Jung Kim	KAIST and Institute for Basic Science, South Korea
Yasuaki Kobayashi	Hokkaido University, Japan
Jakub Kozik (Co-chair)	Jagiellonian University in Kraków, Poland
Rastislav Kralovic	Comenius University Bratislava, Slovakia
Matthias Krause	University of Mannheim, Germany
Piotr Krysta	Augusta University, USA and University of Liverpool, UK
Francis Lazarus	Université Grenoble Alpes, France
Borut Lužar	Faculty of Information Studies in Novo mesto, Slovenia
Tomáš Masařík	University of Warsaw, Poland
Neeldhara Misra	Indian Institute of Technology, Gandhinagar, India
Krzysztof Pietrzak	IST Austria, Austria
Adam Polak	Bocconi University, Italy
Paweł Prałat	Toronto Metropolitan University, Canada
Eric Rivals	University of Montpellier, France
Liam Roditty	Bar-Ilan University, Israel
Paweł Rzążewski	Warsaw University of Technology, Poland
Christian Scheideler	Paderborn University, Germany
Joachim Spoerhase	University of Liverpool, UK
Frank Stephan	National University of Singapore, Singapore
Sabine Storandt	Universität Konstanz, Germany,
Jacek Tabor	Jagiellonian University in Kraków, Poland
Alessandra Tappini	University of Perugia, Italy
Torsten Ueckerdt	Karlsruhe Institute of Technology, Germany

Erik Jan van Leeuwen	Utrecht University, the Netherlands
Jan Volec	Czech Technical University in Prague, Czech Republic
Alexander Wolff (Co-chair)	Universität Würzburg, Germany
Ryo Yoshinaka	Tohoku University, Japan
Meirav Zehavi	Ben-Gurion University of the Negev, Israel
Johannes Zink	Technische Universität München, Germany

Steering Committee

Henning Fernau (Chair)	Trier University, Germany
Serge Gaspers	University of New South Wales, Australia
Ralf Klasing	CNRS and University of Bordeaux, France
Jakub Kozik	Jagiellonian University, Kraków, Poland
Rastislav Královič	Comenius University Bratislava, Slovakia
Věra Kůrková	Czech Academy of Sciences, Czech Republic
Mirosław Kutyłowski	NASK – National Research Institute, Poland
Jan van Leeuwen	Utrecht University, the Netherlands
Tiziana Margaria	University of Limerick, Ireland
Branislav Rovan	Comenius University Bratislava, Slovakia
Július Štuller	Czech Academy of Sciences, Slovakia
Alexander Wolff	Universität Würzburg, Germany

Additional Reviewers

Abu-Khzam, Faisal	Bonerath, Annika
Araujo-Pardo, Gabriela	Bordewich, Magnus
Arora, Pragya	Brand, Timo
Artmann, Matthias	Buchem, Moritz
Arya, Paras	Chekan, Vera
Avin, Chen	Cherif, Sami
Barish, Robert	Chionas, Georgios
Batko, Bogdan	Chistikov, Dmitry
Berthe, Gaétan	Cáceres, Manuel
Bešter Štorgel, Kenny	Dabrowski, Konrad K.
Biondi, Giulio	Das, Bireswar
Blauth, Jannis	de Castro Mendes Gomes, Guilherme
Blažej, Václav	Defrain, Oscar
Bodlaender, Hans L.	DeWolfe, Ryan
Bojko, Dominik	Diatzko, Gregor

Dietl, Guido
DĘbski, Michał
Echeverría, Henry
Elias, Marek
Epstein, Leah
Fleischmann, Pamela
Fujita, Shinya
Förster, Henry
Gahlawat, Harmender
Ghazawi, Samah
Golin, Mordecai J.
Gottlieb, Eric
Gouleakis, Themis
Gusain, Rachana
Gutowski, Grzegorz
Herold, Martin
Hirvensalo, Mika
Hlineny, Petr
Huang, Shenwei
Jakoby, Andreas
Janusz, Szymon
Jiamjitrak, Wanchote Po
Jurdzinski, Tomasz
Kadria, Avi
Kamali, Shahin
Kamiński, Bogumił
Kavšek, Branko
Kawamura, Akitoshi
Kempa, Dominik
Kindermann, Philipp
Klawitter, Jonathan
Koana, Tomohiro
Kobayashi, Yasuaki
Kociumaka, Tomasz
Kowalski, Dariusz
Krekelberg, Bob
Kubaty, Piotr
Lafond, Manuel
Lagerkvist, Victor
Lamprou, Ioannis
Langedal, Kenneth
Lassota, Alexandra
Lauerbach, Antonio
Li, Ruoying
Liedtke, David

Luckner, Marcin
Mann, Kevin
Manne, Fredrik
Marcilon, Thiago
Markou, Euripides
Melissinos, Nikolaos
Mikšaník, David
Miller, Avery
Miller, Jacob
Molter, Hendrik
Mouawad, Amer
Movsum, Ulvi
Mumey, Brendan
Münch, Miriam
Nadara, Wojciech
Nagarajan, Viswanath
Neuenkirch, Andreas
Nogler, Jakob
Olkowski, JĘdrzej
Osipov, George
Otachi, Yota
Ouchi, Katsuhisa
Papasotiropoulos, Georgios
Pardubska, Dana
Peng, Junqiang
Perz, Daniel
Radoszewski, Jakub
Reidl, Felix
Rohwedder, Lars
Roth, Marc
Saarela, Aleksi
Saladi, Rahul
Sapir, Ariel
Sau, Ignasi
Say, A.C. Cem
Scheu-Hachtel, Linda
Schierreich, Šimon
Schmidt, Jonas
Sharma, Roohani
Sieper, Marie Diana
Sikora, Florian
Smid, Michiel
Smith, Ben
Soos, Mate
Struski, Łukasz

Suzuki, Takahiro
Szilágyi, Krisztina
Szufel, Przemyslaw
Taufique, Zain
Tinarrage, Raphaël
Torán, Jacobo
TrĘdowicz, Magdalena
Tsuchiya, Shoichi
Upadhyay, Jalaj
van Ee, Martijn
Vanhatalo, Aleksi
Vihrovs, Jevgēnijs
Vu, Tung Anh
Waleń, Tomasz

Wallinger, Markus
Wang, Qisheng
Watrigant, Rémi
Werthmann, Julian
Wicke, Kristina
Willhalm, Daniel
Wolf, Samuel
Wrona, Michał
Yang, Yongjie
Yang, Zhengyi
Zalonis, Jasmin
Zhang, Louxin
Černá, Daniela

Invited Talks

Semirandom Planted Clique

Jarosław Błasiok

Bocconi University
`jaroslaw.blasiok@unibocconi.it`

The *planted clique problem* is a classical problem in *average-case complexity*, which studies efficient algorithms that succeed on typical inputs drawn from natural distributions. An instance is generated by sampling an *Erdős–Rényi* random graph on n vertices, where each edge appears independently with probability $1/2$, then selecting a random set S of k vertices and adding all edges among them — planting a clique. Given such a graph, the goal is to recover the set S.

By analyzing the expected number of k-cliques in an Erdős–Rényi graph, one finds that such graphs are unlikely to contain cliques much larger than $2 \log n$. Similarly, when $k \gg 2 \log n$, the planted clique is typically the unique largest clique, and an inefficient algorithm can recover S with high probability. Indeed, enumerating all candidate sets yields a quasi-polynomial-time algorithm running in $n^{O(\log n)}$, showing that the planted clique is *information-theoretically identifiable* in this regime.

This motivates the question of when a *polynomial-time* algorithm exists. For $k \geq \Omega(\sqrt{n \log n})$, the clique can be found by inspecting vertex degrees. A celebrated result of Alon, Krivelevich, and Sudakov [1] shows that even for $k = \Theta(\sqrt{n})$, the clique can be recovered using spectral methods based on the adjacency matrix. The famous planted clique conjecture posits an information-theoretic/computational gap: for k between $2 \log n$ and $o(\sqrt{n})$, no polynomial-time algorithm can recover the planted clique with constant probability, despite the existence of inefficient algorithms that can do so.

Semi-random Problems The existence of efficient average-case algorithms for problems whose worst-case versions are NP-hard may help explain why many NP-complete problems arising in practice are easier than worst-case theory suggests. However, this explanation is sometimes unsatisfying: average-case algorithms often rely on fragile properties of the specific input distribution. Real-world instances may contain randomness, but rarely follow exactly the distributions assumed in the analysis.

This motivates the study of a hybrid between worst-case and average-case analysis: the *semi-random model*. In this model, an instance is first drawn from a probability distribution and then modified in a bounded way by an *adversary* with full knowledge of the algorithm. The goal is to design algorithms that succeed with high probability on such instances.

Semi-random Planted Clique In the case of a planted clique, a semi-random instance is prepared in the following way

1. Nature generates an Erdős–Rényi random graph $G(n, 1/2)$ and a set $S \subset [n]$ of size k, and plants a clique on the set S.
2. Then, the adversary can arbitrarily alter the induced subgraph on vertices $[n]\backslash S$.

For a graph like that, no algorithm can succeed in recovering the set S with high probability: an adversary, in their part of a graph, can plant n/k cliques that are (information-theoretically) indistinguishable from S. The more reasonable goal is to ask for a *short* list of candidate sets that with high probability contain the set S. As it turns out, as opposed to the standard planted clique problem, here even in the range where k is relatively large (say, $n^{0.99}$), there is no "trivial" algorithm (similar to just filtering vertices based on their degree). The original algorithm of [4] worked for $k \geq n^{2/3}$, applying a semidefinite-programming relaxation; this was followed by a Sum-of-Squares based approach of [3], applying when $k \geq n^{1/2+\varepsilon}$.

In this talk, we discuss the approach from [2], which proposed a simple greedy algorithm working when $k \geq \Omega(\sqrt{n}\,\text{polylog}(n))$ – an analysis of which, quite surprisingly is heavily reliant on a *restricted isometry property* of a specific matrix — an important property studied in the superficially distant area of compressed sensing.

References

1. Alon, N., Krivelevich, M., Sudakov, B.: Finding a large hidden clique in a random graph. In: Random Struct. Algorithms **13**(3-4), 457–466 (1998)
2. Błasiok, J., Buhai, R.D., Kothari, P.K., Steurer, D.: Semirandom planted clique and the restricted isometry wproperty. In: 2024 IEEE 65th Annual Symposium on Foundations of Computer Science (FOCS), pp. 959–969. IEEE (2024)
3. Buhai, R.D., Kothari, P.K., Steurer, D.: Algorithms approaching the threshold for semi-random planted clique. In: STOC'23—Proceedings of the 55th Annual ACM Symposium on Theory of Computing. ACM, New York, 2023, pp. 1918–1926 (2023)
4. Charikar, M., Steinhardt, J., Valiant, G.: Learning from untrusted data. In: Proceedings of the 49th Annual ACM SIGACT Symposium on Theory of Computing, pp. 47–60 (2017)

On the Bidirected Cut Relaxations for Steiner Tree and Steiner Forest

Jarosław Byrka 🆔

University of Wrocław
`jby@cs.uni.wroc.pl`

The Steiner tree problem is one of the most prominent problems in network design. Given an edge-weighted undirected graph and a subset of the vertices, called terminals, the task is to compute a minimum-weight tree containing all terminals (and possibly further vertices). The best-known approximation algorithms for Steiner tree involve enumeration of a (polynomial but) very large number of candidate components and are therefore slow in practice.

A promising ingredient for the design of fast and accurate approximation algorithms for Steiner tree is the bidirected cut relaxation (BCR): bidirect all edges, choose an arbitrary terminal as a root, and enforce that each cut containing some terminal but not the root has one unit of fractional edges leaving it. BCR is known to be integral in the spanning tree case [3], i.e., when all the vertices are terminals. For general instances, however, it was not even known whether the integrality gap of BCR is better than the integrality gap of the natural undirected relaxation, which is exactly 2. We resolve this question by proving an upper bound of 1.9988 on the integrality gap of BCR.

In the second part of the talk I discuss an LP relaxation for Steiner Forest that generalizes BCR for Steiner Tree. We prove that this relaxation has several promising properties. Among them, it is possible to round any half-integral LP solution to a Steiner Forest instance while increasing the cost by at most a factor 16/9. To prove this result we introduce a novel recursive densest-subgraph contraction algorithm.

The talk is based on joint works with Fabrizio Grandoni and Vera Traub that appeared as [1] and [2].

Keywords: LP relaxation · Steiner tree · Steiner forest

References

1. Byrka, J., Grandoni, F., Traub, V.: The bidirected cut relaxation for Steiner tree has integrality gap smaller than 2. In: 2024 IEEE 65th Annual Symposium on Foundations of Computer Science (FOCS), pp. 730–753. IEEE (2024)
2. Byrka, J., Grandoni, F., Traub, V.: On the bidirected cut relaxation for Steiner forest. In Megow, N., Basu, A. (eds.) Integer Programming and Combinatorial Optimization.

IPCO 2025. LNCS. vol. 15620, pp. 114–127. Springer, Cham (2025). https://doi.org/
10.1007/978-3-031-93112-3_9
3. Edmonds, J.: Optimum branchings. J. Res. Nat. Bur. Stan. Eng. Instrum. C **71**, 233
(1967)

Long-Refinement Graphs

Sandra Kiefer

University of Oxford, UK
sandra.kiefer@cs.ox.ac.uk

Colour Refinement, sometimes also called Naïve Vertex Classification, is an iterative combinatorial method that distinguishes vertices in graphs based on the local structure around them, with the purpose of uncovering asymmetries in the graphs. The structure around each vertex is encoded in a *colour*, and the vertex colours in one iteration are used to refine the structural information in the next iteration. The most prominent application of the algorithm is in approaches to the graph isomorphism problem: if vertices receive distinct colours in some iteration, this implies that they cannot be mapped onto each other via any isomorphism. A generalisation of the Naïve Vertex Classification approach is used in Babai's quasipolynomial-time algorithm for deciding graph isomorphism [2], and it is well understood for which pairs of graphs it suffices to apply Colour Refinement to detect non-isomorphism [1, 7].

In each iteration, the algorithm assigns to two vertices the same colour if and only if they had the same colour and also in each colour the same number of neighbours in the preceding (or, equivalently, in every previous) iteration. The induced partition of the vertex set into colour classes hence becomes gradually finer until it stabilises. The number of iterations until that point is reached is the central complexity parameter of the algorithm, with tight links to parameters in various other areas of computer science. For example, the number of iterations needed to obtain distinct colourings on two input graphs G and H corresponds to the quantifier depth of a distinguishing formula in a well-known counting logic [3] and to the minimal depth of a forest whose homomorphism counts into G and H are distinct [4].

For n-vertex graphs, the trivial upper bound on the number of Colour Refinement iterations until stabilisation is $n-1$. For a long time, it was open whether graphs that attain this number of iterations actually exist. My talk gives an overview of the history of this class of *long-refinement graphs*, as well as a report on recent results. By restricting ourselves to small degrees, we constructed infinite families of long-refinement graphs, thereby showing that the trivial upper bound on the iteration number of Colour Refinement on n-vertex graphs is tight [5]. Recently, via reverse-engineering the connections within and between colour classes in the graphs, we obtained a complete characterisation of the long-refinement graphs with small (or, equivalently, large) degrees [6]. The analysis provides us with a deep understanding of the dynamics of Colour Refinement on long-refinement graphs, and it yields that all long-refinement graphs with maximum degree 3 can be described via compact strings over a very small alphabet.

The work initiated a search for long-refinement graphs that are only distinguished in the last iteration of Colour Refinement before termination. In the talk, I will present a simple argument to show that such graphs do not exist.

Keywords: Colour Refinement · Naïve Vertex Classification · Graph Isomorphism

References

1. Arvind, V., Köbler, J., Rattan, G., Verbitsky, O.: Graph isomorphism, color refinement, and compactness. Comput. Complex. **26**(3), 627–685 (2017). https://doi.org/10.1007/s00037-016-0147-6
2. Babai, L.: Graph isomorphism in quasipolynomial time [extended abstract]. In: Proceedings of the 48th Annual ACM SIGACT Symposium on Theory of Computing, STOC 2016, pp. 684–697 (2016). https://doi.org/10.1145/2897518.2897542
3. Cai, J.Y., Fürer, M., Immerman, N.: An optimal lower bound on the number of variables for graph identification. Combinatorica **12**(4), 389–410 (1992), https://link.springer.com/article/10.1007/BF01305232
4. Fluck, E., Seppelt, T., Spitzer, G.L.: Going deep and going wide: counting logic and homomorphism indistinguishability over graphs of bounded treedepth and treewidth. In: Proceedings of the 32nd EACSL Annual Conference on Computer Science Logic, CSL 2024. LIPIcs. vol. 288, pp. 27:1–27:17 (2024). https://doi.org/10.4230/LIPICS.CSL.2024.27, https://doi.org/10.4230/LIPIcs.CSL.2024.27
5. Kiefer, S., McKay, B.D.: The iteration number of colour refinement. In: Proceedings of the 47th International Colloquium on Automata, Languages, and Programming, ICALP 2020. Leibniz International Proceedings in Informatics (LIPIcs). vol. 168, pp. 73:1–73:19. Schloss Dagstuhl – Leibniz-Zentrum für Informatik, Dagstuhl, Germany (2020). https://doi.org/10.4230/LIPIcs.ICALP.2020.73
6. Kiefer, S., de Mel, T.D.: A classification of long-refinement graphs for Colour Refinement. In: Proceedings of the 2026 Annual ACM-SIAM Symposium on Discrete Algorithms, SODA 2026, pp. 388–407. https://doi.org/10.1137/1.9781611978971.18
7. Kiefer, S., Schweitzer, P., Selman, E.: Graphs identified by logics with counting. ACM Trans. Comput. Log. **23**(1), 1:1–1:31 (2022). https://doi.org/10.1145/3417515

Snapshots of Reconfiguration

Linda Kleist ⓘ

Universität Hamburg, Germany
`linda.kleist@uni-hamburg.de`

In dynamic or interactive environments, a common challenge is to transform one configuration into another through a sequence of small, local modifications. Interesting settings range from robot motion planning, over games such as the 15-puzzle or the Rubik's cube, to the rotation of binary trees.

These reconfiguration processes raise many natural algorithmic questions: Can any two – or two given – configurations be transformed into each other? If so, it is of interest to bound the length of a shortest reconfiguration sequence between (any) two configurations, to compute the length of an optimal sequence, and to efficiently determine a short(est) sequence.

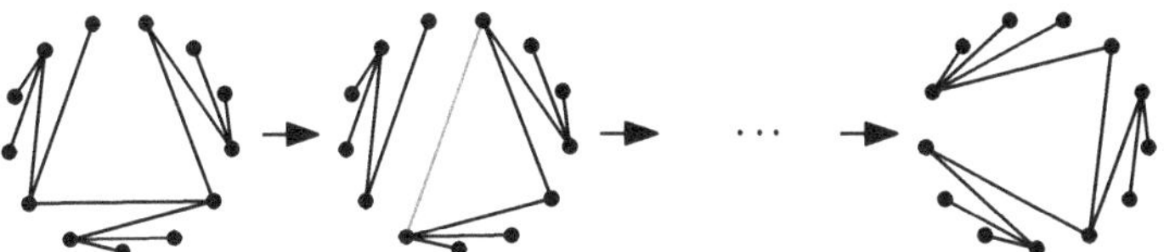

Fig. 1. Can you transform the left non-crossing spanning tree the right by a sequence of edge exchanges? How many do you need?

These questions can be nicely rephrased in terms of a so-called *flip graph*. The flip graph of a discrete reconfiguration problem has a vertex for each configuration and an edge between any two configurations that can be transformed into one another by a basic operation, called a *flip*. In terms of the flip graph, the questions then read as follows: Is the flip graph connected? What is the diameter of the flip graph? Can a shortest path (flip sequence) between two vertices of the flip graph be computed efficiently?

The possibly most famous flip graph is the flip graph of triangulations on convex point sets with edge exchange. This graph is the 1-skeleton of the associahedron and corresponds to rotations of binary trees. For n-point sets in general position, the connectivity of the flip graph [8] and its diameter of $\Theta(n^2)$ [6] are classic results. Moreover, computing the flip distance is known to be NP-hard [9] and -hard [10]. For triangulations of convex point sets, the diameter of the flip graph is at most $2n - 10$, which is tight for $n > 13$ [12, 11]. The computational complexity of computing the flip distance of two triangulations on a convex point set is a tantalizing open problem.

In this talk, we discuss several settings including the reconfiguration of non-crossing graphs on points sets (e.g. triangulations, spanning trees, ...), as well as the reconfiguration of geometric objects in polygonal domains. For examples, consider fig:intro1,fig:intro. We include some insights of recent works [1, 2, 4, 3, 5, 7].

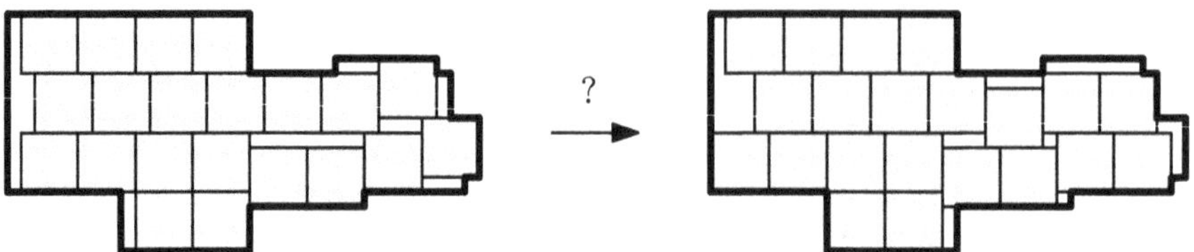

Fig. 2. Can the left configuration of unit squares in a simple polygon be reconfigured to the right configuration of unit squares? Taken from [1].

References

1. Abrahamsen, M., et al.: Reconfiguration of unit squares and disks: Pspace-hardness in simple settings. In: Symposium on Computational Geometry (SoCG). LIPIcs. vol. 332, pp. 1:1–1:18 (2025). https://doi.org/10.4230/LIPICS.SOCG.2025.1

2. Akitaya, H.A., Cardinal, J., Felsner, S., Kleist, L., Lauff, R.: Facet-hamiltonicity. In: Symposium on Discrete Algorithms (SODA), pp. 5051–5064 (2025). https://doi.org/10.1137/1.9781611978322.171

3. Bjerkevik, H.B., Dorfer, J., Kleist, L., Ueckerdt, T., Vogtenhuber, B.: Flip distance of non-crossing spanning trees: Np-hardness and improved bounds (2026). in prep

4. Bjerkevik, H.B., Kleist, L., Ueckerdt, T., Vogtenhuber, B.: Flipping non-crossing spanning trees. In: Symposium on Discrete Algorithms (SODA), pp. 2313–2325. https://doi.org/10.1137/1.9781611978322.77

5. Felsner, S., Kleist, L., Mütze, T., Sering, L.: Rainbow cycles in flip graphs. SIAM J. Discrete Math. **34**(1), 1–39 (2020). https://doi.org/10.1137/18M1216456

6. Hurtado, F., Noy, M., Urrutia, J.: Flipping edges in triangulations. Discrete Comput. Geom. **22**(3), 333–346 (1999). https://doi.org/10.1007/PL00009464

7. Kleist, L., Kramer, P., Rieck, C.: On the connectivity of the flip graph of plane spanning paths. In: Král', D., Milanič, M. (eds) Graph-Theoretic Concepts in Computer Science. WG 2024. LNCS. vol. 14760, pp. 327–342 (2024). https://doi.org/10.1007/978-3-031-75409-8_23

8. Lawson, C.L.: Transforming triangulations. Discrete Math. **3**(4), 365–372 (1972). https://doi.org/10.1016/0012-365X(72)90093-3

9. Lubiw, A., Pathak, V.: Flip distance between two triangulations of a point set is NP-complete. Comput. Geom. Theory Appl. **49**, 17–23 (2015). https://doi.org/10.1016/J.COMGEO.2014.11.001

10. Pilz, A.: Flip distance between triangulations of a planar point set is APX-hard. Comput. Geom. **47**(5), 589–604 (2014). https://doi.org/10.1016/j.comgeo.2014.01.001

11. Pournin, L.: The diameter of associahedra. Adv. Math **259**, 13–42 (2014). https://doi.org/10.1016/j.aim.2014.02.035
12. Sleator, D.D., Tarjan, R.E., Thurston, W.P.: Rotation distance, triangulations, and hyperbolic geometry. In: Symposium on Theory of Computing (STOC), pp. 122–135 (1986). https://doi.org/10.1145/12130.12143

Distributed Quantum Advantage

Jukka Suomela

Aalto University, P.O. Box 15400, 00076 Aalto, Finland
jukka.suomela@aalto.fi
https://jukkasuomela.fi

Modern computer systems are fundamentally *distributed*; our everyday life relies on global communication systems such as the Internet and large-scale data centers that house thousands of interconnected computers. At the same time, *quantum computing* is advancing rapidly, boosted by tens of billions of euros in investment from governments and companies worldwide. But we still do not know whether quantum computing and communication can truly help solve practically relevant tasks in the distributed setting!

In this talk, I will discuss three fundamentally different flavors of distributed quantum advantage. The first one is purely *computational advantage*: if one quantum computer can outperform one classical computer, a network of many quantum computers can outperform a network of many classical computers. But what if the bottleneck is related to communication, not computation?

The second flavor is advantage in *communication bandwidth*: sending one qubit can be more useful than sending one classical bit; such an advantage has been studied in the framework of the quantum-CONGEST model [7]. However, sending one qubit can also be much more expensive than sending one classical bit, and hence it is unclear whether this can ever lead to a practically relevant performance boost in computer networks.

The third flavor is advantage in *communication rounds*: a network of quantum computers can solve computational tasks in a smaller number of communication rounds than any classical system; a message with one qubit can be more powerful than a message with *any* number of classical bits. This kind of advantage has been studied in the framework of the quantum-LOCAL model [2, 8], and if we could realize such an advantage for a practically relevant task, it would result in a convincing, indisputable distributed quantum advantage.

A priori, it is not at all clear that any savings in the number of communication rounds are possible, and there are numerous *negative results* that put limits on the amount of quantum advantage [1, 2, 5, 6, 8]. But it turns out that there is hope! In a 2019 breakthrough result, Le Gall, Nishimura, and Rosmanis [9] presented the first distributed problem with a strict quantum advantage in the quantum-LOCAL model, and our recent work has provided more examples of problems that admit an asymptotic distributed quantum advantage [3, 4].

Yet all these examples of quantum advantage have been purely *artificial* problems, engineered to demonstrate quantum advantage, but serving no practical purpose. Is there any hope of realizing quantum advantage in practice? I will explore the current frontier of what is known and what is unknown, and what are the most promising avenues for overcoming the current barriers.

Keywords: Distributed graph algorithms · Quantum computing

References

1. Akbari, A., et al.: Online locality meets distributed quantum computing. In: Koucký, M., Bansal, N. (eds.) Proceedings of the 57th Annual ACM Symposium on Theory of Computing, STOC 2025, Prague, Czechia, June 23–27, 2025, pp. 1295–1306. ACM (2025). https://doi.org/10.1145/3717823.3718211
2. Arfaoui, H., Fraigniaud, P.: What can be computed without communications? SIGACT News **45**(3), 82–104 (2014). https://doi.org/10.1145/2670418.2670440
3. Balliu, A., et al.: Distributed quantum advantage for local problems. In: Koucký, M., Bansal, N. (eds.) Proceedings of the 57th Annual ACM Symposium on Theory of Computing, STOC 2025, Prague, Czechia, June 23–27, 2025, pp. 451–462. ACM (2025). https://doi.org/10.1145/3717823.3718233
4. Balliu, A., et al.: Distributed quantum advantage in locally checkable labeling problems. In: Proceedings of the 37th ACM-SIAM Symposium on Discrete Algorithms, SODA 2026, Vancouver, Canada (2026). https://doi.org/10.48550/ARXIV.2504.05191
5. Balliu, A., et al.: New limits on distributed quantum advantage: Dequantizing linear programs. In: Kowalski, D.R. (ed.) 39th International Symposium on Distributed Computing, DISC 2025, Berlin, Germany, October 27–31, 2025. LIPIcs, vol. 356, pp. 11:1–11:22. Schloss Dagstuhl - Leibniz-Zentrum für Informatik (2025). https://doi.org/10.4230/LIPICS.DISC.2025.11
6. Coiteux-Roy, X., et al: No distributed quantum advantage for approximate graph coloring. In: Mohar, B., Shinkar, I., O'Donnell, R. (eds.) Proceedings of the 56th Annual ACM Symposium on Theory of Computing, STOC 2024, Vancouver, BC, Canada, June 24–28, 2024, pp. 1901–1910. ACM (2024). https://doi.org/10.1145/3618260.3649679
7. Elkin, M., Klauck, H., Nanongkai, D., Pandurangan, G.: Can quantum communication speed up distributed computation? In: Halldórsson, M.M., Dolev, S. (eds.) ACM Symposium on Principles of Distributed Computing, PODC'14, Paris, France, July 15–18, 2014, pp. 166–175. ACM (2014). https://doi.org/10.1145/2611462.2611488
8. Gavoille, C., Kosowski, A., Markiewicz, M.: What can be observed locally? In: Keidar, I. (ed.) Distributed Computing. DISC 2009. LNCS. vol. 5805, pp. 243–257. Springer, Cham (2009). https://doi.org/10.1007/978-3-642-04355-0_26
9. Le Gall, F., Nishimura, H., Rosmanis, A.: Quantum advantage for the LOCAL model in distributed computing. In: Niedermeier, R., Paul, C. (eds.) 36th International Symposium on Theoretical Aspects of Computer Science, STACS 2019, Berlin, Germany, March 13–16, 2019. LIPIcs, vol. 126, pp. 49:1–49:14. Schloss Dagstuhl - Leibniz-Zentrum für Informatik (2019). https://doi.org/10.4230/LIPICS.STACS.2019.49

Propositions as Types

Philip Wadler

Powerful insights arise from linking two fields of study previously thought separate. Examples include Descartes's coordinates, which links geometry to algebra, Planck's Quantum Theory, which links particles to waves, and Shannon's Information Theory, which links thermodynamics to communication. Such a synthesis is offered by the principle of Propositions as Types, which links logic to computation. At first sight it appears to be a simple coincidence—almost a pun—but it turns out to be remarkably robust, inspiring the design of automated proof assistants and programming languages, and continuing to influence the forefronts of computing.

Propositions as Types is a notion with many names and many origins. It is closely related to the BHK Interpretation, a view of logic developed by the intuitionists Brouwer, Heyting, and Kolmogorov in the 1930s. It is often referred to as the Curry-Howard Isomorphism, referring to a correspondence observed by Curry in 1934 and refined by Howard in 1969. Others draw attention to significant contributions from de Bruijn's Automath and Martin-Löf's Type Theory in the 1970s.

Propositions as Types is a notion with depth. It describes a correspondence between a given logic and a given programming language. At the surface, it says that for each proposition in the logic there is a corresponding type in the programming language—and vice versa. Thus we have

$$\textit{propositions} \text{ as } \textit{types}.$$

It goes deeper, in that for each proof of a given proposition, there is a program of the corresponding type—and vice versa. Thus we also have

$$\textit{proofs} \text{ as } \textit{programs}.$$

And it goes deeper still, in that for each way to simplify a proof there is a corresponding way to evaluate a program—and vice versa. Thus we further have

$$\textit{simplification of proofs}, \text{ as } \textit{evaluation of programs}.$$

Hence, we have not merely a shallow bijection between propositions and types, but a true isomorphism preserving the deep structure of proofs and programs, simplification and evaluation.

Propositions as Types is a notion with breadth. It applies to a range of logics including propositional, predicate, second-order, intuitionistic, classical, modal, and linear. It underpins the foundations of functional programming, explaining features including functions, records, variants, parametric polymorphism, data abstraction, continuations, linear types, and session types. It has inspired automated proof assistants and programming languages including Agda, Automath, Coq, Epigram, $F^{\#}$, $F^{\star}$, Haskell, Lean, LF, ML, NuPRL, Scala, Singularity, and Trellys.

Propositions as Types is a notion with mystery. Why should it be the case that intuitionistic natural deduction, as developed by Gentzen in the 1930s, and simply-typed lambda calculus, as developed by Church around the same time for an unrelated purpose, should be discovered thirty years later to be essentially identical? And why should it be the case that the same correspondence arises again and again? The logician Girard and the computer scientist Reynolds independently developed the same calculus, now dubbed Girard-Reynolds. The logician Hindley and the computer scientist Milner independently developed the same type system, now dubbed Hindley-Milner. Curry-Howard is a double-barrelled name that ensures the existence of other double-barrelled names. Those of us that design and use programming languages may often feel they are arbitrary, but Propositions as Types assures us some aspects of programming are absolute.

The talk is based on a paper that appeared in *Communications of the ACM* [1], and the abstract above is taken from that paper.

References

1. Wadler, P.: Propositions as types. Commun. ACM **58**(12), 75–84 (2015). https://doi. org/10.1145/2699407.

Contents

Posters

On the Complexity of Constrained Reconfiguration and Motion Planning

Nicolas Bousquet[1,2], Remy El Sabeh[3(✉)], Amer E. Mouawad[3,4],
and Naomi Nishimura[3]

[1] CNRS, INSA Lyon, UCBL, LIRIS, UMR5205, 69622 Villeurbanne, France
nicolas.bousquet@cnrs.fr
[2] CNRS - Université de Montréal CRM - CNRS, Montreal, Canada
[3] David R. Cheriton School of Computer Science, University of Waterloo,
Waterloo, Canada
{relsabeh,nishi}@uwaterloo.ca, aa368@aub.edu.lb
[4] Department of Computer Science, American University of Beirut, Beirut, Lebanon
https://perso.liris.cnrs.fr/nbousquet/ ,
https://www.aub.edu.lb/pages/profile.aspx?MemberId=aa368

Abstract. Coordinating the motion of multiple agents in constrained environments is a fundamental challenge in robotics, motion planning, and scheduling. A motivating example involves n robotic arms, each represented as a line segment. The objective is to rotate each arm to its vertical orientation, one at a time, without collisions. This scenario is an example of the more general k-COMPATIBLE ORDERING problem, where n agents, each capable of k state-changing actions, must transition to specific target states under constraints encoded as a set $\mathcal{G}$ of k pairs of directed graphs. We show that k-COMPATIBLE ORDERING is NP-complete, even when $\mathcal{G}$ is planar, degenerate, or acyclic. On the positive side, we provide polynomial-time algorithms for cases such as when $k = 1$ or $\mathcal{G}$ has bounded treewidth. We also introduce generalized variants supporting multiple state-changing actions per agent. These results extend to a wide range of scheduling, reconfiguration, and motion planning applications.

Keywords: search · optimization · robotics · parameterized complexity · combinatorial reconfiguration

1 Introduction

Motion planning of multiple robots sharing a common workspace is a well-studied problem that has been tackled from both centralized [15] and decentralized [8,22]

R. El Sabeh and N. Nishimura—Research supported by the Natural Sciences and Engineering Research Council of Canada.

N. Bousquet—Research partly supported by ANR project ENEDISC (ANR-24-CE48-7768-01).

© The Author(s), under exclusive license to Springer Nature Switzerland AG 2026
J. Kozik and A. Wolff (Eds.): SOFSEM 2026, LNCS 16448, pp. 1–15, 2026.
https://doi.org/10.1007/978-3-032-17801-5_1

perspectives. The main goal of robot motion planning is typically to coordinate the actions of robots, such that the robots attain a particular common goal, while guaranteeing that the interference between the robots is either minimized or eliminated. In other words, the movements of a robot should not cause a collision with any of the other robots, which may be moving either synchronously or asynchronously.

The system can abstractly be defined by a starting position for each robot, a set of potential moves for each robot, a measure we would like to minimize (such as time), and a target position (or positions) for each robot. Many state-of-the-art algorithms for motion planning in the literature are based on sampling from reasonable local choices. While these algorithms may perform well in practice, they are not deterministic and can usually only be assessed empirically against each other without concrete guarantees. In this paper, and inspired by the abstract definition of the motion planning problem, as well as its accompanying geometric constraints, we tackle motion planning from the reconfiguration angle.

The reconfiguration framework studies the transformation of combinatorial objects via a sequence of operations, each involving a single object, while maintaining feasibility throughout the transformation. The objects under consideration typically belong to the solution space of a well-defined problem. Given source and target configurations, we often ask whether there exists a sequence of valid steps to transform the former into the latter. This can be modeled as a reachability question in a graph of configurations, where nodes represent feasible solutions, and edges correspond to possible transformations. However, the solution space is usually exponential, so a polynomial-time solution cannot rely on exhaustive search.

The framework of reconfiguration was formally introduced by Hearn and Demaine with the NONDETERMINISTIC CONSTRAINT LOGIC problem, which established PSPACE-completeness results for many naturally occurring puzzles [6,12,13]. Since then, reconfiguration has extended to graph problems [16]. In the context of graph reconfiguration problems, configurations are transformed by adding, removing, or substituting vertices. While some reconfiguration problems admit polynomial-time algorithms on restricted graph classes (e.g., planar graphs [2], graphs of bounded treewidth [18], or interval graphs [5]), many remain PSPACE-complete or hard to approximate [10,20,21]. For more on reconfiguration problems, we refer the reader to surveys on the topic [3,14,19].

Going back to motion planning, in this paper, we have chosen to restrict the power of robots to study the impact of the number and relative positions of robots on the complexity of the problem. We also focus on centralized algorithms, avoiding synchronization issues that might arise in multi-agent systems.

The ROBOTIC ARM MOTION PLANNING (RAMP) problems are meant to provide context for and a motivation for the paper's two main problems, namely the k-COMPATIBLE ORDERING problem and the k-COMPATIBLE SET ARRANGEMENT problem, with the included results for these two problems being used to refine what we know about the tractability of RAMP problems. Problems on robotic arms will therefore be recurrently used as tangible examples of k-compatible ordering and arrangement problems.

The RAMP problems are defined in the plane $\mathbb{R}^2$, equipped with the Cartesian coordinate system, where a *robotic arm* is a unit-length line segment. An arm can be identified by its *center*, *i.e.*, the midpoint (x, y) of its line segment, and the *angle* $a \in [0, \pi)$ that the line segment forms with the horizontal line passing through its center. A collection of arms is *conflict-free* if the line segments are pairwise non-intersecting. We are particularly interested in arms that are *vertical* (angle $\pi/2$). A reconfiguration operation on a set of arms is a *rotation* of a single arm about its center; a *counterclockwise rotation* (resp. *clockwise rotation*) of a robotic arm updates its angle a to $a' = (a + \alpha) \bmod \pi$, where α is a positive (resp. negative) number in $(-\pi, \pi)$ while rotating clockwise (resp. counterclockwise). If a robotic arm r_1 is rotating, and another arm r_2 intersects one of the two sectors swept by the rotation of r_1, we say that the rotation of r_1 *hits* r_2.

We start by presenting the simplest RAMP problem:

> SINGLE MOVE ROBOTIC ARM MOTION PLANNING (SM-RAMP)
> **Input:** A sequence of robotic arms $\mathcal{R} = r_1, \ldots, r_n$, with distinct fixed centers.
> **Output:** Whether there exists a sequence of conflict-free robotic arm configurations $\mathcal{R}_0, \ldots, \mathcal{R}_n$ such that $\mathcal{R} = \mathcal{R}_0$ and all robotic arms are vertical in $\mathcal{R}_n$. Moreover, for every $0 \leq j \leq n - 1$, there exists a robotic arm r_i satisfying the following:
>
> - $\mathcal{R}_j$ and $\mathcal{R}_{j+1}$ differ only in the angle of r_i;
> - r_i is rotated from its initial angle in $\mathcal{R}_j$ to $\pi/2$ (vertical orientation) in $\mathcal{R}_{j+1}$;
> - The rotation of r_i does not hit any other robotic arm in $\mathcal{R}_j$; and
> - r_i is rotated at most once in the sequence $\mathcal{R}_0, \ldots, \mathcal{R}_n$.

In order to use the machinery of graph reconfiguration to design hardness results or algorithms, we aim to define SM-RAMP in graph theoretic terms. Therefore, we consider a problem on graphs that generalizes it, the k-COMPATIBLE ORDERING problem, defined below:

> k-COMPATIBLE ORDERING
> **Input:** A vertex set V, a collection $\mathcal{G}$ of k pairs of directed graphs $(A_1, B_1), \ldots, (A_k, B_k)$, such that $V(A_i) = V(B_i) = V$ for $1 \leq i \leq k$.
> **Output:** Whether there exists a pair $(\mathcal{S}, \mathcal{L})$, where $\mathcal{S} = s_1, \ldots, s_n$ is an ordering of V, $\mathcal{L} = \ell_1, \ldots, \ell_n$ is a sequence of labels, and vertex $s_i \in \mathcal{S}$ is assigned label $\ell_i \in [k]$, for $1 \leq i \leq n$, such that the following two constraints are satisfied:
>
> - s_i is a sink in $A_{\ell_i}[s_i, \ldots, s_n]$; and
> - s_i is a source in $B_{\ell_i}[s_1, \ldots, s_i]$.

****Lemma 1.** *There exists a polynomial-time reduction from the* SM-RAMP *problem to the* 2-COMPATIBLE ORDERING *problem.*

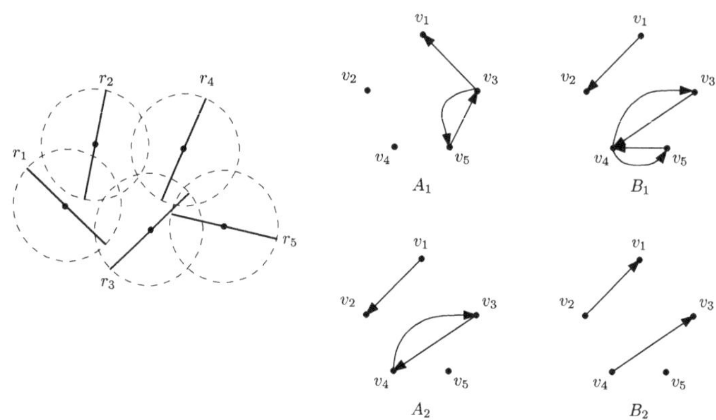

Fig. 1. An SM-RAMP instance (left) and its equivalent 2-COMPATIBLE ORDERING instance (right). Arms can be rotated in the order r_1, r_5, r_3, r_4, r_2, where r_1, r_3 rotate clockwise and r_2, r_4, r_5 rotate counterclockwise. The arc $v_3 v_5$ in A_1 means that a clockwise rotation of r_3 to its final orientation would hit r_5 if r_5 has not rotated. The absence of the arc $v_3 v_5$ from B_2 means that r_3 does not block the counterclockwise rotation of r_5 if r_3 is already vertical. The solution of the 2-COMPATIBLE ORDERING instance corresponding to the solution of SM-RAMP is $\mathcal{S} = v_1, v_5, v_3, v_4, v_2$ and $\mathcal{L} = 1, 2, 1, 2, 2$.

Due to space constraints, the proof of Lemma 1, missing definitions, as well as the complete proofs of all starred results, can be found in the full version of the paper [4]. Figure 1 illustrates an instance of SM-RAMP and the corresponding 2-COMPATIBLE ORDERING instance and gives the main flavor of the reduction.

While the SM-RAMP problem, which we initially set out to solve, inspired the k-COMPATIBLE ORDERING problem, the latter's applications extend beyond problems on robotic arms, as it can capture constraints that are not inherently geometric. The k-COMPATIBLE ORDERING problem models a class of element-ordering problems in which each element can be inserted in one of k modes in the ordering, where the chosen mode determines the set of elements that must precede it and the set that must follow it, independently of the insertion modes of these other elements.

We use the TRUCK UNLOADING problem as a natural example of a 6-COMPATIBLE ORDERING problem. Let $b_1, \ldots, b_n$ be a set of boxes in a truck, where each box is a rectangular cuboid. We are looking to lift the boxes out of the truck, and then place them in a single stack inside a warehouse. When a box is lifted, one of its six faces, labeled 1 to 6 arbitrarily, is oriented downward; it is then placed on the stack in the same orientation (we assume that before lifting, each box can be freely rotated so that any of its six faces may be chosen as the downward face, allowing up to six lifting orientations). Choosing label ℓ for box

b_i means: "lift box b_i in orientation ℓ from the truck onto the stack". The graph A_ℓ has an arc $b_i b_j$ whenever moving b_i outside of the truck in orientation ℓ must wait until b_j has already been moved onto the stack (e.g., b_i sits deeper in the truck than b_j, and cannot be carried out of the truck in orientation ℓ without hitting b_j). The graph B_ℓ has an arc $b_i b_j$ whenever b_i being added to the stack in any orientation before b_j in orientation ℓ would cause the stack to topple, due to the weight distribution in b_i prohibiting b_j from being placed on top of it in orientation ℓ, irrespective of b_i's orientation. Thus, when a box b_i is assigned label ℓ in a solution's ordering, we simultaneously enforce that all its A_ℓ-successors appear earlier than it in the ordering of V in a solution, ensuring that no box prevents box b_i from being moved out of the truck, and all its B_ℓ-predecessors appear later than it in the ordering of V in a solution, ensuring that no box positioned under b_i in the stack causes the stack to topple.

While both the SM-RAMP problem and the TRUCK UNLOADING problem can be modeled as k-COMPATIBLE ORDERING problems ($k = 2$ for SM-RAMP, $k = 6$ for TRUCK UNLOADING), the value of k is fixed and is intrinsic to the definition of both problems. With that being said, TRUCK UNLOADING can be generalized by replacing boxes with polyhedrons that can be oriented in one of k ways. We have therefore decided to include the parameter k in the formulation of k-COMPATIBLE ORDERING not only to capture a wider array of problems, but also to analyze how the value of k affects the problem's tractability.

We also consider a related problem that is interesting in its own right. In the MULTIPLE MOVE ROBOTIC ARM MOTION PLANNING (MM-RAMP) problem, instead of restricting each arm to a single rotation, we seek a transformation where each arm can take on one of multiple predetermined orientations and can rotate multiple times. Moreover, the final desired orientations for all arms need not be vertical. The formal definition of the MM-RAMP problem can be found in the full version of the paper [4].

One then wonders whether we can generalize MM-RAMP as we generalized SM-RAMP. The answer is positive. The k-COMPATIBLE SET ARRANGEMENT problem generalizes MM-RAMP, and is presented and studied in [4].

1.1   Our Contributions

The main technical contributions of the paper are efficient algorithms to find k-compatible orderings, and hardness results for k-COMPATIBLE ORDERING and k-COMPATIBLE SET ARRANGEMENT. We first prove the following dichotomy:

Theorem 1. k-COMPATIBLE ORDERING *can be decided in polynomial time if* $k = 1$ *and is* NP-*complete for every* $k \geq 2$.

The theorem ensures, in particular, that if all the robotic arms are allowed to rotate only in one direction (clockwise for instance), then we can decide the problem in polynomial time. Since robotic arms are able to rotate in two directions in SM-RAMP, and since k-COMPATIBLE ORDERING is NP-complete for $k = 2$, if SM-RAMP can be decided in polynomial time, it must be due to some topological or geometric structure of the constraints.

In that direction, we also give some evidence that only very specific topologies can help. Namely, we prove that 2-COMPATIBLE ORDERING remains NP-complete even when the union of all the graphs in $\mathcal{G}$ is planar, or more formally:

Theorem 2. *k-COMPATIBLE ORDERING is NP-complete for $k \geq 2$ even when restricted to instances where the union of all graphs in $\mathcal{G}$ is planar and has degeneracy 3.*

The theorem above shows that the topology and having few constraints locally (bounded degeneracy) do not help to obtain polynomial-time algorithms. Intuitively, planarity of the union graph corresponds to a non-crossing (geometrically simple) arrangement of constraints, e.g., arms or boxes on a plane. The theorem shows that even this simplified topology does not make the problem tractable. The proof has a drawback: some of the graphs in $\mathcal{G}$ contain directed cycles. However, such cycles do not appear in k-COMPATIBLE ORDERING instances encoding SM-RAMP instances. One may then wonder what happens when we restrict graphs in $\mathcal{G}$ to be acyclic; we prove that the problem remains NP-complete.

Theorem 3. *k-COMPATIBLE ORDERING is NP-complete for $k \geq 2$ even when restricted to instances where none of the graphs in $\mathcal{G}$ contains a directed cycle.*

We then prove that if we strengthen the acyclicity assumption a bit, the problem becomes polynomial-time solvable. Namely, if there exists an integer x such that the union of the pair (A_x, B_x) forms an acyclic graph, then k-COMPATIBLE ORDERING can be decided in polynomial time; unfortunately, in the case of robotic arms, a cycle is possible in the union of the two directed graphs representing constraints on the same rotation direction.

Next, we investigate which parameters make the problem tractable. Our first result proves that the problem is polynomial-time solvable if the union of all (directed) graphs has bounded (undirected) treewidth.

***Theorem 4.** *k-COMPATIBLE ORDERING can be solved in time $\mathcal{O}(|V| \cdot (t^2 \cdot (t + 1)! \cdot k^{t+1}))$, where t is the treewidth of the undirected union of the graphs in $\mathcal{G}$.*

Since robotic arms only have constraints related to nearby robotic arms, one should be able to iteratively encode all the constraints by considering sets of robotic arms, rather than individual arms. These sets are such that two robotic arms belonging to the same set have the same relationship with respect to the robotic arms outside of the set. Moreover, the partitioning is recursive, and robotic arms are never partitioned into too many sets (arms in the same set may have different constraints involving each other). This idea is tightly coupled with the notions of modular decomposition and modular width, which we modify slightly to get them to work in the context of k-COMPATIBLE ORDERING, with the existence of labels and directions on the edges. We prove that if the union of the graphs has bounded labeled modular width (we refer the reader to [4] for a formal definition), then the problem can be decided in polynomial time. For example, instances of TRUCK UNLOADING modeled as 6-COMPATIBLE

ORDERING instances with bounded labeled modular width are those where the boxes are clustered within the truck, and where a box's balancing on the stack is only affected by other boxes in the same cluster.

*__Theorem 5.__ k-COMPATIBLE ORDERING can be solved in time $f(k + mw) \cdot |V|^{\mathcal{O}(1)}$, where f is a computable function and mw is the labeled modular width, assuming a labeled modular decomposition of the labeled union of $\mathcal{G}$ is given. Thus, the problem is solvable in polynomial time when $k + mw$ is a constant._

Unlike the k-COMPATIBLE ORDERING problem, which allows only one move per agent, the k-COMPATIBLE SET ARRANGEMENT problem allows multiple moves per agent and is therefore not surprisingly "harder" to solve. Our next result confirms this by showing that the problem is indeed PSPACE-complete.

*__Theorem 6.__ k-COMPATIBLE SET ARRANGEMENT is PSPACE-complete, even when restricted to instances where the union of the graphs in $\mathcal{G}$ is planar, has maximum degree 6, and has bounded bandwidth._

2   Preliminaries

Graph-theoretic terminology and notation used throughout the paper are as in standard references [1,11].

For $k \in \mathbb{N}$, we denote the set $\{1,\ldots,k\}$ by $[k]$. Given a graph G and a subset $X \subseteq V(G)$, we denote the induced subgraph of G on the vertices in X by $G[X]$. Let the _union_ of a set of graphs that have the same vertex set $G_1 = (V, E_1), G_2 = (V, E_2), \ldots, G_k = (V, E_k)$ be the graph $G' = (V, \{v_i v_j \mid v_i v_j \in \bigcup_{i=1}^{k} E_i\})$. If $G_1, \ldots, G_k$ are directed graphs, and we intend G' to be undirected, we call the operation _undirected union_ instead. For brevity, whenever we refer to the union of $\mathcal{G} = \{(A_1, B_1), \ldots, (A_k, B_k)\}$, we are referring to the union of the graphs in $\mathcal{G}$.

We direct the reader to the literature for definitions of the undirected graph parameters we rely on in the paper, namely degeneracy [11], treewidth [9], modular width [17] and bandwidth [7]. For digraphs, these parameters are understood as those of the underlying undirected graph.

3   Finding k-Compatible Orderings

Recall that in the definition of k-COMPATIBLE ORDERING, we are given constraints represented by pairs of graphs in $\mathcal{G}$ and we seek a pair $(\mathcal{S}, \mathcal{L})$, where $\mathcal{S} = s_1, \ldots, s_n$ is an ordering of V, $\mathcal{L} = \ell_1, \ldots, \ell_n$ is a sequence of labels, and vertex $s_i \in \mathcal{S}$ is assigned label $\ell_i \in [k]$ such that s_i satisfies the following two constraints: (i) s_i is a sink in $A_{\ell_i}[s_i, \ldots, s_n]$, and (ii) s_i is a source in $B_{\ell_i}[s_1, \ldots, s_i]$.

When s_i satisfies the first constraint, we say that s_i satisfies its _sink constraint_ and when it satisfies the second, we say that it satisfies its _source constraint_.

At times, we will refer to the sequence of labels $\mathcal{L} = \ell_1, \ldots, \ell_n$ as a *labeling*, and we say pair $(\mathcal{S}, \mathcal{L})$ is a *labeled ordering*. When we want to highlight that $\mathcal{S}$ together with the labeling $\mathcal{L}$ satisfies all the constraints, we say that $\mathcal{S}$ is *valid* with labeling $\mathcal{L}$, or that $(\mathcal{S}, \mathcal{L})$ is a *valid labeled ordering*. Given instance $(V, \{(A_1, B_1), \ldots, (A_k, B_k)\})$ of k-COMPATIBLE ORDERING, $(V', \{(A_1[V'], B_1[V']), \ldots, (A_k[V'], B_k[V'])\})$ is an *induced instance* of it on $V' \subseteq V$.

The source and sink constraints lead to the following characterizations of any solution to the k-COMPATIBLE ORDERING problem, for any label ℓ and vertices v_i and v_j.

Observation 1. *If $v_i v_j$ is an arc in A_ℓ (resp. B_ℓ), and v_i (resp. v_j) is assigned label ℓ, then v_j must precede v_i in the ordering.*

Observation 2. *For a set of vertices $V' \subseteq V$ forming a directed cycle in a graph A_ℓ or B_ℓ, the vertices in V' cannot all be labeled ℓ.*

Observation 3. *If A_ℓ contains the arc $v_i v_j$, and B_ℓ contains the arc $v_j v_i$, then v_i cannot be labeled ℓ.*

4   Positive Results

It may not be evident why encoding an SM-RAMP instance as a 2-COMPATIBLE ORDERING instance uses two directed graphs per label, and one may be inclined to replace each label's graph pair with their union, treating the arcs in the resulting graphs as source (or, symmetrically, sink) constraints. In fact, one can prove that if substituting the A graph and the B graph of every pair with their union does not preserve solutions, it is because one of the resulting graphs contains a cycle (we refer the reader to [4] for a concrete example). In the case of 1-COMPATIBLE ORDERING instances, checking for cycles in the union of the graphs is found to be both necessary and sufficient to determine yes-instances.

Lemma 2. *1-COMPATIBLE ORDERING can be solved in time $\mathcal{O}(n + m)$, where m is the number of edges in the union graph.*

Proof. We prove that an instance $(V, \{(A_1, B_1)\})$ of 1-COMPATIBLE ORDERING is a yes-instance if and only if the union of A_1 and B_1 is a directed acyclic graph.

For the forward direction, towards a contradiction, suppose that the instance is positive, and that the union of A_1 and B_1 contains at least one cycle. Consider an arbitrary solution $(\mathcal{S} = s_1, \ldots, s_n, \mathcal{L})$, where $\mathcal{L}$ trivially assigns label 1 to all vertices. Since the union of A_1 and B_1 contains at least one cycle, there will exist a vertex s_x in $\mathcal{S}$ that has an out-neighbor s_w in $s_{x+1}, \ldots, s_n$; otherwise, the union of A_1 and B_1 would have a topological ordering and would therefore be acyclic, a contradiction. Whether the arc $s_x s_w$ is in A_1 or B_1, it forces the solution to include s_w before s_x, which contradicts $(\mathcal{S}, \mathcal{L})$ being a solution.

As for the backward direction, let $\mathcal{L}$ be the labeling that assigns label 1 to all vertices. We claim that we can find an ordering $\mathcal{S}$ of the vertices such that $(\mathcal{S}, \mathcal{L})$ is a valid labeled ordering. We first compute a topological ordering for the union of A_1 and B_1 in time $\mathcal{O}(n+m)$ (m being the number of edges in the union of A_1 and B_1). Let $\mathcal{S} = s_1, \ldots, s_n$ be the topological ordering in reverse order. To show that $(\mathcal{S}, \mathcal{L})$ is a solution, suppose, for the sake of contradiction, that there exists a vertex s_x in $\mathcal{S}$ that is assigned label 1 and that violates either the sink or the source constraint. If s_x is not a sink in $A_1[s_x, \ldots, s_n]$, then there exists an out-neighbor of s_x in $s_{x+1}, \ldots, s_n$. However, all the out-neighbors of s_x in A_i are in $s_1, \ldots, s_{x-1}$, given that $\mathcal{S}$ is a reversed topological ordering. An analogous argument ensures that s_x is a source in $B_1[s_1, \ldots, s_x]$. Since the choice of s_x is arbitrary, there is no vertex for which a source or sink constraint is violated, and $(\mathcal{S}, \mathcal{L})$ is a valid labeled ordering, as needed. $\qquad\square$

A generalization of Lemma 2 can be used to characterize trivial yes-instances of k-COMPATIBLE ORDERING as follows:

Corollary 1. *Given an instance $(V, \mathcal{G})$ of k-COMPATIBLE ORDERING, if there exists a pair of directed graphs (A_i, B_i) in $\mathcal{G}$ whose union is a directed acyclic graph, $(V, \mathcal{G})$ is a yes-instance of k-COMPATIBLE ORDERING.*

An algorithm that can be used to check whether a pair $(\mathcal{S}, \mathcal{L})$ solves an instance of k-COMPATIBLE ORDERING follows almost immediately:

Observation 4. *A pair $(\mathcal{S} = s_1, \ldots, s_n, \mathcal{L} = \ell_1, \ldots, \ell_n)$ solves a k-COMPATIBLE ORDERING instance $(V, \mathcal{G})$ if the graph $(V, \{s_i s_j \mid s_i s_j \in A_{\ell_i} \text{ or } s_i s_j \in B_{\ell_j}\})$ is a directed acyclic graph.*

$\star$**Theorem 4.** *k-COMPATIBLE ORDERING can be solved in time $\mathcal{O}(|V| \cdot (t^2 \cdot (t+1)! \cdot k^{t+1}))$, where t is the treewidth of the undirected union of the graphs in $\mathcal{G}$.*

Proof (Sketch). Let $(T, \mathcal{X})$ be a nice tree decomposition of the union graph H of width t. For each bag $X \in \mathcal{X}$, let $\mathrm{ord}(X)$ be an ordering of X and $\lambda : X \to [k]$ a labeling. A dynamic programming (DP) state for a bag X stores whether there exists a partial ordering π of all vertices introduced in the subtree rooted at X such that: (i) π restricted to X equals $\mathrm{ord}(X)$, (ii) each vertex of X is assigned the label λ, (iii) all precedence constraints induced by incoming A_j-arcs from "future" vertices and outgoing B_j-arcs to "past" vertices are respected.

A key observation (proved in the full version) is that all constraints relevant to extending π outside the subtree are already visible inside X: an A_j-arc (u, v) can be violated only if u and v appear together in some bag, and similarly for B_j-arcs. Thus each DP transition needs only to verify compatibility of local orderings and labels inside the bag, and that the relative positions of introduced and forgotten vertices respect all A_j/B_j edges whose endpoints lie in the bag.

For a bag of size at most $t+1$, the number of possible orderings and labelings is at most $(t+1)! \cdot k^{t+1}$. Each transition (introduce, forget, join) checks consistency of constraints encoded by at most $\mathcal{O}(t^2)$ arcs, which can be done in $\mathcal{O}(t^2)$ time for each candidate state.

Since a nice decomposition has $\mathcal{O}(|V|)$ bags, the total running time is: $\mathcal{O}(|V| \cdot (t+1)! \cdot k^{t+1} \cdot t^2)$, which is FPT parameterized by $t + k$. A full description and correctness proof appear in [4]. $\qquad\square$

***Theorem 5.** *k-COMPATIBLE ORDERING can be solved in time $f(k + mw) \cdot |V|^{\mathcal{O}(1)}$, where f is a computable function and mw is the labeled modular width, assuming a labeled modular decomposition of the labeled union of $\mathcal{G}$ is given. Thus, the problem is solvable in polynomial time when $k + mw$ is a constant.*

Proof (Sketch). We begin with brief intuition for the labeled modular decomposition. A *module* in a graph is a set of vertices that behave identically toward the rest of the graph, in the sense that every vertex outside the module either has edges to all of them or to none of them. In our setting we use the *labeled* version: two vertices are indistinguishable not only with respect to adjacencies in the union graph H but also with respect to which A_j- and B_j-edges they participate in. Intuitively, each module is a "super-vertex" whose internal structure can be summarized once its label is chosen, and the entire graph is recursively built by composing such super-vertices via series, parallel, or substitution operations. The labeled modular width mw bounds the size of every prime node in this decomposition.

Let $\mathcal{M}$ be the labeled modular decomposition tree of H. For each node N of $\mathcal{M}$ with children $C_1, \ldots, C_r$ ($r \leq mw$), the DP stores, for every possible labeling of the children, whether N admits a k-compatible ordering extending the children's summaries. A summary for a child records only the label chosen for that module and the feasible ranges of positions that vertices of that label may occupy inside N.

In a prime node, once the labels of the $r \leq mw$ children are fixed, the precedence constraints between modules (arising from A_j- and B_j-arcs) determine whether there exists a consistent ordering; this reduces to testing whether a small precedence graph on at most mw vertices admits a topological ordering.

For a substitution node N, the DP considers all k^r ways to assign labels to its r children. For each assignment, it combines the children's summaries and checks whether the inter-module constraints arising from A_j-/B_j-arcs can be satisfied. Since each module has only $g(k)$ possible summaries for some function g, combining them requires considering only $g(k) \cdot h(mw)$ possibilities for appropriate functions g and h.

Because the modular decomposition has $|V|^{\mathcal{O}(1)}$ nodes and each node explores at most $f(mw, k)$ combinations for some function f depending only on mw and k, we obtain fixed-parameter tractability for parameter $mw + k$. Full details and correctness proof appear in [4]. $\qquad\square$

5   Hardness Results

The instances of k-COMPATIBLE ORDERING that we know how to efficiently solve as a result of Corollary 1 exhibit acyclicity within their input. In this section, we restrict the appearance of cycles to be representative of cyclical constraints that naturally arise in the context of the RAMP problems. We prove that k-COMPATIBLE ORDERING is hard on each of the considered cycle structures.

The hardness results covered in this section stem from a restricted version of the 3-SAT problem, namely the PLANAR 4-BOUNDED 3-SAT problem. Let

φ be a 3-SAT instance, and let $G(\varphi)$ be a graph representation of φ, defined as follows: for each variable and clause in φ, we add a vertex to the vertex set of $G(\varphi)$. As for the edge set of $G(\varphi)$, we have an edge between a clause vertex and a variable vertex if and only if there exists a literal of that variable in the clause. If $G(\varphi)$ is planar, and each variable occurs at most 4 times in φ, irrespective of the polarity of its literals, we say that φ is a PLANAR 4-BOUNDED 3-SAT instance. To prove that k-COMPATIBLE ORDERING is NP-hard, we reduce from PLANAR 4-BOUNDED 3-SAT, which is known to be NP-hard, even on instances where each clause contains exactly 3 literals [23].

Theorem 2. k-COMPATIBLE ORDERING *is* NP-*complete for* $k \geq 2$ *even when restricted to instances where the union of all graphs in* $\mathcal{G}$ *is planar and has degeneracy* 3.

Proof (Sketch). We show that 2-COMPATIBLE ORDERING is NP-complete under the stated conditions; this suffices since 2-COMPATIBLE ORDERING is a special case of k-COMPATIBLE ORDERING.

Membership in NP is straightforward: a labeled ordering $(\mathcal{S}, \mathcal{L})$ serves as a certificate, and constraints can be verified in linear time (Observation 4).

Let $\varphi = C_1 \wedge \cdots \wedge C_p$ be a PLANAR 4-BOUNDED 3-SAT instance on variables $X = \{x_1, \ldots, x_q\}$. We construct a 2-COMPATIBLE ORDERING instance as follows:

Vertices in V:

- For each clause C_i, add three *clause vertices* c_i^1, c_i^2, c_i^3 to V, representing the literals of C_i.
- For each variable $x_j \in X$, add a *variable vertex* to V. Thus, $|V| = 3p + q$.
- Assigning label 1 (resp. 2) to a variable vertex corresponds to setting that variable to *true* (resp. *false*). Assigning label 1 to a clause vertex means its literal is *true*. At least one literal per clause must be labeled 1, in agreement with the label of its associated variable.

Arcs in A_1, B_1, A_2, B_2:

- **Clause cycles in** A_2: For each clause C_i, create a directed 3-cycle among c_i^1, c_i^2, c_i^3 in A_2; at least one of the three will be labeled 1 (Observation 2).
- **Positive clause-variable arcs**: If c^+ is a positive clause vertex for variable x, add c^+x to A_1 and xc^+ to A_2. These arcs ensure that c^+ can be labeled 1 only if x is labeled 1 and appears before c^+ in the ordering (Observation 1).
- **Negative clause-variable arcs**: If c^- is a negative clause vertex for variable x, add c^-x and xc^- in B_1. These arcs ensure c^- can be labeled 1 only if x is labeled 2 and appears after c^- in the ordering (Observations 1 and 2).
- There are no arcs in B_2, and variable vertices form an independent set in all four graphs.

We claim $(V, \mathcal{G} = \{(A_1, B_1), (A_2, B_2)\})$ is a yes-instance of 2-COMPATIBLE ORDERING if and only if φ is satisfiable.

($\Rightarrow$) If φ has a satisfying assignment, label each variable vertex 1 (resp. 2) if it is *true* (resp. *false*) in that assignment. For each clause C_i, pick one *true* literal c_i^a to label 1; any remaining clause vertices of C_i get label 2. Arrange:

$$\underbrace{\text{true variables}}_{\text{label 1}}, \underbrace{\text{satisfied clause vertices}}_{\text{label 1}}, \underbrace{\text{remaining vertices}}_{\text{label 2}}.$$

All arc constraints are satisfied: (1) a clause vertex with label 1 appears after the variable vertex it neighbors in A_1 if the variable is set to true, (2) a clause vertex with label 1 appears before the variable vertex that neighbors it in B_1 if the variable is set to false, and (3) every clause has at least one literal labeled 1. ($\Leftarrow$) Conversely, if there is a valid labeled ordering in $(V, \mathcal{G})$, then each clause cycle in A_2 forces at least one of c_i^1, c_i^2, c_i^3 to be labeled 1. By inspecting the arcs between clause vertices and their associated variables, one sees that a positive (resp. negative) clause vertex labeled 1 requires the corresponding variable to be labeled 1 (resp. 2) and come earlier (resp. later) in the ordering. This yields a truth assignment satisfying all clauses.

Since φ is planar 4-bounded, $G(\varphi)$ has a planar embedding. Replacing each clause vertex with three closely spaced points (for c_i^1, c_i^2, c_i^3) introduces no crossings, so the union remains planar. For degeneracy, each clause vertex has degree at most 3 in the underlying undirected graph without parallel edges (two edges in the cycle and one edge to the corresponding variable). Variable vertices form independent sets, implying any induced subgraph has a vertex of degree at most 3. Hence the union of A_1, A_2, B_1, B_2 has degeneracy at most 3.

Thus, 2-COMPATIBLE ORDERING (and hence k-COMPATIBLE ORDERING) is NP-complete under planarity and degeneracy 3. $\qquad\square$

The construction covered in Theorem 2 goes against the spirit of the problem, because of the existence of cycles (Observation 2). One might therefore wonder whether forbidding directed cycles within the directed graphs makes the problem easier to solve. Our next result shows that the answer is negative.

Theorem 3. k-*COMPATIBLE ORDERING is* NP-*complete for $k \geq 2$ even when restricted to instances where none of the graphs in $\mathcal{G}$ contains a directed cycle.*

Proof (Sketch). Membership of the problem in NP is immediate. We show hardness via a reduction from PLANAR 4-BOUNDED 3-SAT, modifying the construction used in the proof of Theorem 2 to eliminate directed cycles in A_2 and B_1.

Let $\varphi = C_1 \wedge \cdots \wedge C_p$ be a PLANAR 4-BOUNDED 3-SAT instance on variables $X = \{x_1, \ldots, x_q\}$. Starting with the vertex set V in Theorem 2 (which includes *variable vertices* and *clause vertices*), we add:

- A *dummy clause vertex* d_i for each clause C_i,
- A *dummy negative clause vertex* d_i^a for each negative literal c_i^a, and
- A single *global vertex* g.

Hence, $|V| = 4p + q + o + 1$, where o is the total number of negative literals in φ.

We then define arcs in A_1, B_1, A_2, B_2, while preserving the logical structure of the previous reduction:

- **Clause arcs in A_2:** For each clause C_i, label its three clause vertices c_i^1, c_i^2, c_i^3. Add $c_i^1 c_i^2, c_i^2 c_i^3, c_i^3 d_i$ to A_2, and $d_i c_i^1$ to A_1. These arcs replace the cycle in A_2 in the construction of Theorem 2 with a path ending at d_i.
- **Positive clause-variable arcs:** As in Theorem 2, for a positive literal c_i^+ of variable x, add $c_i^+ x$ to A_1 and $x c_i^+$ to A_2.
- **Negative clause-variable arcs:** For each negative literal c_i^a with variable x, add $d_i^a c_i^a, c_i^a x$ to B_1 and $x d_i^a$ to B_2. The dummy vertex d_i^a ensures no cycle arises in B_1.
- **Arcs involving the global vertex g:** For each dummy clause vertex d_i, add $d_i g$ to A_2 and $g d_i$ to B_2. For each dummy negative clause vertex d_i^a, add $d_i^a g$ to A_1 and $g d_i^a$ to B_1. These arcs guarantee dummy vertices obtain particular labels (Observation 3) and prevent cycles involving g.

We claim that $(V, \mathcal{G} = \{(A_1, B_1), (A_2, B_2)\})$ is a yes-instance of 2-COMPATIBLE ORDERING if and only if φ is satisfiable.
($\Rightarrow$) If φ is satisfiable, we begin with the labeled ordering from Theorem 2 that arranges variable vertices and clause vertices consistently. We then insert:

- Each d_i with label 1 immediately after (or before) the associated clause vertices $c_i^1, \ldots, c_i^3$ to satisfy the constraints from the new arcs in A_1 and A_2.
- Each d_i^a with label 2 immediately after its negative literal c_i^a or before its variable x to satisfy the constraints from the new arcs in B_1 and B_2.
- The global vertex g, labeled 1, at the end of the ordering.

These insertions do not create cycles and maintain the constraints of the new vertices and those in Theorem 2's construction, so the extended ordering is valid. ($\Leftarrow$) Conversely, any valid labeled ordering must assign the label 1 to at least one clause vertex in each C_i, as constraints of d_i would be otherwise violated. A satisfying assignment for φ can be derived accordingly, as in Theorem 2.

Hence 2-COMPATIBLE ORDERING (and therefore k-COMPATIBLE ORDERING) remains NP-complete when all graphs are acyclic. $\qquad\square$

6   Conclusion and Open Problems

In this paper, we introduced two generalizations of the geometric RAMP problems, and we settled their complexity under constraints related to acyclicity, planarity, maximum degree and degeneracy. We also presented two efficient algorithms that solve k-COMPATIBLE ORDERING instances with bounded treewidth and labeled modular width respectively, which we believe encapsulate most instances encoding SM-RAMP instances.

We propose several open problems. The complexity of SM-RAMP itself remains unresolved. It would be valuable to determine whether this specific problem admits a polynomial-time solution or if it is NP-complete. One might also explore what happens when the problem is relaxed or generalized further. For instance, what is the complexity of the problem if we remove the assumption that all robotic arms have the same length?

Another intriguing direction concerns the complexity of k-COMPATIBLE ORDERING under geometric constraints, as opposed to purely topological ones like planarity. For example, what is the complexity of k-COMPATIBLE ORDERING when the graphs in $\mathcal{G}$ are unit disk graphs, which arise naturally in scenarios involving spatial or wireless network constraints?

References

1. Bollobás, B.: Modern Graph Theory. GTM, vol. 184. Springer, New York (1998). https://doi.org/10.1007/978-1-4612-0619-4
2. Bousquet, N., Hommelsheim, F., Kobayashi, Y., Mühlenthaler, M., Suzuki, A.: Feedback vertex set reconfiguration in planar graphs. Theor. Comput. Sci. **979**, 114,188 (2023)
3. Bousquet, N., Mouawad, A.E., Nishimura, N., Siebertz, S.: A survey on the parameterized complexity of reconfiguration problems. Comput. Sci. Rev. **53**, 100,663 (2024). https://doi.org/10.1016/J.COSREV.2024.100663
4. Bousquet, N., Sabeh, R.E., Mouawad, A.E., Nishimura, N.: On the complexity of constrained reconfiguration and motion planning (2025). https://arxiv.org/abs/2508.13032
5. Brianski, M., Felsner, S., Hodor, J., Micek, P.: Reconfiguring independent sets on interval graphs. In: Bonchi, F., Puglisi, S.J. (eds.) 46th International Symposium on Mathematical Foundations of Computer Science, MFCS 2021, Tallinn, Estonia, 23–27 August 2021. LIPIcs, vol. 202, pp. 23:1–23:14. Schloss Dagstuhl - Leibniz-Zentrum für Informatik (2021). https://doi.org/10.4230/LIPICS.MFCS.2021.23
6. Buchin, K., Hagedoorn, M., Kostitsyna, I., van Mulken, M.: Dots & boxes is PSPACE-complete. In: Bonchi, F., Puglisi, S.J. (eds.) 46th International Symposium on Mathematical Foundations of Computer Science, MFCS 2021, Tallinn, Estonia, 23–27 August 2021. LIPIcs, vol. 202, pp. 25:1–25:18. Schloss Dagstuhl - Leibniz-Zentrum für Informatik (2021). https://doi.org/10.4230/LIPICS.MFCS.2021.25
7. Chinn, P.Z., Chvátalová, J., Dewdney, A.K., Gibbs, N.E.: The bandwidth problem for graphs and matrices–a survey. J. Graph Theory **6**(3), 223–254 (1982)
8. Ćurković, P., Jerbić, B.: Dual-arm robot motion planning based on cooperative coevolution. In: Camarinha-Matos, L.M., Pereira, P., Ribeiro, L. (eds.) DoCEIS 2010. IAICT, vol. 314, pp. 169–178. Springer, Heidelberg (2010). https://doi.org/10.1007/978-3-642-11628-5_18
9. Cygan, M., et al.: Parameterized Algorithms, vol. 5. Springer, Cham (2015)
10. Demaine, E.D., Hohenberger, S., Liben-Nowell, D.: Tetris is hard, even to approximate. In: Warnow, T., Zhu, B. (eds.) COCOON 2003. LNCS, vol. 2697, pp. 351–363. Springer, Heidelberg (2003). https://doi.org/10.1007/3-540-45071-8_36
11. Diestel, R.: Graph Theory, vol. 173. Springer, Cham (2025)
12. Hearn, R.A., Demaine, E.D.: PSPACE-completeness of sliding-block puzzles and other problems through the nondeterministic constraint logic model of computation. Theoret. Comput. Sci. **343**(1–2), 72–96 (2005)
13. Hearn, R.A., Demaine, E.D.: Games, Puzzles, and Computation. CRC Press (2009)
14. van den Heuvel, J.: The Complexity of change, p. 409. Part of London Mathematical Society Lecture Note Series (2013)

15. Hue, L.T., Anh, N.P.T.: Planning the optimal trajectory for a dual-arm robot system using a genetic algorithm considering the controller. In: 2022 International Conference on Advanced Technologies for Communications (ATC), pp. 292–297 (2022). https://doi.org/10.1109/ATC55345.2022.9942975
16. Ito, T., et al.: On the complexity of reconfiguration problems. Theoret. Comput. Sci. **412**(12–14), 1054–1065 (2011)
17. McConnell, R.M., Spinrad, J.P.: Modular decomposition and transitive orientation. Discret. Math. **201**(1–3), 189–241 (1999)
18. Mouawad, A.E., Nishimura, N., Raman, V., Wrochna, M.: Reconfiguration over tree decompositions. In: Cygan, M., Heggernes, P. (eds.) IPEC 2014. LNCS, vol. 8894, pp. 246–257. Springer, Cham (2014). https://doi.org/10.1007/978-3-319-13524-3_21
19. Nishimura, N.: Introduction to reconfiguration. Algorithms **11**(4), 52 (2018). https://doi.org/10.3390/a11040052
20. Ohsaka, N.: Gap preserving reductions between reconfiguration problems. In: Berenbrink, P., Bouyer, P., Dawar, A., Kanté, M.M. (eds.) 40th International Symposium on Theoretical Aspects of Computer Science, STACS 2023, Hamburg, Germany, 7–9 March 2023. LIPIcs, vol. 254, pp. 49:1–49:18. Schloss Dagstuhl - Leibniz-Zentrum für Informatik (2023). https://doi.org/10.4230/LIPICS.STACS.2023.49
21. Ohsaka, N.: Gap amplification for reconfiguration problems. In: Proceedings of the 2024 Annual ACM-SIAM Symposium on Discrete Algorithms, SODA 2024, pp. 1345–1366. SIAM (2024)
22. Salehian, S.S.M., Figueroa, N., Billard, A.: A unified framework for coordinated multi-arm motion planning. Int. J. Robot. Res. **37**(10), 1205–1232 (2018). https://doi.org/10.1177/0278364918765952
23. Tippenhauer, S., Muzler, W.: On planar 3-SAT and its variants. Fachbereich Mathematik und Informatik der Freien Universitat Berlin (2016)

Efficient Solutions to Variants of Inversion Problems of Range Minimum Queries

Souta Kobayashi[1]([✉])[iD], Dominik Köppl[2][iD], Ryo Yoshinaka[1][iD], and Ayumi Shinohara[1][iD]

[1] Tohoku University, Sendai, Japan
`kobayashi.souta.t8@dc.tohoku.ac.jp`, `{ryoshinaka,ayumis}@tohoku.ac.jp`
[2] University of Yamanashi, Kofu, Japan
`dkppl@yamanashi.ac.jp`

Abstract. Given a set of results from range minimum queries (RMQs), our task is to construct a sequence that is consistent with the results of the queries. We study two types of RMQs: a *value-based* RMQ returns the minimum value and an *index-based* RMQ returns the index of the minimum. While the value-based version has been discussed informally in the context of competitive programming, the index-based version appears to be unexplored. In this paper, we provide a survey and unified analysis of the value-based version, and we propose efficient algorithms for the index-based version. These include algorithms for computing the lexicographically smallest consistent sequence and permutation, as well as an enumeration algorithm that outputs all consistent permutations with constant delay.

1 Problem Definition

Given an integer array $A[1 : n]$ of n integers, a *range minimum query* (RMQ) asks, for a given interval $[l, r]$ to report the smallest entry in the sub-array $A[l : r]$, i.e., $\min_{k \in [l,r]} A[k]$. Literature gives two types of data structures that deal with this problem differently: encoding and indexing data structures. Encoding data structures [3,5,7,11], mostly based on the equivalence with the lowest common ancestor problem, need A only during their construction – at query time they can report the answer directly. The other type, the indexing data structures [9,10,15], needs to consult A even during query time. While the latter type seems to be weaker, it allows squeezing the constructed data structure in only a few bits of space. Indeed, indexing data structures that report only the index i of the minimum need $2n + o(n)$ bits of space, which was reached by Fischer and Heun [10]. In particular, when only the index of the minimum is needed, indexing data structures for RMQs are the way to go. Therefore, it has become natural to think about RMQs as either of two types: returning the value of the minimum or the index at which the minimum is stored. We call the former type *value-based* and the latter *index-based*.

Now suppose that instead of having A or an RMQ data structure on A, we are given some answers to either value-based or index-based RMQs and try to

© The Author(s), under exclusive license to Springer Nature Switzerland AG 2026
J. Kozik and A. Wolff (Eds.): SOFSEM 2026, LNCS 16448, pp. 16–30, 2026.
https://doi.org/10.1007/978-3-032-17801-5_2

reverse-engineer A. We should find an array $A'[1:n]$ that is consistent with the original A, i.e., answering each of the RMQs on A or A' leads to the same results. While there can be infinitely many arrays satisfying the requirements, this paper focuses on reconstructing specific types of arrays as representatives: namely, lexicographically smallest arrays and permutations. Those variants of the value-based version of this problem have occasionally appeared in competitive programming contests [18,19,21,23,25], with several informal explanatory articles available online [20,22,24,26]. In contrast, to the best of the authors' knowledge, the index-based version has not been investigated.

Regarding the reverse engineering of index-based RMQs, we present linear-time algorithms for computing the smallest positive integer sequences and permutations, together with an algorithm that enumerates all consistent permutations with constant delay, in Sect. 3. Since index-based RMQs do not retain the actual array values, consistent permutations can be viewed as representing all consistent arrays without duplicated values. The value-based counterparts of these problems are discussed in Sect. 4. After surveying the existing results, we present a new linear-time algorithm for computing the smallest consistent permutation.

2   Preliminaries

Let $\mathbb{N}$ and $\mathbb{N}_+$ denote the sets of non-negative integers and positive integers, respectively. The interval of two non-negative integers l and r is denoted by $[l,r] = \{\, i \in \mathbb{N} \mid l \leq i \leq r \,\}$. For a sequence $A = (a_1, \ldots, a_n)$ of integers, $A[i] = a_i$ and $A[i:j] = (a_i, \ldots, a_j)$ if $i \leq j$. In the case where $i > j$, $A[i:j]$ is the empty sequence. For a set J of indices of A, we write $A(J) = \{\, A[i] \mid i \in J \,\}$. Thus, $A[i:j]$ is a sequence and $A([i,j])$ is a set. Since this paper is concerned with sequences of positive integers only, an *(integer) sequence* refers to a sequence of positive integers, unless otherwise noted. A sequence A of length n is *(lexicographically) smaller* than another B of the same length if there is $i \in [1,n]$ such that $A[1:i-1] = B[1:i-1]$ and $A[i] < B[i]$.

2.1   Useful Queries and Data Structures

Our algorithms need to solve two types of problems, for which we use data structures from literature.

Interval Union-Find Problem

Consider manipulating an interval partition $\mathcal{I}$ of the set $[1,n]$. Initially, $\mathcal{I}$ is the discrete partition $\mathcal{I} = \{\, [i,i] \mid i \in [1,n] \,\}$. We allow two kinds of operations:

1. $\mathcal{I}.link(i)$ with $i \in [1,n]$ merges the intervals $[l,i], [i+1,r] \in \mathcal{I}$ into $[l,r]$ if such intervals exist in $\mathcal{I}$;
2. $\mathcal{I}.find(i)$ with $i \in [1,n]$ returns $r \in [1,n]$ such that $i \in [l,r] \in \mathcal{I}$.

The above is a specialization of the union-find problem commonly studied in the literature. Our specialization allows a more efficient solution.

Theorem 1. ([12,16]). *One can perform m operations of link and find over $[1,n]$ given in an online manner in $O(n + m \cdot \alpha(m + n, n))$ time, where α is the inverse of Ackermann's function. Moreover, if the machine word size is $\Omega(\log n)$ bits, the factor $\alpha(m + n, n)$ can be replaced by a constant.*

Here, by an *online manner*, we mean that we perform each operation without knowing the succeeding operations.

OFFLINE MINIMUM PROBLEM

Consider manipulating a semi-dynamic set $S \subseteq [1,n]$ of integers. Initially, S is empty ($S = \emptyset$). We allow two kinds of operations:

1. $S.insert(i)$ with $i \in [1,n]$ updates the set S to $S \cup \{i\}$;
2. $S.extract()$ returns and deletes the minimum of S, unless S is empty.

Theorem 2. ([1,12]). *One can perform (at most $2n$) operations of insert and extract over $[1,n]$ given in an offline manner in $O(n \cdot \alpha(n, n))$ time if $insert(i)$ is called at most once for each $i \in [1,n]$. Moreover, if the machine word size is $\Omega(\log n)$ bits, the factor $\alpha(n, n)$ can be replaced by a constant.*

Here, by an *offline manner*, we mean that we are given the sequence of operations before processing the first operation.

3 Index-Based RMQs

This section is concerned with integer sequences consistent with given index-based RMQs and their answers. An *index-based RMQ with answer (iRMQA)* is a triple of positive integers (l, r, k) such that $l, r, k \in [1, n]$ with $l \le k \le r$ for some fixed n. We formally define our problem as follows.

IRMQA REVERSAL PROBLEM

Input: a finite set $Q \subseteq \{\, (l, r, k) \in \mathbb{N}_+^3 \mid 1 \le l \le k \le r \le n \,\}$ of iRMQAs;
Output: a sequence $A[1 : n]$ such that $\min\{\, i \in [l, r] \mid A[i] = \min A[l : r] \,\} = k$ for all $(l, r, k) \in Q$ or a report that no such sequence exists.

In this model, when there are multiple indices that achieve the minimum value within the queried range, the query result is the smallest such index.

Throughout this section, we fix the length n of the underlying sequences and a finite set Q of iRMQAs, and let $q = |Q|$.

Example 1. Given $n = 9$ and $Q_1 = \{(1, 2, 1), (1, 3, 3), (4, 7, 5), (6, 8, 8), (5, 9, 9)\}$, one of the solutions is $B = (7, 7, 5, 9, 5, 5, 5, 3, 1)$. In fact,

$$\min\{\, i \in [1, 2] \mid B[i] = \min B[1 : 2] = 7 \,\} = \min\{1, 2\} = 1,$$
$$\min\{\, i \in [1, 3] \mid B[i] = \min B[1 : 3] = 5 \,\} = \min\{3\} = 3,$$
$$\min\{\, i \in [4, 7] \mid B[i] = \min B[4 : 7] = 5 \,\} = \min\{5, 6, 7\} = 5,$$
$$\min\{\, i \in [6, 8] \mid B[i] = \min B[6 : 8] = 3 \,\} = \min\{8\} = 8,$$
$$\min\{\, i \in [5, 9] \mid B[i] = \min B[5 : 9] = 1 \,\} = \min\{9\} = 9.$$

If two iRMQAs (l, r, k) and (l', r', k) with the same k are given, they can be merged into $(\min(l, l'), \max(r, r'), k)$ without changing the admissible consistent sequences. We call an iRMQA set *canonical* if it has just one iRMQA of the form (l_k, r_k, k) for each k. One can convert Q into canonical form by letting

$$l_k = \min(\{k\} \cup \{\, l \mid (l, r, k) \in Q \text{ for some } r \,\}),$$
$$r_k = \max(\{k\} \cup \{\, r \mid (l, r, k) \in Q \text{ for some } l \,\}).$$

We store those values in two arrays of length n by a preprocessing taking $O(q+n)$ time. Hereafter, we assume that the given iRMQA set is canonical.

3.1 Permutations Consistent with iRMQAs

This subsection is concerned with the problems of enumerating all permutations and finding the smallest permutation consistent with the given iRMQA set Q.

We represent the conditions defined by the iRMQAs of Q with a directed graph $G_Q = ([1, n], E_Q)$, where

$$E_Q = \{\, (i, j) \mid i \neq j \text{ and } l_i \leq j \leq r_i \text{ for } (l_i, r_i, i) \in Q \,\}.$$

Proposition 1. *A permutation P is consistent with an iRMQA set Q if and only if its inverse permutation P^{-1} is a topological ordering of the graph G_Q.*

Thus, finding a permutation consistent with the given iRMQAs is equivalent to finding a topological sort on the graph G_Q. In particular, Q has a consistent permutation if and only if G_Q is acyclic. However, constructing the graph G_Q can take quadratic time. Instead, we compute a subgraph G'_Q, which has less than $2n$ edges and retains the same admissible topological orderings as G_Q. We first observe some properties of the graph G_Q. Let $\widehat{E}_Q$ be the transitive closure of E_Q; that is, $(i, j) \in \widehat{E}_Q$ if and only if there is a path from i to j in G_Q.

Lemma 1. *If $i < j < k$ and $(i, k) \in E_Q$, then $(i, j) \in E_Q$. As a corollary, if $i < j < k$ and $(i, k) \in \widehat{E}_Q$, then $(i, j) \in \widehat{E}_Q$. The same holds when $i > j > k$.*

We call an edge (i, j) *leftward* if $i > j$ and *rightward* if $i < j$. Lemma 1 implies that if we have $(i, k), (j, k) \in E_Q$ with $i < j < k$, deletion of (i, k) does not affect the reachability from i to k since $(i, j) \subset E_Q$. Therefore, it is enough to keep the closest predecessors $par_{\mathsf{L}}[k]$ and $par_{\mathsf{R}}[k]$ on the left and the right, respectively, for each vertex k. We call $par_{\mathsf{L}}[k]$ and $par_{\mathsf{R}}[k]$, if they exist, the *left* and the *right* parents of k, respectively. Formally, define $G'_Q = ([1, n], E'_Q)$ whose edge set $E'_Q \subseteq E_Q$ keeps only edges of the forms $(par_{\mathsf{L}}[k], k)$ and $(par_{\mathsf{R}}[k], k)$ where

$$par_{\mathsf{L}}[k] = \max(\{\, i \mid i < k \text{ and } (i, k) \in E_Q \,\} \cup \{0\}),$$
$$par_{\mathsf{R}}[k] = \min(\{\, i \mid k < i \text{ and } (i, k) \in E_Q \,\} \cup \{n + 1\}).$$

If k has no left (resp. right) predecessors, $par_{\mathsf{L}}[k]$ (resp. $par_{\mathsf{R}}[k]$) is the dummy value 0 (resp. $n + 1$). Let us partition E'_Q into E_{L} and E_{R} that cons of leftward edges of the form $(par_{\mathsf{R}}[k], k)$ and rightward edges of the form $(par_{\mathsf{L}}[k], k)$, respectively. The edge sets E_{L} and E_{R} respectively form forests. Let $\widehat{E}_{\mathsf{L}}$ and $\widehat{E}_{\mathsf{R}}$ be the transitive closures of E_{L} and E_{R}, respectively.

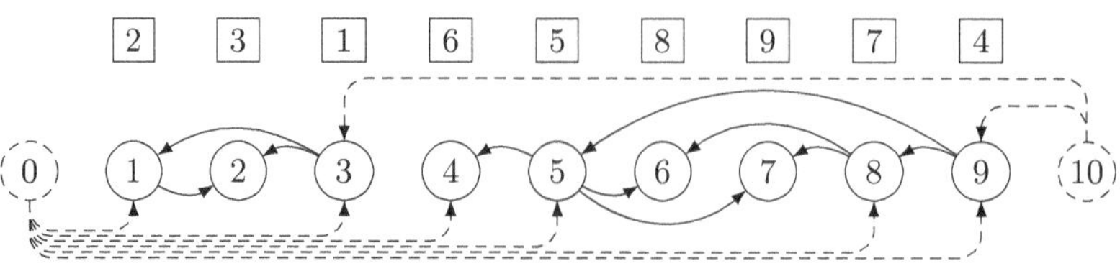

Fig. 1. Graph G'_{Q_1} for Q_1 in Example 1. Edges of E_{L} and E_{R} are drawn by solid leftward and rightward arrows, respectively. The vertex 10 and its outgoing edges drawn with dashed lines appear in T_{L} only. Similarly, 0 appears in T_{R}. The boxed numbers show the preorder numbering in left-to-right depth-first search of the tree T_{L}, which gives the smallest permutation consistent with Q_1.

Lemma 2. *For $(i,j) \in E_Q$, if $i > j$, then $(i,j) \in \widehat{E}_{\mathsf{L}}$. If $i < j$, then $(i,j) \in \widehat{E}_{\mathsf{R}}$.*

Therefore, the graph G'_Q admits exactly the same topological orderings as G_Q. Since $|E'_Q| = |E_{\mathsf{L}}| + |E_{\mathsf{R}}| \le 2(n-1)$, the size of E'_Q is less than $2n$.

Algorithm 1 computes the right parent $par_{\mathsf{R}}[k]$ and the leftmost reachable vertex $reach_{\mathsf{L}}[k] = \min(\{\, j \mid (j,k) \in \widehat{E}_{\mathsf{L}} \,\} \cup \{k\})$ for each k. The array par_{R} can be seen as a tree with the dummy root $n+1$, where we set $l_{n+1} = 1$. That is, we claim that Algorithm 1 constructs the tree $T_{\mathsf{L}} = ([1, n+1], E_{\mathsf{L}}^+)$ where

$$E_{\mathsf{L}}^+ = E_{\mathsf{L}} \cup \{\, (n+1, k) \mid k \text{ has no incoming edges in } E_{\mathsf{L}} \,\}.$$

The function $\mathtt{subtree}(j)$ recursively constructs the subtree rooted at j in the right-to-left depth-first search manner and returns the leftmost descendant minus one. Figure 1 illustrates an example of G'_Q together with T_{L} and T_{R} where T_{R} is defined symmetrically to T_{L}.

Algorithm 1: Computing the tree T_{L}

Input: Canonical iRMQA set $Q = \{\, (l_k, r_k, k) \mid k \in [1, n] \,\}$
Output: arrays par_{R} and $reach_{\mathsf{L}}$ such that $par_{\mathsf{R}}[i]$ is the smallest j satisfying $l_j \le i < j$ and $reach_{\mathsf{L}}[i]$ is the leftmost reachable vertex from i

```
 1 begin
 2 │   par_R, reach_L ← arrays of length n;
 3 │   subtree(n + 1);
 4 │   return par_R and reach_L;

 5 function subtree(j):
 6 │   i ← j − 1;
 7 │   while i ≥ l_j do
 8 │   │   par_R[i] ← j;
 9 │   │   i ← subtree(i);
10 │   reach_L[j] ← i + 1;
11 │   return i;
```

Lemma 3. *Algorithm 1 computes the right parent $par_R[i]$ and the leftmost reachable vertex $reach_L[i]$ of all i in $O(n)$ time.*

Computation of $par_L[k]$ is symmetric to Algorithm 1, so we can construct the graph G'_Q in linear time by the same techniques. If the graph G'_Q contains a cycle, which can be decided in time linear in the graph size, Q admits no consistent permutations. Suppose G'_Q is acyclic. Ono and Nakano [14] have shown that we can enumerate all linear extensions of a given poset with constant delay time. The edge set E'_Q itself is not a poset, but one can recognize its transitive closure in constant time, since $(i, j) \in \widehat{E}_Q$ if and only if $j \in [reach_L[i], reach_R[i]]$, where $reach_R$ is defined and computed in the way exactly symmetric to $reach_L$.

Theorem 3. *One can enumerate all permutations consistent with q given iRMQAs with constant delay time after processing the input in $O(q + n)$ time.*

Now we turn our attention to finding the smallest consistent permutation. We assume G_Q has no cycle. The left-to-right depth-first search of the tree T_L finds the smallest consistent permutation. This might appear surprising, because not every topological sort on T_L gives a topological sort on G'_Q. However, the numbering based on the "left-to-right" depth-first search on T_L always respects the "rightward" edges of E_R, as long as G'_Q has no cycle.

Lemma 4. *The preorder numbering P_0 in left-to-right depth-first search of the tree T_L gives the smallest permutation consistent with Q, where the dummy vertex $n + 1$ is numbered 0.*

Proof. We note the fact that if $i < j$ and j is visited before i, then j is an ancestor of i in T_L.

We first show the consistency of P_0, i.e., for any edge (i, j) of G'_Q, it holds $P_0[i] < P_0[j]$. If $(i, j) \in E_L$, obviously $P_0[i] < P_0[j]$. Suppose $(i, j) \in E_R$. Since G'_Q has no cycle, $(j, i) \notin \widehat{E}_L$. Thus, j is not an ancestor of i in T_L. Since $i < j$, we visit i before j, i.e., $P_0[i] < P_0[j]$. Therefore, P_0 is consistent with Q.

To establish the minimality of P_0, we show that if a permutation P' is consistent with Q and $P'[1 : i - 1] = P_0[1 : i - 1]$, then $P'[i] \geq P_0[i]$. The vertices visited before i belong to either

$$J_i = \{ j \in [i + 1, n] \mid j \text{ is an ancestor of } i \} \text{ or}$$
$$K_i = \{ j \in [1, i - 1] \mid j \text{ is not a descendant of } i \} .$$

That is, $P_0(J_i \cup K_i) = \{1, \ldots, P_0[i] - 1\}$. The assumption $P'[1 : i - 1] = P_0[1 : i - 1]$ immediately implies $P'(K_i) = P_0(K_i)$ by $K_i \subseteq [1, i - 1]$. Thus,

$$P'(J_i) \subseteq [1, n] \setminus P'(K_i) = [1, n] \setminus P_0(K_i) = [P_0[i], n] \cup P_0(J_i) .$$

If $P'(J_i) = P_0(J_i)$, then $P'[i] \in [P_0[i], n]$, so $P'[i] \geq P_0[i]$. If $P'(J_i) \neq P_0(J_i)$, there is $j \in J_i$ such that $P'[j] \in [P_0[i], n]$, i.e., $P'[j] \geq P_0[i]$. Since P' is consistent with Q, we have $P'[i] > P'[j] \geq P_0[i]$. $\square$

Corollary 1. (Smallest Permutation for iRMQAs). *One can find the smallest permutation of length n consistent with q given iRMQAs in $O(q + n)$ time.*

3.2   Smallest Sequence Consistent with iRMQAs

In this subsection, we discuss finding the smallest integer sequence consistent with the given iRMQAs. Similarly to the case of permutations, the conditions defined by the iRMQAs can be represented by the graph $G_Q = ([1,n], E_Q)$ defined in Subsect. 3.1 with an edge weighting $w\colon E_Q \to \{0,1\}$ defined by

$$w((i,j)) = \begin{cases} 0 & \text{if } i < j, \\ 1 & \text{if } i > j. \end{cases}$$

Recall that for any sequence A consistent with Q, if $(l,r,k) \in Q$, then $A[k] < A[i]$ for all $i \in [l, k-1]$ and $A[k] \le A[i]$ for all $i \in [k+1, r]$. Thus, A is consistent with Q if and only if for all $(i,j) \in E_Q$,

$$A[i] + w((i,j)) \le A[j]. \tag{1}$$

The *length of a path* is the sum of the weights of the constituting edges.

Lemma 5. Q *admits a consistent sequence if and only if G_Q is acyclic.*

Hereafter, we write $w(i,j)$ rather than $w((i,j))$ for readability, and assume that G_Q is acyclic. We define a graph $\widetilde{G}_Q$ by adding a vertex $n+1$ to G_Q and edges $(n+1, i)$ for all $i \in [1,n]$, with $w(n+1, i) = 1$ for all $i \in [1,n]$. We then define an array $MaxDist_{G_Q}$ of length n so that $MaxDist_{G_Q}[i]$ is the maximum length of a path from vertex $n+1$ to i in $\widetilde{G}_Q$.

Lemma 6. $MaxDist_{G_Q}$ *is the smallest integer sequence consistent with Q.*

Similarly to the case of permutations in Subsect. 3.1, we use a smaller graph for computing $MaxDist_{G_Q}$.

Lemma 7. *Let T_{L} be the graph defined in Subsect. 3.1. Then, $MaxDist_{G_Q}[i]$ coincides with the depth of i in T_{L} for any vertex $i \in [1,n]$.*
Proof. If a longest path p from $n+1$ to i in E_Q consists of leftward edges only, Lemma 2 implies that there is a path p' in E_{L} that is at least as long as p. Therefore, it suffices to show that for any path from $n+1$ to $i \in [1,n]$ in E_Q, there is a path of the same weight consisting of leftward edges only.

Suppose a path p from $n+1$ in E_Q includes a rightward edge. Since the first edge must be leftward, the first rightward edge (j,k) in p must follow a leftward edge $(i,j) \in E_Q$ with $j < i$ and $j < k$. If $i < k$, then $(j,k) \in E_Q$ implies $(j,i) \in E_Q$. We have a cycle of i and j. So, $i > k$. By Lemma 1, E_Q has a leftward edge (i,k). Replacing the sub-path (i,j,k) with (i,k) in p does not change the length. □

Concerning the iRMQA set Q_1 in Example 1, the depth list of the vertices $1, \ldots, n$ of the tree T_{L} is $(2, 2, 1, 3, 2, 3, 3, 2, 1)$ (see Fig. 1). This is the smallest sequence consistent with Q_1.

Having already discussed linear-time construction of T_{L} (Algorithm 1), we obtain the following theorem.

Theorem 4. *We can compute the smallest sequence of length n consistent with q given iRMQAs in $O(q + n)$ time.*

4   Value-Based RMQs

This section is concerned with integer sequences consistent with given value-based RMQs and their answers. A *value-based RMQ with answer (vRMQA)* is a triple (l, r, v) of positive integers such that $l, r \in [1, n]$ with $l \leq r$ and $v \in \mathbb{N}_+$ for some fixed n. We formally define our problem as follows.

---

vRMQA Reversal Problem
Input: a finite set $Q \subseteq \{ (l, r, v) \in \mathbb{N}_+^3 \mid 1 \leq l \leq r \leq n \}$ of vRMQAs;
Output: a positive integer sequence $A[1 : n]$ such that $\min A[l : r] = v$ for all $(l, r, v) \in Q$ or a report that no such sequence exists.

Throughout this section, we fix the length n of the underlying sequences and a finite set Q of vRMQAs and let $q = |Q|$.

Example 2. Given $n = 9$ and $Q_2 = \{(1, 5, 3), (2, 3, 6), (4, 6, 3), (6, 8, 1)\}$, the sequence $A = (3, 9, 6, 8, 9, 3, 5, 1, 2)$ is one of the solutions. In fact,

$$\min A[1 : 5] = \min(3, 9, 6, 8, 9) = 3\,, \qquad \min A[2 : 3] = \min(9, 6) = 6\,,$$
$$\min A[4 : 6] = \min(8, 9, 3) = 3\,, \qquad \min A[6 : 8] = \min(3, 5, 1) = 1\,.$$

4.1   Smallest Sequence Consistent with vRMQAs

The problem of constructing a sequence consistent with a given set of vRMQAs has been previously addressed in competitive programming contests like [18, 19]. This subsection reviews the editorial of [19] on this problem.

We call a sequence A *consistent with* a vRMQA set Q if, for every $(l, r, v) \in Q$,

$$\forall i \in [l, r],\ A[i] \geq v, \tag{2}$$
$$\exists i \in [l, r],\ A[i] = v. \tag{3}$$

We refer to the first condition (2) as a *lower-bound condition (LBC)* and denote it by $\mathsf{LB}(l, r, v)$. Similarly, the second condition (3) is referred to as an *existential condition (EXC)* and is denoted by $\mathsf{EX}(l, r, v)$.

We define the *lower-bound array LB* of length n by

$$LB[i] = \max \left(\{1\} \cup \{ v \mid \exists (l, r, v) \in Q,\ i \in [l, r] \} \right), \tag{4}$$

which represents the logical conjunction of the LBCs on each position i.

For the instance of Example 2, we have

$$LB = (3, 6, 6, 3, 3, 3, 1, 1, 1)\,.$$

Proposition 2. *If Q admits consistent integer sequences, LB is the smallest such sequence.*

Theorem 5. ([19]). *In $O(n + q \log n)$ time, one can decide whether given vRMQAs admit a consistent sequence and if so, compute the smallest consistent sequence.*

Proof. The array LB can be constructed in $O(n + q \log n)$ time by employing a segment tree with range updates and point queries as follows. For each vRMQA $(l, r, v) \in Q$ (the order is not of importance), we (lazily) update $LB[i]$ for all $i \in [l, r]$ to be v if v is larger than the current value. Furthermore, the EXCs can also be verified in $O(n + q \log n)$ time using a segment tree, with which we can verify $\min LB[l : r] = v$ for each $(l, r, v) \in Q$. Therefore, the problem can be solved in $O(n + q \log n)$ time in total. □

When the maximum value v of $(l, r, v) \in Q$ is $O(n)$, we can compute LB more efficiently using the union-find structure like Algorithm 2 shown later.

4.2   Permutations Consistent with vRMQAs

This subsection discusses counting the permutations and finding the smallest permutation consistent with the given vRMQA set. These problems were previously addressed, respectively, at Nordic Olympiad in Informatics 2018 [21] and Singapore National Olympiad in Informatics 2017 [23]. In addition, a variant problem was addressed at USA Computing Olympiad 2008 [25]. Several unofficial web articles [22,24,26] explain these problems, but they are informal and do not necessarily pursue minimizing the theoretical time complexity. In this subsection, we consolidate these ideas together with our own, and present a formal and comprehensive analysis.

Since we are concerned with permutations, we assume $v \in [1, n]$ for all $(l, r, v) \in Q$. In addition, since there must be a position with value 1 in any permutation, we may assume without loss of generality $(1, n, 1) \in Q$. Furthermore, if there exist two vRMQAs $(l_1, r_1, v), (l_2, r_2, v) \in Q$ with the same value $v \in [1, n]$, since any permutation P over $[1, n]$ has just one position i such that $P[i] = v$, those intervals $[l_1, r_1]$ and $[l_2, r_2]$ must overlap in order to admit a consistent permutation. So, the following two facts hold:

- P satisfies both $\mathsf{LB}(l_1, r_1, v)$ and $\mathsf{LB}(l_2, r_2, v)$ if and only if
  it satisfies $\mathsf{LB}(\min(l_1, l_2), \max(r_1, r_2), v)$;
- P satisfies both $\mathsf{EX}(l_1, r_1, v)$ and $\mathsf{EX}(l_2, r_2, v)$ if and only if
  it satisfies $\mathsf{EX}(\max(l_1, l_2), \min(r_1, r_2), v)$.

Therefore, the information represented in Q is equivalent to Q' defined as follows.

$$Q' = \{\, (l_v^{\mathsf{LB}}, r_v^{\mathsf{LB}}, v) \mid v \in V_Q \,\} \cup \{\, (l_v^{\mathsf{EX}}, r_v^{\mathsf{EX}}, v) \mid v \in V_Q \,\}, \text{ where}$$

$$l_v^{\mathsf{LB}} = \min\{\, l \in [1, n] \mid (l, r, v) \in Q \,\}, \quad r_v^{\mathsf{LB}} = \max\{\, r \in [1, n] \mid (l, r, v) \in Q \,\},$$

$$l_v^{\mathsf{EX}} = \max\{\, l \in [1, n] \mid (l, r, v) \in Q \,\}, \quad r_v^{\mathsf{EX}} = \min\{\, r \in [1, n] \mid (l, r, v) \in Q \,\},$$

and $V_Q = \{\, v \in [1, n] \mid (l, r, v) \in Q \text{ for some } l, r \,\}$ is the set of values that appear in Q. We say that Q' is the *canonical form* of Q. From now on, we assume Q is given in its canonical form and one can obtain $l_v^{\mathsf{LB}}, r_v^{\mathsf{LB}}, l_v^{\mathsf{EX}}, r_v^{\mathsf{EX}}$ in constant time for each $v \in V_Q$. Just by scanning the vRMQA set Q once, we can store those values in four arrays of size n in $O(q + n)$ time. We note that $1 \in V_Q$, $l_1^{\mathsf{LB}} = 1$, and $r_1^{\mathsf{LB}} = n$ by the assumption.

Let $I_v^{\mathsf{LB}} = [l_v^{\mathsf{LB}}, r_v^{\mathsf{LB}}]$ and $I_v^{\mathsf{EX}} = [l_v^{\mathsf{EX}}, r_v^{\mathsf{EX}}]$. Then, a permutation P is consistent with Q if and only if for all $v \in V_Q$,

$$\forall i \in I_v^{\mathsf{LB}}, \ P[i] \geq v, \quad \text{and} \quad \exists i \in I_v^{\mathsf{EX}}, \ P[i] = v.$$

As in Eq. 4 in Sect. 4.1, we define the lower-bound array LB accordingly. Using $I_v^{\mathsf{LB}} = [l_v^{\mathsf{LB}}, r_v^{\mathsf{LB}}]$, the array is expressed as $LB[i] = \max\{\, v \in V_Q \mid i \in I_v^{\mathsf{LB}} \,\}$.

Example 3. For the instance Q_2 in Example 2, we have $V_{Q_2} = \{1, 3, 6\}$ and

$$I_1^{\mathsf{LB}} = [1, 9], \ I_3^{\mathsf{LB}} = [1, 6], \ I_6^{\mathsf{LB}} = [2, 3], \ I_1^{\mathsf{EX}} = [6, 8], \ I_3^{\mathsf{EX}} = [4, 5], \ I_6^{\mathsf{EX}} = [2, 3].$$

Taking advantage of the restriction that $v \in [1, n]$ for any $(l, r, v) \in Q$, Algorithm 2 computes the array LB more efficiently than the method in the proof of Theorem 5. The array LB is initialized with ones. In the **for** loop, decrementing v from n to 2, we set $LB[i] = v$ for all $i \in I_v^{\mathsf{LB}}$ unless $LB[i]$ has already been updated (i.e., $LB[i] \neq 1$). This procedure gives the correct lower-bound array. If we check whether $LB[i] = 1$ for every $i \in I_v^{\mathsf{LB}}$, the computation costs quadratic time, but we can accelerate this by using Theorem 1, following the approach of [26]. We partition the set $[1, n{+}1]$ into intervals with the invariant property that $LB[j] = 1$ if and only if j is the right end of an interval for any $j \in [1, n{+}1]$, assuming $LB[n{+}1] = 1$. This allows us to skip consecutive indices where the values of LB are not 1 and to find the least $j \geq i$ such that $LB[j] = 1$ by $find(i)$. This guarantees that LB is updated at most once on each index i and the total number of iterations of the **while** loop is bounded by n. Therefore, Algorithm 2 computes the lower-bound array LB in $O(n \cdot \alpha(n, n))$ time. It will be $O(n)$ time if the machine word size is $\Omega(\log n)$ bits.

Algorithm 2: Computing the lower bound array

Input: Canonical vRMQA set $Q = \{\, (l_v^{\mathsf{LB}}, r_v^{\mathsf{LB}}, v), (l_v^{\mathsf{EX}}, r_v^{\mathsf{EX}}, v) \mid v \in V_Q \,\}$
Output: The lower-bound array LB

1 **begin**
2     $LB \leftarrow$ an array of length n initialized with ones;
3     $\mathcal{I} \leftarrow$ the discrete partition of $[1, n + 1]$;
4   **for** $v = n, n - 1, \ldots, 2$ **do**
5   **if** $v \in V_Q$ **then**
6             $i \leftarrow \mathcal{I}.find(l_v^{\mathsf{LB}})$;
7   **while** $i \leq r_v^{\mathsf{LB}}$ **do**
8                 $LB[i] \leftarrow v$;
9                 $\mathcal{I}.link(i)$;
10                 $i \leftarrow \mathcal{I}.find(i)$;
11   **return** LB;

Now, we determine all valid positions of values $v \in [1, n]$ in a consistent permutation using LB. Define

$$X_v = \{\, i \in [1, n] \mid LB[i] = v \,\},$$
$$X_{\leq v} = \bigcup_{u=1}^{v} X_u,$$
$$X_v^{\mathsf{EX}} = X_v \cap I_v^{\mathsf{EX}} \text{ for } v \in V_Q.$$

Note that $X_v = \emptyset$ for $v \notin V_Q$ and $X_v^{\mathsf{EX}} \subseteq X_{\leq v}$ for $v \in V_Q$.

Lemma 8. *A permutation P is consistent with Q if and only if*

$$P^{-1}[v] \in \begin{cases} X_v^{\mathsf{EX}} & \text{for all } v \in V_Q, \\ X_{\leq v} & \text{for all } v \notin V_Q. \end{cases}$$

Counting Permutations Consistent with vRMQAs. A strategy for counting the number of permutations consistent with Q is presented in [22]. We determine $P^{-1}[v]$ for $v = 1, \ldots, n$ in this order based on Lemma 8. If $v \in V_Q$, we have $|X_v^{\mathsf{EX}}|$ choices, since none has been chosen from X_v^{EX} earlier. If $v \notin V_Q$, we have $|X_{\leq v}| - (v - 1)$ choices, since we have already picked $v - 1$ from $X_{\leq v}$. Therefore, the number C_Q of consistent permutations is

$$C_Q = \prod_{v \in [1,n]} \max(0, c_v) \text{ where } c_v = \begin{cases} |X_v^{\mathsf{EX}}| & \text{if } v \in V_Q, \\ |X_{\leq v}| - v + 1 & \text{if } v \notin V_Q. \end{cases} \tag{5}$$

Counting $|X_v^{\mathsf{EX}}|$ and $|X_{\leq v}|$ is easy using LB and I_v^{EX}.

Example 4. For the instance Q_2 in Example 2, we have

$$X_1 = \{7, 8, 9\}, \qquad X_3 = \{1, 4, 5, 6\}, \qquad X_6 = \{2, 3\},$$
$$X_1^{\mathsf{EX}} = \{7, 8\}, \qquad X_3^{\mathsf{EX}} = \{4, 5\}, \qquad X_6^{\mathsf{EX}} = \{2, 3\},$$

and

$$C_{Q_2} = 2 \cdot (3 - 1) \cdot 2 \cdot (7 - 3) \cdot (7 - 4) \cdot 2 \cdot (9 - 6) \cdot (9 - 7) \cdot (9 - 8) = 1152.$$

Theorem 6. *We can count the number of permutations of length n consistent with q given vRMQAs in $O(q + n \cdot \alpha(n, n))$ time. Moreover, if the machine word size is $\Omega(\log n)$ bits, the factor $\alpha(n, n)$ can be replaced by a constant.*

By expanding the counting arguments above, we can rank all consistent permutations so that, given a rank r in $[1, C_Q]$, we can return the unique consistent permutation with rank r in $O(n)$ time, and vice versa. This immediately yields an enumeration of consistent permutations with an $O(n)$-time delay. We remark that by applying Uno's technique [17], it is also possible to enumerate all consistent permutations with an $O(\log n)$-time delay after $O(n^{2.5})$ time preprocessing.

Algorithm 3: Computing the smallest consistent permutation

Input: Canonical vRMQA set Q and the lower-bound array LB

Output: The smallest consistent permutation P_0

1 begin

2 $\quad P_0, F \leftarrow$ arrays of length n initialized with zeros;

3 $\quad Z_v \leftarrow \emptyset$ for all $v \in V_Q$;

4 $\quad$ **for** $i = 1, 2, \ldots, n$ **do**

5 $\quad\quad v \leftarrow LB[i]$;

6 $\quad\quad$ **if** $i \in I_v^{\mathsf{EX}}$ and $F[v] = 0$ **then** $P_0[i] \leftarrow v$ and $F[v] \leftarrow 1$;

7 $\quad\quad$ **else** add i to Z_v;

8 $\quad S \leftarrow \emptyset$;

9 $\quad$ **for** $v = 1, 2, \ldots, n$ **do**

10 $\quad\quad$ **if** $v \in V_Q$ **then** $S.insert(i)$ for all i in Z_v ;

11 $\quad\quad$ **else** $P_0[S.extract()] \leftarrow v$;

12 $\quad$ **return** P_0;

Smallest Permutation Consistent with vRMQAs. Assuming that Q admits at least one consistent permutation ($C_Q \neq 0$), we compute the lexicographically smallest such permutation. The strategy in [24] greedily assigns each $v \in V_Q$ to the earliest position satisfying the conditions of Lemma 8. Formally, we define a permutation P_0 via its inverse $P_0^{-1}[v]$ for $v = 1, \ldots, n$ in this order by

$$P_0^{-1}[v] = \min Y_v \text{ for } Y_v = \begin{cases} X_v^{\mathsf{EX}} & \text{if } v \in V_Q, \\ X_{\leq v} \setminus P_0^{-1}([1, v-1]) & \text{if } v \notin V_Q. \end{cases} \quad (6)$$

By the counting argument for Theorem 6, Y_v is never empty and $\min Y_v$ differs from any value of $P_0^{-1}[1 : v-1]$. Thus, P_0 is well-defined and consistent with Q.

Example 5. Let us compare the smallest consistent permutation P_0 and the lower bound array LB for the instance Q_2 in Example 2:

$$P_0 = (4, 6, 7, 3, 5, 8, 1, 2, 9) ,$$
$$LB = (3, \underline{6}, \underline{6}, \underline{3}, 3, 3, \underline{1}, \underline{1}, 1) ,$$

where the positions in X_v^{EX} for $v \in V_{Q_2}$ are underlined in LB.

Lemma 9. P_0 *is the smallest permutation consistent with* Q.

Proof. (Sketch). One can show that if a permutation P is consistent with Q and $P \neq P_0$, then the permutation obtained from P by swapping the values v and u is smaller than P and consistent with Q, where $v \in [1, n]$ is the smallest such that $P^{-1}[v] \neq P_0^{-1}[v]$ and $u = P[P_0^{-1}[v]]$. $\qquad\square$

Algorithm 3 computes P_0 based on the definition (Eq. 6). While the underlying greedy logic follows [24], this is slightly modified to enable efficient computation, improving the $O(n \log n)$ time complexity to $O(n \cdot \alpha(n, n))$. The **for** loop of Line 4 computes $P_0^{-1}[v]$ for $v \in V_Q$ straightforwardly, where $F[v]$ indicates whether $P_0^{-1}[v]$ has already been computed. At the same time, we prepare the sets $Z_v = X_v \setminus \{P_0^{-1}[v]\}$ for all $v \in V_Q$. These sets store the positions where values greater than v and not belonging to V_Q can be placed, and are used in the latter half of the algorithm. We determine the positions $P_0^{-1}[v]$ of $v \notin V_Q$ in the **for** loop of Line 9. We perform the following operations on a semi-dynamic set S, which is initially empty, for each $v \in [1, n]$ in ascending order:

1. if $v \in V_Q$, insert every index $i \in Z_v = X_v \setminus \{P_0^{-1}[v]\}$ into S,
2. if $v \notin V_Q$, set $P_0[\min S] = v$ and remove $\min S$ from S.

These operations maintain the loop invariant that after the vth iteration of the second **for** loop, we have $S = X_{\leq v} \setminus P_0^{-1}([1, v])$. Thus, our construction is faithful to the definition of P_0 (Eq. 6). Since we can fix those operations over S before Line 8, we can apply Theorem 2.

Theorem 7. *We can compute the smallest permutation of length n consistent with the q given vRMQAs in $O(q + n \cdot \alpha(n, n))$ time. Moreover, if the machine word size is $\Omega(\log n)$ bits, the factor $\alpha(n, n)$ can be replaced by a constant.*

5   Concluding Remarks

We presented efficient algorithms for reconstructing sequences and permutations consistent with RMQ answers, addressing various natural variations such as computing the lexicographically smallest solution, enumerating all consistent permutations, and counting their total number, for both value-based and index-based settings.

While we showed that counting consistent permutations is tractable for vRMQAs, the complexity of the analogous counting problem for iRMQAs is still open. Counting the number of permutations consistent with a given iRMQA set Q is equivalent to counting that of linear extensions of the poset generated by the directed acyclic graph G'_Q in Sect. 3. Whilst counting the number of linear extensions of a given poset is #P-complete in general [8], some special cases allow polynomial-time counting (e.g., [4,13]). Since the graph G'_Q is highly restrictive, it might be possible to design a polynomial-time algorithm for counting the consistent permutations.

It is also interesting future work to study the inversion problems of other types of range queries, such as range sum queries and range mode queries.

Acknowledgment. The authors are grateful to the anonymous reviewers for their helpful comments. In particular, we thank the reviewer who pointed out the connection to competitive programming, which encouraged us to examine the existing online results in depth and made us aware that some of the results in our submission were not entirely original. This work was supported in part by JSPS KAKENHI Grant Numbers 25K21150, 23H04378 (DK), 23K11325, 24H00697, 24K14827, 25K14981 (RY), and 25K14979 (AS).

References

1. Aho, A.V., Hopcroft, J.E., Ullman, J.D.: The design and analysis of computer algorithms. Addison-Wesley Series in Computer Science and Information Processing, Addison-Wesley, Reading, Massachusetts (1974)
2. Alba, D.E., Leal, J.A.R.: Algorithms for competitive programming. DND Learning (2023). https://books.google.co.jp/books?id=Uhz50AEACAAJ
3. Alstrup, S., Gavoille, C., Kaplan, H., Rauhe, T.: Nearest common ancestors: a survey and a new algorithm for a distributed environment. Theory Comput. Syst. **37**(3), 441–456 (2004). https://doi.org/10.1007/S00224-004-1155-5
4. Atkinson, M.D.: On computing the number of linear extensions of a tree. Order **7**, 23–25 (1990). https://doi.org/10.1007/BF00383170
5. Bender, M.A., Farach-Colton, M., Pemmasani, G., Skiena, S., Sumazin, P.: Lowest common ancestors in trees and directed acyclic graphs. J. Algorithms **57**(2), 75–94 (2005). https://doi.org/10.1016/j.jalgor.2005.08.001
6. de Berg, M., Cheong, O., van Kreveld, M., Overmars, M.: Computational geometry: algorithms and applications. Springer (2008). https://books.google.de/books?id=tkyG8W2163YC
7. Berkman, O., Vishkin, U.: Recursive star-tree parallel data structure. SIAM J. Comput. **22**(2), 221–242 (1993). https://doi.org/10.1137/0222017
8. Brightwell, G., Winkler, P.: Counting linear extensions. Order **8**, 225–242 (1991). https://doi.org/10.1007/BF00383444
9. Fischer, J., Heun, V.: Theoretical and practical improvements on the RMQ-problem, with applications to LCA and LCE. In: Lewenstein, M., Valiente, G. (eds.) Combinatorial Pattern Matching, pp. 36–48. Springer Berlin Heidelberg, Berlin, Heidelberg (2006). https://doi.org/10.1007/11780441_5
10. Fischer, J., Heun, V.: A new succinct representation of RMQ-information and improvements in the enhanced suffix array. In: Chen, B., Paterson, M., Zhang, G. (eds.) ESCAPE 2007. LNCS, vol. 4614, pp. 459–470. Springer, Heidelberg (2007). https://doi.org/10.1007/978-3-540-74450-4_41
11. Gabow, H.N., Bentley, J.L., Tarjan, R.E.: Scaling and related techniques for geometry problems. In: DeMillo, R.A. (ed.) Proc. STOC, pp. 135–143. ACM (1984). https://doi.org/10.1145/800057.808675
12. Gabow, H.N., Tarjan, R.E.: A linear-time algorithm for a special case of disjoint set union. J. Comput. Syst. Sci. **30**(2), 209–221 (1985). https://doi.org/10.1016/0022-0000(85)90014-5
13. Möhring, R.H.: Computationally tractable classes of ordered sets. In: Rival, I. (ed.) Algorithms and Order, NATO ASI Series C: Mathematical and Physical Sciences, vol. 255, pp. 105–193. Springer, Dordrecht (1989). https://doi.org/10.1007/978-94-009-2639-4_4
14. Ono, A., Nakano, S.i.: Constant time generation of linear extensions. In: Liśkiewicz, M., Reischuk, R. (eds.) Fundamentals of Computation Theory, pp. 445–453. Springer, Berlin, Heidelberg (2005). https://doi.org/10.1007/11537311_39
15. Sadakane, K.: Succinct data structures for flexible text retrieval systems. J. Discrete Algorithms **5**(1), 12–22 (2007). https://doi.org/10.1016/J.JDA.2006.03.011

16. Tarjan, R.E.: Efficiency of a good but not linear set union algorithm. J. ACM **22**(2), 215–225 (1975). https://doi.org/10.1145/321879.321884
17. Uno, T.: A fast algorithm for enumerating bipartite perfect matchings. In: Eades, P., Takaoka, T. (eds.) Algorithms and Computation (ISAAC 2001), pp. 367–379. Springer, Berlin, Heidelberg (2001). https://doi.org/10.1007/3-540-45678-3_32
18. Inverse range minimum query. https://eolymp.com/en/problems/3080, Accessed 01 Dec 2025
19. No.1675 Strange Minimum Query. https://yukicoder.me/problems/no/1675, Accessed 27 Nov 2025
20. No.1675 Strange Minimum Query kaisetsu (editorial). https://yukicoder.me/problems/no/1675/editorial, Accessed 27 Nov 2025
21. Nordic Olympiad in Informatics. https://nordic.progolymp.se/, Accessed 27 Nov 2025
22. xiaoliebao1115: P11505 [nordicoi 2018] mysterious array. https://www.luogu.com.cn/article/j9714zmg (2025), Accessed 27 Nov 2025
23. The Singapore National Olympiad in Informatics. https://noisg.comp.nus.edu.sg/, problem archive available at https://github.com/noisg/sg_noi_archive, Accessed 27 Nov 2025
24. https://github.com/noisg/sg_noi_archive/blob/master/2017/solution_writeup/NOI2017_solutions.pdf , Accessed 01 Dec 2025
25. Haybale guessing. https://www.acmicpc.net/problem/6153, originally USACO 2008 January Contest, Gold Division; Accessed 27 Nov 2025
26. [P2898][USACO08JAN] Haybale Guessing Haybale Guessing. https://www.programmersought.com/article/60757109923/, Accessed 01 Dec 2025

Sublinear Time Algorithms for Abelian Group Isomorphism and Basis Construction

Nader H. Bshouty[(✉)] [iD]

Department of Computer Science, Technion, Haifa, Israel
bshouty@cs.technion.ac.il

Abstract. In this paper, we study the problems of Abelian group isomorphism and basis construction in two models. In the *partially specified model* (PS-model), the algorithm does not know the group size but can access randomly chosen elements of the group along with the Cayley table of those elements, which provides the result of the binary operation for every pair of selected elements. In the stronger *fully specified model* (FS-model), the algorithm knows the size of the group and has access to its elements and Cayley table.

Given two Abelian groups, G, and H, we present an algorithm in the PS-model (and hence in the FS-model) that runs in time $\tilde{O}(\sqrt{|G|})$ and decides if they are isomorphic. This improves on Kavitha's linear-time algorithm and gives the first sublinear-time solution for this problem. We then prove the lower bound $\Omega(|G|^{1/4})$ for the FS-model and the tight bound $\Omega(\sqrt{|G|})$ for the PS-model. This is the first known lower bound for this problem. We obtain similar results for finding a basis for Abelian groups. For deterministic algorithms, a simple $\Omega(|G|)$ lower bound is given.

Keywords: Abelian Group · Isomorphic · Sublinear-time Algorithm

1 Introduction

In this paper, we study the problems of Abelian group isomorphism and constructing a basis in two models. In the *fully specified model*[1] (FS-model), the size of the groups is known, and the algorithm has access to the elements of the group and their Cayley tables. In the *partially specified model* (PS-model), the size of the group is unknown; the algorithm can receive uniform random elements of the groups and access the Cayley table of elements observed so far.

To the best of our knowledge, we provide the first sublinear-time algorithm (in the size of the group) for isomorphism testing and basis construction problems of Abelian groups in these models while also establishing the first tight lower bounds.

All results in this paper are stated for the PS-model, unless explicitly indicated otherwise for the FS-model.

---

[1] Also known as the Cayley or multiplication table model.

© The Author(s), under exclusive license to Springer Nature Switzerland AG 2026

J. Kozik and A. Wolff (Eds.): SOFSEM 2026, LNCS 16448, pp. 31–45, 2026.
https://doi.org/10.1007/978-3-032-17801-5_3

1.1 Generators For Abelian Groups

To address these problems, we first study the problem of constructing a set of generators for Abelian groups G using Oracle access to the Cayley table of G. Our algorithm is randomized and runs in time $\tilde{O}(\sqrt{|G|})$ and constructs a set of generators $A = \{a_i\}_{i \in [t]}$ for G with size at most $t \leq \log |G|$, satisfying the *triangular relations*[2]. We then use these relations to construct an Abelian group Γ_G isomorphic to G, that is a subset of some quotient ring, where each operation in Γ_G can be performed in time $poly(\log |G|)$.

Our algorithm operates within the PS-model. We also show that this algorithm is optimal, i.e., any randomized algorithm in the PS-model that produces a set of generators with any set of relations of size at most $|G|^{o(1)}$ must run in time $\Omega(\sqrt{|G|})$. For the FS-model, we obtain the lower bound $\Omega(|G|^{1/4})$.

We then apply this to isomorphism testing and basis construction.

1.2 Isomorphism Testing

Group isomorphism is a fundamental problem in group theory and computation [1–4,9,10,12–14,18–20,22–25,28].

In this paper, we focus on the Abelian group isomorphism problem [12,18–20,22,23,25,28]: Given two finite Abelian groups H and G with access to their Cayley tables. Decide if H is isomorphic to G.

Lipton et al. [23] showed that Abelian group isomorphism could be solved in polynomial time. Savage [25] gave an algorithm that runs in time $O(|G|^2)$. Vikas [28] improved this bound and gave an $O(|G| \log |G|)$ time algorithm. Then Kavitha [20] and Karagiorgos and Paulakis [18,19] improved it to $O(|G|)$. All the algorithms listed above are deterministic. In this paper, we show.

Theorem 1. *Any deterministic algorithm that decides whether an Abelian group is isomorphic to a given group G must make at least $\Omega(|G|)$ accesses to the elements of the groups and to the Cayley table of G.*

Therefore, for deterministic algorithms, the algorithms of Kavitha [20] and Karagiorgos and Paulakis [18,19] are optimal.

In this paper, we give a randomized algorithm for this problem. We show.

Theorem 2. *The Abelian group isomorphism problem can be solved in time*[3] $\tilde{O}(\sqrt{|G|})$. *The algorithm also computes an explicit isomorphism.*

We also establish tight lower bounds for this problem.

Theorem 3. *Any Abelian group isomorphism algorithm must make at least $\Omega(\sqrt{|G|})$ accesses to the elements of the group and the Cayley table of G.*

---

[2] A *Relation* is an equation for elements of the group. For example, $a^2 b = c^3 d$ for some elements a, b, c, d in a group. For a set of generators $\{a_1, a_2, \ldots, a_t\}$, triangular relations are relations of the form $a_i^{k_i} = a_1^{\lambda_{i,1}} \cdots a_{i-1}^{\lambda_{i,i-1}}$ for all $i \in [t]$.

[3] All results in this paper are stated for the PS-model, unless explicitly indicated otherwise for the FS-model.

To the best of our knowledge, this is the first sublinear upper bound and the first tight lower bound for this problem.

In the FS-model, we establish the following lower bound.

Theorem 4. *In the FS-model, any Abelian group isomorphism algorithm must make at least $\Omega(|G|^{1/4})$ access to the elements of the groups and the Cayley table of G.*

We also note that the problem of determining the existence of a homomorphism between two Abelian groups G and H can also be solved with the same time complexity. Our algorithm also finds the homomorphism and isomorphism.

1.3   Basis Construction

Computing a basis for a finite Abelian group is a fundamental problem in computational group theory, with applications in cryptography, coding theory, and algebraic computations [5,7,8,12,15,16,18,19,21,26].

Every finite Abelian group G can be represented as a direct product of cyclic groups $G_1 \times G_2 \times \cdots \times G_t$. If a_i generates G_i, then the elements $a_1, a_2, \ldots, a_t$ are called *basis* for G.

All known sublinear time algorithms that find a basis for an Abelian group run in time $\tilde{O}(\sqrt{|G|})$ and assume that the group is given by a set of generators along with their orders and the Cayley table [7,8,15,18,26]. Chen and Fu [8], and Karagiorgos and Paulakis [18,19] gave an $O(|G|)$ time deterministic algorithm that accesses only the Cayley table. We show that this algorithm is optimal.

Theorem 5. *Any deterministic algorithm that finds a basis for an Abelian group G must access the elements of the group and its Cayley table at least $\Omega(|G|)$ times.*

In this paper, we give a randomized algorithm for this problem. We show.

Theorem 6. *There exists a randomized algorithm that accesses the Cayley table of an Abelian group G at most $\tilde{O}(\sqrt{|G|})$ time and finds a basis for G.*

To the best of our knowledge, this result provides the first sublinear algorithm for finding the basis of an Abelian group using only the Cayley table.

We then establish the following tight lower bound.

Theorem 7. *Any randomized algorithm that finds a basis for an Abelian group G must access the elements of the group and its Cayley table at least $\Omega(\sqrt{|G|})$ times.*

This is the first tight lower bound for the problem.

Furthermore, in the FS-model, we show.

Theorem 8. *Any randomized algorithm that finds a basis for an Abelian group G in the FS-model must access the elements of the group and its Cayley table at least $\Omega(|G|^{1/4})$ times.*

2 Our Technique

In this section, we outline the techniques used in our algorithms, beginning with those applied to establish the upper bounds, followed by the methods for deriving the lower bounds.

2.1 Generators with Triangular Relation

We first provide an algorithm that, for every Abelian group G, runs in time $\tilde{O}(\sqrt{|G|})$ and constructs a set of generators $A = \{a_1, \ldots, a_t\}$ of size at most $t = \log|G|$ that satisfies *triangular relations*. That is, for every $i \in [t]$, there exists a relation of the form $a_i^{k_i} = a_1^{\lambda_{i,1}} \cdots a_{i-1}^{\lambda_{i,i-1}}$, where $k_i \geq 2$.

We present two algorithms. A deterministic algorithm that runs in time $O(|G|)$ and a randomized algorithm that runs in time $\tilde{O}(\sqrt{|G|})$. In both algorithms, the main idea is to construct a sequence of subgroups $\{e\} < G_1 < G_2 < \cdots < G_t = G$, where G_i is generated by $\{a_1, \ldots, a_i\}$. To construct G_{i+1}, we choose an element a_{i+1} in $G\backslash G_i$, find the order k_{i+1} of a_{i+1} in the quotient group G/G_i, and construct the elements of all the disjoint cosets of G_i to form G_{i+1} and establish the relation for a_{i+1}.

In the deterministic algorithm, we leverage the fact that elements of different cosets are disjoint, which guarantees that each group element is generated exactly once. As a result, the algorithm enumerates all elements of the group without repetition and runs in time $\tilde{O}(|G|)$.

In the randomized algorithm, we use the following simple idea. Let $\mathcal{F}$ be a homomorphism $\mathcal{F} : G \to F$ with a readily computable inverse $\mathcal{F}^{-1}$. For $a \in G$, to determine $\mathcal{F}(a)$, we choose a set of $\tilde{O}(\sqrt{|G|})$ elements R_1 in G with known $\mathcal{F}$ values, then choose another set R_2 of $\tilde{O}(\sqrt{|G|})$ elements in G uniformly at random, also with known $\mathcal{F}$ values. This can be done by choosing $f \in F$ uniformly at random, computing $\chi = \mathcal{F}^{-1}(f)$ and then choosing $x \in \chi$ uniformly at random. Since $aG = G$, with high probability, $R_1 \cap aR_2$ contains an element $b = ac$, where $b \in R_1$ and $c \in R_2$. Then $\mathcal{F}(a) = \mathcal{F}(bc^{-1}) = \mathcal{F}(b)\mathcal{F}(c)^{-1}$.

We use this technique to find an element a_{i+1} in $G\backslash G_i$, determine its order in the quotient group G/G_i, and compute the relation for a_{i+1} in time $\tilde{O}(\sqrt{|G|})$.

2.2 Isomorphism and Basis Construction

Given an Abelian group G, we use the algorithm from the previous section to find a set of generators A with triangular relations in time $\tilde{O}(\sqrt{|G|})$. Using these relations, $\mathcal{R}$, we construct a group $\Gamma(\mathcal{R})$ that is isomorphic to G, which we call the monomial Abelian group derived from $\mathcal{R}$. This is because the elements of this group are monomials over $t = |A|$ formal elements $x_1, \ldots, x_t$, where multiplication in this group is modulo polynomials corresponds to the relations $\mathcal{R}$. The elements x_i, satisfy the same relations as a_i in G. We also show that the multiplication and inversion operation in $\Gamma(\mathcal{R})$ can be performed in time $poly(t)$.

Then, one can apply the Smith normal form to convert these generators and relations into a basis. The time complexity of this method is $poly(\log|G|)$ [17].

This addresses both the basis construction problem and the isomorphism problem.

2.3   Our Technique for the Lower Bounds

In this section, we present the technique used to establish two foundational lower bounds from which all other lower bounds are derived. Our approach is inspired by the technique introduced in [11].

For the lower bound in the FS-model, we first define the class of all Abelian groups, $\mathcal{G}$ on the ground set $[p^{2m}]$ that are isomorphic to either $H_1 = \mathbb{Z}_{p^2}^m$ or $H_2 = \mathbb{Z}_{p^2}^{m-1} \times \mathbb{Z}_p^2$. We then show that any algorithm which, with probability at least $2/3$, determines whether $G \in \mathcal{G}$ is isomorphic to H_1 or H_2, must access the elements of G and its Cayley table at least $\Omega(|G|^{1/4})$ times.

We apply Yao's Minimax Principle [27,29] and show that, for any deterministic algorithm $\mathcal{A}$ that decides whether a given group G is isomorphic to H_1 or H_2 (by outputting 1 or 2), if $\mathcal{A}$ is executed on a uniformly random group $G \in \mathcal{G}$, then with probability at least $2/3$, $\mathcal{A}$ must access elements of G and entries of its Cayley table at least $\Omega(|G|^{1/4})$ times.

To this end, consider such an algorithm $\mathcal{A}$. We map the elements of the group $G \in \mathcal{G}$ to abstract elements $G' = \{\sigma_1, \ldots, \sigma_{|G|}\}$ through a bijective function $\phi : G \to G'$ chosen uniformly at random. Under this mapping, G' is isomorphic to G with the product $\sigma_i \sigma_j = \phi(\phi^{-1}(\sigma_i)\phi^{-1}(\sigma_j))$. Next, we map G' to the free $\mathbb{Z}_{p^2}$-module $L = \{\alpha_1 x_1 + \cdots + \alpha_w x_w | \alpha_i \in \mathbb{Z}_{p^2}, w \in [|G|]\}$ with coefficients in $\mathbb{Z}_{p^2}$, where $\{x_i\}_{i=1}^{\infty}$ are formal elements. In this mapping, each σ_i is mapped to x_i. We then run the algorithm $\mathcal{A}$ on this mapped group, i.e., we run the algorithm and replace each multiplication in the group G' with $+$ in L and each σ_i with x_i.

Since in $\mathcal{A}$ each σ_i is replaced with an element in L, and comparisons in the IF commands (i.e., the command "If $x = y$ Then ... Else ...") involve elements in L, the execution follows a single, well-defined path P in the algorithm $\mathcal{A}$. Assume that this path results in an output of 1.

We then show that if the path P contains at most $|G|^{1/4}$ group operations and elements of the group, then, for ϕ chosen uniformly at random, running the algorithm on G', with high probability, the logical values of the comparisons in the IF commands in the algorithm $\mathcal{A}$ are consistent with the path P. That is, with high probability, the execution on G' will follow the same path P. The intuition here is that both H_1 and H_2 behave similarly to L under a randomly chosen ϕ. Since this path outputs 1, the failure probability of $\mathcal{A}$ is close to $1/2$.

In the PS-model, we define the class of all Abelian groups $\mathcal{G}$ that are isomorphic to either $G_1 = \mathbb{Z}_p^m$ or $G_2 = \mathbb{Z}_p^{m-1}$. We show that if there exists an algorithm running in time T that, with probability at least $2/3$, distinguishes between these groups, then there is also an algorithm running in time T that, with probability at least $2/3$, distinguishes between H_1 and H_2. This result follows from the fact that pH_1 is isomorphic to G_1 and pH_2 is isomorphic to

G_2. Consequently, any algorithm for G_i gives an algorithm for H_i. Hence, we establish that $T = \Omega(|H_1|^{1/4}) = \Omega(p^{m/2}) = \Omega(\sqrt{|G_1|})$.

For deterministic algorithms, consider the two Abelian groups $D_1 = \mathbb{Z}_p^m$ and $D_2 = \mathbb{Z}_p^{m-2} \times \mathbb{Z}_{p^2}$. An adversary can provide elements from $W = \mathbb{Z}_p^{m-2} \times \{(0,0)\}$ for each access to an element σ_i. The sum of any two elements in W remains in W, preventing the algorithm from distinguishing between the Abelian groups until the $(p^{m-2}+1)$-th element is requested. At this point, the adversary must reveal whether the hidden group is D_1 or D_2. Thus, the complexity of the algorithm is at least $\Omega(p^{m-2}) = \Omega(n)$ for any constant p.

Using these two lower bounds, we get the lower bounds for isomorphism testing and basis construction.

3   Generators for Abelian Group Randomized Algorithm

In this section, we prove

Theorem 9. *Given a Cayley multiplication table of an Abelian group G, where each entry in the table can be accessed in constant time. There is a randomized algorithm (in the PS-model and therefore in the FS-model) that runs in time $\tilde{O}(\sqrt{|G|})$ and, with probability $1 - 1/poly(|G|)$, finds*

1. *A set of generators $A = \{a_1, \ldots, a_t\}$ for G of size at most $t \leq \log |G|$.*
2. *Integers k_i and $0 \leq \lambda_{i,j} \leq k_i - 1$ that satisfy the relation $a_i^{k_i} = a_1^{\lambda_{i,1}} \cdots a_{i-1}^{\lambda_{i,i-1}}$.*

Additionally, the algorithm generates subsets $A_i \subset G$ and subgroups G_i of G that satisfy:

1. *$G_i = \langle a_1, a_2, \ldots, a_i \rangle$ and $G = G_t = \langle a_1, \ldots, a_t \rangle$.*
2. *$a_i \notin G_{i-1} = \langle a_1, a_2, \ldots, a_{i-1} \rangle$, and k_i is the smallest integer in $\mathbb{N}$ such that $a_i^{k_i} \in \langle a_1, a_2, \ldots, a_{i-1} \rangle$.*
3. *Every element in G has a unique representation of the form $a_1^{j_1} \cdots a_t^{j_t}$ for some $j_i \in \{0, 1, \ldots, k_i - 1\}$.*
4. *k_i divides $|G|/|G_{i-1}|$.*
5. *$|G_i| = k_1 k_2 \cdots k_i$ and $|G| = k_1 k_2 \cdots k_t$.*

For some basic facts in Groups see Subsect. 3.1 in the full paper [6].

We will first present the proof for the FS-model, and then for the PS-model.

3.1   The Algorithm in the FS-Model

Consider the following algorithm.

Algorithm 1. Random Generators(G)

1: $A_0 = \{\}$; $G_0 = \{e\}$; $i = 0$; $k_0 = 1$.
2: **while** $k_0 k_1 k_2 \cdots k_i \neq |G|$ **do**
3:    $i \leftarrow i + 1$
4:    $a_i \leftarrow$ **Choose**(G, G_{i-1}) /* Choose $a_i \in G \setminus G_{i-1}$ where $G_{i-1} :=$ $\langle a_1, a_2, \ldots, a_{i-1} \rangle$.
5:    $A_i = A_{i-1} \cup \{a_i\}$.
6:    $k_i \leftarrow$ **FindMin**(a_i, G_{i-1}) /*Find the minimal k_i such that $a_i^{k_i} \in G_{i-1}$
7:    $(\lambda_{i,j}) \leftarrow$ **FindExp**$(a_i^{k_i}, G_{i-1})$ /*Find $0 \leq \lambda_{i,j} \leq k_i - 1$ such that $a_i^{k_i} = a_1^{\lambda_{i,1}} \cdots a_{i-1}^{\lambda_{i,i-1}}$.
8: **end while**
9: Output $t := i$, $A := A_i$, $(k_1, k_2, \ldots, k_t)$ and $\lambda_{m,j}$.

In the randomized algorithm, **Random Generators**, at each iteration r, we do not store the elements of G_i for $i \in [r]$ or their unique representation. For all $i \in [t]$, we store only the integer k_i, the set of generators $A_i = \{a_1, \ldots, a_i\}$ for G_i, and the relation $a_i^{k_i} = a_1^{\lambda_{i,1}} \cdots a_{i-1}^{\lambda_{i,i-1}}$. This is the same algorithm as **Generators** in Sect. 4 in the full paper [6], where its correctness is also proved. We will show that each one of the procedures **Choose**, **FindMin** and **FindExp** can be executed in time $\tilde{O}(\sqrt{|G|})$ with a success probability of at least $1 - 1/poly(|G|)$. The result follows by union bound and since the **while** loop runs at most $O(\log |G|)$ time.

We now prove

Lemma 1. *Given $k_1, \ldots, k_r$ and $a_1, \ldots, a_r$, a uniformly random element in $G_r = \langle a_1, a_2, \ldots, a_r \rangle$ can be generated in time $O(\log |G|)$.*

Proof. Every element in G_r has a unique representation $a_1^{j_1} \cdots a_r^{j_r}$, where $0 \leq j_i \leq k_i - 1$. To choose an element uniformly at random, we first choose $0 \leq j_i \leq k_i - 1$ uniformly at random for each $i \in [r]$, and then compute $a_1^{j_1} \cdots a_r^{j_r}$ in time

$$r + \lceil \log j_1 \rceil + \cdots + \lceil \log j_r \rceil \leq 2r + \log(k_1 \cdots k_r) \leq 3 \log |G|.$$

$\square$

In the next section, we will discuss the case when $|G|$ is unknown.

We now show how to choose an element $a \in G \backslash G_{i-1}$ in step 4 in time $\tilde{O}(\sqrt{|G|})$. We prove the following.

Lemma 2. *Let $H < G$ be a subgroup. Suppose we can choose an element uniformly at random from both H and G, and multiply two elements in G in constant time. Then, there exists an algorithm that runs in time $\tilde{O}(\sqrt{|G|})$ and, with probability at least $1 - 1/poly(|G|),$*

1. *Decides, given $a \in G$, whether $a \in H$.*
2. *Finds an element $a \in G \backslash H$.*

Proof. Let G be a group with n elements. Given $a \in G$. If $a \in H$, then $aH = H$. If $a \notin H$, then $aH \cap H = \emptyset$. To test if $aH \cap H = \emptyset$, we choose any $t = O(\sqrt{n \log n})$ elements R_1 in aH and t elements R_2 in H uniformly at random. If $R_1 \cap R_2 = \emptyset$, we conclude that $a \notin H$; otherwise, we conclude $a \in H$. If $aH = H$, the probability that the algorithm fails is at most

$$\left(1 - \frac{O(\sqrt{n \log n})}{|H|}\right)^{O(\sqrt{n \log n})} \leq \left(1 - \frac{O(\sqrt{\log n})}{\sqrt{n}}\right)^{O(\sqrt{n \log n})} = \frac{1}{poly(n)}.$$

We now prove item 2. Since $H < G$, we have $|H|$ divides $|G|$, so if we choose an element $a \in G$ uniformly at random, with probability at least $1/2$, $a \in G \backslash H$. Thus, the algorithm can choose $O(\log n)$ elements uniformly at random from G and run the algorithm from item 1 for each one. With probability at least $1 - 1/poly(n)$, one of these will lie in $G \backslash H$. $\square$

The next lemma shows how to find the smallest $k_i > 1$ such that $a_i^{k_i} \in G_{i-1}$ in step 6 in time $\tilde{O}(\sqrt{|G|})$.

Lemma 3. *Let $H < G$ be a subgroup of G. Suppose we can choose an element uniformly at random from both H and G, and can multiply two elements in G in constant time. Given $a \in G$, the smallest k such that $a^k \in H$ can be found in time $\tilde{O}(\sqrt{|G|})$.*

Proof. To find the smallest k such that $a^k \in H$, we use the fact that k divides $|G|/|H|$ and therefore divides[4] $n = |G|$. Also, if $k|k'$, then $a^{k'} \in H$.

The algorithm to find k is as follows: Let $m = n$. For every prime p that divides m, while $a^{m/p}$ is in H, set $m \leftarrow m/p$.

The final m is the required k. The number of iterations in this algorithm is at most $O(\log n)$. The time complexity is $\tilde{O}(\sqrt{n})$, as checking if $a^{m/p}$ is in H, uses Lemma 2. $\square$

Now we show how to find $\lambda_{i,j}$ in step 7 in time $\tilde{O}(\sqrt{|G|})$.

Lemma 4. *Let $A = \{a_1, \ldots, a_t\}$ be a set of generators of G. The elements $0 \leq \lambda_{i,j} \leq k_i - 1$ that satisfy $a_i^{k_i} = a_1^{\lambda_{i,1}} \cdots a_{i-1}^{\lambda_{i,i-1}}$ can be found in $\tilde{O}(\sqrt{n})$ time.*

Proof. For a_1, we have $a_1^{k_1} = 1$. Suppose we know the representations of $a_i^{k_i} = a_1^{\lambda_{i,1}} \cdots a_{i-1}^{\lambda_{i,i-1}}$ for $i = 1, \ldots, j-1$. Now we have $a_j^{k_j} \in G_{j-1} = \langle a_1, a_2, \ldots, a_{j-1} \rangle$. To find the representation of $a_j^{k_j}$, we choose any $O(\sqrt{n \log n})$ distinct elements in G_{j-1} with known representations. This is achieved by choosing $O(\sqrt{n \log n})$ distinct elements $(w_1, \ldots, w_{j-1}) \in \prod_{h=1}^{j-1}\{0, 1, \ldots, k_h - 1\}$ and computing $a_1^{w_1} \cdots$

---

[4] Factoring $n = |G|$ can be done in time $\tilde{O}(\sqrt{|G|})$.

$a_{j-1}^{w_{j-1}}$. Using Lemma 1, we choose $O(\sqrt{n \log n})$ elements uniformly at random with known representation and multiply each by $a_j^{k_j}$. If a common element appears, we obtain $(a_1^{\beta_1} \cdots a_{j-1}^{\beta_{j-1}})a_j^{k_j} = a_1^{\alpha_1} \cdots a_{j-1}^{\alpha_{j-1}}$. Thus, we find that $a_j^{k_j} = (a_1^{\alpha_1} \cdots a_{j-1}^{\alpha_{j-1}})(a_1^{\beta_1} \cdots a_{j-1}^{\beta_{j-1}})^{-1}$, which can be computed in time $\tilde{O}(\log^2 n)$.

As in the proof of Lemma 2, the algorithm succeeds with probability at least $1 - 1/poly(n)$. $\qquad\square$

3.2   The Algorithm in the PS-Model

In this section, we provide a sketch of how to adapt the algorithm **Random Generators** to work within the PS-model.

Recall that in this model, the size of the group is unknown; the algorithm can receive random, uniformly distributed elements from the group and access the Cayley table of elements that have been observed so far.

In the full paper [6], we show how to estimate $|G|$ within a $poly(\log |G|)$ factor.

Lemma 5. *There exists a randomized algorithm that runs in expected time $\tilde{O}(\sqrt{|G|})$ and, with probability at least $1 - 1/poly(|G|)$, returns a value q satisfying $|G| \leq q \leq \Theta(|G| \log |G|)$. The probability that this algorithm requires more than $\Theta(\sqrt{L|G|} \log^2(L|G|))$ elements of G decreases exponentially with L.*

Lemma 2 still applies here if we replace n with q. The running time of the algorithm is then $\tilde{O}(\sqrt{q}) = \tilde{O}(\sqrt{|G|})$.

The next lemma shows how to find the smallest $k_i > 1$ such that $a_i^{k_i} \in G_{i-1}$ in time $\tilde{O}(\sqrt{|G|})$. This replaces Lemma 3.

Lemma 6. *Let G be an Abelian group with an unknown number of elements and let $H < G$ be a subgroup of G. Suppose we can choose an element uniformly at random from both H and G, and can multiply two elements in G in constant time. There is a randomized algorithm that, given $a \in G$, runs in time $\tilde{O}(\sqrt{|G|})$ and, with probability $1 - 1/poly|G|$, finds the smallest k such that $a^k \in H$.*

Proof. The algorithm samples $m = O(\sqrt{q \log q}) = \tilde{O}(\sqrt{|G|})$ integers $j_1, \ldots, j_m$ in $[2q]$ uniformly at random and computes a^{j_i} for each $i \in [m]$. By the birthday paradox, with probability at least $1 - 1/poly(q) = 1 - 1/poly(|G|)$, there are two distinct indices $j_{i_1} < j_{i_2}$ such that $a^{j_{i_1}} = a^{j_{i_2}}$. Consequently, $a^{j_{i_2} - j_{i_1}} = e$. Next, we apply the algorithm from the proof of Lemma 3, using $m = j_{i_2} - j_{i_1}$ to find k. $\qquad\square$

Finally, Lemma 4 also applies in the PS-model.

4   Isomorphism and Basis for Abelian Groups

In this section, we prove the following.

Theorem 10. *There is a randomized algorithm that accesses the Cayley table of an Abelian group G at most $\tilde{O}(\sqrt{|G|})$ times and finds a basis for G.*

Proof. Given a group G, we start by running the procedure **Random Generators** for G. By Theorem 9, this algorithm runs in time $\tilde{O}(\sqrt{|G|})$ and gives a set of generators A_G for G with triangular relations.

Next, we apply Lemma 12 in [6] to construct the monomial Abelian group Γ_G that is isomorphic to G.

By Lemma 13 in [6], both the multiplication of two elements and computing the inverses in Γ_G can be performed in time $\tilde{O}(\log^2 |G|)$. We then use the algorithm by Kannan and Bachem [17] to find a basis for Γ_G. By Lemma 14 in [6], this algorithm runs in time $poly(\log |G|)$. Once we have a basis for Γ_G, we use the isomorphism between Γ_G and G in Lemma 12 in [6] to get the basis of G. $\square$

We now prove the following.

Theorem 11. *The Abelian group isomorphism problem can be solved in time $\tilde{O}(\sqrt{|G|})$.*

Proof. Given two groups G and H, by Theorem 10, we can find the basis for both G and H in time[5] $\tilde{O}(\sqrt{|G|})$. Using the Smith normal form, we can find groups G' and H' of the form $\mathbb{Z}_{m_1} \times \cdots \times \mathbb{Z}_{m_t}$ that are isomorphic to G and H, respectively. Then, checking whether G' is isomorphic to H' is straightforward. $\square$

5   Lower Bounds

In this section, we give all the lower bounds. We begin with some preliminary results, followed by two lemmas from which we derive the lower bounds.

5.1   Preliminary Results

Let $H_1 = \mathbb{Z}_{p^2}^m$ and $H_2 = \mathbb{Z}_{p^2}^{m-1} \times \mathbb{Z}_p^2$. The following is Lemma 27 in the full paper [6].

Lemma 7. *For uniformly random elements $h_1, \ldots, h_r \in H_2$ and $h'_1, \ldots, h'_r \in H_1$, and any $w_1, \ldots, w_r \in \mathbb{Z}_{p^2}$ not all zero*

$$\mathbf{Pr}[w_1 h'_1 + \cdots + w_r h'_r = 0^m] \leq \mathbf{Pr}[w_1 h_1 + \cdots + w_r h_r = 0^{m+1}] \leq \frac{1}{p^{m-1}}. \quad (1)$$

For uniformly random distinct elements $h_1, \ldots, h_r \in H_2$ and $h'_1, \ldots, h'_r \in H_1$, and for any $w_1, \ldots, w_r \in \mathbb{Z}_{p^2}$ not all zero

$$\mathbf{Pr}[w_1 h'_1 + \cdots + w_r h'_r = 0^m] \leq \mathbf{Pr}[w_1 h_1 + \cdots + w_r h_r = 0^{m+1}] \leq \frac{1}{p^{m-1}} + \frac{r^2}{2p^{2m}}. \quad (2)$$

---

[5] Here we assume that $|H| = \Theta(|G|)$. Otherwise, we can run the algorithm on both G and H in parallel and halt when the number of bases of one exceeds that of the other.

5.2   Three Lower Bounds Proofs

We first prove the following lower bound.

Lemma 8. *Let $\mathcal{G}$ be the set of all groups[6] that are isomorphic to either $H_1 = \mathbb{Z}_{p^2}^m$ or $H_2 = \mathbb{Z}_{p^2}^{m-1} \times \mathbb{Z}_p^2$. Any algorithm that, with probability at least $2/3$, determines whether $G \in \mathcal{G}$ is isomorphic to H_1 or H_2, must access the elements of G and its Cayley table at least $\Omega(p^{m/2-1/2}) = \Omega(|G|^{1/4}/\sqrt{p})$ times.*

Proof. We denote the group sum in H_r, for $r = 1, 2$, by $+_r$. Consider a set of abstract elements $\Sigma = \{\sigma_1, \ldots, \sigma_t\}$ with $t = p^{2m}$ and a bijective map $\phi : \Sigma \to H_r$ where $r = 1, 2$. Define the operation $\sigma_i +_{r,\phi} \sigma_j = \phi^{-1}(\phi(\sigma_i) +_r \phi(\sigma_j))$. Then $(\Sigma, +_{r,\phi})$ is a group isomorphic to H_r. Let $\mathcal{G}' = \{(\Sigma, +_{r,\phi}) | \phi \text{ bijective}, r = 1, 2\}$. Note that the algorithm does not know ϕ or r, but can compute $\sigma_i +_{r,\phi} \sigma_j$ by accessing the Cayley table.

We will use Yao's minimax principle [27,29]. We will show that for any deterministic algorithm $\mathcal{A}$ that decides whether a given group $G \in \mathcal{G}'$ is isomorphic to H_1 or H_2 (by outputting 1 or 2), if $\mathcal{A}$ is executed on a uniformly random group $G \in \mathcal{G}'$, then with probability at least $2/3$, $\mathcal{A}$ must access elements of G and entries of its Cayley table at least $\Omega(|G|^{1/4}/\sqrt{p})$ times.

Let $\mathcal{A}$ be a deterministic algorithm that accesses the elements of a random uniform group $G \in \mathcal{G}'$ and its Cayley table at most $T = p^{(m-1)/2}/20$ times and, with probability at least $2/3$, outputs $r \in \{1, 2\}$ if G is isomorphic to H_r. We assume that the commands of the algorithm are labeled with numbers $\{1, 2, \ldots\}$ and each command is one of the following types:

Type 1: $z_i \leftarrow \sigma_j$.   **Type 2**: $z_i \leftarrow z_j +_{r,\phi} z_k$.
Type 3: If $z_i = z_j$ Goto line w.   **Type 4**: Output z_i

Notice that commands of type 1 and 2 are used at most T times, while the command of type 3 can be used any number of times. This allows the algorithm to search in a table of already accessed elements without any additional cost.

Define the free $\mathbb{Z}_{p^2}$-module $L = \{\alpha_1 x_1 + \cdots + \alpha_w x_w | \alpha_i \in \mathbb{Z}_{p^2}, w \in [|G|]\}$ with coefficients in $\mathbb{Z}_{p^2}$, where $\{x_i\}_{i=1}^{\infty}$ are formal elements. Consider the same algorithm $\mathcal{A}$, modified so that each σ_j in the algorithm is replaced with the formal element x_j, each z_j with the new variable z_j', and $+_{r,\phi}$ with the group sum $+$ of L. Denote this modified algorithm as $\mathcal{A}'$. The elements created by $\mathcal{A}'$ are thus elements of the group L.

We then execute the algorithm $\mathcal{A}'$ until it terminates, and outputs $r_0 \in \{1, 2\}$. Note that, since in $\mathcal{A}'$ each σ_i is replaced with elements in L, $+_{r,\phi}$ is replaced with $+$ of L, and comparisons in the IF commands involve elements of L, the execution proceeds along a single, well-defined path P in the algorithm $\mathcal{A}'$.

In this execution (path P in $\mathcal{A}'$), the algorithm $\mathcal{A}'$ creates at most T variables z_i' where each variable is a linear combination of at most T formal elements x_i.

---

[6] Here and in all the lemmas below, we assume that the set of elements of the group is a subset of a fixed countable set. We make this assumption because, in classical set theory (ZFC), the class (or category) of all groups isomorphic to a given group does not form a set.

We will assume, without loss of generality, that the variables are $z'_1, z'_2, \ldots, z'_\ell$ with $\ell \leq T$ and the formal elements are $x_1, \ldots, x_T$. Thus, for every $i \in [\ell]$, we have $z'_i = z'_{i,1} x_1 + \cdots + z'_{i,T} x_T$, $z'_{i,j} \in \mathbb{Z}_{p^2}$, which is an element in L. If we follow the same execution path P in $\mathcal{A}$, disregarding the If command, we obtain $z_i = z'_{i,1} \sigma_1 +_{r,\phi} \cdots +_{r,\phi} z'_{i,T} \sigma_T$. This holds because, in the groups L, H_1, and H_2, the sum is taken modulo p^2.

Therefore, in all the "If $z_i = z_j$ Goto line w" commands along path P, if $z'_i = z'_j$, then $z_i = z_j$. If $z'_i \neq z'_j$, then by Lemma 7,

$$\mathbf{Pr}_{r,\phi}[z_i \neq z_j] = \mathbf{Pr}_{r,\phi}[z'_{i,1}\sigma_1 +_{r,\phi} \cdots +_{r,\phi} z'_{i,T}\sigma_T \neq z'_{j,1}\sigma_1 +_{r,\phi} \cdots +_{r,\phi} z'_{j,T}\sigma_T]$$

$$= \mathbf{Pr}_{r,\phi}\left[\phi^{-1}\left(\sum_{r=1}^{T}(z'_{i,r} - z_{j,r})\phi(\sigma_i)\right) \neq 0\right] \quad \text{Here the sum is } +_r$$

$$\geq \mathbf{Pr}_{r,\phi}\left[\sum_{r=1}^{T}(z'_{i,r} - z_{j,r})\phi(\sigma_i) \neq 0^{m+1}\right] \quad \text{Here the sum is } +_2$$

$$\geq 1 - \left(\frac{1}{p^{m-1}} + \frac{T^2}{2p^{2m}}\right).$$

Therefore, the probability that for all $i, j \in [T]$, if $z'_i \neq z'_j$, then $z_i \neq z_j$ is at least

$$1 - \left(\frac{T(T-1)}{2}\left(\frac{1}{p^m} + \frac{T^2}{2p^{2m}}\right)\right) \geq 1 - \frac{T^2}{p^{m-1}} - \frac{T^4}{p^{2m}} \geq \frac{99}{100}.$$

Thus, with probability, at least $99/100$, algorithm $\mathcal{A}$ follows the same path of execution P as $\mathcal{A}'$ and output r_0. So, if G is isomorphic to H_r, and $r \neq r_0$, the algorithm fails with probability at least $99/100$. Therefore, with probability at least $(1/2)(99/100) > 1/3$, the algorithm $\mathcal{A}$ fails. A contradiction. $\quad\square$

Lemma 9. *Let $\mathcal{G}$ be the set of all groups isomorphic to either $H_1 = \mathbb{Z}_p^{m-1}$ or $H_2 = \mathbb{Z}_p^m$. Any algorithm that, for $G \in \mathcal{G}$ of unknown size, decides with probability at least $2/3$ whether G is isomorphic to H_1 or H_2 must access an oracle that selects uniformly random elements of G and access the Cayley table of G at least $\Omega(p^{m/2-1/2}/\log p) = \Omega(|G|^{1/2}/(\sqrt{p}\log p))$ times.*

Proof. Since $p(\mathbb{Z}_{p^2}^{m-1} \times \mathbb{Z}_p^2)$ is isomorphic to $\mathbb{Z}_p^{m-1}$ and $p\mathbb{Z}_{p^2}^m$ is isomorphic to $\mathbb{Z}_p^m$, if there exists an algorithm that can distinguish between $H_1 = \mathbb{Z}_p^{m-1}$ and $H_2 = \mathbb{Z}_p^m$ in time T, then we can solve the problem in Lemma 8 in time $T\log p$. Since by Lemma 8, $T\log p = \Omega(p^{m/2-1/2})$, the result follows. $\quad\square$

Lemma 10. *Let $\mathcal{G}$ be the set of all groups that are isomorphic to either $D_1 = \mathbb{Z}_p^m$ or $D_2 = \mathbb{Z}_p^{m-2} \times \mathbb{Z}_{p^2}$. Any deterministic algorithm that determines whether $G \in \mathcal{G}$ is isomorphic to D_1 or D_2, must access the elements of G and its Cayley table at least $\Omega(p^{m-2}) = \Omega(|G|/p^2)$ times.*

Proof. An adversary can provide elements from $W = \mathbb{Z}_p^{m-2} \times \{(0,0)\}$ for each access to an element σ_i. The sum of any two elements in W remains in W, preventing the algorithm from distinguishing between the Abelian groups D_1 and D_2 until the $(p^{m-2}+1)$-th element is requested. At this point, the adversary must reveal whether the hidden group is D_1 or D_2. Thus, the complexity of the algorithm is at least $\Omega(p^{m-2}) = \Omega(n)$ for any constant p. $\quad\square$

5.3 Lower Bounds for Isomorphism and Basis

In this section, we prove the lower bounds. We first show

Theorem 12. *In the FS-model, the following problems cannot be solved in time less than $\Omega(n^{1/4})$.*

1. *Given two Abelian groups G and H of size n, decide if G is isomorphic to H.*
2. *Given an Abelian group G of size n, find a basis for G.*
3. *Given an Abelian group G of size n, find a set of generators for G with relations of size at most $n^{o(1)}$.*

Proof. If there exists an algorithm that solves the isomorphism problem in time T, then the same algorithm can solve the problem in Lemma 8 for $p = O(1)$ in time T. Therefore, $T = \Omega(|G|^{1/4})$. This proves item 1.

If there exists an algorithm that finds a basis for G in time T, then the problem in Lemma 8 can be solved in time $O(T)$ as follows: Given H that is either isomorphic to $H_1 = \mathbb{Z}_{p^2}^m$ or $H_2 = \mathbb{Z}_{p^2}^{m-1} \times \mathbb{Z}_p^2$, find the basis $\{a_1, \ldots, a_t\}$ for H. If $t = m$, then H is isomorphic to H_1; otherwise, $t = m + 1$ and H is isomorphic to H_2. This proves item 2.

If there is an algorithm that finds generators for G with relations of size at most $t = n^{o(1)}$, then by the result of Kannan and Bachem [17] result, one can find the basis of G in time $poly(t, \log n) = n^{o(1)}$. This proves item 3. □

Theorem 13. *In the PS-model, the following problems cannot be solved in time less than $\Omega(|G|^{1/2})$.*

1. *Given two Abelian groups G and H, decide if G is isomorphic to H.*
2. *Given an Abelian group G, find a basis for G.*
3. *Given an Abelian group G, find generators for G with relations of size at most $n^{o(1)}$.*

Proof. We use the same reductions as in the proof of Theorem 12, combined with Lemma 9. □

For deterministic algorithms, we have.

Theorem 14. *For deterministic algorithms, the following problems cannot be solved in time less than $\Omega(|G|)$.*

1. *Given two Abelian groups G and H, decide if G is isomorphic to H.*
2. *Given an Abelian group G, find a basis for G.*
3. *Given an Abelian group G, find generators for G with relations of size at most $n^{o(1)}$.*

Proof. We use the same reductions as in the proof of Theorem 12, combined with Lemma 10. □

References

1. Babai, L.: Graph isomorphism in quasipolynomial time [extended abstract]. In: Proceedings of the 48th Annual ACM SIGACT Symposium on Theory of Computing, STOC 2016, Cambridge, MA, USA, 18–21 June 2016, pp. 684–697 (2016). https://doi.org/10.1145/2897518.2897542
2. Babai, L., Codenotti, P., Grochow, J.A., Qiao, Y.: Code equivalence and group isomorphism. In: Proceedings of the Twenty-Second Annual ACM-SIAM Symposium on Discrete Algorithms, SODA 2011, San Francisco, California, USA, 23–25 January 2011, pp. 1395–1408 (2011). https://doi.org/10.1137/1.9781611973082.107
3. Babai, L., Qiao, Y.: Polynomial-time isomorphism test for groups with abelian sylow towers. In: 29th International Symposium on Theoretical Aspects of Computer Science, STACS 2012, Paris, France, 29th February–3rd March 2012, pp. 453–464 (2012). https://doi.org/10.4230/LIPICS.STACS.2012.453
4. Babai, L., Szemerédi, E.: On the complexity of matrix group problems I. In: 25th Annual Symposium on Foundations of Computer Science, West Palm Beach, Florida, USA, 24–26 October 1984, pp. 229–240 (1984). https://doi.org/10.1109/SFCS.1984.715919
5. Borges-Quintana, M., Borges-Trenard, M.A., Martínez-Moro, E.: On the use of Gröbner bases for computing the structure of finite abelian groups. In: Ganzha, V.G., Mayr, E.W., Vorozhtsov, E.V. (eds.) CASC 2005. LNCS, vol. 3718, pp. 52–64. Springer, Heidelberg (2005). https://doi.org/10.1007/11555964_5
6. Bshouty, N.H.: Sublinear time algorithms for abelian group isomorphism and basis construction. CoRR abs/2504.02387 (2025). https://doi.org/10.48550/ARXIV.2504.02387
7. Buchmann, J., Schmidt, A.: Computing the structure of a finite abelian group. Math. Comput. **74**(252), 2017–2026 (2005). https://doi.org/10.1090/S0025-5718-05-01740-0
8. Chen, L., Fu, B.: Linear and sublinear time algorithms for the basis of abelian groups. Theor. Comput. Sci. **412**(32), 4110–4122 (2011). https://doi.org/10.1016/J.TCS.2010.06.011
9. Chen, Z., Grochow, J.A., Qiao, Y., Tang, G., Zhang, C.: On the complexity of isomorphism problems for tensors, groups, and polynomials III: actions by classical groups. In: 15th Innovations in Theoretical Computer Science Conference, ITCS 2024, Berkeley, CA, USA, 30 January–2 February 2024, pp. 31:1–31:23 (2024). https://doi.org/10.4230/LIPICS.ITCS.2024.31
10. Dietrich, H., Wilson, J.B.: Group isomorphism is nearly-linear time for most orders. In: 62nd IEEE Annual Symposium on Foundations of Computer Science, FOCS 2021, Denver, CO, USA, 7–10 February 2022, pp. 457–467 (2021). https://doi.org/10.1109/FOCS52979.2021.00053
11. Gall, F.L., Yoshida, Y.: Property testing for cyclic groups and beyond. J. Comb. Optim. **26**(4), 636–654 (2013). https://doi.org/10.1007/S10878-011-9445-8
12. Garzon, M.H., Zalcstein, Y.: On isomorphism testing of a class of 2-nilpotent groups. J. Comput. Syst. Sci. **42**(2), 237–248 (1991). https://doi.org/10.1016/0022-0000(91)90012-T
13. Grochow, J.A., Qiao, Y.: Algorithms for group isomorphism via group extensions and cohomology. SIAM J. Comput. **46**(4), 1153–1216 (2017). https://doi.org/10.1137/15M1009767
14. Grochow, J.A., Qiao, Y.: On p-group isomorphism: search-to-decision, counting-to-decision, and nilpotency class reductions via tensors. ACM Trans. Comput. Theory **16**(1), 2:1–2:39 (2024). https://doi.org/10.1145/3625308

15. Iliopoulos, C.S.: Analysis of algorithms on problems in general abelian groups. Inf. Process. Lett. **20**(4), 215–220 (1985). https://doi.org/10.1016/0020-0190(85)90052-3

16. Iliopoulos, C.S.: Worst-case complexity bounds on algorithms for computing the canonical structure of finite abelian groups and the hermite and smith normal forms of an integer matrix. SIAM J. Comput. **18**(4), 658–669 (1989). https://doi.org/10.1137/0218045

17. Kannan, R., Bachem, A.: Polynomial algorithms for computing the smith and hermite normal forms of an integer matrix. SIAM J. Comput. **8**(4), 499–507 (1979). https://doi.org/10.1137/0208040

18. Karagiorgos, G., Poulakis, D.: Efficient algorithms for the basis of finite abelian groups. Discret. Math. Algorithms Appl. **3**(4), 537–552 (2011). https://doi.org/10.1142/S1793830911001401

19. Karagiorgos, G., Poulakis, D.: Linear time algorithms for the basis of abelian groups. In: Fu, B., Du, D.-Z. (eds.) COCOON 2011. LNCS, vol. 6842, pp. 456–466. Springer, Heidelberg (2011). https://doi.org/10.1007/978-3-642-22685-4_40

20. Kavitha, T.: Linear time algorithms for abelian group isomorphism and related problems. J. Comput. Syst. Sci. **73**(6), 986–996 (2007). https://doi.org/10.1016/j.jcss.2007.03.013

21. Koblitz, N., Menezes, A.: A survey of public-key cryptosystems. SIAM Rev. **46**(4), 599–634 (2004). https://doi.org/10.1137/S0036144503439190

22. Miller, G.L.: On the n log n isomorphism technique: a preliminary report. In: Proceedings of the 10th Annual ACM Symposium on Theory of Computing, San Diego, California, USA, 1–3 May 1978, pp. 51–58 (1978). https://doi.org/10.1145/800133.804331

23. Lipton, R.J., Snyder, L., Zalcstein, Y.: The Complexity of Word and Isomorphism Problems for Finite Groups. John Hopkins (1976)

24. Rosenbaum, D.J.: Bidirectional collision detection and faster deterministic isomorphism testing. CoRR abs/1304.3935 (2013). http://arxiv.org/abs/1304.3935

25. Savage, C.: An $o(n^2)$ algorithm for abelian group isomorphism. Technical report, North Carolina State University (1980)

26. Teske, E.: The Pohlig-Hellman method generalized for group structure computation. J. Symb. Comput. **27**(6), 521–534 (1999)

27. Vazirani, V.V.: Approximation Algorithms. Springer, Heidelberg (2001), see Section 2.4, "Yao's Minimax Principle", pp. 24–26

28. Vikas, N.: An $o(n)$ algorithm for abelian p-group isomorphism and an $o(n \log n)$ algorithm for abelian group isomorphism. J. Comput. Syst. Sci. **53**(1), 1–9 (1996). https://doi.org/10.1006/JCSS.1996.0045

29. Yao, A.C.: Probabilistic computations: toward a unified measure of complexity. SIAM J. Comput. **10**(2), 281–290 (1981)

Towards an Algebraic Approach to the Reconfiguration CSP

Kei Kimura[(✉)]

Kyushu University, 744, Motooka, Nishi-ku, Fukuoka, Japan
`kimura.kei.993@m.kyushu-u.ac.jp`

Abstract. This paper investigates the reconfiguration variant of the Constraint Satisfaction Problem (CSP), referred to as the Reconfiguration CSP (RCSP). Given a CSP instance and two of its solutions, RCSP asks whether one solution can be transformed into the other via a sequence of intermediate solutions, each differing by the assignment of a single variable. RCSP has attracted growing interest in theoretical computer science, and when the variable domain is Boolean, the computational complexity of RCSP exhibits a dichotomy depending on the allowed constraint types. A notable special case is the reconfiguration of graph homomorphisms—also known as graph recoloring—which has been studied using topological methods. We propose a novel algebraic approach to RCSP, inspired by techniques used in classical CSP complexity analysis. Unlike traditional methods based on total operations, our framework employs partial operations to capture a reduction involving equality constraints. This perspective facilitates the extension of complexity results from Boolean domains to more general settings, demonstrating the versatility of partial operations in identifying tractable RCSP instances.

Keywords: Constraint satisfaction · Combinatorial reconfiguration · algebraic approach

1 Introduction

The Constraint Satisfaction Problem (CSP) asks whether there exists an assignment of values to variables that satisfies all given constraints. Owing to its expressive power, CSP has been widely studied across mathematics, artificial intelligence, and computer science [1,9]. Numerous variants have been explored, including optimization [35], counting [11], and reconfiguration [15].

A central approach to analyzing CSP complexity involves restricting the constraint language, enabling the study of specific problems such as the Boolean satisfiability problem (SAT) and graph coloring. Algebraic methods have been particularly successful in this context, characterizing complexity via operations preserving constraint languages. These techniques have led to dichotomy theorems [6,33,34] and have been extended to optimization [35] and fine-grained complexity analyses [19,22].

© The Author(s), under exclusive license to Springer Nature Switzerland AG 2026
J. Kozik and A. Wolff (Eds.): SOFSEM 2026, LNCS 16448, pp. 46–60, 2026.
https://doi.org/10.1007/978-3-032-17801-5_4

This work focuses on the Reconfiguration CSP (RCSP), which asks whether one solution can be transformed into another through a sequence of intermediate solutions. RCSP has been actively studied as part of combinatorial reconfiguration [17], with connections to SAT algorithm behavior and worst-case analysis [15,31]. Graph recoloring, a reconfiguration variant of graph homomorphism, is a notable special case and has been analyzed under fixed codomain graphs [2,4,8], aligning with constraint language restrictions.

A seminal work by Gopalan et al. [15] established a dichotomy for the Boolean RCSP with constants[1]: the problem is solvable in polynomial time under a certain condition and PSPACE-complete otherwise, with the criterion later refined by Schwerdtfeger [31]. Subsequent studies have explored structural restrictions [16], as well as connectivity properties of CSP solution spaces [5].

When the constraint language consists of a single binary relation, RCSP coincides with the digraph recoloring problem, whose complexity has been widely studied. For instance, K_3-recoloring is polynomial-time solvable [8], while K_k-recoloring is PSPACE-complete for $k \geq 4$ [2]. Brewster et al. [4] extended this dichotomy to circular coloring, showing that $C_{p,q}$-recoloring is tractable when $2 \leq p/q < 4$ and PSPACE-complete otherwise. As a major result concerning the polynomial-time solvability of graph recoloring, Wrochna [32] introduced a topological approach proving polynomial-time solvability for square-free graphs. This method inspired further results for reflexive and loopless digraphs under various girth conditions [3,23,24].

Despite its relevance, no systematic method exists for analyzing RCSP complexity under constraint language restrictions, and algebraic approaches remain unexplored. Nevertheless, it is noteworthy that approaches based on topological methods have yielded meaningful results for the graph recoloring problem [3,23,24,32].

Our Contribution. In this study, we introduce an algebraic framework for analyzing the computational complexity of the Reconfiguration CSP (RCSP) under constraint language restrictions, using partial operations—operations undefined for some inputs. We focus on partial operations that preserve constraint languages, meaning that if the operation is defined on a set of solutions, the result remains a solution.

We first show that partial operations can capture a reduction between RCSPs when equality constraints are allowed, implying that the complexity of such RCSPs is governed by the set of partial operations under which these are invariant.

Next, we analyze tractable RCSPs in the Boolean domain. Prior work by Gopalan et al. [15] and Schwerdtfeger [31] identifies three conditions—safe OR-freeness, safe NAND-freeness, and safe componentwise bijunctivity—under which RCSPs are solvable in polynomial time. We demonstrate that safe OR-freeness (resp., safe NAND-freeness) is captured by a single partial operation derived from the Maltsev operation, a well-studied polymorphism in the litera-

---

[1] Referred to as the *st*-connectivity problem in [15].

ture on CSPs. We further extend this operation to larger domains, yielding new tractable RCSP classes.

We also show that safe componentwise bijunctivity can be characterized by partial operations, but unlike safe OR-freeness, it cannot be captured by any finite set of them, highlighting a fundamental difference in their algebraic structure.

Organization. The structure of this paper is as follows. Section 2 introduces the formal definition of RCSPs and reviews previously known results. Section 3 focuses on safe OR-freeness, while Sect. 4 discusses safe componentwise bijunctivity. Finally, Sect. 5 concludes the paper.

From this section onwards, due to page limitations, many proofs are presented only in sketch form. For complete proofs, please refer to the full version [20] of this paper.

2   Preliminaries

In this section, we first define the Reconfiguration CSP (RCSP). Then, in Sect. 2.2, we introduce partial operations and explain the reduction relationships between RCSPs. Furthermore, in Sect. 2.3, we summarize known results in the context of Boolean RCSPs.

2.1   Definitions

For a positive integer r we denote $\{1, \ldots, r\}$ by $[r]$.

Throughout the paper, D denotes a finite domain whose size is equal or greater than two.

Definition 1. *An r-ary relation on D is a subset of D^r, where $r \geq 1$.*

Example 1. The *diagonal (or equality) relation* Δ_D on D is a binary relation $\Delta_D = \{(d, d) \in D^2 \mid d \in D\} \subseteq D^2$. The *inequality relation* $\neq_D$ on D is a binary relation $\neq_D = \{(c, d) \in D^2 \mid c \neq d\} \subseteq D^2$. For each $d \in D$, let C_d denote the singleton unary relation $C_d = \{d\} \subseteq D$. The r-ary *empty relation* $\emptyset^{(r)}$ is the empty set $\emptyset \subseteq D^r$, where $r \geq 1$.

Definition 2. *A* constraint language *is a finite set of non-empty, finitary relations on D.*

Definition 3. *For a (not necessarily finite) set of relations Γ, a Γ_C-formula I is a conjunction of the form*

$$I = \bigwedge_{i=1}^{m} R_i(\xi_1^i, \ldots, \xi_{r_i}^i), \tag{1}$$

where m is a positive integer, each R_i is an r_i-ary relation from Γ, and the ξ_j^i are (not necessarily distinct) variables or elements of D (also called constants*). If all ξ_j^i are variables, then I is called a Γ-formula.*

We denote the set of variables that occur in I by $\mathrm{VAR}(I)$. An assignment $t : \mathrm{VAR}(I) \to D$ satisfies I *or is a* solution *of I if for all $i \in [m]$ it holds that $(t(\xi_1^i), \ldots, t(\xi_{r_i}^i)) \in R_i$, where we define $t(d) = d$ for every $d \in D$.*

Note that each Γ_C-formula defines the relation of its satisfying assignments (or solutions). Indeed, we assume that $\mathrm{VAR}(I) = \{x_1, \ldots, x_n\}$ for a Γ_C-formula I, and identify an assignment t with a tuple of $\boldsymbol{t}$ in D^n such that $t_j = t(x_j)$ for all $j \in [n]$. We also denote by $s(I)$ the set of solutions of I and regard it as an n-ary relation, i.e., $s(I) \subseteq D^n$. We say that $s(I)$ is *expressed* by a Γ_C-formula.

Example 2. Let $D = \{0, 1\}$ and $\Gamma_{2\mathrm{SAT}} = \{R_{ij} \mid i, j \in \{0, 1\}\}$, where $R_{ij} = \{0, 1\}^2 \setminus \{(i, j)\}$. Consider an instance (2-CNF) φ of 2-SAT $\varphi(x_1, x_2, x_3) = (x_1 \vee x_2) \wedge \overline{x_2} \wedge x_3$. Then φ can be represented as a Γ_C-formula $I_\varphi = R_{00}(x_1, x_2) \wedge R_{01}(0, x_2) \wedge R_{00}(x_3, x_3)$. The set $s(I_\varphi) = \{(1, 0, 1)\} \subseteq \{0, 1\}^3$.

Following Gopalan et al. [15], we consider constraint satisfaction problems (CSPs) that allow constant assignments.

Definition 4 (CSP). *Let Γ be a constraint language on D. The* constraint satisfaction problem (CSP) *on Γ with constants, denoted by $\mathrm{CSP}_C(\Gamma)$, is a problem of determining if a given Γ_C-formula has a solution or not.*

Example 3. The 2-SAT problem is a CSP with constants over the Boolean domain $\{0, 1\}$. In addition, the graph k-coloring problem (with prescribed colors) is equivalent to $\mathrm{CSP}_C(\{\neq_D\})$, where $D = \{0, 1, \ldots, k - 1\}$.

Now, we define the reconfiguration CSP (RCSP).

Definition 5 (Solution graph). *For a relation $R \subseteq D^r$, the* solution graph $G(R) = (V(R), E(R))$ *associated with R is an undirected graph defined as follows. Its vertex set is R, i.e., $V(R) = R$. Moreover, for $\boldsymbol{x}, \boldsymbol{y} \in R$, we have $\{\boldsymbol{x}, \boldsymbol{y}\} \in E(R)$ if and only if $\mathrm{dist}(\boldsymbol{x}, \boldsymbol{y}) = 1$, where $\mathrm{dist}(\boldsymbol{x}, \boldsymbol{y}) = |\{j \mid x_j \neq y_j\}|$ is the Hamming distance of $\boldsymbol{x}$ and $\boldsymbol{y}$.*

For a CSP instance I, the solution graph $G(s(I))$ of $s(I)$ is denoted by $G(I)$ for brevity.

Definition 6 (RCSP). *Let Γ be a constraint language on D. The* reconfiguration constraint satisfaction problem (RCSP) *with constants on Γ, denoted by $\mathrm{RCSP}_C(\Gamma)$, consists of the instances of the form $(I, \boldsymbol{s}, \boldsymbol{t})$, where I is an instance of $\mathrm{CSP}_C(\Gamma)$ and $\boldsymbol{s}, \boldsymbol{t}$ are solutions of I. An instance $(I, \boldsymbol{s}, \boldsymbol{t})$ is a* yes *instance if $\boldsymbol{s}$ and $\boldsymbol{t}$ are in the same connected component in $G(I)$, and a* no *instance otherwise.*

For CSPs, the computational complexity of $\mathrm{CSP}(\Gamma)$ remains unchanged when Γ is retracted to its core and all constant relations are added. Therefore, considering CSPs with constants does not impose any restriction in the context of complexity classification. In contrast, for RCSPs, retracting to the core may alter the yes/no status of instances, so it is not clear whether considering RCSPs with constants is without loss of generality.

2.2 Partial Polymorphism and Logical Expression (reduction)

In the algebraic approach to the CSP, the computational complexity of the CSP is characterized using total operations. In this subsection, we show that the computational complexity of the RCSP is characterized using *partial* operations.

Definition 7 (Partial operation). *A k-ary partial operation $f : D^k \rightharpoonup D$ on D is a mapping $f : D' \to D$ for some $D' \subseteq D^k$. Here, D' is called the domain of f and is denoted by $\mathrm{dom}(f)$. If $\boldsymbol{x} \in D'$, then we say that $f(\boldsymbol{x})$ is defined and otherwise (that is, if $\boldsymbol{x} \notin D'$) undefined. If $\mathrm{dom}(f) = D^k$, then f is a total operation.*

Example 4 ([21, Example 3.7], [22, Definition 3.19]). The partial Maltsev operation $M_p : D^3 \rightharpoonup D$ on D is defined as follows. Firstly, its domain $\mathrm{dom}(M_p) = \{(x,y,y) \mid x,y \in D\} \cup \{(y,y,x) \mid x,y \in D\} \subseteq D^3$. Then, for all $x,y \in D$, $M_p(x,y,y) = M_p(y,y,x) = x$.

Definition 8. *Let f and g be two partial operations on D with the same arity. We say that f is a subfunction of g if $\mathrm{dom}(f) \subseteq \mathrm{dom}(g)$ and $f(\boldsymbol{x}) = g(\boldsymbol{x})$ for every $\boldsymbol{x} \in \mathrm{dom}(f)$. We write $f \leq g$ to denote this relation.*

The partial Maltsev operation is a partial version (more precisely, a subfunction) of a Maltsev operation, where Maltsev operations have been extensively studied in the algebraic approach to constraint satisfaction problems (CSPs). Here, a total operation $M : D^3 \to D$ is called *Maltsev* if for all $x,y \in D$, it holds that $M(x,y,y) = M(y,y,x) = x$. A Maltsev operation is a generalization of the affine operation $f(x,y,z) = x - y + z$, under which the solution set of linear equations is invariant.

A partial operation f on D is called *idempotent* if $f(x,\ldots,x) = x$ holds for every $x \in D$. Note that the partial Maltsev operation is idempotent.

Now, we define the notion of partial polymorphism (e.g., [22]), a key concept in the algebraic approach to RCSPs. Intuitively, a partial operation is said to be a partial polymorphism of a constraint language Γ if, whenever it is defined on several tuples from a relation in Γ, the result of applying the operation to those tuples also belongs to the same relation.

For a k-ary partial operation f on D and tuples $\boldsymbol{x}^{(1)}, \ldots, \boldsymbol{x}^{(k)}$ in D^r, define $f(\boldsymbol{x}^{(1)}, \ldots, \boldsymbol{x}^{(k)})$

as $f(\boldsymbol{x}^{(1)}, \ldots, \boldsymbol{x}^{(k)}) = \left(f(x_1^{(1)}, \ldots, x_1^{(k)}), \ldots, f(x_r^{(1)}, \ldots, x_r^{(k)}) \right) \in D^r$, where $\boldsymbol{x}^{(i)} = (x_1^{(i)}, \ldots, x_r^{(i)})$ for each $i \in [k]$. If $f(x_j^{(1)}, \ldots, x_j^{(k)})$ is defined for all $j \in [r]$, then we say $f(\boldsymbol{x}^{(1)}, \ldots, \boldsymbol{x}^{(k)})$ is *defined*; otherwise it is *undefined*.

Definition 9 (Partial polymorphism).

- *A k-ary partial operation f on D is a partial polymorphism of an r-ary relation $R \subseteq D^r$ if for any $\boldsymbol{x}^{(1)}, \ldots, \boldsymbol{x}^{(k)} \in R$ such that $f(\boldsymbol{x}^{(1)}, \ldots, \boldsymbol{x}^{(k)})$ is defined, we have $f(\boldsymbol{x}^{(1)}, \ldots, \boldsymbol{x}^{(k)}) \in R$.*
- *If f is a partial polymorphism of R, R is* invariant *under f.*

- *For a set Γ of relations, f is called a partial polymorphism of Γ if it is a polymorphism of every relation in Γ. In this case, Γ is* invariant *under f.*

Definition 10 (pPol and Inv). *Let Γ be a set of relations on D. The set $\mathrm{pPol}(\Gamma)$ denotes the set of partial operations that are a partial polymorphism of Γ, i.e.,*

$$\mathrm{pPol}(\Gamma) = f \mid \text{for every } R \in \Gamma, f \text{ is a partial polymorphism of } R.$$

Let F be a set of partial operations on D. The set $\mathrm{Inv}(F)$ denotes the set of relations that are invariant under every partial operation in F, i.e.,

$$\mathrm{Inv}(F) = R \mid \text{for every } f \in F, R \text{ is invariant under } f.$$

The fact that the set of partial polymorphisms is closed under taking subfunctions is well established in the literature and follows directly from the definition.

Lemma 1. *Let Γ be a set of relations. If $f \leq g \in \mathrm{pPol}(\Gamma)$, then $f \in \mathrm{pPol}(\Gamma)$.*

Proposition 1 below states that the relations that are invariant under partial polymorphisms can be characterized using a logical expression.

Definition 11. *Let Γ be a set of relations. $\langle \Gamma \rangle_{\wedge,=,\mathbf{f}}$ is the set of relations that can be expressed as a $(\Gamma \cup \{\Delta_D\} \cup \{\emptyset^{(1)}\})$-formula.*

Proposition 1 ([14,28]). *Let Γ be a constraint language on D. Then $\mathrm{Inv}(\mathrm{pPol}(\Gamma)) = \langle \Gamma \rangle_{\wedge,=,\mathbf{f}}$.*

Consequently, Γ can be characterized by partial operations if and only if $\Gamma = \langle \Gamma \rangle_{\wedge,=,\mathbf{f}}$. Moreover, the idempotent partial polymorphisms of a constraint language Γ are inherited by the set of solutions of an instance of $\mathrm{CSP_C}(\Gamma)$ as shown below.

Corollary 1. *Let Γ be a constraint language on D. Let I be an instance of $\mathrm{CSP_C}(\Gamma)$ and f be an idempotent partial polymorphism of Γ. Then $s(I)$ is invariant under f.*

Proof. Note that f is a partial polymorphism of $\Gamma \cup \bigcup_{d \in D}\{C_d\}$, since it is idempotent. Then the statement follows from Proposition 1, since $s(I)$ can be represented as a Γ_C-formula.

The following theorem shows that the inclusion relation of partial polymorphisms leads to a reduction between RCSPs provided that they have the equality relation. This allows us to characterize the computational complexity of RCSPs having the equality relation by the set of partial polymorphisms. The following proof is similar to that of Theorem 10 in [19].

Theorem 1. *Let Γ_1 and Γ_2 be constraint languages on D. Assume that the equality relation Δ_D is contained in Γ_1 and that $\mathrm{pPol}(\Gamma_1) \subseteq \mathrm{pPol}(\Gamma_2)$. Then $\mathrm{RCSP_C}(\Gamma_2)$ is polynomial-time reducible to $\mathrm{RCSP_C}(\Gamma_1)$.*

Proof. Given an instance $(I, \boldsymbol{s}, \boldsymbol{t})$ of $\mathrm{RCSP_C}(\Gamma_2)$ with n variables, we transform it into an equivalent instance $(I', \boldsymbol{s}', \boldsymbol{t}')$ of $\mathrm{RCSP_C}(\Gamma_1)$. From Proposition 1, every constraint $R(\xi_1, \ldots, \xi_r)$ in I can be replaced with constraints

$$R_1(\xi_{11}, \ldots, \xi_{1r_1}) \wedge \cdots \wedge R_\ell(\xi_{\ell 1}, \ldots, \xi_{\ell r_\ell}), \tag{2}$$

where $R_1, \ldots, R_\ell \in \Gamma_1$ and $\xi_{11}, \ldots, \xi_{\ell r_\ell} \in \{\xi_1, \ldots, \xi_r\} \cup D$. The resulting instance I' has the same set of solutions as that of I. By setting $\boldsymbol{s}' = \boldsymbol{s}$ and $\boldsymbol{t}' = \boldsymbol{t}$, we have that $(I, \boldsymbol{s}, \boldsymbol{t})$ is a yes instance if and only if $(I', \boldsymbol{s}', \boldsymbol{t}')$ is a yes instance. Since each constraint in I is replaced in constant time, the above reduction can be done in polynomial time.

2.3 Known Results for the Boolean RCSP

We summarize the results by Gopalan et al. [15] and Schwerdtfeger [31] for the RCSP on the Boolean domain, i.e., the case of $D = \{0, 1\}$.

For an r-ary relation R on D and $0 < k < r$, we can define an $(r - k)$-ary relation $R'(x_1, \ldots, x_{r-k}) = R(\xi_1, \ldots, \xi_n)$, where each $\xi_j \in \{x_1, \ldots, x_{r-k}\} \cup D$. If each variable x_j occurs at most once in $(\xi_1, \ldots, \xi_n)$, we say that R' is obtained from R by *substitution of constants*. If each ξ_j is a variable, we say that R' is obtained from R by *identification of variables*.

Example 5. Recall the example in Example 2. There, $R'_{01}(x_2) := R_{01}(0, x_2)$ is obtained from R_{01} by substitution of constants. Moreover, $R'_{00}(x_3) := R_{00}(x_3, x_3)$ is obtained from R_{00} by identification of variables.

Definition 12. *Let R be a relation on $\{0, 1\}$.*

- *R is OR-free (resp., NAND-free) if the binary relation OR $= \{(0, 1), (1, 0), (1, 1)\}$ (resp., NAND $= \{(0, 0), (0, 1), (1, 0)\}$) cannot be obtained from R by substitution of constants.*
- *R is safely OR-free (resp., safely NAND-free) if R and every relation R' obtained from R by identification of variables is OR-free (resp., NAND-free).*

We denote the set of safely OR-free (resp., safely NAND-free) relations by Γ_{SOF} (resp., Γ_{SNF}).

Definition 13. *Let R be a relation on $\{0, 1\}$.*

- *R is bijunctive if it is the set of solutions of a 2-CNF-formula.*
- *R is componentwise bijunctive if every connected component $G(R)$ is a bijunctive relation.*
- *R is safely componentwise bijunctive if R and every relation R' obtained from R by identification of variables is componentwise bijunctive.*

We denote the set of safely componentwise bijunctive relations by Γ_{SCB}.

Now, we introduce the relations investigated in [15, 31].

Definition 14. *A set Γ of relations on $\{0,1\}$ is* tight *(resp., safely tight) if at least one of the following conditions holds:*

- *every relation in Γ is componentwise bijunctive (resp., safely componentwise bijunctive).*
- *every relation in Γ is OR-free (resp., safely OR-free).*
- *every relation in Γ is NAND-free (resp., safely NAND-free).*

The following dichotomy for the computational complexity of $\mathrm{RCSP_C}$ on the Boolean domain is essentially shown by Gopalan et al. [15]. However, the result is later corrected by Schwerdtfeger [31] by adding "safely" to the condition for polynomial-time solvability.

Theorem 2 ([15, Theorem 2.9] and [31]). *Let Γ be a constraint language on $\{0,1\}$. If Γ is safely tight, then $\mathrm{RCSP_C}(\Gamma)$ is in P; otherwise, $\mathrm{RCSP_C}(\Gamma)$ is PSPACE-complete.*

3 Safely OR-Free (NAND-Free) Relations and Their Extension

In this section, we characterize the sets of safely OR-free and safely NAND-free relations, denoted by Γ_{SOF} and Γ_{SNF} respectively, using a single partial operation on $\{0,1\}$. We then extend this operation to larger domains, yielding a class of RCSPs solvable in polynomial time. Finally, in Subsect. 3.3, we show that this class encompasses several known CSP and graph homomorphism cases, including instances where the tractability of RCSPs was previously unknown.

3.1 Ordered Partial Maltsev Operation and Characterization of Γ_{SOF}

First, we introduce a new partial operation based on the Maltsev operation.

Definition 15. *Let D be a finite set equipped with a total order $\leq$. The $(D,\leq)$-Maltsev operation $M_{D,\leq}$ is the ternary partial operation $M : D^3 \rightharpoonup D$ that satisfies $M(x,y,y) = M(y,y,x) = x$ for any $x,y \in D$ with $x \leq y$ and $M(x,y,z)$ being undefined for the other inputs.*

We call a partial operation ordered partial Maltsev *if it coincides with $M_{D,\leq}$ for some totally ordered finite set $(D,\leq)$.*

By definition, an ordered partial Maltsev operation is a subfunction of the partial Maltsev operation. Note that by definition ordered partial Maltsev operations are idempotent.

Example 6. When $D = \{0,1\}$ and $0 < 1$, the $(D,\leq)$-*Maltsev* operation $M_{D,\leq}$ is a ternary partial operation $M_{D,\leq} : \{0,1\}^3 \rightharpoonup \{0,1\}$ such that $M(0,0,0) = M(0,1,1) = M(1,1,0) = 0$, $M(1,1,1) = 1$, and undefined for the other inputs.

Example 7. When $D = \{0,1\}$ and $1 < 0$, the $(D, \leq)$-*Maltsev* operation $M_{D,\leq}$ is a ternary partial operation $M_{D,\leq} : \{0,1\}^3 \rightharpoonup \{0,1\}$ such that $M(0,0,0) = 0$, $M(1,0,0) = M(0,0,1) = M(1,1,1) = 1$, and undefined for the other inputs.

We will show that the partial operations in Examples 6 and 7 characterize the sets of safely OR-free and safely NAND-free relations, respectively.

Lemma 2. *Let* $D = \{0,1\}$ *and* $0 < 1$. *Let* r *be a positive integer and* R *be an* r-*ary relation* $R \subseteq \{0,1\}^r$. *Then,* R *is safely OR-free if and only if it is invariant under the* $(D, \leq)$-*Maltsev operation.*

Proof. (sketch). We only show the only-if part, which is proven by contradiction. Assume that R is invariant under $M_{D,\leq}$ but not safely OR-free. Then there exists $\xi \in \{x_1, x_2, 0, 1\}^r$ such that $R'(x_1, x_2) := R(\xi_1, \ldots, \xi_r)$ is OR$(= \{(0,1), (1,0), (1,1)\})$. It follows that, up to permutation, R contains tuples

$$t^1 = (\overbrace{0, \ldots, 0}^{\xi_i = x_1}, \overbrace{1, \ldots, 1}^{\xi_i = x_2}, \overbrace{0, \ldots, 0}^{\xi_i = 0}, \overbrace{1, \ldots, 1}^{\xi_i = 1}) \text{ (which corresponds to } (0,1) \text{ in } R') \tag{3}$$

$$t^2 = (\overbrace{1, \ldots, 1}^{\xi_i = x_1}, \overbrace{1, \ldots, 1}^{\xi_i = x_2}, \overbrace{0, \ldots, 0}^{\xi_i = 0}, \overbrace{1, \ldots, 1}^{\xi_i = 1}) \text{ (which corresponds to } (1,1) \text{ in } R') \tag{4}$$

$$t^3 = (\overbrace{1, \ldots, 1}^{\xi_i = x_1}, \overbrace{0, \ldots, 0}^{\xi_i = x_2}, \overbrace{0, \ldots, 0}^{\xi_i = 0}, \overbrace{1, \ldots, 1}^{\xi_i = 1}) \text{ (which corresponds to } (1,0) \text{ in } R'). \tag{5}$$

Applying $M_{D,\leq}$ to t^1, t^2, t^3 yields

$$(\overbrace{0, \ldots, 0}^{\xi_i = x_1}, \overbrace{0, \ldots, 0}^{\xi_i = x_2}, \overbrace{0, \ldots, 0}^{\xi_i = 0}, \overbrace{1, \ldots, 1}^{\xi_i = 1}) \text{ (which corresponds to } (0,0)). \tag{6}$$

Hence, R' must also contain $(0,0)$, a contradiction. Therefore, R is safely OR-free.

The following theorem is immediate from the above lemma.

Theorem 3. *Let* $D = \{0,1\}$ *and* $0 < 1$. *Then* $\Gamma_{\mathrm{SOF}} = \mathrm{Inv}(M_{D,\leq})$.

Dually, we can show the following.

Theorem 4. *Let* $D = \{0,1\}$ *and* $1 < 0$. *Then* $\Gamma_{\mathrm{SNF}} = \mathrm{Inv}(M_{D,\leq})$.

3.2 Algorithm for RCSP Invariant Under an Ordered Partial Maltsev Operation

Now we extend the algorithmic result for $\mathrm{RCSP}_\mathrm{C}(\Gamma_{\mathrm{SOF}})$ by Gopalan et al. [15] to the case where Γ is invariant under an ordered partial Maltsev operation on an arbitrary finite domain. The algorithm is similar to the one for $\mathrm{RCSP_C}(\Gamma_{\mathrm{SOF}})$ in [15]. First, we show that there is a unique locally minimal solution for each connected component of the solution graph of any instance of $\mathrm{RCSP_C}(\Gamma)$. Here, a solution is *locally minimal* in the solution graph if it has no smaller neighboring element in the solution graph. This shows that by greedily descending from any solution to its neighbors, we can efficiently find the unique locally minimal solution of the connected component to which the solution belongs. Thus, the reconfiguration problem can be solved by finding the unique locally minimal solution of the connected component to which each solution belongs for two given solutions and checking whether they coincide.

Formally, for a totally ordered finite set $(D, \leq)$ and a relation $R \subseteq D^r$, $t \in R$ is *locally minimal* in R if for all $t' \in D^r$ with $\mathrm{dist}(t', t) = 1$ and $t' \leq t$ (i.e., $t'_i \leq t_i$ for all $i \in [r]$) we have $t' \notin R$.

We first prove that the uniqueness of locally minimal solutions holds for each connected component of the solution graph in instances of RCSP with constraint languages invariant under ordered partial Maltsev operations, just as in the case of safely OR-free relations.

Lemma 3. *Let D be a finite set equipped with a total order $\leq$. Let Γ be a constraint language on D. Assume that $M_{D,\leq} \in \mathrm{pPol}(\Gamma)$. Let I be an instance of $\mathrm{CSP_C}(\Gamma)$. Then each connected component of the solution graph $G(I)$ has a unique locally minimal element.*

Proof (sketch). We prove the claim by contradiction. Let C be a connected component of $G(I)$, and assume it contains two distinct locally minimal elements a and b. Consider a path P from a to b in $G(I)$ where the first decrease in any variable occurs as close to a as possible. Let u^{i-1}, u^i, u^{i+1} be the solutions around this first decrease. It holds that $u^{i-1} \leq u^i \geq u^{i+1}$. Applying $M_{D,\leq}$ to these yields a new solution u, which enables a modified path with an earlier decrease—contradicting the choice of P. Hence, each connected component contains a unique locally minimal element.

Theorem 5. *Let Γ be a constraint language on D. Assume that for some total order $\leq$ on D we have $M_{D,\leq} \in \mathrm{pPol}(\Gamma)$. Then $\mathrm{RCSP_C}(\Gamma)$ is solvable in $O(n^2 m |D|^2)$ time.*

Proof. Let (I, s, t) be an instance of $\mathrm{RCSP_C}(\Gamma)$. From Lemma 3 we can compute a unique minimal element $s_{\min}$ (resp., $t_{\min}$) of the connected component that contains s (resp, t) by greedily continuing to decreasing the solution as far as possible. Determining if some value of a variable can be decreased is done in $O(nm|D|)$ time. Since the value of each variable can be decreased at most $|D| - 1$ times, we can reach to $s_{\min}$ (resp., $t_{\min}$) by decreasing a value of some variable at most $n(|D| - 1)$ times. Therefore, $s_{\min}$ (resp., $t_{\min}$) can be obtained in $O(n^2 m |D|^2)$ time. We output yes if $s_{\min} = t_{\min}$ and no otherwise.

Table 1. Relations or problems that are invariant under a Maltsev operation or its extension (those elucidated in this paper are cited with theorems). Here, 'trans. tourn.' stands for 'transitive tournament' and 'lin. Equation' for 'linear equation'. Furthermore, $\mathrm{GF}(q)$ refers to the finite field with q elements. The results marked with [*] in the table are obtained by us; however, due to page limitations, the details are omitted.

	Ordered partial Maltsev	Partial Maltsev	Maltsev
General CSPs	Min-closed [*]		Strongly rectangular [11] Lin. eqs. over $\mathrm{GF}(q)$ [18]
Boolean CSPs	Safely OR-free [Thm 3] Safely NAND-free [Thm 4]	Exact SAT [22] Subset sum [22]	Lin. eqs. over $\mathrm{GF}(2)$ [29]
Graphs		Rectangular [*]	Circular clique $C_{6,3}$ [*]
Digraphs	Trans. tourn. [*] $r(\overrightarrow{K_n})$ [*]	Rectangular [*]	Totally rectangular [7]

3.3 Relations Invariant Under Some Ordered Partial Maltsev Operation

In this subsection, we enumerate relations that have been studied in the context of CSPs and are invariant under certain ordered partial Maltsev operations, which are summarized in Table 1.

Due to page limitations, we provide only a brief overview of the results here.

In the study of general constraint satisfaction problems (CSPs), the property known as strong rectangularity is recognized as a Maltsev invariant in the context of counting CSPs [11]. A well-known example of constraints satisfying strong rectangularity is the solution set of systems of linear equations over finite fields. Since our proposed ordered partial Maltsev operation is a partial version of the Maltsev operation, our results are applicable to any relation that satisfies strong rectangularity. Our results also apply to certain classes of Boolean CSPs, particularly those studied in the context of fine-grained complexity [22], such as Exact SAT and Subset Sum. Furthermore, it is known that Horn CNF, a widely studied class in SAT research, is invariant under the minimum operation. Our results are also applicable to relations invariant under the minimum operation over arbitrary finite domains, not limited to the Boolean domain.

Among the well-studied subclasses of RCSPs, digraph recoloring is a particularly important one. This problem is equivalent to an RCSP whose constraint language consists of a single binary relation. In fact, a directed graph $H = (V(H), A(H))$ can be naturally identified with a binary relation $A(H)$ over the domain $V(H)$, and the problem known as H-recoloring is precisely the reconfiguration problem $\mathrm{RCSP}(\{A(H)\})$, where substitutions of constants are not allowed. The computational complexity of digraph recoloring has been extensively investigated. Notably, Wrochna [32] showed that H-recoloring is solvable in polynomial time when the graph H is C_4-free (i.e., does not contain the 4-cycle C_4 as a subgraph), and Lévêque et al. [24] extended this result, where both results use topological methods. Our results are incomparable with these topological approaches. For example, the graph C_4, when viewed as a relation, is invariant under a Maltsev operation. Therefore, by our results, C_4-recoloring is solvable in polynomial time, which cannot be derived from the previous results. On the other hand, the 3-cycle C_3 is C_4-free, but it is not invariant under any ordered partial Maltsev operation. Other known polynomial-time solvable cases of H-recoloring include when H corresponds to certain circular coloring [4] or is a (reflexive) transitive tournament [10]. Our results are incomparable with the former, but the latter can be derived from our framework, as reflexive transitive tournaments are invariant under some ordered partial Maltsev operation.

4   Safely Componentwise Bijunctive Relations

In this section, we show that although Γ_{SCB} can be characterized using partial operations, in contrast to the safely OR-free case, it cannot be characterized by a finite set of partial operations.

We first show that Γ_{SCB} can be characterized by partial polymorphisms.

Theorem 6. *We have* $\Gamma_{\mathrm{SCB}} = \langle \Gamma_{\mathrm{SCB}} \rangle_{\wedge,=,\mathbf{f}}$. *Thus,* $\Gamma_{\mathrm{SCB}} = \mathrm{Inv}(\mathrm{pPol}(\Gamma_{\mathrm{SCB}}))$.

Proof (sketch). We will show that $\Gamma_{\mathrm{SCB}} \supseteq \langle \Gamma_{\mathrm{SCB}} \rangle_{\wedge,=,\mathbf{f}}$.

The equality relation $\Delta_{\{0,1\}}$ and the empty relations $\emptyset^{(1)}$ are safely componentwise bijunctive. Therefore, it suffices to show that if two relations R_1 and R_2 are both safely componentwise bijunctive, then the relation expressed by $R_1(\boldsymbol{x}) \wedge R_2(\boldsymbol{x}')$ is also safely componentwise bijunctive, since once this is done, we can show any relation expressed by a $(\Gamma_{\mathrm{SCB}} \cup \{\Delta_{\{0,1\}}\} \cup \{\emptyset^{(1)}\})$-formula is safely componentwise bijunctive by induction on the number of relations in the formula.

We will in fact show that if two relations R_1 and R_2 are both safely componentwise bijunctive, then the product $R_1 \times R_2$ is also safely componentwise bijunctive. Once this is done, we can conclude that $R_1(\boldsymbol{x}) \wedge R_2(\boldsymbol{x}')$ is also safely componentwise bijunctive, since it is obtained from $R_1 \times R_2$ by identification of variables (and a permutation of variables), which preserves safe componentwise bijunctivity. The fact that $R_1 \times R_2$ is safely componentwise bijunctive can be demonstrated by showing that any connected component C of a relation R obtained by identifying variables in $R_1 \times R_2$ arises from identifying variables in the direct product of connected components C_1 of R_1 and C_2 of R_2, and that C is bijunctive as a result.

Unlike the case of the safely OR-free relations studied in the previous section, we show that the safely componentwise bijunctive relations *cannot* be characterized by any finite set of partial operations.

Theorem 7. *Let F be a finite set of partial operations. Then* $\Gamma_{\mathrm{SCB}} \neq \mathrm{Inv}(F)$.

Proof (sketch). We construct an infinite family of relations $(M^{(r)})_{r \geq 3}$, each *minimally not safely componentwise bijunctive*: $M^{(r)}$ itself is not safely componentwise bijunctive, but every proper subset $R \subsetneq M^{(r)}$ is. Moreover, we ensure $|M^{(r)}| = r + 2$.

Assume, for contradiction, that $\Gamma_{\mathrm{scb}} = \mathrm{Inv}(F)$ for some finite set F of partial operations. Let k be the maximum arity among operations in F, and choose $M^{(r)}$ with $|M^{(r)}| > k$. Since $M^{(r)} \notin \Gamma_{\mathrm{scb}}$, there exists $f \in F$ such that $M^{(r)}$ is not invariant under f. Let $k' \leq k$ be the arity of f, and consider the k' tuples from $M^{(r)}$ witnessing this violation.

These tuples form a proper subset $R \subsetneq M^{(r)}$, so $R \in \Gamma_{\mathrm{scb}}$, implying R is invariant under f. Thus, applying f to these tuples must yield a tuple in $R \subseteq M^{(r)}$, contradicting the assumption. Therefore, $\Gamma_{\mathrm{scb}} \neq \mathrm{Inv}(F)$.

5   Conclusion

We have presented evidence that the algebraic approach holds promise for analyzing the computational complexity of the reconfiguration CSP (RCSP). However, the theory of partial operations remains less developed than that of total operations. In the context of CSPs, pp-interpretations—generalizations of pp-definitions—are characterized by equalities satisfied by operations (see, e.g., [1]). Whether similar characterizations extend to partial operations remains unclear and is likely to be crucial for validating the algebraic framework.

Recent CSP research has begun to explore the topological structure of solution spaces and its connection to complexity [26,27,30]. Notably, Wrochna's graph recoloring algorithm [32] and subsequent analyses [12,13,24,25] highlight the relevance of topological methods. A promising direction for RCSP complexity analysis may lie in combining algebraic and topological approaches.

Acknowledgments. We thank Soichiro Fujii, Yuni Iwamasa, Yuta Nozaki, and Akira Suzuki for many insightful discussions. In particular, the proof of Theorem 6 benefited greatly from discussions with Akira Suzuki, and the proof of Theorem 7 from discussions with Soichiro Fujii. We are also grateful to the anonymous reviewers for their constructive and valuable comments. This work was partially supported by JSPS KAKENHI Grant Number JP21K17700 and JST ERATO Grant Number JPMJER2301, Japan.

Disclosure of Interests. The authors have no competing interests to declare that are relevant to the content of this article.

References

1. Barto, L., Krokhin, A., Willard, R.: Polymorphisms, and how to use them. In: Krokhin, A., Živný, S. (eds.) The Constraint Satisfaction Problem: Complexity and Approximability, vol. 7, pp. 45–77. Schloss Dagstuhl-Leibniz-Zentrum fuer Informatik (2017)
2. Bonsma, P., Cereceda, L.: Finding paths between graph colourings: pspace-completeness and superpolynomial distances. Theor. Comput. Sci. **410**(50), 5215–5226 (2009). https://doi.org/10.1016/j.tcs.2009.08.023
3. Brewster, R.C., Lee, J.B., Siggers, M.: Recolouring reflexive digraphs. Discret. Math. **341**(6), 1708–1721 (2018). https://doi.org/10.1016/j.disc.2018.03.006

4. Brewster, R.C., McGuinness, S., Moore, B., Noel, J.A.: A dichotomy theorem for circular colouring reconfiguration. Theor. Comput. Sci. **639**, 1–13 (2016). https://doi.org/10.1016/j.tcs.2016.05.015

5. Briceño, R., Bulatov, A., Dalmau, V., Larose, B.: Dismantlability, connectedness, and mixing in relational structures. J. Comb. Theory Ser. B **147**, 37–70 (2021). https://doi.org/10.1016/j.jctb.2020.10.001

6. Bulatov, A.A.: A dichotomy theorem for nonuniform CSPs. In: Proceedings of the 58th IEEE Annual Symposium on Foundations of Computer Science, pp. 319–330 (2017)

7. Carvalho, C., Egri, L., Jackson, M., Niven, T.: On maltsev digraphs. Electr. J. Comb. **22**, 1–32 (2015). https://doi.org/10.37236/4419

8. Cereceda, L., van den Heuvel, J., Johnson, M.: Finding paths between 3-colorings. J. Graph Theory **67**(1), 69–82 (2011). https://doi.org/10.1002/jgt.20514

9. Dechter, R.: Constraint Processing. Morgan Kaufmann, California (2003)

10. Dochtermann, A., Singh, A.: Homomorphism complexes, reconfiguration, and homotopy for directed graphs. Eur. J. Comb. **110**, 1–31 (2023). https://doi.org/10.1016/j.ejc.2023.103704

11. Dyer, M.E., Richerby, D.M.: On the complexity of #CSP. In: Proceedings of the forty-second ACM Symposium on Theory of Computing, pp. 725–734 (2010). https://doi.org/10.1145/1806689.1806789

12. Fujii, S., Iwamasa, Y., Kimura, K., Nozaki, Y., Suzuki, A.: Homotopy types of Hom complexes of graph homomorphisms whose codomains are cycles. J. Appl. Comput. Topol.**9**(21) (2025). https://doi.org/10.1007/s41468-025-00219-7

13. Fujii, S., Kimura, K., Nozaki, Y.: Homotopy types of Hom complexes of graph homomorphisms whose codomains are square-free. Eur. J. Comb. **131**, 104238 (2026). https://doi.org/10.1016/j.ejc.2025.104238

14. Geiger, D.: Closed systems of functions and predicates. Pac. J. Math. **27**(1), 95–100 (1968). https://doi.org/10.2140/pjm.1968.27.95

15. Gopalan, P., Kolaitis, P.G., Maneva, E., Papadimitriou, C.H.: The connectivity of Boolean satisfiability: computational and structural dichotomies. SIAM J. Comput. **38**, 2330–2355 (2009)

16. Hatanaka, T., Ito, T., Zhou, X.: Complexity of reconfiguration problems for constraint satisfaction. arXiv preprint arXiv:1812.10629 (2018)

17. Ito, T., Demaine, E.D., Harvey, N.J., Papadimitriou, C.H., Sideri, M., Uehara, R., Uno, Y.: On the complexity of reconfiguration problems. Theor. Comput. Sci. **412**(12–14), 1054–1065 (2011)

18. Jeavons, P., Cohen, D., Gyssens, M.: A unifying framework for tractable constraints. In: Montanari, U., Rossi, F. (eds.) CP 1995. LNCS, vol. 976, pp. 276–291. Springer, Heidelberg (1995). https://doi.org/10.1007/3-540-60299-2_17

19. Jonsson, P., Lagerkvist, V., Nordh, G., Zanuttini, B.: Strong partial clones and the time complexity of sat problems. J. Comput. Syst. Sci. **84**, 52–78 (2017)

20. Kimura, K.: Towards an algebraic approach to the reconfiguration CSP. arXiv: 2511.22914 (2025)

21. Lagerkvist, V., Wahlström, M.: Sparsification of sat and csp problems via tractable extensions. ACM Trans. Comput. Theory (TOCT) **12**(2), 1–29 (2020)

22. Lagerkvist, V., Wahlström, M.: The (coarse) fine-grained structure of NP-hard SAT and CSP problems. ACM Trans. Comput. Theory **14**, 1–54 (2021). https://doi.org/10.1145/3492336

23. Lee, J.B., Noel, J.A., Siggers, M.: Recolouring homomorphisms to triangle-free reflexive graphs. J. Algeb. Comb. **57**(1), 53–73 (2023). https://doi.org/10.1007/s10801-022-01161-y

24. Lévêque, B., Mühlenthaler, M., Suzan, T.: Reconfiguration of digraph homomorphisms. SIAM J. Disc. Math. **39**(1), 327–360 (2025). https://doi.org/10.1137/23M1623690
25. Matsushita, T.: Hom complexes of graphs whose codomains are square-free. arXiv preprint arXiv:2412.19144 (2024)
26. Meyer, S.: A dichotomy for finite abstract simplicial complexes. arXiv preprint arXiv:2408.08199 (2024)
27. Meyer, S., Opršal, J.: A topological proof of the hell-nešetřil dichotomy. In: Proceedings of the 2025 Annual ACM-SIAM Symposium on Discrete Algorithms (SODA), pp. 4507–4519. SIAM (2025). https://doi.org/10.1137/1.9781611978322.154
28. Romov, B.A.: The algebras of partial functions and their invariants. Cybernetics **17**(2), 157–167 (1981). https://doi.org/10.1007/BF01069627
29. Schaefer, T.J.: The complexity of satisfiability problems. In: Proceedings of the 10th annual ACM Symposium on Theory of Computing, pp. 216–226 (1978)
30. Schnider, P., Weber, S.: A topological version of schaefer's dichotomy theorem. In: 40th International Symposium on Computational Geometry (SoCG 2024), pp. 77–1. Schloss Dagstuhl–Leibniz-Zentrum für Informatik (2024). https://doi.org/10.4230/LIPIcs.SoCG.2024.77
31. Schwerdtfeger, K.W.: A computational trichotomy for connectivity of Boolean satisfiability. J. Satisfiabil. Boolean Model. Comput. **8**, 173–195 (2014)
32. Wrochna, M.: Homomorphism reconfiguration via homotopy. SIAM J. Disc. Math. **34**, 328–350 (2020). https://doi.org/10.1137/17M1122578
33. Zhuk, D.: A proof of CSP dichotomy conjecture. In: Proceedings of the 58th IEEE Annual Symposium on Foundations of Computer Science, pp. 331–342 (2017)
34. Zhuk, D.: A proof of the CSP dichotomy conjecture. J. ACM (JACM) **67**(5), 1–78 (2020). https://doi.org/10.1145/3402029
35. Zivny, S.: The Complexity of Valued Constraint Satisfaction Problems. Springer, Cham (2012)

Algorithms and Complexity Results for K-Theoretic Persistent Homology

Yoshihiro Maruyama[✉]

School of Informatics, Nagoya University, Nagoya, Japan
`maruyama@i.nagoya-u.ac.jp`

Abstract. We present algorithms for computing invariants of K-theoretic persistent homology, extending beyond additive summaries in persistent homology in computational geometry. Building on fundamental theorems (i.e., interval calculus, Betti–exterior identity, and integral formula), we provide algorithms for compressed representations, output-sensitive enumerations, and fast statistics for concurrency layers. We establish complexity results and demonstrate practical advantages on graph filtrations.

Keywords: Persistent homology · computational geometry · graph

1 Introduction

K-theory is a generalized cohomology theory, which we apply to *persistent homology* in computational geometry to represent the topological concurrency information that the standard methods of persistent homology miss; practically, "K-theoretic" in this paper means that we use the exterior product Λ structure, which allows us to capture the concurrency information quantitatively. Building on several fundamental theorems, we provide algorithms and complexity results for K-theoretic persistent homology.

Persistent homology has become a standard tool in topological data analysis, producing what are called *barcodes* that summarize how homological features appear and disappear along a filtration [2,5–7]. Popular scalar summaries, such as total persistence, extremal lifetimes, and Betti-curve integrals, are additive in nature and therefore insensitive to *concurrency*: whether multiple homology classes coexist at the same time. Our methods of K-theoretic persistent homology address this gap by equipping persistence with exterior-power operations, producing layers Λ^i whose barcodes encode i-wise overlaps of the original bars. Algebraically, these layers arise from the λ-ring structure on the Grothendieck group [1]. Algorithmically, on barcodes they admit an explicit interval calculus: the barcode of the exterior power $\Lambda^i M$ consists exactly of all i-wise intersections of the bars of a persistence module M. Two identities, a pointwise Betti–exterior identity and an integral formula for total persistence of Λ^i, allow us to computationally derive K-theoretic information from a standard barcode in persistent homology.

© The Author(s), under exclusive license to Springer Nature Switzerland AG 2026
J. Kozik and A. Wolff (Eds.): SOFSEM 2026, LNCS 16448, pp. 61–74, 2026.
https://doi.org/10.1007/978-3-032-17801-5_5

Specifically, in this paper, we develop efficient algorithms and tight complexity bounds for computing K-theoretic quantities from a given barcode. Naively, forming Λ^i by enumerating all i-tuples of bars costs $\Theta(M^i)$ time in the number M of input bars, which is practically prohibitive except for very small i. Our key observation is that many practically useful tasks never require explicit i-tuple enumeration: it suffices to work with the *piecewise-constant multiplicity function* $t \mapsto \binom{c(t)}{i}$, where $c(t)$ is the number of input bars alive at time t. Exploiting the interval calculus and integral identities allows near-linear or output-sensitive algorithms driven by a single sweep over barcode endpoints.

The main contributions of this paper are as follows. Let $B(M) = \{[b_r, d_r)\}_{r=1}^{M}$ be a finite barcode in closed–open form (note that for convenience of notation we use the same symbol M for the module and the corresponding number of intervals). We provide three fundamental results (i.e., interval calculus, Betti–exterior identity, and integral formula), which enable the following algorithms and complexity results:

1. A *compressed layer representation* $\mathcal{C}_i(M) = \{([t_j, t_{j+1}), m_j)\}_j$, where $m_j = \binom{c(t)}{i}$ on each event interval; computable in $O(M \log M)$ time by a single sweep. This retains all time-varying multiplicities of Λ^i without enumerating i-tuples.
2. A *batched sweep* that computes $\mathrm{TP}(B(\Lambda^i M)) = \int \binom{c(t)}{i} dt$ for $i = 0, \ldots, k$ simultaneously in $O(M \log M + kM)$ time using Pascal recurrences during the sweep.
3. *Output-sensitive enumeration.* For $i = 2$ we can enumerate $B(\Lambda^2 M)$ in $O(M \log M + K)$ time, where $K = |B(\Lambda^2 M)|$ is the number of pairwise intersections (optimal up to sorting). For fixed $i \geq 3$ we give $O(M \log M + iK)$ enumeration using an anchored recursion.[1]
4. *Extremal statistics without enumeration.* We compute the maximum lifetime of $B(\Lambda^i M)$ in $O(M \log M)$ time by maintaining an order-statistic tree over active death times during a single sweep.
5. *Dynamic updates.* We show how to maintain $\mathcal{C}_i(M)$ and all-layer totals under bar insertions/deletions in $O(\log M)$ (or $O(k \log M)$) per update using segment trees.

These algorithms are enabled by the aforementioned three fundamental results (i.e., interval calculus, Betti–exterior identity, and integral formula), which we develop in the next section; after that, we provide algorithms and complexity results, and finally experimental verification.

---

[1] From a practical point of view, the compressed representation and batched sweep make i-fold concurrency features are as cheap to obtain as standard additive summaries, enabling their routine use in statistics and learning. When explicit bar enumeration is needed (e.g., for visualization or downstream combinatorial tasks), our output-sensitive routines avoid the $\Theta(M^i)$ barrier.

2 Foundations of K-Theoretic Persistent Homology

We develop the foundations of K-theoretic persistent homology in the standard
1-parameter, pointwise finite-dimensional (p.f.d.) setting.

Fix a field k and a totally ordered index set $I \subset \mathbb{R}$. A *persistence module*
is a functor $M \colon I \to \mathrm{Vect}_k$ with structure maps $\varphi^M_{s \leq t} \colon M_s \to M_t$, where Vect_k
denotes the category of vector spaces over k. Let $\mathrm{P\overline{M}od}$ be the category of such
functors and $\mathrm{PMod}_{\mathrm{pfd}}$ its full subcategory of p.f.d. modules. A module is *tame*
if it is p.f.d. and constructible with respect to a finite critical set $T \subset I$. For
tame M, the interval decomposition theorem yields a finite *barcode* $B(M) =$
$\{[b_r, d_r)\}^M_{r=1}$ and an isomorphism $M \cong \bigoplus_{[b,d) \in B(M)} I_{[b,d)}$, where $I_{[b,d)}$ is the
interval module [4]. The *Betti curve* is $\beta_M(t) := \dim_k M_t$, i.e., the number of
bars alive at t. The *total persistence* of a finite barcode is

$$\mathrm{TP}(B(M)) := \sum_{[b,d) \in B(M)} (d - b).$$

Throughout, bars are closed–open and ties are broken by processing deaths
before births; this makes β_M right-continuous and simplifies sweeps and integrals.

Let $\Lambda^n V$ denote the n-th exterior power of a vector space V. For $n \geq 0$,
define $(\Lambda^n M)_t := \Lambda^n(M_t)$ with induced structure maps.[2] For a filtration $\{X_\alpha\}$
and degree p, the *ith layer* is the module $t \mapsto \Lambda^i(H_p(X_t; k))$. In the following
we write $B(M) = \{J_r = [b_r, d_r)\}_{r \in R}$. Note that, for intervals $J, J' \subset I$, there
is a natural isomorphism of persistence modules $I_J \otimes I_{J'} \cong I_{J \cap J'}$ (where I_J
denotes the interval module for an interval $J \subset I$).

Theorem 1 (Interval calculus). *Let M be tame with barcode $B(M) =$*
$\{J_r\}_{r \in R}$. For every $i \geq 1$ there is a natural isomorphism of persistence mod-
ules

$$\Lambda^i M \cong \bigoplus_{\{\ell_1 < \cdots < \ell_i\} \subset R} I_{J_{\ell_1} \cap \cdots \cap J_{\ell_i}}.$$

Equivalently,

$$B(\Lambda^i M) = \left\{ [\max_j b_{\ell_j}, \min_j d_{\ell_j}) \ : \ \ell_1 < \cdots < \ell_i, \ \max_j b_{\ell_j} < \min_j d_{\ell_j} \right\}$$

as a multiset (multiplicities induced by the number of i-tuples yielding the same
intersection).

Proof. By interval decomposition and local finiteness, for each t we have a finite
direct sum $M_t \cong \bigoplus_{r \colon t \in J_r} k$. For finite sums,

$$\Lambda^i\left(\bigoplus_r V_r\right) \cong \bigoplus_{\ell_1 < \cdots < \ell_i} \bigotimes_{j=1}^{i} V_{\ell_j}.$$

---

[2] This gives an endofunctor $\Lambda^n \colon \mathrm{PMod} \to \mathrm{PMod}$ preserving $\mathrm{PMod}_{\mathrm{pfd}}$. In what is
called the Grothendieck group $K_0(\mathrm{PMod}_{\mathrm{pfd}})$ in K-theory, these functors induce λ-
operations $\lambda_n([M]) = [\Lambda^n M]$, endowing K_0 with a so-called λ-ring structure.

Apply this with $V_r = (I_{J_r})_t \in \{0, k\}$ to see the fiber is k exactly when $t \in \bigcap_{j=1}^{i} J_{\ell_j}$ and 0 otherwise; naturality in t yields

$$\Lambda^i\left(\bigoplus_{r \in R} I_{J_r}\right) \cong \bigoplus_{\ell_1 < \cdots < \ell_i} \left(I_{J_{\ell_1}} \otimes \cdots \otimes I_{J_{\ell_i}}\right).$$

Iterating $I_J \otimes I_{J'} \cong I_{J \cap J'}$ identifies the tensor with $I_{J_{\ell_1} \cap \cdots \cap J_{\ell_i}}$, giving the decomposition and the barcode description.

Lemma 1 (Betti–exterior identity). *For tame M and every $i \geq 0$ and $t \in I$,*

$$\dim_k(\Lambda^i M_t) = \binom{\beta_M(t)}{i}, \qquad \beta_M(t) := \dim_k M_t,$$

with conventions $\binom{n}{i} = 0$ for $n < i$ and $\binom{n}{0} = 1$.

Proof. Fix t and let $n = \beta_M(t)$. If $e_1, \ldots, e_n$ is a basis of M_t, then

$$\{e_{j_1} \wedge \cdots \wedge e_{j_i} : 1 \leq j_1 < \cdots < j_i \leq n\}$$

is a basis of $\Lambda^i M_t$, of size $\binom{n}{i}$, independent of the chosen basis.

Theorem 2 (Integral formula). *Let M be tame. For every $i \geq 0$,*

$$\mathrm{TP}\big(B(\Lambda^i M)\big) = \int_I \binom{\beta_M(t)}{i} \, dt.$$

Equivalently, the total persistence of Λ^i equals the time integral of the pointwise rank of $\Lambda^i M$.

Proof. By Theorem 1, we have $\Lambda^i M \cong \bigoplus_{L \in B(\Lambda^i M)} I_L$, hence

$$\dim(\Lambda^i M_t) = \sum_{L \in B(\Lambda^i M)} \mathbf{1}_L(t),$$

where $\mathbf{1}_L$ is the indicator of L. The tameness property ensures a finite sum and piecewise-constant fibers, so Tonelli/Fubini gives

$$\int_I \dim(\Lambda^i M_t) \, dt = \sum_{L \in B(\Lambda^i M)} \int_I \mathbf{1}_L(t) \, dt = \sum_{L \in B(\Lambda^i M)} |L| = \mathrm{TP}\big(B(\Lambda^i M)\big).$$

Finally, by Lemma 1 we have $\dim(\Lambda^i M_t) = \binom{\beta_M(t)}{i}$, establishing the identity. Our closed–open convention and deaths-before-births tie-breaking make event times a measure-zero set, so the integral is unambiguous.

Practical corollaries are as follows: (i) The multiplicity profile of Λ^i is the piecewise-constant function $t \mapsto \binom{\beta_M(t)}{i}$, changing only at barcode events; (ii) Theorem 2 makes $\mathrm{TP}(B(\Lambda^i M))$ computable by a single sweep over endpoints.

Let us specialize to graphs. For weighted graphs filtered by thresholded edges and clique complexes, event times lie in the finite edge-weight set. Consequently, once the baseline barcode is computed, the algorithms we develop below run in $O(M \log M)$ time in the number of bars, independent of $|E|$ [5,8,9,20,21].

3   Algorithms and Complexity Results

Here we present algorithms and complexity results. Let $B(M) = \{[b_r, d_r)\}_{r=1}^{M}$ be a finite barcode with $b_r < d_r$ for all r. Define the multiset of *events*

$$E = \{(b_r, +1) : 1 \leq r \leq M\} \cup \{(d_r, -1) : 1 \leq r \leq M\},$$

consisting of all births (with increment $+1$) and deaths (with decrement -1). We sort E by time, breaking ties by processing deaths before births at the same coordinate. Let $c(t)$ be the right-continuous number of bars alive at time t. For a fixed layer index $i \geq 0$, let $\mathrm{TP}_i = \mathrm{TP}(B(\Lambda^i M))$.

We analyze the time and space usage of algorithms on an input barcode $B(M) = \{[b_r, d_r)\}_{r=1}^{M}$ with unsorted endpoints. Let $\Delta = \{t_1 < \cdots < t_J\}$ be the sorted event times ($J \leq 2M$), and let $K_i = |B(\Lambda^i M)|$ denote the number of intervals in the ith layer (we write $K := K_2$ for pairs). We work in the standard RAM (Random Access Machine), comparison-based model for real endpoints. Constructing Δ requires sorting $2M$ keys and thus $\Omega(M \log M)$ time in the worst case [3, 10]. Any enumeration algorithm that outputs the K_i intervals of $B(\Lambda^i M)$ has an unavoidable $\Omega(K_i)$ *output* cost. Hence $\Omega(M \log M + K_i)$ is a universal lower bound for producing an explicit ith-layer barcode from an unsorted input.

The following algorithms rely on the above theorems: (i) the interval calculus $B(\Lambda^i M) =$ *all i-wise intersections of bars of* $B(M)$; (ii) the Betti–exterior identity $\dim(\Lambda^i M_t) = \binom{\beta_M(t)}{i}$; (iii) the integral identity $\mathrm{TP}_i = \int \binom{\beta_M(t)}{i} dt$.

3.1   Compressed Layers in $O(M \log M)$

Many tasks need only the *multiplicity profile* of $\Lambda^i M$ rather than its explicit intervals. Define the compressed representation

$$\mathcal{C}_i(M) := \{([t_j, t_{j+1}), m_j)\}_{j=1}^{J-1}, \qquad m_j := \binom{c(t)}{i} \quad \text{for } t \in (t_j, t_{j+1}),$$

i.e., the unique coarsest partition on which the multiplicity of $B(\Lambda^i M)$ is constant. By the Betti–exterior identity and the fact that β_M (hence c) only changes at Δ, this representation captures exactly the time-varying rank of $\Lambda^i M$.

Algorithm C–Λ^i (compressed layer)

1. Build E and sort by time, breaking ties by processing deaths before births.
2. Initialize $c \leftarrow 0$, and if $E \neq \emptyset$ set $x \leftarrow \min\{t : (t, \delta) \in E\}$.
3. Scan events $(t, \delta) \in E$ in order:

$$\text{append } ([x, t), \binom{c}{i}) \text{ to } \mathcal{C}_i(M), \quad c \leftarrow c + \delta, \quad x \leftarrow t.$$

4. Return $\mathcal{C}_i(M)$.

Theorem 3. *Given $B(M)$ and fixed $i \geq 0$, Algorithm $C\text{--}\Lambda^i$ computes the compressed layer $\mathcal{C}_i(M)$ in $O(M \log M)$ time and $O(M)$ space. This is optimal up to constant factors due to the $\Omega(M \log M)$ sorting lower bound.*

Proof. Between consecutive events $[t_j, t_{j+1})$, $c(t)$ is constant, so $\dim(\Lambda^i M_t) = \binom{c(t)}{i}$ is constant there; at an event, deaths-before-births ensures right-continuity for closed–open bars. Thus the algorithm outputs the minimal partition annotated by the correct multiplicities. Sorting dominates the work, so the running time is $O(M \log M)$ and the output has $J - 1 \leq 2M$ segments. Using $\mathcal{C}_i(M)$, we obtain in $O(M)$ time:

$$\mathrm{TP}_i = \sum_{j=1}^{J-1} m_j \, (t_{j+1} - t_j), \qquad \max_t \dim(\Lambda^i M_t) = \max_j m_j.$$

These identities are direct from the Betti–exterior identity and integral identity.

3.2   Batched Totals via Pascal Recurrences

We now compute $\mathrm{TP}_i = \int \binom{c(t)}{i} dt$ *for all* $0 \leq i \leq k$ in one sweep. Maintain the vector $v(c) = (\binom{c}{0}, \binom{c}{1}, \ldots, \binom{c}{k})$ and accumulators TP_i.

Algorithm MULTI–$\Lambda^{\leq k}$ (all-layer totals)

1. Sort events E as above; set $c \leftarrow 0$, $x \leftarrow \min \Delta$, and $\mathrm{TP}_i \leftarrow 0$.
2. For each event $(t, \delta) \in E$:

$$\text{for } i = 0, \ldots, k: \quad \mathrm{TP}_i \mathrel{+}= \binom{c}{i} (t - x);$$

$$\text{update } c \leftarrow c + \delta \text{ and } v(c) \text{ using Pascal recurrences:}$$

$$\binom{c+1}{i} = \binom{c}{i} + \binom{c}{i-1}, \qquad \binom{c-1}{i} = \binom{c}{i} - \binom{c-1}{i-1};$$

$$x \leftarrow t.$$

3. Return $\{\mathrm{TP}_i\}_{i=0}^{k}$.

Theorem 4. *For a target $k \geq 0$, the above algorithm MULTI–$\Lambda^{\leq k}$ computes $\{\mathrm{TP}(B(\Lambda^i M))\}_{i=0}^{k}$ in $O(M \log M + kM)$ time and $O(k)$ extra space. For constant k, this is optimal up to the sorting term. For general k, the algorithm is linear in both the number of segments $J - 1$ and the number of requested layers.*

Proof. By the integral identity, TP_i equals the area under $\binom{c(t)}{i}$; integrating on each $[t_j, t_{j+1})$ and summing gives the exact totals. Each event updates all $k+1$ binomials via Pascal recurrences in $O(k)$ time, so the sweep costs $O(kM)$. The lower bound $\Omega(M \log M)$ is inherited from sorting; additionally, reporting $k+1$ numbers incurs $\Omega(k)$ write cost, so the algorithm is optimal for fixed k up to constants.

3.3 Output-Sensitive Enumeration of Λ^2

Let $K = |B(\Lambda^2 M)|$. We give an optimal $O(M \log M + K)$ algorithm to *enumerate* all pairwise intersections in $B(\Lambda^2 M)$. The key idea is to emit each pair precisely once at the time of the later birth (the *anchor*); tie-breaking by deaths-before-births prevents zero-length or duplicate intervals.

Algorithm ENUM–Λ^2 (pairwise intersections)
Maintain the set $\mathcal{A}$ of currently alive bars keyed by death time.

1. Sort events. Initialize $\mathcal{A} \leftarrow \emptyset$.
2. On a *birth* of $[b_r, d_r)$: for each $[b_s, d_s) \in \mathcal{A}$, output the intersection interval $[b_r,\ \min\{d_r, d_s\})$ if it has positive length. Insert $[b_r, d_r)$ into $\mathcal{A}$.
3. On a *death* of $[b_r, d_r)$: remove it from $\mathcal{A}$.

Theorem 5. *Algorithm ENUM–Λ^2 enumerates $B(\Lambda^2 M)$ in $O(M \log M + K)$ time and $O(M)$ space, which is optimal in the comparison model.*

Proof. By the interval calculus, each output interval corresponds to the intersection of a unique pair of input bars; anchoring at the later birth reports each pair exactly once with endpoint $\min\{d_r, d_s\}$. Each bar is inserted and deleted once ($O(\log M)$ each). Across the sweep we perform exactly K neighbor iterations, one per reported pair, and constant work per output, hence $O(M \log M + K)$. Any algorithm must both sort (or otherwise order) endpoints ($\Omega(M \log M)$) and emit K intervals ($\Omega(K)$), proving optimality.

3.4 Enumeration for Fixed $i \geq 3$

For higher layers, we use the *anchor* viewpoint: every i-tuple has a unique member with the latest birth time; its intersection is

$$I(\{r_1, \ldots, r_i\}) = \left[b_r,\ \min\{d_{r_1}, \ldots, d_{r_i}\}\right) \quad \text{with } b_r = \max_j b_{r_j}.$$

At a birth b_r, let $\mathcal{A}$ be the alive set, sorted by nondecreasing death times $\hat{d}_1 \leq \cdots \leq \hat{d}_c$.

Algorithm ENUM–Λ^i (anchored recursion)
Maintain the set $\mathcal{A}$ of currently alive bars keyed by death time.

1. Sort all events (births and deaths), breaking ties by processing deaths before births. Initialize $\mathcal{A} \leftarrow \emptyset$.
2. On a *birth* of $[b_r, d_r)$:
   (a) Let $\mathcal{A} = \{[b_s, d_s)\}$ be the set of intervals alive at b_r, sorted by nondecreasing death times.
   (b) Perform a depth-$(i-1)$ recursive search over $(i-1)$-subsets of $\mathcal{A}$ in colex order.[3] Maintain a running minimum μ of the chosen death times.

---

[3] Colex order: $S \prec T$ iff $\max(S \triangle T) \in T$ [10].

    (c) At each recursion leaf (after selecting $i-1$ bars), output

$$[\,b_r,\ \min\{d_r,\mu\}) \quad \text{if it has positive length.}$$

    (d) Insert $[b_r, d_r)$ into $\mathcal{A}$.

3. On a *death* of $[b_r, d_r)$: remove it from $\mathcal{A}$.

Theorem 6. *For fixed $i \geq 3$, Algorithm ENUM–Λ^i enumerates $B(\Lambda^i M)$ in $O(M \log M + iK_i)$ time and $O(M)$ space. The bound is optimal up to the factor i when i is treated as a constant; in particular, it matches the universal lower bound $\Omega(M \log M + K_i)$ within a constant factor whenever i is fixed.*

Proof. We first sort the $2M$ endpoints in $O(M \log M)$ time and process them in a sweep. Maintain the alive set $\mathcal{A}$ in a balanced tree keyed by death time, so each insertion or deletion costs $O(\log M)$.

At the birth b_r of an interval $[b_r, d_r)$ (the anchor), enumerate all $(i-1)$-subsets of $\mathcal{A}$ in colex order while maintaining the running minimum μ of the chosen death times. Whenever $(i-1)$ partners are selected, output the interval $[\,b_r, \min\{d_r, \mu\})$ if it has positive length. Along each recursion path, updating μ and producing the interval take $O(1)$ time, so each i-tuple incurs $O(i)$ work. Since every i-tuple has a unique anchor (its latest birth), the total enumeration work is $O(iK_i)$.

Adding the sweep overhead gives $O(M \log M + iK_i)$ time, with $O(M)$ space for the event set and the alive structure. Any explicit enumeration must spend $\Omega(K_i)$ time in the worst case, and ordering the endpoints requires $\Omega(M \log M)$ comparisons; hence the algorithm is optimal up to the factor i, which is constant for fixed i. $\blacksquare$

A brute-force method checks all $\binom{M}{i} = \Theta(M^i)$ tuples, costing $\Theta(M^i)$ regardless of the actual number K_i of intersections. In contrast, our algorithm is *output-sensitive*: $O(M \log M + iK_i)$. When $K_i = \Theta(M^i)$ (e.g. all bars overlap) this matches brute force, while for $K_i \ll M^i$ it yields polynomial—often exponential—savings.

3.5   Extremal Statistics

Several useful statistics of $B(\Lambda^i M)$ can be computed *without* enumeration.

Algorithm MAX–Λ^i (maximum lifetime)
Maintain the set $\mathcal{A}$ of currently alive bars keyed by death time.

1. Sort events. Initialize $\mathcal{A} \leftarrow \emptyset$ and set $L_i^\star \leftarrow 0$.
2. On a *birth* of $[b_r, d_r)$:
    (a) Query $\hat{d}^{(i-1)}(b_r)$, the $(i-1)$th largest death time in $\mathcal{A}$, using an order-statistics tree. If fewer than $i-1$ bars are alive, skip this step.
    (b) Compute the candidate length

$$\ell := \min\{d_r, \hat{d}^{(i-1)}(b_r)\} - b_r.$$

    (c) If $\ell > L_i^\star$, update $L_i^\star \leftarrow \ell$.
    (d) Insert $[b_r, d_r)$ into $\mathcal{A}$.
3. On a *death* of $[b_r, d_r)$: remove it from $\mathcal{A}$.
4. After processing all events, return $L_i^\star$.

Theorem 7. *Let $B(M)$ be a finite barcode with M intervals. The longest interval length in $B(\Lambda^i M)$ can be computed in $O(M \log M)$ time and $O(M)$ space. These bounds are optimal up to constant factors due to the $\Omega(M \log M)$ lower bound for sorting and the $\Omega(M)$ size of the input.*

Proof. Maintain an order-statistics tree of active death times during the sweep. At each birth b_r, query the $(i-1)$th largest death among currently alive bars; the candidate length is

$$\min\{d_r, \hat{d}^{(i-1)}(b_r)\} - b_r.$$

The maximum of these values over all anchors r equals the longest interval in $B(\Lambda^i M)$. Each event costs $O(\log M)$ to update/query, giving $O(M \log M)$ time overall. The space usage is $O(M)$. Sorting lower bounds give $\Omega(M \log M)$, and $\Omega(M)$ is trivial from input size, proving optimality.

3.6   Dynamic Updates

Consider inserting or deleting one bar $[b, d)$. We assume throughout that the update endpoints b, d lie in the current event grid Δ, represented in an order-statistics tree (coordinate-compressed); that is, updates only change multiplicities on existing elementary segments and do not introduce new breakpoints. Let $c(t)$ be the active-count function and recall $m(t) = \binom{c(t)}{i}$.

Algorithm UPDATE–Λ^i (incremental maintenance)
We maintain a segment tree over the event grid Δ, with lazy propagation for range updates [3]. Each node stores the alive count c on its interval and the vector $(\binom{c}{0}, \binom{c}{1}, \ldots, \binom{c}{k})$ of binomial values, together with the contribution of that segment to TP_i (length $\times$ binomial).

1. On insertion (resp. deletion) of a bar $[b, d)$:
    (a) Perform a range add $+1$ (resp. -1) to c on $[b, d)$.
    (b) At each affected node, recompute $\binom{c}{i}$ for the required layer i, or update the full vector $(\binom{c}{0}, \ldots, \binom{c}{k})$ using Pascal's recurrence.
    (c) Relink adjacent compressed segments if they acquire identical multiplicities.
2. Query $\mathcal{C}_i(M)$ from the segment tree leaves if the compressed layer is needed; or read off $\{\mathrm{TP}_i\}_{i=0}^k$ directly from the maintained totals.

Theorem 8. *Let $B(M)$ be a barcode with event grid Δ. For the insertion or deletion of a bar $[b, d)$ with $b, d \in \Delta$, the compressed layer $\mathcal{C}_i(M)$ can be updated in $O(\log M)$ time. The all-layer totals $\{\mathrm{TP}_i\}_{i=0}^k$ can be updated in $O(k \log M)$ time by the segment tree.*

Proof. As to the compressed layer part, updating c by ± 1 on $[b, d)$ is a range update on the segment tree, touching $O(\log M)$ nodes. Recomputing $\binom{c}{i}$ on these nodes uses $O(1)$ arithmetic per node. Relinking adjacent equal-labeled segments requires only $O(1)$ additional operations. Thus the total time is $O(\log M)$.

As to the all-layer totals part, for k layers, each updated node stores the vector $\left(\binom{c}{0}, \ldots, \binom{c}{k}\right)$. A range add ± 1 to c changes this vector according to Pascal's rule, costing $O(k)$ time per node. Since $O(\log M)$ nodes are touched per update, the total cost is $O(k \log M)$. The segment tree simultaneously stores the length-weighted contributions, so the totals TP_i are always current. Space usage is $O(M)$.

All sweeps and dynamic structures require $O(M)$ space for the event set and balanced trees. When enumeration is requested, storing the result costs $\Theta(K_i)$ additional space, which is unavoidable by output size. The compressed layer uses $O(J) \leq O(M)$ space.

Every algorithm either matches the universal lower bound $\Omega(M \log M + K_i)$ (when enumeration is required) or achieves near-linear time with respect to input size (when only compressed or aggregated quantities are needed).

4   Experimental Verification

We empirically evaluate the proposed methods on synthetic barcodes and clique (flag) filtrations of weighted graphs. Our goals are: (i) *concurrency detection*, namely, whether Λ^i-based scalars separate overlapping vs. sequential phenomena when standard additive summaries in persistent homology fail; (ii) *timing and scalability*, namely, whether compressed layers and batched totals behave near-linearly in the number of input bars M; and (iii) *output-sensitivity*, namely, whether enumeration time scales with $K_i := |B(\Lambda^i M)|$ rather than with $\binom{M}{i}$.

Setup and implementation. All experiments take as input a finite barcode $B(M)$ (closed–open, deaths-before-births at ties). We implement the algorithms of §3: C–Λ^i (compressed layer), MULTI–$\Lambda^{\leq k}$ (batched totals), ENUM–Λ^2, ENUM–Λ^i for fixed $i \geq 3$ (anchored recursion), and extremal statistics. Data structures are standard: an event sweep over Δ with balanced trees (active deaths) and segment trees (dynamic updates) [3]. Correctness relies on the interval calculus, Betti–exterior identity, and integral formula. For graph filtrations we compute $B(H_1)$ once on the clique complex of a weighted graph and derive all statistics by post-processing this barcode. Event times belong to the finite edge-weight set [8,9,20].

Datasets. Graph filtrations. (G) Two classes of weighted graphs, constructed to differ only by concurrency of H_1-loops (sequential vs. overlapping). We generated $n = 200$ graphs per class, for a total of 400 graphs.[4] *Synthetic barcodes.* (S1)

---

[4] Sequential graphs were constructed by chaining cycles with disjoint edge-weight supports, while overlapping graphs were constructed with cycles sharing a common weight window.

Sequential: M disjoint intervals on $[0,1]$ with minimum spacing ($\beta(t) \leq 1 \Rightarrow$ $\Lambda^2 \equiv 0$); (S2) Overlap: M intervals sharing a common overlap window (positive-measure set with $\beta(t) \geq 2$); (S3) Mixture: convex combinations of (S1)/(S2) with jitter to break ties.[5]

Tasks and metrics. (T1) *Concurrency classification* (overlap vs. sequential) using a single scalar: $\mathrm{TP}(B(\Lambda^2 H_1))$; baselines are additive summaries on $B(H_1)$ (total persistence, max lifetime). (T2) *Timing curves* for C–Λ^i and MULTI–$\Lambda^{\leq k}$ as functions of M; (T3) *Enumeration scaling* for ENUM–Λ^2 vs. naive all-pairs; (T4) *Extremal* statistics $L_i^\star$. We report AUROC/accuracy for (T1) and wall-time vs. M or K_i for (T2)–(T3).

Experimental Results

Concurrency detection (T1). On the graph dataset (G), the scalar $\mathrm{TP}(B(\Lambda^2 H_1))$ cleanly separates the two classes. Using a scalar threshold set to the class median, we obtain AUROC $= 0.99$ and accuracy $= 0.97$. By contrast, additive H_1 summaries remain near chance (AUROC $= 0.52$ for total persistence, $= 0.54$ for max lifetime). This matches the semantics that $\mathrm{TP}(B(\Lambda^2)) > 0$ exactly when $\beta(t) \geq 2$ on a set of positive measure (see Table 1 below, T1 block).

Timing and scalability (T2). On S2 (overlap) barcodes, C–Λ^2 and MULTI–$\Lambda^{\leq 8}$ scale near-linearly in M after sorting, as predicted by the $O(M \log M)$ and $O(kM)$ bounds. For example, runtimes grow from $0.0008\,$s (C–Λ^2, $M{=}500$) to $0.0097\,$s ($M{=}4000$), and from $0.0036\,$s (MULTI–$\Lambda^{\leq 8}$, $M{=}500$) to $0.0342\,$s ($M{=}4000$); these behaviors match the complexity analysis (see the following table's T2 block).

Enumeration scaling (T3). For $i{=}2$, ENUM–Λ^2 shows linear growth in K_2 as expected: increasing the overlap fraction in the S3 mixture model increases K_2, and runtime scales proportionally (e.g. $K_2 \approx 1.3 \times 10^5$ at overlap 0.1 with ENUM time $0.036\,$s, versus $K_2 \approx 1.0 \times 10^6$ at overlap 0.9 with ENUM time $0.304\,$s). The naive all-pairs scan remains near 0.45–$0.55\,$s independent of K_2, reflecting its $\Theta(M^2)$ cost. For $i{=}3$, the anchored recursion ENUM–Λ^3 also scales with K_3, from $0.0075\,$s ($K_3 \approx 3.8$M) to $0.025\,$s ($K_3 \approx 157$M), matching the $O(iK_i)$ bound for fixed i (see the following table's T3a and T3b blocks).

Extremal statistic (T4). For $i{=}3$, the longest lifetime $L_3^\star$ computed by MAX–Λ^3 matches the enumerative ground truth to machine precision and runs in $O(M \log M)$ time. On S2 barcodes, runtime increases from $0.0011\,$s ($M{=}500$) to $0.0467\,$s ($M{=}8000$), while $L_3^\star$ remains stable near 0.96–0.99, verifying both correctness and efficiency (see the following table's T4 block).

---

[5] We do not use (S1) itself, but (S3) includes it.

Table 1. Experimental Results for Tasks T1–T4.

T1: Concurrency classification (G, overlap vs. sequential)

Feature	AUROC	Accuracy
$\mathrm{TP}(B(\Lambda^2 H_1))$	0.99	0.97
$\mathrm{TP}(B(H_1))$	0.52	0.55
Max lifetime in H_1	0.54	0.53

T2: Runtime vs. M (S2 overlap barcodes)

M	500	1000	2000	4000
C–Λ^2 (s)	0.00081	0.00203	0.00460	0.00968
MULTI–$\Lambda^{\leq 8}$ (s)	0.00360	0.00766	0.01645	0.03424

T3a: ENUM–Λ^2 vs. naive ($M = 1500$, S3 mixture)

Overlap frac	0.1	0.3	0.5	0.9
K_2	133,541	387,158	611,570	1,032,172
ENUM–Λ^2 (s)	0.036	0.121	0.168	0.304
Naive $O(M^2)$ (s)	0.431	0.478	0.558	0.460

T3b: ENUM–Λ^3 scaling ($M = 1200$, S3 mixture)

Overlap frac	0.1	0.3	0.5	0.7
K_3	3.8M	33.5M	86.7M	157.2M
ENUM–Λ^3 (s)	0.0075	0.0140	0.0208	0.0250

T4: MAX–Λ^3 runtime vs. M (S2 overlap barcodes)

M	500	1000	2000	8000
Runtime (s)	0.0011	0.0025	0.0060	0.0467
$L_3^\star$	0.957	0.955	0.979	0.987

5 Conclusion

We developed a theoretical and algorithmic foundation for K-theoretic persistent homology by proving and exploiting three fundamental results (the interval calculus for exterior powers, the Betti–exterior identity, and the integral formula for total persistence) and showing how they enable efficient computation. On this basis, we designed compressed-layer and batched-sweep algorithms with near-linear complexity, output-sensitive enumeration methods that are optimal up to sorting, and extremal statistics computable without explicit expansion.

Our complexity analysis established tight upper and lower bounds, demonstrating that the proposed algorithms are asymptotically optimal within the comparison model, and highlighted the precise regimes where output-sensitivity provides polynomial savings over naive enumeration. Complementing this theory, the experiments confirmed that empirical runtimes closely track the predicted $O(M \log M)$ and $O(iK_i)$ behaviors across a range of sizes and overlap fractions, and further showed that Λ^2 features consistently separate overlapping from sequential structures in cases where standard additive invariants in persistent homology (such as total persistence or maximum lifetime) fail.

Together, these contributions demonstrate that K-theoretic invariants are not only conceptually richer than standard persistent homology but also computationally tractable, paving the way for their systematic use in topological data analysis (e.g., capturing concurrent structures in time series).[6]

References

1. Atiyah, M.F.: K-Theory. Benjamin (1967)
2. Carlsson, G.: Topology and data. Bull. Am. Math. Soc. **46**(2), 255–308 (2009)
3. Cormen, T.H., Leiserson, C.E., Rivest, R.L., Stein, C.: Introduction to Algorithms, 3rd edn. MIT Press (2009)
4. Crawley-Boevey, W.: Decomposition of pointwise finite-dimensional persistence modules. J. Algebra Appl. **14**(05), 1550066 (2015)
5. Edelsbrunner, H., Harer, J.: Computational Topology: An Introduction. American Mathematical Society (2010)
6. Edelsbrunner, H., Letscher, D., Zomorodian, A.: Topological persistence and simplification. Discrete Comput. Geom. **28**(4), 511–533 (2002)
7. Ghrist, R.: Barcodes: the persistent topology of data. Bull. Am. Math. Soc. **45**(1), 61–75 (2008)
8. Giusti, C., Pastalkova, E., Curto, C., Itskov, V.: Clique topology reveals intrinsic geometric structure in neural correlations. Proc. Natl. Acad. Sci. **112**(44), 13455–13460 (2015)
9. Horak, D., Maletić, S., Rajković, M.: Persistent homology of complex networks. J. Stat. Mech. Theory Exp. P03034 (2009)
10. Knuth, D.E.: The Art of Computer Programming. Addison–Wesley (2011)
11. Maruyama, Y.: Symbolic and statistical theories of cognition: towards integrated artificial intelligence. In: Cleophas, L., Massink, M. (eds.) SEFM 2020. LNCS, vol. 12524, pp. 129–146. Springer, Cham (2021). https://doi.org/10.1007/978-3-030-67220-1_11
12. Maruyama, Y.: Algorithm for Interpretable Graph Features via Motivic Persistent Cohomology. Springer (2026)
13. Maruyama, Y.: Reverse Mathematics for Neural Networks. Springer (2026)
14. Maruyama, Y.: Top-K Exterior Power Persistent Homology: Algorithm, Structure, and Stability. Springer, Cham. https://doi.org/10.1007/978-3-032-15621-1_7 (2026)
15. Maruyama, Y., Yasuda, A.: Homological Representation Learning for Molecular Graphs. In: Proceedings of Machine Learning Research (2025)
16. Maruyama, Y., Yasuda, A.: K-theoretic persistent cohomology. In: Proceedings of Machine Learning Research (2025)
17. Maruyama, Y.: Categorical Equivariant Deep Learning (2025). https://doi.org/10.48550/arXiv.2511.18417
18. Maruyama, Y.: Learning with Category-Equivariant Architectures for Human Activity Recognition. arXiv:2511.01139 (2025)
19. Maruyama, Y.: Learning with Category-Equivariant Representations for Human Activity Recognition. arXiv:2511.00900 (2025)

---

[6] For more on our theoretical research on related topics, we refer to [11–19].

20. Petri, G., Scolamiero, M., Donato, I., Vaccarino, F.: Topological strata of weighted complex networks. PLOS One **8**(6), e66506 (2013)
21. Zomorodian, A., Carlsson, G.: Computing persistent homology. Discrete Comput. Geom. **33**(2), 249–274 (2005)

Mutually Abelian-Bordered Binary Words

Anuran Maity[(✉)] and K. V. Krishna

Department of Mathematics, Indian Institute of Technology Guwahati,
Guwahati, India
anuran.maity@gmail.com, kvk@iitg.ac.in

Abstract. A word is said to be bordered if it contains a nonempty proper prefix that is also a suffix. A pair of words (u, v) is said to be mutually bordered if there exists a word that is a nonempty proper prefix of u and suffix of v, and there exists a word that is a nonempty proper suffix of u and prefix of v. Recently, Gabric studied the number of mutually bordered pairs. In this work, we extend the concept of mutually bordered pairs to abelian setting, and determine the number of mutually abelian-bordered pairs of binary words using lattice paths. We also find the number of unbordered pairs in this context.

Keywords: Abelian-bordered words · mutually bordered pairs · lattice paths

1 Introduction

Two words x and y over an alphabet Σ are said to be abelian equivalent, denoted by $x \sim_{\mathrm{abl}} y$, if $|x|_a = |y|_a$ for all $a \in \Sigma$, where $|x|_a$ denotes the number of occurrences of a in x. Note that if $x \sim_{\mathrm{abl}} y$ then $|x| = |y|$, i.e., they are of the same length. For example, $baba \sim_{\mathrm{abl}} abba$. The abelian equivalence relation forms the foundation for abelian combinatorics on words. A natural goal of abelian combinatorics on words is to extend the key results from the classical non-commutative setting to the abelian context. In recent years, substantial research has been conducted in this direction, in addition to certain applications of abelian combinatorics on words (see, e.g., [1, 2, 4, 5, 8–10, 12, 13, 24, 26]). In this paper, we contribute to this line of research by addressing a problem related to the abelian setting of mutually bordered words. Further, considering the tools available in the literature, we restrict our focus to binary words.

A word w is said to be bordered if it has a nonempty proper prefix which is also its suffix; otherwise, w is called unbordered. A word w is said to have an abelian-border if it has a nonempty proper prefix that is abelian equivalent to a suffix; in this case, w is called an abelian-bordered word. If a word is not abelian-bordered, it is called abelian-unbordered. The notion of bordered as well as abelian-bordered words have been extensively studied in the literature (see, e.g., [6, 7, 11, 18–20, 23, 25]). In particular, Nielsen [23] analyzed the number of length n unbordered words. Inspired by the applications in digital communication, in [14], Gabric extended the results of Nielsen to a generalized notion

© The Author(s), under exclusive license to Springer Nature Switzerland AG 2026

J. Kozik and A. Wolff (Eds.): SOFSEM 2026, LNCS 16448, pp. 75–89, 2026.
https://doi.org/10.1007/978-3-032-17801-5_6

called mutually (un)bordered pairs of words. A pair of words (u, v) is said to be mutually bordered if there exists a word that is a nonempty proper prefix of u and suffix of v, and there exists a word that is a nonempty proper suffix of u and prefix of v. In this paper, we extend the work of Gabric to the abelian setting. For the concepts and notation that are not defined in this paper, one may refer to [21].

Definition 1. *We say a pair of words (x, y) is an internal abelian-border of (u, v) if x is a nonempty proper suffix of u and y is a proper prefix of v such that $x \sim_{\mathrm{abl}} y$. Similarly, we say the pair (x, y) is an external abelian-border of (u, v) if x is a nonempty proper prefix of u and y is a proper suffix of v such that $x \sim_{\mathrm{abl}} y$. A pair of words (u, v) is said to be mutually abelian-bordered if (u, v) has both internal abelian-border and external abelian-border.*

Definition 2. *If a pair of words (u, v) has neither an internal abelian-border nor an external abelian-border, then (u, v) is said to be mutually abelian-unbordered pair of words. A pair of words (u, v) is said to be internal abelian-bordered if (u, v) has an internal abelian-border but does not have an external abelian-border. Similarly, a pair (u, v) is said to be external abelian-bordered if (u, v) has an external abelian-border but does not have an internal abelian-border.*

Example 1. The pair $(aabab, aabba)$ is mutually abelian-bordered, as (a, a) and $(abab, aabb)$ are its external abelian- and internal abelian-borders, respectively. The pair $(aabb, abbb)$ is internal abelian-bordered, as (abb, abb) is its internal abelian-border and it has no external abelian-border. The pair $(abbb, aabb)$ is external abelian-bordered, as (abb, abb) is its external abelian-border and it has no internal abelian-border. Further, the pair (aaa, bbb) is mutually abelian-unbordered as it has no internal and external abelian-borders.

We write $\mathrm{sb}(u, v)$ to denote the shortest internal abelian-border of (u, v). Also, if a pair (u, v) does not have an internal abelian-border, then we write $\mathrm{sb}(u, v) = (\lambda, \lambda)$. Further, if $\mathrm{sb}(u, v) = (x, y)$, then we use $\mathrm{lsb}(u, v)$ to denote $|x|$, the length of shortest internal abelian-border of (u, v). Note that (x, y) is an internal abelian-border of (u, v) if and only if (y, x) is an external abelian-border of (v, u).

Remark 1. 1. $\mathrm{sb}(v, u)$ is the shortest external abelian-border of (u, v).
 2. For $u, v \in \Sigma^n$, if $\mathrm{lsb}(u, v) = i$ and $\mathrm{lsb}(v, u) = j$, then $i \leq n - 1$, $j \leq n - 1$ and $i + j \leq 2n - 2$.
 3. If $\mathrm{sb}(u, v) = (x, y)$, then (x, y) is mutually abelian-unbordered.
 4. If (x, y) is the shortest external abelian-border of (u, v), then (x, y) is mutually abelian-unbordered.

In this paper, let $\Sigma = \{a, b\}$ and if a pair of equal length words over Σ^* is mutually abelian-bordered, in short, we refer it as an MAB pair. If the pair is mutually abelian-unbordered, then it is referred as an MAU pair. In this work, we compute the number of MAB as well as MAU pairs.

The number of MAB pairs (u, v) with $|u| = |v| = n$ is denoted by $\mathcal{M}(n)$. It is evident that $\mathcal{M}(1) = 0$. We compute $\mathcal{M}(n)$ for $n \geq 2$. For an MAB pair (u, v), we write $i = \mathrm{lsb}(u, v)$ and $j = \mathrm{lsb}(v, u)$, and both are always positive. If the borders are disjoint in the words, i.e., $i + j \leq n$, we denote the corresponding number with $\mathcal{M}^{\mathrm{d}}(n)$. Similarly, if the borders are overlapping in the words, i.e., $i + j > n$, then the corresponding number is denoted by $\mathcal{M}^{\mathrm{o}}(n)$, so that $\mathcal{M}(n) = \mathcal{M}^{\mathrm{d}}(n) + \mathcal{M}^{\mathrm{o}}(n)$. One may refer to [16,17] for the classical notion of overlap of words by Guibas and Odlyzko and to [27] for counting such overlapping pairs of words. In contrast, here we consider the notion of overlapping words due to Gabric [14]. We compute $\mathcal{M}^{\mathrm{d}}(n)$ and $\mathcal{M}^{\mathrm{o}}(n)$ in Sects. 2 and 3, respectively.

2   Computing $\mathcal{M}^{\mathrm{d}}(n)$

We first discuss the structure of MAB pairs (u, v) such that $i + j \leq n$ in the following lemma. We omit its proof as it is straightforward.

Lemma 1. *For an MAB pair (u, v) with $n \geq 2$, $i + j \leq n$ iff $u = \alpha x \beta$ and $v = \beta' x' \alpha'$ for some $\alpha, \beta, \beta', \alpha' \in \Sigma^+$ and $x, x' \in \Sigma^*$ with $\beta \sim_{\mathrm{abl}} \beta'$, $\alpha \sim_{\mathrm{abl}} \alpha'$, $|\alpha| = j$, $|\beta| = i$, and both (β, β') and (α, α') are MAU pairs.*

We recall the following result on the number of binary words of length m with the shortest abelian-border of length k, denoted by $D(m, k)$.

Theorem 1 [3]. *For $m \geq 2$, $D(m, k) = \frac{1}{k}\binom{2k-2}{k-1}2^{m-2k+1}$.*

Theorem 2. *For $n \geq 2$, $\mathcal{M}^{\mathrm{d}}(n) = \sum_{i=1}^{n-1}\sum_{j=1}^{n-i} D(2n - 2i, j)D(2i, i).$*

Proof. Let (u, v) be an MAB pair with $i + j \leq n$. In view of Lemma 1, $uv = \alpha x \beta \beta' x' \alpha'$ and note that $\beta \beta'$ is an abelian-bordered word with the shortest abelian-border of length $|\beta|$. The number of such words in uv is $D(2i, i)$. Also, $\alpha x x' \alpha'$ is an abelian-bordered word with the shortest abelian-border of length $|\alpha|$ in uv. Now the number of words of the form $\alpha x x' \alpha'$ is $D(2n - 2i, j)$. Hence, the total number of desired pairs (u, v) is $\sum_{i=1}^{n-1}\sum_{j=1}^{n-i} D(2n - 2i, j)D(2i, i)$. $\square$

As per Table 1, we have verified Theorem 2 for some initial values through a computer program.

3   Computing $\mathcal{M}^{\mathrm{o}}(n)$

We first present a straightforward property on the structure of MAB pairs (u, v) such that $i + j > n$ in the following lemma.

Lemma 2. *For an MAB pair (u, v) with $n \geq 3$, $i + j > n$ iff $u = \alpha \gamma \beta$ and $v = \beta' \gamma' \alpha'$ for some $\alpha, \gamma, \beta, \beta', \gamma', \alpha' \in \Sigma^+$ with $\alpha \gamma \sim_{\mathrm{abl}} \gamma' \alpha'$, $\gamma \beta \sim_{\mathrm{abl}} \beta' \gamma'$, $|\alpha \gamma| = j$, $|\gamma \beta| = i$, and both $(\gamma \beta, \beta' \gamma')$ and $(\alpha \gamma, \gamma' \alpha')$ are MAU pairs.*

To calculate $\mathcal{M}^\circ(n)$, we use the existing relationship between binary words and the lattice paths. A *lattice path* is a sequence of one unit steps between two points in $\mathbb{Z}^2$ such that each step is towards right (east) or up (north). We depict lattice paths in XY-plane by considering the lattice points, i.e., points with integer coordinates. Representing each right move by one letter and a move upward by another letter, and vice versa, we can see that there is a bijection between lattice paths and binary words. Figure 2 gives an example of lattice paths and corresponding words. We use the following notation throughout the text.

1. For a binary word w, we denote its corresponding lattice path which starts at a point A by $p_A(w)$. Given a lattice path p starting from a point A, the corresponding binary word is denoted by $w_A(p)$; if the end point B is also specified, then the word is denoted by $w_A^B(p)$.
2. Let p and q be two lattice paths. The set of all lattice points common to them is denoted by $p \cap q$. If q starts where p ends, then pq denotes the path which starts at the starting point of p and ends at the ending point of q.
3. If A is point in XY-plane, then $X(A)$ and $Y(A)$ denote the X- and Y-coordinates of A, respectively.
4. For $A = (X_1, Y_1), B = (X_2, Y_2) \in \mathbb{Z}^2$, the number of lattice paths from A to B is denoted by N_A^B. Let $\Delta(X) = X_2 - X_1$ and $\Delta(Y) = Y_2 - Y_1$. If both $\Delta(X), \Delta(Y) \geq 0$, then $N_A^B = \binom{X_2 - X_1 + Y_2 - Y_1}{X_2 - X_1}$; else, $N_A^B = 0$.
5. For distinct lattice points A, B and C, let $N_{A,B}^C$ denote the number of pairs of lattice paths (p, q) such that p is from A to C, q is from B to C, and $p \cap q = \{C\}$. Also, we write $N_A^{B,C}$ to denote the number of pairs of lattice paths (p', q') such that p' is from A to B, q' is from A to C, and $p' \cap q' = \{A\}$.
6. Let A, B and C be distinct lattice points such that $X(A) < X(B)$. We write ${}_A\mathcal{N}_B^C$ to denote the set of all triplets of lattice paths (p, q, p') such that p is from A to C, q is from B to C, p' is from B corresponding to the word $w_A^C(p)$, $p \cap q = \{C\}$, and $q \cap p' = \{B\}$. Further, let $|{}_A\mathcal{N}_B^C| = {}_A N_B^C$.
7. Let L be a straight line intersecting the two distinct lattice paths p and p'. We define the *internal points* on L with respect to p and p' as those lattice points on L which lie in between p and p', but not on either p or p'. We denote the set of all such points by $\mathtt{IP}_{p,p'}(L)$. Also, we write $\mathtt{IP}_{p,p'}(L \,\&\, X = m)$ to denote the set of all internal points (X_1, Y_1) on L such that $X_1 \leq m$.

We recall the relation between abelian-border of a word and lattice paths.

Theorem 3 [25]. *A binary word w has an abelian-border of length k iff for any point A in $\mathbb{Z}^2$, the lattice paths $p_A(w)$ and $p_A(w^R)$ intersect after k steps.*

Remark 2. We have the following straightforward observations from Theorem 3.

1. The length of the shortest abelian-border of a word w is k iff for any lattice point A, the paths $p_A(w)$ and $p_A(w^R)$ intersect again only after k steps.
2. Let $u, v \in \Sigma^+$ and $A \in \mathbb{Z}^2$. The lattice paths $p_A(u)$ and $p_A(v)$ intersect after k steps iff there exist $u' = \mathrm{Pref}_k(u)$ and $v' = \mathrm{Pref}_k(v)$ such that $u' \sim_{\mathrm{abl}} v'$.

Example 2. The word $w = ababaaabb$ has abelian-borders of lengths 4 and 5. As shown in Fig. 2, $p_O(w)$ and $p_O(w^R)$ intersect at A and B (except the end points) after 4 and 5 steps, respectively. Note that these paths intersect for second time at A after 4 steps, and length of the shortest abelian-border of w is 4.

Remark 3. The function $f : \Sigma^n \to P_A(n)$ defined by $f(w) = p_A(w)$ is a bijection, where $P_A(n)$ is the set of all lattice paths from a point $A = (k, l)$ to the line $X + Y = k + l + n$.

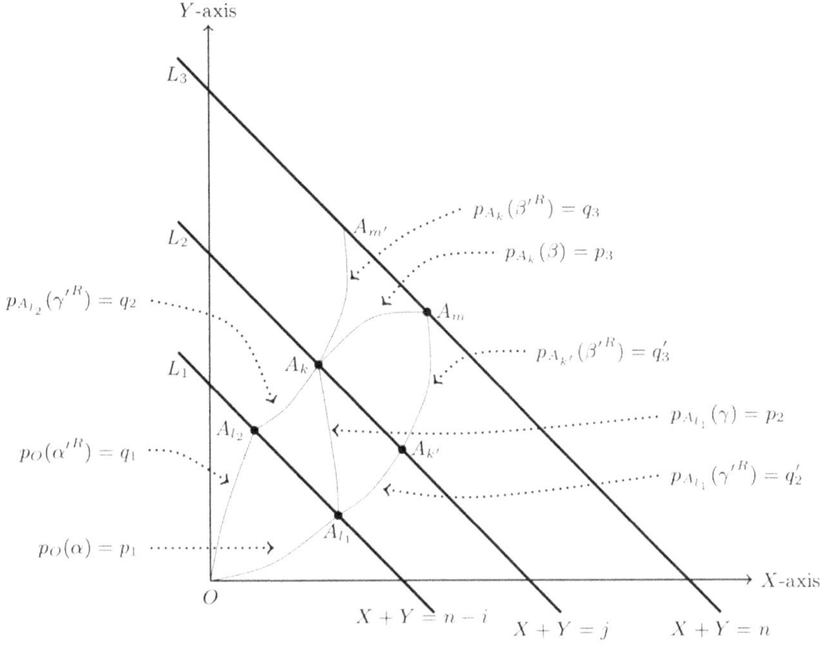

Fig. 1. Lattice paths $p_O(\alpha\gamma\beta)$ and $p_O(\alpha'^R\gamma'^R\beta'^R)$.

3.1   Strategy of Computation

We now describe the strategy of counting $\mathcal{M}^\circ(n)$. Let (u, v) be an arbitrary MAB pair with $|u| = |v| = n$. In view of Lemma 2, we have the following structure of (u, v): $u = \alpha\gamma\beta$ and $v = \beta'\gamma'\alpha'$, for some $\alpha, \gamma, \beta, \beta', \gamma', \alpha' \in \Sigma^+$, and $\text{sb}(u, v) = (\gamma\beta, \beta'\gamma')$, $\text{sb}(v, u) = (\gamma'\alpha', \alpha\gamma)$ with $\text{lsb}(u, v) = i$, $\text{lsb}(v, u) = j$ so that $|\gamma| = |\gamma'| = i + j - n$, $|\alpha| = |\alpha'| = n - i$, and $|\beta| = |\beta'| = n - j$. Throughout Sect. 3, we use these notation along with the corresponding geometric elements described in the following.

Consider the lines $L_1 : X + Y = n - i$, $L_2 : X + Y = j$ and $L_3 : X + Y = n$, and the lattice paths $p = p_O(\alpha\gamma\beta)$ and $q = p_O(\alpha'^R\gamma'^R\beta'^R)$ from the origin $O = (0, 0)$

in XY-plane (For instance, see Fig. 1). Suppose p intersects L_1, L_2 and L_3 at the points A_{l_1}, A_k, and A_m, respectively. Note that, since $\alpha\gamma \sim_{abl} \alpha'^R\gamma'^R$, both the paths p and q intersect the line L_2 at the same point A_k. Suppose q intersects L_1 and L_3 at the points A_{l_2} and $A_{m'}$, respectively.

Note that the path p has three parts, say p_1, p_2 and p_3, corresponding to α, γ, and β, respectively, so that $p = p_1 p_2 p_3$. Also, the path q has three parts, say q_1, q_2 and q_3, corresponding to α'^R, γ'^R, and β'^R, respectively, so that $q = q_1 q_2 q_3$.

Let $A_{l_1} = (l_1, n - i - l_1)$, $A_{l_2} = (l_2, n - i - l_2)$, $A_k = (k, j - k)$, $A_m = (m, n - m)$ and $A_{m'} = (m', n - m')$, where $l_1 \neq l_2$, $0 \leq l_1, l_2 \leq n - i$. Observe that $m' = l_2 + |\gamma'^R\beta'^R|_a$, $n - m' = n - i - l_2 + |\gamma'^R\beta'^R|_b$, $m = l_1 + |\gamma\beta|_a$, and $n - m = n - i - l_1 + |\gamma\beta|_b$. Since $\gamma\beta \sim_{abl} \gamma'^R\beta'^R$, $X(A_m) - X(A_{m'}) = m - m' = l_1 - l_2 = X(A_{l_1}) - X(A_{l_2})$ and $Y(A_{m'}) - Y(A_m) = Y(A_{l_2}) - Y(A_{l_1})$. Thus, the lattice path $p_{A_{l_1}}(\gamma'^R\beta'^R)$ intersects the line L_3 at the point A_m. Suppose $p_{A_{l_1}}(\gamma'^R\beta'^R)$ intersects L_2 at the point $A_{k'}$, say $(k', j - k')$. Let $q_2' = p_{A_{l_1}}(\gamma'^R)$ and $q_3' = p_{A_{k'}}(\beta'^R)$. Since $\gamma\beta \sim_{abl} \beta'\gamma'$ and $(\gamma\beta, \beta'\gamma')$ is an MAU pair, $p_2 p_3 \cap q_2' q_3' = \{A_{l_1}, A_m\}$. Similarly, $p_1 p_2 \cap q_1 q_2 = \{O, A_k\}$. Now by Remark 3, the number of pairs of words $(\alpha\gamma\beta, \alpha'^R\gamma'^R\beta'^R)$ equals the number of pairs of lattice paths (p, q). To count the number of pairs of lattice paths (p, q), we first observe the following:

- for each pair of paths (p, q), there exist pairs $(p_1 p_2, q_1 q_2)$ and $(p_2 p_3, q_2' q_3')$ such that $p_1 p_2 \cap q_1 q_2 = \{O, A_k\}$ and $p_2 p_3 \cap q_2' q_3' = \{A_{l_1}, A_m\}$.
- for pairs of paths $(p_1 p_2, q_1 q_2)$ and $(p_2 p_3, q_2' q_3')$ such that $p_1 p_2 \cap q_1 q_2 = \{O, A_k\}$ and $p_2 p_3 \cap q_2' q_3' = \{A_{l_1}, A_m\}$, there exists a pair of paths (p, q) such that $p_1 p_2$ is an initial part of p, $p_2 p_3$ is a terminal part of p, $q_1 q_2$ is an initial part of q, and $q_2 q_3$ is a terminal part of q.

Thus, $\mathcal{M}^\circ(n)$ is the number of pairs of paths $(p_1 p_2, q_1 q_2)$ and $(p_2 p_3, q_2' q_3')$ such that $p_1 p_2 \cap q_1 q_2 = \{O, A_k\}$ and $p_2 p_3 \cap q_2' q_3' = \{A_{l_1}, A_m\}$.

First consider $l_1 > l_2$ (see Fig. 1). Since we can move only horizontally or vertically along a lattice path, $l_1 \leq k \leq l_2 + i + j - n$ and $k' \leq m \leq k + n - j$. Now we calculate the following: (i) The number of pairs of paths (p_1, q_1) such that $p_1 \cap q_1 = \{O\}$, i.e., $N_O^{A_{l_1}, A_{l_2}}$, (ii) the number of triplets of paths (q_2, p_2, q_2') such that $q_2 \cap p_2 = \{A_k\}$, and $p_2 \cap q_2' = \{A_{l_1}\}$, i.e., $_{A_{l_2}}N_{A_{l_1}}^{A_k}$, and (iii) the number of pairs of paths (p_3, q_3') such that $p_3 \cap q_3' = \{A_m\}$, i.e., $N_{A_k, A_{k'}}^{A_m}$.

Then the number $(N_O^{A_{l_1}, A_{l_2}})(_{A_{l_2}}N_{A_{l_1}}^{A_k})(N_{A_k, A_{k'}}^{A_m})$ gives us the required number of paths for fixed i, j, l_1, l_2, k, m. Thus, for fixed i, j, and for all $0 \leq l_2 < l_1 \leq n - i$, the number of required paths is

$$S = \sum_{l_2} \sum_{l_1 > l_2} \sum_{k = l_1}^{(l_2 + i + j - n)} \sum_{m = k'}^{(k + n - j)} (N_O^{A_{l_1}, A_{l_2}})(_{A_{l_2}}N_{A_{l_1}}^{A_k})(N_{A_k, A_{k'}}^{A_m}).$$

For fixed i, j, when $l_1 < l_2$, the required number of paths is also S by symmetry. Hence, we have the following theorem.

Theorem 4. *For $n \geq 3$,*

$$\mathcal{M}^o(n) = 2 \sum_{j} \sum_{i} \sum_{l_2} \sum_{l_1 > l_2} \sum_{k=l_1}^{(l_2+i+j-n)} \sum_{m=k'}^{(k+n-j)} (N_O^{A_{l_1},A_{l_2}})(_{A_{l_2}}N_{A_{l_1}}^{A_k})(N_{A_k,A_{k'}}^{A_m}).$$

3.2   Computation

We first compute $N_O^{A_{l_1},A_{l_2}}$ and $N_{A_k,A_{k'}}^{A_m}$. Then we compute $\mathcal{M}^o(n)$ for each of cases $|\gamma| = 1$, $|\gamma| = 2$, $|\gamma| = 3$, $|\gamma| = 4$, and $|\gamma| \geq 5$. Let $\mathcal{M}_r^o(n)$ be the number of n-length MAB pairs with overlapped portion of length r so that $\mathcal{M}^o(n) = \sum_{r \geq 1} \mathcal{M}_r^o(n)$. In these calculations, we need the points $E_1 = (l_1, n - i - l_1 + 1)$, the one just above the point A_{l_1}, and $E_2 = (k, j - k - 1)$, the one just below the point A_k (see Fig. 5 for reference).

We directly obtain the values of $N_O^{A_{l_1},A_{l_2}}$ and $N_{A_k,A_{k'}}^{A_m}$ from [15, Theorem 2], as stated in the following proposition.

Proposition 1. *For $l_1 - l_2 \geq 1$,*

$$N_O^{A_{l_1},A_{l_2}} = \frac{l_1-l_2}{n-i}\binom{n-i}{l_1}\binom{n-i}{l_2} \quad and \quad N_{A_k,A_{k'}}^{A_m} = \frac{l_1-l_2}{n-j}\binom{n-j}{m-k}\binom{n-j}{m-k'}.$$

The following remark is evident on a binary alphabet.

Remark 4. For $u, v \in \{a,b\}^+$ with $|u| = |v|$, $u \sim_{\text{abl}} v$ iff $|u|_c = |v|_c$ for some $c \in \{a,b\}$.

Lemma 3. $||\alpha|_c - |\alpha'|_c| = 1 \; \forall \; c \in \{a,b\}$ *iff* $i + j - n = 1$.

Proof. Suppose $||\alpha|_c - |\alpha'|_c| = 1 \; \forall c \in \{a,b\}$. Since $\alpha\gamma \sim_{\text{abl}} \gamma'\alpha'$, we have $||\gamma'|_c - |\gamma|_c| = 1 \; \forall c \in \{a,b\}$. Without loss of generality, suppose $|\alpha|_a - |\alpha'|_a = 1$. If possible, let $i + j - n = |\gamma| = |\gamma'| \geq 2$. Then we have the following cases:

Case 1: Let $\gamma = xa$ and $\gamma'^R = yb$ for some $x, y \in \Sigma^+$. Then, as $|\alpha|_a > |\alpha'|_a$, $a \in \text{Suff}(\gamma)$ and $b \in \text{Suff}(\gamma'^R)$, it can be observed that the paths $p_O(\alpha\gamma)$ and $p_O(\alpha'^R\gamma'^R)$ intersect before the point A_k; which is a contradiction as $|\alpha\gamma|$ is shortest external abelian-border of (u,v).

Case 2: Let $\gamma = xb$ and $\gamma'^R = ya$ for some $x, y \in \Sigma^+$. Now, $\gamma\beta \sim_{\text{abl}} \beta'\gamma'$ implies $|\beta|_a - |\beta'|_a = 1$. Then $|b\beta|_a = |\beta|_a = 1 + |\beta'|_a = |\beta'a|_a$ and $(u,v) = (\alpha xb\beta, \beta'ay^R\alpha')$. Thus $b\beta \sim_{\text{abl}} \beta'a$, i.e., $\text{lsb}(u,v) \leq |b\beta| < |\gamma\beta|$, a contradiction.

Case 3: Let $\gamma = xa_1$ and $\gamma'^R = ya_1$ for some $x, y \in \Sigma^+$, $a_1 \in \{a,b\}$. Then, $\alpha\gamma \sim_{\text{abl}} \gamma'\alpha'$ implies $\alpha x \sim_{\text{abl}} y^R\alpha'$ which is a contradiction as $\text{lsb}(v,u) = |\alpha\gamma|$.

Therefore, our assumption $i + j - n \geq 2$ was wrong. Hence as $i + j - n \geq 1$, $i + j - n = 1$.

Conversely, let $i + j - n = 1$, i.e., $|\gamma| = |\gamma'| = 1$. Since $\gamma \neq \gamma'$, consider $\gamma = a_1$ and $\gamma' = b_1$ where $a_1 \neq b_1$, and $a_1, b_1 \in \{a,b\}$. Now, $\alpha\gamma \sim_{\text{abl}} \gamma'\alpha'$ implies $|\alpha a_1|_{a_1} = |b_1\alpha'|_{a_1}$, i.e., $|\alpha'|_{a_1} - |\alpha|_{a_1} = 1$. Similarly, $|\alpha a_1|_{b_1} = |b_1\alpha'|_{b_1}$ implies $|\alpha|_{b_1} - |\alpha'|_{b_1} = 1$. Therefore $||\alpha|_c - |\alpha'|_c| = 1 \; \forall c \in \{a,b\}$. $\square$

Corollary 1.   *1.* $||\alpha|_c - |\alpha'|_c| = 1 \; \forall c \in \{a,b\} \Leftrightarrow ||\beta|_c - |\beta'|_c| = 1 \; \forall c \in \{a,b\}$.

2. $||\beta|_c - |\beta'|_c| = 1 \ \forall c \in \{a, b\} \Leftrightarrow i + j - n = 1$.

Using Lemma 3 and Corollary 1, we have the following remark:

Remark 5. If $i + j = 2n - 2$, then $|\alpha| = |\gamma| = |\beta| = 1$ so that $n = 3$ and $(u, v) \in \{(aba, bab), (bab, aba)\}$.

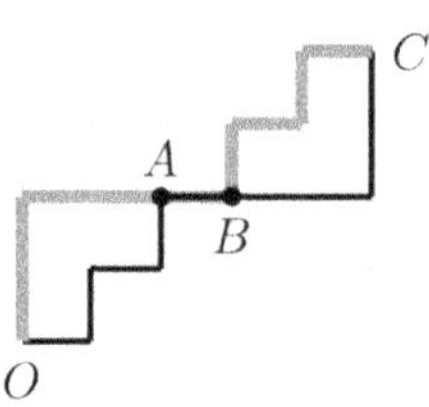

Fig. 2. Lattice paths $p_O(w)$ (thin) and $p_O(w^R)$ (thick), for $w = ababaaabb$.

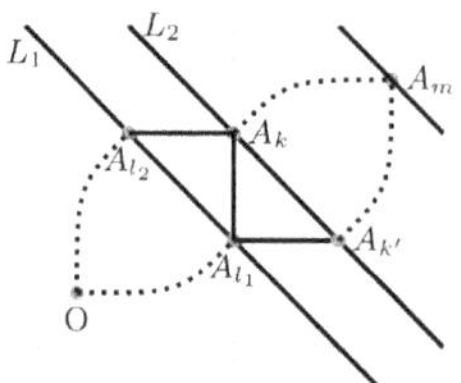

Fig. 3. Illustration to the proof of Proposition 2 when $r = 1$.

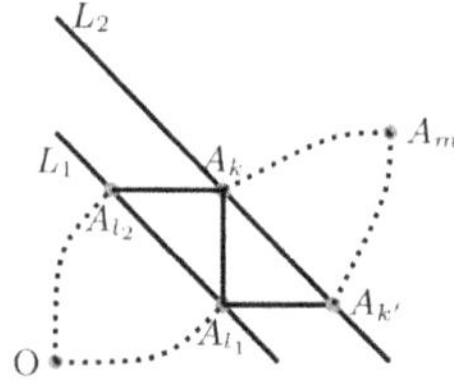

Fig. 4. Illustration to the proof of Proposition 2 when $r = 2$.

Lemma 4. *For $|\gamma| \geq 2$, if $(p, q, p') \in {}_{A_{l_2}}\mathcal{N}_{A_{l_1}}^{A_k}$, then $w_{A_{l_2}}^{A_k}(p) = aza$ for some $z \in \Sigma^*$ and consequently, $w_{A_{l_1}}^{A_k}(q) = bz'b$ for some $z' \in \Sigma^*$.*

Proof. Let $w_{A_{l_2}}^{A_k}(p) = w_p$ and $w_{A_{l_1}}^{A_k}(q) = w_q$. Then $|w_p| = |w_q| = |\gamma|$.

Case 1: Let $b \in \mathrm{Suff}(w_p)$. Then p must pass through the point E_2. Since $p \cap q = \{A_k\}$, q reaches the point A_k through $(k - 1, j - k)$. Then as $l_1 > l_2$ and $j - k \geq Y(A_{l_2}) > Y(A_{l_1})$, p and q must intersect before A_k, which is a contradiction.

Case 2: Let $b \in \mathrm{Pref}(w_p)$. Then as $w_p = w_{A_{l_1}}(p')$, p' pass through the point E_1. Since $q \cap p' = \{A_{l_1}\}$, q must pass through the point $(l_1 + 1, n - i - l_1)$. Since $Y(A_k) - Y(A_{k'}) = Y(A_{l_2}) - Y(A_{l_1}) > 0$, path p' and q must intersect after the point A_{l_1}, which is a contradiction.

Hence, w_p is of the form aza for some $z \in \Sigma^*$. Since $p \cap q = \{A_k\}$ and $q \cap p' = \{A_{l_1}\}$, w_q must be of the form $bz'b$ for some $z' \in \Sigma^*$. $\square$

Remark 6. By Lemma 4, we have $w_{A_{l_2}}^{A_k}(p) = a^{n_1} b^{n_2} \cdots a^{n_{2r-1}}$ and $w_{A_{l_1}}^{A_k}(q) = b^{m_1} a^{m_2} \cdots b^{m_{2r'-1}}$ for some $r, r', n_1, \cdots n_{2r-1}, m_1, \cdots m_{2r'-1} \in \mathbb{N}$.

Proposition 2. *For $1 \leq r = i + j - n \leq 4$, the values of $\mathcal{M}_r^\circ(n)$ are as per the following:*

$$\mathcal{M}_1^\circ(n) = \sum_{i=2}^{n-1} \sum_{l_1=1}^{n-i} \sum_{m=l_1+1}^{i+l_1-1} \frac{2}{(i-1)(n-i)} \binom{n-i}{l_1} \binom{n-i}{l_1-1} \binom{i-1}{m-l_1} \binom{i-1}{m-l_1-1}.$$

$$\mathcal{M}_2^o(n) = \sum_{i=3}^{n-2} \sum_{l_1=2}^{n-i} \sum_{m=l_1+2}^{i+l_1-2} \frac{8}{(i-2)(n-i)} \binom{n-i}{l_1} \binom{n-i}{l_1-2} \binom{i-2}{m-l_1} \binom{i-2}{m-l_1-2}.$$

$$\mathcal{M}_3^o(n) = \sum_{i=5}^{n-2} \sum_{l_2=0}^{n-i-2} \sum_{l_1=2+l_2}^{3+l_2} \sum_{k=l_1}^{l_2+3} \sum_{m=k+l_1-l_2}^{n-j+k} \frac{2(l_1-l_2)^2}{(n-i)(n-j)} \binom{n-i}{l_1} \times$$

$$\binom{n-i}{l_2} \binom{n-j}{m-k} \binom{n-j}{m-k-l_1+l_2} {}_{A_{l_2}}N_{A_{l_1}}^{A_k}$$

where

$$_{A_{l_2}}N_{A_{l_1}}^{A_k} = \begin{cases} 1 & \text{if } l_1 = 2 + l_2 \ \& \ k = l_1 \text{ or } l_1 + 1, \\ 1 & \text{if } l_1 = 3 + l_2 \ \& \ k = l_1, \\ 0 & \text{otherwise.} \end{cases}$$

$$\mathcal{M}_4^o(n) = \sum_{i=6}^{n-2} \sum_{l_2=0}^{n-i-2} \sum_{l_1=2+l_2}^{4+l_2} \sum_{k=l_1}^{l_2+4} \sum_{m=k+l_1-l_2}^{n-j+k} \frac{2(l_1-l_2)^2}{(n-i)(n-j)} \binom{n-i}{l_1} \times$$

$$\binom{n-i}{l_2} \binom{n-j}{m-k} \binom{n-j}{m-k-l_1+l_2} {}_{A_{l_2}}N_{A_{l_1}}^{A_k} \ \text{where}$$

$$_{A_{l_2}}N_{A_{l_1}}^{A_k} = \begin{cases} 1 & \text{if } (l_1 = 2+l_2 \ \& \ k = l_1 \text{ or } l_1+2) \text{ or } (l_1 = 4+l_2 \ \& \ k = l_1), \\ 2 & \text{if } (l_1 = 2+l_2 \ \& \ k = l_1+1) \text{ or } (l_1 = 3+l_2 \ \& \ k = l_1 \text{ or } l_1+1), \\ 0 & \text{otherwise.} \end{cases}$$

Proof. Since we can move only horizontally or vertically along a lattice path, $\max\{l_1, l_2\} \le k \le \min\{l_1, l_2\} + |\gamma|$ and $\max\{k, k'\} \le m \le \min\{k, k'\} + n - j$ (see Fig. 1). We now compute the value of $_{A_{l_2}}N_{A_{l_1}}^{A_k}$ for $1 \le |\gamma| \le 4$.

Let $|\gamma| = 1$, i.e., $|\gamma'| = i + j - n = 1$. Then by Lemma 3, $||\alpha|_c - |\alpha'|_c| = 1 \ \forall c \in \{a, b\}$. Let $|\alpha|_a - |\alpha'|_a = 1$. Then $l_1 - l_2 = 1$, $A_{l_2} = (l_1 - 1, n - i - l_1 + 1)$ and $l_1 \le k \le l_1$, i.e., $k = l_1$ (see Fig. 3). Then $Y(A_k) = j - k = n - i + 1 - l_1$ and $A_k = (l_1, n - i - l_1 + 1)$. This implies $\gamma = b$ and $\gamma'^R = a$. Then $A_{k'} = (l_1 + 1, n - i - l_1)$. This implies $_{A_{l_2}}N_{A_{l_1}}^{A_k} = 1$. Now $l_1 + 1 \le m \le l_1 + n - j$, i.e., $l_1 + 1 \le m \le l_1 + i - 1$. Thus, the value of $\mathcal{M}_1^o(n)$ is as desired.

Let $|\gamma| = 2$, i.e., $|\gamma'| = i + j - n = 2$. Then by Lemma 3, $||\alpha|_c - |\alpha'|_c| \ge 2 \ \forall c \in \{a, b\}$. Let $|\alpha|_a - |\alpha'|_a \ge 2$. Then $l_1 - l_2 \ge 2$. Now, $l_1 \le k \le l_2 + 2$. This implies $l_1 - l_2 \le 2$. Thus $l_1 - l_2 = 2$. Thus $A_{l_2} = (l_1 - 2, n - i - l_1 + 2)$. Now $l_1 \le k \le l_1 - 2 + 2$, i.e., $k = l_1$ (see Fig. 4). Then $Y(A_k) = j - k = n - i + 2 - l_1$ and $A_k = (k, n - i - l_1 + 2)$. This implies $\gamma = bb$ and $\gamma'^R = aa$. Then $A_{k'} = (l_1 + 2, n - i - l_1)$. Then $_{A_{l_2}}N_{A_{l_1}}^{A_k} = 1$. Now $k' \le m \le k + n - j$, i.e., $l_1 + 2 \le m \le l_1 + i - 2$. Thus the value of $\mathcal{M}_2^o(n)$ is as desired.

For $|\gamma| = 3$ and 4, proofs are similar to the above. For details of the proof, one may refer to the arXiv version of the paper [22]. $\qquad\square$

We have verified Proposition 2 through a computer program and the results are given in Table 1.

Table 1. Values of $\mathcal{M}^{\mathrm{d}}(n)$ and $\mathcal{M}_r^{\mathrm{o}}(n)$, for $1 \leq r \leq 4$ and $1 \leq n \leq 12$.

n	1	2	3	4	5	6	7	8	9	10	11	12
$\mathcal{M}^{\mathrm{d}}(n)$	0	4	24	116	520	2248	9520	39796	164904	679064	2783440	11368904
$\mathcal{M}_1^{\mathrm{o}}(n)$	0	0	2	8	28	96	330	1144	4004	14144	50388	180880
$\mathcal{M}_2^{\mathrm{o}}(n)$	0	0	0	0	0	2	16	88	416	1820	7616	31008
$\mathcal{M}_3^{\mathrm{o}}(n)$	0	0	0	0	0	0	4	32	178	856	3820	16320
$\mathcal{M}_4^{\mathrm{o}}(n)$	0	0	0	0	0	0	0	8	64	360	1760	8002

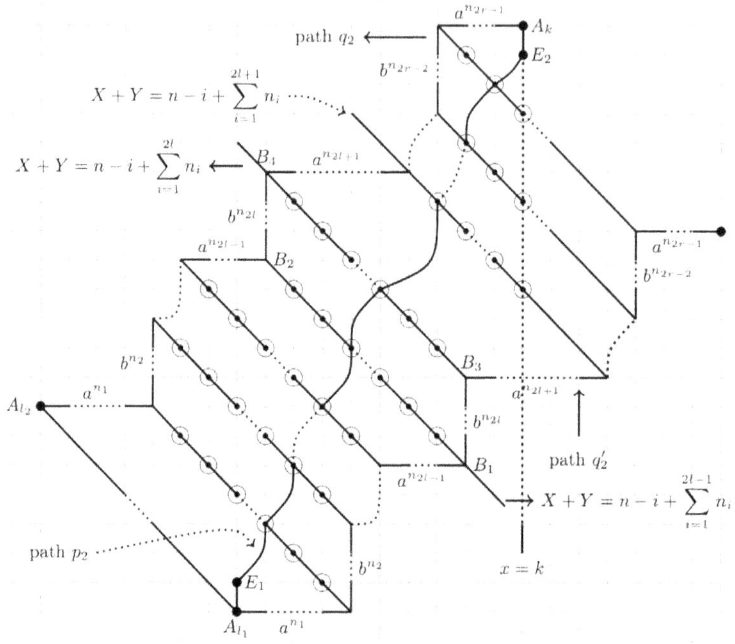

Fig. 5. Illustration to the proof of Lemma 5 and Proposition 3. Here, big dots surrounded by circles denote internal points.

For $r \in \mathbb{N}$, consider $L_r' : X + Y = n - i + \sum_{t=1}^r n_t$ and $\mathrm{IP}_{q_2,q_2'}(L_r')$, in short we write $\mathrm{IP}_{q_2,q_2'}(r)$. We use these notation only in Lemma 5.

Lemma 5. *Let* $z = a^{n_1}b^{n_2}a^{n_3}\cdots a^{n_{2r-1}}$, $q_2 = p_{A_{l_2}}(z)$ *and* $q_2' = p_{A_{l_1}}(z)$ *where* $l_1 - l_2 \geq 2$, $r \in \mathbb{N}$ *and each* $n_t \geq 1$. *Then for any* $1 \leq s \leq 2r - 2$, *the lattice paths between the points of* $\mathrm{IP}_{q_2,q_2'}(s)$ *and* $\mathrm{IP}_{q_2,q_2'}(s+1)$ *do not intersect* q_2 *and* q_2'. *Also, lattice paths from the point* E_1 *to* $\mathrm{IP}_{q_2,q_2'}(1)$, *and that from the points of* $\mathrm{IP}_{q_2,q_2'}(2r-2)$ *to the point* E_2 *do not intersect* q_2 *and* q_2'.

Proof. For $1 \leq l \leq r - 1$, let L_{2l-1}' intersect q_2' and q_2 at the points B_1 and B_2, respectively (see Fig. 5). Also, let L_{2l}' intersects q_2' and q_2 at the points

B_3 and B_4, respectively (see Fig. 5). By geometric argument, $X(B_2) = X(B_4)$ and $X(B_1) = X(B_3)$. If $(X_1, Y_1) \in \mathrm{IP}_{q_2, q_2'}(2l - 1)$ and $(X_2, Y_2) \in \mathrm{IP}_{q_2, q_2'}(2l)$, then $X(B_2) + 1 \leq X_1 \leq X(B_1) - 1$ and $X(B_4) + 1 \leq X_2 \leq X(B_3) - 1$. If p' is any lattice path between (X_1, Y_1) and (X_2, Y_2) and C is any point on p', then as we can move only horizontally or vertically along a lattice path, $X(B_2) + 1 \leq X(C) \leq X(B_1) - 1$. Thus p' never intersects q_2 and q_2'. Therefore any lattice path (if it exists) between (X_1, Y_1) and (X_2, Y_2) do not intersect q_2' and q_2. Since (X_1, Y_1) and (X_2, Y_2) are arbitrary, this holds between any two points of $\mathrm{IP}_{q_2, q_2'}(2l - 1)$ and $\mathrm{IP}_{q_2, q_2'}(2l)$. Similar things happen between the points of $\mathrm{IP}_{q_2, q_2'}(2l)$ and $\mathrm{IP}_{q_2, q_2'}(2l + 1)$. This proves the first part of the statement. Taking clue from Fig. 5, one can prove the remaining parts of the statement. $\square$

We now compute $_{A_{l_2}} N_{A_{l_1}}^{A_k}$ when $|\gamma| \geq 5$. Because of limited space, we only present a sketch of the proof with necessary details.

Proposition 3. *For $|\gamma| \geq 5$ and $l_1 - l_2 \geq 2$, the value of $_{A_{l_2}} N_{A_{l_1}}^{A_k}$ is*

$$\sum_E \sum_D \sum_C \binom{n_1 - 1}{l_2 - l_1 + n_1 + s_1} \binom{n_2}{t_2 - s_1} \times$$

$$\binom{n_{2r-1} - 1}{n_{2r-1} - t_{2r-2}} \prod_{r''=2}^{r-1} \binom{n_{2r''-1} - 1}{n_{2r''-1} + s_{2r''-1} - t_{2r''-2}} \binom{n_{2r''}}{t_{2r''} - s_{2r''-1}},$$

in which $E = \{ r \in \mathbb{N} : 1 \leq r \leq \lfloor \frac{i+j-n+1}{2} \rfloor \}$; D consists of tuples $(n_1, n_2, \cdots, n_{2r-1})$ satisfying $\sum_{i'=1}^{2r-1} n_{i'} = |\gamma|$, where $\sum_{j'=1}^{r} n_{2j'-1} = k - l_2$ and $\sum_{j'=1}^{r-1} n_{2j'} = |\gamma| - k + l_2$; and C is the collection of the following conditions:

(i) $1 \leq s_1, t_2, s_3, t_4, \cdots, s_{2l-1}, t_{2l} \leq l_1 - l_2 - 1$
(ii) $1 \leq s_{2l+1}, t_{2l+2} \leq l_1 - l_2 - n_{2l+1}''$
(iii) $1 \leq s_{2l+s}, t_{2l+s+1} \leq l_1 - l_2 - n_{2l+1}'' - n_{2l+3} - \cdots - n_{2l+s}$ for $3 \leq s \leq 2r - 2l - 3$

where $0 \leq l \leq r - 1$, $n_{2l+1} = n_{2l+1}' + n_{2l+1}''$ with $n_{2l+1}' \geq 0$ and $n_{2l+1}'' > 0$, and $l_1 - l_2 = n_{2l+1}'' + n_{2l+3} + \cdots + n_{2r-1}$.

Proof. For $r \in \mathbb{N}$ and integers $n_1, n_2, \cdots, n_{2r-1} \geq 1$, let $P = (n_1, n_2, \cdots, n_{2r-1})$ be a $(2r - 1)$-partition of $i + j - n$ such that $w_{A_{l_2}}^{A_k}(q_2) = a^{n_1} b^{n_2} a^{n_3} \cdots b^{n_{2r-2}} a^{n_{2r-1}}$ for some lattice path q_2 from the point A_{l_2} to the point A_k. Let q_2' be a lattice path from A_{l_1} corresponding to the word $w_{A_{l_2}}^{A_k}(q_2)$ (see Fig. 5). Using $2D$ geometry, we now compute $\mathrm{IP}_{q_2, q_2'}(X + Y = n - i + \sum_{i'=1}^{s'} n_{i'}, X = k)$ for each $1 \leq s' \leq 2r - 2$ (see Fig. 5). In short, we denote this set as $\mathrm{IP}_{q_2, q_2'}^{P}(s')$ where $1 \leq s' \leq 2r - 2$. Furthermore, by geometric arguments, we observe that $\mathrm{IP}_{q_2, q_2'}^{P}(s')$ is nonempty for every $1 \leq s' \leq 2r - 2$.

Now we calculate all possible lattice paths p_2 that begin at the point A_{l_1}, proceed to the point E_1, then move sequentially through the points of $\mathrm{IP}_{q_2, q_2'}^{P}(1)$, $\mathrm{IP}_{q_2, q_2'}^{P}(2)$, ..., $\mathrm{IP}_{q_2, q_2'}^{P}(2r - 2)$, followed by the point E_2, and finally terminate at A_k (see Fig. 5). From Lemma 5, these paths p_2 intersect the paths q_2' and q_2

only at A_{l_1} and A_k, respectively. From these paths p_2, we now construct the set of all possible triplets (q_2, p_2, q_2'). Since this set depends on q_2, and q_2 depends on the partition P, we denote it by $G(P)$.

Since $w^{A_k}_{A_{l_2}}(q_2) = a^{n_1} b^{n_2} a^{n_3} \cdots b^{n_{2r-2}} a^{n_{2r-1}}$, we have $n_1 + n_2 + \cdots + n_{2r-1} = i + j - n$ where $n_1 + n_3 + \cdots + n_{2r-1} = k - l_2$ and $n_2 + n_4 + \cdots + n_{2r-2} = (i + j - n) - (k - l_2)$.

Then there are $\binom{k-l_2-1}{r-1} \binom{i+j-n-k+l_2-1}{r-2}$ distinct $(2r-1)$-partitions of the integer $i + j - n$ where sum of odd-positioned terms is $k - l_2$ and sum of even-positioned terms is $i+j-n-k+l_2$. Let $\mathscr{P}$ be the set of all such $(2r-1)$-partitions of $i+j-n$. Then for any $P = (n_1, n_2, \cdots, n_{2r-1}) \in \mathscr{P}$, we consider a path q_2 from A_{l_2} to A_k such that $w^{A_k}_{A_{l_2}}(q_2) = a^{n_1} b^{n_2} \cdots a^{n_{2r-1}}$ and compute $G(P)$. Consider $H(2r - 1) = \bigcup_{P \in \mathscr{P}} G(P)$. Now, $1 \leq 2r - 1 \leq i + j - n$, i.e., $1 \leq r \leq \lfloor \frac{i+j-n+1}{2} \rfloor$. Let $T = \bigcup_{r=1}^{\lfloor \frac{i+j-n+1}{2} \rfloor} H(2r - 1)$. We now show that $_{A_{l_2}}\mathcal{N}^{A_k}_{A_{l_1}} = T$.

Let $(q_0, p_0, q_0') \in {}_{A_{l_2}}\mathcal{N}^{A_k}_{A_{l_1}}$. Then q_0 is a path from A_{l_2} to A_k, p_0 is a path from A_{l_1} to A_k, q_0' is a path from A_{l_1} corresponding to the word $w^{A_k}_{A_{l_2}}(q_0)$, $q_0 \cap p_0 = \{A_k\}$, and $p_0 \cap q_0' = \{A_{l_1}\}$. Now, $|w^{A_k}_{A_{l_2}}(q_0)| = |w^{A_k}_{A_{l_1}}(p_0)| = i+j-n \geq 5$. Then by Remark 6, there exists a partition $P_0 = (n_1', n_2', \cdots, n_{2r'-1}')$ of $i + j - n$ such that $w^{A_k}_{A_{l_2}}(q_0) = a^{n_1'} b^{n_2'} a^{n_3'} \cdots a^{n_{2r'-1}'}$ and $w^{A_k}_{A_{l_1}}(p_0)$ starts and ends with b. Thus, the path p_0 starts at $A_{l_1} = (l_1, n - i - l_1)$, then it moves to E_1. Since $q_0 \cap p_0 = \{A_k\}$, and $p_0 \cap q_0' = \{A_{l_1}\}$, the path p_0 subsequently passes through some points of $\mathbf{IP}^{P_0}_{q_0,q_0'}(1), \mathbf{IP}^{P_0}_{q_0,q_0'}(2), \cdots, \mathbf{IP}^{P_0}_{q_0,q_0'}(2r' - 2)$ and then E_2 and finally reaches $(k, j - k)$. Since $\mathbf{IP}^{P_0}_{q_0,q_0'}(1), \mathbf{IP}^{P_0}_{q_0,q_0'}(2), \cdots, \mathbf{IP}^{P_0}_{q_0,q_0'}(2r' - 2)$ are nonempty, there always exists a route for p_0. Thus, (q_0, p_0, q_0') is an element of $G(P_0)$. This implies $(q_0, p_0, q_0') \in T$. Thus $_{A_{l_2}}\mathcal{N}^{A_k}_{A_{l_1}} \subseteq T$.

Let $(q_2, p_2, q_2') \in T$. Then for some $1 \leq 2r'' - 1 \leq i + j - n$, $(q_2, p_2, q_2') \in H(2r'' - 1)$. Then for some $(2r'' - 1)$-partition $P'' = (n_1'', n_2'', \cdots, n_{2r''-1}'')$ of $i + j - n$ where $n_1'' + n_3'' + \cdots + n_{2r''-1}'' = k - l_2$ and $n_2'' + n_4'' + \cdots + n_{2r''-2}'' = i + j - n - k + l_2$, we have $(q_2, p_2, q_2') \in G(P'')$. So, q_2 is a path from A_{l_2} to A_k, q_2' is a path from A_{l_1} corresponding to word $w^{A_k}_{A_{l_2}}(q_2)$, p_2 is a path from A_{l_1} to A_k such that $q_2 \cap p_2 = \{A_k\}$ and $p_2 \cap q_2' = \{A_{l_1}\}$. Thus $(q_2, p_2, q_2') \in {}_{A_{l_2}}\mathcal{N}^{A_k}_{A_{l_1}}$. Hence, $T \subseteq {}_{A_{l_2}}\mathcal{N}^{A_k}_{A_{l_1}}$ so that $T = {}_{A_{l_2}}\mathcal{N}^{A_k}_{A_{l_1}}$.

Now we calculate the cardinality of T. As a first step, we compute the cardinality of $G(P)$. This is straightforward since the coordinates of the internal points are known, and the formula for the number of lattice paths between two points is available. This gives $|G(P)| =$

$$\sum_C \binom{n_1 - 1}{l_2 - l_1 + n_1 + s_1} \binom{n_2}{t_2 - s_1} \times$$

$$\binom{n_{2r-1} - 1}{n_{2r-1} - t_{2r-2}} \prod_{r'''=2}^{r-1} \binom{n_{2r'''-1} - 1}{n_{2r'''-1} + s_{2r'''-1} - t_{2r'''-2}} \binom{n_{2r'''}}{t_{2r''''} - s_{2r'''-1}}.$$

Here C is the collection of the following conditions that arise in the determination of the coordinates of the internal points: $1 \leq s_1, t_2, s_3, t_4, \cdots, s_{2l-1}, t_{2l} \leq l_1 - l_2 - 1$, $\quad 1 \leq s_{2l+1}, t_{2l+2} \leq l_1 - l_2 - n''_{2l+1}$, $1 \leq s_{2l+s}, t_{2l+s+1} \leq l_1 - l_2 - n''_{2l+1} - n_{2l+3} - \cdots - n_{2l+s}$, for odd integer s $(3 \leq s \leq 2r - 2l - 3)$ where $n_{2l+1} = n'_{2l+1} + n''_{2l+1}$ for some $l \geq 0$ with $n'_{2l+1} \geq 0$, $n''_{2l+1} > 0$ and $l_1 - l_2 = n''_{2l+1} + n_{2l+3} + \cdots + n_{2r-1}$.

Then we vary P over all partitions in $\mathscr{P}$ to obtain the cardinality of $H(2r-1)$. This gives $|H(2r-1)| = \sum_{P \in \mathscr{P}} |G(P)|$. Finally, we vary the value of r from 1 to $\lfloor \frac{i+j-n+1}{2} \rfloor$ and compute $|T|$. This gives $|T| = \sum_{r=1}^{\lfloor \frac{i+j-n+1}{2} \rfloor} |H(2r-1)|$. $\square$

Theorem 5. *For $|\gamma| = i + j - n \geq 5$, the value of $\mathcal{M}^{\circ}_{|\gamma|}(n)$ is*

$$2 \sum_{j} \sum_{i} \sum_{l_2=0}^{n-i-2} \sum_{l_1=l_2+2}^{n-i} \sum_{k=l_1}^{i+j-n+l_2} \sum_{m=k'}^{n-j+k} \left(N_O^{A_{l_1}, A_{l_2}} \right) \left({}_{A_{l_2}} N_{A_{l_1}}^{A_k} \right) \left(N_{A_k, A_{k'}}^{A_m} \right)$$

where $N_O^{A_{l_1}, A_{l_2}}$, ${}_{A_{l_2}} N_{A_{l_1}}^{A_k}$ and $N_{A_k, A_{k'}}^{A_m}$ are given by propositions 1 and 3.

4   Conclusion

In this paper, we have computed the value of $\mathcal{M}(n)$ for $n \geq 1$. Using Theorem 2, Proposition 2 and Theorem 5, we obtained a formula for $\mathcal{M}(n)$. Further, let $\overline{\mathcal{M}}(n)$ denote the number of MAU pairs (u, v) with $u, v \in \Sigma^n$. By definition, $\overline{\mathcal{M}}(1) = 4$. For $n \geq 2$, the following result for $\overline{\mathcal{M}}(n)$ can be obtained.

Theorem 6. $\overline{\mathcal{M}}(n) = 2 \sum_{i=1}^{n-1} \frac{1}{n-1} \binom{n-1}{i} \binom{n-1}{i-1} + 2 \sum_{r_2=0}^{n-2} \sum_{r_1=r_2+2}^{n} {}_A N_O^{B_{r_2}}$
where $2 \leq r_1 - r_2 \leq n$, $A = (r_2 - r_1, r_1 - r_2)$, $O = (0, 0)$, $B_{r_2} = (r_2, n - r_2)$ and the value of ${}_A N_O^{B_{r_2}}$ follows from propositions 2 and 3.

For $u, v \in \Sigma^n$, it is clear that if (u, v) is an external abelian-bordered pair, then (v, u) is internal abelian-bordered pair, and vice versa. Since these are the pairs which are neither MAB pairs nor MAU pairs, the number of such pairs can be computed as consequence of this work as stated below:

Proposition 4. *The number of external/internal abelian-bordered pairs (u, v) with $|u| = |v| = n$ is $\frac{2^{2n} - \mathcal{M}(n) - \overline{\mathcal{M}}(n)}{2}$.*

We also pose the following questions that arise naturally in this context:

1. Do the limits $\lim_{n \to \infty} \frac{\mathcal{M}(n)}{2^{2n}}$ and $\lim_{n \to \infty} \frac{\overline{\mathcal{M}}(n)}{2^{2n}}$ exist?
2. What is the number of mutually abelian-bordered pairs of binary words (u, v) when $|u| \neq |v|$?
3. What is the number of mutually abelian-bordered pairs of non-binary words?

References

1. Aberkane, A., Currie, J.D., Rampersad, N.: The number of ternary words avoiding abelian cubes grows exponentially. J. Integer Seq. **7**(2), 04.2.7, 13 (2004)
2. Avgustinovich, S., Puzynina, S.: Weak abelian periodicity of infinite words. Theory Comput. Syst. **59**(2), 161–179 (2016)
3. Blanchet-Sadri, F., Chen, K., Hawes, K.: Dyck words, lattice paths, and abelian borders. Int. J. Found. Comput. Sci. **33**(03–04), 203–226 (2022)
4. Carpi, A.: On the number of abelian square-free words on four letters. Disc. Appl. Math. **81**(1–3), 155–167 (1998)
5. Cassaigne, J., Richomme, G., Saari, K., Zamboni, L.Q.: Avoiding abelian powers in binary words with bounded abelian complexity. Int. J. Found. Comput. Sci. **22**(04), 905–920 (2011)
6. Charlier, E., Harju, T., Puzynina, S., Zamboni, L.Q.: Abelian bordered factors and periodicity. Eur. J. Comb. **51**, 407–418 (2016)
7. Christodoulakis, M., Christou, M., Crochemore, M., Iliopoulos, C.S.: Abelian borders in binary words. Disc. Appl. Math. **171**, 141–146 (2014)
8. Currie, J., Linek, V.: Avoiding patterns in the abelian sense. Can. J. Math. **53**(4), 696–714 (2001)
9. Currie, J.D., Rampersad, N.: Fixed points avoiding abelian k-powers. J. Comb. Theory Ser. A **119**(5), 942–948 (2012)
10. Currie, J.D., Visentin, T.I.: Long binary patterns are abelian 2-avoidable. Theor. Comput. Sci. **409**(3), 432–437 (2008)
11. Ehrenfeucht, A., Silberger, D.: Periodicity and unbordered segments of words. Disc. Math. **26**(2), 101–109 (1979)
12. Fici, G., Mignosi, F., Shallit, J.: Abelian-square-rich words. Theor. Comput. Sci. **684**, 29–42 (2017)
13. Fici, G., Postic, M., Silva, M.: Abelian antipowers in infinite words. Adv. Appl. Math. **108**, 67–78 (2019)
14. Gabric, D.: Mutual borders and overlaps. IEEE Trans. Inf. Theory **68**(10), 6888–6893 (2022)
15. Gessel, I., Goddard, W., Shur, W., Wilf, H.S., Yen, L.: Counting pairs of lattice paths by intersections. J. Comb. Theory Ser. A **74**(2), 173–187 (1996)
16. Guibas, L.J., Odlyzko, A.M.: Periods in strings. J. Comb. Theory Ser. A **30**(1), 19–42 (1981)
17. Guibas, L.J., Odlyzko, A.M.: String overlaps, pattern matching, and nontransitive games. J. Comb. Theory Ser. A **30**(2), 183–208 (1981)
18. Harju, T., Nowotka, D.: Border correlation of binary words. J. Comb. Theory Ser. A **108**(2), 331–341 (2004)
19. Harju, T., Nowotka, D.: Periodicity and unbordered words. In: Diekert, V., Habib, M. (eds.) STACS 2004. LNCS, vol. 2996, pp. 294–304. Springer, Heidelberg (2004). https://doi.org/10.1007/978-3-540-24749-4_26
20. Holub, Š., Müller, M.: Fully bordered words. Theor. Comput. Sci. **684**, 53–58 (2017)
21. Lothaire, M.: Combinatorics on words. Cambridge Mathematical Library, Cambridge University Press, Cambridge (1997)
22. Maity, A., Krishna, K.V.: Mutually abelian-bordered binary words. arXiv preprint arXiv:2509.20773 (2025)
23. Nielsen, P.: A note on bifix-free sequences (corresp.). IEEE Trans. Inf. Theory **19**(5), 704–706 (1973)

24. Peltomäki, J., Whiteland, M.A.: Avoiding abelian powers cyclically. Adv. Appl. Math. **121**, 102095 (2020)
25. Rampersad, N., Rigo, M., Salimov, P.: On the number of abelian bordered words. In: Béal, M.-P., Carton, O. (eds.) DLT 2013. LNCS, vol. 7907, pp. 420–432. Springer, Heidelberg (2013). https://doi.org/10.1007/978-3-642-38771-5_37
26. Richmond, L.B., Shallit, J.: Counting abelian squares. Electron. J. Comb. **16**(1) (2009)
27. Rivals, E., Wang, P.: Counting overlapping pairs of words. In: International Computing and Combinatorics Conference, pp. 381–395. Springer, Heidelberg (2025). https://doi.org/10.1007/978-981-95-0218-9_28

Minimum Length Word-Representants of Treelike Permutation Graphs

Tithi Dwary$^{(\boxtimes)}$ and K. V. Krishna

Department of Mathematics, Indian Institute of Technology Guwahati,
Guwahati, India
`{tithi.dwary,kvk}@iitg.ac.in`

Abstract. A simple graph is called a word-representable graph if there is a word over its vertex set such that any two vertices are adjacent in the graph if and only if they alternate in the word; such a word is called a word-representant of the graph. Srinivasan and Hariharasubramanian proved that a minimum length word-representant for a non-complete triangle-free circle graph on n vertices with at least one edge is of $2n - 2$ length. Further, they posed an open problem to find classes of word-representable graphs whose minimum length word-representants are of $2n - \kappa$ length, where n is the number of vertices and κ is the clique number of a graph.

A graph is called a treelike comparability graph if it admits a transitive orientation such that the transitive reduction is a tree. When such a graph is also a permutation graph, it is called a treelike permutation graph. Recently, a subclass of treelike permutation graphs, viz., the class of double-arborescences, was established to be the first example satisfying the criterion given in the above-mentioned open problem. In this work, we devise a polynomial-time algorithm to construct a minimum length word-representant for a given treelike permutation graph and show that all treelike permutation graphs satisfy the criterion of the open problem.

Keywords: Word-representable graph $\cdot$ minimum length
word-representant $\cdot$ treelike permutation graph $\cdot$ double-arborescence

1 Introduction and Preliminaries

First, we recall the necessary background material on treelike comparability graphs from [1] and word-representable graphs from [8], and fix the notation used in this paper. One may refer to [8], for the notions which are used but not defined in this paper.

A graph G is a pair (V, E), where V is a finite set and E is a set of 2-element subsets of V. The elements of V and E are called vertices and edges of the graph G, respectively. If there is an edge between two vertices, we say they are adjacent. Every graph has a depiction in which vertices are denoted by thick dots and adjacent vertices are connected by a line segment or a curve.

© The Author(s), under exclusive license to Springer Nature Switzerland AG 2026
J. Kozik and A. Wolff (Eds.): SOFSEM 2026, LNCS 16448, pp. 90–103, 2026.
https://doi.org/10.1007/978-3-032-17801-5_7

Definition 1. *An orientation of a graph $G = (V, E)$ is an assignment of direction to each edge, i.e., fixing an order for the vertices of each edge. Thus, the resultant edge set, say T, is a subset of $V \times V$. In this case, we say G is oriented and that T is the orientation of G. For $\{a, b\} \in E$, if $(a, b) \in T$, we also write $\overrightarrow{ab}$ indicating the orientation/direction assigned to the edge.*

A path in a graph $G = (V, E)$ is a sequence of vertices $a_1, a_2, \ldots, a_k$, written $a_1 - a_2 - \cdots - a_k$, such that $\{a_i, a_{i+1}\} \in E$ for $1 \leq i < k$. If G is oriented, we also indicate the direction of each edge in the path. For example, $a_1 \to a_2 \leftarrow a_3 \to a_4$ is a oriented path between a_1 and a_4. If all edges on a path are in the same direction, then we say it is a directed path. For example, $a_1 \to a_2 \to a_3 \to a_4$ is a directed path from a_1 to a_4. A graph is said to be connected if there is a path between any two vertices of the graph. In this work, we deal with only connected graphs.

Definition 2. *A graph $G = (V, E)$ is said to be a comparability graph if it admits a transitive orientation, i.e., an orientation T in which if $\overrightarrow{ab} \in T$ and $\overrightarrow{bc} \in T$, then the transitive edge $\overrightarrow{ac} \in T$. The transitive reduction, say T', of the comparability graph G with respect to T is the oriented graph obtained by deleting the transitive edges from T.*

Remark 1. Let G be a comparability graph with the transitive reduction T' of a transitive orientation T of G. For any two vertices a and b of G, there is a directed path from a to b in T' if and only if $\overrightarrow{ab} \in T$.

Definition 3. *A graph G is called a treelike comparability graph if it admits a transitive orientation such that the underlying graph of its transitive reduction is a tree; in this case, the corresponding orientation is called a treelike orientation. If a treelike comparability graph is also a permutation graph, i.e., an intersection graph of segments whose endpoints lie on two parallel lines, it is called a treelike permutation graph.*

It was shown in [1] that a treelike orientation of a comparability graph, if it exists, is unique up to isomorphism and reversing the whole orientation.

Definition 4. *A treelike comparability graph $G = (V, E)$ is called a double-arborescence if there is an all-adjacent vertex in G, i.e., there is a vertex r such that $V = \{r\} \cup N_G(r)$, where $N_G(r)$ is the set of all vertices adjacent to r. We consider r as the root of G. In addition, under the treelike orientation, if the root r is a source (or a sink), i.e., the indegree (respectively, outdegree) of r is zero, then G is called an arborescence.*

The transitive reduction of an arborescence is a directed rooted tree. Figure 1 depicts these types of graphs by their transitive reductions.

Definition 5. *A treelike comparability graph G is called a path of m double-arborescences (or simply, a path of double-arborescences) if G (precisely) consists of m double-arborescences and a path (ignoring the directions) connecting their roots; such a path is called a root-path of G.*

It was proved in [1, Theorem 6] that a graph is a treelike permutation graph if and only if it is a path of double-arborescences. For a detailed information on these graphs, one may refer to [6, 12] and the references thereof.

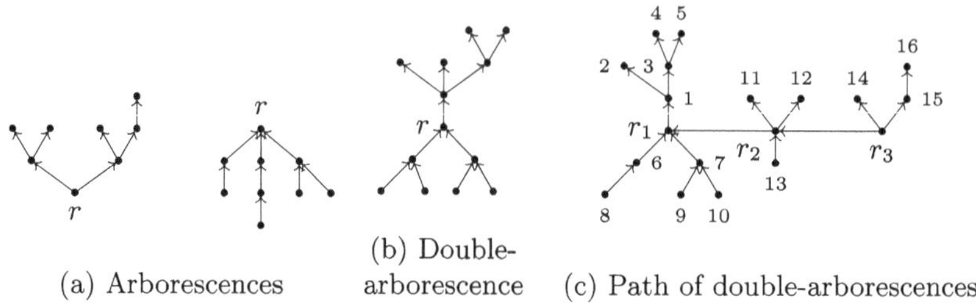

(a) Arborescences   (b) Double-arborescence   (c) Path of double-arborescences

Fig. 1. Examples of treelike comparability graphs in terms of transitive reductions.

Let X be a finite set of letters. A word over X is a finite sequence of letters of X written by juxtaposing them. A subword u of a word w is a subsequence of the sequence w, denoted by $u \ll w$. A subword u is a factor of w if $w = xuy$, for some words x and y, possibly empty words, denoted by $u \ll_f w$. For example, $ababbcac$ is a word over $\{a, b, c\}$, $aabcc \ll ababbcac$ and $abbc \ll_f ababbcac$. For a word $w = a_1 a_2 \cdots a_k$, the reversal $w^R = a_k a_{k-1} \cdots a_1$, where a_i's are letters. Let w be a word over a set X and $Y \subseteq X$. We write $w|_Y$ to denote the subword of w that is obtained by deleting all the letters belonging to $X \setminus Y$. For instance, if $w = ababbcac$, then $w|_{\{a,b\}} = ababba$. Let w be a word over a set containing the letters a, b. We say a and b alternate in w if $w|_{\{a,b\}}$ is in one of the following forms: $(ab)^k, (ba)^k, (ab)^k a, (ba)^k b$, for some k, where x^k denotes the word $xx \cdots x$ for k times. A word w is called a permutation on the letters of w if each letter appears exactly once in w.

Definition 6. *A graph $G = (V, E)$ is called a word-representable graph if there is a word w over the vertex set V such that, for all $a, b \in V$, a and b are adjacent in G if and only if a and b alternate in w; such a word w is called a word-representant of G and we say w represents G. The length of a minimum length word-representant of G is denoted by $\ell(G)$.*

The class of word-representable graphs includes several important classes of graphs such as comparability graphs, circle graphs, and 3-colorable graphs. A word-representable graph has infinitely many word-representants [9]. In this work, we study minimum length word-representants of word-representable graphs. Let G be a word-representable graph on n vertices. Then, it is evident that $\ell(G) \geq n$. A circle graph is an intersection graph of chords of a circle. It was shown in [7] that the class of circle graphs characterizes the word-representable graphs represented by the words in which every letter appears exactly twice. Thus, if G is a circle graph, $\ell(G) \leq 2n$. In the seminal work [4], Gaetz and Ji considered the subclasses, viz., cycles and trees, of circle graphs and provided explicit formulae for both the length and the number of minimum length word-representants. In [11], Srinivasan and Hariharasubramanian proved that there is no circle graph G with $\ell(G) = 2n$ and an edgeless graph G is the only circle graph having $\ell(G) = 2n - 1$. Also, they showed that $\ell(G) = 2n - 2$ for a non-complete triangle-free circle graph G containing at least one edge. Further, they proved that $\ell(G) \geq 2n - \kappa$ for any word-representable graph G with clique number[1] κ. In this connection, they posed an open problem to find classes of word-representable graphs G such that $\ell(G) = 2n - \kappa$. Recently, in [2], a subclass of treelike permutation graphs, viz., the class of double-arborescences, was proved to be the first example satisfying the criterion of the above-mentioned open problem.

In this work, we devise a polynomial-time algorithm to construct a minimum length word-representant of a given treelike permutation graph. Contributing to the above-mentioned open problem, we prove that if G is a treelike permutation graph on n vertices with clique number κ, then $\ell(G) = 2n - \kappa$.

2   Double-Arborescences

In what follows, for a comparability graph $G = (V, E)$ with a transitive orientation T, we use the following notation. For $a \in V$, the upper set of a, $U(a) = \{b \in V \mid \overrightarrow{ab} \in T\}$ and $p_{\overline{a}}, p'_{\overline{a}}, q_{\overline{a}}, q'_{\overline{a}}$ denote permutations on $U(a)$. Further, the lower set of a, $L(a) = \{b \in V \mid \overrightarrow{ba} \in T\}$ and $p_{\underline{a}}, p'_{\underline{a}}, q_{\underline{a}}, q'_{\underline{a}}$ denote permutations on $L(a)$. A permutation on the empty set is considered to be the empty word. For $A \subseteq V$, $G[A]$ denotes the subgraph of G induced by A.

Let T' be the transitive reduction of the treelike orientation T of an arborescence or a double-arborescence $G = (V, E)$ with root r. In [2], we have algorithms to construct minimum length word-representants of G by considering a longest directed path C in T' as input. In this section, we present some structural properties of these minimum length word-representants, which are useful in this work.

First, we consider an arborescence G, and observe the following properties of a minimum length word-representant w of G constructed in [2].

Remark 2. (i) If r is the source in T, then $V = \{r\} \cup U(r)$ and the word w is of the form $p_{\overline{r}} r u$, where u is a subword of w such that no element of C appears in u. Moreover, for any vertex $a \in U(r)$, $p_{\overline{a}} a \ll_f p_{\overline{r}} \ll_f w$.

(ii) If r is the sink in T, then $V = \{r\} \cup L(r)$ and the word w is of the form $q_{\underline{r}} r v$, where v is a subword of w such that no element of C appears in v. Moreover, for any vertex $a \in L(r)$, $q_{\underline{a}} a \ll_f q_{\underline{r}} \ll_f w$.

Proposition 1. *Let r be the source in T and $(r =)a_1 \rightarrow a_2 \rightarrow \cdots \rightarrow a_s$ be a directed path in T', for some integer $s \geq 2$. Then, for all $1 \leq i \leq s - 1$,*
$$p_{\overline{a_{i+1}}} a_{i+1} \ll_f p_{\overline{a_i}} \ll_f p_{\overline{a_1}} \ll_f w.$$

---

[1] A clique is a set of vertices that induces a complete subgraph. The size of a maximum clique in a graph is its clique number.

Proof. In view of Remark 2(i), we have $p_{\overline{a_2}} a_2 \ll_f p_{\overline{a_1}} \ll_f w$. Thus, the result is true for $i = 1$. Suppose for $2 \leq t < s-1$, we have $p_{\overline{a_{t+1}}} a_{t+1} \ll_f p_{\overline{a_t}} \ll_f p_{\overline{a_1}} \ll_f w$. We show that the result is true for $i = t+1$. Since $\{a_{t+2}\} \cup U(a_{t+2}) \subseteq U(a_{t+1})$, exactly one occurrence of each element of $U(a_{t+2}) \cup \{a_{t+2}\}$ appears in $p_{\overline{a_{t+1}}}$. Further, since $p_{\overline{a_{t+2}}} a_{t+2} \ll_f p_{\overline{a_1}}$ (by Remark 2(i)) and $p_{\overline{a_{t+1}}} \ll_f p_{\overline{a_1}}$, we have $p_{\overline{a_{t+2}}} a_{t+2} \ll_f p_{\overline{a_{t+1}}} \ll_f p_{\overline{a_1}}$. Thus, the result is true for $i = t+1$. Hence, for all $1 \leq i \leq s - 1$, the result follows. $\square$

Similarly, if r is the sink in T, we have the following result.

Proposition 2. *Let r be the sink in T and $a_s \to a_{s-1} \to \cdots a_2 \to a_1 (= r)$ be a directed path in T', for some integer $s \geq 2$. Then, for all $1 \leq i \leq s - 1$,*
$$q_{\underline{a_{i+1}}} a_{i+1} \ll_f q_{\underline{a_i}} \ll_f q_{\underline{a_1}} \ll_f w.$$

Now let $G = (V, E)$ be a double-arborescence. Consider the restrictions[2] T_A, T_B of T and T'_A, T'_B of T' on the vertex sets A and B, respectively, where $A = \{r\} \cup L(r)$ and $B = \{r\} \cup U(r)$. Then note that r is the sink in T_A and r is the source in T_B and $A \cup B = V$. Further, both $G[A]$ and $G[B]$ of G are arborescences with root r. Considering the longest directed paths $C \cap A$ and $C \cap B$ in T'_A and T'_B, construct minimum length word-representants $q_{\underline{r}} rv$ and $p_{\overline{r}} ru$ (cf. Remark 2) of $G[A]$ and $G[B]$, respectively. Then, by [2, Theorem 4.6], the word $w' = v^R p_{\overline{r}} r q_{\underline{r}}^R u$ is a minimum length word-representant of G of length $2n - \kappa$, where z^R denotes the reversal of a word z. Further, based on the construction, we observe the following properties of w'.

Remark 3. The word w' is of the form $v_1 p_{\overline{r}} r p_{\underline{r}} u$, where $v_1 = v^R$ and $p_{\underline{r}} = q_{\underline{r}}^R$ is a permutation on $L(r)$. Moreover, we have the following properties of $p_{\overline{r}}$ and $p_{\underline{r}}$.

(i) Since $p_{\overline{r}} ru$ is a minimum length word-representant of the arborescence $G[B]$, by Remark 2(i), if $a \in U(r)$, then $p_{\overline{a}} a \ll_f p_{\overline{r}} \ll_f w'$.
(ii) Since $q_{\underline{r}} rv$ is a minimum length word-representant of the arborescence $G[A]$, by Remark 2(ii), if $a \in L(r)$, then $q_{\underline{a}} a \ll_f q_{\underline{r}}$ so that $a p_{\underline{a}} \ll_f p_{\underline{r}} \ll_f w'$, where $(q_{\underline{a}} a)^R = a p_{\underline{a}}$.

Remark 4. Let a and b be two distinct vertices of G such that $\overrightarrow{ra} \in T'$ and $\overrightarrow{rb} \in T'$. Thus, a and b are non-adjacent in G and $a, b \in U(r)$. Then, in view of Remark 3(i), we have $p_{\overline{a}} a \ll_f p_{\overline{r}}$ and $p_{\overline{b}} b \ll_f p_{\overline{r}}$. Since $U(a) \cap U(b) = \varnothing$, $p_{\overline{r}} = x_1 p_{\overline{a}} a x_2 p_{\overline{b}} b x_3$ (if a appears before b in $p_{\overline{r}}$) or $p_{\overline{r}} = x_1 p_{\overline{b}} b x_2 p_{\overline{a}} a x_3$ (if b appears before a in $p_{\overline{r}}$), for some words x_1, x_2, x_3.

Remark 5. Let a and b be two distinct vertices of G such that $\overrightarrow{ar} \in T'$ and $\overrightarrow{br} \in T'$. Then, in view of Remark 3(ii), we have $p_{\underline{r}} = y_1 a p_{\underline{a}} y_2 b p_{\underline{b}} y_3$ (if a appears before b in $p_{\underline{r}}$) or $p_{\underline{r}} = y_1 b p_{\underline{b}} y_2 a p_{\underline{a}} y_3$ (if b appears before a in $p_{\underline{r}}$), for some words y_1, y_2, y_3.

---

[2] For $A \subseteq V$, the restriction T_A of an orientation T is given by $T_A = T \cap (A \times A)$.

Remark 6. In view of propositions 1, 2 and Remark 3, we have the following properties of the word w'.

(i) Let $(r =)a_1 \rightarrow a_2 \rightarrow \cdots \rightarrow a_s$ be a directed path in T', for some integer $s \geq 2$. Then, for all $1 \leq i \leq s - 1$, $p_{\overline{a_{i+1}}}a_{i+1} \ll_f p_{\overline{a_i}} \ll_f p_{\overline{a_1}} \ll_f w'$.

(ii) Let $a_s \rightarrow a_{s-1} \cdots \rightarrow a_2 \rightarrow a_1(= r)$ be a directed path in T', for some integer $s \geq 2$. Then, for all $1 \leq i \leq s - 1$, $a_{i+1}p_{\underline{a_{i+1}}} \ll_f p_{\underline{a_i}} \ll_f p_{\underline{a_1}} \ll_f w'$.

3 Treelike Permutation Graphs

This section consists of two subsections. In Subsect. 3.2, we give a polynomial-time algorithm to construct a minimum length word-representant of a treelike permutation graph, which uses permutations representing arborescences. While the literature has various methods for constructing permutations representing a permutation graph (cf. [5,10]), we construct a specific type of permutations for arborescences in Subsect. 3.1, which are useful in Subsect. 3.2.

3.1 Permutations for an Arborescence

Let $G = (V, E)$ be an arborescence with root r in the treelike orientation T of G. Let T' be the transitive reduction of T. As G is a treelike comparability graph, both T and T' can be found in linear time [1, Theorem 5]. In Algorithm 1, we give a procedure based on breadth-first search (BFS) on the tree T' to construct two permutations q and q' such that the word qq' represents G.

Algorithm 1: Constructing two permutations for an arborescence.

    Input: An arborescence $G = (V, E)$ and its transitive reduction T' of T.
    Output: Two permutations q and q' on V.
1 Initialize $q = r$, $q' = r$ and Q with r (where Q is a queue as in BFS).
2 **while** Q *is not empty* **do**
3   Remove the first element of Q, say a.
4   **if** *a is not a leaf in T'* **then**
5   Let $c_1, c_2, \ldots, c_m$ be the children of a in T' and append them in Q.
6   Replace a in q by $c_1 c_2 \cdots c_m a$ and in q' by $c_m c_{m-1} \cdots c_1 a$.
7   **end**
8 **end**
9 **return** q and q'.

Remark 7. Algorithm 1 runs in $O(n^2)$ time, where n is the number of vertices of G. Note that, in each iteration of total n iterations, testing condition in step 4 and updating q and q' in step 6 can be done in $O(n)$ time.

Remark 8. We have the following properties of q and q':

(i) If r is the sink in T, the permutations q and q' are of the form $q = q_{\underline{r}}r$ and $q' = q'_{\underline{r}}r$ such that the word $q_{\underline{r}}q'_{\underline{r}}$ represents the induced subgraph $G[L(r)]$.

(ii) Moreover, if $a \in L(r)$, then $p_{\underline{a}}a \ll_f q_{\underline{r}}$ and $p'_{\underline{a}}a \ll_f q'_{\underline{r}}$ such that $p_{\underline{a}}p'_{\underline{a}}$ represents the induced subgraph $G[L(a)]$.

(iii) If r is the source in T, the permutations q and q' are of the form $q = q_{\underline{\bar{r}}}r$ and $q' = q'_{\underline{\bar{r}}}r$, and for $a \in U(r)$, we have $p_{\underline{a}}a \ll_f q_{\underline{\bar{r}}}$ and $p'_{\underline{a}}a \ll_f q'_{\underline{\bar{r}}}$ such that $q_{\underline{\bar{r}}}q'_{\underline{\bar{r}}}$ and $p_{\underline{a}}p'_{\underline{a}}$ represent $G[U(r)]$ and $G[U(a)]$, respectively.

Remark 9. In view of Remark 8, we further observe the following.

(i) Let r be the sink in T and $a_s \to a_{s-1} \to \cdots a_2 \to a_1(= r)$ be a directed path in T', for some integer $s \geq 2$. Then, for all $1 \leq i \leq s - 1$, $p_{\underline{a_{i+1}}}a_{i+1} \ll_f p_{\underline{a_i}} \ll_f q_{\underline{a_1}}$ and $p'_{\underline{a_{i+1}}}a_{i+1} \ll_f p'_{\underline{a_i}} \ll_f q'_{\underline{a_1}}$, where for all $1 \leq i \leq s$, the word $p_{\underline{a_i}}p'_{\underline{a_i}}$ represents the induced subgraph $G[L(a_i)]$.

(ii) Let r be the source in T and $(r =)a_1 \to a_2 \to \cdots \to a_s$ be a directed path in T', for some integer $s \geq 2$. Then, for all $1 \leq i \leq s-1$, $p_{\underline{\overline{a_{i+1}}}}a_{i+1} \ll_f p_{\underline{\overline{a_i}}} \ll_f q_{\underline{\overline{a_1}}}$ and $p'_{\underline{\overline{a_{i+1}}}}a_{i+1} \ll_f p'_{\underline{\overline{a_i}}} \ll_f q'_{\underline{\overline{a_1}}}$, where for all $1 \leq i \leq s$, the word $p_{\underline{\overline{a_i}}}p'_{\underline{\overline{a_i}}}$ represents the induced subgraph $G[U(a_i)]$.

Lemma 1. *For $a, b \in V$, if a and b are adjacent in G, then a and b alternate in the word qq'.*

Proof. Without loss of generality, suppose $\overrightarrow{ba} \in T$. Let $(b =)a_s \to a_{s-1} \to \cdots \to a_0(= a)$ be the directed path from b to a in T'. Suppose the root r is the sink in T (if r is the source in T, the proof will be similar). As per Algorithm 1, a is visited first, then $a_1, a_2, \ldots, a_{s-1}$ and lastly b is visited. For $1 \leq i \leq s$, using induction on i, we show that $a_i a \ll q$ and $a_i a \ll q'$. Then, in particular for $i = s$, we have $ba \ll q$ and $ba \ll q'$ so that a and b alternate in qq'.

For $i = 1$, since a_1 is a child of a, as per step 6, we have $a_1 a \ll q$ and $a_1 a \ll q'$. Hence, a_1 and a alternate in the word qq'. For $i = t$ suppose $a_t a \ll q$ and $a_t a \ll q'$. For $i = t + 1$, since a_{t+1} is a child of a_t, a_t is replaced in both q and q' by a word containing $a_{t+1}a_t$ as a subword. Thus, we have $a_{t+1}a_t a \ll q$ and $a_{t+1}a_t a \ll q'$ so that $a_{t+1}a \ll q$ and $a_{t+1}a \ll q'$. Thus, $qq'|_{\{a,b\}} = baba$ so that a and b alternate in qq'. $\square$

Lemma 2. *For $a, b \in V$, if a and b are not adjacent in G, then they do not alternate in the word qq'.*

Proof. Suppose the root r is the sink in T (if r is the source in T, the proof will be similar). Since a and b are non-adjacent in T', and r is the sink in T', there exists a vertex c in G such that there are directed paths from a to c, and b to c in T'. Let $(a =)a_s \to a_{s-1} \to \cdots \to a_1 \to c$, and $(b =)b_{s'} \to b_{s'-1} \to \cdots \to b_1 \to c$ be the directed paths in T'. Note that a_1 and b_1 are children of c.

Without loss of generality, suppose $a_1 b_1 \ll q$. For $1 \leq i \leq s$, using induction on i, we show that $a_t b_1 \ll q$ and $b_1 a_t \ll q'$. For $i = 1$, since a_1, b_1 are children of c, by step 6, we have $a_1 b_1 \ll a_1 b_1 c \ll q$ and $b_1 a_1 \ll b_1 a_1 c \ll q'$. For $i = t$, suppose $a_t b_1 \ll q$ and $b_1 a_t \ll q'$. For $i = t + 1$, since a_{t+1} is a child of a_t, a_t is replaced in both q and q' by a word containing $a_{t+1}a_t$ as a subword. Thus, we have $a_{t+1}a_t b_1 \ll q$ and $b_1 a_{t+1}a_t \ll q'$. Hence, $a_{t+1}b_1 \ll q$ and $b_1 a_{t+1} \ll q'$ so that in particular for $i = s$ we have, $ab_1 \ll q$ and $b_1 a \ll q'$.

Now for $1 \leq i \leq s'$, using induction on i, we can observe that $ab_t \ll q$ and $b_t a \ll q'$ so that $ab \ll q$ and $ba \ll q'$. Hence, a and b do not alternate in qq'. $\square$

Thus, in view of lemmas 1 and 2, we have the following theorem establishing the correctness of Algorithm 1.

Theorem 1. *The word qq' represents the arborescence G.*

3.2 Minimum Length Word-Representant

Let $G = (V, E)$ be a treelike permutation graph (hence, a path of double-arborescences) and T be the treelike orientation of G. Suppose C is a longest directed path in the transitive reduction T' of T. Note that C can be found in quadratic time [3]. In this section, we construct a minimum length word-representant of G through Algorithm 2.

In Algorithm 2, we consider a smallest possible root-path in T' and the double-arborescence $G_{r_\tau} = G[\{r_\tau\} \cup U(r_\tau) \cup L(r_\tau)]$, where r_τ is the first amongst the roots that appear in C. Using the construction given in [2], we first construct a minimum length word-representant of G_{r_τ} and extend it to that of G iteratively. In each iteration, in a specific order of the roots, we include two permutations on the lower or on the upper set of each root that is already included in the word. Similar to the notation of permutations on $U(a)$ and $L(a)$, in Algorithm 2, we denote the permutations on $U^e(b)$ by $q_{\overline{b^e}}$ or $q'_{\overline{b^e}}$, and the permutations on $L^e(b)$ by $q_{\underline{b^e}}$ or $q'_{\underline{b^e}}$. A demonstration of Algorithm 2 is given in Table 1, in which we provide stepwise results by indicating the step number from Algorithm 2.

We establish the correctness of Algorithm 2 through a series of results, as per the following. In the first iteration of Algorithm 2, the availability of the factor $p_{\overline{r_{\tau-1}}}r_{\tau-1}$ (in line 11) or the factor $r_{\tau-1}p_{\underline{r_{\tau-1}}}$ (in line 18) of the word v_{r_τ} is ensured by the following remark.

Remark 10. If $\overrightarrow{r_\tau r_{\tau-1}} \in T'$, by Remark 3(i), we have $p_{\overline{r_{\tau-1}}}r_{\tau-1} \ll_f p_{\overline{r_\tau}} \ll_f v_{r_\tau}$. Also, in view of Remark 3(ii), if $\overrightarrow{r_{\tau-1}r_\tau} \in T'$, then $r_{\tau-1}p_{\underline{r_{\tau-1}}} \ll_f p_{\underline{r_\tau}} \ll_f v_{r_\tau}$.

Further, from the second iteration onwards, the presence of the factors $p_{\overline{b}}b$, $p_{\underline{b}}b$, or $bp_{\underline{b}}$ in the word v_a, if $b \neq r_{\tau+1}$, is ensured by the following propositions.

Proposition 3. *Suppose v_a is obtained at the end of the ith iteration and $b = head(Q)$ in the $(i+1)^{th}$ iteration such that $b \neq r_{\tau+1}$. If $\overrightarrow{ab} \in T'$, then $p_{\overline{b}}b \ll_f v_a$.*

Proof. Note that $b \in U(a)$ and $a \neq r_\tau$. We prove that $p_{\overline{b}}b \ll_f v_a$ in various cases as per the following. If $a \in V_{r_\tau} \setminus \{r_\tau\}$, we consider the following cases.

Case 1: $a \in U(r_\tau)$. Then, we have the portion of the root-path $r_\tau \to \cdots \to a' \to a \to b$ in T'. Hence, as per line 11 of Algorithm 2, in the ith iteration, the

---

Algorithm 2: Construction of minimum length word-representant of a treelike permutation graph.

Input: A treelike permutation graph $G = (V, E)$, the transitive reduction T' of the treelike orientation T of G, and a longest directed path C in T'.

Output: A minimum length word-representant of G.

1   Let $r_1 - r_2 - \cdots - r_m$ be the root-path in T' such that m is the smallest possible, and r_τ be the first amongst the roots that appear in C.

2   Let $G_{r_\tau} = (V_{r_\tau}, E_{r_\tau})$ be the double-arborescence $G[\{r_\tau\} \cup U(r_\tau) \cup L(r_\tau)]$.

3   Let $w_{r_\tau} = u v_{r_\tau} u'$ be the minimum length word-representant of G_{r_τ}, constructed using C and the restriction of T' to G_{r_τ}, where $v_{r_\tau} = p_{\overline{r_\tau}} r_\tau p_{r_\tau}$.

4   Set $a = r_\tau$ and consider a queue $Q = \langle r_{\tau-1}, r_{\tau-2}, \ldots, r_1, r_{\tau+1}, r_{\tau+2}, \ldots, r_m \rangle$.

5   **while** Q *is not empty* **do**

6   Let $b = \text{head}(Q)$. If $b = r_{\tau+1}$, then set $a = r_\tau$, $v_a = v_{r_1}$ and $V_a = V_{r_1}$.

7   **if** $\overrightarrow{ab} \in T'$ **then**

8   Let $L^e(b) = L(b) \setminus (L(a) \cup \{a\})$.

9   Using Algorithm 1, construct two permutations q and q' for the arborescence $G[\{b\} \cup L^e(b)]$ with T' restricted to the arborescence.

10   Let $q_{\underline{b}e}$ and $q'_{\underline{b}e}$ be the permutations obtained from q and q' by deleting b so that $q_{\underline{b}e} q'_{\underline{b}e}$ represents $G[L^e(b)]$.

11   Let v_b be the word obtained by replacing a factor $p_{\overline{b}} b$ in v_a with the word $q_{\underline{b}e} p_{\overline{b}} b q'_{\underline{b}e}$.

12   Set $w_b = u v_b u'$ and $V_b = V_a \cup L^e(b)$.

13   **end**

14   **else**

15   Let $U^e(b) = U(b) \setminus (U(a) \cup \{a\})$.

16   Using Algorithm 1, construct two permutations q and q' for the arborescence $G[\{b\} \cup U^e(b)]$ with T' restricted to the arborescence.

17   Let $q_{\overline{b}e}$ and $q'_{\overline{b}e}$ be the permutations obtained from q and q' by deleting b so that $q_{\overline{b}e} q'_{\overline{b}e}$ represents $G[U^e(b)]$.

18   Let v_b be the word obtained by replacing a factor $p_{\underline{b}} b$ or $b p_{\underline{b}}$ (whichever is available) in v_a with the word $q_{\overline{b}e} p_{\underline{b}} b q'_{\overline{b}e}$ or $q_{\overline{b}e} b p_{\underline{b}} q'_{\overline{b}e}$, respectively, as appropriate.

19   Set $w_b = u v_b u'$ and $V_b = V_a \cup U^e(b)$.

20   **end**

21   Remove $\text{head}(Q)$ from Q; and update a by b.

22 **end**

23 **return** w_{r_m}

---

word v_a is obtained from the word $v_{a'}$ (or from v_{r_1}, if $a = r_{\tau+1}$) by replacing the factor $p_{\overline{a}} a$ of $v_{a'}$ (or of v_{r_1}) with the word $q_{\underline{a}e} p_{\overline{a}} a q'_{\underline{a}e}$. Thus, $p_{\overline{a}} \ll_f v_a$. Further, in view of Remark 6(i), we have $p_{\overline{b}} b \ll_f p_{\overline{a}} \ll_f v_a$.

Case 2: $a \in L(r_\tau)$. Then, we have the portion of the root-path $b \leftarrow a \rightarrow a' \rightarrow \cdots \rightarrow r_\tau$ in T'. Hence, as per line 18 of Algorithm 2, in the ith iteration, the word v_a is obtained from the word $v_{a'}$ (or from v_{r_1}, if $a = r_{\tau+1}$) by replacing the factor $a p_{\underline{a}}$ or $p_{\underline{a}} a$ of $v_{a'}$ (or of v_{r_1}) with the word $q_{\overline{a}e} a p_{\underline{a}} q'_{\overline{a}e}$ or $q_{\overline{a}e} p_{\underline{a}} a q'_{\overline{a}e}$, respectively,

Table 1. Stepwise details of implementation of Algorithm 2 on Fig. 1(c)

Input: The transitive reduction T' depicted in Fig. 1(c) of a graph G and a longest directed path $C: 13 \to r_2 \to r_1 \to 1 \to 3 \to 4$ in T'.

1. $r_1 \leftarrow r_2 \leftarrow r_3$ is the root-path in T' and $r_\tau = r_2$.
2. $G_{r_2} = G[\{r_2\} \cup U(r_2) \cup L(r_2)]$.
3. $w_{r_2} = uv_{r_2}u'$, where $u = r_3$, $u' = (12)(11)25$ and
$v_{r_2} = p_{\overline{r_2}}r_2 p_{\underline{r_2}}$, where $p_{\overline{r_2}} = 45321r_1(11)(12)$ and $p_{\underline{r_2}} = r_3(13)$.
4. $a = r_2$ and $Q = \langle r_1, r_3 \rangle$.

First iteration	Second iteration
6. $b = r_1$	6. $b = r_3$, $a = r_2$, $v_a = v_{r_1}$
7. $\overrightarrow{r_2 r_1} \in T'$	14. $\overrightarrow{r_3 r_2} \in T'$
8. $L^e(r_1) = \{6, 7, 8, 9, 10\}$	15. $U^e(r_3) = \{14, 15, 16\}$
9. $q = 869(10)7r_1, q' = (10)9786r_1$	16. $q = (14)(16)(15)r_3, q' = (16)(15)(14)r_3$
10. $q_{r_1^e} = 869(10)7, q'_{r_1^e} = (10)9786$	17. $q_{r_3^e} = (14)(16)(15), q'_{r_3^e} = (16)(15)(14)$
11. $v_{r_1} = 869(10)745321r_1(10)9786$ $(11)(12)r_2r_3(13)$	18. $v_{r_3} = 869(10)745321r_1(10)9786(11)$ $(12)r_2(14)(16)(15)r_3(16)(15)(14)(13)$
12. $w_{r_1} = r_3 v_{r_1}(12)(11)25$	19. Output $w_{r_3} = r_3 v_{r_3}(12)(11)25$

as appropriate. Thus, in any case, $q_{\overline{a^e}} \ll_f v_a$. Further, since $\{b\} \cup U(b) \subseteq U^e(a)$, in view of Remark 8(iii), we have $p_{\overline{b}}b \ll_f q_{\overline{a^e}} \ll_f v_a$.

Suppose $a \notin V_{r_\tau}$. Then $a \notin \{r_{\tau-1}, r_{\tau+1}\}$; otherwise, a would belong to either $U(r_\tau)$ or $L(r_\tau)$. Also, note that $a \notin \{r_1, r_m\}$. Further, observe that $a \in L^e(r_s)$ or $a \in U^e(r_s)$, where $s \notin \{1, \tau, m\}$ and r_s appears before a in the queue Q. We consider the following cases.

Case 1: $a \in L^e(r_s)$. Then, we have the portion of the root-path $b \leftarrow a \to a' \to \cdots \to r_s$ in T'. Since r_s appears before a in Q, r_s is visited first, then a', a are visited in the algorithm. Hence, as per line 18 of Algorithm 2, in the ith iteration, the word v_a is obtained from the word $v_{a'}$ by replacing the factor $p_a a$ or ap_a of $v_{a'}$ with the word $q_{\overline{a^e}}p_a a q'_{\overline{a^e}}$ or $q_{\overline{a^e}}ap_a q'_{\overline{a^e}}$, respectively. Thus, in any case, $q_{\overline{a^e}} \ll_f v_a$. Since $\{b\} \cup U(b) \subseteq U^e(a)$, in view of Remark 8(iii), we have $p_{\overline{b}}b \ll_f q_{\overline{a^e}} \ll_f v_a$.

Case 2: $a \in U^e(r_s)$. Then, we have the portion of the root-path $r_s \to \cdots \to a' \to a \to b$ in T'. Thus, as per line 11 of Algorithm 2, in the ith iteration, the word v_a is obtained from the word $v_{a'}$ by replacing the factor $p_{\overline{a}}a$ of $v_{a'}$ with the word $q_{\underline{a^e}}p_{\overline{a}}a q'_{\underline{a^e}}$. In view of Remark 9(ii), we have $p_{\overline{b}}b \ll_f p_{\overline{a}} \ll_f v_a$. $\qquad\square$

Proposition 4. *Suppose v_a is obtained at the end of the ith iteration. If $(r_{\tau+1} \neq)b = head(Q)$ in the $(i+1)^{th}$ iteration and $\overrightarrow{ba} \in T'$, then $p_{\underline{b}}b \ll_f v_a$ or $bp_{\underline{b}} \ll_f v_a$.*

Proof. Note that $b \in L(a)$ and $a \neq r_\tau$. We prove that either $p_{\underline{b}}b$ or $bp_{\underline{b}}$ is a factor of v_a in various cases. If $a \in V_{r_\tau} \setminus \{r_\tau\}$, we consider the following cases.

Case 1: $a \in U(r_\tau)$. Then, we have the portion of the root-path $r_\tau \to \cdots \to a' \to a \leftarrow b$ in T'. Thus, as per line 11 of Algorithm 2, in the ith iteration, the word v_a is obtained from the word $v_{a'}$ (or from v_{r_1}, if $a = r_{\tau+1}$) by replacing the factor $p_{\bar{a}}a$ of $v_{a'}$ (or of v_{r_1}) with the word $q_{\underline{a}^e}p_{\bar{a}}aq'_{\underline{a}^e}$ so that $q_{\underline{a}^e} \ll_f v_a$. Since $\{b\} \cup L(b) \subseteq L^e(a)$, by Remark 8(ii), we have $p_{\underline{b}}b \ll_f q_{\underline{a}^e} \ll_f v_a$.

Case 2: $a \in L(r_\tau)$. Then, we have the portion of the root-path $b \to a \to a' \to \cdots \to r_\tau$ in T'. Hence, as per line 18 of Algorithm 2, in the ith iteration, the word v_a is obtained from the word $v_{a'}$ (or from v_{r_1}, if $a = r_{\tau+1}$) by replacing the factor $ap_{\underline{a}}$ or $p_{\underline{a}}a$ of $v_{a'}$ (or of v_{r_1}) with the word $q_{\overline{a}^e}ap_{\underline{a}}q'_{\overline{a}^e}$ or $q_{\overline{a}^e}p_{\underline{a}}aq'_{\overline{a}^e}$, respectively, as appropriate. Note that in any case, $p_{\underline{a}} \ll_f v_a$. Further, in view of Remark 6(ii), $bp_{\underline{b}} \ll_f p_{\underline{a}} \ll_f v_a$.

Suppose $a \notin V_{r_\tau}$. Then $a \notin \{r_{\tau-1}, r_{\tau+1}\}$; otherwise, a would belong to either $U(r_\tau)$ or $L(r_\tau)$. Also, note that $a \notin \{r_1, r_m\}$. Further, observe that $a \in L^e(r_s)$ or $a \in U^e(r_s)$, where $s \notin \{1, \tau, m\}$ and r_s appears before a in the queue Q. We now consider the following cases.

Case 1: $a \in L^e(r_s)$. Then, we have the portion of the root-path $b \to a \to a' \to \cdots \to r_s$ in T'. Since r_s appears before a in Q, r_s is visited first, then a', a are visited in the algorithm. Hence, as per line 18 of Algorithm 2, in the ith iteration, the word v_a is obtained from the word $v_{a'}$ by replacing the factor $p_{\underline{a}}a$ or $ap_{\underline{a}}$ of $v_{a'}$ with the word $q_{\overline{a}^e}p_{\underline{a}}aq'_{\overline{a}^e}$ or $q_{\overline{a}^e}ap_{\underline{a}}q'_{\overline{a}^e}$, respectively, as appropriate. Further, in view of Remark 9(i), we have $p_{\underline{b}}b \ll_f p_{\underline{a}} \ll_f v_a$.

Case 2: $a \in U^e(r_s)$. Then, we have the portion of the root-path $r_s \to \cdots \to a' \to a \leftarrow b$ in T'. Thus, as per line 11 of Algorithm 2, in the ith iteration, the word v_a is obtained from the word $v_{a'}$ by replacing the factor $p_{\bar{a}}a$ of $v_{a'}$ with the word $q_{\underline{a}^e}p_{\bar{a}}aq'_{\underline{a}^e}$. Since $\{b\} \cup L(b) \subseteq L^e(a)$, by Remark 8, $p_{\underline{b}}b \ll_f q_{\underline{a}^e} \ll_f v_a$.
$\square$

If $b = r_{\tau+1}$, we show that either $p_{\overline{r_{\tau+1}}}r_{\tau+1}$ or $r_{\tau+1}p_{\underline{r_{\tau+1}}}$ is a factor of v_{r_1}.

Proposition 5. *If $\overrightarrow{r_\tau r_{\tau+1}} \in T'$, then $p_{\overline{r_{\tau+1}}}r_{\tau+1}$ is a factor of v_{r_1}. Otherwise, $r_{\tau+1}p_{\underline{r_{\tau+1}}}$ is a factor of v_{r_1}.*

Proof. Suppose $\overrightarrow{r_\tau r_{\tau+1}} \in T'$. Recall that $v_{r_\tau} = p_{\overline{r_\tau}}r_\tau p_{\underline{r_\tau}}$. If $\overrightarrow{r_\tau r_{\tau-1}} \in T'$, then both $r_{\tau-1}, r_{\tau+1} \in U(r_\tau)$. Thus, in view of Remark 3(i), $p_{\overline{r_{\tau-1}}}r_{\tau-1}$ and $p_{\overline{r_{\tau+1}}}r_{\tau+1}$ are factors of $p_{\overline{r_\tau}}$. Since $U(r_{\tau-1}) \cap U(r_{\tau+1}) = \varnothing$, no element of $U(r_{\tau-1})$ appears in $p_{\overline{r_{\tau+1}}}$ and no element of $U(r_{\tau+1})$ appears in $p_{\overline{r_{\tau-1}}}$. Hence, if $r_{\tau-1}$ appears before $r_{\tau+1}$ in $p_{\overline{r_\tau}}$, then $v_{r_\tau} = u_1 p_{\overline{r_{\tau-1}}}r_{\tau-1}u_2 p_{\overline{r_{\tau+1}}}r_{\tau+1}u_3 r_\tau p_{\underline{r_\tau}}$; and if $r_{\tau+1}r_{\tau-1} \ll p_{\overline{r_\tau}}$, then $v_{r_\tau} = u_1 p_{\overline{r_{\tau+1}}}r_{\tau+1}u_2 p_{\overline{r_{\tau-1}}}r_{\tau-1}u_3 r_\tau p_{\underline{r_\tau}}$. In any case, $p_{\overline{r_{\tau+1}}}r_{\tau+1} \ll_f v_{r_\tau}$.

Recall that the algorithm obtains v_{r_1} by iteratively updating v_{r_i} to $v_{r_{i-1}}$ (for $\tau \geq i \geq 2$) through concatenation of two permutations on the either side of the factor $p_{\overline{r_{i-1}}}r_{i-1}$ or $r_{i-1}p_{\underline{r_{i-1}}}$ or $p_{\underline{r_{i-1}}}r_{i-1}$. Observe that the factor that is updated in any step is either $p_{\overline{r_{\tau-1}}}r_{\tau-1}$ or it appears in $p_{\overline{r_{\tau-1}}}$ or it appears in the permutations that are included in one of the previous iterations. Thus, $p_{\overline{r_{\tau+1}}}r_{\tau+1}$ remains a factor of v_{r_τ} in any iteration so that we have $p_{\overline{r_{\tau+1}}}r_{\tau+1} \ll_f v_{r_1}$.

If $\overrightarrow{r_{\tau-1}r_\tau} \in T'$, then $r_{\tau-1} \in L(r_\tau)$ and $r_{\tau+1} \in U(r_\tau)$. Thus, by Remark 3, $v_{r_\tau} = x_1 p_{\overline{r_{\tau+1}}} r_{\tau+1} x_2 r_\tau x_3 r_{\tau-1} p_{\overline{r_{\tau-1}}} x_4$, for some words x_i, so that $p_{\overline{r_{\tau+1}}} r_{\tau+1}$ is a factor of v_{r_τ}. Hence, as in the previous case, we have $p_{\overline{r_{\tau+1}}} r_{\tau+1} \ll_f v_{r_1}$.

Along the similar lines, one can prove that $r_{\tau+1} p_{\overline{r_{\tau+1}}}$ is a factor of v_{r_1} if $\overrightarrow{r_{\tau+1}r_\tau} \in T'$. $\qquad\square$

We now prove that w_{r_m} represents the graph G using the following lemmas.

Lemma 3. *For $i \geq 0$, if $\overrightarrow{ab} \in T'$, let V_a and $V_b = V_a \cup L^e(b)$ be the subsets of V obtained at the end of ith and $(i+1)^{th}$ iterations of Algorithm 2, respectively. For $a_1 \in L^e(b)$ and $b_1 \in V_a$, a_1 and b_1 are adjacent in $G[V_b]$ if and only if either $b_1 = b$ or $b_1 \in U(b)$.*

Proof. Note that for $i = 0$, $a = r_\tau$ and $V_a = V_{r_\tau}$. Further, note that $\{b\} \cup U(b) \subseteq V_a$. If $b_1 = b$ or $b_1 \in U(b)$, then clearly b_1 is adjacent to a_1. For the converse, suppose a_1 and b_1 are adjacent in $G[V_b]$. Since $a_1 \in L^e(b)$, we have $\overrightarrow{a_1b} \in T$. Thus, if $b_1 = b$, we are done. Suppose $b_1 \neq b$. Note that $\overrightarrow{a_1b_1} \in T$; otherwise, $b_1 \in L(a_1) \subseteq L^e(b)$, a contradiction that $b_1 \in V_a$. Thus, there is a directed path from a_1 to b_1 in T'. Also, since $\overrightarrow{a_1b} \in T$, there is a directed path from a_1 to b in T'. Moreover, observe that the directed path from a_1 to b_1 is via b in T'. Hence, $\overrightarrow{bb_1} \in T$ so that $b_1 \in U(b)$. Note that the above argument also works if $b = r_{\tau+1}$, $a = r_\tau$, $v_a = v_{r_1}$ and $V_a = V_{r_1}$. $\qquad\square$

Along the lines of Lemma 3, the following lemma can be obtained.

Lemma 4. *For $i \geq 0$, if $\overrightarrow{ba} \in T'$, suppose V_a and $V_b = V_a \cup U^e(b)$ are the subsets of V obtained at the end of ith and $(i+1)^{th}$ iterations of Algorithm 2, respectively. For $a_1 \in U^e(b)$ and $b_1 \in V_a$, a_1 and b_1 are adjacent in $G[V_b]$ if and only if either $b_1 = b$ or $b_1 \in L(b)$.*

Lemma 5. *Let A and B be two disjoint subsets of a set X, p_A be a permutation on A and q_B, q'_B be permutations on B. For some words x_1 and x_2, let $w = x_1 q_B p_A q'_B x_2$ be a word over X such that each letter of X appears at most twice in w. Then, $b_1 \in B$ alternates with $b_2 \in X \setminus B$ in w if and only if $b_2 \in A$.*

Proof. Suppose an element $b_1 \in B$ alternates with an element $b_2 \in X \setminus B$ in w. Note that if $b_2 \notin A$, b_2 cannot appear in p_A. Further, since $b_2 \notin B$ and b_1 appears in q_B as well as in q'_B, $w|_{\{b_1,b_2\}}$ has a factor b_1b_1, a contradiction.

Conversely, let $b_2 \in A$. Then, by the construction of w if b_2 occurs exactly once in w, we have $w|_{\{b_1,b_2\}} = b_1b_2b_1$, and if b_2 occurs exactly twice in w, we have $w|_{\{b_1,b_2\}} = b_1b_2b_1b_2$ or $b_2b_1b_2b_1$. Thus, b_1 and b_2 alternate in w. $\qquad\square$

Theorem 2. *For $i \geq 0$, let w_a and w_b be the words obtained at the end of ith and $(i+1)^{th}$ iterations of Algorithm 2, respectively. If w_a represents $G[V_a]$, then w_b represents $G[V_b]$. Hence, the output w_{r_m} represents G.*

Proof. Note that, for $i = 0$, $w_a = w_{r_\tau}$ represents G_{r_τ} by [2, Theorem 4.6]. Further, observe that $w_a = uv_a u'$. As per line 6 of Algorithm 2, if $b = r_{\tau+1}$, we consider $a = r_\tau$, $w_a = w_{r_1}$, $v_a = v_{r_1}$ and $V_a = V_{r_1}$ in the following argument. We prove the result in the following cases.

Case 1: $\overrightarrow{ab} \in T'$. Then, as per line 12 of Algorithm 2, we have $V_b = V_a \cup L^e(b)$ and $w_b = uv_b u'$, where the word v_b is obtained from the word v_a by replacing the factor $p_{\overline{b}}b$ of v_a with the word $q_{\underline{be}}p_{\overline{b}}bq'_{\underline{be}}$. Observe that $w_b = xq_{\underline{be}}p_{\overline{b}}bq'_{\underline{be}}x'$, for some words x, x' over V_b, in which $w_a = xp_{\overline{b}}bx'$. Thus, $w_b|_{V_a} = w_a$ and $w_b|_{L^e(b)} = q_{\underline{be}}q'_{\underline{be}}$. Since the subwords w_a and $q_{\underline{be}}q'_{\underline{be}}$ of w_b represent $G[V_a]$ and $G[L^e(b)]$, respectively, any two vertices of V_a (or any two vertices of $L^e(b)$) are adjacent in $G[V_b]$ if and only if they alternate in the word w_b. Suppose $a_1 \in L^e(b)$ and $b_1 \in V_a$. Then, by Lemma 3, a_1 and b_1 are adjacent in $G[V_b]$ if and only if either $b_1 = b$ or $b_1 \in U(b)$. Further, in view of Lemma 5, a_1 alternates with b_1 in w_b if and only if either $b_1 = b$ or $b_1 \in U(b)$. Thus, w_b represents $G[V_b]$.

Case 2: $\overrightarrow{ba} \in T'$. Then, as per line 19 of Algorithm 2, we have $V_b = V_a \cup U^e(b)$ and $w_b = uv_b u'$, where v_b is obtained from v_a by replacing the factor $p_{\underline{b}}b$ or $bp_{\underline{b}}$ of v_a with the word $q_{\overline{be}}p_{\underline{b}}bq'_{\overline{be}}$ or $q_{\overline{be}}bp_{\underline{b}}q'_{\overline{be}}$, respectively. Observe that $w_b = yq_{\overline{be}}p_{\underline{b}}bq'_{\overline{be}}y'$ or $yq_{\overline{be}}bp_{\underline{b}}q'_{\overline{be}}y'$, for some words y, y' over V_b. Also, $w_b|_{V_a} = w_a$ and $w_b|_{U^e(b)} = q_{\overline{be}}q'_{\overline{be}}$. Since the subwords w_a and $q_{\overline{be}}q'_{\overline{be}}$ of w_b represent $G[V_a]$ and $G[U^e(b)]$, respectively, any two vertices of V_a (or any two vertices of $U^e(b)$) are adjacent in $G[V_b]$ if and only if they alternate in w_b. Suppose $a_1 \in U^e(b)$ and $b_1 \in V_a$. Then, by Lemma 4, a_1 and b_1 are adjacent in $G[V_b]$ if and only if either $b_1 = b$ or $b_1 \in L(b)$. Further, by Lemma 5, a_1 alternates with b_1 in w_b if and only if either $b_1 = b$ or $b_1 \in L(b)$. Thus, w_b represents $G[V_b]$. $\square$

We now conclude that the output w_{r_m} of Algorithm 2 is a minimum length word-representant of G.

Theorem 3. *If G is a treelike permutation graph on n vertices with clique number κ, then $\ell(G) = 2n - \kappa$.*

Proof. Note that the number of vertices in the longest directed path C is κ, as κ is the clique number of G. Since w_{r_τ} is a minimum length word-representant of the double-arborescence G_{r_τ}, by [2, Theorem 4.6], the length of w_{r_τ} is $2|V_{r_\tau}| - \kappa$. In each iteration of Algorithm 2, exactly two vertices of $L^e(b)$ or of $U^e(b)$ are inserted in the word w_a. Thus, the length of the output w_{r_m} (a word-representant of G, by Theorem 2) is $2|V| - \kappa$ so that $\ell(G) \leq 2n - \kappa$. Further, in view of [11, Theorem 2.9], we have $2n - \kappa \leq \ell(G)$. Hence, w_{r_m} is a minimum length word-representant of G and $\ell(G) = 2n - \kappa$. $\square$

Remark 11. The time complexity of Algorithm 2 is a polynomial in the size of the graph. For instance, note that the root-path in step 1 can be found in linear time (by [1, Theorem 7]). Further, step 3 can be done in $O(n^2)$ time, where n is the number of vertices of G. Also, construction of permutations in steps 9 and 16 representing an arborescence can be performed in $O(n^2)$ time (by Remark 7). It can be observed that each step inside the while loop can be executed in $O(n^2)$ time. Hence, Algorithm 2 runs in $O(n^3)$ time, as the number of iterations is at most n.

In this work, we proved that a minimum length word-representant of a treelike permutation graph on n vertices with clique number κ is of $2n - \kappa$ length. This contributes a class to the open problem posed in [11] and also extends the corresponding result of double-arborescences from [2]. In this context, it is natural to ask the following question. Can this work be extended to a more general class of treelike comparability graphs?

References

1. Cornelsen, S., Di Stefano, G.: Treelike comparability graphs. Disc. Appl. Math. **157**(8), 1711–1722 (2009)
2. Dwary, T., Krishna, K.V.: Characterization of double-arborescences and their minimum length word-representants. Disc. Appl. Math. **380**, 599–611 (2026)
3. Even, S., Pnueli, A., Lempel, A.: Permutation graphs and transitive graphs. J. Assoc. Comput. Mach. **19**, 400–410 (1972)
4. Gaetz, M., Ji, C.: Enumeration and extensions of word-representants. Disc. Appl. Math. **284**, 423–433 (2020)
5. Golumbic, M.C.: Algorithmic graph theory and perfect graphs. Ann. Disc. Math. 57, 2nd edn (2004)
6. Golumbic, M.C.: Why are they called trivially perfect graphs? Cadernos do IME-Série Informática **47**, 40–45 (2022)
7. Halldórsson, M.M., Kitaev, S., Pyatkin, A.: Alternation graphs. In: Kolman, P., Kratochvíl, J. (eds.) WG 2011. LNCS, vol. 6986, pp. 191–202. Springer, Heidelberg (2011). https://doi.org/10.1007/978-3-642-25870-1_18
8. Kitaev, S., Lozin, V.: Words and graphs. In: Monographs in Theoretical Computer Science. An EATCS Series. Springer, Heidelberg (2015)
9. Kitaev, S., Pyatkin, A.: On representable graphs. J. Autom. Lang. Comb. **13**(1), 45–54 (2008)
10. McConnell, R.M., Spinrad, J.P.: Modular decomposition and transitive orientation. Disc. Math. **201**(1–3), 189–241 (1999)
11. Srinivasan, E., Hariharasubramanian, R.: Minimum length word-representants of word-representable graphs. Disc. Appl. Math. **343**, 149–158 (2024)
12. Wolk, E.S.: The comparability graph of a tree. Proc. Amer. Math. Soc. **13**, 789–795 (1962)

A Quadratic Lower Bound for 2DFAS Against One-Way Liveness

Kehinde Adeogun and Christos Kapoutsis[(✉)]

Carnegie Mellon University in Qatar, Education City, Doha, Qatar
kadeogun@andrew.cmu.edu, cak@cmu.edu

Abstract. We show that every *two-way deterministic finite automaton* (2DFA) that solves *one-way liveness* on height h has $\Omega(h^2)$ states. This implies a quadratic lower bound for converting *one-way nondeterministic finite automata* to 2DFAS, which asymptotically matches Chrobak's well-known lower bound for this conversion on unary languages. In contrast to Chrobak's simple proof, which relies on a 2DFA's inability to differentiate between any two sufficiently distant locations in a unary input, our argument works on alphabets of arbitrary size and is structured around a main lemma that is general enough to potentially be reused elsewhere.

Keywords: two-way deterministic finite automata · Sakoda-Sipser conjecture · state complexity · minicomplexity · descriptional complexity

1 Introduction

Back in the 1970s, Sakoda and Sipser [1] (but see also [2]) asked a fundamental question, which to this day remains an important open problem in automata and complexity theory: Can every *two-way nondeterministic finite automaton* (2NFA) with some number of states s be simulated by a *two-way deterministic finite automaton* (2DFA) with a number of states $f(s)$ which is polynomial in s?

A standard way to approach this question is to study variant cases: cases where either the 2DFA is enhanced or the 2NFA is restricted, if we want to prove an upper bound for $f(s)$; or cases where either the 2DFA is restricted or the 2NFA is enhanced, if we want to prove a lower bound.

Indeed, a *polynomial* upper bound for $f(s)$ is known when the simulating 2DFA is enhanced into some restricted single-tape Turing machine: a *weight-reducing* one (i.e., one that overwrites symbols only by "lighter" ones) or a *bounded & linear-time* one (i.e., a linear-time one that writes only within the

---

K. Adeogun—Supported by Qatar Foundation, via CMUQ's Seed Research program and project *Minicomplexity*. All statements made herein are the responsibility of the authors.

© The Author(s), under exclusive license to Springer Nature Switzerland AG 2026
J. Kozik and A. Wolff (Eds.): SOFSEM 2026, LNCS 16448, pp. 104–116, 2026.
https://doi.org/10.1007/978-3-032-17801-5_8

input) [3]. Likewise, a *quasi-polynomial* upper bound is known when the simulated 2NFA is restricted to *unary* (i.e., accepting only unary inputs) [4]; or *letter-bounded* (i.e., accepting only inputs of the form $a_1^* a_2^* \cdots a_k^*$ for fixed k) [5]; or *outer-nondeterministic* (i.e., using nondeterminism only on the end-markers) [6].

On the flip side, an *exponential* lower bound for $f(s)$ is known when the simulating 2DFA is restricted to *single-pass* (i.e., halting upon reaching an end-marker) [2]; or *sweeping* (i.e., reversing only on the endmarkers) [7]; or *(almost) oblivious* (i.e., following sub-linearly many trajectories per input length) [8]; or *few-reversal* (i.e., reversing sub-linearly often) [9]; or *logical* (i.e., satisfying some formula on "reachability variables" in each state) [10–12]; and an *infinite* lower bound is known when the 2DFA is a *mole* (i.e., it only explores the configuration graph of the 2NFA) [13]. Interestingly enough, we know of no lower bound attempt for cases where the simulated 2NFA is somehow enhanced.

In this landscape, a natural question is: *What about lower bounds for unrestricted* 2DFAS? Unfortunately, there are only two results of this kind and, quite counterintuitively, both are obtained in scenaria where the simulated 2NFA is (not in any way enhanced, but instead) severely restricted!

The first one is Chrobak's well-known bound $f(s) = \Omega(s^2)$ for when the simulated 2NFA is both *unary* (as above) and *one-way* (i.e., moving its input head only forward; namely, a standard *one-way nondeterministic finite automaton* or 1NFA) [14, Thm. 6.3] (cf. also [15]). Expectedly, the simple proof of this result relies heavily on the inability of the simulating 2DFA to differentiate between distant locations on the unary input. Moreover, a matching upper bound [14, Thm. 6.2] shows that no better lower bound is possible in this special case.

The second result is the bound $f(s) = \Omega(s^2 / \log s)$ of [16, Lemma 7] for the case where the simulated 2NFA is both *one-way* (again) and *accepting only inputs of length* 3.[1] The (counting) argument for this second result is also quite simple. Moreover, here too, the possibility of a better lower bound is removed by a (quite non-trivial) matching upper bound [16, Lemma 8].

Our present study offers a third lower bound for unrestricted 2DFAS. As in the two existing results, we focus on 2NFAs that are *one-way*; and, like Chrobak, we prove $f(s) = \Omega(s^2)$. Unlike the two existing results, however, our 2NFAs work on alphabets of arbitrary size and inputs of arbitrary length.

We express our argument in terms of *one-way liveness* (OWL), the complete language (family) for the conversion of 1NFAS to 2DFAS [1, Thm. 2.3]. Specifically, we prove that: *every* 2DFA *that solves* OWL *on height h has* $\Omega(h^2)$ *states*, which clearly implies that $f(s) = \Omega(s^2)$, since OWL on height h is solved by a 1NFA with h states [1, Thm. 2.3i].

After some preliminaries (Sect. 2), we recall (from [13]) the basic definitions and facts about *generic strings* (Sect. 3.1); and show how these force a 2DFA to use its states on their outer boundaries (Sect. 3.2). We then prove our main lemma (Lemma 10), that a suitably selected pair of properties can force a 2DFA to spend ≥ 1 state just to distinguish between them. Finally, we identify a

---

[1] This is (easily) an equivalent restatement of [16]'s lower bound for 3OWL.

quadratically-long sequence of properties where, every two successive ones are as described in the lemma (Sect. 4). This then implies the promised lower bound.

2   Preparation

The set Σ^* consists of all finite strings over Σ. If $z \in \Sigma^*$, we let $|z|$, z_i, z^i be its length, its ith symbol ($1 \leq i \leq |z|$), and its concatenation with itself $i \geq 0$ times. If $f : S \to S$, then f^i is the composition of $i \geq 1$ copies of f.

2.1   Two-Way Finite Automata

A *two-way deterministic finite automaton* (2DFA) is any tuple of the form $M = (Q, \Sigma, \delta, q_\mathrm{s}, q_\mathrm{a}, q_\mathrm{r})$, where Q is a set of *states*, Σ is an *alphabet*, $q_\mathrm{s}, q_\mathrm{a}, q_\mathrm{r} \in Q$ are the *start, accept,* and *reject* states, and $\delta : Q \times (\Sigma \cup \{\vdash, \dashv\}) \to Q \times \{\mathrm{L,R}\}$ is a total *transition function*, for $\vdash, \dashv \notin \Sigma$ the left and right endmarkers and L,R the two directions. An input $z \in \Sigma^*$ is presented to M between the endmarkers, as $\vdash z \dashv$. The computation starts at q_s on the left endmarker. In each step, the next state and head move are derived from δ and the current state and symbol in the standard way. Endmarkers are never violated, except for $\dashv$ when the next state is q_a or q_r. Hence, the computation can either loop; or fall off $\dashv$ into one of $q_\mathrm{a}, q_\mathrm{r}$.

Formally, the *computation* of M from state p on the jth symbol of string z, denoted by $\mathrm{COMP}_{M,p,j}(z)$, is the unique sequence $c = (q_t, j_t)_{0 \leq t \leq m}$ of pairs, called *configurations*, where $0 \leq m \leq \infty$; $q_0 := p$ and $j_0 := j$; and every next state q_{t+1} and position j_{t+1} are derived from δ and the current state q_t and symbol z_{j_t}. If $m = \infty$, we say the computation *loops*. Otherwise, m is finite and j_m is either $|z|+1$ or 0, in which cases we say the computation *hits right* or *hits left* into q_m, respectively. If $j = 1$ or $j = |z|$, we get the *left computation* and *right computation* of M from p on z, respectively, namely $\mathrm{LCOMP}_{M,p}(z) := \mathrm{COMP}_{M,p,1}(z)$ and $\mathrm{RCOMP}_{M,p}(z) := \mathrm{COMP}_{M,p,|z|}(z)$. Then the *computation of M on z* is its left computation from q_s on $\vdash z \dashv$, in symbols $\mathrm{COMP}_M(z) := \mathrm{LCOMP}_{M,q_\mathrm{s}}(\vdash z \dashv)$; and M *accepts* z if this computation hits right into q_a. As usual, M *solves* a language $L \subseteq \Sigma^*$ if it accepts exactly the strings in L.

2.2   One-Way Liveness

Let $h \geq 1$. The alphabet $\Sigma_h := \mathcal{P}([h] \times [h])$ consists of every two-column directed graph with h nodes per column and only rightward arrows (Fig. 1). An n-long string $z \in \Sigma_h^*$ is naturally viewed as a graph of $n+1$ columns, indexed from 0 to n, where edges connect successive columns and, for simplicity, are undirected. If there exists an n-long path from some node of column 0 to some node of column n, then z is *live*; otherwise, it is *dead*. Checking that a given $z \in \Sigma_h^*$ is live is the *one-way liveness* problem for height h, denoted by OWL_h.

Given any $z \in \Sigma_h^*$, its *connectivity* $C(z)$ is the $h \times h$ Boolean matrix where cell (i, j) is 1 iff z contains an n-long path from node i of column 0 to node j of column n (Fig. 1(R)). Easily, concatenation of strings corresponds to *Boolean*

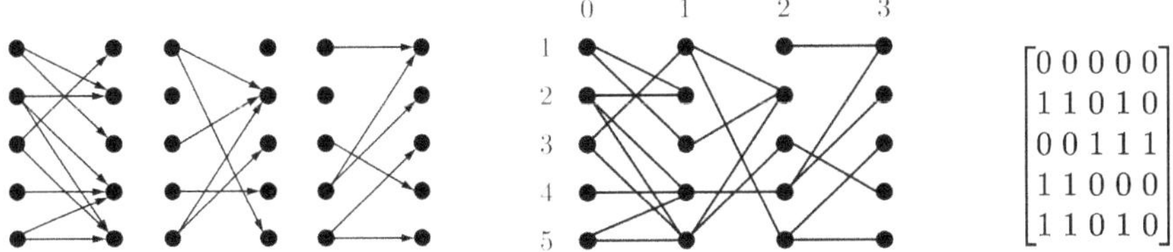

Fig. 1. Three symbols in Σ_5 (L); their string (M); and its connectivity (R).

multiplication of connectivities, namely $C(xy) = C(x)C(y)$, where every $+$ and $\cdot$ on Boolean values is $\vee$ and $\wedge$, respectively. So, equivalently, OWL$_h$ is defined over the alphabet of all $h \times h$ Boolean matrices and consists in checking that a given string of them has non-zero Boolean product.

The problem is important in studying how easily 2DFAS can simulate *one-way nondeterministic finite automata* (1NFAS, as typically defined [17,18]). Formally, the family OWL $= ($OWL$_h)_{h \geq 1}$ is complete, under homomorphic reductions, for the class 1N of all language families $L = (L_h)_{h \geq 1}$ that are recognized by 2DFA families $M = (M_h)_{h \geq 1}$ where the number of states grows at most polynomially in h [1]. Intuitively, OWL$_h$ captures in its instances all computations of all h-state 1NFAS. So, if a 2DFA can solve OWL$_h$ with $\leq s(h)$ states, then every h-state 1NFA can be simulated by a 2DFA with $\leq s(h)$ states; and, conversely, if every 2DFA for OWL$_h$ needs $\geq s(h)$ states, then some h-state 1NFA requires $\geq s(h)$ states in every 2DFA that simulates it.

2.3   Properties

A *property* (of the strings) over an alphabet Σ is any $P \subseteq \Sigma^*$.

We call P *smooth*, if it is "closed under concatenation with infix", meaning that every two strings in it have "an infix that keeps their concatenation in the property": $(\forall x,z \in P)(\exists y)(xyz \in P)$. We say that properties P, P' are *separated* by language $L \subseteq \Sigma^*$ if there is always a context where the choice between P and P' affects membership in L: $(\forall x \in P)(\forall z \in P')(\exists u, v)(uxv \in L \iff uzv \notin L)$.

Properties can also relate via extensions. We say that P LR-*extends* to P' if every string in it can be LR-extended into P': $(\forall x \in P)(\exists y)(xy \in P')$. Sometimes, properties depend so much on how the strings end that a single suffix y can push into P' every x from P or P': $(\exists y)(\forall x \in P \cup P')(xy \in P')$. We then say that P *has a suffix into* P'. In even more extreme cases, that suffix y can be chosen to end in any way $v \in P$ that we want: $(\forall v \in P)(\exists u)(\forall x \in P \cup P')(xuv \in P')$. We then say that P *has suffix of choice into* P'.

Finally, in the symmetric variants of the above relations P RL-*extends to* P': $(\forall x \in P)(\exists y)(yx \in P')$; or *has a prefix into* P': $(\exists y)(\forall x \in P \cup P')(yx \in P')$; which could be *of choice*: $(\forall v \in P)(\exists u)(\forall x \in P \cup P')(vux \in P')$.

3   A Lower-Bound Lemma

Let L be a language over some alphabet Σ. This section proves a general lemma for lower-bounding the number of states in a 2DFA solving L. Section 4 uses this lemma for $\Sigma = \Sigma_h$ and $L = \text{OWL}_h$ to show a lower bound of $\Omega(h^2)$.

So, fix a 2DFA M that solves L with states Q. (Since it is fixed, we drop unnecessary subscripts: e.g., for $\text{LCOMP}_{M,p}(z)$, we just write $\text{LCOMP}_p(z)$.)

3.1   Generic Strings

Introduced in [19], generic strings continue to be the main building block that we have for constructing hard inputs for 2DFAS of various kinds [9,13,20–22]. Here, we are using the flavor of [13, Sect. 3.2]. This subsection briefly recalls concepts and facts from that work that we need in our arguments.

For any $y \in \Sigma^*$, consider the set of *"exit states"* that M hits right into after a left computation on y:

$$Q_{\text{LR}}(y) := \{q \in Q \mid (\exists p \in Q)(\text{LCOMP}_p(y) \text{ hits right into } q)\}.$$

When we LR-extend y to some yz, the size of this set cannot possibly increase, because the function $\alpha_{y,z}$ that maps every $q \in Q_{\text{LR}}(y)$ to the state (if any) that is hit right by the computation $\text{COMP}_{q,|y|+1}(yz)$ is easily seen to be a partial surjection from $Q_{\text{LR}}(y)$ onto $Q_{\text{LR}}(yz)$.

Lemma 1 ([13, Fact 3]). *For all y,z, the function $\alpha_{y,z}$ partially surjects $Q_{\text{LR}}(y)$ onto $Q_{\text{LR}}(yz)$; hence $|Q_{\text{LR}}(y)| \geq |Q_{\text{LR}}(yz)|$.*

Consequently, if we start with a string y in a property P and keep LR-extending it inside P, we will at some point have minimized the size of the set of exit states. When that happens, the string has become "LR-generic".

Definition 1 ([13, Def. 3]). *A string y is LR-generic for property P if $y \in P$ and, for all $yz \in P$, we have $|Q_{\text{LR}}(y)| = |Q_{\text{LR}}(yz)|$.*

Lemma 2 ([13, Lemma 3]). *Every property $P \neq \emptyset$ has LR-generic strings.*

Proof. Pick $y \in P$. If y is LR-generic, we are done. Otherwise, there is $yz_1 \in P$ such that $|Q_{\text{LR}}(y)| > |Q_{\text{LR}}(yz_1)|$. So, we replace y with yz_1 and repeat: either yz_1 is LR-generic and we are done; or there is $yz_1z_2 \in P$ such that $|Q_{\text{LR}}(yz_1z_2)|$ is even smaller, and we replace yz_1 with yz_1z_2 and repeat. Clearly, this eventually terminates, as M has finitely many states and every step decreases $|Q_{\text{LR}}(\cdot)|$. $\square$

In addition, every further forward extension of a LR-generic string inside the property continues to be LR-generic for the property:

Lemma 3. *If y is LR-generic for P, then so is every $yz \in P$.*

Proof. Towards a contradiction, say y is LR-generic for P but some $yz \in P$ is not. Then there is $yzz' \in P$ with $|Q_{\text{LR}}(yz)| > |Q_{\text{LR}}(yzz')|$. Then $\tilde{z} := zz'$ is such that $y\tilde{z} \in P$ but $|Q_{\text{LR}}(y)| > |Q_{\text{LR}}(y\tilde{z})|$, contrary to the LR-genericity of y. $\square$

Finally, the size $|Q_{\mathrm{LR}}(yz)|$ is upper bounded, not only by the respective size $|Q_{\mathrm{LR}}(y)|$ for the prefix y (the *"prefix-monotonicity"* given by Lemma 1), but also by the respective size $|Q_{\mathrm{LR}}(z)|$ for the suffix z (a *"suffix-monotonicity"*).

Lemma 4 ([13, Fact 4]). *For all y,z: $Q_{\mathrm{LR}}(yz) \subseteq Q_{\mathrm{LR}}(z)$; so $|Q_{\mathrm{LR}}(yz)| \leq |Q_{\mathrm{LR}}(z)|$.*

Symmetrically to the above, we can also define the set $Q_{\mathrm{RL}}(y)$ of *exit states* that M hits left into after right computations on y; note the partial surjection $\beta_{z,y} : Q_{\mathrm{RL}}(y) \rightarrow Q_{\mathrm{RL}}(zy)$, mapping every p to the state (if any) hit left by $\mathrm{COMP}_{p,|z|}(zy)$; conclude that $|Q_{\mathrm{RL}}(y)|$ cannot increase when we RL-extend y (*suffix-monotonicity*); define RL-*generic strings* and show that they exist for every $P \neq \emptyset$; show that RL-extensions preserve RL-genericity; and argue that $Q_{\mathrm{RL}}(z) \supseteq Q_{\mathrm{RL}}(zy)$ and thus *prefix-monotonicity* also holds.

In the end, we are interested in strings that are generic in both directions, and we are guaranteed to have them for non-empty properties that are also smooth.

Definition 2 ([13, Def. 3]). *A string is generic if it is both LR- and RL-generic.*

Lemma 5 ([13, Lemma 3]). *Every smooth property $P \neq \emptyset$ has generic strings.*

Proof. Since $P \neq \emptyset$, there exist $x, z \in P$ that are respectively LR- and RL-generic for P (by Lemma 2 and its symmetric variant). Since P is smooth, there exists y such that $w := xyz \in P$. Then w is both LR- and RL-generic for P, as a LR-extension of x and a RL-extension of z (Lemma 3 and its symmetric variant). Therefore, w is generic for P. $\qquad\square$

3.2   Exit Sizes

The previous section studied how the size of the set $Q_{\mathrm{LR}}(y)$ changes as $y \in P$ is LR-extended within P. Call this number $a(y) := |Q_{\mathrm{LR}}(y)|$ the "LR-*size*" of y. We now study how it changes when y is replaced by other strings.

We first note that switching between LR-generic strings for P preserves the LR-size, provided that P is smooth. Hence, this number is a feature of P

Lemma 6. *Let x,z be LR-generic strings for a smooth P. Then $a(x) = a(z)$.*

Proof. Suppose P is smooth and strings x, z are LR-generic for it. Since P is smooth, there exists y such that $xyz \in P$. Then $a(x) = |Q_{\mathrm{LR}}(x)| = |Q_{\mathrm{LR}}(xyz)|$, by the definition of $a(x)$ and since $xyz \in P$ LR-extends the LR-generic x. But also $|Q_{\mathrm{LR}}(xyz)| \leq |Q_{\mathrm{LR}}(z)| = a(z)$, by Lemma 4 and the definition of $a(z)$. Overall, $a(x) \leq a(z)$. Finally, $a(z) \leq a(x)$ is also true, by a symmetric argument. $\qquad\square$

Definition 3. *The LR-size $a(P)$ of a smooth $P \neq \emptyset$ is the common LR-size of all LR-generic strings of P.*

Moreover, $a(P)$ is a lower bound for the LR-size of any string in P, and is reached by exactly the LR-generic strings of P.

Lemma 7. *Let $P \neq \emptyset$ be smooth. Every $x \in P$ satisfies $a(x) \geq a(P)$; and the equality holds iff x is LR-generic for P.*

Proof. Let $x \in P$. As in the proof of Lemma 2, x can be LR-extended to some LR-generic string xz for P. Then $a(x) = |Q_{\mathrm{LR}}(x)| \geq |Q_{\mathrm{LR}}(xz)| = a(xz) = a(P)$ (by Lemma 1 and the definitions of $a(\cdot), a(P)$), hence $a(x) \geq a(P)$.

Now let x be LR-generic for P. Then $a(x) = a(P)$, by Def. 3. Conversely, assume $a(x) = a(P)$. Pick any forward extension $xz \in P$. We know $|Q_{\mathrm{LR}}(x)| \geq |Q_{\mathrm{LR}}(xz)|$, by Lemma 1. We also know $|Q_{\mathrm{LR}}(x)| \leq |Q_{\mathrm{LR}}(xz)|$, as $|Q_{\mathrm{LR}}(x)| = a(x) = a(P)$ (by definition of $a(x)$ and our assumption) and $|Q_{\mathrm{LR}}(xz)| \geq a(P)$ (by the first part of our proof, and since $xz \in P$). Overall, $|Q_{\mathrm{LR}}(x)| = |Q_{\mathrm{LR}}(xz)|$ for the arbitrary LR-extension of x in P. So, x is LR-generic for P. □

Finally, we note that the LR-sizes of distinct smooth properties can be compared, if one extends to the other.

Lemma 8. *If $P, P' \neq \emptyset$ are smooth and P LR-extends to P', then $a(P) \geq a(P')$.*

Proof. Pick any LR-generic string x for P. Since P LR-extends to P', there exists y such that $xy \in P'$. Then $a(P) = a(x) = |Q_{\mathrm{LR}}(x)| \geq |Q_{\mathrm{LR}}(xy)| \geq a(P')$, where the equalities are by the definitions of $a(P), a(x)$ (and since x is LR-generic) and the two inequalities hold by Lemmas 1 and 7, respectively. □

Symmetrically to the above, we can also define the RL-*size* $b(y) := |Q_{\mathrm{RL}}(y)|$ of a string y; prove that it is common for all RL-generic strings of a smooth P; define it as the RL-*size* $b(P)$ of P; show that it lower bounds the RL-size of all $x \in P$ and is met iff x is RL-generic; and prove that $b(P') \leq b(P)$ whenever a smooth P RL-extends a smooth P'. We can then also prove the following.

Lemma 9. *If $P, P' \neq \emptyset$ are smooth and P has suffix of choice into P', then both $a(P) \geq a(P')$ and $b(P) \geq b(P')$.*

Proof. Since P has suffix of choice into P', it clearly LR-extends to P', therefore $a(P) \geq a(P')$ (by Lemma 8), and we just need to prove that $b(P) \geq b(P')$.

Let v be any generic string for P (Lemma 5). Then $v \in P$ and, since P has suffix of choice into P', some u is such that the suffix uv forces every string xuv into P', no matter which $x \in P \cup P'$ we pick. Let us pick $x = v$. Then $vuv \in P'$, and it suffices to prove that $b(P) \geq b(vuv) \geq b(P')$.

The second inequality follows by the symmetric variant of Lemma 7. The first one follows from the fact that $b(P) = |Q_{\mathrm{RL}}(v)| \geq |Q_{\mathrm{RL}}(vuv)| = b(vuv)$, where we are using the definitions of $b(P)$ (since v is RL-generic for P) and $b(vuv)$, and prefix-monotonicity (the symmetric variant of Lemma 4). □

3.3   The Lemma

We are now ready to state and prove our main lemma.

Lemma 10 (Main Lemma). *Let $P, P' \neq \emptyset$ be two smooth properties separated by L. If P has suffix of choice into P', then $a(P) \geq a(P')$ and $b(P) \geq b(P')$, and at least one of these inequalities is strict.*

Proof. Inequalities $a(P) \geq a(P')$ and $b(P) \geq b(P')$ follow from Lemma 9. So we just need to prove that at least one of the two is strict. Toward a contradiction, suppose that neither is, namely $a(P) = a(P')$ and $b(P) = b(P')$.

Let ϑ be generic for P (Lemma 5). Since P has suffix of choice into P', there exists x such that the suffix $x\vartheta$ forces every string $wx\vartheta$ into P', for all $w \in P \cup P'$:

$$(\forall w \in P \cup P')(wx\vartheta \in P') \tag{1}$$

Claim. For all $t \geq 1$, we have $\vartheta(x\vartheta)^t \in P'$.

Proof. By induction on $t \geq 1$. For $t = 1$, apply (1) for $w := \vartheta$ (which is in P as generic string), to conclude that P' contains $wx\vartheta = \vartheta x\vartheta$. For $t \geq 2$, apply (1) for $w := \vartheta(x\vartheta)^{t-1}$ (which is in P' by the inductive hypothesis), to conclude that P' also contains $wx\vartheta = \vartheta(x\vartheta)^{t-1}x\vartheta = \vartheta(x\vartheta)^t$.   □

So, for every $t \geq 1$ (and since $\vartheta \in P$, $\vartheta(x\vartheta)^t \in P'$, and P, P' are separated by L), we know that there exists a context u_t, v_t such that exactly one of $u_t\vartheta v_t$ and $u_t\vartheta(x\vartheta)^t v_t$ is in L. In what follows, we find $t \geq 1$ such that M decides identically on $u_t\vartheta v_t$ and $u_t\vartheta(x\vartheta)^t v_t$, contradicting the assumption that M solves L.

We first focus on the string $\vartheta x\vartheta$ and the function $\alpha_{\vartheta,x\vartheta}$ that partially surjects $Q_{\mathrm{LR}}(\vartheta)$ to $Q_{\mathrm{LR}}(\vartheta x\vartheta)$ (Lemma 1). Letting $\alpha := \alpha_{\vartheta,x\vartheta}$ and $A := Q_{\mathrm{LR}}(\vartheta)$, we prove:

Claim. α is a permutation of A.

Proof. We know $Q_{\mathrm{LR}}(\vartheta x\vartheta)$ is a subset of $Q_{\mathrm{LR}}(\vartheta)$ (Lemma 4); and at least as big, because $|Q_{\mathrm{LR}}(\vartheta x\vartheta)| \geq a(P')$ (by Lemma 7 and since $\vartheta x\vartheta \in P'$) and $a(P') = a(P)$ (our assumption) and $a(P) = |Q_{\mathrm{LR}}(\vartheta)|$ (by definition of $a(P)$ and since ϑ is LR-generic). So, it must be $Q_{\mathrm{LR}}(\vartheta x\vartheta) = Q_{\mathrm{LR}}(\vartheta) = A$, which implies that $\alpha = \alpha_{\vartheta,x\vartheta}$ is total and injective. Since it is also surjective, it bijects A to A. (Fig. 2z.)   □

Moreover, by standard arguments (see, e.g., [13, Facts 7,8]), we know that the same holds for the respective function on other strings $\vartheta(x\vartheta)^t$ (Fig. 2z).

Claim. For all $t \geq 1$, the function $\alpha_{\vartheta,(x\vartheta)^t}$ is the permutation α^t of A.

Finally, by symmetric arguments, the partial surjection $\beta := \beta_{\vartheta x,\vartheta}$ (cf. comments after Lemma 4) of $B := Q_{\mathrm{RL}}(\vartheta)$ to $Q_{\mathrm{RL}}(\vartheta x\vartheta)$ is also a permutation of B; and for every $t \geq 1$, the respective mapping $\beta_{(\vartheta x)^t,\vartheta}$ is the permutation β^t of B.

Now, let $t_{\mathrm{LR}}, t_{\mathrm{RL}} \geq 1$ be such that $\alpha^{t_{\mathrm{LR}}}$ and $\beta^{t_{\mathrm{RL}}}$ become the identity mappings id_A on A and id_B on B, respectively. Then for $t := t_{\mathrm{LR}} \cdot t_{\mathrm{RL}}$ and for the string $\vartheta(x\vartheta)^t$, we have $\alpha_{\vartheta,(x\vartheta)^t} = \alpha^t = (\alpha^{t_{\mathrm{LR}}})^{t_{\mathrm{RL}}} = (\mathrm{id}_A)^{t_{\mathrm{RL}}} = \mathrm{id}_A$ and likewise $\beta_{(\vartheta x)^t,\vartheta} = \beta^t = \mathrm{id}_B$. Intuitively, this means that no LR-traversal of $\vartheta(x\vartheta)^t$ by M notices the t copies of $x\vartheta$ to the right of the leftmost ϑ; and, likewise, no RL-traversal notices the t copies of ϑx to the left of the rightmost ϑ. This is an issue when we place the two strings ϑ and $\vartheta(x\vartheta)^t$ in the context of $u := u_t$ and $v := v_t$.

Claim. M decides identically on $u\vartheta v$ and $u\vartheta(x\vartheta)^t v$.

Proof. Consider the computation c of M on $u\vartheta v$ (Fig. 2a). Let $0 \le m \le \infty$ be the number of times c fully traverses ϑ (i.e., crosses into ϑ from one side and, after computing inside ϑ, crosses out of ϑ from the opposite side). For $i = 1, 2, \ldots, m$, let the ith *crucial point* of c be the configuration (p_i, k_i) at the end of the ith full traversal, so that k_i is either $|u\vartheta|+1$ (odd i) or $|u|$ (even i); let the 0th crucial point (p_0, k_0) be the start configuration $(q_s, 0)$; and the $m + 1$st crucial point (p_{m+1}, k_{m+1}) be the last configuration of c (if any). Then call c_i the infix of c from the $i - 1$st to the ith crucial point (inclusive); so that it lies entirely inside either $\vdash u\vartheta$ (odd i) or $\vartheta v \dashv$ (even i), except only for its last configuration.

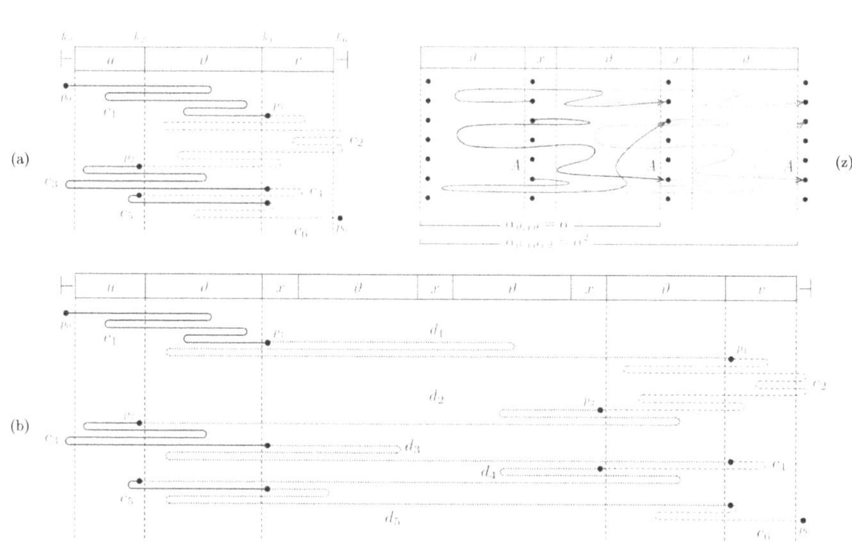

Fig. 2. (a) A computation on $u\vartheta v$ with $m+2 = 7$ crucial points; and the corresponding $m + 1 = 6$ computation infixes $c_1, \ldots, c_6$, odd (bold) and even (dashed). (b) The computation c' on $u\vartheta(x\vartheta)^t v$ (here $t = 3$), created by joining the concatenations c'_i of the c_i and d_i. (z) How $\alpha := \alpha_{\vartheta, x\vartheta}$ (solid lines) permutes $A := Q_{\text{LR}}(\vartheta)$; and $\alpha_{\vartheta,(x\vartheta)^t}$ (solid & dotted) permutes A, too, sometimes as identity (here $t = 2$).

Now let d_i be the computation of M that justifies why $\alpha_{\vartheta,(x\vartheta)^t}(p_i) = \alpha^t(p_i) = \text{id}_A(p_i) = p_i$, if i is odd; or why $\beta_{(\vartheta x)^t, \vartheta}(p_i) = \beta^t(p_i) = \text{id}_B(p_i) = p_i$, if i is even. Form c'_i by "joining" c_i and d_i at the shared configuration (p_i, k_i) (Fig. 2b). Then "join" all c'_i to obtain the full computation c' of M on $u\vartheta(x\vartheta)^t v$. By construction, c' loops or halts exactly as c, so M decides identically on both inputs.   □

Since exactly one of $u\vartheta v$ and $u\vartheta(x\vartheta)^t v$ is in L, we conclude that M does not solve L, a contradiction.   □

4   The Lower Bound

To prove the promised lower bound, we introduce a sequence of $1+N$ non-empty properties of strings over Σ_h

$$\emptyset \neq P_0, P_1, \dots, P_N \subseteq \Sigma_h^* \tag{2}$$

such that $N = \binom{h}{2}$ and every two successive ones, P_{t-1}, P_t $(1 \leq t \leq N)$, satisfy the conditions of the Main Lemma: they are smooth (Lemma 13), separated by OWL_h (Lemma 14), and such that the earlier one has suffix of choice into the later one (Lemma 15). Then the lemma implies that the exit sizes of P_{t-1} upper bound those of P_t, and strictly so on at least one of the two sides. Therefore, in the sequence of the corresponding pairs of exit sizes

$$(a(P_0), b(P_0)),\ (a(P_1), b(P_1)),\ \dots,\ (a(P_N), b(P_N)),$$

no step increases a component; and each step decreases at least one. This is a total of N decrements, to be divided between the two nonincreasing sequences

$$a(P_0), a(P_1), \dots, a(P_N) \qquad \text{and} \qquad b(P_0), b(P_1), \dots, b(P_N)\,.$$

So, at least one of them decreases $\geq N/2$ times. Since all numbers are obviously between $|Q|$ and 0, it follows that $|Q| \geq N/2 = \frac{1}{2}\binom{h}{2} = \frac{1}{4}h^2 - \frac{1}{4}h = \Omega(h^2)$.

To define the properties of (2), we first define a respective sequence of connectivities $C_0, C_1 \dots, C_N$ and then let each P_t be the property of having connectivity C_t, namely:

$$P_t := \{z \in \Sigma_h^* \mid C(z) = C_t\}\,. \tag{3}$$

So, in the rest of this section, we introduce the connectivities; prove a list of facts about them (Lemmas 11 and 12); then use these facts to prove the necessary facts about our properties (Lemmas 13–15).

To construct the connectivities $C_0, C_1, \dots, C_N$, we take the identity matrix I_h and start flipping the $N = \binom{h}{2}$ bits of its strict upper triangle, from 0 to 1, one at a time, from column h to column 2, and from bottom to top in each column: that is, we start with cells $(h{-}1, h), (h{-}2, h), \dots, (1, h)$, continue with $(h{-}2, h{-}1), \dots, (1, h{-}1)$, etc., all the way to $(2, 3), (1, 3)$, and finally $(1, 2)$. (Fig. 3.) More carefully, if we let "*the t-th cell*" refer to the tth cell (i_t, j_t) in this sequence, then

$$C_0 := I_h \qquad \text{and} \qquad C_t := C_{t-1} + E_t, \tag{4}$$

where E_t is the $h \times h$ Boolean matrix with only the t-th cell (i_t, j_t) set to 1.

Two important facts about these matrices is that they are idempotent; and that the product of every two successive ones (in either order) produces the later one (Lemma 12). To prove these facts, we must first prove some other, more elementary facts about how the C_ts interact with the E_ts. In the process, it helps to also consider the matrix E_t' obtained from E_t by also placing 1s in the cells to the right of the t-th cell in the same row (so that the 1s in E_t' are exactly at $(i_t, j_t), (i_t, j_t + 1), \dots, (i_t, h)$) (Fig. 3).

$$C_0 = \begin{bmatrix} 1&0&0&0&0 \\ 0&1&0&0&0 \\ 0&0&1&0&0 \\ 0&0&0&1&0 \\ 0&0&0&0&1 \end{bmatrix} \quad C_7 = \begin{bmatrix} 1&0&0&1&1 \\ 0&1&0&1&1 \\ 0&0&1&1&1 \\ 0&0&0&1&1 \\ 0&0&0&0&1 \end{bmatrix} \quad C_{10} = \begin{bmatrix} 1&1&1&1&1 \\ 0&1&1&1&1 \\ 0&0&1&1&1 \\ 0&0&0&1&1 \\ 0&0&0&0&1 \end{bmatrix} \quad E_8 = \begin{bmatrix} 0&0&0&0&0 \\ 0&0&1&0&0 \\ 0&0&0&0&0 \\ 0&0&0&0&0 \\ 0&0&0&0&0 \end{bmatrix} \quad E_8' = \begin{bmatrix} 0&0&0&0&0 \\ 0&0&1&1&1 \\ 0&0&0&0&0 \\ 0&0&0&0&0 \\ 0&0&0&0&0 \end{bmatrix}$$

Fig. 3. Some connectivities for $h = 5$.

Lemma 11. *For all $t = 1, 2, \ldots, N$:*

(a) $C_{t-1} + E_t' = C_t + E_t' = C_t$ (c) $E_t C_{t-1} = E_t'$

(b) $C_{t-1} E_t = E_t$ (d) $E_t^2 = 0$

Proof. Fix t and let $(i, j) := (i_t, j_t)$ be the t-th cell. In some of the parts, we write E_t as $e_i e_j^T$, where e_k stands for the h-long column vector with a single 1 at cell k, and its transposition e_k^T is the corresponding row vector.

(a) C_{t-1} and C_t already have 1s in the cells to the right of the tth cell in the same row. So, adding E_t' does not affect these cells (since the addition is Boolean). For the t-th cell, this is still 0 in C_{t-1}, so the addition makes it 1 and we get C_t; and is already 1 in C_t, so the addition does not cause any change.

(b) It is $C_{t-1} E_t = C_{t-1}(e_i e_j^T) = (C_{t-1} e_i) e_j^T = e_i e_j^T = E_t$, where $C_{t-1} e_i = e_i$ because only the ith column of C_{t-1} matters, and is just e_i (because $i < j$).

(c) Let $X := E_t C_{t-1}$. Since every row $i' \neq i$ in E_t is zero, the corresponding row i' in X is also zero. So, we just study row i. Every cell (i, r) in it is the product of the ith row of E_t, namely e_i, with the r-th column c_r of C_{t-1}; so, it is 1 iff c_r has 1 in its j-th cell. By the structure of C_{t-1}, this holds only for $c_j, c_{j+1}, \ldots, c_h$. So, the cells of X with 1 are $(i, j), (i, j+1), \ldots, (i, h)$. So, $X = E_t'$.

(d) We have $E_t^2 = (e_i e_j^T)(e_i e_j^T) = e_i(e_j^T e_i) e_j^T = e_i 0 e_j^T = 0$, where $e_j^T e_i = 0$ because $i \neq j$ (since $i < j$). $\qquad\square$

Lemma 12. *For all $t = 1, 2, \ldots, N$:* $C_t^2 = C_t$ *and* $C_{t-1} C_t = C_t = C_t C_{t-1}$.

Proof. For the idempotence of C_t, we induct on $t \geq 0$. If $t = 0$, then $C_0^2 = I_h^2 = I_h = C_0$. If $t \geq 1$, then $C_t^2 = (C_{t-1} + E_t)^2 = C_{t-1}^2 + C_{t-1} E_t + E_t C_{t-1} + E_t^2$. By the inductive hypothesis and Lemma 11bcd, this becomes $C_{t-1} + E_t + E_t' + 0$, where the first two terms give C_t. So, it is just $C_t + E_t'$; which is C_t, by Lemma 11a.

For the next equality, the definition implies $C_{t-1} C_t = C_{t-1}(C_{t-1} + E_t) = C_{t-1}^2 + C_{t-1} E_t$. By idempotence and Lemma 11b, this is $C_{t-1} + E_t$, which is just C_t (by the definition again). Hence, $C_{t-1} C_t = C_t$.

For the last equality, the definition implies $C_t C_{t-1} = (C_{t-1} + E_t) C_{t-1} = C_{t-1}^2 + E_t C_{t-1}$. By idempotence and Lemma 11c, this is just $C_{t-1} + E_t'$, which is C_t, by Lemma 11a. Hence, $C_t C_{t-1} = C_t$. $\qquad\square$

We are now ready to prove the three facts that we promised about the P_t.

Lemma 13. *For all $t = 0, \ldots, N$, property P_t is smooth.*

Proof. Pick any $x, z \in P_t$. Let y be any string with identity connectivity (e.g., the empty string). Then the connectivity of xyz is $C(xyz) = (C(x)C(y))C(z) = (C_t I_h)C_t = C_t^2 = C_t$, where we used idempotence (Lemma 12). So, $xyz \in P_t$. $\square$

Lemma 14. *For all $t = 1, \ldots, N$, properties P_{t-1}, P_t are separated by* OWL$_h$.

Proof. Pick any $x \in P_{t-1}$ and $z \in P_t$. Then $C(x) = C_{t-1}$ and $C(z) = C_t$. By the definitions of C_{t-1}, C_t, it follows that the tth cell (i_t, j_t) is 0 in $C(x)$; and 1 in $C(z)$. Hence, x contains no path from the i_tth node of the leftmost column to the j_tth node of the rightmost column, but z does. Consider the symbols $u := \{(1, i_t)\}, v := \{(j_t, 1)\} \in \Sigma_h$. Clearly then, uxv is dead and uzv is live. $\square$

Lemma 15. *For all $t = 1, \ldots, N$, property P_{t-1} has suffix of choice into P_t.*

Proof. Pick any $v \in P_{t-1}$. Let u be arbitrary in P_t. To show $xuv \in P_t$, for all $x \in P_{t-1} \cup P_t$, we start with arbitrary such x and find the connectivity of xuv: $C(xuv) = C(x)\big(C(u)C(v)\big) = C(x)(C_t C_{t-1}) = C(x)C_t$, where the last equality uses Lemma 12. Now, since x is in P_{t-1} or P_t, this connectivity is $C_{t-1}C_t$ or C_t^2, which are both equal to C_t, again by Lemma 12. So, $xuv \in P_t$. $\square$

5   Conclusion

We proved that every 2DFA solving OWL$_h$ needs $\Omega(h^2)$ states and thus, in general, the conversion of 1NFAS to 2DFAS incurs a quadratic blow-up in the number of states. Although this simply matches the known lower bound by Chrobak [14,15] and only slightly exceeds that of [16], it results from an analysis of 2DFA computations that are clearly richer than those realized on unary or 3-long inputs.

A more careful construction in Sect. 4 could, in fact, result in a slightly longer sequence of properties, $\binom{h+1}{2}$ instead of only $\binom{h}{2}$, which we conjecture is also the maximum achievable in the current context. The full version of this paper will present this better construction and discuss its maximality.

A major deficiency in the study of 2DFA computations is the lack of general lower-bounding tools. In this work, we have very consciously tried to extract from our syllogisms the statement of a lemma that could potentially serve as such a tool. It would be great if the result of this effort, our main lemma (Lemma 10), could indeed be used to also lower bound other 2DFAS.

References

1. Sakoda, W.J., Sipser, M.: Nondeterminism and the size of two-way finite automata. In: Proceedings of STOC, pp. 275–286 (1978)
2. Seiferas, J.I.: Untitled, manuscript communicated to M. Sipser (1973)
3. Guillon, B., Pighizzini, G., Prigioniero, L., Průša, D.: Converting nondeterministic two-way automata into small deterministic linear-time machines. Inf. Comput. **289** (2022)
4. Geffert, V., Mereghetti, C., Pighizzini, G.: Converting two-way nondeterministic unary automata into simpler automata. Theor. Comput. Sci. **295**, 189–203 (2003)

5. Clerici, A., Pighizzini, G., Prigioniero, L.: Two-way automata and bounded languages. In: Proceedings of CIAA, pp. 73–85 (2025)
6. Geffert, V., Guillon, B., Pighizzini, G.: Two-way automata making choices only at the endmarkers. Inf. Comput. **239**, 71–86 (2014)
7. Sipser, M.: Lower bounds on the size of sweeping automata. J. Comput. Syst. Sci. **21**(2), 195–202 (1980)
8. Hromkovič, J., Schnitger, G.: Nondeterminism versus determinism for two-way finite automata: generalizations of Sipser's separation. In: Proceedings of ICALP, pp. 439–451 (2003)
9. Kapoutsis, C.: Nondeterminism is essential in small two-way finite automata with few reversals. Inf. Comput. **222**, 208–227 (2013)
10. Hromkovič, J., Královič, R., Královič, R., Štefanec, R.: Determinism vs. nondeterminism for two-way automata: representing the meaning of states by logical formulæ. Int. J. Found. Comput. Sci. **24**(7), 955–978 (2013)
11. Bianchi, M.P., Hromkovič, J., Kováč, I.: On the size of two-way reasonable automata for the liveness problem. In: Proceedings of DLT, pp. 120–131 (2015)
12. Raszyk, M.: On the size of logical automata. In: Proceedings of SOFSEM, pp. 447–460 (2019)
13. Kapoutsis, C.: Deterministic moles cannot solve liveness. J. Autom. Lang. Comb. **12**(1–2), 215–235 (2007)
14. Chrobak, M.: Finite automata and unary languages. Theor. Comput. Sci. **47**, 149–158 (1986)
15. To, A.W.: Unary finite automata vs. arithmetic progressions. Inf. Process. Lett. **109**, 1010–1014 (2009)
16. Kapoutsis, C.: Optimal 2DFA algorithms for one-way liveness on two and three symbols. In: Böckenhauer et al. [23], pp. 33–48 (2018)
17. Rabin, M.O., Scott, D.: Finite automata and their decision problems. IBM J. Res. Dev. **3**, 114–125 (1959)
18. Sipser, M.: Introduction to the Theory of Computation, 3rd edn. Cengage Learning (2012)
19. Sipser, M.: Halting space-bounded computations. Theor. Comput. Sci. **10**, 335–338 (1980)
20. Kapoutsis, C.: Small sweeping 2NFAs are not closed under complement. In: Proceedings of ICALP, pp. 144–156 (2006)
21. Kapoutsis, C., Královič, R., Mömke, T.: Size complexity of rotating and sweeping automata. J. Comput. Syst. Sci. **78**(2), 537–558 (2012)
22. Kapoutsis, C., Pighizzini, G.: Reversal hierarchies for small 2DFAs. In: Proceedings of MFCS, pp. 554–565 (2012)
23. Böckenhauer, H.-J., Komm, D., Unger, W. (eds.): Adventures between lower bounds and higher altitudes – Essays dedicated to Juraj Hromkovič on the occasion of his 60th birthday, vol. 11011 of Lecture Notes in Computer Science. Springer, Heidelberg (2018)

Vertical-Horizontal Full Compatibility
of One-Dimensional Subshifts

Arthur Mittelstaedt[(✉)] and Gaétan Richard

Normandie Univ, UNICAEN, ENSICAEN, CNRS, GREYC, 14000 Caen, France
`{arthur.mittelstaedt,gaetan.richard}@unicaen.fr`

Abstract. We study the notion of full-compatibility: given two one-dimensional subshifts H and V, is there a two-dimensional subshift X such that $H = \{c_{|\mathbb{Z} \times \{0\}} \mid c \in X\}$ and $V = \{c_{|\{0\} \times \mathbb{Z}} \mid c \in X\}$? We show that this problem is decidable when both X and Y are nearest-neighbor SFTs but undecidable when at least one is allowed to be of slightly higher complexity. We also prove that the problem is undecidable for three nearest-neighbor SFTs combined in a three-dimensional subshift.

Keywords: Subshifts · Wang tiles · Undecidability · Domino Problem · Tilings · Subshifts of Finite Type

1   Introduction

In science, local constraints or rules are often well understood; however, their global implications are often difficult or impossible to determine. To better study the link between locality and globality, one can look at a formal simplified and regular case. One example of such a context consists of using a regular line (or grid) where each position is endowed with a state chosen among a finite set, and look at the implication of local constraints given by a set of forbidden words.

On the discrete line (dimension 1), this framework corresponds to an infinite word version of formal languages and has been heavily studied and its basics are now well understood [11]. Most problems relating locality and globality are decidable.

On the grid (dimension 2), this framework corresponds to *tilings*. In this case, one surprising thing is that simple local constraints can lead to very complex constructions. A first example is the existence of an aperiodic set of Wang tiles [3]. This kind of "strange" object has been proven fruitful when physicists have encountered quasicrystals. Further studies have found and constructed many other variants of such elements using self-similarity [5,10], minimizing the size of constraints [4,7], or using geometry [9,12]. Depending on the construction, the understanding of the global structure is more or less complete.

With advances in dimension two, one current axis of research is to explore its links with the one dimensional case. One recent example of such result is the one from N. Aubrun and M. Sablik [2] that proves that any one dimensional infinite language from a large "complex" class (formally, whose forbidden patterns set is

© The Author(s), under exclusive license to Springer Nature Switzerland AG 2026
J. Kozik and A. Wolff (Eds.): SOFSEM 2026, LNCS 16448, pp. 117–127, 2026.
https://doi.org/10.1007/978-3-032-17801-5_9

semi-computable), can be seen as the projection of a "simple" two dimensional one (with a finite set of forbidden patterns). One other example is the work of S. Esnay [6] and in particular [1] on the computational complexity of two-dimensional tilings when horizontal constraints are fixed.

In this paper, we study the following question: given **two** one-dimensional infinite languages, can both of them appear in a two dimensional tiling (horizontally and vertically). We also add one extra demand that **every** word appears: we call this problem *full compatibility*. The paper is divided as follows: In Sect. 2, we give the formal definition of full compatibility and several basic and meaningful examples. Then, we present one case where full compatibility is decidable (Sect. 3) and several cases where it is undecidable (Sect. 4).

2   The Full Compatibility Problem

2.1   Subshift and Traces

The elementary objects studied are the members of the set $\Sigma^{\mathbb{Z}^d}$, given d a dimension and Σ a finite alphabet. We call this set the full-shift over Σ and its elements are called *configurations*. When $d = 1$ we alternately call them bi-infinite words or just words. The elements of $\mathbb{Z}^d$ are called *positions* or vectors.

If c is a 2D configuration ($d = 2$), and (x, y) is a position, then $c_{(x,y)} = c(x, y)$ refers to the letter at position (x, y). The space of these configurations is endowed with the distance defined by $d(c, c') = 2^{-\min(\{|i| \ |c_i \neq c'_i\})}$, using the L_∞ norm. The resulting space is metric and compact. The *shift* is the action of an elementary vector $v : \sigma_v(c)_p = c_{p+v}$.

These objects are intuitively considered as tilings, and since they are formally defined using functions, we will use some notations usually employed for functions. Especially, we often use the restriction notation to refer to a precise part of a configuration. Let us say that c is a 2D configuration, $c_{|\mathbb{Z}\times\{0\}}$ corresponds to the tiled horizontal line at height 0 (resp. $c_{|\{0\}\times\mathbb{Z}}$ is the central column).

A *pattern* is a function of Σ^P where P is a finite part of $\mathbb{Z}^d$, called the supporting set of the pattern. A pattern m *appears* in a configuration c if $\exists v \in \mathbb{Z}^d, \sigma_v(c)_{|P} = m$, a configuration avoids a set of forbidden patterns if none of its elements appears in it. For convenience, in dimension 2, we identify patterns and tuples.

Definition 1 (Subshift). *A* subshift *is the subset of configurations of* $\Sigma^{\mathbb{Z}^d}$ *avoiding* $\mathcal{F}$, *a certain set of forbidden patterns.*

The name subshift comes from the fact that these sets are stable under the shift operation. In fact subshifts are exactly the subsets of $\Sigma^{\mathbb{Z}^d}$ which are stable under the shift and topologically closed. Note that there is always several sets of forbidden patterns to describe a single subshift.

We also introduce some elementary notations to describe words. We use the symbol ω as an exponent on the left or on the right of a letter to symbolize an infinite repetition. For example if a and b are letters and w is some finite word then ${}^{\omega}awb^{\omega}$ stands for a word with an infinite number of successive a followed by w, again followed by an infinite repetition of the letter b. Note that even if w is fixed, this description formally corresponds to an infinity of words that are just shifted versions of one another. We often abusively write $c = {}^{\omega}awb^{\omega}$ to say that c is one of the words described by ${}^{\omega}awb^{\omega}$. Note that the set of words corresponding to ${}^{\omega}awb^{\omega}$, despite being stable under the shift, is not a subshift as it is not necessarily closed, we should add the words ${}^{\omega}a^{\omega}$ and ${}^{\omega}b^{\omega}$ to obtain a subshift.

Definition 2. *We can distinguish several types of subshifts according to the complexity of their set of forbidden patterns:*

- *A subshift is* of finite type *(**SFT** for short) if it is defined by a finite set of forbidden patterns;*
- *Such a subshift is* nearest-neighbor SFT *(**NNSFT**) if it can be described by a set of forbidden patterns whose supporting sets consists of exactly two adjacent positions. For one-dimensional subshifts this is equivalent to $\mathcal{F} \subseteq \Sigma^{\{0,1\}}$.*
- *A subshift is* sofic *if it is a letter-to-letter projection of an SFT.*
- *A subshift is* effective *if its set of forbidden patterns is recursively enumerable.*

All these definitions are usual and widespread ones [8]. The nearest-neighbor notion is a bit more specific case but appears naturally [1]: on the line, NNSFT correspond to (bi-infinite) traces of states encountered marching along a finite graph (and also Rauzy graph of order one); on the plane, *Wang* tilings [13] are a classical example of NNSFT. In many cases, the additional size restriction of the size of patterns is not significant as any SFT is equivalent up to conjugacy to a NNSFT (up to scaling). In this paper, this scaling operation is significant and will impact results.

2.2   Full-Compatibility

One classical question on subshifts is the following: "Given a finite set of forbidden patterns, is the subshift avoiding it non-empty?". This problem is computable on the line but undecidable on the plane [3]. Due to the separation between these two cases, one can study the link between these two dimensions. A result by N. Aubrun and M. Sablik [2] states that any effective 1D subshift can be seen as the set of lines of a 2D sofic subshift. More recently, S. Esnay has studied the decidability of 2D subshift given horizontal constraints.

In this paper, our point of view is to give **both** the horizontal and the vertical languages and see if they can be achieved by a 2D subshift simultaneously. More formally, given two one-dimensional subshifts H and V, is there a two-dimensional subshift X whose set of lines is exactly H and whose set of columns is exactly V?

Definition 3 (Fully-Compatible subshift).

Two one-dimensional subshifts H and V with common alphabet are fully-compatible if there exists a two-dimensional subshift X such that $H = \{c_{|\mathbb{Z}\times\{0\}} \mid c \in X\}$ and $V = \{c_{|\{0\}\times\mathbb{Z}} \mid c \in X\}$.

Example 1 (Two not fully-compatible subshifts). Let H be the set of configurations of the form ${}^{\omega}(1^{\omega}2^{\omega}3)^{\omega}$ and V be the set of configurations of the form ${}^{\omega}(12^{\omega}3^{\omega})^{\omega}$. One can verify that H and V are two NNSFTs. Let X be their *combiner.* X is not empty, as configurations of the form depicted in Fig. 1 are valid in it. But H and V are not fully-compatible, as we will immediately show. Take a configuration of H that contains two consecutive 1s, it cannot be a row of X since the row just below would have to contain two consecutive 3 which is forbidden in H.

3	1	2	2	3	1	2	2	2
2	3	1	2	2	3	1	2	2
2	2	3	1	2	2	3	1	2
2	2	2	3	1	2	2	3	1
2	2	2	2	3	1	2	2	3
2	2	2	2	2	3	1	2	2
3	2	2	2	2	2	3	1	2
2	3	2	2	2	2	2	3	1
1	2	3	2	2	2	2	2	3

Fig. 1. A representative configuration of X: diagonals of 1, 2, and 3; the number of consecutive diagonals of 2 can be arbitrary

For following sections, where the decidability of the problem will be studied we will consider the necessary condition that all letters that appears in some configuration of H also appear in some configuration of V.

2.3   Basic Properties and Examples

This first formulation motivates the use of "compatibility" in the name of the problem and show the goal of the study but it hides a clearer way to present this question. Indeed, it is easy to verify that the set $X = \{c \in \Sigma^{\mathbb{Z}^2} \mid \forall i, c_{|\mathbb{Z}\times\{i\}} \in H, c_{|\{i\}\times\mathbb{Z}} \in V\}$ is a subshift, as it can be obtained by combining forbidden patterns from H and verticalized forbidden patterns from from V. We call X the *combiner* of H and V. The compatibility problem can then be stated as : given H and V, can all their configurations be found as lines and respectively columns of configurations from X. Note that X could also be empty.

Definition 4 (combiner of H and V).

Given H and V two one-dimensional subshifts with alphabet Σ, the combiner of H and V is the subshift X defined by :

$$X = \{c \in \Sigma^{\mathbb{Z}^2} \mid \forall i, (c_{(j,i)})_{j \in \mathbb{Z}} \in H \text{ and } (c_{(i,j)})_{j \in \mathbb{Z}} \in V\}$$

Lemma 1. *Let H and V be both SFT (resp. nearest-neighbor SFT, resp. sofic-subshift, resp. effective subshift), their combiner is a subshift and can be chosen SFT (resp. nearest-neighbor SFT, resp. sofic, resp. effective)*

Proof. It is sufficient to take for the combiner the disjoint union of the horizontal forbidden patterns of H and the vertical ones of V.

Let us give some simple examples. The $H = V = {}^\omega(01)^\omega$ are fully-compatible (corresponding to the 2D checkboard). More generally, any subshift V is fully-compatible with $H = {}^\omega(01)^\omega$ if and only if V is closed under complement.

On the other hand, given a set of Wang tiles (2D nearest-neighbor SFT), it is undecidable to determine the set of its horizontal lines. In the case of [4], the set of horizontal lines is not sofic and partially understood (encodings of irrational numbers).

2.4   General Direction

It can be noted that our paper deals with two orthogonal directions. Most of the results can be extended to cases where those directions are arbitrary.

Lemma 2. *Let U and V be two subshifts, u and v two non-colinear vectors. The following properties are equivalent:*

- *There is a two-dimensional subshift X such that: $U = \{c_{|\mathbb{Z}u} \mid c \in X\}$ and $V = \{c_{|\mathbb{Z}v} \mid c \in X\}$.*
- *U and V are fully-compatible.*

Proof. Let us assume the first property, We take the subshift X and construct C extracting point on the grid defined by (u, v) by:

$$C = \left\{c(ku + lv)_{(k,l) \in \mathbb{Z}^2} \mid c \in X\right\}$$

It is clear that C is a combiner of U and V.

Conversely, let C be the combiner of U and V. As u and v are non-colinear, the regular grid can been seen as a finite disjoint union of (u, v) subgrids: $\mathbb{Z}^2 = \biguplus_{0 \le i \le N} \{ku + lv + d_i \mid k, l \in \mathbb{Z}\}$ with a fixed family of vectors $d_0, d_1, \cdots, d_N \in \mathbb{Z}^2$. Let us define the subshift obtained by taking arbitrary configuration of C on each subgrid:

$$X = \{c(x,y) = c_i(k,l) \text{ with } ku + lv + d_i = (x,y) \mid c_0, c_1, \ldots, c_N \in C\}$$

C satisfies the first property.

It can be noted that, in the previous proof X is a sofic subshift if and only if C is. However, the property is not true for SFT. In particular, the subshift X realising U and V along (u, v) could be of finite type while U and V are not.

3 Decidability of the Compatibility of Two Nearest-Neighbor Subshifts

In this section, we show that the compatibility problem is decidable for the case where H and V are nearest-neighbor SFT. Let H and V be one-dimensional subshifts of finite type described by the finite sets of forbidden patterns $\mathcal{F}_H$ and $\mathcal{F}_V$. As H and V are nearest-neighbor SFT we can consider that all their forbidden patterns only consists of two adjacent letters.

Lemma 3. *Let H and V be two nearest-neighbor SFT, they are compatible if and only if the 4 following propositions are verified:*

$$1)\ \forall a \in H\ \exists b \in H\ \forall i, (a_i, b_i) \notin \mathcal{F}_V$$
$$2)\ \forall a \in H\ \exists b \in H\ \forall i, (b_i, a_i) \notin \mathcal{F}_V$$
$$3)\ \forall a \in V\ \exists b \in V\ \forall i, (a_i, b_i) \notin \mathcal{F}_H$$
$$4)\ \forall a \in V\ \exists b \in V\ \forall i, (b_i, a_i) \notin \mathcal{F}_H$$

Theses formulas state that for any configuration a from H (resp. V), you can find configurations b, b' of H (resp. V) such that stacking b, a and b' on top (resp. at the right) of one another does not create a forbidden pattern of V (resp.H). Note that this lemma strongly use the fact that H and V are nearest-neighbor SFT and is not true for SFT.

Proof. Indeed, it is clear that these propositions are necessary for H and V to be compatible. By using compactness, we can also show that it is sufficient : suppose that the propositions are verified, then for any configuration a of H you can construct a valid strip of any width with a in the middle. by alternately applying the first two formulas to stack enough lines to obtain a strip of any width with a in the middle. Similarly you can stack columns for a configuration of V.

We now should show how to computationally verify these four propositions.

Proposition 1. *Given two nearest-neighbor SFT U and V, it is decidable whether U and V are fully-compatible.*

Proof. H can also be seen as the set of bi-infinite walks on some digraph. Let Σ be the set of vertices with an edge between l and l' if and only if $(l, l') \notin \mathcal{F}_H$. Configurations of H then correspond to bi-infinite walks in this graph. Same goes for V. We can construct the graph (Σ, A_H) and (Σ, A_V) where: $A_H = \{(l_1, l_2) \notin \mathcal{F}_H\}, A_V = \{(l_1, l_2) \notin \mathcal{F}_V\}$.

Let us show how to verify the proposition 1), the other ones being similar.

Take the digraph $(\Sigma^2 \setminus \mathcal{F}_V, A)$ where $A = \{((l_1, l_1'), (l_2, l_2')) \mid (l_1, l_2) \in A_H,\ (l_1', l_2') \in A_H\}$ (see example in Fig. 2). It is now sufficient to study the bi-infinite walks of this new graph, omitting the second element of each state. More explicitly, we are interested in $\{(l_i)_{i \in \mathbb{Z}} \mid \exists (l_i')_{i \in \mathbb{Z}}, (l_i, l_i')_{i \in \mathbb{Z}}$ is a bi-infinite walk $\}$. We can now directly identify this set with $\{a \in H \mid \exists b \in H\ \forall i, (a_i, b_i) \notin \mathcal{F}_V\}$.

All that is left is to verify that this set is H, we reduced the problem to testing the equality of the languages of two graphs (or automata), which is decidable.

The algorithm for the compatibility problem can be summed up as:

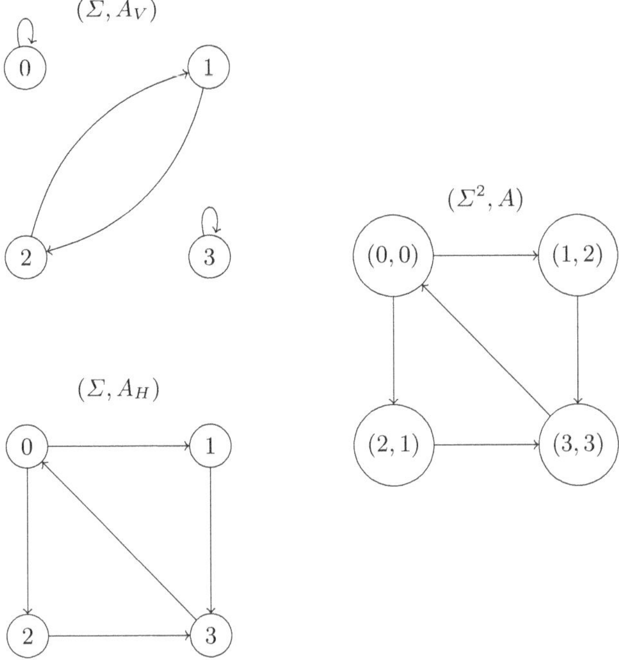

Fig. 2. Example of Construction of the digraph.

- Build (Σ, A_H) and (Σ, A_V)
- For each one of the four propositions :
  - Build (Σ^2, A)
  - Switch to (Σ, A) by erasing the second elements
  - Verify the equivalence of (Σ, A) and (Σ, A_H) (or (Σ, A_V) depending on the current proposition)
- H and V are compatible if all the propositions were found to be true.

4   Undecidability Result

We have shown that for the class of pairs of nearest-neighbor SFTs, the problem is decidable. In this section we show that this problem is undecidable for the complementary class with respect to the class of pairs of SFT. In other words there is no algorithm to decide if two subshifts are compatible when at least one of them is allowed not to be a nearest-neighbor SFT.

In the previous section, we assumed that $\mathcal{F}_H$ and $\mathcal{F}_V$ contain only patterns consisting of two adjacent letters. But in the case of a subshift that is not a nearest-neighbor SFT, there are larger forbidden patterns, the propositions 1), 2) are no longer sufficient to build a configuration with any configuration of H as one of its lines.

We will proceed by reduction from the well-known Domino problem.

Definition 5. *Domino problem Given Σ an alphabet and $\mathcal{F}$ a finite set of forbidden patterns on Σ. Is the subshift defined by $\mathcal{F}$ empty ?*

This problem is known to be undecidable [13], even when the patterns are those of a set of wang tiles, a special case of NNSFT. In the following part we will consider that the Domino problem instances are NNSFT described by forbidden patterns whose supporting sets consists of two adjacent cells.

Theorem 1. *When V is a nearest-neighbor SFT and H is SFT, there is no algorithm to decide whether they are compatible.*

Proof. Let X be a two-dimensional NNSFT on the alphabet Σ, described by its finite set of forbidden patterns $\mathcal{F}$. Write $\mathcal{F} = \mathcal{F}_h \cup \mathcal{F}_v$ by separating vertical and horizontal constraints. Without loss of generality, we can assume that there are words in $\Sigma^{\mathbb{Z}}$ avoiding $\mathcal{F}_h$, as well as words avoiding $\mathcal{F}_v$. Indeed, this can be computationally verified and X would be trivially empty otherwise.

Let $\Sigma' = \Sigma \cup \{\#\}$ with $\# \notin \Sigma$. Now let H be the *SFT* with alphabet Σ' and forbidden pattern set $\{(a, \#) \mid a \in \Sigma\} \cup \{(a, b, c) \mid a \in \Sigma, (b, c) \in \mathcal{F}_h\}$. Thus a word of H is either the word ${}^{\omega}\#^{\omega}$, or a word on Σ avoiding $\mathcal{F}_h$, or word of the form ${}^{\omega}\#aw$ where a is any letter from Σ and w is the half of a word on Σ avoiding $\mathcal{F}_h$. Intuitively, in H, consecutive letters from Σ must respect constraints of X except when they are preceded by the special symbol $\#$. We choose V to be the nearest-neighbor subshift on Σ' avoiding $\mathcal{F}_v$. It's the set of configurations consisting of sequences of letters from Σ avoiding $\mathcal{F}_v$, interrupted by sequences of $\#$.

The claim is that H and V are compatible if and only if X is not empty.

Suppose that H and V are compatible. Let w be a word in $\Sigma^{\mathbb{Z}}$ avoiding $\mathcal{F}_v$, w is in V and there must be a configuration c of the combiner with its column 0 being w. As patterns of the set $\{(a, \#) \mid a \in \Sigma\}$ are forbidden in H, all the columns on the left of column 0 do not contain any occurrence of the symbol $\#$. Then the half plane $c \mid_{\mathbb{Z} \times \mathbb{N}^*}$ avoids both $\mathcal{F}_h$ and $\mathcal{F}_v$ so by compactness we can extract a valid configuration of X from it.

Reciprocally suppose that X is not empty, and let w be one of its configurations. We must find configurations of the combiner for each of the words of H and V. We first deal with H. Let h be a word of H, define c such that $c_{(i,0)} = h_i$ and $c_{(i,j)} = \#$ otherwise, then c is a configuration of the combiner. Now let v be a configuration of V. v consists of an alternation of (possibly infinite) sequences of $\#$ and of letters from Σ avoiding $\mathcal{F}_v$. We construct the 2D configuration c as depicted in Fig. 3 :

- $c_{(0,j)} = v_j$ for all j
- $c_{(i,j)} = \#$ for all $i < 0$ and all j
- $c_{(i,j)} = \#$ for all if $i > 0$ all all j such that $c_{(0,j)} = \#$
- $c_{(i,j)} = w_{(i,j)}$ for all if $i > 0$ all all j such that $c_{(0,j)} \in \Sigma$

c is indeed a configuration of the combiner as adjacent Σ letters from column 0 and column 1 do not have to avoid $\mathcal{F}_v$ as they are always preceded by a $\#$ on column -1.

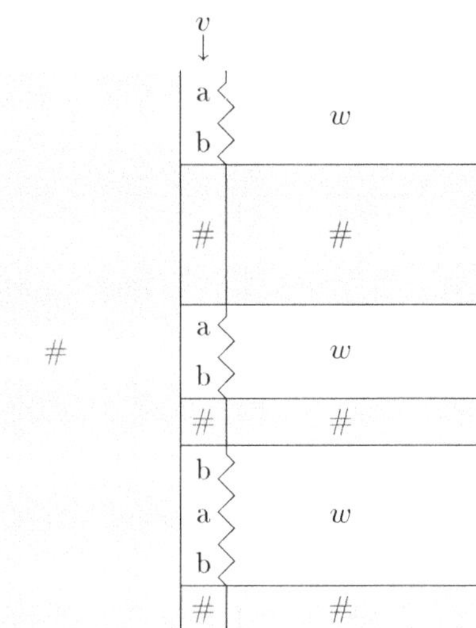

Fig. 3. Construction of a configuration with a given v as one of its columns

The undecidability extends trivially to the general case (SFT, sofic, or effective subshift).

5   Undecidability of 3D Compatibility

The 3D compatibility of a 2D SFT and 1D SFT is undecidable, as the emptiness of 2D SFT already is , A better question is whether three 1D subshifts are fully-compatible—the definition extends naturally requiring each subshift being the projection along any orthogonal direction.

Proposition 2. *Given X, Y and Z three one-dimensional nearest-neighbor SFT, it is undecidable to know whether X, Y, Z are fully compatible.*

Proof. The reduction is done again starting from the domino problem. However, this reduction is one-to-many. More formally, we reduce one instance of the domino problem to several instances of 3D compatibility instances.

Given X an instance of the domino problem, the intuitive idea is to build a 3D subshift where the domino appear as a diagonal plane, this requires to thicken the plane to ensure transmission and achieving that along any orthogonal direction, we do cross at most two cells of the plane. As drawing and thinking in 3D is not so easy, we first give the sketch of the proof in 2D (which by itself does not prove anything as 1D-SFT emptiness is decidable).

Let us take a 1D nearest-neighbor SFT language L over alphabet Σ defined by the set of forbidden words $\mathcal{F}$. We want to construct two 1D-SFTs X and Z such that each non-uniform 2D configuration in the combiner corresponds exactly to one configuration of L (up to a shift). For this, we construct the alphabet $\Sigma \cup \Sigma' \cup \{\#, -\}$ where Σ' is a copy of Σ. The encoding is as follows (see Fig. 4): for a word $(a_i)_{i \in \mathbb{Z}}$, we places a_i at position (i, i), a copy a'_i at position $(i + 1, i)$, $-$ at position (i, j) for $j < i$ and $\#$ at position (i, j) for $j > i$.

$-$	$-$	$-$	$-$	$-$	$-$	$-$	$-$	a_8
$-$	$-$	$-$	$-$	$-$	$-$	$-$	a_7	a'_7
$-$	$-$	$-$	$-$	$-$	$-$	a_6	a'_6	$\#$
$-$	$-$	$-$	$-$	$-$	a_5	a'_5	$\#$	$\#$
$-$	$-$	$-$	$-$	a_4	a'_4	$\#$	$\#$	$\#$
$-$	$-$	$-$	a_3	a'_3	$\#$	$\#$	$\#$	$\#$
$-$	$-$	a_2	a'_2	$\#$	$\#$	$\#$	$\#$	$\#$
$-$	a_1	a'_1	$\#$	$\#$	$\#$	$\#$	$\#$	$\#$
a_0	a'_0	$\#$	$\#$	$\#$	$\#$	$\#$	$\#$	$\#$

Fig. 4. Encoding of a 1D-nearest-neighbor SFT in two 1D-nearest-neighbor SFT

To achieve this encoding, we define the (copy) 1D-SFT X as the closure of the set $\{{}^\omega\!- a\, a'\, \#^\omega \mid \forall a \in \Sigma\}$. And the (check) 1D-SFT Z as the closure of $\{{}^\omega\!- a'\, b\, \#^\omega \mid \forall (a, b) \text{ admissible pattern of size two}\}$.

It is clear that any encoding of a word of $a \in L$ satisfy that every line is in X and every column is in Z. It is also clear that any configuration of the combiner is either composed only of $-$, only of $\#$, or is an encoding of a word of L.

The last point is to note that X and Z may not be fully compatible even though L is non-empty (that is some case of consecutive letters could be allowed but never occur). If they are not compatible, there is still two fully compatible subshifts $X' \subset X$ and $Z' \subset Z$ such that their combiner is exactly the combiner of X and Z. The key point is that since both X and Z are finite up to shift, there is only a finite number of possibilities for X' and Z' and if ever L is not empty, X' and Z' are distinct from ${}^\omega\!-^\omega \cup {}^\omega\!\#^\omega$

Thus, the algorithm for deciding if L is empty is the following:

- Construct X and Z
- For any $S_X \subsetneq \Sigma$ and any $S_Y \subsetneq \Sigma' \times \Sigma$, take X' be X with S_X as additional forbidden patterns and Z' be Z with S_Z as additional forbidden patterns, test whether X' and Z' are fully compatible.
- If there at least one pair verifying this, L is not empty.

The real 3D case is very similar to the previous one except we add one additional subshift Y. Let us take a 2D-nearest-neighbor SFT S and a configuration $a_{(i,j)} \in \Sigma^{\mathbb{Z}}$. The encoding is done by putting: $a_{(i,j)}$ at position $(i,j,i+j)$, $a'_{(i,j)}$ at position $(i+1,j,i+j)$, $a''_{(i,j)}$ at position $(i,j+1,i+j)$, and as previously, preceded by $^{\omega}-$ and followed by $+^{\omega}$. It can be noted that this encoding places two symbols $a'_{(i-1,j)}$ and $a''_{(i,j-1)}$ at position $(i,j,i+j-1)$.

We construct similarly over the set $\Sigma \cup (\Sigma' \times \Sigma'') \cup \{\#, -\}$ the 1D-SFT:

$$X = \{^{\omega}- \; a \; (a',b'') \; \#^{\omega} \mid a,b \in \Sigma\}$$
$$Y = \{^{\omega}- \; b \; (a',b'') \; \#^{\omega} \mid a,b \in \Sigma\}$$
$$Z = \{^{\omega}- \; (a',b'') \; c \; \#^{\omega} \mid (a,c) \text{ admissible horizontal pattern}, (b,c) \text{ vertical one }\}$$

The properties and algorithm are the same as in the 1D case but here, since tiling by 2D-nearest-neighbor SFT is undecidable, it implies that checking full compatibility of three 1D-nearest-neighbor SFT X, Y and Z is undecidable.

References

1. Aubrun, N., Esnay, S., Sablik, M.: Domino problem under horizontal constraints. In: Paul, C., Bläser, M. (eds.) STACS 2020. vol. 154, pp. 26:1–26:15. Schloss Dagstuhl – Leibniz-Zentrum für Informatik, Dagstuhl (2020). https://doi.org/10.4230/LIPIcs.STACS.2020.26
2. Aubrun, N., Sablik, M.: Simulation of effective subshifts by two-dimensional subshifts of finite type. Acta Appl. Math. **126**(1), 35–63 (2013). https://doi.org/10.1007/s10440-013-9808-5
3. Berger, R.: The undecidability of the domino problem. Ph.D. thesis, Harvard University (1964)
4. Culik, K., Kari, J.: An aperiodic set of wang cubes. In: Puech, C., Reischuk, R. (eds.) STACS 1996. LNCS, vol. 1046, pp. 137–146. Springer, Heidelberg (1996). https://doi.org/10.1007/3-540-60922-9_12
5. Durand, B., Levin, L.A., Shen, A.: Complex tilings. In: STOC, pp. 732–739 (2001)
6. Esnay, S.: Limitation de la complexité de certains invariants des sous-décalages par contraintes dynamiques et structurelles. Ph.D. thesis, Université Toulouse III, Paul Sabatier (2022)
7. Jeandel, E., Rao, M.: An aperiodic set of 11 wang tiles. In: Advances in Combinatorics (2021). https://doi.org/10.19086/aic.18614
8. Lind, D., Marcus, B.H.: An Introduction to Symbolic Dynamics and Coding, 2nd edn. Cambridge University Press, Cambridge (2021)
9. Penrose, R.: The role of aesthetics in pure and applied mathematical research. Bull. Inst. Math. Appl. **10**(2), 266–271 (1974)
10. Robinson, R.: Undecidability and nonperiodicity for tilings of the plane. Inventiones Mathematicae **12** (1971)
11. Rozenberg, G., Salomaa, A. (eds.): Handbook of Formal Languages, Volume 3: Beyond Words. Springer, Heidelberg (1997). https://doi.org/10.1007/978-3-642-59126-6
12. Smith, D., Myers, J.S., Kaplan, C.S., Goodman-Strauss, C.: An aperiodic monotile. CoRR arxiv:2303.10798 (2023). https://doi.org/10.48550/ARXIV.2303.10798
13. Wang, H.: Proving theorems by pattern recognition - ii. Bell Syst. Techn. J. **40**(1), 1–41 (1961). https://doi.org/10.1002/j.1538-7305.1961.tb03975.x

Asymptotically Optimal Representation
of Palindromic Structure

Michael Itzhaki[(✉)]

Department of Computer Science, Bar-Ilan Universicty, Ramat Gan, Israel
michaelitzhaki@gmail.com

Abstract. We introduce an asymptotically optimal representation of the Manacher array of a string that supports constant-time access. The approach relies on the combinatorial properties of palindromes, yielding a compact yet efficient structure. This work fits within the broader study of compressed text indexing and highlights structural aspects of palindromic substrings that may inspire further algorithmic applications.

1 Introduction

Palindromes are sequences of characters that read the same forward and backward and have fascinated mathematicians and linguists for centuries due to their unique properties. Since the inception of modern stringology, identifying palindromes in a string has been an active field of research [7,22,28]. Palindromes represent a simple yet profound form of symmetry found not only in language but also in number theory [4] and DNA sequences [36], and have applications ranging from error detection in coding theory [26] to the study of self-replicating structures in bioinformatics [6,10,27].

The *Manacher array* is a fundamental structure for palindrome processing. For each center c (integer or half-integer), the Manacher array stores the radius of the longest palindrome mirrored around c. The array can be computed in linear time [7,28] and has become pervasive in string algorithms [2,19,23,34].

Storing the Manacher array naively requires $\mathcal{O}(n \log n)$ bits, even though $\Theta(n)$ bits are necessary (folklore; for example, see [16]). This raises a natural and previously unresolved question: *Can the Manacher array be encoded in $\mathcal{O}(n)$ bits while still supporting efficient access?*

Classical compression algorithms, such as Huffman coding [12], the Lempel-Ziv family [24,25,30,37], and succinct data structures like bitvectors [3,18], wavelet trees [11,32], and FIDs [33], provide a general framework for compact sequence representation. However, these do not exploit the specific combinatorial structure of palindromes.

In this work, we study the compression of the Manacher array in the setting of compressed pattern matching, a line of research that has received significant recent attention [8,9,20,21]. Whereas prior work has focused on compressing

A preliminary poster version of this work appeared at DCC 2025 [17].

© The Author(s), under exclusive license to Springer Nature Switzerland AG 2026

J. Kozik and A. Wolff (Eds.): SOFSEM 2026, LNCS 16448, pp. 128–143, 2026.
https://doi.org/10.1007/978-3-032-17801-5_10

the strings or patterns themselves, we instead target a derived combinatorial structure, aiming to compress it without sacrificing access efficiency. A very recent preprint [29] studies a related problem and obtains results of a similar flavor.

Our Contributions. This paper studies the compression of the Manacher array of a string; we show a compact representation of the Manacher array that supports constant-time access. Our contribution is summarized in the following theorem:

Theorem 1. *Let S be a string of length n, and let A denote its Manacher array. The array A admits a lossless encoding of size $\mathcal{O}(n)$ bits that supports element access in constant time.*

This paper advances the well-established line of research on the combinatorial properties of palindromes, providing refined structural insights and extending the current understanding of their behavior [1,13,16,23,31,35].

Due to space constraints, several technical proofs and additional examples are only provided in the full version of this work [15].

2   Preliminaries

Computation Model. We assume the Word RAM model with a word length of size $\Omega(\log n)$. All indices in this paper are 1-based, and logarithms are taken to base two. In this model, logarithms can be computed in constant time.

Basic Definitions. A *string* is an ordered sequence of characters. We denote the length of S as $|S|$, its i-th character by $S[i]$ and its factor $S[i]S[i+1]\ldots S[j]$ by $S[i\mathbin{..}j]$. The empty string of length 0 is denoted by ε. When $j < i$, we denote $S[i\mathbin{..}j]$ as the empty string ε. A factor $S[1\mathbin{..}i]$ is called a *prefix*, and a factor $S[i\mathbin{..}|S|]$ is called a *suffix*. If $1 \leq i < |S|$ (resp. $|S| \geq j > 1$) the prefix (resp. suffix) is said to be *proper*. The *concatenation* of strings S and T is $S\,T: \ S[1]S[2]\ldots S[|S|]T[1]T[2]\ldots T[|T|]$. When clear from context, we simply write ST. For a positive integer i, we denote S^i as the concatenation of S to itself i times, i.e., $S^1 = S$, and $S^i = S^{i-1} \cdot S$. To refer to a specific occurrence of a factor in a string, we use the interval notation $[i\mathbin{..}j] = \{t \mid t \in \mathbb{N}, \ i \leq t \leq j\}$. An interval $[i\mathbin{..}t]$ is called a *prefix* of $[i\mathbin{..}j]$ if $t \leq j$, and a suffix is defined symmetrically. The center of an interval $\mathcal{P} = [i\mathbin{..}j]$ is denoted as $c_{\mathcal{P}} :- \frac{i+j}{2}$.

Periods and Repetitions. A string S has *period* p if p is a positive integer and $S[i] = S[i+p]$ for every $1 \leq i \leq |S| - p$. The *period* of S is the smallest such value p, and is denoted by $\mathrm{period}(S)$. A string S is said to be *periodic* if $\mathrm{period}(S) \leq |S|/2$. A *repetition* in S is an interval $[i\mathbin{..}j] \subseteq [1\mathbin{..}|S|]$ for which the corresponding factor $S[i\mathbin{..}j]$ is periodic. If $p = \mathrm{period}(S)$, we refer to the factor $S[1\mathbin{..}p]$ as the *root* of S. The root of a repetition is defined as the root of its corresponding factor. A *run* (or *maximal repetition*) in a string is a repetition that cannot be extended to the left or to the right with the same period. For instance, consider $S = \texttt{cababab}$. The repetition $[3 \mathinner{.\,.} 7]$ corresponds to the factor $\texttt{babab}$ with period 2. However, it is not a run, as it can be extended by one character to the left, resulting in the repetition $[2 \mathinner{.\,.} 7]$ which has the same period. In contrast, the repetition $[2 \mathinner{.\,.} 7]$ is a run. Note that any repetition in a string can be extended (possibly by zero characters) to a unique run of the same period. We refer to the result of the extension as the *periodic extension* of the repetition.

Palindromes. A string S is called a *palindrome* or *palindromic* if S equals its reversal, i.e., $S[i] = S[|S| - i + 1]$ for every $1 \leq i \leq |S|$. For example, the strings $\texttt{abcba}$ and $\texttt{abccba}$ are palindromes, while $\texttt{abcbaa}$ is not. An interval $\mathcal{P} = [i \mathinner{.\,.} j]$ is called a *subpalindrome* of S if its corresponding factor $S[i \mathinner{.\,.} j]$ is palindromic. We say that $\mathcal{P}$ is *centered* at $c_{\mathcal{P}}$, and that its *radius* is $\lceil \frac{j-i}{2} \rceil$. A palindromic prefix (resp. suffix) is the subpalindrome $[1 \mathinner{.\,.} i]$ (resp. $[i \mathinner{.\,.} |S|]$). The interval $\mathcal{P}$ is called a *maximal subpalindrome* if $\mathcal{P}$ is a subpalindrome and $S[i - 1 \mathinner{.\,.} j + 1]$ is either not defined or not a palindrome. We regard the empty prefix $S[1 \mathinner{.\,.} 0]$ and empty suffix $S[|S| + 1 \mathinner{.\,.} |S|]$ as subpalindromes of radius 0 centered at 0.5 and $|S| + 0.5$, respectively. For every center $c \in \{1, 1.5, \ldots, |S|\}$ there is a corresponding unique maximal subpalindrome $\mathcal{P}$ with $c_{\mathcal{P}} = c$. The Manacher array records for every center the radius of its corresponding maximal subpalindrome.

Definition 1 (Manacher array). *Let S be a string of length n. The* Manacher *array of S is the array $\mathsf{A}[1 \mathinner{.\,.} 2n + 1]$ where each position i corresponds to a center of S, and is defined as follows:*

- *If i is even, $i = 2k$, then $\mathsf{A}[i]$ is the maximum integer $r \geq 0$ such that $S[k - r \mathinner{.\,.} k + r]$ is a palindrome, and $1 \leq k - r \leq k + r \leq n$.*
- *If i is odd, $i = 2k + 1$, then $\mathsf{A}[i]$ is the maximum integer $r \geq 0$ such that $S[k - r + 1 \mathinner{.\,.} k + r]$ is a palindrome, and $1 \leq k - r + 1$, $k + r \leq n$.*

The following lemma correlates periodicity with palindromes:

Lemma 1 (Folklore, [14]). *Let P be a palindrome. If the palindrome P has a proper palindromic suffix of length p then $|P| - p$ is a period of P.*

Additional Notation. When S is clear from context, we denote the maximal subpalindrome of S at center c as $\mathcal{P}_c$. When $\mathcal{P}_c$ is periodic, we denote by $\mathcal{R}_c$ its periodic extension and by r_c the root of $\mathcal{R}_c$. Notice that $\mathcal{P}_c$ and $\mathcal{R}_c$ are intervals, while r_c is a string. Additionally, note that the root of the factor corresponding to $\mathcal{P}_c$ is not necessarily r_c.

Bitvector Operations. We use the standard *rank* and *select* operations on bitvectors. Given $v \in \{0, 1\}^n$, $\mathrm{rank}_1(v, i)$ returns the number of 1-bits in $v[1 \mathinner{.\,.} i]$, and $\mathrm{select}_1(v, i)$ returns the position of the i-th 1 in v. The bitvector can be preprocessed into a structure of size $n + o(n)$ bits that supports both operations in constant time [3,18].

Standard bitwise shift operations scale indices by powers of 2. To adapt these to 1-based indexing, for an index i and an integer $m \geq 0$, we define the 1-based right shift as $\mathrm{shr}_m(i) := ((i - 1) \gg m) + 1$, and the 1-based left shift as $\mathrm{shl}_m(i) := ((i - 1) \ll m) + 1$. Recall that $i \ll m = i \cdot 2^m$ and $i \gg m = \lfloor i \cdot 2^{-m} \rfloor$.

SDC and Smooth Arrays. The *Simple Dense Coding* (SDC) representation [5] binary-encodes all elements in an array A of non-negative integers, concatenates the codes, and marks the beginning of each codeword in an auxiliary bitvector. Supporting *select* on the auxiliary bitvector yields constant-time access to any element, while requiring $\mathcal{O}\left(\sum_{i=1}^{n} \log(A[i] + 1)\right)$ bits of space. We call the latter expression the *log-sum* of A. In particular, an array of non-negative integers that admits a linear log-sum can be SDC-encoded using $\mathcal{O}(n)$.

Remark 1. In general, Manacher arrays do not admit a linear log-sum.

For instance, consider the string $S = a^n$. The log-sum of its Manacher array is $\mathcal{O}(n \log n)$, hence the SDC-encoding of a Manacher array requires $\omega(n)$ bits of space in the worst case.

Our ultimate goal is to encode a Manacher array with $2n + 1$ elements using $\mathcal{O}(n)$ bits while still supporting constant-time access. To do so, we construct arrays with linear log-sum and encode them using SDC. To show that an array has linear log-sum we impose a stronger condition which is more natural for Manacher arrays and their derivatives, and better captures their structure. We proceed to define the condition:

Definition 2 (Smooth array). *Let A be an array with ℓ integral elements from the range $[0 \mathinner{.\,.} \ell - 1]$. We say that the array is* smooth *if it satisfies that*

$$\forall i < j \quad \min(A[i], A[j]) \leq j - i.$$

Observation 2. *Any smooth array admits a linear log-sum.*

This observation follows from the definition of smooth arrays, which requires values greater than 2^m to be stored at least 2^m indices apart. Consequently, there are at most $\ell \cdot 2^{-m}$ such values, and the log-sum satisfies:

$$\sum_{i=1}^{\ell} \log(A[i] + 1) \leq \ell \sum_{m=1}^{\lceil \log \ell \rceil} m \cdot 2^{-m} = \mathcal{O}(\ell).$$

Corollary 1. *It is possible to SDC-encode any constant number of smooth arrays, each with ℓ elements, using $\mathcal{O}(\ell)$ bits in total.*

3   Framework and Further Definitions

In this section, we introduce the terminology and notation for periodic palindromes, palindromic runs, and their centers. These definitions will be used later in the construction of the compressed Manacher array.

A *periodic palindrome* is a string that is both periodic and palindromic. Such objects play a central role in our construction. We begin by citing a useful structural property:

Lemma 3 (Canonical decomposition of periodic palindromes [23]**).** *Any periodic palindrome P admits a unique decomposition $(q_0 q_1)^k q_0$, where:*

1. *q_0 and q_1 are palindromes,*
2. *$r = q_0 q_1$ is the root of P,*
3. *$P \bmod |r| = |q_0|$,*
4. *$k \geq 2$.*

In particular, condition (3) implies that $|q_1| \geq 1$.
We call this decomposition the canonical decomposition *of P, and $q_0 q_1$ is the* canonical decomposition *of r.*

Although well defined, periodic palindromes pose challenges for our objective. To simplify our framework, we make use of a different, more robust notion: *palindromic run*, which is the periodic extension of a maximal periodic subpalindrome.

Definition 3 (Palindromic Run). *Let S be a string, and let $\mathcal{R}$ be a run of S with root r. The run $\mathcal{R}$ is said to be a* palindromic run *if $r = q_0 q_1$ for some palindromes q_0 and q_1.*

Lemma 1 in [23] guarantees that the decomposition $r = q_0 q_1$ with $|q_1| > 0$ is unique when it exists. We say that two palindromic runs are the same (denoted by $\mathcal{R} = \mathcal{R}'$) if their underlying intervals are equal.

In this work, we encode palindromic runs as *palindromic run descriptor* (PRD).

Definition 4 (Palindromic run descriptor (PRD)). *Let $\mathcal{R} = [i \mathinner{.\,.} j]$ be a palindromic run of S with root $r = q_0 q_1$. The* palindromic run descriptor *(PRD) of $\mathcal{R}$ is the tuple $(i, |q_0|, |q_1|, k, e)$, where:*

- *i is the starting index of the interval $\mathcal{R}$.*
- *q_0, q_1 are the palindromic factors from the decomposition of r (Lemma 3).*
- *k, e are the unique integers satisfying $S[i \mathinner{.\,.} j] = r^k r[1 \mathinner{.\,.} e]$, and $0 \leq e < |r|$.*

The value $\exp(\mathcal{R}) :- k + \frac{e}{|r|}$ is commonly referred to as the *exponent* of $\mathcal{R}$.

Next, we categorize the centers in a given string. Recall from the preliminaries that the maximal subpalindrome centered at c is $\mathcal{P}_c$, the palindromic extension of $\mathcal{P}_c$ (when exists) is $\mathcal{R}_c$, and the root of $\mathcal{R}_c$ is r_c. First, we define a *periodic center* as a center whose associated maximal subpalindrome $\mathcal{P}_c$ is a repetition.

Among all centers inside a palindromic run, those aligned with occurrences of the palindromic factors of the root (i.e., q_0, q_1) are particularly interesting as they satisfy several useful combinatorial properties. Let $\mathcal{R} = [i \mathinner{.\,.} j]$ be a palindromic run with root $r = q_0 q_1$. If a center c coincides with the center of an occurrence of q_0 or q_1, we say that c is a *repeat center of $\mathcal{R}$*. Set $c_1 :- i + \frac{|q_0| - 1}{2}$ to be the first repeat center of $\mathcal{R}$. Formally, the set of repeat centers of $\mathcal{R}$ is

$$C_{\mathrm{rep}}(\mathcal{R}) = \left\{ c_1 + t \cdot \frac{|r|}{2} \mid t \in \mathbb{Z}_{\geq 0}, \; i - \frac{1}{2} \leq c_1 + t \cdot \frac{|r|}{2} \leq j + \frac{1}{2} \right\}.$$

More generally, we say that a center c is an *internal center of* $\mathcal{R}$ if it coincides with the center of an occurrence of q_0 or q_1 *other than* the first or last occurrences of each. The number of internal centers of $\mathcal{R}$ is exactly $|C_{\mathrm{rep}}(\mathcal{R})| - 4$, and since for every palindromic run $|C_{\mathrm{rep}}(\mathcal{R})| \geq 5$, at least one internal center exists. Notice that for every internal center c of $\mathcal{R}$, the length of $\mathcal{P}_c$ satisfies $|\mathcal{P}_c| \geq 2|r|$. Lastly, we observe that if c is a periodic center, then it is an internal center of $\mathcal{R}_c$.

Example 1. Consider the string $S = \mathtt{baaab}$, with the palindromic run $\mathcal{R} = [2 .. 4]$. The repeat centers of $\mathcal{R}$ are $\{1.5, 2, \ldots, 4, 4.5\}$, for a total of seven centers. Of them, $\{2.5, 3, 3.5\}$ are the internal centers of $\mathcal{R}$. Although $c = 3$ is an internal center of $\mathcal{R}$, it is not a periodic center as the maximal subpalindrome $\mathcal{P}_3 = [1 .. 5]$ is not a repetition. Therefore, the periodic centers of S are $\{2.5, 3.5\}$.

The last definition in this section is the center-period array, an array that stores for every periodic center its associated period.

Definition 5 (Center-period array). *Let S be a string. The* center-period *array* L *assigns to (i) each periodic center c the period of its associated palindromic run, i.e. $|r_c|$, and (ii) zero to all other centers. Formally, the center-period array* L *is defined as:*

$$\mathsf{L}[2c] = \begin{cases} 0 & \text{if } c \text{ is not a periodic center,} \\ |r_c| & \text{otherwise.} \end{cases}$$

The following example summarizes all the definitions of this section.

Example 2. We demonstrate the main definitions in this section: palindromic run, palindromic run descriptor, periodic centers and internal centers (of a run), and the center-period array. Let S be a string of length 29:

$$S = \mathtt{xxababababacddcababacddcabababy}.$$

There are five palindromic runs in S:

$$\mathcal{R} = [5 .. 27], \; \mathcal{R}^{(1)} = [3 .. 9], \; \mathcal{R}^{(2)} = [14 .. 18], \; \mathcal{R}^{(3)} = [23 .. 28], \; \mathcal{R}^{(4)} = [1 .. 2].$$

The runs $\mathcal{R}$, $\mathcal{R}^{(2)}$ and $\mathcal{R}^{(4)}$ all have five repeat centers, hence one internal center. These centers are 16, 16 and 1.5, respectively. The internal centers of $\mathcal{R}^{(1)}$ are $\{5, 6, 7\}$ and the internal centers of $\mathcal{R}^{(3)}$ are $\{25, 26\}$. The runs and their respective roots, repeat centers and internal centers are drawn in Fig. 1.

Although 16 is an internal center of both $\mathcal{R}^{(2)}$ and $\mathcal{R}$, it is not a periodic center, as $\mathcal{P}_{16} = [4 .. 28]$ is not a repetition. The other internal centers are all periodic. An illustration of the center-period array of S together with the Manacher array can be found in Fig. 2.

We conclude by showing the PRD of the palindromic runs:

$$\mathtt{PRD} = (i, |q_0|, |q_1|, k, e) \quad \mathtt{PRD}(\mathcal{R}) = (5, 5, 4, 2, 5) \quad \mathtt{PRD}(\mathcal{R}^{(4)}) = (1, 0, 1, 2, 0)$$

$$\mathtt{PRD}(\mathcal{R}^{(1)}) = (3, 1, 1, 3, 1) \quad \mathtt{PRD}(\mathcal{R}^{(2)}) = (14, 1, 1, 2, 1) \quad \mathtt{PRD}(\mathcal{R}^{(3)}) = (23, 1, 1, 3, 0)$$

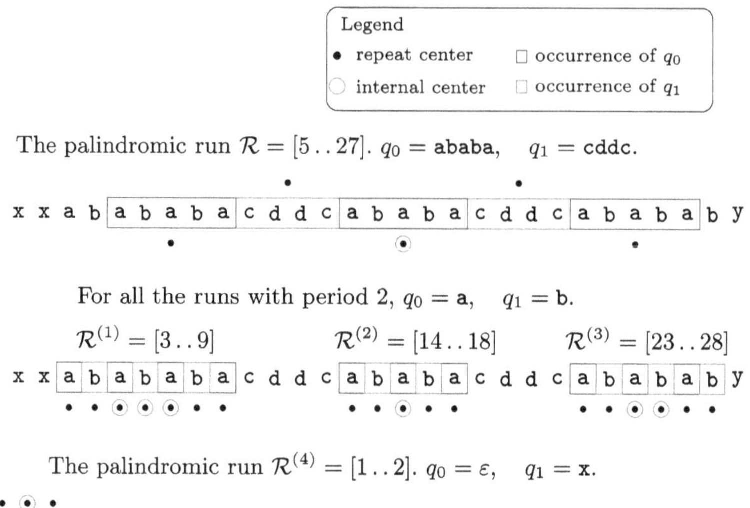

The palindromic run $\mathcal{R} = [5 \mathinner{.\,.} 27]$. $q_0 = \mathtt{ababa}$, $q_1 = \mathtt{cddc}$.

For all the runs with period 2, $q_0 = \mathtt{a}$, $q_1 = \mathtt{b}$.

$$\mathcal{R}^{(1)} = [3 \mathinner{.\,.} 9] \qquad \mathcal{R}^{(2)} = [14 \mathinner{.\,.} 18] \qquad \mathcal{R}^{(3)} = [23 \mathinner{.\,.} 28]$$

The palindromic run $\mathcal{R}^{(4)} = [1 \mathinner{.\,.} 2]$. $q_0 = \varepsilon$, $q_1 = \mathtt{x}$.

Fig. 1. Visual demonstration of repeat centers and internal centers. At the top, the palindromic run with period nine is described, below the three runs with period two, and lastly the run with period one.

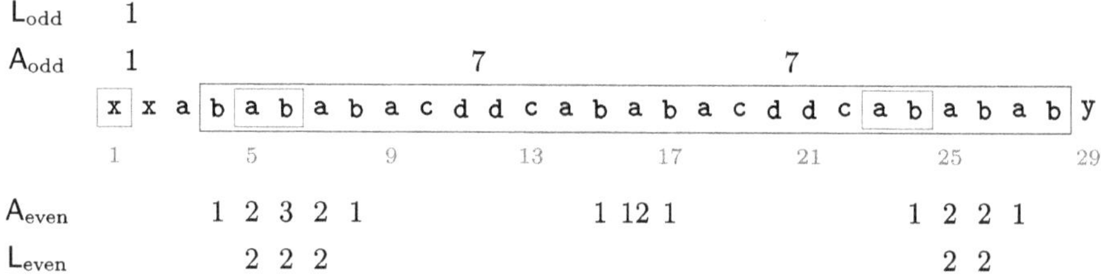

Fig. 2. A visualization of the center-period array L and the Manacher array A. For each periodic center, the root r_c is highlighted in red. The maximal palindrome is highlighted in blue. For clarity, each array is split into two subarrays – the values in odd and in even indices. (Color figure online)

4   Properties of Periodic Palindromes

Using the terminology introduced in the previous section, we develop the structural properties of palindromic runs and periodic palindromes. The proofs in this sections are sketched, and are formally provided in the full version [15]. Our starting point is the following lemma:

Lemma 4 (Lemma 3 in [23]). *Let P be a periodic palindrome and suppose $(q_0 q_1)^k q_0$ is the canonical decomposition of P. If $\mathcal{U} = [i \mathinner{.\,.} j]$ is a subpalindrome of P such that $|\mathcal{U}| \geq |q_0 q_1| - 1$, then the center of $\mathcal{U}$ coincides with the center of some q_0 or q_1 from the decomposition.*

Next, we prove a vital structural property: a maximal periodic subpalindrome (i.e., a maximal subpalindrome that is periodic) is a prefix or a suffix of its periodic extension.

Lemma 5. *Let S be a string, $\mathcal{P} = [i \mathinner{..} j]$ be a maximal periodic subpalindrome, and $\mathcal{R} = [i' \mathinner{..} j']$ be the periodic extension of $\mathcal{P}$. Then $i = i'$ or $j = j'$.*

This lemma is proved by establishing that if $\mathcal{P}$ does not border with $\mathcal{R}$, the properties of a maximal subpalindrome force a mismatch between two characters, while the properties of a repetition force these characters to be equal, hence resulting in a contradiction. A corollary of this lemma is that given a periodic center c, the radius of $\mathcal{P}_c$ is the distance to the closer border of $\mathcal{R}_c$.

We prove another key structural fact, showing that if a palindrome has a sufficiently long periodic factor, then the palindrome itself is periodic.

Lemma 6. *Let P be a palindrome, and $\mathcal{T} = [2 \mathinner{..} |P|]$ be a repetition with root r. The period of P is $|r|$.*

We prove this lemma by showing that $S[1] = r[|r|]$. We split into two cases: where $|\mathcal{T}| = 2|r|$, and where $|\mathcal{T}| > 2|r|$. The first case is proven directly by the palindromic property, and the second case is proven using the properties of the periodic palindrome $S[2 \mathinner{..} |P| - 1]$.

We conclude the section by proving that every internal center c of a palindromic run $\mathcal{R}$ yields $\mathcal{R}_c = \mathcal{R}$, with the sole exception of $c = c_{\mathcal{R}}$.

Lemma 7. *Let S be a string, $\mathcal{R}$ be a palindromic run, and c be an internal center of $\mathcal{R}$. At least one of the following is satisfied:*

  i The maximal subpalindrome $\mathcal{P}_c$ is a prefix or a suffix of $\mathcal{R}$, or
  ii $\mathcal{P}_c$ and $\mathcal{R}$ share the same center (i.e., $c_{\mathcal{P}_c} = c_{\mathcal{R}}$).

The idea of the proof is to invoke Lemma 6 on a subpalindrome of $\mathcal{P}_c$. By definition, if c is an internal center, the maximal subpalindrome $\mathcal{P}_c$ is either a prefix, a suffix, or not contained in $\mathcal{R}$. The first two options satisfy the lemma's conditions, whereas the last contradicts Lemma 6 when considering the subpalindrome centered at c that extends the interval $\mathcal{R}$ by exactly one character. The latter construction is possible for every internal center $c \neq c_{\mathcal{R}}$.

Corollary 2. *Let c_1, c_2 be two internal centers of a palindromic run $\mathcal{R}$. At least one of the centers is periodic, and its palindromic run is $\mathcal{R}$, i.e., $\mathcal{R}_c = \mathcal{R}$.*

5   Compact Manacher Array

This section introduces a compressed data structure for the Manacher array, ultimately proving Theorem 1. Throughout the section, we denote by S the input string of length n, by A its Manacher array, and by L its center-period array.

Remark 2. For clarity, we assume throughout this section that S has no even-length subpalindromes. This simplifying assumption *does not affect the generality of the result*, but ensures consistent and clearer notation. Specifically, a subpalindrome $\mathcal{P}$ of S with radius $\mathrm{rad}(\mathcal{P})$ and center $c_\mathcal{P}$ corresponds to the factor $S[c_\mathcal{P} - \mathrm{rad}(S) \mathinner{..} c_\mathcal{P} + \mathrm{rad}(S)]$, and $c_\mathcal{P}$ is guaranteed to be an integer.

5.1 High-Level Idea

The idea underlying the data structure is to store information about periodic and non-periodic centers in separate, smooth arrays, and encode those arrays using SDC, supporting the required time and space bounds.

1. Initially, we show that setting the radii at periodic centers in A to zero guarantees that the resulting array $\tilde{\mathsf{A}}$ is smooth (Lemma 8).
2. We proceed by showing that the center-period array L is smooth (Lemma 9).
3. We then describe a data structure $\mathbf{I}$, that given an arbitrary periodic center c locates the first internal center of $\mathcal{R}_c$ in constant time (Lemma 10).
4. Lastly, we show that all of these structures suffice to compute the radius of the maximal subpalindrome at any given center in constant time (Lemma 11).

5.2 Description of the Structures

The first data structure employed is the *sparse Manacher array*.

Definition 6. (Sparse Manacher array). *Given a string S and its Manacher array* A*, the sparse Manacher array* $\tilde{\mathsf{A}}$ *is defined by:*

$$\tilde{\mathsf{A}}[2c] = \begin{cases} 0 & \textit{if } c \textit{ is a periodic center,} \\ \mathsf{A}[2c] & \textit{otherwise.} \end{cases}$$

Lemma 8. *The sparse Manacher array of S is smooth.*

Proof. We prove the claim by contradiction. Let $c_1 < c_2$ be two centers satisfying:

$$\tilde{\mathsf{A}}[2c_1] \leq \tilde{\mathsf{A}}[2c_2] \quad \text{and} \quad \tilde{\mathsf{A}}[2c_1] > (2c_2 - 2c_1).$$

Denote $d := c_2 - c_1$. Substituting d into the expression, the condition $\tilde{\mathsf{A}}[2c_1] > 2d$ implies a significant overlap between the maximal subpalindromes at centers c_1, c_2, i.e., $\mathcal{P}_{c_1}, \mathcal{P}_{c_2}$. More precisely:

  i $[c_1 - 2d \mathinner{..} c_1 + 2d]$ is a subpalindrome centered at c_1, and
  ii $[c_2 - d \mathinner{..} c_2 + d]$ is its palindromic suffix centered at c_2.

Consequently (Lemma 1), this overlap implies that the factor $S[c_1 - 2d \mathinner{..} c_1 + 2d]$ is periodic with period $(4d + 1) - (2d + 1) = 2d$.

By invoking Lemma 4, both centers c_1, c_2 are internal centers of some palindromic run. However, by Corollary 2, at least one of them is a periodic center and would have been set to zero, contradicting the definition of $\tilde{\mathsf{A}}$. $\square$

Next, we show how to compute values from the Manacher array at periodic centers. We begin by claiming that the center-period array is smooth.

Lemma 9. *The center-period array L of a string S is smooth.*

Proof (Sketch). The high-level idea supporting the proof is that internal centers of different palindromic runs that occur "too close" to one another force a common period for both runs.

The proof proceeds by contradiction. We assume two centers c_1, c_2 violate the smoothness condition. We analyze their palindromic runs $\mathcal{R}_{c_1}$ and $\mathcal{R}_{c_2}$, dividing the problem into three cases based on alignment and containment.

Case 1: $\mathcal{R}_{c_1} = \mathcal{R}_{c_2}$: Periodic centers of the same run inherently satisfy the smoothness condition.

Case 2 (Containment): If c_1 is an internal center of $\mathcal{R}_{c_2}$, it forces a synchronization of periods, and if c_1 is not an internal center of $\mathcal{R}_{c_2}$ then its period is sufficiently small and cannot violate the smoothness condition with c_2.

Case 3 (Partial or no overlap). We assume that $|r_{c_1}| \leq |r_{c_2}|$ and $c_1 < c_2$. Consider c_1, c_2 and their adjacent repeat centers along the runs,

$$c_1^{(+1)},\ c_1^{(+2)},\ c_2^{(-1)},\ c_2^{(-2)}.$$

Using the fact that c_1, c_2 form a counterexample and that the runs are not nested, it follows that all four of these centers lie inside both $\mathcal{R}_1$ and $\mathcal{R}_2$. In particular, there is a subpalindrome of length $|r_{c_1}|$ centered at $c_2^{(-1)}$, which together with Lemma 4 yields a subpalindrome of length $|r_{c_2}|$ centered at $c' :- c_2^{(-1)} + \frac{1}{2}|r_{c_1}|$. Thus $c_2^{(-1)}$ and c' are repeat centers shared by $\mathcal{R}_{c_1}$ and $\mathcal{R}_{c_2}$; in particular, both runs have the same period, and the runs synchronize, which was covered by Case 1. The above argument also rules out the disjoint configuration. The full derivation is provided in the full version.

So far, we have shown that the arrays storing (i) radii at non-periodic centers and (ii) periods at periodic centers are both smooth, and can be encoded using SDC, supporting constant-time access. The missing piece for the complete data structure is to compute the radius at a periodic center. The primary obstacle to doing so is locating the borders of the palindromic run of the given center. The following lemma resolves this structural challenge by introducing an auxiliary data structure $\mathbf{I}$ that identifies the first internal center of a run given any internal center of the same run. This constitutes the final non-trivial construction; the derivation of the radius from this center is mechanical and is detailed subsequently.

Lemma 10. *Given a string S of length n, there exists a data structure $\mathbf{I}$ occupying $\mathcal{O}(n)$ bits that supports constant-time queries for the first (lowest) internal center of $\mathcal{R}_c$ for any periodic center c.*

Proof. The proof presents the construction of $\mathbf{I}$ and the corresponding query procedure, followed by a correctness proof.

Construction. We construct a hierarchy of bitvectors $\mathbf{I}$, where each bitvector in the hierarchy is preprocessed for `rank/select` queries.

The first bitvector $\mathbf{I}[1]$ has $2n$ bits, the second bitvector $\mathbf{I}[2]$ has n bits, and each new level is half the size of the preceding; the total space is $\mathcal{O}(n)$ bits. Since the largest possible period is $n/2$, the number of layers is $\lceil \log(n/2 + 1) \rceil$.

The bitvector at layer m (i.e., $\mathbf{I}[m]$) stores information for periodic centers whose period falls within the interval $[2^{m-1}, 2^m)$. In the bitvectors, a set bit ('1') indicates the first internal center of a palindromic run, and the other bits are unset ('0'). Accessing the bit corresponding to center c at level m requires scaling. We denote the scaling operation as scl, defined as $\mathrm{scl}_m(c) :- \mathrm{shr}_{m-1}(2c)$.

Formally, given a palindromic run $\mathcal{R}$ with period p:

1. Determine the layer m that satisfies $2^{m-1} \le p < 2^m$.
2. Denote the set of internal centers of $\mathcal{R}$ as C_{internal}.
3. Set the first internal center of $\mathcal{R}$; $c_1 :- \min\{C_{\text{internal}}\}$.
4. Set $c_m^{(1)} \leftarrow \mathrm{scl}_m(c_1)$.
5. Assign $\mathbf{I}[m][c_m^{(1)}] \leftarrow 1$.

Given a hierarchy $\mathbf{I}$ where all bitvectors initially contain only zeroes, we perform the above procedure for every palindromic run in S. Then, we preprocess all bitvectors in the hierarchy to support `rank/select` queries in constant time. The resulting hierarchy $\mathbf{I}$ is our data structure.

Query. The query procedure computes the first internal center c_1 of $\mathcal{R}_c$ given a periodic center c. The procedure consists of five steps:

1. Retrieve the period of $\mathcal{R}_c$ from L; $p_c \leftarrow \mathsf{L}[2c]$.
2. Determine the appropriate layer m that satisfies $2^{m-1} \le p_c < 2^m$.
3. Compute the scaled index for center c; $c_m \leftarrow \mathrm{scl}_m(c)$.
4. Perform a predecessor query on $\mathbf{I}[m]$ at position c_m via `rank/select` queries to get an approximation of the required center:

$$\hat{c}_1 \leftarrow \mathrm{select}_1(\mathbf{I}[m], \ \mathrm{rank}_1(\mathbf{I}[m], \ c_m)).$$

5. Compute the resulting index c_1. The resulting index satisfies:

$$(i) \quad \mathrm{shl}_{m-1}(\hat{c}_1) \le 2c_1 < \mathrm{shl}_{m-1}(\hat{c}_1 + 1), \qquad (ii) \quad 2(c - c_1) = 0 \quad (\mathrm{mod} \ p_c).$$

Since $2^{m-1} \le p_c$, at most one possible center satisfies the above conditions, as two periodic centers are separated by $p_c \ge 2^{m-1}$ array positions. Therefore, given that $\hat{c}_1 = \mathrm{scl}_m(c_1)$, the correct result will always be returned.

Correctness. The correctness of the data structure relies on the smoothness of the center-period array L. We denote $\mathcal{R}^{(1)} :- \mathcal{R}_c$.

For the query to fail, the bitvector $\mathbf{I}[m]$ must identify an invalid internal center $c_1' > c_1$ as the first internal center of $\mathcal{R}^{(1)}$. Denote $\mathcal{R}^{(2)} :- \mathcal{R}_{c_1'}$. Since c_1' is indicated in level m, it follows that $|r^{(2)}| \in [2^{m-1}, 2^m)$.

The collision implies that the scaled indices satisfy $\mathrm{scl}_m(c_1) < \mathrm{scl}_m(c_1') \leq \mathrm{scl}_m(c)$. Because $2^{m-1} \leq |r^{(1)}| < 2^m$, a center $\hat{c}$ exists such that $2|\hat{c} - c_1'| < 2^{m-1}$. However, $\min(|r^{(1)}|, |r^{(2)}|) \geq 2^{m-1}$, which implies $\min(\mathsf{L}[2\hat{c}], \mathsf{L}[2c_1']) \geq 2^{m-1} > 2|\hat{c} - c_1'|$, violating the smoothness condition of L, hence contradicting Lemma 9. $\qquad\square$

We now have all the required pieces to compute the radius of the maximal subpalindrome at any given center. We formally present the final data structures:

Definition 7 (Enriched center-period array). *The* Enriched Center-Period Array $\tilde{\mathsf{L}}$ *is an extension to the center-period array. For each periodic center c with* PRD $= (i, |q_0|, |q_1|, k, e)$*, the array $\tilde{\mathsf{L}}$ stores the triplet $(|r_c|, e, |q_0|)$.*

Since $e, |q_0| < |r_c|$, these values can be stored as three different smooth arrays[1].

The final data structure is the boundary locator.

Definition 8 (Boundary locator). *The* boundary locator $\tilde{\mathbf{I}}$ *is a composite data structure consisting of the layered bitvector $\mathbf{I}$ (from Lemma 10) and a symmetric data structure $\mathbf{I}_{last}$, that given an arbitrary periodic center c locates the last (highest) internal center c_m of the run $\mathcal{R}_c$.*

The following lemma assembles the pieces and explains how to find the radius given a periodic center.

Lemma 11. *Let c be a periodic center of S. The radius of the maximal subpalindrome centered at c can be found in constant time, using $\mathcal{O}(n)$ bits of space.*

Proof. To compute the radius, we first reconstruct the PRD of the palindromic run $\mathcal{R}_c$, using $\tilde{\mathsf{L}}$ and $\tilde{\mathbf{I}}$.

Algorithm. The retrieval proceeds in three steps:

1. **Find Boundaries:** Query $\tilde{\mathbf{I}}$ with c to obtain c_1 and c_m.
2. **Retrieve Parameters:** Query $\tilde{\mathsf{L}}[2c]$ to retrieve the period $|r_c|$, the excess e, and $|q_0|$. From these, we compute the suffix length $|q_1| = |r_c| - |q_0|$.
3. **Compute** PRD**:** Using the retrieved boundaries and period length, we compute the number of repetitions k:

$$k = 1 + \left\lceil \frac{c_m - c_1}{|r_c|} \right\rceil + \begin{cases} 1 & \text{if } e < |r_c| - \frac{|q_1|}{2}, \\ 0 & \text{otherwise.} \end{cases}$$

The computation of k diverges by values of e, as the suffix $r[1\,..\,e]$ introduces two repeat centers to the run $\mathcal{R}$ when $e \geq |r_c| - \frac{1}{2}|q_1|$.

The starting index i of the run is computed relative to the first internal center:

$$i = c_1 - |r_c| - \frac{|q_0| - 1}{2}.$$

---

[1] Strictly speaking, storing the triplet for every periodic center is suboptimal but does not affect the asymptotic space bounds.

Radius Calculation. With the tuple $(i, |q_0|, |q_1|, k, e)$ fully determined, the boundaries of the palindromic run $\mathcal{R}_c$ are known. Thus, the radius is simply the distance from c to the nearest run boundary. (Recall Lemma 5).

Since both $\tilde{\mathsf{I}}$ and $\tilde{\mathsf{L}}$ consume $\mathcal{O}(n)$ bits and support constant time query, and all arithmetic operations are constant time, the lemma holds. $\square$

5.3   The Complete Data Structure

We can now combine the components described in the previous subsections to prove our main result. We restate the main theorem:

Theorem 1. *Let S be a string of length n, and let A denote its Manacher array. The array A admits a lossless encoding of size $\mathcal{O}(n)$ bits that supports element access in constant time.*

Proof. The data structure consists of three components:

1. $\tilde{\mathsf{A}}$: The sparse Manacher array, storing radii for all non-periodic centers, and zero for periodic centers (Definition 6).
2. $\tilde{\mathsf{L}}$: The enriched center-period array, storing run details $(|r|, |q_0|, e)$ for periodic centers (Definition 7).
3. $\tilde{\mathsf{I}}$: The boundary locator, computing the first/last internal center of a palindromic run given one of its internal centers (Definition 8).

As established in the respective lemmas, each component occupies $\mathcal{O}(n)$ bits. Thus, the total space usage is $\mathcal{O}(n)$ bits.

To retrieve the radius at an arbitrary center c, the query proceeds as follows: First, we check $\tilde{\mathsf{A}}[c]$. If the value is non-zero, we return it directly. Otherwise, c is a periodic center. We query $\tilde{\mathsf{L}}$ to retrieve the triplet $(|r_c|, |q_0|, e)$, and query $\tilde{\mathsf{I}}$ to locate the run boundaries c_1 and c_m. By Lemma 11, these parameters allow us to compute the radius of the maximal subpalindrome at c in constant time.
$\square$

6   Future Work

While our data structure achieves asymptotically optimal space and constant access time, several questions remain. First and foremost: a full and formal construction algorithm running in linear time is a natural next step, potentially using $\mathcal{O}(n)$ bits of working space. Another open problem is determining the exact entropy of the Manacher array. Finally, investigating the adequacy of our techniques for other combinatorial structures in string processing may further advance compressed indexing and retrieval.

References

1. Amir, A., Itzhaki, M.: Reconstructing general matching graphs. In: Inenaga, S., Puglisi, S.J. (eds.) 35th Annual Symposium on Combinatorial Pattern Matching (CPM 2024). Leibniz International Proceedings in Informatics (LIPIcs), vol. 296, pp. 2:1–2:15 (2024). https://doi.org/10.4230/LIPIcs.CPM.2024.2
2. Charalampopoulos, P., Pissis, S.P., Radoszewski, J.: Longest palindromic substring in sublinear time. In: Bannai, H., Holub, J. (eds.) 33rd Annual Symposium on Combinatorial Pattern Matching (CPM 2022). LIPIcs, vol. 223, pp. 20:1–20:9. Schloss Dagstuhl – Leibniz-Zentrum für Informatik (2022). https://doi.org/10.4230/LIPIcs.CPM.2022.20
3. Clark, D.: Compact PAT Trees. Ph.D. thesis, University of Waterloo, Waterloo, Canada (1996). https://dl.acm.org/doi/10.5555/287799
4. Dvořáková, L., Kruml, S., Ryzák, D.: Antipalindromic numbers. Acta Polytechnica 61(3), 428–434 (2021). https://doi.org/10.14311/AP.2021.61.0428
5. Fredriksson, K., Nikitin, F.: Simple compression code supporting random access and fast string matching. In: Demetrescu, C. (ed.) WEA 2007. LNCS, vol. 4525, pp. 203–216. Springer, Heidelberg (2007). https://doi.org/10.1007/978-3-540-72845-0_16
6. Fuglsang, A.: Distribution of potential type II restriction sites (palindromes) in prokaryotes. Biochem. Biophys. Res. Commun. 310(2), 280–285 (2003). https://doi.org/10.1016/j.bbrc.2003.09.014
7. Galil, Z., Seiferas, J.: Recognizing certain repetitions and reversals within strings. In: Proceedings of the 17th Annual Symposium on Foundations of Computer Science (FOCS 1976), pp. 236–252. IEEE Computer Society (1976). https://doi.org/10.1109/SFCS.1976.25
8. Ganardi, M., Gawrychowski, P.: Pattern matching on grammar-compressed strings in linear time. In: Proceedings of the 2022 Annual ACM-SIAM Symposium on Discrete Algorithms (SODA 2022), pp. 2833–2846. Society for Industrial and Applied Mathematics (2022). https://doi.org/10.1137/1.9781611977073.110
9. Gawrychowski, P., Gourdel, G., Starikovskaya, T., Steiner, T.A.: Compressed consecutive pattern matching. In: Proceedings of the 2024 Data Compression Conference (DCC 2024), pp. 163–172. IEEE (2024). https://doi.org/10.1109/DCC58796.2024.00024
10. Gelfand, M.S., Koonin, E.V.: Avoidance of palindromic words in bacterial and archaeal genomes: a close connection with restriction enzymes. Nucleic Acids Res. 25, 2430–2439 (1997). https://doi.org/10.1093/nar/25.12.2430
11. Grossi, R., Gupta, A., Vitter, J.S.: High-order entropy-compressed text indexes. In: Proceedings of the 14th Annual ACM-SIAM Symposium on Discrete Algorithms (SODA 2003), pp. 841–850. Society for Industrial and Applied Mathematics (2003). https://doi.org/10.5555/644108.644250
12. Huffman, D.A.: A method for the construction of minimum-redundancy codes. Proc. IRE 40, 1098–1101 (1952). https://doi.org/10.1109/JRPROC.1952.273898
13. I, T., Inenaga, S., Bannai, H., Takeda, M.: Counting and verifying maximal palindromes. In: Chavez, E., Lonardi, S. (eds.) String Processing and Information Retrieval. LNCS, vol. 6393, pp. 135–146. Springer, Heidelberg (2010). https://doi.org/10.1007/978-3-642-16321-0_13
14. I, T., Sugimoto, S., Inenaga, S., Bannai, H., Takeda, M.: Computing palindromic factorizations and palindromic covers on-line. In: Kulikov, A.S., Kuznetsov, S.O., Pevzner, P. (eds.) CPM 2014. LNCS, vol. 8486, pp. 150–161. Springer, Cham (2014). https://doi.org/10.1007/978-3-319-07566-2_16

15. Itzhaki, M.: Asymptotically optimal representation of palindromic structure (2025). https://arxiv.org/abs/2410.09984
16. Itzhaki, M.: Combinatorics of palindromes. In: Jeż, A., Otop, J. (eds.) Fundamentals of Computation Theory. LNCS, vol. 16106, pp. 252–266. Springer, Heidelberg (2025). https://doi.org/10.1007/978-3-032-04700-7_19
17. Itzhaki, M.: Palindromes compression and retrieval. In: 2025 Data Compression Conference (DCC), p. 375 (2025). https://doi.org/10.1109/DCC62719.2025.00062
18. Jacobson, G.: Space-efficient static trees and graphs. In: Proceedings of the 30th Annual Symposium on Foundations of Computer Science (FOCS 1989), pp. 549–554. IEEE Computer Society (1989). https://doi.org/10.1109/SFCS.1989.63533
19. Jeuring, J.: The derivation of on-line algorithms, with an application to finding palindromes. Algorithmica **11**(2), 146–184 (1994). https://doi.org/10.1007/BF01182773
20. Kempa, D., Kociumaka, T.: Breaking the $\mathcal{O}(n)$-barrier in the construction of compressed suffix arrays and suffix trees. In: Proceedings of the 2023 ACM-SIAM Symposium on Discrete Algorithms (SODA 2023), pp. 5122–5202. Society for Industrial and Applied Mathematics (2023). https://doi.org/10.1137/1.9781611977554.ch187
21. Kempa, D., Kociumaka, T.: Collapsing the hierarchy of compressed data structures: suffix arrays in optimal compressed space. In: Proceedings of the 64th Annual IEEE Symposium on Foundations of Computer Science (FOCS 2023), pp. 1877–1886. IEEE Computer Society (2023). https://doi.org/10.1109/FOCS57990.2023.00114
22. Knuth, D.E., Morris, J.H., Pratt, V.R.: Fast pattern matching in strings. SIAM J. Comput. **6**, 323–350 (1977). https://doi.org/10.1137/0206024
23. Kosolobov, D., Rubinchik, M., Shur, A.M.: Pal^k is linear recognizable online. In: Italiano, G.F., Margaria-Steffen, T., Pokorný, J., Quisquater, J.J., Wattenhofer, R. (eds.) SOFSEM 2015: Theory and Practice of Computer Science. LNCS, vol. 8939, pp. 289–301. Springer, Heidelberg (2015).https://doi.org/10.1007/978-3-662-46078-8_24
24. Lempel, A., Ziv, J.: On the complexity of finite sequences. IEEE Trans. Inf. Theory **22**, 75–81 (1976). https://doi.org/10.1109/TIT.1976.1055501
25. Lempel, A., Ziv, J.: Compression of two-dimensional images. In: Apostolico, A., Galil, Z. (eds.) Combinatorial Algorithms on Words, NATO ASI Series F, vol. 12, pp. 141–154. Springer, Heidelberg (1985). https://doi.org/10.1007/978-3-642-82456-2_10
26. Lenz, A., Wachter-Zeh, A., Yaakobi, E.: Duplication-correcting codes. Des. Codes Cryptogr. **87**(2), 277–298 (2019). https://doi.org/10.1007/s10623-018-0523-0
27. Lisnic, B., Svetec, I.K., Saric, H., Nikolic, I., Zgaga, Z.: Palindrome content of the yeast *Saccharomyces cerevisiae* genome. Curr. Genetics **47**, 289–297 (2005). https://doi.org/10.1007/s00294-005-0573-5
28. Manacher, G.: A new linear-time "on-line" algorithm for finding the smallest initial palindrome of a string. J. ACM **22**(3), 346–351 (1975). https://doi.org/10.1145/321892.321896
29. Mieno, T.: Compact representation of maximal palindromes (2025). https://arxiv.org/abs/2508.14384
30. Miller, V.S., Wegman, M.N.: Variations on a theme by Ziv and Lempel. In: Apostolico, A., Galil, Z. (eds.) Combinatorial Algorithms on Words, NATO ASI Series F, vol. 12, pp. 131–140. Springer, Heidelberg (1985).https://doi.org/10.1007/978-3-642-82456-2_9
31. Nagashita, S., I, T.: PalFM-index: FM-index for palindrome pattern matching. In: Bulteau, L., Lipták, Z. (eds.) 34th Annual Symposium on Combinatorial Pattern

Matching (CPM 2023). LIPIcs, vol. 259, pp. 23:1–23:15. Schloss Dagstuhl – Leibniz-Zentrum für Informatik (2023). https://doi.org/10.4230/LIPIcs.CPM.2023.23

32. Navarro, G.: Wavelet trees for all. In: Kärkkäinen, J., Stoye, J. (eds.) CPM 2012. LNCS, vol. 7354, pp. 2–26. Springer, Heidelberg (2012). https://doi.org/10.1007/978-3-642-31265-6_2

33. Raman, R., Raman, V., Satti, S.R.: Succinct indexable dictionaries with applications to encoding k-ary trees, prefix sums and multisets. ACM Trans. Algor. **3**(4), 43 (2007). https://doi.org/10.1145/1290672.1290680

34. Rubinchik, M., Shur, A.M.: EERTREE: an efficient data structure for processing palindromes in strings. Eur. J. Comb. **68**, 249–265 (2018). https://doi.org/10.1016/j.ejc.2017.07.021

35. Rubinchik, M., Shur, A.M.: Palindromic k-factorization in pure linear time. In: Esparza, J., Král', D. (eds.) 45th International Symposium on Mathematical Foundations of Computer Science (MFCS 2020). Leibniz International Proceedings in Informatics (LIPIcs), vol. 170, pp. 81:1–81:14. Schloss Dagstuhl – Leibniz-Zentrum für Informatik, Dagstuhl (2020). https://doi.org/10.4230/LIPIcs.MFCS.2020.81

36. Srivastava, S.K., Robins, H.S.: Palindromic nucleotide analysis in human T cell receptor rearrangements. PLoS ONE **7**(12), e52250 (2012). https://doi.org/10.1371/journal.pone.0052250

37. Ziv, J., Lempel, A.: A universal algorithm for sequential data compression. IEEE Trans. Inf. Theory **23**(3), 337–343 (1977). https://doi.org/10.1109/TIT.1977.1055714

On the Order-Diameter Ratio of Girth-Diameter Cages

Stijn Cambie[1] , Jan Goedgebeur[1,2] , Jorik Jooken[1] ,
and Tibo Van den Eede[1(✉)]

[1] Department of Computer Science, KU Leuven Campus Kulak-Kortrijk, Kortrijk 8500, Belgium
{stijn.cambie,jan.goedgebeur,jorik.jooken,tibo.vandeneede}@kuleuven.be
[2] Department of Mathematics, Computer Science and Statistics, Ghent University, Ghent 9000, Belgium

Abstract. For integers k, g, d, a $(k; g, d)$-cage (or simply girth-diameter cage) is a smallest k-regular graph of girth g and diameter d (if it exists). The order of a $(k; g, d)$-cage is denoted by $n(k; g, d)$. We determine asymptotic lower and upper bounds for the ratio between the order and the diameter of girth-diameter cages as the diameter goes to infinity. We also prove that this ratio can be computed in constant time for fixed k and g.

We theoretically determine the exact values $n(3; g, d)$, and count the number of corresponding girth-diameter cages, for $g \in \{4, 5\}$. Moreover, we design and implement an exhaustive graph generation algorithm and use it to determine the exact order of several open cases and obtain – often exhaustive – sets of the corresponding girth-diameter cages. The largest case we generated and settled with our algorithm is a $(3; 7, 35)$-cage of order 136.

Keywords: cage problem · degree diameter problem · extremal graphs · generation algorithm

1 Introduction

In this paper, we study girth-diameter cages, a concept recently introduced by Araujo-Pardo et al. [1], which is closely related to the Cage Problem and also has connections to the Degree Diameter Problem. These are two important problems in the area of extremal graph theory. We first introduce these two problems before discussing girth-diameter cages.

Given integers $k \geq 2$, $g \geq 3$, the *Cage Problem* asks for the minimum order of a k-regular graph of girth g (denoted by $n(k, g)$) and the corresponding graphs (called (k, g)-*cages*). This problem started receiving a lot of attention after Sachs [14] proved that k-regular graphs with girth g (called (k, g)-*graphs*) exist for every value of $k \geq 2$ and $g \geq 3$. The exact value of $n(k, g)$ has only been determined for a few infinite families and for some specific small pairs (k, g)

© The Author(s), under exclusive license to Springer Nature Switzerland AG 2026
J. Kozik and A. Wolff (Eds.): SOFSEM 2026, LNCS 16448, pp. 144–156, 2026.
https://doi.org/10.1007/978-3-032-17801-5_11

(mainly using clever algorithmic searches [6]), but remains wide open for many pairs (k, g). Apart from some exceptional cases, the best upper bounds on $n(k, g)$ are given by an explicit construction due to Lazebnik, Ustimenko and Woldar [9]. The state of the art as such tells us that $M(k, g) \leq n(k, g) \leq C \cdot M(k, g)^{3/2}$ for some constant C, where $M(k, g)$ denotes the well-known Moore bound (also presented later in (1)) for k-regular graphs of girth g.

The Degree Diameter Problem asks for the order of the largest graph of given maximum degree Δ and diameter d (denoted by $n'(\Delta, d)$) as well as the corresponding graphs. The Cage Problem and the Degree Diameter Problem are connected through the Moore bound, which simultaneously gives a lower bound for $n(k, g)$ as well as an upper bound for $n'(\Delta, d)$.

For more detailed information, we refer the interested reader to the Dynamic Cage Survey by Exoo and Jajcay [6] on the Cage Problem and to the excellent survey by Miller and Širáň [13] on the Degree Diameter Problem.

A $(k; g, d)$-*graph* is a k-regular graph of girth g and diameter d. Analogously to the classical cage problem, a smallest $(k; g, d)$-graph is called a $(k; g, d)$-*cage* (or simply a *girth-diameter cage* if the values of k, g and d are not specified) and its order is denoted by $n(k; g, d)$. The authors of [1] mainly studied the case $n(k; 5, 4)$. We will be interested in the exact determination of $n(k; g, d)$ for certain specific tuples (k, g, d) as well as determining the smallest real number $f(k, g)$ such that, as the diameter d goes to infinity, we have $n(k; g, d) \leq f(k, g)d + O_{k,g}(1)$. Here, $O_{k,g}(1)$ is a constant depending only on k and g. In cases where the exact determination of $f(k, g)$ or $n(k; g, d)$ is not possible, we will be interested in lower and upper bounds instead.

Our main theorem can be stated as follows and is proven in Sect. 2.

Theorem 1. *For all integers* $k, g \geq 3$, *we have* $\frac{M(k,g)}{g} \leq f(k, g)$ *and* $f(k, g) \leq \frac{n(k,g)}{g}$. *Moreover, neither of the two bounds are sharp for all pairs* (k, g).

The exact value of $n(k; g, d)$ is open for most cases except for some notable examples from the literature that were already known (and are implied by our result). When the girth equals 4, using $n(k, 4) = M(k, 4) = 2k$, Theorem 1 implies that $n(k; 4, d) = \frac{k}{2}d + O_k(1)$, which is already known by [5, Thm. 2] (it is stated for K_3-free graphs). Also without girth condition, up to the $O_k(1)$ term, sharpness of the ratio $\frac{M(k,3)}{3} = \frac{k+1}{3}$ was already known by classical results in [5], and made precise by [8].

Note that if a (k, g)-Moore graph exists, i.e., $M(k, g) = n(k, g)$, then $n(k; g, d)$ is determined up to a constant by Theorem 1. This is the case when $g \in \{3, 4\}$, $g = 5$ and $k \in \{2, 3, 7\}$ and possibly $k = 57$ (the existence of such a Moore graph is a famous open problem [6]), and $g \in \{6, 8, 12\}$ for some values of k (e.g. when $k - 1$ is a prime power). So the smallest cases (for k resp. g) for which the bounds are not asymptotically determined, are $(k, g) \in \{(3, 7), (4, 5)\}$. We will study these pairs in Sect. 3.

We also prove that for all integers $k, g \geq 3$, determining $f(k, g)$ is a task that requires only finitely many computations[1]. Due to space constraints we had to

---

[1] We remark that this is mainly of theoretical importance, since for many (k, g) this finite number is extremely large.

omit several proofs and details from this text, as is the case for the proof of Theorem 2. A full-length version of this article containing these omissions can be found on arXiv [3]. We explicitly indicate with a ♠ that a proof is omitted.

Theorem 2 (♠). *For all integers $k, g \geq 3$, one can compute $f(k, g)$ in $O_{k,g}(1)$ time.*

In Sect. 3, we compute several previously unknown values of $n(k; g, d)$, starting with the theoretical determination of $n(3; 4, d)$ and $n(3; 5, d)$, as well as the number of corresponding girth-diameter cages. Additionally, we design and implement an exhaustive generation algorithm for $(k; g, d)$-graphs. This allowed us to determine the values of $n(k; g, d)$ for 107 new triples (k, g, d) and the corresponding $(k; g, d)$-cages (often exhaustive). The full results derived from the computations are omitted here, but are listed in [3].

Finally, in Sect. 4 we conclude with some further observations and open questions, in particular a question relating $(k; g, d)$-cages to the conjecture that every even girth (k, g)-cage is bipartite.

2   Proof of Theorem 1

In this section, we prove Theorem 1 in three parts separately. First, we prove a lower bound $M'(k; g, d)$ on $n(k; g, d)$ in Proposition 1 which implies that $\frac{M(k,g)}{g} \leq f(k, g)$. As a side result, we also prove a better lower bound $M''(k; g, d)$ for $d \leq g$ where g is even, i.e., $g = 2t \geq 4$ for some t, in Proposition 2. The proofs of these lower bounds contain the underlying ideas for the generation algorithm described in Subsect. 3.2. Second, we prove $f(k, g) \leq \frac{n(k,g)}{g}$ in Proposition 3. Third, we prove that neither of the two bounds on $f(k, g)$ are sharp for all pairs (k, g) in Proposition 4.

Recall the Moore bound $M(k, g)$, defined by

$$n(k, g) \geq M(k, g) = \begin{cases} 1 + k \sum_{i=0}^{t-1}(k-1)^i & \text{for } g = 2t+1 \\ 2 \sum_{i=0}^{t-1}(k-1)^i & \text{for } g = 2t. \end{cases} \tag{1}$$

This bound is obtained by realising that for any vertex u in a (k, g)-graph, there cannot be an edge between two vertices v and w if the sum of the distance between u and v and between u and w is strictly smaller than $g - 1$. Hence, locally around every vertex u in a (k, g)-graph, the graph looks like a tree of order $M(k, g)$ (and we call this tree a Moore tree). We will furthermore split $M(k, 2t + 1)$ in the sum of the sizes of even and odd neighbourhoods towards the center of the Moore tree; $M(k, 2t + 1) = M_0(k, 2t + 1) + M_1(k, 2t + 1)$, where

$$M_0(k, 2t + 1) = 1 + k \sum_{i=0}^{\lfloor \frac{t-2}{2} \rfloor} (k-1)^{2i+1} \text{ and } M_1(k, 2t + 1) = k \sum_{i=0}^{\lfloor \frac{t-1}{2} \rfloor} (k-1)^{2i}.$$

Note that it is usually only defined for $k \geq 2$ and $g \geq 3$, but we allow the formula to be interpreted for any $g > 0$ and further define $M(k,g) = 0$ for every $g \leq 0$.

In Proposition 1, we give a lower bound for $n(k; g, d)$, which we denote by $M'(k; g, d)$ in analogy to this. For certain values, this lower bound will turn out to be exact (see Sect. 3). A direct corollary of this theorem is that $f(k,g) \geq \frac{M(k,g)}{g}$.

Proposition 1. *Suppose $k \geq 3, g \geq 3, d \geq t := \lfloor g/2 \rfloor$, then $n(k; g, d) \geq M'(k; g, d)$, where*

$$M'(k; g, d) = \begin{cases} M(k,g) + M(k, 2d - 2t - 1) & \text{for } t \leq d \leq 2t, \\ \\ (r+2)M(k,g) + M(k,s) & \text{for } \begin{array}{l} d = 2t + 1 + rg + s \text{ with} \\ r \geq 0, \ s \in \{0, \ldots, g-1\} \end{array} \end{cases}$$

Proof. Let G be a $(k; g, d)$-graph of order $n = n(k; g, d)$ and let u, v be two vertices for which $d(u,v) = d = diam(G)$.

We define the neighbourhoods $N_i := N_i(u) = \{w \in V(G) \mid d(u,w) = i\}$. It holds that $|V(G)| = \sum_{i=0}^{d} |N_i|$. Since G is a (k,g)-graph, we claim that

$$\sum_{i=0}^{t} |N_i| \geq M(k,g). \tag{2}$$

To see this, note that if g is odd, $\bigcup_{0 \leq i \leq t} N_i$ spans a (supergraph of a) k-ary tree of order $M(k,g)$, as one can see by taking a rooted tree starting from u with height t, i.e., by considering a (k,g)-Moore tree rooted at the vertex u. If g is even, the claim follows by considering $\bigcup_{1 \leq i \leq t-1} (N_i(u) \cup N_i(z)) \subseteq \bigcup_{0 \leq i \leq t} N_i(u)$ for a neighbour z of u, i.e., by considering a (k,g)-Moore tree rooted at the edge uz.

We now deal with the case $t \leq d \leq 2t$. Similarly, by considering a rooted subtree in v of height $d - t - 1$, one concludes that

$$\sum_{i=0}^{d-t-1} |N_{d-i}| \geq M(k, 2(d-t) - 1).$$

Hence, for $t \leq d \leq 2t$, we obtain $|V(G)| \geq M(k,g) + M(k, 2(d-t) - 1)$.

Henceforth, we focus on the other case. Let G' be a $(k; g, d)$-graph with $d = 2t + 1 + rg + s$ where $r \geq 0, s \in \{0, \ldots, g-1\}$. For every $1 \leq j \leq g$ and for every $0 \leq q \leq d-j$, similarly as before, we conclude that $\sum_{i=1}^{j} |N_{q+i}| \geq M(k,j)$, by considering rooted subtree(s) from a vertex $w \in N_{q+\lceil \frac{j}{2} \rceil}$ if j is odd, or edge $xy \in N_{q+\frac{j}{2}} \times N_{q+\frac{j}{2}+1}$ if j is even. Hence, the theorem follows since

$$|V(G')| = \sum_{i=0}^{t} |N_i| + \sum_{l=0}^{r-1} \sum_{i=1}^{g} |N_{t+lg+i}| + \sum_{i=1}^{s} |N_{t+rg+i}| + \sum_{i=0}^{t} |N_{d-i}|$$
$$\geq M(k,g) + rM(k,g) + M(k,s) + M(k,g)$$
$$= (r+2)M(k,g) + M(k,s).$$

$\square$

When the girth g is even, and $d \leq g$, the above bounds can be sharpened.

Proposition 2. *Suppose* $k \geq 3$ *and* $d \leq g = 2t \geq 4$, *then* $n(k; g, d) \geq M''(k; g, d)$, *where*

$$M''(k; g, d) = M(k, g) + 2\max\{M_0(k, 2d - 2t - 1), M_1(k, 2d - 2t - 1)\}.$$

Furthermore, this bound can only be attained by bipartite graphs.

Proof. Let u, v be vertices satisfying $d(u, v) = d$. Since $\bigcup_{i=0}^{t-1} N_i(u)$ and $\bigcup_{i=0}^{t-1} N_i(v)$ span a tree, we can deduce that in each case (depending on $d \pmod 2$ and $t \pmod 2$) the following neighbourhoods form an independent set (and so does their union, since distances between vertices of different neighbourhoods are at least two)

$$\begin{aligned}
&(N_1(u) \cup N_3(u) \cup \ldots \cup N_{t-1}(u)) \cup (N_1(v) \cup N_3(v) \cup \ldots \cup N_{d-t-1}(v)), \\
&(N_1(u) \cup N_3(u) \cup \ldots \cup N_{t-1}(u)) \cup (N_0(v) \cup N_2(v) \cup \ldots \cup N_{d-t-1}(v)), \\
&(N_0(u) \cup N_2(u) \cup \ldots \cup N_{t-1}(u)) \cup (N_0(v) \cup N_2(v) \cup \ldots \cup N_{d-t-1}(v)), \\
&(N_0(u) \cup N_2(u) \cup \ldots \cup N_{t-1}(u)) \cup (N_1(v) \cup N_3(v) \cup \ldots \cup N_{d-t-1}(v)).
\end{aligned}$$

Since the graph needs to be k-regular, its order is at least two times the independence number (and equality implies the bipartiteness), which is at least the sum of sizes of the corresponding neighbourhoods listed above. Here $M(k, g) = 2M_\varepsilon(k, 2t - 1)$ for $\varepsilon \in \{0, 1\}$ satisfying $2 \nmid t + \varepsilon$ ($\varepsilon \equiv t - 1 \pmod 2$). This can be concluded from the fact that if $M(k, g)$ would be attained, the extremal graph is bipartite and regular, but also from direct verification where one uses $k = 1 + (k - 1)$. $\square$

Corollary 1. *Suppose* $k \geq 3, g \geq 3, d \geq t := \lfloor g/2 \rfloor$, *then* $n(k; g, d) \geq M(k; g, d)$, *where*

$$M(k; g, d) = \begin{cases} M''(k; g, d) & \text{for } d \leq g = 2t, \\ M'(k; g, d) & \text{otherwise.} \end{cases}$$

Note here that $M''(k; g, d) \geq M'(k; g, d)$ for all $t \leq d \leq g = 2t \geq 4$, since $M(k, 2d - 2t - 1) = M_0(k, 2d - 2t - 1) + M_1(k, 2d - 2t - 1) \leq 2\max\{M_0(k, 2d - 2t - 1), M_1(k, 2d - 2t - 1)\}$.

We now discuss a construction leading to an infinite family of (k, g)-graphs for which the ratio between the order and the diameter tends to $\frac{n(k,g)}{g}$ as the diameter of the graph tends to infinity.

Proposition 3. *For all integers* $k, g \geq 3$, *we have* $f(k, g) \leq \frac{n(k,g)}{g}$.

Proof. Let M be a (k, g)-cage of order $n(k, g)$.

If k is even, let M' be the graph obtained by removing a vertex u (disjoint from at least one shortest cycle of M), adding two vertices v and w and adding edges between v and 2 of the initial neighbours of u and edges between w and the $k - 2$ remaining initial neighbours of u. Let $r \geq 1$ be an integer and consider the graph M'_r obtained by taking r copies of M' (where vertices in copy i receive the subscript i) and identifying v_i with w_{i-1} for each $2 \leq i \leq r$. Note that $|V(M'_r)| = r|V(M)| + 1$, that M'_r has girth g, that the distance between v_1 and w_r is at least gr and that every vertex in M'_r has degree k, except for v_1 (having degree 2) and w_r (having degree $k - 2$). Let G' be the graph obtained by taking a $(k, g + 2)$-cage, removing a vertex x, removing an edge yz where y was initially a neighbour of x and adding a matching of size $k/2$ between the vertices of degree $k - 1$. Note that G' has girth at least g and all vertices have degree k, except for y (having degree $k - 2$). We can complete M'_r to a (k, g)-graph G'' by identifying y with v_1 and adding another copy of M with one edge subdivided, and identifying this additional vertex (of degree 2) with w_r. Now G'' is k-regular, the diameter of G'' is at least as large as the diameter of M'_r, G'' has girth g and for r tending to infinity, the ratio of its order and diameter tends to $\frac{n(k,g)}{g}$.

If k is odd, let M' be the graph obtained by removing an edge vw (disjoint from at least one shortest cycle) from M. Let $r \geq 1$ be an integer and consider the graph M'_r obtained by taking r copies of M' and adding an edge between v_i and w_{i-1} for each $2 \leq i \leq r$. Now $|V(M'_r)| = r|V(M)|$, M'_r has girth equal to g, the distance between v_1 and w_r is at least $gr - 1$ and each vertex in M'_r has degree k, except for v_1 and w_r, which each have degree $k - 1$. Let G' be the graph obtained by taking a $(k, g + 1)$-cage, removing a vertex x and adding a matching of size $\frac{k-1}{2}$ between $k - 1$ of the initial neighbours of x (so G' has girth at least g and all vertices, except for an initial neighbour y of x, have degree k). Let G'' be the graph obtained by taking M'_r, taking two copies G'_1 and G'_2 of G', adding an edge between y_1 and v_1 and an edge between y_2 and w_r. Again, G'' is a k-regular graph with girth g for which the ratio between its order and diameter tends to $\frac{n(k,g)}{g}$ as r tends to infinity. An illustration of this is presented in Fig. 1. Here the matching of order $(3 - 1)/2 = 1$ in G' is the dashed line, and dotted lines are used for the edge connecting M'_r and G', as well as the edges $v_i w_{i-1}$. ⊔

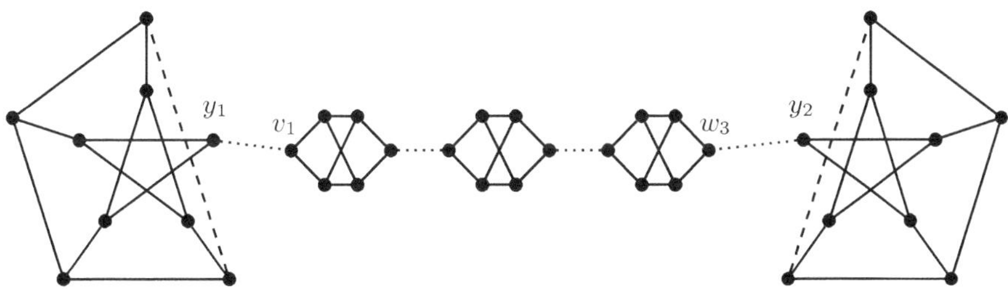

Fig. 1. Illustration of the construction in Proposition 3 for $(k, g) = (3, 4)$ and $r = 3$.

Next, it is natural to wonder whether for all integers $k, g \geq 3$, we have $f(k, g) = \frac{M(k,g)}{g}$ or $f(k, g) = \frac{n(k,g)}{g}$. We will show that this is not the case.

Proposition 4 ($\spadesuit$). *Neither of the two bounds $\frac{M(k,g)}{g} \leq f(k, g)$ and $f(k, g) \leq \frac{n(k,g)}{g}$ are sharp for all pairs (k, g).*

Due to space constraints we omit the proof. See [3].
Together, Propositions 1 to 4 yield Theorem 1.

3   Determining $n(k; g, d)$ for Several Open Cases

In this section, we first theoretically determine the values of $n(k; g, d)$, and count the number of girth-diameter cages for $(k, g) = (3, 4)$ and $(k, g) = (3, 5)$ in Sect. 3.1. Then, in Sect. 3.2, we present an exhaustive generation algorithm for $(k; g, d)$-graphs and use it to computationally determine several more values of $n(k; g, d)$ and the corresponding $(k; g, d)$-cages.

3.1   Theoretical Determination of $n(3; g, d)$ for $g \in \{4, 5\}$

In what follows, we determine $n(k; g, d)$ when $(k, g) \in \{(3, 4), (3, 5)\}$. Hence the $O_{k,g}(1)$ term for these values is determined exactly. The fact that this is in general a hard task, can be concluded from the computational results in Appendix B of [3], e.g. observe the values for $(k, g) = (3, 7)$ when $7 \mid d$; for $d \in \{7, 14, 21\}$, the lower bound $M(k; g, d)$ is sharp, but for $d \in \{28, 35\}$ it is not. The data seems to suggest that $n(k; g, d) = n(k; g, d-7)+24$ for $d \geq 23$ and all further results may be analogous to the cases where $(k, g) \in \{(3, 4), (3, 5)\}$.

Proposition 5 ($\spadesuit$). *Let $d = 5 + 4k + j$, where $0 \leq j \leq 3$ and $k \geq 0$. A $(3; 4, d)$-cage G has at least k bridges and the order of G is $n = 14 + 2j + 6k$.*

Due to space constraints we omit the proof. See [3].
With a careful analysis of components, we can also characterise the $(3; 4, d)$-cages, and in particular count them.

Proposition 6 ($\spadesuit$). *The number of $(3; 4, d)$-cages for $d \geq 9$ equals*

$$\begin{cases} 1 & \text{for } d \equiv 1 \pmod 4 \\ 4 & \text{for } d \equiv 2 \pmod 4 \\ 17 + \lfloor \frac{d}{8} \rfloor & \text{for } d \equiv 3 \pmod 4 \\ 27 + d + \lfloor \frac{d-4}{8} \rfloor & \text{for } d \equiv 0 \pmod 4. \end{cases}$$

Due to space constraints we omit the proof. See [3].

Proposition 7 ($\spadesuit$). *For $d \geq 5$, $n(3; 5, d) = 2d + 10$ if $d \equiv 0, 1 \pmod 5$ and $n(3; 5, d) = 2d + 8$ if $d \equiv 2, 3, 4 \pmod 5$. For $2 \leq d \leq 4$, $n(3; 5, d) = 2d + 6$.*

Due to space constraints we omit the proof. See [3].

By a careful analysis, one can also count the number of $(3; 5, d)$-cages.

Proposition 8 (♠). *The number of $(3; 5, d)$-cages for $d \geq 6$ equals*

$$
\begin{cases}
d - \left\lfloor \frac{d}{10} \right\rfloor & \text{for } d \equiv 1 \pmod 5 \\
1 & \text{for } d \equiv 2 \pmod 5 \\
4 & \text{for } d \equiv 3 \pmod 5 \\
10 & \text{for } d \equiv 4 \pmod 5 \\
128 + 11.3d & \text{for } d \equiv 0 \pmod{10} \\
138.5 + 11.3d & \text{for } d \equiv 5 \pmod{10}.
\end{cases}
$$

Due to space constraints we omit the proof. See [3].

3.2   Further Values of $n(k; g, d)$ with Exhaustive Generation

Using the exhaustive generation algorithm for $(k; g, d)$-graphs which we will describe in this section, we could determine the value of $n(k; g, d)$ for 177 different triples (k, g, d) of which 107 triples are "new" (in the sense that they have not yet been determined in Sect. 3.1 of this paper or in the literature). We considered triples with k, g, d in the following sets: $k \in \{3, \ldots, 7\}$, $g \in \{4, \ldots, 8, 10\}$ and $d \in \{2, \ldots, 40\}$. For all 177 triples we determine at least one $(k; g, d)$-cage and we determine the exhaustive set for 123 triples. The $(k; g, d)$-cages we obtained with the algorithm can be downloaded from [16] and a selection of these cages can be found on the House of Graphs [4] by searching for the keyword "girth-diameter cages".

In the rest of this subsection we will describe our generation algorithm and explain the sanity checks that we performed to verify the correctness of our implementation.

Our algorithm, whose pseudocode can be found in Algorithm 1 and 2, generates all $(k; g, d)$-graphs of order n. It is based on the generation algorithm for (k, g)-graphs by McKay, Myrvold and Nadon [10] complemented with an extra pruning rule by the same authors and Exoo [7]. Their algorithm receives as input k, g and the order n. It starts from a (k, g)-Moore tree, adds $n - M(k, g)$ isolated vertices, and recursively adds edges one by one. To limit the search space, several pruning rules are applied. Additionally, a heuristic is applied to choose the next edge to add. In particular, the algorithm first adds the edge incident to the vertex x that has the fewest available incident edges that can be added without violating the degree or girth constraint. As mentioned in [10], this heuristic was one of the factors for the significant speedup compared to the previous best algorithm for (k, g)-graphs described in [2].

Our algorithm extends theirs with new optimisations to incorporate the diameter constraint. Contrary to their algorithm, we start from the tree T of order $M'(k; g, d)$ corresponding to Proposition 1. However, when $g = 2t$ and $d = 2t-1$, we start from the tree of girth $g - 1$ (instead of g) since this tree contains more vertices, i.e., $M'(k; 2t-1, 2t-1) > M'(k; 2t, 2t-1)$, and is still a subgraph of any $(k; 2t, 2t-1)$-graph. After the construction of the tree, we proceed in a similar way: add isolated vertices until there are n vertices and recursively add edges with an analogous heuristic that also incorporates the diameter. In addition to the pruning rules of [7, 10], we also prune the search space by disallowing an edge that would decrease the distance between the two vertices u, v at distance d in T. All of this is described in more detailed pseudocode in Algorithm 1 and 2.

Algorithm 1: $genGirthDiamGraphs(n, k, g, d)$

---

1 **if** $n < M(k; g, d)$ or $d < \lfloor g/2 \rfloor$ **then**

    `// If the order n is smaller than` $M(k; g, d)$ `or if` $d < \lfloor g/2 \rfloor$`,`
      `there are definitely no graphs.`

2   **return**

  `// Make tree corresponding to Proposition 1.`

3 $T \leftarrow makeTree(n, k, g, d)$

  `// Add` $n - M(k; g, d)$ `isolated vertices to` T`.`

4 $G_{\text{start}} \leftarrow addIsoVertices(T, n - M(k; g, d))$

  `// Determine which edges can be added without exceeding the degree`
    k `and without violating the girth` g `and diameter` d

5 $E_{\text{add}} \leftarrow calcValidAddableEdges(G_{\text{start}}, k, g, d)$

  `// Heuristic: edges will be added incident with the vertex` x `that`
    `has the fewest options in` E_{add}

6 $x \leftarrow \arg\min_{w \in V(G), \deg(w) < r}(|\{e : e \in E_{\text{add}} \text{ and } w \in e\}|)$

7 $recursivelyAddEdges(G_{\text{start}}, E_{\text{add}}, k, g, d, x)$ `// See Algorithm 2`

The graph isomorphism detection is done with **nauty** [11] by computing and comparing the canonical form of two graphs. Here it is important to notice that u and v are two distinguished vertices in our algorithm (i.e., the pruning rules related to the diameter treat u and v different from other vertices in the graph). Therefore, one has to only consider isomorphisms between two graphs that map these distinguished vertices to each other. To achieve this, we compute the canonical form of the graph G' that is obtained by adding two vertices u', v' and two edges uu', vv' to the original graph G. This leads u and v to be of degree $k + 1$. Since no other vertex can be of degree $k + 1$, u and v will never be mapped to other vertices. The source code of our implementation of this algorithm can be found in the GitHub repository https://github.com/AGT-Kulak/girthDiamGen [16].

We ran the algorithm for various small values of n, k, g, d, leading to the determination of $n(k; g, d)$ and the corresponding cages. Table 1 and the tables in Appendix B of [3] show the results we obtained using our algorithm, i.e. the values of $n(k; g, d)$, the lower bound $M(k; g, d)$, (a lower bound on) the number of pairwise non-isomorphic $(k; g, d)$-cages and whether all generated $(k; g, d)$-cages are bipartite. In case $M(k; g, d) = n(k; g, d)$, both values are marked in bold.

Algorithm 2: $recursivelyAddEdges(G, E_{\mathrm{add}}, k, g, d, x)$

```
// x: current vertex to add edges to
```
1 **if** method was called with graph that is isomorphic with G (where the isomorphism maps the two distinguished vertices whose distance is d to each other) **then**

2 $\quad$ | **return**

3 **if** $\forall v \in V(G)\colon \deg(v) = k$ **then** `// G is a k-regular graph.`

4 $\quad$ | **if** $g(G) = g$ and $diam(G) = d$ **then**

5 $\quad$ | $\quad$ | output G

6 $\quad$ | **return**

```
// Choose new x, if the current vertex x has been completed to
   degree k.
```
7 **if** $\deg(x) = k$ **then**

8 $\quad$ | $x \leftarrow \arg\min_{w \in V(G), \deg(w) < k}(|\{e\colon e \in E_{\mathrm{add}} \text{ and } w \in e\}|)$

```
// Apply pruning rule from [7] on all E_add.
```
9 $E_{\mathrm{add}} \leftarrow pruneValidEdgesExoo(G, E_{\mathrm{add}})$

10 **if** $\neg enoughValidEdgesLeft(G, E_{\mathrm{add}})$ **then**

11 $\quad$ | **return**

```
// Try adding each valid edge incident to x.
```
12 **foreach** $w \in \{w\colon xw \in E_{\mathrm{add}}\}$ **do**

```
      // Copy E_add in prevE_add to restore it after removing xw again.
```
13 $\quad$ | $prevE_{\mathrm{add}} \leftarrow E_{\mathrm{add}}$

14 $\quad$ | $E(G) \leftarrow E(G) \cup \{xw\}$

```
      // Remove edges from E_add resulting from adding edge xw.
```
15 $\quad$ | $E_{\mathrm{add}} \leftarrow updateValidAddableEdges(G, E_{\mathrm{add}}, k, g, d)$

```
      // Check if we can prune xw.
```
16 $\quad$ | **if** none of the pruning rules can be applied **then**

17 $\quad$ | $\quad$ | $recursivelyAddEdges(G, E_{\mathrm{add}}, k, g, d, x)$

18 $\quad$ | $E(G) \leftarrow E(G) \setminus \{xw\}$

```
      // Restore E_add and remove xw from it.
```
19 $\quad$ | $E_{\mathrm{add}} \leftarrow prevE_{\mathrm{add}} \setminus \{xw\}$

20 $\quad$ | **if** $\neg enoughValidEdgesLeft(G, E_{\mathrm{add}})$ **then**

21 $\quad$ | $\quad$ | **return**

The generated $(k; g, d)$-cages can also be found in `graph6` format on the GitHub repository [16].

Lastly, we explain which extra steps we took to ensure the correctness of the implementation of our exhaustive generation algorithm. We used the generator `GENREG` [12] to generate (k, g)-graphs of order n (for small values of n, k, g) and split the graphs based on the diameter d, to obtain all $(k; g, d)$-graphs. More specifically, we checked for $k = 3$ and $g \in \{3, \ldots, 8\}$ and for $k \in \{4, 5\}$ and $g \in \{3, 4, 5\}$ for orders $n(k, g), n(k, g) + 1, \ldots, n(k, g) + C$, where C was chosen such that `GENREG` was able to finish the computation within 100 CPU hours for the given k, g and n. For each n, k, g, d, we obtained exactly the same set of $(k; g, d)$-graphs with our generator as `GENREG`.

Table 1. Values of $M(k;g,d)$, $n(k;g,d)$, number of $(k;g,d)$-cages and bipartiteness for $(3;6,d)$ with $d \in \{3,\ldots,20\}$. In case $M(k;g,d) = n(k;g,d)$, both values are marked in bold.

k	g	d	$M(k;g,d)$	$n(k;g,d)$	Number of cages	All bipartite
3	6	3	**14**	**14**	1	Yes
3	6	4	**16**	**16**	1	Yes
3	6	5	**20**	**20**	6	Yes
3	6	6	**28**	**28**	3016	Yes
3	6	7	28	30	8	No
3	6	8	29	32	7	No
3	6	9	30	34	9	No
3	6	10	32	36	6	No
3	6	11	34	38	6	No
3	6	12	38	44	13953	No
3	6	13	42	44	1	No
3	6	14	43	46	2	No
3	6	15	44	48	6	No
3	6	16	46	50	6	No
3	6	17	48	52	6	No
3	6	18	52	58	13987	No
3	6	19	56	58	1	No
3	6	20	57	60	2	No

Besides that, we checked that we obtained all (k,g)-cages (for small k,g) with our generator by running it for the corresponding diameter of these (k,g)-cages. Moreover, Araujo-Pardo et al. found $(k;5,4)$-cages for $k \in \{3,4,5,6\}$ [1]. For each k, we obtained the same number of $(k;5,4)$-cages with our algorithm as Araujo-Pardo et al. Finally, we also verified that the results of our generator agree with the theoretical results for $(3;g,d)$ with $g \in \{4,5\}$ in Sect. 3.1 (both for the order and number of girth-diameter cages).

4   Further Observations on $(k;g,d)$-Cages

4.1   Remark on $(k;3,3)$

Knor [8] considered the question of finding the minimum order of a k-regular graph of diameter d, which we will denote by $n(k;d)$. He determined $n(k;d)$ for all $k \geq 2$ and $d \geq 1$. He also describes k-regular graphs with diameter d of order $n(k;d)$. For $k \geq 3$ the described families have girth 3, besides for $d = 3$, where the family has girth 4. Hence, we know $n(k;3,d)$ for all $d \neq 3$. In Proposition 9 we "complete" the determination of $n(k;3,d)$ by proving the order for $d = 3$.

Proposition 9 (♠). *For all $k \geq 3$, it holds that*

$$n(k; 3, 3) = 2(k + 1)$$

Due to space constraints we omit the proof. See [3].

4.2   Behaviour of $(k; g, d)$-Cages

Despite the existence of non-bipartite $(k; g, d)$-cages with even g, we can observe in Table 1 and in the tables in Appendix B of [3] that all generated even girth $(k; g, d)$-cages with $d \leq g$ are bipartite. In other words, we only found non-bipartite $(k; g, d)$-cages for $d > g$. Often this is the case exactly when G has a cutvertex (a regular bipartite graph cannot have a cutvertex), as is the case for the $(3; 6, 12)$-cages, but e.g. the graph depicted in Fig. 2 (also added as https:// houseofgraphs.org/graphs/54022 on the House of Graphs [4]) is a 2-connected non-bipartite $(4; 4, 5)$-cage.

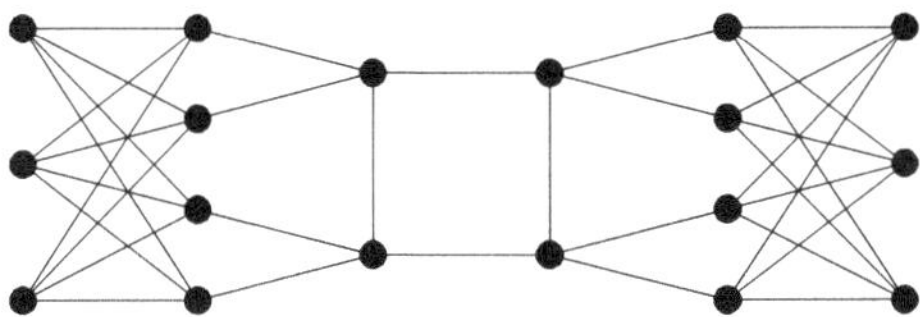

Fig. 2. A 2-connected non-bipartite $(4; 4, 5)$-cage.

Moreover, $(k; g, d)$-cages with $d > g$ can never be (k, g)-cages since Sauer proved that every (k, g)-cage G satisfies $diam(G) \leq g$ [15]. Hence, if it holds that every $(k; g, d)$-cage with g even and $d \leq g$ is bipartite, then this proves the longstanding conjecture that every even girth (k, g)-cage is bipartite [6, 17].

Therefore, the following question can be of interest.

Question 1. For even g and $d \leq g$, is every $(k; g, d)$-cage bipartite?

Notice that Proposition 2 is sharp for the smallest case, $g = 4$, by observing the extremal graphs $K_{k,k}$, $K_{k+1,k+1} \backslash M$ (a perfect matching M removed), and the graph formed from two copies of $K_{k,k}^{-}$ ($K_{k,k}$ minus one edge) connected with edges between the degree $(k-1)$-vertices. Equality in the latter case can only be attained by a k-regular (connected) bipartite graph of order $4k$ for which there are two vertices u, v in the same bipartition class with $N(u) \cap N(v) = \emptyset$.

Also for $g = 6, 8$ and 12 one can conclude that the answer for Question 1 is positive for $d \in \{g/2, g\}$ whenever a (k, g)-Moore graph exists.

Further investigating $n(k; g, d)$-graphs for $\lfloor \frac{g}{2} \rfloor < d \leq g$ may lead to improvements on the gap between the Moore bound and the known upper bound on the order of cages. In particular, a sharp bound on $n(k; g, g - 1)$ might possibly be obtained by finding a good construction of a set of perfect matchings. The latter would already be interesting to think about for $g = 10$.

Acknowledgments. Stijn Cambie and Jorik Jooken are supported by a Postdoctoral Fellowship of the Research Foundation Flanders (FWO) with grant numbers 1225224N and 1222524N, respectively. Jan Goedgebeur and Tibo Van den Eede are supported by Internal Funds of KU Leuven and an FWO grant with grant number G0AGX24N.

Several of the computational resources used in this work were provided by the VSC (Flemish Supercomputer Center), funded by the Research Foundation - Flanders (FWO) and the Flemish Government. We are grateful to the anonymous referees of SOFSEM2026 for their careful reading and suggestions that helped to improve the presentation of the paper.

Disclosure of Interests. The authors have no competing interests to declare that are relevant to the content of this article.

References

1. Araujo-Pardo, G., Conder, M., García-Colín, N., Kiss, G., Leemans, D.: A note on girth-diameter cages. Art Discrete Appl. Math. **8**(3), Paper No. 3.06, 7 (2025)
2. Brinkmann, G., McKay, B.D., Saager, C.: The smallest cubic graphs of girth nine. Comb. Probab. Comput. **4**, 317–329 (1995)
3. Cambie, S., Goedgebeur, J., Jooken, J., Van den Eede, T.: On the order-diameter ratio of girth-diameter cages. arXiv preprint arXiv:2511.21144 (2025)
4. Coolsaet, K., D'hondt, S., Goedgebeur, J.: House of graphs 2.0: a database of interesting graphs and more. Discrete Appl. Math. **325**, 97–107 (2023). https://houseofgraphs.org/
5. Erdős, P., Pach, J., Pollack, R., Tuza, Z.: Radius, diameter, and minimum degree. J. Comb. Theor. Ser. B **47**(1), 73–79 (1989)
6. Exoo, G., Jajcay, R.: Dynamic cage survey. Electron. J. Comb. **DS16**, 48 (2013)
7. Exoo, G., McKay, B.D., Myrvold, W., Nadon, J.: Computational determination of (3,11) and (4,7) cages. J. Discrete Algorithms **9**(2), 166–169 (2011)
8. Knor, M.: Smallest regular graphs of given degree and diameter. Discussiones Mathematicae Graph Theory **34**(1), 187–191 (2014)
9. Lazebnik, F., Ustimenko, V.A., Woldar, A.J.: New upper bounds on the order of cages. Electron. J. Combin. **4**(2), R13 (1997)
10. McKay, B.D., Myrvold, W., Nadon, J.: Fast backtracking principles applied to find new cages. In: 9th Annual ACM-SIAM Symposium on Discrete Algorithms, pp. 188–191 (1998)
11. McKay, B.D., Piperno, A.: Practical graph isomorphism, II. J. Symb. Comput. **60**, 94–112 (2014)
12. Meringer, M.: Fast generation of regular graphs and construction of cages. J. Graph Theory **30**(2), 137–146 (1999)
13. Miller, M., Širáň, J.: Moore graphs and beyond: a survey of the degree/diameter problem. Electron. J. Comb. **DS14** (2012)
14. Sachs, H.: Regular graphs with given girth and restricted circuits. J. Lond. Math. Soc. **1**(1), 423–429 (1963)
15. Sauer, N.: Extremaleigenschaften regulärer Graphen gegebener Taillenweite I & II. Sitzungsberichte Österreich. Acad. Wiss. Math. Natur. Kl. s-B II, 176:9–25, 27-43 (1967)
16. Van den Eede, T.: Source code of the program `girthDiamGen`. https://github.com/AGT-Kulak/girthDiamGen (2025), GitHub repository
17. Wong, P.K.: Cages–a survey. J. Graph Theory **6**(1), 1–22 (1982)

Enumeration With Nice Roman Domination Properties

Kevin Mann[(✉)][iD]

Universität Trier, Fachbereich 4 – Abteilung Informatikwissenschaften,
54286 Trier, Germany
`mann@uni-trier.de`

Abstract. Although EXTENSION PERFECT ROMAN DOMINATION is NP-complete, all minimal (with respect to the pointwise order) perfect Roman dominating functions can be enumerated with polynomial delay. This algorithm uses a bijection between minimal perfect Roman dominating functions and Roman dominating functions and the fact that all minimal Roman dominating functions can be enumerated with polynomial delay. This bijection considers the set of vertices with value 2 under the functions. In this paper, we will generalize this idea by defining so called *nice Roman Domination properties* for which we can employ this method. With this idea, we can show that all minimal *maximal Roman Dominating functions* can be enumerated with polynomial delay in $\mathcal{O}(1.9332^n)$ time. Furthermore, we prove that enumerating all minimal connected/total Roman dominating functions on cobipartite graphs can be achieved with polynomial delay. Additionally, we show the existence of a polynomial-delay algorithm for enumerating all minimal connected Roman dominating function on interval graphs. We show some downsides to this method as well.

1 Introduction

Enumeration problems have many different applications, as for example data mining, machine learning [10] or biology/chemistry [19]. One possible property of enumeration algorithms is polynomial delay. An enumeration algorithm has polynomial delay if the time between two new outputs is polynomially bounded. This is interesting for multi-agent systems, where one agent enumerates and the other processes the output. In this scenario, we know that the second agent does not need to wait for too long and use unnecessary resources. One example for polynomial delay enumeration is enumerating minimal Rdfs [2]. This is interesting, since it is open for more than 4 decades if there exists an enumeration algorithm for minimal dominating sets with polynomial delay (or even output polynomial time).

ROMAN DOMINATION is a well-studied variation of DOMINATING SET, see [5,7,15,16], being motivated by a defense strategy of the Roman Empire [23]. In this scenario, you want to put up to 2 legions per region such that each region has at least one legion on itself or there is a neighbored region with two legions

© The Author(s), under exclusive license to Springer Nature Switzerland AG 2026
J. Kozik and A. Wolff (Eds.): SOFSEM 2026, LNCS 16448, pp. 157–171, 2026.
https://doi.org/10.1007/978-3-032-17801-5_12

on it. This can be modeled on graphs where each region is represented by its own vertex and vertices are connected iff the corresponding regions are neighbors. The distribution of the legions can be represented by a function $f\colon V \to \{0,1,2\}$ for the set of regions V. $f\colon V \to \{0,1,2\}$ is called a *Roman dominating function* (Rdf for short) on a graph $G = (V,E)$ if, for all $v \in V$ with $f(v) = 0$, there is a $u \in V$ with $f(u) = 2$ and $\{v,u\} \in E$. A function $f\colon V \to \{0,1,2\}$ can also be represented as a partition of V into $(V_0(f), V_1(f), V_2(f))$, where $V_i(f) := \{v \in V \mid f(v) = i\}$ for all $i \in \{0,1,2\}$.

The polynomial delay algorithm in [2] uses EXT RD, which is polynomial time solvable. Using the extension version of a problem is a usual strategy to obtain a polynomial-delay algorithm. In this paper, we will only consider monotonically increasing problems. Therefore, we will explain the idea of extension problems with respect to these. For more information, we refer to [4]. The idea of an extension problem is to find a minimal solution which is greater than or equal to a given presolution with respect to the given partial order. Then, EXTENSION X ROMAN DOMINATION (where X specifies the variant, which is empty in the classical version) is defined as follows:

Problem name: EXTENSION X ROMAN DOMINATION, or EXT X RD for short.

Given: A graph $G = (V,E)$ and a function $f \in \{0,1,2\}^V$.

Question: Is there a minimal X Rdf g on G with $f \le g$?

Here, $f \le g$ is understood pointwisely: for all $v \in V$, $f(v) \le g(v)$. In [18], it is proven that enumerating minimal perfect Roman dominating functions (or pRdfs for short; a pRdf is a Rdf f such that all vertices in $V_0(f)$ have exactly one neighbor in $V_2(f)$) can be done with polynomial delay even though the extension version is NP-complete. The enumeration algorithm is based on a bijection B between minimal Rdfs and minimal pRdf on the same graph. For a minimal Rdf f, $V_2(f) = V_2(B(f))$. In this paper, we want to generalize this idea for more Roman domination variations like total, connected or maximal. To this end, we will separate the choice of vertices with value 1 and value 2. Therefore, we define nice Roman domination properties in Sect. 3. We also prove how to make use of this by enumerating all minimal functions of a variation with a nice Roman domination property and a fixed set of vertices with value 2. From this situation, some enumeration properties can be inherited to enumerating all minimal functions of this variation. In Sect. 4, we use these results to prove that all minimal mRdfs can be enumerated with polynomial delay and in $\mathcal{O}(1.9332^n)$. We use this technique also for enumerating minimal tRdfs/cRdfs on cobipartite in Sect. 5. This section also includes a polynomial delay algorithm for enumerating minimal cRdfs in interval graphs. We also show that this approach sometimes fails unless $\mathsf{P} = \mathsf{NP}$ in Sect. 6.

Due to space restrictions, some proofs are omitted. These can be found in the long version [17]. The corresponding results are marked by $(\star)$.

2   Definitions and Notation

By $\mathbb{N} = \{0, 1, 2, \ldots\}$ we denote the non-negative integers. For all $n \in \mathbb{N}$, $[n] :=$ $\{1, \ldots, n\}$. $\binom{A}{k}$ denotes all subsets of a set A of size exactly $k \in \mathbb{N}$. For two sets A, B, the set of all functions $f : A \to B$ is denoted by B^A. A special function is the *characteristic function* $\chi_A : B \to \mathbb{N}$ such that for all $v \in B$, $\chi_A(v) = 1$ iff $v \in A$ and $\chi_A(v) = 0$ otherwise.

In this paper, we only consider simple undirected graphs $G = (V, E)$, where V denotes the vertex set and $E \subseteq \{\{v, u\} \mid v, u \in V\}$ is the set of edges. The by $A \subseteq V$ induced subgraph of G is defined as $G[A] = (A, E \cap \binom{A}{2})$. For all $v \in V$, $N(v) := \{u \mid \{v, u\} \in E\}$ is called the (open) neighborhood of v with respect to G. The closed neighborhood of $v \in V$ is denoted by $N[v] := \{v\} \cup N(v)$. For a set $A \subseteq V$, we define the neighborhood by the union $N(A) = \bigcup_{v \in A} N(v)$ or $N[A] = \bigcup_{v \in A} N[v]$. $P_{G,A}(v) = N[v] \setminus N[A \setminus \{v\}]$ is the set of private neighbors of $v \in A \subseteq V$ with respect to A on G. We also consider hypergraphs $H = (X, S)$, where X is the universe and $S \subseteq 2^X$ is the set of hyperedges.

We call a set $D \subseteq V$ a dominating set of the graph $G = (V, E)$ if, for all $v \in V$, $N[v] \cap D \neq \emptyset$. This can be viewed as a special case of hitting sets. A hitting set $D \subseteq X$ of the hypergraph $H = (X, S)$ if for all $s \in S$, $s \cap D \neq \emptyset$. There are many different variations of dominating sets on a graph $G = (V, E)$, as connected dominating set, cds for short, (D has to be connected), or total dominating set (D is a hitting set of $(V, \{N(v) \mid v \in V\})$). Similar to dominating set, Roman domination has also many variations. Some of these are mentioned in the appendix. In this paper, we will only consider maximal/total/connected Rdf, extensively.

Further, we consider some notation of enumeration. As for decision problems, we need an alphabet Σ. Let $A \subseteq \Sigma^* \times \Sigma^*$ be a binary relation. An enumeration problem ENUM A is defined as enumerating all elements in $A(x) := \{y \in \Sigma^* \mid A(x, y)\}$ for a given instance $x \in \Sigma^*$ without repetitions.

The analysis of the running time for enumeration is divided into input-sensitive (running time with respect to the input size $|x|$) and output-sensitive (running time with respect to the input and output size $|x| + |A(x)|$) We mentioned already polynomial delay which is a output-sensitive property. For a polynomial delay algorithm there exists a polynomial $p \colon \mathbb{N} \to \mathbb{N}$ such that the running time to the first output, between two consecutive outputs and after the last is bounded in $p(|x|)$ for the input x. Another possible property of an enumeration algorithm is output-polynomial time where there exists a polynomial $p \colon \mathbb{N} \times \mathbb{N} \to \mathbb{N}$ such that the running time is bounded in $p(|x|, |A(x)|)$ for the input x. It should be mentioned that a polynomial delay algorithm is also output polynomial. Enumerating minimal hitting sets on a hypergraph, also known as minimal transversal enumeration problem (denoted by ENUM TR), is a very famous enumeration problem, for which it is open for more than four decades whether there is an output-polynomial time algorithm.

As for classical complexity, there are also reductions for enumeration problems. For example *parsimonious reduction* [24] and *e-reduction* [8]. The definition to both can be found in the appendix.

3   Nice Roman Domination Property

In this section, we will set the theoretical basis of this paper. The main goal is to prove Theorem 1. For this paper, we assume that a property $\mathcal{P}$ is a set of functions mapping the vertex set of a graph to $\{0, 1, 2\}$. Instead of $f \in \mathcal{P}$, we say f has property $\mathcal{P}$.[1] To this end, we start by defining nice Roman domination properties. We call a property $\mathcal{P}$ a *nice Roman domination property* (*nrdp* for short) if for every $G = (V, E)$ and every $f \in \{0, 1, 2\}^V$ with property $\mathcal{P}$ fulfills the following constraints:

1. f is a Rdf.
2. For all $v \in V_0(f)$, $f + \chi_{\{v\}}$ has property $\mathcal{P}$.
3. For all $v \in V_2(f)$, $f - \chi_{\{v\}}$ has property $\mathcal{P}$ iff $P_{G[V \setminus V_1(f)], V_2(f)}(v) \subseteq \{v\}$, i.e. $f - \chi_{\{v\}}$ does not have property $\mathcal{P}$ iff v has a private neighbor, but itself, on the graph $G[V_0(f) \cup V_2(f)]$.

For a nrdp $\mathcal{P}$, a graph G and an $A \subseteq V$, define $C_{\mathcal{P},G}[A]$ as the set of minimal $f \in \{0, 1, 2\}^V$ with respect to $\leq$ and the property $\mathcal{P}$ such that $V_2(f) = A$. Now, we are ready to formulate our main theorem.

Theorem 1. *Let $\mathcal{P}$ be a nrdp and $\mathcal{G}$ be a class of graphs. If there exists an algorithm $\mathcal{A}$ that enumerates all elements in $C_{\mathcal{P},G}[A]$ in output-polynomial time (resp. with polynomial delay) for all $G = (V, E) \in \mathcal{G}$ and $A \subseteq V$, then there exists an algorithm to enumerate all minimal elements with property $\mathcal{P}$ on any $G \in \mathcal{G}$ in output-polynomial time (resp. with polynomial delay).*

Theorem 1 is a modification of $\leq_e$-reductions [8] for nrdp to get stronger results and even output-polynomial time. For the proof, we first show a couple of lemmas. The next lemma follows from [2, Lemma 7].

Lemma 1. *Let $G = (V, E)$ be a graph. $F(A) := 2 \cdot \chi_A + \chi_{V \setminus N[A]}$ is a bijection between all sets $A \subseteq V$ such that for all $v \in V$, $P_{G[N[A]], A}(v) \not\subseteq \{v\}$ and all minimal Rdfs of G.*

Lemma 2. *Let $\mathcal{P}$ be a nrdp, $G = (V, E)$ be graph and $A \subseteq B \subseteq V$. If $C_{\mathcal{P},G}[B] \neq \emptyset$ then $C_{\mathcal{P},G}[A] \neq \emptyset$.*

Proof. Let $g \in C_{\mathcal{P},G}[B]$ and for each $v \in B$ let $Y_v = P_{G,B}(v) \cap V_0(g)$. Define $Y_A := \bigcup_{v \in A} Y_v$ and $f := \chi_{V \setminus Y_A} + \chi_A$. We want to show that f has property $\mathcal{P}$. Observe $V_0(f) = Y_A \subseteq N(A)$. Hence, f is a Rdf. By Constraint 2 of nrdp and adding $\chi_{\{u\}}$ for $u \in V_0(g) \setminus Y_A$ to g, $\widetilde{g} := g + \chi_{V_0(g) \setminus Y_A}$ has property $\mathcal{P}$. Furthermore, $\widetilde{g} = f + \chi_{B \setminus A} \geq f$, as $V_2(\widetilde{g}) = B = V_2(f) \cup B \setminus A$, $V_0(\widetilde{g}) = Y_A = V_0(f)$ and $B \setminus A \subseteq V_1(f)$ as they are no private neighbors of elements in A with respect to B. By definition of f and Y_A, $N(u) \cap V_0(f) = \emptyset$ for all $u \in B \setminus A$. By an inductive argument, Constraint 3 of nrdp implies $\widetilde{g} - \chi_{B \setminus A} = f$ has property $\mathcal{P}$.

---

[1] Actually, we have to say f has property $\mathcal{P}$ with respect to a graph G, but the graph is known due to the context.

We now show that from f, we can construct some $h \in C_{\mathcal{P},G}[A]$. If f is minimal, set $h = f$. Otherwise, assume there is a $h \in \{0,1,2\}^V$ with property $\mathcal{P}$ on G, $h \leq f$ and a $u \in V$ such that $h(u) < f(u)$. If $f(u) = 2$ then the elements in Y_u are not dominated by h, as they private neighbors of u. This is not possible and, therefore, $V_2(h) = V_2(f)$. This implies $h = f - \chi_M$ for an $M \subseteq V_1(f)$, $M \neq \emptyset$, and $h \in C_{\mathcal{P},G}[A]$. Thus, $C_{\mathcal{P},G}[A] \neq \emptyset$. $\qquad\square$

Now we will show a connection between $C_{\mathcal{P},G}$ and extension problems.

Lemma 3. *Let $G = (V,E)$ be a graph and $A \subseteq V$. $C_{\mathcal{P},G}[A] \neq \emptyset$ iff there is a minimal $g \in \{0,1,2\}^V$ with property $\mathcal{P}$ on G such that $2 \cdot \chi_A + \chi_{V \setminus N[A]} \leq g$. Further, for all $h \in C_{\mathcal{P},G}[A]$, $2 \cdot \chi_A + \chi_{V \setminus N[A]} \leq h$.*

Proof. Let $C_{\mathcal{P},G}[A] \neq \emptyset$. By Lemma 1, $f := 2 \cdot \chi_A + \chi_{V \setminus N[A]}$ is the only minimal Rdf with $V_2(f) = A$. Thus, every Rdf g on G with $V_2(g) = A$ fulfills $f \leq g$. Conversely, let $g \in \{0,1,2\}^V$ be minimal with the property $\mathcal{P}$ on G with $f \leq g$. This implies $V_0(g) \subseteq V_0(f) = N[A] \setminus A$. If there exists an $x \in V_2(g) \setminus A$, x has no private neighbor in $V_0(g)$ and g is not minimal (see Constraint 3 of nrdps). Hence, $C_{\mathcal{P},G}[A] \neq \emptyset$ and for all $g \in C_{\mathcal{P},G}[A]$, $f \leq g$. $\qquad\square$

Let us shortly follow a corollary before proving Theorem 1. We use this result to discuss the why our approach is helpful.

Corollary 1. $(\star)$ *Let $\mathcal{P}$ be a nrdp, $G = (V,E)$ a graph and $f \in \{0,1,2\}^V$ with $N(V_1(f)) \cap V_2(f) = \emptyset$. Then, there is an $h \in \{0,1,2\}$ minimal with the property $\mathcal{P}$ and $f \leq h$ iff $C_{\mathcal{P},G}[V_2(f)] \neq \emptyset$.*

Now we can turn our attention back to Theorem 1.

Proof (of Theorem 1). We will only consider the output-polynomial time case. The proof for the polynomial-delay case is analogous. Hence, let $\mathcal{A}$ be the algorithm for enumerating $C_{\mathcal{P},G}[A]$ for given $G = (V,E) \in \mathcal{G}$ and $A \subseteq V$ and p the polynomial such that $p(|V|+|E|, |C_{\mathcal{P},G}[A]|)$ is the running time of the algorithm. W.l.o.g., all the multiplicative constants of the monomials are positive. Otherwise, we could use the polynomial with the absolute values of the multiplicative constants instead.

The idea of the algorithm for this proof is based on a branching algorithm. We branch on each vertex if it is in A. Every time we add a vertex to A, we compute $C_{\mathcal{P},G}[A]$. If $C_{\mathcal{P},G}[A] = \emptyset$, we can skip this branch. This case can be checked in polynomial time, as we can enumerate all elements in output-polynomial time. If $C_{\mathcal{P},G}[A] \neq \emptyset$, we enumerate the elements and go on with the next vertex.

Now we want to prove that this algorithm works. So, let $G = (V,E) \in \mathcal{G}$. Clearly, $\bigcup_{A \subseteq V} C_{\mathcal{P},G}[A]$ is exactly the set of all minimal elements with property $\mathcal{P}$ on G. By Lemma 2, we can skip the branches with $C_{\mathcal{P},G}[A] = \emptyset$. Therefore, we only need to show that the algorithm runs in output-polynomial time. To do this, we will prove the following claim:

Claim. The algorithm considers at most $|V|^2$ many consecutive sets $A \subseteq V$ with $C_{\mathcal{P},G}[A] = \emptyset$ before the algorithm either stops or finds a $B \subseteq V$ with $C_{\mathcal{P},G}[B] \neq \emptyset$.

Proof. If $C_{\mathcal{P},G}[\emptyset] = \emptyset$ then the algorithm stops directly by Lemma 2. So, assume there is a $B' \subseteq V$ such that $C_{\mathcal{P},G}[B']$ is not empty. In this case, the algorithm would now go through at most $|V|$ vertices which it did not consider so far to obtain B'. For each of these vertices $v \in V \setminus B'$, the algorithm either has found $C_{\mathcal{P},G}[B' \cup \{v\}] \neq \emptyset$ or would cut this branch and go on with the next vertex.

After this, the algorithm goes through all of the following cases: it deletes an element from B' and checks as before. It will go on with this process until it either finds a set B with $C_{\mathcal{P},G}[B] \neq \emptyset$ or it stops. As the algorithm can only delete at most $|B'| \leq |V|$ vertices from B' and after each deletion step it has to consider at most $|V|$ before deleting the next vertex, the algorithm considers at most $|V|^2$ consecutive sets $A \subseteq V$ with $C_{\mathcal{P},G}[A] = \emptyset$ before stopping.

Let $\mathcal{B} := \{B \subseteq V \mid C_{P,G}[B] \neq \emptyset\}$. The claim implies that we consider at most $|V|^2 \cdot |\mathcal{B}|$ sets $A \subseteq V$ with $C_{\mathcal{P},G}[A] = \emptyset$. Hence, the running time is bounded by

$$\left(\sum_{A \in \mathcal{B}} p(|V| + |E|, |C_{\mathcal{P},G}[A]|) \right) + |V|^2 \cdot |\mathcal{B}| \cdot p(|V| + |E|, 0)$$

$$\leq \left(\sum_{B \in \mathcal{B}} |C_{\mathcal{P},G}[B]| \right) \cdot \left(p(|V| + |E|, \sum_{B \in \mathcal{B}} |C_{\mathcal{P},G}[B]|) + |V|^2 \right) \cdot p(|V| + |E|, 0).$$

This is a polynomial in $|V| + |E|$ and the output size. $\quad\square$

It is important for the polynomial delay case that in the claim of Theorem 1 the sets are consecutive. Otherwise, the time between two outputs could be too long. This theorem only gives us an output-sensitive results. Now, we prove that if we can bound $C_{\mathcal{P},G}$ polynomially, we can bound the input-sensitive running time.

Lemma 4. *Let $\mathcal{P}$ be a nrdp and $\mathcal{G}$ a class of graphs such that there is an algorithm enumerating all minimal Rdfs on $G = (V, E) \in \mathcal{G}$ in $\mathcal{O}^*(r^{|V|})$ for $r \in \{x \in \mathbb{R} \mid x \geq 1\}$. If there is an algorithm $\mathcal{A}$ that enumerates all elements in $C_{\mathcal{P},G}[A]$ with polynomial delay and if there is polynomial p such that $|C_{\mathcal{P},G}[A]| \leq p(|V|)$ for each $G = (V, E) \in \mathcal{G}$ and $A \subseteq V$, then there is an algorithm to enumerate all minimal $g \in \{0, 1, 2\}^V$ with property $\mathcal{P}$ on $G = (V, E) \in \mathcal{G}$ in $\mathcal{O}^*(r^{|V|})$.*

Proof. Let $G = (V, E) \in \mathcal{G}$ be a graph. By Theorem 1 there exists a polynomial-delay algorithm to enumerate all minimal $g \in \{0, 1, 2\}^V$ with property $\mathcal{P}$ on $G = (V, E) \in \mathcal{G}$. By [2, Proposition 8], a minimal Rdf is uniquely defined by the vertices of value 2. Together with Lemma 1 there is a bijection between all minimal Rdfs on G and subsets $A \subseteq V$ such that for all $v \in A$, $P_{G,A}(v) \not\subseteq \{v\}$. Hence, the number of elements with property $\mathcal{P}$ on G is bounded by $r^{|V|} \cdot p(|V|)$ for the given polynomial p. As the delay is bounded by a polynomial q, the running time of the algorithm is $q(|V|)p(|V|)r^{|V|}$ or $\mathcal{O}^*(r^n)$. $\quad\square$

Fig. 1. Non-trivial no-instance of EXT MAX RD.

4   Maximal Roman Dominating Functions

In this section, we want to make use of the results in Sect. 3 by looking at maximal Roman dominating functions. For a graph $G = (V, E)$, a function $f \in \{0, 1, 2\}^V$ is a *maximal Roman dominating function* (or mRdf for short) on G iff $N(v) \cap V_2(f) \neq \emptyset$ for all $v \in V_0(f)$ and $N[V_0(f)] \neq V$; see [22]. First, we will show a characterization result for minimal mRdfs.

Theorem 2. $(\star)$ *Let $G = (V, E)$ be a graph and $f \in \{0, 1, 2\}^V$. f is a minimal mRdf iff the following constraints hold.*

1. *f is an mRdf.*
2. *For all $v \in V_1(f)$, either $N(v) \cap V_2(f) = \emptyset$ or $V \setminus N[V_0(f)] \subseteq N[v]$.*
3. *For all $v \in V_2(f)$, $P_{G[V_0(f) \cup V_2(f)], V_2(f)}(v) \not\subseteq \{v\}$.*

To show that our approach is really helpful, we prove hardness for the extension problem. First, let us consider a non-trivial no-instance. Let $G = (\{1, 2, 3, 4\}, \{\{1, 2\}, \{2, 3\}, \{3, 4\}\})$ be a path of 4 vertices with the function $f := 2 \cdot \chi_{\{1,4\}}$. This instance is visualized by Fig. 1. f is a Rdf, but $V_0(f)$ is a ds. Furthermore, each vertex in $V_0(f)$ is a unique private neighbor of a vertex in $V_2(f)$. Theorem 2 implies, that for any $g \in \{0, 1, 2\}^{\{1,2,3,4\}}$ with $f \leq g$, g is not a minimal mRdf.

Theorem 3. $(\star)$ EXT MAX RD *is* **NP**-*complete even when* $|V_1(f)| = 2$.

Let $\mathcal{MRDF}$ denote the property of mRdfs. To apply earlier results, we now show that $\mathcal{MRDF}$ is nice.

Lemma 5. $\mathcal{MRDF}$ *is a nrdp.*

Proof. Clearly, every mRdf is also a Rdf. Let $G = (V, E)$ be a graph and let f be an mRdf on G. Since domination and Roman domination are monotonically increasing (with respect to $\subseteq$ and $\leq$), $f + \chi_{\{v\}}$ is still an mRdf for all $v \in V_0(f)$. Therefore, we only need to consider the Constraint 3.

Let $v \in V_2(f)$ and $f_v := f - \chi_{\{v\}}$. Clearly, $V_0(f_v) = V_0(f)$. Therefore, if f_v is no mRdf, there is a $u \in V_0(f_v)$ which is not dominated by $V_2(f_v) = V_2(f) \setminus \{v\}$. Hence, $u \in N(v)$ is a private neighbor of v. If $P_{G[V \setminus V_1(f)], V_2(f)}(v) \subseteq \{v\}$, then for each $w \in V_0(f) = V_0(f_v)$, there exists a $u \in (V_2(f) \setminus \{v\}) \cap N(w) = V_2(f_v) \cap N(w)$. Therefore, $V_0(f) = V_0(f_v)$ is dominated by $V_2(f) \setminus \{v\} = V_2(f_v)$. □

Lemma 6. *Let $G = (V, E)$ be a graph and let $A \subseteq V$. Then, $|C_{\mathcal{MRDF},G}[A]| \leq n$. Furthermore, $C_{\mathcal{MRDF},G}[A]$ can be computed in polynomial time.*

Proof. We can assume that each $v \in A$ has a private neighbor but itself. Otherwise, $C_{\mathcal{MRDF},G}[A] = \emptyset$ by Constraint 3 of Theorem 2. This property can be checked in polynomial time.

Define $f \in \{0,1,2\}^V$ with $f := 2 \cdot \chi_A + \chi_{V \setminus N[A]}$. As $V_0(f) \subseteq N(A) = N(V_2(f))$, f is a (minimal) Rdf. Since each $v \in V \setminus N[A]$ has no neighbor in V_2, $f \leq g$ for all $g \in C_{\mathcal{MRDF},G}[A]$. Therefore, if f is an mRdf, then $C_{\mathcal{MRDF},G}[A] = \{f\}$. So, assume there is no $v \in V_1(f)$ such that $N(v) \cap V_0(f) \neq \emptyset$. Then define for all $v \in V \setminus A$, $g_v \in \{0,1,2\}^V$ with $g_v = f + \chi_{V_0(f) \cap N[v]}$. Clearly, $f \leq g_v$ and g_v is a Rdf. Since $N[v] \cap V_0(g_v) = N[v] \cap (V_0(f) \setminus N[v])$, g_v is an mRdf.

Assume there is a minimal mRdf h on G such that $V_2(h) = A$, but there is no $v \in V \setminus A$ with $h = g_v$. As h is a Rdf, $f \leq h$. By Theorem 2, there is a $v \in V_1(h) \subseteq V \setminus A$, such that $v \notin N[V_0(h)]$. Hence, for each $v \in V \setminus A$, $g_v = f + \chi_{N[v] \cap V_0(f)} \leq h$. Thus, h is not minimal in the first place and $C_{\mathcal{MRDF},G}[A] \subseteq \{g_v \mid v \in V \setminus A\}$. Therefore, $|C_{\mathcal{MRDF},G}[A]| \leq |V \setminus A| \leq n$.

It can be checked in polynomial time if f and g_v (for all $v \in V \setminus V_2$) are minimal mRdfs. Thus $C_{\mathcal{MRDF},G}[A]$ can be constructed in polynomial time. $\square$

Theorem 1 and Lemma 4 imply with Theorem 11 of [2] the following:

Corollary 2. *There is a polynomial-space algorithm that enumerates all minimal mRdfs of a given graph of order n with polynomial delay, in time $\mathcal{O}(1.9332^n)$.*

The results in [1] imply the following:

Corollary 3. *There is a polynomial-space algorithm that enumerates all minimal mRdfs of a given graph of order n with polynomial delay and in time*

- $\mathcal{O}(1.8940^n)$ *on chordal graphs;*
- $\mathcal{O}(\sqrt{3}^n)$ *on interval graphs or on trees;*
- $\mathcal{O}(\sqrt[3]{3}^n)$ *on split or on cobipartite graphs.*

The next corollary follows by Corollary 1 and Lemma 6.

Corollary 4. EXT MAX RD *is polynomial time solvable for the instance (G, f) if $N(V_1(f)) \cap V_2(f) = \emptyset$.*

Remark 1. In the proof of Theorem 3 we construct a graph $\widetilde{G} = (\widetilde{V}, \widetilde{E})$. For the hardness we makes use of a $t \in \widetilde{V}$ of value 1 for which a $s \in \widetilde{V}$ of value 2 is the only neighbor. Because of Constraint 2 of Theorem 2, there is no $w \in \widetilde{V} \setminus \{t\}$ of value 1 without a neighbor of value 0.

This is interesting since the instance would be polynomial time solvable, if the value of t would be 0 (see Corollary 4). Therefore, just one wrongly placed 1 is enough to make the extension problem hard. This is the motivation for considering the vertices of value 1 after choosing the vertices with the value 2.

5   Extension Total/Connected Roman Domination

For this chapter we want to consider total and connected Roman Domination. The so-called total Roman Domination parameter was introduced in [16]. A Rdf $f \in \{0,1,2\}^V$ is a *total Roman dominating function* (or *tRdf* for short) with respect to G if, for each $v \in V_1(f) \cup V_2(f)$, $N(v) \cap (V_1(f) \cup V_2(f))$ is not empty. A *connected Roman dominating function* (or *cRdf* for short) with respect to G is a Rdf such that $V_1(f) \cup V_2(f)$ is connected in G.[2] References [14, 16] provide proofs for the NP-completeness for the natural decision problem of TOTAL or CONNECTED ROMAN DOMINATION (given a graph $G = (V, E)$ and $k \in \mathbb{N}$, is there a tRdf or cRdf f with $\sum_{v \in V} f(v) \leq k$). We start by proving a characterization result of a minimal tRdf/cRdf:

Theorem 4. $(\star)$ *Let $G = (V, E)$ be a graph and $f \in \{0,1,2\}^V$. f is a minimal tRdf (or cRdf, resp.) iff the following conditions hold:*

1. *f is a tRdf (or cRdf, resp.),*
2. *for all $v \in V_1(f)$, (a) $N(v) \cap V_2(f) = \emptyset$, or (b) $G[(V_1(f) \cup V_2(f)) \setminus \{v\}]$ has an isolated vertex (or (b) $G[(V_1(f) \cup V_2(f)) \setminus \{v\}]$ is not connected, resp.) and*
3. *for all $v \in V_2(f)$, $P_{G[V_0(f) \cup V_2(f)], V_2(f)}(v) \nsubseteq \{v\}$.*

This result is important for the remaining section, especially for the next result:

Theorem 5. $(\star)$ $\mathcal{CRDF}$ *and* $\mathcal{TRDF}$ *are nrdps.*

5.1   Minimal cRdf/tRdf on Cobipartite Graphs

A graph $G = (V, E)$ is called *cobipartite* if there is a partition of V into two cliques C_1, C_2. It is known by [12] that enumerating minimal connected dominating sets is as hard as enumerating minimal hitting sets. The goal of this subsection is the following theorem:

Theorem 6. $(\star)$ *For all cobipartite graphs $G = (V, E)$ and $A \subseteq V$, $C_{\mathcal{CRDF},G}[A]$ and $C_{\mathcal{TRDF},G}[A]$ can be enumerated with polynomial delay and $|C_{\mathcal{CRDF},G}[A]| \in \mathcal{O}(|V|^2)$ and $|C_{\mathcal{TRDF},G}[A]| \in \mathcal{O}(|V|^2)$. Furthermore, there is a polynomial-delay algorithm enumerating all minimal cRdfs (or tRdfs) on cobipartite graphs of order n within time $\mathcal{O}^*(\sqrt[3]{3}^{\,n})$.*

We will shortly describe the idea of the proof. Details can be found in [17]. Let $G = (V, E)$ be a cobipartite graph with the two cliques C_1, C_2, $A \subseteq V$, $f_A := 2 \cdot \chi_A + \chi_{V \setminus N[A]}$ and $g \in C_{\mathcal{X RDF},G}[A]$, where $\mathcal{X} \in \{\mathcal{T}, \mathcal{C}\}$. We can conclude $|(V_1(g) \cap C_i) \setminus V_1(f_A)| \leq 1$ for $i \in \{1, 2\}$. Therefore, $|C_{\mathcal{X RDF},G}[A]| \leq \mathcal{O}(|V|^2)$ and we can enumerate all elements in $C_{\mathcal{X RDF},G}[A]$ with polynomial delay.

---

[2] There is a different definition in [20]. In our opinion, the given definition fits better than the original, military motivation of distributing legions.

5.2   Interval Graphs

In this subsection, we will consider cRdfs on interval graphs. A graph $G = (V, E)$ is an *interval graph* if for a each $v \in V$ there is an interval $I_v = [\ell_v, r_v]$ such that $\{u, v\} \in E$ iff $I_v \cap I_u \neq \emptyset$. We make use of *locally definable properties* from [3].

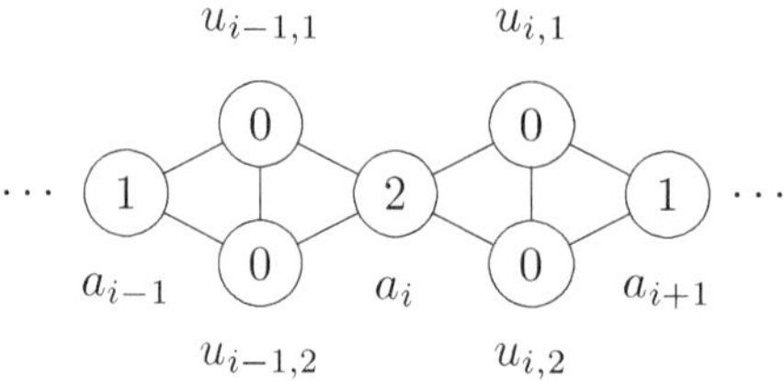

Fig. 2. Construction of G_n for $i \in \{2i \mid i \in \left[\lfloor \frac{n}{2} \rfloor\right]\}$.

Before going into details, we show that we cannot apply Lemma 4 in this case:

Example 1. We consider the graph $G_n := (V_n, E_n)$ with

$$V_n := \{a_i \mid i \in [n]\} \cup \{u_{i,j} \mid i \in [n-1], j \in [2]\},$$
$$E_n := \{\{a_i, u_{i,j}\} \mid i \in [n-1], j \in [2]\} \cup \{\{a_i, u_{i-1,j}\} \mid i \in [n] \setminus \{1\}, j \in [2]\}.$$

G_n is an interval graph with the representation $I_{a_i} = [i - 0.3, i + 0.3]$ and $I_{u_{k,j}} = [k + 0.2, k + 0.8]$ for $i \in [n], k \in [n-1], j \in [2]$. Furthermore, G_n is planar which can be observed in Fig. 2. We define $A := \{a_{2i} \mid i \in \left[\lfloor \frac{n}{2} \rfloor\right]\}$. Hence, $N(A) = \{u_{i,j} \mid i \in [n-1], j \in [2]\}$. For $f_A := 2\chi_A + \chi_{V \setminus N[A]}$, $V_1(f_A) \cup V_2(f_A)$ is an independent set. To extend f_A to minimal cRdf, we must add for all $i \in [n-1]$ exactly one of $u_{i,1}$ and $u_{i,2}$ to $V_1(f_A)$. Therefore, $|C_{G_n, \mathcal{CRDF}}[A]| \geq \sqrt[3]{2}^n$.

For our enumeration algorithm, we order the vertices by ℓ_v (for all $v, u \in V$, $v \leq u$ iff $\ell_v \leq \ell_u$). Let $A \subseteq V$. Define $f_A := 2\chi_A + \chi_{V \setminus N[A]}$. We call a set $X \subseteq N(A) \setminus A$, f_A-*connected* iff $f_A + \chi_X$ is a minimal cRdf.

As f_A-connectivity is a locally definable property on interval graphs $(\star)$, with [3, Theorem 3.2], we can deduce:

Corollary 5. *Let $G = (V, E)$ be an interval graph and $A \subseteq V$. The set of all f_A-connected sets can be enumerated with polynomial delay.*

Since there is a bijection between the f_A-connected sets on G and $C_{G, \mathcal{CRDF}}[A]$ which can be calculated in polynomial time, we obtain:

Theorem 7. *There is a polynomial-delay algorithm for enumerating all minimal cRdfs on interval graphs.*

Remark 2. The approach described in [3] use of an existing polynomial delay algorithm for enumerating paths on directed graphs. Therefore, in this subsection we consider the combination of our (extension) approach together with this one. This is another strength of the nrdp approach. By considering the vertices of value 1 after fixing the vertices of value 2, we can combine different approaches. Even if we do not consider further such cases it would be interesting to see more such combinations (for example with the supergraph method [12,13]).

6   Downsides of the Approach

Is this section, we show that using this method naively is not helpful every time. To this end, we define the following problem for any nrdp $\mathcal{P}$:

> **Problem name:** $\mathcal{P}$-Non-Emptiness
> **Given:** Graph $G = (V, E)$ and $A \subseteq V$.
> **Question:** Is $C_{\mathcal{P},G}[A] \neq \emptyset$?

Theorem 8. $\mathcal{CRDF}$-Non-Emptiness *is* **NP**-*complete even on bipartite 2-degenerate graphs of maximum degree 4.*

Proof. For the hardness we use Monotone 3-Sat-(2,2). This is a variation of 3-Sat, where each variable appears twice positively and twice negatively. [9] provides an NP-completeness proof for this problem. So, let $X := \{x_1, \ldots, x_n\}$ be the set of variables and $C := \{c_1, \ldots, c_m\}$ the set of clauses. To simplify the notation, denote $X' = \{x_i, \overline{x_i} \mid i \in [n]\}$. We can assume that there is no pair $(i, j) \in [n] \times [m]$ such that $\{x_i, \overline{x_i}\} \subseteq c_j$. Otherwise, we can delete c_j from C. For each $j \in [m]$, let $c_j = \{l_{j,1}, l_{j,2}, l_{j,3}\} \in C$ with $l_{j,1}, l_{j,2}, l_{j,3} \in X'$.

Define $G = (V, E)$ with

$$V := \{v_i, \overline{v_i}, w_i \mid i \in [n]\} \cup \{p_j \mid j \in [m]\} \cup \{u_i, u_i' \mid i \in [n-1]\},$$
$$E := \{\{v_i, w_i\}, \{\overline{v_i}, w_i\} \mid i \in [n]\} \cup \{\{u_i, w_i\}, \{u_i, w_{i+1}\}, \{u_i, u_i'\} \mid i \in [n-1]\} \cup$$
$$\{\{v_i, p_j\} \mid j \in [m], i \in [n], \exists k \in [3] : x_i = l_{j,k}\} \cup$$
$$\{\{\overline{v_i}, p_j\} \mid j \in [m], i \in [n], \exists k \in [3] : \overline{x_i} = l_{j,k}\}.$$

This graph has the partition classes $\{v_i, \overline{v_i} \mid i \in [n]\} \cup \{u_i \mid i \in [n-1]\}$ and $\{w_i \mid i \in [n]\} \cup \{u_i' \mid i \in [n-1]\} \cup \{p_j \mid j \in [m]\}$. Thus, G is a bipartite graph. Now we will show that G is 2-degenerate. At first, for all $i \in [n-1]$, $\deg(u_i') = 1$. After deleting these vertices from G, u_i has degree 2 for all $i \in [n-1]$ (u_i has degree 3 in G). Let G' be the graph after deleting also these vertices. Now let $i \in [n]$. While w_i has degree at most 4 in G, the degree in G' is 2. After deleting $\{w_r \mid r \in [n]\}$, $v_i, \overline{v_i}$ have degree 2. In G, $v_i, \overline{v_i}$ have degree 3. Therefore, we only need consider p_j for $j \in [m]$. But $G[\{p_j \mid j \in [m]\}]$ has no edges. Hence, G is a 2-degenerate graph with maximum degree 4.

Let $A := \{w_i \mid i \in [n]\} \cup \{u_i \mid i \in [n-1]\}$. The gadget is also depicted in Fig. 3.

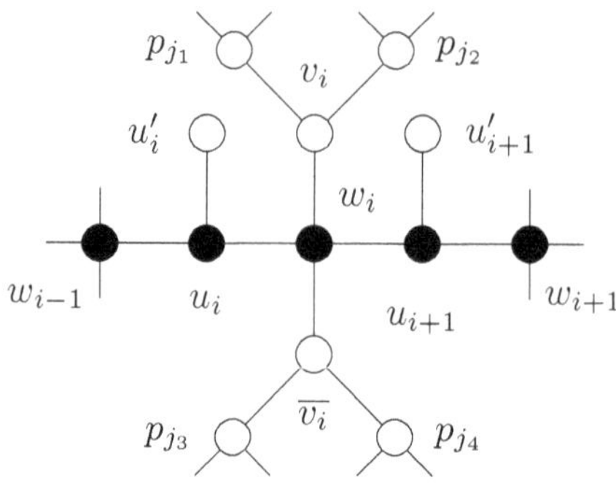

Fig. 3. Construction of Theorem 8 where the black vertices are in A.

Claim. There is an assignment ϕ satisfying C iff $C_{\mathcal{CRDF},G}[A] \neq \emptyset$.

Proof. Let ϕ be a satisfying assignment. Let $I = [n]$. If there is an $i \in I$, such that $\phi - (1 - \phi(x_i))\chi_{\{x_i\}}$ is still satisfying C, delete i from I. Since there is no pair $(i, j) \in [n] \times [m]$ with $\{x_i, \overline{x}_i\} \subseteq C_j$, ϕ restricted to I satisfies C. Define $h := 2 \cdot \chi_A + \chi_{\{v_i | i \in I, \phi(x_i)=1\} \cup \{\overline{v_i} | i \in I, \phi(x_i)=0\}} + \chi_{\{p_1,\ldots,p_m\}}$. Clearly, h is a Rdf, as $V \setminus N(A) = \{p_1, \ldots, p_m\}$. As ϕ satisfies C, for each $j \in [m]$ there is an $i \in I$ such that w_i is connected to p_j in $V_1(h) \cup V_2(h)$. Furthermore, A is connected. Hence, $V_1(h) \cup V_2(h)$ is connected and h a cRdf. Next, we want to prove that each element in $V_2(h) = \{w_1, u_1, w_2, \ldots, u_{n-1}, w_n\}$ has a private neighbor. Since ϕ is assignment, $\emptyset \neq \{v_i, \overline{v}_i\} \cap V_0(h) = N(w_i) \cap V_0(h)$ for all $i \in [n]$. Further, u_i' is the private neighbor of u_i for $i \in [n-1]$. This implies Property 3 of Theorem 4. For $j \in [m]$, $N(p_j) \cap V_2(h) = \emptyset$. This leaves only to show that $G[(V_1(h) \cup V_2(h)) \setminus \{t\}]$ is not connected for any $t \in \{v_i, \overline{v}_i \mid i \in I\} \cap V_1(h)$. Let $i \in I$ and w.l.o.g $h(v_i) = 1$ (for $h(\overline{v}_i) = 1$ works analogously). By definition of I there is a $j \in [m]$ such that C_j is only satisfied by x_i. Hence, p_j has no neighbor in $V_1(h - \chi_{\{v_i\}}) \cup V_2(h - \chi_{\{v_i\}})$. Therefore, h is a minimal cRdf.

Conversely, assume there is a minimal cRdf $g \in \{0, 1, 2\}^V$ with $f \leq g$. Define $\phi := \chi_{\{x_i | i \in [n], g(v_i) \neq 0\}}$. Let $i \in [n]$. Since $w_i \in V_2(f) \subseteq V_2(g)$, it needs a private neighbor and $\{v_i, \overline{v}_i\} \cap V_0(g) \neq \emptyset$. Let $j \in [m]$. W.l.o.g., we consider C_j includes only negative literals. Since g is a cRdf, there is an $i \in [n]$ with $\overline{v}_i \in N(p_j) \cap (V_1(g) \cup V_2(g))$. Hence, ϕ satisfies C_j.

Since the graph has order $3n + m$ and $6n$ edges, G and f can be constructed in polynomial time. $\square$

There are simular results for tRdfs:

Theorem 9. $(\star)$ *$\mathcal{TRDF}$-*NON-EMPTINESS *is* NP-*complete even on bipartite graphs of maximum degree 3.*

We have even some hardness results for split graphs. For the proof we use the following lemma:

Lemma 7. ($\star$) *Let $G = (V, E)$ be a connected split graph without universal vertices. Then $f \in \{0, 1, 2\}^V$ is a tRdf on G iff f is a cRdf.*

Theorem 10. ($\star$) ENUM $C_{\mathcal{CRDF}}$ *and* ENUM $C_{\mathcal{TRDF}}$ *are at least as hard as* ENUM TR *even on split graphs.*

To clarify, why these hardness result are limitations to our approach, we present the construction of the proof. The complete prove can be found in [17].

Proof. Let $H = (X, S)$ be the hypergraph with $X := \{x_1, \ldots, x_n\}$ and $S := \{s_1, \ldots, s_m\}$. We can assume that there is no $i \in [n]$ such that $x_i \in s_j$ for all $j \in [m]$. Otherwise, $Tr(G) = \{\{x_i\}\} \cup Tr((X \setminus \{x_i\}, S))$. Define $G = (V, E)$ with

$$V := \{a, b\} \cup \{u_i \mid i \in [n]\} \cup \{w_j \mid j \in [m]\}$$

$$E := \binom{\{a, b\} \cup \{u_i \mid i \in [n]\}}{2} \cup \{\{u_i, w_j\} \mid i \in [n], j \in [m], x_i \in s_j\}.$$

Claim. $F : C_{\mathcal{CRDF}, G}[\{a\}] \to Tr(H), g \mapsto \{x_i \mid i \in [n], g(u_i) = 1\}$ is a bijection.

$\square$

Remark 3. The hardness result in this section imply that this approach is not helpful in every case. For the proof of Theorem 1, we need a output-polynomial enumeration algorithm for $C_{\mathcal{P}}[A]$. This is not possible, if it is NP-complete to decide if there is element in $C_{\mathcal{P}}[A]$.

The algorithm of Theorem 1 does not consider $C_{\mathcal{P}, G}[B]$ for $B \subseteq A$. The supergraph method, for example, uses already existing solutions to compute new once. This implies the question, whether we could use different approaches (besides extension) for results like Theorem 1.

By Theorem 10, we know, that even the idea of using $C_{\mathcal{CRDF}, G}[B]$ for $C_{\mathcal{CRDF}, G}[A]$, when $B \subseteq A$, is not helpful in every case. In this proof $A = \{a\}$. So, the only proper subset is $\emptyset$. By the definition, $C_{\mathcal{CRDF}, G}[\emptyset] = \{\chi_V\}$. Hence, using proper subsets is not helpful in every case.

7   Conclusion

We introduced the notion of nrdp and used it to obtain some enumeration results for different Roman domination variations like mRdf, tRdf, cRdf. Furthermore, we proved that our approach is not always helpful unless $\mathsf{P} = \mathsf{NP}$. In these cases, it does not prove that the enumeration is hard. Thus, it would be interesting to prove some results for enumerating minimal cRdfs/tRdfs on general graphs. There could be even more graph classes with good enumeration results. Here, it could be interesting to consider to use different approaches/methods as mentioned in Remark Remark 2 and Remark 3. If we have hardness results we could also use the approach to obtain FPT delay algorithms. Improving the input-sensitive bounds for enumerating minimal Rdfs could strengthen results implied by Lemma 4.

Furthermore, not all Roman domination variations are nrdp. For example, unique response Roman domination (see [21]) or Italian domination (see [6,11]) fail to be. Finding more nrdps, or even a generalization of nrdp which also handles some of these examples could be a future research direction. Another question is if there are results beyond enumeration which could use nrdp.

References

1. Abu-Khzam, F.N., Fernau, H., Mann, K.: Roman census: enumerating and counting Roman dominating functions on graph classes. In: Leroux, J., Lombardy, S., Peleg, D. (eds.) 48th International Symposium on Mathematical Foundations of Computer Science, MFCS. Leibniz International Proceedings in Informatics (LIPIcs), vol. 272, pp. 6:1–6:15. Schloss Dagstuhl – Leibniz-Zentrum für Informatik (2023)
2. Abu-Khzam, F.N., Fernau, H., Mann, K.: Minimal Roman dominating functions: extensions and enumeration. Algorithmica **86**, 1862–1887 (2024)
3. Blind, S., Creignou, N., Olive, F.: Locally definable vertex set properties are efficiently enumerable. Discret. Appl. Math. **303**, 186–202 (2021)
4. Casel, K., Fernau, H., Ghadikolaei, M.K., Monnot, J., Sikora, F.: On the complexity of solution extension of optimization problems. Theoret. Comput. Sci. **904**, 48–65 (2022)
5. Chapelle, M., Cochefert, M., Couturier, J.F., Kratsch, D., Liedloff, M., Perez, A.: Exact algorithms for weak Roman domination. In: Lecroq, T., Mouchard, L. (eds.) IWOCA 2013. LNCS, vol. 8288, pp. 81–93. Springer, Heidelberg (2013). https://doi.org/10.1007/978-3-642-45278-9_8
6. Chellali, M., Haynes, T.W., Hedetniemi, S.T., McRae, A.A.: Roman 2-domination. Discret. Appl. Math. **204**, 22–28 (2016)
7. Cockayne, E.J., Dreyer, P., Jr., Hedetniemi, S.M., Hedetniemi, S.T.: Roman domination in graphs. Discret. Math. **278**, 11–22 (2004)
8. Creignou, N., Kröll, M., Pichler, R., Skritek, S., Vollmer, H.: A complexity theory for hard enumeration problems. Discret. Appl. Math. **268**, 191–209 (2019)
9. Darmann, A., Döcker, J.: On simplified NP-complete variants of monotone 3-SAT. Discret. Appl. Math. **292**, 45–58 (2021)
10. Gunopulos, D., Khardon, R., Mannila, H., Toivonen, H.: Data mining, hypergraph transversals, and machine learning. In: Mendelzon, A.O., Özsoyoglu, Z.M. (eds.) Proceedings of the Sixteenth ACM SIGACT-SIGMOD-SIGART Symposium on Principles of Database Systems, PODS. pp. 209–216. ACM Press (1997)
11. Henning, M.A., Klostermeyer, W.F.: Italian domination in trees. Discret. Appl. Math. **217**, 557–564 (2017)
12. Kobayashi, Y., Kurita, K., Mann, K., Matsuo, Y., Ono, H.: Enumerating minimal vertex covers and dominating sets with capacity and/or connectivity constraints. Algorithms **18**(2) (2025)
13. Kurita, K., Wasa, K.: An approximation algorithm for K-best enumeration of minimal connected edge dominating sets with cardinality constraints. Theoret. Comput. Sci. **1005**, 114628 (2024)
14. Li, K., Ran, Y., Zhang, Z., Du, D.: Nearly tight approximation algorithm for (connected) Roman dominating set. Optim. Lett. **16**(8), 2261–2276 (2022)
15. Liedloff, M., Kloks, T., Liu, J., Peng, S.L.: Efficient algorithms for Roman domination on some classes of graphs. Discret. Appl. Math. **156**(18), 3400–3415 (2008)

16. Liu, C.H., Chang, G.J.: Roman domination on strongly chordal graphs. J. Comb. Optim. **26**(3), 608–619 (2013)
17. Mann, K.: Enumeration with nice Roman domination properties. Technical. Rep. 2511.20367, ArXiv, Cornell University, USA (2025)
18. Mann, K., Fernau, H.: Perfect Roman domination: aspects of enumeration and parameterization. Algorithms **17**(12:576) (2024)
19. Marino, A.: Analysis and Enumeration. Algorithms for Biological Graphs, Atlantis Studies in Computing, vol. 6. Atlantis Press, Paris (2015)
20. Muddebihal, M.H.: Sumangaladevi: connected Roman domination in graphs. Int. J. Res. Eng. Technol. **2**, 333–340 (2013)
21. Rubalcaba, R.R., Slater, P.J.: Roman dominating influence parameters. Discret. Math. **307**(24), 3194–3200 (2007)
22. Shao, Z., Song, Y., Liu, Q., Duan, Z., Jiang, H.: On maximal Roman domination in graphs: complexity and algorithms. RAIRO Informatique théorique et Appl./Theor. Inf. Appl. **58**(4), 2709–2731 (2024)
23. Stewart, I.: Defend the Roman empire. Scientific American **281**(6), 136–139 (1999)
24. Strozecki, Y.: Enumeration complexity. EATCS Bull. **129** (2019)

Enumeration Kernels of Polynomial Size
for Cuts of Bounded Degree

Christian Komusiewicz[1] and Diptapriyo Majumdar[2]($\boxtimes$)

[1] Friedrich-Schiller-Universität Jena, Jena, Germany
c.komusiewicz@uni-jena.de
[2] Indraprastha Institute of Information Technology Delhi, New Delhi, India
diptapriyo@iiitd.ac.in

Abstract. We consider the NP-complete d-CUT problem in the context of enumeration kernelization. The decision version of d-CUT asks if a given undirected graph G has a cut (A, B) such that every $u \in A$ has at most d neighbors in B and every $v \in B$ has at most d neighbors in A. We study various structural parameterizations of three different enumeration variants of the problem–ENUM d-CUT, ENUM MIN-d-CUT and ENUM MAX-d-CUT in which one aims to enumerate all d-cuts, all inclusion-wise *minimal* d-cuts, and all inclusion-wise *maximal* d-cuts, respectively. For parameterization by vertex cover number (vc) and neighborhood diversity (nd) we provide polynomial-delay enumeration kernels of polynomial size for ENUM d-CUT and ENUM MAX-d-CUT and fully-polynomial enumeration kernels of polynomial size for ENUM MIN-d-CUT. For parameterization by the clique partition number (pc), we provide bijective enumeration kernels for all three problems.

Keywords: Parameterized Complexity · Kernelization · Enumeration Algorithms · d-Cut · Polynomial-Delay Enumeration Kernelization

1 Introduction

Kernelization [13,15,16] is a central subfield of parameterized complexity that was initially explored for decision problems. The objective of a kernelization algorithm is to shrink the input instance in polynomial time into an equivalent instance the size of which is dependent only on the parameter. The effectiveness of a kernelization is determined by the size bound of the kernel. Later, several extensions of kernelization have been adopted for optimization problems [27], counting problems [19,26], and enumeration problems [12,14,17]. To make the kernelization framework flexible enough to handle enumeration problems with many solutions, Creignou et al. [10–12] introduced a notion which asks for two algorithms: The first algorithm is the *kernelization algorithm* that shrinks the

Research of Diptapriyo Majumdar has been supported by Anusandhan National Research Foundation (ANRF) research grant SRG/2023/001592.

© The Author(s), under exclusive license to Springer Nature Switzerland AG 2026
J. Kozik and A. Wolff (Eds.): SOFSEM 2026, LNCS 16448, pp. 172–186, 2026.
https://doi.org/10.1007/978-3-032-17801-5_13

input to an instance (the kernel) whose size is bounded by the parameter. The second algorithm is the *solution-lifting algorithm* that given a solution of the kernel, outputs a collection of solutions to the input instance. The definition of Creignou et al. [10–12] allows the solution-lifting algorithm to run in FPT-delay.

Golovach et al. [17] observed that allowing FPT-delay is, however, too weak: In this case, the solution-lifting algorithm may *ignore* the kernel when outputting all solutions to the input instance which implies that all problems with FPT enumeration algorithms have kernels of constant size. Consequently, the instance produced by the kernel is not necessarily a representation of the original instance. To remedy this, Golovach et al. [17] introduced two *refined* notions of enumeration kernels: (i) *fully-polynomial enumeration kernels* and (ii) *polynomial-delay enumeration kernels* where the solution-lifting algorithms must run in polynomial time and polynomial delay, respectively (see Definitions 2.1). This definition has several desirable consequences [17]: First an enumeration problem admits a polynomial-delay (or polynomial-time) enumeration algorithm if and only if it admits a polynomial-delay (fully-polynomial) enumeration kernel of constant size. This is analogous to the fact that a problem admits a polynomial-time algorithm if and only if it admits a kernel of constant size. Second, a parameterized enumeration problem admits an FPT-delay algorithm if and only if it admits a polynomial-delay enumeration kernel. This automatic enumeration kernel however has exponential size with respect to the parameter. As for decision problems, this begs the following question: if a parameterized enumeration problem admits an FPT-delay algorithm, does it admit a polynomial-delay enumeration kernel of *polynomial size*? Such a kernel gives us a *compact* representation of all solutions of the original instance, and a guarantee that a collection of 'equivalent' solutions for each kernel solutions can be enumerated very efficiently. In this work, we aim to extend the set of enumeration problems for which such kernelizations are known.

Motivated by the work on polynomial-delay enumeration kernelization for MATCHING CUT [17] and the work on standard kernelization for the more general d-CUT problem [7], we initiate a systematic study of d-CUT from the perspective of enumeration kernelization, with the focus of providing fully polynomial enumeration kernelizations and polynomial-delay enumeration kernelizations. The d-CUT problem can be formalized as follows. A *cut* of a graph G is a partition (A, B) of $V(G)$ into two nonempty sets A and B. A cut (A, B) is a *d-cut* of G if every vertex of A has at most d neighbors in B and every vertex of B has at most d neighbors in A. For a fixed integer $d \geq 1$, the d-CUT problem asks whether a given graph G has a d-cut.

For the enumeration variants of d-CUT, it is more meaningful to characterize d-cuts via *edge cuts*. For two disjoint vertex sets A and B in G, let $E_G(A, B)$ denote the set of edges with one endpoint in A and the other endpoint in B. If (A, B) is a cut, then $E_G(A, B)$ is an *edge cut* and an edge cut $E_G(A, B)$ is a d-cut if every vertex of $V(G)$ is an endpoint of at most d edges of $E_G(A, B)$. This viewpoint allows us to reason about minimality and maximality of cuts: A d-cut (A, B) is *(inclusion) minimal* if there is no d-cut (A', B') such that

$E_G(A', B') \subsetneq E_G(A, B)$. Similarly, a d-cut (A, B) of G is *(inclusion) maximal* if there is no d-cut (A', B') of G such that $E_G(A, B) \subsetneq E_G(A', B')$. This gives rise to the following three enumeration variants of d-CUT.

ENUM d-CUT/ENUM MIN-d-CUT/ENUM MAX-d-CUT
Input: A simple undirected graph $G = (V, E)$.
Objective: Enumerate all/all minimal/all maximal d-cuts of G.

We underline that in our definition, two d-cuts (A, B) and (A', B') are *identical* if their edge cuts are the same, that is, if $E_G(A, B) = E_G(A', B')$, and *distinct* otherwise. The 1-CUT problem is precisely the MATCHING CUT problem. Both MATCHING CUT and d-CUT are NP-complete even in special classes of graphs [5,8,9,22,24,25,28,29] and have been explored from the perspective of parameterized complexity [1–3,7,20].

Our Contributions. We apply both types of enumeration kernelizations to ENUM MIN-d-CUT, ENUM d-CUT, and ENUM MAX-d-CUT. As d-CUT is NP-hard [9], it is unlikely that ENUM d-CUT and the other two variants admit polynomial-delay enumeration algorithms. Moreover, none of these problems admits an enumeration kernelization of polynomial size when parameterized by a combined parameter of *solution size, maximum degree*, and *treewidth of the input graph* [7] and polynomial-size kernels are also excluded for parameterization by the vertex-integrity of the input graph (see Proposition 2.1). Hence, it is interesting to study structurally larger parameters such as minimum vertex cover size (vc) and neighborhood diversity (nd).

Our main results for all three variants are the following kernelizations for parameterization by the vertex cover number (vc) of the input graph.

Theorem 1. *For every fixed $d \geq 1$, ENUM MIN-d-CUT parameterized by vc admits a fully-polynomial enumeration kernel with $\mathcal{O}(d\mathsf{vc}^2)$ vertices.*

Theorem 2. *For every fixed $d \geq 1$, ENUM d-CUT parameterized by vc admits a polynomial-delay enumeration kernel with $\mathcal{O}(d\mathsf{vc}^2)$ vertices.*

Theorem 3. *For every fixed $d \geq 1$, ENUM MAX-d-CUT parameterized by vc admits a polynomial-delay enumeration kernel with $\mathcal{O}(d^3\mathsf{vc}^{d+1})$ vertices.*

Next, we consider the parameters neighborhood diversity (nd) and clique partition number (pc).

Theorem 4. *For every fixed $d \geq 1$, ENUM MIN-d-CUT parameterized by nd admits a fully-polynomial enumeration kernel with $\mathcal{O}(d^2\mathsf{nd})$ vertices. Moreover, ENUM d-CUT and ENUM MAX-d-CUT admit polynomial-delay enumeration kernels with $\mathcal{O}(d^2\mathsf{nd})$ vertices.*

Theorem 5. *For every fixed $d \geq 1$, ENUM d-CUT, ENUM MIN-d-CUT, and ENUM MAX-d-CUT parameterized by pc admits a bijective enumeration kernels with $\mathcal{O}(\mathsf{pc}^{d+2})$ vertices.*

Due to lack of space, the proofs of Theorems 3 and 5 and the proofs of several lemmas can be found in the full version available in arXiv:2308.01286.

Further Related Work. We refer to a survey by Wasa [31] for a detailed overview of enumeration algorithms. Bentert et al. [4] introduced a notion of *advice enumeration kernels*, an extension to the enumeration problem by Creignou et al. [12] where the solution-lifting algorithm does not need to know the whole input but rather only some possibly smaller advice. Another application of polynomial-delay enumeration kernels was given recently for structural parameterizations of the LONG PATH problem [21]. Additionally, there have been some recent works on FPT-delay enumeration algorithms for MATCHING MULTICUT [18] and polynomial-delay enumeration kernels for VERTEX COVER and FEEDBACK VERTEX SET [6].

2 Preliminaries

We use $[r]$ to denote $\{1, \ldots, r\}$ for some $r \in \mathbb{N}$ and $A \uplus B$ to denote the disjoint union of the sets A and B. For a set A, we use 2^A to denote the *power set* of A. For a finite set A and a nonnegative integer $d \geq 0$, we use $\binom{A}{d}$, $\binom{A}{\leq d}$, and $\binom{A}{\geq d}$ to denote the collection of all subsets of A of size equal to d, at most d, and at least d, respectively. Throughout this paper, we consider simple undirected graphs. For a subset $X \subseteq V(G)$, $G[X]$ denotes the subgraph induced by the vertex subset X. Similarly, $G - X$ denotes the graph obtained after deleting the vertex set X, that is, $G - X = G[V(G) \setminus X]$. For a vertex u, we write $G - u$ for $G - \{u\}$. Similarly, for a set of edges $F \subseteq E(G)$ we write $G - F$ to denote the graph obtained after deleting the edges of F. For an edge e, we write $G - e$ for $G - \{e\}$. For a vertex v, we denote by $N_G(v)$ the *open neighborhood* of v in G, i.e. the set of vertices that are adjacent to v in G and by $N_G[v] := N_G(v) \cup \{v\}$ the *closed neighborhood* of v in G. If the graph is clear from the context, we omit the subscript.

A set of vertices $S \subseteq V(G)$ is called *vertex cover* if for every $uv \in E(G)$, $u \in S$ or $v \in S$ (or both). The size of a minimum vertex cover of a graph is known as the *vertex cover number* and is denoted by vc. Given a graph G, a *neighborhood decomposition* of G is a partition $\mathcal{U} = \{X_1, \ldots, X_k\}$ of $V(G)$ such that every set X_i is a clique or an independent set and all the vertices of X_i have the same neighborhood outside of X_i. The *neighborhood diversity* of G, denoted by $\mathsf{nd}(G)$ is the size of a smallest neighborhood decomposition of G. A partition $\mathcal{C} = \{C_1, \ldots, C_k\}$ of $V(G)$ is a *clique partition* of G if for all $1 \leq i \leq k$, C_i is a clique. The *clique partition number* of G is denoted by $\mathsf{pc}(G)$ is the minimum number k such that G has a clique partition with k cliques. The *vertex-integrity* of G, denoted by $\mathsf{vi}(G)$ is the minimum number k such that G has a vertex set X of size $k' \leq k$ with the largest connected component in $G - X$ having size $k - k'$.

We make use of the following graph operation. Let G be an undirected graph such that $u \in V(G)$. We define the operation of *attaching an r-vertex pendant clique to u* as follows: add a clique C of r vertices disjoint from $V(G)$ to G such that for every $v \in C$, $N(v) \setminus C = \{u\}$.

Some Fundamental Properties of d-Cuts. A set $T \subseteq V(G)$ is *d-monochromatic* if for every d-cut (A, B) of G, either $T \subseteq A$ or $T \subseteq B$; since d is usually clear from context, we will write monochromatic instead. The following lemma gives fundamental structural characterizations of d-cuts of G. Some of them were already observed by Gomes and Sau [7].

Lemma 2.1. *The following statements hold true for every undirected graph G and every positive integer d.*

   (i) *If C is a clique of G with at least $2d+1$ vertices, then C is monochromatic.*
  (ii) *If X and Y are monochromatic sets with a nonempty intersection, then $X \cup Y$ is monochromatic.*
 (iii) *If a vertex u has more than d neighbors in a monochromatic set X, then $X \cup \{v\}$ is monochromatic.*
  (iv) *Let T be a subset of $V(G)$ such that there are more than $2d$ vertices in $V(G) \setminus T$ whose neighborhoods contain T. Then, T is monochromatic.*

We can also observe the following for d-cuts without edges.

Observation 2.1. *A graph G is disconnected if and only if the unique minimal d-cut of G is $\emptyset$.*

Parameterized Complexity and Kernelization. An instance of a parameterized problem $L \subseteq \Sigma^* \times \mathbb{N}$ consists of an instance $x \in \Sigma^*$ and a natural number $k \in \mathbb{N}$ in which k is called the *parameter*. A parameterized problem L is *fixed-parameter tractable (FPT)* if there exists an algorithm $\mathcal{A}$, a computable function $f : \mathbb{N} \to \mathbb{N}$ such that given an instance $(x, k) \in \Sigma^* \times \mathbb{N}$, the algorithm $\mathcal{A}$ correctly decides whether $(x, k) \in L$ or not in $f(k) \cdot |x|^{\mathcal{O}(1)}$-time. We say that the algorithm $\mathcal{A}$ is a *fixed-parameter algorithm* (or *FPT algorithm*) for L. Observe that we allow exponential blow-up in k while the algorithm runs in polynomial in $|x|$. Considering a parameterized problem L, kernelization (or preprocessing) is often used as an important part to design FPT algorithm. Informally, it solves the easier parts of the instance in polynomial-time and reduces the instance size. Formally, a parameterized problem L admits a *kernelization* if there exists an algorithm $\mathcal{B}$ that given an instance $(x, k) \in \Sigma^* \times \mathbb{N}$, runs in time polynomial in $|x| + k$ and outputs an instance $(x, k) \in \Sigma^* \times \mathbb{N}$ such that

   1. $(x, k) \in L$ if and only if $(x', k') \in L$, and
   2. $|x'| + k' \leq g(k)$ for some computable function $g : \mathbb{N} \to \mathbb{N}$.

In this definition, the output instance (x', k') is called a *kernel* and g is the *size* of the kernel. It is well-known due to [13,15] that a decidable parameterized problem is FPT if and only if it admits a kernelization. For more details related to parameterized complexity and kernelization, we refer to [13,15,16].

Parameterized Enumeration and Enumeration Kernelization. We use the framework for parameterized enumeration that was proposed by Creignou et al. [12]. An *enumeration problem* over a finite alphabet Σ is a tuple (L, Sol) such that

   (i) $L \subseteq \Sigma^*$ is a decidable problem, and

(ii) $\mathsf{Sol} : \Sigma^* \to 2^{\Sigma^*}$ is a computable function such that $\mathsf{Sol}(x)$, the *set of solutions of* x, is nonempty if and only if $x \in L$.

Observe that L is decidable since Sol is a computable function. A *parameterized enumeration problem* is defined as a triple $\Pi = (L, \mathsf{Sol}, \kappa)$ such that (L, Sol) is an enumeration problem and $\kappa : \Sigma^* \to \mathbb{N}$ is the *parameter*. We define here the parameter as a computable function $\kappa(x)$; it is natural to assume that the parameter is given with the input or $\kappa(x)$ can be computed in polynomial time. An *enumeration algorithm* $\mathcal{A}$ for a parameterized enumeration problem Π is a deterministic algorithm that given $x \in \Sigma^*$, outputs $\mathsf{Sol}(x)$ exactly without duplicates and terminates after a finite number of steps. If $\mathcal{A}$ terminates in $f(\kappa(x)) \cdot |x|^{\mathcal{O}(1)}$-time, then $\mathcal{A}$ is called an *FPT-enumeration algorithm*. For $x \in L$ and $1 \le i < |\mathsf{Sol}(x)|$, the i-th *delay* of $\mathcal{A}$ is the time taken between outputting the i-th and $(i+1)$-th solution of $\mathsf{Sol}(x)$. The time from the start of the algorithm until the first output is the 0-*th delay*, also called *precalculation* time. The time from the last output to the termination of $\mathcal{A}$ is the $|\mathsf{Sol}(x)|$-*th delay*, also called *postcalculation* time. If every delay of $\mathcal{A}$ is in $f(\kappa(x)) \cdot |x|^{\mathcal{O}(1)}$, then $\mathcal{A}$ is called an *FPT-delay enumeration algorithm*.

Definition 2.1. *Let $\Pi = (L, \mathsf{Sol}, \kappa)$ be a parameterized enumeration problem. A* polynomial-delay enumeration kernel(ization) *for Π is a pair of algorithms $\mathcal{A}$ and $\mathcal{A}'$ such that:*

(i) For every instance x of Π, the kernelization algorithm $\mathcal{A}$ *computes in time polynomial in $|x| + \kappa(x)$ an instance y of Π such that $|y| + \kappa(y) \le f(\kappa(x))$ for a computable function f.*

(ii) For every $s \in \mathsf{Sol}(y)$, the solution-lifting algorithm $\mathcal{A}'$ *computes with delay polynomial in $|x| + |y| + \kappa(x) + \kappa(y)$ a nonempty set of solutions $S_s \subseteq \mathsf{Sol}(x)$ such that $\{S_s \mid s \in \mathsf{Sol}(y)\}$ is a partition of $\mathsf{Sol}(x)$.*

We call f the size *of the kernel(ization). If f is a polynomial function, then Π is said to admit a polynomial-delay enumeration kernel(ization) of* polynomial size.

Observe that Property (ii) of the definition above implies that $\mathsf{Sol}(x) \ne \emptyset$ if and only if $\mathsf{Sol}(y) \ne \emptyset$.

Fully-polynomial enumeration kernel(ization) is defined as in Definition 2.1, the only difference is that condition (ii) is replaced by the following condition.

(ii*) For every $s \in \mathsf{Sol}(y)$, the *solution-lifting algorithm* $\mathcal{A}'$ computes in time polynomial in $|x| + |y| + \kappa(x) + \kappa(y)$ a nonempty set of solutions $S_s \subseteq \mathsf{Sol}(x)$ such that $\{S_s \mid s \in \mathsf{Sol}(y)\}$ is a partition of $\mathsf{Sol}(x)$.

An enumeration kernel is *bijective* if for every $s \in \mathsf{Sol}(y)$, the solution-lifting algorithm produces in polynomial time a unique solution $\hat{s} \in \mathsf{Sol}(x)$ giving a bijection between $\mathsf{Sol}(y)$ and $\mathsf{Sol}(x)$. Bijective enumeration kernelizations are fully-polynomial enumeration kernelizations which in turn are polynomial-delay enumeration kernelizations [17].

Nonexistence of Enumeration Kernels for d-Cut. To put our main positive results in context, we observe that for some parameters, certain kernelizations of polynomial size can be easily excluded. First, we show that when parameterized by the vertex-integrity, none of the desired kernels is achievable.

Proposition 2.1. *For every fixed $d \geq 1$, ENUM d-CUT, ENUM MAX-d-CUT, and ENUM MIN-d-CUT when parameterized by the vertex integrity (vi) or by the treedepth (td) do not admit polynomial-delay enumeration kernels of polynomial size unless* $NP \subseteq coNP/poly$.

For the enumeration of all or all maximal d-cuts, we can exclude some fully-polynomial enumeration kernels by showing that there are too many solutions.

Proposition 2.2. ENUM d-CUT *and* ENUM MAX-d-CUT *parameterized by* vc *or* nd *do not admit fully-polynomial enumeration kernels of any size.*

3 Parameterization by the Vertex Cover Number

In this section, we provide a fully-polynomial enumeration kernel for ENUM MIN-d-CUT and a polynomial-delay enumeration kernel for ENUM d-CUT when parameterized by the vertex cover number (vc) of the input graph. For both kernels, we assume that the input consists of the graph G together with a vertex cover S of size $k \leq 2vc(G)$. We can make this assumption without loss of generality since such a vertex cover can be computed in linear time [30]. We let $I = V(G) \setminus S$ denote the corresponding independent set.

3.1 Enumerating Minimal d-Cuts

Marking Scheme and Kernelization Algorithm. If the input graph G is disconnected, then our kernelization algorithm outputs $H = 2K_1$ as the output graph. Otherwise, we proceed as follows. As S is a vertex cover of G, every vertex of I has at least one neighbor in S since G is connected. We partition $I = I_1 \uplus I_2$ where I_1 denotes the vertices of degree exactly one and I_2 denotes the vertices of degree at least two. Our marking scheme (inspired by the approach of Golovach et al. [17] for MATCHING CUT) works as follows.

(i) For every $x \in S$, mark an arbitrary vertex of $N_G(x) \cap I_1$ if such a vertex exists.

(ii) For every $\{x, y\} \in \binom{S}{2}$, choose $\min\{2d + 1, |N_G(x) \cap N_G(y) \cap I_2|\}$ arbitrary vertices from I_2 that are adjacent to x and y and mark them for the pair $\{x, y\}$.

(iii) If I contains at least one vertex of degree at most d, then do the following. Mark an arbitrary such vertex u, attach a pendant clique C of $2d + 2$ vertices to u and mark the vertices of C.

Let Z be the set of all marked vertices. Output the graph $H = G[S \cup Z]$. This completes the kernelization algorithm. Clearly, the algorithm has a polynomial running time. The main idea of the kernelization algorithm is that we remove vertices of I as long as we guarantee that their neighborhood is monochromatic. This is formalized as follows.

Lemma 3.1. *Assume that G is connected and let u be a vertex in $V(G) \setminus V(H)$, then $N_G(u)$ is monochromatic in G and in H.*

Proof. If $u \in I_1$, then $N_G(u)$ contains only one vertex which is trivially monochromatic. Hence, we assume that u has at least two neighbors. We show that for every pair of vertices v and w in $N_G(u)$, the set $\{v, w\}$ is monochromatic. Since u is not in $V(H)$, the vertices v and w have at least $2d + 1$ common neighbors in G and, by the construction of the kernel, also in H. Thus, by item (iv) of Lemma 2.1, $\{v, w\}$ is monochromatic in G and in H. $\qquad\square$

We may observe the following bound on the size of the kernel.

Observation 3.1. *If G is disconnected, then H is isomorphic to $2K_1$. If G is connected, then H is connected and has $\mathcal{O}(d \cdot \mathsf{vc}^2)$ vertices.*

Equivalence Classes of Minimal d-Cuts. It is not hard to see that H does not keep the information of all minimal d-cuts of G. For example, when some vertex v of I_1 is not contained in H, then the kernel does not contain a cut directly corresponding to the minimal d-cut $(\{v\}, V \setminus \{v\})$. To establish a relation between the minimal d-cuts of G and the minimal d-cuts of H, we define an equivalence relation as follows.

Let F_1 and F_2 be edge sets of the connected graph G. We say that F_1 and F_2 are *equivalent* if $F_1 = F_2$ or there are two distinct vertices $u, v \in I$ of degree at most d such that $F_1 = \{uw \in E(G) \mid w \in N_G(u)\}$ and $F_2 = \{vw \in E(G) \mid w \in N_G(v)\}$. Observe that the above-defined relation is an equivalence relation. The same notion of equivalence relation can be defined between an edge set $F_1 \subseteq E(G)$ and an edge set $F_2 \subseteq E(H)$ as well. The following observation is true by the definition of the equivalence relation.

Observation 3.2. *Let F_1 and F_2 be two edge sets of G (respectively, two edge sets of H, or $F_1 \subseteq E(G)$ and $F_2 \subseteq E(H)$) that are equivalent to each other. Then, F_1 is a d-cut of G (respectively, of H), if and only if F_2 is a d-cut of G (respectively, of H).*

Avoiding Duplicate Enumeration. To ensure that our solution-lifting algorithm does not enumerate duplicate d-cuts, we define the notion of distinguished minimal d-cut of H as follows. Let $F' = E_H(A', B')$ be a minimal d-cut of H. We say that (A', B') is *distinguished* if $A' = C \cup \{v\}$ for some $v \in I$, and $B' = V(H) \setminus A'$ such that C is a pendant clique of $2d + 1$ vertices attached to v and is disjoint from G. The polynomial-time solution-lifting algorithm now works as follows. If the given minimal d-cut $F' = E_H(A', B')$ of H is distinguished, then the solution-lifting algorithm outputs F' and all minimal d-cuts of G that are not minimal d-cuts of H. For every other minimal d-cut $F' = E_H(A', B')$ of H that is not distinguished, the solution-lifting algorithm only outputs F'. In the next lemma, we formally prove that we can design such an enumeration algorithm when a distinguished minimal d-cut of H is given as part of the input.

Lemma 3.2. *There exists a polynomial-time algorithm that outputs all minimal d-cuts of G that are not minimal d-cuts of H.*

Proof. We consider again the case that G is connected as, otherwise, $\emptyset$ is the unique minimal d-cut of G and H. We show the following characterization from which a polynomial-time algorithm can be easily derived.

> A d-cut $F = E_G(A, B)$ is a minimal d-cut of G that is not a minimal d-cut of H if and only if $F = E_G(\{u\}, V(G) \setminus \{u\})$ for some vertex $u \in I \setminus Z$ that has degree at most d.

First, let F be such a minimal d-cut. Since F is not a d-cut of H, it contains at least one edge incident with some vertex $u \in I \setminus Z$. By Lemma 3.1, $N_G(u)$ is monochromatic. Thus, F contains all edges incident with u. Consequently, u has degree at most d. Moreover, since F is minimal it contains no further edges, that is, $F = E_G(\{u\}, V(G) \setminus \{u\})$ as claimed. Conversely, let u be a vertex in $I \setminus Z$ that has degree at most d. Then, $F = E_G(\{u\}, V(G) \setminus \{u\})$ is a d-cut of G and since u is not contained in H it is not a d-cut of H. Since, again by Lemma 3.1, $N_G(u)$ is monochromatic, F is a minimal d-cut of G. $\qquad\square$

Using the above lemma, we are ready to prove our result.

Theorem 1. *For every fixed $d \geq 1$, ENUM MIN-d-CUT parameterized by vc admits a fully-polynomial enumeration kernel with $\mathcal{O}(d\mathsf{vc}^2)$ vertices.*

Proof. Our enumeration kernel has two parts. The first part is the kernelization algorithm and the second part is the solution-lifting algorithm. Let G be the input graph and S be a vertex cover of G such that $|S| = k \leq 2\mathsf{vc}(G)$. The kernelization algorithm invokes the *marking scheme* and outputs the graph H. It follows from Observation 3.1 that H has $\mathcal{O}(d|S|^2) = \mathcal{O}(d\mathsf{vc}^2)$ vertices.

Our solution-lifting algorithm works as follows. If H is a disconnected graph, then due to Observation 3.1, H is isomorphic to $2K_1$ and G is disconnected as well. Clearly, $\emptyset$ is the unique minimal d-cut of both G and H. Hence, given a minimal d-cut $\emptyset$ of H, the solution-lifting algorithm outputs $\emptyset$.

If G is connected, then due to Observation 3.1, H is also connected. Therefore, every minimal d-cut of H is nonempty. Let (A', B') be a given minimal d-cut. First, we output $E_G(A', B')$. Afterwards, we check whether (A', B') is the distinguished cut. More precisely, we check whether $A' = C \cup \{v\}$ and $B' = V(H) \setminus A'$ such that C is the pendant clique of $2d + 2$ vertices attached to v. If this is the case, then we invoke Lemma 3.2 to output all minimal d-cuts of G that are not minimal d-cuts of H in polynomial time.

The correctness is implied directly by the fact that every minimal d-cut of H is also a minimal d-cut of G and that the minimal d-cuts of G that are not minimal d-cuts of H are output only for the distinguished minimal cut of H. This completes the proof of a fully-polynomial enumeration kernel with $\mathcal{O}(d\mathsf{vc}^2)$ vertices for ENUM MIN-d-CUT parameterized by the vertex cover number. $\square$

3.2   Enumerating All d-Cuts

Next, we describe a polynomial-delay enumeration kernelization for ENUM d-CUT. While the kernelization is almost the same, the solution-lifting algorithm will be very different from the one in ENUM MIN-d-CUT. Recall that we may assume that the input consists of G and a vertex cover S with $|S| \leq 2\mathsf{vc}(G)$. We partition the independent set $I = V(G) \setminus S$ into $I_0 \uplus I_1 \uplus I_2$ where I_0 denotes the set of isolated vertices, I_1 denotes the vertices with degree 1, and I_2 denotes the vertices with degree at least 2 in I. In the following, we assume that G contains at least one edge, otherwise the kernelization is trivial.

Kernelization Algorithm. The marking scheme now works as follows.

1. Mark an arbitrary vertex from I_0.
2. For every $x \in S$, mark an arbitrary vertex from $N_G(x) \cap I_1$.
3. For every $\{x, y\} \in \binom{S}{2}$, choose an arbitrary set of $\min\{2d + 1, |N_G(x) \cap N_G(y) \cap I_2|\}$ vertices from I_2 that are adjacent to x and y, and mark them for the pair $\{x, y\} \in \binom{S}{2}$.

Let Z be the set of all marked vertices from I. The kernelization outputs $H = G[S \cup Z]$. As above, the removed vertices have a monochromatic neighborhood.

Lemma 3.3. *Let u be a vertex in $V(G) \setminus V(H)$, then $N_G(u)$ is monochromatic in G and in H.*

We have the following bound on the kernel size.

Observation 3.3. *Let H be the graph obtained from G after invoking the above marking scheme. Then, H has $\mathcal{O}(d \cdot |S|^2)$ vertices and at most one isolated vertex.*

Equivalence Classes of d-Cuts. Let F' be the edge set of a d-cut (A', B') of H. Let $S'_A = S \cap A'$, $S'_B = S \cap B'$, $I'_A = Z \cap A'$, and $I'_B = Z \cap B'$. By Lemma 3.3, for every $u \in I \setminus Z$, it holds that either $N_G(u) \subseteq A'$ or $N_G(u) \subseteq B'$. Let $J_A = \{u \in I \setminus Z \mid N_G(u) \subseteq A', N_G(u) \neq \emptyset\}$ and $J_B = \{u \in I \setminus Z \mid N_G(u) \subseteq B', N_G(u) \neq \emptyset\}$. Observe that $(A' \cup J_A, B' \cup J_B)$ is a d-cut of G. In particular, the edge cut of (A', B') and $(A' \cup J_A, B' \cup J_B)$ are the same. Hence, if the edge cut of some d-cut (A, B) of G is a proper superset of the edge cut of (A', B'), then there exist two sets $Q \subseteq J_A$ and $P \subseteq J_B$ such that $(A' \cup P \cup (J_A \setminus Q), B' \cup Q \cup (J_B \setminus P))$ is a d-cut and at least one of P and Q is nonempty. Informally, J_A and J_B are those vertices outside of the kernel that will create additional cut edges by being on a different side of the cut than their monochromatic neighborhood.

We now characterize those sets $Q \subseteq J_A$ and $P \subseteq J_B$ that give a d-cut. To this end, for every $u \in S'_A$, let $h(u)$ be the number of neighbors of u that are in B'. Analogously, for every $u \in S'_B$, let $h(u)$ be the number of neighbors of u that are in A'. We now say that a pair of sets $(P, Q) \in 2^{J_A} \times 2^{J_B}$ is *legal* with respect to a d-cut (A', B') of H if the following conditions are satisfied:

(i) every vertex of P has at most d neighbors in B',
(ii) every vertex of Q has at most d neighbors in A',
(iii) every $u \in S'_A$ has at most $d - h(u)$ neighbors in Q, and
(iv) every $u \in S'_B$ has at most $d - h(u)$ neighbors in P.

Finally, we call a nonempty cut (A, B) of G a *legal extension* of a nonempty d-cut (A', B') of H if $A = A' \cup P \cup (J_A \setminus Q)$ and $B = B' \cup Q \cup (J_B \setminus P)$ such that (P, Q) is a legal pair with respect to (A', B'). Observe that $E(A', B') \subseteq E(A, B)$ and that every edge of $E(A, B) \setminus E(A', B')$ is incident with at least one vertex that is not in $A' \cup B'$. The following lemma shows how legal extensions relate d-cuts of G with d-cuts of H.

Lemma 3.4. *Let (A', B') be a d-cut of H and (A, B) be a cut of G that is a legal extension of (A', B'). Then, (A, B) is a d-cut of G. Conversely, for every d-cut (A, B) of G, there exists a unique d-cut (A', B') of H such that (A, B) is a legal extension of (A', B').*

The Solution-Lifting Algorithm. Lemma 3.4 now provides the basis for the solution-lifting algorithm: we only need to enumerate for each d-cut of H all the legal extensions. The following lemma states that we can do this efficiently.

Lemma 3.5. *Let (A^*, B^*) be a d-cut of H. Then, there is an enumeration algorithm that enumerates all d-cuts of G that are legal extensions of (A^*, B^*) without duplicates with polynomial delay.*

Proof. Our enumeration algorithm ENUMEQVT is a recursive backtracking algorithm which maintains three disjoint sets P, Q, and D, each containing only vertices from $I \setminus V(H)$. In a call ENUMEQVT(P, Q, D), we search for all legal extensions (P', Q') of (A^*, B^*) such that $P \subseteq P'$, $Q \subseteq Q'$, and P' and Q' avoid D. We initialize D to be the set of all vertices from $I \setminus V(H)$ that have degree more than d. In addition, we initialize $P = Q = \emptyset$. In other words, the initial call is ENUMEQVT$(\emptyset, \emptyset, D)$.

Observe that the initial definition of D is correct in the sense that every legal extension must avoid D: Every vertex u of D has a monochromatic neighborhood of size at least $d + 1$. Hence, u cannot be added to A^* (respectively, to B^*) if $N_G(u) \subseteq B^*$ (respectively, if $N_G(u) \subseteq A^*$).

As the first step of ENUMEQVT(P, Q, D), we output the edge cut of the current d-cut $(A^* \cup P \cup (J_A \setminus Q), B^* \cup Q \cup (J_B \setminus P))$. Next, we consider one-by-one the vertices $v \in I \setminus (Z \cup D \cup P \cup Q)$. For each such vertex v, we check whether it can be added to P or Q, that is, we check whether

(i) $v \in J_A$ and $(P, Q \cup \{v\})$ is legal with respect to (A^*, B^*), or

(ii) $v \in J_B$ and $(P \cup \{v\}, Q)$ is legal with respect to (A^*, B^*).

If (i) is satisfied, then $(A^* \cup P \cup \{v\} \cup (J_A \setminus Q), B^* \cup Q \cup (J_B \setminus P))$ is a legal extension of (A^*, B^*). Then, the enumeration algorithm recursively calls itself with $(P \cup \{v\}, Q, D)$. If (ii) is satisfied, then $(A^* \cup P \cup (J_A \setminus Q), B^* \cup Q \cup \{v\} \cup (J_B \setminus P))$ is a legal extension of (A^*, B^*). Then, the enumeration algorithm recursively calls itself with $(P, Q \cup \{v\}, D)$. If neither of the above situations occur but a vertex $v \notin D \cup P \cup Q$ exists, then the algorithm recursively calls itself with $(P, Q, D \cup \{v\})$. When there does not exist any vertex outside $P \cup Q \cup D$ but in $J_A \cup J_B$ that are of degree at most d, then the algorithm terminates.

Delay Bound: First, observe that at each node of the tree defined by the recursive calls, the total time spent at that node (without the time for the recursive calls) is polynomial in n. Moreover, for each recursive call, one d-cut is output. Thus, to obtain a bound on the delay, it is sufficient to bound the number of nodes visited between two consecutive outputs. Now, after a recursive call, the algorithm visits at most n nodes before either spawning a new recursive call or terminating, since the tree has depth at most n. Therefore, the algorithm has polynomial delay.

Enumeration without duplicates: Observe that every legal extension (A, B) of (A^*, B^*) satisfies that $A = A^* \cup P \cup (J_A \setminus Q)$ and $B = B^* \cup Q \cup (J_B \setminus P)$. At every node of the tree, a legal extension (A, B) of (A^*, B^*) is output for a different pair (P, Q). Hence, no legal extension of (A^*, B^*) is output twice. $\square$

We now have all the ingredients to show the kernel. In particular, the solution-lifting algorithm is essentially a direct application of Lemma 3.5.

Theorem 2. *For every fixed $d \geq 1$,* ENUM d-CUT *parameterized by* vc *admits a polynomial-delay enumeration kernel with $\mathcal{O}(d\mathsf{vc}^2)$ vertices.*

4   Parameterization by the Neighborhood Diversity

Next, we sketch the proof of Theorem 4.

Theorem 4. *For every fixed $d \geq 1$,* ENUM MIN-d-CUT *parameterized by* nd *admits a fully-polynomial enumeration kernel with $\mathcal{O}(d^2\mathsf{nd})$ vertices. Moreover,* ENUM d-CUT *and* ENUM MAX-d-CUT *admit polynomial-delay enumeration kernels with $\mathcal{O}(d^2\mathsf{nd})$ vertices.*

Proof sketch. Let G be the input graph. First, we invoke a polynomial-time algorithm [23] to compute a neighborhood decomposition $\mathcal{U} = (X_1, \ldots, X_k)$ with the minimum module number $k = \mathsf{nd}(G)$. Now, if G has at most $(3d + 2)k$ vertices, then we output $H = G$ which gives a bijective enumeration kernelization. Otherwise, our kernelization algorithm and solution-lifting algorithm work as follows. Recall that every module X_i is a clique or an independent set. If X_i is a clique, then we call it a *clique module,* and if X_i is an independent set, we call it an *independent module*. We say that an independent module X_i is *nice* if $1 \leq |N(X_i)| \leq d$. Clearly, a nice module X_i is adjacent to at most d other modules that together span at most d vertices.

Kernelization. For each of the three problems, our kernelization algorithm is the same. If a module X_i is independent and not adjacent to any other module, then mark an arbitrary vertex from X_i. For every nice module $X_i \in \mathcal{U}$, mark $\min\{3d + 2, |X_i|\}$ vertices. If X_i is a nice module that is larger than $3d + 2$, that is, X_i has some unmarked vertex, then pick d arbitrary marked vertices $u_1, \ldots, u_d$ from X_i and, for every $i \in [d]$, attach a pendant clique C_i with $2d + i$ vertices into u_i. For every other module X_i, mark $\min\{2d + 1, |X_i|\}$ vertices. After this procedure is complete, remove the unmarked vertices of G to obtain the graph H, and output H. It is not hard to observe that $\mathsf{nd}(H) \leq (2d + 1)k$ and $|V(H)| \leq (3d^2 + 3d + 2)k = \mathcal{O}(d^2\mathsf{nd}(G))$.

Solution-Lifting. We describe the solution-lifting algorithm for ENUM d-CUT; the solution-lifting algorithms for ENUM MIN-d-CUT and ENUM MAX-d-CUT have similar flavors. Consider a d-cut (A', B') of H. Observe that for every nice module X_i of G with at least $3d + 3$ vertices, exactly $3d + 2$ vertices of X_i are in H. Out of these $3d + 2$ vertices, there are exactly d vertices to which pendant cliques of size $2d + 1, \ldots, 3d$ are attached. We consider the edge cut $F' = E_H(A', B')$, and initialize $A := A', B := B'$. As X_i is a nice module, $T = N_G(X_i)$ has at most d vertices. Due to item (iv) of Lemma 2.1, $T \subseteq A'$ or $T \subseteq B'$. We assume without loss of generality that $T \subseteq A'$. We call T an *r-suitable* set with respect to (A', B') if $X_i \cap B'$ contains exactly those r vertices of X_i to which the pendant cliques of size $\{2d + 1, \ldots, 2d + r\}$ have been attached. We look at every nice module X_i that has more than $3d + 2$ vertices in G one by one. If $N_G(X_i)$ is not r-suitable with respect to (A', B'), then we keep $X_i \cap B$ and $X_i \cap A$ unchanged. If $N_G(X_i)$ is r-suitable with respect to (A', B'), then we first enumerate all possible sets $Z \in \binom{X_i}{r}$ such that Z contains at least one unmarked vertex from X_i. For each set Z, we remove the vertices of $X_i \cap B$ from B, add the verticez of Z to B, and modify A accordingly. We repeat this for every nice module that has more than $3d + 2$ vertices in G. As there are at most k nice modules, and this enumeration has at most $\binom{n}{d}$ branches for each of the nice modules, the depth of the recursion is at most n. At every leaf, we output a d-cut (A, B). Observe that the computation at each branching node runs in polynomial time, and we output a d-cut in every leaf node. Therefore, this solution-lifting algorithm has polynomial delay. $\qquad\square$

5   Conclusion

It would be interesting to explore structurally smaller parameters such as feedback vertex set size or cluster vertex deletion set size. The existence of (polynomial-delay) enumeration kernels for these parameters is open for $d = 1$, that is, for the enumeration variants of MATCHING CUT. Another interesting research direction would be to identify natural parameterized problems whose decision versions admit a polynomial kernel but whose enumeration versions do not admit polynomial-delay enumeration kernelizations of polynomial size. Obtaining such a negative result would likely need a new lower bound framework.

Acknowledgments. We would like to thank the reviewers of SOFSEM for their valuable feedbacks and suggestions that have helped improve the presentation of our results.

References

1. Aravind, N.R., Kalyanasundaram, S., Kare, A.S.: On structural parameterizations of the matching cut problem. In: Gao, X., Du, H., Han, M. (eds.) COCOA 2017. LNCS, vol. 10628, pp. 475–482. Springer, Cham (2017). https://doi.org/10.1007/978-3-319-71147-8_34
2. Aravind, N.R., Kalyanasundaram, S., Kare, A.S.: Vertex partitioning problems on graphs with bounded tree width. Discret. Appl. Math. **319**, 254–270 (2022)
3. Aravind, N.R., Saxena, R.: An FPT algorithm for matching cut and d-cut. In: Flocchini, P., Moura, L. (eds.) IWOCA 2021. LNCS, vol. 12757, pp. 531–543. Springer, Cham (2021). https://doi.org/10.1007/978-3-030-79987-8_37
4. Bentert, M., Fluschnik, T., Nichterlein, A., Niedermeier, R.: Parameterized aspects of triangle enumeration. J. Comput. Syst. Sci. **103**, 61–77 (2019)
5. Bonsma, P.S.: The complexity of the matching-cut problem for planar graphs and other graph classes. J. Graph Theory **62**(2), 109–126 (2009)
6. Bougeret, M., Gomes, G.C.M., dos Santos, V.F., Sau, I.: Enumeration kernels for vertex cover and feedback vertex set. In: IPEC (2025). https://arxiv.org/abs/2509.08475
7. de C. M. Gomes, G., Sau, I.: Finding cuts of bounded degree: complexity, FPT and exact algorithms, and kernelization. Algorithmica **83**(6), 1677–1706 (2021)
8. Chen, C., Hsieh, S., Le, H., Le, V.B., Peng, S.: Matching cut in graphs with large minimum degree. Algorithmica **83**(5), 1238–1255 (2021)
9. Chvátal, V.: Recognizing decomposable graphs. J. Graph. Theor. **8**(1), 51–53 (1984)
10. Creignou, N., Kröll, M., Pichler, R., Skritek, S., Vollmer, H.: A complexity theory for hard enumeration problems. Discret. Appl. Math. **268**, 191–209 (2019)
11. Creignou, N., Ktari, R., Meier, A., Müller, J., Olive, F., Vollmer, H.: Parameterised enumeration for modification problems. Algorithms **12**(9), 189 (2019)
12. Creignou, N., Meier, A., Müller, J., Schmidt, J., Vollmer, H.: Paradigms for parameterized enumeration. Theor. Comput. Syst. **60**(4), 737–758 (2017)
13. Cygan, M., Fomin, F.V., Kowalik, L., Lokshtanov, D., Marx, D., Pilipczuk, M., Pilipczuk, M., Saurabh, S.: Parameterized Algorithms. Springer (2015)
14. Damaschke, P.: Parameterized enumeration, transversals, and imperfect phylogeny reconstruction. Theor. Comput. Sci. **351**(3), 337–350 (2006)
15. Downey, R.G., Fellows, M.R.: Fundamentals of Parameterized Complexity. Texts in Computer Science, Springer (2013)
16. Fomin, F.V., Lokshtanov, D., Saurabh, S., Zehavi, M.: Kernelization: Theory of Parameterized Preprocessing. Cambridge University Press (2019)
17. Golovach, P.A., Komusiewicz, C., Kratsch, D., Le, V.B.: Refined notions of parameterized enumeration kernels with applications to matching cut enumeration. J. Comput. Syst. Sci. **123**, 76–102 (2022)

18. Gomes, G.C.M., Juliano, E., Martins, G., dos Santos, V.F.: Matching (multi)cut: algorithms, complexity, and enumeration. In: Bonnet, É., Rzazewski, P. (eds.) 19th International Symposium on Parameterized and Exact Computation, IPEC 2024, September 4-6, 2024, Royal Holloway, University of London, Egham, United Kingdom. LIPIcs, vol. 321, pp. 25:1–25:15. Schloss Dagstuhl - Leibniz-Zentrum für Informatik (2024)
19. Jansen, B.M.P., van der Steenhoven, B.: Kernelization for counting problems on graphs: preserving the number of minimum solutions. In: Misra, N., Wahlström, M. (eds.) 18th International Symposium on Parameterized and Exact Computation, IPEC 2023, September 6-8, 2023, Amsterdam, The Netherlands. LIPIcs, vol. 285, pp. 27:1–27:15. Schloss Dagstuhl - Leibniz-Zentrum für Informatik (2023)
20. Komusiewicz, C., Kratsch, D., Le, V.B.: Matching cut: kernelization, single-exponential time FPT, and exact exponential algorithms. Discret. Appl. Math. **283**, 44–58 (2020)
21. Komusiewicz, C., Majumdar, D., Sommer, F.: Polynomial-size enumeration kernelizations for long path enumeration. In: Graph-Theoretic Concepts in Computer Science, 27th International Workshop, WG 2025, Otzenhausen, Germany, 2025, Proceedings (2025). https://doi.org/10.1007/978-3-032-11835-6_24
22. Kratsch, D., Le, V.B.: Algorithms solving the matching cut problem. Theor. Comput. Sci. **609**, 328–335 (2016)
23. Lampis, M.: Algorithmic meta-theorems for restrictions of treewidth. Algorithmica **64**(1), 19–37 (2012)
24. Le, H., Le, V.B.: A complexity dichotomy for matching cut in (bipartite) graphs of fixed diameter. Theor. Comput. Sci. **770**, 69–78 (2019)
25. Le, V.B., Telle, J.A.: The perfect matching cut problem revisited. Theor. Comput. Sci. **931**, 117–130 (2022)
26. Lokshtanov, D., Misra, P., Saurabh, S., Zehavi, M.: Kernelization of counting problems. In: Guruswami, V. (ed.) 15th Innovations in Theoretical Computer Science Conference, ITCS 2024, January 30 to February 2, 2024, Berkeley, CA, USA. LIPIcs, vol. 287, pp. 77:1–77:23. Schloss Dagstuhl - Leibniz-Zentrum für Informatik (2024)
27. Lokshtanov, D., Panolan, F., Ramanujan, M.S., Saurabh, S.: Lossy kernelization. In: Hatami, H., McKenzie, P., King, V. (eds.) Proceedings of the 49th Annual ACM SIGACT Symposium on Theory of Computing, STOC 2017, Montreal, QC, Canada, June 19-23, 2017. pp. 224–237. ACM (2017)
28. Lucke, F., Paulusma, D., Ries, B.: Finding matching cuts in H-free graphs. In: Bae, S.W., Park, H. (eds.) 33rd International Symposium on Algorithms and Computation, ISAAC 2022, December 19-21, 2022, Seoul, Korea. LIPIcs, vol. 248, pp. 22:1–22:16. Schloss Dagstuhl - Leibniz-Zentrum für Informatik (2022)
29. Patrignani, M., Pizzonia, M.: The complexity of the matching-cut problem. In: Brandstädt, A., Le, V.B. (eds.) Graph-Theoretic Concepts in Computer Science, 27th International Workshop, WG 2001, Boltenhagen, Germany, June 14-16, 2001, Proceedings. LNCS, vol. 2204, pp. 284–295. Springer (2001)
30. Savage, C.D.: Depth-first search and the vertex cover problem. Inf. Process. Lett. **14**(5), 233–237 (1982)
31. Wasa, K.: Enumeration of enumeration algorithms (2016). https://doi.org/10.48550/arXiv.1605.05102

Weighted Food Webs Make Computing Phylogenetic Diversity So Much Harder

Jannik Schestag[✉] [iD]

TU Delft, Delft, The Netherlands
`j.t.schestag@tudelft.nl`

Abstract. In a phylogenetic tree, present-day species are leaves and an edge from u to v indicates that u is an ancestor of v. Weights on these edges indicate the phylogenetic distance. The phylogenetic diversity (PD) of a set of species A is the total weight of edges that are on any path between the root of the phylogenetic tree and a species in A.

Selecting a small set of species that maximizes phylogenetic diversity for a given phylogenetic tree is an essential task in preservation planning, where limited resources naturally prevent saving all species. An optimal solution can be found with a greedy algorithm [Steel, Systematic Biology, 2005; Pardi and Goldman, PLoS Genetics, 2005]. However, when a food web representing predator-prey relationships is given, finding a set of species that optimizes phylogenetic diversity subject to the condition that each saved species should be able to find food among the preserved species is NP-hard [Spillner et al., IEEE/ACM, 2008].

We present a generalization of this problem, where, inspired by biological considerations, the food web has weighted edges to represent the importance of predator-prey relationships. We show that this version is NP-hard even when both structures, the food web and the phylogenetic tree, are stars. To cope with this intractability, we proceed in two directions. Firstly, we study special cases where a species can only survive if a given fraction of its prey is preserved. Secondly, we analyze these problems through the lens of parameterized complexity. Our results include that finding a solution is fixed-parameter tractable with respect to the vertex cover number of the food web, assuming the phylogenetic tree is a star.

1 Introduction

The ongoing *sixth mass extinction* [1,5] presents a significant challenge to humanity. There is a moral imperative to preserve species [21]; moreover, maintaining biodiversity is also critical for human well-being [4,28].

However, conservation efforts are constrained by limited political will, funding, and other resources, making it impossible to protect every species that is on the edge of extinction. As a result, strategic decisions are made about which species to prioritize. To provide biological evidence on how relevant the protection of a certain set of species (taxa) is, biologists developed the *phylogenetic*

Funded by the Dutch Research Council (NWO), OCENW.GROOT.2019.015.

© The Author(s), under exclusive license to Springer Nature Switzerland AG 2026
J. Kozik and A. Wolff (Eds.): SOFSEM 2026, LNCS 16448, pp. 187–202, 2026.
https://doi.org/10.1007/978-3-032-17801-5_14

diversity (PD) measure [10]. Given a phylogenetic tree—a directed tree where today's species are leaves and edges weights describe related a species is to its genetic parent—the phylogenetic diversity of a set of species A is the total weight of edges on paths from the root to species in A. Although phylogenetic diversity is not a perfect proxy for biological diversity [18], it is the best approach to capturing the number of unique features represented in a species set [11] and has become the most widely used biodiversity measures [39]. In the MAXIMIZE PHYLOGENETIC DIVERSITY (MAX-PD) problem, one is given a phylogenetic tree and a budget k, and the goal is to select k species that maximize phylogenetic diversity [10]. A greedy algorithm optimally solves MAX-PD [10,25,36].

One important extension is the problem OPTIMIZING PD WITH DEPENDENCIES (ε-PDD), introduced in [24], where a food web encodes predator-prey relationships. Here, the goal is to select k species that maximize phylogenetic diversity, with the constraint that each selected species must either be a food source of the ecological system or have at least one prey among the selected species. Food webs are key ecological models that describe species' roles in their environments and the flow of energy through ecosystems [27]. Introducing weights to these interactions–reflecting their ecological importance–gives further insight into the function of the system and has become increasingly common for food webs [14,23,42]. In fact, it has been noted that "weighting ecological interactions is especially important in case of food webs" [33]. However, ε-PDD assumes unweighted food webs, limiting its capacity to represent interaction significance.

Our Contribution. We close this gap by introducing WEIGHTED-PDD, a generalization of ε-PDD in which the food web is edge-weighted. We are tasked to select k species that maximize phylogenetic diversity under the constraint that each selected species is either a source or receives a total incoming weight of at least 1 from other selected species. We prove that WEIGHTED-PDD is NP-hard to solve, even on elementary instances, such as if the food web is a clique or a star.

To address this computational hardness, we pursue two directions. First, we define and study the RESTRICTED WEIGHTED PDD (RW-PDD) problem, where species require that a predefined fraction of their prey also be preserved. This problem is a special case of WEIGHTED-PDD and generalizes the following.

- ε-PDD: A selected species must have at least one preserved prey;
- $1/2$-PDD: At least half of the prey of a selected species must be preserved;
- 1-PDD: All prey of selected species must be preserved.

Second, we perform a detailed analysis within the framework of parameterized complexity, where we ask whether instances $\mathcal{I}$ of a problem Π can be solved in $f(\kappa) \cdot |\mathcal{I}|^{\mathcal{O}(1)}$ time (FPT) or $|\mathcal{I}|^{f(\kappa)}$ time (XP), where f is a computable function, κ is problem-specific parameter, and $|\mathcal{I}|$ the size of the instance. W[1]-hardness with respect to p provides evidence that no FPT-algorithm exists.

We examine RW-PDD and 1-PDD with respect to parameters categorizing the structure of the food web. We focus on the vertex cover number of instances of RW-PDD, where we provide an XP-algorithm in the general case and, for the case that the phylogenetic tree is replaced with a vertex-weighting, called RW-PDD$_s$, an FPT-algorithm. A comprehensive overview of the complexity results for RW-PDD and RW-PDD$_s$ with respect to the main structural parameters is provided in Fig. 3 and for 1-PDD and 1-PDD$_s$ in Fig. 5.

Structure of the Paper. In the next section, we prove first observations and the NP-hardness of WEIGHTED-PDD. In Sects. 3 and 4, we, respectively, analyze RW-PDD and 1-PDD with respect to structural parameters of the food web. Finally, in Sect. 5, we discuss our results and present future research ideas.

2   Preliminaries

2.1   Definitions

For a positive integer $a \in \mathbb{N}$, by $[a]$ we denote the set $\{1, 2, \ldots, a\}$, and by $[a]_0$ the set $\{0\} \cup [a]$. For functions $f, f' : A \to \mathbb{R}$, we define $f(A') := \sum_{a \in A'} f(a)$ for subsets A' of A, and we write $f' \leq f$ if $f'(a) \leq f(a)$ for all $a \in A$. For a condition Φ, the *Kronecker delta* δ_Φ is 1 if Φ holds and otherwise δ_Φ is 0.

We consider, unless stated otherwise, simple directed graphs $G = (V, E)$ with vertex-set $V(G) := V$ and edge-set $E(G) := E$. The *underlying undirected graph of* G is obtained by omitting edge directions. We write uv for directed edges from u to v and $\{u, v\}$ for an undirected edge between u and v. The *in-degree* $\deg^-(v)$ of a vertex v is the number of incoming edges at v. For a graph G and a vertex set $V' \subseteq V(G)$, the *subgraph of* G induced by V' is denoted with $G[V'] := (V', \{uv \in E(G) \mid u, v \in V'\})$. We define $G - V' := G[V \setminus V']$. A *star* with center v is a connected graph in which every edge is incident with v.

Phylogenetic Trees and Phylogenetic Diversity. A *tree* $T = (V, E)$ is a directed, connected, cycle-free graph, where the *root*, often denoted with ρ, is the only vertex with an in-degree of zero, and each other vertex has an in-degree of one. Vertices that have an out-degree of zero are called *leaves*.

For a given set X, a *phylogenetic X tree* $\mathcal{T} = (V, E, \omega)$ is a tree $T = (V, E)$ in which each non-leaf vertex has an out-degree of at least two, with an *edge-weight* function $\omega : E \to \mathbb{N}_{>0}$, and X is the set of leaves. In biological applications, X is a set of *taxa* (or species), all other vertices of $\mathcal{T}$ correspond to biological ancestors of these taxa and edge weight $\omega(uv)$ describes the *phylogenetic distance* between u and v. As u and v correspond to distinct, possibly extinct taxa, we assume this distance to be positive. For an edge $uv \in E$ in a tree, v is a *child* of u.

Given a phylogenetic tree $\mathcal{T}$ and set $A \subseteq X$, let $E_\mathcal{T}(A)$ denote the set of edges on a path to a leaf in A. The *phylogenetic diversity* $PD_\mathcal{T}(A)$ of A is

$$PD_\mathcal{T}(A) := \sum_{e \in E_\mathcal{T}(A)} \omega(e). \tag{1}$$

Informally, the phylogenetic diversity of a set A is the total weight of edges to A.

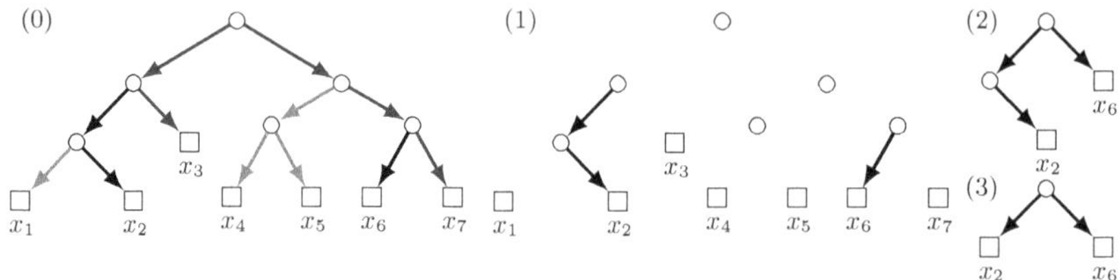

Fig. 1. (0): A hypothetical phylogenetic tree $\mathcal{T}$. For $A = \{x_3, x_7\}$ and $B = \{x_1, x_4, x_5\}$, blue edges are in $E_{\mathcal{T}}(A)$ and red edges are in $E_{\mathcal{T}}^+(B)$. For $i \in [3]$, (i) shows the (A, B)-contraction of $\mathcal{T}$ after Step i. To increase readability, edge weights are omitted.

A degree-2 vertex v with incident edges uv and vw is *contracted* if an edge uw with weight $\omega(uv) + \omega(vw)$ is added and v is removed. A vertex v *is identified with the root* ρ if all children of v become children of the root ρ and v is removed. Let $A, B \subseteq X$ be taxa sets. Let $E_{\mathcal{T}}^+(B)$ denote the set of edges uv for which only taxa in B can be reached from v. (See Fig. 1.) The (A, B)-*contraction* of a phylogenetic tree $\mathcal{T}$ results from applying exhaustively: 1) Remove edges in $E_{\mathcal{T}}(A)$ and in $E_{\mathcal{T}}^+(B)$. 2) Identify vertices that became in-degree zero vertices after Step 1 with the root. 3) Contract vertices with an in- and out-degree of 1.

We always consider (A, B)-contractions in the context of subtracting $PD_{\mathcal{T}}(A)$ from the threshold of diversity. Therefore, intuitively, the (A, B)-contraction of a tree is the tree resulting from saving taxa in A and letting taxa in B die out.

Food-Webs. For a set X of taxa, a *food web* $\mathcal{F} = (X, E)$ *on* X is a directed, acyclic graph with an edge-weight function $\gamma : E \to (0, 1]$. For each edge xy, we say x is *prey* of y and y is a *predator* of x. The set of prey and predators of x are $N_<(x)$ and $N_>(x)$, respectively. A taxon x without prey is a *source*.

For a food web $\mathcal{F}$, a set $A \subseteq X$ of taxa is γ-*viable* if $\sum_{e \in A_v} \gamma(e) \geq 1$ for each non-source $v \in A$, where A_v is the set of edges $uv \in E(\mathcal{F})$ with $u \in A$. In other words, each $v \in A$ is either a source of $\mathcal{F}$, or the total weight of edges incoming from another vertex in A is at least 1. If for each taxon all incoming edges have the same weight, then we say that γ is *restricted*. We observe that if γ is restricted and A is γ-viable, then for any non-source $v \in A$, at least $\gamma_v := \lceil \gamma(uv)^{-1} \rceil$ prey of v are in A, where uv is an arbitrary incoming edge of v.

Problem Definitions and Parameterizations. We define the following problem.

Weighted-PDD

Input: A phylogenetic X-tree $\mathcal{T}$, a food web $\mathcal{F}$ on X with edge-weights γ, and integers k and D.

Question: Is there a γ-viable set $S \subseteq X$ of size at most k such that $PD_{\mathcal{T}}(S) \geq D$?

The set S is called a *solution* of the instance. We adopt the convention that n is the number of taxa, $|X|$, and m is the number of edges of the food web, $|E(\mathcal{F})|$. Observe that $\mathcal{T}$ has $\mathcal{O}(n)$ edges. In RESTRICTED WEIGHTED PDD (RW-PDD), γ has to be restricted. The problems 1-PDD, ½-PDD, and ε-PDD are special cases of RW-PDD where, respectively, $\gamma(e)$ is $1/\deg^-(v)$, $2/\deg^-(v)$, and 1 for each edge e incoming at $v \in X$. Thus, a taxon can be saved only if all, half, or at least one of its prey are also preserved.

In the respective special cases RW-PDD$_\mathrm{s}$, 1-PDD$_\mathrm{s}$, ½-PDD$_\mathrm{s}$, and ε-PDD$_\mathrm{s}$, we require $\mathcal{T}$ to be a star. It is noted that such an instance can be viewed as only containing a vertex-weighted food web and no phylogenetic tree [12].

For an instance of RW-PDD, W_max as the maximum γ_v over taxa v.

2.2   Related Work

ε-PDD has been defined by Moulton et al. [24]. The conjecture that ε-PDD is NP-hard [35] has been proven in [12] even for the case that the food web is a directed tree—a spider graph to be more precise. Further, ε-PDD$_\mathrm{s}$ is NP-hard even if the food web is bipartite [12] but can be solved in polynomial time if the food web is a directed tree [12]. ε-PDD can be approximated with a constant factor if the longest path in the food web has a constant length [8].

It has been shown that ε-PDD is FPT when parameterized by the budget k plus the height of the phylogenetic tree [20]. Further, 1-PDD and ½-PDD are W[1]-hard and in XP, when parameterized by the budget k or the threshold of diversity D [16]. ½-PDD is W[1]-hard with respect to the treewidth of the food web, but FPT when parameterized with the food web's node scanwidth [32]. None of the three problems, 1-PDD, ½-PDD, and ε-PDD, admits a polynomial kernel with respect to vc $+D$, where vc is the vertex cover of the food web [16].

2.3   Preliminary Observations

Lemma 1. *Given an instance $\mathcal{I} = (\mathcal{T}, \mathcal{F}, k, D)$ of* WEIGHTED-PDD *and a set $A \subseteq X$, one can check whether A is a solution of $\mathcal{I}$ in $\mathcal{O}(n + m)$ time.*

Proof. We can compute whether $PD_\mathcal{T}(A) \geq D$ in $\mathcal{O}(n)$ time, by summing the weight of edges in $E_\mathcal{T}(A)$. One can check $|A| \leq k$ in $\mathcal{O}(k)$ time. To check whether A is γ-viable, we need to iterate over the set of prey for each taxon and check the weight of edges coming from A, which takes $\mathcal{O}(m)$ time.   □

Lemma 2. *Let $\mathcal{I} = (\mathcal{T}, \mathcal{F}, k, D)$ be a* yes-*instance of* WEIGHTED-PDD. *A solution of size of exactly k exists, subject to $k \leq |X|$.*

Proof. Let S be a solution for $\mathcal{I}$ with $|S| < k$. Assume $S \neq X$ and let x be a taxon in $X \setminus S$ being a source or having all prey in S. Such a taxon exists as $\mathcal{F}$ is a directed acyclic graph and has a topological order. Because S is γ-viable, also $S \cup \{x\}$ is γ-viable. Observe $PD_\mathcal{T}(S \cup \{x\}) \geq PD_\mathcal{T}(S)$ for each taxon $x \in X$. So $S \cup \{x\}$ is a solution and consequently, there is a solution of size k.   □

Fig. 2. An illustration of the Lemma 3. Here, all edges are directed towards the right.

Lemma 3 ($\star$). *Let $\mathcal{F}$ be a food web with $\gamma(uv) = 1/|N_<(v)|$. Let R and Q be sets of taxa, such that, in $\mathcal{F}$, taxa of $X \setminus R$ can not reach any taxon of R and taxa of Q can not reach any taxon of $X \setminus Q$. If S is γ-viable in $\mathcal{F} - (R \cup Q)$, then $S \cup R$ is γ-viable in $\mathcal{F}$.*

Proofs of theorems marked with $\star$ are partly or fully deferred to Arxiv [31]. For a visualization, consider Fig. 2.

2.4   Hardness of Weighted PDD

Now, we prove that solving WEIGHTED-PDD is NP-hard, even on instances that can be considered as containing only elementary information.

Theorem 1. WEIGHTED-PDD *is weakly NP-hard in general and W[1]-hard when parameterized by the solution size k, even if*

- *the phylogenetic tree is a star and the food web is a star, or*
- *the phylogenetic tree is a star and the food web is a clique.*

These cases become strongly NP-hard, if rationals are allowed as edge weights in the phylogenetic tree.

Note that ε-PDD is strongly NP-hard. However, if the food web is a star or a clique, then solving the problem can be done in polynomial time, because after compulsorily saving the source, all taxa can be selected without further conditions and the instance can be reduced to MAX-PD and solved with Faith's greedy algorithm [10]. Consequently, this theorem shows NP-hardness of cases that are computationally easy for ε-PDD and even for RW-PDD.

Proof. We reduce from KNAPSACK, in which a set of items $A = \{a_1, \ldots, a_n\}$, a cost-function $c : A \to \mathbb{N}$, a value-function $\nu : A \to \mathbb{N}$, and two integers $B, D \in \mathbb{N}$ are given. It is asked whether a set $A' \subseteq A$ with $c(A') \leq B$ and $\nu(A') \geq D$ exists. KNAPSACK is NP-hard [19] and W[1]-hard with respect to the solution size k [7]. Allowing rational costs and values makes KNAPSACK strongly NP-hard [41].

Observe that after multiplying $c(a)$ and B with $k + 1$ for each $a \in A$ and adding k items of cost 1 and value 0, we may assume that if there is a solution, then there is also one of size k.

Reduction. Given an instance $\mathcal{I} := (A = \{a_1, \ldots, a_n\}, c, \nu, B, D)$ of KNAPSACK, we construct an instance $\mathcal{I}' := (\mathcal{T}, \mathcal{F}, k', D')$ of WEIGHTED-PDD as follows.

Define $X := A \cup \{\star, \overline{a}\}$ and let N and M be big integers. Let $\mathcal{T}$ be a star with root ρ, leaves X, and edge weights $\omega(\rho a) := \nu(a)$ for each $a \in A$ and $\omega(\rho\star) := \omega(\rho\overline{a}) := N$. Let $\mathcal{F}$ contain edges $a\star$ for each $a \in A \cup \{\overline{a}\}$ of weight $\gamma(a\star) := (M - c(a))/(M(k+1) - B)$ and $\gamma(\overline{a}\star) := M/(M(k+1) - B)$.

As constructed so far, $\mathcal{F}$ is a star. To obtain a clique, we add edges $\overline{a}a_i$ and $a_p a_q$, all of weight 1, for each $i \in [n]$ and each combination $1 \leq p < q \leq n$.

Finally, we set $k' := k + 2$ and $D' := 2N + D$.

Intuition. The construction ensures that $c(A') \leq B$ if and only if $A'' := A' \cup \{\star, \overline{a}\}$ is γ-viable in $\mathcal{F}$ and $\nu(A') \geq D$ if and only if $PD_{\mathcal{T}}(A'') \geq D'$ for any set $A' \subseteq A$. $\qquad\square$

Correctness. The reduction is computed in polynomial time. We only consider the correctness when $\mathcal{F}$ is a star and omit the equivalent case of $\mathcal{F}$ being a clique.

Let A' be a solution of $\mathcal{I}$ of size k. We show that $S := A' \cup \{\star, \overline{a}\}$ is a solution of $\mathcal{I}'$. It is $PD_{\mathcal{T}}(S) = 2N + \nu(A') \geq 2N + D = D'$ and the size of S is clearly $|A'| + 2 = k + 2$. It remains to show that S is γ-viable. Since $A' \cup \{\overline{a}\}$ are sources, it is sufficient to check that the incoming weight of $\star$ is at least 1. It is

$$\gamma(\overline{a}\star) + \sum_{a \in A'} \psi(a\star) = \frac{M + \sum_{a \in A'} M - c(a)}{M(k+1) - B} \tag{2}$$

$$= \frac{(k+1)M - \sum_{a \in A'} c(a)}{M(k+1) - B} \tag{3}$$

$$\geq \frac{(k+1)M - B}{M(k+1) - B} = 1. \tag{4}$$

Consequently, S is γ-viable and a solution for instance $\mathcal{I}'$ of WEIGHTED-PDD.

Conversely, let S be a solution for instance $\mathcal{I}'$. For N big enough, we may assume $\star, \overline{a} \in S$. We define $A' := S \setminus \{\star, \overline{a}\}$ and show that A' is a solution for $\mathcal{I}$. It is $\nu(A') = PD_{\mathcal{T}}(S) - 2N \geq D$. Because S is γ-viable, $\gamma(\overline{a}\star) + \sum_{a \in A'} \psi(a\star) \geq 1$. Further, we may assume by Lemma 2 that $|S| = k'$. Consequently,

$$\frac{M + \sum_{a \in A'} M - c(a)}{M(k+1) - B} \geq 1 \tag{5}$$

$$\Longleftrightarrow \quad M + \sum_{a \in A'} M - c(a) \geq M(k+1) - B \tag{6}$$

$$\Longleftrightarrow \quad \sum_{a \in A'} c(a) \leq B \tag{7}$$

Thus, A' is a solution of $\mathcal{I}$. $\qquad\square$

3   Structural Parameters of the Food-Web for Rw-PDD

In this section, we consider RW-PDD with respect to parameters categorizing the food web's structure. A comprehensive overview of the complexity results for

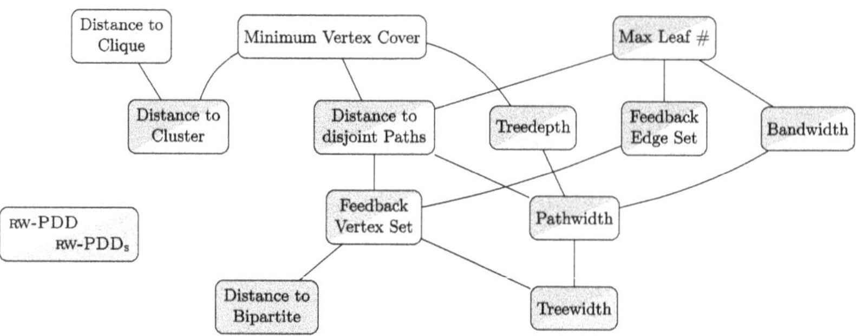

Fig. 3. In this figure, the complexity of RW-PDD and RW-PDD$_s$ with respect to several structural parameters of the food web is presented. The complexity of RW-PDD is in the top left of each box, and the complexity of RW-PDD$_s$ is in the bottom right. A parameter p is marked in red (◉) if RW-PDD / RW-PDD$_s$ is NP-hard for constant values of p, or in amber (○) or green (○) if RW-PDD$_s$/RW-PDD$_s$ admits an XP-, or, respectively, an FPT-algorithm with respect to p. Classifying RW-PDD parameterized by distance to clique remains open. RW-PDD$_s$ with respect to treewidth is W[1]-hard [32] and in XP. Two parameters p_1 and p_2 are connected with an edge if in every graph the parameter p_1 further up is bounded by a function in p_2. A more in-depth look into the hierarchy of graph parameters can be found in [34]. (Color figure online)

the main structural parameters are provided in Fig. 3. All three described XP-algorithms are FPT-algorithms if $W_{\max}$ is added to the parameter.

The hardness results are direct implications of results of [12] or [20].

3.1   Minimum Vertex Cover

In this section, we parameterize RW-PDD with the minimum vertex cover number (vc) of the food web $\mathcal{F}$. We start with a useful pre-processing step.

Lemma 4 (⋆). *Given an instance $\mathcal{I} = (\mathcal{T}, \mathcal{F}, k, D)$ of RW-PDD and a vertex cover $C \subseteq X$ of $\mathcal{F}$ of size vc, in $\mathcal{O}(2^{\mathrm{vc}} \cdot (n + m))$ time, one can compute 2^{vc} instances $\mathcal{I}_A = (\mathcal{T}_A, \mathcal{F}_A, k_A, D_A)$ of RW-PDD, one for each $A \subseteq C$, such that $\mathcal{I}$ is a yes-instance of RW-PDD, if and only if $\mathcal{I}_A$ is a yes-instance of RW-PDD for some $A \subseteq C$ and (1) the taxa in A are children of the root of $\mathcal{T}_A$, (2) the height of $\mathcal{T}_A$ is at most the height of $\mathcal{T}$, (3) $\mathcal{T}_A$ contains $\mathcal{O}(n)$ vertices, (4) $u \notin A$ and $v \in A$ for each edge $uv \in E(\mathcal{F}_A)$, (5) γ remains unchanged on edges that are in both instances, and (6) A is a subset of each solution S of $\mathcal{I}_A$.*

Intuitively, $A' := C \setminus A$ and some taxa in $X \setminus C$ can not survive, after fixing A. Lemma 4 is used in both of the following theorems.

Theorem 2 (⋆). *Instances $\mathcal{I} = (\mathcal{T}, \mathcal{F}, k, D)$ of RW-PDD can be solved in $\mathcal{O}((1 + W_{\max})^{2\,\mathrm{vc}} \cdot (n + m)k)$ time, if a vertex cover $C \subseteq X$ of $\mathcal{F}$ of size vc is given.*

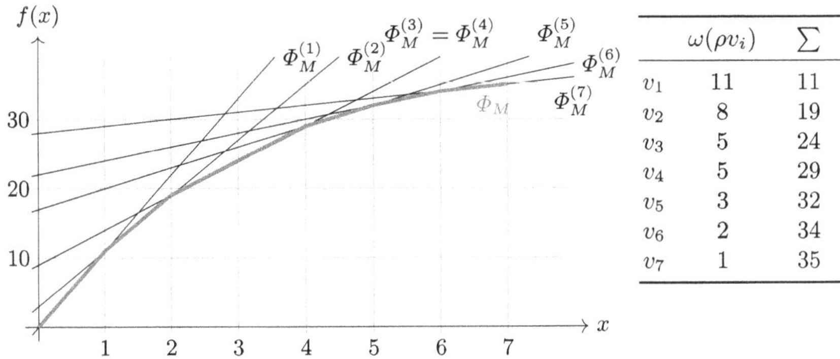

Fig. 4. An illustrative example of how to compute the function Φ_M for values of $\omega(\rho v_i)$.

In the following, we show how to reduce instances of RW-PDD$_s$ to instances of integer linear programming feasibility (ILP-FEASIBILITY) with 2^{vc} variables. ILP-FEASIBILITY on n variables can be solved using $n^{2.5n+o(n)} \cdot |\mathcal{I}|$ arithmetic operations, where $|\mathcal{I}|$ is the input length [13,22]. Using a randomized algorithm even a running time of $\log(2n)^{\mathcal{O}(n)}$ is possible [29]. It follows that RW-PDD$_s$ is FPT when parameterized with the vertex cover number.

Theorem 3. *Instances* $\mathcal{I} = (\mathcal{T}, \mathcal{F}, k, D)$ *of* RW-PDD$_s$ *can be solved in* $vc^{\mathcal{O}(2^{vc})} \cdot (n \log n + m)$ *time, if a vertex cover* $C \subseteq X$ *of* $\mathcal{F}$ *of size* vc *is given.*

Proof. Algorithm and Correctness. Apply Lemma 4 and iterate over the instances $\mathcal{I}_A = (\mathcal{T}_A, \mathcal{F}_A, k_A, D_A)$ of RW-PDD$_s$. We provide a reduction from $\mathcal{I}_A$ to an instance of ILP-FEASIBILITY with $2^{|A|}$ variables.

For subsets M of A, define $[M]_\sim$ as the set of taxa $v \in X \setminus A$ that have M as predators. For each $a \in A$, define A_a to be the family of sets $S \subseteq A$ containing a. We define an instance of ILP-FEASIBILITY, with variables x_M, upper bounded by $q_M := |[M]_\sim|$, indicating how many taxa are chosen from $[M]_\sim$. Recall that γ_a is the number of prey of a taxon a that have to be saved to save $a \in A$.

$$\sum_{M \subseteq A} x_M \leq k_A - |A| \tag{8}$$

$$\sum_{M \in A_a} x_M \geq \gamma_a \qquad \forall a \in A \tag{9}$$

$$\sum_{M \subseteq A} \Phi_M(x_M) \geq D_A - PD_{\mathcal{T}_A}(A) \tag{10}$$

$$\sum_{M \subseteq A} x_M \leq q_M \tag{11}$$

Recall, we have to save all taxa in A, by Lemma 4. Inequality (8) ensures that at most k_A taxa are saved. Inequality (9) ensures that for each taxon $a \in A$

the necessary number of prey are saved so that the solution is γ-viable. Inequality (11) provides the (logical) upper bound of x_M. With $\Phi_M(x_M)$, the best phylogenetic diversity that can be achieved when x_M taxa are saved from M is given. Since all taxa in A have to be saved, $D_A - PD_{T_A}(A)$ diversity has to be contributed overall from the taxa $X \setminus A$. Thus, Inequality (10) ensures the diversity threshold is met. It remains to show how to compute $\Phi_M(x_M)$. We do this with an approach similar to the one used to show that KNAPSACK is FPT when parameterized by the number of numbers [9]. An example is given in Fig. 4.

For each $M \subseteq A$, order the taxa $v_1, \ldots, v_{q_M}$ of $[M]_\sim$, such that $\omega(\rho v_i) \geq \omega(\rho v_{i+1})$, for each $i \in [q_M]$, where ρ is the root of T_A. For $i \in [q_M]$, define linear functions $\Phi_M^{(i)}$ with $\Phi_M^{(i)}(i-1) = \sum_{j=1}^{i-1} \omega(\rho v_j)$ and $\Phi_M^{(i)}(i) = \sum_{j=1}^{i} \omega(\rho v_j)$. Define $\Phi_M(j) := \min_{i \in [q_M - 1]} \Phi_M^{(i)}(j)$. This completes the algorithm. The correctness follows from the correct definition of the ILP-FEASIBILITY instance.

Running Time. The algorithm in Lemma 4 returns 2^{vc} instances in $\mathcal{O}(2^{vc} \cdot (n + m))$ time. The sets $[M]_\sim$ can be computed in time $\mathcal{O}(2^{vc} + n + m)$ by an iteration over X and computing the predators. All functions Φ_M are computed in $\mathcal{O}(2^{vc} \cdot n \log n)$ time. Then, the overall running time is dominated by the running time of ILP-FEASIBILITY, which is $\log(2 \cdot 2^{vc})^{\mathcal{O}(2^{vc})} = (vc + 1)^{\mathcal{O}(2^{vc})} = vc^{\mathcal{O}(2^{vc})}$.

$\square$

3.2 Distance to Cluster

In this section, we consider RW-PDD on instances where the food web is almost a cluster graph. In a cluster graph, every connected component is a clique. Cluster graphs generalize cliques and independent sets.

It remains open whether $1/2$-PDD—and therefore RW-PDD—can be solved in polynomial time on cliques. In ε-PDD, it is sufficient to save the source and reduce to MAX-PD. In 1-PDD, we have to save taxa in topological order.

In the following, we observe that $1/2$-PDD is NP-hard if the food web is a cluster graph and show that RW-PDD$_s$ admits an XP-algorithm when parameterized by the number of taxa that need to be removed to obtain a cluster. The hardness result follows from a result in [12].

Corollary 1. $1/2$-PDD *is NP-hard, even if the food web is a cluster graph and each connected component contains four taxa.*

In the following, we show that RW-PDD$_s$ is XP with respect to the distance to clusterand FPT when adding $W_{\max}$ to the parameter.

Theorem 4 ($\star$). *Instances* $\mathcal{I} = (T, F, k, D)$ *of* RW-PDD$_s$ *can be solved in* $\mathcal{O}((W_{\max} + 1)^{2\,cvd} \cdot n^2 k)$ *time, if a set* $M \subseteq X$ *of size* cvd *such that* $F - M$ *is a cluster graph is given.*

3.3   Treewidth

Finally, we show that RW-PDD$_s$ is XP with respect to the treewidth τ of the food web $\mathcal{F}$ and FPT when adding $W_{\max}$ to the parameter.

Theorem 5 ($\star$). *Instances* $\mathcal{I} = (\mathcal{T}, \mathcal{F}, k, D)$ *of* RW-PDD$_s$ *can be solved in* $\mathcal{O}(W_{\max}^{2\tau} \cdot \tau \cdot nk^2)$ *time, if a nice tree-decomposition* T *of* $\mathcal{F}$ *with treewidth* τ *is given.*

4   Structural Parameters of the Food-Web for 1-PDD

In this section, we analyze the complexity of 1-PDD with respect to parameters that categorize the food web of an instance. A detailed overview of these results is provided in Fig. 5. It is somewhat remarkable that for all these parameterizations, 1-PDD seemingly has the same tractability result as ε-PDD [20].

4.1   Distance to Cluster

In this section, we consider how difficult 1-PDD is to solve on instances, where the food web almost is a cluster graph.

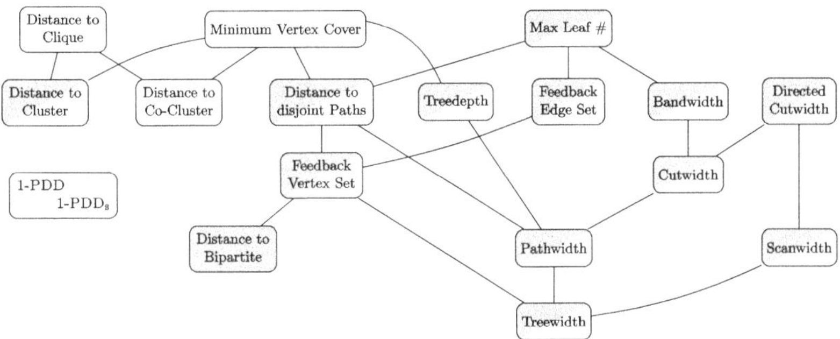

Fig. 5. This figure, similar to Fig. 3, shows the complexity of 1-PDD and 1-PDD$_s$ with respect to the main structural parameter of the food-web.

Recall that in a cluster graph, every connected component is a clique. In 1-PDD, every clique is essentially a path, as every vertex that appears earlier in the topological orientation has to be saved first. Consequently, with [12, Theorem 5.1.], we can conclude the following for 1-PDD.

Corollary 2. 1-PDD *is* NP-*hard, even if the food web is a cluster graph and each connected component contains three taxa.*

Next, we show that 1-PDD$_s$ is polynomial-time solvable when the food web is a cluster graph. Afterward, we generalize this result and show that 1-PDD$_s$ is FPT when parameterized by the size of a given cluster vertex deletion set.

Lemma 5. *Instances of* 1-PDD$_s$ *can be solved in* $\mathcal{O}((n+m) \cdot k^2)$ *time, if the food web in the input is a cluster graph.*

Proof. Algorithm. Let an instance $\mathcal{I} := (\mathcal{T}, \mathcal{F}, k, D)$ of 1-PDD$_s$ be given, where $\mathcal{F}$ is a cluster graph. Let $C_1, \ldots, C_q$ be the connected components of $\mathcal{F}$. The topological order of C_i, for each $i \in [q]$, directly indicates which set of taxa $S_{i,j}$ will be saved if $j \in [k]$ taxa can be saved from C_i. Define $\omega_{i,j} := PD_{\mathcal{T}}(S_{i,j})$.

Define a dynamic programming algorithm with table DP. In DP$[i, k']$, store the maximum phylogenetic diversity when k' taxa can be saved from $C_1, \ldots, C_i$.

As a base case, for each $j \in [\min\{k, |C_1|\}]_0$, store DP$[1, j] = \omega_{1,j}$.

To compute further values, we use the recurrence

$$\mathrm{DP}[i+1, j] := \max_{\ell \in [j]_0} \mathrm{DP}[i, \ell] + \omega_{i+1, j-\ell}. \tag{12}$$

Return yes if DP$[q, k] \geq D$. Otherwise, return no.

Correctness. Since the phylogenetic tree is a star, the only dependence of the taxa is given by the food web. Therefore, the sets $S_{i,j}$ are well-defined. The rest of the proof is straight-forward.

Running Time. By iterating over the edges, we can compute the in-degree of every vertex, which defines the topological order. Then, all values of $\omega_{i,j}$ can be computed in $\mathcal{O}(n)$ time. The table DP has $\mathcal{O}(q \cdot k)$ entries which can be computed in $\mathcal{O}(k)$ time, each. Thus, the overall running time is $\mathcal{O}(m \cdot k^2)$. $\quad\square$

Theorem 6. *Instances* $\mathcal{I} := (\mathcal{T}, \mathcal{F}, k, D)$ *of* 1-PDD$_s$ *can be solved in* $\mathcal{O}(2^{|M|} \cdot (n+m) \cdot k^2))$ *time if a set* $M \subseteq X$ *is given such that* $\mathcal{F} - M$ *is a cluster graph.*

Proof. Algorithm. Iterate over subsets $Y \subseteq M$. We want that Y are the taxa in M that are being saved and $M \setminus Y$ should die out. Let R_Y be the set of taxa which can reach Y in $\mathcal{F}$ and let Q_Y be the set of taxa which can be reached from $M \setminus Y$ in $\mathcal{F}$. If $R_Y \cap Q_Y \neq \emptyset$, then continue with the next set Y. Otherwise, compute whether $\mathcal{I}' := (\mathcal{T} - (R_Y \cup Q_Y), \mathcal{F} - (R_Y \cup Q_Y), k - |R_Y|, D - PD_{\mathcal{T}}(R_Y))$ is a yes instance of 1-PDD$_s$ with Lemma 5 and return yes if so. Otherwise, continue with the next set Y. Return no after the iteration.

Correctness. Let S be a solution of $\mathcal{I}$ and define $Y := S \cap M$. By Lemma 3, $R_Y \subseteq S$ and $Q_Y \cap S = \emptyset$. We conclude that $S \setminus R_Y$ is a solution of $\mathcal{I}'$. As Y is considered in the iteration, the algorithm returns yes.

Conversely, assume that the algorithm returns yes on Y. Because Y is to be saved, each taxon which can reach Y needs to be saved. Similarly, each taxon that can be reached from $M \setminus Y$ will go extinct when $M \setminus Y$ does. Assume now that S is a solution for $\mathcal{I}'$. By Lemma 3, $S \cup R_Y$ is valid in $\mathcal{F}$. Further, $|S \cup R_Y| = |S| + |R_Y| \leq k$ and $PD_{\mathcal{T}}(S \cup R_Y) = PD_{\mathcal{T}}(S) + PD_{\mathcal{T}}(R_Y) \geq D$.

Running Time. For a given Y, the sets R and Q can be computed in $\mathcal{O}(n+m)$ time. By Lemma 5, we can compute a solution for $\mathcal{I}''$ in $\mathcal{O}((n+m) \cdot k^2)$ time. $\square$

4.2   Distance to Co-Cluster

Now, we show that 1-PDD is FPT with respect to the distance to co-cluster. Recall, a co-cluster graph is the complement of a cluster graph. Similar as in the last section, we show that 1-PDD is polynomial-time solvable on co-clusters, first.

Lemma 6. *Instances of* 1-PDD *can be solved in* $\mathcal{O}(nk \cdot (n + m))$ *time, if the food web in the input is a co-cluster graph.*

Proof. Algorithm. Let an instance $\mathcal{I} := (\mathcal{T}, \mathcal{F}, k, D)$ of 1-PDD be given, where $\mathcal{F}$ is a co-cluster graph. Compute a topological order $x_1, \ldots, x_n$ of $\mathcal{F}$. Iterate over taxa $x_i \in X$. We want x_i to be the first taxon to die out. By definition, the set $A_i = \{x_1, \ldots, x_{i-1}\}$ survives and the set Q_i of taxa reachable from x_i dies out. Observe that $X_i := X \setminus (A_i \cup Q_i)$ are not neighbors of x_i in $\mathcal{F}$ and so, as $\mathcal{F}$ is a co-cluster, $\mathcal{F}[X_i]$ is an independent set. Let $\mathcal{T}_i$ be the (A_i, Q_i)-contraction of $\mathcal{T}$.

Return yes, if $\mathcal{I}_i := (\mathcal{T}_i, k - |A_i|, D - PD_{\mathcal{T}}(A_i))$ is a yes instance of MAX-PD. Otherwise, continue with the next taxon. After the iteration, return no.

Correctness. Let S be a solution for $\mathcal{I}$ and consider the computed topology. Let x_i be the taxon of $X \setminus S$ such that $A_i \subseteq S$. As $x_i \notin X \setminus S$ and S is γ-viable (for $\gamma = 1/|N_<(v)|$) if and only if $X_{\leq x} \subseteq S$ for each $x \in S$ [16], $Q_i \cap S = \emptyset$. Define $S' := S \setminus A_i \subseteq X_i$ and observe $|S'| = |S| - |A_i| \leq k - |A_i|$ and $PD_{\mathcal{T}_i}(S') = PD_{\mathcal{T}}(S) - PD_{\mathcal{T}}(A_i) \geq D - PD_{\mathcal{T}}(A_i)$. Thus, S' is a solution for $\mathcal{I}_i$ and the algorithm returns yes.

Conversely, if there is a taxon x_i such that $\mathcal{I}_i$ is a yes-instance of MAX-PD with solution S_i, then by analogous argument, $S_i \cup A_i$ is a solution for $\mathcal{I}$.

Running Time. For each taxon x_i, the sets A_i and Q_i can be computed in time $\mathcal{O}(n + m)$. Computing MAX-PD with Faith's Algorithm takes $\mathcal{O}(n \cdot k)$ time [25,36]. So, the overall running time is $\mathcal{O}(n \cdot (n + m) \cdot k)$. □

Theorem 7. *Instances* $\mathcal{I} := (\mathcal{T}, \mathcal{F}, k, D)$ *of* 1-PDD *can be solved in* $\mathcal{O}(2^{|M|} \cdot nk \cdot (n + m))$ *time if a set* $M \subseteq X$ *is given such that* $\mathcal{F} - M$ *is a co-cluster graph.*

Theorem 7 is proven similar to Theorem 6. We iterate over subsets Y of M and want that Y are the taxa that are surviving, while $M \setminus Y$ do not survive. After removing the taxa which can reach Y or which can be reached from $M \setminus Y$, the food web is a co-cluster and a solution can be found with Lemma 6.

4.3   Treewidth

In the following, we show that 1-PDD$_s$ is FPT with respect to the treewidth τ of $\mathcal{F}$. We use a coloring on the vertices to indicate whether a taxon is saved or not. This approach is similar to the one used in [20] to show that ε-PDD$_s$ is FPT when parameterized with τ. Common definitions can be found in [6,30].

Theorem 8 ($\star$). *Instances* $\mathcal{I} := (\mathcal{T}, \mathcal{F}, k, D)$ *of* 1-PDD$_s$ *can be solved in* $\mathcal{O}(2^{\tau} \tau \cdot nk^2)$ *time if a nice tree-decomposition* T *of* $\mathcal{F}$ *with treewidth* τ *is given.*

5 Discussion

In this paper, we defined WEIGHTED-PDD, a problem considering weighted food webs in the context of phylogenetic diversity maximization, as well as three special cases, RW-PDD, 1-PDD, and $\frac{1}{2}$-PDD. We analyzed these problems in the light of parameterized complexity for structural parameters of the food web and presented several XP-algorithms for RW-PDD$_s$ and several FPT-algorithms for 1-PDD$_s$. It is a somewhat surprising observation that for these parameters, 1-PDD and 1-PDD$_s$ have the same complexity as ε-PDD and ε-PDD$_s$.

It remains open whether $\frac{1}{2}$-PDD can be solved in polynomial time on instances where the food web is a clique and whether some of the presented XP-algorithms for the vertex cover number, distance to cluster, or treewidth of the food web can be improved to FPT-algorithms.

Some biological applications consider species interaction that generalizes one-on-one interactions [2], which may be represented with a hypergraph [15]. We wonder how such interactions could be modeled in the context of maximization of phylogenetic diversity and whether such problems can be solved efficiently.

Another recent line of research is defining phylogenetic diversity in phylogenetic networks [3,17,37,38,40]. So far, these concepts are considered without considering biological interactions. We expect a combination of these concepts to result in very hard problems, as ε-PDD is already hard if the phylogenetic tree and the food web are elementary trees and most definitions of phylogenetic diversity for networks are already hard on easy network structures. Yet, future research may identify special cases where efficient algorithms are feasible.

References

1. Barnosky, A.D., Matzke, N., Tomiya, S., et al.: Has the Earth's sixth mass extinction already arrived? Nature **471**(7336), 51–57 (2011)
2. Battiston, F., Cencetti, G., Iacopini, I., et al.: Networks beyond pairwise interactions: structure and dynamics. Phys. Rep. **874**, 1–92 (2020)
3. Bordewich, M., Semple, C., Wicke, K.: On the complexity of optimising variants of phylogenetic diversity on phylogenetic networks. Theoret. Comput. Sci. **917**, 66–80 (2022)
4. Cardinale, B.J., Duffy, J.E., Gonzalez, A., et al.: Biodiversity loss and its impact on humanity. Nature **486**(7401), 59–67 (2012)
5. Cowie, R.H., Bouchet, P., Fontaine, B.: The Sixth Mass Extinction: fact, fiction or speculation? Biol. Rev. **97**(2), 640–663 (2022)
6. Cygan, M., et al.: Parameterized Algorithms. Springer, Cham (2015)
7. Downey, R.G., Fellows, M.R.: Fixed-parameter tractability and completeness II: On completeness for W[1]. Theoret. Comput. Sci. **141**(1–2), 109–131 (1995)
8. Dvorák, W., Henzinger, M., Williamson, D.P.: Maximizing a submodular function with viability constraints. Algorithmica **77**(1), 152–172 (2017)
9. Etscheid, M., Kratsch, S., Mnich, M., Röglin, H.: Polynomial kernels for weighted problems. J. Comput. Syst. Sci. **84**, 1–10 (2017)
10. Faith, D.P.: Conservation evaluation and phylogenetic diversity. Biol. Cons. **61**(1), 1–10 (1992)

11. Faith, D.P.: The PD phylogenetic diversity framework: linking evolutionary history to feature diversity for biodiversity conservation. In: Pellens, R., Grandcolas, P. (eds.) Biodiversity Conservation and Phylogenetic Systematics. TBC, vol. 14, pp. 39–56. Springer, Cham (2016). https://doi.org/10.1007/978-3-319-22461-9_3

12. Faller, B., Semple, C., Welsh, D.: Optimizing phylogenetic diversity with ecological constraints. Ann. Comb. **15**(2), 255–266 (2011)

13. Frank, A., Tardos, É.: An application of simultaneous diophantine approximation in combinatorial optimization. Combinatorica **7**(1), 49–65 (1987)

14. Girardin, V., Grente, T., Niquil, N., Regnault, P.: Analysis of ecological networks: linear inverse modeling and information theory tools. Phys. Sci. Forum **9**(1), 24 (2024)

15. Golubski, A.J., Westlund, E.E., Vandermeer, J., Pascual, M.: Ecological networks over the edge: hypergraph trait-mediated indirect interaction (TMII) structure. Trends Ecol. Evol. **31**(5), 344–354 (2016)

16. Holtgrefe, N., Schestag, J., Zeh, N.: Limits of Kernelization and Parametrization for Phylogenetic Diversity with Dependencies (2025, Manuscript in Preparation)

17. Jones, M., Schestag, J.: How can we maximize phylogenetic diversity? Parameterized approaches for networks. In: Proceedings of the 18th International Symposium on Parameterized and Exact Computation (IPEC 2023), pp. 30:1–30:12. Schloss-Dagstuhl-Leibniz Zentrum für Informatik (2023)

18. Karanth, K.P., Gautam, S., Arekar, K., Divya, B.: Phylogenetic diversity as a measure of biodiversity: pros and cons. J. Bombay Nat. Hist. Soc. **116**, 53–61 (2019)

19. Karp, R.M.: Reducibility Among Combinatorial Problems. Springer, Boston (2010)

20. Komusiewicz, C., Schestag, J.: Maximizing phylogenetic diversity under ecological constraints: a parameterized complexity study. In: Proceedings of the 44th IARCS Annual Conference on Foundations of Software Technology and Theoretical Computer Science (FSTTCS 2024), pp. 28:1–28:18. Schloss-Dagstuhl-Leibniz Zentrum für Informatik (2024)

21. Kopnina, H.: Half the earth for people (or more)? Addressing ethical questions in conservation. Biol. Cons. **203**, 176–185 (2016)

22. Lenstra, H.W., Jr.: Integer programming with a fixed number of variables. Math. Oper. Res. **8**(4), 538–548 (1983)

23. Lieberman, E., Hauert, C., Nowak, M.A.: Evolutionary dynamics on graphs. Nature **433**(7023), 312–316 (2005)

24. Moulton, V., Semple, C., Steel, M.: Optimizing phylogenetic diversity under constraints. J. Theor. Biol. **246**(1), 186–194 (2007)

25. Pardi, F., Goldman, N.: Species Choice for Comparative Genomics: Being Greedy Works. PLoS Genet. **1** (2005)

26. Pardi, F., Goldman, N.: Resource-aware taxon selection for maximizing phylogenetic diversity. Syst. Biol. **56**(3), 431–444 (2007)

27. Pimm, S.L.: Food Webs. Springer, New York (1982)

28. Rands, M.R., Adams, W.M., Bennun, L., et al.: Biodiversity conservation: challenges beyond. Science **329**(5997), 1298–1303 (2010)

29. Reis, V., Rothvoss, T.: The subspace flatness conjecture and faster integer programming. In: Proceedings of the 64th Annual Symposium on Foundations of Computer Science (FOCS 2023), pp. 974–988. IEEE (2023)

30. Robertson, N., Seymour, P.D.: Graph Minors. X. Obstructions to tree-decomposition. J. Comb. Theory Ser. B **52**(2), 153–190 (1991)

31. Schestag, J.: Weighted food webs make computing phylogenetic diversity so much harder. arXiv preprint arXiv:2510.05911 (2025)

32. Schestag, J., Zeh, N.: The First Known Problem That Is FPT with Respect to Node Scanwidth but Not Treewidth (2025, Manuscript in Preparation)
33. Scotti, M., Podani, J., Jordán, F.: Weighting, scale dependence and indirect effects in ecological networks: a comparative study. Ecol. Complex. 4(3), 148–159 (2007)
34. Sorge, M., et al.: The Graph Parameter Hierarchy (2020). https://manyu.pro/assets/parameter-hierarchy.pdf
35. Spillner, A., Nguyen, B.T., Moulton, V.: Computing phylogenetic diversity for split systems. IEEE/ACM Trans. Comput. Biol. Bioinf. 5(2), 235–244 (2008)
36. Steel, M.: Phylogenetic Diversity and the greedy algorithm. Syst. Biol. 54(4), 527–529 (2005)
37. van Iersel, L., Jones, M., Schestag, J., Scornavacca, C., Weller, M.: Phylogenetic network diversity parameterized by reticulation number and beyond. In: Proceedings of the 22nd RECOMB International Workshop on Comparative Genomics (RECOMB-CG 2025), pp. 107–130. Springer (2025)
38. van Iersel, L., Schestag, J., Jones, M., Scornavacca, C., Weller, M.: Average-tree phylogenetic diversity of networks. In: Proceedings of the 25th International Workshop on Algorithms in Bioinformatics (WABI 2025), pp. 14:1–14:21. Schloss Dagstuhl–Leibniz-Zentrum für Informatik (2025)
39. Vellend, M., Cornwell, W.K., Magnuson-Ford, K., Mooers, A.Ø.: Measuring phylogenetic biodiversity (2011)
40. Wicke, K., Fischer, M.: Phylogenetic diversity and biodiversity indices on phylogenetic networks. Math. Biosci. 298, 80–90 (2018)
41. Wojtczak, D.: On strong NP-completeness of rational problems. In: Fomin, F.V., Podolskii, V.V. (eds.) CSR 2018. LNCS, vol. 10846, pp. 308–320. Springer, Cham (2018). https://doi.org/10.1007/978-3-319-90530-3_26
42. Yang, R., Feng, M., Liu, Z., Wang, X., Qu, Z.: Analysis of keystone species in a quantitative network perspective based on stable isotopes. Ecol. Complex. 59, 101092 (2024)

A Practical Algorithm for 3-Admissibility

Christine Awofeso[iD], Patrick Greaves[iD], Oded Lachish[iD], and Felix Reidl[(✉)][iD]

Birkbeck, University of London, London, UK
{cawofe01,pgreav01}@student.bbk.ac.uk, {o.lachish,f.reidl}@bbk.ac.uk

Abstract. The 3-admissibility of a graph is a promising measure to identify real-world networks that have an algorithmically favourable structure.

We design an algorithm that decides whether the 3-admissibility of an input graph G is at most p in time $O(mp^7)$ and space $O(np^3)$, where m is the number of edges in G and n the number of vertices. To the best of our knowledge, this is the first explicit algorithm to compute the 3-admissibility.

The linear dependence on the input size in both time and space complexity, coupled with an 'optimistic' design philosophy for the algorithm itself, makes this algorithm practicable, as we demonstrate with an experimental evaluation on a corpus of 217 real-world networks.

Our experimental results show, surprisingly, that the 3-admissibility of most real-world networks is not much larger than the 2-admissibility, despite the fact that the former has better algorithmic properties than the latter.

1   Introduction

Our work here is motivated by efforts to apply algorithms from sparse graph theory to real-world graph data, in particular algorithms that work efficiently if certain *sparseness measures* of the input graph are small. In algorithm theory, specifically from the purview of parametrized algorithms, this approach has been highly successful: by designing algorithms around sparseness measures like treewidth [1,3,13], maximum degree [15], the size of an excluded minor [4], or the size of a 'shallow' excluded minor [8,11], many hard problems allow the design of approximation or parametrized algorithms with some dependence on these measures.

For real-world applications, many of the algorithmically very useful measures turn out to be too restrictive, that is, the measures will likely be too large for most practical instances. Other graph measures, such as degeneracy, might be bounded in practice but provide only a limited benefit for algorithm design. We therefore aim to identify measures that strike a balance: we would like to find measures that are small on many real-world networks *and* providing an algorithmic benefit. Additionally, we would like to be able to compute such measures efficiently.

© The Author(s), under exclusive license to Springer Nature Switzerland AG 2026
J. Kozik and A. Wolff (Eds.): SOFSEM 2026, LNCS 16448, pp. 203–215, 2026.
https://doi.org/10.1007/978-3-032-17801-5_15

A good starting point here is the *degeneracy* measure, which captures the maximum density of all subgraphs. Recall that a graph is d-degenerate if its vertices can be ordered in such a way that every vertex x has at most d neighbours that are smaller than x. Such orderings cannot exist for *e.g.* graphs that have a high minimum degree or contain large cliques as subgraphs. In a survey of 206 networks from various domains by Drange *et al.* [5], it was shown that the degeneracy of most real-world networks is indeed small: it averaged about 23 with a median of 9.

Awofeso *et al.* identified the 2-admissibility [2] as a promising measure since it provides more structure than degeneracy and therefore better algorithmic properties (see their paper for a list of results). The 2-admissibility is part of a family of measures called *r-admissibility* which we define further below. The family includes degeneracy for $r = 1$, intuitively the larger the value r the 'deeper' into the network structure we look. Awofeso *et al.* designed a practical algorithm to compute the 2-admissibility and experimentally showed that this measure is still small for many real-world networks, specifically for most networks with degeneracy d, the 2-admissibility is around $d^{1.25}$.

Motivated by theoretical results [7–9] which imply that graphs with bounded 3-admissibility allow stronger algorithmic results than graphs with bounded 2-admissibility[1], we set out to design a practicable algorithm to compute the 3-admissibility of real-world networks.

Theoretical Contribution. We design and implement an algorithm that decides whether the 3-admissibility of an input graph G is at most p in time $O(mp^7)$ using $O(np^3)$ space. This improves on a previous algorithm described by Dvořák [6] with running time $O(n^{3p+5})$ and also beats the 3-approximation with running time $O(pn^3)$ described in the same paper when $p < n^{2/7}$. Dvořák also provides a theoretical linear-time algorithm for r-admissibility in *bounded expansion* classes (which include *e.g.* planar graphs, bounded-degree graphs and classes excluding a minor or topological minor); however, this algorithm relies on a data structure for dynamic first-order model checking [9], which certainly is not practical. Our algorithm runs in linear time as long as the 3-admissibility is a constant; this includes graph classes where *e.g.* the 4-admissibility is unbounded.

For reasons of space, we have relegated most theoretical results and their proofs as well as the detailed pseudocode of our algorithm to the full version of the paper[2]. Our main result is the following:

Theorem 1. *There exists an algorithm that, given a graph G and an integer p, decides whether* $\mathrm{adm}_3(G) \leq p$ *in time* $O(mp^7)$ *and space* $O(np^3)$.

---

[1] It is hard to quantify from these algorithmic meta-results how much more 'tractable' problems become, though from works like [12,14] it is clear that *e.g.* some graphs H can be counted in linear time in graphs of bounded 3-admissibility, while the same is not possible (modulo a famous conjecture) in graphs of bounded 2-admissibility.

[2] Please find the full paper under https://arxiv.org/abs/2512.01121.

Implementation and Experiments

We implemented the algorithm in Rust and ran experiments on a dedicated machine with modest resources (2.60Ghz, 16 GB of RAM) on a corpus of used by Awofeso *et al.*[3]. Our algorithm was able to compute the 3-admissibility for all but the largest few networks in this data set (209 completed, largest completed network is `teams` with 935K nodes). For more than half of the networks, the computation takes less than a second, and for 89% of these the computation took less than ten minutes. As such, the program is even practicable on higher-end laptops.

Surprisingly, we find that the 3-admissibility for many networks (93) is *equal* to the 2-admissibility, and for the remaining network, the 3-admissibility is never larger than twice the 2-admissibility. We discuss why this is indeed surprising, possible explanations, and exciting potential consequences in Sect. 5. Detailed experimental results for all networks can be found in the full version of this paper.

2   Preliminaries

In this paper, all graphs are simple unless explicitly mentioned otherwise. For a graph G we use $V(G)$ and $E(G)$ to refer to its vertex set and edge set, respectively. We use the shorthands $|G| := |V(G)|$ and $\|G\| := |E(G)|$. The degree of a vertex v in a graph G, denoted $\deg_G(v)$, is the number of neighbours v has in G.

For sequences of vertices $x_1, x_2, \ldots, x_\ell$, in particular paths, we use notation like $x_1 P x_\ell$, $x_1 Q$ and $R x_\ell$ to denote the subsequences $P = x_2, \ldots, x_{\ell-1}$, $Q = x_2, \ldots, x_\ell$ and $R = x_1, \ldots, x_{\ell-1}$, respectively. Note that further on, we sometimes refer to paths as having a *start-point* and an *end-point*, despite the fact that they are not directed. We do so to simplify the reference to the vertices involved. A path P *avoids* a vertex set L if no innerr vertex of P is contained in L. Note that we allow both endpoints to be in L. The length of a path P is the number of edges it has and is denoted by $\text{length}(P)$. The distance between two vertices in a graph G, denoted $\text{dist}_G(u, v)$, is the length of the shortest path in G having u and v as endpoints.

An *ordered qraph* is a pair $\mathbb{G} = (G, <)$ where G is a graph and $<$ is a total order relation on $V(G)$. We write $\leq_\mathbb{G}$ to denote the ordering of $\mathbb{G}$ and extend this notation to derive the relations $<_\mathbb{G}, >_\mathbb{G}, \geq_\mathbb{G}$.

3   Graph Admissibility

To define *r-admissibility* we need the following ideas and notation. Let $\mathbb{G} = (G, \leq)$, $L \subseteq V(G)$ and $v \in V(G)$. A path vPx is (r, L)-*admissible* if its length $\text{length}(vPx) \leq r$ and it avoids L. For every $v \in V(G)$, set $L \subseteq V(G)$ and integer $i > 0$ we let $\text{Target}_L^i(v)$ be the set of all vertices in $x \in L$ such that x is reachable from v via an (i, L)-admissible path. Note that $\text{Target}_L^i(v) \subseteq \text{Target}_L^{i+1}(v)$.

---

[3]   Please find the implementation and all data sets used in our experiments under https://github.com/chrisateen/three-admissibility.

Fig. 1. On the left, a maximal $(3, L)$-admissible packing rooted at u as well as the sets $T_1 := \mathrm{Target}_L^1(u)$, $T_2 := \mathrm{Target}_L^2(u) \setminus T_1$, and $T_3 := \mathrm{Target}_L^3(u) \setminus (T_1 \cup T_2)$. The sets S_i contain vertices in R whose shortest $(2, L)$-admissible path to u has length i. On the right, the same local subgraph but embedded in a tree of height 3.

An (r, L)-*admissible packing* is a collection of paths $\{vP_ix_i\}_i$ with v referred to as the *root* v such that every path vP_ix_i is (r, L)-admissible, the paths P_ix_i are pairwise vertex-disjoint, and each endpoint $x_i \in \mathrm{Target}_L^r(v)$. Note that in particular, all endpoints $\{x_i\}_i$ are distinct. We call such a packing *maximum* if there is no larger (r, L)-packing with the same root and *maximal* if the packing cannot be increased by adding a (r, L)-admissible path from v to an unused vertex in $\mathrm{Target}_L^r(v)$. We often treat (r, L)-admissible packings as trees rooted at v and use terms such as 'parent', 'child' or 'leaf'.

An example of a 3-admissible packing is depicted in Fig. 1. We write $\mathrm{pp}_L^r(v)$ to denote the maximum size of any (r, L)-admissible packing rooted at v.

Given an ordered graph $\mathbb{G}$, we define $\mathrm{pp}_{\mathbb{G}}^r(v)$ to be $\mathrm{pp}_L^r(v)$, where $L = \{u \in V(G) \mid u \leq_{\mathbb{G}} v\}$. The r-*admissibility* of an ordered graph $\mathbb{G}$, denoted $\mathrm{adm}_r(\mathbb{G})$ and the admissibility of an unordered graph G, denoted $\mathrm{adm}_r(G)$ are[4]

$$\mathrm{adm}_r(\mathbb{G}) := \max_{v \in \mathbb{G}} \mathrm{pp}_{\mathbb{G}}^r(v)$$

$$\text{and} \quad \mathrm{adm}_r(G) := \min_{\mathbb{G} \in \pi(G)} \mathrm{adm}_r(\mathbb{G}),$$

where $\pi(G)$ is the set of all possible orderings of G.

If $\mathbb{G}$ is an ordering of G such that $\mathrm{adm}_r(\mathbb{G}) = \mathrm{adm}_r(G)$, then we call $\mathbb{G}$ an *admissibility ordering* of G. The 1-admissibility of a graph coincides with its degeneracy. For $r \geq 2$, an optimal ordering can also be computed in linear time in n if the class has *bounded expansion*, *i.e.* if the graph class has bounded admissibility for *every* r (see [6]).

The following fact is a simple consequence of the fact that every (p, r)-admissible ordering is, in particular, a p-degeneracy ordering.

Fact 2. *If G is (p, r)-admissible, then $|E(G)| \leq p \cdot |V(G)|$.*

The following well-known facts about r-admissibility are important for the algorithm we present. These facts hold for all values of r, although we will only need

---

[4] Note that some authors choose to define the admissibility as $1 + \max_{v \in \mathbb{G}} \mathrm{pp}_{\mathbb{G}}^r(v)$ as this matches some other related measures.

them for $r \le 3$. Recall that in a d-degenerate graph, we can always find a vertex of degree $\le d$. The first lemma shows that a similar property holds in graphs of bounded r-admissibility:

Lemma 1. *A graph G is (p, r)-admissible if and only if, for every nonempty $L \subseteq V(G)$, there exists a vertex $u \in V(G) \setminus L$ such that $\mathrm{pp}_L^r(u) \le p$.*

Proof. Suppose first that for every nonempty $L \subseteq V$, there exists a vertex $u \in L$ such that $\mathrm{pp}_{V \setminus L}^r(u) \le p$. Then, an r-admissible order of G can be found as follows: first initialise the set L to be equal to V and i to $|G|$, and then repeat the following two steps until L is empty: (1) find a vertex $u \in L$ such that $\mathrm{pp}_{V \setminus L}^r(u) \le p$ removing it from L and adding it in the i'th place of the order (2) decrease i by 1.

We note that by construction, the r-admissibility of the order we got is at most p. Thus, G has r-admissibility p.

Suppose that G has r-admissibility p. Then, there exists an ordering, $\mathbb{G}$ of $V(G)$ such that $\mathrm{adm}_r(\mathbb{G}) \le p$. Let $u_1, u_2, \ldots u_{|G|}$ be the vertices of $\mathbb{G}$ in order. Let $u_k \in U$ be the vertex with the maximum index in L and define $U := \{u_1, \ldots, u_k\}$. We note that $L \subseteq U$.

Assume for the sake of contradiction that $\mathrm{pp}_L^r(u_k) > p$. Then, there exists an (r, L)-admissible packing H rooted at u_k of size $p + 1$. By definition, every path in H starts in u_k and ends in a vertex in L.

Thus, since $L \subseteq U$, every path P in H either already avoids U or includes a vertex from U. If the second case holds for such a path P then it has the form $u_k P_1 y P_2$ with $y \in U$ and $P_1 \cap U = \emptyset$. Replacing P by $u_k P_1 y$ and repeating this step for each path that contains a vertex from U results in a (r, U)-admissible packing H' with $|H'| = |H|$. Thus $\mathrm{pp}_U^r(u_k) > p$, contradicting our assumption that $\mathbb{G}$ has $\mathrm{adm}_r(\mathbb{G}) \le p$. $\square$

The second lemma allows us to conclude that if during the algorithm run we find a vertex that does not have a large $(3, L)$-path packing for some set L, then we know that this property will hold even if the set L shrinks in the future:

Lemma 2. *Let G be a graph, and let $L' \subseteq L \subseteq V(G)$. Then for every $v \in V(G)$, we have $\mathrm{pp}_{L'}^r(v) \le \mathrm{pp}_L^r(v)$.*

Proof. Fix a non-empty set $L \subseteq V(G)$ and an arbitrary vertex $u \in L$, let $L' = L \setminus \{u\}$. We show that in this case $\mathrm{pp}_{L'}^r(v) \le \mathrm{pp}_L^r(v)$ for all $v \in V(G)$, the claim then follows by induction.

Assume towards a contradiction that $\mathcal{P}'$ is a (r, L')-path packing rooted at v of size $s > \mathrm{pp}_L^r(v)$. If u does not appear in $\mathcal{P}'$, then $\mathcal{P}'$ is a (r, L)-path packing of size s, a contradiction. The same is true if $u = v$. Assume therefore that $P \in \mathcal{P}'$ contains u and $u \ne v$. Let Q be the segment of P that starts in v and ends in u, clearly $|Q| \le |P| \le r$ and Q avoids L. Then $\mathcal{P} = \mathcal{P}' \setminus P \cup \{Q\}$ is a (r, L)-path packing of size s, again contradicting that $s > \mathrm{pp}_L^r(v)$. We conclude that $\mathrm{pp}_{L'}^r(v) \le \mathrm{pp}_L^r(v)$ and the claim follows. $\square$

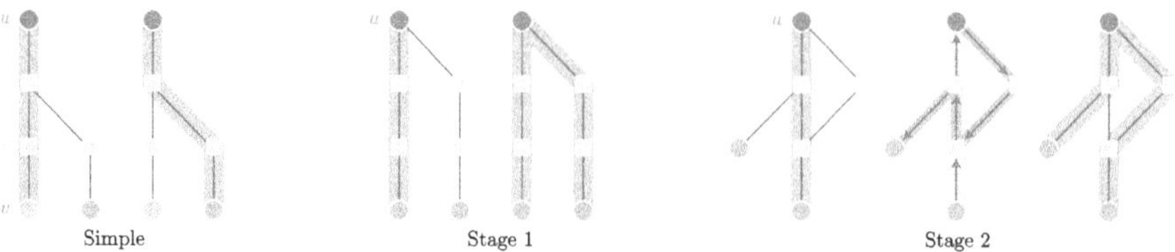

Fig. 2. The three 'escalations' of increasing path packings. During a Simple update (left), the path $uwxv$ is lost since v moved to R, and the algorithm attempts to find a replacement. For small packings, Stage 1 attempts to find disjoint paths rooted at u. If this does not increase the packing size, Stage 2 constructs a suitable flow network to either increase the packing size, or prove that the current packing is maximum.

The final lemma is a well-known bound between the sizes of path-packing and the size of the target sets, slightly adapted for our purposes here:

Lemma 3. *Fix integers $p, r \geq 1$. Let G be a (p, r)-admissible graph and L, R a partition of $V(G)$ such that for all $u \in R$, $\mathrm{pp}_L^r(u) \leq p$. Then for $u \in R$, $|\mathrm{Target}_L^r(u)| \leq p^r$ and for $v \in L$, $|\mathrm{Target}_L^r(v)| \leq |N_R(v)|(p-1)^{r-1}$.*

Proof. Consider $u \in R$ first and let Γ be a tree constructed from the shortest (r, L)-admissible paths from each vertex in $|\mathrm{Target}_L^r(u)|$ to u. For every interior vertex $x \in \Gamma$, we can construct a (r, L)-admissible packing by taking one path through each child of x to a leaf of Γ. Since $\mathrm{pp}_L^r(x) \leq p$, x cannot have more than p children in Γ. The same logic applies to the root, thus Γ has at most p^r leaves, which therefore bounds $|\mathrm{Target}_L^r(u)|$.

We apply the same trick to $v \in L$, except that for each interior vertex $x \in \Gamma$, we can also add a path from x to v to the packing, hence x has at most $p - 1$ children. Consequently, $|\mathrm{Target}_L^r(v)|$ contains at most $|N_R(v)|(p-1)^{r-1}$ vertices. $\qquad\square$

4   Algorithm Overview

We provide here a description of how the Algorithm works. The `Main` algorithm uses an Oracle that iteratively returns the next vertex in the admissibility order (starting at the last vertex and ending with the first). Due to space constraints, we only provide a high-level description of this Oracle here. A formal description, proof of correctness, and analysis of complexity can be found in the full version of this paper.

The input to the `Main` algorithm is a graph G and an integer p. We assume that $\|G\| \leq pn$, since this can easily be checked, and if it does not hold, then by Fact 2, the 3-admissibility number of G is strictly greater than p.

Let us for now assume that we have access to an Oracle that, given a subset L of $V(G)$, can provide us with a vertex $v \in L$ such that $\mathrm{pp}_L^3(v) \leq p$ if such a vertex exists. With the help of this Oracle, the following greedy algorithm

---

Algorithm `Update`: Returns a $(p,3)$-admissible ordering of G if one exists and otherwise returns FALSE.

Input: A graph G and a parameter $p \in \|G\|$

1 Initialise $L := V(G)$, $R := \emptyset$, $i := |G|$, and the Oracle.
2 **while** $L \neq \emptyset$ **do**
3   Ask the Oracle to return a vertex $v \in L$ such that $\mathrm{pp}_L^3(v) \leq p$.
4   **if** *the Oracle returned FALSE instead of a vertex* **then**
5   return FALSE
6   set $v_i := v$
7   remove v from L
8   prepend v to R
9   set $i := i - 1$
10 Return R.

(Algorithm `Update`) returns an ordering $\mathbb{G}$ such that $\mathrm{adm}_3(\mathbb{G}) \leq p$ if such an order exists and otherwise returns FALSE:
This greedy strategy works because of Lemma 1.

Given the Oracle, implementing the above algorithm is straightforward. The running time of the above algorithm is $O(n)$ plus the overall running time used by the oracle. The same applies for space complexity. Thus, the problem of finding a $(p,3)$-admissible ordering of a graph G is reduced to the problem of implementing the Oracle, which we outline below after making some structural observations on admissible packings.

4.1 The Structure of $(3, L)$-Admissible Packings

We need two properties of $(3, L)$-admissible packings that play a central role in the algorithm. Let G be a graph and L, R a partition of $V(G)$ and let H be a $(3, L)$-admissible packing rooted at $v \in L$. Then we call H *chordless* if for every path $vwP \in H$ (with P potentially empty) there is no edge between v and P.

We call H *covering* if for every vertex $x \in \mathrm{Target}_L^3(v)$ either $x \in N(v)$ and the path xv is in H, or there exists a $(3, L)$-admissible path from x to v which intersects $V(H) \setminus \{v\}$. In other words, every target vertex of v is either in the packing or has at least one $(3, L)$-admissible path to v which intersects H in a vertex other than v.

The important observation here is that if for $v \in L$ there exists a $(3, L)$-packing rooted at v of size k, then it also has a chordless packing of the same size. If $vwP \in H$ has a cord, then we can replace vwP by a shorter path with the same endpoint that is completely contained within vwP. This observation is already enough to show that one can implement an Oracle for 3-admissibility that works in polynomial time; the following construction provides important intuition and is key in our proof for a much faster Oracle.

Definition 1 (Packing Flow Network). *Let L, R be a partition of $V(G)$. For $u \in L$, the* packing flow network Π *is a directed flow network constructed* *as follows: start a BFS from u which stops whenever it encounters a vertex in L (except u) and stop the process after three steps. Remove all vertices discovered in the third step that are in R.*

Call the vertices discovered in the ith step S_i and T_i ($i \in \{0, \ldots, 3\}$), where $S_i \subseteq R$ to $T_i \subseteq L$, with $T_0 = \{u\}$ and $S_0 = S_3 = \emptyset$. The arcs of Π are the edges of G from layer T_0 to $T_1 \cup S_1$, S_1 to $S_2 \cup T_2$, and from S_2 to $T_2 \cup T_3$.

The source of the flow network is u and the sinks are $T_1 \cup T_2 \cup T_3$. We set the capacities to one for all arcs as well as all vertices, with the exception of u which has infinite capacity.

Lemma 4. *Let L, R be a partition of $V(G)$ and let Π be the packing flow network for $u \in L$. Then there is a one-to-one correspondence between integral flows on Π and chordless $(3, L)$-admissible packings rooted at u.*

Proof. To see that an integral flow corresponds to a packing, note that since the vertices have a capacity of one, every vertex except u has at most one incoming and one outgoing unit of flow. The saturated arcs therefore form a collection of paths all starting at u and ending at $T_1 \cup T_2 \cup T_3$. The internal vertices of these paths all lie in $S_1 \cup S_2$ and since the maximum distance from u is three, we conclude that all paths are indeed $(3, L)$-admissible and thus form a $(3, L)$-admissible packing rooted at u. Finally, since there are no arcs within S_1 and no arcs from u to $T_2 \cup S_2 \cup T_3$, we conclude that the packing is chordless.

In the other direction, we can convert any chordless $(3, L)$-admissible packing rooted at v into an integral flow by sending one unit of flow along each path. Since the packing is chordless, all edges are present as arcs in Π. $\qquad\square$

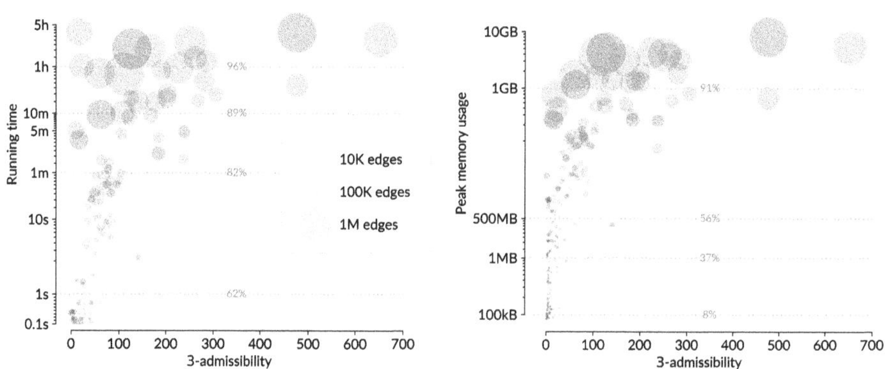

Fig. 3. Running time (left) and peak memory consumption (right) of our experiments. Networks where $\mathrm{adm}_2 = \mathrm{adm}_3$ are coloured purple, all other teal. The marker size indicates the number of edges in the network, the horizontal lines show the number of data points below certain interesting thresholds. (Color figure online)

4.2   The Oracle

The Oracle has access to the input graph G and the value of p.

As already mentioned on every call to the Oracle except for possibly the last, the Oracle returns a vertex. In the following, the set L is a variable of the oracle that contains all the vertices that the Oracle has yet to return and the set R is a variable that contains all the vertices that the oracle has already returned, *i.e.* $R = V(G) \setminus L$. We note that the set L has the same role as previously defined and it can be assumed that the oracle maintains this set and not the algorithm that calls it.

When queried, the Oracle must return a vertex v such that $\mathrm{pp}_L^3(v) \leq p$ or FALSE if no such vertex exists. To do so, the Oracle maintains a set Cand and ensures that before every query this set contains exactly the vertices $v \in L$ with $\mathrm{pp}_L^3(v) \leq p$. Hence, the Oracle returns FALSE if Cand is empty and otherwise an arbitrary vertex from this set. In the latter case, the returned vertex is removed from Cand and L and then added to R. The Oracle further updates all its data structures to be consistent with the new value of L and R.

The challenging part for the Oracle is to ensure that the correct vertices are added to Cand. Note that a vertex will only be removed from Cand when it is returned by the Oracle; this is because of the following property: when a vertex $v \in L$ is added to Cand, we have $\mathrm{pp}_L^3(v) \leq p$. Since vertices are only removed from the set L, Lemma 2 ensures that from then on $\mathrm{pp}_L^3(v) \leq p$ will hold in the future. Next, we explain how the oracle ensures that it adds the correct vertices to Cand.

In the initialisation stage, before the Oracle receives its first query, L contains all vertices and therefore $\mathrm{pp}_L^3(v) = \mathrm{pp}_{V(G)}^3(v) = \deg_G(v)$. Indeed, initially, the Oracle adds to Cand all the vertices $v \in L$ such that $\deg_G(v) \leq p$. We conclude that the first answer provided by the Oracle is always correct and uses at most $O(n)$ time and space.

The rest of this section is dedicated to explaining if the contents of Cand were correct before some call to the Oracle; the Oracle can efficiently ensure that the contents of Cand are correct before the next call. Taken together with the observation that the state of the Oracle is initially correct, this implies inductively that the Oracle works correctly.

For every vertex $v \in L \setminus \mathsf{Cand}$, the oracle maintains a $(3, L)$-admissible packing $\mathsf{Pack}(v)$, which is updated every time a vertex is removed from L. For these packings, we maintain two invariants, namely that they are all *covering* and *chordless*.

Once a vertex is added to Cand, the oracle stops maintaining its packing until it is removed from Cand and added to R. At this stage, the existing packing is discarded and a new maximal $(2, L)$-packing is computed for the vertex and maintained. The Oracle guarantees that these packings are *chordless*, and the *covering* property is implied by their maximality. As these packings only decrease in size over time and initially contain at most p paths, operations on these packings are cheap.

Let us now discuss how the Oracle decides when to add a vertex to Cand. Every time a vertex v is moved from L to R, the oracle updates all the packings of vertices in R and $L \setminus$ Cand which include v. This includes trying to add 'replacement' paths in case the move of v invalidated a path; these replacements ensure that the packing invariants are maintained. If for a vertex $u \in L$ the packing size could not be increased in this way, and the packing size has reached p, then the Oracle 'escalates' by attempting to add further paths in more computationally expensive ways (see Fig. 2). First, it attempts to find a path that intersects the current packing only in u. If it finds such a path, it adds it to Pack(u) which increases its size to $p + 1$ and u is not added to Cand in this round. If such a path does not exist, then Pack(u) is a *maximal* packing. The Oracle then attempts to increase the packing size by constructing a small auxiliary flow graph and augmenting a flow corresponding to the current packing. Here, our theoretical contribution is to show how to construct a small flow graph that mimics the properties of the complete packing flow network (Definition 1). If the flow increases, the Oracle recovers a $p + 1$ packing for u and does not add u to Cand in this round. If the flow cannot be increased further, the packing Pack(u) is already maxim*um*—no $(3, L)$-admissible packing rooted at u of size larger than p exists; hence u is added to Cand.

This approach ensures that the Oracle only resorts to performing the costly flow computation if the current packing is already small. The invariants of chordless and covering $(3, L)$-admissible packings are central here, since it brings the following advantages:

i. Given a covering $(3, L)$-admissible packing rooted at $v \in L$, when moving v from L to R the vertices in L whose packing need to be updated can be efficiently found with the help of a maximal $(2, L)$-admissible packings stored for vertices in R.
ii. Updating a chordless, covering $(3, L)$-admissible packing can be done efficiently when the size of the packing is strictly larger than $p + 1$.

For the pseudocode of the various parts of the algorithms, proof of correctness, and running time, we refer the reader to the full version of this paper.

5   Experimental Evaluation

We implemented the algorithm in Rust (2024 edition, `rustc` version 1.88.0) and ran experiments on a dedicated machine with 2.60Ghz AMD Ryzen R1600 CPUs and 16 GB of RAM. To optimise performance, we used the compile flag `target-cpu=native` and settings `codegen-units = 1`, `lto = "fat"`, `panic = "abort"`, and `strip = "symbols"` to minimize the final binary size.

Of the 217 networks in the corpus, our experiments were completed on 209.

Computing the 3-Admissibility is Practicable

The 'optimistic' design of the algorithm, which avoids expensive computations like the flow augmentation as much as possible, resulted in a practical implementation that computed the 3-admissibility on networks with up to 592K nodes.

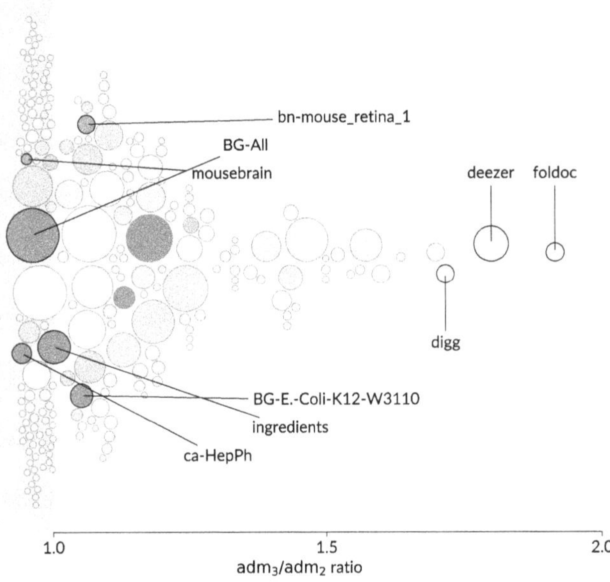

Fig. 4. The horizontal position indicates the ratio adm_3 / adm_2, networks in the grey strip all satisfy $adm_3 = adm_2$ and are distributed horizontally to keep the plot area small. The colour indicates the network degeneracy (as a measure of density), higher values are darker. Networks with the highest ratio are labelled, as well as those networks which have a high density but a low adm_3 / adm_2 ratio.

Figure 3 summarises the results: more than half of the networks finished in less than a second, 82% in under a minute (including a network on 33K nodes), and 89% in under ten minutes (including a network on 225K nodes). Memory usage is also modest by modern standards, with more than half of the networks needing less than 500 MB and 91% needing less than 1 GB.

This means that our implementation can run even on a modest laptop for quite large networks. As a point of comparison, the computation for the `offshore` network (278K nodes, 505K edges) ended in about 40 min on a laptop with a 2.40 GHz Intel i5-1135G7 processor while using about 500 MB RAM.

The 3-Admissibility is Surprisingly Small

To our surprise, the 3-admissibility for all tested networks is not much larger than the 2-admissibility. In fact, for 90 of the networks, both values are the same, and for the remaining 110 networks, the 3-admissibility is less than twice the 2-admissibility (the largest factor in the data set is 1.91). Figure 4 visualizes these ratios, Fig. 3 shows the absolute values for adm_3.

We find these results surprising for two reasons. First, experimental measurements of a related measure, the *weak r-colouring number* $wcol_r$, performed by Nadara *et al.* [10] showed that $wcol_3$ was substantially larger than $wcol_2$ across most instances. Second, Awofeso *et al.* [2] showed that the 2-admissibility is about $d^{1.25}$, where d is the degeneracy (the 1-admissibility) of the network, so we expected to see a similar relation going from $r = 2$ to $r = 3$.

There are two plausible explanations for this observation. For some networks, we could be seeing a plateau at $r = 3$, *e.g.*; for some $r \geq 4$, we would see an increase again. This would be plausible in *e.g.* road networks or other infrastructure networks which contain longer paths connecting hub vertices; however, then we would expect to see a similar effect in the experiments by Nadara *et al.*

The second explanation is that this is indeed the maximum value for *any r*, which would have quite significant implications about the structure of such networks: graphs with bounded '∞-admissibility' can be thought of as gluing together graphs of bounded degree with a constant number of high-degree vertices added in (see the recent survey by Siebertz [16] for a good overview). This is consistent with the observations by Nadara *et al.*, since graphs of bounded degree will have wcol_r increasing with r, while adm_r would be bounded by a universal constant since a path-packing rooted at some vertex v is always limited by the degree of v.

6   Conclusion

We demonstrated that a careful algorithm design and an 'optimistic' approach resulted in a resource-efficient implementation to compute the 3-admissibility of real-world networks.

Our experiments not only demonstrate that the implementation is of practical use, but also that the 3-admissibility of all 209 networks was surprisingly low; for almost half, it was even equal to the 2-admissibility. As we outlined above, it is likely that the r-admissibility of many real-world networks is already maximal for $r = 2$, which has interesting implications for the structure of such networks.

In the future, we intend on investigating whether the structure theorem of graphs with bounded '∞-admissibility' indeed applies to real-world networks in a meaningful way, as suggested by these experimental results. An important step will be the design of a comparable algorithm to compute the 4-admissibility, using the lessons learned in this work.

References

1. Arnborg, S., Lagergren, J., Seese, D.: Easy problems for tree-decomposable graphs. J. Algorithms **12**, 308–340 (1991)
2. Awofeso, C., Greaves, P., Lachish, O., Reidl, F.: A practical algorithm for 2-admissibility. In: Mutzel, P., and Prezza, N. (eds.) 23rd International Symposium on Experimental Algorithms, SEA 2025, July 22–24, Venice, Italy, vol. 338 of LIPIcs, Schloss Dagstuhl - Leibniz-Zentrum für Informatik, 2025, pp. 3:1–3:19 (2025)
3. Courcelle, B.: The monadic second-order logic of graphs. I. Recognizable sets of finite graphs. Inf. Comput., **85**, 12–75 (1990)

4. Demaine, E., Hajiaghayi, M., Kawarabayashi, K.: Algorithmic graph minor theory: Decomposition, approximation, and coloring. In: 46th Annual IEEE Symposium on Foundations of Computer Science (FOCS'05), pp. 637–646 (2025)
5. Drange, P.G., Greaves, P., Muzi, I., Reidl, F.: Computing Complexity Measures of Degenerate Graphs. In: 18th International Symposium on Parameterized and Exact Computation (IPEC 2023), vol. 285 of Leibniz International Proceedings in Informatics (LIPIcs), Dagstuhl, Germany, Schloss Dagstuhl – Leibniz-Zentrum für Informatik, pp. 14:1–14:21 (2023)
6. Dvořák, Z.: Constant-factor approximation of the domination number in sparse graphs. Eur. J. Comb. **34**, 833–840 (2013)
7. Dvořák, Z.: Approximation metatheorems for classes with bounded expansion. In: 18th Scandinavian Symposium and Workshops on Algorithm Theory, SWAT 2022, June 27-29, 2022, Tórshavn, Faroe Islands, vol. 227 of LIPIcs, Schloss Dagstuhl - Leibniz-Zentrum fur Informatik, pp. 22:1–22:17 (2022)
8. Dvořák, Z., Král, D., Thomas, R.: Deciding first-order properties for sparse graphs. In: IEEE 51st Annual Symposium on Foundations of Computer Science. IEEE **2010**, 133–142 (2010)
9. Dvořák, Z., Král, D., Thomas, R.: Testing first-order properties for subclasses of sparse graphs. J. ACM, **60**, 36:1–36:24 (2013)
10. Nadara, W., Pilipczuk, M., Rabinovich, R., Reidl, F., Siebertz, S.: Empirical evaluation of approximation algorithms for generalized graph coloring and uniform quasi-wideness. ACM J. Exp. Algorithmics, **24**, 2.6:1–2.6:34 (2019)
11. Nešetřil, J., Ossona de Mendez, P.: Grad and classes with bounded expansion II. Algorithmic aspects. Eur. J. Comb., **29**, 777–791 (2008)
12. Paul-Pena, D., Seshadhri, C.: A dichotomy hierarchy for linear time subgraph counting in bounded degeneracy graphs. In: Azar, Y., Panigrahi, D. (eds.) Proceedings of the 2025 Annual ACM-SIAM Symposium on Discrete Algorithms, SODA 2025, New Orleans, LA, USA, January 12–15, 2025, SIAM, pp. 48–87 (2025)
13. Reed, B.A.: Algorithmic Aspects of Tree Width. In: Reed, B.A., Sales, C.L. (eds.) Recent Advances in Algorithms and Combinatorics, pp. 85–107. Springer, New York, NY (2003)
14. Reidl, F., Sullivan, B.D.: A color-avoiding approach to subgraph counting in bounded expansion classes. Algorithmica **85**, 2318–2347 (2023)
15. Seese, D.: Linear time computable problems and first-order descriptions. Math. Struct. Comput. Sci. **0**, 505–526 (1990)
16. Siebertz, S.: On the generalized coloring numbers. arXiv preprint arXiv:2501.08698 (2025)

Using Ray-Shooting Queries for Sublinear Algorithms for Dominating Sets in RDV Graphs

Therese Biedl[1]([envelope]) [ORCID] and Prashant Gokhale[2]

[1] David R. Cheriton School of Computer Science, University of Waterloo, Waterloo, ON, Canada
biedl@uwaterloo.ca
[2] University of Wisconsin-Madison, Madison, WI, USA
prashant.gokhale@wisc.edu

Abstract. In this paper, we study the dominating set problem in *RDV graphs*, a graph class that lies between interval graphs and chordal graphs and defined as the **v**ertex-intersection graphs of **d**ownward paths in a **r**ooted tree. It was shown in a previous paper that adjacency queries in an RDV graph can be reduced to the question whether a horizontal segment intersects a vertical segment. This was then used to find a maximum matching in an n-vertex RDV graph, using priority search trees, in $O(n \log n)$ time, i.e., without even looking at all edges. In this paper, we show that if additionally we also use a ray shooting data structure, we can also find a minimum dominating set in an RDV graph $O(n \log n)$ time (presuming a linear-sized representation of the graph is given).. The same idea can also be used for a new proof to find a minimum dominating set in an interval graph in $O(n)$ time.

1 Introduction

In a graph G, a subset S of vertices is called *dominating* if every vertex in the graph is either included in S or adjacent to a vertex in S. Finding a minimum dominating set is one of the oldest problems in graph theory and graph algorithms, with applications in routing problems in wireless networks and facility location problems. Finding a minimum dominating set is NP-hard in general [13], as there exists a straightforward reduction from the set cover problem. On the other hand, an easy $O(\log n)$-approximation algorithm is well known [7].

In this paper, we study the problem of finding a minimum dominating set in graph classes whose structure makes it feasible to find a minimum dominating set very efficiently. The minimum dominating set problem remains NP-hard for chordal graphs [3] but polynomial time algorithms exist for permutation graphs

Research of TB supported by NSERC, RGPIN-2020-03958. Research of PG supported by a MITACS Globalink Graduate Fellowship. The results also appeared as part of the second author's Master's thesis [17] at the University of Waterloo.

© The Author(s), under exclusive license to Springer Nature Switzerland AG 2026
J. Kozik and A. Wolff (Eds.): SOFSEM 2026, LNCS 16448, pp. 216–230, 2026.
https://doi.org/10.1007/978-3-032-17801-5_16

[12], cographs [8], and strongly chordal graphs [11]. (Section 2 will review those definitions that are needed later again.) For more information about the complexity of minimum dominating set in various subclasses of perfect graphs, the interested reader can look at an overview paper by Corneil and Stewart [9].

This paper is concerned with finding *sub-linear* time algorithms for minimum dominating set, i.e., we are not even allowed to look at all edges. The main ingredient here is that the graph is given implicitly, typically via an intersection representation. For example in an *interval graph* (i.e., the intersection graph of n 1-dimensional intervals), a minimum dominating set can be found in $O(n)$ time if a suitable interval representation is given, even though the graph may have $\Theta(n^2)$ edges.Other problems can also be solved in sub-linear time in interval graphs, see e.g. [5] for path cover, Hamiltonian cycle and domatic partition, or [20] for maximum matching. Cheng, Kaminski, and Zaks [6] extended the algorithm for dominating set in interval graphs to a graph class slightly bigger than interval graphs, specifically, intersection graphs of downward paths in a *spider* (a tree consisting of multiple paths sharing a common node).

In this paper, we study a graph class called *RDV graphs*, which strictly contains the graph class of Cheng, Kaminski and Zaks (and hence the interval graphs). RDV graphs are the intersection graphs of downward paths in an arbitrary rooted tree. These are known to be strongly chordal graphs, and as such we can find a minimum dominating set in them in linear time [11]. In fact, this was known even earlier: Booth and Johnson showed that a straightforward linear-time greedy algorithm performs optimally if applied to an RDV graph [3].

The goal of this paper is to modify the greedy-algorithm by Booth and Johnson to achieve run-time $O(|T| + n \log n)$, where $|T|$ is the size of the host-tree in a given intersection representation of the RDV graph. (We may assume that $|T| \in O(n)$ [2], and will therefore typically state the bound as $O(n \log n)$ time.) While this is not quite as fast as the algorithm for interval graphs, it is still sub-linear since RDV graphs may have quadratically many edges. The key ingredient is the insight (first published in a companion paper [2]) that adjacency queries in an RDV graph can be reduced to an intersection problem between horizontal and vertical segments. In [2], this insight was combined with range minimum queries (implemented with a priority search tree) to obtain sublinear algorithms for maximum matching in RDV graphs. In this paper, we show that if we *additionally* also use a ray-shooting data structure, then more complicated neighbourhood queries can also be solved efficiently without looking at all edges, and use this to solve the minimum dominating set in sublinear time. (The technique should work more broadly; we briefly show that we can also find a minimum dominating set in an interval graph with it.)

The paper is structured as follows. In Sect. 2, after reviewing some definitions, we briefly explain results from the companion paper that we need. In Sect. 3 we give the greedy-algorithm and explain which of its operations cannot easily be sped up with the known techniques. Next we explain how to use a ray-shooting data structure to implement this operation. This gives us the faster run-time for the greedy-algorithm for dominating set. We discuss an $O(n)$-time algorithm for dominating set in interval graphs in Sect. 4 and end in Sect. 5 with further thoughts.

2   Background

We assume familiarity with graphs, and refer e.g. to Diestel's book for basic terminology [10]. For a vertex z in a graph, we write $N[z]$ for the *closed neighbourhood*, i.e., the vertex z and all its neighbours. Often we will fix a vertex order $z_1, \ldots, z_n$. We then write $N_{\geq}[z_i]$ for those vertices z_q in the closed neighbourhood of z_i that satisfy $q \geq i$ (and similarly define $N_<[z_i], N_{\leq}[z_i], N_>[z_i]$).

In this paper we only study a graph G that has a *representation as a vertex-intersection graph of subtrees of a tree* (or *tree-intersection representation* for short). This consists of a *host-tree* T and, for each vertex z of G, a subtree $T(z)$ of T such that (z, w) is an edge of G if and only if $T(z)$ and $T(w)$ share at least one node of T. As convention, we use the term 'node' for the vertices of T, to distinguish them from the vertices of G. It is well-known [14] that a graph has a tree-intersection representation if and only if it is *chordal*, i.e., it has no induced cycle of length 4 or more.

Numerous subclasses of chordal graphs can be defined by restricting the host-tree and/or the subtrees $T(z)$ in some way. For example, we mentioned the (unnamed) class of graphs studied by Cheng, Kaminski, and Zaks [6] where the host-tree is restricted to be a spider, and the subtrees are restricted to be paths that do not contain the center of the spider as interior vertex. The graph class we study here was named by Monma and Wei [23], though studied even earlier by Gavril [15]:

Definition 1. *A* rooted directed path graph *(or* RDV *graph [23]) is a graph that has an* RDV *representation, i.e., a tree-intersection representation with a rooted host-tree T where for every vertex z the subtree $T(z)$ is a* downward path, *i.e., a path that begins at some node and then always goes downwards in T.*

We may assume that T has $O(n)$ nodes [2]. We assume throughout the paper that an RDV graph G is given implicitly via an RDV representation (T, t, b), i.e., we are given a rooted tree T, and for every vertex z of G, a *top node $t(z)$* and a *bottom node $b(z)$* that is a descendant of $t(z)$, with the convention that $T(z)$ is the path between $t(z)$ and $b(z)$. A *bottom-up enumeration* of G is a vertex order obtained by sorting vertices by decreasing distance of $t(z)$ to the root, breaking ties arbitrarily. This can be computed in $O(|T| + n)$ time using bucket-sort.

2.1   From RDV Representation to Segments

We briefly review some crucial insights taken from our companion paper [2]. Let G be a graph with an RDV representation with host tree T. Assign points to the nodes of the host-tree T as follows: For each node ν, define y-coordinate $y(\nu)$ to be the distance of node ν from the root of T. To define the x-coordinate, first fix an arbitrary order of children at each node, then enumerate the leaves of T

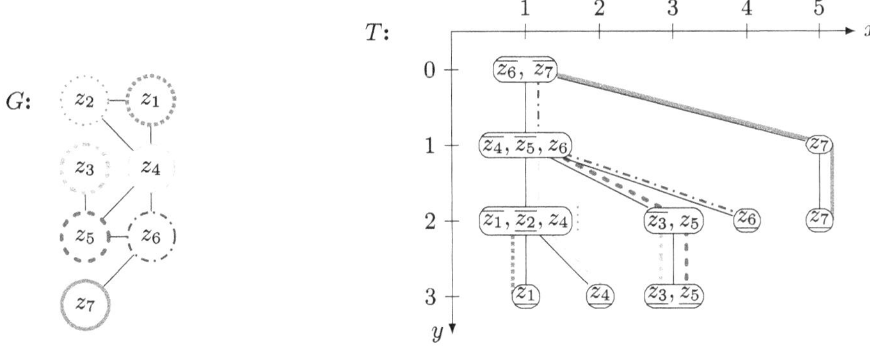

Fig. 1. An RDV graph with an RDV representation. Each node ν lists those vertices z with $\nu \in T(z)$; we write $\underline{z}$ if $\nu = b(z)$ and $\overline{z}$ if $\nu = t(z)$. We also show the paths as poly-lines, with colors/dash-pattern matching the vertices. Nodes are drawn at their coordinates, and indices correspond to a bottom-up enumeration order.

as $L_1, \ldots, L_\ell$ from left to right, and then for each node ν set $x(\nu)$ to be the index of the leftmost leaf that is a descendant of ν. Figure 1 shows each node ν drawn at point $(x(\nu), y(\nu))$ (where y-coordinates increase top-to-bottom). We can compute these coordinates in $O(|T|)$ time.

Definition 2. *Let G be a graph with an RDV representation. For each vertex z, define the following (see also Fig. 2):*

- *The* horizontal segment $\mathbf{h}(z)$ *of z is the segment that extends rightward from the point of $t(z)$ until the x-coordinate of the rightmost descendant of $t(z)$.*
- *The* vertical segment $\mathbf{v}(z)$ *of z is the vertical segment that extends upward from the point of $b(z)$ until the y-coordinate of of $t(z)$ (i.e., it hits $\mathbf{h}(z)$).*

Crucial to developing efficient algorithms is that edges in the graph can be characterized via intersections of these segments.

Theorem 1. *[2] Let G be a graph with an RDV representation and let $z_1, \ldots, z_n$ be a bottom-up enumeration of vertices. Then for any $i < j$, edge (z_i, z_j) exists if and only if the vertical segment $\mathbf{v}(z_j)$ intersects the horizontal segment $\mathbf{h}(z_i)$.*

2.2 Pre-processing the RDV Representation

For our data structures to come, it will be helpful if each set of segments defined by vertices are disjoint. This could be achieved with perturbing the coordinates, but we find it easier to modify instead the RDV representation to have some properties that may be of independent interest. The following is easy to show (see Fig. 2 for an example).

Lemma 1. *Let G be a graph with an RDV representation (T, t, b). Then, in $O(|T| + n)$ time, we can convert (T, t, b) into an RDV representation (T', t', b') such that $t'(z) \neq t'(w)$ and $b'(z) \neq b'(w)$ for all pairs of vertices $z \neq w$. Furthermore, all bottom nodes are leaves and no top node is a leaf.*

Proof. We assume that each node ν of T knows all vertices for which it is the top node, and all vertices for which it is the bottom node. (We can compute this in $O(|T| + n)$ time.) Now perform the following at each node ν of T. First, if there are d vertices $z_1, \ldots, z_d$ for which ν is the bottom node, then add d new leaves as children of ν, and make these the new bottom nodes for $z_1, \ldots, z_d$. (With this, no top node is a leaf anymore, since the corresponding bottom node became a strict descendant.) Second, if there are $k \geq 2$ vertices $w_1, \ldots, w_k$ for which ν is the top node, then let ρ be the parent of ν (temporarily insert a new node as parent ρ if ν was the root). Subdivide the link from ν to ρ with $k - 1$ new nodes, and make these the new top nodes of $w_2, \ldots, w_k$. One easily verifies that the newly defined paths intersect in a node if and only if the old paths did. We added at most two new nodes for each vertex of G, and so $|T'| \leq |T| + O(n)$.

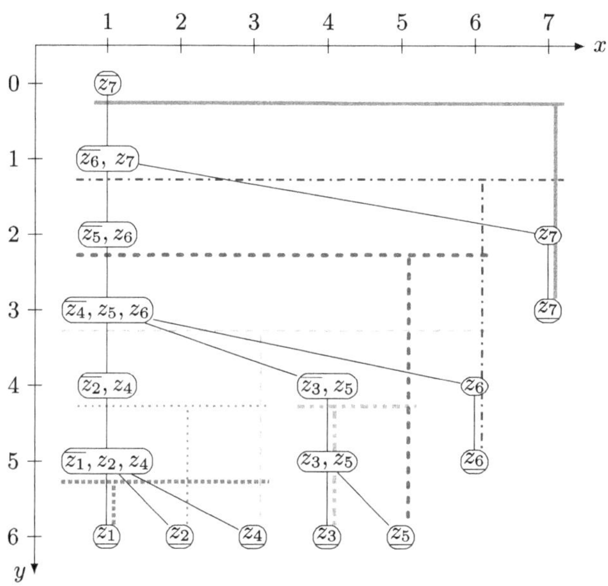

Fig. 2. The modified RDV representation, and the segments corresponding to vertices (drawn slightly offset for legibility).

From now on, we assume that the given RDV representation satisfies the conditions of Lemma 1. With this, $\mathbf{v}(z)$ is disjoint from $\mathbf{v}(w)$ for any $z \neq w$, since $b(z) \neq b(w)$ are both leaves, hence obtain distinct x-coordinates, and this determines the x-coordinates of the vertical segments. Likewise, $\mathbf{h}(z)$ is disjoint from $\mathbf{h}(w)$: If $t(z), t(w)$ have distinct y-coordinates, then so do the two horizontal segments. If $t(z)$ and $t(w)$ have the same y-coordinate, then they are on the same level of the host tree, and by $t(z) \neq t(w)$ they have disjoint sets of descendants. Since the x-coordinates of the descendants determine the x-range of the horizontal segments, this makes $\mathbf{h}(z)$ and $\mathbf{h}(w)$ disjoint.

3   Dominating Sets in RDV Graphs

To find a minimum dominating set in an RDV graph efficiently, we use a well-known greedy-algorithm [3]. Assume that we are given a vertex order $z_1, \ldots, z_n$. Initialize the dominating set D as empty and mark all vertices as "undominated". Then, for $i = 1, \ldots, n$, proceed as follows:

If z_i is "undominated", then select the vertex $z_q \in N[z_i]$ that maximizes the index q in the vertex order, add z_q to D, and mark all neighbours of z_q as "dominated".

Clearly this gives a dominating set, and as Booth and Johnson showed [3], it gives a minimum dominating set if run on an RDV graph using a bottom-up elimination order as the vertex order. Also, the algorithm clearly can be implemented in linear time since we spend $O(\deg(z))$ time whenever a vertex z becomes dominated for the first time, or is added to D, and everything else takes $O(n)$ time in total.

To reduce the run-time to sublinear in the number of edges (and specifically, $O(n \log n)$ after computing the coordinates of host-tree T), we separate out the crucial operations that we need.

A. For vertex z_i, find the vertex $z_q \in N[z_i]$ that has the maximal index q.
B. Given a vertex z_q, ensure that all vertices in $N[z_q]$ are marked "dominated".

Note that each operation is performed at most once per iteration, hence n times in total, so it would be sufficient to our run time to show how to implement them in $O(\log n)$ time. (We will achieve this bound only in an amortized sense: We use $O(n \log n)$ time over all iterations together.)

Operation A is very similar to an operation that we needed in our companion paper [2] (the only difference is that we use the closed neighbourhood, rather than the open one, and we maximize rather than minimize). As shown there, it can therefore be implemented in $O(\log n)$ time per operation using a priority search tree; we review this briefly below.

However, this implementation works efficiently only for the "current vertex" v_i, and therefore cannot be used for Operation B, where the input-vertex z_q is arbitrary. (We know that $q \geq i$ since $z_i \in N[z_i]$ and we maximize the index, but $q > i$ is possible.) As such, we need to develop a new and entirely different approach to support Operation B.

3.1 Implementing Operation A in $O(\log n)$ Time per Iteration

Define for each vertex z of G a *tuple* consisting of a *value* and a *weight*. We set the value to be $x(b(z))$, i.e., the x-coordinate of $\mathbf{v}(z)$. We set the weight to be the index that z has in the vertex order, i.e., it is q if $z = z_q$. To support Operation A, we store some of these tuples in a *priority search tree* $\mathcal{P}$, a data structure for 2-dimensional points that was introduced by McCreight [21]. This supports (among others) the operation *range-maximum-query*, which receives as input a range $[x, x']$ of values, and it returns, among all tuples that are stored in $\mathcal{P}$ and whose value falls into the range, the one that maximizes the weight (breaking ties arbitrarily, and returning "null" if there is no such tuple). It can do so in $O(\log |\mathcal{P}|)$ time [21]. (Our priority search tree always stores at most n tuples, so this is $O(\log n)$ time.) A priority search tree also supports the addition or deletion of a tuple in $O(\log n)$ time. With this, we have all tools that are needed for Operation A.

Lemma 2. *We can implement Operation A in amortized $O(\log n)$ time.*

Proof. We maintain a priority search tree $\mathcal{P}$, initially empty. We will ensure that throughout the following holds for $\mathcal{P}$:

Invariant: In iteration i, the priority search tree $\mathcal{P}$ stores exactly those tuples of vertices z_q for which $q \geq i$ and $y(b(z_q)) \geq y(t(z_i)) \geq y(t(z_q))$, i.e., the y-range of $\mathbf{v}(z_q)$ includes $y(t(z_i))$.

So at the beginning of iteration i, we add to $\mathcal{P}$ all those tuples of vertices z_p where we newly have $y(b(z_p)) \geq y(t(z_i))$, which implies $i = 1$ or $y(b(z_p)) < y(t(z_{i-1}))$. (We can find these vertices efficiently by pre-sorting a list L of all vertices by the y-coordinate of the bottom-node, and transferring a vertex from L to $\mathcal{P}$ once this coordinate is big enough.) A vertex z_p that was newly added to $\mathcal{P}$ satisfies $p \geq i$, since either $i = 1$ or $y(t(z_p)) \leq y(b(z_p)) < y(t(z_{i-1}))$. Hence $p > i-1$ since we use a bottom-up elimination order. At the end of iteration i, we remove z_i's tuple from $\mathcal{P}$, hence all tuples left in $\mathcal{P}$ have index bigger than i and the invariant holds for the next iteration. Since $y(t(z_i))$ can only decrease as i increases, every vertex is added to $\mathcal{P}$ exactly once. It also is deleted exactly once, and the total time to maintain $\mathcal{P}$ over all iterations is $O(n \log n)$.

Recall that Operation A means finding (in iteration i) the vertex $z_q \in N[z_i]$ that maximizes index q. Since $z_i \in N[z_i]$, we have $q \geq i$, so we only need to search among vertices in $N_{\geq}[z_i]$. We claim that doing so is the same as to perform a range-maximum query in $\mathcal{P}$ with the x-range of $\mathbf{h}(z_i)$. Assume first that the range-maximum query returns the tuple of z_q. Then the x-range of $\mathbf{h}(z_i)$ contains $x(b(z_q))$, the x-coordinate of $\mathbf{v}(z_q)$. Also $q \geq i$ and the y-coordinate of $\mathbf{h}(z_i)$ falls into the y-range of $\mathbf{v}(z_q)$ by the invariant. Hence $\mathbf{v}(z_q)$ intersects $\mathbf{h}(z_i)$, and $z_q \in N_{\geq}[z_i]$. Similarly one argues that any vertex $z_q \in N_{\geq}[z_i]$ gives an element of $\mathcal{P}$ whose value falls into the range. Therefore range-maximum query considers exactly the (tuples of) vertices of $N_{\geq}[z_i]$ and returns the one with the largest index, as required. So performing Operation A means one range-minimum query which takes $O(\log n)$ time.

In our example of Fig. 2, in the first iteration we have $y(t(z_1)) = 5$ and $\mathcal{P}$ stores the tuples of $z_1, \ldots, z_6$ since their vertical segments span across y-coordinate 5. We search for x-range $[2,4]$, and the vertical segments of z_1, z_2, z_4 fall into this range. Of these vertices, z_4 has the largest weight and we add it to the dominating set.

3.2 Implementing Operation B in $O(\log n)$ Time per Iteration

In Operation B (in iteration i), we are given a vertex z_q with $q \geq i$, and we must ensure that its closed neighbourhood is marked dominated. We break this up into two steps, considering $N_\leq[z_q]$ and $N_>[z_q]$ separately. As discussed earlier, we cannot use the priority search tree idea for this, since the segments stored in the priority search tree depend on the y-coordinate of $t(z_i)$, and so it can only be used for a search within $N[z_i]$. We instead store two different data structures (one for horizontal and one for vertical segments) that support *ray shooting queries*.

We need some background material first. Let S be a set of disjoint horizontal segments. A *ray shooting query* receives as input a vertical ray $\overrightarrow{q}$ specified by its endpoint q and the direction in which it emanates. The desired output is the first segment of S that is *hit* by the ray, i.e., it contains a point p of $\overrightarrow{q}$, and the open segment $\overline{qp}$ intersects no segment of S. (We return "null" if $\overrightarrow{q}$ hits no segments of S.) Many data structures have been developed for answering ray shooting queries efficiently, and that permit the insertion or deletion of segments. See for example [19,22] for older results with slower processing time. The current best run-time bounds are by Giyora and Kaplan [16]; their data structure uses $O(|S|)$ space and supports updates and queries in $O(\log |S|)$ time. (In our setting, we will always have $|S| \leq n$.) It was shown later that these run-time bounds can be achieved even when segments and rays are not orthogonal to each other [24], though we do not need this for our application.

Lemma 3. *We can implement Operation B in amortized $O(\log n)$ time.*

Proof. We explain here first how to mark all undominated vertices in $N_\leq[z_q]$ and then briefly sketch how to handle $N_>[z_q]$ symmetrically. We maintain throughout the algorithm a set S_H that stores the horizontal segments of all those vertices that are *not* dominated. (It is very important to remove segments of dominated vertices from this set for the amortized analysis to work out.) Initially S_H simply stores $\mathbf{h}(z)$ for all vertices z. (Recall that these segments are disjoint.) We remove $\mathbf{h}(z)$ from S_H when we first mark z as "dominated". Clearly this takes $O(n \log n)$ time in total, and hence $O(\log n)$ amortized time per iteration.

We define for every vertex z a ray $\overrightarrow{z}$ that starts at the top end of $\mathbf{v}(z)$ and shoots along it (i.e., towards ∞ in y-coordinates). Assume that in iteration i we have fixed vertex z_q, and perform a ray-shooting query with ray $\overrightarrow{z_q}$ within the segment-set S_H. If the query returns $\mathbf{h}(z_p)$, then we explicitly check whether $z_p \in N[z_q]$; since we can argue that $p \leq q$ (see below) this takes constant time using Theorem 1. If we indeed have $z_p \in N_\leq(z_q)$ then we mark z_p as dominated (it was "undominated" before since its segment was in S_H). We remove its segment from S_H (and also from the "other" data structure S_V that we define below). Then repeat the ray-query.

To see that we must have $p \leq q$, recall that we shoot the ray downwards (towards larger y-coordinates), so $y(t(z_p)) \geq y(t(z_q))$. So we could have $p > q$ only if $y(t(z_p)) = y(t(z_q))$, but since top nodes are distinct this then means that the x-range of $\mathbf{h}(z_p)$ is disjoint from the one of $\mathbf{h}(z_q)$. The latter range contains the x-coordinate of the ray, so then $\overrightarrow{z_q}$ cannot possibly have hit $\mathbf{h}(z_p)$.

Assume now that the query returns $\mathbf{h}(z_p)$ for some vertex z_p where $p \leq q$ but $z_p \notin N[z_q]$. We claim that then we are done searching for neighbours. (For the search in $N_{\leq}[z_q]$ this case actually cannot happen, but for the search in $N_{>}[z_q]$ this may happen.) Namely, the intersection point between the ray and $\mathbf{h}(z_p)$ then is *outside* the y-range of $\mathbf{v}(z_q)$ (otherwise there would be an edge by Theorem 1). Further queries can only find segments where the y-coordinate is even bigger, hence also outside the y-range of $\mathbf{v}(z_q)$, and so will not reveal any more elements of $N_{\leq}[z_q]$.

So if we perform ray-shooting queries until we either receive "null" or find a vertex not in the neighbourhood of z_q, then all found vertices are in $N_{\leq}[z_q]$. Vice versa, any undominated vertex in $N_{\leq}[z_q]$ will be found, since its horizontal segment intersects $\mathbf{v}(z_q)$ and so it will be hit by the ray. On the example of Fig. 2, in the first iteration (where we had $z_q = z_4$), we shoot a vertical downward ray from point $(3, 3)$, which hits the horizontal segments of z_4, z_2, z_1 (in this order) and then returns "null".

If we found k vertices in $N_{\leq}[z_q]$ (for some $k \geq 0$), then we had to perform $k+1$ ray-shooting queries, which took $O((k+1)\log n)$ time. But we also marked k vertices as dominated. We can hence attribute one of these log-factors to the iteration i that triggered Operation B, and all others to the vertices that are for the first time marked as "dominated". This happens once per vertex, and hence is $O(\log n)$ amortized overall.

In a symmetric way we can ensure that all vertices in $N_{>}[z_q]$ are marked as dominated: Store a data structure S_V that stores $\mathbf{v}(z)$ for every undominated vertex z. (Recall that these segments are disjoint.) To find $N_{>}[z_q]$, shoot a ray from the left endpoint of $\mathbf{h}(z_q)$ rightward, and find the first segment in S_V that it intersects. If this intersects, say, $\mathbf{v}(z_p)$, then first test whether the x-coordinate of the intersection point lies within $\mathbf{h}(z_q)$. If not, then by Theorem 1 (and because we return the first segment that was hit) we have found all undominated vertices in $N_{>}[z_q]$. If the x-coordinate is within range, then z_p is an undominated neighbour of z_q (and necessarily in $N_{>}[z_q]$ since we found all undominated vertices in $N_{\leq}[z_q]$ earlier already and removed their segments from S_V). We mark it as dominated and repeat the ray-shooting until it returns "null". In our example, we would shoot a rightward ray from point $(1, 3)$. The first segment hit by the ray belongs to z_5 (remember that z_4 was marked dominated earlier, and its segment removed from S_V in consequence). We mark z_5 as dominated, and likewise z_6 (whose segment is hit next). Next the ray hits the vertical segment of z_7, but here the x-coordinate of the intersection-point is outside the range of $\mathbf{h}(z_4)$; we stop the search and have found all of $N_>[z_4]$.

Algorithm 1: Dominating set in RDV graphs

Input: A graph G with an RDV representation T

1 Compute the coordinates of nodes of T.
2 Bucket-sort the vertices by decreasing $y(t(\cdot))$ to obtain vertex order $z_1, \ldots, z_n$.
3 Bucket-sort the vertices by decreasing of $y(b(\cdot))$ to obtain a list L.
4 Initialize dominating set $D = \emptyset$ and empty priority search tree $\mathcal{P}$
5 Initialize orthogonal ray shooting structures S_H and S_V
6 **for** $i = 1, 2, \cdots, n$ **do**
7 $\quad\lfloor$ S_H.insert($\mathbf{h}(z_i)$), S_V.insert($\mathbf{v}(z_i)$), mark z_i as "undominated".

8 **for** $i = 1, 2, \cdots, n$ **do**
9 $\quad$ **while** *the first vertex z_p in L satisfies $y(b(y_p)) \geq y(t(z_i))$* **do** // Add to $\mathcal{P}$
10 $\quad\quad\lfloor$ delete z_p from L and add its tuple $[x(b(z_q)), q]$ to $\mathcal{P}$

11 $\quad$ **if** z_i *is "undominated"* **then**
12 $\quad\quad$ $[\cdot, q] \leftarrow \mathcal{P}$.rangemax($x$-range of $\mathbf{h}(z_i)$)   // Operation A
13 $\quad\quad$ $D \leftarrow D \cup \{z_q\}$
14 $\quad\quad$ $\overrightarrow{z_q} \leftarrow$ downward ray from top end of $\mathbf{v}(z_q)$ // Operation B, $N_\leq$
15 $\quad\quad$ $s = S_H$.orthrayshoot($\overrightarrow{z_q}$)
16 $\quad\quad$ **while** *s is not null (say $s = \mathbf{h}(z_p)$)* **do**
17 $\quad\quad\quad$ **if** *$y(t(z_p))$ lies in the y-range of $\mathbf{v}(z_q)$* **then**
18 $\quad\quad\quad\quad$ Mark z_p as "dominated", S_H.delete $(\mathbf{h}(z_p))$, S_V.delete $(\mathbf{v}(z_p))$
19 $\quad\quad\quad\quad$ $s \leftarrow S_H$.orthrayshoot($\overrightarrow{z_q}$)
20 $\quad\quad\quad$ **else** $s \leftarrow$ null

21 $\quad\quad$ $\overrightarrow{z_q} \leftarrow$ rightward ray from left end of $\mathbf{h}(z_q)$ // Operation B, $N_>$
22 $\quad\quad$ $s = S_V$.orthrayshoot($\overrightarrow{z_q}$)
23 $\quad\quad$ **while** *s is not null (say $s = \mathbf{v}(z_p)$)* **do**
24 $\quad\quad\quad$ **if** *$x(b(z_p))$ lies in the x-range of $\mathbf{h}(z_q)$* **then**
25 $\quad\quad\quad\quad$ Mark z_p as "dominated", S_H.delete $(\mathbf{h}(z_p))$, S_V.delete $(\mathbf{v}(z_p))$
26 $\quad\quad\quad\quad$ $s \leftarrow S_V$.orthrayshoot($\overrightarrow{z_q}$)
27 $\quad\quad\quad$ **else** $s \leftarrow$ null

28 $\quad\lfloor$ delete tuple of z_i from $\mathcal{P}$
29 **return** D

This ends the description of how to implement the operations efficiently, and gives us our main result (we also summarize the algorithm with the pseudocode in Algorithm 1).

Theorem 2. *Given an n-vertex graph G with an $O(n)$-sized RDV representation T, a minimum dominating set of G can be found in $O(n \log n)$ time.*

The $O(\log n)$ amortized run-time is achieved with the ray shooting data structure of Giyora and Kaplan [16]. This uses dynamic fractional cascading and a generalized version of van Emde Boas trees, and hence is quite complex. One could instead use an older ray-shooting data structure of Overmars [25] which is considerably simpler. However, this uses run-time $O(\log^2 n)$ for updates and queries, so we would incur an extra $\log n$ factor in the runtime bounds.

4 $O(n)$-Time Algorithm for Interval Graphs

As mentioned in the introduction, an algorithm to find a minimum dominating set in an interval graph with run-time $O(n)$ was known to Cheng et al. [6], but we have not been able to find a description of it in the literature. Since it is very easy to state such an algorithm using the ideas of the previous section, we therefore give an independent proof of this result here.

Every interval graph is an RDV-graph, because we can view its representation by intersecting intervals as an RDV-representation where the host tree T is a path (rooted at one end), and the interval of vertex v becomes a path $T(z)$ connecting its top end $t(z)$ to its bottom end $b(z)$. With operations similar to Lemma 1, we can achieve that $t(z) \neq t(w)$ for any two vertices $z \neq w$ while keeping the host tree a path of size $O(|T| + n)$. Since T is a path, *all* nodes of T have the same x-coordinate, and only the y-coordinates are relevant for intersections. For ease of description, we will identify nodes of T with their y-coordinates (i.e., distance from the root), and so for example write $[t(z), b(z)]$ for the interval of vertex v, and statements such as $b(w) \leq t(z)$.

To find a minimum dominating set in an interval graph, we use the same greedy-algorithm, but implement the two required operations differently.

Operation A: We maintain a set $\mathcal{P}$ with the same invariant, but can use a simpler data structure since all x-coordinates are the same: We store $\mathcal{P}$ as an unsorted list of vertices. Clearly we can add to this and delete from it in $O(1)$ time, presuming every vertex keeps track of where it is stored in $\mathcal{P}$.

In iteration i, list $\mathcal{P}$ stores exactly those vertices z_p with $t(z_p) \leq t(z_i) \leq b(z_p)$. All these vertices intersect the interval of z_i (at $t(z_i)$), and $p \geq i$, so $\mathcal{P}$ stores exactly $N_\geq(z_i)$. To find the vertex $z_q \in N_\geq(z_i)$ that maximizes index q, it hence suffices to find the vertex with maximum index in $\mathcal{P}$. We do this via brute-force search in $O(|\mathcal{P}|)$ time; we will see that this is $O(1)$ amortized time.

We add z_q to the dominating set D, and claim that with this *all* vertices in $z_1, \ldots, z_q$ are dominated. (Here it will be crucial that we have an interval graph.) For if $1 \leq p \leq q$, then either $p < i$ (then we made z_p dominated in an earlier iteration), or $t(z_q) \leq t(z_p) \leq t(z_i) \leq b(z_q)$ (since $q \geq p \geq i$ and $z_q \in \mathcal{P}$) and so the intervals of z_q and z_p intersect.

Therefore Operation A (performed only when z_i was undominated) will be followed by $q - i$ iterations where Operation A will *not* be performed. Also, $|N_\geq(z_i)| \leq q - i + 1$ by choice of q. So these "cheap" iterations pay for the cost of performing Operation A, which hence becomes $O(1)$ time amortized.

Operation B: We have to ensure that all vertices in $N_<(z_q) \cup N_\geq(z_q)$ are marked "dominated". This is easy for $N_<(z_q)$, a subset of $\{z_1, \ldots, z_{q-1}\}$, because we simply mark *all* of $z_i, \ldots, z_{q-1}$ as dominated (correctness was argued above). The time for this is $O(q-i)$, which as above is $O(1)$ time amortized.

It remains to find all vertices in $N_{\geq}(z_q)$ that were undominated before iteration i. Observe that *any* vertex z_p with $p > q$ is undominated at this time: We have $z_p \notin N(z_i)$ (by choice of q), and since $t(z_p) < t(z_i)$ by $p > q > i$ we therefore have $t(z_p) \leq b(z_p) < t(z_i)$. On the other hand, any vertex z_d that is in D before iteration i satisfies $d < i$ (since z_i is undominated), so $t(z_i) < t(z_d) \leq b(z_d)$ and $z_p \notin N[z_d]$. So all we have to compute is $N_{\geq}(z_q)$. But we saw the method for this with Operation A: This is the same as the contents of $\mathcal{P}$ at the beginning of iteration q. So rather than implementing Operation B, we instead wait until iteration q (all iterations inbetween are trivial anyway since those vertices are dominated by z_q), and in that iteration, mark the corresponding neighbours as "dominated". Since they were undominated before, this is $O(1)$ amortized time.

We summarize the algorithm in the pseudocode in Algorithm 2.

Theorem 3. *Given an n-vertex interval graph G in form of an $O(n)$-sized RDV representation T where the host tree is a path, a minimum dominating set of G can be found in $O(n)$ time.*

Algorithm 2: Dominating set in interval graphs

    Input: Graph G with RDV representation where the host-tree T is a path

1 Compute the coordinates of nodes of T.

2 Bucket-sort the vertices by decreasing $y(t(\cdot))$ to obtain vertex order $z_1, \ldots, z_n$.

3 Bucket-sort the vertices by decreasing of $y(b(\cdot))$ to obtain a list L.

4 Initialize dominating set $D = \emptyset$ and an empty list $\mathcal{P}$

5 Mark all vertices as "undominated"

6 **for** $i = 1, 2, \cdots, n$ **do**

7   **while** *the first vertex z_q in L satisfies $y(b(z_q)) \geq y(t(z_i))$* **do**

8   delete z_q from L and add it to $\mathcal{P}$

9   **if** z_i *is "undominated"* **then**

10   find vertex z_q with maximum index in $\mathcal{P}$ `// Operation A`

11        $D \leftarrow D \cup \{z_q\}$

12   **for** $j = i$ *to* $q-1$ **do** mark z_j as "dominated";   `// Operation B, $N_<$`

13   **if** z_i *belongs* D **then**   `// `z_i` was `z_q` during last Operation A`

14   mark all vertices in $\mathcal{P}$ as "dominated"   `// Operation B, $N_{\geq}$`

15   delete z_i from $\mathcal{P}$

16 **return** D

---

5 Further Thoughts

In this paper, we gave a sublinear time algorithm to find the minimum dominating set in an RDV graph. The main ingredient is to implement the needed operations (in a natural greedy-algorithm) efficiently via geometric data structures involving priority search trees and ray shooting.

Our techniques can be applied for sublinear time algorithms for other problems. As an example, consider the independent set problem, i.e., we want to find a maximum set I of vertices such that no two of them are adjacent. There is a natural greedy algorithm that finds the correct answer in a chordal graph [18]: Fix a so-called perfect elimination order $z_1, \ldots, z_n$ (for an RDV graph, a bottom-up elimination order is a perfect elimination order). Initialize I as an empty set and mark all vertices as "not covered". For $i = 1, \ldots, n$, if z_i has not yet been covered, then add z_i to I and mark all vertices in $N[z_i]$ as "covered". With Operation B, we can perform in an RDV graph the operation of marking all neighbours as covered in $O(k \log n)$ time, where k is the number of vertices that were not previously covered. Since this happens to every vertex only once, the run-time becomes $O(n \log n)$ in an RDV graph.

We end with an open question. Can the runtime of Theorem 2 be improved to $O(n \log \log n)$ or even $O(n)$? One possible approach is to improve the runtime of priority search trees and orthogonal ray shooting data structures, at least in the special situation when all coordinates are integers in $O(n)$. But it is also conceivable that one could generalize the algorithm for interval graphs, i.e., exploit the structure of RDV graphs further to argue that simpler data structures suffice for the operations.

A   Missing Proof

Lemma 1. *Let G be a graph with an RDV representation (T, t, b). Then, in $O(|T| + n)$ time, we can convert (T, t, b) into an RDV representation (T', t', b') such that $t'(z) \neq t'(w)$ and $b'(z) \neq b'(w)$ for all pairs of vertices $z \neq w$. Furthermore, all bottom nodes are leaves and no top node is a leaf.*

Proof. We assume that each node ν of T knows all vertices for which it is the top node, and all vertices for which it is the bottom node. (We can compute this in $O(|T| + n)$ time.) Now perform the following at each node ν of T. First, if there are d vertices $z_1, \ldots, z_d$ for which ν is the bottom node, then add d new leaves as children of ν, and make these the new bottom nodes for $z_1, \ldots, z_d$. (With this, no top node is a leaf anymore, since the corresponding bottom node became a strict descendant.) Second, if there are $k \geq 2$ vertices $w_1, \ldots, w_k$ for which ν is the top node, then let ρ be the parent of ν (temporarily insert a new node as parent ρ if ν was the root). Subdivide the link from ν to ρ with $k - 1$ new nodes, and make these the new top nodes of $w_2, \ldots, w_k$. One easily verifies that the newly defined paths intersect in a node if and only if the old paths did. We added at most two new nodes for each vertex of G, and so $|T'| \leq |T| + O(n)$.

References

1. Bertossi, A.A., Gori, A.: Total domination and irredundance in weighted interval graphs. SIAM J. Discret. Math. **1**(3), 317–327 (1988). https://doi.org/10.1137/0401032
2. Biedl, T., Gokhale, P.: Finding maximum matchings in RDV graphs efficiently. Comput. Geom. Topology (2025). Submitted to special issue for papers from CCCG'24
3. Booth, K., Johnson, J.H.: Dominating sets in chordal graphs. SIAM J. Comput. **11**(1), 191–199 (1982). https://doi.org/10.1137/0211015
4. Brandstädt, A.: The computational complexity of feedback vertex set, hamiltonian circuit, dominating set, steiner tree, and bandwidth on special perfect graphs. J. Inf. Process. Cybern. **23**(8/9), 471–477 (1987)
5. Chang, M.-S., Peng, S.-L., Liaw, J.-L.: Deferred-query: an efficient approach for some problems on interval graphs. Networks **34**(1), 1–10 (1999). https://doi.org/10.1002/(SICI)1097-0037(199908)34:1
6. Cheng, S.W., Kaminski, M., Zaks, S.: Minimum dominating sets of intervals on lines. Algorithmica **20**(3), 294–308 (1998). https://doi.org/10.1007/PL00009197
7. Chvatal, V.: A greedy heuristic for the set-covering problem. Math. Oper. Res. **4**(3), 233–235 (1979). https://doi.org/10.1287/moor.4.3.233
8. Corneil, D.G., Perl, Y.: Clustering and domination in perfect graphs. Discrete Appl. Math. **9**(1), 27–39 (1984). https://doi.org/10.1016/0166-218X(84)90088-X
9. Corneil, D.G., Stewart, L.K.: Dominating sets in perfect graphs. Discrete Math. **86**(1), 145–164 (1990). https://doi.org/10.1016/0012-365X(90)90357-N
10. Diestel, R.: Graph Theory, 4th Edition, volume 173 of Graduate Texts in Mathematics. Springer, 2012. https://doi.org/10.1007/978-3-642-14279-6
11. Farber, M.: Domination, independent domination, and duality in strongly chordal graphs. Discrete Appl. Math. **7**(2), 115–130 (1984). https://doi.org/10.1016/0166-218X(84)90061-1
12. Farber, M., Keil, J.M.: Domination in permutation graphs. J. Algorithms **6**(3), 309–321 (1985). https://doi.org/10.1016/0196-6774(85)90001-X
13. Garey, M.R., Johnson, D.S.: Computers and Intractability: A Guide to the Theory of NP-Completeness. Freeman (1979)
14. Gavril, F.: The intersection graphs of subtrees in trees are exactly the chordal graphs. J. Combi. Theory Ser. B **16**, 47–56 (1974). https://doi.org/10.1016/0095-8956(74)90094-X
15. Gavril, F.: A recognition algorithm for the intersection graphs of directed paths in directed trees. Discret. Math. **13**(3), 237–249 (1975). https://doi.org/10.1016/0012-365X(75)90021-7
16. Giyora, Y., Kaplan, H.: Optimal dynamic vertical ray shooting in rectilinear planar subdivisions. ACM Trans. Algorithms, 5(3), pp. 28:1–28:51, 2009. https://doi.org/10.1145/1541885.1541889
17. Gokhale, P.: Efficient algorithms for RDV graphs. In: Master's thesis, David R. Cheriton School of Computer Science, University of Waterloo, 2025
18. Golumbic, M.C.: Algorithmic graph theory and perfect graphs, 1st edn. Academic Press, New York (1980)
19. Kaplan, H., Molad, E., Tarjan, R.: Dynamic rectangular intersection with priorities. In: Proceedings of the Thirty-Fifth Annual ACM Symposium on Theory of Computing, STOC '03, pp. 639–648, New York (2003). Association for Computing Machinery. https://doi.org/10.1145/780542.780635

20. Liang, Y., Rhee, C.: Finding a maximum matching in a circular-arc graph. Inf. Process. Lett. **45**(4), 185–190 (1993). https://doi.org/10.1016/0020-0190(93)90117-R
21. McCreight, E.M.: Priority search trees. SIAM J. Comput. **14**(2), 257–276 (1985). https://doi.org/10.1137/0214021
22. Mehlhorn, K., Näher, S.: Dynamic fractional cascading. Algorithmica **5**(1–4), 215–241 (2023). https://doi.org/10.1007/BF01840386
23. Monma, C., Wei, V.: Intersection graphs of paths in a tree. J. Comb. Theor. Ser. B **41**(2), 141–181 (1986). https://doi.org/10.1016/0095-8956(86)90042-0
24. Nekrich, Y.: Dynamic planar point location in optimal time. In: Proceedings of the 53rd Annual ACM SIGACT Symposium on Theory of Computing, STOC 2021, pp. 1003–1014 (2021). https://doi.org/10.1145/3406325.3451100
25. Overmars, M.H.: Range searching in a set of line segments. In: Proceedings of the First Annual Symposium on Computational Geometry, SCG '85, pp. 177–185, New York (1985). Association for Computing Machinery. https://doi.org/10.1145/323233.323257
26. Ramalingam, G., Pandu Rangan, C.: A unified approach to domination problems on interval graphs. Inf. Process. Lett. **27**(5), 271–274 (1988). https://doi.org/10.1016/0020-0190(88)90091-9

Sublinear Work Parallel Quantum Algorithms for Computational Geometry

Shion Fukuzawa$^{(\boxtimes)}$ [ID], Michael Goodrich [ID], and Sandy Irani [ID]

University of California, Irvine, Irvine, CA 92697, USA
{fukuzaws,goodrich}@uci.edu, ssirani@ics.uci.edu

Abstract. Previous work on parallel quantum algorithms has focused on minimizing the length of the critical path of the circuit (i.e., query span), motivated by the small decoherence times of qubits. In this paper, we study similar settings where queries to a quantum circuit can be made in parallel in each timestep. Rather than focus exclusively on query span, however, we are interested in also minimizing the total number of queries made across all parallel processes (i.e., query work). We give parallel quantum algorithms that solve the maxima set and convex hull problems for n lexicographically sorted points in the plane in $\tilde{O}(\sqrt{n})$ query span and $\tilde{O}(\sqrt{nh})$ query work, where h is the size of the output. These results therefore resolve a natural question of whether we can use quantum computers in parallel to improve the depth of a quantum algorithm without requiring more work than a sequential algorithm. Also, note that our work bounds are sublinear when h is $O(n^{1-\epsilon})$, for a small constant, $\epsilon > 0$.

Keywords: quantum computing · parallel computing · convex hulls · maxima sets · span · work

1 Introduction

Quantum computing is implicitly parallel, in that the state of a quantum circuit can exist in multiple states defined by a superposition of quantum bits, or **qubits** [18]. This form of parallelism has significant limitations, however, because information in quantum superposition can only be accessed by a quantum measurement, causing the quantum state to collapse, which results in a loss of information. Nonetheless, quantum superposition can be carefully exploited to solve certain problems, like factoring [29], with the right kind of mathematical structure. See, e.g., Nielsen and Chuang [22]. This notion of parallelism is different from the parallelism used in classical computing, however, where independent processors cooperate to solve a problem together.

We believe it is important to consider explicit parallelism in quantum computing, where multiple quantum computers work in concert, possibly with parallel classical computers, to solve a computational problem. Since quantum circuits must be reversible, the traditional quantum circuit model naturally captures the

© The Author(s), under exclusive license to Springer Nature Switzerland AG 2026
J. Kozik and A. Wolff (Eds.): SOFSEM 2026, LNCS 16448, pp. 231–245, 2026.
https://doi.org/10.1007/978-3-032-17801-5_17

notion of parallel quantum processors. In this setting, some unit of qubits can be thought of as a single processor, and actions on this processor act as gates applied in this register. We can also model classical communication between quantum processors via controlled gates.

In classical parallel algorithm design, one defines the **span** of a computation to be the length of the longest sequence of operations that cannot be executed in parallel, and the **work** of a computation to be the total number of operations that the computation executes. Intuitively, the span of a parallel algorithm is its running time if it were implemented on an unbounded number of processors and its work is its running time if it were implemented on a single processor. In this paper we introduce and motivate analogues of span and work for parallel models of quantum computing.

It is important to focus on minimizing the time used to run the quantum processors, but we find that only focusing on such measures can neglect complexities that correspond to work. For example, if the question of quantum query span emerges from concerns over hardware limitations, it is also important to consider the total quantum work being performed, since gates applied in parallel must be managed by classically controlled devices. Using this lens, we can observe that several existing quantum parallel algorithms have non-trivial improvements in span, but they end up costing more work than a sequential algorithm would achieve. See Jeffery, Magniez, and Wolf [15] for an example.

This phenomenon can be captured by a simple thought experiment. Consider a parallel quantum algorithm that uses p processors in parallel to achieve a good span but it ignores work. Such an algorithm could, for example, allow for an improvement in span from $O(n^r)$ to $O((n/p)^r)$, for some $r < 1$. However, by ignoring the work bound, each of the p processes runs the same quantum algorithm, implying a total work bound of $O(n^r p^{1-r})$; hence, the work is minimized when using only one processor! This leads to the natural question of whether we can design a parallel quantum algorithm that achieves an improvement in the span without incurring a penalty in work. We show that this is possible, by describing quantum algorithms for computational geometry problems that can be parallelized so that the work matches the optimal sequential runtime.

We believe studying parallel quantum computing in terms of both span and work can help better exploit near term quantum devices. That is, we believe utilizing such circuits in parallel can be a way of better leveraging advances that can be achieved via quantum computing in the near term. In such instances, it is desirable to improve the span of quantum algorithms through the use of explicit parallelism while also achieving good work bounds.

1.1   Related Prior Work

Cleve and Watrous [6] show how to implement the quantum Fourier Transform in a parallel fashion, which implies that Shor's factoring algorithm [29] can be parallelized. More recently, Regev [26] showed that under certain number theoretic conjectures, factoring can be parallelized to improve the span of the algorithm.

Beyond factoring, several works have analyzed the improvement in query complexity when allowing parallelization. Jeffery, Magniez, and Wolf [15] explore the analogue to span and show that the element distinctness and k-sum problem can be parallelized to go from $\Theta(n^{2/3})$ and $\Theta(n^{k/(k+1)})$ query complexity sequentially to $\Theta((n/p)^{2/3})$ and $\Theta((n/p)^{k/(k+1)})$ query span respectively. Girish, Sinha, Tal and Wu [11] showed that there are problems where one extra parallel query can provide an exponential improvement in the query span.

On the other hand, there have been several works that explore lower bounds and limitations of parallelization in the quantum setting [2,4,17,19,23,30,32].

Many parallel models assume certain restrictions and properties we have seen from early quantum hardware, and these limitations have initiated exploration into distributed quantum computing, which is discussed in depth in a recent survey by Caleffi, Amoretti, Ferrari, Illiano, Manzalini, and Cacciapuoti [3].

There is considerable interest in quantum algorithms for solving computational geometry problems, but previous work has focused exclusively on sequential algorithms. For example, Sadakane, Sugawara, and Tokuyama [27] give a output sensitive quantum convex hull algorithm that runs in $\tilde{O}(h\sqrt{n})$ time,[1] and this time bound is also achieved by Wang and Zhou [31], where h is the size of the output. Note that if $h = n$, as can occur, for instance, with points distributed on a circle or parabola, then these algorithms run in $\tilde{O}(n^{3/2})$ time, which is significantly slower than the best classical algorithms which run in $O(n\log h)$ time [5,7,16]. Recently, in prior work [9], the authors show how to improve this performance for sequential quantum algorithms, giving algorithms for lexicographically sorted inputs that run in $\tilde{O}(\sqrt{nh})$ time. This divide-and-conquer algorithm can be explicitly parallelized by parallelizing the divide step, but it then sequentially solves h smaller instances of size n/h. Unfortunately, doing this leads to a total query work of $\tilde{O}(\sqrt{n}h^{3/2})$, which is not work efficient.

1.2   Our Results

In this paper, we extend a model that parallelizes the query complexity analysis of quantum algorithms, focusing on a *quantum parallel query* model [15], where an algorithm can make quantum queries in parallel in each timestep. In particular we are interested in seeing if there is a way to minimize the number of queries along a computational path, without having to asymptotically increase the total number of queries made across all paths. We define these ideas as *query span* and *query work*, respectively, from the analogous classical-computing ideas they are based on. Sublinear improvements in the query span often occur at the expense of the total query work, so we explore the possibility of interesting computational problems that can have improved query span while maintaining the same asymptotic query work as the optimal sequential quantum algorithm.

We apply this model to some output sensitive geometric problems, using an assumption that the input point set is sorted. The general framework of the algorithms can be thought of as parallelized divide-and-conquer algorithms,

---

[1] We use $\tilde{O}(f(n))$ to denote $O(f(n)\log^c n)$, for some constant $c \geq 0$.

where each processor handles a recursive call. To this end, we assume we can access up to $O(n)$ quantum processors, where we will only use $O(h)$ of them to store our outputs. The recursion relies on being able to find a property related to the median point that can be used by recursive calls maintaining $O(1)$ bits of information about the points they are not assigned to.

We also improve the query span over the best known sequential quantum algorithm for computing convex hulls on sorted input data [9]; here the query span is $\tilde{O}(\sqrt{nh})$ with a query work of $\tilde{O}(\sqrt{n}h^{3/2})$ due to the fact that blocks of size n/h are searched simultaneously by h separate subprocesses. In the two-dimensional version of this problem, one is given a set, S, of n lexicographically sorted points in the plane and asked to output a representation of the smallest convex polygon that contains the points in S. (See Fig. 1.) In this work we introduce $\tilde{O}(\sqrt{n})$ query span and $\tilde{O}(\sqrt{nh})$ query work algorithms for the maxima set and convex hull problems. This means that we can remove the dependency on h in the runtime by increasing the number of processors. The work bounds are, of course, sublinear when h is $O(n^{1-\epsilon})$, for constant $\epsilon > 0$.

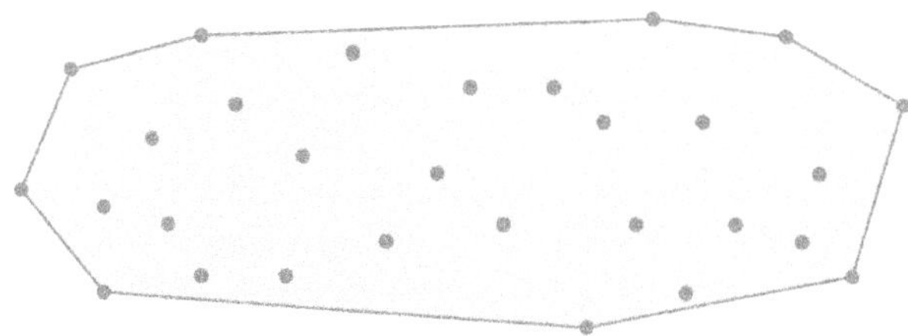

Fig. 1. A two-dimensional convex hull.

The assumption of pre-sorted inputs for the problems rise from the facts that the sorting problem can be reduced to computing the convex hull problem, and there is no quantum speed-up for sorting [14]; hence, we don't expect quantum algorithms to outperform classical algorithms for this problem in the general case, so we apply this assumption to see see what parallel queries can achieve for pre-sorted inputs. This assumption is also common in prior work on classical parallel algorithms; see, e.g., Ghouse and Goodrich [10], Hayashi, Nakano, and Olarlu [13], Berkman, Schieber, and Vishkin [1], Nakano [21], Goodrich [12], and Nakagawa, Man, Ito, and Nakano [20].

2 Preliminaries

Basics of Quantum Computing. Following the approach from our previous work [9], we describe algorithms in this paper so that a reader unfamiliar with quantum computing can appreciate the results by treating quantum subroutines as blackboxes, and we provide a brief overview of quantum computing and the subroutines we use in this section. For a more detailed introduction to the field, we refer the reader to texts such as by Nielsen and Chuang [22].

The basic unit of computation in quantum computing is a **quantum state** over n qubits, defined as a unit vector in $\mathbb{C}^{2^n}$. Among these quantum states, of particular interest are the **computational basis states** $\{|j\rangle\}$ for $j \in [0, \ldots, 2^n - 1]$ defined as the basis vectors containing a 1 in the j-th index and 0 elsewhere. We will describe other vectors in this space using the computational basis,

$$|\Psi\rangle = \sum_{j=0}^{2^n - 1} \alpha_j |j\rangle. \tag{1}$$

If at least two distinct α_j are nonzero, we say that the state $|\Psi\rangle$ is in a superposition of computational basis states. We refer to quantum algorithms that map between two states as **unitaries**.

A key feature in quantum algorithm design is the requirement of a **measurement**. In this work, we assume measurements are done in the computational basis. When a state $|\Psi\rangle$ is measured, we say it **collapses** to the state $|j\rangle$ with probability $|\alpha_j|^2$. Once the state collapses, there is no general way to return to the pre-measurement state $|\Psi\rangle$ without rerunning the circuit that prepares it. We will use the state where all computational basis states are equally likely to be measured, which can be prepared by a circuit of depth 1 by applying a **Hadamard gate** to each qubit.

We assume that the input data $S = [p_0, p_1, \ldots, p_{n-1}]$ is accessible via a **data loading unitary** U_S that maps an index i to the corresponding element p_i. Since quantum operations must be reversible, it is standard to model the action of this unitary by using an index and data register as $|i\rangle|0\rangle \xrightarrow{U_S} |i\rangle|p_i\rangle$. We can use linearity to examine the action of this unitary to a state in superposition,

$$U_S \left(\sum_{j=0}^{n-1} \frac{1}{\sqrt{n}} |j\rangle|0\rangle \right) = \sum_{j=0}^{n-1} \frac{1}{\sqrt{n}} (U_S |j\rangle|0\rangle) = \frac{1}{\sqrt{n}} \sum_{j=0}^{n-1} |j\rangle|p_j\rangle, \tag{2}$$

from which the data point p_j can be retrieved with probability $|\frac{1}{\sqrt{n}}|^2$ by measuring the index register. In summary, we assume a model where a quantum state representing a uniform distribution of our inputs can be prepared by a single query to the data loading unitary. The complexity of data loading unitaries is often high, and so we focus on the total number of queries required to U_S.

Quantum Subroutines. The quantum computing components of our algorithms are primarily invocations of two subroutines, and the ideas of the algorithms should be clear to a reader new to quantum computing. When a quantum subroutine returns a value, the value resides in a register to be read at the end of the computation. We discuss this in more detail when we discuss our computational model.

The first subroutine is for preparing superposition states of subsets of points, when the size of the subset is a power of 2. Recall that the data

$S = [p_0, p_1, \ldots, p_{n-1}]$ is encoded as a sorted list in U_S, and each parallel thread will prepare a uniform superposition of equally sized blocks of the data. We define the state,

$$|S_{t,j}\rangle := \frac{1}{\sqrt{n/2^t}} \sum_{k=0}^{n/2^t - 1} |jn/2^t + k\rangle |p_{jn/2^t+k}\rangle, \tag{3}$$

as the superposition of all the states from $jn/2^t$ to $(j+1)n/2^t - 1$. That is, given a subdivision of the input set into 2^t equally sized blocks, we are preparing all the points contained in block j. To construct $|S_{t,j}\rangle$, we begin with a register storing a $\log n$-bit zero string $|0^{\log n}\rangle$, and another register large enough to store points in the dataset. We call these the index and data register respectively. The first step is to set the first t bits of the index register to j. Next, take the last $\log n - t$ bits in the index register and apply Hadamard gates to each of these qubits to prepare a uniform superposition state over the bitstrings with the prefix j. Finally, apply U_S to this state and the data register to prepare $|S_{t,j}\rangle$. We summarize this process in Algorithm 1.

Algorithm 1. qPrep(t, j)

Require: t: This parameter tells us that a block will have 2^t elements.
Require: j: The index of the block to prepare.
Require: Global access to the unitary U_S that encodes the sorted data S
 1: Prepare two registers, initialized to $|0^{\log n}\rangle_I |0\rangle_D$.
 2: Set the first t bits of I to j
 3: Apply a Hadamard gate to each remaining qubit in I.
 4: Apply U_S.
 5: **return** $|S_{t,j}\rangle$, the state described in equation 3.

The second subroutine we will use is a quantum max/min finding algorithm due to Durr and Hoyer [8]. The input to this algorithm is a Boolean function $f : D \to \{0, 1\}$, a comparator, $\preceq$, to maximize (or minimize) over, the depth of recursion this function is applied t, and the block we are searching in j. The superposition state described in Theorem 1 is prepared by qPrep(t, j). This has the effect of applying the algorithm only to the points in block j, and allows us to search the block in $\tilde{O}(\sqrt{n/2^t})$ time. We use the signature, qMax/qMin$(f, \preceq, t, j)$, when calling this function, and the output is an element of D or null if there are no elements such that $f(d) = 1$. The function f can be passed as a classical circuit that can be converted to a quantum circuit.

Theorem 1 (Quantum Maximum/Minimum Finding [8]). *Let* $D = [d_0, \ldots, d_{n-1}]$ *be a list of* n *elements represented by* w *bits, and let* S *be the time required to prepare the state,*

$$|\psi\rangle = \frac{1}{\sqrt{n}} \sum_{i=0}^{n-1} |i\rangle |d_i\rangle, \tag{4}$$

and Q the time it takes to query a boolean function $f : D \to \{0, 1\}$. Let M be the subset of D containing all $x \in D$ such that $f(x) = 1$. Also, let $\preceq$ be some ordering of data values in the data register such that comparisons according to this ordering can be performed in $O(\log w)$ time. Then we can find the maximum (or minimum) value of M under the specified ordering in time $S \cdot Q \cdot \tilde{O}(\sqrt{n})$ with success probability $2/3$ or $Q \cdot \tilde{O}(\sqrt{n})$ calls to qPrep.

3   Model

The Quantum Parallel Query Model. Our model is a ***quantum parallel query*** model, where up to p quantum processors have access to a set of n points via a unitary U encoding the data of interest with the action

$$U|j\rangle|0\rangle \to |j\rangle|x_j\rangle. \tag{5}$$

We analyze our algorithms by counting the **total** number of queries made to this oracle. We define the minimum number of queries on a critical path of computation as the **query span** and the total number of queries to the oracle over all processors as the **query work**. Similar models (for example, see Jeffery *et al.* [15]) consider a unitary that is queried by p processors simultaneously and are interested in the required number of simultaneous calls made to such a unitary. These models consider a simultaneous call to the unitary as one call, which is useful for analyzing the query span of an algorithm. Our models are equivalent when only considering query span, but we require that each processor pay 1 unit of work to the total to use it. For the algorithms considered here, we are not primarily interested in how the span changes with the number of processors, and assume we have access to up to $O(n)$ processors.

A general description of our model is illustrated in Fig. 2, where the algorithm begins with a call to A over the full point set of size n. The output is classical data, which can be copied around in a quantum circuit. The figure illustrates this process via the vertical wires. Given these pieces of data from the previous iteration, now two processors each call A on subsets of size $n/2$.

Both the maxima set and convex hull problems are output sensitive algorithms. We define the recursive calls such that once there is no need to search a block, a new subprocess does not spawn. Furthermore, each processor will store one of the output elements at the end of a terminating subprocess. Our model makes sure that each new function call is allocated to the next idle processor in an array, as shown in Fig. 2. Because of this, once the algorithm terminates, it is possible to retrieve all h elements in $O(h)$ extra time.

Another important feature of counting the total work is that it allows us to recursively run bounded-error algorithms as we can now determine the number of runs required for all subprocesses to succeed.

4   Parallel Quantum Convex Hull Algorithm

We now apply the model to describe a quantum parallel algorithm to construct the convex hull of n points in the plane in $\tilde{O}(\sqrt{n})$ query span and $\tilde{O}(\sqrt{nh})$ query

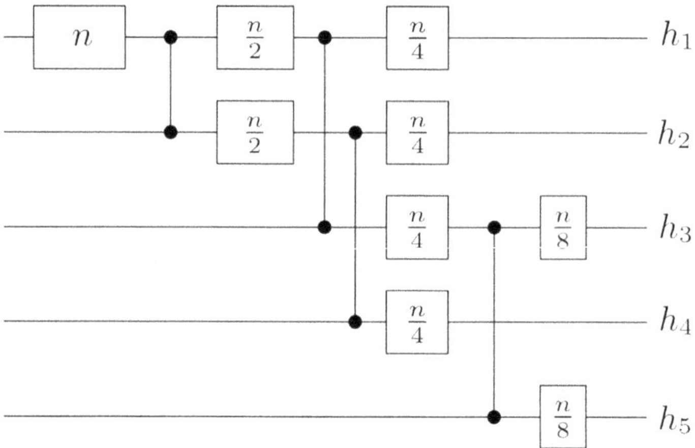

Fig. 2. An example demonstrating how the model uses extra processors for solving subprocesses. Each box is a call to a recursive algorithm A, and the labels of each box describe the size of the input to that call to A. Subprocesses that have solved a search space can maintain a value in the output set, and the solution can be read out in $O(h)$ postprocessing time.

work, where h is the size of the output. The query span improves upon existing quantum query algorithms for the same problem by removing its dependency on h. In each call, we search for the edge of the convex hull that lies above the median point x. We call such an edge a **bridge edge** over x. After finding this edge, we recurse by dividing the input in halves.

4.1 Algorithm

We briefly review a well-known classical (sequential) divide-and-conquer algorithm for 2D convex hulls (see, e.g., [7,24,25,28]). The convex hull $CH(S)$ of a set, S, of n points in the plane can be described as a union of the polygonal chains that form the upper hull $UH(S)$ and lower hull $LH(S)$ of the points, where $UH(S)$ (resp., $LH(S)$) is the set of edges of $CH(S)$ with positive (negative) normals. At a high level, we can divide S into a left set, S_1, and right set, S_2, and recursively find $UH(S_1)$ and $UH(S_2)$. Then we look for the tangent edge above $UH(S_1)$ and $UH(S_2)$ and prune away the points under the bridge, resulting in an algorithm running in $O(n \log n)$ time.

It can be shown that the bridge edge lies on the solution to a two-dimensional linear program using a point-line duality [7,24,28]. This fact is summarized in Lemma 1. In this duality, each point p with coordinate (p_x, p_y) in the primal plane maps to a line $y = p_x x - p_y$ in the dual plane and vice versa. See Fig. 3.

Lemma 1 (see, e.g., [7,24,28]). *Given a vertical line L, and a set S of n points in a plane, one can determine the edge e of $UH(S)$ that intersects L (or*

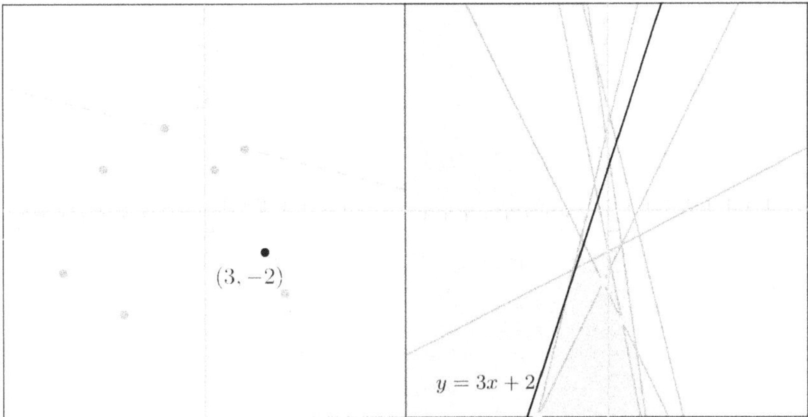

Fig. 3. An example of point-line duality for the convex hull. The left shows the primal plane showing the input set in purple, and the lines containing the upper hull in green. The solid edge is the bridge edge crossing the y-axis. Each point $p = (p_x, p_y)$ in the primal plane maps to a line $y = p_x x - p_y$ in the dual plane on the right and vice versa. The green lines in the primal plane map to the lower envelope in the dual plane. (Color figure online)

that no such edge exists) by finding the highest point of intersection between $O(n)$ halfplanes.

We also use the following lemma due to Sadakane et al. showing a quantum algorithm for computing the highest point in the intersection of n halfplanes. Their result follows from a quantum algorithm to solve linear programs.

Lemma 2 (Sadakane et al. [27]). *The highest point in the intersection of n lower-halfplanes can be computed in $O(\sqrt{n} \log^2 n)$ time, using a quantum computer, with probability $1 - n^{-c}$ for any constant c given the linear constraints in superposition.*

The linear constraints in superposition required in the lemma refers to the following state, which we can prepare using qPrep (Algorithm 1):

$$|\psi\rangle := \frac{1}{\sqrt{n}} \sum_{i=0}^{n-1} |i\rangle |p_i\rangle, \tag{6}$$

Here p_i are the coordinates of each of the n points. By point-line duality, the coordinates fully describe the halfplanes in the dual plane, so a single call to qPrep is sufficient to use Lemma 2.

The output of Lemma 2 is a point in the dual plane, which can be mapped to a line in the primal plane containing the bridge edge. To find it, we can find the two endpoints by minimizing the distance to the line using qMin. Algorithm 2 summarizes this process, and returns the endpoints $p_s, p_f \in S$ of the bridge edge that crosses the median coordinate in the t-th level of recursion in branch j. The query complexity of this algorithm is dominated by the use of Lemma 2, giving a total query complexity of $\tilde{O}(\sqrt{n/2^t})$.

Algorithm 2. Bridge(t, j)

Require: t: The layer of recursion. This implies that the size of blocks to search over are $n/2^t$.
Require: j: The index of the block we are searching in.
1: Call `qPrep(t, j)` to load the points.
2: Let L be a vertical line drawn at the x coordinate of the median point in the j-th block of size 2^t. This can be computed by querying U_S at index $j2^t + 2^t/2$.
3: Use Lemmas 1 and 2 to find the bridge edge in the j-th block of size $n/2^t$ with p_s, p_f being the two endpoints.
4: Return the two endpoints $p_s \in S_i$ and $p_f \in S_j$ of the bridge edge.

We now describe the full algorithm. As described in Sect. 3, we start with one processor that computes the bridge edge over the median point of the input set S. It then claims the first half of the point set for the recursive call, and spawns a new subprocess to handle the other half. Each subprocess maintains up to two bridge edges that overlap with the set of points they are considering. They then recursively perform the same task, with the only change being that they will not spawn a new subprocess in the case that the median point lies under a previously found bridge edge. In this case, the processor will start a search on the half of its assigned search space that is not covered by an existing upper hull edge (See Fig. 4). The full algorithm is given in Algorithm 3.

Algorithm 3. qParallelConvexHull(t, j, l, r, P)

Require: t: The level of recursion. The top call is $t = 0$.
Require: j: The block to search in.
Require: l, r: The outgoing left and right bridge edges l, r (if any) as a tuple of points $(l_s, l_e), (r_s, r_e) \in S^2$. By definition, l_s, r_e are outside the block of interest, and l_e, r_s are in this block.
Require: P: The processor assigned to this task.
1: `qPrep(t, j)`
2: Let p_m be the point with median coordinate in block j at level t.
3: **if** p_m lies under l **then return** qParallelConvexHull$(t + 1, 2j + 1, l, r, P)$.
4: **if** p_m lies under r **then return** qParallelConvexHull$(t + 1, 2j, l, r, P)$.
5: $(p_s, p_f) = $ Bridge(t, j)
6: **if** p_s and p_f connect to the neighboring bridge points, **then return** (No need to recurse as all relevant points are found.)
7: qParallelConvexHull$(t + 1, 2j, l, (p_s, p_f), P)$.
8: qParallelConvexHull$(t + 1, 2j + 1, (p_s, p_f), r,$ new processor$)$.

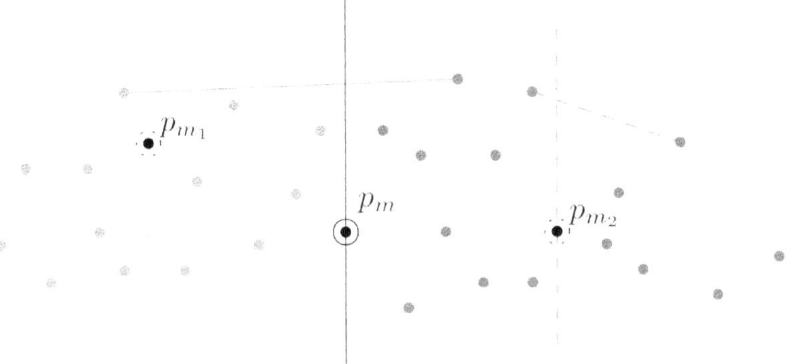

Fig. 4. An illustration of the distribution of work in the quantum parallel convex hull algorithm. The median points are circled. At the top level, p_m is selected as a median and the algorithm searches for the edge on the upper hull that lies above p_m. The algorithm then recurses to the orange and purple point sets respectively, where the median points in each group is labeled $p_{m_1}(Color figure online)$ and p_{m_2}. In the left half, p_{m_1} intersects a known upper hull, so the right half can be safely ignored.

Lemma 3. *Algorithm 3 has a total query span of $\tilde{O}(\sqrt{n})$.*

Proof. The query span of this algorithm can be characterized by the following recurrence relation:

$$QS(n) \leq QS(n/2) + \tilde{O}(\sqrt{n}). \tag{7}$$

The total number of points in the output set only affects the branches that get explored so does not appear in this recurrence. The $\tilde{O}(\sqrt{n})$ term comes from finding the bridge edge over the median point, as described in Lemmas 1 and 2. Since the work is distributed in a recursive manner, at each level of recursion an individual processor only needs to manage half of the previous problem's size. The recurrence solves to $QS(n) = \tilde{O}(\sqrt{n})$.

Once completed, what remains is to reconstruct the upper hull from the bridge edges that were found. This incurs an additional additive $O(h)$ post-processing time, without any queries to the oracle. Thus, the query span of this algorithm given n processors is $\tilde{O}(\sqrt{n})$.

Lemma 4. *Algorithm 3 has a total query work of $\tilde{O}(\sqrt{nh})$ where h is the size of the convex hull and the tilde suppresses logarithmic terms in n.*

Proof. The recurrence and the solution are stated below. The $\tilde{O}(\sqrt{n})$ work being done per process comes from computing the bridge edge using Lemmas 1 and 2.

$$QW(n,h) \leq QW(n/2, h_1) + QW(n/2, h_2) + \tilde{O}(\sqrt{n}) \leq \tilde{O}(\sqrt{nh}). \tag{8}$$

The recurrence can be solved by strong induction, using the inductive hypothesis that $QW(k, h) \leq cl\sqrt{kh}$ for all $k < n$. Here l is a polylogarithmic factor in

n. The terms h_1 and h_2 corresponding to the size of the solution in each of the two halves, and satisfy $h_1 + h_2 = h$. The work performed in each level requires $\tilde{O}(\sqrt{n})$ queries during the call to qMax. The polylogarithmic factors from the calls to qMax emerge from sampling. We now solve this recurrence.

$$QW(n, h) \leq QW(n/2, h_1) + QW(n/2, h_2) + \tilde{O}(\sqrt{n}) \tag{9}$$

$$\leq cl\sqrt{(n/2)h_1} + cl\sqrt{(n/2)h_2} + \tilde{O}(\sqrt{n}) \tag{10}$$

$$= cl\sqrt{\frac{nh}{2(h/h_1)}} + cl\sqrt{\frac{nh}{2(h/h_2)}} + \tilde{O}(\sqrt{n}) \tag{11}$$

$$= cl\sqrt{nh}\left(\frac{1}{\sqrt{2h/h_1}} + \frac{1}{\sqrt{2h/h_2}}\right) + \tilde{O}(\sqrt{n}) \tag{12}$$

$$= cl\sqrt{nh}\left(\frac{\sqrt{h_1} + \sqrt{h_2}}{\sqrt{2h}}\right) + \tilde{O}(\sqrt{n}) \tag{13}$$

$$\leq cl\sqrt{nh} + \tilde{O}(\sqrt{n}) = \tilde{O}(\sqrt{nh}) \tag{14}$$

The inequality in the last line is due to the fact that $\sqrt{h_1} + \sqrt{h_2}$ is maximized when $h_1 = h_2 = h/2$. In summary, the total query work required for this algorithm is $\tilde{O}(\sqrt{nh})$, which matches the best known sequential quantum query algorithm to solve the same problem [9].

Lemma 5. *Algorithm 3 succeeds with probability* $1 - \frac{3}{n}$.

Proof. Minimum finding (Theorem 1) and Bridge (Algorithm 2) are bounded-error probabilistic quantum algorithms with success probabilities of $1 - 1/3$ and $1 - \frac{1}{n^k}$ respectively for some constant k. We can decrease the probability of failure to an inverse exponential by repeating $c\log n$ times. We only need to see the target one time since these are search problems, so the probability of failure for each is $\frac{1}{3^{c\log n}} = O(1/n^c)$ and $\frac{1}{n^{ck\log n}}$ respectively. The probability that at least one of these fail during the same call can be determined using the union bound:

$$\mathbb{P}[\text{qMin fails}] \cup \mathbb{P}[\text{qMin fails}] \cup \mathbb{P}[\text{Lemma 2 fails}] \leq \frac{2}{n^c} + \frac{1}{n^{ck\log n}} \leq \frac{3}{n^c}. \tag{15}$$

In the worst case we spawn $O(n)$ processors at a depth of $O(\log n)$. At each call, we are running two instances of minimum finding and one instance of Bridge. In total, we have at most $O(n\log n)$ calls to the qParallelConvexHull subroutine that has at most a $\frac{3}{n^c}$ probability of failure assuming each bounded-error subroutine is run $\log n$ times. We can apply the union bound again to show that the probability that at least one of these calls fail is at most

$$\frac{3n\log n}{n^c} \leq \frac{3n^2}{n^c} = \frac{3}{n^{c-2}}. \tag{16}$$

Since c is a constant we can choose when rerunning the bounded error subroutines to reduce error, we use $c \geq 3$ for a total success probability of at least $1 - \frac{3}{n}$ for one iteration of the algorithm.

We summarize our results into the following theorem.

Theorem 2. *There is a quantum parallel algorithm for finding the convex hull of a presorted set of n points in the plane accessed via a unitary encoding the data with $\tilde{O}(\sqrt{n})$ query span, $\tilde{O}(\sqrt{nh})$ query work, and $1 - \frac{3}{n}$ success probability where h is the size of the output.*

5   Discussion

We considered a new approach and model for quantum parallel algorithms, prioritizing keeping query work low, while improving query span. We showed that for the maxima set and convex hull problems, we can improve the query span by a factor of $\sqrt{h}$ with respect to the leading sequential quantum algorithms [9] while maintaining the same level of query work. We did not consider whether restricting the number of processors could affect these parameters, however, and we believe a different model of quantum parallelism could analyze algorithms in terms of reducing the query span, query work, and the width (number of processors allowed). A deeper understanding of the tradeoff between these three parameters could serve as an important step in classifying parallel quantum algorithms. We leave this problem as an open question, and believe that the problems proposed in this paper are good starting points.

References

1. Berkman, O., Schieber, B., Vishkin, U.: A fast parallel algorithm for finding the convex hull of a sorted point set. Int. J. Comput. Geometry Appl. **06**(02), 231–241 (1996). https://doi.org/10.1142/S0218195996000162
2. Burchard, P.: Lower bounds for parallel quantum counting (2019). https://arxiv.org/abs/1910.04555
3. Caleffi, M., Amoretti, M., Ferrari, D., Illiano, J., Manzalini, A., Cacciapuoti, A.S.: Distributed quantum computing: a survey. Comput. Netw. **254**, 110672 (2024). https://www.sciencedirect.com/science/article/pii/S1389128624005048,   https://doi.org/https://doi.org/10.1016/j.comnet.2024.110672
4. Carolan, J., Gilani, A.S., Vempati, M.: Quantum advantage and lower bounds in parallel query complexity (2025). https://arxiv.org/abs/2410.02665
5. Chan, T.M., Snoeyink, J., Yap, C.-K.: Output-sensitive construction of polytopes in four dimensions and clipped Voronoi diagrams in three. In: 6th ACM-SIAM Symposium on Discrete Algorithms (SODA), pp. 282–291 (1995)
6. Cleve, R., Watrous, J.: Fast parallel circuits for the quantum fourier transform. In: Proceedings 41st Annual Symposium on Foundations of Computer Science, pp. 526–536 (2000). https://doi.org/10.1109/SFCS.2000.892140
7. de Berg, M., Cheong, O., van Kreveld, M., Overmars, M.: Computational Geometry: Algorithms and Applications. Springer, 3rd edn. (2008). https://doi.org/10.1007/978-3-540-77974-2

8. Durr, C., Hoyer, P.: A quantum algorithm for finding the minimum (1999). http://arxiv.org/abs/quant-ph/9607014
9. Fukuzawa, S., Goodrich, M.T., Irani, S.: Quantum combine and conquer and its applications to sublinear quantum convex hull and maxima set construction. In: 41th International Symposium on Computational Geometry (SoCG 2025) (2025)
10. Ghouse, M.R., Goodrich, M.T.: In-place techniques for parallel convex hull algorithms. In: 3rd ACM Symposium on Parallel Algorithms and Architectures (SPAA), pp. 192–203 (1991)
11. Girish, U., Sinha, M., Tal, A., Wu, K.: The power of adaptivity in quantum query algorithms. In: Proceedings of the 56th Annual ACM Symposium on Theory of Computing, STOC 2024, pp. 1488–1497, New York, NY, USA (2024). Association for Computing Machinery. https://doi.org/10.1145/3618260.3649621
12. Goodrich, M.T.: Constructing the convex hull of a partially sorted set of points. Comput. Geometry 2(5), 267–278 (1993). https://www.sciencedirect.com/science/article/pii/092577219390023Y, https://doi.org/https://doi.org/10.1016/0925-7721(93)90023-Y
13. Hayashi, T., Nakano, K., Olarlu, S.: An $O((\log\log n)^2)$ time algorithm to compute the convex hull of sorted points on reconfigurable meshes. IEEE Trans. Parallel Distrib. Syst. 9(12), 1167–1179 (1998). https://doi.org/10.1109/71.737694
14. Høyer, P., Neerbek, J., Shi, Y.: Quantum complexities of ordered searching, sorting, and element distinctness, pp. 346–357. Springer, Heidelberg (2001). http://dx.doi.org/10.1007/3-540-48224-5_29
15. Jeffery, S., Magniez, F., de Wolf, R.: Optimal parallel quantum query algorithms. Algorithmica 79(2), 509–529 (2016). https://doi.org/10.1007/s00453-016-0206-z
16. Kirkpatrick, D.G., Seidel, R.: Output-size sensitive algorithms for finding maximal vectors. In: O'Rourke, J. (ed.) Proceedings of the First Annual Symposium on Computational Geometry, Baltimore, Maryland, USA, 5-7 June 1985, pp. 89–96. ACM, 1985. https://doi.org/10.1145/323233.323246
17. Koiran, P., Landes, J., Portier, N., Yao, P.: Adversary lower bounds for nonadaptive quantum algorithms. J. Comput. Syst. Sci. 76(5), 347–355 (2010). Workshop on Logic, Language, Information and Computation. https://www.sciencedirect.com/science/article/pii/S0022000009000968, https://doi.org/https://doi.org/10.1016/j.jcss.2009.10.007
18. Margolus, N.: Parallel quantum computation. Complexity, Entropy Phys. Inf. Santa Fe Inst. Stud. Sci. Complexity 8, 273–287 (1990)
19. Moore, C., Nilsson, M.: Parallel quantum computation and quantum codes. SIAM J. Comput. 31(3), 799–815 (2001). https://doi.org/10.1137/S00097539799355053
20. Nakagawa, M., Man, D., Ito, Y., Nakano, K.: A simple parallel convex hulls algorithm for sorted points and the performance evaluation on the multicore processors. In: 2009 International Conference on Parallel and Distributed Computing, Applications and Technologies, pp. 506–511 (2009). https://doi.org/10.1109/PDCAT.2009.56
21. Nakano, K.: Computation of the convex hull for sorted points on a reconfigurable mesh. Parallel Algorithms Appl. 8(3-4), 243–250 (1996). https://doi.org/10.1080/10637199608915555
22. Nielsen, M.A., Chuang, I.L.: Quantum Computation and Quantum Information. Cambridge University Press, 10th anniversary edn. (2010)
23. Nishimura, H., Yamakami, T.: An algorithmic argument for nonadaptive query complexity lower bounds on advised quantum computation. In: Fiala, J., Koubek, V., Kratochvíl, J. (eds.) MFCS 2004. LNCS, vol. 3153, pp. 827–838. Springer, Heidelberg (2004). https://doi.org/10.1007/978-3-540-28629-5_65

24. O'Rourke, J.: Computational Geometry in C. Cambridge University Press (1998)
25. Preparata, F.P., Shamos, M.I.: Computational Geometry: An Introduction. Springer (2012). https://doi.org/10.1007/978-1-4612-1098-6
26. Regev, O.: An efficient quantum factoring algorithm. arXiv:2308.06572 (2024)
27. Sadakane, K., Sugawara, N., Tokuyama, T.: Quantum computation in computational geometry. Interdiscip. Inf. Sci. **8**(2), 129–136 (2002)
28. Seidel, R.: Linear programming and convex hulls made easy. In: 6th Symposium on Computational Geometry (SoCG), pp. 211–215 (1990)
29. Shor, P.W.: Polynomial-time algorithms for prime factorization and discrete logarithms on a quantum computer. SIAM Rev. **41**(2), 303–332 (1999). https://doi.org/10.1137/S0036144598347011
30. van Dam, W.: Quantum oracle interrogation: getting all information for almost half the price. In: Proceedings 39th Annual Symposium on Foundations of Computer Science (Cat. No.98CB36280), pp. 362–367 (1998). https://doi.org/10.1109/SFCS.1998.743486
31. Wang, C., Zhou, R.G.: A quantum search algorithm of two-dimensional convex hull. Commun. Theor. Phys. **73**(11), 115102 (2021). https://doi.org/10.1088/1572-9494/ac1da0
32. Zalka, C.: Grover's quantum searching algorithm is optimal.: Phys. Rev. A **60**, 2746–2751 (1999). https://doi.org/10.1103/PhysRevA.60.2746

On Strictly Output-Sensitive Color Frequency Reporting

Erwin Glazenburg$^{(\boxtimes)}$ and Frank Staals

Utrecht University, Utrecht, The Netherlands
`{e.p.glazenburg,f.staals}@uu.nl`

Abstract. Given a set of n colored points $P \subset \mathbb{R}^d$ we wish to store P such that, given some query region Q, we can efficiently report the colors of the points appearing in the query region, along with their frequencies. This is the *color frequency reporting* problem. We study the case where query regions Q are axis-aligned boxes or dominance ranges. If Q contains k colors, the main goal is to achieve "strictly output-sensitive" query time $O(f(n) + k)$. We show that, for every $s \in \{2, \ldots, n\}$, there exists a simple $O(ns \log_s n)$-size data structure for points in $\mathbb{R}^2$ that allows frequency reporting queries in $O(\log n + k \log_s n)$ time. Furthermore, we give a lower bound for the weighted version of the problem in the arithmetic model, proving that with $O(m)$ space one can not achieve query times better than $\Omega\left(\phi \frac{\log(n/\phi)}{\log(2m/n)}\right)$, where ϕ is the number of possible colors. This means that our data structure is near-optimal. We extend these results to higher dimensions as well. Finally, we give an $O(n^{1+\varepsilon} + m \log n + K)$-time algorithm that can answer m dominance queries $\mathbb{R}^2$ with total output complexity K, while using only linear working space.

Keywords: Range Searching · Data structure · Lower bound

1 Introduction

Let P be a set of n points in $\mathbb{R}^d$, with $d = O(1)$, each of which has a color from $[\phi] = \{1, \ldots, \phi\}$ (with $\phi \leq n$), and let P_c denote the subset of points of color c. We want to store P so that given an axis aligned query box Q we can efficiently report the colors of the points appearing in the query range, together with their frequency. That is, we wish to report a set of k color, frequency pairs (c, f) so that $f = |P_c \cap Q| > 0$ and the sum of the frequencies is $|P \cap Q|$. See Fig. 1. We call this *color frequency reporting*, but it is also known by the rather vague name of *type-2* color counting [12]. More generally, we may associate every point $p \in P$ with a weight $w(p)$ from some appropriate semigroup, and instead report, for each color c, the total weight $w(P_c \cap Q) = \sum_{p \in P_c \cap Q} w(p)$ in the query range,

E. Glazenburg was supported by the Netherlands Organisation for Scientific Research (NWO) under project OCENW.M20.135.

© The Author(s), under exclusive license to Springer Nature Switzerland AG 2026
J. Kozik and A. Wolff (Eds.): SOFSEM 2026, LNCS 16448, pp. 246–259, 2026.
https://doi.org/10.1007/978-3-032-17801-5_18

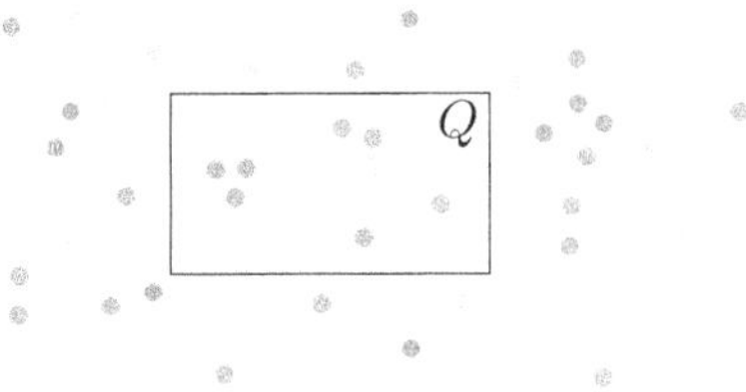

Fig. 1. A set of colored points P in $\mathbb{R}^2$ and a query box Q with color frequencies (red,4) and (blue,3). There are no green or purple points in Q, hence they do not appear in the output. (Color figure online)

provided that this weight is non-zero. Our main goal is to obtain small, ideally (near) linear space data structures for these problems that achieve a *strictly output-sensitive* query time of the form $O(q(n) + k)$, i.e. a query time strictly linear in the output size k (and whose dependency $q(n)$ on n is of course also sublinear, and preferably even (poly-)logarithmic). We are generally interested in data structures in the pointer machine model, as this tends to give relatively simple and flexible data structures.

Related Work. Color frequency reporting is a form of *generalized intersection searching* [11,12,14]. For points in $\mathbb{R}^1$ the problem was already solved optimally in 1995: Bozanis et al. [4] gave a data structure of linear size and query time $O(\log n + k)$. Similarly, there is an optimal data structure for the dual problem of reporting the frequencies of a set of colored intervals stabbed by a query point [12].

In $\mathbb{R}^2$, it is relatively simple to obtain a data structure of size $O(n \log n)$ with query time $O(\log n + k \log n)$: we first report all intersected colors in $O(\log n + k)$ time using a colored range reporting query [16], and then for all k colors separately perform a standard uncolored range counting query in $O(\log n)$ time each using range trees [3].

Rahul, Gupta, and Rajan [15] describe an $O(n^{1+\varepsilon})$ space data structure that with $O(\log n + k)$ query time, based on bootstrapping. Here, and throughout the rest of the paper $\varepsilon > 0$ is an arbitrarily small constant. To achieve query time $O(\text{polylog } n + k)$, we need at least $\Omega(n(\log n / \log \log n)^{d-1})$ space, as Chazelle's lower bound for range reporting [7] applies (just assign each point a unique color), however, there are no near-linear-space structures that achieve strictly output sensitive query times. In particular, it is open whether one can achieve even a query time of $O(n^{0.99} + k)$ using near-linear space.

In the more powerful word-RAM model, some recent progress has been made towards near-linear-space structures for points in $\mathbb{R}^2$. Chan et al. [5] achieve query time $O(\frac{\log n}{\log \log n} + k \log \log n)$ using $O(n \log^{1+\varepsilon} n)$ space using a recursive grid approach and several "bit-tricks" that allow them to sum the counts of multiple colors simultaneously. Afshani et al. [1] present a linear-space data structure for simplex color frequency reporting queries. They achieve query time $O(n^{1-1/d} + n^{1-1/d}\phi^{1/d}/w^{\alpha})$, where $\alpha > 0$ is some constant and w is the word size (which is typically $\Theta(\log n)$). Furthermore, they solve approximate color frequency counting with an *additive* error εn for dominance queries with $O(n)$ space and $O(\log n + \varepsilon^{-1} n + k)$ query time.

In external memory with block size B, Ganguly et al. [10] give a data structure of size $O(n)$ that answers *constant-factor approximate* frequency reporting queries in $O(\log_B n + \log^* B + k/B)$ time (here, $\log^* B$ is the iterated logarithm of B, i.e. the smallest value i for which $\log^{(i)} B = \log(\log^{(i-1)} B) \leq 2$ (with $\log^{(1)} B = \log B$)). In the pointer machine model, their techniques can be used to answer $(1 + \varepsilon)$-approximate frequency reporting queries using $O(n/\varepsilon)$ space in $O(\log n + k)$ time.

Challenges. Achieving our goal of strict output sensitivity using near-linear space turns out to be extremely challenging. Most data structuring approaches (including the techniques used in the above results) follow a divide-and-conquer approach that partitions the space (and thereby the input point set). However, this may cause the points $P_c \cap Q$ of color c to be stored in many, say $g(n)$, different places in the data structure. This means that we have to aggregate their frequencies at query time. If this happens for $\Omega(k)$ colors, we get an $\Omega(kg(n))$ cost in the query time. If we instead store the points per color, we may have to spend $\Theta(\log n)$ time per color to actually count the subset that lies in Q. Filtering search [6] is a common technique used in reporting queries which allows us to charge some of the "searching" costs to the output. However, as our main challenge is combining the color counts, filtering search is of limited use.

Results and Organization. We present the following results:

- When P is a set of n points in $\mathbb{R}^2$, and we are given a parameter $s \in \{2, \ldots, n\}$, we present a simple $O(ns \log_s n)$ size data structure storing P that allows frequency reporting queries for dominance ranges in $O(\log n + k \log_s n)$ time. See Sect. 3. For ease of exposition, we present our data structure for color frequency reporting, but our results apply also to the weighted version of the problem. Setting $s = n^\varepsilon$, this then yields an $O(n^{1+\varepsilon})$ space structure with strictly output-sensitive $O(\log n + k)$ query time. This matches, and we believe simplifies, the result of Rahul et al. [15]. Setting e.g. $s = 2^{\sqrt{\log n}}$ reduces the space to $O(n 2^{\sqrt{\log n}})$ space, while still answering queries at a cost of $\sqrt{\log n}$ per reported color.
- We extend the above results to queries with arbitrary axis aligned boxes in $\mathbb{R}^d$. Every additional dimension comes at the cost of an $(s \log_s n)$ factor in space, and a $(\log_s n)$ factor per color (and an additive $O(\log^{d-1} n)$ term) in query time. Every additional side we allow in the query range (from d-sided dominance ranges to arbitrary $2d$-sided axis-aligned boxes) comes at the cost of a factor $\log n$ in space, and a factor 2 in query time. With $s \approx n^\varepsilon$ this still leads to strictly output-sensitive query time $O(\log n + k)$ using $O(n^{1+\varepsilon})$ space.

- We prove a lower bound for the weighted case with dominance queries in the arithmetic model; see Sect. 4. In this model, computation is essentially free: we just count the number of pre-computed weights (of subsets of points) stored by the data structure, and the number of these weights we have to combine to obtain the answer to a query. Building on Chazelle's lower bounds for counting queries in this model [8], we prove that using $O(m)$ space one can not achieve query time faster than $\Omega\left(\phi\left(\frac{\log(n/\phi)}{\log(m/n)}\right)^{d-1}\right)$.

  In particular, compare this to our trade-off data structure from Sect. 3, which uses $O(ns^{d-1}\operatorname{polylog} n)$ space for $O(\operatorname{polylog} n + k\log_s^{d-1} n)$ query time; the lower bound says that using $O(ns^{d-1})$ space, one cannot achieve query time better than $\Omega\left(\phi\left(\frac{\log(n/\phi)}{\log((ns^{d-1})/n)}\right)^{d-1}\right) = \Omega\left(\phi\left(\frac{\log(n/\phi)}{\log(s^{d-1})}\right)^{d-1}\right) =$ $\Omega\left(\phi\left(\frac{\log(n/\phi)}{\log s}\right)^{d-1}\right) = \Omega(\phi\log_s^{d-1}(n/\phi))$. So for $\phi \leq n^c$ with $c < 1$, our data structure is worst-case optimal w.r.t. the query dependency on k, up to $\log n$ factors in space. (Note that for larger values of ϕ, the problem becomes less interesting.)

  This suggests that it will be hard to get strictly output sensitive query times for (unweighted) color frequency reporting queries using near linear space, in the pointer machine model. At the very least, one has to use properties specific to counting (e.g. that we can subtract frequencies) to achieve this goal. Note however that even in the uncolored case no such results are known.
- We present a transformation that, for some values of s and ϕ, allows us to reduce the space for 2D dominance queries to $O(ns^\varepsilon \phi^\varepsilon)$, by answering several 1D dominance queries simultaneously; see Sect. 5. Hence, this allows us to shave some logarithmic factor in space in certain scenarios.
- Lastly, we present a result on the algorithmic version of the problem in which we are given a set of n colored points $P \subset \mathbb{R}^2$ as well as a set $\mathcal{Q}$ of m 3-sided color frequency reporting queries that we *all* have to answer. See Fig. 2. Our algorithm in Sect. 6 uses linear, i.e. $O(n + m)$, working space, and answers all m queries in $O(n^{1+\varepsilon} + m\log n + K)$ total time, where K is the total output complexity of all queries. Note that e.g. for $m = \Theta(n)$ we thus essentially obtain $O(n^\varepsilon + k)$ time per query, which compares favorable with the $O(n^\varepsilon k)$ cost per query for the naive solution that uses only $O(n)$ space.

2 Preliminaries

We focus on range searching where the query region $Q = [a_1, b_1] \times [a_2, b_2] \times \ldots \times [a_d, b_d]$ is an axis-aligned box in d-dimensional space. A box in $\mathbb{R}^1$ is an interval, and a box in $\mathbb{R}^2$ is a rectangle. In the simplest case, the box is unbounded in one direction for every dimension, e.g. towards $-\infty$. A query $Q = (-\infty, x] \times (-\infty, y] \times (-\infty, z] \times \ldots$ of this type is a *dominance* query; see Fig. 2. We will often identify the query Q with its corner point $q = (x, y, z, \ldots)$. A dominance

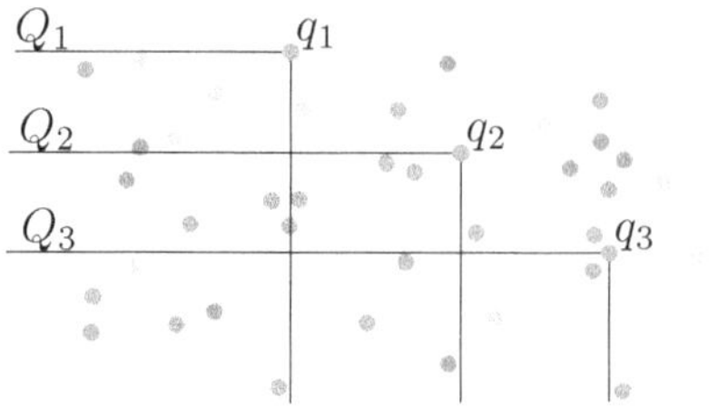

Fig. 2. A set of dominance queries. In the offline version, we wish to answer them all at once.

query can be seen as a *d-sided box*: it is bounded on one side in each of the d dimensions. Similarly an i-sided box $B_{d,i}$, for $d \leq i \leq 2d$, is bounded on two sides in $i - d$ dimensions, and on one side in the remaining $2d - i$ dimensions. General axis-aligned boxes are thus $2d$-sided boxes.

Color Frequency Reporting Queries in $\mathbb{R}^1$. We briefly review the approach of Bozanis et al. [4] (which itself uses ideas from Gupta et al. [12]) that solves frequency reporting queries in $\mathbb{R}^1$. We focus on the case of dominance queries $(-\infty, q]$. Fix a color $c \in [\phi]$, and let x_i be the ith smallest point in P_c. They then map x_i to the point (x_i, x_{i+1}) in $\mathbb{R}^2$ (or (x_i, ∞) if no successor exists). Let $\tilde{P}_c$ denote the resulting set of mapped points, and let $\tilde{P} = \bigcup_c \tilde{P}_c$. Bozanis et al. observe that the query quadrant $\tilde{Q} = (-\infty, q] \times [q, \infty)$ contains at most one point $(x_i, x_{i+1}) \in \tilde{P}$ of each color c, namely the rightmost point of color c in $(-\infty, q]$. Thus if for color c query $\tilde{Q}$ contains point (x_i, x_{i+1}), then the original query Q contains exactly i points of color c. They store the set $\tilde{P}$ in a priority search tree in order to quickly report (the ranks of) all k points contained in query $\tilde{Q}$ in $O(\log n + k)$ time. We show that we can instead report the points using a binary search and $O(k)$ additional time.

Lemma 1. *Let P be a set of n points in $\mathbb{R}^1$ with colors from $[\phi]$. In $O(n \log n)$ time we can build an $O(n)$ space data structure, so that we can answer dominance color frequency reporting queries $Q = (-\infty, q]$ using a binary search and additional $O(k)$ time.*

We can answer weighted queries as well simply by setting the weight of the point (x_i, x_{i+1}) to be $\sum_{h \leq i} w(x_h)$.

3 A Strictly Output-Sensitive Data Structure

We first describe a generic data structure for dominance queries in $\mathbb{R}^2$. We then extend the result to axis-aligned query boxes in $\mathbb{R}^d$.

Lemma 2. *Let P be a set of n points in $\mathbb{R}^2$, each of which has a color from $[\phi]$, and let $2 \leq s \leq n$ be a parameter. In $O(ns \log n \log_s n)$ time we can construct a data structure of size $O(ns \log_s n)$ that can answer dominance frequency reporting queries $Q = (-\infty, x] \times (-\infty, y]$ in $O(\log n + k \log_s n)$ time.*

Proof. We build a balanced s-ary tree on the x-coordinates of the points, of height $O(\log_s n)$. See Fig. 3. Every node ν corresponds to a vertical strip S_ν and the subset of points $P_\nu = P \cap S_\nu$ in S_ν; the root node r corresponds to a singular 'strip' covering the whole plane, and $P_r = P$. Each strip S_ν is then horizontally partitioned into s vertical strips $S_1, \ldots, S_s$, one strip S_i for each child c_i of ν, each of which contains a $1/s$ fraction of the points.

Let $L_i = \bigcup_{h<i} P_{c_h}$ be the subset of points in P_ν that lie left of S_i, and for a point $p \in L_i$ let p' be the y-coordinate of p, and let $L_i' = \{p' \mid p \in L_i\}$ denote the resulting set of 1D points (y-coordinates). We essentially regard p' as the projection of p onto the left boundary of strip S_i, see Fig. 4. We store each set L_i' in a one-dimensional color frequency reporting data structure $\mathcal{L}_i$ as described in Lemma 1. We do this for every node ν and every strip S_i in ν.

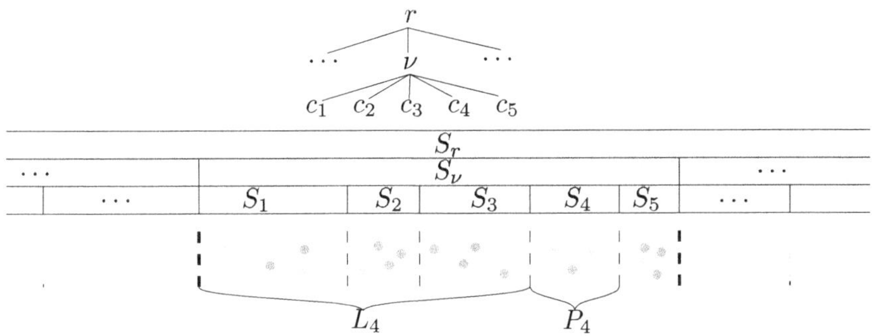

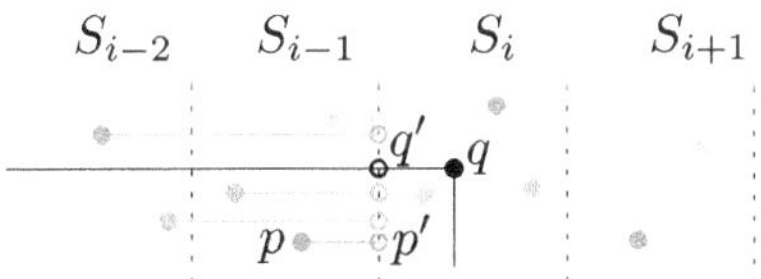

Fig. 3. A set of points in $\mathbb{R}^2$ with an s-ary strip-tree on its x-coordinates (with $s = 5$). The points P_4 and L_4 for strip S_4 are indicated.

Fig. 4. A dominance query whose corner point $q = (x, y)$ lies in strip S_i.

We use fractional cascading [9] to link all of these $\mathcal{L}_i$ data structures in the entire tree together, so that given a query value y we can perform the binary searches for y in all $O(\log_s n)$ nodes along a root-to-leaf path in $O(\log n)$ time in total.

The total space usage is $O(ns)$ per level, and thus $O(ns \log_s n)$ in total. Furthermore, constructing all these data structures takes $O(ns \log n)$ time per level, and thus $O(ns \log n \log_s n)$ time in total.

The Query Algorithm. To answer a query $Q = (-\infty, x] \times (-\infty, y]$ we follow a root-to-leaf path towards x while maintaining the frequency of each (relevant) color. At node ν, let S_i be the (sub)strip of S_ν containing x. We query $\mathcal{L}_i$ with $(-\infty, y]$ and update the frequencies of the at most k reported colors. Indeed observe that a point $p \in L_i$ left of strip S_i appears in Q if and only if $p' \in (-\infty, y]$, that none of the points right of S_i lies in Q, and that the points in S_i are accounted for in the node corresponding to S_i.

At each node of the search path we have to update the frequencies for at most k reported colors. Note that this is not as trivial as it might sound, especially in the pointer machine model, as the data structures of Lemma 1 do not report the colors in a fixed order. Hence, we may have to sort these colors in $O(k \log k)$ time before merging the two lists of color counts. To update the color frequencies in $O(k)$ time per node, we instead use a mechanism similar to Gupta et al. [13]: our data structure will also store a linked list of length ϕ, whose c^{th} node μ_c (with $c \in [\phi]$) will maintain the current total count of color c. If a query on one of the subsets returns some tuple (c, k_c) in an answer, we increment the value at μ_c by k_c. For each point of color c stored in one of the Lemma 1 data structures, we also store a pointer to node μ_c. This allows us to update the answer in $O(1)$ time per color, at no extra space cost. We additionally maintain a list of pointers to all non-zero nodes to report all final counts in $O(k)$ time (whenever we increment the count of a color c and the previous count of that color was 0, we add a pointer to μ_c to this list). We appropriately reset these linked lists after answering a query.

Query Time. We can find the root-to-leaf path towards x in $O(\log n)$ total time by doing a single binary search on the leaves, and walking up towards the root. This path has length $O(\log_s n)$, and at each node we perform one 1D query in $O(\log n + k)$ time; note that each of the k output colors may appear in every node. Using the above procedure of updating frequencies, the total query time is thus $O(\log n + k \log_s n)$ (recall that the 1D structures are linked using fractional cascading). $\qquad\square$

We can extend the above approach to the case of dominance queries in $\mathbb{R}^d$, for $d \geq 2$, costing us a factor $s \log_s n$ per dimension in space, and a factor $\log_s n$ per dimension in query time; the techniques are essentially the same.

Theorem 1. *Let P be a set of n points in $\mathbb{R}^d$, with $d \geq 2$, each of which has a color from $[\phi]$, and let $2 \leq s \leq n$ be a parameter. In $O(n \log n(s \log_s n)^{d-1})$ time, we can build a data structure of size $O(n(s \log_s n)^{d-1})$ that can answer a frequency reporting dominance query Q in $O(\log n \log_s^{d-2} n + k \log_s^{d-1} n)$ time.*

Lastly, we can extend this to general axis-aligned query boxes. Recall a dominance query is a d-sided box, and general axis-aligned boxes are $2d$-sided boxes. We pay a factor $\log n$ in space for every side we add, and a factor 2 in query time, resulting in the following theorem:

Theorem 2. *Let P be a set of n points in $\mathbb{R}^d$, each of which has a color from $[\phi]$, and let $2 \leq s \leq n$ be a parameter. In $O(n \log^{d+1} n (s \log_s n)^{d-1})$ time, we can build a data structure of size $O(n \log^d n (s \log_s n)^{d-1})$ that can answer axis-aligned frequency reporting queries in $O(\log n \log_s^{d-2} n + k \log_s^{d-1} n)$ time.*

This result again also applies to the weighted case.

Our main goal was to achieve strictly output-sensitive query time. We can achieve this by using Theorem 2 with $s = n^{\varepsilon/d}$, for any constant $\varepsilon > 0$, resulting in a datastructure with $O(n \log^d n (n^{\varepsilon/d} d/\varepsilon)^{d-1}) \leq O(n^{1+\varepsilon})$ space, $O(n \log^{d+1} n (n^{\varepsilon/d} d/\varepsilon)^{d-1}) \leq O(n^{1+\varepsilon})$ preprocessing time, and $O(\log n (d/\varepsilon)^{d-2} + k(d/\varepsilon)^{d-1}) = O(\log n + k)$ query time (using that $\log_{n^{(\varepsilon/d)}}(n) = d/\varepsilon = O(1)$). The overall result matches that of Gupta et al. [13], however we believe that our approach is simpler.

4   A Lower Bound in the Arithmetic Model

In this section we focus on the weighted version of the problem, where each point in $P \subset \mathbb{R}^d$ also has a weight from some semigroup. Given a (dominance) query Q we now wish to report, for every color c appearing in Q, the weight $w(P_c \cap Q)$.

Our lower bound holds in the arithmetic model, and builds on a result by Chazelle [8] on the uncolored version of this problem. We describe the model here briefly, see Sect. 2 of Chazelle's article for more detail. We then state Chazelle's result, and afterwards show how to apply it to obtain a lower bound for our colored problem.

The Arithmetic Model. In this model we can store *generators*, which consist of a subset of points, and the total sum of their weights. A collection of generators is called a *storage scheme Γ* if, for any query Q, we can find a subset of generators containing exactly the points in $P \cap Q$. The answer to an (uncolored) query is then simply the sum of the values of these generators. The size of Γ is the number of generators stored, and the cost of answering a query is the minimum number of generators required to compute the answer. Importantly, the query cost does not include the time required to actually find the correct generator(s), and as such the arithmetic model is very suitable for proving lower bounds.

As a simple example, consider the 1D colored weight reporting problem in the arithmetic model. The data structure from Lemma 1 stores n generators: each point p of color c stores the total weight of all points of color c left of p. For any query, the correct weight for each color is thus stored in a single generator. In the arithmetic model, the query cost is thus simply k, while in the pointer machine model we still need extra techniques to find the correct generators in $O(\log n)$ additional time.

We require that the used semi-group $(\mathcal{S}, +)$ is *faithful*; see Sect. 2.3 of Chazelle's paper for technical details. Most semigroups arising in practice, e.g. $(\mathbb{N}, +)$, $(\mathbb{N}, \max)$, $(\{0, 1\}, \text{or})$, have this property. Essentially, this means that a storage scheme must work for *any* weight assignment: if we change the weight assignment, then the subset of points contained in a generator may not change, but of course the total sum of their weights may.

The Uncolored Lower Bound. For a point set P in $\mathbb{R}^d$ of a single color, storage scheme Γ, and dominance query Q, let $t(P, \Gamma, Q)$ be the cost of answering query Q, i.e. the smallest number of generators from Γ needed to answer Q. Chazelle shows the following:

Lemma 3 (Chazelle's uncolored lower bound [8]). *Let $\varepsilon > 0$, let P be a random set of n points in $\mathbb{R}^d$, each with a weight from some faithful semigroup, let Γ be any storage scheme for P of size m, and let Q be a random dominance query. There exists a constant r (depending on ε) such that, with probability $(1 - \varepsilon)$, it holds that $t(P, \Gamma, Q) \geq r \left(\frac{\log n}{\log(2\,m/n)} \right)^{d-1}$.*

Essentially, there is a high probability that a query Q is "difficult" to answer, meaning that we have to take the union of at least $r \left(\frac{\log n}{\log(2m/n)} \right)^{d-1}$ generators.

The Colored Lower Bound. We now show how to extend this lower bound to the colored version of the problem. The input point set now contains points of multiple colors, and thus in principle we could have generators consisting of points of multiple colors too. However, it turns out that an optimal storage scheme can never use such multi-colored generators:

Lemma 4. *Let P be a set of n colored points, each with a weight from some faithful semigroup, and let Γ be any optimal storage scheme for P. Each generator in Γ contains points of only a single color.*

Since each generator contains points of only a single color, we can view the generators as being colored too. If we group the generators by color, we observe that a colored storage scheme for points of ϕ colors is simply a collection of ϕ uncolored storage schemes. Similar to the uncolored case, let $t(P, \Gamma, Q)$ be the cost of answering query Q using storage scheme Γ on the colored point set P. The above lemma implies that $t(P, \Gamma, Q) = \sum_c t(P_c, \Gamma_c, Q)$; the weights for each color c must be summed independently, using the generators Γ_c of color c, so the total query cost is simply the sum of the query cost for all colors. Essentially, the problems are entirely separate, and thus Lemma 3 applies to each color separately. This leads to the following result:

Theorem 3. *Let $\varepsilon > 0$, let P be a random set of n points in $\mathbb{R}^d$ of ϕ colors such that there are n/ϕ points of each color, each with a weight from some faithful semigroup, let Γ be any storage scheme for P of size m, and let Q be a random dominance query. There exists a constant R (depending on ε) such that, with probability at least $(1 - \varepsilon)$, it holds that $t(P, \Gamma, Q) \geq R\phi \left(\frac{\log n}{\log(2\,m/n)} \right)^{d-1}$.*

Since our bound is in the arithmetic model, it essentially gives a lower bound on the number of values that we need to add together to compute an answer. The pointer machine model is weaker than the arithmetic model, so the lower bound directly holds there. In contrast, in the word-RAM model we can encode multiple values in a single word, and sum them up using a single operation, so the lower bound does not hold there.

5   A Transformation for Reducing Space in 2D

For certain values of ϕ and s we can reduce the space usage in Lemma 2 for 2D dominance queries using one level of "bootstrapping". The main observation is that using the structure for weighted color frequency queries in $\mathbb{R}^3$, we can actually support simultaneously answering a dominance query in $\mathbb{R}^1$ on a prefix $P_1, .., P_j$ of t point sets in $\mathbb{R}^1$. This allows us to reduce the space used at every node in the structure of Lemma 2.

Querying Multiple Sets Simultaneously. Let $P_1, .., P_t$ be t sets of points in $\mathbb{R}^1$, and let n denote the total number of points in $P = P_{\leq t} = \bigcup_{j \leq t} P_j$. Following the same strategy as Bozanis et al. [4], consider a point set P_j and a color c, and let x_i be the ith point of color c in P_j. We map point x_i to the point $(x_i, x_{i+1}, j) \in \mathbb{R}^3$, and associate it with weight i. We do this for every point set P_j and every color c. Let $\tilde{P}$ be the resulting set of points. An octant of the form $(-\infty, x] \times [y, \infty) \times (-\infty, i]$ is k-*shallow* with respect to a set $\tilde{P}$ if it contains at most k points of $\tilde{P}$. We then observe:

Lemma 5. *For any $1 \leq i \leq t$ and any $x \geq y$, the query $\tilde{Q} = (-\infty, x] \times [y, \infty) \times (-\infty, i]$ is $i\phi$-shallow with respect to $\tilde{P}$.*

Proof. For every $j \leq i$, $\tilde{Q}$ contains at most one point from $\tilde{P}_{j,c}$ for every color c; for $j > i$, $\tilde{Q}$ contains no points. $\square$

Set $r = t\phi$ and build an r-shallow cutting for such quadrants on P [2]. An r-shallow cutting Ξ is a set of $O(n/r)$ quadrants (of the above form) such that (i) every $k \leq r$-shallow quadrant is contained in a quadrant in Ξ, and (ii) every quadrant $\gamma \in \Xi$ contains at most $O(r)$ points $\tilde{P}_\gamma$ from $\tilde{P}$. This set $\tilde{P}_\gamma$ is the *conflict list* of γ. One can compute an r-shallow cutting of size $O(n/r)$ and process it in $O(n \log n)$ time, so that given a query point q we can find a quadrant of Ξ that contains it (if such a quadrant exists) in $O(\log(n/r)$ time [2].

We build a data structure for 3D weighted color frequency reporting queries on the conflict list $\tilde{P}_\gamma$ for every quadrant $\gamma \in \Xi$; say this uses $S(n)$ space for n points, with $O(Q(n)+k)$ query time. Hence, the total space used is $O((n/r)S(r))$.

Now consider a query, consisting of an integer $j \in 1..t$ and a range $(-\infty, q]$. Lemma 5 guarantees that the quadrant $\tilde{Q} = (-\infty, q] \times [q, \infty) \times (-\infty, j]$ is contained in a quadrant $\gamma \in \Xi$. Hence, querying with the corner point (q, q, j) we find such a quadrant γ in $O(\log(n/r))$ time. We then query its associated data structure in $O(Q(r,k))$ time. Since γ contains $\tilde{Q}$, any point contained in $\tilde{Q}$ is contained in γ, and none of the points outside of γ can be contained in $\tilde{Q}$. Moreover, for each color c, the total number of points in $P_{<j}$ is exactly the total weight of the points in $\tilde{Q} \cap \tilde{P}_{\leq j,c}$.

Lemma 6. *Given an $S(n)$ space data structure that supports weighted color frequency queries for dominance ranges in $\mathbb{R}^3$ in $O(Q(n)+k)$ time, we can build a data structure of size $O((n/t\phi)S(t\phi))$ that stores t sets of points $P_1,..,P_t$ in $\mathbb{R}^1$ of total size n. We can answer color frequency reporting queries for dominance ranges $(-\infty,q]$ in any $P_{\leq j} = \bigcup_{i\leq j} P_i$ in $O(\log \frac{n}{t\phi} + Q(t\phi) + k)$ time.*

Note that unfortunately we cannot easily extend this idea to simultaneously answer dominance queries on a prefix $P_1,..,P_j$ of points in $\mathbb{R}^2$, as shallow cuttings in $\mathbb{R}^4$ use quadratic space.

Reducing the Space for Dominance Queries in $\mathbb{R}^2$. For points in $\mathbb{R}^2$ and parameter s, we now follow the same recursive approach as in Lemma 2, recursively partitioning the set of points using $t = s$ vertical strips $S_1,..,S_s$. However, we now build the data structure from Lemma 6 on (the y-coordinates of) the sets $P_j = S_j \cap P$, instead of constructing a 1D structure on (the y-coordinates of) each of the sets $L_j = \bigcup_{i<j} P_i$. At every level we thus use $O\left(\frac{S(s\phi)n}{s\phi}\right)$ space, and thus $O\left(\frac{S(s\phi)n \log_s n}{s\phi}\right)$ space in total.

Consider a dominance query whose corner point lies in strip S_j. We can now compute the contribution of the points in strips $S_1,..,S_{j-1}$ in $O(\log(n/(s\phi)) + Q(s\phi)+k)$ time by querying the data structure of Lemma 6. Therefore, by doing this on all $O(\log_s n)$ levels of the tree, the total query time becomes $O(\log n + \log_s n(k + \log(n/(s\phi)) + Q(s\phi)))$.

Lemma 7. *Let P be a set of n points in $\mathbb{R}^2$, each of which has a color from $[\phi]$, and let $2 \leq s \leq n$ be a parameter. Given an $S(n)$-space data structure that supports weighted color frequency queries for dominance ranges in $\mathbb{R}^3$ in $O(Q(n) + k)$ time, we can build a data structure of size $O\left(\frac{S(s\phi)n \log_s n}{s\phi}\right)$ that can answer a color frequency reporting query for a dominance range in $O(\log n + \log_s n(k + \log(n/(s\phi)) + Q(s\phi)))$ time.*

In particular, starting with the data structure from Theorem 1 (set to use $O(r^{1+\varepsilon})$ space for r points) gives an $O(ns^\varepsilon \phi^\varepsilon \log_s n)$ space structure with query time $O(\log_s n(\log n + k))$. Hence, it allows us to reduce the space usage from $O(ns)$ to $O(ns^\varepsilon \phi^\varepsilon)$ while not losing anything in the query time when k and ϕ and s are sufficiently large.

Consider for example the case when $\phi = \Theta(\log n)$, for a query where $k = \Omega(\phi)$, and we choose $s = \phi$. Lemma 2 yields a data structure of size $O(ns \log_s n) = O(n\frac{\log^2 n}{\log\log n})$ with $O(\log n + k\log_s n) = O(\frac{\log^2 n}{\log\log n})$ query time, and Lemma 7 yields a data structure of size $O(n(s\phi)^\varepsilon \log_s n) = O(n\frac{\log^{1+2\varepsilon} n}{\log\log n})$ with $O(\log_s n(\log n + k)) = O(\frac{\log^2 n}{\log\log n})$ query time; the latter improves the space usage by almost a factor $\log n$.

Corollary 1. *When $\phi = \Theta(\log n)$, we can build an $O(n \log^{2\varepsilon} n \frac{\log n}{\log \log n})$ space data structure that can answer 2D color frequency reporting queries with $O(\frac{\log^2 n}{\log \log n})$ query time.*

6   Low-Space Algorithms in 2D

In this section we turn to the algorithmic problem. Here we are given a set $\mathcal{Q}$ of m dominance queries, and we wish to answer them all. For every query $Q \in \mathcal{Q}$ we must output a list of k_Q (color,count) tuples, where k_Q is the number of colors contained in query Q. These lists do not have to be in any specific order, as long as all color counts of a specific query are together. Ideally, we provide these answers in a stream so that we do not have to store the entire output of size $K = \sum_Q k_Q$.

Note that for point sets in $\mathbb{R}^1$ we can simply build and query the linear space solution from Bozanis et al. [4] in $O(n \log n + m \log n + K)$ time. In higher dimensions we could similarly build the data structure from Theorem 1 and use it to answer each query. This would take $O(n \log n (s \log_s n)^{d-1})$ time to build the datastructure plus $O(m \log n \log_s^{d-2} n + K \log_s^{d-1})$ to answer all the queries, using $O(n(s \log_s n)^{d-1})$ working space. We can lower the space usage by observing that we do not have to build every $(d-1)$-dimensional datastructure $\mathcal{L}_i$ simultaneously. If we answer the queries $\mathcal{Q}$ in order of x-coordinate, we can build each $(d-1)$-dimensional datastructure once, use it to answer the relevant $(d-1)$-dimensional queries, and destroy it again. As such, each point will only be contained in a single $(d-1)$-dimensional datastructure at once, yielding the following result:

Theorem 4. *Let P be a set of n points in $\mathbb{R}^d$, with $d \geq 2$, each of which has a color from $[\phi]$, and let $\mathcal{Q}$ be a set of m dominance queries, and let $2 \leq s \leq n$ be a parameter. We can answer all queries in $O(n \log n (s \log_s n)^{d-1} + m \log n \log_s^{d-2} n + K \log_s^{d-1} n)$ time, where K is the total output size, using $O(n(\varepsilon \log_s(n))^{d-2} + m)$ working space*

Note that this yields a linear space algorithm in 2D.

We would like to reduce the space usage even further, preferably all the way down to $O(n+m)$ for any dimension d, but this may not be as easy as it sounds. A logical approach would be to use the fact that we are given all queries in advance, and rather than building a $(d-1)$-dimensional data structure on the set L'_i and querying it one by one, instead performing all queries on L'_i at once using a $(d-1)$-dimensional algorithm. The issue is that we then obtain a list of up to K sub-answers that we need to combine with other sub-answers (obtained from other $(d-1)$-dimensional data structures) to compute the full answer to each query. However, these lists will come in arbitrary query-order, and as such combining them might take up to $O(K \log m)$ time. Note that the trick used in Theorem 1 is not enough in this case.

For the same reason, it is difficult to extend this result to general more-sided queries using standard techniques; this would also require combining many subqueries. We can however increase the number of sides by 1, using the fact that we can answer dominance queries in order of y-coordinate, allowing us to combine all sub-answers in $O(K)$ total time.

Theorem 5. *Let P be a set of n points in $\mathbb{R}^2$, each of which has a color from $[\phi]$, and let $\mathcal{Q}$ be a set of 3-sided queries of the form $[x_1, x_2] \times (-\infty, y)$, and let $2 \leq s \leq n$ be a parameter. We can answer all queries in $O(n \log^2 ns \log_s n + m \log n + K \log_s n)$ time, using $O(n + m)$ space.*

References

1. Afshani, P., Cheng, P., Roy, A.B., Wei, Z.: On range summary queries. In: 50th International Colloquium on Automata, Languages, and Programming, ICALP 2023. LIPIcs, vol. 261, pp. 7:1–7:17. Schloss Dagstuhl - Leibniz-Zentrum für Informatik (2023). https://doi.org/10.4230/LIPICS.ICALP.2023.7
2. Afshani, P., Tsakalidis, K.: Optimal deterministic shallow cuttings for 3-d dominance ranges. Algorithmica **80**(11), 3192–3206 (2018). https://doi.org/10.1007/S00453-017-0376-3
3. de Berg, M., Cheong, O., van Kreveld, M.J., Overmars, M.H.: Computational Geometry: Algorithms and Applications, 3rd edn. Springer, Heidelberg (2008). https://doi.org/10.1007/978-3-540-77974-2
4. Bozanis, P., Kitsios, N., Makris, C., Tsakalidis, A.: New upper bounds for generalized intersection searching problems. In: Fülöp, Z., Gécseg, F. (eds.) ICALP 1995. LNCS, vol. 944, pp. 464–474. Springer, Heidelberg (1995). https://doi.org/10.1007/3-540-60084-1_97
5. Chan, T.M., He, Q., Nekrich, Y.: Further results on colored range searching. In: 36th International Symposium on Computational Geometry, SoCG 2020. LIPIcs, vol. 164, pp. 28:1–28:15. Schloss Dagstuhl - Leibniz-Zentrum für Informatik (2020). https://doi.org/10.4230/LIPICS.SOCG.2020.28
6. Chazelle, B.: Filtering search: a new approach to query-answering. SIAM J. Comput. **15**(3), 703–724 (1986). https://doi.org/10.1137/0215051
7. Chazelle, B.: Lower bounds for orthogonal range searching: I. the reporting case. J. ACM **37**(2), 200–212 (1990). https://doi.org/10.1145/77600.77614
8. Chazelle, B.: Lower bounds for orthogonal range searching II. the arithmetic model. J. ACM **37**(3), 439–463 (1990). https://doi.org/10.1145/79147.79149
9. Chazelle, B., Guibas, L.J.: Fractional cascading: I. A data structuring technique. Algorithmica **1**(2), 133–162 (1986). https://doi.org/10.1007/BF01840440
10. Ganguly, A., Munro, J.I., Nekrich, Y., Shah, R., Thankachan, S.V.: Categorical range reporting with frequencies. In: 22nd International Conference on Database Theory, ICDT 2019. LIPIcs, vol. 127, pp. 9:1–9:19. Schloss Dagstuhl - Leibniz-Zentrum für Informatik (2019). https://doi.org/10.4230/LIPICS.ICDT.2019.9
11. Gupta, P., Janardan, R., Rahul, S., Smid, M.: Computational geometry: generalized (or colored) intersection searching. In: Handbook of Data Structures and Applications, pp. 1043–1058. Chapman and Hall/CRC (2004). https://doi.org/10.1201/9781420035179

12. Gupta, P., Janardan, R., Smid, M.H.M.: Further results on generalized intersection searching problems: counting, reporting, and dynamization. J. Algorithms **19**(2), 282–317 (1995). https://doi.org/10.1006/JAGM.1995.1038
13. Gupta, P., Janardan, R., Smid, M.H.M.: A technique for adding range restrictions to generalized searching problems. Inf. Process. Lett. **64**(5), 263–269 (1997). https://doi.org/10.1016/S0020-0190(97)00183-X
14. Janardan, R., López, M.A.: Generalized intersection searching problems. Int. J. Comput. Geom. Appl. **3**(1), 39–69 (1993). https://doi.org/10.1142/S021819599300004X
15. Rahul, S., Gupta, P., Rajan, K.S.: Data structures for range aggregation by categories. In: Proceedings of the 21st Annual Canadian Conference on Computational Geometry, Vancouver, British Columbia, Canada, 17–19 August 2009, pp. 133–136 (2009). http://cccg.ca/proceedings/2009/cccg09_35.pdf
16. Shi, Q., JáJá, J.F.: Optimal and near-optimal algorithms for generalized intersection reporting on pointer machines. Inf. Process. Lett. **95**(3), 382–388 (2005). https://doi.org/10.1016/J.IPL.2005.04.008

On the Complexity of Capacitated Vehicle Routing with Order Restrictions

Steven Miltenburg[ID], Tim Oosterwijk[ID], and René Sitters[(✉)][ID]

Vrije Universiteit Amsterdam, De Boelelaan 1105, 1081 HV Amsterdam, The Netherlands
{s.j.g.miltenburg,t.oosterwijk,r.a.sitters}@vu.nl

Abstract. The capacitated vehicle routing problem (c-VRP), which is also known as the k-tour cover problem, has been widely studied in many settings. In the Euclidean plane, a PTAS for constant capacity c was given more than 40 years ago, which has been improved and extended ever since. Here, we present arguably the easiest form of a PTAS possible. The algorithm simply chooses between greedy or complete enumeration. A direct corollary of this simplification is that the PTAS extends to basically any kind of order restriction, as long as the objective is to minimize total tour length.

Furthermore, we show that when the metric is a tree, there is a clear separation in complexity between the c-VRP with and without order restrictions. We show that even for a fixed order, the c-VRP is APX-hard for arbitrary (non-constant) capacity c. This in contrast to the polynomial time approximation scheme for the standard c-VRP on trees.

Keywords: Capacitated vehicle routing · PTAS · Precedence constraints · APX-hardness · Euclidean plane

1 Introduction

Efficient routing of capacitated vehicles from a central depot is a fundamental challenge in logistics and transportation, with applications ranging from courier services and waste collection to supply distribution in urban and rural settings. In many practical scenarios, the order in which requests are served is subject to precedence constraints: Certain deliveries or pickups must occur before others due to physical, temporal, or logical dependencies. Such constraints arise naturally in contexts like inventory restocking, production systems, and multistage service chains (e.g., in health care or maintenance). These applications motivate the study of the Capacitated Vehicle Routing Problem (c-VRP) with precedence constraints, where the goal is to find a set of feasible vehicle routes that satisfy both capacity and precedence requirements while minimizing total travel cost.

In this problem, we are given a metric space with a distinguished root (or depot), a capacity c, and a set of requests, where each request is a point in the metric space. Points of the metric space may contain more than one request.

© The Author(s), under exclusive license to Springer Nature Switzerland AG 2026
J. Kozik and A. Wolff (Eds.): SOFSEM 2026, LNCS 16448, pp. 260–273, 2026.
https://doi.org/10.1007/978-3-032-17801-5_19

The goal is to partition the requests into subsets of size at most c and for each subset to find a tour through its requested points and such that each tour starts and ends at the root and respects a given partial order on the requests, while minimizing the total length of all tours. Without the partial order, the problem coincides with the classic Capacitated Vehicle Routing Problem, also known as the k-Tour Cover problem. If, on the other hand, the precedence constraints form a total ordering, the problem is also referred to as the capacitated Fixed Order Routing problem [18]. Here, we study the variant where precedence constraints only take effect between requests on the *same* tour. Although this might feel odd at first, this setting does appear naturally in many application. For example, in logistics, where the order in which items are loaded in and unloaded from a truck (or ship) is constrained by the different weights and shapes. Clearly, these precedence constraints are not present between items assigned to different trucks. To the best of our knowledge, the approximability of this type of precedence constraints has not been studied before.

Related Literature. c-VRP is a fundamental combinatorial optimization problem of significant theoretical and practical interest. It is arguably one of the most important problems in operations research and NP-hard in general metric spaces [15]. The seminal work of Haimovich and Rinnooy Kan [10] introduced the iterated tour partitioning heuristic, which was shown to be a $1 + (1 - 1/c)\alpha_{\text{TSP}}$-approximation [3], where α_{TSP} is the approximation guarantee of a TSP algorithm. Recently, Blauth et al. [7] improved this to $1 + \alpha_{\text{TSP}} - \varepsilon$ for some small constant $\varepsilon > 0$.

In $\mathbb{R}^2$, Adamaszek et al. [2] give a PTAS for $c \leq 2^{\log^\delta n}$, where $\delta = \delta(\varepsilon)$. Their algorithm reduces an instance of c-VRP to a small number of independent instances with a small number of points and then uses the QPTAS of Das and Mathieu [8] in the reduced instance. A generalization to higher dimensions was given by Khachay and Dubinin [13]. For trees, there exists a 4/3-approximation by Becker [6]. Asano et al. [4] proved it is not possible to achieve a better approximation guarantee using the lower bound by Hamaguchi and Katoh [11]. Rather recently, Mathieu and Zhou [17] obtained a PTAS.

Adding precedence constraints to the c-VRP makes the problem significantly more difficult. Most of the literature for c-VRP with precedence constraints focuses on exact methods with super-polynomial running time or heuristics that do not provide hard bounds on the approximation guarantee. We refer the interested reader to the survey by Liu et al. [16].

Our Contribution. In this paper, we consider the capacitated vehicle routing problem with precedence constraints in Euclidean spaces and on trees. First, we provide a very simple PTAS for constant capacity c in $\mathbb{R}^D$ for fixed dimension D. To our knowledge, this is the first PTAS that works for any partial order. The algorithm first decomposes an instance into bounded instances and then constructs a grid on each of them. It then solves each bounded instance separately, either by full enumeration or by a trivial solution on the instance where all requests are moved to the nearest grid point. We then show how this PTAS extends to $c = \tilde{\mathcal{O}}\left(\log^{1/3} n\right)$ and to c-VRP with time windows, if we allow the time windows to be violated by a factor $1 + \mathcal{O}(\varepsilon)$.

Second, we prove APX-hardness on tree metrics where the precedence constraints form a total ordering. In contrast, without precedence constraints, there is a PTAS by Mathieu and Zhou [17] for arbitrary capacity c. Our result implies APX-hardness for general total orderings.

Model and Notation. An instance of c-VRP is given by a set R of n requests. Each request r_i is associated with its location, also denoted by r_i. The metric space is endowed with a distance function $d(\cdot, \cdot)$. In this paper, we consider the Euclidean space $\mathbb{R}^D$ for fixed dimension $D \geq 1$, and trees, where each edge has an associated length, and the distance between two vertices is the length of the shortest path between them. We denote the set of locations in the metric space by V, with a special location r_0 called the root. Note that there might be multiple requests at the same location. We often identify a request by its location, in which case R can be viewed as a multi-set of locations.

By a *tour* we mean a path from r_0 to r_0 serving at most c requests, and we define its length as the sum of distances between the adjacent locations on the tour. A feasible solution to an instance of c-VRP is a collection of tours such that every request is on exactly one tour. The objective is to find a solution that minimizes the sum of the lengths of its tours.

In the c-VRP with precedence constraints, we are additionally given a partial order $\preceq$ on the requests. In this problem, the sequence of request visits needs to respect the partial order for each tour; there are no order constraints between different tours.

In the c-VRP with time windows, each request r_i is additionally given a time interval $[a_i, b_i]$ in which it must be visited (all tours start at r_0 at time 0). We measure time by distance plus waiting time (in practical terms, we assume without loss of generality that vehicles traversing the tours have speed 1). However, the objective function still minimizes the traveled distance, and not the time. We assume that $b_i \geq d(r_0, r_i)$ for every request r_i, since otherwise there is no feasible solution.

Although the c-VRP with precedence constraints and the c-VRP with time windows are related, it is not the case that one of the two is a special case of the other; observe that there are precedence constraint structures that cannot be modeled by time windows. Fixed order routing is a special case of both problems, though: $\preceq$ can be a total ordering, and choosing sequential time windows of appropriate width can enforce the fixed order.

For convenience, we denote $d_v = d(r_0, v)$ for any point v in the metric space. When we abuse notation by identifying a request with its location, we write $d_v = d(r_0, v)$ for a request v as well; no confusion shall arise. For a given instance, we write $d_{\min} = \min_{v \in V} d_v$ and $d_{\max} = \max_{v \in V} d_v$. Without loss of generality, assume that $d_v \geq 1$ for all $v \in V$.

Paper Outline. Section 2 presents the PTAS in $\mathbb{R}^D$ for precedence constraints and for time windows. Section 3 proves the APX-hardness on trees for a total ordering.

2   A Simple PTAS for c-VRP Under Arbitrary Precedence Constraints or Time Windows

The polynomial time approximation scheme that we present here is arguably the easiest PTAS possible for capacitated VRP with constant capacity c, and even holds for any precedence relation represented by a directed acyclic graph (DAG) or arbitrary time window constraints. First, we reduce to a set of so-called *bounded instances* and then simply take a trivial solution for each of these instances, or, if the number of points is small, apply complete enumeration.

A reduction to bounded instances has been applied before (cf. [1,9,17]). All these papers use a Baker's type approach [5], where the plane is partitioned into geometrically increasing rings that partition the point set V into subsets. Then, every $1/\varepsilon$-th ring is 'marked'. Removing the marked rings leaves a collection of bounded instances (separated by the marked rings). The instances on the marked rings are not bounded, but (when randomized) contribute only an $\mathcal{O}(\varepsilon)$ fraction of the optimal value and are approximated using a simple algorithm. For example, Adamaszek et al. [1] use the iterated tour partitioning algorithm for the marked subsets. Our approach is similar to that of Sitters [19] and is simpler and more general than previous approaches. We show that there is no need to mark every $1/\varepsilon$-th ring and partition the set V into subsets V_i, all of which define a bounded instance. Hence, there is no need to deal with marked (unbounded) instances in a less efficient way. The proof is similar to [19] and we defer the proof to the full version of the paper. We shall first present the analysis for arbitrary precedence constraints and consider time windows in Sect. 2.3. To simplify the exposition, we shall first present the PTAS in $\mathbb{R}^2$ and then discuss its extension to $\mathbb{R}^D$.

2.1   Reduction to Bounded Instances

Let I be an instance of c-VRP in $\mathbb{R}^2$ under precedence constraints. Let $\varepsilon > 0$. We show that we can split I into a set of instances I_i for which $d_{\max}/d_{\min}$ is bounded by some constant Q depending on ε only. We call these instances *bounded*. A geometric interpretation is as follows: We draw circles with center r_0 and randomly chosen radii $x_1 < x_2 < \cdots < x_q$ and define V_i as the set of points enclosed by the circles with radii x_{i-1} and x_i. We can assume that no location is exactly on one of the circles, since this happens with probability 0.

Since there is no bound on the number of tours in the solution, the union of the solutions for the bounded instances I_i is a feasible solution for I. The next lemma shows that a PTAS for bounded instances yields a PTAS for arbitrary, unbounded instances. As described above, similar reductions have been applied before. The proof is deferred to the full version of this conference paper.

Algorithm 1: Reducing to bounded instances

---

1   $a \leftarrow 2/\varepsilon$ and let $b \sim U[0,1]$

2   $x_0 \leftarrow 0$ and $x_i \leftarrow e^{a(i-1+b)}$ for $i = 1, 2, \ldots, q$ where q is s.t. $x_q > d_{\max}$

3   Partition V into $V_i = \{v \mid x_{i-1} \leq d_v < x_i\}$ for $i = 1, 2, \ldots, q$, and let I_i be the instance defined by the requests on V_i

4   $\Gamma_i \leftarrow$ a solution to I_i for all i

5   **return** $\Gamma_1 \cup \ldots \cup \Gamma_q$

Lemma 1. *Let $\varepsilon > 0$, let* Opt *be the optimal value for I and let* $\textsc{Opt}_i$ *be the optimal value for I_i as defined in Algorithm 1. Then*

$$\sum_i \mathbb{E}[\textsc{Opt}_i] \leq (1 + \varepsilon)\textsc{Opt}.$$

Note that the reduction is easy to derandomize since the random choice of b will only generate $\mathcal{O}(n^2)$ different possible partitions of V. From now on we consider an instance I to be a bounded instance, i.e., the ratio between the maximum and minimum distance to the root of any point is bounded by a constant $Q \leq e^a = e^{2/\varepsilon}$.

2.2   A Simple PTAS for Bounded Instances

We shall drop the index i used for the bounded instances in the previous section and consider here some bounded instance I. The PTAS for bounded instances is extremely simple. We shall define a threshold value N, which is a constant depending on ε and c, and apply complete enumeration if $n \leq N$. In the other case, we define a grid and cluster the points by shifting each to the nearest grid point and then take a trivial solution where each tour only visits one grid point.

One of the standard lower bounds, also used in [10], is the following:

$$\textsc{Opt} \geq \sum_i (2/c)d(r_0, r_i). \tag{1}$$

This lower bound can be understood as the minimal path from the root to a location being the direct distance, and since every tour has at most capacity c and the tour needs to both visit a location and go back, the contribution of every request r_i is at least $(2/c)d(r_0, r_i)$.

Rounding the Instance by Shifting to Grid Points. We create a rounded instance I' as follows. Starting with I, we take a grid where the distance between two consecutive grid lines is equal to

$$\sigma = \varepsilon d_{\min}/(\sqrt{2}c),$$

and move every request to a grid point at the intersection of two grid lines. To enhance the computation, this can be any of the four corner points of the grid cell containing the request. In this way, by choosing the corner point appropriately, we can assume that the minimum and maximum distances to the origin are still, respectively, at least $d_{\min}$ and at most $d_{\max}$. Let us bound the error made by the rounding.

Lemma 2. *Let* OPT *and* OPT' *be the optimal values for* I *and* I', *respectively. Then,* $\mathrm{OPT}' \leq (1 + \varepsilon)\mathrm{OPT}$. *Furthermore, given any solution to the rounded instance of cost* Z, *we can find a solution to the unrounded instance of cost at most* $Z + \varepsilon\mathrm{OPT}$.

Proof. An obvious lower bound resulting from Eq. (1) on the optimal value is

$$\mathrm{OPT} \geq (2/c)nd_{\min} .$$

For each request, the detour caused by the rounding is at most $2\sqrt{2}\sigma$. Hence, the rounded solution has a cost of at most

$$\mathrm{OPT} + n2\sqrt{2}\sigma = \mathrm{OPT} + 2n\varepsilon d_{\min}/c \leq (1 + \varepsilon)\mathrm{OPT}.$$

Similarly, for any solution for I' of value Z, the corresponding solution for I has cost at most $Z + n2\sqrt{2}\sigma \leq Z + \varepsilon\mathrm{OPT}$. $\qquad\qquad\square$

The PTAS: Apply Complete Enumeration or Take a Trivial Solution. After reducing to bounded instances and moving points to grid points, the PTAS becomes extremely simple. Consider the rounded instance I'. Let $n(p)$ be the number of requests in a grid point p. We define the *trivial solution* as the solution that sends $\lceil n(p)/c \rceil$ vehicles to each grid point p.

We shall choose N such that the error made by the trivial algorithm is small. The number of grid lines is $\mathcal{O}(d_{\max}/\sigma) = \mathcal{O}(d_{\max}/(\varepsilon d_{min}/c)) = \mathcal{O}(cQ/\varepsilon)$. Let Δ be the number of grid points. Then, $\Delta = \mathcal{O}(c^2 Q^2/\varepsilon^2)$. We choose the threshold value $N = \Delta Qc/\varepsilon$.

Algorithm 2: A simple PTAS for bounded instances

---

1 **if** $n \leq \Delta Qc/\varepsilon$ **then**
2   **return** the optimal solution by complete enumeration
3 **else**
4   **return** the trivial solution for the rounded instance

We now prove the approximation guarantee and running time of our PTAS.

Lemma 3. *Algorithm 2 returns a* $(1+3\varepsilon)$-*approximation and runs in polynomial time for any constant* $\varepsilon > 0$.

Proof. The solution is optimal for the rounded instance by definition if the algorithm applies complete enumeration. So, assume the other case, i.e., $n > \Delta Qc/\varepsilon$. Again, we use the lower bound from Eq. (1):

$$\textsc{Opt} \geq 2nd_{\min}/c > \Delta 2Qcd_{\min}/(c\varepsilon) = 2\Delta Q/\varepsilon d_{\min} \geq 2\Delta d_{\max}/\varepsilon.$$

Let $\textsc{Opt}'$ be the optimal value for the rounded instance. Then, applying Eq. (1) to the rounded instance, we can bound the optimal value by

$$\textsc{Opt}' \geq \sum_i (2/c)d(r_0, r_i') = 2\sum_{p=1}^{\Delta}(n(p)/c)d_p.$$

Using these two lower bounds, we see that the cost of the trivial solution is at most

$$\sum_p 2\lceil n(p)/c\rceil d_p \leq \textsc{Opt}' + 2\sum_p d_p \leq \textsc{Opt}' + 2\sum_p d_{\max} = \textsc{Opt}' + 2\Delta d_{\max}$$

$$\leq \textsc{Opt}' + \varepsilon\textsc{Opt} \leq (1 + 2\varepsilon)\textsc{Opt}.$$

By Lemma 2, the cost for the corresponding solution for the original instance I is at most $(1 + 2\varepsilon)\textsc{Opt} + \varepsilon\textsc{Opt} = (1 + 3\varepsilon)\textsc{Opt}$.

For the running time, it is clear that the trivial solution can be found in polynomial time and the complete enumeration is done in constant time since $\Delta = \mathcal{O}(c^2Q^2/\varepsilon^2)$ and Q is a constant. $\qquad\square$

Increasing the capacity to $c = \tilde{\mathcal{O}}\left(\log^{1/3}\right) n$. The complete enumeration on the instance of $n \leq N$ points can be done by a dynamic program. Let $\mathcal{S}$ be the set of all subsets of requests. For each $S \in \mathcal{S}$, let $Z(S)$ be the optimal value for the instance restricted to S. For $|S| \leq c$ this can be computed in $c! = \mathcal{O}(c^c)$ time, and we have at most N^c of these sets. In general, we can compute $Z(S)$ by enumerating all sets $S' \subset S$ with $|S'| \leq c$ and minimize over $Z(S \setminus S') + Z(S')$. Hence, the DP runs in time

$$\mathcal{O}(2^N 2^c + N^c c^c).$$

Recall that $N = \tilde{\mathcal{O}}(c^3)$, where we use $\tilde{\mathcal{O}}(\cdot)$ to hide the polynomial dependency of the hidden constant on ε. Hence, the running time is

$$2^{\tilde{\mathcal{O}}(c^3)}2^c + \tilde{\mathcal{O}}(c^3)^c c^c = 2^{\tilde{\mathcal{O}}(c^3)}, \tag{2}$$

which is polynomial for $c = \tilde{\mathcal{O}}\left(\log^{1/3} n\right)$.

N.B. Note that our bound on c is much smaller (worse) than the bound of $c = 2^{\log^{\delta} n}$ by Adamaszek et al. [2]. However, our approach does not rely on the QPTAS for c-VRP by Das and Mathieu [8] and includes arbitrary precedence constraints.

Extension to $\mathbb{R}^D$. While we presented our algorithm and analysis for $\mathbb{R}^2$, we can extend it to $\mathbb{R}^D$ in a straightforward way. The circles around r_0 translate into D-dimensional balls in general, which still form bounded instances. The grid becomes D-dimensional, which gives $\Delta = \mathcal{O}\big(c^D Q^D / \varepsilon^D\big)$ grid points. This increases the value of N to $\tilde{\mathcal{O}}\big(c^{D+1}\big)$. The running time of the PTAS in Eq. (2) remains polynomial for $c = \tilde{\mathcal{O}}\big(\log^{1/(D+1)} n\big)$.

Theorem 1. *There exists a PTAS for c-VRP with arbitrary precedence constraints in $\mathbb{R}^D$ for fixed $D \geq 1$ for $c = \tilde{\mathcal{O}}\big(\log^{1/(D+1)} n\big)$.*

2.3   Time Windows

Approximation schemes for the capacitated VRP with time windows have been studied in [14] and [20]. Both papers use a model where time windows are disjoint and servers can travel at any speed. Hence, these models are similar to fixed order routing, i.e., c-VRP with a total ordering. Here, we consider arbitrary time windows $[a_i, b_i]$ for the requests r_i and assume the vehicle either moves at a constant, unit speed or holds still. The polynomial time approximation scheme derived above applies to c-VRP with time windows in almost the same way. However, the computed solution may violate the time window by a factor $1 + \mathcal{O}(\varepsilon)$.

Clearly, our PTAS does not work for hard time constraints (where no violation is allowed) because of the shifting of location of requests to grid points. To our knowledge, APX hardness of the Euclidean c-VRP with constant c and hard time window constraints is open. Here, we show that a PTAS is possible for arbitrary time windows if these can be violated by a factor $1 + \mathcal{O}(\varepsilon)$. The algorithm and proof are almost the same as for precedence constraints, and we will briefly address the small modifications.

Reducing to Bounded Instances. In the reduction, we split any tour from an optimal solution into separate tours that each visit points from only one of the subinstances I_i. The ordering of the points remains the same, and we never arrive later at a point. Hence, when the server waits in case it is early, all time constraints are met exactly.

Rounding to Grid Points and Returning the Trivial Solution. For the rounded instance, the algorithm returns the trivial solution, i.e., any tour only contains requests from a single grid point. We then visit these points in any order that meets the time windows; for example, we can visit them in non-decreasing order of starting time a_i of the window. Now consider some grid point p and a tour for p in the trivial solution. For the unrounded solution, the last request on the tour is served latest at time $d_p + 2\sqrt{2}\sigma c = d_p + \sqrt{2}\varepsilon d_{\min} \leq (1 + \sqrt{2}\varepsilon)d_p$. Since we may assume that b_i is at least the distance to the origin for every request r_i, we see that each time window is violated by at most a factor $(1 + \mathcal{O}(\varepsilon))$.

Increasing c to $c = \tilde{\mathcal{O}}\left(\log^{1/(D+1)} n\right)$ in $\mathbb{R}^D$. For the complete enumeration, we can apply exactly the same dynamic program structure. Instances with at most c requests can be solved in $\mathcal{O}(c^c)$ time, and in general, the value of an instance on n requests is found by enumerating single-tour solutions in combination with the solution of the remaining requests, both stored in the DP table. Hence, the upper bound on the running time as given in Eq. (2) still holds, and the arguments for the extension to $\mathbb{R}^D$ carry over as before.

Theorem 2. *There exists a PTAS for c-VRP with arbitrary time windows in $\mathbb{R}^D$ for fixed $D \geq 1$ for $c = \tilde{\mathcal{O}}\left(\log^{1/(D+1)} n\right)$ if the time windows are allowed to be violated by at most a factor $1 + \mathcal{O}(\varepsilon)$.*

3 APX-Hardness for *c*-VRP with a Total Ordering on Trees

In this section, we consider the c-VRP problem with precedence constraints on trees, where the precedence constraints $\preceq$ form a total ordering. We will assume that the set of requests is indexed according to $\preceq$: for requests r_i and r_j, $i < j$ implies $i \prec j$. In the previous section, we showed that a PTAS exists in the Euclidean plane by reducing an instance to bounded subproblems and solving these separately. Although a similar reduction works for trees as well, the resulting subproblems are not as trivial to solve. Below we give some intuition as to why a PTAS on trees is nontrivial even for bounded instances. We further support the intuition that ordered routing is harder on trees than in the plane by showing that for arbitrary capacity the problem is APX-hard for totally ordered requests. In contrast, without precedence constraints, a PTAS for arbitrary capacity was recently given by Mathieu and Zhou [17]. In fact, this is a special case where the order is defined by a depth-first-search in the tree, since DFS is optimal for any tour in the solution.

For trees, it is possible to reduce every instance to bounded instance as in Lemma 1 in a similar way. However, for trees, when we cluster points to $\mathcal{O}(1/\varepsilon)$ locations, then in general, each cluster is still of non-constant diameter and may still contain $\mathcal{O}(n)$ locations. Solving a bounded subproblem is still non-trivial.

In contrast to the classic version, c-VRP with precedence constraints is already non-trivial on trees of depth 2. Figure 1 presents an instance where it is suboptimal to bundle c requests from a single vertex. The optimal solution requires splitting the 30 requests in v_A into sets of 10 combined with a set of 5 from one of the other vertices. In the absence of precedence constraints, it would be optimal to send 2 vehicles to V_A and use one for the other vertices. In general, adding precedence constraints to the VRP on trees makes the problem non-trivial.

In the remainder of this section we show that c-VRP with a total ordering on trees is APX-hard. Since a total ordering of the requests is a special case of both partial orders and time windows, the APX-hardness extends to c-VRP with precedence constraints and to c-VRP with time windows. We use

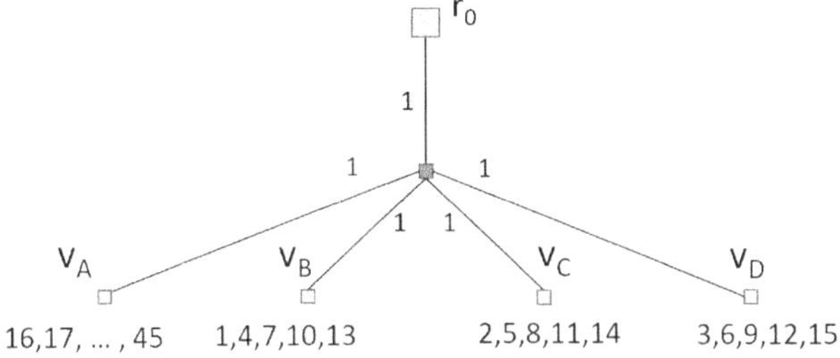

Fig. 1. An instance of c-VRP with a total ordering on a tree with $c = 15$. The optimal solution requires splitting the 30 requests in v_A in sets of 10 combined with a set of 5 from one of the other vertices. In the absence of precedence constraints, it would be optimal to send 2 vehicles to V_A and use one for the other vertices.

an approximation-preserving reduction from the following problem. (To avoid notational confusion, we use A, B, E instead of A, B, C in the definition.)

Maximum Bounded 3-Dimensional Matching (Max-3DM-B)

Instance:	Sets $A = \{a_1, a_2, \ldots, a_n\}$, $B = \{b_1, b_2, \ldots, b_n\}$, and $E = \{e_1, e_2, \ldots, e_n\}$. A subset T of $A \times B \times E$ such that any element of A, B, and E occurs in exactly one, two or three triples in T.
Solution:	A subset T' of T such that no two triples of T' agree in any coordinate.
Goal:	Maximize the cardinality of T'.

Kann [12] proved the maximum bounded 3-dimensional matching problem to be APX-complete. This holds true in particular if we assume that every element occurs in exactly three triples. The remainder of this section proves the following theorem.

Theorem 3. *The c-VRP on a tree is APX-hard under precedence constraints forming a total ordering, and c being part of the input.*

From the above theorem, c-VRP with arbitrary precedence constraints and c-VRP with time windows are APX-hard on a tree, since they both contain the total order as a special case. Without precedence constraints, a PTAS for c-VRP on trees, for arbitrary c, was given only quite recently by Mathieu and Zhou [17]. Our result implies that the PTAS cannot be extended with a total order constraint, assuming $P \neq NP$. For constant c and precedence constraints, determining the complexity remains an open problem for now.

3.1   The Reduction

Let I be an instance of the maximum bounded 3-dimensional matching problem. Our instance I' of c-VRP on a tree consists of an edge-weighted tree, a multiset of requests on this tree, and a total ordering on these requests. The tree is

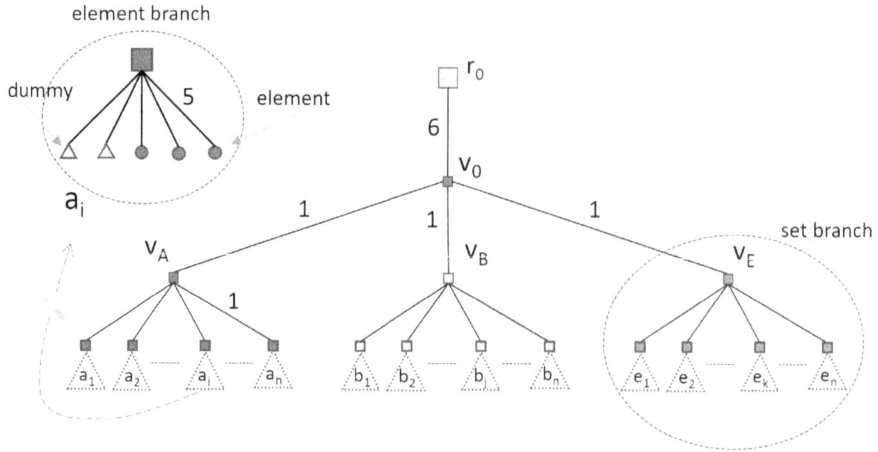

Fig. 2. The instance I' for c-VRP with a total ordering, created from the Max-3DM-B instance I.

illustrated in Fig. 2. For each element $a_i \in A$ in I we create a vertex labeled a_i in I', and we do the same for the elements of B and E. These $3n$ vertices are called *element-vertices*. We attach five leaves to each of these element-vertices as follows. First, create a leaf for each of the three triples in which the corresponding element occurs in I. These are called the *element-leaves*. Then create two other leaves, referred to as *dummy leaves*. The total number of leaves in the tree is thus $5 \cdot 3n = 15n$. We also create a vertex v_A that connects to each vertex a_i and define v_B and v_E in a similar way. Finally, we connect these three vertices to a vertex v_0, which is connected to the root r_0. We refer to the 3 branches rooted at v_A, v_B, v_E as *set-branches* and to the $3n$ branches corresponding to the elements as *element-branches*. The length of the edge (r_0, v_0) is 6 and the length of any edge connected to a leaf is 5. All other edges have length 1.

The description of the graph in our instance I' is now complete, so we turn our attention to the requests and their ordering. Both element-leaves and dummy leaves are given a large number of requests, all other vertices will have no requests.

The requests for the element-leaves are the colored squares depicted in Fig. 3, where we define $K = 7n$ in our reduction. Each of these requests is defined by a unique pair $(x, y) \in \mathbb{N}^2$, where x corresponds to the row and y to the column. We use the following order on the element requests, column by column:

$$(x, y) \prec (x', y') \text{ iff either } y < y' \text{ or } y = y' \text{ and } x < x'. \tag{3}$$

We create the requests for these element-leaves based on the triples they occur in. Let $t_1, t_2, \ldots, t_{3n}$ be the triples in arbitrary order. All requests with index x (row x in Fig. 3) correspond to triple t_x ($1 \leq x \leq 3n$). These requests are divided over its 3 corresponding leaves.

- For the a-leaf, the y-indices of the requests are $2x-1, 2x, 2x+1, \ldots, K+x-1$.

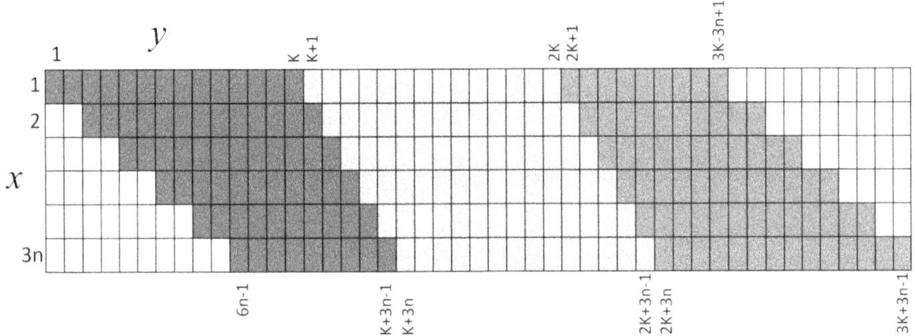

Fig. 3. The requests for element-leaves. In this illustration, $n = 2$.

- For the b-leaf, the y-indices of the requests are $K + x, K + x + 1, \ldots, 2K + x - 1$.
- For the e-leaf, the y-indices of the requests are $2K + x, 2K + x + 1, \ldots, 3K - 3n + 2x - 1$.

So we see that for triple t_x (row x), the y-indices are $2x - 1, 2x, \ldots, 3K - 3n + 2x - 1$, giving a total of $3K - 3n + 1$, which is independent of x – so every triple generates the same number of requests. We define the capacity as $c = 3K - 3n + 1$.

Finally, each dummy leaf is given $n_D := c - K$ requests, and these dummy requests precede all other requests. The order among the dummy requests is arbitrary. This completes the description of the instance I' created from I.

3.2 The Analysis

Suppose that there exists a perfect 3-dimensional matching $\mathcal{S}$ in I, then we create the following solution $\mathcal{S}'$ for I'. For each triple in $\mathcal{S}$, create a tour in $\mathcal{S}'$ that visits the 3 corresponding element-leaves. These tours have length 54. For each element-vertex, there are two additional tours that each visit one dummy leaf plus one element leaf and these tours have length 36. We denote the cost of this solution by

$$\text{OPT}_{3DM} := 54n + 36 \cdot 6n = 270n \,.$$

Lemma 4. *If there exists a solution for our c-VRP instance of cost $Z \leq (1 + \varepsilon)\text{OPT}_{3DM}$, for any $\varepsilon > 0$. Then, there is a matching of size at least $(1 - 90\varepsilon)n$.*

The proof of Lemma 4 is deferred to the full version of this conference paper. The lemma implies that any PTAS for c-VRP with a total ordering on trees implies a PTAS for the bounded 3-dimensional matching problem. However, no such PTAS exists, assuming $P \neq NP$ [12].

Acknowledgements. The authors would like to thank Leen Stougie for fruitful discussions.

References

1. Adamaszek, A., Czumaj, A., Englert, M., Räcke, H.: Almost tight bounds for reordering buffer management. SIAM J. Comput. **51**(3), 701–722 (2022)
2. Adamaszek, A., Czumaj, A., Lingas, A.: PTAS for k-tour cover problem on the plane for moderately large values of k. Int. J. Found. Comput. Sci. **21**(06), 893–904 (2010)
3. Altinkemer, K., Gavish, B.: Heuristics for delivery problems with constant error guarantees. Transp. Sci. **24**(4), 294–297 (1990)
4. Asano, T., Katoh, N., Kawashima, K.: A new approximation algorithm for the capacitated vehicle routing problem on a tree. J. Comb. Optim. **5**(2), 213–231 (2001)
5. Baker, B.S.: Approximation algorithms for NP-complete problems on planar graphs. J. ACM **41**(1), 153–180 (1994)
6. Becker, A.: A tight 4/3 approximation for capacitated vehicle routing in trees. In: Blais, E., Jansen, K., Rolim, J.D.P., Steurer, D. (eds.) Approximation, Randomization, and Combinatorial Optimization. Algorithms and Techniques (APPROX/RANDOM 2018). Leibniz International Proceedings in Informatics (LIPIcs), vol. 116, pp. 3:1–3:15. Schloss Dagstuhl – Leibniz-Zentrum für Informatik, Dagstuhl (2018)
7. Blauth, J., Traub, V., Vygen, J.: Improving the approximation ratio for capacitated vehicle routing. Math. Program. **197**(2), 451–497 (2023)
8. Das, A., Mathieu, C.: A quasi-polynomial time approximation scheme for Euclidean capacitated vehicle routing. In: Proceedings of the Twenty-First Annual ACM-SIAM Symposium on Discrete Algorithms, SODA 2010, pp. 390–403. Society for Industrial and Applied Mathematics (2010)
9. Grandoni, F., Mathieu, C., Zhou, H.: Unsplittable Euclidean capacitated vehicle routing: a $(2 + \varepsilon)$-approximation algorithm (2022)
10. Haimovich, M., Rinnooy Kan, A.: Bounds and heuristics for capacitated routing problems. Math. Oper. Res. **10**(4), 527–542 (1985)
11. Hamaguchi, S., Katoh, N.: A capacitated vehicle routing problem on a tree. In: Chwa, K.-Y., Ibarra, O.H. (eds.) Algorithms and Computation, pp. 399–407. Springer, Heidelberg (1998)
12. Kann, V.: Maximum bounded 3-dimensional matching is MAX SNP-complete. Inf. Process. Lett. **37**(1), 27–35 (1991)
13. Khachay, M., Dubinin, R.: Approximability of the d-dimensional Euclidean capacitated vehicle routing problem. In: AIP Conference Proceedings, vol. 1776, p. 10 (2016)
14. Khachay, M., Ogorodnikov, Y.: PTAS for the Euclidean capacitated vehicle routing problem with time windows. In: International Conference on Learning and Intelligent Optimization (2019)
15. Lenstra, J.K., Kan, A.R.: Complexity of vehicle routing and scheduling problems. Networks **11**(2), 221–227 (1981)
16. Liu, F., Lu, C., Gui, L., Zhang, Q., Tong, X., Yuan, M.: Heuristics for vehicle routing problem: a survey and recent advances. CoRR, abs/2303.04147 (2023)
17. Mathieu, C., Zhou, H.: A PTAS for capacitated vehicle routing on trees. ACM Trans. Algorithms **19**(2), 1–28 (2023)
18. Miltenburg, S., Oosterwijk, T., Sitters, R.: Complexity of fixed order routing. In: Approximation and Online Algorithms: 22nd International Workshop, WAOA 2024, Egham, UK, 5–6 September 2024, pp. 183–197. Springer, Heidelberg (2025)

19. Sitters, R.: An $n^{O(\log \log n)}$ time approximation scheme for capacitated VRP in the Euclidean plane. CoRR, abs/2507.15549 (2025)
20. Song, L., Huang, H., Hongwei, D.: Approximation schemes for Euclidean vehicle routing problems with time windows. J. Comb. Optim. **32**(4), 1217–1231 (2016)

Exact Matching and Top-k Perfect Matching Parameterized by Neighborhood Diversity or Bandwidth

Nicolas El Maalouly and Kostas Lakis[✉]

ETH Zurich, Zurich, Switzerland
`{nicolas.elmaalouly,konstantinos.lakis}@inf.ethz.ch`

Abstract. The Exact Matching (EM) problem asks whether there exists a perfect matching which uses a prescribed number of red edges in a red/blue edge-colored graph. While there exists a randomized polynomial-time algorithm for the problem, only some special cases admit a deterministic one so far, making it a natural candidate for testing the P=RP hypothesis. A polynomial-time equivalent problem, Top-k Perfect Matching (TkPM), asks for a perfect matching maximizing the weight of the k heaviest edges.

We study the above problems, mainly the latter, in the scenario where the input is a blown-up graph, meaning a graph which had its vertices replaced by cliques or independent sets. We describe an FPT algorithm for TkPM parameterized by k and the neighborhood diversity of the input graph, which is essentially the size of the graph before the blow-up; this graph is also called the prototype. We extend this algorithm into an approximation scheme with a much softer dependency on the aforementioned parameters, time-complexity wise. Moreover, for prototypes with bounded bandwidth but unbounded size, we develop a recursive algorithm that runs in subexponential time. Utilizing another algorithm for EM on bounded neighborhood diversity graphs, we adapt this recursive subexponential algorithm to EM.

Our approach is similar to the use of dynamic programming on e.g. bounded treewidth instances for various problems. The main point is that the existence of many disjoint separators is utilized to avoid including in the separator any of a set of "bad" vertices during the split phase.

Keywords: Exact Matching · Top-k Perfect Matching · Parameterized Complexity · Dynamic Programming · Neighborhood Diversity · Bandwidth

1 Introduction

The concept of Perfect Matchings (PM) is central to the study of a long catalog of graph-theoretic algorithmic problems. By deciding the existence of a PM or finding one which maximizes (or minimizes) the total edge weight, one can design

© The Author(s), under exclusive license to Springer Nature Switzerland AG 2026
J. Kozik and A. Wolff (Eds.): SOFSEM 2026, LNCS 16448, pp. 274–287, 2026.
https://doi.org/10.1007/978-3-032-17801-5_21

efficient algorithms for such problems, since the former tasks are solvable in polynomial time [3].

An interesting twist one can impose on the PM problem is the following. Given an edge-coloring of the input graph G using two colors (say red and blue), one can ask whether there exists a PM of G which includes precisely k red edges, where k is also part of the input. We call this problem **Exact Matching (EM)**.

Exact Matching (EM)
Input: A graph G with edges colored red/blue and an integer k.
Task: Decide if there exists a PM M of G such that M has exactly k red edges.

This seemingly benign modification to the PM problem was conceived by Papadimitriou and Yannakakis [13] whilst trying to investigate the effect of restricting the shape of the output tree in the minimum weight spanning tree problem. Specifically, this restriction can be expressed as a graph class which the output tree must fall into. For example, if this class is the one containing all paths, then the problem is at least as hard as deciding the existence of a Hamiltonian path, which is **NP**-hard. Interestingly, Papadimitriou and Yannakakis showed that for a specific class (called *double 2-stars*) the resulting constrained spanning tree problem is polynomial-time equivalent to EM.

At the time of its conception, EM was conjectured to be **NP**-hard. A few years later, though, Mulmuley, Vazirani and Vazirani [12] used a polynomial identity testing approach based on the Schwartz-Zippel Lemma [15,17] along with the *isolation lemma* they developed to solve the problem in randomized polynomial time. Therefore, it was shown that it is quite unlikely for EM to be **NP**-hard after all. Even after so many years however, no deterministic algorithm is known for the general case of EM. This borderline placement of the problem in terms of computational difficulty is what makes it interesting. Specifically, it pertains to the long standing question of whether **P**=**RP**, being one of the very few natural problems suitable to test this question.

In an effort to study EM from a different angle, El Maalouly introduced the **Top-k Perfect Matching (TkPM)** problem [4], which is the main focus of this work. In TkPM, an edge-weighted (as opposed to edge-colored) graph G is given as input along with a positive integer k. The task is to compute a matching M which maximizes the total weight of the k heaviest edges.

Top-k Perfect Matching (TkPM)
Input: A graph G, a weight function w over the edges and an integer k.
Task: Compute a PM M of G which maximizes $|M|_k$ among all PMs of G, where $|M|_k$ refers to the total weight of M's k heaviest edges.

For $k = \frac{n}{2}$, the problem reduces to the classical maximum weight perfect matching problem. The investigation in question [4,10] showed that TkPM (with polynomial weights, which we also assume) is in fact polynomial-time equivalent to EM, making TkPM another valid candidate for the **P=RP** question.[1] The main difference between EM and TkPM is that the latter can be naturally tackled as an approximation task. Conceptually similar optimization objectives had already been studied for problems such as k-clustering and load balancing [2].

Previous Work on TkPM. El Maalouly [4] studied TkPM both from an exact and an approximation viewpoint. All algorithms mentioned below are deterministic. For general graphs, an algorithm achieving a 0.5 approximation ratio was given. This was specialized for bipartite graphs, yielding an approximation ratio that tends to 0.8 as the number of vertices increases.

For the exact algorithms, the independence number α and bipartite independence number β of the graph was utilized. The bipartite independence number is defined in a similar way to the independence number. For a given bipartition A, B of $V(G)$, the bipartite independence number β is the largest number such that there exists an independent set in G using β vertices from A and another β vertices from B, yielding a total of 2β vertices. Given these parameters, exact algorithms were proposed that run in *Fixed Parameter Tractable (FPT)* time with respect to both k, α in the general case and k, β in the bipartite case.

Our Contributions. In this work we mainly concern ourselves with blown-up graphs. By blowing up a graph we mean the process of repeatedly taking one original vertex and replacing it with a clique or an independent set (which we call a *blob*) of arbitrary size. These newly created vertices inherit the same neighborhood relationships as the original vertex for vertices outside the blob.

In Sect. 3, we focus on TkPM for graphs which are blow-ups of constant size graphs. More accurately, we design an exact algorithm (Algorithm 1) that runs in FPT time with respect to k and the *neighborhood diversity* parameter. Intuitively, a graph has neighborhood diversity γ, if its vertices can be partitioned into γ sets such that the neighborhood of a vertex only depends on which set of the partition includes it. We formally define this in Definition 1.

Theorem 1. *Algorithm 1 runs in time* $\mathcal{O}\left(\binom{2k+\gamma-1}{\gamma-1} f(n)\right)$, *where $f(n)$ is a polynomial function. In other words, Algorithm 1 runs in FPT time when parameterized by both γ and k.*

We modify the previous exact algorithm slightly and derive an approximation scheme (Algorithm 2) whose complexity is governed by the accuracy required, k and γ. However, the dependency on the latter two is much lighter.

Theorem 2. *Algorithm 2 runs in time* $\mathcal{O}\left(\left(\frac{\log_2 k}{\log_2(1/(1-\epsilon))}\right)^{\gamma^2} f(n)\right)$, *where $f(n)$ is a polynomial function depending only on n and $1 - \epsilon$ is the desired approximation ratio.*

---

[1] However, the reduction does not preserve the graph structure.

Going back to exact algorithms, in Sect. 4 we study how the structure of the initial graph (before the blow-up) can help us solve the problem efficiently, resulting in Algorithm 3. We show that if the initial graph has bounded *bandwidth*, then we can solve TkPM exactly in the blown-up graph in subexponential time. The bandwidth of a graph has been treated as both a quantity to compute and as a parameter to derive efficient algorithms as we do here [1,6,7,11,14,16]. We do so by exploiting the fact that bounded-bandwidth graphs have many *disjoint*, balanced and small separators. This is important because with our recursive approach there are blobs which we cannot afford to use to split the initial graph for the recursion step, so we must find a separator which contains none of them. We are fortunately always able to find one such "free" separator, as long as the graph is large. Otherwise we may simply use Algorithm 1.

Theorem 3. *theorem Algorithm 3 runs in time at most* $2^{\mathcal{O}\left(\phi^2 \sqrt{n} \log^2 n\right)}$ *when the input prototype $\mathcal{P}$ has bandwidth ϕ.*

In Sect. 5, we discuss how one can adapt this algorithm to work for EM. We employ an existing algorithm for EM in bounded neighborhood diversity graphs [5] as a base case. Since this algorithm runs in XP time and not FPT as our (base case) algorithm for TkPM, we have to slightly modify our recursive approach details (the definition of which blobs we can/cannot use for splitting) and that results in a larger, but still subexponential runtime.

Theorem 4. *There exists an algorithm that solves EM and runs in time at most* $2^{\mathcal{O}\left(\phi^2 n^{12/13} \log^2 n\right)}$ *when the input prototype $\mathcal{P}$ has bandwidth ϕ.*

Some proofs are omitted due to space constraints. The full version contains these proofs and is available online [9].

2 Preliminaries

An edge from u to v with weight w is denoted by $(\{u, v\}, w)$. For a matching M, we use $|M|$ to denote the sum of the weights of the edges in M. Similarly, we use $|M|_k$ to refer to the sum of the k highest weight edges in M, i.e. the value of the top-k objective for M.

We almost exclusively consider blowups of graphs in this work. By blowup we specifically mean the following process. First, each vertex is replaced by an empty or complete graph on some arbitrary number of new vertices. Then, for each edge (u, v) of the original graph we add all edges between pairs of vertices u', v', where u' belongs to the empty/complete subgraph that replaced u and v' belongs to the subgraph that replaced v. The weights of the new edges can be chosen arbitrarily.

We generally use $\mathcal{P}$ to refer to the initial graph and G to refer to the resulting graph of the blowup. We also use the term *prototype* to refer to the initial graph $\mathcal{P}$. Moreover, we use the term *blob* to refer to a vertex of a prototype $\mathcal{P}$ or to the set of vertices that replaced it during the blowup, interchangeably. Similarly, we use the term *band* to refer to an edge of $\mathcal{P}$ or to the set of new edges created from it in the resulting graph. We may also use the term band for the set of edges created while blowing-up a vertex into a clique.

Notationally, to keep things clear, we use $E(B)$ to refer to the set of edges created from a band B. Moreover, given a set of edges E, we use $V(E)$ to refer to the set of vertices incident to least one edge of E. We remark that $V(\mathcal{P})$ and $E(\mathcal{P})$ still carry the standard meaning they would for any graph, i.e. they refer to the set of vertices and edges of prototype $\mathcal{P}$ viewed as a regular graph before the blowup. Hence, to formally refer to the set of edges in G created by some band in $\mathcal{P}$ we will use $E(B)$, where $B \in E(\mathcal{P})$.

In general, when we say that we *guess* some sort of information (for example a set of edges or a vertex-separator), we mean that we try all combinations, possibly under some constraints which will be mentioned or clear from the context. Finally, since we focus on perfect matchings, we will assume that the input graphs (not the prototypes) have $2n$ vertices, i.e. n refers to the size of the PM and not the number of vertices as usual.

3 Top-k Perfect Matching in Graphs of Bounded Neighbourhood Diversity

We define the neighborhood diversity of a graph below.

Definition 1 (Neighborhood diversity). *We say that a graph G has neighbourhood diversity number γ if we can partition its vertices into γ sets $V_1, V_2, \ldots, V_\gamma$ such that for any $u, u' \in V_i$ and any $v \neq u, u'$ we have $(u, v) \in E(G) \iff (u', v) \in E(G)$ and γ is the minimum integer for which this partitioning is possible.*

In other words, we can think of the vertices of G as each belonging to one of γ *types*, and the existence of an edge only depends on the types of the endpoints. Note here that this is only with regard to the *existence* of edges; their weights can be arbitrarily chosen. For example, r-partite complete graphs have $\gamma \leq r$. Moreover, a blowup G' of a graph $G = (V, E)$ has $\gamma \leq |V|$, regardless of the size of G' and the inclusion or not of the edges "inside" the blown-up vertices.

It is important to mention that the neighborhood diversity of a graph can be computed in polynomial time. Additionally, an optimal partitioning of the vertices in different sets according to Definition 1 can also be computed in polynomial time [8]. Thus, we may freely assume from now on that such a partition is given.

Designing the Algorithm via Reductions. We describe an algorithm (Algorithm 1) that runs in FPT time parameterized by both k and γ (The running time is polynomial if γ is bounded by a constant). The idea is that since all vertices of one type are equivalent in terms of their neighbourhoods, any two matchings that use a fixed number of vertices per type are equivalent as far as extendibility goes. Thus, if one knew the precise number of vertices per type which are incident to the top-k edges in the optimal solution, then they would only need to optimize under this constraint (number per type), since any solution (and thus also the best) would be extendible, as the top-k edges of the optimal solution are extendible.

It turns out that replacing the perfect matching constraint with this "type" constraint allows one to solve the optimization problem easily. We describe here how to solve this new problem, which we call *Type-Constrained Maximum Weight Matching*, or *TC-MWM*. The idea is quite simple, we simply add some "killer" vertices which we fully connect to each type in order to enforce the type constraints and then run Maximum Weight Perfect Matching.

The TC-MWM Problem. In more detail, let $G = (V_1 \cup V_2 \cup \cdots \cup V_\gamma, E)$ be a graph of bounded neighbourhood diversity and $(c_1, c_2, \ldots, c_\gamma)$ be a tuple of type constraints. The problem of TC-MWM asks us to find a set of edges which are incident to precisely c_i vertices in V_i and is of maximum total weight under these constraints.

Type-Constrained Maximum Weight Matching (TC-MWM)
Input: An edge-weighted graph $G = (V_1 \cup V_2 \cup \cdots \cup V_\gamma, E)$ of neighborhood diversity γ and a tuple $(c_1, c_2, \ldots, c_\gamma)$ with $c_i \in [0, |V_i|]$, for all i.
Task: Out of all matchings of G which cover *exactly* c_i vertices of V_i for all i, compute one which maximizes the total weight.

We solve TC-MWM by defining the graph G' where we add vertex sets K_i of cardinality $|V_i| - c_i$, fully connected to V_i, with each new edge getting weight 0. We run MWPM on G' and then drop any edges not in G from the result. The construction of G' is exemplified in Fig. 1.

Note that since the sets K_i must be matched, exactly $|K_i|$ vertices of V_i are reserved for that and the rest, $|V_i| - |K_i| = c_i$, are matched to vertices of G. So, any candidate solution to MWPM for G' respects the type constraints. We now need to show that it is also part of a perfect matching in G. This follows because we can extend M^* to a perfect matching using the sets K_i, as there are precisely $|K_i|$ vertices of V_i unmatched in M^*.

Using TC-MWM to Solve TkPM. The final piece of the puzzle is: how does one know the number of vertices per type that the optimal TkPM solution uses, i.e. the vector $(c_1, \ldots, c_\gamma)$? The answer is that we can "guess" it, i.e. try all possible options. But how many are there? Note that the top k edges of the optimal TkPM solution will "touch" precisely $2k$ vertices. These are distributed in *some* way between the γ types. Thus, it suffices to enumerate all solutions of the equation $c_1 + c_2 + \cdots + c_\gamma = 2k$, which are bounded by a function of k and γ. Namely, this number equals the number of ways to distribute $2k$ balls into γ distinguishable bins, which equals $\binom{2k+\gamma-1}{\gamma-1}$. Importantly, this number does not depend on n. Note, however, that not all solutions of this equation are relevant;

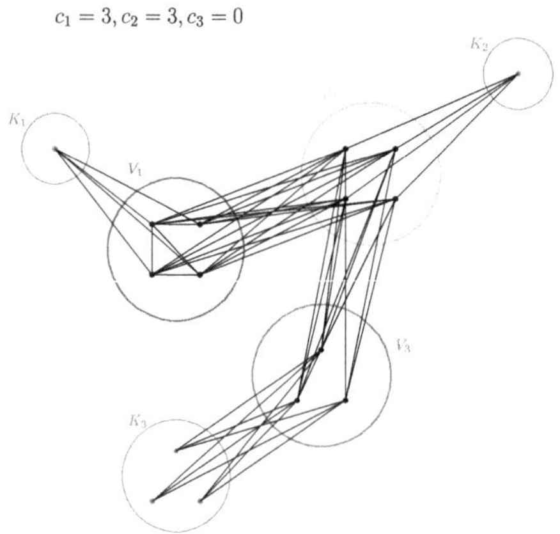

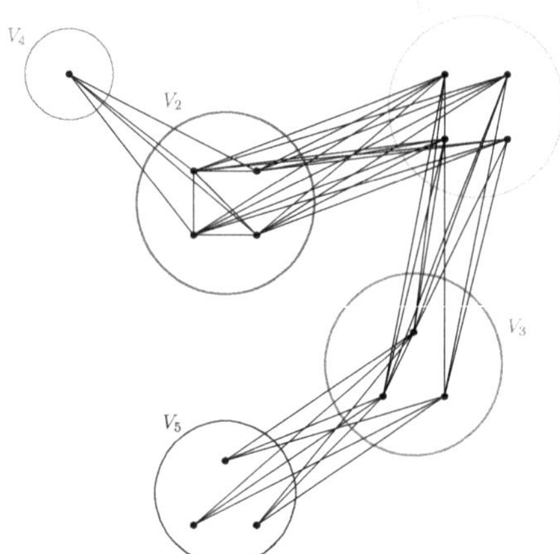

Fig. 1. The graph used for the reduction of TC-MWM to MWPM.

Fig. 2. A case where a small separator (here $\{V_1\}$) in the prototype blows up into a big separator in the graph, both in terms of vertices and adjacent edges.

Algorithm 1: TkPM for Bounded Neighborhood Diversity

Input: k, $G = (V_1 \cup V_2 \cup \cdots \cup V_\gamma, E)$
Output: A PM of G maximizing the top-k objective, $\emptyset$ if none exist
$M_{\text{best}} \leftarrow \emptyset$
for $(c_1, c_2, \ldots, c_\gamma) \in [0, k]^\gamma$ with $\sum_{i=1}^{\gamma} c_i = 2k$ **do**
    for $i \leftarrow 1$ **to** γ **do**
        $K_i \leftarrow \{v_{i_1}, v_{i_2}, \ldots, v_{i_{|V_i|-c_i}}\}$ $E_i' \leftarrow \{(\{u, v\}, 0) \mid u \in V_i, v \in K_i\}$
    $G' \leftarrow (V(G) \cup \bigcup_{i=1}^{\gamma} K_i, E(G) \cup \bigcup_{i=1}^{\gamma} E_i')$
    $M' \leftarrow \textsc{MaximumWeightPerfectMatching}(G') \cap E(G)$
    $G'' \leftarrow G[V(G) \setminus V(M')]$
    if G'' has a Perfect Matching M'' **and** $|M'|_k > |M_{\text{best}}|_k$ **then**
        $M_{\text{best}} \leftarrow M' \cup M''$
return M_{best}

---

some may produce matchings that are not extendible to a PM in G. We simply discard such cases. The pseudocode of the overall algorithm that solves TkPM is given in Algorithm 1.

3.1 An Approximation Scheme

Algorithm 1 essentially implements a brute force search throughout *all* the options that the optimal solution M^* has for distributing its $2k$ contributing edge endpoints into the blobs of the graph. Hence, for the correct guess of this distribution, we retrieve an optimal solution. What if instead of an optimal solution, we aim for an approximate one? This is in fact possible with slight changes to the algorithm, resulting in an approximation scheme for it. That is, for any desired $\epsilon > 0$, we can output a PM which achieves a $(1 - \epsilon)$ approximation for the top-k objective. We describe Algorithm 2 below.

Exponentially Growing Band Counts. The first main difference from Algorithm 1 is that instead of considering the number of vertices touched per blob (the c_i values), we focus on the number of edges utilized per band and also within blobs (we shall use b_i to refer to these numbers). We might slightly abuse terminology (as warned) and use the term band below to also refer to a set of edges staying within a blob. Note that now we may have up to $\binom{\gamma}{2} + \gamma$ different b_i values, whose sum must not exceed k. The trick we utilize to achieve the aforementioned approximation scheme is to consider the values $A = \{0, 1, \lceil \alpha \rceil, \lceil \alpha^2 \rceil, \ldots, k\}$ for each b_i instead of all values in $[0, k]$, for some appropriate $\alpha > 1$ that depends on ϵ. We additionally replace the constraint $\sum_{i=1}^{\gamma} c_i = 2k$ with $\sum_{i=1}^{\gamma} b_i \le k$. Therefore, we might not get the distribution of M^*'s edges (per band/within blobs) exactly right, but we get one which is close enough for our purposes (and which is also extendible). More details on this are found in the formal proof.

Runtime. For such a choice of α, we see that each c_i is endowed with roughly $\log_\alpha k$ different choices. Hence, we can have at most

$$\mathcal{O}\left(\left(\log_{\frac{1}{1-\epsilon}} k\right)^{\gamma^2}\right) = \mathcal{O}\left(\left(\frac{\log_2 k}{\log_2\left(\frac{1}{1-\epsilon}\right)}\right)^{\gamma^2}\right)$$

different combinations for these numbers, which governs the time complexity of our approximation scheme. Notice that we have thus greatly reduced the effect of k on the runtime, introducing an ϵ-dependent overhead instead. In fact, for $\gamma = \mathcal{O}\left(\sqrt{\frac{\log n}{\log \log n}}\right)$ we get a *Polynomial-Time Approximation Scheme (PTAS)*, since then for any fixed ϵ we can bound the overall runtime by a polynomial (whose exponent is roughly the hidden constant in the asymptotic bound for γ).

4 Going Further: A Recursive Approach

Discovering the Algorithm. Before presenting the new class of graphs we will treat and our algorithm, let us go through the thought process by which one decides upon them. Recall that the graphs of bounded neighborhood diversity that we considered can be simply viewed as arbitrary blowups of "small" graphs. Let us use the word *prototype* to refer to these initial graphs. We would like to leverage some sort of structural property (not the property of being small) about these prototypes in order to get an algorithm that runs faster than Algorithm 1. The first thought that comes to mind is to consider prototypes of some bounded graph parameter, for example treewidth. Under this assumption, it initially looks like one could repeatedly decompose the problem into smaller subproblems in

---

Algorithm 2: Approximation Scheme for TkPM.

Input: k, $G = (V_1 \cup V_2 \cup \cdots \cup V_\gamma, E)$, approximation slack $\epsilon \in (0, 1)$.
Output: A PM of G approximating the top-k objective, $\emptyset$ if none exist
$M_{\text{best}} \leftarrow \emptyset$
$\alpha \leftarrow \frac{1}{1-\epsilon}$
$A \leftarrow \{0, 1, \lceil \alpha \rceil, \lceil \alpha^2 \rceil, \ldots, k\}$
for $\left(b_1, b_2, \ldots, b_{\binom{\gamma}{2}+\gamma}\right) \in A^{\binom{\gamma}{2}+\gamma}$ with $\sum_{i=1}^{\binom{\gamma}{2}+\gamma} b_i \leq k$ **do**
    Compute the corresponding tuple $(c_1, c_2, \ldots, c_\gamma)$
    for $i \leftarrow 1$ **to** γ **do**
        $K_i \leftarrow \{v_{i_1}, v_{i_2}, \ldots, v_{i_{|V_i|-c_i}}\}$
        $E_i' \leftarrow \{(\{u, v\}, 0) \mid u \in V_i, v \in K_i\}$
    $G' \leftarrow (V(G) \cup \bigcup_{i=1}^{\gamma} K_i, E(G) \cup \bigcup_{i=1}^{\gamma} E_i')$
    $M' \leftarrow \textsc{MaximumWeightPerfectMatching}(G') \cap E(G)$
    $G'' \leftarrow G[V(G) \setminus V(M')]$
    if G'' has a Perfect Matching M'' **and** $|M'|_k > |M_{\text{best}}|_k$ **then**
        $M_{\text{best}} \leftarrow M' \cup M''$
return M_{best}

---

a Dynamic Programming fashion. However, as will be made clear later, most commonly used parameters such as treewidth or pathwidth fail to give rise to a decomposition scheme for the blown-up graphs. This is due to some very specific demands about the separators of the prototype.

Why Blowing-Up Makes Decomposition Difficult. Consider some prototype graph $\mathcal{P}$. Let us refer to its vertices as *blobs* and to its edges as *bands*. In the resulting graph G, the blobs turn into cliques or independent sets and the bands turn into edge sets of a complete bipartite graph. Suppose that we are guaranteed the existence of a small (in terms of number of blobs) separator by some graph parameter. The problem now becomes; how do we actually recurse in order to build an overall solution? It could happen that the small blob-separator of $\mathcal{P}$ is actually a huge set of vertices in G which also connects to arbitrarily many edges (for example, see Fig. 2). Hence, there would be too many different ways to recurse depending on the set of edges chosen to be included in the matching (within the edges separating the graph), defeating the purpose of the decomposition. It seems logical therefore to attempt to avoid this kind of phenomenon and guarantee in some manner that the number of recursion choices is relatively controlled in the blown-up graph.

Avoiding the Blow-Up Side Effect. How can we mitigate the difficulty explained in the previous paragraph? Our approach is to find a separator which touches a controlled number of (matching) edges in the blown-up graph, hence making the recursion workload less demanding. This motivates the following definitions. Let M^* be the matching of the best solution. The edges of M^* are distributed in *some* way between the bands of the prototype (edges staying within blobs also exist in

M^*, but that is irrelevant for the following discussion). Some of the bands will contain "many" edges, while some will contain "few". We characterize the former as *tight* bands and the latter as *loose* bands. The threshold for the number of edges contained in a tight/loose band is set to $\sqrt{n}$. An immediate observation using these definitions is that there can be at most $\sqrt{n}$ tight bands, otherwise there would be more than n edges in M^*, a contradiction. An illustration of these definitions is given in Fig. 3. Note that since we are aiming for a subexponential algorithm only, we can assume that we have the information of which bands are tight and which are loose, since we can try every possible combination, of which there are subexponentially many.

Using the Tightness/Looseness of Bands. We now have a tool that we can use to combat the aforementioned blow-up side effect. Specifically, to see why this tool helps, consider the "lucky" case where we have a small blob-separator which is incident only to loose bands. Let us call such a separator a *loose separator*. This allows us to make the following optimization in the task of recursing to find a global solution. Instead of considering all subsets of edges within the loose bands, we can simply consider these subsets which respect the bound of $\sqrt{n}$ edges per band, since the optimal solution M^* is guaranteed to fall within such a selection (for the correct choice of tight bands). Of course, we still need to assure that there are few bands also; this is guaranteed by the graph parameter we will use.

Finding Loose Separators. We have described how in the fortunate case of having a loose separator we can proceed with the decomposition of the problem. The question remains; how do we actually find loose separators? If our graph parameter guarantees the existence of a single small separator (as most commonly used graph parameters do), then the situation looks grim, as it could very well be that some of the blobs are connected to tight bands. A natural idea then is to ask for *many* disjoint small separators. Indeed, if we have this sort of (strong enough) guarantee, we can find one separator which is touched by no tight band (i.e. a loose separator), as each tight band only touches 2 blobs. Therefore, the task is now to find a graph parameter which will guarantee that our prototype graphs have many small separators.

Graphs of Bounded Bandwidth Have Many Small Separators. A parameter such as the one we are looking for turns out to be the bandwidth. The bandwidth of a graph G is defined as follows.

Definition 2. *The bandwidth $\phi(G)$ of a graph G is the minimum integer such that there exists an ordering $v_1, v_2, \ldots, v_n$ of the vertices of G such that $\{v_i, v_j\} \in E(G) \implies |i - j| \leq \phi(G)$.*

It is not very hard to see that any consecutive $\phi(G)$ vertices in such an ordering constitute a separator. Therefore, one can find many disjoint separators of G using that fact. Of course, the separators need to be balanced, in the sense that at least some constant fraction of vertices are preserved in the resulting two parts of the graph after the separation. This can be achieved by e.g. only considering separators of vertices v_i with $i \in [n/4, 3n/4]$.

Interestingly, the bandwidth serves as an approximate upper bound for the maximum degree. Indeed, there cannot be a vertex u with more than $2\phi(G)$ neighbors, since then in any linear ordering of the vertices there must exist a neighbor v of u which is more than $\phi(G)$ distance apart from u in said ordering, a contradiction. This implies that a blob-separator of small size is also adjacent to a correspondingly small number of bands, which is crucial for our algorithm.

When to Stop Separating. To find a loose separator, we need to be able to find many small separators compared to the number of tight bands. If the number of separators is more than twice the number of tight bands, there has to exist one which is loose. What if that is no longer true after some steps of decomposition? In such a case, we can show that the number of blobs remaining is of the same order of magnitude as the number of tight bands, i.e. we have $\mathcal{O}(\phi\sqrt{n})$ blobs. Fortunately, we can now simply use Algorithm 1 to solve the problem as a base case of the recursion.

The Overall Algorithm. We described in the previous paragraphs the various ingredients used to make our recursive algorithm. Let us now collect them. First, if the number of tight bands surpasses a certain fraction (depending on ϕ) of the number of blobs (or if the number of blobs is bounded by some constant), we simply use Algorithm 1, since then we would have that the number of blobs is conveniently $\mathcal{O}(\phi\sqrt{n})$. Otherwise, the existence of a loose blob-separator of size at most ϕ is guaranteed. We find such a separator $\mathcal{S}$ and proceed as follows.[2]

Let $\mathcal{P}_1$ and $\mathcal{P}_2$ be the separated parts of $\mathcal{P}$ and G_1, G_2 the corresponding subgraphs of G. We loop through every selection E' of at most $\sqrt{n}$ edges within each band touching $\mathcal{S}$ such that all edges in E' have at least one endpoint within the vertices in the blobs of $\mathcal{S}$. For each such E' and for every $(k', k_{\text{blobs}}, k_1, k_2) \in [0, k]^4$ such that $k' + k_{\text{blobs}} + k_1 + k_2 = k$, we recursively call our algorithm on the inputs $\mathcal{P}_1, G_1, k_1$ and $\mathcal{P}_2, G_2, k_2$ and join the matchings found with E', as well as a top-k_{blobs} PM calculated for the subgraph of G induced by vertices of blobs of $\mathcal{S}$ not touched by E' using Algorithm 1. If no PM of G can be built in this way, we ignore this specific choice of E'. We return the best valid matching found over all such recursive calls. A visual explanation is given in Fig. 4. The pseudocode is given in Algorithm 3.

---

[2] It suffices to try all possibilities and this does not affect the runtime. We do not require that the ordering of Definition 2 is given nor do we need to compute it.

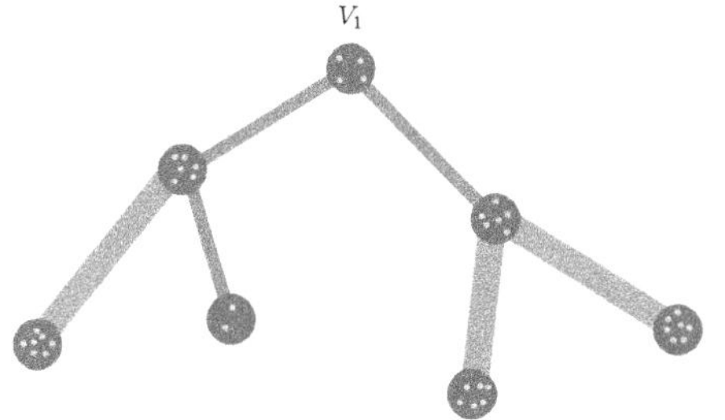

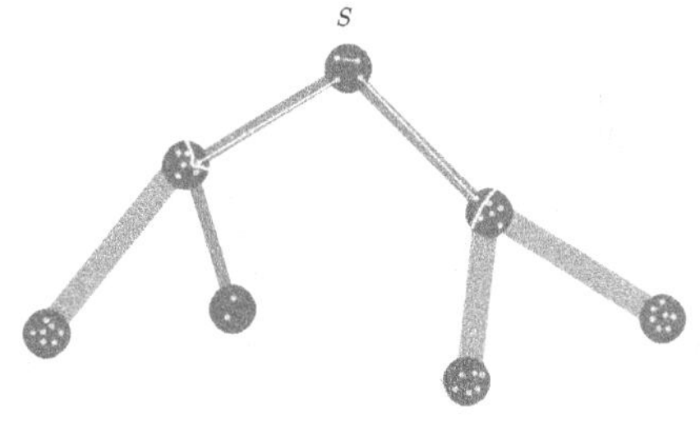

Fig. 3. An illustration of loose/tight bands. Loose bands (blue) contain less than $\sqrt{n}$ edges of the solution PM M and tight ones (red) contain at least that many. The set $\{V_1\}$ is a loose separator, since it splits the prototype in a balanced way, contains at most ϕ blobs and is only adjacent to loose bands. (Color figure online)

Fig. 4. An illustration of the recursive solution to the problem. The complete edge set E' (in yellow) leaving blobs of the loose separator $\mathcal{S}$ is guessed. Then, the problem decomposes into the matchings M_1, M_2 of the smaller subproblems (in orange) as well as the matching M_{blobs} which is comprised of edges which stay within a single blob of $\mathcal{S}$ (in pink). (Color figure online)

Algorithm 3: TkPM for Blowups of Bounded Bandwidth Graphs *(BBB)*

Input: k, prototype $\mathcal{P}$, bandwidth ϕ, G, subset of tight bands $\mathcal{T} \subseteq E(\mathcal{P})$, n
Output: A PM of G maximizing the top-k objective, $\emptyset$ if none exist
if G has no Perfect Matching **then**
    $\lfloor$ **return** $\emptyset$
$M_{\text{best}} \leftarrow \emptyset$
if $2|\mathcal{T}| \geq \left\lfloor \left(\frac{|V(\mathcal{P})|}{2} - 3 \right) / \phi \right\rfloor$ **or** $|V(\mathcal{P})| \leq C$ for some constant C **then**
    $\lfloor$ Call Algorithm 1 on (k, G) and **return** its output.

else
    Find a separator $\mathcal{S}$ of $\mathcal{P}$ with $V(\mathcal{T}) \cap \mathcal{S} = \emptyset$ separating $\mathcal{P}$ into $\mathcal{P}_1$ and $\mathcal{P}_2$
    with $|V(\mathcal{P}_i)| \in \left[\frac{1}{4}|V(\mathcal{P})|, \frac{3}{4}|V(\mathcal{P})| \right]$ for $i \in \{1, 2\}$, such that $|\mathcal{S}| \leq \phi$.
    Let V_3 be the set of vertices in blobs of $\mathcal{S}$.
    for $E' \subseteq E(G)$ s.t. $E' \subseteq V_{\mathcal{S}} \times V(G)$ **and** $|E' \cap E(B)| \leq \sqrt{n}$ $\forall B \in E(\mathcal{P})$ **do**
        for $(k', k_{\text{blobs}}, k_1, k_2) \in [0, k]^4$ with $k' + k_{\text{blobs}} + k_1 + k_2 = k$ **do**
            $G_i \leftarrow$ the subgraph of G induced by the vertices in blobs of $\mathcal{P}_i$ not
            touched by E', for $i \in \{1, 2\}$
            $M_i \leftarrow BBB(k_i, \mathcal{P}_i, \phi, G_i, \mathcal{T} \cap E(\mathcal{P}_i), n)$, for $i \in \{1, 2\}$
            if $M' = E' \cup M_1 \cup M_2$ extends to a PM of G **then**
                Call Algorithm 1 on $(k_{\text{blobs}}, G_{\text{blobs}})$ to get M_{blobs}, where G_{blobs}
                consists only of $V_{\mathcal{S}}$ and the edges within single blobs of $\mathcal{S}$.
                $M'' \leftarrow M' \cup M_{\text{blobs}}$
                $M_{\text{best}} \leftarrow \max_{|\cdot|_k} \{M'', M_{\text{best}}\}$

return M_{best}

---

5 Adapting the Recursive Algorithm to Exact Matching

Taking a step back, one notices that our recursive approach for TkPM is not very closely related to it but instead exhibits a slight generality. Disregarding the base case, all we did was decompose the problem based on a characterization of the bands using the number of their edges utilized in the solution. Then, we tried all possible combinations of the edges in the loose bands recursively.

The EM Algorithm. In fact, we can use the same approach to derive a subexponential time algorithm for the same class of graphs for EM also. Specifically, we keep roughly the same decomposition procedure and use a different algorithm for the base case. To do this, we rely on the existence of an algorithm solving EM on instances with neighborhood diversity γ in time $n^{\mathcal{O}(\gamma^{12})}$ from [5]. Note however that this algorithm does not run in FPT time, in contrast to our base case algorithm for TkPM. Instead, it runs in XP time, i.e. the exponent of the polynomial depends on γ. We can deal with the consequences of this by tweaking the definition of tight bands to balance the costs of the base case and the recursive decomposition costs.

As we said, we will change the definition of a tight band, specifically by modifying the threshold from $\sqrt{n}$ to n^{α}, for some $\alpha < 1$. Note that now this means that the set of tight bands $\mathcal{T}$ can have at most $|\mathcal{T}| \leq n^{1-\alpha}$ elements. This modification affects the runtime of our algorithm in the following two ways:

- In the base case, we have that the number of blobs n' is $\mathcal{O}(|\mathcal{T}|) = \mathcal{O}(n^{1-\alpha})$. So, the runtime using the algorithm mentioned for EM is $n^{\mathcal{O}(n^{12(1-\alpha)})}$.
- For the recursive decomposition step, we still get the same results about existence of loose separators. However, since loose bands are now allowed to have a different number of edges, we now see that there are at most $\binom{n^2}{n^\alpha}^{\mathcal{O}(\phi^2)} = n^{\mathcal{O}(\phi^2 n^\alpha)}$ different ways to recurse instead of $n^{\mathcal{O}(\phi^2 \sqrt{n})}$.

To balance the costs of these two steps, we want to have the same power of n in the exponent, therefore we set $12(1 - \alpha) = \alpha$ and get $\alpha = \frac{12}{13}$. Mimicking the proof of Theorem 3, one sees that with this value of α we get an algorithm that runs in time $2^{\mathcal{O}(\phi^2 n^{12/13} \log^2 n)}$, which is subexponential.

Theorem 4. *There exists an algorithm that solves EM and runs in time at most $2^{\mathcal{O}(\phi^2 n^{12/13} \log^2 n)}$ when the input prototype $\mathcal{P}$ has bandwidth ϕ.*

Acknowledgments. The authors wish to thank Andor Vári-Kakas for his insightful help during the initial stages of the project.

References

1. Beisegel, J., Klost, K., Knorr, K., Ratajczak, F., Scheffler, R.: A graph width perspective on partially ordered Hamiltonian paths. arXiv preprint arXiv:2503.03553 (2025)

2. Chakrabarty, D., Swamy, C.: Approximation algorithms for minimum norm and ordered optimization problems. In: Proceedings of the 51st Annual ACM SIGACT Symposium on Theory of Computing, pp. 126–137 (2019)
3. Edmonds, J.: Paths, trees, and flowers. Can. J. Math. **17**, 449–467 (1965)
4. El Maalouly, N.: Exact Matching: algorithms and Related Problems. In: 40th International Symposium on Theoretical Aspects of Computer Science (STACS), pp. 29:1–29:17 (2023)
5. El Maalouly, N., Haslebacher, S., Wulf, L.: On the exact matching problem in dense graphs. In: 41st International Symposium on Theoretical Aspects of Computer Science (STACS), pp. 33:1–33:17 (2024)
6. Golovach, P., Heggernes, P., Kratsch, D., Lokshtanov, D., Meister, D., Saurabh, S.: Bandwidth on at-free graphs. Theoret. Comput. Sci. **412**(50), 7001–7008 (2011)
7. Kaplan, H., Shamir, R.: Pathwidth, bandwidth, and completion problems to proper interval graphs with small cliques. SIAM J. Comput. **25**(3), 540–561 (1996)
8. Lampis, M.: Algorithmic meta-theorems for restrictions of treewidth. Algorithmica **64**(1), 19–37 (2012)
9. Maalouly, N.E., Lakis, K.: Exact matching and top-k perfect matching parameterized by neighborhood diversity or bandwidth (2025). https://arxiv.org/abs/2510.12552
10. Maalouly, N.E., Wulf, L.: Exact matching and the top-k perfect matching problem. arXiv preprint arXiv:2209.09661 (2022)
11. Monien, B., Sudborough, I.H.: Bounding the bandwidth of NP-complete problems. In: Noltemeier, H. (ed.) WG 1980. LNCS, vol. 100, pp. 279–292. Springer, Heidelberg (1981). https://doi.org/10.1007/3-540-10291-4_20
12. Mulmuley, K., Vazirani, U.V., Vazirani, V.V.: Matching is as easy as matrix inversion. In: Proceedings of the Nineteenth Annual ACM Symposium on Theory of Computing, pp. 345–354 (1987)
13. Papadimitriou, C.H., Yannakakis, M.: The complexity of restricted spanning tree problems. J. ACM (JACM) **29**(2), 285–309 (1982)
14. Saxe, J.B.: Dynamic-programming algorithms for recognizing small-bandwidth graphs in polynomial time. SIAM J. Algebraic Discr. Methods **1**(4), 363–369 (1980)
15. Schwartz, J.T.: Fast probabilistic algorithms for verification of polynomial identities. J. ACM (JACM) **27**(4), 701–717 (1980)
16. Zabih, R.: Some applications of graph bandwidth to constraint satisfaction problems. In: AAAI, pp. 46–51 (1990)
17. Zippel, R.: Probabilistic algorithms for sparse polynomials. In: Ng, E.W. (ed.) Symbolic and Algebraic Computation. LNCS, vol. 72, pp. 216–226. Springer, Heidelberg (1979). https://doi.org/10.1007/3-540-09519-5_73

Overlapping Biclustering

Matthias Bentert[1,2] (iD), Pål Grønås Drange[2(✉)] (iD), and Erlend Haugen[2]

[1] Technische Universität Berlin, Berlin, Germany
{Pal.Drange,Erlend.H}@uib.no
[2] University of Bergen, Bergen, Norway
bentert@tu-berlin.de

Abstract. We study the problem of transforming bipartite graphs into bicluster graphs. Abu-Khzam, Isenmann, and Merchad [IWOCA '25] introduced two variants of this problem. In both problems, the goal is to transform a bipartite graph into a bicluster graph with at most k operations, where the allowed operations are inserting an edge, deleting an edge, and splitting a vertex. Splitting a vertex v refers to replacing v by two new vertices whose combined neighborhood equals the neighborhood of v. The latter models overlapping clusters, that is, vertices belonging to multiple clusters, and is motivated by several real-world applications. The versions differ in that one variant allows splitting any vertex, while the second variant only allows vertex splits on one side of the bipartition.

Regarding computational complexity, they showed APX-hardness for both variants and a polynomial kernel (with $O(k^5)$ vertices) for the one-sided variant. They asked as open problems whether the polynomial kernel can be improved and whether it can also be extended for the other variant. We answer both questions in the affirmative and give kernels with $O(k^2)$ vertices for both variants. We also show that both problems can be solved in $O(k^{11k}+n+m)$ time, where n and m denote the number of vertices and edges in the input graph, respectively.

Keywords: Biclustering · Clustering · Editing · Graph Modification · Vertex Splitting

1 Introduction

Clustering is a fundamental task in data analysis, aiming to group similar objects based on a given similarity measure. In *correlation clustering*, the input is a set of objects with pairwise similarity or dissimilarity information, and the goal is to partition the objects into clusters of mutually similar entities [7]. This approach is

Part of this work was achieved by Erlend Haugen during his Master's thesis. This work received financial support from the European Research Council (ERC) under the European Union's Horizon 2020 research and innovation programme (grant agreement No. 819416) and the German Research Foundation (DFG, grant 1-5003036: SPP 2378 – ReNO-2). The full version of the article can be found on arXiv [12].

© The Author(s), under exclusive license to Springer Nature Switzerland AG 2026

J. Kozik and A. Wolff (Eds.): SOFSEM 2026, LNCS 16448, pp. 288–302, 2026.
https://doi.org/10.1007/978-3-032-17801-5_22

widely used in machine learning, document classification, gene expression analysis, image segmentation, and social network analysis. Clustering is often modeled as the graph problem CLUSTER EDITING [15,27]. Here, objects are vertices and an edge between two vertices indicates similarity. The task is then to add or delete the fewest number of edges to obtain a *cluster graph*—a disjoint union of cliques.

However, many datasets do not naturally express similarity in terms of pairwise relationships. Instead, they are *dichotomous*, representing binary associations between two distinct types of objects. These datasets are common in a variety of applications: politicians and the bills they support, movie enjoyers and the films they watch, or students and the courses they take [26]. Such data can be modeled as bipartite graphs, where one part corresponds to decision-makers and the other to their choices. Similar binary structures arise in gene-disease associations [9] and chemical reactions [14].

To cluster such dichotomous data, cliques are no longer the appropriate target. Instead, the goal becomes transforming the input into a *bicluster graph*—a disjoint union of bicliques (complete bipartite graphs). This leads to the problem of BICLUSTER EDITING, which asks for the minimum number of edge modifications needed to achieve such a structure. BICLUSTER EDITING has numerous applications, including gene expression analysis, community detection, and in recommendation systems [8,13,30,32,34,35].

In certain cases, some objects naturally belong to multiple clusters. In document classification, where we assign keywords to documents, a single document often fits several keywords, and the resulting groups do not form disjoint clusters. In community detection within social network analysis, we typically assume each person belongs to only one community—an overly strong assumption. Similarly, in sentiment analysis, where we classify the emotional tone of a message, a sentence may naturally belong to more than one category. For example *"I love this example, but I hate how long it took me to come up with it."*

In the bipartite case, similar challenges arise. Consider movie recommender systems. On the one side, we have the users and on the other side, we have the movies. Edges indicate that a particular user likes a particular movie. One can imagine a group of users all liking the same set of action movies—forming a biclique and another group doing the same for romantic films. But full separation is unlikely since some users will enjoy both action movies and romances. Naturally, such users should belong to both clusters, that is, clusters should allow for some amount of *overlap* [19]. Such overlapping (bi)clusters are modeled in the context of graph problems by *vertex splitting*: a vertex v with neighborhood $N(v)$ is split into (replaced by) two vertices v_1 and v_2 such that $N(v_1) \cup N(v_2) = N(v)$, where each new vertex represents the involvement of v in one (bi)cluster (or is split further). An example of such a vertex split is given in Fig. 1. Allowing this operation in addition to the insertion and removal of edges leads to the following problem, first studied by Abu-Khzam et al. [3].

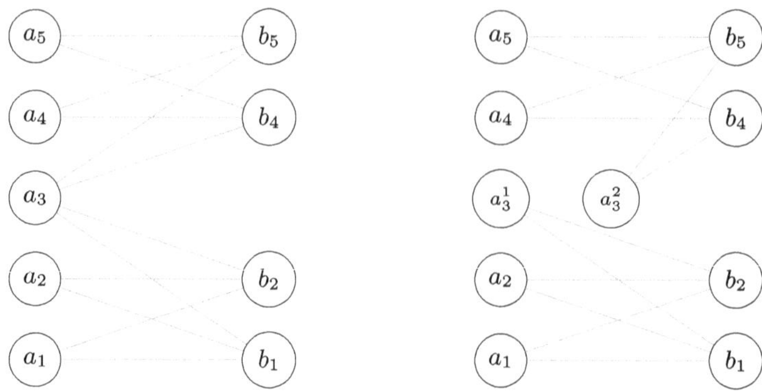

Fig. 1. An example instance on the left where two operations are needed when only edge insertions and edge removals are allowed. On the right side a solution resulting from a single vertex split is shown.

BICLUSTER EDITING WITH VERTEX SPLITTING

Input: A bipartite graph $G = (V_1 \uplus V_2, E)$ and an integer $k \geq 0$.
Task: Decide whether there exists a bicluster graph which is the result of at most k edit operations on G, where each operation adds an edge, removes an edge, or splits a vertex.

We mention that there are two natural variants of the BICLUSTER EDITING problem and it makes sense to study both with the additional vertex splitting operation. In one variant, the input is an arbitrary graph, and the goal is to transform it into a bicluster graph without constraints on which vertices can belong to either side of the resulting bipartite graph. This model fits applications such as ownership or trading networks, where entities may simultaneously act on both sides (e.g., as buyers and sellers). In the second variant, the input graph is bipartite with a fixed bipartition. This assumption naturally occurs in the aforementioned application in recommender systems, but also in the a context of phylogenetics and other biological datasets. We focus on the second variant in this paper as we believe that it has the potential to improve recommender systems, especially in terms of explainability. While many current systems rely on statistical approaches or machine learning methods and operate as black boxes, clustering-based approaches can reveal groupings transparently, providing both users and providers with insight into why certain recommendations are made.

We also briefly mention the following problem, introduced by Abu-Khzam, Isenmann, and Merchad [3], as both our kernel and FPT algorithms work for this:

BICLUSTER EDITING WITH ONE-SIDED VERTEX SPLITTING

Input: A bipartite graph $G = (V_1 \uplus V_2, E)$ and an integer $k \geq 0$.

Task: Decide whether there exists a bicluster graph which is the result of at most k edit operations on G, where each operation adds an edge, removes an edge, or splits a vertex in V_1.

Related Work. We start with related work on vertex splitting. One of the first uses of vertex splitting as a graph modification operation appeared in the context of graph drawing as a way to make graphs planar. Eades and de Mendonça Neto [20] introduced this problem under the name PLANAR VERTEX SPLITTING. Faria et al. [21] showed that it is NP-complete, even on cubic graphs. Ahmed et al. [5] showed that vertex splitting to bipartite graphs is FPT and used this result to develop graph drawing algorithms [4]. Vertex splitting was also studied in the context of graph classes of bounded treewidth: Baumann et al. [10] showed that for MSO_2 testable graph classes Π with bounded treewidth, vertex splitting to Π is FPT.

Abu-Khzam et al. [1] used vertex splitting as an added operation in CLUSTER EDITING to model overlapping clustering. Here, the idea is that a vertex that belongs to two different clusters can be split into two, one copy for each of the two clusters it belongs to. The resulting problem as well as different variants of CLUSTER EDITING with this additional vertex splitting operation have since been studied [2, 11, 22].

We continue with related work on BICLUSTER EDITING. The problem is NP-complete [6], even on subcubic graphs [18], and has been extensively studied due to its applications in computational biology [30]. The problem has seen steady progress in both kernelization and fixed-parameter tractability: Protti et al. [33] gave an $O(k^2)$ vertex kernel and a branching algorithm running in $O(4^k + n + m)$ time. This was improved by Guo et al. [24] to $O(3.24^k + n + m)$, along with a smaller kernel, which was later found to contain a minor flaw [28]. Subsequent works further advanced the understanding and efficiency of the problem [18, 34, 37]. These efforts culminated in the current state-of-the-art results by Tsur [36], who presented an FPT algorithm running in $O^*(2.22^k)$ time, and Lafond [29] who gave a refined kernel containing at most $4.5k$ vertices. For a comprehensive overview of early developments and biological motivations, we refer to the survey by Madeira and Oliveira [30].

Abu-Khzam, Isenmann, and Merchad [3] recently introduced the problems BICLUSTER EDITING WITH VERTEX SPLITTING and its one-sided variant BICLUSTER EDITING WITH ONE-SIDED VERTEX SPLITTING. They proved both problems to be NP-complete, APX-hard, and intractable under ETH even on bipartite planar graphs of maximum degree three. On the positive side, they showed that BICLUSTER EDITING WITH ONE-SIDED VERTEX SPLITTING admits a kernel with $O(k^5)$ vertices that can be computed in $O(n^2)$ time and developed an algorithm running in $O(5^{3k} k^{2k} n^2)$ for it.

Our Contribution. We show that both problems admit kernels with $O(k^2)$ vertices that can be computed in $O(n + m)$ time. We also show that the algorithm by Abu-Khzam, Isenmann, and Merchad for BICLUSTER EDITING WITH ONE-SIDED VERTEX SPLITTING contains a flaw and design an algorithm running in $O(k^{11k} + n + m)$ time for both BICLUSTER EDITING WITH VERTEX SPLITTING and BICLUSTER EDITING WITH ONE-SIDED VERTEX SPLITTING. These two results solve two open problems raised by Abu-Khzam, Isenmann, and Merchad [3].

2   Preliminaries

For a positive integer k, we use $[k]$ to denote the set $\{1, 2, \ldots, k\}$ of all positive integers up to k. All logarithms in this paper use 2 as their base. We use standard graph-theoretic notation and refer the reader to the textbook by Diestel [17] for standard definitions. For $G = (V, E)$, we will denote by n_G and m_G the number of vertices and edges, respectively. All graphs in this work are simple, unweighted, and undirected. We denote the open neighborhood of a vertex v by $N_G(v)$. When G is clear from context, we will drop the subscripts. For an introduction to parameterized complexity, fixed-parameter tractability, and kernelization, we refer the reader to the textbooks by Flum and Grohe [23], Niedermeier [31], and Cygan et al. [16].

A graph $G = (V_1 \uplus V_2, E)$ is bipartite if each edge in E has one endpoint in V_1 and one endpoint in V_2—note that any one of those can be empty. A biclique is a complete bipartite graph, that is, each vertex in V_1 is adjacent to each vertex in V_2. If $G = (V_1 \uplus V_2, E)$ is a biclique, then we will also refer to $G = (V_1, V_2)$ as a biclique for notational convenience. A *bicluster graph* is a graph in which each connected component is a biclique. The class of bicluster graphs is exactly the graphs that have neither C_3 nor P_4 as an induced subgraph. An (inclusive) vertex split replaces a vertex v by two vertices v_1, v_2 such that $N(v_1) \cup N(v_2) = N(v)$. An example of the splitting operation is given in Fig. 2 and we call the two new vertices v_1 and v_2 *copies* of the original vertex v (and if v_1 or v_2 are further split in the future, then the resulting vertices are also called copies of v). For the sake of notational convenience, we also say that v is a copy of itself.

An edit sequence is a sequence of operations, where each operation is (i) the addition of an edge, (ii) the removal of an edge, or (iii) the inclusive split of a vertex. We denote the graph resulting from applying an edit sequence σ to a graph G by $G_{|\sigma}$.

A *critical independent set* is a maximal set I of vertices such that $N(u) = N(v)$ for all $u, v \in I$. Equivalently, two vertices u and v belong to the same critical independent set if and only if they are false twins. It is known that critical independent sets form a partition of the vertex set of a graph and that all critical independent sets can be computed in linear time [25]. The *critical independent set quotient graph* $\mathcal{I}$ of G contains a node for each critical independent set in G and two nodes are adjacent if and only if the two respective critical independent sets I_1 and I_2 are adjacent, that is, there is an edge between each vertex in I_1 and

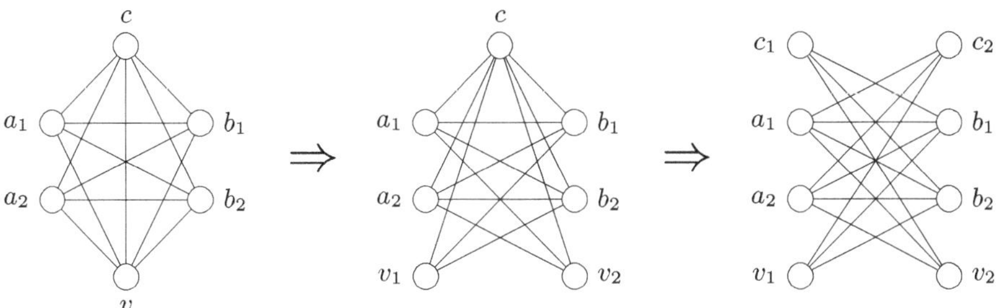

Fig. 2. An illustration of two (inclusive) vertex splits (in a non-bipartite graph). In the first split, the vertex v is replaced by v_1 and v_2, with the vertices a_1, a_2, b_1 and b_2 being adjacent to exactly one of the two vertices and c being adjacent to both. In the second, c is split into c_1 and c_2.

each vertex in I_2. Note that by the definition of critical independent sets, this is equivalent to the condition that at least one edge $\{u, v\}$ with $u \in I_1$ and $v \in I_2$ exists. To avoid confusion, we will call the vertices in $\mathcal{I}$ *nodes* and in G *vertices*.

3   Structural Results

In this section, we prove two structural results about optimal solutions to BICLUSTER EDITING WITH VERTEX SPLITTING and BICLUSTER EDITING WITH ONE-SIDED VERTEX SPLITTING that are used in later proofs. The first state that the bipartitions of the final bicluster graph follow the bipartition of the input graph, that is, no edges between copies of vertices in V_1 and V_2 are added.

Lemma 1. *Let $(G = (V_1 \uplus V_2, E), k)$ be a yes-instance of* BICLUSTER EDITING *WITH* VERTEX SPLITTING *or* BICLUSTER EDITING *WITH* ONE-SIDED VERTEX SPLITTING. *Then, there exists a solution σ of length at most k such that for each biclique $H_j = (A_j, B_j)$ in $G_{|\sigma}$ it holds that $A'_j \subseteq V_1$ and $B'_j \subseteq V_2$, where A'_j and B'_j consist of all vertices $v \in V_1 \uplus V_2$ such that A_j and B_j contain a single copy of v, respectively.*

Proof. Let σ be a solution of size $k' \leq k$. We will construct a solution σ' of size $k'' \leq k'$ that satisfies the listed requirements. If σ already satisfies the requirements, then we are done. So assume that it does not and let $H_j = (A_j, B_j)$ be a biclique in $G_{|\sigma}$ that does not satisfy the requirement. Then, we assume without loss of generality that $A'_j \cap V_1 \neq \emptyset$, $B'_j \cap V_2 \neq \emptyset$, and $B'_j \cap V_1 \neq \emptyset$ (the case where $A'_j \cap V_2 \neq \emptyset$ is symmetric even for BICLUSTER EDITING WITH ONE-SIDED VERTEX SPLITTING). We partition B_j into two sets B_1 and B_2 such that B_1 contains all copies of vertices in V_1 and B_2 contains all copies of vertices in V_2. We also partition A_j into A_1 and A_2, where A_2 is potentially empty. Note that all edges between vertices in A_1 and B_1 and all edges between vertices in A_2 and B_2 are added in σ since V_1 and V_2 induce independent sets in G. Consider the sequence σ' that is equal to σ except for the fact that it does not create these edges. Note that σ' is strictly shorter than σ so it only remains to show that σ' is a valid solution.

To this end, consider the effect on $H_j = (A_j, B_j)$. All edges between vertices in A_1 and in B_2 are still there. So are all edges between vertices in A_2 and in B_1. Since H_j is a biclique, there are no edges between vertices in A_1 and in A_2 or between vertices in B_1 and in B_2. Moreover, by construction, there are no edges between vertices in A_1 and in B_1 or between vertices in A_2 and in B_2 in $G_{|\sigma'}$. Thus, the vertices in H_j now form the two bicliques $H_j^1 = (A_1, B_2)$ and $H_j^2 = (B_1, A_2)$. Since H_j was chosen arbitrarily, the same holds for all bicliques that did not initially satisfy the requirement. Moreover, since $A_1', B_1' \subseteq V_1$ and $A_2', B_2' \subseteq V_2$ by construction, all new bicliques also satisfy the requirement of the lemma. This concludes the proof. $\square$

The following lemma is a generalization of a lemma due to Guo et al. [24] in the context of vertex splitting. It states that two copies of the same vertex are always contained in two different bicliques in (the graph obtained by) any optimal solution.

Lemma 2. *Let $(G = (V_1 \uplus V_2, E), k)$ be a yes-instance of* BICLUSTER EDITING WITH VERTEX SPLITTING *or* BICLUSTER EDITING WITH ONE-SIDED VERTEX SPLITTING. *Then, there exists a solution σ of length at most k such that for each critical independent set $I_i \subseteq V_1$ in G and each biclique $H_j = (A_j, B_j)$ in $G_{|\sigma}$ it holds that B_j does not contain a copy of any vertex in I_i and A_j contains exactly one copy of each vertex in I_i or does not contain any copy of a vertex in I_i. The same applies to critical independent sets $I_p \subseteq V_2$ with the roles of A_j and B_j swapped.*

Proof. Let $\hat{\sigma}$ be a solution for (G, k) that satisfies Lemma 1. For each critical independent set I_i, we select a representative vertex $r_i \in I_i$ by picking any vertex in I_i with the fewest number of appearances in $\hat{\sigma}$, that is, a vertex which minimizes the sum of incident added edges, incident removed edges, and vertex splits in $\hat{\sigma}$. Note that due to the assumption that $\hat{\sigma}$ satisfies Lemma 1, it holds for any critical independent set $I_i \subseteq V_1$, any critical independent set $I_{i'}$, and each biclique $H = (A_j, B_j)$ in $G_{|\hat{\sigma}}$ that A_j does not contain a copy of $I_{i'}$ and B_j does not contain a copy of a vertex in I_i. If it also holds for each critical independent set $I_i \subseteq V_1$, each critical independent set $I_{i'} \subseteq V_2$, and each biclique $H = (A_j, B_j)$ in $G_{|\hat{\sigma}}$ that (i) A_j does not contain any copy of a vertex in I_i or A_j contains exactly one copy of each vertex in I_i and (ii) B_j does not contain any copy of a vertex in $I_{i'}$ or B_j contains exactly one copy of each vertex in $I_{i'}$, then $\hat{\sigma}$ satisfies all requirements of the lemma statement, and we are done.

Otherwise, there exists a biclique $H_j = (A_j, B_j)$ in $G_{|\hat{\sigma}}$ such that (i) A_j contains two copies of some vertex $v \in I_i$ or (ii) there exists a critical independent set I_i (or $I_{i'}$) and two vertices u, v in it such that A_j (B_j) contains a copy of u but no copy of v. In case (i), we remove all operations from $\hat{\sigma}$ involving one of the two copies. Note that this results in a solution of strictly shorter length as we do not need to create this additional copy and bicliques are closed under vertex deletion. In case (ii), we assume without loss of generality that A_j contains a copy of u but no copy of v as the case where B_j contains a copy of u but no copy of v is analogous. Then, there also exists such a pair of vertices where one of u and v is r_i: if A_j contains a copy of r_i, then we can take the pair (r_i, v) and if A_j does not contain a copy of r_i, then we can take the pair (u, r_i). Let $w \in \{u, v\}$ be the other vertex.

We find a new optimal solution by removing all operations involving w and copying all operations including r_i and replacing r_i by w in the copy. For the sake of notational convenience, we say that if some operation e_p involves the jth copy of r_i, then the operation e_{p+1} is a copy of e_p and it uses the jth copy of w. Moreover, when splitting another vertex (not r_i or w), then we treat each copy of w the same as the corresponding copy of r_i. Note that since $\hat{\sigma}$ satisfies Lemma 1, it does not create any edge between a copy of w and a copy of r_i. Our modifications to $\hat{\sigma}$ also do not add such edges so the resulting sequence σ' still satisfies Lemma 1. Since w is involved in at least as many operations as r_i (recall that r_i was picked as having the fewest number of appearances in $\hat{\sigma}$), it holds that σ' will be at most as long as $\hat{\sigma}$.

We next show that σ' is also a solution. Note that the graph $G_{|\sigma'}$ is the graph obtained from $G_{|\hat{\sigma}}$ by deleting all copies of w, and then adding a false twin for each copy of r_i. Since the class of bicluster graphs is closed under vertex deletion and under adding false twins, it follows that $G_{|\sigma'}$ is also a bicluster graph. Moreover, the number of vertices that behave differently than their representative is one smaller in σ' compared to $\hat{\sigma}$. Hence, repeating the above procedure at most n times results in an optimal solution σ as stated in the lemma. $\qquad\square$

In the following section, we will use Lemma 2 to develop a polynomial kernel, as well as a $2^{O(k \log k)}(n + m)$-time algorithm for both problem variants, and importantly, the kernelization algorithms run in linear time in the input.

4   A Polynomial Kernel

In this section, we prove that BICLUSTER EDITING WITH VERTEX SPLITTING and BICLUSTER EDITING WITH ONE-SIDED VERTEX SPLITTING admit kernels with $O(k^2)$ vertices which can be computed in linear time. This improves upon the kernel with $O(k^5)$ vertices computable in $O(n^2)$ time by Abu-Khzam, Isenmann, and Merchad [3]. We note that Lafond recently proved a $4.5k$-vertex kernel for BICLUSTER EDITING [29]. However, the kernel does not generalize to BICLUSTER EDITING WITH VERTEX SPLITTING or BICLUSTER EDITING WITH ONE-SIDED VERTEX SPLITTING. The kernel consists of three rules and then an argument that if the resulting graph contains more than $4.5k$ vertices, then it is a no-instance. The graph in Fig. 3a does not allow for the application of any of the three rules of Lafond for $k = 2$ but it is a yes-instance since the two vertices u, v can each be split once to obtain a bicluster graph.

Moreover, a common approach for linear kernels for BICLUSTER EDITING is to argue that if the second neighborhood[1] of a critical independent set I_i is

---

[1] The second neighborhood of v, $N^2(v)$ is the set of vertices at distance 2 from v.

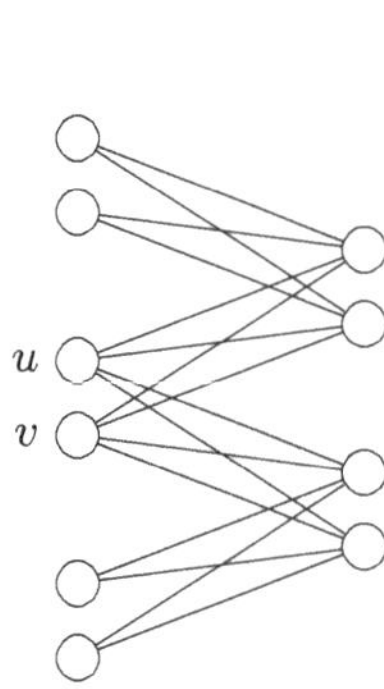

(a) A graph in which none of the reduction rules used by Lafond [29] apply which is a yes-instance for $k = 2$ but contains more than 9 vertices.

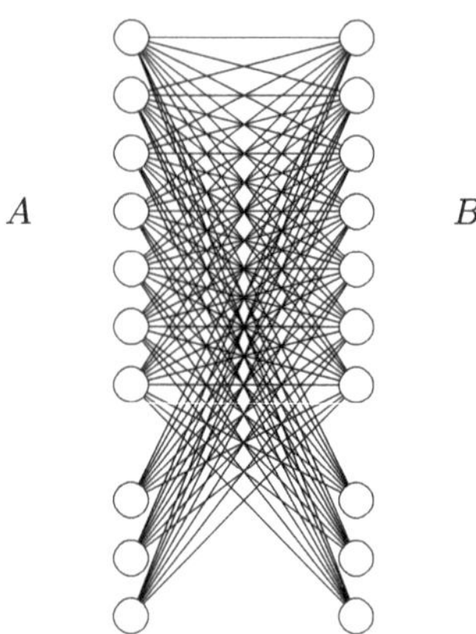

(b) A graph in which removing a vertex from either set A or B reduces the size of an optimal solution. An optimal solution splits each vertex in A once and has therefore size 7. Removing one vertex from A or B drops the solution size to 6 as only the vertices in the smaller set need to be split.

Fig. 3. Examples showing that previous kernelization techniques for BICLUSTER EDITING do not extend to our setting.

smaller than $|I_i|$, then one can safely remove a vertex from I_i. The graph in Fig. 3b shows that such an approach does not work for BICLUSTER EDITING WITH VERTEX SPLITTING.

However, a quadratic kernel for BICLUSTER EDITING WITH VERTEX SPLITTING (or BICLUSTER EDITING WITH ONE-SIDED VERTEX SPLITTING) can be achieved relatively easily.

Theorem 1. *BICLUSTER EDITING WITH VERTEX SPLITTING and BICLUSTER EDITING WITH ONE-SIDED VERTEX SPLITTING each admit a kernel with at most $6k(k + 1)$ vertices and at most $6k$ critical independent sets, which can be computed in linear time.*

Proof. We first compute all critical independent sets and the critical independent set quotient graph $\mathcal{I}$ of G. Next, if a connected component in $\mathcal{I}$ is an isolated vertex or two vertices connected by an edge, then we remove all vertices in the corresponding critical independent sets.

We formalize this as the following reduction rule:

Reduction rule 1 *Remove all vertices contained in critical independent sets corresponding to isolated nodes or isolated edges in $\mathcal{I}$.*

Note that the former corresponds to a set of isolated vertices in G and the latter corresponds to a connected component that is a complete bipartite graph. Since both are bicliques and not connected to the rest of the graph, we can safely remove them.

Next, we reduce the size of each critical independent set.

Reduction rule 2 *If $|I_i| > k + 1$ for some critical independent set I_i, then remove an arbitrary vertex in I_i from G.*

We next prove that the reduction rule is safe. To this end, let G' be the graph obtained after removing a vertex u as described in the reduction rule. First assume that (G, k) is a yes-instance (of either problem). Since removing a vertex can never increase the distance from being a bicluster graph, (G', k) is clearly also a yes-instance. Now assume that (G', k') is a yes-instance. Combining Lemma 2 with the observation that $|I_i|$ remains strictly greater than k after the removal of u, we can assume without loss of generality that there is an optimal solution σ, which does not add or remove any edges incident to vertices in I_i and also does not split any vertex in I_i. Hence, adding the vertex u back does not change the fact that σ still results in a bicluster graph.

We next analyze the running time. Computing all critical independent sets and the critical independent set quotient graph $\mathcal{I}$ of G takes linear time [25]. Finding isolated vertices and edges in $\mathcal{I}$ also takes linear time. Iterating over all critical independent sets and removing vertices also takes linear time. Finally, checking whether $\mathcal{I}$ contains more than $6k$ nodes after performing the above reduction rules exhaustively and returning a trivial no-instance if that is the case takes constant time. Since each step takes linear time, the entire kernel can be computed in linear time.

It remains to show that if a graph G does not allow for the application of either of the two above reduction rules and contains more than $6k$ critical independent sets, then (G, k) is a no-instance of Bicluster Editing with Vertex Splitting and of Bicluster Editing with One-Sided Vertex Splitting. We show that if a graph does not allow for an application of either Reduction rules 1 or 2 and contains more than $6k$ critical independent set, then we are dealing with a no-instance. Note that since Reduction rule 2 does not apply, this also shows that the number of vertices is upper bounded by $6k(k+1)$. Afterwards, we analyze the running time of computing the kernel.

Assume that there is a solution sequence σ of length at most k resulting in a bicluster graph $H = (V', E')$. We assume without loss of generality that σ satisfies Lemmas 1 and 2. We will partition the vertices in V' into two sets. The set X contains all vertices that are *touched* by σ, that is, vertices that are the result of a vertex-split operation or vertices that are incident to an edge that is added or removed by σ. Note that $|X| \leq 2k$ as each edge is incident to two vertices and each vertex split adds two new vertices to the graph. The set Y contains all other vertices in V'. Since vertices in Y are not split, they are also part of the original graph. Consider a connected component (a biclique) $H_j = (A_j, B_j)$ in H. Since Reduction rule 1 does not apply, at least one vertex in H_j is contained in X. Moreover, each side of H_j contains vertices in Y from at most one critical independent set as shown next. Assume towards a contradiction that A_j contains vertices $a_1, a_2 \in Y$ belonging to different critical independent sets. Since $a_1, a_2 \in Y$, no edges incident to these two vertices are added or removed during σ. Hence, B_j consists of exactly one copy of vertices in $N(a_1)$ and the same for a_2. However, these two neighborhoods are different as a_1 and a_2 belong to different critical independent sets (and are both contained in V_1). Thus, A_j only contains vertices in Y from one critical independent set and the same is true for B_j by the same argument. Hence, the number of critical independent sets with vertices in Y is at most $2|X| \leq 4k$. Combined with the fact that $|X| \leq 2k$ and therefore at most $2k$ critical independent sets (in the original graph) contain vertices with copies in X, this shows that the number of critical independent sets in the graph before applying σ is at most $6k$. This concludes the proof. $\square$

We mention that the question whether BICLUSTER EDITING WITH VERTEX SPLITTING and/or BICLUSTER EDITING WITH ONE-SIDED VERTEX SPLITTING admit kernels with $O(k)$ vertices is an interesting open problem for the future.

5   An FPT Algorithm

Theorem 1 implies that BICLUSTER EDITING WITH VERTEX SPLITTING is fixed-parameter tractable since we can compute the kernel in linear time and then solve the kernel using brute force. For BICLUSTER EDITING WITH ONE-SIDED VERTEX SPLITTING, Abu-Khzam et al. [3] developed an algorithm running in time $O(5^{3k} k^{2k} n^2)$. Unfortunately, we show that the algorithm contains a flaw and show a counter example where the algorithm does not compute an optimal solution. Afterwards, we will present an algorithm that solves BICLUSTER EDITING WITH VERTEX SPLITTING and BICLUSTER EDITING WITH ONE-SIDED VERTEX SPLITTING in $O(k^{11k} + n + m)$ time. This algorithm is an adaptation of an algorithm with similar running time due to Abu-Khzam et al. [1].

The counter example for the algorithm by Abu-Khzam et al. [3] is given in the full version [12]. In short, the algorithm iteratively finds an induced $P_4 = (a, b, c, d)$, where $a, c \in V_1$ and c is not marked. Then, it branches into five different cases: deleting any of the three edges, adding the edge $\{a, d\}$, or marking c. Marking corresponds to splitting the vertex c but the authors delay the decision on how to split c until all edge modifications are guessed and then present an algorithm that optimally splits all marked vertices. The problem is that marking vertices and only searching for P_4s in which c is unmarked can make it impossible for the algorithm to delete an edge that has to be removed in any optimal solution. We develop a different algorithm with a slightly worse running time in the parameter but better dependency on the input size for both BICLUSTER EDITING WITH VERTEX SPLITTING and BICLUSTER EDITING WITH ONE-SIDED VERTEX SPLITTING.

Theorem 2. *BICLUSTER EDITING WITH VERTEX SPLITTING and BICLUSTER EDITING WITH ONE-SIDED VERTEX SPLITTING are solvable in* $O(k^{11k} + n + m)$ *time.*

Proof. First, we compute the kernel G' from Theorem 1, all critical independent sets, and the critical independent set quotient graph $\mathcal{I}$ of the kernel in linear time [25]. Note that $\mathcal{I}$ contains at most $6k$ vertices. Let σ be a solution that satisfies Lemmas 1 and 2. Let $\mathcal{X} = \{H_1, H_2, \ldots, H_\ell\}$, where each $H_j = (A_j, B_j)$ is a biclique in $G'_{|\sigma}$. Note also that $\mathcal{X}$ contains $\ell \leq 2k$ bicliques as each operation can complete at most two bicliques of the solution (removing an edge between two bicliques or splitting a vertex contained in both bicliques) and Reduction rule 1 removed all isolated bicliques (bicliques with no further connections in the input graph). Hence, if there are more than $2k$ bicliques in the solution, then we cannot reach the solution with k operations. To streamline the following argumentation, we will cover the nodes in $\mathcal{I}$ by bicliques $H_1, H_2, \ldots, H_{\ell=2k}$ and assume that an optimal solution contains exactly $2k$ bicliques by allowing some of the bicliques to be empty. Next, let $a_1, a_2, \ldots, a_\ell, b_1, b_2, \ldots, b_\ell, c$ be $2\ell + 1$ colors. Next, we iterate over all possible colorings of the nodes in $\mathcal{I}$ such that each critical independent set $I_i \subseteq V_1$ gets a color in $\{c, a_1, a_2, \ldots, a_\ell\}$ and each critical independent set $I_{i'} \subseteq V_2$ gets a color in $\{c, b_1, b_2, \ldots, b_\ell\}$ for BICLUSTER EDITING WITH VERTEX SPLITTING and a color in $\{b_1, b_2, \ldots, b_\ell\}$ for BICLUSTER EDITING WITH ONE-SIDED VERTEX SPLITTING. Note that there are at most $6k$ critical independent sets in G and hence there are at most $(\ell + 1)^{6k} \in O((2k + 1)^{6k})$ such colorings.

The idea behind the coloring is the following. We will try to find a solution satisfying Lemmas 1 and 2. Each color a_j corresponds to the "left side" A_j of a biclique $H_j = (A_j, B_j)$ in the solution graph. The color b_j corresponds to set B_j. That is, we try to find a solution where all (vertices in critical independent sets corresponding to) nodes of the same color (except for color c) belong to the side of the same biclique in the solution. The color c indicates that the node will belong to multiple bicliques in the solution, that is, all vertices in the respective critical independent set will be split. Since each such split operation reduces k by one, we can reject any coloring in which the number of vertices in critical independent sets corresponding to nodes with color c is more than k. In particular, we can reject any coloring in which more than k nodes have color c.

Next, we guess two indices $i \in [k], j \in [\ell]$ and assume that the ith node of color c belongs[2] to A_j or B_j—or that all nodes of color c have been assigned to all bicliques they belong to. Note that each guess is over $k\ell + 1$ possibilities. Moreover, at most $k+1$ guesses do not reduce k by at least one (the last guess and the first time each index $i \in [k]$ is guessed) Hence, we can make at most $2k + 1$ guesses. Thus, the number of such guesses is at most

$$(k\ell + 1)^{2k+1} = (2k^2 + 1)^{2k+1} \in O((2k + 1)^{4k+1}).$$

It remains to compute the best solution corresponding to each possible set of guesses. To this end, we first iterate over each pair of vertices and add an edge between them if this edge does not already exist and we guess that there

---

[2] We only consider B_j in the case of BICLUSTER EDITING WITH VERTEX SPLITTING and the choice depends on whether the vertices in the critical independent set are contained in V_1 or V_2.

is a biclique H_j which contains a copy of each of the two vertices. Moreover, we remove an existing edge between them if we guessed that the two vertices do not appear in a common biclique. Finally, we perform all split operations. Therein, we iteratively split one vertex v into two vertices u_1 and u_2 where u_1 will be the vertex in some set A_j or B_j and u_2 might be split further in the future. The vertex u_1 is adjacent to all vertices that are guessed to belong to B_j or A_j. The vertex u_2 is adjacent to all vertices that u was adjacent to, except for vertices that are adjacent to u_1 and not guessed to also belong to some other biclique $H_{j'}$ which (some copy of) u_2 belongs to.

Since our algorithm performs an exhaustive search of all possible solutions satisfying Lemmas 1 and 2, it will find a solution if one exists. It only remains to analyze the running time. We first compute the kernel in $O(n + m)$ time. We then try $O((2k + 1)^{6k})$ possible colorings of $\mathcal{I}$ and for each coloring $O((2k + 1)^{4k+1})$ guesses. Afterwards, we compute the solution in $O(k^2)$ time as $n \in O(k)$ by Theorem 1. Thus, the overall running time is in

$$O((2k + 1)^{10k+1} \cdot k^2 + n + m) \subseteq O(k^{11k} + n + m).$$

$\square$

6   Conclusion

We showed that both BICLUSTER EDITING WITH VERTEX SPLITTING and BICLUSTER EDITING WITH ONE-SIDED VERTEX SPLITTING admit polynomial kernels with $O(k^2)$ vertices computable in linear time. This resolves two open problems by Abu-Khzam et al. [3]. Moreover, we show that their FPT algorithm for BICLUSTER EDITING WITH ONE-SIDED VERTEX SPLITTING contains a flaw and present a different algorithm solving both problem variants in $O(k^{11k} + n + m)$ time.

We highlight several directions for future work. One direction is approximation: is there a factor-c approximation that can be computed in polynomial time for any constant c for either problem? Another direction is to study BICLUSTER EDITING WITH VERTEX SPLITTING on non-bipartite input graphs. A key question is whether a variant of Lemma 2 still holds in that setting. Next, we leave it open whether the polynomial kernels can be improved to $O(k)$ vertices. Finally, we ask whether an algorithm running in $2^{O(k)}\mathrm{poly}(n)$ time exists for BICLUSTER EDITING WITH VERTEX SPLITTING or BICLUSTER EDITING WITH ONE-SIDED VERTEX SPLITTING.

References

1. Abu-Khzam, F.N., et al.: Cluster editing with vertex splitting. Discrete Appl. Math. **371**, 185–195 (2025)
2. Abu-Khzam, F.N., Davot, T., Isenmann, L., Thoumi, S.: On the complexity of 2-club cluster editing with vertex splitting. In: Proceedings of the 31st International Computing and Combinatorics Conference (COCOON), pp. 3–14. Springer Nature (2026)

3. Abu-Khzam, F.N., Isenmann, L., Merchad, Z.: Bicluster editing with overlaps: a vertex splitting approach. In: Proceedings of the 36th International Workshop on Combinatorial Algorithms (IWOCA), pp. 146–159. Springer Nature (2025)

4. Ahmed, A.R., et al.: Splitting vertices in 2-layer graph drawings. IEEE Comput. Graph. Appl. **43**(3), 24–35 (2023)

5. Ahmed, R., Kobourov, S., Kryven, M.: An FPT algorithm for bipartite vertex splitting. In: Proceeding of the 30th International Symposium on Graph Drawing and Network Visualization (GD), pp. 261–268. Springer (2022)

6. Amit, N.: The bicluster graph editing problem. PhD Dissertation, Tel Aviv University (2004)

7. Bansal, N., Blum, A., Chawla, S.: Correlation clustering. Mach. Learn. **56**(1–3), 89–113 (2004)

8. Barber, M.J.: Modularity and community detection in bipartite networks. Phys. Rev. E **76**(6), 066102 (2007)

9. Bauer-Mehren, A., Bundschus, M., Rautschka, M., Mayer, M.A., Sanz, F., Furlong, L.I.: Gene-disease network analysis reveals functional modules in mendelian, complex and environmental diseases. PLoS ONE **6**(6), e20284 (2011)

10. Baumann, J., Pfretzschner, M., Rutter, I.: Parameterized complexity of vertex splitting to pathwidth at most 1. Theoret. Comput. Sci. **1021**, 114928 (2024)

11. Bentert, M., Crane, A., Drange, P.G., Reidl, F., Sullivan, B.D.: Correlation clustering with vertex splitting. In: Proceedings of the 19th Scandinavian Symposium and Workshops on Algorithm Theory (SWAT), pp. 8:1–8:17. Schloss Dagstuhl - Leibniz-Zentrum für Informatik (2024)

12. Bentert, M., Drange, P.G., Haugen, E.: Overlapping biclustering (2025)

13. Cheng, Y., Church, G.M.: Biclustering of expression data. In: Proceedings of the 8th International Conference on Intelligent Systems for Molecular Biology (ISMB), pp. 93–103. AAAI Press (1999)

14. Craciun, G., Feinberg, M.: Multiple equilibria in complex chemical reaction networks: II. The species-reaction graph. SIAM J. Appl. Math. **66**(4), 1321–1338 (2006)

15. Crespelle, C., Drange, P.G., Fomin, F.V., Golovach, P.: A survey of parameterized algorithms and the complexity of edge modification. Comput. Sci. Rev. **48**, 100556 (2023)

16. Cygan, M., et al.: Parameterized Algorithms. Springer, Marcin Pilipczuk (2015)

17. Diestel, R.: Graph Theory. Springer (2005)

18. Drange, P.G., Reidl, F., Villaamil, F.S., Sikdar, S.: Fast biclustering by dual parameterization. In: Proceedings of the 10th International Symposium on Parameterized and Exact Computation (IPEC), pp. 402–413. Schloss Dagstuhl — Leibniz-Zentrum für Informatik (2015)

19. Du, N., Wang, B., Wu, B., Wang, Y.: Overlapping community detection in bipartite networks. In: Proceedings of the 2008 IEEE/WIC/ACM International Conference on Web Intelligence (WI), pp. 176–179. IEEE Computer Society (2008)

20. Eades, P., de Mendonça Neto, C.F.X.: Vertex splitting and tension-free layout. In: Proceedings of the 3rd International Symposium on Graph Drawing (GD), pp. 202–211. Springer (1995)

21. Faria, L., de Figueiredo, C.M.H., de Mendonça Neto, C.F.X.: Splitting number is NP-complete. Discrete Appl. Math. **108**(1-2), 65–83 (2001)

22. Firbas, A., Dobler, A., Holzer, F., Schafellner, J., Sorge, M., Villedieu, A., Wißmann, M.: The complexity of cluster vertex splitting and company. Discret. Appl. Math. **365**, 190–207 (2025)

23. Flum, J., Grohe, M.: Parameterized Complexity Theory. Springer (2006)
24. Guo, J., Hüffner, F., Komusiewicz, C., Zhang, Y.: Improved algorithms for bicluster editing. In: Proceedings of the 5th International Conference on Theory and Applications of Models of Computation (TAMC), pp. 445–456. Springer (2008)
25. Hsu, W.-L., Ma, T.-H.: Substitution decomposition on chordal graphs and applications. In: Proceedings of the 2nd International Symposium on Algorithms (ISA), pp. 52–60. Springer (1991)
26. Jiang, Y., Skufca, J.D., Sun, J.: BiFold visualization of bipartite datasets. EPJ Data Sci. **6**(1), 2 (2017)
27. Komusiewicz, C., Uhlmann, J.: Cluster editing with locally bounded modifications. Discret. Appl. Math. **160**(15), 2259–2270 (2012)
28. Lafond, M.: Even better fixed-parameter algorithms for bicluster editing. In: Computing and Combinatorics, pp. 578–590. Springer International Publishing (2020)
29. Lafond, M.: Improved kernelization and fixed-parameter algorithms for bicluster editing. J. Comb. Optim. **47**(5), 90 (2024)
30. Madeira, S.C., Oliveira, A.L.: Biclustering algorithms for biological data analysis: a survey. IEEE/ACM Trans. Comput. Biol. Bioinf. **1**(1), 24–45 (2004)
31. Niedermeier, R.: An Invitation to Fixed-Parameter Algorithms. Oxford University Press (2006)
32. Pontes, B., Giráldez, R., Aguilar-Ruiz, J.S.: Biclustering on expression data: a review. J. Biomed. Inform. **57**, 163–180 (2015)
33. Protti, F., da Silva, M.D., Szwarcfiter, J.L.: Applying modular decomposition to parameterized cluster editing problems. Theory Comput. Syst. **44**(1), 91–104 (2009)
34. Sun, P., Guo, J., Baumbach, J.: Efficient large-scale bicluster editing. In: Proceedings of the German Conference on Bioinformatics 2014, pp. 54–60. GI (2014)
35. Tanay, A., Sharan, R., Shamir, R.: Biclustering algorithms: a survey. Handbook Comput. Mol. Biol. **9**(1–20), 122–124 (2005)
36. Tsur, D.: Faster parameterized algorithms for bicluster editing and flip consensus tree. Theoret. Comput. Sci. **953**, 113796 (2023)
37. Xiao, M., Kou, S.: A simple and improved parameterized algorithm for bicluster editing. Inf. Process. Lett. **174**, 106193 (2022)

Finding a HIST: Chordality, Structural Parameters, and Diameter

Tesshu Hanaka[1] , Hironori Kiya[3(✉)] , and Hirotaka Ono[2]

[1] Kyushu University, Fukuoka 819-0395, Japan
hanaka@inf.kyushu-u.ac.jp
[2] Nagoya University, Nagoya 464-8601, Japan
ono@nagoya-u.ac.jp
[3] Osaka Metropolitan University, Osaka 599-8531, Japan
kiya@omu.ac.jp

Abstract. A *homeomorphically irreducible spanning tree* (HIST) is a spanning tree with no degree-2 vertices, serving as a structurally minimal backbone of a graph. While the existence of HISTs has been widely studied from a structural perspective, the algorithmic complexity of finding them remains less understood. In this paper, we provide a comprehensive investigation of the HIST problem from both structural and algorithmic viewpoints. We present a simple characterization that precisely describes which chordal graphs of diameter at most 3 admit a HIST, leading to a polynomial-time decision procedure for this class. In contrast, we show that the problem is NP-complete for strongly chordal graphs of diameter 4. From the perspective of parameterized complexity, we establish that the HIST problem is W[1]-hard when parameterized by clique-width, indicating that the problem is unlikely to be efficiently solvable in general dense graphs. On the other hand, we present fixed-parameter tractable (FPT) algorithms when parameterized by treewidth, modular-width, or cluster vertex deletion number. Specifically, we develop an $O^*(4^k)$-time algorithm parameterized by modular-width k, and an FPT algorithm parameterized by the cluster vertex deletion number based on kernelization techniques that bound clique sizes while preserving the existence of a HIST. These results together provide a clearer understanding of the structural and computational boundaries of the HIST problem.

Keywords: HIST · chordal graph · diameter · modular width

1 Introduction

Let G be a connected graph. A *homeomorphically irreducible spanning tree* (HIST) of G is a spanning tree of G that contains no vertex with degree two [14]. Because a HIST has no degree-2 vertex, it cannot be obtained by subdividing

Supported by KAKENHI JP20H05967, JP22H00513, JP23H04388, JP24K02898, and JP25K03077, and JST CRONOS Grant Number JPMJCS24K2.

© The Author(s), under exclusive license to Springer Nature Switzerland AG 2026
J. Kozik and A. Wolff (Eds.): SOFSEM 2026, LNCS 16448, pp. 303–316, 2026.
https://doi.org/10.1007/978-3-032-17801-5_23

the edges of a smaller tree; hence, it is called "homeomorphically irreducible." This exclusion imposes a particular structural constraint compared to general spanning trees, often resulting in a more branching or less linear configuration. In trees, degree-2 vertices typically function as internal points along paths. By excluding them, the definition enforces that all internal vertices must have a degree of at least 3. Such trees or trees with a few degree-2 vertices have applications in the design of efficient communication networks with minimal redundancy [5,21]. Beyond their practical relevance, HISTs are also of significant theoretical interest, as they play a key role in major conjectures in graph theory, such as the 3-decomposition conjecture [3,15]. For these reasons, HISTs have been extensively studied from both applied and theoretical perspectives.

HISTs have been studied extensively in graph theory, particularly with respect to their existence. For example, the paper by Albertson et al. [1] shows that if a graph with n vertices has minimum degree at least $4\sqrt{2n}$, then it admits a HIST, and that this bound is nearly tight. Very recently, this bound was improved to $4\sqrt{n}$ by Furuya et al. [10]. The paper by Albertson et al. [1] also demonstrates that some graphs with diameter two do not admit a HIST, but proves that every such graph has a spanning tree with at most three degree-2 vertices. Later, Shan and Tsuchiya [19] completely characterized diameter-2 graphs that contain a HIST. Albertson et al. [1] also resolved a conjecture by Hill [14] by confirming that every triangulation of the plane has a HIST. They extended this to near-triangulations, showing that the structure of planar embeddings can guarantee the existence of a HIST under certain face and edge conditions. Importantly, in 2013, Chen and Shan [4] resolved an open problem that had remained unsolved for more than 20 years [1,17], proving that any graph in which every edge is contained in at least two triangles contains a HIST.

However, from a practical standpoint, the existence of a HIST is not the only aspect of interest. In many applications, constructing a HIST, rather than merely proving its existence, is equally or even more critical. Yet the algorithmic task of constructing a HIST has received little attention in the literature. Deciding whether a given graph admits a HIST is NP-complete [1], even for planar graphs with maximum degree at most 3 [9]. However, as far as the authors know, few additional results are known beyond these.

In this paper, we address algorithms and computational complexity for finding HISTs from two complementary perspectives. The first perspective investigates the relationship between the existence of HISTs and structural properties of graphs, such as graph diameter and triangulated structures, which have been central themes in previous studies. Regarding graph diameter, small diameters, such as 3 or 4, are of particular interest based on the series of prior works. On the side of triangulated structures, we focus on *chordal graphs*, also known as *triangulated graphs*. Chordal graphs, characterized by the presence of chords in every cycle, form a graph class where many problems that are otherwise computationally hard can be solved efficiently. In particular, their strong relationship with tree decompositions plays a key role in algorithm design. Furthermore, the class of chordal graphs includes several well-known subclasses, such as split graphs, block graphs, and interval graphs, which have been widely studied. Based on these, we examine the existence of HISTs and the computational complexity of finding them in chordal graphs with small diameters.

The second perspective focuses on algorithmic aspects, particularly on the design of efficient algorithms for constructing HISTs in general graphs. To circumvent the NP-hardness of the problem, two primary approaches are commonly considered: the design of approximation algorithms or parameterized algorithms. Since the HIST problem is also a decision problem, it is natural to focus on the latter, namely, the design of parameterized algorithms. In this study, we investigate whether the HIST problem is fixed-parameter tractable (FPT) with respect to structural graph parameters such as treewidth and modular-width. The detailed results of this study are presented in Sect. 1.1.

1.1 Our Contribution

Chordal Graphs with a Small Diameter. As seen in the previous section, prior studies on the existence of HISTs have primarily focused on the role of triangle structures and graph diameter. Shan and Tsuchiya [19] provide a complete characterization of graphs with diameter 2 that admit a HIST. Based on this result, we first confirm that such graphs can be recognized in polynomial time.

We then investigate split graphs and block-split graphs, which are representative examples of chordal graphs with diameter 3. In this paper, we first characterize block-split graphs that admit a HIST (Theorem 2) and then extend this characterization to split graphs (Theorem 4) and chordal graphs with diameter 3 (Theorem 5) that admits a HIST. Using these characterizations, we can find a HIST of a given chordal graph with diameter 3 in polynomial time. On the other hand, we prove that deciding whether a strongly chordal graph with diameter 4 admits a HIST is NP-complete, which provides a sharp boundary. The results can be summarized as follows:

- For graphs with diameter at most 2, the existence of a HIST can be decided in polynomial time [Theorem 1]
- In contrast, deciding whether a graph with diameter 4 admits a HIST is NP-complete, even when restricted to strongly chordal graphs [Theorem 6].
- For chordal graphs with diameter 3, we give a simple characterization of graphs that admit a HIST, which enables us to determine whether it has a HIST in polynomial time [Theorem 5]. However, the complexity of deciding whether a given graph with diameter 3 contains a HIST remains open.

Algorithms for Structural Graph Parameters. As a basic result, we first present a $4^n n^{O(1)}$-time algorithm that decides whether a given graph of order n admits a HIST. We then investigate its parameterized complexity. Since the reduction in the proof of Theorem 6 implies that the problem is W[1]-hard when parameterized by clique-width, we turn our attention to more specific parameters: treewidth, modular-width, and cluster-vertex-deletion number.

These parameters provide upper bounds on clique-width and characterize more restricted graph classes (see e.g.,[20]). Moreover, they capture distinct structural properties: treewidth reflects global tree-likeness, modular-width measures recursive modularity, and cluster-vertex-deletion number quantifies proximity to clustered structures. By studying the problem for these parameters, we aim to clarify the structural conditions under which the problem becomes tractable, providing a more fine-grained perspective beyond clique-width.

Figure 1 summarizes the current landscape of parameterized complexity with respect to these structural graph parameters. These results collectively help clarify the structural boundaries that govern the computational complexity of finding a HIST.

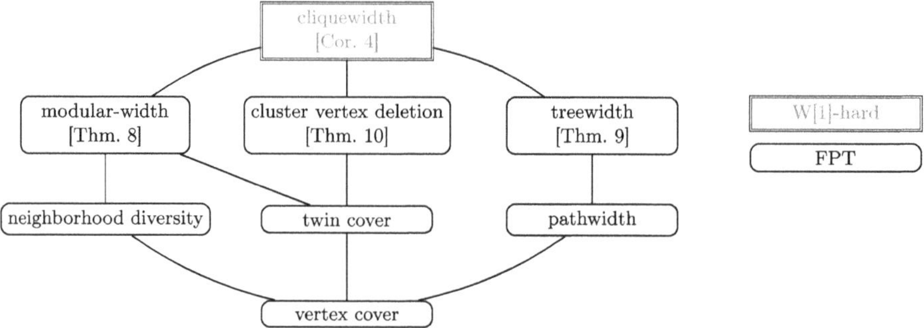

Fig. 1. Parameterized complexity of HIST with respect to structural graph parameters. The connection between two parameters means that the upper parameter p is bounded by some computable function $f(\cdot)$ of the lower parameter q, i.e., $p \leq f(q)$. The double and rounded rectangles indicate that the problem is W[1]-hard and fixed-parameter tractable, respectively.

The remainder of this paper is organized as follows. Section 2 introduces fundamental definitions and basic results. Section 3 provides structural characterizations of chordal graphs with diameter at most 3, including a simple necessary and sufficient condition for the existence of a HIST. Section 4 presents hardness results, showing that deciding whether a strongly chordal graph with diameter 4 admits a HIST is NP-complete. Section 5 develops an exact exponential-time algorithm for the general case. Section 6 focuses on parameterized algorithms, presenting fixed-parameter tractability results with respect to treewidth, modular-width, and cluster vertex deletion number. Due to space limitations, many proofs are omitted; proofs for theorems marked by an asterisk appear in the full version [12].

2 Preliminaries

2.1 Notations and Terminology

We assume basic knowledge of graph theory and discrete algorithms. All graphs considered in this paper are finite, simple, and undirected. Let $G = (V(G), E(G))$

be a simple undirected graph, where $V(G)$ and $E(G)$ respectively denote the set of vertices and the set of edges of G. We sometimes simply write V and E to refer to $V(G)$ and $E(G)$, respectively. A graph G is said to be *connected* if a path exists between every pair of vertices in V. For a graph $G = (V, E)$ and a vertex $v \in V$, the *neighborhood* of v, denoted by $N_G(v)$, is the set of vertices adjacent to v, that is, $N_G(v) = \{u \in V \mid \{u, v\} \in E\}$. The *degree* of a vertex $v \in V$ in G, denoted by $d_G(v)$, is the number of vertices in $N_G(v)$, i.e., $|N_G(v)|$. Both $N_G(v)$ and $d_G(v)$ may be simply denoted respectively by $N(v)$ and $d(v)$, if the considered graph G is clear from context. An edge $\{u, v\} \in E$ is called a *pendant edge* if one of its endpoints has degree one. A vertex of degree one is called a *pendant vertex*, the endpoint of a pendant edge. *Twins* are a pair of vertices that share the same neighbors, excluding each other. When twins are adjacent, they are called *true twins*; when they are non-adjacent, they are called *false twins*. Formally, vertices u and v are twins if $N(u) \setminus \{v\} = N(v) \setminus \{u\}$. They are *true twins* if they are adjacent, and *false twins* if they are not adjacent and satisfy $N(u) = N(v)$. A *path* in G is a sequence of distinct vertices $(v_0, v_1, \ldots, v_k)$ such that $\{v_{i-1}, v_i\} \in E$ for all $i = 1, \ldots, k$. The length of the path is defined by the number of its edges, i.e., k. The *distance* between two vertices u and v in G is the length of the shortest path between u and v. The *diameter* of G is the maximum distance between any pair of vertices in V. A *spanning tree* of a connected graph $G = (V, E)$ is a subgraph $T = (V, E')$ such that T is a tree and $E' \subseteq E$.

A graph is called *chordal* if every cycle of length at least four has a *chord*, an edge connecting two non-consecutive vertices in the cycle. A graph is called *strongly chordal* if it is chordal and every even cycle of length at least six has an odd chord, that is, a chord that divides the cycle into two paths of odd length. A graph $G = (V, E)$ is called a *split graph* if its vertex set V can be partitioned into two disjoint sets C and I, where C induces a clique and I induces an independent set. A *block-split graph* is a split graph $G = (C, I, E)$ such that each vertex in the independent set I has degree at most one. Every vertex in I is isolated or adjacent to exactly one vertex in C. Equivalently, the subgraph induced by $C \cup I$ contains only pendant edges between I and C, and all remaining edges lie within the clique C.

Given a graph $G = (V, E)$, a *Hamiltonian path* is a path that visits each vertex of V exactly once. In particular, we consider the *s-t Hamiltonian path*, which is a Hamiltonian path that starts at a designated vertex s and ends at another designated vertex t. The *Hamiltonian path problem* asks whether a given graph contains a Hamiltonian path between two specified vertices. It is well known that the $(s\text{-}t)$ Hamiltonian path problem is NP-complete in general graphs. Moreover, the hardness persists even when the input graph is restricted to bipartite, chordal, or split graphs.

A set $V' \subseteq V$ is called a *dominating set* of G if for all $u \in V \setminus V'$ there is a $v \in V'$ such that $u \in N(v)$. If we require, in addition, that the subgraph induced by V' in G be a clique, then the corresponding set is called a *dominating clique*.

The following proposition is obvious, but useful.

Proposition 1. *For a graph $G = (V, E)$, G contains a HIST if there exists a spanning subgraph G' of G such that G' has a HIST.*

In the following sections, we repeatedly use the argument, based on Proposition 1, that to prove G admits a HIST, it suffices to find an appropriate subgraph G' that has a HIST. In some cases, we explicitly mention this step, while in others, we simply present a subgraph with a HIST without further explanation.

2.2   Finding a HIST of a Graph with Diameter 2

As a basic result, we here see that finding a HIST of a graph with diameter at most two is done in polynomial time. The case of diameter one is trivial. Since a graph with diameter one is a complete graph, it is easy to see that a HIST exists if and only if its order is not three.

We now consider the case of diameter two. For graphs with diameter two and at least 10 vertices, Shan and Tsuchiya [19] presented a complete characterization of graphs having a HIST, which is explained below: Let $(p_1, \ldots, p_k)$ be a vector consisting of k positive integers, and for $(p_1, \ldots, p_k)$, we define a graph $A(p_1, \ldots, p_k)$ as follows:

$$V(A(p_1, \ldots, p_k)) = \{x, y_1, \ldots, y_k\} \cup \bigcup_{i=1}^{k} U_i, \text{ where } U_i = \{u_j^{(i)} \mid j = 1, \ldots, p_i\},$$

$$E(A(p_1, \ldots, p_k)) = \bigcup_{i=1}^{k} \{\{x, u\}, \{y_i, u\} \mid u \in U_i\} \cup \left\{ \{y_a, y_b\} \in \binom{\{y_1, \ldots, y_k\}}{2} \right\}.$$

Namely, in $A(p_1, \ldots, p_k)$, $\{y_1, \ldots, y_k\}$ forms a clique with order k, and each $G[U_i \cup \{x, y_i\}]$ forms a complete bipartite graph, and each vertices in U_i for each i are twins. Let $\mathcal{A} = \bigcup_{k \in \mathbb{Z}^+} \{A(p_1, \ldots, p_k) \mid p_i \in \mathbb{Z}^+ \text{for each } i\}$. Furthermore, we define B_n by $V(B_n) = V(A(2, n-5))$ and $E(B_n) = E(A(2, n-5)) \cup \{\{u_1^{(1)}, u_2^{(1)}\}\}$. The following lemma is known.

Lemma 1 ([19, Theorem 1]). *Let G be a graph of order $n \geq 10$ and diameter 2. Then G contains a HIST if and only if $G \notin \mathcal{A} \cup \{B_n\}$.*

By this lemma, we have the following theorem.

Theorem 1. *For a graph G with diameter at most 2, the existence of a HIST can be determined in polynomial time.*

3   Chordal Graphs with Diameter 3

In this section, we provide a characterization of chordal graphs of diameter 3 with a HIST. To this end, we first characterize block split graphs that admit a HIST and then extend this characterization to include split graphs and chordal graphs with diameter 3 that admit a HIST. Note that split graphs are one of the most well-studied subclasses of chordal graphs, and they have a diameter of at most 3. We start with block-split graphs.

3.1 Block-Split Graph Having a HIST

Let $G = (C, I, E)$ be a block-split graph. A vertex $u \in C$ is called *good* if its degree satisfies $d(u) \neq |C|$, *bad* otherwise. A good vertex has either 0 or at least 2 neighbors in I, each of which is a pendant vertex (i.e., has degree 1). A bad vertex has exactly one pendant neighbor in I. Note that every vertex in C is categorized as good or bad, and the terms "good" and "bad" are used only for vertices in C. The following characterizes block-split graphs containing a HIST.

Theorem 2. *Let $G = (C, I, E)$ be a connected block-split graph that is not a tree and not isomorphic to K_3. Then G admits a HIST if and only if it contains at least two good vertices.*

This characterization is not hard to show. Actually, Diemunsch et al. provide a characterization of P_5-free graphs with a HIST, which also leads to Theorem 2 as a corollary of [8, Theorem 1.1].

Corollary 1. *For a block split graph $G = (C, I, E)$, we can determine whether G has a HIST in linear time.*

In a graph $G = (V, E)$, a subgraph $F = (V, E')$ is called a *homeomorphically irreducible spanning forest* (HISF) if F contains neither a cycle nor a vertex with degree two. Then the necessary condition of G having a HIST in Theorem 2 is easily extended to a HISF with k connected components.

Theorem 3. *Let $G = (C, I, E)$ be a block-split graph that is not a tree and not isomorphic to K_3. If G admits a HISF with k connected components, then it contains at least $2k$ good vertices.*

3.2 Split Graph Having a HIST

Let $G = (C, I, E)$ be a split graph. By fully utilizing Theorems 2 and 3, we can prove the following theorem, which characterizes split graphs without a HIST (and thus, those with one). Again, the characterization of P_5-free graphs with a HIST by Diemunsch et al. also leads to Theorem 4 as a corollary of [8, Theorem 1.1].

Theorem 4. *Let $G = (C, I, E)$ be a connected split graph that is not a block-split graph. Then, G does not admit a HIST if and only if one of the following holds:*

1. *For all $u \in C$, $|N(u) \cap I| = 1$ and $|C| - |I| = 1$.*
2. *Let $U = \{u \in C \mid |N(u) \cap I| = 2\}$. All of the following hold:*
   (a) For all $u \in C$, $|N(u) \cap I| \in \{1, 2\}$ and $|U| \geq 2$,
   (b) For each $u \in U$, the neighborhoods $N(u) \cap I$ are distinct,
   (c) For each $u \in C \setminus U$, the vertex in $N(u)$ is a pendant vertex,
   (d) $\left| \bigcap_{u \in U} (N(u) \cap I) \right| = 1$.

Since the conditions of Theorem 4 can be checked in linear time, we have the following corollary.

Corollary 2. *Given a split graph $G = (C, I, E)$, we can determine whether G has a HIST in linear time.*

3.3   Chordal Graphs with Diameter 3 Having a HIST

We are now ready to give a characterization of chordal graphs with diameter 3 that admit a HIST. We first see a property of chordal graphs with diameter 3.

Lemma 2 ([16, Theorem 2.1]). *A chordal graph G has a dominating clique if and only if it has a diameter at most 3.*

Let G be a chordal graph with diameter 3 that is not a split graph. By Lemma 2, G has a (maximal) dominating clique, say C. Let $G = (C, \bar{C}, E)$, where $\bar{C} = V \setminus C$. Here, $G[\bar{C}]$ contains at least one edge, because otherwise it becomes a split graph.

Theorem 5. *Let $G = (C, \bar{C}, E)$ be a chordal graph with diameter 3 that is not a split graph, and let $U = \{u \in C \mid |N(u) \cap \bar{C}| = 2\}$. Then, G does not admit a HIST if and only if one of the following holds:*

1. *$\exists u^* \in C : |N(u^*) \cap \bar{C}| \geq 3$, and $\forall u \in C \setminus \{u^*\} : |N(u) \cap \bar{C}| = 1$ and the neighbor of u in $\bar{C}$ is a pendant vertex.*
2. *$\forall u \in C : |N(u) \cap \bar{C}| \in \{1, 2\}$, every vertex in $\bigcup_{u \in C \setminus U} N(u)$ is a pendant vertex, and*
   (a) $|U| = 1$, or
   (b) $|U| = 2$ and $\left|\bigcap_{u \in U}(N(u) \cap \bar{C})\right| = 1$, or
   (c) $|U| \geq 3$, for each $u \in U$, the neighborhoods $N(u) \cap \bar{C}$ are distinct, $\left|\bigcap_{u \in U}(N(u) \cap \bar{C})\right| = 1$, and $G[\bar{C}]$ contains 1 edge.

Note that finding a dominating clique of a chordal graph can be done in polynomial time, because we can enumerate all maximal cliques of a chordal graph in linear time [11]. Given a maximal dominating clique, the conditions of Theorem 5 can be easily checked, which implies the following corollary.

Corollary 3. *Given a chordal graph G with diameter at most 3, we can determine whether G has a HIST in polynomial time.*

4   Hardness Results

This section gives an NP-hardness proof, which yields several hardness results.

What we show in this section is the NP-hardness of deciding whether a given strongly chordal graph with diameter at most 4 admits a HIST. Although we omit the details, we start from the *s-t* Hamiltonian path problem, which is NP-complete for chordal bipartite graphs with two pendant vertices. We then make an instance graph a split graph after adding a small number of vertices, which plays a role in making the diameter short. The chordal bipartiteness guarantees that the resulting graph is strongly chordal. This argument leads to the following.

Lemma 3. *For a given strongly chordal split graph G, $s, t \in V(G)$, determining whether G admits an s-t Hamiltonian path is NP-complete.*

Finally, we give a reduction from the *s-t* Hamiltonian path problem to the decision problem of the existence of a HIST.

Theorem 6 (*). *For a strongly chordal graph G of diameter at most 4, determining whether G has a HIST is NP-complete.*

It is shown that *s-t* Hamiltonian Path is NP-complete even on chordal bipartite graphs [13] and planar graphs of maximum degree 3 [18]. Moreover, it is W[1]-hard when parameterized by cliquewidth [13]. Since our reduction (with a small modification) in Theorem 6 is only attaching a pendant vertex' to each vertex, the resulting graph keeps chordal bipartiteness, or planarity. Also, it increases the maximum degree by one and clique-width by at most one. Thus, we have the following theorem.

Corollary 4. *Determining whether G has a HIST is NP-complete even on chordal bipartite graphs and planar graphs of maximum degree 4. Moreover, it is W[1]-hard when parameterized by cliquewidth.*

5   Exact Exponential-Time Algorithm

We can design a dynamic programming algorithm that decides whether a HIST exists in G. The basic idea is to record, for every vertex subset $S \subseteq V$ and for every pair of subsets $S_1, S_2 \subseteq S$, a boolean entry $C[S][S_1][S_2]$, which is true exactly when the induced subgraph $G[S]$ admits a spanning tree whose set of degree-1 vertices is S_1 and whose set of degree-2 vertices is S_2. In this bookkeeping, an omitted vertex of S corresponds to a tree vertex whose degree is at least three or whose degree has not yet been fixed by the partial construction. Every spanning tree on S has at least one leaf, so we can build larger feasible configurations from smaller ones by peeling off a leaf. Concretely, to update the table for a given triple (S, S_1, S_2) we choose a candidate leaf $j \in S_1$ and one of its neighbors $k \in N(j) \cap S$, then consider the effect of removing j: the degree of k in the remaining tree is one less, so the status of k must be adjusted in the smaller subproblem (for example, a vertex that was degree two may become degree one after removal). By enumerating all such local removals and the corresponding transfers between the role-sets S_1, S_2 in the smaller set $S \setminus j$, and by filling the table in increasing order of $|S|$, the algorithm systematically tests all spanning-tree degree patterns. Since we iterate over all choices of S and all choices of $S_1, S_2 \subseteq S$, the number of table entries and the transitions to check gives a running time of the form $4^n n^{O(1)}$. A pseudocode of this algorithm is found in the full version of this paper [12].

Theorem 7. *For an n-vertex graph, the existence of a HIST can be decided in $4^n n^{O(1)}$ time.*

6   FPT Algorithms by Structural Graph Parameters

In this section, we investigate the parameterized complexity of the HIST problem with respect to several structural graph parameters.

Parameterization by Modular-Width. The modular-width of a graph is a structural parameter based on the concept of a module. A *module* in a graph is a vertex subset such that every vertex outside the module is either adjacent to all vertices in the module or to none of them. The *modular decomposition* recursively partitions a graph into modules, forming a decomposition tree. Each node in the tree is classified as either a *parallel node* (an independent set), a *series node* (a clique), or a *prime node* (neither). The *modular-width* is defined as the maximum size of a prime node in the modular decomposition tree.

Suppose that a graph G is decomposed into modules $M_1, \ldots, M_k \subseteq V$. That is, for all $u, v \in M_i$ and $w \in V \setminus M_i$, if $\{u, w\} \in E$, then $\{v, w\} \in E$, and if $\{u, w\} \notin E$, then $\{v, w\} \notin E$. We assume that there exists a module consisting of a single vertex. If no such module exists, we can arbitrarily split one of the modules with at least two vertices into a singleton and the remainder. Although this increases the number of modules by one, as will be shown later, this has no impact on the computational complexity.

Concerning a HIST, we can show the following lemma.

Lemma 4 (*). *Suppose that the graph G is decomposed into modules $M_0, \ldots,$ $M_k \subseteq V$, where $|M_0| = 1$. If G admits a HIST in which the vertex in M_0 is not a leaf, then there exists a HIST T satisfying the following constraints: For each $i = 1, \ldots, k$, one of the following holds:*

1. *All vertices in M_i are leaves in T, i.e., $\forall v \in M_i : d_T(v) = 1$.*
2. *There exists a vertex $u \in M_i$ such that $d_T(u) \geq 3$, and all other vertices in $M_i \setminus \{u\}$ are leaves in T, i.e., $\forall v \in M_i \setminus \{u\} : d_T(v) = 1$. In this case, one of the following holds:*
   (a) $d_T(u) = 3$ and the degree of u within $T[M_i]$ is 1.
   (b) $d_T(u) \geq 3$ and the degree of u within $T[M_i]$ is 0.

Based on this lemma, we further see how a HIST may form in the *quotient graph*. Suppose that G is decomposed into a set of modules $M_0, \ldots, M_k$ such that $|M_0| = 1$. If G admits a HIST T, then from Lemma 4, each module contains at most one vertex of degree at least 3 in T. We refer to such a vertex as the *representative vertex* of the module. Consider the quotient graph $H = (\mathcal{M}, \mathcal{E})$ where $\mathcal{M} = \{M_i \mid i = 0, \ldots, k\}$ represents the modules of G. Let $\mathcal{M}'$ denote the set of modules that contain a representative vertex. The subgraph defined by the connections between representative vertices forms a tree.

Note that, in this structure, the degrees of the representative vertices themselves do not reach three if we consider only the connections between representatives. The HIST is completed by connecting pendant vertices to the representative vertices. Furthermore, within each module, it suffices to assume that exactly one vertex is connected as a pendant vertex.

To formalize these, we partition the set $\mathcal{M}'$ in the quotient graph $H = (\mathcal{M}, \mathcal{E})$ (where $\mathcal{M} = \{M_i \mid i = 0, \ldots, k\}$) as follows. For a spanning tree T of the subgraph $H[\mathcal{M}']$, we divide $\mathcal{M}'$ into: $\mathcal{M}_1 := \{M \in \mathcal{M}' \mid d_T(M) = 1\}, \mathcal{M}_2 :=$

$\{M \in \mathcal{M}' \mid d_T(M) = 2\}, \mathcal{M}_3 := \{M \in \mathcal{M}' \mid d_T(M) \geq 3\}$. Let $\mathcal{M}^{\perp}$ denote the set of modules that form independent sets. The following must be satisfied:

$$\forall M_i \in \mathcal{M}_1 : \sum_{M_j : \{M_i, M_j\} \in \mathcal{E}} x_{ji} + x_{ii} \geq 2, \tag{1}$$

$$\forall M_i \in \mathcal{M}_2 : \sum_{M_j : \{M_i, M_j\} \in \mathcal{E}} x_{ji} + x_{ii} \geq 1 \tag{2}$$

$$\forall M_i \in \mathcal{M}' : \sum_{M_j : \{M_i, M_j\} \in \mathcal{E}} x_{ij} + x_{ii} = |M_i| - 1 \tag{3}$$

$$\forall M_i \in \mathcal{M} \setminus \mathcal{M}' : \sum_{M_j : \{M_i, M_j\} \in \mathcal{E}} x_{ij} = |M_i| \tag{4}$$

$$\forall M_i \in \mathcal{M}' : x_{ii} \in \begin{cases} \{0\} & \text{if } M_i \in \mathcal{M}^{\perp} \\ \{0, 1\} & \text{otherwise} \end{cases}. \tag{5}$$

Here, x_{ij} denotes the number of vertices in module M_i that are connected as pendant vertices to the representative vertex of module M_j. Also, x_{ii} represents the number of pendant vertices connected from within module M_i to its own representative vertex. From Lemma 4, we know that x_{ii} is at most one.

Note that if the module itself is not an independent set, such an assignment is always feasible. This feasibility is precisely expressed by condition (5).

This assignment problem determines whether a HIST of G can be constructed based on the spanning tree T of $H[\mathcal{M}']$. In the desired HIST, the representative vertices of $\mathcal{M}'$ must have degree 3. However, the representative vertices in the modules of $\mathcal{M}_1$ and $\mathcal{M}_2$ can only attain degrees of 1 or 2 by using the edges present in T alone. Therefore, this formulation verifies whether it is possible to assign pendant vertices to achieve the required degrees appropriately.

The right-hand sides of constraints (1) and (2) represent the required number of pendant vertices for each module, while the left-hand sides specify from which modules these pendant vertices can be assigned. Constraints (3) and (4) ensure that the total number of assigned pendant vertices equals $|M_i| - 1$ (excluding the representative vertex) or $|M_i|$ (when all vertices are assigned as pendants), respectively. If this assignment problem has a feasible solution, then by choosing the representative vertices as the internal vertices of the tree T and connecting the remaining vertices according to the assignment solution as pendant vertices, we can construct a HIST of G. It should be noted that the above discussion focuses solely on the fact that T is a spanning tree of H and on the degrees within each module. From this observation, the following result holds.

Lemma 5. *Suppose that the graph G is decomposed into a set of modules $\mathcal{M} = \{M_i \subseteq V \mid i = 0, \ldots, k\}$, where $|M_0| = 1$. If there exists a subset of modules $\mathcal{M}' \subseteq \mathcal{M}$ and a partition $(\mathcal{M}_1, \mathcal{M}_2, \mathcal{M}_3)$ of $\mathcal{M}'$ satisfying the following two conditions, then G admits a HIST:*

1. *There exists a spanning tree T of the subgraph $H[\mathcal{M}']$ of the quotient graph $H = (\mathcal{M}, \mathcal{E})$ of G such that $\mathcal{M}_1 := \{M \in \mathcal{M}' \mid d_T(M) = 1\}$, $\mathcal{M}_2 := \{M \in \mathcal{M}' \mid d_T(M) = 2\}$, and $\mathcal{M}_3 := \{M \in \mathcal{M}' \mid d_T(M) \geq 3\}$.*
2. *The assignment problem defined by (1)–(5) has a feasible solution.*

Based on Lemma 5, we can determine the existence of a HIST in G by exhaustively enumerating all subsets $\mathcal{M}' \subseteq \mathcal{M}$, considering all possible partitions $(\mathcal{M}_1, \mathcal{M}_2, \mathcal{M}_3)$, verifying the existence of a spanning tree T in $H[\mathcal{M}']$ that satisfies condition (1), and solving the corresponding assignment problem to check condition (2). The former can be verified using a slight modification of exact exponential-time algorithm. The latter can be solved in polynomial time, which leads to the following theorem.

Theorem 8. *Given an n-vertex graph with modular-width k, the existence of a HIST can be determined in $O^*(4^k)$ time, where the O^* notation suppresses polynomial factors.*

Parameterization by Treewidth. In this section, we demonstrate the fixed-parameter tractability of computing a HIST for treewidth. To this end, it is sufficient to give an appropriate MSO_2 formulation by Courcelle's theorem [2,6]. It is well-known that verifying an edge subset F is a tree can be expressed as an MSO_2 formula $\mathtt{tree}(F)$ [7]. Then, a constant-length MSO_2 formula verifying F is a HIST can be expressed as follows:

$$\varphi(F) = \mathtt{tree}(F) \wedge \big(\forall v \in V.\ \exists e_1 \in F.\ \mathtt{inc}(v, e_1)$$
$$\wedge\ (\exists e_1, e_2 \in F.\ (e_1 \neq e_2) \wedge \mathtt{inc}(v, e_1) \wedge \mathtt{inc}(v, e_2)))$$
$$\implies \exists e_1, e_2, e_3 \in F.\ (e_1 \neq e_2) \wedge (e_2 \neq e_3) \wedge (e_3 \neq e_1)$$
$$\wedge\ \mathtt{inc}(v, e_1) \wedge \mathtt{inc}(v, e_2) \wedge \mathtt{inc}(v, e_3)).$$

Theorem 9. *Finding a HIST is fixed-parameter tractable when parameterized by treewidth.*

Parameterization by Cluster Deletion Number. A graph in which each connected component is a complete graph (clique) is called a *cluster graph*. The *cluster vertex deletion number* of a graph $G = (V, E)$ is the minimum number of vertices that need to be deleted to make the graph a cluster graph. Formally, the cluster vertex deletion number $\mathsf{cvd}(G)$ is defined as follows: $\mathsf{cvd}(G) = \min_{S \subseteq V}\{|S| \mid G \setminus S$ is a cluster graph$\}$.

Let S be a vertex subset such that $G \setminus S$ is a cluster graph. Suppose that $G \setminus S$ consists of a set of connected cliques $\mathcal{C}$.

Lemma 6 (*). *Let G, S, and $\mathcal{C}$ be as defined above, and suppose that G admits a HIST. Then, there exists a HIST T of G satisfying the following: For each clique $C \in \mathcal{C}$ and for each partition of C into true twin vertex sets $(M_1, \ldots, M_l)$, each M_i contains at most one vertex of degree at least 3 in T; that is, $|\{u \in M_i \mid d_T(u) \geq 3\}| \leq 1$.*

From Lemma 6, in each clique C, for a module $M \subseteq C$, if $|M| \geq l + |S| + 3$, we can safely delete vertices from M until $|M| = l + |S| + 1$. We can show the following: Let $D \subseteq M \subseteq C$ be the set of deleted vertices, and let $G' = G \setminus D$ denote the graph after deleting D. Then, G admits a HIST if and only if G' admits a HIST. Since the number of modules in each clique C of $G \setminus S$ is at most 2^{cvd}, we obtain the following.

Lemma 7. *By applying the above reduction, the size of the maximum clique in G can be reduced to at most $2^{\mathsf{cvd}}(2^{\mathsf{cvd}} + 3 + \mathsf{cvd}) + \mathsf{cvd}$.*

From Lemma 7, it follows that in graphs where the cluster vertex deletion number cvd is bounded, the size of the graph can be reduced while preserving the existence of a HIST, such that its treewidth is at most $2^{\mathsf{cvd}+1}(2^{\mathsf{cvd}+1} + 2 + 2\mathsf{cvd}) + \mathsf{cvd}$. Therefore, by Theorem 9, the problem is fixed-parameter tractable (FPT) with respect to cvd.

Theorem 10. *Finding a HIST is fixed-parameter tractable when parameterized by cluster deletion number.*

Acknowledgement. We thank the anonymous reviewers for helpful comments that improved the paper's readability, and in particular, we are grateful to one reviewer for bringing reference [8] to our attention.

References

1. Albertson, M.O., Berman, D.M., Hutchinson, J.P., Thomassen, C.: Graphs with homeomorphically irreducible spanning trees. J. Graph Theory **14**(2), 247–258 (1990)
2. Arnborg, S., Lagergren, J., Seese, D.: Easy problems for tree-decomposable graphs. J. Algorithms **12**(2), 308–340 (1991)
3. Bachtler, O.: On Algorithmic Certification of Graph Structures. doctoralthesis, Rheinland-Pfälzische Technische Universität Kaiserslautern-Landau (2023)
4. Chen, G., Shan, S.: Homeomorphically irreducible spanning trees. J. Comb. Theory B **103**(4), 409–414 (2013)
5. Cong, J., He, L., Koh, C., Madden, P.H.: Performance optimization of VLSI interconnect layout. Integr. **21**(1–2), 1–94 (1996)
6. Courcelle, B.: The monadic second-order logic of graphs. i. recognizable sets of finite graphs. Inf. Comput. **85**(1), 12–75 (1990)
7. Courcelle, B.: On the expression of graph properties in some fragments of monadic second-order logic. In: Immerman, N., Kolaitis, P.G. (eds.) Descriptive Complexity and Finite Models, Proceedings of a DIMACS Workshop 1996, Princeton, New Jersey, USA, 14–17 January 1996. DIMACS Series in Discrete Mathematics and Theoretical Computer Science, vol. 31, pp. 33–62. DIMACS/AMS (1996)
8. Diemunsch, J., et al.: A characterization of P_5-free graphs with a homeomorphically irreducible spanning tree. Discret. Appl. Math. **185**, 71–78 (2015)
9. Douglas, R.J.: NP-completeness and degree restricted spanning trees. Discret. Math. **105**(1–3), 41–47 (1992)

10. Furuya, M., Saito, A., Tsuchiya, S.: Refinements of degree conditions for the existence of a spanning tree without small degree stems. Discret. Math. **348**(2), 114307 (2025)
11. Gavril, F.: Algorithms for minimum coloring, maximum clique, minimum covering by cliques, and maximum independent set of a chordal graph. SIAM J. Comput. **1**(2), 180–187 (1972)
12. Hanaka, T., Kiya, H., Ono, H.: Finding a hist: Chordality, structural parameters, and diameter (2025). https://arxiv.org/abs/2510.04418
13. Hanaka, T., Kobayashi, Y.: Finding a minimum spanning tree with a small non-terminal set. Theor. Comput. Sci. **1033**, 115092 (2025)
14. Hill, A.: Graphs with homeomorphically irreducible spanning trees. In: Combinatorics (Proceedings of British Combinatorial Conference, University of College Wales, Aberystwyth, 1973), pp. 61–68 (1974)
15. Hoffmann-Ostenhof, A.: Nowhere-Zero Flows and Structures in Cubic Graphs. Ph.D. thesis, University of Vienna (2011)
16. Kratsch, D., Damaschke, P., Lubiw, A.: Dominating cliques in chordal graphs. Discret. Math. **128**(1–3), 269–275 (1994)
17. Malkevitch, J.: Spanning trees in polytopal graphs. Ann. N. Y. Acad. Sci. **319**(1), 362–367 (1979)
18. de Melo, A.A., de Figueiredo, C.M.H., Souza, U.S.: On the terminal connection problem. In: Bures, T., et al. (eds.) SOFSEM 2021: Theory and Practice of Computer Science - 47th International Conference on Current Trends in Theory and Practice of Computer Science, SOFSEM 2021, Bolzano-Bozen, Italy, 25–29 January 2021, Proceedings. LNCS, vol. 12607, pp. 278–292. Springer (2021)
19. Shan, S., Tsuchiya, S.: Characterization of graphs of diameter 2 containing a homeomorphically irreducible spanning tree. J. Graph Theory **104**(4), 886–903 (2023)
20. Tran, D.L.: Expanding the graph parameter hierarchy. Bachelor thesis, Technische Universität Berlin, Institute of Software Engineering and Theoretical Computer Science (AKT), September 2022, https://fpt.akt.tu-berlin.de/publications/theses/BA-Duc-Long-Tran.pdf, supervisor: Dr. André Nichterlein; Second reviewer: Prof. Dr. Stephan Kreutzer
21. Zhalechian, M., Torabi, S.A., Mohammadi, M.: Hub-and-spoke network design under operational and disruption risks. Transp. Res. Part E: Logist. Transp. Rev. **109**, 20–43 (2018)

Parameterized Algorithms for Locally Minimal Defensive Alliance

Ajinkya Gaikwad[1] , Soumen Maity[1(✉)], and Saket Saurabh[2,3]

[1] Indian Institute of Science Education and Research, Pune, India
ajinkya.gaikwad@students.iiserpune.ac.in, soumen@iiserpune.ac.in
[2] Institute of Mathematical Sciences, Chennai, India
saket@imsc.res.in
[3] University of Bergen, Bergen, Norway

Abstract. A set D of vertices of a graph is a *defensive alliance* if, for each element of D, the majority of its neighbors are in D. We consider the notion of local minimality in this paper. A defensive alliance D is called a *locally minimal defensive alliance* if removing any vertex $v \in D$ destroys the defensive alliance property, i.e., $D \setminus \{v\}$ is no longer a defensive alliance [1]. Given an undirected graph $G = (V, E)$ and an integer $k \in \mathbb{N}$, we study Locally Minimal Defensive Alliance, where the goal is to check whether G has a locally minimal defensive alliance of size at least k. This problem is known to be NP-hard, but its parameterized complexity remains open until now. We enhance our understanding of the problem from the viewpoint of parameterized complexity by showing that (1) the problem admits a fixed-parameter tractable (FPT) algorithm on general graphs when parameterized by the solution size k, and (2) we also present a subexponential algorithm on planar graphs of minimum degree at least two using the tool of bidimensionality.

Keywords: Parameterized Complexity · FPT · Locally Minimal Defensive Alliance

1 Introduction

Throughout history, humans have formed communities, guilds, faiths, etc. in the hope of coming together with a group of people having similar requirements, visions, and goals. Their reasons to do so usually rest on the fact that any group with common interests often provides added mutual benefits to the union in fields of trade, culture, defense, etc., as compared to the individual. Such activities are commonly seen today in geopolitics, cultures, trades, economics, unions, etc., and are popularly termed as *alliances*.

Based on an alliance structure, formation, and goals, many variations of the problem exist in graph theory. A defensive alliance is usually formed to defends its members against non-members; hence, it is natural to ask that each member

Full version with complete proofs is available at https://arxiv.org/abs/2208.03491.

© The Author(s), under exclusive license to Springer Nature Switzerland AG 2026
J. Kozik and A. Wolff (Eds.): SOFSEM 2026, LNCS 16448, pp. 317–331, 2026.
https://doi.org/10.1007/978-3-032-17801-5_24

of the alliance should have more friends within the alliance (including oneself) than outside. Similarly, an offensive alliance is formed with the inverse goal of offending or attacking non-members of the alliance. These lead to algorithmic problems in the formation of alliances. The main topic of this paper is the study of one such problem from the perspective of parameterized complexity.

A non-empty set D of vertices of a graph is a *defensive alliance* if, for each element of D, the majority of its neighbors are in D. In the DEFENSIVE ALLIANCE problem, given a graph G and a positive integer k, the objective is to decide whether there exists a vertex subset of size at most k, that forms a defensive alliance. During the last 20 years, the DEFENSIVE ALLIANCE problem, and several of its variants, have been studied extensively from both the combinatorial and computational perspective [1–3, 3–5, 8, 9, 11, 15–18, 20–22]. As expected, the problems of finding small defensive and offensive alliances are NP-complete.

The focus has been chiefly on finding small alliances, although studying large alliances make a lot of sense from the original motivation of these notions and were actually also delineated in the very first papers on alliances [16]. To remedy this situation, researchers have considered the notion of minimal defensive alliance problems. We first caution that being a defensive alliance is not a hereditary property, that is, a superset or subset of a defensive alliance is not necessarily a defensive alliance. Shafique [19] called an alliance a *locally minimal alliance* if the set obtained by removing any vertex of the alliance is not an alliance. Bazgan et al. [1] considered another notion of alliance that they called a *globally minimal alliance*, which has the property that no proper subset is an alliance. Bazgan et al. [1] proved that deciding if a graph contains a locally minimal defensive alliance of size at least k is NP-complete, even when restricted to bipartite graphs with an average degree less than 5.6. This naturally leads to the question, what is the tractability of this problem when considered from the algorithmic paradigms meant to cope with NP-hardness, such as parameterized complexity?

In this paper, we take up such a study and consider the LOCALLY MINIMAL DEFENSIVE ALLIANCE problem from the perspective of parameterized complexity. In this problem, given an undirected graph $G = (V, E)$ and an integer $k \in \mathbb{N}$, the goal is to check whether G has a locally minimal defensive alliance of size at least k, say D. Such a vertex set D is called a *solution*. The parameterized complexity of alliance problems is well-studied with both natural and structural parameters. The problem of finding a defensive alliance of size at most k admits an FPT algorithm when parameterized by solution size, see [7]. We complement this positive result by showing that LOCALLY MINIMAL DEFENSIVE ALLIANCE also admits an FPT algorithm, which uses a similar approach of extreme combinatorics mentioned in [6].

Our Results: Our main contribution is that LOCALLY MINIMAL DEFENSIVE ALLIANCE is FPT when parameterized by solution size.

Theorem 1. LOCALLY MINIMAL DEFENSIVE ALLIANCE *is FPT.*

The proof of Theorem 1 is based on a win–win strategy. We first compute $\mathrm{diam}(G)$ in polynomial time. If $\mathrm{diam}(G) \geq 4k^2$, then (G, k) is a yes-instance

(Theorem 3), so we henceforth assume $\mathrm{diam}(G) < 4k^2$. Under this assumption, we analyze the vertices of G according to their degrees. Let $f_0(k) = k^{k^{c_0 k}}$ for a sufficiently large constant c_0, and call a vertex *high-degree* if $\deg(v) > 2f_0(k)$. Every vertex of G is therefore either high-degree or not, and so the following three cases cover all possibilities.

1. *High-degree vertices with many neighbours of degree at least 2.* If there exists a high-degree vertex v with $|N_2(v)| \geq f_0(k)$, then Lemma 8 implies that G contains a locally minimal defensive alliance of size at least k.
2. *High-degree vertices with few neighbours of degree at least 2.* If v is high-degree but $|N_2(v)| < f_0(k)$, then we can in FPT-time determine whether there exists a locally minimal defensive alliance containing v (Lemma 10). If such an alliance exists, we are done.
3. *No high-degree vertices.* If $\Delta(G) \leq 2f_0(k)$, then because $\mathrm{diam}(G) < 4k^2$, the number of vertices in G is bounded by a function of k only. In this case the problem can be solved by brute force on the bounded-size graph.

Our algorithm uses these structural cases together with the transformation procedures of Algorithms 1 and 2. These procedures start with an initial defensive alliance (in particular the trivial one, $V(G)$) and greedily eliminate vertices that violate local minimality. Either the greedy process directly yields a locally minimal defensive alliance of size at least k, or the failure of the process identifies structural conditions that place the instance in one of the cases above. The main technical difficulty is to maintain suitable witnesses of marginal protection during this transformation. This is highly non-trivial and uses several interesting ideas.

Theorem 2. LOCALLY MINIMAL DEFENSIVE ALLIANCE *on the planar graphs of minimum degree at least 2, admits an FPT algorithm with running time* $2^{\mathcal{O}(\sqrt{k}\log k)}n^{\mathcal{O}(1)}$.

The proof of this theorem relies on a win/win argument that exploits the relation between treewidth and the linear grid theorem [10]. For an integer $t \geq 1$, we define Γ_t to be the planar graph obtained as follows. Start with the $t \times t$ square grid, add exactly one diagonal in each unit square with all diagonals oriented in the same direction (thus forming the $t \times t$ triangular grid), and then choose one corner vertex r (say the bottom-right corner) and add an edge between r and every vertex on the outer boundary of the grid We show that the existence of $\Gamma_{10\sqrt{k}+4}$ as a minor obtained by only edge contractions guarantees the existence of locally minimal defensive alliance of size at least k. When this case does not arise, we use a dynamic programming algorithm on graphs of treewidth at most $\mathcal{O}(\sqrt{k})$ and maximum degree Δ with running time $\Delta^{\mathcal{O}(\sqrt{k})}n^{\mathcal{O}(1)}$. Finally, to get the required time complexity mentioned in Theorem 2, we show that if the maximum degree of the input graph is at least $\mathcal{O}(k^5)$, then there always exists a locally minimal defensive alliance of size at least k.

2   Definitions and Preliminaries

Throughout this article, $G = (V, E)$ denotes a finite, simple and undirected graph of order $|V| = n$. The *(open) neighborhood* $N_G(v)$ of a vertex $v \in V(G)$ is the set $\{u \mid (u, v) \in E(G)\}$. The *closed neighborhood* $N_G[v]$ of a vertex $v \in V(G)$ is the set $\{v\} \cup N_G(v)$. The *degree* of $v \in V(G)$ is $|N_G(v)|$ and denoted by $\deg_G(v)$. The subgraph induced by $D \subseteq V(G)$ is denoted by $G[D]$. We use $\deg_D(v)$ to denote the number of neighbors of vertex v in D. The complement of the vertex set D in V is denoted by D^c. The minimum degree of graph G is denoted by $\delta(G)$.

Definition 1. A non-empty set $D \subseteq V$ is a *defensive alliance* in G if for each $v \in D$, $\deg_D(v) + 1 \geq \deg_{D^c}(v)$.

A vertex $v \in D$ is said to be *protected* if $\deg_D(v) + 1 \geq \deg_{D^c}(v)$. Here v has $\deg_D(v) + 1$ defenders and $\deg_{D^c}(v)$ attackers in G. A set $D \subseteq V$ is a defensive alliance if every vertex in D is protected.

Definition 2. A vertex $v \in D$ is said to be *marginally protected* if it becomes unprotected when any of its neighbors in D is moved from D to $V \setminus D$. A vertex $v \in D$ is said to be *overprotected* if it remains protected even when any of its neighbors is moved from D to $V \setminus D$.

Definition 3. [1] A defensive alliance D is called a *locally minimal defensive alliance* if for any $v \in D$, $D \setminus \{v\}$ is not a defensive alliance.

It is important to note that if D is a locally minimal defensive alliance, then for every vertex $v \in D$, at least one of its neighbors in D is marginally protected.

Definition 4. [1] A defensive alliance D is a *globally minimal defensive alliance* or shorter *minimal alliance* if no proper subset is a defensive alliance.

It is easy to see that every globally minimal defensive alliance is also locally minimal, but the converse need not hold. A defensive alliance D is called *connected* if the subgraph induced by D is connected. An alliance D is called a *connected locally minimal alliance* if for any $v \in D$, $D \setminus \{v\}$ is not an alliance and the graph induced by vertices in D is a connected graph. Notice that any globally minimal alliance is always connected. Furthermore, it has been shown that deciding whether a graph contains a globally minimal defensive alliance of size at least k is NP-complete even when $k = 3$ [14]. In this paper, we consider LOCALLY MINIMAL DEFENSIVE ALLIANCE. We define the problem as follows:

LOCALLY MINIMAL DEFENSIVE ALLIANCE
Input: An undirected graph $G = (V, E)$ and an integer k.
Question: Does G have a locally minimal defensive alliance D with $|D| \geq k$?

---

3   FPT Algorithm Parameterized by solution Size

We first introduce the notion of crucial vertices in a defensive alliance.

Definition 5. Let D be a defensive alliance in G. A vertex $u \in D$ is said to be a *crucial vertex* of D if u has a marginally protected neighbor u_1 in D and u_1 also has a marginally protected neighbor u_2 in D (u_2 and u are not necessarily distinct vertices).

Lemma 1. *A defensive alliance D of size at least two is a locally minimal defensive alliance in G if and only if all the vertices of D are crucial.*

Proof. Suppose D is a locally minimal defensive alliance of size at least two in G. For any $u \in D$, $D \setminus \{u\}$ is not a defensive alliance. This implies that u must have a marginally protected neighbor u_1 in D. Due to the same reason u_1 must also have a marginally protected neighbor u_2 in D. Therefore u is a crucial vertex in D. On the other hand, suppose every vertex of D is crucial. Therefore, for any $u \in D$, $D \setminus \{u\}$ cannot form a defensive alliance as u is adjacent to a marginally protected vertex in D. $\qquad\square$

We now propose a simple greedy algorithm that takes as input a defensive alliance and returns a locally minimal defensive alliance. A vertex in D is said to be a *bad* vertex of D if it has no marginally protected neighbor in D. Similarly, a vertex $u \in D$ is said to be a *good* vertex if it has at least one marginally protected neighbor. Every crucial vertex is also a good vertex, but the converse is untrue.

Algorithm 1: Make a defensive alliance locally minimal

Input: A graph $G = (V, E)$ and a defensive alliance D_{in} in G
Output: A locally minimal defensive alliance D
$D \leftarrow D_{in}$;
while *D contains a bad vertex* **do**
    Identify a bad vertex $u \in D$;
    $D \leftarrow D \setminus \{u\}$;
return D;

During the execution of Algorithm 1, A good vertex may become a bad vertex and finally get removed from D. Similarly, during the execution of Algorithm 1, a good or bad vertex may become crucial and then is never deleted from D.

Lemma 2. *The locally minimal defensive alliance D obtained by Algorithm 1 on (G, D_{in}) contains all crucial vertices of D_{in}.*

Proof. Suppose u is an arbitrary crucial vertex in D_{in}. That is, u has a marginally protected neighbor u_1 in D_{in} and u_1 also has a marginally protected neighbor u_2 in D_{in}. As u_1 has a marginally protected neighbor u_2 in D_{in} and similarly u_2 has a marginally protected neighbor u_1 in D_{in}, u_1, u_2 will never be removed during the execution of Algorithm 1. As u has a marginally protected neighbor u_1 which will never be removed, u will also never be removed from D_{in}. Thus the locally minimal defensive alliance D obtained by Algorithm 1 contains all crucial vertices of D_{in}. $\square$

We now modify Algorithm 1. Note that in Algorithm 1, the bad vertices of D were getting removed from D_{in} in an arbitrary order. In the following algorithm, we restrict the order in which the bad vertices will be removed from D_{in}. First, the bad vertices of $D_{in} \setminus C$ will be removed, and then the bad vertices of C will be removed. Note that due to Lemma 2, we know that if there exists a defensive alliance D_{in} that contains k crucial vertices then we have a locally minimal defensive alliance D of size at least k.

Algorithm 2: Make a defensive alliance locally minimal given a fixed set C

Input: A graph $G = (V, E)$, a defensive alliance $D \subseteq V(G)$, and a set $C \subseteq D$
Output: A locally minimal defensive alliance
while $D \setminus C$ *contains a bad vertex* **do**
    Identify a bad vertex $u \in D \setminus C$;
    $D \leftarrow D \setminus \{u\}$;
Run Algorithm 1 on (G, D);

Lemma 3. *Suppose D_0 is the defensive alliance obtained at the end of* **while** *loop in Algorithm 2. Then every vertex in $D_0 \setminus N[C]$ is a crucial vertex of D_0. Furthermore, all the vertices in $D_0 \setminus N[C]$ will be there in the locally minimal defensive alliance obtained at the end of Algorithm 2.*

Proof. Let u be an arbitrary vertex in $D_0 \setminus N[C]$. Since every vertex in $D_0 \setminus C$ has a marginally protected neighbor in D_0, u also has a marginally protected neighbor u_1 in D_0. Clearly, u_1 is in $D_0 \setminus C$. Therefore u_1 also has a marginally protected neighbor u_2 (u_2 can be u) in D_0. By Definition 5, u is crucial in D_0. Therefore every vertex in $D_0 \setminus N[C]$ is a crucial vertex. Due to Lemma 2, since every vertex in the set $D_0 \setminus N[C]$ is crucial in D_0, applying Algorithm 1 will return a locally minimal defensive alliance which contains every vertex in the set $D_0 \setminus N[C]$. $\square$

Next, we show that choosing a set C correctly in Algorithm 2 from the given graph G can generate a locally minimal defensive alliance of size at least k.

3.1 Graphs with Diameter $\geq 4k^2$

Let G be a graph with $diam(G) \geq 4k^2$. In this section, we prove that G contains a locally minimal defensive alliance of size at least k. In particular, we prove the following theorem.

Theorem 3. *Let k be a positive integer and S_{in} be a defensive alliance in G. If $diam(G[S_{in}]) \geq 4k^2$, then G has a locally minimal defensive alliance of size at least k. Moreover, such a locally minimal defensive alliance can be obtained in polynomial time.*

We begin by applying Algorithm 3 on defensive alliance S_{in}.

Algorithm 3: Modify a defensive alliance by removing certain bad vertices

---

    Input: A graph $G = (V, E)$ and a defensive alliance S_{in} in G
    Output: A modified defensive alliance S
    $S \leftarrow S_{in}$;
    while *S has either a degree-one bad vertex or a bad vertex with a degree-one neighbor in S* **do**
        for *each $u \in S$* **do**
            if *u is a bad vertex and $\deg_G(u) = 1$* **then**
                $S \leftarrow S \setminus \{u\}$;
        for *each $v \in S$* **do**
            if *v is a bad vertex and v has a neighbor $u \in S$ with $\deg_G(u) = 1$* **then**
                $S \leftarrow S \setminus \{v\}$;
    return S;

Lemma 4. *Let S be the defensive alliance obtained at the end of Algorithm 3. Let $u, v \in S$ be two adjacent vertices such that $deg_G(u) = 1$ and v be the only neighbor of u in G. Then both u and v are crucial.*

Let S be the defensive alliance obtained at the end of Algorithm 3. Note that $G[S]$ may be disconnected. Suppose $S_1, S_2, \ldots, S_l$ are the connected defensive alliances such that $S = \bigcup_{i=1}^{l} S_i$.

Lemma 5. *For each $i \in [l]$, Algorithm 1 on (G, S_i) produces a connected locally minimal defensive alliance of size of at least two.*

Lemma 6. *Suppose S_{in} is the input defensive alliance in Algorithm 3. Suppose the diameter of graph $G[S_{in}]$ is at least $4k^2$, that is, there exists a pair of vertices $u, v \in S_{in}$ such that $d(u, v) \geq 4k^2$ in $G[S_{in}]$. Let $v_0, v_1, \ldots, v_{4k^2}$ be a shortest u–v path in $G[S_{in}]$, where $v_0 = u$ and $v_{4k^2} = v$. Then for each $1 \leq i \leq 4k^2 - 1$, either $v_i \in S$ or $v_{i+1} \in S$ where S is the defensive alliance obtained at the end of Algorithm 3.*

Due to Lemma 6, it is clear that if $diam(G[S_{in}]) \geq 4k^2$ either S contains a connected defensive alliance S_i of diameter at least $8k$ or a collection of at least $\frac{k}{2}$ connected defensive alliances $S_1, S_2, \ldots, S_l$. In the latter case, it is easy to see that one can obtain a locally minimal defensive alliance of size at least two by simply applying Algorithm 1 on (G, S_i) for each i. This is true due to Lemma 5.

Lemma 7. *If there is a connected defensive alliance S_i of diameter at least $8k$, then there exists a locally minimal defensive alliance of size at least k.*

Proof. Since $\mathrm{diam}(G[S_i]) \geq 8k$, there exist two vertices $w_0, w_{8k} \in S_i$ such that $d_{G[S_i]}(w_0, w_{8k}) = 8k$. Let $w_0, w_1, \ldots, w_{8k}$ be a shortest w_0–w_{8k} path in $G[S_i]$. Define the set $C = \{w_{4\ell} \mid 0 \leq \ell \leq 2k\}$, so $|C| = 2k + 1$.

Run Algorithm 2 on (G, S_i, C), and let $S_i(C)$ denote the defensive alliance obtained at the end of the **while**-loop of Lines 1–4 of Algorithm 2. The set $S_i(C)$ need not be connected. Let $X_1, X_2, \ldots, X_p$ be the connected components of $S_i(C)$. For each $j \in [p]$, define $C_j = C \cap X_j$. Thus $\{C_1, \ldots, C_p\}$ is a partition of C, and every C_j is nonempty because Algorithm 2 maintains $C \subseteq S_i(C)$. Line 5 of Algorithm 2 applies Algorithm 1 to each pair (G, X_j). Let S_{ij} denote the connected locally minimal defensive alliance returned by Algorithm 1 on (G, X_j).

Claim. For each $j \in [p]$, the set S_{ij} is a connected locally minimal defensive alliance and $|S_{ij}| \geq \max\{2, |C_j| - 1\}$.

Proof. By Lemma 5, Algorithm 1 always outputs a defensive alliance of size at least 2 because $X_j \subseteq S_i$. Let $C_j = \{c_1, \ldots, c_q\}$ with $q \geq 2$. Along the original path $w_0, w_1, \ldots, w_{8k}$, the vertices $c_1, \ldots, c_q$ occur in increasing order of distance from w_0. For each $1 \leq \ell < q$, any shortest path between c_ℓ and $c_{\ell+1}$ inside $G[X_j]$ must contain a vertex $x \in X_j \setminus N[C_j]$. By Lemma 3, such a vertex x is crucial in X_j. Since there are $q - 1$ such pairs, X_j contains at least $q - 1$ crucial vertices. By Lemma 2, Algorithm 1 preserves all crucial vertices when transforming X_j into S_{ij}. Hence $|S_{ij}| \geq q - 1 = |C_j| - 1$, proving the claim.

Since the connected components $X_1, \ldots, X_p$ are disjoint and their closed neighborhoods do not intersect, the sets $S_{i1}, S_{i2}, \ldots, S_{ip}$ are vertex-disjoint. Moreover, $|C| = \sum_{j=1}^{p} |C_j| \geq 2k + 1$. By the above claim, we obtain that $\sum_{j=1}^{p} |S_{ij}| \geq \sum_{j=1}^{p}(|C_j| - 1) = |C| - p \geq (2k + 1) - p \geq k$, because $p \leq k + 1$. Thus $\bigcup_{j=1}^{p} S_{ij}$ is a (not necessarily connected) locally minimal defensive alliance of size at least k. This completes the proof of the lemma.

Proof of Theorem 3. We present a polynomial-time algorithm to obtain a locally minimal defensive alliance of size at least k. We begin by applying Algorithm 3 on (G, S_{in}). Let $S = \bigcup_{i=1}^{l} S_i$ be the defensive alliance obtained at the end of Algorithm 3 where $S_1, S_2, \ldots, S_l$ are connected defensive alliances such that $N[S_i] \cap S_j = \emptyset$ for all $i, j \in [l]$ and $i \neq j$. Due to Lemma 5, we know that applying Algorithm 1 on each S_i returns a locally minimal defensive alliance of size at least 2. Therefore we can assume that $l < \frac{k}{2}$, otherwise it is a yes instance.

As diameter of $G[S] \geq 4k^2$ and $l < \frac{k}{2}$, due to Lemma 6 we observe that there exists $i \in [l]$ such that $diam(G[S_i]) \geq 8k$. Due to Lemma 7, one can obtain a locally minimal defensive alliance of size at least k by applying Algorithm 2 on (G, S_i). As Algorithm 1, 2 and 3 run in polynomial time, the above algorithm runs in polynomial time as well. This completes the proof of Theorem 3.   □

3.2 Graphs with Diameter $< 4k^2$

Let G be a graph with $diam(G) < 4k^2$. We show that G contains a locally minimal defensive alliance of size at least k. Consider two cases: $\Delta(G)$ large and small.

$\Delta(G)$ Large: Partition $V(G)$ into $H = \{v \in V \mid \deg(v) > 2f_0(k)\}$ and $L = \{v \in V \mid \deg(v) \leq 2f_0(k)\}$, where $f_0(k) = k^{k^{c_0 k}}$ for a large constant c_0. Vertices in H are *high-degree*. For $v \in V$, let $N_2(v) = \{u \in N(v) \mid \deg(u) \geq 2\}$ and $N_1(v) = \{u \in N(v) \mid \deg(u) = 1\}$. Further partition H as $H_1 = \{v \in H \mid |N_2(v)| < f_0(k)\}$ and $H_2 = \{v \in H \mid |N_2(v)| \geq f_0(k)\}$.

Lemma 8. *If $H_2 \neq \emptyset$ then there exists a locally minimal defensive alliance of size at least k.*

Proof. Let us assume that $u_0 \in H_2$. We will show that there exists a locally minimal defensive alliance of size at least k.

Case 1. Let us assume that $|N_1(u_0)| \leq |N_2(u_0)|$. This implies that $V(G) \backslash N_1(u_0)$ is a defensive alliance. This is true because the number of neighbors of u_0 in $V(G) \setminus N_1(u_0)$ is greater than or equal to the number of neighbors of u_0 in $N_1(u_0)$. So u_0 is protected; the other vertices are clearly protected. Now run Algorithm 2 on $(G, V(G) \setminus N_1(u_0), \{u_0\})$ and suppose it returns D_0.

Subcase 1.1 If $u_0 \in D_0$ then at least $\frac{\deg(u_0)}{2} > k$ neighbors of u_0 are in D_0 for its protection. That means, $|D_0| \geq k$.

Subcase 1.2 Suppose u_0 is not in D_0. Let D_0' be the defensive alliance obtained at the end of the **while** loop of lines 1–4 in Algorithm 2 and all neighbors of u_0 in D_0' are overprotected. This is why u_0 is deleted from D_0' in line 5. We partition the vertices of defensive alliance $D_0' \setminus \{u_0\}$ into two parts: part $P_1^{u_0} \subseteq N_2(v)$ contains vertices adjacent to u_0 and part $P_2^{u_0}$ contains vertices not adjacent to u_0. Note that all the vertices in $P_1^{u_0}$ have a degree of at least two. We emphasize that the sets $P_1^{u_0}$ and $P_2^{u_0}$ are unrelated to the previous notation N_1, N_2 and H_1, H_2; here the subscript only indicates whether a vertex is adjacent to u_0. Due to Lemma 3, the vertices of $P_2^{u_0}$ will be there in a final locally minimal defensive alliance as they are crucial in D_0' because for any two consecutive vertices of C the shortest path between them inside D_0' contains an internal vertex from $P_2^{u_0}$. Thus we have $|P_2^{u_0}| \leq k - 1$, otherwise we can obtain a solution of size at least k by applying Algorithm 1 on D_0'. Furthermore, every vertex in $P_2^{u_0}$ has degree at most $2k$. Indeed, if some $x \in P_2^{u_0}$ had $\deg_G(x) \geq 2k + 1$, then by Lemma 3 the vertex x must appear in every locally minimal defensive alliance contained in D_0'. Since x has at least $2k+1$ neighbours, its closed neighbourhood already contains at least k vertices, implying the existence of a locally minimal defensive alliance of size at least k. As every vertex in $P_2^{u_0}$ has degree at most $2k$, we prove that $P_1^{u_0}$ contains at least one vertex u_1 with degree at least $f_1(k)$ where $f_1(k) = k^{k^{c_1 k}}$ for some large constant c_1. Note that $G[D_0' \setminus \{u_0\}]$ contains at most $\frac{k}{2} - 1$ connected components of size at least two, otherwise by applying Algorithm 1 on each of these defensive alliances, we will obtain a locally minimal defensive alliance of size at least k. Moreover, every component of $G[D_0' \setminus \{u_0\}]$ has diameter at most $4k^2$, otherwise by Theorem 3, it is a yes instance. Therefore, if the maximum degree of $G[D_0' \setminus \{u_0\}]$ is $f_1(k)$ then we know that it contains at most $\mathcal{O}(k.f_1(k)^{k^2})$ vertices. We also know that $G[D_0' \setminus \{u_0\}]$ contains at least $\mathcal{O}(f_0(k))$ vertices. Therefore, we get $cf_0(k) \leq c'k.f_1(k)^{k^2}$ where c and c' are two positive real numbers. This implies

$$f_1(k) \geq \frac{c}{c'}\left(\frac{f_0(k)}{k}\right)^{\frac{1}{k^2}}.$$

Since we have $f_0(k) = k^{k^{c_0 k}}$, we can assume that $f_1(k) = k^{k^{c_1 k}}$ for some large constant $c_1 \leq c_0$.

We again run Algorithm 2 on $(G, D_0' \setminus \{u_0\}, \{u_1\})$. Let D_1' is a defensive alliance obtained at the end of the **while** loop of lines 1–4 in Algorithm 2 on $(G, D_0' \setminus \{u_0\}, \{u_1\})$. We can apply the same argument as before and either obtain a locally minimal defensive alliance of size at least k containing u_1 or a defensive alliance $D_1' \setminus \{u_1\}$ containing a vertex u_2 of degree $f_2(k) = k^{k^{c_2 k}}$ for some large constant c_2.

We repeat the same procedure k times. Let u_k be the vertex obtained by the same procedure with a degree at least $f_k(k) = k^{k^{c_k k}}$ for some large constant c_k. Call Algorithm 2 on $(G, D_{k-1}' \setminus \{u_{k-1}\}, u_k)$ and suppose it returns D_k. If u_k is in D_k then at least $\frac{\deg(u_k)}{2} > k$ neighbors of u_k are in D_k for its protection. That means we have a locally minimal defensive alliance D_k of size at least k; so we are done. Suppose u_k is not in D_k. Let D_k' be the defensive alliance obtained at the end of **while** loop of Algorithm 2. We assume that all neighbors of u_k in D_k' are overprotected and this is why u_k is deleted. We partition the vertices of defensive alliance $D_k' \setminus \{u_k\}$ into two parts: $P_1^{u_k}$ and $P_2^{u_k}$. A vertex x is in part $P_1^{u_k}$ if it is adjacent to all u_i's for $i \in [k]$ and a vertex x is in part $P_2^{u_k}$ if it is not adjacent to at least one u_i for some $i \in [k]$. By Lemma 3, the vertices of $P_2^{u_k}$ are crucial, and therefore these crucial vertices will be there in the final locally minimal defensive alliance. If $|P_2^{u_k}| \geq k$, then we have a locally minimal defensive alliance of size at least k, and we are done. Thus we assume $|P_2^{u_k}| < k$. Consider a vertex $x \in P_1^k$. By the definition of $P_1^{u_k}$, x is adjacent to $k+1$ vertices $u_0, u_1, \ldots, u_k$ which are outside the defensive alliance $D_k' \setminus \{u_k\}$. As x is protected in $D_k' \setminus \{u_k\}$ it must have at least k neighbors in $D_k' \setminus \{u_k\}$. Thus $\deg_G(x) \geq 2k+1$. Line 6 of Algorithm 2 executes Algorithm 1 on $(G, D_k' \setminus \{u_k\})$; suppose it outputs S. We prove that at least one vertex of $P_1^{u_k}$ will survive in S. For the sake of contradiction, assume that this is false, that is, no vertex survive in S. Suppose vertex x in $P_1^{u_k}$ is deleted last. It has at most

k neighbors, including itself, in the current defensive alliance as $P_2^{u_k}$ contains at most $k-1$ vertices and $k+1$ neighbors outside the current defensive alliance. That means, x is not protected, a contradiction to the fact that x is protected in $D'_k \setminus \{u_k\}$. Therefore, the locally minimal defensive alliance S obtained by Algorithm 2 contains a vertex of degree at least $2k+1$; hence S is of size at least k.

Case 2. Let us assume that $|N_1(u_0)| \geq |N_2(u_0)|$. Suppose $N(u_0) = \{u_1, \ldots, u_l\}$. As we know that $|N_1(u_0)| \geq |N_2(u_0)|$, without loss of generality, we can assume that $\deg(u_i) = 1$ for $i \in \left[\lfloor \frac{l-1}{2} \rfloor\right]$. Let us denote $N_1^*(u_0) = \{u_1, u_2, \ldots, u_{\lfloor \frac{l-1}{2} \rfloor}\} \subseteq N_1(u_0)$. Clearly, $V(G) \setminus N_1^*(u_0)$ is a defensive alliance. Now run Algorithm 2 on $(G, V(G) \setminus N_1^*(u_0), \{u_0\})$. Suppose it returns D_0.

Subcase 2.1 If $u_0 \in D_0$, then there exists a locally minimal defensive alliance of size at least k.

Subcase 2.2 Suppose $u_0 \notin D_0$. Let D'_0 be the defensive alliance obtained at the end of lines 1–4 in Algorithm 2 and all neighbors of u_0 in D'_0 are overprotected. This is why we deleted u_0 from D'_{u_0}. As u_0 is marginally protected in $V(G) \setminus N_1^*(u_0)$ and $N_2(u_0) \subseteq V(G) \setminus N_1^*(u_0)$ implies that $N_2(u_0) \subseteq D'_0$. This is true because we cannot remove the neighbors of u_0 anymore in $V(G) \setminus N_1^*(u_0)$ as u_0 is marginally protected in $V(G) \setminus N_1^*(u_0)$. Next, we consider the defensive alliance $D''_0 = D'_0 \setminus (N_1(u_0) \cup \{u_0\})$ where we get rid of degree one neighbors of u_0 from D'_0. Just as in Subcase 1.2, we can partition the vertices of D''_0 into two parts $P_{u_0}^1 = \{w \in N(u_0) \cap D''_0\}$ and $P_{u_0}^2 = \{w \notin N(u_0) \mid w \in D''_0)\}$. Clearly, all the vertices in $P_{u_0}^1$ are of degree at least 2, in fact, $P_{u_0}^1 = N_2(u_0)$. Also, we know that $|N_2(u_0)| = |P_{u_0}^1| \geq f_0(k)$. From here on, we can apply the same argument as in Subcase 1.2 and obtain a locally minimal defensive alliance of size at least k. This proves Lemma 8. $\qquad \square$

Now let us focus on vertices in the set H_1. First, we make some simple observations about vertices in H_1.

Observation 1. Suppose $u \in H_1$ is contained in a locally minimal defensive alliance S and u is overprotected. Then u can be made marginally protected in S by moving some degree one neighbors of u from S to $V \setminus S$.

Observation 2. A graph G has a defensive alliance D with a pair of adjacent marginally protected vertices $u, v \in D$ if and only if G has a locally minimal defensive alliance S in which u is marginally protected.

Lemma 9. *Given a partition of $V(G)$ into three parts A, B, C, we can check in polynomial time if there is a defensive alliance D in G such that $A \subseteq D$ and $B \cap D = \emptyset$. Also if such a defensive alliance exists then it can be obtained in polynomial time.*

Lemma 10. *Given a vertex $v \in H_1$ and $H_2 = \emptyset$, we can determine if there exists a locally minimal defensive alliance containing v in FPT time.*

Given a graph G, we first check whether $H_2 \neq \emptyset$. If yes, then due to Lemma 8 conclude that we are dealing with a yes-instance. Hence, we assume that $H_2 = \emptyset$. Next, for each $v \in H_1$, we check in FPT time whether there exists a locally minimal defensive alliance containing v. If there exists a locally minimal defensive alliance containing v then conclude that we are dealing with a yes-instance as $\deg(v) > 2k$. Therefore, we can assume that there does not exist any locally minimal defensive alliance containing a vertex from H_1. So we remove the vertices of H_1 from graph G. To do this, we create a weighted graph G' from G in the following way: $G' = G - H_1$ and $w : V(G') \to \mathbb{Z}$ is a weight function where $w(v)$ is equal to the number of neighbors of v in H_1. Note that G' can be a disconnected graph. Suppose $G'_1, G'_2, \ldots, G'_l$ are the connected components of G'. We observe that $D \subseteq V(G'_i)$ is a defensive alliance in G if for all $v \in D$, we have $\deg_D(v) + 1 \geq \deg_{V(G'_i) \setminus D}(v) + w(v)$. Given a connected component G'_i, if there exists a vertex $v \in V(G'_i)$ such that $\deg_{G'_i}(v) + 1 < w(v)$ then we update G'_i to $G'_i - \{v\}$ and update the weight of the vertices in $G'_i - \{v\}$ to $w(u) = w(u) + 1$ for all $u \in N(v)$, otherwise we keep the same weight. This is true because such a vertex v cannot be part of any locally minimal defensive alliance in G. Now let us assume that after this reduction, we get $G''_1, G''_2, \ldots, G''_{l'}$, $l' \leq l$, as nonempty connected components.

Claim. For each $i \in [l']$, $V(G''_i)$ forms a defensive alliance in G.

Proof. Let $v \in V(G''_i)$. We know that $\deg_{V(G''_i)}(v) + 1 \geq w(v)$, otherwise v must have been deleted from G''_i. Note that $w(v) = \deg_{V(G) \setminus V(G''_i)}(v)$ by construction. Therefore, we have $\deg_{V(G''_i)}(v) + 1 \geq \deg_{V(G) \setminus V(G''_i)}(v)$ for all $v \in V(G''_i)$. This completes the proof of the claim. $\square$

Note that we have deleted the vertices v for which we know that there is no locally minimal defensive alliance containing v. Now, one can check if $diam(G[V(G''_i)]) \geq 4k^2$ for some i. If yes then conclude that we are dealing with a yes-instance due to Theorem 3. Hence, we assume that $diam(G[V(G''_i)]) < 4k^2$ for all i. We also know that $\Delta(G''_i) \leq f_0(k)$. This means that we have a weighted graph where the size of each connected component is bounded by a function of k. From here, one can do brute force to find a largest locally minimal defensive alliance in each G''_i. Finally, we add the sizes of largest locally minimal defensive alliances from different connected components, and if the sum is at least k then conclude that we are dealing with a yes-instance.

$\boldsymbol{\Delta(G)}$ **is small:** Given a graph G, if we know that $\Delta(G) \leq k^{k^{c_0 k}}$ for some constant c_0 and also that the diameter of each connected component is at most $4k^2$ then clearly each connected component size is bounded by a function of k only. As explained in previous paragraph, we can easily get a locally minimal defensive alliance of size at least k in FPT time.

Proof of Theorem 1. Given a graph G, we first check if the diameter of G is at least $4k^2$. If yes, then due to Theorem 3 it is a yes instance. Hence, we assume that the diameter of G is less than $4k^2$. We find a vertex v in G with maximum degree $\Delta(G)$. As explained in Sect. 3.2, if $\Delta(G) > k^{k^{c_0 k}}$ for some constant c_0 then we can solve the problem in FPT time. Finally, if $\Delta(G) \leq k^{k^{c_0 k}}$ then due to Sect. 3.2, we know how to find a locally minimal defensive alliance of size at least k in FPT time. This proves Theorem 1.

It is known that LOCALLY MINIMAL DEFENSIVE ALLIANCE is NP-complete even on planar graphs [12]. Next, we show that the problem admits subexponential time algorithm when parameterized by the solution size k on planar graphs with minimum degree at least 2.

Theorem 4. *If a connected planar graph G satisfies $tw(G) > 9t + 5$, then G contains Γ_t as a contraction; moreover an $\mathcal{O}(n^2)$ algorithm either outputs a decomposition of width $(9 + \epsilon)t + 5$ or a contraction to Γ_t.*

If G contains $\Gamma_{10\sqrt{k}+4}$ as a contraction (obtained only via edge contractions), we choose a $2k$-sized 4-scattered set $C \subseteq V(G)$ from the contracted grid, run Algorithm 2 on $(G, V(G), C)$, and apply Algorithm 1 to each resulting component. Spacing guarantees yield $\sum_i |S_i'| \geq k$, giving:

Lemma 11. *If a planar graph of minimum degree at least 2 contains $\Gamma_{10\sqrt{k}+4}$ as a contraction, then it has a locally minimal defensive alliance of size at least k.*

If the above case does not occur, Theorem 4 returns a tree decomposition of width $\mathcal{O}(\sqrt{k})$. We then apply:

Theorem 5 ([13]). *Given a nice decomposition of width w, a maximum locally minimal defensive alliance can be computed in $\mathcal{O}^*(9^w \Delta^{\mathcal{O}(w)})$.*

Large-diameter defensive alliances also force large locally minimal ones:

Lemma 12. *If $\delta(G) \geq 2$ and a connected defensive alliance S satisfies that $\mathrm{diam}(G[S]) \geq 8k$, then G has a locally minimal defensive alliance of size at least k.*

Finally, we show that a vertex with large degree immediately implies a yes instance.

Lemma 13. *Let G be planar with minimum degree at least 2. If $\Delta(G) \geq (16k)^5$, then G has a locally minimal defensive alliance of size at least k.*

Combining the above: (i) if $\Delta(G) \geq (16k)^5$, use Lemma 13; (ii) else use Theorem 4; (iii) if $\Gamma_{10\sqrt{k}+4}$ occurs as a contraction, apply Lemma 11; (iv) otherwise the treewidth is $\mathcal{O}(\sqrt{k})$ and Theorem 5 solves the problem in $2^{\mathcal{O}(\sqrt{k}\log k)} n^{\mathcal{O}(1)}$ time.

Theorem 6. LOCALLY MINIMAL DEFENSIVE ALLIANCE *on planar graphs with $\delta(G) \geq 2$ parameterized by k can be solved in time $2^{\mathcal{O}(\sqrt{k}\log k)} n^{\mathcal{O}(1)}$.*

References

1. Bazgan, C., Fernau, H., Tuza, Z.: Aspects of upper defensive alliances. Discret. Appl. Math. **266**, 111–120 (2019)
2. Bliem, B., Woltran, S.: Defensive alliances in graphs of bounded treewidth. Discret. Appl. Math. **251**, 334–339 (2018)
3. Cami, A., Balakrishnan, H., Deo, N., Dutton, R.: On the complexity of finding optimal global alliances. J. Combin. Math. Combin. Comput. **58**, 23–31 (2006)
4. Enciso, R.: Alliances in graphs: parameterized algorithms and on partitioning series -parallel graphs. Ph.D. thesis, University of Central Florida, USA (2009)
5. Fernau, H., Raible, D.: Alliances in graphs: a complexity-theoretic study. In: Proceeding Volume II of the 33rd International Conference on Current Trends in Theory and Practice of Computer Science (2007)
6. Fernau, H.: Extremal kernelization: a commemorative paper. In: Brankovic, L., Ryan, J., Smyth, W.F. (eds.) Combinatorial Algorithms, pp. 24–36. Springer International Publishing, Cham (2018)
7. Fernau, H., Raible, D.: Alliances in graphs: a complexity-theoretic study. In: van Leeuwen, J., Italiano, G.F., van der Hoek, W., Meinel, C., Sack, H., Plasil, F., Bieliková, M. (eds.) SOFSEM 2007: Theory and Practice of Computer Science, 33rd Conference on Current Trends in Theory and Practice of Computer Science, Harrachov, Czech Republic, January 20–26, 2007, Proceedings, vol. II, pp. 61–70. Institute of Computer Science AS CR, Prague (2007)
8. Fernau, H., Rodriguez-Velazquez, J.A.: A survey on alliances and related parameters in graphs. Electronic J. Graph Theory Appl. **2**(1) (2014). https://doi.org/10.5614/ejgta.2014.2.1.7
9. Fernau, H., Rodríguez, J.A., Sigarreta, J.M.: Offensive r-alliances in graphs. Discret. Appl. Math. **157**(1), 177–182 (2009)
10. Fomin, F.V., Golovach, P., Thilikos, D.M.: Contraction obstructions for treewidth. J. Comb. Theory Ser. B **101**(5), 302–314 (2011). https://doi.org/10.1016/j.jctb.2011.02.008, https://www.sciencedirect.com/science/article/pii/S0095895611000256
11. Fricke, G., Lawson, L., Haynes, T., Hedetniemi, M., Hedetniemi, S.: A note on defensive alliances in graphs. Bull. Inst. Combin. Appl. **38**, 37–41 (2003)
12. Gaikwad, A., Maity, S., Tripathi, S.K.: Parameterized complexity of locally minimal defensive alliances. CoRR **abs/2105.10742** (2021), https://arxiv.org/abs/2105.10742
13. Gaikwad, A., Maity, S., Tripathi, S.K.: Parameterized complexity of locally minimal defensive alliances. In: Mudgal, A., Subramanian, C.R. (eds.) Algorithms and Discrete Applied Mathematics, pp. 135–148. Springer International Publishing, Cham (2021)
14. Horn, B.: Parameterized complexity of defensive alliances. Master's thesis, Utrecht University (2023), https://studenttheses.uu.nl/handle/20.500.12932/43668, master's thesis
15. Jamieson, L.H., Hedetniemi, S.T., McRae, A.A.: The algorithmic complexity of alliances in graphs. J. Comb. Math. Comb. Comput. **68**, 137–150 (2009)
16. Kristiansen, P., Hedetniemi, M., Hedetniemi, S.: Alliances in graphs. J. Comb. Math. Comb. Comput. **48**, 157–177 (2004)
17. Carvajal, R., Matamala, M., Rapaport, I., Schabanel, N.: Small alliances in graphs. In: Kučera, L., Kučera, A. (eds.) Mathematical Foundations of Computer Science, MFCS 2007, LNCS. vol. 4708. Springer Berlin Heidelberg, Berlin, Heidelberg (2007)

18. Rodríguez-Velázquez, J., Sigarreta, J.: Global offensive alliances in graphs. Electron. Notes Discrete Math. **25**, 157–164 (2006)
19. Shafique, K.H.: Partitioning a graph in alliances and its application to data clustering. Ph.D. thesis, University of Central Florida (2004)
20. Sigarreta, J., Bermudo, S., Fernau, H.: On the complement graph and defensive k-alliances. Discret. Appl. Math. **157**(8), 1687–1695 (2009)
21. Sigarreta, J., Rodríguez, J.: On defensive alliances and line graphs. Appl. Math. Lett. **19**(12), 1345–1350 (2006)
22. Sigarreta, J., Rodríguez, J.: On the global offensive alliance number of a graph. Discret. Appl. Math. **157**(2), 219–226 (2009)

Exploiting Low Scanwidth to Resolve Soft Polytomies

Sebastian Bruchhold[1(✉)] and Mathias Weller[2]

[1] Technische Universität Berlin, Berlin, Germany
bruchhold@math.tu-berlin.de
[2] LIGM, CNRS, Univ Gustave Eiffel, 77454 Marne-la-Vallée, France
mathias.weller@cnrs.fr

Abstract. Phylogenetic networks allow modeling reticulate evolution, capturing events such as hybridization and horizontal gene transfer. A fundamental computational problem in this context is the TREE CONTAINMENT problem, which asks whether a given phylogenetic network is compatible with a given phylogenetic tree. However, the classical statement of the problem is not robust to poorly supported branches in biological data, possibly leading to false negatives. In an effort to address this, a relaxed version that accounts for uncertainty, called SOFT TREE CONTAINMENT, has been introduced by Bentert, Malík, and Weller [SWAT'18]. We present an algorithm that solves SOFT TREE CONTAINMENT in $2^{\mathcal{O}(\Delta_T \cdot k \cdot \log(k))} \cdot n^{\mathcal{O}(1)}$ time, where $k := \mathrm{sw}(\Gamma) + \Delta_N$, with Δ_T and Δ_N denoting the maximum out-degrees in the tree and the network, respectively, and $\mathrm{sw}(\Gamma)$ denoting the "scanwidth" [Berry, Scornavacca, and Weller, SOFSEM'20] of a given tree extension of the network, while n is the input size. Our approach leverages the fact that phylogenetic networks encountered in practice often exhibit low scanwidth, making the problem more tractable.

Keywords: Phylogenetic networks · Containment relations · Directed width measures

1 Introduction

Evolutionary processes are not always represented well by tree-like models, as hybridization, horizontal gene transfer, and other reticulate events often give rise to more complex structures [7,25]. To capture these evolutionary relationships, researchers use phylogenetic networks, which generalize phylogenetic trees by allowing for "reticulations" that represent non-tree-like events (see the monographs by Gusfield [13] and Huson, Rupp, and Scornavacca [15]).

This paper presents a result from the first author's master's thesis (available at https://doi.org/10.14279/depositonce-21448), which was supervised by the second author.

© The Author(s), under exclusive license to Springer Nature Switzerland AG 2026
J. Kozik and A. Wolff (Eds.): SOFSEM 2026, LNCS 16448, pp. 332–346, 2026.
https://doi.org/10.1007/978-3-032-17801-5_25

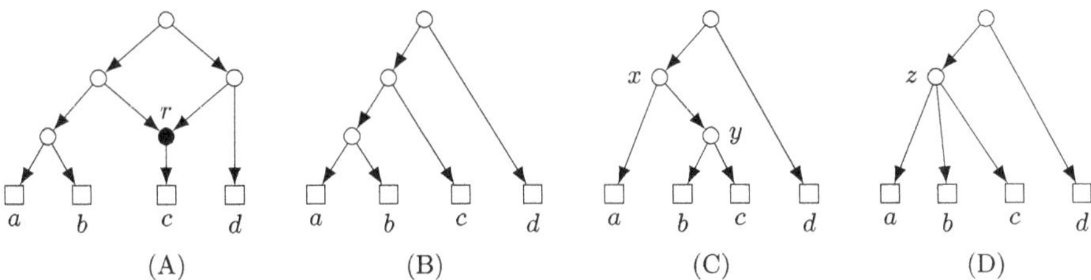

Fig. 1. (A) Example of a phylogenetic network N with one reticulation r, and (B) a tree T_B that is (firmly) displayed by N ("compatible" with N). The tree T_C depicted in (C) is *not* firmly displayed by N. However, if the branch xy in T_C has low support in the biological data used to construct T_C, it is possible that T_C does not reflect the true evolutionary history between taxa a, b, and c. To avoid such artifacts, branches with low support are usually contracted, resulting in the tree T_D, depicted in (D), which now represents exactly the information that is well supported in the data. The information represented by T_D is now consistent with the information represented by N, so we would like to say that T_D is "compatible" with N, even though N does not firmly display T_D. This motivates the formulation of "soft containment". Indeed, T_D is softly displayed by N (since T_B is a binary resolution of T_D that is firmly displayed by N).

A fundamental computational problem arising in this context is the TREE CONTAINMENT problem, which asks whether a given phylogenetic network is "compatible" with ("firmly displays" (see Definition 5)) a given phylogenetic tree, that is, whether the network contains a subdivision of the tree (as a subdigraph respecting leaf-labels). This question is important for reconstruction methods and in the estimation of distances between evolutionary scenarios.

While the TREE CONTAINMENT problem is known to be NP-hard [18,20], numerous special cases of practical interest have been shown to admit polynomial-time solutions [5,9–12,17,18,20,26]. However, the classical formulation of TREE CONTAINMENT is not robust against branches that are only poorly supported by the biological data. Indeed, to be compatible, all branches must be represented as-is, even when they are weakly supported (see Fig. 1 for an example).

To address this limitation, a relaxed version of TREE CONTAINMENT, called SOFT TREE CONTAINMENT, was proposed by Bentert and Weller [2]. This relaxed framework allows "undoing" contractions of poorly supported branches (contracting low-support branches is a common preprocessing step in network analysis), effectively allowing uncertainty in the form of so-called "soft polytomies". The corresponding notion of "soft display" (see Definition 5) permits us to resolve each high-degree vertex into a binary tree. Then, the resulting binary resolution of the network ought to firmly display the resulting binary resolution of the tree.

Although both versions of TREE CONTAINMENT remain NP-hard [2], we aim to solve even large instances in practice. To achieve this, we leverage the fact that phylogenetic networks occurring in empirical studies are expected to exhibit a certain degree of tree-likeness, which can be effectively quantified using the recently introduced measure of tree-likeness called "scanwidth" [3]. Our algorithm solves SOFT TREE CONTAINMENT in $2^{\mathcal{O}(\Delta_T \cdot k \cdot \log(k))} \cdot n^{\mathcal{O}(1)}$ time, where

$k := \mathrm{sw}(\Gamma) + \Delta_N$, with Δ_T and Δ_N denoting the maximum out-degrees in the tree and the network, respectively, and $\mathrm{sw}(\Gamma)$ denoting the scanwidth of a tree extension that is given as part of the input. The algorithm first makes the network binary through "stretching" and "in-splitting". This special case (where the network is binary but the tree need not be) is then solved via bottom-up dynamic programming along the given tree extension.

If a claim is marked with a star ($\star$), then its proof is given in the full version of this paper [6].

2   Related Work

On a binary network N and a binary tree T, the TREE CONTAINMENT problem can be solved in $2^{\mathcal{O}(t^2)} \cdot |A(N)|$ time, where t denotes the treewidth of N [16]. This special case of TREE CONTAINMENT is also a special case of SOFT TREE CONTAINMENT, as the notions of "firm" and "soft" display coincide when N and T are both binary [1]. Additionally, SOFT TREE CONTAINMENT is a generalization of the CLUSTER CONTAINMENT problem [1]. Van Iersel, Semple, and Steel [18] show that the latter problem can be solved in polynomial time on binary networks of constant "level". Note that level-c networks have scanwidth at most $c + 1$ [14]. A comparison of the level, scanwidth, and treewidth of real-world phylogenetic networks was conducted by Holtgrefe [14], concluding that treewidth is not much smaller than scanwidth in practice. He further points out that "edge-treewidth" by Magne et al. [21] is the undirected analogue of scanwidth. We also remark that SOFT TREE CONTAINMENT is conceptually similar to MINOR CONTAINMENT: The question of whether a binary network N softly displays a possibly non-binary tree T is equivalent to asking if T can be obtained from a subtree of N through arc contraction (while preserving leaf-labels). In fact, the reduction of general SOFT TREE CONTAINMENT to this special case is the subject of Sect. 4.2. Janssen and Murakami [19] study a similar containment notion based on arc contraction.

3   Preliminaries

Directed Graphs. A *directed graph* (or *digraph*) is an ordered pair $D = (V, A)$ that consists of a non-empty, finite set V of *vertices* and a set $A \subseteq \{(u, v) \in V^2 \mid u \neq v\}$ of *arcs*. We also use the notation $V(D) := V$ and $A(D) := A$. The in-neighbors of a vertex $v \in V$ are referred to as its *parents* (denoted $V_D^-(v)$), and its out-neighbors are referred to as its *children* (denoted $V_D^+(v)$). The set of incoming arcs of v is denoted $A_D^-(v)$, and the set of outgoing arcs of v is denoted $A_D^+(v)$. The *in-degree* of v is $\deg_D^-(v) := |V_D^-(v)|$, the *out-degree* of v is $\deg_D^+(v) := |V_D^+(v)|$, and the *degree* of v is $\deg_D(v) := \deg_D^-(v) + \deg_D^+(v)$. The maximum out-degree of any vertex in D is denoted Δ_D. If the maximum in-degree and the maximum out-degree of D are both at most 2, then D is called *binary*. A *root* of D is a vertex with in-degree 0, and D is called *rooted* if it has exactly one root (denoted $\mathrm{root}(D)$). A *leaf* of D is a vertex with out-degree 0, and the set of all leaves of D is denoted $L(D)$. We use $\geq_D \subseteq V^2$ to denote the reachability relation of D (so $u \geq_D v$ means that there is a directed path from u to v, possibly having length 0). The irreflexive kernel of $\geq_D$ is denoted $>_D$. We say that D is *weakly connected* if the underlying graph of D is connected. The result of removing v from D is denoted $D - v$. We also use the notation $D - U$ when removing a set of vertices U from D. If a digraph D' can be obtained by removing vertices and arcs from D, then D' is a *subdigraph* of D, denoted $D' \subseteq D$, and D' is called *leaf-monotone* if, additionally, $L(D') \subseteq L(D)$. The subdigraph *induced* by a non-empty set $U \subseteq V$ is $D[U] := (U, A \cap U^2)$. When *subdividing* an arc $(u, v) \in A$ with a new vertex $w \notin V$, the arc (u, v) is removed from D, while the vertex w and the arcs (u, w) and (w, v) are added. The inverse operation, removing w with its incident arcs and adding (u, v), is called *suppression* of w. When *contracting* (u, v), the parents of v, except for u, become parents of u, and the children of v, except for possibly u, become children of u; the vertex v and its incident arcs are then removed. If D can be obtained from a digraph D' by subdividing arcs, then D is a *subdivision* of D'. The *union* of two digraphs D and D' is $D \cup D' := (V(D) \cup V(D'), A(D) \cup A(D'))$. A *directed acyclic graph* (DAG) is a digraph without directed cycles. An *out-tree* is a rooted DAG in which every vertex has in-degree at most 1. Note that rooted DAGs, and thus out-trees, are always weakly connected.

Phylogenetic Networks. A *(phylogenetic) network* is a rooted DAG whose root has out-degree at least 2 and in which each non-root vertex either has in-degree 1 (and is called a *tree vertex*) or has out-degree 1 (and is called a *reticulation*), but not both (see Fig. 1(A)). The root of a network is also considered to be a tree vertex. If a network has no reticulations (that is, it is an out-tree), then it is called a *(phylogenetic) tree.*

Scanwidth. Undoubtedly, one of the most successful concepts in algorithmic graph theory is the notion of *treewidth* [4]. While some problems in phylogenetics yield efficient algorithms on networks of bounded treewidth, the "undirected nature" of treewidth often makes dynamic programming prohibitively complicated or outright unimplementable [24]. Since phylogenetic networks exhibit a "natural flow of information" along the direction of the arcs, Berry, Scornavacca, and Weller [3] introduced a width measure similar to treewidth, but respecting the direction of the arcs (see [22,23] for previous efforts to port the notion of treewidth to directed graphs), called "scanwidth". The idea is to "inscribe" the arcs of the network N into an out-tree Γ (called a "tree extension") on $V(N)$, such that all network arcs are directed away from the root of Γ (see Fig. 2(middle) on page 8). This structure allows a bottom-up dynamic-programming approach whose exponential part depends on the size of the set $\mathrm{GW}_v(\Gamma)$ of arcs of N that would be cut when slicing just above any vertex v of Γ.

Formally: Let D be a DAG. A *tree extension* of D is an out-tree Γ on $V(\Gamma) = V(D)$ with $A(D) \subseteq >_\Gamma$. For a tree extension Γ of D, define $\mathrm{GW}_t^D(\Gamma) := \{(u, v) \in A(D) \mid u >_\Gamma t \geq_\Gamma v\}$, for every $t \in V(\Gamma)$, and $\mathrm{sw}_D(\Gamma) := \max_{t \in V(\Gamma)} |\mathrm{GW}_t^D(\Gamma)|$. The *scanwidth* of D, denoted $\mathrm{sw}(D)$, is the minimum $\mathrm{sw}_D(\Gamma)$ achievable by any tree extension Γ of D. If D is rooted, then a *canonical* tree extension Γ of D is such that $D[\{v \in V(D) \mid t \geq_\Gamma v\}]$ is weakly connected for all $t \in V(\Gamma)$ [14]; further, $L(\Gamma) = L(D)$ and $\deg_\Gamma^+(v) \leq \deg_D^+(v)$ for each $v \in V(D)$ [3]. For every tree extension Γ of a rooted DAG D, a canonical tree extension Γ' of D with $\mathrm{sw}_D(\Gamma') \leq \mathrm{sw}_D(\Gamma)$ can be constructed in polynomial time [14]. Scanwidth upper bounds maximum in-degree [14] but is incomparable with maximum out-degree, as witnessed by in-stars and out-stars.

Assuming that no ambiguity arises, we may omit sub- or superscripts.

4   Deciding Soft Tree Containment

In this paper, we develop a parameterized algorithm for SOFT TREE CONTAINMENT. Informally speaking, this problem, first considered by Bentert, Malík, and Weller [1], asks us to decide if a given phylogenetic tree can be embedded in a given phylogenetic network while taking into account uncertainty in the form of soft polytomies (high-degree vertices to be resolved into a binary tree). We build up to its formal definition (Problem 1) in a way that differs slightly from the original. First, we adapt the notion of *splitting* [2, 19].

Definition 1. *Let N be any network, let $v \in V(N)$ have out-degree at least 3, and let $u, w \in V_N^+(v)$ be two distinct children of v. Then,* out-splitting u and w *from v produces the network $N' := (V(N) \uplus \{x\}, A)$, where $x \notin V(N)$ is a new vertex and $A := (A(N) \backslash \{(v, u), (v, w)\}) \cup \{(v, x), (x, u), (x, w)\}$.*

Exhaustive out-splitting enables us to resolve a high-out-degree tree vertex into a binary out-tree. Its dual for high-in-degree reticulations is defined below.

Definition 2. *Let N be any network, let $v \in V(N)$ have in-degree at least 3, and let $u, w \in V_N^-(v)$ be two distinct parents of v. Then,* in-splitting u and w *from v produces the network $N' := (V(N) \uplus \{x\}, A)$, where $x \notin V(N)$ is a new vertex and $A := (A(N) \backslash \{(u, v), (w, v)\}) \cup \{(u, x), (w, x), (x, v)\}$.*

We now introduce terminology for partially and fully resolved networks. Note that we use a different definition of *binary resolution* than Bentert, Malík, and Weller [1]. This avoids corner cases violating the intuition behind soft polytomies.

Definition 3. *Let N be a network. Any network that is obtained from N through exhaustive out-splitting is called a* binary out-resolution *of N, and any network that is obtained from N through exhaustive in-splitting is called a* binary in-resolution *of N. A* binary resolution *of N is a binary out-resolution of a binary in-resolution of N.*

When discussing firm and soft display, we want to treat leaves as labeled and all other vertices as unlabeled. This is achieved via the following notion of an isomorphism that agrees with the identity function on all leaves.

Definition 4. *Let D and D' be DAGs with $L(D) = L(D')$. Let $\iota \colon V(D) \to V(D')$ be an isomorphism between D and D' with $\iota|_{L(D)} = \mathrm{id}_{L(D)}$, where $\iota|_{L(D)} \colon L(D) \to V(D')$ denotes the restriction of ι to $L(D)$. Then, ι is called* leaf-respecting.

At this point, we have everything needed to define the central notions of display for this work. Recall that, when taking a leaf-monotone subdigraph, no new leaves can be created.

Definition 5 ([1]). *Let N be any network and let T be any tree. We say that N firmly displays T if there is a leaf-respecting isomorphism between a subdivision of T and a leaf-monotone subdigraph of N. We say that N softly displays T if there is a binary resolution N' of N as well as a binary resolution T' of T such that N' firmly displays T'.*

We are now ready to define the problem under consideration. Note that, for technical reasons, we restrict the input slightly more than Bentert, Malík, and Weller [1]; in particular, networks in our sense cannot contain any vertex that has both in-degree and out-degree 1.

Problem 1 ([1]) Soft Tree Containment
Input: A phylogenetic network N and a phylogenetic tree T with $L(T) \subseteq L(N)$.
Question: Does N softly display T?

We solve this problem in two steps, corresponding to the next two sections. In Sect. 4.1, we consider the special case where the given network N is binary. In Sect. 4.2, we describe a reduction mapping an arbitrary network to a binary one.

4.1   Soft Display by Binary Networks

Van Iersel, Jones, and Weller [17] present a dynamic-programming algorithm that uses tree extensions, thereby leveraging low scanwidth, to solve the Tree Containment problem, which asks for firm rather than soft containment. In the following, we adapt their algorithm for our purposes. Although our approach closely follows theirs, access to the unpublished manuscript is *not* required to understand our presentation. We will, however, highlight both similarities and differences between the two approaches. To start with, we use their definition of a *downward-closed subforest* and of *top arcs* in T, which are mapped into the "bags" of a given tree extension of N.

In contrast to the definition of a phylogenetic network/tree, it will be more convenient to work with N and T whose roots have out-degree 1. This is for technical reasons and does not limit practical applicability. We also assume to be given a canonical tree extension, which might be more problematic in practice, but it allows us to separate the problem of computing/approximating low-scanwidth tree extensions from the problem of deciding soft containment. The following assumption formally introduces these objects, but its scope is restricted to this section and we drop it for Sect. 4.2.

Assumption 1. *Let N^* be any binary phylogenetic network, and let T^* be an arbitrary phylogenetic tree with $L(T^*) = L(N^*)$. We construct the rooted DAG $N := (V(N^*) \uplus \{\rho_N\}, A(N^*) \cup \{(\rho_N, \mathrm{root}(N^*))\})$ by attaching a new root $\rho_N \notin V(N^*)$, and we construct the out-tree $T := (V(T^*) \uplus \{\rho_T\}, A(T^*) \cup \{(\rho_T, \mathrm{root}(T^*))\})$ by attaching a new root $\rho_T \notin V(T^*)$. Also, let Γ be a canonical tree extension of N.*

We generalize the strict reachability relation defined earlier in order to define reachability between arcs and vertices/arcs. The new reachability relation agrees with the old relation when comparing vertices with vertices.

Definition 6 ([17]). *Let D be a DAG, $u, v \in V(D)$ vertices, and $(w, x), (y, z) \in A(D)$ arcs. Then, the relation $>_D \subseteq (V(D) \uplus A(D))^2$ is defined as follows:*

1. *We have $u >_D v$ if and only if $u \geq_D v$ and $u \neq v$,*
2. *we have $(w, x) >_D (y, z)$ if and only if $x \geq_D y$,*
3. *we have $u >_D (y, z)$ if and only if $u \geq_D y$, and*
4. *we have $(w, x) >_D v$ if and only if $x \geq_D v$ (read v is reachable from (w, x)).*

This relation lets us describe parts of N or T that are "below" a specified set of arcs. In particular, our dynamic-programming table will indicate whether the parts of T that are below an anti-chain of arcs (called "top arcs") can be embedded below a set of arcs in N.

Definition 7 ([17]). *A downward-closed subforest of T is a subdigraph F of T that does not contain isolated vertices and whose arc set is such that, for all arcs $a \in A(F)$ and $b \in A(T)$, if $a >_T b$, then $b \in A(F)$. An arc $(x, y) \in A(F)$ is called a* top arc *of F if $\deg_F^-(x) = 0$. The set $S \subseteq A(F)$ that contains exactly the top arcs of F is called a* top-arc set *for $L(F)$, and F is called the* forest below S.

In contrast to firm containment, soft containment allows embedding distinct arcs (x, y) and (x, z) of T as paths in N that may start with the same arcs but separate further down. We call such paths "eventually arc-disjoint".

Definition 8. *Two distinct paths $P, Q \subseteq N$ are called* eventually arc-disjoint *if they have length at least 1 and (a) P and Q are arc-disjoint or (b) P and Q start at the same vertex $v \in V(N)$ and $P - v$ and $Q - v$ are eventually arc-disjoint.*

We remark that the union of several pairwise eventually arc-disjoint paths that all start with the same vertex can be used to form an out-tree. To witness soft containment, we associate each vertex of T with such a collection of paths in N. Formalizing this intuition, we adapt the notion of a *pseudo-embedding* [17] to work with eventually arc-disjoint paths. Then, we show that such an embedding of T into N witnesses N^* softly displaying T^*.

Definition 9. *A* soft pseudo-embedding *of a downward-closed subforest F of T into the rooted DAG N is a function ϕ that maps every arc $(x, y) \in A(F)$ to some directed path $\phi(x, y) \subseteq N$ of length at least 1 in N and satisfies the following:*

(SPE1) *For any two arcs of the form $(x, y), (y, z) \in A(F)$, the last vertex of $\phi(x, y)$ is the first vertex of $\phi(y, z)$.*
(SPE2) *For all arcs $(x, y), (x', y') \in A(F)$ with $x \neq x'$, the two directed paths $\phi(x, y)$ and $\phi(x', y')$ are arc-disjoint.*
(SPE3) *For all arcs $(x, y), (x, y') \in A(F)$ with $y \neq y'$, the two directed paths $\phi(x, y)$ and $\phi(x, y')$ are eventually arc-disjoint.*

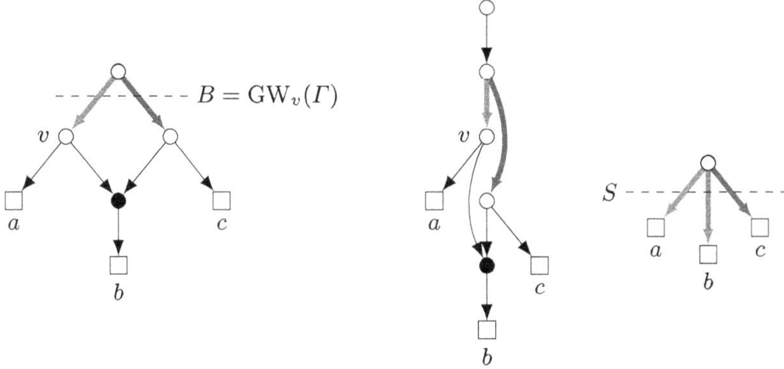

Fig. 2. A valid signature $[B, S, \psi]$, adapted from [17]. **Right:** A top-arc set S in an out-tree T. The three arcs of S have distinct colors. **Left:** A "bag" B of arcs in N. The colors indicate which arc of S is mapped to which arc in B by ψ. Note that ψ maps both the magenta and the green arc of S to the same magenta–green arc in B. **Middle:** A canonical tree extension Γ of N (arcs of Γ depicted in gray) with the arcs of N drawn in. The set $\mathrm{GW}_v(\Gamma)$ is the current choice for B.

(SPE4) For every arc $(x, \ell) \in A(F)$ with $\ell \in L(T)$, the last vertex of $\phi(x, \ell)$ is
        ℓ.

Also, for any arc $(x, y) \in A(F)$, we use $\phi(y)$ to denote the last vertex of $\phi(x, y)$.

Lemma 1 ($\star$)**.** *The binary network N^* softly displays the tree T^* if and only if there is a soft pseudo-embedding of T into N.*

Recall that we assume to be given a canonical tree extension Γ of N (see Assumption 1). We define $\mathrm{HW}_t(\Gamma) := \{(u, v) \in A(N) \mid u \geq_\Gamma t >_\Gamma v\}$ for every $t \in V(\Gamma)$. When visualizing scanwidth via scanner lines [3], the set $\mathrm{HW}_t(\Gamma)$ may be thought of as containing the arcs of N cut right below $t \in V(\Gamma)$, whereas $\mathrm{GW}_t(\Gamma)$ contains the arcs cut right above t.

In the following, we develop a dynamic-programming algorithm that computes its table entries ("signatures") bottom-up along Γ. Intuitively, for each $t \in V(\Gamma)$, each top-arc set S in T, and each $\psi \colon S \to \mathrm{GW}_t(\Gamma)$, we set the table entry / signature $[\mathrm{GW}_t(\Gamma), S, \psi]$ to "valid" if the forest below S in T is softly displayed by the network below $\mathrm{GW}_t(\Gamma)$ in such a way that each arc $a \in S$ is embedded in a path starting with $\psi(a) \in \mathrm{GW}_t(\Gamma)$. An example of a valid signature is depicted in Fig. 2. Note that ψ is not required to be injective.

Definition 10. *A* signature *is an ordered triple $[B, S, \psi]$ containing a set $B \subseteq A(N)$ of arcs of N, a top-arc set $S \subseteq A(T)$ for the set of leaves reachable from B in N, and a function $\psi \colon S \to B$. Such a signature $[B, S, \psi]$ is called* valid *if there is a soft pseudo-embedding ϕ of the forest below S into N such that, for every arc $(x, y) \in S$, the first arc of the path $\phi(x, y)$ is $\psi(x, y)$; then, ϕ is a* witness *for the valid signature $[B, S, \psi]$.*

We now prove several lemmas leading up to Proposition 1, bounding the running time and establishing the correctness of Algorithm 1, which decides if N^* softly displays T^* by determining whether there is a soft pseudo-embedding of T into N. This can be phrased as a question about the existence of a valid signature involving the top arc of T (Lemma 6), and Algorithm 1 answers this question by generating all valid signatures of interest along Γ. To bound the running time, it is then important to bound the number of valid signatures, which we do next (Lemma 2). After that, we derive three lemmas that describe how valid signatures are to be generated, starting from the leaves of Γ. The lemmas correspond to steps of Algorithm 1 and, informally speaking, capture how the validity of signatures propagates along vertices $v \in V(\Gamma)$ of the canonical tree extension, alternating between signatures that use $\mathrm{GW}_v(\Gamma)$ and signatures that use $\mathrm{HW}_v(\Gamma)$.

As indicated, we first give a bound on the number of valid signatures, utilizing a bound on the number of *important separators* due to Chitnis, Hajiaghayi, and Marx [8, Lemma 4.2] that implies a bound on the number of bounded-size top-arc sets for any fixed set of leaves [17, Lemma 6].

Lemma 2 ($\star$). *For each set $B \subseteq A(N)$, there are at most $(4 \cdot |B|)^{\Delta_T \cdot |B|}$ valid signatures of the form $[B, S, \psi]$.*

For every non-root $v \in V(\Gamma)$, Algorithm 1 constructs the set GWS_v of pairs (S, ψ) for which $[\mathrm{GW}_v(\Gamma), S, \psi]$ is a valid signature (and analogously for HWS_v). The construction has three steps: First, handling the case where v is a leaf. Second, constructing HWS_v when v is not a leaf. Third, constructing GWS_v when v is not a leaf. The next three lemmas imply the steps' correctness.

Lemma 3 ($\star$). *For each leaf $v \in L(\Gamma)$, there is exactly one ordered pair (S, ψ) such that $[\mathrm{GW}_v(\Gamma), S, \psi]$ is a valid signature.*

For every non-root, non-leaf $v \in V(\Gamma)$, Algorithm 1 constructs HWS_v by combining pairs from GWS_{q_1} and GWS_{q_2}, where q_1 and q_2 are the children of v in Γ. We define $D_u := D[\{w \in V(D) \mid u \geq_D w\}]$ for each $u \in V(D)$ in a DAG D, to be able to refer to subtrees of T and Γ rooted at a particular vertex. Recall that $\psi|_Z$ is the restriction of ψ to Z.

Lemma 4 ($\star$). *Let $v \in (V(\Gamma)\backslash\{\mathrm{root}(\Gamma)\})\backslash L(\Gamma)$ have children $V_\Gamma^+(v) = \{q_1, q_2\}$ in Γ, where q_1 and q_2 are not necessarily distinct, and let $[\mathrm{HW}_v(\Gamma), S, \psi]$ be a signature. Further, for each $i \in \{1, 2\}$, let $S_i := \{(x, y) \in S \mid L(T_y) \subseteq L(\Gamma_{q_i})\}$. Then, the signature $[\mathrm{HW}_v(\Gamma), S, \psi]$ is valid if and only if (a) $S = S_1 \cup S_2$ and (b) $[\mathrm{GW}_{q_i}(\Gamma), S_i, \psi|_{S_i}]$ is a valid signature for each $i \in \{1, 2\}$.*

The next lemma concerns the correctness of the procedure PopulateGWS. Given a non-root, non-leaf vertex $v \in V(\Gamma)$, this procedure constructs the set GWS_v based on the elements of HWS_v. Informally speaking, this step from $\mathrm{HW}_v(\Gamma)$ to $\mathrm{GW}_v(\Gamma)$ unlocks the incoming arcs of v in N. There are three ways in which these arcs may be used: (1) not at all, (2) to elongate paths of an existing embedding, or (3) to enlarge the downward-closed subforest to be embedded.

Algorithm 1. Solve SOFT TREE CONTAINMENT on N^* and T^*.

Input: N (binary), T (not necessarily binary), and Γ; see Assumption 1.
Output: If N^* softly displays T^*, then **YES**, otherwise **NO**.

1: **procedure** SOLVESTC(N, T, Γ)
2:    **for** $v \in V(\Gamma) \setminus \{\mathrm{root}(\Gamma)\}$ in a bottom-up traversal of Γ **do**
3:       $GWS_v \leftarrow \varnothing$;
4:       $HWS_v \leftarrow \varnothing$;
5:       **if** $v \in L(\Gamma)$ **then**
6:          $p \leftarrow$ parent of v in T;
7:          $u \leftarrow$ parent of v in N;
8:          $GWS_v \leftarrow \{(\{(p, v)\}, \psi)\}$, where $\psi(p, v) := (u, v)$;
9:       **else**
10:          $\{q_1, q_2\} \leftarrow V_\Gamma^+(v)$; $\triangleright$ *Note that q_1 and q_2 are not necessarily distinct.*
11:          **if** $q_1 = q_2$ **then**
12:             $HWS_v \leftarrow GWS_{q_1}$;
13:          **else**
14:             **for** $((S_1, \psi_1), (S_2, \psi_2)) \in GWS_{q_1} \times GWS_{q_2}$ **do**
15:                Add $(S_1 \cup S_2, \psi_1 \cup \psi_2)$ to HWS_v;
16:          POPULATEGWS(N, T, Γ, v); $\triangleright$ *Populate GWS_v using HWS_v.*
17:    $v \leftarrow$ child of the root of N;
18:    $a \leftarrow$ top arc of T;
19:    **if** $\exists (S, \psi) \in GWS_v \colon S = \{a\}$ **then**
20:       **return** YES;
21:    **else**
22:       **return** NO;

23: **procedure** POPULATEGWS(N, T, Γ, v)
24:    **for** $(S, \psi) \in HWS_v$ **do**
25:       **if** $\psi^{-1}(A_N^+(v)) = \varnothing$ **then**
26:          Add (S, ψ) to GWS_v;
27:       **else**
28:          **if** $\exists x \in V(T) \colon \psi^{-1}(A_N^+(v)) \subseteq A_T^+(x)$ **then**
29:             **for** $u \in V_N^-(v)$ **do**
30:                Define $\psi' \colon S \to \mathrm{GW}_v(\Gamma)$ such that
31:                $\psi'(x, y) := (u, v)$ for every $(x, y) \in \psi^{-1}(A_N^+(v))$ and
32:                $\psi'(x', y') := \psi(x', y')$ for every $(x', y') \in S \setminus \psi^{-1}(A_N^+(v))$;
33:                Add (S, ψ') to GWS_v;
34:          **if** $\exists y \in V(T) \setminus \{\mathrm{root}(T)\} \colon \psi^{-1}(A_N^+(v)) = A_T^+(y)$ **then**
35:             $x \leftarrow$ parent of y in T;
36:             $S' \leftarrow (S \setminus \psi^{-1}(A_N^+(v))) \cup \{(x, y)\}$;
37:             **for** $u \in V_N^-(v)$ **do**
38:                Define $\psi' \colon S' \to \mathrm{GW}_v(\Gamma)$ such that $\psi'(x, y) := (u, v)$ and
39:                $\psi'(x', y') := \psi(x', y')$ for every $(x', y') \in S \setminus \psi^{-1}(A_N^+(v))$;
40:                Add (S', ψ') to GWS_v;

Lemma 5 ($\star$). *Let $v \in (V(\Gamma) \backslash \{\mathrm{root}(\Gamma)\}) \backslash L(\Gamma)$ be a vertex and $[\mathrm{GW}_v(\Gamma), S, \psi]$ a signature. Then, $[\mathrm{GW}_v(\Gamma), S, \psi]$ is valid if and only if at least one of the three conditions below is met:*

(1) $\psi(x, y) \notin A_N^-(v)$ for every $(x, y) \in S$, and $[\mathrm{HW}_v(\Gamma), S, \psi]$ is a valid signature.
(2) $(u, v) \in \psi(S)$ for exactly one vertex $u \in V_N^-(v)$; further, it holds that
  (2.1) $X := \psi^{-1}(u, v) \subseteq A_T^+(x)$ for some $x \in V(T)$, and
  (2.2) $[\mathrm{HW}_v(\Gamma), S, \psi']$ is a valid signature, where $\psi' \colon S \to \mathrm{HW}_v(\Gamma)$ is some function with $\psi'(X) \subseteq A_N^+(v)$ and, for all $(x', y') \in S \backslash X$, we have $\psi'(x', y') = \psi(x', y')$.
(3) $\psi(x, y) \in A_N^-(v)$ for exactly one $(x, y) \in S$; further, it holds that
  (3.1) $y \notin L(T)$, and
  (3.2) $[\mathrm{HW}_v(\Gamma), S_{x,y}, \psi']$ is a valid signature, where $S_{x,y} := (S \backslash \{(x, y)\}) \cup A_T^+(y)$ and $\psi' \colon S_{x,y} \to \mathrm{HW}_v(\Gamma)$ is some function with $\psi'(A_T^+(y)) \subseteq A_N^+(v)$ and, for all $(x', y') \in S \backslash \{(x, y)\}$, we have $\psi'(x', y') = \psi(x', y')$.

While the previous lemmas established that Algorithm 1 constructs its valid signatures correctly, the next lemma implies that its final answer is correct.

Lemma 6 ($\star$). *Let $v \in V(N)$ be the child of the root of N, and let $a \in A(T)$ be the top arc of T. Then, N^* softly displays T^* if and only if there exists a valid signature of the form $[\mathrm{GW}_v(\Gamma), \{a\}, \psi]$.*

We now piece the previous lemmas together to conclude this section.

Proposition 1. *Given N, T, and Γ fulfilling Assumption 1, Algorithm 1 decides whether N^* softly displays T^* in $\mathcal{O}^*(2^{\Delta_T \cdot (\mathrm{sw}(\Gamma)+1) \cdot \log_2(4 \cdot \mathrm{sw}(\Gamma)+4)})$ time.*

Proof. Let $v \in V(\Gamma) \backslash \{\mathrm{root}(\Gamma)\}$ be a vertex. Whenever Algorithm 1 adds a pair (S, ψ) to GWS_v (or HWS_v), it is true that $[\mathrm{GW}_v(\Gamma), S, \psi]$ (or $[\mathrm{HW}_v(\Gamma), S, \psi]$) is a signature. From Lemmas 3 to 5, it follows that Algorithm 1 generates precisely every such signature that is valid. Then, the correctness of Algorithm 1 is a consequence of Lemma 6. The bound on the running time may be deduced as follows: Because N is binary, we have $|\mathrm{HW}_v(\Gamma)| \leq |\mathrm{GW}_v(\Gamma)| + 1 \leq \mathrm{sw}(\Gamma) + 1$. Hence, by Lemma 2, the sets GWS_v and HWS_v contain at most $2^{\Delta_T \cdot (\mathrm{sw}(\Gamma)+1) \cdot \log_2(4 \cdot \mathrm{sw}(\Gamma)+4)}$ elements each. This also bounds the number of iterations of the loop in Line 14, as each iteration adds a new pair to HWS_v.

4.2 Soft Display by Arbitrary Networks

In this section, we devise a reduction that maps any network to a binary one while preserving the trees that the network softly displays. The reduction can be performed in polynomial time, and the resulting increase in the scanwidth of the network can be bounded. Relying on the considerations of the previous section, we then obtain our main result (Theorem 1). The reduction consists of two steps:

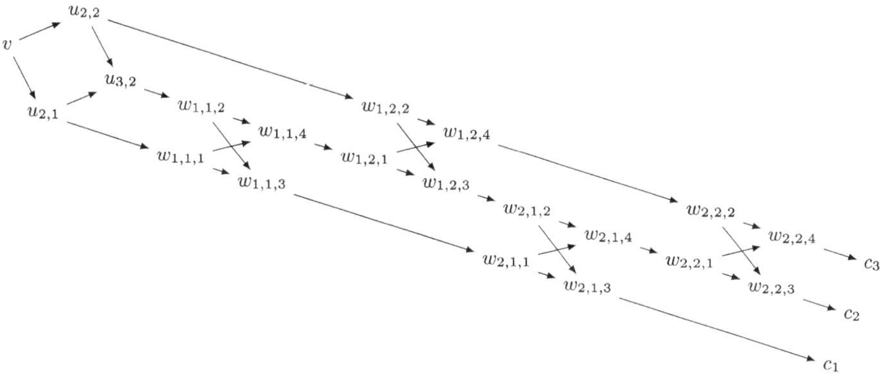

Fig. 3. Stretching a vertex v with three children c_1, c_2, and c_3 gives this stretch gadget.

1. We *stretch* the network. This means that we modify the network using gadgets that are conceptually very similar to the *universal networks* by Zhang [27] (see Fig. 3, also compare [27, Fig. 4]). Like these universal networks, our gadgets consist of a triangular upper part and a lower part that is based on a sorting network.

   The triangular upper part is essentially one half of a grid that was slightly modified to obtain a phylogenetic network. In it, any binary phylogenetic tree (up to some number of leaves) can be embedded. As each high-out-degree vertex is replaced by a gadget when stretching, the stretched network represents all possible binary out-resolutions.

   However, embedding a binary tree in the triangular upper part artificially enforces an order on the leaves of the binary tree. The lower part allows us to undo this. We choose a different sorting network than Zhang [27].
2. We compute an arbitrary binary in-resolution of the stretched network through exhaustive in-splitting. This does not affect which trees are firmly displayed.

Definition 11. *Let N be any arbitrary network, and let $v \in V(N)$ be a vertex with an out-degree of at least 3 in N. Then, stretching v produces the network $N' := (V, A)$ formally specified in the full version [6]; see Fig. 3 for an example. Further, stretching N refers to stretching every vertex of N that has an out-degree of at least 3, and it produces the network $\mathrm{str}(N)$.*

The proof of Theorem 1 relies on four key claims: Lemmas 7 and 8, which state that stretching and in-splitting preserve the collection of trees softly displayed a network, and Lemmas 9 and 10, which state that the increase in scanwidth when stretching and in-splitting is bounded, using the network's maximum out-degree.

Lemma 7. ($\star$). *Let N be any network and let T be any tree. Then, N softly displays T if and only if $\mathrm{str}(N)$ softly displays T.*

We argue that a stretched network can be made binary through in-splitting without altering the collection of softly displayed trees; the collection of firmly displayed trees also stays unaltered.

Lemma 8. $(\star)$. *Let N be any network such that $\Delta_N = 2$, let N' be any binary (in-)resolution of N, and let T be any tree. Then, N firmly/softly displays T if and only if N' firmly/softly displays T.*

Finally, we show that neither stretching nor in-splitting to obtain a binary in-resolution can blow up the scanwidth too much, in order to prove Theorem 1.

Lemma 9. $(\star)$. *Let N be any network. Then, $\mathrm{sw}(\mathrm{str}(N)) \leq \mathrm{sw}(N) + 2 \cdot \Delta_N$.*

Lemma 10. $(\star)$. *Let N' arise from any network N through in-splitting. Then, $\mathrm{sw}(N') \leq \mathrm{sw}(N)$.*

Theorem 1. *Given a network N, a tree T, and a tree extension Γ of N, we can decide whether N softly displays T in $\mathcal{O}^*(2^{\mathcal{O}(\Delta_T \cdot k \cdot \log(k))})$ time, where $k := \mathrm{sw}_N(\Gamma) + \Delta_N$.*

Proof. Let N' be a binary (in-)resolution of $\mathrm{str}(N)$. By Lemmas 7 and 8, we know that N softly displays T if and only if N' softly displays T. Based on Γ, we construct a tree extension Γ' of N' such that $\mathrm{sw}_{N'}(\Gamma')$ is bounded from above by some linear function of $\mathrm{sw}_N(\Gamma) + \Delta_N$, as described in the proofs of Lemmas 9 and 10. It is easy to modify N' and Γ' so as to fulfill the requirements of Assumption 1, since we reject if $L(T) \nsubseteq L(N)$. This involves removing excess leaves from N', updating Γ' accordingly, and making it canonical. Finally, we invoke Algorithm 1, from which we derive the promised running time; see Proposition 1.

5 Conclusion

We devised a parameterized algorithm for the SOFT TREE CONTAINMENT problem, where it has to be decided whether a phylogenetic network softly displays a phylogenetic tree. This algorithm consists of two steps: solving the problem via dynamic programming when the network is binary and reducing any non-binary network to an equivalent binary one. The running time of this algorithm depends exponentially on a function of the network's and the tree's maximum out-degree and of the scanwidth attained by a given tree extension of the network. The dependence on the input size is polynomial.

Future work can attempt to eliminate the superpolynomial dependence on the maximum out-degrees or, alternatively, rule out the existence of such an algorithm. Because of the dependence on scanwidth, it is essential to be able to obtain low-scanwidth tree extensions in a reasonable amount of time. This further motivates the search for exact or approximate algorithms to construct such tree extensions. Additionally, it is of interest to tackle SOFT TREE CONTAINMENT with parameters stronger than scanwidth, such as node-scanwidth [14], edge-treewidth [21], or treewidth. Even weaker parameters could be of interest, when they give rise to algorithms that perform better in practice for example. Finally, it is worth exploring whether our techniques can be adapted to approach other containment problems, for instance, NETWORK CONTAINMENT in the sense of Janssen and Murakami [19].

Disclosure of Interests. The authors have no competing interests to declare.

References

1. Bentert, M., Malík, J., Weller, M.: Tree containment with soft polytomies. In: 16th Scandinavian Symposium and Workshops on Algorithm Theory (SWAT 2018). Leibniz International Proceedings in Informatics (LIPIcs), vol. 101, pp. 9:1–9:14. Schloss Dagstuhl – Leibniz-Zentrum für Informatik (2018). https://doi.org/10.4230/LIPIcs.SWAT.2018.9
2. Bentert, M., Weller, M.: Tree containment with soft polytomies. J. Graph Algorithms Appl. **25**(1), 417–436 (2021). https://doi.org/10.7155/jgaa.00565
3. Berry, V., Scornavacca, C., Weller, M.: Scanning phylogenetic networks is NP-hard. In: Chatzigeorgiou, A., et al. (eds.) SOFSEM 2020. LNCS, vol. 12011, pp. 519–530. Springer, Cham (2020). https://doi.org/10.1007/978-3-030-38919-2_42 https://hal.science/hal-02353161v2
4. Bodlaender, H.L.: A tourist guide through treewidth. Acta Cybernetica **11**(1–2), 1–21 (1993). https://cyber.bibl.u-szeged.hu/index.php/actcybern/article/view/3417
5. Bordewich, M., Semple, C.: Reticulation-visible networks. Adv. Appl. Math. **78**, 114–141 (2016). https://doi.org/10.1016/j.aam.2016.04.004
6. Bruchhold, S., Weller, M.: Exploiting low scanwidth to resolve soft polytomies (2025). https://doi.org/10.48550/arXiv.2511.20771
7. Chan, J.M., Carlsson, G., Rabadan, R.: Topology of viral evolution. Proc. Natl. Acad. Sci. **110**(46), 18566–18571 (2013). https://doi.org/10.1073/pnas.1313480110
8. Chitnis, R., Hajiaghayi, M., Marx, D.: Fixed-parameter tractability of directed multiway cut parameterized by the size of the cutset. SIAM J. Comput. **42**(4), 1674–1696 (2013). https://doi.org/10.1137/12086217X
9. Fakcharoenphol, J., Kumpijit, T., Putwattana, A.: A faster algorithm for the tree containment problem for binary nearly stable phylogenetic networks. In: 12th International Joint Conference on Computer Science and Software Engineering (JCSSE'15), pp. 337–342. IEEE (2015). https://doi.org/10.1109/JCSSE.2015.7219820
10. Gambette, P., Gunawan, A.D.M., Labarre, A., Vialette, S., Zhang, L.: Locating a tree in a phylogenetic network in quadratic time. In: Przytycka, T.M. (ed.) RECOMB 2015. LNCS, vol. 9029, pp. 96–107. Springer, Cham (2015). https://doi.org/10.1007/978-3-319-16706-0_12
11. Gunawan, A.D.M.: Solving tree containment problem for reticulation-visible networks with optimal running time (2017). https://doi.org/10.48550/arXiv.1702.04088
12. Gunawan, A.D.M., DasGupta, B., Zhang, L.: A decomposition theorem and two algorithms for reticulation-visible networks. Inf. Comput. **252**, 161–175 (2017). https://doi.org/10.1016/j.ic.2016.11.001
13. Gusfield, D.: ReCombinatorics: The Algorithmics of Ancestral Recombination Graphs and Explicit Phylogenetic Networks. MIT Press, Cambridge (2014). https://doi.org/10.7551/mitpress/9432.001.0001
14. Holtgrefe, N.: Computing the Scanwidth of Directed Acyclic Graphs. Master's thesis, Delft University of Technology (2023). http://resolver.tudelft.nl/uuid:9c82fd2a-5841-4aac-8e40-d4d22542cdf5
15. Huson, D.H., Rupp, R., Scornavacca, C.: Phylogenetic Networks: Concepts, Algorithms and Applications. Cambridge University Press, Cambridge (2010). https://doi.org/10.1017/CBO9780511974076

16. van Iersel, L., Jones, M., Weller, M.: Embedding phylogenetic trees in networks of low treewidth. Discrete Math. Theor. Comput. Sci. **25**(2), 4 (2023). https://doi.org/10.46298/dmtcs.10116
17. van Iersel, L., Jones, M., Weller, M.: Tree and network containment parameterized by scanwidth (2025). unpublished manuscript
18. van Iersel, L., Semple, C., Steel, M.: Locating a tree in a phylogenetic network. Inf. Process. Lett. **110**(23), 1037–1043 (2010). https://doi.org/10.1016/j.ipl.2010.07.027
19. Janssen, R., Murakami, Y.: On cherry-picking and network containment. Theor. Comput. Sci. **856**, 121–150 (2021). https://doi.org/10.1016/j.tcs.2020.12.031
20. Kanj, I.A., Nakhleh, L., Than, C., Xia, G.: Seeing the trees and their branches in the network is hard. Theoret. Comput. Sci. **401**(1–3), 153–164 (2008). https://doi.org/10.1016/j.tcs.2008.04.019
21. Magne, L., Paul, C., Sharma, A., Thilikos, D.M.: Edge-treewidth: algorithmic and combinatorial properties. Discr. Appl. Math. **341**, 40–54 (2023). https://doi.org/10.1016/j.dam.2023.07.023
22. Reed, B.: Introducing directed tree width. Electronic Notes in Discrete Mathematics **3**, 222–229 (1999). https://doi.org/10.1016/S1571-0653(05)80061-7. 6th Twente Workshop on Graphs and Combinatorial Optimization
23. Safari, M.A.: D-width: A more natural measure for directed tree width. In: Mathematical Foundations of Computer Science 2005, pp. 745–756. Springer, Berlin, Heidelberg (2005). https://doi.org/10.1007/11549345_64
24. Scornavacca, C., Weller, M.: Treewidth-based algorithms for the small parsimony problem on networks. Algorithms Mol. Biol. **17**, 15 (2022). https://doi.org/10.1186/s13015-022-00216-w
25. Treangen, T.J., Rocha, E.P.C.: Horizontal transfer, not duplication, drives the expansion of protein families in prokaryotes. PLoS Genet. **7**(1), e1001284 (2011). https://doi.org/10.1371/journal.pgen.1001284
26. Weller, M.: Linear-time tree containment in phylogenetic networks. In: Blanchette, M., Ouangraoua, A. (eds.) RECOMB-CG 2018. LNCS, vol. 11183, pp. 309–323. Springer, Cham (2018). https://doi.org/10.1007/978-3-030-00834-5_18
27. Zhang, L.: On tree-based phylogenetic networks. J. Comput. Biol. **23**(7), 553–565 (2016). https://doi.org/10.1089/cmb.2015.0228

Maintaining a Kingdom in a Tournament

Oren Weimann[(⊠)][iD] and Raphael Yuster[iD]

University of Haifa, Haifa, Israel
`oren@cs.haifa.ac.il`, `raphy@math.haifa.ac.il`

Abstract. A king in an n-vertex tournament graph G is a vertex that can reach any other vertex v with a path of length at most two. A kingdom is a data structure that given any vertex v returns such a path from a king to v in $O(1)$ time. In this paper, we show how to maintain a kingdom while the tournament graph G undergoes updates. We consider both edge updates that flip the direction of an edge, and vertex updates that insert/delete vertices by activating/deactivating rows and columns of the graph's adjacency matrix.

For a *single* edge-flip, we show that after $O(n^{3/2})$ preprocessing time, we can maintain a kingdom in $O(1)$ time following the edge-flip. With $\tilde{O}(n^2)$ preprocessing time, we can support any *constant* number of edge-flips, vertex insertions, and vertex deletions in $O(\log n)$ time per operation. For an *arbitrary* number of edge-flips, we present a randomized algorithm that maintains a kingdom in $O(\log n)$ expected time following every edge-flip, and another algorithm that supports vertex insertions in $O(\sqrt{n})$ amortized time per insertion.

Keywords: Tournament · king · dynamic graphs · fault tolerance

1   Introduction

A tournament is a directed graph $G = (V(G), E(G))$ with exactly one edge between every pair of vertices. Already in 1953, Landau [5] observed that every tournament has a *king* – a vertex that can reach any other vertex with a (directed) path of length at most two. Indeed, it is not difficult to see that any vertex of a tournament with maximum out-degree is a king. Alternatively, the source of every maximal transitive subtournament is a king. See Maurer [8] for many basic mathematical properties of kings.

Both Landau and Maurer used kings in tournaments in order to analyze the mathematical structure of dominance hierarchies in animals. In such applications (as well as others such as sports competitions or elections), not only the identity of the king is important, but also the dominance hierarchy that the king induces (which we call a *kingdom*). A kingdom, given any vertex v, reveals the path (of length at most two) from the king to v. Formally, a kingdom is a directed spanning tree of depth at most 2 rooted at a king x. I.e., for every child u of x there is an edge $(x, u) \in E(G)$, and for every child v of u there is an edge

© The Author(s), under exclusive license to Springer Nature Switzerland AG 2026
J. Kozik and A. Wolff (Eds.): SOFSEM 2026, LNCS 16448, pp. 347–360, 2026.
https://doi.org/10.1007/978-3-032-17801-5_26

$(u, v) \in E(G)$. Notice that not all out-neighbors of x in G must be children of x in the kingdom (it is possible that some of them are grandchildren).

In this paper, we are interested in maintaining a kingdom in a *dynamic* tournament G. That is, a tournament that undergoes edge-flips, and vertex insertions/deletions. Maintaining a king itself is actually trivial by simply maintaining the out-degree of every vertex (and reporting the maximal one as the king). Maintaining a kingdom however, is much more challenging.

The Kingdom Structure. There are several combinatorial and algorithmic questions that naturally arise regarding kingdoms. We call a king plus its children a *dominating out-star*. Notice that a dominating out-star D is a dominating set in the tournament. Namely, for every vertex $v \in V(G) \backslash V(D)$, there is a vertex $u \in V(D)$ such that $(u, v) \in E(G)$; we say that u *dominates* v. A natural question is to determine the smallest size of a dominating out-star. Let $f(n)$ denote the maximum over all n-vertex tournaments, of the minimum size of a dominating out-star. It is not difficult to show that $f(n) = (1+o(1)) \log n$ (see Proposition 1) although an exact formula seems elusive. This problem is naturally related to the well-known problem of Erdős and Moser [4] on determining $h(n)$, the minimum over all n-vertex tournaments, of the maximum order of a transitive subtournament. Since every maximal transitive subtournament contains a dominating out-star, we have $f(n) \leq h(n)$. It is well known that $2 \log n \geq h(n) \geq \log n$ [4,11]; unlike $f(n)$, the exact asymptotics of $h(n)$ is not known. We mention also that it is known that $g(n)$, the maximum over all n-vertex tournaments, of the minimum size of a dominating set (so clearly $g(n) \leq f(n)$) also satisfies $g(n) = (1 + o(1)) \log n$ [6,12].

As for the algorithmic side, there are several results on king detection which also compute a corresponding kingdom [2,3,7,10]. In particular, an $O(n^{3/2})$ time algorithm of Shen, Sheng, and Wu [10] computes a king and is easily seen to also compute a dominating out-star and a kingdom. It is also shown in [10] that any *deterministic* algorithm for finding a king requires $\Omega(n^{4/3})$ queries to the tournament adjacency matrix, hence no deterministic algorithm can run in $o(n^{4/3})$ time. Yet, it was observed by Abboud, Grossman, Naor, and Solomon [1] that there is an $O(n)$ randomized Las Vegas algorithm for detecting a king and computing a corresponding kingdom.

The main results in this paper extend these algorithms to the *dynamic* and *fault-tolerant* setting. That is, we are allowed to preprocess the graph, such that we can efficiently (either worst case or amortized) maintain the kingdom after any vertex/edge change. In a tournament, changing an edge means flipping its direction, inserting a vertex v additionally inserts an edge between v and every other vertex, and deleting a vertex v deletes all its incident edges. In a fault tolerant algorithm, we wish to preprocess the graph efficiently, such that after a single change (or perhaps constantly many changes) we can maintain the kingdom efficiently.

It is not difficult to see that a single edge-flip or vertex insertion or vertex deletion might cause a vertex to no longer be a king and can cause significant change to the kingdom. As we do not want to compute the new kingdom explicitly following every change, we will maintain the kingdom implicitly with a data structure that stores all the structural information of the resulting kingdom. That is, given a vertex v, our data structure supports a constant time query $q(v)$ which outputs a path of length at most two from the newly computed king x to v. Formally, for a tournament G, a *Kingdom Structure* (KS) is a triple (x, Y, S) such that x is a king of G, Y is a list of (not necessarily all) out-neighbors of x so that $\{x\} \cup Y$ forms a dominating out-star, and S is a data structure that can be queried in *constant time* to return $q(v)$. Notice that a KS encodes the entire information of a kingdom.

Our Results. Our first result is a fault tolerant algorithm that handles any constant number of changes efficiently. As an edge-flip is a special case of two vertex operations (deletion of a vertex v and reinsertion of v with one incident edge changed compared to v's original adjacencies), it suffices to describe the algorithm following only vertex operations. Formally, we assume that the tournament G is represented by its (binary) adjacency matrix M (indexed by $V(G)$), and that M includes additional rows and columns that are initially marked as *inactive*. A change (A, R) is given by a list of vertices R (active rows/columns of M that should be marked as inactive) and a list of vertices A (inactive rows/columns of M that should be marked as active). We never access an inactive entry in M, and accessing an active entry in M is done in constant time.

Theorem 1. *Given an n-vertex tournament G, for every constant k, we can preprocess G in $\tilde{O}(n^2)$ deterministic time or $\tilde{O}(n)$ expected time, so that following any set (A, R) of changes with $|A \cup R| \leq k$, we can compute a KS for the new tournament deterministically in $O(\log n)$ time. If $A = \emptyset$ (only removals are allowed) then the new KS can be computed in $O(1)$ time.*

Our second result is a dynamic algorithm for handling vertex insertions only. Each insertion is presented by activating a new row/column at the bottom/right of the current adjacency matrix M. Denoting by n the current number of vertices, each insertion is performed in amortized $O(\sqrt{n})$ time (i.e., if we start with a graph that has n vertices and perform $t = O(n)$ subsequent insertions, then the total time of all insertions is worst case $O(t\sqrt{n})$).

Theorem 2. *Given an n-vertex tournament G, there is a dynamic algorithm that, after preprocessing G in $O(n^{3/2})$ deterministic time or $\tilde{O}(n)$ expected time, can compute a KS in $O(\sqrt{n})$ amortized time following every new vertex insertion.*

We then move on to consider a dynamic algorithm for handling edge-flips. We present an algorithm that can handle any number of edge-flips in (nearly optimal) $O(\log n)$ expected time per flip. We assume that the sequence of edge-flips is chosen in advance by an oblivious adversary as he wishes but only according to the structure of the initial tournament graph G. The oblivious-adversary model is standard for measuring the performance of randomized dynamic algorithms. The sequence of edge-flips is oblivious to the coin tosses made by the algorithm, and the algorithm is oblivious to the sequence of edge-flips. Handling an adaptive-adversary is left as an open problem.

Theorem 3. *Given an n-vertex tournament G, there is a dynamic algorithm that, after preprocessing G in $\tilde{O}(n)$ expected time, can compute a KS in $O(\log n)$ expected time following every edge-flip.*

Finally, we show how to improve the above $O(\log n)$ bound to (optimal) $O(1)$ deterministic time in the special case where only a single edge-flip is allowed.

Theorem 4. *Given an n-vertex tournament G, we can preprocess G in $O(n^{3/2})$ time, so that following any single edge-flip, a KS of the resulting tournament can be computed in $O(1)$ time.*

Roadmap. In Sect. 2 we describe a very simple $O(n^2)$ time deterministic algorithm to obtain a KS for a tournament in which the dominating out-star has $O(\log n)$ vertices. This algorithm is used as an ingredient in the preprocessing part of Theorem 1 (which is presented in Sect. 3), and in the dynamic algorithm of Theorem 2 (which is presented in Sect. 4). The dynamic algorithm of Theorem 3 for handling edge-flips is presented in Sect. 5 and the single edge-flip algorithm of Theorem 4 is presented in Sect. 6.

2   Computing a KS with a Small Dominating Out-Star

Recall that $f(n)$ denotes the maximum over all n-vertex tournaments, of the minimum size of a dominating out-star. It is easy to see that $f(n) \leq \log(n+1)$ by repeatedly finding a maximum out-degree vertex and eliminating its out-neighbors. Formally,

Proposition 1. $f(n) \leq \log(n+1)$.

Proof. It suffices to prove that every tournament with $2^k - 2$ vertices has a dominating out-star of order at most $k - 1$. This clearly holds for $k = 2$. Proceeding inductively, let G be a tournament with $n = 2^k - 2$ vertices and let v be a vertex with maximum out-degree. Since the maximum out-degree in a tournament is at least $(n - 1)/2$, the out-degree of v is at least $2^{k-1} - 1$. Remove v and its out-neighbors from G to obtain a tournament on at most $2^{k-1} - 2$ vertices which, by induction, has a dominating out-star of order at most $k - 2$. Such an out-star, together with v, is a dominating out-star for G. $\square$

Corollary 1. *Given an n-vertex tournament G, we can construct a KS (x, Y, S) for G with $|Y| = O(\log n)$ in $O(n^2)$ deterministic time or $O(n)$ expected time.*

Proof. Suppose that M is the (binary) adjacency matrix of G (so $M(u, v) = 1$ if and only if $(u, v) \in E(G)$). Initially, we set all vertices of G as *unmarked*. Also we set an array P indexed by $V(G)$ where $P(v)$ is either $\emptyset$ or else $P(v) = u$ such that $(u, v) \in E(G)$. Finally, let Y be an (initially empty) list and let x be a variable that at the end will contain a king.

Notice that at any given time we can compute a vertex of maximum out-degree in the subtournament of G induced by the unmarked vertices in $O(nd)$

time where d is the number of unmarked vertices. We repeatedly proceed as in the proof of Proposition 1. We find a vertex of maximum out-degree in the subtournament of the yet unmarked vertices. Suppose this vertex is y. Set y as marked. For each unmarked out-neighbor v of y, set $P(v) = y$ and set v as marked. If all vertices $V(G)$ are now marked, set $x = y$ and halt. Otherwise, add y to Y and repeat.

As at each round, the number of unmarked vertices reduces by a factor of at least 2, so the overall runtime is $O(n^2)$ as claimed. Alternatively, this can be done in $O(n)$ expected time, if instead of the maximum-degree vertex we randomly choose an unmarked vertex (whose degree is expected to be at least half of the maximum-degree vertex). Notice that (x, Y, P) is indeed a KS, as given $v \in V(G)$, either $(x, v) \in E(G)$ and the query $q(v)$ returns the single edge (x, v) or $(v, x) \in E(G)$ and the query $q(v)$ returns the path $(x, P(v), v)$ where $P(v) \in Y$. $\qquad\square$

Recall from the introduction that the algorithm of Corollary 1 is not optimal. The algorithm from [10] computes a KS in $O(n^{3/2})$ time. However, that algorithm only guarantees that $|Y| \leq \sqrt{n}$ while the algorithm from Corollary 1 guarantees $|Y| = O(\log n)$; which will be important for us later. The algorithm from [1] computes a KS in expected $O(n)$ time, but it only guarantees that $|Y|$ has expected size $O(\log n)$. We can use the latter as an alternative for Corollary 1 if we settle for a randomized algorithm. However, in our applications, the running time of Corollary 1 will not be a bottleneck as we shall use it as a subroutine for relatively small subtournaments.

Another noteworthy observation is that in all of these algorithms, the computed dominating out-star is a *transitive* subtournament of G, and that the king is the source of that transitive subtournament. One may wonder whether it is always the case that a smallest dominating out-star can be taken to be a transitive subtournament. This, however, is false as the following example demonstrates.

Let G be tournament consisting of n vertices, with four vertices designated a, x_1, x_2, x_3 and the remaining $n - 4$ vertices are partitioned into three (say equal) parts V_1, V_2, V_3. Define the edges to be (a, x_1), (a, x_2), (a, x_3), (x_1, x_2), (x_2, x_3), (x_3, x_1). Further, for $i \in \{1, 2, 3\}$, all vertices of V_i are in-neighbors of a and of x_j for $j \neq i$ and are out-neighbors of x_i. Finally, orient the edges of the subtournament induced by $V_1 \cup V_2 \cup V_3$ at random. Clearly, $\{a, x_1, x_2, x_3\}$ is a dominating out-star where a is the king, and $\{a, x_1, x_2, x_3\}$ is not transitive. It is easily checked that with high probability (as n grows), there is no dominating out-star with fewer than four vertices and that $\{a, x_1, x_2, x_3\}$ is the only dominating out-star with four vertices.

Finally, it is easy to see that $f(3) = f(4) = f(5) = f(6) = 2$, and that $f(n) = 3$ for all $n = 7, \ldots, 14$. To see, say, the latter, notice that $f(14) \leq 3$ by Proposition 1, while $f(7) \geq 3$ by considering the so-called Paley tournament [9] on seven vertices. This is the (unique) regular tournament on seven vertices (i.e., the in-degree and out-degree of each vertex is 3) for which the out-neighbors of every vertex form a directed triangle. Clearly, it cannot be dominated by two vertices. Thus, for all $2 \leq n \leq 14$, we have that $g(n) = f(n)$ (recall that $g(n) \leq f(n)$ is the maximum over all n-vertex tournaments of the minimum size of a dominating set). A simple probabilistic argument shows that $g(n) = (1 + o(1)) \log n$, hence so does $f(n)$. It seems interesting to determine whether $f(n) = g(n)$ for all n.

3   A Fault Tolerant Algorithm

Proof of Theorem 1. The idea is to repeatedly apply Corollary 1 and store the information in a rooted tree $\mathcal{T}$ of depth k. The nodes of $\mathcal{T}$ correspond to subgraphs obtained from G by removing at most k vertices. The root of $\mathcal{T}$ corresponds to the entire graph G, and stores a KS (x, Y, S) for G obtained using Corollary 1. The children of the root are subgraphs of the form $G \setminus \{v\}$. The crucial observation is that we only need to consider vertices v that belong to Y (other vertices do not affect the KS), and by Corollary 1 we have $|Y| = O(\log n)$. I.e., the root has only $O(\log n)$ children. We repeat this process until depth k.

Formally, the nodes of $\mathcal{T}$ are pairs of the form $(Q, (x_Q, Y_Q, S_Q))$ where Q is a subtournament of G and (x_Q, Y_Q, S_Q) is a KS for Q obtained using Corollary 1. The root of $\mathcal{T}$ is $(G, (x_G, Y_G, S_G))$. For a node $(Q, (x_Q, Y_Q, S_Q))$ of $\mathcal{T}$ its children are defined as follows: For each $v \in \{x_Q\} \cup Y_Q$ there is a child $(Q_v, (x_{Q_v}, Y_{Q_v}, S_{Q_v}))$ where $Q_v = Q \setminus \{v\}$ is the tournament obtained from Q after removing v. We construct $\mathcal{T}$ only until level k (the level of the root is 0). By Corollary 1, the number of children of a vertex of $\mathcal{T}$ at level r is at most $\log(n + 1 - r)$ as every tree vertex at level r corresponds to a tournament with $n - r$ vertices, whose corresponding KS is constructed in $O((n - r)^2) = O(n^2)$ deterministic time or $O(n)$ expected time. Hence, the entire tree $\mathcal{T}$ is of size $O(\log^k n)$ and is constructed in $O(n^2 \cdot \log^k n) = \tilde{O}(n^2)$ deterministic time or $\tilde{O}(n)$ expected time, as required.

We next describe how changes are handled. Suppose (A, R) is a set of changes where A is a set of newly added vertices, $R \subset V(G)$ is a set of removed vertices, and $|A \cup R| \leq k$. We shall first handle the removed vertices R. To this end, we locate in $\mathcal{T}$ a node $(Q, (x_Q, Y_Q, S_Q))$ where $V(G) \setminus V(Q) \subseteq R$ and $(\{x_Q\} \cup Y_Q) \cap R = \emptyset$. This can be done by traversing $\mathcal{T}$ from the root as we next describe: Initialize $R^* = R$ and proceed as follows. If the root $(G, (x_G, Y_G, S_G))$ already has the required property, we halt at the root. Otherwise, let $v \in (\{x_G\} \cup Y_G) \cap R^*$. Set $R^* = R^* \setminus \{v\}$ and traverse to the child $(Q, (x_Q, Y_Q, S_Q))$ where $V(G) \setminus V(Q) = \{v\}$. Now continue in the same manner. Suppose we are now at some tree node $(Q, (x_Q, Y_Q, S_Q))$. If $(\{x_Q\} \cup Y_Q) \cap R^* = \emptyset$ we halt at this tree vertex. Otherwise, let $v \in (\{x_Q\} \cup Y_Q) \cap R^*$. Set $R^* = R^* \setminus \{v\}$ and continue to the child $(Q_v, (x_{Q_v}, Y_{Q_v}, S_{Q_v}))$ where $V(Q) \setminus V(Q_v) = \{v\}$. Notice that since $|R| \leq k$ and since $|R^*|$ decreases by 1 as we traverse to the next child, the procedure must halt at some tree node $(Q, (x_Q, Y_Q, S_Q))$ after k steps as required. Also note that the time required to traverse the tree is $O(\log n)$ as each list Y_Q has $O(\log n)$ elements. In fact, we may further reduce the traversal time to $O(1)$ since recall that in the KS of Corollary 1, S_Q is just an array $P = P_Q$, where $P(v) = \emptyset$ if

$v \in x_Q \cup Y_Q$. So we just need to query $P(v)$ for every $v \in R^*$ and see if it is empty.

Once we arrive at our desired tree node $(Q, (x_Q, Y_Q, S_Q))$ we have the property that $V(G) \setminus V(Q) \subseteq R$ and $(\{x_Q\} \cup Y_Q) \cap R = \emptyset$. Recall also that S_Q is just an array $P = P_Q$ such that $P(v) = \emptyset$ for $v \in \{x_Q\} \cup Y_Q$ and otherwise $P(v)$ is a vertex of $x_Q \cup Y_Q$ which dominates v. Let $G^* = G \setminus R$, namely the subtournament obtained from G after removing R. Define $S_{G^*} = (P, R)$ (namely, the data structure has two components, the array P and the set R of removed vertices). We claim that $(G^*, (x_Q, Y_Q, S_{G^*}))$ is a KS for G^*. Indeed, given $v \in V(G)$ we wish to return $q(v)$. If $v \in R$ then we return that the query $q(v)$ is invalid. Otherwise, if $v = x_Q$ we return $q(v) = x_Q$, if $(x_Q, v) \in E(G)$ then $(x_Q, v) \in E(G^*)$ and we return $q(v) = (x_Q, v)$. Otherwise, we return $q(v) = (x_Q, P(v), v)$ (notice that this is a valid path of length 2 in G^*). We have proved that a KS for G^* can be constructed in constant time, as required.

We next need to handle the vertex addition set A. Starting from G^* (which is just G with $|R|$ rows and columns of the adjacency matrix of G marked inactive) we add the vertices of A (recall, this means that we activate a bottom row and a rightmost column of the adjacency matrix) one by one. We will show how this can be done in $O(\log n)$ time for each addition. To simplify notation, suppose that $A = \{a_1, \ldots, a_t\}$ where the row/column corresponding to a_i is the $n + i$ row/column of the adjacency matrix. Recall that the KS for G^* is the $(x_Q, Y_Q, S_{G^*}) = (x_Q, Y_Q, (P, R))$ obtained by the deletion procedure above. Starting with a_1, we check whether (y, a_1) is an edge for some $y \in \{x_Q\} \cup Y_Q$. If so, simply set $P(a_1) = y$. This takes $O(\log n)$ time since $|\{x_Q\} \cup Y_Q| = O(\log n)$. Otherwise, a_1 dominates all vertices of the dominating star $\{x_Q\} \cup Y_Q$ of G^* so it is a king of $G^* \cup \{a_1\}$. Therefore, we just modify the KS by setting $Y_Q := Y_Q \cup \{x_Q\}$ and $x_Q := a_1$. Similarly, when adding a_i, if (y, a_i) is an edge for some $y \in \{x_Q\} \cup Y_Q$ then we set $P(a_i) = y$. This search takes $O(\log n)$ time since $|Y_Q| = O(i + \log n) = O(\log n)$ since $i = O(1)$. Otherwise, a_i dominates all vertices of the dominating star of $G^* \cup \{a_1, \ldots, a_{i-1}\}$, hence it is a king of $G^* \cup \{a_1, \ldots, a_{i-1}\}$, and we modify the KS by setting $Y_Q := Y_Q \cup \{x_Q\}$ and $x_Q := a_i$. As $|A| \leq k$ is constant, the entire procedure takes $O(\log n)$ time and we arrive at a KS for $G^* \cup A$, as required. $\qquad\square$

4   A Dynamic Insertion Algorithm

Proof of Theorem 2. Recall that we start from an n-vertex tournament G given by its adjacency matrix, and we want to support every new insertion by exposing a new row at the bottom of the matrix and a new column at the right of the matrix. Our goal is to preprocess G and obtain a KS for it, and then show that following $\Theta(n)$ subsequent insertions, the total number of operations required for maintaining a KS during t insertions is $O(t\sqrt{n})$.

We initially preprocess G in $O(n^{3/2})$ time using the algorithm from [10]. Recall that this computes a KS of the form (x, Y, P) where $|Y| \leq \sqrt{n}$ and P is an array indexed by $V(G)$ such that $P(v) = \emptyset$ if $v \in \{x\} \cup Y$ and otherwise $P(v) \in \{x\} \cup Y$ and $(P(v), v) \in E(G)$.

We insert new vertices one by one. In most cases, an insertion takes only $O(\sqrt{n})$ time, but from time to time it will take longer, yet still not violate the claimed amortized bound. Let $Y_0 := \{x\} \cup Y$ (as we will modify Y and x, we need to remember the original dominating out-star). We shall also maintain a list Z consisting of "worrisome" newly inserted vertices. Initially, $Z = \emptyset$. We use G to refer to the *current* tournament and its adjacency matrix.

Consider the first vertex v to be inserted. We check in $O(\sqrt{n})$ time whether there is some $y \in \{x\} \cup Y = Y_0$ such that (y, v) is an edge. If so, then set $P(v) = y$. Otherwise, v dominates the dominating out-star of G, so v is a king of $G \cup \{v\}$. We then modify $Y := \{x\} \cup Y$, $x := v$, and $P(v) := \emptyset$ to obtain the new KS for $G \cup \{v\}$. We also add v to Z. We continue in this manner for the first $\sqrt{n}$ insertions. I.e., when inserting v, if there is some $y \in \{x\} \cup Y$ where (y, v) is an edge then we set $P(v) = y$. If $y \in (\{x\} \cup Y) \setminus Y_0$ then we add v to Z. Otherwise, we modify $Y := \{x\} \cup Y$, $x := v$, and $P(v) := \emptyset$ and add v to Z. After doing $\sqrt{n}$ insertions, we have $|Y| \leq 2\sqrt{n}$ and $|Z| \leq \sqrt{n}$, so it takes $O(\sqrt{n})$ time to do each of the first $\sqrt{n}$ insertions.

Once we arrive at insertion number $\sqrt{n}$ we "rearrange" our data structure. Notice that $(\{x\} \cup Y) \setminus Y_0 \subseteq Z$ and that $Y \setminus Y_0$ is a prefix of Y. Using the algorithm of Corollary 1, we compute a KS for the subtournament $G[Z]$ induced by Z. Notice that $G[Z]$ is accessible in the bottom right $\sqrt{n} \times \sqrt{n}$ of the present adjacency matrix (so we have $O(1)$ access to its edges). The time required for computing the KS of $G[Z]$ is therefore $O(\sqrt{n}^2) = O(n)$ and the resulting dominating out-star is of size at most $O(\log \sqrt{n}) = O(\log n)$. Let the obtained KS of $G[Z]$ be (x_Z, Y_Z, P_Z). We merge it with our current KS (x, Y, P) of G as follows. We set $Y := Y_0 \cup Y_z$, $x := x_Z$ and $P(v) := P_Z(v)$ for each $v \in Z$ (for $v \notin Z$, we keep $P(v)$ intact). Observe that the entire rearranging operation requires only $O(n)$ time, which is amortized $O(\sqrt{n})$ time over the previous $\sqrt{n}$ "cheap" insertions. Crucially, however, is that now $|Y| \leq \sqrt{n} + \log n$.

We may now perform $\sqrt{n}/\log n$ iterations of $\sqrt{n}$ insertions each, where at the end of each iteration, we perform the rearrangement procedure just described for the Z vertices inserted during the iteration. Each such iteration extends the current Y by at most $\log n$ vertices, so each insertion still requires $O(\sqrt{n})$ amortized time. Note that we have now already added $n/\log n$ vertices. We could have, in fact, performed $\sqrt{n}$ iterations to account for n insertions. This is perfectly fine, but if we do so, the size of Y could be as large as $\sqrt{n} \log n$, so this yields an insertion algorithm with amortized time $O(\sqrt{n} \log n)$ which is only slightly larger than what we are claiming. Let us see how we can further avoid this $\log n$ factor.

Once we perform the $\sqrt{n}/\log n$ iterations described above (call these "level 1" iterations), we can rearrange the union of the sets Z of all iterations. Notice that such union of Z's has at most $n/\log n$ vertices. We can now use the algorithm of [10] to obtain a KS with dominating out-star of size $(n/\log n)^{1/2}$ and the time to construct it is $(n/\log n)^{3/2}$ which is only $O((n/\log n)^{1/2})$ when amortized over all $n/\log n$ insertions. We then merge it as described earlier with the current KS to obtain a KS for the current tournament whose Y is only of size $\sqrt{n} + (n/\log n)^{1/2}$.

We can now perform $\sqrt{\log n}$ "level 2" iterations where each iteration invokes a "level 1 iteration". Once done, we have inserted $n/\sqrt{\log n}$ vertices, and use [10] to rearrange, in $(n/\sqrt{\log n})^{3/2}$ time for a KS of the current tournament whose Y is of size only $\sqrt{n} + (n/\sqrt{\log n})^{1/2}$. Continuing in this way, we perform "level t" iterations, to obtain a KS whose dominating out-star is of size at most $\sqrt{n} + n^{1/2}/(\log n)^{1/2^t}$ and the time for rearranging using [10] is $(n/(\log n)^{1/2^{t-1}})^{3/2}$. Once we have $t = O(\log \log \log n)$, we have inserted $\Theta(n)$ vertices and the current Y is still of size $O(\sqrt{n})$. This is because $\sum_{t=1}^{\log \log \log n} (1/\log n)^{1/2^t} = O(1)$. The amortized time for each insertion is therefore at most $O(\sqrt{n})$. $\qquad\square$

5   A Dynamic Edge-Flip Algorithm

Proof of Theorem 3. We will use the following simple claim.

Claim 1. *Given any n-vertex tournament, we can find a vertex with out-degree at least $n/4$ in expected $O(n)$ time.*

Proof. We claim that in every tournament there are at least $(n-1)/2$ vertices whose out-degree is at least $n/4$. Suppose not, and let X be the set of vertices whose out-degree is less than $n/4$, so we have that $|X| \geq n/2 + 1$. Consider the subtournament induced by X and a vertex with maximum out-degree in it. Already in that subtournament, the out-degree of this vertex is at least $(|X| - 1)/2 \geq n/4$, a contradiction. We therefore conclude that a randomly chosen vertex v has out-degree at least $n/4$ with probability at least $1/2 - 1/2n \geq 1/3$. The lemma follows since checking v's out-degree is done in $O(n)$ time. $\qquad\square$

Let $F = \{f_1, \ldots, f_m\}$ be any sequence of flips. Let G^i be the tournament obtained from G after performing the flips $f_1, \ldots, f_i$ (so that $G^0 = G$). I.e., if $f_i = \{x, y\}$ then G^i is obtained from G^{i-1} by flipping the edge between x and y.

We first define the components of our data structure that contains the KS following the i'th flip. Initially, the data structure will contain the KS for $G^0 = G$.

- The adjacency matrix A of the current tournament G^i.
- A variable K holding a king of G^i.
- An array D of vertices forming a dominating out-star of G^i. We call the vertices of D *backbone vertices*. It will always hold that if $D = \{v_1^i, \ldots, v_{t_i}^i\}$ (i.e., $D[j] = v_j^i$), then $t_i = O(\log n)$ and $(v_j^i, v_{j'}^i) \in E(G^i)$ if and only if $j > j'$. Furthermore, the last vertex $v_{t_i}^i$ is K.
- An array B of size n such that if $v = v_j^i$ is a backbone vertex, then $B[v] = -j$. Otherwise, $B[v]$ is the index of a backbone vertex such that $(D[B[v]], v) \in E(G^i)$. We call the set of all vertices v for which $B[v] = j$, the *bag of v_j^i*.
- An array L of size n such that the first segment of L contains all vertices in the first bag, as well as the first backbone vertex. The second segment of L contains all vertices in the second bag, as well as the second backbone vertex and so on. The ordering within each segment is arbitrary.

- An array P such that $P[j]$ is the index in L of the last vertex of the j'th segment. If j is larger than the number of backbone vertices, then $P[j]$ is irrelevant (and will not be used).
- An array U of *unmarked vertices*. These can be in any order. We will only use U when updating the data structure after a flip. Once the current update is complete, U is empty.
- A variable M holding the number of marked vertices (i.e., $M = n - |U|$).

Let us observe that a query can be performed in $O(1)$ time. Given a vertex v, we check $B[v]$. If $B[v]$ is negative (so v is a backbone vertex), then if $v \neq K$ we return the path (K, v), and otherwise we return the path (K). If $B[v]$ is positive (i.e., it is equal to the index of the bag containing v), then let $D[B[v]]$ be its dominator. If $D[B[v]] \neq K$, then return the path $(K, D[B[v]], v)$. Otherwise, return the path (K, v).

Let us next see how we set our data structure for the initial tournament $G = G^0$ whose adjacency matrix A is given. We use Claim 1 to locate (in expected $O(n)$ time) a vertex v (so that $v = v_1^0$ will be the first backbone vertex of G^0) with out-degree at least $n/4$. For each out-neighbor u of v, we set $B[u] = 1$. We also set $B[v] = -1$ and set $D[1] = v$. All out-neighbors of v and v itself are assigned consecutively to the initial segment of L. We set M to be the number of out-neighbors of v plus one. We set $P[1] = M$ (since this is the index in L of the end of the first segment). We set U to be the list of all in-neighbors of v.

Locating the j'th Backbone Vertex of G^0. More generally, we describe how to locate the j'th backbone vertex v of G^0. We will later, during updates, use this procedure to determine a j''th backbone vertex of G^i, if necessary. We repeatedly choose an index p at random in $\{1, \ldots, n - M\}$ (as we must choose an unmarked vertex, and the number of unmarked vertices is $n - M$), and let $v = U[p]$ (so v is a randomly-chosen unmarked vertex). For each vertex $u \in U$ we query $A[v, u]$ to find the out-degree of v in the remaining subtournament on the unmarked vertices. If this number is larger than $(n - M)/4$ we are happy with v, otherwise we choose a different v. By Claim 1, the expected time to find v is $O(n - M)$. Once v is found we continue as follows: For each unmarked out-neighbor u of v (i.e., in $O(n - M)$ time), we set $B[u] = j$. We also set $B[v] = -j$ and set $D[j] = v$. All unmarked out-neighbors of v and v itself are assigned consecutively to the next segment of L. Note that we know that this segment starts at index $P[j - 1] + 1$ of L. We increase M by the number of unmarked out-neighbors of v plus one. We set $P[j] = M$ (since this is now the index in L of the last vertex of the j'th segment). We set U to be the list of all unmarked in-neighbors of v (as these now remain unmarked). Note that the total expected time to locate the j'th backbone vertex and perform the data structure modifications is $O(n - M)$. Also observe that the case $j = 1$ described in the previous paragraph can be merged to the general case if we just initialize $U = [n]$, $M = 0$ and $P[0] = 0$.

We repeatedly apply the above procedure increasing j by one each time, halting when $M = n$ and set K to be the backbone vertex v in the final iteration.

Notice that after this preprocessing stage, the data structure satisfies all required properties listed earlier. Furthermore, as in each iteration the number of unmarked vertices decreases by a fraction of at least $3/4$, we have that the size of the backbone is at most $O(\log n)$. Also, the expected runtime of the j'th iteration is $O((3/4)^j n)$ and hence expected $O(n)$ in total.

The Update Procedure. Let B_j^i be the *bag* of v_j^i, and let G_j^i be the subtournament consisting of all backbone vertices $v_j^i, v_{j+1}^i, \ldots, v_{t_i}^i$ and their bags. In particular, $G_1^i = G^i$. We shall maintain the following additional property for our data structure: if $x \in B_j^i \cup \{v_j^i\}$, then there is an edge from x to every backbone vertex v_k^i with $k < j$. Clearly, this property holds after the preprocessing stage (i.e., for $i = 0$). Let us now describe the update procedure. Suppose we arrive at flip $f_{i+1} = \{x, y\}$ where the current tournament is G^i. We consider three cases:

1. Both x and y are not backbone vertices of G^i. In this case we only need to modify $A[x, y]$ (in constant time) to obtain the adjacency matrix of G^{i+1}. Note that we can tell if a vertex v is backbone or not by just checking if $B[v]$ is positive or negative.
2. One of $\{x, y\}$ (say, w.l.o.g x) is a backbone vertex and the other is not, and it also holds that $x = v_j^i$, $y \in B_k^i$, and $k < j$. Again, we modify $A[x, y]$ to obtain the adjacency matrix of G^{i+1}. Note that we can determine j as $j = -B[x]$ and determine k as $k = B[y]$.
3. Otherwise, let k be the smallest index such that v_k^i is a backbone vertex and $v_k^i \in \{x, y\}$. Again, k is found by examining $B[x]$ and $B[y]$. Notice that both x and y are vertices of G_k^i. We modify $A[x, y]$ to obtain the adjacency matrix of G^{i+1} (in particular, G_k^i becomes G_k^{i+1}). Next, we recompute a KS and a backbone of G_k^{i+1} to obtain a backbone and corresponding bags. Informally, this is done as in the aforementioned preprocessing algorithm, only we start at its k'th iteration (see details below). Finally, we concatenate the obtained backbone and bags to the end of the prefix backbone $\{v_1^i, ..., v_{k-1}^i\}$ and their bags (this is the prefix backbone of G^i which remains intact) to obtain a backbone and bags for G^{i+1}.

It is easy to see that cases 1 and 2 require $O(1)$ time and maintain all the data structure properties. Let us now be more formal regarding case 3. Notice that we need to "reset" our data structure such that it looks as if it now starts the k'th iteration (as the first $k - 1$ bags and backbone vertices remain intact). Let $w = P[k-1]+1$ and let U be all the vertices of $L[w, \ldots, n]$ in U (so we effectively "unmark" all vertices of G_k^i). Also set $M = P[k-1]$ to be the number of marked vertices. Now set $j = k$ and use the aforementioned procedure *locating the j'th backbone vertex*, performing it repeatedly until no unmarked vertices are left. This will yield the backbone vertices $v_k^{i+1}, v_{k+1}^{i+1}, \ldots, v_{t_{i+1}}^{i+1}$ and their bags, and modify the data structure accordingly, maintaining its properties.

It remains to analyze the running time of case 3. Note that the running time is (in expectation) linear in the number of vertices of G_k^i (i.e., $O(|V(G_k^i)|)$ time). In particular, the larger k is, the less time the update is expected to take, as the sizes of the bags geometrically decrease (in the worst case, if $k = 1$, we actually run the entire preprocessing algorithm from scratch in $O(n)$ time). We next prove that case 3 is not expected to occur frequently.

For a vertex v, a flip stage i, and an index j, Let $p(v, i, j)$ be the probability that v is the j'th backbone vertex of G^i, i.e., $v = v_j^i$. Recall that $1 \leq j \leq O(\log n)$. Consider the last flip-stage that changed v_j^i (this could be at flip-stage i or earlier). Then, by Claim 1, v_j^i was chosen at random among a pool of at least $|V(G_j^i)|/4$ vertices and the probability that v was chosen to be v_j^i is therefore at most $4/|V(G_j^i)|$. Thus, $p(v, i, j) \leq 4/|V(G_j^i)|$.

Fix $1 \leq k \leq O(\log n)$. Consider the $(i+1)$'th flip $f_{i+1} = \{x, y\}$. The probability that $x = v_k^i$ or $y = v_k^i$ is therefore at most $4/|V(G_k^i)| + 4/|V(G_k^i)| = 8/|V(G_k^i)|$. Hence, the probability that case 3 occurred and the corresponding value of k in it is our chosen k is at most $8/|V(G_k^i)|$. The expected runtime in this case is $O(|V(G_k^i)|)$. Hence, the a priori expected runtime is only $O(|V(G_k^i)| \cdot 8/|V(G_k^i)|) = O(1)$. As there are $O(\log n)$ possible choices for k, the expected runtime of an update is $O(\log n)$, as required. $\qquad\square$

6   A Single Edge-Flip Algorithm

Proof of Theorem 4. Let (x, Y, S) be a KS for G. Recall that x is a king, Y is a list of (some) out-neighbors of x, and S is a data structure such that given $v \in V(G)$ where $(v, x) \in E(G)$, S may be queried in constant time to yield a vertex $y \in Y$ such that (x, y, v) is a path of length 2. We next define the components (and properties) of S. These properties hold after preprocessing and, when specified, also hold after a single edge-flip.

1. A variable m (meaning *marked*). After preprocessing, $m = empty$, but following an edge-flip, m may be assigned a vertex.
2. An array P indexed by $V(G)$, such that $P(v)$ is a list of at most two vertices. After preprocessing, it holds for each $v \in \{x\} \cup Y$ that $P(v)$ is empty, and it holds for $v \notin \{x\} \cup Y$ that $P(v)$ contains at most two vertices, the first of which, say y, is in $\{x\} \cup Y$ such that $(y, v) \in E(G)$. After an edge-flip, it will be the case that if v is not a vertex of the current dominating out-star, then the first *unmarked* vertex of $P(v)$ (there will always be such a vertex) is a vertex of the current dominating out-star.
3. After preprocessing, suppose that $Y = \{y_1, \ldots, y_t\}$. Then, if $v \notin \{x\} \cup Y$ and the first vertex of $P(v)$ is y_j, it shall always be the case that $(v, y_i) \in E(G)$ for $i < j$. Also, if the first vertex of $P(v)$ is x, then $(v, y_i) \in E(G)$ for all $1 \leq i \leq t$.
4. An array W indexed by $V(G)$ such that after preprocessing, $W(v)$ is the list of all vertices in $\{x\} \cup Y$ that are in-neighbors of v. Furthermore, if $v \notin \{x\} \cup Y$ then the first vertex of $W(v)$ is also the first vertex of $P(v)$.
5. After preprocessing, let Z be the set of all vertices v in $V(G) \setminus (\{x\} \cup Y)$ such that the first vertex of $P(v)$ is x (in particular these vertices are directly dominated by the original king x, but notice that there may also be other vertices not in Z which are dominated by the king). Assuming $Z \neq \emptyset$, we

compute a king z for $G[Z]$ (the subtournament induced by Z), a list Y_Z such that $\{z\} \cup Y_Z$ is a dominating out-star of $G[Z]$, and an array P_Z such that for each $v \in Z \setminus (\{z\} \cup Y_Z)$, we have that $(P_Z(v), v) \in E(G)$ and $P_Z(v) \in (\{z\} \cup Y_Z)$. In particular, (z, Y_Z, P_Z) is a KS for $G[Z]$. Now, for each $v \in Z \setminus (\{z\} \cup Y_Z)$, we will have that the *second* vertex of $P(v)$ is $P_Z(v)$ (recall that the first vertex of $P(v)$ is x).

We invoke the algorithm of [10] twice to construct the KS specified above on the original G. We first invoke it for G to obtain x, Y, and the array P where $P(v) = \emptyset$ if $v \in \{x\} \cup Y$ and otherwise $P(v)$ is a single vertex in $\{x\} \cup Y$ which dominates v. Recall that Item 2 above is an artifact of the algorithm (and that $\{x\} \cup Y$ is a transitive subtournament). We then gather all vertices of Z as in Item 5 and invoke [10] again on $G[Z]$ to obtain z, Y_Z, P_Z and the modified two-element lists $P(v)$ for $v \in Z \setminus (\{z\} \cup Y_Z)$. Both invocations require $O(n^{3/2})$ time, and the construction of W in Item 4 is done in $O(|Y|n) = O(n^{3/2})$ time, since $|Y| \le \sqrt{n}$. Altogether, the preprocessing takes $O(n^{3/2})$ time, as claimed.

Let us next be formal about our query procedure that may be used either on the original graph after preprocessing, or after a single edge-flip (where the correctness after an edge-flip is proved below). Here we refer to G, x, Y, S as either the original tournament and its KS or the tournament and its modified KS after a single flip. Given $v \in V$, if $v = x$, then return $q(v) = (x)$. Else, if $(x, v) \in E(G)$ return $q(v) = (x, v)$, else return (x, y, v) where y is the first unmarked vertex of $P(v)$. Clearly this takes $O(1)$ time and is correct for the original tournament.

Now consider an edge-flip. There are several cases:

- If (u, v) is flipped to (v, u) where both $u, v \notin \{x\} \cup Y$. This can be verified in constant time as for such vertices we have that $P(v)$ and $P(u)$ are both not empty. In this case we do nothing. We keep the same KS and this is valid since the dominating out-star has not changed.
- If (v, y) is flipped to (y, v) for $v \notin \{x\} \cup Y$ and $y \in \{x\} \cup Y$. Here again we know that $y \in \{x\} \cup Y$ since $P(y)$ is empty and again we do nothing. This is valid since v is still dominated by the first vertex of $P(v)$ (which can't be y).
- If (y, y') is flipped to (y', y) for $y, y' \in Y$. Once again, we do nothing, as $\{x\} \cup Y$ is still a dominating out-star.
- If (y, v) is flipped to (v, y) for $v \notin \{x\} \cup Y$ and for $y \in \{x\} \cup Y$. If y is not the first vertex of $P(v)$, then we do nothing. If y is the first vertex of $P(v)$, we must search for another vertex to dominate v. Since the first vertex of $W(v)$ is also y, we look for the next vertex. If y' is such, then just set $P(v) = y'$. If, however, there is no next vertex, then v dominates all of $\{x\} \cup Y$. Hence, v is a new king, so we can modify the KS to be $Y := \{x\} \cup Y$, $x := v$, and we do not need to change the array P at all (all vertices except v keep their dominators).
- If (x, y) is flipped for $y \in Y$. Then x is no longer guaranteed to be king. There are now two possibilities. If $Z = \emptyset$ then we can just define $P(x) := y$, set the new king as y_t (the last vertex of Y; recall that Y still induces a transitive subtournament) and modify $Y = Y \setminus \{y_t\}$. If, however, Z is not empty, we

mark $m = x$, add $Y := Y \cup Y_Z$ (so that the order of Y is $(y_1, \ldots, y_t, Y_Z)$ and note that this takes $O(1)$ time by concatenating lists), and set the new king to be z. Notice that queries work correctly even for vertices that were originally in Z, as when scanning their list $P(v)$, the first vertex (i.e., x) is marked so is not returned, and their second vertex is unmarked and returned correctly. $\square$

References

1. Abboud, A., Grossman, T., Naor, M., Solomon, T.: From donkeys to kings in tournaments. In: 32nd ESA, vol.308, pp. 3:1–3:14 (2024)
2. Ajtai, M., Feldman, V., Hassidim, A., Nelson, J.: Sorting and selection with imprecise comparisons. ACM Trans. Algorithms **12**(2), 19:1–19:19 (2016)
3. Biswas, A., Jayapaul, V., Raman, V., Satti, S.R.: Finding kings in tournaments. Discret. Appl. Math. **322**, 240–252 (2022)
4. Erdős, P., Moser, L.: On the representation of directed graphs as unions of orderings. Math. Inst. Hung. Acad. Sci **9**, 125–132 (1964)
5. Landau, H.G.: On dominance relations and the structure of animal societies: III the condition for a score structure. Bull. Math. Biophys. **15**, 143–148 (1953)
6. Lu, X., Wang, D.-W., Wong, C.K.: On the bounded domination number of tournaments. Discret. Math. **220**(1–3), 257–261 (2000)
7. Mande, N.S., Paraashar, M., Saurabh, N.: Randomized and quantum query complexities of finding a king in a tournament. In: Bouyer, P., Srinivasan, S. (eds.) 43rd FSTTCS, pp. 30:1–30:19 (2023)
8. Maurer, S.B.: The king chicken theorems. Math. Mag. **53**(2), 67–80 (1980)
9. R. E. A. C. Paley. J. math. and phys. Bull. Math. Biophys. **12**, 311–320 (1933)
10. Shen, J., Sheng, L., Wu, J.: Searching for sorted sequences of kings in tournaments. SIAM J. Comput. **32**(5), 1201–1209 (2003)
11. Stearns, R.: The voting problem. Am. Math. Mon. **66**(9), 761–763 (1959)
12. Szekeres, E., Szekeres, G.: On a problem of Schütte and Erdös. Math. Gazette 290–293 (1965)

Learning-Augmented Online Bipartite Matching in the Random Arrival Order Model

Kunanon Burathep[(✉)] , Thomas Erlebach , and William K. Moses Jr.

Department of Computer Science, Durham University, Durham, UK
{kunanon.burathep,thomas.erlebach,william.k.moses-jr}@durham.ac.uk

Abstract. We study the online unweighted bipartite matching problem in the random arrival order model, with n offline and n online vertices, in the learning-augmented setting: The algorithm is provided with untrusted predictions of the types (neighborhoods) of the online vertices. We build upon the work of Choo et al. (ICML 2024, pp. 8762–8781) who proposed an approach that uses a prefix of the arrival sequence as a sample to determine whether the predictions are close to the true arrival sequence and then either follows the predictions or uses a known baseline algorithm that ignores the predictions and is β-competitive. Their analysis is limited to the case that the optimal matching has size n, i.e., every online vertex can be matched. We generalize their approach and analysis by removing any assumptions on the size of the optimal matching while only requiring that the size of the predicted matching is at least αn for any constant $0 < \alpha \leq 1$. Our learning-augmented algorithm achieves $(1 - o(1))$-consistency and $(\beta - o(1))$-robustness. Additionally, we show that the competitive ratio degrades smoothly between consistency and robustness with increasing prediction error.

Keywords: Learning-Augmented Algorithms · Untrusted Predictions · Competitive Analysis · Consistency · Robustness

1 Introduction

Online bipartite matching is a fundamental problem in theoretical computer science with significant applications in areas such as online advertising, ride-sharing platforms, and resource allocation. This problem involves a bipartite graph where one side of the graph (the offline vertices) is known in advance, while the other side (the online vertices) arrives sequentially. When an online vertex arrives, it must be permanently matched to an offline vertex or left unmatched, without knowledge of future arrivals. The goal is to maximize the size of the matching.

Traditional approaches to online bipartite matching consider the worst-case guarantees in the adversarial arrival model, where an adversary controls the graph's structure and the arrival sequence of online vertices. It is easy to see

© The Author(s), under exclusive license to Springer Nature Switzerland AG 2026
J. Kozik and A. Wolff (Eds.): SOFSEM 2026, LNCS 16448, pp. 361–375, 2026.
https://doi.org/10.1007/978-3-032-17801-5_27

that the greedy algorithm, which matches each online vertex to an arbitrary unmatched neighbor if one exists, is $\frac{1}{2}$-competitive, and it is known that randomization allows one to obtain an improved competitive ratio of $1-1/e \approx 0.632$ [19]. A setting that makes it possible to achieve an even better competitive ratio, at least $\beta \approx 0.696$, is the random arrival order model, where the sequence of online vertices is assumed to arrive in a uniformly random order [13,18,23].

Recently, learning-augmented algorithms have emerged as a means to enhance online algorithms by incorporating (untrusted) predictions about future inputs. These predictions are also referred to as advice. Learning-augmented algorithms aim to achieve near-optimal performance when the predictions are accurate while maintaining a competitive ratio close to that of online algorithms without predictions when the predictions are inaccurate. For learning-augmented online bipartite matching, various types of predictions have been studied, such as degree information [1], advice derived from reinforcement learning [20], and predictions of the types of online vertices [9]. In this paper, we build upon the work of Choo et al. [9] who considered the latter type of predictions in the random arrival order model. They proposed an approach that uses a prefix of the arrival sequence as a sample to determine whether the predictions are close to the true arrival sequence and then either follows the predictions or uses a known baseline algorithm that ignores the predictions and is β-competitive. Here, β is the best-known competitive ratio of an online algorithm for bipartite matching without predictions in the random arrival order model, currently $\beta \approx 0.696$ [23]. The analysis by Choo et al. is limited to the case that the optimal matching has size n, i.e., every online vertex can be matched.

We generalize their approach and analysis by removing any assumptions on the size of the optimal matching while only requiring that the size of the predicted matching is at least αn for an arbitrary constant $0 < \alpha \leq 1$. Our learning-augmented algorithm achieves $(1 - o(1))$-consistency and $(\beta - o(1))$-robustness. Additionally, we show that the competitive ratio degrades smoothly between consistency and robustness with increasing prediction error: We show that the competitive ratio is at least $1 - \frac{2L_1(p,q)}{2\alpha + L_1(p,q)} - o(1)$, where $L_1(p,q)$ is the L_1-distance between the distribution of vertex types in the predicted arrival sequence and the true arrival sequence.

Our result addresses the following key limitations of the algorithm and analysis proposed by Choo et al. [9]. In cases where the predicted matching size is smaller than βn, their approach does not achieve $(1 - o(1))$-consistency even if the predictions are perfectly accurate: Their algorithm runs the baseline algorithm whenever the predicted matching size is smaller than βn. Furthermore, their analysis relies on the assumption that the optimal matching size equals n.

Although Choo et al. do not explicitly state the assumption that the optimal matching has size $n^* = n$, this assumption underlies several parts of their analysis, including Theorem 4.1 and Lemmas 4.3 and 4.4. For a detailed discussion, see Sect. 1.3. Consequently, their analysis fails to establish $(1 - o(1))$-consistency when the optimal matching size is smaller than n, even if the predicted matching size exceeds βn. In the more general setting with arbitrary optimal matching size, the analysis in the random arrival order model requires a refined argument:

after switching to the baseline algorithm, its performance depends on how many matches are still possible once the sampling phase has passed. Our analysis addresses this using Lemma 3 to bound the expected number of such remaining optimal matches and Corollary 1 to bound the algorithm's expected total number of matches.

The paper is structured as follows. Section 1.1 introduces preliminaries, Sect. 1.2 reviews related work, and Sect. 1.3 compares our result to that by Choo et al. [9]. Section 2 presents our algorithm and its analysis: we first outline the algorithm, then provide the key components of the analysis in Sects. 2.1, 2.2 and 2.3, and combine these results to prove our main theorem in Sect. 2.4. Section 3 concludes the paper. Proofs of statements marked with $(\star)$ are omitted due to the page limit. For the full version of this paper, see [8].

1.1   Preliminaries

A *matching* in a graph with edge set E is a set $M \subseteq E$ of edges such that each vertex of the graph has at most one incident edge in M. We consider the problem of online unweighted bipartite matching in a graph $G = (U \cup V, E)$, where U and V are the sets of n offline and n online vertices, respectively. The online vertices arrive sequentially. Upon the arrival of an online vertex, the algorithm must either match it to one of its available neighbors (i.e., to a neighbor that has not yet been matched) or leave it unmatched. The objective value is the number of online vertices that have been matched, also referred to as the number of matches made by the algorithm.

We focus on the random arrival order model, where the arrival order of the online vertices $v \in V$ is a uniformly random permutation. The performance of a (possibly randomized) algorithm is measured using the *competitive ratio*, defined as

$$\inf_{G=(U \cup V, E)} \frac{E_{V\text{'s arrival seq.}}\left[E_{\text{random choices of alg.}}\left[\#\text{matches by algorithm}\right]\right]}{\#\text{matches by optimal solution}}$$

The size of the optimal matching does not depend on the arrival order. The size of the matching produced by an algorithm depends on the order of the arrival sequence and on the random choices made by the algorithm, hence we take the expectation over both of these. The competitive ratio is a value ≤ 1, and the closer to 1 it is, the better.

Learning-augmented algorithms are algorithms that are provided with (untrusted) predictions, and the goal is to design algorithms that achieve performance close to the optimal solution when the predictions are accurate, but still maintain a competitive ratio close to that of online algorithms without predictions if the predictions are arbitrarily wrong. These properties are formalized as *consistency* and *robustness*: An algorithm is γ-*consistent* if its competitive ratio is at least γ when the predictions are entirely correct, and it is ρ-*robust* if its competitive ratio is at least ρ no matter how wrong the predictions are. In addition, one is interested in *smoothness*, i.e., one aims to show that the competitive ratio of an algorithm degrades gracefully between consistency and robustness, as a function of the prediction error.

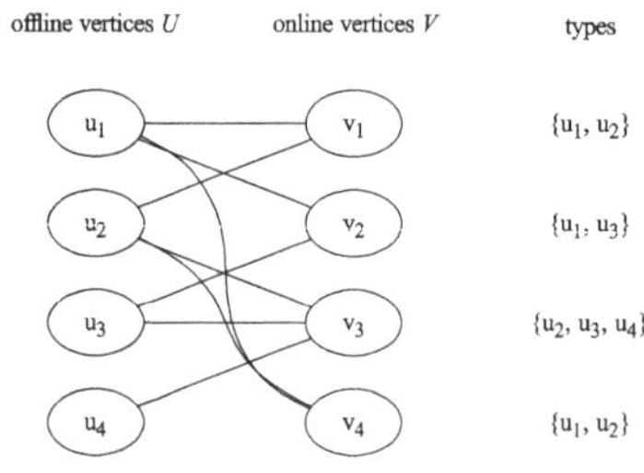

(a) Bipartite graph and types of online vertices, e.g., v_1 and v_4 both have type $\{u_1, u_2\}$ as they have edges to u_1 and u_2

	Type t	Count $c^*(t)$
	$\{u_1, u_2\}$	2
T^*	$\{u_1, u_3\}$	1
	$\{u_2, u_3, u_4\}$	1
$2^U \setminus T^*$	$\ldots$	0

(b) Table with number $c^*(t)$ of type t vertices, e.g., $c^*(\{u_1, u_2\}) = 2$

Fig. 1. Illustration of vertex types.

The *type* of an online vertex $v \in V$ is the set of offline vertices in U that are adjacent to v, i.e., the set $\{u \in U \mid \{u, v\} \in E\}$. There are 2^n possible types, but since there are only n online vertices, the number of different types that occur in any given arrival sequence is at most n. The idea of representing online vertices as types was first introduced by Borodin et al. [5]. The set of online vertices can be represented by a function c^* that maps each type t to the number of type t vertices contained among the online vertices. We denote by T^* the set of types t with $c^*(t) > 0$, and let $r^* = |T^*|$. See Fig. 1 for an illustration. We assume that the algorithm is provided with a function $\hat{c}$ that maps each type t to the number of online vertices of type t that are predicted to arrive, and we denote the set of types t with $\hat{c}(t) > 0$ by $\hat{T}$ and its cardinality by $\hat{r}$. It is clear that $\sum_{t \in T^*} c^*(t) = n$, and we assume also that $\sum_{t \in \hat{T}} \hat{c}(t) = n$. We denote the L_1-distance between c^* and $\hat{c}$ by $L_1(c^*, \hat{c}) = \sum_t |c^*(t) - \hat{c}(t)|$. The type prediction $\hat{c}$ determines a predicted graph $\hat{G} = (U \cup \hat{V}, \hat{E})$ in which $\hat{V}$ contains exactly $\hat{c}(t)$ vertices of type t. We use n^* to denote the size of an optimal matching in G and $\hat{n}$ to denote the size of an optimal matching in $\hat{G}$.

If p and q are two probability distributions over a domain T, their L_1-distance $L_1(p, q)$ is defined as $L_1(p, q) = \sum_{t \in T} |p(t) - q(t)|$. We can identify the online vertices with a probability distribution p where type t has probability $c^*(t)/n$, and the predicted online vertices with a probability distribution q where type t has probability $\hat{c}(t)/n$. Note that $L_1(c^*, \hat{c}) = n \cdot L_1(p, q)$ holds in this case. In order to estimate the L_1-distance between the probability distributions representing the predicted online vertices and the true online vertices based on a sample, we use the following theorem stated by Choo et al. [9], which in turn is based on results by Jiao et al. [16].

Theorem 1 (Choo et al.[9]). *For any $\varepsilon > 0$ and $\delta' \in (0,1)$, let q be a reference distribution over a domain T of size r. There exists an even integer $s \in \Theta\left(\frac{r \log(1/\delta')}{\varepsilon^2 \log(r)}\right)$ such that an algorithm drawing $s_1 + s_2$ i.i.d samples from an unknown distribution p over T, where $s_1, s_2 \sim Poisson(s/2)$, outputs an estimate $\hat{L}_1(p,q)$ satisfying $|\hat{L}_1(p,q) - L_1(p,q)| \leq \varepsilon$ with probability at least $1 - \delta'$.*

1.2 Related Work

Online Bipartite Matching Without Predictions. Karp et al. [19] first introduced the online bipartite matching problem and presented a randomized RANKING algorithm, which achieves a competitive ratio of $1 - 1/e$ in the adversarial arrival model. Goel and Mehta [13] studied the random arrival order model and presented a deterministic algorithm with a competitive ratio of $1 - 1/e$. They further showed that no online algorithms in the random order model can achieve a competitive ratio better than 0.83. Karande et al. [18] demonstrated that RANKING achieves a competitive ratio of at least 0.653 in the random arrival order model. They also described a family of graphs that show an upper bound for the competitive ratio of RANKING of 0.727 in the random order model. Mahdian and Yan [23] showed that RANKING achieves competitive ratio at least 0.696.

Another model is the online stochastic model, where each online vertex $v \in V$ arrives according to some known probability distribution. We are given a probability distribution $\mathcal{D}$ over the elements of V such that each time a vertex v arrives, it is drawn independently from the distribution $\mathcal{D}$. This model is also referred to as the known i.i.d. arrival model. Feldman et al. [11], who first studied the known i.i.d. arrival model, presented a 0.67-competitive algorithm. Manshadi et al. [24] established a hardness bound of 0.823 for randomized algorithms in the known i.i.d. model and also presented an algorithm achieving a competitive ratio of 0.702. Jaillet and Lu [15] presented a 0.725-competitive algorithm, and Brubach et al. [6,7] further improved the competitive ratio to 0.7299.

Learning-Augmented Algorithms for Matching. Learning-augmented algorithms have been studied across various fields, including the ski rental problem [2,3, 14,26,29], non-clairvoyant scheduling [26,29], and caching [22,27,28]. For further information about this area, see the survey by Mitzenmacher and Vassilvitskii [25], and for further references, refer to the website managed by Lindermayr and Megow [21].

For online bipartite matching, various prediction models have been explored. Aamand et al. [1] studied adversarial arrival models using predicted degree information as advice. Their algorithm is optimal for graphs drawn from the Chung-Lu-Vu random graph model [10], where the expected values of offline degrees are known and provided as advice.

Feng et al. [12] proposed a two-stage vertex-weighted variant in which the advice is a proposed matching for the online vertices arriving in the first stage. Jin and Ma [17] further established a tight robustness-consistency trade-off for the two-stage model, presenting an algorithm with $(1 - (1 - \sqrt{1 - R})^2)$-consistency and R-robustness, where $R \in [0, 3/4]$.

Antoniadis et al. [4] studied online weighted bipartite matching in a random arrival model. In their model, the predictions provide information about the maximum edge weights adjacent to each offline vertex. Additionally, their analysis is based on a hyperparameter of prediction confidence. Consequently, the consistency and robustness of their algorithm depend on the trade-off hyperparameter.

Li et al. [20] studied online weighted capacity bipartite matching using pre-trained reinforcement learning as advice. In their setting, the offline vertices have capacities, denoted by c_u, that limit the number of matches they can make with online vertices. The setting is referred to as *no-free disposal*. They also consider *free disposal*, where each offline vertex can match with more than c_u online vertices, but only the top c_u rewards are counted. Their algorithm achieves $(1 - \rho)$-consistency and $(\rho \cdot \beta)$-robustness, where ρ is the hedging ratio that balances robustness and consistency, and β is the best competitive ratio of online algorithms without predictions in their setting.

As mentioned earlier, Choo et al. [9] studied unweighted bipartite matching with random-order arrivals. The predictions specify the types of the online vertices, and the number of online vertices of each type. In this paper, we modify their algorithm and its analysis to address several limitations. A detailed comparison of our result with their work is provided in the following section.

1.3   Comparison with the Result of Choo et al. [9]

In this section, we discuss our interpretation of the analysis of Choo et al. [9] and clarify how our results differ. Their main result is the following:

Theorem 4.1 (in Choo et al. [9]). *For any advice $\hat{c}$ with $|\hat{T}| = \hat{r}, \varepsilon > 0$ and $\delta > \frac{1}{poly(\hat{r})}$, let $\hat{L}_1$ be the estimate of $L_1(p,q)$ obtained from $k = s_{\hat{r},\varepsilon,\delta} \cdot \sqrt{\log(\hat{r}+1)}$ IID samples of p. TESTANDMATCH produces a matching of size m with competitive ratio of at least $\frac{\hat{n}}{n} - \frac{L_1(p,q)}{2} \geq \beta$ when $\hat{L}_1 \leq 2(\frac{\hat{n}}{n} - \beta) - \varepsilon$, and at least $\beta \cdot (1 - \frac{k}{n})$ otherwise, with success probability $1 - \delta$.*

Although their Theorem 4.1 does not state it explicitly, their proof relies on the assumption $n^* = n$, and so the theorem is only proven for that case (see below for details). Hence, the theorem shows only for $k = o(n)$ and $n^* = n$ that their algorithm achieves 1-consistency and $(\beta - o(1))$-robustness with probability $1 - \delta$. They mention that the guarantees on the competitive ratio that are obtained with probability $1 - \delta$ can be converted into guarantees on the expected competitive ratio, giving $(1 - \delta)$-consistency and $\beta(1 - o(1))(1 - \delta)$-robustness in expectation (and they suggest $\delta = 0.001$ as a possible choice). We state our results directly in terms of the expected competitive ratio instead.

Differences to Our Result. The algorithm TESTANDMATCH by Choo et al. runs the β-competitive baseline algorithm on all arrivals whenever $\frac{\hat{n}}{n} \leq \beta$. In contrast, our algorithm replaces this requirement by $\frac{\hat{n}}{n} \leq \alpha$ for an arbitrarily small constant $\alpha > 0$. As a result, our algorithm obtains both $(1 - o(1))$-consistency and competitive ratio greater than β over a larger set of inputs, as we explain in the following.

First, consider consistency. Assume that the optimal matching has size n^* with $\alpha n \leq n^* < \beta n$ and that the predictions are accurate: $\hat{n} = n^*$ and $L_1(c^*, \hat{c}) = 0$. Our algorithm is $(1-o(1))$-competitive for such inputs. The algorithm of Choo et al. [9], however, does not achieve $(1 - o(1))$-consistency because the condition $\hat{n} = n^* < \beta n$ causes it to run the baseline algorithm on all arrivals, yielding only competitive ratio β. Choo et al. mention the possibility of modifying the predictions by adding a complete bipartite graph between unmatched offline and unmatched predicted vertices (see Sect. 5.3 and Appendix D.3 of [9]), thus ensuring that the maximum predicted matching has size n for the modified predictions $\hat{c}'$. Note that this causes the prediction error to increase from 0 to $L_1(c^*, \hat{c}') = 2(n - \hat{n}) > 2(1 - \beta)n$, which implies $\hat{L}_1(p, q) > 2(1 - \beta) - \varepsilon$ with high probability, exceeding the threshold. Consequently, their algorithm still switches to the baseline algorithm after the sampling phase, achieving only competitive ratio $\beta - o(1)$.

Second, consider inputs with non-zero prediction error. The analysis by Choo et al. establishes competitive ratio better than β only if $n^* = n$ and $\hat{L}_1(p, q) < 2(\frac{\hat{n}}{n} - \beta) - \varepsilon$. On the other hand, our analysis establishes competitive ratio better than β whenever $\hat{n} \geq \alpha n$ and $\hat{L}_1(p, q) < \frac{2\hat{n}}{n} \cdot \frac{1-\beta}{1+\beta} - \varepsilon$. In particular, we obtain competitive ratio better than β for all inputs where both n^* and $\hat{n}$ lie in $[\alpha n, n]$ and the prediction error is small.

Implicit Perfect Matching Assumption in Choo et al.'s Analysis. The analysis of Choo et al. [9] implicitly assumes $n^* = n$. One of their key lemmas [9, Lemma 4.3] states that, if an algorithm makes j matches during the first k arrivals and then switches to the baseline algorithm, the expected total number of matches is at least $j + \beta(n - k - j)$. The proof uses the argument that the number of matches that are still possible during the remaining $n - k$ arrivals is at least $n - k - j$, and this argument clearly holds only under the assumption that the size of the optimal matching is equal to n. Lemma 4.4 in [9] then builds directly on Lemma 4.3, and their proof of Theorem 4.1 in turn relies on Lemma 4.4. Thus, their proofs of all these statements require the assumption that $n^* = n$.

2   Learning-Augmented Algorithm and Analysis

In this section, we present a learning-augmented algorithm that is $(1 - o(1))$-consistent and $(\beta - o(1))$-robust provided that $\hat{n} \geq \alpha n$ for an arbitrary constant $\alpha \in (0, 1]$, where β is the best-known competitive ratio for online bipartite matching in the random arrival order model without predictions, currently $\beta = 0.696$ [23]. Additionally, we provide a smoothness analysis for our algorithm.

The high-level idea of the algorithm by Choo et al. [9] is to use an initial prefix of the arrival sequence as a sample to assess the predictions and, based on this assessment, either continue with the prediction-based algorithm or switch to the online bipartite matching algorithm without predictions.

Our algorithm (see Algorithm 1) follows the approach proposed by Choo et al. [9] of using a prefix of the arrival sequence as a sample, estimating the L_1-distance between the predicted and true distribution of input vertices via Theorem 1, and then either following the predictions (if the estimated L_1-distance is at most $\tau - \varepsilon$ for $\tau = \frac{2\hat{n}}{n} \cdot \frac{1-\beta}{1+\beta}$) or switching to the baseline algorithm for the remaining input. Here, the baseline algorithm is chosen as the best-known algorithm for online bipartite matching in the random arrival order model without predictions, and its competitive ratio is denoted by β. We use BASELINE to refer to that algorithm. The algorithm that follows the predictions is called MIMIC (see Algorithm 2): Whenever an online vertex arrives, the algorithm matches it according to a pre-computed maximum matching $\hat{M}$ of the predicted input graph $\hat{G}$. During the sampling phase, the arriving vertices are also processed by MIMIC. More precisely, our algorithm evaluates the accuracy of the predictions by comparing the predicted distribution ($q = \hat{c}/n$) with the actual input distribution ($p = c^*/n$) based on a sample with expected size $s_{n,\varepsilon,\delta'} = \Theta\left(\frac{(n+1)\log(1/\delta')}{\varepsilon^2 \log(n+1)}\right)$. As done by Choo et al. [9], the actual sample size is chosen as $s_1 + s_2$, where s_1 and s_2 are drawn independently from a Poisson distribution with parameter $\frac{1}{2}s_{n,\varepsilon,\delta'}$. If the sample size is too large, i.e., if $s_1 + s_2 > s_{n,\varepsilon,\delta'}(1 + \sqrt{\log(n+1)})$, which happens with probability $o(1)$, we do not sample and run BASELINE from the beginning. We refer to $s_{n,\varepsilon,\delta'}(1 + \sqrt{\log(n+1)})$ as the *sample size limit*. Otherwise, we compute a sample T_p^s of $s = s_1 + s_2$ input vertices with replacement (lines 10–19 of Algorithm 1 implement sampling with replacement in the same way as done by Choo et al. [9]) and use the algorithm of Theorem 1 to determine an estimate $\hat{L}_1(p,q)$ of the L_1-distance $L_1(p,q)$ between p and q.

We apply that algorithm using domain size $n+1$ (as opposed to $\hat{r}+1$ used by Choo et al. [9]) using a similar approach as Choo et al. [9]: We consider only the $\hat{r} \leq n$ types t with $\hat{c}(t) > 0$, a single dummy type t' that represents all types that appear in the true arrival sequence but not in the predicted arrival sequence, and $n - \hat{r}$ arbitrary other types. If $\hat{L}_1(p,q)$ is smaller than the threshold $\tau - \varepsilon$ for $\tau = \frac{2\hat{n}}{n} \cdot \frac{(1-\beta)}{(1+\beta)}$, we consider the prediction error small and continue to run MIMIC for the remainder of the input. Otherwise, we run BASELINE on the remainder of the input (while disregarding edges of newly arriving online vertices to offline vertices that have already been matched during the sampling phase).

The main differences between our algorithm TEST-AND-MATCH+ and the original algorithm TEST-AND-MATCH are: Our algorithm only requires $\hat{n} \geq \alpha n$ for an arbitrary constant $\alpha \in (0,1]$, as opposed to the condition that $\hat{n} \geq \beta n$ for $\beta = 0.696$ required by Choo et al. [9]. We choose the expected sample size as $s_{n,\varepsilon,\delta'} = \Theta\left(\frac{(n+1)\log(1/\delta')}{\varepsilon^2 \log(n+1)}\right)$ as opposed to $\Theta\left(\frac{(\hat{r}+1)\log(1/\delta')}{\varepsilon^2 \log(\hat{r}+1)}\right)$, and multiply by $1 + \sqrt{\log(n+1)}$ instead of $\sqrt{\log(\hat{r}+1)}$ to determine the sample size limit. This has the advantage that the probability that the sample size exceeds the sample size limit can be bounded by $1/\text{poly}(n)$ instead of $1/\text{poly}(\hat{r})$ and is thus guaranteed to be $o(1)$ even if $\hat{r}$ is a small constant. Furthermore, the threshold on the estimated L_1-distance below which we run MIMIC on the remaining input is set to $\frac{2\hat{n}}{n} \cdot \frac{1-\beta}{1+\beta} - \varepsilon$, as opposed to $2(\frac{\hat{n}}{n} - \beta) - \varepsilon$ by Choo et al. [9]. This makes the threshold useful also for situations where $n^* < n$. The main changes in the analysis include: Instead of assuming $n^* = n$, we need to bound n^* based on $\hat{n}$ and the prediction error. The analysis of the expected size of an optimal matching on the remaining input becomes significantly more challenging without the assumption that $n^* = n$.

Our main result about the performance of the TEST-AND-MATCH+ algorithm (executed with $\delta' = \frac{1}{\log\log\log n}$) can be stated as follows:

Theorem 2. *Let α be a constant in the range $(0,1]$, $\varepsilon \in (0, \alpha\frac{1-\beta}{1+\beta}]$ be another constant. Consider any instance with predicted input $\hat{c}$ and predicted matching size $\hat{n} \geq \alpha n$, and let $L_1(p,q)$ be the prediction error. If $L_1(p,q) \leq \frac{2\hat{n}}{n} \cdot \frac{(1-\beta)}{(1+\beta)} - 2\varepsilon$, TEST-AND-MATCH+ achieves competitive ratio at least $1 - \frac{2L_1(p,q)}{2\alpha + L_1(p,q)} - o(1)$ (and at least $\beta - o(1)$). Otherwise, the algorithm achieves competitive ratio at least $\beta - o(1)$.*

Note that if the predicted matching size $\hat{n}$ is less than αn, we run the BASELINE algorithm right away, which guarantees a competitive ratio of at least β. This ensures the robustness of the solution.

The following three subsections provide the building blocks for the proof of Theorem 2. In Sect. 2.1, we bound the number of matches achieved when running MIMIC for the entire input as well as the number of matches obtained if we switch to BASELINE after we obtain the estimate $\hat{L}_1(p,q)$. We also provide an upper bound on the size of the optimal matching. In Sect. 2.2, we justify the choice of the threshold τ to ensure that our algorithm achieves a competitive ratio better than β if we continue with MIMIC. In Sect. 2.3, we bound the probability that the sampling fails (because the sample size $s_1 + s_2$ exceeds the limit we set) or that the estimate returned when calling the algorithm from Theorem 1 has additive error greater than ε. Finally, in Sect. 2.4, we combine these building blocks to complete the proof of Theorem 2.

2.1   Bounds on the Number of Matches

The following two lemmas establish bounds on the size of the matching produced by MIMIC and on the size of the optimal matching in terms of $\hat{n}$ and the prediction error.

Lemma 1. ($\star$) *Assume that we run MIMIC for the entire input sequence. Then the number of matches we obtain is at least $\hat{n} - \frac{n}{2}L_1(p,q)$, where $L_1(p,q)$ is the L_1-distance between the input distribution $p = c^*/n$ and the predicted distribution $q = \hat{c}/n$.*

Lemma 2. ($\star$) *The size n^* of the optimal matching is at most $\hat{n} + \frac{n}{2}L_1(p,q)$.*

Furthermore, if we run BASELINE instead of MIMIC after we determine $\hat{L}_1(p,q)$, the following lemma gives the expected number of optimal matches in the remaining input. Afterwards, we use this to give a lower bound on the expected total number of matches we obtain after we switch to run BASELINE for the remaining input.

Algorithm 1. TEST-AND-MATCH+

Input: Set of predicted input types $\hat{c}$ with $\hat{r} = |\hat{T}|$, desired error bounds $\delta' > 0$ and $\varepsilon > 0$

Initialization: $i \leftarrow 0$, $T_p^s \leftarrow \emptyset$, $A \leftarrow \emptyset$, $\tau \leftarrow \frac{2\hat{n}}{n} \cdot \frac{(1-\beta)}{(1+\beta)}$

1: **if** $\frac{\hat{n}}{n} < \alpha$ **then**
2:     Run BASELINE for the whole input.
3: **end if**
4: Define $s_{n,\varepsilon,\delta'} = \Theta\left(\frac{(n+1)\log(1/\delta')}{\varepsilon^2 \log(n+1)}\right)$
5: Draw the numbers $s_1, s_2 \sim \text{Poisson}(\frac{1}{2}s_{n,\varepsilon,\delta'})$
6: **if** $s_1 + s_2 > s_{n,\varepsilon,\delta'} \cdot \left(1 + \sqrt{\log(n+1)}\right)$ **then**
7:     Run BASELINE for all remaining input   ▷ Happens with probability
    $\delta_{poi} = \mathcal{O}\left(\frac{1}{\text{poly}(n+1)}\right)$
8: **end if**
9: Compute a maximum matching $\hat{M}$ using $\hat{c}$ as input
10: **while** the size of T_p^s is smaller than $s_1 + s_2$ **do**   ▷ Sample $s_1 + s_2$ input vertices with replacement
11:     Flip a coin that lands on heads with probability i/n and tails otherwise
12:     **if** the coin flip yields heads **then**
13:         Pick an element uniformly at random from the set $\{A[0], \ldots, A[i-1]\}$ and add it to T_p^s
14:     **else**
15:         Add the $(i+1)$-th arriving online vertex to T_p^s and store it in $A[i]$
16:         $i \leftarrow i + 1$
17:         Run MIMIC (Algorithm 2) to process the online vertex
18:     **end if**
19: **end while**
20: Run the algorithm of Theorem 1 on the distribution $q = \frac{\hat{c}}{n}$ and the sample T_p^s from distribution $p = \frac{c^*}{n}$ with domain size n to obtain an estimate $\hat{L}_1(p,q)$ such that $|\hat{L}_1(p,q) - L_1(p,q)| \leq \varepsilon$ with probability $1 - \delta'$
21: **if** $\hat{L}_1(p,q) \leq \tau - \varepsilon$ **then**
22:     Run MIMIC (Algorithm 2) on the remaining input
23: **else**
24:     Run BASELINE on the remaining input
25: **end if**

Algorithm 2. MIMIC

Input: Let c be a copy of $\hat{c}$ (initialized before the first online vertex arrives)
1: When an input vertex v with type t arrives:
2: **if** $c(t) > 0$ **then**
3:     Match v according to an arbitrary unused type t match in $\hat{M}$ if such a match exists
4:     Decrease $c(t)$ by 1
5: **else**
6:     Leave v unmatched
7: **end if**

Lemma 3. $(\star)$ *Let α be an arbitrary constant in the range $(0, 1]$, $\varepsilon \in (0, \alpha\frac{1-\beta}{1+\beta}]$ be another constant, $\delta' = \frac{1}{\log\log\log n}$, and sample size $k = s_1 + s_2 \leq s_{n,\varepsilon,\delta'} \cdot \left(1 + \sqrt{\log(n+1)}\right)$. Let $k' \leq k$ be the number of vertex arrivals during the sampling phase. Then the number of matches made by the optimal matching among the remaining $n - k' \geq n - k$ arrivals is at least $\left(\frac{n-k}{n} - o(1)\right) \cdot n^*$ with probability $1 - o(1)$.*

Proof (Sketch). Fix an optimal matching M^* of size n^*, and let O_1 and O_2 denote the number of vertices from M^* appearing in the sampling phase and in the remainder of the input, respectively. The analysis reduces to determining an upper bound on O_1 since $O_2 = n^* - O_1$. Using Chernoff bounds, we show that when n^* is large, O_1 remains close to its expectation $(k/n)n^*$, while for small n^*, the probability of sampling even one matched vertex is negligible. In both cases, it follows that $O_2 \geq \frac{n-k}{n}n^* - o(n^*)$ with high probability. The full proof is provided in [8]. $\qquad\square$

Corollary 1. *Assume that the algorithm has matched j vertices during the sampling phase and then switches to* BASELINE *for the remaining input. The total expected number of matches made by the algorithm is at least $\beta\frac{n-k}{n} \cdot n^* - o(n^*)$, where β is the competitive ratio of* BASELINE *and k is the number of arrivals before we switch to* BASELINE.

Proof. By Lemma 3, with probability $1 - o(1)$ the remaining input contains a set V^* of at least $\frac{n-k}{n}n^* - o(n^*)$ vertices that are matched by the optimal matching. If the algorithm has already made j matches at the point of switching to BASELINE, j of these $\frac{n-k}{n}n^* - o(n^*)$ vertices in the remaining input may no longer be matchable. At least $\frac{n-k}{n}n^* - j - o(n^*)$ vertices will still be matchable, however, and therefore BASELINE will make at least $\beta\left(\frac{n-k}{n}n^* - j - o(n^*)\right)$ matches. Thus, the total expected number of matches made by the algorithm is at least $j + (1 - o(1))\beta\left(\frac{n-k}{n}n^* - j - o(n^*)\right) \geq \beta \cdot \frac{n-k}{n} \cdot n^* - o(n^*)$. $\qquad\square$

2.2   Choice of Threshold

Corollary 1 shows that the competitive ratio of our algorithm is at least $\beta - o(1)$ if the sampling phase processes $k = o(n)$ vertices and we switch to BASELINE for the remainder of the input. In this section, we justify the choice of the threshold $\tau - \varepsilon$ to ensure that the algorithm is also $(\beta - o(1))$-competitive if the algorithm decides to run MIMIC on the whole input.

Lemma 4. $(\star)$ *If $L_1(p, q) \leq \tau$, where $\tau = \frac{2\hat{n}}{n} \cdot \frac{(1-\beta)}{(1+\beta)}$ and our algorithm runs* MIMIC *on the whole input, then the competitive ratio is at least β.*

If the estimated L_1-distance $\hat{L}(p, q)$ is at most $\tau - \varepsilon$ and the estimation error is at most ε, it follows that $L(p, q) \leq \tau$. Thus, we obtain the following corollary.

Corollary 2. *If $\hat{L}_1(p, q) \leq \tau - \varepsilon$ and $\hat{L}_1(p, q)$ satisfies the property $|\hat{L}_1(p, q) - L_1(p, q)| \leq \varepsilon$, then our algorithm runs* MIMIC *on the whole input and has competitive ratio at least β.*

2.3   Bounding the Probability of Sampling Failures

In Algorithm 1, two cases may prevent the algorithm from carrying out the sampling phase and obtaining a good estimate of $L_1(p,q)$: First, the sum $s_1 + s_2$ of two numbers drawn from a Poisson distribution might exceed $s_{n,\varepsilon,\delta'} \cdot \left(1 + \sqrt{\log(n+1)}\right)$. In this case, we run BASELINE for the whole input. Second, the estimated value of $L_1(p,q)$, obtained via the algorithm of Theorem 1, may not satisfy the property $|\hat{L}_1(p,q) - L_1(p,q)| \leq \varepsilon$. This can cause the algorithm to make an incorrect decision regarding whether to run MIMIC or BASELINE on the remaining input. We show that the probability for these cases to happen is $o(1)$, thus reducing the competitive ratio only by $o(1)$. First, the following lemma, which can be shown by applying standard tail bounds of the Poisson distribution, bounds the probability that the sample size $s_1 + s_2$ is too large.

Lemma 5. $(\star)$ *Let α be any constant in the range $(0,1]$, $\varepsilon \in (0, \alpha\frac{1-\beta}{1+\beta}]$, $\delta' = \frac{1}{\log\log\log n}$, and $s_{n,\varepsilon,\delta'} = \Theta\left(\frac{(n+1)\cdot\log(1/\delta')}{\varepsilon^2\cdot\log(n+1)}\right)$. The probability that two numbers s_1, s_2 drawn independently from a Poisson distribution with parameter $s_{n,\varepsilon,\delta'}/2$ satisfy $s_1 + s_2 > s_{n,\varepsilon,\delta'} \cdot \left(1 + \sqrt{\log(n+1)}\right)$ is at most $\delta_{poi} = \mathcal{O}\left(\frac{1}{\mathrm{poly}(n+1)}\right)$.*

Lemma 6. $(\star)$ *The probability that the two following events both occur is at least $1 - \delta$, where $\delta = \delta_{poi} + \delta'$. First, the sum of the two numbers s_1, s_2, drawn independently from a Poisson distribution with parameter $s_{n,\varepsilon,\delta'}/2$, is at most $s_{n,\varepsilon,\delta'} \cdot \left(1 + \sqrt{\log(n+1)}\right)$. Second, the estimate $\hat{L}_1(p,q)$ computed by the algorithm of Theorem 1 satisfies $|\hat{L}_1(p,q) - L_1(p,q)| \leq \varepsilon$.*

2.4   Proof of Theorem 2

We are now ready to prove Theorem 2.

Proof (of Theorem 2). We distinguish three cases regarding the relationship between $L_1(p,q)$ and τ as follows.

Case 1: $L_1(p,q) \leq \tau - 2\varepsilon$

In this case, if the estimate $\hat{L}_1(p,q)$ satisfies the property $|\hat{L}_1(p,q) - L_1(p,q)| \leq \varepsilon$, our algorithm continues to run MIMIC for the whole input as $\hat{L}_1(p,q) \leq L_1(p,q) + \varepsilon \leq (\tau - 2\varepsilon) + \varepsilon = \tau - \varepsilon$. By Lemma 1, MIMIC produces a matching of size $\hat{n} - \frac{n}{2}L_1(p,q)$ in this case. Furthermore, by Lemma 6, the probability that the sample size $s_1 + s_2$ does not exceed the sample size limit $s_{n,\varepsilon,\delta'} \cdot \left(1 + \sqrt{\log(n+1)}\right)$ and $|\hat{L}_1(p,q) - L_1(p,q)| \leq \varepsilon$ is at least $1 - \delta$ with $\delta = \delta_{poi} + \delta' = \mathcal{O}\left(\frac{1}{\mathrm{poly}(n+1)}\right) + \frac{1}{\log\log\log n} = o(1)$, where δ_{poi} denotes the probability bound obtained in Lemma 5 and δ' follows the definition in Lemmas 3 and 5. Thus, the expected size of the computed matching is at least $(1 - \delta) \cdot \left(\hat{n} - \frac{n}{2}L_1(p,q)\right) = \hat{n} - \frac{n}{2}L_1(p,q) - o(\hat{n})$. As the optimal matching size is at most $\hat{n} + \frac{n}{2}L_1(p,q)$ by Lemma 2, the competitive ratio is at least $\frac{\hat{n} - \frac{n}{2}L_1(p,q) - o(\hat{n})}{\hat{n} + \frac{n}{2}L_1(p,q)} = 1 - \frac{nL_1(p,q)}{\hat{n} + \frac{n}{2}L_1(p,q)} - o(1) \geq 1 - \frac{2L_1(p,q)}{2\alpha + L_1(p,q)} - o(1)$, where the final inequality holds as $\frac{\hat{n}}{n} \geq \alpha$. Furthermore, by Lemma 4, the competitive ratio is also at least $\beta - o(1)$. In summary, for $L_1(p,q) \leq \tau - 2\varepsilon$, our algorithm obtains a competitive ratio of $1 - \frac{2L_1(p,q)}{2\alpha + L_1(p,q)} - o(1)$, and the competitive ratio is also at least $\beta - o(1)$.

Case 2: $L_1(p,q) \geq \tau$ If the algorithm switches to BASELINE after observing $k \leq s_{n,\varepsilon,\delta'} \cdot \left(1 + \sqrt{\log(n+1)}\right)$ arrivals, then, by Corollary 1, the expected matching size is at least $\beta \frac{n-k}{n} \cdot n^* - o(n^*)$. The probability that the sample size $s_1 + s_2$ does not exceed the sample size limit and the estimate $\hat{L}_1(p,q)$ satisfies $|\hat{L}_1(p,q) - L_1(p,q)| \leq \varepsilon$ is at least $1 - \delta$. Thus, the expected matching size is at least $(1 - \delta)\left(\beta \frac{n-k}{n} \cdot n^* - o(n^*)\right) = \beta \frac{n-k}{n} \cdot n^* - o(n^*)$.

Therefore, the competitive ratio is $\frac{\left(\beta \cdot \frac{n-k}{n} \cdot n^*\right) - o(n^*)}{n^*} = \beta \cdot \frac{n-k}{n} - o(1)$, which is at least $\beta \cdot \left(1 - \frac{s_{n,\varepsilon,\delta'} \cdot \left(1 + \sqrt{\log(n+1)}\right)}{n}\right) - o(1) = \beta - o(1)$. Thus, for $L_1(p,q) \geq \tau$, our algorithm achieves competitive ratio at least $\beta - o(1)$.

Case 3: $L_1(p,q) \in (\tau - 2\varepsilon, \tau)$

In this case, we show that our algorithm is $(\beta - o(1))$-competitive no matter whether it runs MIMIC or BASELINE. If the algorithm runs MIMIC on the whole input, then by Lemma 4, since $L_1(p,q) < \tau$, the competitive ratio is at least β. If the algorithm runs BASELINE on the whole input, it is β-competitive as BASELINE is β-competitive. If the algorithm switches to BASELINE after the sampling phase, we can apply the same justification as in **Case 2** to show that it achieves competitive ratio at least $\beta - o(1)$. $\qquad\square$

3   Conclusions

We have studied online bipartite matching with predictions in the random arrival order model. By extending the algorithm and analysis by Choo et al. [9], we have shown that one can achieve $(1 - o(1))$-consistency and $(\beta - o(1))$-robustness provided that the predicted matching size $\hat{n}$ is at least αn for an arbitrary constant $\alpha \in (0, 1]$, no matter what the size n^* of the optimal matching is. In contrast, the analysis by Choo et al. required $n^* = n$ and $\hat{n} \geq \beta n$. Therefore, it does not establish $(1 - o(1))$-consistency when $n^* < n$, even under perfect predictions. In addition, we have shown that our algorithm has competitive ratio at least $1 - \frac{2L_1(p,q)}{2\alpha + L_1(p,q)} - o(1)$, where $L_1(p,q)$ is the prediction error, thus establishing smoothness. Regarding future work, it would be interesting to investigate whether the results can be generalized to the case where the predicted and optimal matching size are sublinear in n. It seems difficult to adapt the sampling-based approach to this case as an error of ε (for constant ε) in the estimate of the L_1-distance corresponds to a linear number of mispredicted vertices, which can be larger than the optimal matching size.

References

1. Aamand, A., Chen, J., Indyk, P.: (Optimal) online bipartite matching with degree information. Adv. Neural. Inf. Process. Syst. **35**, 5724–5737 (2022)
2. Anand, K., Ge, R., Panigrahi, D.: Customizing ml predictions for online algorithms. In: International Conference on Machine Learning, pp. 303–313. PMLR (2020)
3. Angelopoulos, S., Dürr, C., Jin, S., Kamali, S., Renault, M.: Online computation with untrusted advice. J. Comput. Syst. Sci. **144**, 103545 (2024)
4. Antoniadis, A., Gouleakis, T., Kleer, P., Kolev, P.: Secretary and online matching problems with machine learned advice. In: Larochelle, H., Ranzato, M., Hadsell, R., Balcan, M., Lin, H. (eds.) Advances in Neural Information Processing Systems, vol. 33, pp. 7933–7944. Curran Associates, Inc. (2020). https://proceedings.neurips.cc/paper_files/paper/2020/file/5a378f8490c8d6af8647a753812f6e31-Paper.pdf
5. Borodin, A., Karavasilis, C., Pankratov, D.: An experimental study of algorithms for online bipartite matching. J. Exp. Algorithmics (JEA) **25**, 1–37 (2020)
6. Brubach, B., Sankararaman, K.A., Srinivasan, A., Xu, P.: New algorithms, better bounds, and a novel model for online stochastic matching. In: 24th Annual European Symposium on Algorithms (ESA 2016). Schloss Dagstuhl-Leibniz-Zentrum fuer Informatik (2016)
7. Brubach, B., Sankararaman, K.A., Srinivasan, A., Xu, P.: Online stochastic matching: new algorithms and bounds. Algorithmica **82**(10), 2737–2783 (2020)
8. Burathep, K., Erlebach, T., Moses Jr, W.K.: Learning-augmented online bipartite matching in the random arrival order model. arXiv preprint arXiv:2511.23388 (2025)
9. Choo, D., Gouleakis, T., Ling, C.K., Bhattacharyya, A.: Online bipartite matching with imperfect advice. In: Salakhutdinov, R., et al. (eds.) Proceedings of the 41st International Conference on Machine Learning. Proceedings of Machine Learning Research, vol. 235, pp. 8762–8781. PMLR (2024). https://proceedings.mlr.press/v235/choo24a.html
10. Chung, F., Lu, L., Vu, V.: Spectra of random graphs with given expected degrees. Proc. Natl. Acad. Sci. **100**(11), 6313–6318 (2003)
11. Feldman, J., Mehta, A., Mirrokni, V., Muthukrishnan, S.: Online stochastic matching: Beating 1-1/e. In: 2009 50th Annual IEEE Symposium on Foundations of Computer Science, pp. 117–126. IEEE (2009)
12. Feng, Y., Niazadeh, R., Saberi, A.: Two-stage stochastic matching with application to ride hailing. In: Proceedings of the 2021 ACM-SIAM Symposium on Discrete Algorithms (SODA), pp. 2862–2877. SIAM (2021)
13. Goel, G., Mehta, A.: Online budgeted matching in random input models with applications to ADWORDS. In: SODA, vol. 8, pp. 982–991 (2008)
14. Gollapudi, S., Panigrahi, D.: Online algorithms for rent-or-buy with expert advice. In: International Conference on Machine Learning, pp. 2319–2327. PMLR (2019)
15. Jaillet, P., Lu, X.: Online stochastic matching: new algorithms with better bounds. Math. Oper. Res. **39**(3), 624–646 (2014)
16. Jiao, J., Han, Y., Weissman, T.: Minimax estimation of the l_1 distance. IEEE Trans. Inf. Theory **64**(10), 6672–6706 (2018). https://doi.org/10.1109/TIT.2018.2846245
17. Jin, B., Ma, W.: Online bipartite matching with advice: tight robustness-consistency tradeoffs for the two-stage model. Adv. Neural. Inf. Process. Syst. **35**, 14555–14567 (2022)

18. Karande, C., Mehta, A., Tripathi, P.: Online bipartite matching with unknown distributions. In: Proceedings of the Forty-Third Annual ACM Symposium on Theory of Computing, pp. 587–596 (2011)
19. Karp, R.M., Vazirani, U.V., Vazirani, V.V.: An optimal algorithm for on-line bipartite matching. In: Proceedings of the Twenty-Second Annual ACM Symposium on Theory of Computing, pp. 352–358 (1990)
20. Li, P., Yang, J., Ren, S.: Learning for edge-weighted online bipartite matching with robustness guarantees. In: International Conference on Machine Learning, pp. 20276–20295. PMLR (2023)
21. Lindermayr, A., Megow, N.: Algorithms with predictions (2025). https://algorithms-with-predictions.github.io
22. Lykouris, T., Vassilvitskii, S.: Competitive caching with machine learned advice. J. ACM (JACM) **68**(4), 1–25 (2021)
23. Mahdian, M., Yan, Q.: Online bipartite matching with random arrivals: an approach based on strongly factor-revealing LPs. In: Proceedings of the Forty-Third Annual ACM Symposium on Theory of Computing, pp. 597–606 (2011)
24. Manshadi, V.H., Gharan, S.O., Saberi, A.: Online stochastic matching: online actions based on offline statistics. Math. Oper. Res. **37**(4), 559–573 (2012)
25. Mitzenmacher, M., Vassilvitskii, S.: Algorithms with predictions. Commun. ACM **65**(7), 33–35 (2022)
26. Purohit, M., Svitkina, Z., Kumar, R.: Improving online algorithms via ml predictions. In: Advances in Neural Information Processing Systems, vol. 31 (2018)
27. Rohatgi, D.: Near-optimal bounds for online caching with machine learned advice. In: Proceedings of the Fourteenth Annual ACM-SIAM Symposium on Discrete Algorithms, pp. 1834–1845. SIAM (2020)
28. Wei, A.: Better and simpler learning-augmented online caching. arXiv preprint arXiv:2005.13716 (2020)
29. Wei, A., Zhang, F.: Optimal robustness-consistency trade-offs for learning-augmented online algorithms. Adv. Neural. Inf. Process. Syst. **33**, 8042–8053 (2020)

On the Sprague-Grundy Values of Games with a Pass

Hikaru Manabe[1]([✉]) [ID], Ryohei Miyadera[2,6] [ID], and Koki Suetsugu[3,4,5,6] [ID]

[1] University of Tsukuba College of Informatics, 1-1-1 Tennodai, Tsukuba, Japan
urakihebanam@gmail.com
[2] Keimei Gakuin Junior and Senior High School, 9-5-1 Yokoo, Suma-ku, Kobe, Japan
[3] Osaka Metropolitan University, 3-3-138 Sugimoto, Sumiyoshi-ku, Osaka, Japan
[4] Waseda University, 513 Waseda-Tsurumaki-Cho, Sinjuku-ku, Tokyo, Japan
[5] Gifu University, 1-1 Yanagido, Gifu, Japan
[6] 1-7-11 Akabanedai, Kita-ku, Tokyo, Japan

Abstract. In this paper, we consider two-player impartial games with a pass-move. A disjunctive compound of positions is a position in which, on each turn, the current player chooses one of the components and makes a legal move in it. For disjunctive compounds, it is known that the time to determine which player has a winning strategy is bounded by the time to compute the SG-values of the components plus the time for their XOR. However, if we allow a pass-move during the play, the analysis of such games becomes much more difficult. A pass-move allows each player to skip exactly one turn in non-terminal positions during the game, after which neither player may use a pass-move again. We establish a homomorphism on the SG-values of positions with a pass-move. That is, if every component satisfies a condition called one-move position, the SG-value of the disjunctive compound of the components with a pass-move is the same as the SG-value of NIM with a pass-move where the size of every pile is the same as the SG-value of every component of the compound. This guarantees that the time to determine which player has a winning strategy in a disjunctive compound with a pass can be bounded by the sum of the time to determine SG-values of all components without a pass and a position in NIM with a pass. We also show how the homomorphism is used for determining SG-values of some CHOCOLATE GAMES.

Keywords: Combinatorial game theory · Impartial games · Sprague-Grundy value · Pass-move · One-move positions · CHOCOLATE GAME

1 Introduction

In this paper, we consider two-player *impartial games*, that is, two-player games in which, in every position, the sets of options for both players are the same. We assume that the games are under *normal play convention*, that is, the player who has no legal move loses. In addition, we assume that every position is *short*;

© The Author(s), under exclusive license to Springer Nature Switzerland AG 2026
J. Kozik and A. Wolff (Eds.): SOFSEM 2026, LNCS 16448, pp. 376–389, 2026.
https://doi.org/10.1007/978-3-032-17801-5_28

that is, for any position, there is a sufficiently large integer n, and the game must end in fewer than n moves, regardless of the choice of moves. Developing effective methods for determining whether a player has a winning strategy is a central topic in theoretical computer science. For example, NIM is a two-player game with several piles of tokens, and in each turn, the current player chooses one of the piles and removes any positive number of tokens from the pile. Since we assume that games are under normal play convention, the player who has no legal move is the loser. Bouton showed that in NIM, the previous player has a winning strategy if and only if the exclusive OR (XOR) of the numbers of tokens in the piles is zero [4]. This is a much faster way to determine which player has a winning strategy than brute-force algorithms.

Furthermore, for any impartial games under normal play convention, Sprague and Grundy independently generalized Bouton's theory and showed that when a position is regarded as a disjunctive compound of some components, that is, in each turn, the player chooses exactly one of the components and makes a legal move on it, the previous player has a winning strategy if and only if the XOR of the parameter, called *SG-value*, of each component is zero [7,17]. Therefore, if a position is a disjunctive compound, the time for determining which player has a winning strategy is bounded by the sum of the times for calculating SG-values of components and the time for calculating XOR of the SG-values.

On the other hand, it becomes more difficult to analyze games if we allow a pass-move in the play. Under this convention, a pass-move may be used at most once in the game, but not when the position is terminal. Once the pass has been used by either player, it is no longer available. There are some early results considering this convention [5,10,12,14]. Also, this convention is generalized as a sum of positions called "split sum" in [8].

NIM is a good example to explain how the pass-move makes it difficult to analyze games. As mentioned above, it is easy to solve three-pile NIM. However, no mathematical formula is known for the previous player's winning position when a pass-move is allowed in this ruleset and this has been posed as an open problem in the list of unsolved problems in combinatorial game theory [15].

In this paper, we consider disjunctive compounds with a pass-move and show that if the components satisfy a special property, called one-move position, the time to determine which player has a winning strategy can be bounded by the sum of the time to determine SG-values of all components without a pass and a position in NIM with a pass, similarly to the case of standard disjunctive compounds.

The outline of this paper is as follows. In the latter part of this section, we introduce necessary early results for completeness. In Sect. 2, we establish a homomorphism of SG-values on disjunctive compound with a pass, which makes the calculation for determining which player has a winning strategy easier. In Sect. 3, we show how the result can be used for analyzing games by using an example, CHOCOLATE GAME.

1.1 Disjunctive Compound

We briefly review some of the necessary concepts of combinatorial game theory; refer to [1] for more details. Let $\mathbb{Z}_{\geq 0}$ be the set of nonnegative integers.

Definition 1. *A position is referred to as a $\mathcal{P}$-position (resp. an $\mathcal{N}$-position) if the previous (resp. next) player has a winning strategy in the position.*

Note that the player who is to move is traditionally referred to as the "next player."

Definition 2. *1. For any position g of an impartial game, there is a set of positions that can be reached in precisely one move from g, which is denoted by $\mathrm{move}(g)$. An element in $\mathrm{move}(g)$ is called an* option *of g. Let $\mathrm{move}^0(g) = \{g\}$, and $\mathrm{move}^n(g) = \bigcup_{h \in \mathrm{move}^{n-1}(g)} \mathrm{move}(h)$ for any $n \geq 1$. If $g' \in \mathrm{move}^n(g)$ for $n \geq 0$, g' is a* follower *of g.*
2. The minimum excluded value (mex) *of a set S of nonnegative integers is the least nonnegative integer that is not in S. That is, $\mathrm{mex}(S) = \min(\mathbb{Z}_{\geq 0} \setminus S)$.*
3. Each position g of an impartial game has an associated Sprague-Grundy value, *or SG-value, which is denoted by $\mathcal{G}(g)$. The SG-value is obtained recursively as follows. $\mathcal{G}(g) = \mathrm{mex}(\{\mathcal{G}(h) \mid h \in \mathrm{move}(g)\})$.*

Theorem 1. *([17],[7]). For any position g of an impartial game, $\mathcal{G}(g) = 0$ if and only if g is a $\mathcal{P}$-position and $\mathcal{G}(g) \neq 0$ if and only if g is an $\mathcal{N}$-position.*

Definition 3. *Let $g_1, \ldots, g_n$ be positions in impartial games. A* compound *of positions is a function C such that $C(g_1, \ldots, g_n)$ is also a position in an impartial game.*

Definition 4. *Let $g_1, \ldots, g_n$ be positions in impartial games. The* disjunctive compound *of $g_1, \ldots, g_n$, denoted by $C_+(g_1, \ldots, g_n)$, or, $g_1 + \cdots + g_n$, is a position such that the set of options of the position is*

$$\mathrm{move}(C_+(g_1, \ldots, g_n))$$
$$= \bigcup_{i=1}^{n} \{C_+(g_1', \ldots, g_n') \mid g_i' \in \mathrm{move}(g_i) \text{ and } g_j' = g_j \text{ for any } j \neq i\}.$$

Here, if all of $g_1, \ldots, g_n$ are terminal positions, $C_+(g_1, \ldots, g_n)$ is a terminal position since $\mathrm{move}(C_+(g_1, \ldots, g_n)) = \emptyset$.

In other words, in a disjunctive compound of positions, each player chooses exactly one component and makes a legal move on it. Disjunctive compound is one of the most famous compounds.

Let $\oplus$ be the XOR operator for binary notation.

Theorem 2 *([17],[7]). For any positions $g_1, \ldots, g_n$ in impartial games, we have*

$$\mathcal{G}(C_+(g_1, \ldots, g_n)) = \mathcal{G}(g_1) \oplus \cdots \oplus \mathcal{G}(g_n).$$

For any compound C, let $C^{\mathrm{nim}}(m_1, \ldots, m_n)$ be the position obtained by applying C to single-pile NIM positions with m_i tokens. For example, since n-pile NIM is disjunctive compound of positions of single-pile NIM, a position in n-pile NIM can be denoted by $C_+^{\mathrm{nim}}(m_1, \ldots, m_n)$.

The SG-value of n-pile NIM is equal to the XOR of the numbers of tokens in the piles since the SG-value of one-pile NIM with m tokens is m. Thus, we have $\mathcal{G}(C_+^{\mathrm{nim}}(m_1, \ldots, m_n)) = m_1 \oplus \cdots \oplus m_n$ and from Theorem 2, we have a homomorphism

$$\mathcal{G}(C_+(g_1, \ldots, g_n)) = \mathcal{G}(C_+^{\mathrm{nim}}(\mathcal{G}(g_1), \ldots, \mathcal{G}(g_n))).$$

However, when we consider compounds other than disjunctive compound, this kind of homomorphism may not hold. Therefore, it is important to reveal that for what compound C and what kind of positions g_i, the homomorphism

$$\mathcal{G}(C(g_1, \ldots, g_n)) = \mathcal{G}(C^{\mathrm{nim}}(\mathcal{G}(g_1), \ldots, \mathcal{G}(g_n)))$$

holds. We call this homomorphism an *SG-homomorphism*.

1.2   Hypergraph Compound and SG-Decreasing Positions

In [3], the *hypergraph compound* of positions is defined as follows.

Let $\mathcal{H}$ be a hypergraph $\mathcal{H} \subseteq 2^{[n]} \setminus \{\emptyset\}$, where $[n] = \{1, 2, \ldots, n\}$. On each vertex i of $\mathcal{H}$, there is an impartial position g_i. A player, on their turn, chooses a hyperedge $H \in \mathcal{H}$ and makes a move in every position g_i, where $i \in H$.

The hypergraph compound of $g_1, \ldots, g_n$ is denoted by $C_{\mathcal{H}}(g_1, \ldots, g_n)$ for given hypergraph $\mathcal{H}$.

Definition 5. *Let g be a position in an impartial game. g is SG-decreasing if for any nonnegative integer n, every $g' \in \mathrm{move}^n(g)$ satisfies $\mathcal{G}(g') > \mathcal{G}(g'')$ for any $g'' \in \mathrm{move}(g')$.*

Positions in one pile NIM, MINIMAL NIM [9], and EXACT-k NIM [2] with $2k > n$, are examples of SG-decreasing positions.

Theorem 3 ([3]). *Let $\mathcal{H}$ be a hypergraph $\mathcal{H} \subseteq 2^{[n]} \setminus \{\emptyset\}$. Assume that $g_1, \ldots, g_n$ are SG-decreasing positions. Then, we have*

$$\mathcal{G}(C_{\mathcal{H}}(g_1, \ldots, g_n)) = \mathcal{G}(C_{\mathcal{H}}^{\mathrm{nim}}(\mathcal{G}(g_1), \ldots, \mathcal{G}(g_n))).$$

Table 1 summarizes what compound and what kind of positions satisfy SG-homomorphism. In addition to the shown two pairs, we introduce one-pass compound and one-move positions in the next section, as the third example of SG-homomorphism. Actually, the set of one-move positions is a subset of positions in impartial games, and the set of SG-decreasing positions is a subset of one-move positions. Thus, we consider that our result falls between the two results.

Table 1. Pairs of compound and positions which satisfy SG-homomorphism.

Reference	Compound	Positions
Sprague [17], Grundy [7]	disjunctive compound	any positions in impartial games
This paper	one-pass compound	one-move positions
Boros et al. [3]	hypergraph compound	SG-decreasing positions

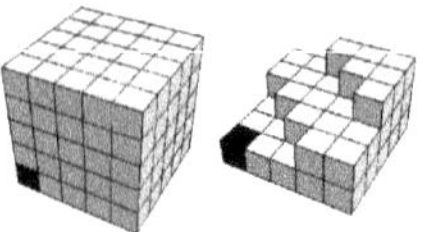

Fig. 1. Two-dimensional chocolate bars.   **Fig. 2.** Three-dimensional chocolate bars.

1.3 CHOCOLATE GAMES

CHOCOLATE GAMES were first presented in [16]. CHOCOLATE GAMES look similar to the CHOMP game that was presented in [6], but how to cut these chocolate bars differs from the cutting rule in CHOMP game. Therefore, CHOCOLATE GAMES differ considerably from CHOMP games. Given that the classical three-pile NIM is mathematically equivalent to a rectangular CHOCOLATE GAME, CHOCOLATE GAMES are generalizations of the classical NIM.

A two- or three-dimensional chocolate bar is a grid of squares or cubes containing at least one bitter cell printed in black. See Figs. 1 and 2. Players alternately break the bar along a groove and take the separated piece. The player who leaves the opponent with only the bitter cell wins. Some examples of cut in three-dimensional chocolate bar are shown in Figures. 3, 4, and 5.

Definition 6. *A single-variable function f of $\mathbb{Z}_{\geq 0}$ onto itself is said to be* monotonically increasing *if $f(u) \leq f(v)$ for $u, v \in \mathbb{Z}_{\geq 0}$, with $u \leq v$, and a two-variable function $F : \mathbb{Z}_{\geq 0} \times \mathbb{Z}_{\geq 0} \to \mathbb{Z}_{\geq 0}$ is said to be* monotonically increasing *if $F(u, v) \leq F(x, z)$ for $x, z, u, v \in \mathbb{Z}_{\geq 0}$ with $u \leq x$ and $v \leq z$.*

Definition 7. *Let F be a monotonically increasing two-variable function. Let $x, y, z \in \mathbb{Z}_{\geq 0}$ such that $y \leq F(x, z)$. The three-dimensional chocolate bar comprises a set of $1 \times 1 \times 1$ boxes. For $u, w \in \mathbb{Z}_{\geq 0}$ such that $u \leq x$ and $w \leq z$, the height of the column at position (u, w) is $\min(F(u, w), y) + 1$. There is a bitter*

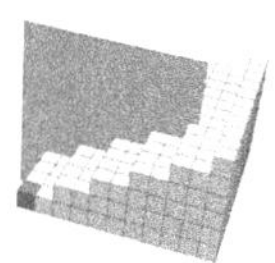

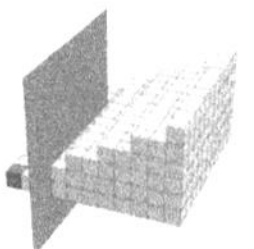

Fig. 3. A vertical cut.   **Fig. 4.** A vertical cut.   **Fig. 5.** A horizontal cut.

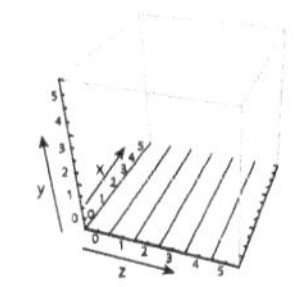

Fig. 6. Coordinate.

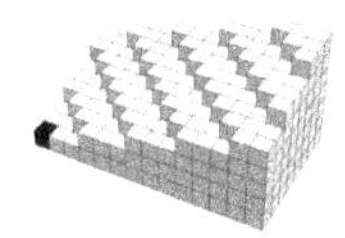

Fig. 7. $CB(F_1, 7, 6, 13)$

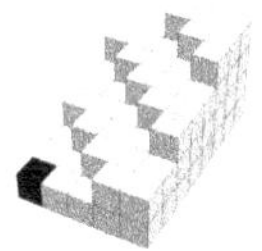

Fig. 8. $CB(F_1, 7, 3, 4)$.

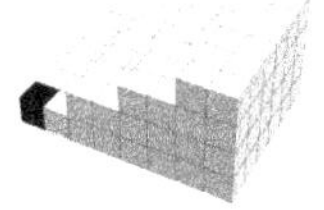

Fig. 9.
$CB(F_2, 5, 3, 7)$.

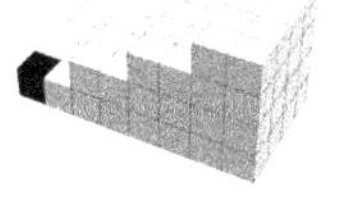

Fig. 10.
$CB(F_2, 3, 3, 7)$.

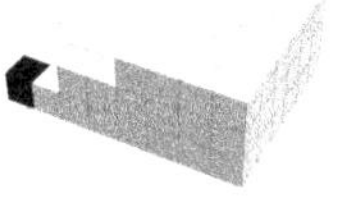

Fig. 11.
$CB(F_2, 5, 2, 7)$.

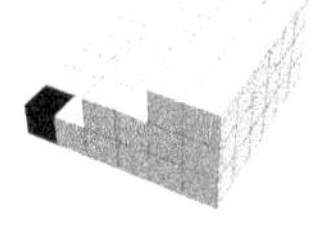

Fig. 12.
$CB(F_2, 5, 2, 5)$.

box at position $(0,0,0)$. We denote this chocolate bar by $CB(F, x, y, z)$. Note that $x + 1, y + 1$, and $z + 1$ are the length, height, and width of the bar, respectively.

Next, we define $\text{move}_F(x, y, z)$ in Definition 8. $\text{move}_F(x, y, z)$ is a set that contains all the positions that can be directly reached from position $CB(F, x, y, z)$ in a single step. That is, $(x', y', z') \in \text{move}_F(x, y, z)$ if and only if $CB(F, x', y', z') \in \text{move}(CB(F, x, y, z))$.

Definition 8. *For $x, y, z \in \mathbb{Z}_{\geq 0}$,*

$$\text{move}_F(x, y, z) = \{(u, \min(F(u, z), y), z) : u < x\} \cup \{(x, v, z) : v < y\}$$

$$\cup \{(x, \min(y, F(x, w)), w) : w < z\},$$

where $u, v, w \in \mathbb{Z}_{\geq 0}$.

The following examples show how the coordinates change when we cut chocolate bars. Figure 6 shows which direction corresponds to $x, y,$ or z. Let $\lfloor\ \rfloor$ be the floor function.

Example 1. Let $F_1(x, z) = \lfloor \frac{x+z}{3} \rfloor$. $CB(F_1, 7, 6, 13)$ and $CB(F_1, 7, 3, 4)$, shown in Figs. 7 and 8, are positions of this CHOCOLATE GAME and $CB(F_1, 7, 3, 4)$ is an option of $CB(F_1, 7, 6, 13)$.

Example 2. Let $F_2(x, z) = \lfloor \frac{z}{2} \rfloor$. Positions $CB(F_2, 3, 3, 7), CB(F_2, 5, 2, 7)$, and $CB(F_2, 5, 2, 5)$, shown in Figs. 10, 11, and 12, are options of $CB(F_2, 5, 3, 7)$, shown in Fig. 9. Note that there are three different directions of cutting chocolates and in some cases, we reduce two coordinates at the same time.

The original two-dimensional chocolate bar shown at left in Fig. 1 and three-dimensional chocolate bar shown at left in Fig. 2 were introduced by Robin [16]. Since the length, height, and width can be reduced independently but not simultaneously, $m \times n \times \ell$ cuboid chocolate bar can be considered as three-pile NIM whose piles have $m - 1$, $n - 1$ and $\ell - 1$ tokens. Since the SG-value of the NIM is $(m - 1) \oplus (n - 1) \oplus (\ell - 1)$, the SG-value of $m \times n \times \ell$ cuboid chocolate bar is $(m - 1) \oplus (n - 1) \oplus (\ell - 1)$. Here, the following question arises naturally: What are the necessary and sufficient conditions for a three-dimensional chocolate bar to have the SG-value $(m - 1) \oplus (n - 1) \oplus (\ell - 1)$, where m, n, and ℓ are the length, height, and width of the bar, respectively? The answer to this question is presented in [13], which is referred to as Theorem 4 in this section.

Definition 9.

(i) *Let $h(z)$ be a single-variable monotonically increasing function. The function h is said to have the NS-property if h satisfies $h(0) = 0$ and the following condition:*
Suppose that $\lfloor \frac{z}{2^i} \rfloor = \lfloor \frac{z'}{2^i} \rfloor$ for some $z, z' \in \mathbb{Z}_{\geq 0}$, and some positive integer i. Then, $\lfloor \frac{h(z)}{2^{i-1}} \rfloor = \lfloor \frac{h(z')}{2^{i-1}} \rfloor$.
(ii) *Let $F(x, z)$ be a double-variable monotonically increasing function. Let $g_n(z) = F(n, z)$ and $h_m(x) = F(x, m)$ for $n, m \in \mathbb{Z}_{\geq 0}$. The function F is said to have the NS-property if for any n and m, g_n and h_m satisfy the NS-property in (i) of this definition.*

The terminology NS-property follows [13], where the notion first appeared. Although no explanation was given there, "NS" is simply short for "necessary and sufficient," used purely as a convenient name.

The following proposition shows examples of functions that satisfy the NS-property in (i) of Definition 9. For the proofs of the proposition, see the full version of this paper [11].

Proposition 1.   *1. Let $h(z) = \lfloor \frac{z}{2k} \rfloor$ for some positive integer k. Then $h(z)$ satisfies NS-property.*
2. Let $h(z) = 2^{\lfloor \log_2 z \rfloor} - 1$ for $z > 0$ and $h(0) = 0$. Then, $h(z)$ satisfies NS-property.

Lemma 1. *Let $h(z)$ be a function which has NS-property. For any positive integer z, we have $h(z) \leq 2^{\lfloor \log_2 z \rfloor} - 1$.*

Proof. Let $i = \lfloor \log_2 z \rfloor$. We have $\lfloor \frac{0}{2^{i+1}} \rfloor = \lfloor \frac{z}{2^{i+1}} \rfloor = 0$. Therefore, from the definition of NS-property, $\lfloor \frac{h(z)}{2^i} \rfloor = \lfloor \frac{h(0)}{2^i} \rfloor = 0$, which means $h(z) \leq 2^i - 1$. $\square$

Theorem 4 ([13]). *Let $F(x, z)$ be a monotonically increasing function. $F(x, z)$ has NS-property if and only if the SG-value of chocolate bar $CB(F, x, y, z)$ is $x \oplus y \oplus z$.*

Table 2. SG-values of CHOCOLATE GAME $CB_2(h, y, z)$

$y\backslash z$	0	1	2	3	4	5	6	7	8	9	10	11	12	13	14	15
0	0	1	2	3	4	5	6	7	8	9	10	11	12	13	14	15
1			3	2	5	4	7	6	9	8	11	10	13	12	15	14
2					6	7	4	5	10	11	8	9	14	15	13	12
3					7	6	5	4	11	10	9	8	15	14	13	12
4									12	13	14	15	8	9	10	11
5									13	12	15	14	9	8	11	10
6									14	15	12	13	10	11	8	9
7									15	14	13	12	11	10	9	8

By Theorem 4, the formula for SG-values in CHOCOLATE GAMES $CB(F, x, y, z)$ with NS-property is the same as that of three-pile NIM, but as shown in Sect. 3, some CHOCOLATE GAMES with NS-property also have a formula for $\mathcal{P}$-positions even if a pass-move is allowed. Therefore, the pass-move was found to have a minimal impact on these CHOCOLATE GAMES. This is remarkable since three-pile NIM does not have any known formula for $\mathcal{P}$-positions when a pass-move is allowed.

Corollary 1. *Let $h(z)$ be a monotonically increasing function. Assume that function $h(z)$ has NS-property. Then, consider two-dimensional* CHOCOLATE *GAME $CB_2(h, y, z) = CB(F, 0, y, z)$, where $F(x, z) = h(z)$ and y is restricted by F as $y \leq F(x, z)$. The SG-value of the chocolate bar $CB_2(h, y, z)$ is $y \oplus z$.*

Proof. When $h(z)$ has NS-property, $F(x, z) = h(z)$ also has NS-property. Therefore, the SG-value of position $CB_2(h, y, z)$ is the same as $CB(F, 0, y, z)$, which is $0 \oplus y \oplus z = y \oplus z$. $\qquad\square$

Table 2 shows the SG-value of CHOCOLATE GAME $CB_2(h, y, z)$ for small bars, where h has NS property. Note that from Lemma 1, whatever the function h is, we do not need to consider the case $y > 2^{\lfloor \log_2 z \rfloor} - 1$. Therefore, the shape of the chocolate bar of a position where $z \leq 15$ must be included in this table.

Lemma 2. *Assume that $h(z)$ has NS-property. For any $z \geq 16$, we have $y \oplus z \geq 16$ if $y \leq h(z)$.*

Proof. Assume that $h(z)$ has NS-property. From Lemma 1, for any integer z, there is an integer i such that $2^i \leq z$ and $h(z) < 2^i$. Thus, when $z \geq 16$, we have $i \geq 4$ and $y \oplus z \geq 2^4 = 16$. $\qquad\square$

Lemma 3. *Assume that h has NS-property. Let y and z be nonnegative integers with $y \leq h(z)$. Then, the following holds.*

- *$CB_2(h, y, z)$ has SG-value 0 if and only if $y = z = 0$.*
- *$CB_2(h, y, z)$ has SG-value 1 if and only if $y = 0$ and $z = 1$.*

- $CB_2(h, y, z)$ *has SG-value 2 if and only if* $(y, z) \in \{(0, 2), (1, 3)\}$.
- $CB_2(h, y, z)$ *has SG-value 3 if and only if* $(y, z) \in \{(0, 3), (1, 2)\}$.
- $CB_2(h, y, z)$ *has SG-value 4 if and only if* $(y, z) \in \{(0, 4), (1, 5), (2, 6), (3, 7)\}$.
- $CB_2(h, y, z)$ *has SG-value 5 if and only if* $(y, z) \in \{(0, 5), (1, 4), (2, 7), (3, 6)\}$.
- $CB_2(h, y, z)$ *has SG-value 6 if and only if* $(y, z) \in \{(0, 6), (1, 7), (2, 4), (3, 5)\}$.
- $CB_2(h, y, z)$ *has SG-value 7 if and only if* $(y, z) \in \{(0, 7), (1, 6), (2, 5), (3, 4)\}$.
- $CB_2(h, y, z)$ *has SG-value 8 if and only if* $(y, z) \in \{(0, 8), (1, 9), (2, 10), (3, 11), (4, 12), (5, 13), (6, 14), (7, 15)\}$.

Proof. From Lemma 2, we only need to consider the case $z \leq 15$. Therefore, from Table 2, we can confirm that the statement is true. $\qquad\square$

2   One-Move Positions with a Pass

In this section, we show that the SG-value of a disjunctive compound of positions with a pass can be calculated by using SG-value of NIM with a pass if every component is a position that can arrive at a terminal position in one move.

Definition 10. *Let* $g_1, \ldots, g_n$ *be positions in impartial games. One-pass compound of* $g_1, \ldots, g_n$ *is denoted by* $C_{(+,\text{pass})}(g_1, \ldots, g_n)$, *and the set of its options is*

$$\bigcup_{i=1}^{n} \{(C_{(+,\text{pass})}(g_1', \ldots, g_n')) \mid g_i' \in \text{move}(g_i) \text{ and } g_j' = g_j \text{ for any } j \neq i\}$$
$$\cup \{C_+(g_1, \ldots, g_n)\},$$

when at least one of g_i *is not a terminal position, and if every* g_i *is a terminal position, then* $\text{move}(C_{(+,\text{pass})}(g_1, \ldots, g_n)) = \emptyset$.

That is, one-pass compound is almost the same as disjunctive compound, but the players can make a pass-move except in a terminal position, and once the pass-move is used, neither player may use it again. Note that when the original position is a $\mathcal{P}$-position, the position changes to an $\mathcal{N}$-position when a pass-move is added since the player can use the pass. However, when the original position is $\mathcal{N}$-position, there are two cases that it changes to a $\mathcal{P}$-position or it remains to be an $\mathcal{N}$-position. Therefore, even if the winner of the original position is determined, it is still hard to determine the winner if a pass-move is added. For example, NIM with three or more piles is known as an unsolved problem if we allow a pass-move even though it is solved when the pass-move is not allowed [15].

Definition 11. *A position* g *is* one-move position *exactly when every follower* h *satisfies*

$$\mathcal{G}(h) = 0 \iff h \text{ is terminal.}$$

Recall that the set of followers of a position includes the position itself. Therefore, if g is a one-move position and $\mathcal{G}(g) = 0$, then g is a terminal position. We also note that if a position is SG-decreasing, then it is also a one-move position, but the converse may not hold. For example, two-dimensional CHOCOLATE GAME where function h satisfies NS-property is a one-move position, but like the move $CB_2(h, 1, 3)$ to $CB_2(h, 1, 2)$, there are moves that increase SG-values (see Table 2). In addition, if a one-move position is considered as a position by itself (i.e., not as a component of a disjunctive compound), then the game ends in at most one move since if a position does not have SG-value 0, then it must have an option to a position whose SG-value is 0, that is, a terminal position. However, when two or more one-move positions are combined by a disjunctive compound, then it is not guaranteed to end in a certain number of moves, since moving a component to a terminal position may not be a winning strategy.

Theorem 5. *Let $g_1, \ldots, g_n$ be one-move positions. Then, we have*

$$\mathcal{G}(C_{(+,\text{pass})}(g_1, \ldots, g_n)) = \mathcal{G}(C^{\text{nim}}_{(+,\text{pass})}(\mathcal{G}(g_1), \ldots, \mathcal{G}(g_n))).$$

Proof. We prove this by induction on the sum of the heights of game trees of $g_1, \ldots, g_n$. When every g_i is terminal position,

$$\mathcal{G}(C_{(+,\text{pass})}(g_1, \ldots, g_n)) = \mathcal{G}(C^{\text{nim}}_{(+,\text{pass})}(\mathcal{G}(g_1), \ldots, \mathcal{G}(g_n))) = 0$$

Assume that at least one of g_i is not a terminal position. Then, from the induction hypothesis,

$$\mathcal{G}(C_{(+,\text{pass})}(g_1, \ldots, g_n))$$
$$= \text{mex}(\bigcup_{i=1}^{n} \{\mathcal{G}(C_{(+,\text{pass})}(g_1', \ldots, g_n')) \mid g_i' \in \text{move}(g_i) \text{ and } g_j' = g_j \text{ for any } j \neq i\}$$
$$\cup \{\mathcal{G}(C_+(g_1, \ldots, g_n))\})$$
$$= \text{mex}(\bigcup_{i=1}^{n} \mathcal{G}(C^{\text{nim}}_{(+,\text{pass})}(\mathcal{G}(g_1'), \ldots, \mathcal{G}(g_n'))) \mid g_i' \in \text{move}(g_i) \text{ and } g_j' = g_j \text{ for any } j \neq i\}$$
$$\cup \{\mathcal{G}(g_1) \oplus \cdots \oplus \mathcal{G}(g_n)\}).$$

Let
$$S_1 = \bigcup_{i=1}^{n} \mathcal{G}(C^{\text{nim}}_{(+,\text{pass})}(\mathcal{G}(g_1'), \ldots, \mathcal{G}(g_n'))) \mid g_i' \in \text{move}(g_i) \text{ and } g_j' = g_j \text{ for any } j \neq i\}$$
and $S_2 = \{\mathcal{G}(g_1) \oplus \cdots \oplus \mathcal{G}(g_n)\}$. In order to prove $\mathcal{G}(C^{\text{nim}}_{(+,\text{pass})}(\mathcal{G}(g_1), \ldots, \mathcal{G}(g_n))) = \text{mex}(S_1 \cup S_2)$, we show that $\mathcal{G}(C^{\text{nim}}_{(+,\text{pass})}(\mathcal{G}(g_1), \ldots, \mathcal{G}(g_n))) \notin S_1 \cup S_2$ and for any $\ell < \mathcal{G}(C^{\text{nim}}_{(+,\text{pass})}(\mathcal{G}(g_1), \ldots, \mathcal{G}(g_n))), \ell \in S_1 \cup S_2$. Here,

$$\mathcal{G}(C^{\text{nim}}_{(+,\text{pass})}(\mathcal{G}(g_1), \ldots, \mathcal{G}(g_n))) \notin S_1$$

because if $m_i \neq m_i'$,

$$\mathcal{G}(C^{\text{nim}}_{(+,\text{pass})}(m_1, \ldots, m_{i-1}, m_i, m_{i+1}, \ldots, m_n))$$
$$\neq \mathcal{G}(C^{\text{nim}}_{(+,\text{pass})}(m_1, \ldots, m_{i-1}, m_i', m_{i+1}, \ldots, m_n)).$$

Furthermore, since every g_i is a one-move position and at least one of g_i is not a terminal position, at least one of $\mathcal{G}(g_i)$ is a positive integer and therefore,

$$\mathcal{G}(C^{\mathrm{nim}}_{(+,\mathrm{pass})}(\mathcal{G}(g_1),\ldots,\mathcal{G}(g_n)) \neq \mathcal{G}(g_1)\oplus\ldots\oplus\mathcal{G}(g_n) = \mathcal{G}(C^{\mathrm{nim}}_{+}(\mathcal{G}(g_1),\ldots,\mathcal{G}(g_n))).$$

Next, assume that $\ell < \mathcal{G}(C^{\mathrm{nim}}_{(+,\mathrm{pass})}(\mathcal{G}(g_1),\ldots,\mathcal{G}(g_n)))$. Since

$$\mathcal{G}(C^{\mathrm{nim}}_{(+,\mathrm{pass})}(\mathcal{G}(g_1),\ldots,\mathcal{G}(g_n))) = \mathrm{mex}(\{\mathcal{G}(G) \mid G \in \mathrm{move}(C^{\mathrm{nim}}_{(+,\mathrm{pass})}(\mathcal{G}(g_1),\ldots,\mathcal{G}(g_n)))\}),$$

there is an integer i such that $\ell = \mathcal{G}(C^{\mathrm{nim}}_{(+,\mathrm{pass})}(\mathcal{G}(g_1),\ldots,\mathcal{G}(g_{i-1}),k,\mathcal{G}(g_{i+1}),\ldots,\mathcal{G}(g_n)))$, where $k < \mathcal{G}(g_i)$, or $\ell = \mathcal{G}(C^{\mathrm{nim}}_{+}(\mathcal{G}(g_1),\ldots,\mathcal{G}(g_n)))$. Then, for the former case, $\ell \in S_1$ and for the latter case, $\ell = \mathcal{G}(g_1) \oplus \cdots \oplus \mathcal{G}(g_n)$. Thus, $\ell \in S_1 \cup S_2$ and from the definition of mex,

$$\mathcal{G}(C^{\mathrm{nim}}_{(+,\mathrm{pass})}(\mathcal{G}(g_1),\ldots,\mathcal{G}(g_n))) = \mathrm{mex}(S_1 \cup S_2)$$

holds, so we have

$$\mathcal{G}(C_{(+,\mathrm{pass})}(g_1,\ldots,g_n)) = \mathcal{G}(C^{\mathrm{nim}}_{(+,\mathrm{pass})}(\mathcal{G}(g_1),\ldots,\mathcal{G}(g_n))).$$

$\square$

Note that when at least one of g_i is not a one-move position, it is not guaranteed that SG-homomorphism holds. Let us assume that one of the components, say, g_1, is not a one-move position and $g_2,\ldots,g_n$ are one-move positions. Then, g_1 has a follower g_1' whose SG-value is zero but not a terminal position. Let $g_2',\ldots,g_n'$ be terminal positions which can be reached from $g_2,\ldots,g_n$ in one move. Then,

$$\mathcal{G}(C^{\mathrm{nim}}_{(+,\mathrm{pass})}(\mathcal{G}(g_1'),\mathcal{G}(g_2')\ldots,\mathcal{G}(g_n'))) = \mathcal{G}(C^{\mathrm{nim}}_{(+,\mathrm{pass})}(0,0,\ldots,0)) = 0,$$

but $\mathcal{G}(C_{(+,\mathrm{pass})}(g_1',g_2'\ldots,g_n')) \neq 0$ since $C_{(+,\mathrm{pass})}(g_1',g_2'\ldots,g_n')$ has an option $C_{+}(g_1',g_2'\ldots,g_n')$, whose SG-value is

$$\mathcal{G}(C_{+}(g_1',g_2'\ldots,g_n')) = \mathcal{G}(g_1') \oplus \mathcal{G}(g_2') \oplus \cdots \oplus \mathcal{G}(g_n') = 0 \oplus 0 \oplus \cdots \oplus 0 = 0.$$

From this theorem, the time needed to compute the SG-value of a one-pass compound of one-move positions is bounded by the total time required to compute the SG-values of all the component positions, plus the time required to compute the SG-value of a NIM position that allows one pass.

Unfortunately, closed formula for SG-values of positions in NIM with a pass is unknown. However, by using this result, we can expect faster analysis than with brute-force algorithms. In particular, we have closed formulas for small SG-values in two-pile NIM with a pass as follows. These results can be used for analyzing games, like some CHOCOLATE GAMES demonstrated in the next section.

Let $\mathcal{G}_{\mathrm{P}}(x,y) = \mathcal{G}(C^{\mathrm{nim}}_{(+,\mathrm{pass})}(x,y))$. Table 3 shows the values of $\mathcal{G}_{\mathrm{P}}(x,y)$. For the cases that $\mathcal{G}_{\mathrm{P}}(x,y)$ is small, we have the following closed formulas. For the proofs of them, see the full version of this paper [11].

Table 3. Values of $\mathcal{G}_P(x, y)$.

$x \backslash y$	0	1	2	3	4	5	6	7	8	9	10	11	12
0	0	2	1	4	3	6	5	8	7	10	9	12	11
1	2	1	0	3	4	5	6	7	8	9	10	11	12
2	1	0	2	5	7	3	8	4	6	12	11	10	9
3	4	3	5	1	0	2	7	6	9	8	12	13	10
4	3	4	7	0	1	8	9	2	5	6	13	14	15
5	6	5	3	2	8	1	0	9	4	7	14	15	13
6	5	6	8	7	9	0	1	3	2	4	15	16	14
7	8	7	4	6	2	9	3	1	0	5	16	17	18
8	7	8	6	9	5	4	2	0	1	3	17	18	16
9	10	9	12	8	6	7	4	5	3	1	0	19	2
10	9	10	11	12	13	14	15	16	17	0	1	2	3
11	12	11	10	13	14	15	16	17	18	19	2	1	0
12	11	12	9	10	15	13	14	18	16	2	3	0	1

Lemma 4. *Let x and y be nonnegative integers.*

(a) $\mathcal{G}_P(x, y) = 0$ if and only if one of the following holds;
 - *$x = y = 0$.*
 - *$x = 2n - 1$ and $y = 2n$ for a positive integer n.*
 - *$x = 2n$ and $y = 2n - 1$ for a positive integer n.*

(b) $\mathcal{G}_P(x, y) = 1$ if and only if one of the following holds;
 - *$(x, y) \in \{(0, 2), (2, 0)\}$.*
 - *$x = y, x \neq 0$, and $x \neq 2$.*

(c) $\mathcal{G}_P(x, y) = 2$ if and only if one of the following holds;
 - *$(x, y) \in \{(0, 1), (1, 0), (2, 2), (3, 5), (4, 7), (5, 3), (6, 8), (7, 4), (8, 6)\}$.*
 - *$(x - 1) \oplus (y - 1) = 3$ and $x, y \geq 9$.*

3   Application for Three-Dimensional CHOCOLATE GAME

In this section, we show how Theorem 5 can be used for analyzing games.

Let us consider a CHOCOLATE GAME in Definition 7 under the condition that $F(x, z) = h(z)$, where h satisfies NS-property. We call this game STAIR CHOCO-LATE GAME. Some examples of positions in the game are shown in Figs. 9, 10, 11, and 12. Then, from Theorem 4, the SG-value of the chocolate bar $CB(F, x, y, z)$ in the game is $x \oplus y \oplus z$.

Note that since $F(x, z) = h(z)$, the value of x is independent from y and z. Therefore, we can consider this ruleset as a disjunctive compound of one-heap NIM and two-dimensional CHOCOLATE GAME. We can use this construction for analyzing STAIR CHOCOLATE GAME with a pass move. The position of this game

388   H. Manabe et al.

is represented by four coordinates (x, y, z, p), where $p = 1$ if the pass is still available; otherwise, $p = 0$.

We already have small SG-values of two-dimensional chocolate games where h has NS-property in Lemma 3. We also have small SG-values of two-pile NIM with a pass in Lemma 4. By combining these results using Theorem 5, we have the following corollaries. Since we do not have any formula for SG-values or $\mathcal{P}$-positions of three-pile NIM with a pass, it is remarkable that we have some formulas for SG-values including $\mathcal{P}$-positions of these three-dimensional CHOCO-LATE GAMES. For the proofs of these corollaries, see the full version of this paper [11].

Corollary 2. *Let*

$$A_0 = \{(x, y, z, 0) \mid x \oplus y \oplus z = 0, y \le h(z)\},$$
$$A_1 = \{(x, y, z, 1) \mid x \text{ is odd}, (x + 1) \oplus y \oplus z = 0, y \le h(z)\},$$
$$A_2 = \{(x, y, z, 1) \mid x \ge 2, x \text{ is even}, (x - 1) \oplus y \oplus z = 0, y \le h(z)\}, \text{ and}$$
$$A = A_0 \cup A_1 \cup A_2 \cup \{(0, 0, 0, 1)\}.$$

A position (x, y, z, p) is in A if and only if the position is a $\mathcal{P}$-position.

Corollary 3. *Let*

$$B_0 = \{(x, y, z, 0) \mid x \oplus y \oplus z = 1, y \le h(z)\},$$
$$B_1 = \{(x, y, z, 1) \mid (x, y, z) \in \{(0, 0, 2), (0, 1, 3), (2, 0, 0)\}, y \le h(z)\},$$
$$B_2 = \{(x, y, z, 1) \mid x \oplus y \oplus z = 0, y \le h(z)\},$$
$$B_3 = \{(x, y, z, 1) \mid (x, y, z) \in \{(0, 0, 0), (2, 0, 2), (2, 1, 3)\}, y \le h(z)\}, \text{ and}$$
$$B = B_0 \cup B_1 \cup B_2 - B_3.$$

A position (x, y, z, p) is in B if and only if the position has SG-value 1.

Corollary 4. *Let*

$$C_0 = \{(x, y, z, 0) \mid x \oplus y \oplus z = 2, y \le h(z)\},$$
$$\begin{aligned}
C_1 = \{(x, y, z, 1) \mid (x, y, z) \in \{&(0, 0, 1), (1, 0, 0), (2, 0, 2), (2, 1, 3), (3, 0, 5), (3, 1, 4), \\
&(3, 2, 7), (3, 3, 6), (4, 0, 7), (4, 1, 6), (4, 2, 5), (4, 3, 4), (5, 0, 3), \ (5, 1, 2), (6, 0, 8), \\
&(6, 1, 9)(6, 2, 10), (6, 3, 11), (6, 4, 12), (6, 5, 13), (6, 6, 14), (6, 7, 15), (7, 0, 4), \\
&(7, 1, 5), (7, 2, 6), (7, 3, 7), (8, 0, 6), (8, 1, 7), (8, 2, 4), (8, 3, 5)\}, y \le h(z)\},
\end{aligned}$$
$$C_2 = \{(x, y, z, 1) \mid (((x - 1) \oplus 3) + 1) \oplus y \oplus z = 0, y \le h(z)\}$$
$$\begin{aligned}
C_3 = \{(x, y, z, 1) \mid (x, y, z) \in \{&(1, 0, 4), (1, 1, 5), (1, 2, 6), (1, 3, 7), (2, 0, 3), (2, 1, 2), \\
&(3, 0, 2), (3, 1, 3), (4, 0, 1), (5, 0, 8), (5, 1, 9), (5, 2, 10), (5, 3, 11), (5, 4, 12), \\
&(5, 5, 13), (5, 6, 14), (5, 7, 15), (6, 0, 7), (6, 1, 6), (6, 2, 5), (6, 3, 4), (7, 0, 6), \\
&(7, 1, 7), (7, 2, 4), (7, 3, 5), (8, 0, 5), (8, 1, 4), (8, 2, 7), (8, 3, 6)\}, y \le h(z)\}, \text{ and}
\end{aligned}$$
$$C = C_0 \cup C_1 \cup C_2 - C_3.$$

A position (x, y, z, p) is in C if and only if the position has SG-value 2.

Acknowledgement. The authors have no competing interests to declare that are relevant to the content of this article. The authors extend their gratitude to Mr. Keito Tanemura, Mr. Yuji Sasaki, and Mr. Yuki Tokuni for their help on developing a calculation program for SG-values. The authors are also grateful to the anonymous reviewers for their valuable comments and suggestions.

References

1. Albert, M.H., Nowakowski, R.J., Wolfe, D.: Lessons In Play: An Introduction to Combinatorial Game Theory, 2nd edn. A K Peters/CRC Press, Boca Raton, Florida, United States (2019)
2. Boros, E., Gurvich, V., Ho, N.B., Makino, K., Mursic, P.: On the Sprague-Grundy function of exact k-nim. Discret. Appl. Math. **239**, 1–14 (2018)
3. Boros, E., Gurvich, V., Ho, N.B., Makino, K., Mursic, P.: Impartial games with decreasing Sprague-Grundy function and their hypergraph compound. Int. J. Game Theory **53**, 1119–1144 (2024)
4. Bouton, C.L.: Nim, a game with a complete mathmatical theory. Ann. Math. **3**, 35–39 (1902)
5. Chan, W.H., Low, R.M., Locke, S.C., Wong, O.L.: A map of the P-positions in 'Nim With a Pass' played on heap sizes of at most four. Discret. Appl. Math. **244**, 44–55 (2018)
6. Gale, D.: A curious Nim-type game. Math. Monthly **81**(8), 876–879 (1974)
7. Grundy, P.M.: Mathematics and games. Eureka **2**, 6–8 (1939)
8. Hirsch, E.: Investigations of impartial games with a pass. preprint **arXiv:2010.10643** (2020)
9. Levine, L.: Fractal sequences and restricted nim. Ars Combinatoria **LXXX**, 113–128 (2006)
10. Low, R.M., Chan, W.H.: An atlas of N- and P-positions in 'Nim with a pass'. Integers **15**(#G2) (2015)
11. Manabe, H., Miyadera, R., Suetsugu, K.: On the Sprague-Grundy values of games with a pass. preprint arXiv:2501.15786 (2025)
12. Miyadera, R., Inoue, M., Fukui, M.: Impartial chocolate bar games with a pass. Integers **16**(#G5) (2016)
13. Miyadera, R., Nakaya, Y.: Grundy numbers of impartial three-dimensional chocolate-bar games. Integers **21B**(#A19) (2021)
14. Morrison, R.E., Friedman, E.J., Landsberg, A.S.: Combinatorial games with a pass: A dynamical systems approach. Chaos, An Interdisciplinary Journal of Nonlinear Science **21**(4) (2011)
15. Nowakowski, R.J.: Unsolved problems in combinatorial games. Games No Chance **V**, 125–168 (2019)
16. Robin, A.C.: A poisoned chocolate problem. Problem Corner, The Mathematical Gazette **73**(466), 341–343 (1989)
17. Sprague, R.P.: Über mathematische Kampfspiele. Tôhoku Mathematical Journal **41**, 438–444 (1935–36)

Reverse Mathematics for Neural Networks

Yoshihiro Maruyama[(✉)]

School of Informatics, Nagoya University, Nagoya, Japan
maruyama@i.nagoya-u.ac.jp

Abstract. We rigorously measure the reverse mathematical strengths of fundamental analytic principles for two neural architectures, Transformer and MLP (MultiLayer Perceptron). We prove that the standard compactness principle for Transformer is equivalent (over RCA_0) to ACA_0. For ReLU MLPs on compact domains, we prove that the standard universal approximation theorem for MLPs (with rational parameters) is equivalent to WKL_0. These results mean that the transformer compactness strictly exceeds the MLP universality in the reverse mathematical hierarchy. This paves the way for a systematic program of reverse mathematical classification of neural networks, bridging proof theory, computable analysis, and machine learning in a novel, interdisciplinary manner.

Keywords: Reverse mathematics $\cdot$ Transformer $\cdot$ MLP

1   Introduction

Analyses of modern neural architectures often invoke two kinds of analytic principles: (i) *approximation* on compact domains (e.g. "universal approximation" for MLPs [2,4,8–10]), and (ii) *compactness/subsequence* arguments for stability or limiting behavior (e.g. in transformer attention patterns [6,28,31,33]). Reverse Mathematics (RM) allows us to calibrate such mathematical principles against the standard subsystems of second–order arithmetic (RCA_0, WKL_0, ACA_0, ATR_0, $\Pi_1^1\text{-}\mathsf{CA}_0$) by asking *which system is exactly needed to prove a given theorem* [3,7,29]. Classical calibrations include that Heine–Cantor on $[0,1]$ (every continuous f is uniformly continuous) is equivalent to WKL_0, while Bolzano–Weierstrass on $[0,1]$ is equivalent to ACA_0 over RCA_0 [29]. Beyond nonuniform implications, their *uniform* counterparts live in the Weihrauch lattice and separate operations like compact choice and subsequence selection even more finely [1].

In this paper, rather than comparing architectures by numerical computational power or runtime, we compare the logical structural strength of mathematical principles underlying them, using the methods of reverse mathematics. For ReLU MLPs, the canonical statement is the universal approximation theorem (UAT) on $[0,1]^d$ with finitely coded rational parameters (cf. [2,4,8–10]). For transformers, a natural compactness principle asserts that, for any uniformly

© The Author(s), under exclusive license to Springer Nature Switzerland AG 2026
J. Kozik and A. Wolff (Eds.): SOFSEM 2026, LNCS 16448, pp. 390–403, 2026.
https://doi.org/10.1007/978-3-032-17801-5_29

bounded sequence of (even very minimal) attention blocks and any finite set of inputs, there is a subsequence along which all attention weights converge coordinatewise. We formulate this principle as $\mathsf{TRANS\text{-}SIM}_{\text{fin}}$ (Definition 9) and the UAT as $\mathsf{MLP\text{-}UAT}_d$ (Definition 10), both in the language of RCA_0.

The main results of this pape are as follows. Our first theorem shows that the transformer compactness principle sits exactly at the ACA_0 level; the second shows that the MLP UAT sits exactly at the WKL_0 level. Over RCA_0, the simultaneous attention–subsequence principle $\mathsf{TRANS\text{-}SIM}_{\text{fin}}$ for a (one–head, one–layer, two–token) transformer block is equivalent to ACA_0. Fix $d \geq 1$. Over RCA_0, the ReLU MLP universal approximation statement $\mathsf{MLP\text{-}UAT}_d$ on $[0,1]^d$ with rational parameters is equivalent to WKL_0. Since $\mathsf{WKL}_0 \not\vdash \mathsf{ACA}_0$ over RCA_0 [29], these theorems yield a strict separation in the reverse mathematical hierarchy between the canonical compactness principle for transformers and the canonical approximation principle for MLPs.

A more technical overview is as follows. For $\mathsf{TRANS\text{-}SIM}_{\text{fin}} \Rightarrow \mathsf{ACA}_0$, we reduce Bolzano–Weierstrass on $[0,1]$ to $\mathsf{TRANS\text{-}SIM}_{\text{fin}}$: given an arbitrary bounded real sequence (r_n), we encode r_n as the first attention weight of a minimal block by a fixed rational choice of tokens such that the weight is a strictly increasing *affine* function of r_n; applying $\mathsf{TRANS\text{-}SIM}_{\text{fin}}$ yields a convergent subsequence of the attention weights and hence, by rational decoding, of (r_n). For $\mathsf{ACA}_0 \Rightarrow \mathsf{TRANS\text{-}SIM}_{\text{fin}}$, we apply Bolzano–Weierstrass coordinatewise and perform a finite diagonal argument to obtain a single subsequence working simultaneously on the finite input set. For $\mathsf{WKL}_0 \Rightarrow \mathsf{MLP\text{-}UAT}_d$, Heine–Cantor provides a uniform modulus, from which we build a multilinear grid interpolant P_δ; each 1D hat is realized *exactly* by a depth-2 ReLU, and d-fold products are approximated by a depth-3 ReLU multiplication gadget derived from a spline approximation to $t \mapsto t^2$ [32] (see also the ReLU–spline links [26] and quantitative results in higher dimension [27]). Summing the blocks yields a rational-parameter network N with $\|f - N\|_\infty < \varepsilon$. For $\mathsf{MLP\text{-}UAT}_d \Rightarrow \mathsf{WKL}_0$, every rational-parameter ReLU MLP is globally Lipschitz with a constant computable from its code; approximating f within $\varepsilon/3$ and propagating the Lipschitz bound gives a rational modulus of uniform continuity for f, i.e. Heine Cantor, hence WKL_0.

Broadly speaking, this means that, in the rigorous reverse mathematical sense, the logical commitments typically used in transformer analyses are strictly richer (ACA_0) than those used in MLP universal approximation (WKL_0). This paves the way for a systematic program of reverse mathematical classification of neural networks, bridging proof theory, computable analysis, and machine learning.

The rest of the paper is organized as follows. In Sect. 2, we give preliminaries. In Sect. 3, we prove the compactness principle for Transformer is equivalent to ACA_0. In Sect. 4, we prove that the universal approximation theorem for MLPs is equivalent to WKL_0.

2   Preliminaries and Formal Setting

We work over the base system RCA_0 (i.e. RCA_0). Throughout, WKL_0 and ACA_0 abbreviate WKL_0 and ACA_0. Vectors carry the sup norm $\|\cdot\|_\infty$; matrices use the induced operator norm $\|W\|_\infty$ (maximum absolute row sum).

Reals and fast Cauchy Names. We represent real numbers in the standard RM way. A *real* is coded by a *fast Cauchy* sequence of rationals $q : \mathbb{N} \to \mathbb{Q}$ with $|q(m) - q(n)| \leq 2^{-n}$ for all $m \geq n$. Given codes q_x, q_y, we set $x = y$ iff $\forall n \, (|q_x(n) - q_y(n)| \leq 2^{-(n-1)})$. Arithmetic operations on reals are defined by primitive recursive transforms on names inside RCA_0.

Fast Cauchy Sequences of Reals. Sequences of reals are handled similarly. A sequence $(x_k)_{k \in \mathbb{N}}$ of reals is *fast Cauchy* if there is $N : \mathbb{N} \to \mathbb{N}$ with $|x_k - x_\ell| \leq 2^{-n}$ whenever $k, \ell \geq N(n)$. A *subsequence* is given by a strictly increasing $g : \mathbb{N} \to \mathbb{N}$.

We also need a coding for continuous functions on compact cubes, standard in reverse mathematics.

Definition 1 (Codes for continuous functions). *A code for $f \in C([0,1]^d)$ consists of:*

(i) a modulus $\mu : \mathbb{N} \to \mathbb{N}$ such that $\|x - y\|_\infty \leq 2^{-\mu(n)} \Rightarrow |f(x) - f(y)| \leq 2^{-n}$;
(ii) a procedure assigning to each rational box $B \subseteq [0,1]^d$ and $n \in \mathbb{N}$ a rational $r(B, n)$ with $|f(x) - r(B, n)| \leq 2^{-n}$ for all $x \in B$.

From the code one computes rational approximations to f at rational points.

In particular, we will often use the dyadic grid of mesh 2^{-m} in $[0,1]^d$ and the multilinear interpolant on that grid; both can be defined within RCA_0 from the code of f and a requested precision.

We next formalize the neural network primitives.

Definition 2 (Affine maps and norms). *An affine map $A(x) = Wx + b$ is coded by rational W, b. Then $|A(x) - A(y)| \leq \|W\|_\infty \|x - y\|_\infty$ for all x, y.*

Affine maps are the linear building blocks of networks. We combine them with the standard ReLU nonlinearity:

Definition 3 (ReLU). *$\sigma(t) = \max\{t, 0\}$ is 1-Lipschitz and piecewise linear with rational breakpoints, hence exactly evaluable at rational inputs in RCA_0 [5].*

From these components we define multilayer perceptrons:

Definition 4 (ReLU MLP). *A (scalar-output) ReLU MLP $N : [0,1]^d \to \mathbb{R}$ of depth L is a finite list $((W_\ell, b_\ell))_{\ell=1}^{L+1}$ of rational matrices/vectors with*

$$x^{(1)} = \sigma(W_1 x + b_1), \quad x^{(2)} = \sigma(W_2 x^{(1)} + b_2), \quad \ldots, \quad N(x) = W_{L+1} x^{(L)} + b_{L+1}.$$

The code is the concatenation of these rationals.

The following shows every such network has a computable Lipschitz constant:

Lemma 1 (Global Lipschitz bound). *If N has layers (W_ℓ, b_ℓ) as above, then for all x, y*

$$|N(x) - N(y)| \leq \left(\prod_{\ell=1}^{L} \|W_\ell\|_\infty\right) \|x - y\|_\infty,$$

and the product is a rational computed from the code of N in RCA_0.

For approximation constructions we also need simple ReLU gadgets:

Definition 5 (Hat functions). *For $t \in \mathbb{Q}$ and $h \in \mathbb{Q}_{>0}$, the 1D hat $\mathrm{hat}_{t,h}(x) := \max\{0, 1 - |x - t|/h\}$ satisfies the exact identity*

$$\mathrm{hat}_{t,h}(x) = \tfrac{1}{h}\big(\sigma(x - (t - h)) - 2\sigma(x - t) + \sigma(x - (t + h))\big),$$

hence is a depth-2 ReLU with rational parameters. Tensor products of such hats over a dyadic grid form the usual multilinear partition of unity on $[0,1]^d$.

Another useful primitive is approximate multiplication:

Definition 6 (Approximate multiplication). *For each rational $\delta > 0$ there is a depth-3 ReLU MLP $M_\times$ with rational parameters such that $\sup_{(x,y) \in [0,1]^2} |M_\times(x,y) - xy| < \delta$. Construction in RCA_0: approximate $t \mapsto t^2$ on $[0,1]$ by a piecewise linear map and use $xy = \tfrac{1}{4}\big((x + y)^2 - (x - y)^2\big)$; all bookkeeping is primitive recursive in δ (cf. [26, 32]).*

We now turn to transformer blocks. First, we define a normalizer that converts arbitrary scores into attention weights in $(0, 1)$:

Definition 7 (Affine attention normalizer). *Fix $c_0 \in \mathbb{Q}$ with $c_0 \geq 1$. For $u, v \in \mathbb{R}$ define*

$$\alpha(u, v) := \frac{u + c_0}{u + v + 2c_0}, \qquad \mathrm{Norm}(u, v) := (\alpha(u, v), 1 - \alpha(u, v)).$$

Convention. *To avoid division by 0 uniformly and keep α total, we apply Norm only to the scaled scores $(\tilde{u}, \tilde{v})$ defined below; with that scaling the denominator satisfies $\tilde{u} + \tilde{v} + 2c_0 \geq 2c_0 - 1 \geq 1$, hence α is always defined.*

With the normalizer in hand we can define minimal attention blocks:

Definition 8 (Minimal transformer block and scaling). *Fix rational tokens $x_{\mathrm{cls}}, x_1, x_2 \in \mathbb{Q}$. A (two–token, one–head) block is specified by reals Q, K (coded by fast Cauchy names). Given a uniform bound certificate $B \in \mathbb{Q}_{>0}$ for a sequence (Q_n, K_n) (i.e. $|Q_n| \leq B$ and $|K_n| \leq B$ for all n), define the raw scores*

$$s_1 = QK\,x_{\mathrm{cls}}x_1, \qquad s_2 = QK\,x_{\mathrm{cls}}x_2,$$

and the scaling factor

$$S := 1 + B^2 \, |x_{\mathrm{cls}}| \, (|x_1| + |x_2|) \; \in \; \mathbb{Q}_{>0}.$$

Set the scaled *scores* $\tilde{s}_i := s_i/S$ *for* $i = 1, 2$. *Then* $|\tilde{s}_1 + \tilde{s}_2| \leq 1$. *The first attention weight is the real*

$$\alpha(Q, K; x_{\mathrm{cls}}, x_1, x_2) \; := \; \alpha(\tilde{s}_1, \tilde{s}_2) \; = \; \frac{\tilde{s}_1 + c_0}{\tilde{s}_1 + \tilde{s}_2 + 2c_0}.$$

A finite input set *is a list* $F = ((x_{\mathrm{cls}}, x_1^{(j)}, x_2^{(j)}))_{j<m}$ *with common* x_{cls}.

For fixed $x_{\mathrm{cls}}, x_1, x_2$ and bound B, the map $(Q, K) \mapsto \alpha(Q, K; x_{\mathrm{cls}}, x_1, x_2)$ is continuous in RCA_0. On instances used in the reversal (Sect. 3) we take $x_2 = -x_1$ so that $\tilde{s}_1 + \tilde{s}_2 \equiv 0$, and then $\alpha = \dfrac{\tilde{s}_1 + c_0}{\tilde{s}_1 + 2c_0}$ is strictly increasing in $\tilde{s}_1$, with a rational inverse on bounded intervals; this is the only monotonicity we need.

We can now state the two central principles studied in the rest of the paper.

Definition 9 (The principle $\mathsf{TRANS\text{-}SIM}_{\mathrm{fin}}$). *Let* $(Q_n, K_n)_{n\in\mathbb{N}}$ *be a sequence together with a rational* $B > 0$ *certifying* $|Q_n|, |K_n| \leq B$ *for all* n, *and let* $F = ((x_{\mathrm{cls}}, x_1^{(j)}, x_2^{(j)}))_{j<m}$ *be a finite input set. We say* $\mathsf{TRANS\text{-}SIM}_{\mathrm{fin}}$ *holds if there exists a strictly increasing* $g : \mathbb{N} \to \mathbb{N}$ *such that, for every* $j < m$, *the sequence of reals*

$$\left(\alpha(Q_{g(k)}, K_{g(k)}; x_{\mathrm{cls}}, x_1^{(j)}, x_2^{(j)}) \right)_{k\in\mathbb{N}}$$

is fast Cauchy (hence convergent). The quantifiers in $\mathsf{TRANS\text{-}SIM}_{\mathrm{fin}}$ *range over all such sequences with their bound certificates and all finite input sets.*

Definition 10 (ReLU MLP universal approximation $\mathsf{MLP\text{-}UAT}_d$). *A ReLU MLP has* rational *parameters as in Definition 4. For* $d \geq 1$, $\mathsf{MLP\text{-}UAT}_d$ *states that for every* $f \in C([0,1]^d)$ *and every rational* $\varepsilon > 0$ *there exists a ReLU MLP* N *with rational parameters such that* $\sup_{x\in[0,1]^d} |f(x) - N(x)| < \varepsilon$.

Finally, we recall two classical calibrations from RM that serve as targets for our reductions. (i) *Heine–Cantor*: Over RCA_0, the statement "every $f \in C([0,1]^d)$ is uniformly continuous on $[0,1]^d$" is equivalent to WKL_0 (for any fixed $d \geq 1$). (ii) *Bolzano–Weierstrass on* $[0,1]$: Over RCA_0, the statement "every sequence $(r_n) \subseteq [0,1]$ has a convergent subsequence" is equivalent to ACA_0. These equivalences are standard (see [29]) and will be used without proof.

3   Transformer Compactness $\Longleftrightarrow$ ACA_0

Here we prove the following theorem.

Theorem 1 (TRANS–ACA Equivalence). *Over* RCA_0, $\mathsf{TRANS\text{-}SIM}_{\mathrm{fin}}$ *is equivalent to* ACA_0.

Proof (sketch; full proof below). ($\Rightarrow$) Given $(r_n) \subseteq [0,1]$, code it as a bounded transformer sequence by $Q_n := 1$, $K_n := r_n$ and tokens $(x_{\text{cls}}, x_1, x_2) = (1, 1, -1)$. With the scaling from Definition 8, we obtain

$$\alpha^{(n)} = \alpha(Q_n, K_n; x_{\text{cls}}, x_1, x_2) = \frac{\tilde{s}_1^{(n)} + c_0}{\tilde{s}_1^{(n)} + \tilde{s}_2^{(n)} + 2c_0} = \frac{\tilde{s}_1^{(n)} + c_0}{2c_0} = \tfrac{1}{2} + \frac{\tilde{s}_1^{(n)}}{2c_0}.$$

Since $\tilde{s}_1^{(n)} = r_n/S$ for a fixed rational $S > 0$, $\alpha^{(n)}$ is an *affine* function of r_n. By TRANS-SIM$_{\text{fin}}$ there is g such that $\alpha^{(g(k))}$ is fast Cauchy; applying the rational linear inverse yields a fast Cauchy subsequence of (r_n), i.e. Bolzano–Weierstrass on $[0, 1]$, and hence ACA$_0$.

($\Leftarrow$) Assuming ACA$_0$, Bolzano–Weierstrass holds on $[0, 1]$. For each input in the finite set F, apply BW to the corresponding attention sequence to obtain a convergent subsequence; a finite diagonalization yields a single subsequence working *simultaneously* for all inputs in F, as required by TRANS-SIM$_{\text{fin}}$.

3.1   Full Proof Details

We now give full proof details in RCA$_0$.

Setup. Fix $c_0 \in \mathbb{Q}$ with $c_0 \geq 1$. For a bounded parameter sequence (Q_n, K_n) with rational certificate $B > 0$ (i.e. $|Q_n|, |K_n| \leq B$ for all n) and fixed rational tokens $(x_{\text{cls}}, x_1, x_2)$, Definition 8 defines the raw scores $s_i^{(n)} = Q_n K_n x_{\text{cls}} x_i$ and the rational scaling

$$S := 1 + B^2 |x_{\text{cls}}| (|x_1| + |x_2|) \in \mathbb{Q}_{>0}, \qquad \tilde{s}_i^{(n)} := s_i^{(n)}/S,$$

so that $\tilde{s}_1^{(n)} + \tilde{s}_2^{(n)} \in [-1, 1]$ for all n. The attention weight is

$$\alpha^{(n)} := \alpha(\tilde{s}_1^{(n)}, \tilde{s}_2^{(n)}) = \frac{\tilde{s}_1^{(n)} + c_0}{\tilde{s}_1^{(n)} + \tilde{s}_2^{(n)} + 2c_0}.$$

Because the denominator is $\geq 2c_0 - 1 \geq 1$, $\alpha^{(n)}$ is a well-defined real in $(0, 1)$ in the sense of RCA$_0$.

Lemma 2 (Encoding). *Over* RCA$_0$, *let* $(r_n) \subseteq [0, 1]$ *be any sequence of reals. Define a transformer sequence by* $Q_n := 1$, $K_n := r_n$, *and* $(x_{\text{cls}}, x_1, x_2) := (1, 1, -1)$. *Then with* $B := 1$ *and the resulting scale* $S = 1 + 2 \in \mathbb{Q}$ *we have, for all* n,

$$\alpha^{(n)} = \tfrac{1}{2} + \frac{r_n}{2c_0 S} = \tfrac{1}{2} + \frac{r_n}{6c_0}.$$

Proof. With $Q_n = 1$, $K_n = r_n$ and $(x_{\text{cls}}, x_1, x_2) = (1, 1, -1)$ we have $s_1^{(n)} = r_n$, $s_2^{(n)} = -r_n$, hence $\tilde{s}_1^{(n)} = r_n/S$, $\tilde{s}_2^{(n)} = -r_n/S$. Therefore

$$\alpha^{(n)} = \frac{\tilde{s}_1^{(n)} + c_0}{\tilde{s}_1^{(n)} + \tilde{s}_2^{(n)} + 2c_0} = \frac{\frac{r_n}{S} + c_0}{0 + 2c_0} = \tfrac{1}{2} + \frac{r_n}{2c_0 S} = \tfrac{1}{2} + \frac{r_n}{6c_0}.$$

All computations are primitive recursive on the given names in RCA$_0$.

Lemma 3 (Linear decoding). *Let* $\mathrm{Dec} : (0,1) \to \mathbb{R}$ *be defined by* $\mathrm{Dec}(\alpha) := 6c_0 \left(\alpha - \frac{1}{2} \right)$. *If* (α_k) *is fast Cauchy, then* $(\mathrm{Dec}(\alpha_k))$ *is fast Cauchy. Moreover, with* $\alpha^{(n)}$ *from Lemma 2,* $\mathrm{Dec}(\alpha^{(n)}) = r_n$ *for all* n.

Proof. Dec is rational affine; hence $|\mathrm{Dec}(\alpha) - \mathrm{Dec}(\beta)| = 6c_0 |\alpha - \beta|$, and fast Cauchiness is preserved with the same rate up to a rational factor. The identity $\mathrm{Dec}(\alpha^{(n)}) = r_n$ follows by direct substitution of $\alpha^{(n)} = \frac{1}{2} + \frac{r_n}{6c_0}$.

Proposition 1 (TRANS-SIM$_{\mathrm{fin}}$ $\Rightarrow$ ACA$_0$). *Over* RCA$_0$, TRANS-SIM$_{\mathrm{fin}}$ *implies Bolzano–Weierstrass on* $[0,1]$ *and hence* ACA$_0$.

Proof. Let $(r_n) \subseteq [0,1]$ be arbitrary. Form the transformer sequence and single-ton input set F as in Lemma 2; this is admissible for TRANS-SIM$_{\mathrm{fin}}$ with bound certificate $B = 1$. By TRANS-SIM$_{\mathrm{fin}}$ there is a strictly increasing $g : \mathbb{N} \to \mathbb{N}$ such that $(\alpha^{(g(k))})$ is fast Cauchy. By Lemma 3, $(r_{g(k)}) = (\mathrm{Dec}(\alpha^{(g(k))}))$ is fast Cauchy, hence convergent as a real sequence in RCA$_0$. Thus BW$([0,1])$ holds. Over RCA$_0$, BW$([0,1])$ is equivalent to ACA$_0$.

Proposition 2 (ACA$_0$ $\Rightarrow$ TRANS-SIM$_{\mathrm{fin}}$). *Over* RCA$_0$, ACA$_0$ *implies* TRANS-SIM$_{\mathrm{fin}}$.

Proof. Fix a bounded parameter sequence (Q_n, K_n) with certificate B and a finite input set $F = ((x_{\mathrm{cls}}, x_1^{(j)}, x_2^{(j)}))_{j<m}$. For each $j < m$, consider the real sequence

$$a_j^{(n)} := \alpha(Q_n, K_n; x_{\mathrm{cls}}, x_1^{(j)}, x_2^{(j)}) \in (0,1).$$

By ACA$_0$, Bolzano–Weierstrass holds on $[0,1]$. Construct a subsequence by finite thinning: apply BW to $(a_0^{(n)})$ to get g_0 with $(a_0^{(g_0(k))})$ convergent (hence fast Cauchy). Assume g_t is built for $t < m-1$ so that $(a_j^{(g_t(k))})$ is fast Cauchy for all $j \le t$. Apply BW to the sequence $(a_{t+1}^{(g_t(k))})_k$ to obtain a strictly increasing h_{t+1} with $(a_{t+1}^{(g_t(h_{t+1}(k)))})$ convergent; let $g_{t+1} := g_t \circ h_{t+1}$. Because a subsequence of a fast Cauchy sequence is fast Cauchy in RCA$_0$, the properties for indices $\le t$ are preserved at each step. After m steps, $g := g_{m-1}$ witnesses that all $j < m$ coordinates are fast Cauchy along $n = g(k)$, as required by TRANS-SIM$_{\mathrm{fin}}$.

Combining Propositions 1 and 2 yields Theorem 1.

3.2   Robustness

The reversal uses only a *single* head, a *single* layer, and *two* tokens; strengthening the architecture (more heads/layers/tokens) only yields a priori stronger principles, hence the equivalence persists. The affine normalizer (Definition 7) is algebraic and avoids developing exponentials; a Softmax variant is also admissible:

Proposition 3 (Softmax variant). *Working in* RCA_0 *on bounded score ranges, define* $\mathrm{softmax}(u, v) := \left(\frac{e^u}{e^u + e^v}, \frac{e^v}{e^u + e^v} \right)$ *where* e^x *is represented on compact intervals by its power series with rational error bounds. Then replacing* α *in* $\mathsf{TRANS\text{-}SIM}_{\mathrm{fin}}$ *by the first Softmax coordinate yields a principle equivalent to* ACA_0 *over* RCA_0.

Proof. On any compact interval $[-M, M]$, e^x has a uniform modulus in RCA_0 via its series, hence so does $u \mapsto \frac{e^u}{e^u + e^c}$ for fixed c. In the reversal, pick tokens with $\tilde{s}_2 \equiv 0$ to obtain $\alpha^{(n)} = \frac{e^{\tilde{s}_1^{(n)}}}{e^{\tilde{s}_1^{(n)}} + 1}$, a strictly increasing bijection $[-M, M] \to [\sigma(-M), \sigma(M)]$ with computable inverse modulus. Decoding preserves fast Cauchy subsequences exactly as in Lemma 3. The $\mathsf{ACA}_0 \Rightarrow \mathsf{TRANS\text{-}SIM}_{\mathrm{fin}}$ direction is identical, using BW and finite thinning.

4   MLP Universal Approximation $\Longleftrightarrow$ WKL$_0$

Here we prove the following theorem; we use some technical lemmas in the proof, which are proven below after the main theorem proof.

Theorem 2 (MLP–WKL Equivalence). *Fix $d \geq 1$. Over* RCA_0, *the statement* $\mathsf{MLP\text{-}UAT}_d$ *is equivalent to* WKL_0.

Proof. ($\mathsf{WKL}_0 \Rightarrow \mathsf{MLP\text{-}UAT}_d$): Over RCA_0, WKL_0 is equivalent to the Heine–Cantor principle on $[0, 1]^d$ (uniform continuity for all continuous $f : [0, 1]^d \to \mathbb{R}$). Fix $f \in C([0, 1]^d)$ and $\varepsilon \in \mathbb{Q}_{>0}$. By Heine–Cantor there is a (uniform) modulus $\mu : \mathbb{N} \to \mathbb{N}$ such that $\|x - y\|_\infty \leq 2^{-\mu(n)}$ implies $|f(x) - f(y)| \leq 2^{-n}$.

Choose $n \in \mathbb{N}$ with $2^{-n} \leq \varepsilon/3$ and let $\delta := 2^{-\mu(n)}$. Partition $[0, 1]^d$ into the rational grid $\mathcal{G}$ of mesh δ. For each grid cube C with vertices $\{v^\lambda\}_{\lambda \in \{0,1\}^d}$, let $\theta_i^\lambda(\cdot)$ be the 1D hat functions along axis i centered at the two vertex coordinates of C (Definition 5). Define the multilinear interpolant

$$P_\delta(x) := \sum_{\lambda \in \{0,1\}^d} \left(\prod_{i=1}^{d} \theta_i^\lambda(x_i) \right) f(v^\lambda).$$

By Lemma 5 below, $\|f - P_\delta\|_\infty \leq 2^{-n} \leq \varepsilon/3$.

We now build a rational-parameter ReLU network N with $\|P_\delta - N\|_\infty \leq \varepsilon/3$.

(A) Exact realization of 1D hats. By Lemma 6, each θ_i^λ is *exactly* a depth-2 ReLU with rational parameters.

(B) Approximate d-fold products. For $\eta > 0$ to be set later, let $M_\times^\eta$ be the depth-3 ReLU multiplication gadget from Lemma 8 (with domain $[-2, 2]$), satisfying $\sup_{(u,v) \in [-2,2]^2} \left| M_\times^\eta(u, v) - uv \right| < \eta$. (We use the general interval because in the identity $xy = \frac{1}{4}\left((x+y)^2 - (x-y)^2\right)$ the arguments $x \pm y$ range over $[-1, 2]$ for $x, y \in [0, 1]$.)

Using a balanced binary tree of copies of $M_\times^\eta$ and the bounds $0 \le \theta_i^\lambda \le 1$, we obtain a ReLU subnetwork $\widetilde{\Phi}_\lambda$ such that

$$\sup_{x \in [0,1]^d} \left| \widetilde{\Phi}_\lambda(x) - \prod_{i=1}^d \theta_i^\lambda(x_i) \right| \le (d-1)\,\eta. \tag{1}$$

This is proved by induction on d (each multiplication introduces at most η absolute error and the factors remain in $[0,1]$).

(C) Rounding vertex weights. Using the code of f, for each vertex v^λ compute a rational $\widehat{f}(v^\lambda)$ with $|f(v^\lambda) - \widehat{f}(v^\lambda)| \le \rho$ (to be chosen). Let $M_f \in \mathbb{Q}_{>0}$ be a rational upper bound for $\max_{x \in [0,1]^d} |f(x)|$, obtained in RCA_0 by dyadic sampling guided by the modulus (see Lemma 4).

(D) Assemble the network. Define the ReLU MLP

$$N(x) \ := \ \sum_{\lambda \in \{0,1\}^d} \widehat{f}(v^\lambda)\, \widetilde{\Phi}_\lambda(x).$$

This is a finite ReLU network with rational parameters: two layers for the hats, $O(\log d)$ layers for the product tree, and a final affine summation.

(E) Error bound. For $x \in [0,1]^d$,

$$|P_\delta(x) - N(x)| \le \sum_\lambda \left| \left(\prod_i \theta_i^\lambda(x_i) \right) f(v^\lambda) - \widehat{f}(v^\lambda)\, \widetilde{\Phi}_\lambda(x) \right|$$

$$\le \sum_\lambda \left| \left(\prod_i \theta_i^\lambda(x_i) \right) (f(v^\lambda) - \widehat{f}(v^\lambda)) \right| + \sum_\lambda |\widehat{f}(v^\lambda)| \left| \prod_i \theta_i^\lambda(x_i) - \widetilde{\Phi}_\lambda(x) \right|$$

$$\le \sum_\lambda \rho \ + \ \sum_\lambda (M_f + \rho)\,(d-1)\eta$$

$$= 2^d\, \rho \ + \ 2^d\,(M_f + \rho)\,(d-1)\eta.$$

Choose rational parameters

$$\rho \ := \ \frac{\varepsilon}{6 \cdot 2^d}, \qquad \eta \ := \ \frac{\varepsilon}{6 \cdot 2^d \cdot (d-1) \cdot (M_f + 1)} \qquad \text{(if } d = 1,\ \eta := 0\text{)},$$

to obtain $\|P_\delta - N\|_\infty \le \varepsilon/3$. Therefore

$$\|f - N\|_\infty \ \le \ \|f - P_\delta\|_\infty + \|P_\delta - N\|_\infty \ < \ \varepsilon/3 + \varepsilon/3 \ < \ \varepsilon.$$

All constructions are primitive recursive in the codes of f and ε within RCA_0, and all network parameters are rational. Thus $\mathsf{MLP\text{-}UAT}_d$ holds over $\mathsf{RCA}_0 + \mathsf{WKL}_0$.

$(\mathsf{MLP\text{-}UAT}_d \Rightarrow \mathsf{WKL}_0)$: Over RCA_0 it suffices to derive Heine–Cantor on $[0,1]^d$ (which is equivalent to WKL_0). Fix $f \in C([0,1]^d)$ and $\varepsilon \in \mathbb{Q}_{>0}$. By $\mathsf{MLP\text{-}UAT}_d$ there exists a rational-parameter ReLU MLP N with $\|f - N\|_\infty < \varepsilon/3$. By Lemma 9, N is globally Lipschitz with rational constant

$$L \ := \ \prod_{\ell=1}^L \|W_\ell\|_\infty \ \in \ \mathbb{Q}_{>0},$$

effectively computed from the code of N. Set $\delta := \varepsilon/(3L) \in \mathbb{Q}_{>0}$. Then for any $x, y \in [0,1]^d$ with $\|x - y\|_\infty < \delta$ we have

$$|f(x)-f(y)| \;\leq\; |f(x)-N(x)|+|N(x)-N(y)|+|N(y)-f(y)| \;<\; \varepsilon/3+L\delta+\varepsilon/3 \;=\; \varepsilon.$$

Hence f is uniformly continuous on $[0,1]^d$ with rational modulus δ. This is precisely Heine–Cantor on $[0,1]^d$, which is equivalent to WKL_0 over RCA_0.

Note that the constructed network N has depth 2 for the exact hat functions, plus $O(\log d)$ for the product tree, and one final affine layer, so the overall depth is $2 + O(\log d) + 1$. All parameters are rational and their magnitudes are polynomially bounded in $1/\delta$ and $1/\eta$. Independent of our RM calibration, such constructions relate to known depth/width expressivity tradeoffs in approximation theory [11,25,30].

We stress that our argument uses the standard coding convention for continuous functions: Heine–Cantor on $[0,1]^d$ is equivalent to WKL_0 over RCA_0 for these pointwise codes, and it is precisely this calibration that underlies the proof.

4.1   Proofs of Lemmas Used Above

We prove the lemmas used in the above proof. All are provable in RCA_0.

Lemma 4 (Bounded sup from a code). *Let $f \in C([0,1]^d)$ be given by a code with uniform modulus μ. For each n one can compute $M_n \in \mathbb{Q}$ with $\|f\|_\infty \leq M_n \leq \|f\|_\infty + 2^{-n}$. In particular, a rational upper bound $M_f \in \mathbb{Q}$ for $\|f\|_\infty$ is computable in RCA_0.*

Proof. Take the dyadic grid of mesh $2^{-\mu(n+1)}$ and evaluate f to precision $2^{-(n+1)}$ on all grid points using the code. Let M_n be the maximum of these rationals plus $2^{-(n+1)}$. By the modulus, every point is within $2^{-\mu(n+1)}$ of a grid point, so the bound holds.

Lemma 5 (Multilinear interpolation error). *Let $f \in C([0,1]^d)$ with modulus μ. For n and $\delta = 2^{-\mu(n)}$, the multilinear interpolant P_δ on the grid of mesh δ satisfies $\|f - P_\delta\|_\infty \leq 2^{-n}$.*

Proof. Fix x in a grid cube C with vertices v^λ. Each x is a convex combination of the vertices with weights $\prod_i \theta_i^\lambda(x_i)$. The diameter of C in $\|\cdot\|_\infty$ is δ, hence $|f(x) - f(v^\lambda)| \leq 2^{-n}$ by the modulus. Therefore

$$|f(x) - P_\delta(x)| \;\leq\; \sum_\lambda \Big(\prod_i \theta_i^\lambda(x_i)\Big) |f(x) - f(v^\lambda)| \;\leq\; 2^{-n}\sum_\lambda \prod_i \theta_i^\lambda(x_i) \;=\; 2^{-n},$$

since the weights form a partition of unity. The argument is formalizable in RCA_0.

Lemma 6 (Exact hats by ReLU). *For $t \in \mathbb{Q}$ and $h \in \mathbb{Q}_{>0}$,*

$$\mathrm{hat}_{t,h}(x) := \max\{0, 1-|x-t|/h\} = \tfrac{1}{h}(\sigma(x - (t - h)) - 2\sigma(x - t) + \sigma(x - (t + h))).$$

Proof. Check the four regions $x \leq t - h$, $t - h \leq x \leq t$, $t \leq x \leq t + h$, and $x \geq t + h$. This is an identity of piecewise linear functions with rational breakpoints in RCA_0.

Lemma 7 (ReLU realization of splines). *Let $a < b$ be rationals. For any continuous piecewise linear $S : [a, b] \to \mathbb{R}$ with rational breakpoints $a = s_0 < s_1 < \cdots < s_m = b$ and rational values $S(s_j)$, there exist rational coefficients $c_0, c_1, \ldots, c_m$ with*

$$S(x) = c_0 + c_1 x + \sum_{j=1}^{m-1} c_{j+1} (x - s_j)_+, \qquad (t)_+ := \sigma(t),$$

so S is exactly a depth-2 ReLU network with rational parameters.

Proof. Let κ_j be the slope of S on (s_{j-1}, s_j). Then $S(x) = S(s_0) + \kappa_1(x - s_0) + \sum_{j=1}^{m-1}(\kappa_{j+1} - \kappa_j)(x - s_j)_+$. All quantities are rational.

Lemma 8 (ReLU multiplication on a bounded interval). *For any $\delta \in \mathbb{Q}_{>0}$ there is a depth-3 ReLU MLP $M_\times^\delta$ with rational parameters such that*

$$\sup_{(x,y)\in[-2,2]^2} \left| M_\times^\delta(x, y) - xy \right| < \delta.$$

Proof. Approximate $t \mapsto t^2$ uniformly on $[-2, 2]$ by a piecewise linear spline S_δ with mesh $O(\delta)$ at rational breakpoints; by Lemma 7, S_δ is exactly a depth-2 ReLU with rational parameters. Define

$$M_\times^\delta(x, y) := \tfrac{1}{4}\big(S_\delta(x + y) - S_\delta(x - y)\big),$$

implemented by composing S_δ with affine maps $(x, y) \mapsto x \pm y$ and summing. Then $\sup |M_\times^\delta(x, y) - xy| = O(\delta)$; refine the mesh to achieve the stated δ. All constructions are primitive recursive in RCA_0.

Lemma 9 (Global Lipschitz constant of MLP). *Let N be a ReLU MLP with layers (W_ℓ, b_ℓ) (rational). Then*

$$|N(x) - N(y)| \leq \left(\prod_{\ell=1}^{L} \|W_\ell\|_\infty\right) \|x - y\|_\infty \qquad \forall x, y \in \mathbb{R}^d,$$

and the product is computable in RCA_0 from the code of N.

Proof. Each affine map $x \mapsto W_\ell x + b_\ell$ is $\|W_\ell\|_\infty$-Lipschitz; σ is 1-Lipschitz; compose and multiply the constants.

5   Conclusions

We established two exact calibrations in Reverse Mathematics (RM) for principles that arise naturally in the analysis of standard neural architectures. For a minimal transformer block, the simultaneous attention–subsequence principle TRANS-SIM$_{\mathrm{fin}}$ (Definition 9) is equivalent over RCA$_0$ to ACA$_0$ (Theorem 1). For ReLU multilayer perceptrons (MLPs) on $[0,1]^d$ with rational parameters, the universal approximation statement MLP-UAT$_d$ (Definition 10) is equivalent over RCA$_0$ to WKL$_0$ (Theorem 2). Since WKL$_0 \not\vdash$ ACA$_0$ over RCA$_0$, these results yield a strict separation of RM strength between the canonical compactness principle used for transformers and the canonical approximation principle used for MLPs.

We emphasize robustness in the following sense. The transformer equivalence TRANS-SIM$_{\mathrm{fin}} \equiv$ ACA$_0$ is proven already for a single head, a single layer, and two tokens; enlarging the architecture (more heads, layers, or tokens) only strengthens the principle. The proof was given for an affine normalizer to keep the development algebraic in RCA$_0$, but a Softmax variant is admissible on bounded score ranges with the same calibration. On the MLP side, MLP-UAT$_d \equiv$ WKL$_0$ is insensitive to architectural detail so long as parameters are rational and the domain is a compact cube; the construction via dyadic grids, exact ReLU hat functions, and a spline–based multiplication gadget is primitive recursive in the input codes. Both theorems are *non–uniform* (there is no single functional that, in RCA$_0$, must produce a witnessing subsequence or network from the input data); uniform (realizer) variants belong to the Weihrauch framework [1] and separate even more finely the operations of compact choice and subsequence selection (more remarks on this below).

Methodological consequences are as follows. Calibrating analysis steps against RCA$_0$, WKL$_0$, and ACA$_0$ clarifies which compactness or comprehension axioms are silently used in common arguments. This can guide proof mining: when a convergence proof crosses to ACA$_0$ via an Ascoli/Bolzano–Weierstrass step, one should not expect a modulus of convergence or an effective selection functional without additional structure; conversely, approximation arguments that remain at WKL$_0$ admit fully explicit, finitary realizations with rational parameters and computable error bounds.

There are some remarks on uniform (Weihrauch) variants. In the Weihrauch lattice one studies the *uniform computational content* of principles, treating each as a multi–valued function between represented spaces. For compact sets, *compact choice* C$_{[0,1]}$ maps a non–empty closed $K \subseteq [0,1]$ (given by negative information) to some $x \in K$, capturing the uniform content of "ERM existence". For sequences, the Bolzano–Weierstrass principle BWT$_{[0,1]}$ maps a bounded sequence $(x_n) \subseteq [0,1]$ to a cluster point, capturing the uniform content of "subsequence selection". It is known [1] that BWT$_{[0,1]} \equiv$ C$'_{[0,1]}$, i.e. Bolzano–Weierstrass is the *jump* of compact choice, hence strictly stronger. Thus even uniformly the operations "choose a near–minimizer" and "select a convergent subsequence" occupy distinct degrees.

In summary, our results clarify where standard neural network principles sit in the RM hierarchy: transformer compactness at ACA$_0$, MLP universality at WKL_0. We hope this promotes a more disciplined use of compactness and selection arguments in learning theory, and provides a gateway to finer uniform classifications of the computational content of such analyses. More broadly, this work may be seen as paving the way for a systematic program of reverse mathematical classification of neural networks, bridging proof theory, computable analysis, and machine learning in an interdisciplinary manner.[1]

References

1. Brattka, V., Gherardi, G., Marcone, A.: The Bolzano-Weierstrass theorem is the jump of weak König's lemma. Ann. Pure Appl. Log. **163**(6), 623–655 (2012)
2. Cybenko, G.: Approximation by superpositions of a sigmoidal function. Math. Control Signals Syst. **2**(4), 303–314 (1989)
3. Eastaugh, B.: Reverse mathematics. In: Zalta, E.N., Nodelman, U. (eds.) SEP. Stanford University (2024)
4. Funahashi, K.: On the approximate realization of continuous mappings by neural networks. Neural Netw. **2**(3), 183–192 (1989)
5. Goodfellow, I., Bengio, Y., Courville, A.: Deep Learning. MIT Press (2016)
6. Hahn, M.: Theoretical limitations of self-attention in neural sequence models. Trans. Assoc. Comput. Linguist. **8**, 156–171 (2020)
7. Hirschfeldt, D.R.: Slicing the Truth: On the Computable and Reverse Mathematics of Combinatorial Principles. World Scientific (2014)
8. Hornik, K.: Approximation capabilities of multilayer feedforward networks. Neural Netw. **4**(2), 251–257 (1991)
9. Hornik, K., Stinchcombe, M., White, H.: Multilayer feedforward networks are universal approximators. Neural Netw. **2**(5), 359–366 (1989)
10. Leshno, M., Lin, V.Y., Pinkus, A., Schocken, S.: Multilayer feedforward networks with a nonpolynomial activation function can approximate any function. Neural Netw. **6**(6), 861–867 (1993)
11. Lu, Z., Pu, H., Wang, F., Hu, Z., Wang, L.: The expressive power of neural networks: a view from the width. In: NeurIPS (2017)
12. Maruyama, Y.: Fundamental results for pointfree convex geometry. Ann. Pure Appl. Logic **161**(12), 1486–1501 (2010)
13. Maruyama, Y.: Reasoning about fuzzy belief and common belief: with emphasis on incomparable beliefs. In: IJCAI (2011)
14. Maruyama, Y.: Natural duality, modality, and coalgebra. J. Pure Appl. Algebra **216**(3), 565–580 (2012)
15. Maruyama, Y.: Categorical harmony and paradoxes in proof-theoretic semantics. Adv. Proof-Theor. Semant. **43**, 95–114 (2016)
16. Maruyama, Y.: Symbolic and statistical theories of cognition: towards integrated artificial intelligence. In: Cleophas, L., Massink, M. (eds.) SEFM 2020. LNCS, vol. 12524, pp. 129–146. Springer, Cham (2021). https://doi.org/10.1007/978-3-030-67220-1_11
17. Maruyama, Y.: Algorithm for interpretable graph features via motivic persistent Cohomology. In: Zaroliagis, C., Bhandari, D., Gupta, P., Das, S. (eds.) ICAA 2026. LNCS, vol. 16423, pp. 87–99. Springer, Cham (2026). https://doi.org/10.1007/978-3-032-15621-1_8

---

[1] For more on our theoretical research on related topics, we refer to [12–24].

18. Maruyama, Y.: Algorithms and complexity results for K-theoretic persistent homology. LNCS. Springer, Cham (2026)
19. Maruyama, Y.: Top-K exterior power persistent homology: algorithm, structure, and stability. In: Zaroliagis, C., Bhandari, D., Gupta, P., Das, S. (eds.) ICAA 2026. LNCS, vol. 16423, pp. 74–86. Springer, Cham (2026). https://doi.org/10.1007/978-3-032-15621-1_7
20. Maruyama, Y., Yasuda, A.: Homological representation learning for molecular graphs. In: Proceedings of Machine Learning Research (2025)
21. Maruyama, Y., Yasuda, A.: K-theoretic persistent Cohomology. In: Proceedings of Machine Learning Research (2025)
22. Maruyama, Y.: Categorical Equivariant deep learning: category-equivariant neural networks and universal approximation theorems. arXiv:2511.18417 (2025)
23. Maruyama, Y.: Learning with category-equivariant architectures for human activity recognition. arXiv:2511.01139 (2025)
24. Maruyama, Y.: Learning with category-equivariant representations for human activity recognition. arXiv:2511.00900 (2025)
25. Montúfar, G.F., Pascanu, R., Cho, K., Bengio, Y.: On the number of linear regions of deep neural networks. In: NeurIPS (2014)
26. Unser, M.: A representer theorem for deep neural networks. J. Mach. Learn. Res. **20**, 1–30 (2019)
27. Petersen, P., Voigtlaender, F.: Optimal approximation of piecewise smooth functions using deep ReLU neural networks. Neural Netw. **108**, 296–330 (2018)
28. Von Oswald, J., et al.: Transformers learn in-context by gradient descent. arXiv:2212.07677 (2022)
29. Simpson, S.G.: Subsystems of Second Order Arithmetic. CUP (2009)
30. Telgarsky, M.: Benefits of depth in neural networks. In: COLT (2016)
31. Vaswani, A., et al.: Attention is all you need. In: NeurIPS (2017)
32. Yarotsky, D.: Error bounds for approximations with deep ReLU networks. Neural Netw. **94**, 103–114 (2017)
33. Yun, C., Bhojanapalli, S., Rawat, A., Reddi, S., Kumar, S.: Are transformers universal approximators of sequence-to-sequence functions? In: ICLR (2020)

Private Graph Colouring with Limited Defectiveness

Aleksander B. G. Christiansen[1], Eva Rotenberg[2],
Teresa Anna Steiner[3], and Juliette Vlieghe[2(✉)]

[1] Technical University of Denmark, Kongens Lyngby, Denmark
[2] IT University of Copenhagen, Copenhagen, Denmark
`{erot,juvl}@itu.dk`
[3] University of Southern Denmark, Odense, Denmark
`steiner@imada.sdu.dk`

Abstract. Differential privacy is the gold standard for privacy preserving data analysis, which is crucial in a wide range of disciplines. Vertex colouring is one of the most fundamental graph problems. In this work, we study vertex colouring in the differentially private setting.

We consider edge-differential privacy, in which the edges of the graph are the private information that should be protected. To satisfy nontrivial privacy guarantees under this notion of privacy, a colouring algorithm needs to be defective: a colouring is d-defective if a vertex can share a colour with at most d of its neighbours. Without defectiveness, any edge-differentially private colouring algorithm needs to assign n different colours to the n different vertices. We show the following lower bound for the defectiveness: any ϵ-edge differentially private algorithm that returns a c-vertex colouring of a graph of maximum degree $\Delta > 0$ with high probability must have defectiveness at least $d \in \Omega\left(\log n / (\epsilon + \log c)\right)$.

We complement our lower bound by presenting an ϵ-differentially private algorithm for $O\left(\Delta / \log n + \epsilon^{-1}\right)$-colouring a graph with defectiveness at most $O\left(\log n\right)$.

Keywords: Differential Privacy · Graph Colouring · Defective Colouring

1 Introduction and Related Work

Graph colouring is a family of fundamental problems with many applications in computer science, including scheduling, routing, register allocation, visualisation, network analysis, and clustering problems. The vertex colouring problem is the following: one wants to assign each vertex a colour such that no two neighbours have the same colour, that is, each colour class is an independent set. In most applications of colouring, the edges of the graph represent some kind of conflict: the two adjacent vertices cannot be scheduled at the same time, cannot be visualised with the same colour, or cannot be in the same cluster or team. In many applications, the conflicts and therefore the edges of the graph

© The Author(s), under exclusive license to Springer Nature Switzerland AG 2026
J. Kozik and A. Wolff (Eds.): SOFSEM 2026, LNCS 16448, pp. 404–415, 2026.
https://doi.org/10.1007/978-3-032-17801-5_30

are sensitive. For example, they may encode physical proximity of devices which gives information about the location or movement pattern of users, or they may encode interpersonal conflicts. Releasing a proper colouring may leak whether two vertices are adjacent, as their consistent separation across colour classes provides evidence about an underlying edge. Edge differential privacy addresses this issue by ensuring that the presence or absence of any single edge has a limited influence on the published colouring. To enforce this notion, we need to tolerate some residual conflict.

In this paper, we consider the hypothesis that a conflict is sensitive information, and ask whether it is possible to somehow approximately colour the vertices, without revealing their exact conflicts and we bound the possible compromises between utility and privacy. This notion of 'approximately' colouring the graph is that of defective colouring: a colouring is (c, d)-defective if it uses a colour palette of size c and any vertex can share a colour with at most d of its neighbours. In other words, we want to partition the set of vertices $V(G)$ into sets $V_1, \ldots, V_c$ such that for every $i \in [c]$, each vertex in V_i has at most d neighbours in V_i. There has been a lot of interesting work on the defective colouring of graphs [1,6,11,18,21,24,28,29,34,41,44,48], also in distributed graph algorithms [3,4,25,36].

There is a large body of work studying graphs under differential privacy, including estimating subgraph counts [5,8,19,23,31–33,35,47,49], the degree distribution of the graph [12,23,27,42,50], and the densest subgraph [15,22,38], as well as approximating the minimum spanning tree and computing clusterings [9,10,15,23,30,37,39,40,46] and cuts and shortest paths [2,7,13,20,26,43,45].

When studying differential privacy, the hope is to give algorithms that disclose some information or analysis of the whole data set (in this case, the graph), while keeping the individual data points (in this case, the edges) private. A graph colouring, and especially one with few colours, gives significant insight into the graph structure, which should intuitively reveal much of the private information. We formalise this intuition by giving a lower bound on the defectiveness of any differentially private colouring.

1.1   Roadmap

In Sect. 2, we prove the following lower bound: any algorithm that c-vertex colours with high probability graphs of n nodes of degree at most Δ must have a defectiveness at least $d \in \Omega \left(\log n / (\epsilon + \log c) \right)$. The proof builds on the following idea: Given a valid (c, d)-defective colouring on a graph G, if there exist $d + 2$ nodes of the same colour, then there is a $(d + 1)$-neighbouring graph to G on which this colouring is invalid: namely, connecting these $d + 2$ nodes into a star. This can be used to upper-bound the probability of any colouring producing $d + 2$ nodes of the same colour. However, if $c < n/(d + 1)$, then for any valid (c, d)-defective colouring there *must* exist $d + 2$ nodes of the same colour. Combining these two observations yields our lower bound.

In Sect. 3, we propose a simple colouring algorithm. With Chernoff bounds and a random colouring, we can get defectiveness $\Omega(\log n)$, so we minimise the number of colours used, which results in a ϵ-edge differentially private $\left(O\left(\Delta/\log n + \epsilon^{-1}\right), O\left(\log n\right)\right)$ colouring.

1.2 Definitions and Notations

We recall some basic definitions and lemmas regarding differential privacy in graphs. A more comprehensive introduction can be found in [17].

Definition 1 (Edge-neighbouring graphs [27]). *Let $n \in \mathbb{N}$. Let $G = (V, E)$ and $G' = (V, E')$ be two graphs on n nodes. We say that G and G' are k-edge-neighbouring if the cardinality of the symmetric difference of E and E' is k. In particular, we say that G is an edge-neighbouring graph of G' if they differ by exactly one edge.*

In this work, we only consider edge-neighbouring (as opposed to *node-neighbouring*, see [27]). We will thus refer to edge-neighbouring graphs as simply *neighbouring* graphs. We focus on edge differential privacy, which protects the privacy of edges.

Definition 2 (Edge differential privacy). *A randomised algorithm A on n-node graphs is ϵ-edge differentially private if for all edge-neighbouring graphs G and G' and for all possible sets of outputs $S \subseteq \mathrm{range}(A)$:*

$$\Pr(A(G) \in S) \leq e^{\epsilon} \Pr(A(G') \in S).$$

The next Lemma follows from the definition of differential privacy.

Lemma 1 (Group privacy [16]). *Let G and G' be k-edge neighboring. Let A be an $(\epsilon, 0)$-edge differentially private on n-node graphs. Then for any possible set of outputs $S \subseteq \mathrm{range}(A)$ we have*

$$\Pr(A(G) \in S) \leq e^{k\epsilon} \Pr(A(G') \in S).$$

Next, we recall the Laplace mechanism and its properties. A common technique to make a function edge differentially private is to add Laplace noise adjusted to the sensitivity of the problem and to the privacy parameter. Informally, the sensitivity of a function states how much its result may differ if we change a single edge. For example, adding or deleting an edge may change the maximum degree Δ by at most 1. Intuitively, the more sensitive a function is, the more noise we need to add to the result to preserve privacy. In Sect. 3, we will add Laplace noise to the maximum degree of the graph: we can then use the noisy maximum degree in the algorithm with the guaranty that differential privacy is respected.

Definition 3 (Laplace distribution). *The Laplace Distribution (centered at 0) with scale b is the distribution with probability density function:*

$$\mathrm{Lap}(x|b) = \frac{1}{2b} \exp\left(-\frac{|x|}{b}\right)$$

Lemma 2 (Laplace Tailbound). *If $Y \sim \mathrm{Lap}(b)$, then: $P(|Y| \geq t \cdot b) = e^{-t}$.*

In the following, we denote the data universe by χ. For example, χ can be the set of all graphs with n vertices.

Definition 4 (L_1-sensitivity). *Let $f : \chi \to \mathbb{R}^k$. The L_1-sensitivity of f is given by $\max_{G,G' \text{ neighbouring}} \|f(G) - f(G')\|_1$.*

Definition 5 (Laplace Mechanism). *Given $f : \chi \to \mathbb{R}^k$ with L_1-sensitivity S_1, the Laplace mechanism is defined as:*

$$M_L(x, f, \epsilon) = f(x) + (Y_1, ..., Y_k),$$

where Y_i are i.i.d random variables drawn from $\mathrm{Lap}(S_1/\epsilon)$.

Lemma 3 ([16]). *The Laplace mechanism preserves ϵ-differential privacy.*

We will use the following Chernoff bound:

Lemma 4 (Additive Chernoff bound). *Let $X_1, \ldots, X_m$ be independent random variables s.t. $0 \leq X_i \leq 1$. Let S denote their sum and $\mu = \mathbb{E}(S)$. Then for any $\eta \geq 0$:*

$$P(S \geq (1 + \eta)\mu) \leq e^{-\frac{\eta^2 \mu}{2 + \eta}}$$

Therefore for any $0 \leq \eta \leq 1$:

$$P(S \geq (1 + \eta)\mu) \leq e^{-\frac{\eta^2 \mu}{3}}$$

2 A Lower Bound on the Defectiveness of Private Colouring with High Probability

In the following, we state our main observation, which is a lower bound on the defectiveness d for any colouring with less than n colours.

Theorem 1. *Let $c, d, \Delta, n \in \mathbb{N}$ with $2 \leq c < n$ and $d < \Delta$, and let $\epsilon > 0$. Consider an ϵ-differentially private algorithm that with high probability returns a (c, d)-defective colouring of any n-node graph with maximum degree at most Δ. It must be that:*

$$d \in \Omega\left(\frac{\log n}{\log c + \epsilon}\right)$$

Proof. If $c(d + 1) \geq n$, then $\frac{\log n}{\log c + \log(d+1)} \in O(1)$. Note that d cannot be 0 as otherwise, $c \geq n$ in contradiction to our assumption, so $d \geq 1$ and therefore it holds that $d \in \Omega\left(\frac{\log n}{\log c + \log(d+1)}\right)$. Either $c \geq d$ and the theorem holds, or $d > c$ and therefore $d \in \Omega(\sqrt{n}) \subseteq \Omega(\log n/(\epsilon + \log c))$.

In the following, we consider the case where $c(d + 1) < n$. Let $\alpha > 0$ be a constant and assume the algorithm returns a valid (c, d)-defective colouring with probability at least $1 - n^{-\alpha}$ for any graph of degree at most Δ. Define

$n_0 = c(d+1) + 1$. Any valid (c,d)-defective colouring on a graph G has to have the following property: In any subset V_0 of n_0 nodes in G, there has to exist a colour such that at least $d+2$ nodes have that colour, by the pigeonhole principle. Since our algorithm returns a valid (c,d)-defective colouring with high probability on any graph G with maximum degree Δ, it has to fulfil for any subgraph V_0 of n_0 nodes:

$$P(\nexists(d+2) \text{ nodes in } V_0 \text{ with the same colour in } G) \le \frac{1}{n^\alpha}. \tag{1}$$

If we consider a second graph on the same set of nodes such that these $d+2$ nodes form a star, this colouring becomes invalid, therefore the probability of it happening in the second graph should be low. We formalise and quantify this constraint.

Consider the empty graph G with n nodes. Pick any subset V_0 with n_0 nodes. Pick any set $U \subseteq V_0$ with $d+2$ nodes. We can find a $d+1$-neighbouring graph G' such that any valid (c,d)-defective colouring needs at least 2 colours within U: We can add $(d+1)$ edges to form a star with $d+1$ leaves. Since the center can have at most d neighbours of the same colour, we need at least two colours to colour the star. Since our algorithm returns a valid (c,d)-defective colouring with high probability on G', it has to fulfill:

$$P(\text{all nodes in } U \text{ have the same colour in } G') \le \frac{1}{n^\alpha}$$

and by group privacy,

$$P(\text{all nodes in } U \text{ have the same colour in } G) \le e^{\epsilon(d+1)}\frac{1}{n^\alpha}.$$

There are $\binom{n_0}{d+2}$ ways of choosing $U \subseteq V_0$, therefore, by the union bound:

$$P(\exists(d+2) \text{ nodes in } V_0 \text{ with the same colour in } G) \le \binom{n_0}{d+2}e^{\epsilon(d+1)}\frac{1}{n^\alpha}$$

Combined with (1), this gives:

$$\frac{1}{n^\alpha} \ge P(\nexists(d+2) \text{ nodes in } V_0 \text{ with the same colour in } G) \ge 1 - \binom{n_0}{d+2}e^{\epsilon(d+1)}\frac{1}{n^\alpha}$$

and further:

$$\frac{1}{n^\alpha} \ge 1 - \binom{n_0}{d+2}e^{\epsilon(d+1)}\frac{1}{n^\alpha}$$

$$\ge 1 - \binom{n_0}{d+2}e^{\epsilon(d+2)}\frac{1}{n^\alpha}.$$

We multiply with n^α and use the following upper bound on binomial coefficients:
$\binom{n_0}{d+2} \leq \left(\frac{n_0 e}{d+2}\right)^{d+2}$. This gives

$$n^\alpha \leq 1 + \binom{n_0}{d+2} e^{\epsilon(d+2)}$$
$$\leq 1 + \left(\frac{n_0 e^{\epsilon+1}}{d+2}\right)^{d+2}$$
$$\in O\left(\left(c e^{\epsilon+1}\right)^{d+2}\right)$$

Taking the logarithm, we get

$$\alpha \log n \in O\left((d+2)(\log(c) + \epsilon + 1)\right)$$
$$\in O\left(d(\log(c) + \epsilon)\right)$$

Therefore:

$$d \in \Omega\left(\frac{\log n}{\log c + \epsilon}\right)$$

which concludes the proof.

Note that if $c \geq n$ we could use a different colour for each vertex, and if $d \geq \Delta$ we could colour the entire graph with one colour, therefore the assumption $c < n$ and $d < \Delta$ is not restrictive. Additionally, if $d = 0$, we know that any two vertices of the same colour do not share an edge, so we cannot have any non-trivial privacy guarantees unless we can use a different colour for each vertex, which is again of no interest, therefore we assume $d \geq 1$.

The bound from Theorem 1 in particular says that for constant ϵ and $c \ll n$, any ϵ-differentially private algorithm which outputs a (c, d)-defective colouring with high probability needs $d \in \Omega(\log n)$. Also, note that by the definition of d, if we want our algorithm to give a (c, d)-defective colouring for any graph of degree Δ, we need $c > \Delta/(d+1)$ (else, there are graphs such that the colouring fails to be (c, d)-defective with probability 1: e.g. the complete graph on $c(d+1)+1$ vertices).

When ϵ grows to infinity, the lower bound on the defectiveness goes to 0 and we find a non-private colouring as expected.

In the following section, we give an algorithm which is gives an $(O(\Delta/\log n),$ $O(\log n))$-defective colouring for constant ϵ with high probability, which, according to the discussion above, is tight in the sense that i) we cannot hope for an asymptotically lower defectiveness if c is subpolynomial in n and ii) given $d = O(\log n)$, we cannot hope for a palette with asymptotically fewer colors.

3   An ϵ-edge differentially private colouring algorithm

Theorem 2. *Let $n, \Delta \in \mathbb{N}$. Let $\epsilon > 0$ and $0 < \eta \leq 1$. There is an ϵ-edge differentially private algorithm that, for any input graph G on n vertices, outputs an* *assignment of colours to the vertices, such that if G has maximum degree Δ, the colour assignment is an $\left(O\left(\eta^2\left(\Delta/\log n + \epsilon^{-1}\right)\right), O\left(\eta^{-2}\log n\right)\right)$-defective colouring with high probability. In particular, when $\epsilon < 1$, choosing $\eta^2 = \epsilon$ gives an $\left(O\left(\Delta\epsilon/\log n + 1\right), O\left(\epsilon^{-1}\log n\right)\right)$ colouring.*

Proof. Given a graph G, we first augment G such that every node in our graph has degree approximately Δ with high probability. We augment G in the following way: First, we compute an ϵ-differentially private estimate of the maximum degree by $\widetilde{\Delta} = \Delta + \mathrm{Lap}(\epsilon^{-1}) + \epsilon^{-1}\alpha\log n$. Since the L_1-sensitivity of the maximum degree is 1, computing $\widetilde{\Delta}$ fulfills ϵ-differential privacy by Definition 5. By Lemma 2, the additive error is bounded by $\epsilon^{-1}\alpha\log n$ with probability at least $1 - n^{-\alpha}$. Thus, we get $\Delta + 2\epsilon^{-1}\alpha\log n \geq \widetilde{\Delta} \geq \Delta$ with high probability. We then add $\widetilde{\Delta}$ dummy nodes to our graph and from every $v \in V$, we add $\widetilde{\Delta}$ edges from v to the dummy nodes. In this augmented graph $\widetilde{G}$, we have $\deg_{\widetilde{G}}(v) = \deg_G(v) + \widetilde{\Delta}$, and thus with probability at least $1 - n^{-\alpha}$, every vertex v in G fulfills $\deg_{\widetilde{G}}(v) \in [\widetilde{\Delta}, 2\widetilde{\Delta}]$.

We now colour the resulting graph as follows: Let $0 < \eta \leq 1$ and $0 \leq \beta \leq 1$. We consider a palette S of size:

$$c = |S| = \frac{\eta^2\widetilde{\Delta}}{3\log\frac{1}{\beta}}.$$

Then, all vertices pick a colour uniformly at random from S. Our colouring of G is now given by this colouring restricted to the vertices in G.

The algorithm fulfills ϵ-differential privacy, since the only time we use information about the edges is to estimate the maximum degree, which we do privately using the Laplace mechanism.

To analyse accuracy, we condition on $\Delta + 2\epsilon^{-1}\alpha\log n \geq \widetilde{\Delta} \geq \Delta$, which is true with probability $1 - n^{-\alpha}$. Then for every node v in the original graph G, we analyse the expected number of neighbours in the augmented graph which are assigned the same colour as v, which we denote by $d_{G_{\mathrm{conflict}}}(v)$. The expectation of $d_{G_{\mathrm{conflict}}}(v)$ is given by the degree of v divided by the number of available colours:

$$E\left(d_{G_{\mathrm{conflict}}}(v)\right) = \frac{\deg_{\widetilde{G}}(v)}{|S|}$$

We have

$$E(d_{G_{\mathrm{conflict}}}(v)) = \frac{3\deg_{\widetilde{G}}(v)\log\frac{1}{\beta}}{\widetilde{\Delta}\eta^2} \geq \frac{3\log\frac{1}{\beta}}{\eta^2}$$

and

$$E(d_{G_{\mathrm{conflict}}}(v)) = \frac{3\deg_{\widetilde{G}}(v)\log\frac{1}{\beta}}{\widetilde{\Delta}\eta^2} \leq \frac{6\log\frac{1}{\beta}}{\eta^2}$$

We can show via Chernoff bound that the resulting colouring is $\left(6\eta^{-2}(1+\eta)\log(1/\beta)\right)$-defective with probability β. Since $\eta \le 1$, the Chernoff bound Lemma 4 gives:

$$P\left(d_{G_{\text{conflict}}}(v) \ge (1+\eta)\mu\right) \le \exp\left(-\frac{\eta^2}{2+\eta}\mu\right)$$

$$\le \exp\left(-\frac{\eta^2}{3}\mu\right)$$

Plugging in the upper and lower bounds on μ gives

$$P\left(d_{G_{\text{conflict}}}(v) \ge \frac{6(1+\eta)\log(1/\beta)}{\eta^2}\right) \le \exp\left(-\frac{\eta^2}{3}\frac{3\log\frac{1}{\beta}}{\eta^2}\right)$$

$$\le \beta$$

To get the defectiveness we use a union bound over all vertices.

$$P(d < x) \ge 1 - \sum_v P(deg_{\widetilde{G}}(v) \ge x)$$

Plugging in the previous result:

$$P\left(d < \frac{6(1+\eta)}{\eta^2}\log(1/\beta)\right) \ge 1 - n\beta$$

We want the algorithm to return a valid (c,d)-colouring with high probability, that is, we want probability of success $1 - n^\alpha$ for any constant α. We plug in $\beta = 1/n^{\alpha+1}$:

$$P\left(d < \frac{6(1+\eta)}{\eta^2}(\alpha+1)\log(n)\right) \ge 1 - \frac{n}{n^{\alpha+1}}$$

This means that with probability at least $1 - n^{-\alpha}$, we have defectiveness

$$d = O\left(\frac{(1+\eta)}{\eta^2}\log(n)\right) = O\left(\frac{1}{\eta^2}\log(n)\right)$$

and

$$c = \Theta\left(\eta^2\left(\frac{\Delta}{\log n} + \epsilon^{-1}\right)\right).$$

Using again union bound, both the condition on the degree and the bound on the defectiveness hold at the same time with probability $1 - 2n^{-\alpha}$. Choosing $\alpha = \alpha' + \log 2$, we get probability $1 - n^{\alpha'}$ for any α'. Clearly the defectiveness in the original graph is smaller than the defectiveness in the augmented graph.

With this strategy we will always get $\Omega(\log n)$ defectiveness, even if we only want the result to hold with constant probability, due to the union bound. However, we can increase the defectiveness and, as a trade off, decrease the size of the palette.

Subsequent Work: Since our original preprint, Dhulipala, Henzinger, Li, Liu, Sricharan, and Zhu [14] have improved this upper bound: they propose algorithms that result in $O(\epsilon^{-1}\log n)$ defectiveness. Thus, the current state-of-the-art gap between lower bound and upper bound has been reduced to a factor that is logarithmic in the number of colours. Dhulipala et al. frame their algorithms in the distributed setting, and given a graph of arboricity α, they either use $O(1 + \alpha\epsilon/\log n)$ colours and $O(n)$ rounds or $O(\alpha\epsilon\log n + \log^2 n)$ colours and $O(\log^2 n)$ rounds.

This leaves open the following question: consider a number of colours that is not subpolynomial in n, e.g. $c = \sqrt{n}$. In all existing algorithms, the upper bounds are no longer close to the lower bound. For any $c < n, c \in O(n^{1-\gamma})$ with $0 < \gamma < 1$, can we find a colouring algorithm such that $d = o(logn)$ or a tighter lower bound on the defectiveness?

4   Conclusion

In this paper, we show that to be ϵ-edge differentially private, a colouring algorithm using c colours needs to have defectiveness $d \in \Omega\left(\log n/(\epsilon + \log c)\right)$ and we propose an ϵ-edge differentially private algorithm using $O\left(\Delta/\log n + \epsilon^{-1}\right)$ colours and defectiveness $O(\log n)$. For any constant ϵ and defectiveness $\Omega(\log n)$, the number of colours is asymptotically tight, and our lower bound shows that we cannot hope for a much smaller defectiveness unless c or Δ are very large.

Acknowledgments. This work was supported by the VILLUM Foundation grants VIL37507 "Efficient Recomputations for Changeful Problems" and VIL51463 "Differential Privacy for String Algorithms and Data Structures".

References

1. Archdeacon, D.: A note on defective colorings of graphs in surfaces. J. Graph Theory **11**(4), 517–519 (1987). https://doi.org/10.1002/jgt.3190110408
2. Arora, R., Upadhyay, J.: On differentially private graph sparsification and applications. In: Proceedings of 32nd NeurIPS 2019, pp. 13378–13389 (2019). https://proceedings.neurips.cc/paper/2019/hash/e44e875c12109e4fa3716c05008048b2-Abstract.html
3. Barenboim, L., Elkin, M.: Distributed (delta+1)-coloring in linear (in delta) time. In: Mitzenmacher, M. (ed.) Proceedings of the 41st Annual ACM Symposium on Theory of Computing, STOC 2009, Bethesda, MD, USA, May 31 - June 2, 2009. pp. 111–120. ACM (2009). https://doi.org/10.1145/1536414.1536432
4. Barenboim, L., Elkin, M.: Deterministic distributed vertex coloring in polylogarithmic time. J. ACM **58**(5), 23:1–23:25 (2011). https://doi.org/10.1145/2027216.2027221
5. Blocki, J., Blum, A., Datta, A., Sheffet, O.: Differentially private data analysis of social networks via restricted sensitivity. In: Proceedings of the 4th ITCS, pp. 87–96 (2013). https://doi.org/10.1145/2422436.2422449

6. Borodin, O.V., Ivanova, A.O., Montassier, M., Ochem, P., Raspaud, A.: Vertex decompositions of sparse graphs into an edgeless subgraph and a subgraph of maximum degree at most k. J. Graph Theory **65**(2), 83–93 (2010). https://doi.org/10.1002/jgt.20467

7. Chen, J.Y., et al.: Differentially private all-pairs shortest path distances: Improved algorithms and lower bounds. In: Proceedings of the 34th SODA, pp. 5040–5067 (2023). https://doi.org/10.1137/1.9781611977554.ch184

8. Chen, S., Zhou, S.: Recursive mechanism: towards node differential privacy and unrestricted joins. In: Proceedings of the ACM SIGMOD International Conference on Management of Data, SIGMOD 2013, New York, NY, USA, 22–27 June 2013, pp. 653–664. ACM (2013). https://doi.org/10.1145/2463676.2465304

9. Cohen, E., Kaplan, H., Mansour, Y., Stemmer, U., Tsfadia, E.: Differentially-private clustering of easy instances. In: Proceedings of the 38th ICML 2021, pp. 2049–2059 (2021). https://proceedings.mlr.press/v139/cohen21c.html

10. Cohen-Addad, V., et al.: Scalable differentially private clustering via hierarchically separated trees. In: Proceedings of the 28th KDD, pp. 221–230 (2022). https://doi.org/10.1145/3534678.3539409

11. Cowen, L.J., Goddard, W., Jesurum, C.E.: Coloring with defect. In: Proceedings of the Eighth Annual ACM-SIAM Symposium on Discrete Algorithms, SODA 1997, pp. 548–557. Society for Industrial and Applied Mathematics, USA (1997)

12. Day, W., Li, N., Lyu, M.: Publishing graph degree distribution with node differential privacy. In: Proceedings of the 2016 ACM SIGMOD, pp. 123–138 (2016), https://doi.org/10.1145/2882903.2926745

13. Deng, C., Gao, J., Upadhyay, J., Wang, C.: Differentially private range query on shortest paths. In: Proceedings of the 18th WADS, pp. 340–370 (2023). https://doi.org/10.1007/978-3-031-38906-1_23

14. Dhulipala, L., Henzinger, M., Li, G.Z., Liu, Q.C., Sricharan, A.R., Zhu, L.: Near-optimal differentially private graph algorithms via the multidimensional abovethreshold mechanism. ESA (2025)

15. Dhulipala, L., Liu, Q.C., Raskhodnikova, S., Shi, J., Shun, J., Yu, S.: Differential privacy from locally adjustable graph algorithms: k-core decomposition, low outdegree ordering, and densest subgraphs. In: Proceedings of the 63rd FOCS, pp. 754–765 (2022). https://doi.org/10.1109/FOCS54457.2022.00077

16. Dwork, C., McSherry, F., Nissim, K., Smith, A.D.: Calibrating noise to sensitivity in private data analysis. In: Proceedings of the 3rd TCC, pp. 265–284 (2006). https://doi.org/10.1007/11681878_14

17. Dwork, C., Roth, A.: The Algorithmic Foundations of Differential Privacy. Now Publishers Inc. (2014)

18. Eaton, N., Hull, T.: Defective list colorings of planar graphs. Bull. Inst. Combin. Appl. **25**(79–87), 40 (1999)

19. Eden, T., Liu, Q.C., Raskhodnikova, S., Smith, A.D.: Triangle counting with local edge differential privacy. In: Proceedings of the 50th ICALP, pp. 52:1–52:21 (2023). https://doi.org/10.4230/LIPIcs.ICALP.2023.52

20. Eliás, M., Kapralov, M., Kulkarni, J., Lee, Y.T.: Differentially private release of synthetic graphs. In: Proceedings of the 31st SODA, pp. 560–578 (2020). https://doi.org/10.1137/1.9781611975994.34

21. Erdős, P., Hajnal, A.: On decomposition of graphs. Acta Math. Acad. Sci. Hungar **18**, 359–377 (1967)

22. Farhadi, A., Hajiaghayi, M., Shi, E.: Differentially private densest subgraph. In: Proceedings of the 25th AISTATS, pp. 11581–11597 (2022). https://proceedings.mlr.press/v151/farhadi22a.html

23. Fichtenberger, H., Henzinger, M., Ost, L.: Differentially private algorithms for graphs under continual observation. In: Proceedings of the 29th ESA, pp. 42:1–42:16 (2021). https://doi.org/10.4230/LIPIcs.ESA.2021.42
24. Frick, M.: A survey of (m, k)-colorings. In: Gimbel, J., Kennedy, J.W., Quintas, L.V. (eds.) Quo Vadis, Graph Theory?, Annals of Discrete Mathematics, vol. 55, pp. 45–57. Elsevier (1993). https://doi.org/10.1016/S0167-5060(08)70374-1
25. Fuchs, M., Kuhn, F.: List defective colorings: distributed algorithms and applications. In: Oshman, R. (ed.) 37th International Symposium on Distributed Computing, DISC 2023, 10–12 October 2023, L'Aquila, Italy. LIPIcs, vol. 281, pp. 22:1–22:23. Schloss Dagstuhl - Leibniz-Zentrum für Informatik (2023). https://doi.org/10.4230/LIPICS.DISC.2023.22
26. Gupta, A., Ligett, K., McSherry, F., Roth, A., Talwar, K.: Differentially private combinatorial optimization. In: Proceedings of the 21st SODA, pp. 1106–1125 (2010). https://doi.org/10.1137/1.9781611973075.90
27. Hay, M., Li, C., Miklau, G., Jensen, D.D.: Accurate estimation of the degree distribution of private networks. In: Proceedings of the 9th ICDM, pp. 169–178 (2009). https://doi.org/10.1109/ICDM.2009.11
28. Hendrey, K., Wood, D.R.: Defective and clustered choosability of sparse graphs. Comb. Probab. Comput. **28**(5), 791–810 (2019). https://doi.org/10.1017/S0963548319000063
29. van den Heuvel, J., Wood, D.R.: Improper colourings inspired by hadwiger's conjecture. J. Lond. Math. Soc. **98**(1), 129–148 (2018). https://doi.org/10.1112/jlms.12127
30. Huang, Z., Liu, J.: Optimal differentially private algorithms for k-means clustering. In: Proceedings of the 37th PODS, pp. 395–408 (2018). https://doi.org/10.1145/3196959.3196977
31. Imola, J., Murakami, T., Chaudhuri, K.: Locally differentially private analysis of graph statistics. In: Proceedings of the 30th USENIX, pp. 983–1000 (2021). www.usenix.org/conference/usenixsecurity21/presentation/imola
32. Imola, J., Murakami, T., Chaudhuri, K.: Communication-efficient triangle counting under local differential privacy. In: Proceedings of the 31st USENIX, pp. 537–554 (2022). www.usenix.org/conference/usenixsecurity22/presentation/imola
33. Imola, J., Murakami, T., Chaudhuri, K.: Differentially private triangle and 4-cycle counting in the shuffle model. In: Proceedings of the 29th CCS, pp. 1505–1519 (2022). https://doi.org/10.1145/3548606.3560659
34. Jing, Y., Kostochka, A., Ma, F., Xu, J.: Defective dp-colorings of sparse simple graphs. Discret. Math. **345**(1), 112637 (2022). https://doi.org/10.1016/j.disc.2021.112637
35. Karwa, V., Raskhodnikova, S., Smith, A.D., Yaroslavtsev, G.: Private analysis of graph structure. ACM Trans. Database Syst. **39**(3), 22:1–22:33 (2014). https://doi.org/10.1145/2611523
36. Kuhn, F.: Weak graph colorings: distributed algorithms and applications. In: auf der Heide, F.M., Bender, M.A. (eds.) SPAA 2009: Proceedings of the 21st Annual ACM Symposium on Parallelism in Algorithms and Architectures, Calgary, Alberta, Canada, 11–13 August 2009, pp. 138–144. ACM (2009). https://doi.org/10.1145/1583991.1584032
37. Lin, Z., Gao, L., Hu, X., Zhang, Y., Liu, W.: Differentially private graph clustering algorithm based on structure similarity. In: Proceedings of the 9th ICCNS, pp. 63–68 (2019). https://doi.org/10.1145/3371676.3371693

38. Nguyen, D., Vullikanti, A.: Differentially private densest subgraph detection. In: Proceedings of the 38th ICML, pp. 8140–8151 (2021). https://proceedings.mlr.press/v139/nguyen21i.html
39. Nissim, K., Raskhodnikova, S., Smith, A.D.: Smooth sensitivity and sampling in private data analysis. In: Proceedings of the 39th STOC, pp. 75–84 (2007). https://doi.org/10.1145/1250790.1250803
40. Nissim, K., Stemmer, U., Vadhan, S.P.: Locating a small cluster privately. In: Proceedings of the 35th PODS, pp. 413–427 (2016). https://doi.org/10.1145/2902251.2902296
41. Ossona De Mendez, P., Oum, S.-I., Wood, D.R.: Defective colouring of graphs excluding a subgraph or minor. Combinatorica **39**(2), 377–410 (2018). https://doi.org/10.1007/s00493-018-3733-1
42. Raskhodnikova, S., Smith, A.D.: Lipschitz extensions for node-private graph statistics and the generalized exponential mechanism. In: Proceedings of the 57th FOCS, pp. 495–504 (2016). https://doi.org/10.1109/FOCS.2016.60
43. Sealfon, A.: Shortest paths and distances with differential privacy. In: Milo, T., Tan, W. (eds.) Proceedings of the 35th PODS, pp. 29–41 (2016), https://doi.org/10.1145/2902251.2902291
44. Škrekovski, R.: List improper colourings of planar graphs. Comb. Probab. Comput. **8**(3), 293–299 (1999)
45. Stausholm, N.M.: Improved differentially private euclidean distance approximation. In: Proceedings of the 40th PODS, pp. 42–56 (2021). https://doi.org/10.1145/3452021.3458328
46. Stemmer, U., Kaplan, H.: Differentially private k-means with constant multiplicative error. In: Proceedings of the 31st NeurIPS, pp. 5436–5446 (2018). https://proceedings.neurips.cc/paper/2018/hash/32b991e5d77ad140559ffb95522992d0-Abstract.html
47. Sun, H., et al.: Analyzing subgraph statistics from extended local views with decentralized differential privacy. In: Proceedings of the 26th CCS, pp. 703–717 (2019). https://doi.org/10.1145/3319535.3354253
48. Wood, D.R.: Defective and clustered graph colouring. arXiv preprint: 1803.07694 (2018)
49. Zhang, J., Cormode, G., Procopiuc, C.M., Srivastava, D., Xiao, X.: Private release of graph statistics using ladder functions. In: Proceedings of the 2015 ACM SIGMOD, pp. 731–745 (2015). https://doi.org/10.1145/2723372.2737785
50. Zhang, S., Ni, W., Fu, N.: Differentially private graph publishing with degree distribution preservation. Comput. Secur. **106**, 102285 (2021). https://doi.org/10.1016/J.COSE.2021.102285

Complexity Aspects of Homomorphisms of Ordered Graphs

Michal Čertík[1]([✉]) [iD], Andreas Emil Feldmann[2] [iD], Jaroslav Nešetřil[1] [iD], and Paweł Rzążewski[3] [iD]

[1] Computer Science Institute, Faculty of Mathematics and Physics Charles University, Prague, Czech Republic
michal.certik@gmail.com
[2] Department of Computer Science, University of Sheffield, Sheffield, UK
[3] Warsaw University of Technology, University of Warsaw, Warsaw, Poland

Abstract. We examine ordered graphs, defined as graphs with linearly ordered vertices, from the perspective of homomorphisms (and colorings) and their complexities. We demonstrate the corresponding computational and parameterized complexities, along with algorithms associated with related problems. These questions are interesting and we show that numerous problems lead to various complexities. The reduction from homomorphisms of unordered structures to homomorphisms of ordered graphs is proved, achieved with the use of ordered bipartite graphs. We then determine the NP-completeness of the problem of finding ordered homomorphisms of ordered graphs and the XP and W[1]-hard nature of this problem parameterized by the number of vertices of the image ordered graph. Classes of ordered graphs for which this problem can be solved in polynomial time are also presented.

Keywords: Computational Complexity · Parameterized Complexity · Algorithms · Ordered Graphs · Homomorphisms

1 Introduction and Motivation

An *ordered graph* is a graph whose vertex set is totally ordered. For two ordered graphs G and H, an *ordered homomorphism* from G to H (denoted by $G \to H$) is a mapping f from $V(G)$ to $V(H)$ that preserves edges and the orderings of vertices, i.e.,

1. for every $uv \in E(G)$ we have $f(u)f(v) \in E(H)$,
2. for $u, v \in V(G)$, if $u \leq v$, then $f(u) \leq f(v)$.

The forthcoming full version of the article [16] will contain the complete list of results along with their proofs on this topic.

P. Rzążewski—Supported by the National Science Centre grant 2024/54/E/ST6/00094.

© The Author(s), under exclusive license to Springer Nature Switzerland AG 2026

J. Kozik and A. Wolff (Eds.): SOFSEM 2026, LNCS 16448, pp. 416–431, 2026.
https://doi.org/10.1007/978-3-032-17801-5_31

Ordered graphs frequently emerge in various contexts: extremal theory [18, 42], category theory [29,40], Ramsey theory [2,3,29,39], model theory [6,7]), among others.

The richness of the field of ordered graphs is reflected not only in its theoretical depth and challenges but also in its numerous applications across science and technology. Related research spans a wide range of domains, including physics [45], medicine and biology [25], large language models [24], neural networks [28], machine learning [25], self-supervised learning [37], data analysis and subspace clustering [47], systems and networks [36], software optimization [43], malware detection [44], business process management [33], workflow models [32], decision making [46], dynamic system call sandboxing [48], fault tolerance [17], blockchains [38], curriculum development [35], multi-linear forms [5], ordered graph grammars [10], rigidity theory [19], shuffle squares [26], and tilings [4], among many others.

In relation to the aforementioned research, homomorphisms of ordered graphs provide both validation and extension: although they impose stricter conditions in comparison to standard homomorphisms (see, for example, [31]), they also exhibit their own unique complexity (see, for instance [1,9,11,12,14,27,41].

The exploration of complexities and parameterized complexities concerning ordered graphs and their homomorphisms has also been examined from multiple perspectives. Recently, [34] has shown that ordering problems for graphs defined by finitely many forbidden ordered subgraphs capture the class NP. In [22], the complexities of decision problems involving ordered graphs and their subgraphs are studied. We address in [15] (parameterized) complexities related to the core of ordered graphs and hypergraphs problems, where a core is defined as an ordered graph that is not homomorphic to a proper ordered subgraph.

In this article, we therefore try to extend the research by results on the complexity and parameterized complexity of fundamental problems related to homomorphisms of ordered graphs.

2 Unordered Structures to Ordered Graphs Reduction

The reduction of unordered homomorphisms to ordered ones is of significant importance and represents one of the major results in our study. We try to undertake this reduction with full generality, ensuring its applicability and utility in the unfolding narrative of our work.

A signature σ consists of a finite set of relation symbols with specified arities. The arity of the relation R is denoted by $\mathsf{ar}(R)$. A *structure* $\mathcal{H}$ with signature $\sigma(\mathcal{H})$ consists of a universe $V(\mathcal{H})$ together with a set of relations $\mathbf{R}(\mathcal{H}) = \{R(\mathcal{H}) \mid R \in \sigma(\mathcal{H})\}$ over the universe $V(\mathcal{H})$. We write $\mathcal{H} = (V(\mathcal{H}), \mathbf{R}(\mathcal{H}))$. By $\|\mathcal{H}\|$ we denote $|\sigma(\mathcal{H})| + |V(\mathcal{H})| + \sum_{R \in \sigma(\mathcal{H})} |R(\mathcal{H})| \cdot \mathsf{ar}(R)$, which is the size of a "reasonable" encoding of $\mathcal{H}$.

Given two structures $\mathcal{G}$ and $\mathcal{H}$ with the same signature σ, a function $f : V(\mathcal{G}) \to V(\mathcal{H})$ *respects* $R \in \sigma$ if, for each $\mathbf{x} \in R(\mathcal{G})$, $f(\mathbf{x}) \in R(\mathcal{H})$, where f is evaluated elementwise. A *homomorphism* from $\mathcal{G}$ to $\mathcal{H}$ is a function $h : V(\mathcal{G}) \to$

$V(\mathcal{H})$ that respects every $R \in \sigma$. If there exists a homomorphism from structure $\mathcal{G}$ to structure $\mathcal{H}$, we denote it by $\mathcal{G} \to \mathcal{H}$.

Theorem 1. *Given two unordered structures $\mathcal{G}, \mathcal{H}$ of the same signature σ, in time polynomial in $||\mathcal{G}||+||\mathcal{H}||$, we can construct a pair of ordered bipartite graphs G, H, such that $\mathcal{G} \to \mathcal{H}$ if and only if $G \to H$.*

Proof. Without loss of generality, we can assume that every $v \in V(\mathcal{G})$ appears in some $\mathbf{x} \in \bigcup_{R \in \sigma} R(\mathcal{G})$, as otherwise we can safely remove v from $\mathcal{G}$ obtaining an equivalent instance of our problem. Fix some arbitrary ordering on the elements of σ. Furthermore, we fix an arbitrary ordering on the elements of $V(\mathcal{G})$, and an arbitrary ordering on the elements of $V(\mathcal{H})$. Finally, for all $R \in \sigma$, we fix an arbitrary ordering of the elements of $R(\mathcal{G})$ and of $R(\mathcal{H})$.

We will perform the construction in two steps: first, we will build a pair of ordered bipartite graphs G, H, such that every vertex v of G is equipped with a list $L(v) \subseteq V(H)$, and G has a homomorphism h to H respecting lists L if and only if $\mathcal{G} \to \mathcal{H}$. By ordered homomorphism $h : G \to H$ *respecting the lists L*, we will mean that for each $v \in V(G)$, it holds that $h(v) \in L(v) \subseteq V(H)$.

Definition of H. The set $V(H)$ is partitioned into sets A_H and B_H which form the bipartition of H. The set A_H contains a vertex (v, u) for every $v \in V(\mathcal{G})$ and $u \in V(\mathcal{H})$. The set B_H contains a vertex $(R, \mathbf{x}, \mathbf{y})$ for every $R \in \sigma$, $\mathbf{x} \in R(\mathcal{G})$ and $\mathbf{y} \in R(\mathcal{H})$. Vertices (v, u) and $(R, \mathbf{x}, \mathbf{y})$ are adjacent in H if and only if there exists $i \in [\mathsf{ar}(R)]$ such that $v = \mathbf{x}_i$ and $u = \mathbf{y}_i$.

The ordering of $V(H)$ is defined as follows. All vertices of A_H precede all vertices of B_H. The vertices within A_H are ordered lexicographically according to the orderings of $V(\mathcal{G})$ and $V(\mathcal{H})$, i.e., $(v, u) < (v', u')$ if and only if $v < v'$ or $v = v'$ and $u < u'$. Similarly, the vertices in B_H are ordered lexicographically according to the orderings of the sets σ, $R(\mathcal{G})$, and $R(\mathcal{H})$.

Definition of G. The graph G is defined as the *incidence graph* of $\mathcal{G}$. More specifically, its vertex set consists of two independent sets A_G and B_G, where A_G is $V(\mathcal{G})$ and B_G is $\bigcup_{R \in \sigma} \bigcup_{\mathbf{x} \in R(\mathcal{G})} \{(R, \mathbf{x})\}$.

In the ordering of vertices, first we have all vertices from A_G ordered according to their ordering in $\mathcal{G}$, and then all the vertices from B_G, ordered lexicographically according to the orderings of σ and $R(\mathcal{G})$, respectively.

Finally, we set the list of every $v \in A_G$ to $\bigcup_{u \in V(\mathcal{H})} \{(v, u)\}$, and the list of every $(R, \mathbf{x}) \in B_G$ to $\bigcup_{\mathbf{y} \in R(\mathcal{H})} \{(R, \mathbf{x}, \mathbf{y})\}$. This completes the definition of G and H.

Note that $|V(G)| = |V(\mathcal{G})| + \sum_{R \in \sigma} |R(\mathcal{G})|$, and $|V(H)| = |V(\mathcal{G})| \cdot |V(\mathcal{H})| + \sum_{R \in \sigma} |R(\mathcal{G})| \cdot |R(\mathcal{H})|$, and these graphs can be constructed in time polynomial in $||\mathcal{G}|| + ||\mathcal{H}||$.

Equivalence of instances. First, suppose that there is a homomorphism $f : \mathcal{G} \to \mathcal{H}$. We define a mapping $h : V(G) \to V(H)$ as follows. For each $v \in A_G$, we set $h(v) = (v, f(v))$. For each $(R, \mathbf{x}) \in B_G$, we set $h((R, \mathbf{x})) = (R, \mathbf{x}, f(\mathbf{x}))$, where $f(\mathbf{x})$ is evaluated element-wise.

It is clear that h respects the lists L and the orderings of $V(G)$ and $V(H)$. Respecting the lists can be seen from the definition of h, and h respecting the ordering follows from the lexicographic ordering in the definition of G and H. Furthermore, h is a homomorphism (it respects (binary) relations), since f is.

On the other hand, suppose that h is an order-preserving homomorphism from G to H that respects the lists L. We define the function $f : V(\mathcal{G}) \to V(\mathcal{H})$ by mapping each $v \in V(\mathcal{G})$ to $u \in V(\mathcal{H})$ such that $h(v) = (v, u)$. We claim that f is a homomorphism from $\mathcal{G}$ to $\mathcal{H}$. Taking some $R \in \sigma$ and $\mathbf{x} \in R(\mathcal{G})$, we aim to show that $f(\mathbf{x}) \in R(\mathcal{H})$. Note that $h((R, \mathbf{x}))$ must be a vertex $(R, \mathbf{x}, \mathbf{y})$ of H which is a common neighbor of $\bigcup_{v \in \mathbf{x}} h(v)$. By construction of H, such a vertex exists if and only if $\mathbf{y} \in R(\mathcal{H})$. Consequently, we have $f(\mathbf{x}) = \mathbf{y} \in R(\mathcal{H})$.

This shows that the constructed instances are indeed equivalent.

Non-list variant. We modify G as follows. Let $p = |V(\mathcal{G})|$ and $q = |\bigcup_{R \in \sigma} R(\mathcal{G})|$. We modify G into G' by adding a separate component, which is a path P on $(p + 1) + 1 + (q + 1)$ vertices denoted consecutively by $x_0, \ldots, x_p, y, z_0, \ldots, z_q$. In the ordering of $V(G')$ the vertex x_0 is the first one, then we insert x_i immediately after the i-th vertex from A_G. The vertex x_p is succeeded by y and z_0, and then we insert z_i after the i-th vertex from B_G.

In an analogous way, we modify H into H': we introduce a new component, which is a path P' with consecutive vertices $x'_0, \ldots, x'_p, y', z'_0, \ldots, z'_q$. The vertex x'_0 is the first vertex of H'. The vertex x'_i for $i \in [p]$ is inserted immediately after the last vertex from $L(v)$, where v is the i-th vertex in A_G. The vertex x'_p is succeeded by y' and then by z'_0. Then we insert each z'_i for $i \in [q]$ after the last vertex of $L((R, \mathbf{x}))$, where $(R, \mathbf{x})$ is the i-th vertex from B_G.

We argue that $G' \to H'$ if and only if G admits a homomorphism to H that preserves lists L. If such a homomorphism from G to H exists, we can easily extend it to the homomorphism from G' to H' by mapping every x_i to x'_i, y to y' and every z_i to z'_i.

Now suppose that h is a homomorphism from G' to H'. We claim that h restricted to $V(G)$ is a homomorphism from G to H that preserves the lists L.

Let us now define a *forward path* as a path whose vertices are ordered in a natural way. Let G_1, G_2 be ordered graphs, and let g be an order-preserving homomorphism from G_1 to G_2. Let P be a forward path in G_1. Then we observe that g is injective on P and the image of P contains a spanning forward path.

Observe that the longest forward path in H has two vertices and P has at least five vertices. Consequently, by the previous observation, h maps each vertex from P to its primed counterpart. Notice that if we now ensure that each vertex of G is mapped to a vertex of H, we know that h must respect lists L.

For contradiction, suppose that some vertex of G was mapped to P'. Notice that since every $v \in V(\mathcal{G})$ appears in some relation $R(\mathcal{G})$, G has no isolated vertices. As the whole connected component of G' must be mapped to the same connected component of H', we conclude that there are some vertices $v \in A_G$ and $(R, \mathbf{x}) \in B_G$ that are adjacent in G that are mapped to P'. Say v is the i-th vertex of A_G and $(R, \mathbf{x})$ is the j-th vertex of B_G. Since P is mapped to P', we conclude that v is mapped either to x'_{i-1} or to x'_i, and $(R, \mathbf{x})$ is mapped either to z'_{j-1} or to z'_j. However, all these vertices are pairwise nonadjacent in P', contradicting the fact that h is a homomorphism. This completes the proof.

You can find the application of this result in Corollary 1.

3   Complexities of Finding Homomorphisms of Ordered Graphs

In this section, we shall focus on what we consider as natural problems arising when considering complexities of ordered homomorphisms.

3.1   Polynomial-Time H-Coloring

Analogously to the H-coloring problem of unordered graphs, let us start with the following problem $\text{HOM}_<(H)$, assuming a fixed ordered graph H.

Problem 1.

$\text{HOM}_<(H)$

Input: Ordered graph G.
Question: Does there exist an ordered homomorphism from G to the fixed ordered graph H?

We consistently denote $n = |V(G)|$ and $h = |V(H)|$. The following complexity then follows easily.

Proposition 2. *The $\text{HOM}_<(H)$ problem can be solved in time $\mathcal{O}(n^{h-1})$. Consequently, for every fixed ordered graph H, $\text{HOM}_<(H)$ is in* P.

Proof. The statement follows since we can simply count all the $\binom{n+h-1}{h-1}$ possible mappings that preserve the ordering (this can be seen as the number of possible solutions to $x_1 + \ldots + x_h = n$ with $x_i \in \{0, \ldots, n\}$), which for fixed h is $\mathcal{O}(n^{h-1})$.

This is, of course, in sharp contrast to the H-coloring problem of unordered graphs, which is NP-complete for any non-bipartite graph H (see [30]).

Similarly to unordered graphs, we can also consider the problem of minimum coloring of G and define *the (ordered) chromatic number* $\chi^<(G)$ to be the minimum k such that $V(G)$ can be partitioned into k disjoint independent intervals. Notice that for ordered graphs this is the size of the smallest homomorphic image and, alternatively, the minimum k such that $G \to K_k$, K_k being a complete graph with fixed linear ordering. We shall also call $\chi^<(G)$ a *coloring* of G.

We have shown the determination of $\chi^<(G)$ using a simple greedy algorithm in [14], and the topic of $\chi^<$-boundedness of ordered graphs is covered in [1,14]. Again, this is in contrast with the similar notion for unordered graphs, which is NP-complete for $K_i, i > 2$.

3.2 NP-Completeness of General Problem

Let us now generalize the $\text{HOM}_<(H)$ problem and start with a definition of the computational problem we refer to as $\text{HOM}_<$, whose input is a pair of ordered graphs G and H and we ask if G admits an ordered homomorphism to H.

Problem 2.

$\text{HOM}_<$

Input: Ordered graphs G and H.
Question: Does there exist ordered homomorphism from G to H?

The following result then follows from Theorem 1.

Corollary 1. *The problem* $\text{HOM}_<$ *is* **NP**-*complete. Furthermore, there is no algorithm that solves every instance* G, H *of* $\text{HOM}_<$ *in time subexponential in* $|V(G)| + |V(H)|$, *unless the ETH fails.*

Proof. Let $\mathcal{G}$ be an instance of 3-COLORING of maximum degree 4. We know from [20], that this problem is **NP**-hard and has ETH lower bound.

We can see this problem as the homomorphism problem of structures with signature $\{E\}$, where the target structure $\mathcal{H}$ is of constant size. We invoke theorem 1 to obtain in polynomial time an equivalent instance (G, H) of $\text{HOM}_<$, which proves that $\text{HOM}_<$ is **NP**-hard.

Note that the number of vertices of G is $|V(\mathcal{G})| + |E(\mathcal{G})| = \mathcal{O}(|V(\mathcal{G})|)$, and the number of vertices of H is $3|V(\mathcal{G})| + 6|E(\mathcal{G})| = \mathcal{O}(|V(\mathcal{G})|)$. Consequently, the lower bound of the ETH holds.

4 Parameterized Complexity

We shall continue by investigating the parameterized complexity of $\text{HOM}_<$.

Similarly to unordered graphs, one of the interesting parameters for exploring the parameterized complexity of $\text{HOM}_<$ is $h = V(H)$. We therefore denote $\text{HOM}_<$ parameterized by h by $\text{HOM}_<(|V(H)|)$.

To justify exploring the parameterized complexity of $\text{HOM}_<(|V(H)|)$, we provide its following parameterized complexity upper bound, following from the proof of Proposition 2.

Proposition 3. *The* $\text{HOM}_<(|V(H)|)$ *problem is in* **W[P]**.

Proof. According to the definition of class **W[P]**, a parameterized problem, parameterized by k, is in **W[P]** if it can be solved by a nondeterministic algorithm that runs in time $f(k) \cdot n^{\mathcal{O}(1)}$ and uses at most $g(k) \cdot \log n$ nondeterministic bits.

We will show that for $\text{HOM}_<(|V(H)|)$, where $k = |V(H)|$, such an algorithm exists.

Indeed, from the proof of Proposition 2, a homomorphism $f : V(G) \to V(H)$ can be represented by the numbers $a_1, a_2, \ldots, a_k$, where a_i is the number of vertices of G mapped to the i-th vertex of H. Each $a_i \in [0, n]$ requires $\log n$ bits, so the entire mapping can be nondeterministically guessed using $k \log n$ bits. We can see that this mapping preserves the ordering.

The preservation of edges can then be verified in polynomial time by checking that each edge in G is mapped to an edge in H (which takes time $\mathcal{O}(|E(G)|)$, which is polynomial in $|V(G)| = n$).

Let us now prove the main parameterized complexity result for $\mathrm{HOM}_<(|V(H)|)$.

Theorem 4. *The* $\mathrm{HOM}_<(|V(H)|)$ *problem is* W[1]*-hard. Furthermore, it cannot be solved in time* $n^{o(h)}$, *unless the ETH fails.*

Proof. Similarly as in the proof of Theorem 1, we will prove this theorem in two steps. First, let us show the hardness in the *list* setting: Each vertex v of G is equipped with a list $L(v) \subseteq V(H)$, and we additionally require that the homomorphism f we are looking for satisfies $f(v) \in L(v)$ for every v.

We reduce from MULTICOLORED INDEPENDENT SET. Let F be the instance graph whose vertex set is partitioned into k subsets $V_1, V_2, \ldots, V_k$. We ask if F has an independent set of size k, intersecting every set V_i. By copying some vertices if necessary, without loss of generality, we may assume that for each $i \in [k]$ we have $|V_i| = \ell$. The problem MULTICOLORED INDEPENDENT SET cannot be solved in time $(k\ell)^{o(k)}$, unless the ETH fails, and the problem is W[1]-hard, parameterized by k (see, e.g., [8,21,23]).

Definition of H. The graph H has $5k$ vertices $\bigcup_{i \in [k]}\{a_i, b_i, x_i, y_i, z_i\}$, ordered as follows: $a_1, b_1, a_2, b_2, \ldots, a_k, b_k, x_1, y_1, z_1, x_2, y_2, z_2, \ldots, x_k, y_k, z_k$.

For each $i \in [k]$, we add edges $a_i x_i, a_i y_i, b_i y_i, b_i z_i$. Furthermore, we add all edges in the set $\bigcup_{i \in [k]}\{x_i, y_i, z_i\}$, except for the edges $y_i y_j$. The vertices $\bigcup_{i \in [k]}\{y_i\}$ form an independent set in H. This completes the definition of H.

Definition of G. Fix $i \in [k]$ and an arbitrary total order on $V_i = \{v_1^i, \ldots, v_\ell^i\}$. We introduce to G a set P^i with vertices $p_0^i, p_1^i, \ldots, p_\ell^i$ (with such an ordering). We set lists $L(p_0^i) = \{a_i\}$, $L(p_\ell^i) = \{b_i\}$, and $L(p_j^i) = \{a_i, b_i\}$ for all $j \in [\ell - 1]$.

Note that in any order-preserving mapping from P^i to H there is exactly one $j \in [\ell]$ such that p_{j-1}^i is mapped to a_i and p_j^i is mapped to b_i. We will interpret choosing such a mapping as choosing v_j^i to the solution (that is, the independent set in F).

Next, we introduce a set Q^i of $\ell + 2$ vertices $q_0^i, q_1^i, \ldots, q_\ell^i, q_{\ell+1}^i$ (with such ordering). We set lists $L(q_0^i) = \{x_i\}, L(q_{\ell+1}^i) = \{z_i\}$, and $L(q_j^i) = \{x_i, y_i, z_i\}$ for $j \in [\ell]$. For $j \in [\ell]$, the vertex q_j^i is adjacent to p_{j-1}^i and p_j^i. We observe that if p_{j-1}^i is mapped to a_i and p_j^i is mapped to b_i (i.e., v_j^i is selected to the independent set), then q_j^i must be mapped to y_i. All vertices $q_{j'}^i$ for $j' < j$ are mapped to x_i or y_i (and it is possible to map them all to x_i). Similarly, all vertices $q_{j'}^i$ for $j' > j$ are mapped to y_i or z_i (and it is possible to map them all to z_i).

Finally, for $j, j' \in [\ell]$ and distinct $i, i' \in [k]$, we add an edge $q_j^i q_{j'}^{i'}$ if an only if v_j^i is adjacent to $v_{j'}^{i'}$.[1]

We order the vertices of G as follows: $P^1, P^2, \ldots, P^k, Q^1, Q^2, \ldots, Q^k$, where the ordering within each set is as specified above. G has $k(\ell+1)+k(\ell+2) = 2k\ell + 3k$ vertices and can clearly be constructed in polynomial time. This completes the definition of G.

Equivalence of Instances. Suppose that F has a yes instance of MULTICOLORED INDEPENDENT SET, that is, for each $i \in [k]$ there is j_i, such that $I = \bigcup_{i \in [k]} \{v_{j_i}^i\}$ is an independent set in F.

We now define $f : V(G) \to V(H)$. Fix $i \in [k]$. We map p_j^i to a_i if $j < j_i$ and to b_i otherwise. We map q_j^i to x_i if $j < j_i$, to y_i if $j = j_i$, and to z_i if $j > j_i$.

Clearly, f respects lists and the ordering of vertices. Let us discuss that it preserves edges. As discussed above, all edges between sets P^i and Q^i are mapped to edges of H. So let us consider an edge $q_j^i q_{j'}^{i'}$. For contradiction, suppose that its image is a non-edge of H, which means that q_j^i is mapped to y_i and $q_{j'}^{i'}$ is mapped to $y_{i'}$. This means that the vertices v_j^i and $v_{j'}^{i'}$ are in I. However, they are adjacent in F as $q_j^i q_{j'}^{i'}$ is an edge in G, a contradiction.

Now consider a homomorphism $f : V(G) \to V(H)$, which respects lists and the ordering of vertices. As observed before, for each $i \in [k]$ there is exactly one $j_i \in [\ell]$ such that $f(p_{j_i-1}^i) = a_i$ and $f(p_{j_i}^i) = b_i$. Let $I = \bigcup_{i \in [k]} \{v_{j_i}^i\}$, we claim that I is independent in F. For contradiction, suppose that $v_{j_i}^i$ is adjacent to $v_{j_{i'}}^{i'}$. This means that $q_{j_i}^i$ is adjacent to $q_{j_{i'}}^{i'}$ by the definition of G. However, $f(q_{j_i}^i) = y_i$ (since no other vertex of $\{x_i, y_i, z_i\}$ is adjacent to both a_i and b_i) and analogously $f(q_{j_{i'}}^{i'}) = y_{i'}$. As y_i is not adjacent to $y_{i'}$ in H, we obtain a contradiction.

This completes the proof of the list version of the theorem.

Hardness in the non-list setting. Let G and H be the graphs obtained so far. We will modify them to obtain G' and H', respectively, such that G' admits a homomorphism to H' if and only if G admits a homomorphism to H which respects lists L.

Observe that the largest clique in H has $2k + 1$ vertices: It consists of all vertices in $\bigcup_{i \in [k]} \{x_i, z_i\}$ and one vertex y_i. We introduce $2k + 2$ new vertices $c_1, c_2, \ldots, c_{2k+2}$ that form a clique. For each $i \in [k]$, we make c_i adjacent to a_i, and c_{i+1} adjacent to b_i. Furthermore, for each $i \in [k]$, we make c_{k+i} adjacent to x_i, and c_{k+i+1} adjacent to z_i. The ordering of vertices of H' is defined as follows:

$$c_1, a_1, b_1, \ldots, c_k, a_k, b_k, c_{k+1}, x_1, y_1, z_1, c_{k+2}, \ldots, c_{2k}, x_k, y_k, z_k, c_{2k+1}, c_{2k+2}.$$

We modify G similarly: we add a clique on the $2k+2$ vertices $r_1, r_2, \ldots, r_{2k+2}$. For each $i \in [k]$, we make r_i adjacent to p_0^i, and r_{i+1} adjacent to p_ℓ^i. Furthermore,

---

[1] Note that we can assume that in our instance of MULTICOLORED INDEPENDENT SET each set V_i is independent. Then the subgraph of G induced by $\bigcup_{i \in [k]} \bigcup_{j \in [\ell]} \{q_j^i\}$ is isomorphic to F.

for $i \in [k]$, we make r_{k+i} adjacent to q_0^i, and r_{k+i+1} adjacent to q_{l+1}^i, where the ordering within the sets is as before.

We need to show that any homomorphism $f : V(G') \to V(H')$, restricted to $V(G)$, respects the lists L. Note that the only $(2k+2)$-clique in H' consists of vertices $\bigcup_{i \in [2k+2]} \{c_i\}$. Consequently, the $(2k+2)$-clique $\bigcup_{i \in [2k+2]} \{r_i\}$ of G' must be mapped by f to this clique. Furthermore, the only way to achieve this while preserving the ordering of the vertices is to map every r_i to c_i (for $i \in [2k+2]$).

Consider a vertex p_0^i for $i \in [k]$. Since f is order-preserving, we have $f(p_0^i) \in \{c_i, a_i, b_i, c_{i+1}\}$. However, p_0^i is adjacent to r_i, and therefore $f(p_0^i) \in \{a_i, c_{i+1}\}$. If $f(p_0^i) = c_{i+1}$, then we also have $f(p_j^i) = c_{i+1}$ for all $j \in [\ell]$. This is a contradiction, as $p_\ell^i r_{i+1} \in E(G')$ and $f(r_{i+1}) = c_{i+1}$. Consequently, $f(p_0^i) = a_i$. Analogously, we can argue that $f(p_\ell^i) = b_i$. Thus, again using that f is order-preserving, we obtain that $f(p_j^i) \in \{a_i, b_i\}$ for all $j \in [\ell - 1]$.

In exactly the same way, we argue that for all $i \in [k]$ we have $f(q_0^i) = x_i$ and $f(q_{\ell+1}^i) = z_i$, and consequently for all $j \in [\ell]$ we have $f(q_j^i) \in \{x_i, y_i, z_i\}$.

As $V(H') = 5k + 2k + 2 = 7k + 2$ and $V(G') = 2k\ell + 3k + 2k + 2 = 2k\ell + 5k + 2$, the claimed hardness follows. This completes the proof.

5   Polynomial-Time Cases of $\mathrm{HOM}_<$

In Sect. 3 we have determined that the complexity of the general $\mathrm{HOM}_<$ problem is NP-complete. In this section, we shall focus on parameters and classes of ordered graphs, for which $\mathrm{HOM}_<$ and its variations are solvable in polynomial time.

5.1 k-Shifted Cliques

We start with a definition of a class of ordered graphs $\mathcal{H}$, for which the $\mathrm{HOM}_<$ problem, where $H \in \mathcal{H}$, is polynomial.

We call an ordered graph H a *k-shifted clique* if:

- $V(H)$ can be partitioned into segments $V_1, \ldots, V_k$; $V_i < V_{i+1}, i \in [k-1]$, each of which induces a clique in H,
- For $i > 1$ and each $u, u' \in V_i$ such that $u < u'$, it holds that $N(u') \cap \bigcup_{j<i} V_j \subseteq N(u) \cap \bigcup_{j<i} V_j$.

In fact, we will derive the result for a more general problem $\mathrm{HOM}_<^\star$.

Problem 3.

$\mathrm{HOM}_<^\star$

Input: Ordered graph H that is a k-shifted clique and an ordered graph G equipped with functions $\mathsf{low}, \mathsf{up} : V(G) \to [h]; h = |V(H)|$.

Question: Does there exist a homomorphism $f : G \to H$ such that for every $v \in V(G)$ we have $\mathsf{low}(v) \le f(v) \le \mathsf{up}(v)$.

We aim to show the following result:

Theorem 5. *If H is a k-shifted clique, then the $\mathrm{HOM}^\star_<$ problem can be solved in time $n^{\mathcal{O}(k)} \cdot h^{\mathcal{O}(1)}$, where $n = |V(G)|$ and $h = |V(H)|$.*

Clearly, $\mathrm{HOM}^\star_<$ is again a list homomorphism problem and setting $\mathsf{low} \equiv 1$ and $\mathsf{up} \equiv h$ gives us a polynomial-time algorithm for $\mathrm{HOM}_<$.

We say that a solution f to $\mathrm{HOM}^\star_<$ (that is, an ordered homomorphism respecting functions low and up) is *minimum* if, for any solution f' and any $v \in V(G)$, it holds that $f(v) \le f'(v)$.

The base case is for $k = 1$, that is, if H is a clique. We will now present two lemmas. We will omit full proofs of these lemmas, which will be included in a forthcoming full version of the article.

Lemma 1. *One can solve an instance of $\mathrm{HOM}^\star_<$ where G has n vertices and $H = K_h$ in time $n^{\mathcal{O}(1)}$. Furthermore, if there is a solution, one can even find a minimum solution f.*

Proof. To prove the statement, we can exhaustively perform the following two update operations. For an edge $vv' \in E(G)$ such that $v < v'$, set $\mathsf{low}(v') = \max(\mathsf{low}(v'), \mathsf{low}(v)+1)$. For pair v and $v' \in V(G)$, such that $v < v'$, set $\mathsf{low}(v') = \max(\mathsf{low}(v'), \mathsf{low}(v))$. We terminate the procedure when no further changes are made. Clearly, the procedure takes time polynomial in n and h (in particular, $\mathcal{O}(n^2)$, since we iterate over all pairs of vertices).

If there exists a vertex $v \in V(G)$ such that $\mathsf{low}(v) > \mathsf{up}(v)$, we terminate the algorithm and answer that no solution exists. Otherwise, we set $f = \mathsf{low}$.

A similar result to the Lemma 1 and an algorithm to find a minimum solution f can also be found in [14].

Lemma 2. *Consider an instance $(G, H, \mathsf{low}, \mathsf{up})$ of $\mathrm{HOM}^\star_<$, where H is a k-shifted clique for $k \ge 2$. Let $R = \{v_p, \ldots, v_h\}$ be the last of the segments that induce a clique (that is, $R = V_k$). Let $L = V(H) \setminus R$.*

Suppose that there exists a solution $\phi : G \to H$ and let $X = \phi^{-1}(R)$. Let f be the minimum solution to $\mathrm{HOM}^\star_<$ on the instance $(G[X], H[R], \mathsf{low}', \mathsf{up}')$, where the functions $\mathsf{low}', \mathsf{up}' : X \to \{p, \ldots, h\}$ are defined as follows, $\mathsf{low}'(u) = \max(\mathsf{low}(u), p), \mathsf{up}'(u) = \mathsf{up}(u)$. Then the function $\phi' : V(G) \to V(H)$ is defined as

$$\phi'(u) = \begin{cases} f(u) & \text{if } u \in X, \\ \phi(u) & \text{if } u \notin X. \end{cases}$$

is a solution to the original instance $(G, H, \mathsf{low}, \mathsf{up})$ of $\mathrm{HOM}^\star_<$.

Now we are ready to prove Theorem 5.

Proof (of Theorem 5.). We proceed by induction on k. If $k = 1$, we are done by Lemma 1. Suppose that $k \ge 2$ and that the result is true for all $k - 1$.

Let $R = V_k = \{v_p, \ldots, v_h\}$, that is, the last segment that induces a clique in H and let $L = V(H) \setminus R$. Note that $H' = H - R$ is a $(k-1)$-shifted clique. We first check if the instance admits a solution whose codomain is L (therefore solution to the instance of G and H'); we do it by calling the algorithm inductively (we need to modify the function up in the obvious way by setting $\mathsf{up}(u) = \min(\mathsf{up}(u), p-1), u \in G$). If so, we return it as a solution we seek and stop. We see that this gives us a complexity $n^{\mathcal{O}(k)}$, by the algorithm of Lemma 1 being performed inductively $\mathcal{O}(k)$ times.

Otherwise, we exhaustively guess the first vertex v of G mapped to a vertex in R, which gives n possibilities. Let X be the subset of $V(G)$ consisting of v and all its successors.

We call the algorithm from the proof of Lemma 1 for $G[X]$ and $H[R]$ with the function low modified as in Lemma 2. If it finds no solution, we terminate the current branch (of the loop over n). Suppose it found minimum solution f.

We call the algorithm inductively for $G[V(G) \setminus X]$ and $H[L]$ and the functions $\mathsf{low}', \mathsf{up}'$ obtained as follows. Consider $u \in V(G) \setminus X$ and $U = X \cap N(u)$.

We set

$$\mathsf{up}'(u) = \min(\mathsf{up}(u), p-1)$$
$$\mathsf{low}'(u) = \max(\mathsf{low}(u), \max_{u' \in U} \min N(f(u'))).$$

This $\mathsf{low}'(u)$ adjustment is necessary because it ensures that $\mathsf{low}'(u)$ is connected to $f(u'), u' \in U$ (since $\mathsf{low}(u)$ could not be connected to all $f(u'), u' \in U$), while keeping $\mathsf{low}'(u)$ minimal (since it is the minimum vertex to which $f(u')$ is connected for all $u' \in U$ (by the properties of the k-shifted cliques)).

If it does not find a solution, we terminate the current branch. If both calls find solutions, we return a mapping that is the union of two mappings found. Note again that it is a solution by the properties of k-shifted cliques.

Since the second call takes time $h^{\mathcal{O}(1)}$, the overall complexity is $n^{\mathcal{O}(k)} h^{\mathcal{O}(1)}$.

On the other hand, as shown in Lemma 2, if there exists any solution, there is a solution of the kind that we seek by Lemma 2.

5.2   Small Generalized Pathwidth

In this subsection, we start with the definition of the parameter $c(H)$.

Let H be an ordered graph with a vertex set $(v_1, \ldots, v_h)$. By $V_{\leq i}$ (resp. $V_{>i}$) we denote the set $\{v_1, \ldots, v_i\}$ (resp. $\{v_{i+1}, \ldots, v_h\}$). Let A_i consist of the vertices of $V_{\leq i}$ that have neighbors in $V_{>i}$. Similarly, let B_i consist of the vertices of $V_{\leq i}$ that have non-neighbors in $V_{>i}$. Then, let $c(H) = \max_i \min(|A_i|, |B_i|)$.

We point out that $\max_i |A_i|$ is roughly equal to the pathwidth of H. We can approximately see this using the following construction. Each ordering of vertices of an unordered graph can define a path decomposition as follows. We start with a bag B_1 consisting of a single vertex v_1. Now suppose that we defined $B_1, \ldots, B_i$ and want to define B_{i+1}. We obtain it from B_i by adding v_{i+1} and removing all vertices whose all edges are already covered.

Optimum path decomposition can be obtained from some (optimal) ordering. Since we deal with a fixed ordering, the parameter we obtain is not the same as the pathwidth of the underlying unordered graph, but it behaves similarly.

Theorem 6. *The* HOM$_<$ *problem can be solved in time* $n^{\mathcal{O}(c(H))} \cdot h^{\mathcal{O}(1)}$, *where* $n = |V(G)|$ *and* $h = |V(H)|$.

Proof. Let $k = c(H)$ and $V(H) = (v_1, \ldots, v_h)$. For i, let i' be the minimum value of $j \geq i$ such that $|A_j| \leq k$ (if such j exists). Similarly, let i'' be the minimum value of $j \geq i$ such that $|B_j| \leq k$ (again, if such j exists). We define $\widetilde{A}_i = A_{i'} \cap V_{\leq i}$ and $\widetilde{B}_i = B_{i''} \cap V_{\leq i}$. If i' (resp. i'') is not defined, we set $\widetilde{A}_i = \emptyset$ (resp. $\widetilde{B}_i = \emptyset$). Observe that for all i we have $|\widetilde{A}_i| \leq k$ and $|\widetilde{B}_i| \leq k$ and for each i we either have $A_i = \widetilde{A}_i$ or $B_i = \widetilde{B}_i$.

We will construct a dynamic programming procedure that computes partial homomorphisms from G to H. The dynamic programming table Tab is indexed by a tuple consisting of:

- a vertex $u \in V(G)$,
- $i \in [h]$,
- a set $X \subseteq V(G)$, such that for each $x \in X$ we have $x \leq u$,
- a function ξ mapping the vertices of G smaller than or equal to u to $\widetilde{A}_i \cup \widetilde{B}_i$.

Note that the number of choices for X and ξ is at most n^{4k}: the size of $|\widetilde{A}_i \cup \widetilde{B}_i|$ is at most $2k$ and the preimage of each element of $\widetilde{A}_i \cup \widetilde{B}_i$ is a segment that can be described by two vertices of G (its first and last element), giving us at most $(n \cdot n)^{2k}$ options.

Thus, the total number of entries in Tab is at most $n^{4k+1} \cdot h$.

We aim to set $\mathsf{Tab}[u, i, X, \xi] = \mathsf{true}$ if and only if there exists a homomorphism $f : G[X] \to H$, $G[X]$ being an induced ordered subgraph of G on vertices X, such that:

- $f(u) = i$,
- $f^{-1}(\widetilde{A}_i \cup \widetilde{B}_i) = X$, and
- $f|_X = \xi$.

Clearly, $G \to H$ if and only if Tab contains a true entry such that the first coordinate of the indexing tuple is the last vertex of G.

Now, let us describe how we fill the table Tab. Let u be the first vertex of G; then the entries for the first vertex of G are straightforward to fill in. For each $i \in [h]$, if $v_i \notin \widetilde{A}_i \cup \widetilde{B}_i$, we set $\mathsf{Tab}[u, i, \emptyset, \emptyset] = \mathsf{true}$. Otherwise, we set $\mathsf{Tab}[u, i, \{u\}, \xi] = \mathsf{true}$, where ξ maps u to v_i.

So, now suppose that we want to fill the entries for some $u \in V(G)$, and the entries for all predecessors of u are already filled. Let u' be the direct predecessor of u.

For each $i \in [h]$ and each tuple (u', i', X', ξ') for which $\mathsf{Tab}[u', i', X', \xi'] = \mathsf{true}$ and $i' \leq i$, we proceed as follows.

- Consider two subcases:

- If $\widetilde{A}_{i'} = A_{i'}$, for every neighbor u'' of u that precedes it, check whether $u'' \in X'$ and $\xi'(u'') \in N(v_i)$ (such $\xi'(u'')$ must be in $A_{i'}$). If not, proceed to the next tuple (since the tuple would not preserve edges).
- If $\widetilde{B}_{i'} = B_{i'}$, for every neighbor u'' of u that precedes it, if $u'' \in X'$, check whether $\xi'(u'') \in N(v_i)$. If not, proceed to the next tuple (since the tuple would not preserve edges).
  - Consider two subcases:
    - If $i \notin \widetilde{A}_i \cup \widetilde{B}_i$, set $X = X' \cap \xi'^{-1}(\widetilde{A}_i \cup \widetilde{B}_i)$ and $\xi = \xi'|_X$ (because then u can be mapped to $i' < i$).
    - If $i \in \widetilde{A}_i \cup \widetilde{B}_i$, set $X = (X' \cap \xi'^{-1}(\widetilde{A}_i \cup \widetilde{B}_i)) \cup \{u\}$ and set ξ such that $\xi|_{X \setminus \{u\}} = \xi'|_{X \setminus \{u\}}$ and $\xi(u) = i$.
  - Set $\mathsf{Tab}[u, i, X, \xi] = \mathsf{true}$.

As mentioned, if Tab contains a true entry such that the first coordinate of the indexing tuple is the last vertex of G, we get an ordered homomorphism $G \to H$. This completes the proof.

Identifying all the classes of ordered graphs for which the $\mathrm{HOM}_<$ problem can be solved in polynomial time remains to be solved. We have shown in [13] that the $\mathrm{HOM}_<$ problem for ordered matchings, defined as ordered graphs where each vertex has exactly one incident edge, is NP-complete.

The same article shows that deciding whether for a given ordered graph G and ordered matching M there exists an ordered homomorphism $G \to M$ is fixed-parameter tractable with respect to $|V(M)|$. We observe that while the ordered graphs G and H in Theorem 4 can be complex, the ordered graph candidates for H, for which this problem is fixed-parameter tractable, are rather restricted in article [13]. Therefore, it remains to fill the gap between these classes of ordered graphs and categorize which of the classes of ordered graphs make the $\mathrm{HOM}_<(|V(H)|)$ problem W[1]-hard and which of them make it fixed-parameter tractable. We shall not address this question in the article at hand.

As mentioned above, determining for which levels i of the W-hierarchy the $\mathrm{HOM}_<$ problem, parameterized by $|V(H)|$, is W[i]-complete, also remains open.

References

1. Axenovich, M., Rollin, J., Ueckerdt, T.: Chromatic number of ordered graphs with forbidden ordered subgraphs. Combinatorica **38**, 1021–1043 (2016). https://api.semanticscholar.org/CorpusID:7776173
2. Balko, M., Cibulka, J., Král, K., Kynčl, J.: Ramsey numbers of ordered graphs. Electron. J. Comb. **27**(1) (2020). https://doi.org/10.37236/7816, http://dx.doi.org/10.37236/7816
3. Balko, M., Poljak, M.: On off-diagonal ordered ramsey numbers of nested matchings (2022)
4. Balogh, J., Li, L., Treglown, A.: Tilings in vertex ordered graphs. J. Comb. Theory, Ser. B **155**, 171–201 (2022). https://doi.org/10.1016/j.jctb.2022.02.006

5. Bhowmik, P., Iosevich, A., Koh, D., Pham, T.: Multi-linear forms, graphs, and lp -improving measures in bbb fqd. Can. J. Math. **77**(1), 208–251 (2023). https://doi.org/10.4153/s0008414x2300086x

6. Bonnet, É., Giocanti, U., de Mendez, P.O., Simon, P., Thomassé, S., Toruńczyk, S.: Twin-width iv: ordered graphs and matrices (2021)

7. Bonnet, É., Nešetřil, J., de Mendez, P.O., Siebertz, S., Thomassé, S.: Twin-width and permutations (2024)

8. Bonnet, É., Bousquet, N., Charbit, P., Thomassé, S., Watrigant, R.: Parameterized complexity of independent set in h-free graphs. Algorithmica **82**(8), 2360–2394 (2020). https://doi.org/10.1007/s00453-020-00730-6

9. Bose, P., Gudmundsson, J., Morin, P.: Ordered theta graphs. Comput. Geom. **28**(1), 11–18 (2004). https://doi.org/10.1016/j.comgeo.2004.01.003

10. Brandenburg, F.J.: Graph-grammars and their application to computer science. In: 3rd International Workshop Warrenton, Virginia, USA, 2–6 December 1986. LNCS, pp. 99–111 (2005). https://doi.org/10.1007/3-540-18771-5_48

11. Braun, B., Browder, J., Klee, S.: Cellular resolutions of ideals defined by nondegenerate simplicial homomorphisms. Israel J. Math. **196**(1), 321–344 (2013). https://doi.org/10.1007/s11856-012-0149-2

12. Čertík, M.: Homomorphisms of Ordered Graphs. Ph.D. thesis, Charles University, Prague (2025), PhD thesis, in preparation

13. Čertík, M., Feldmann, A.E., Nešetřil, J., Rzążewski, P.: On computational aspects of ordered matchings (2025). http://arxiv.org/abs/2511.23093

14. Čertík, M., Nešetřil, J.: Duality, $\chi^<$-boundedness and order density of ordered graphs (2023). https://arxiv.org/abs/2310.00852

15. Čertík, M., Feldmann, A.E., Nešetřil, J., Rzążewski, P.: On computational aspects of cores of ordered graphs (2025). http://arxiv.org/abs/2511.23099

16. Čertík, M., Feldmann, A.E., Nešetřil, J., Rzążewski, P.: Complexity aspects of homomorphisms of ordered graphs (2025). https://arxiv.org/abs/2511.23078

17. Chen, J., Liu, Y., Liu, X., Peng, S.: PGS-BFT: a pipeline-based graph structure byzantine fault tolerance consensus algorithm. In: 2023 3rd International Conference on Computer Science and Blockchain (CCSB), vol. 00, pp. 133–138 (2023). https://doi.org/10.1109/ccsb60789.2023.10398859

18. Conlon, D., Fox, J., Lee, C., Sudakov, B.: Ordered ramsey numbers (2016)

19. Connelly, R., Gortler, S.J., Theran, L.: Reconstruction in one dimension from unlabeled euclidean lengths. Combinatorica **44**(6), 1325–1351 (2024). https://doi.org/10.1007/s00493-024-00119-x

20. Cygan, M., et al.: Tight bounds for graph homomorphism and subgraph isomorphism, pp. 1643–1649. https://doi.org/10.1137/1.9781611974331.ch112, https://epubs.siam.org/doi/abs/10.1137/1.9781611974331.ch112

21. Cygan, M., et al.: Parameterized algorithms, pp. 285–319 (2015). https://doi.org/10.1007/978-3-319-21275-3_9

22. Duffus, D., Ginn, M., Rödl, V.: On the computational complexity of ordered subgraph recognition. Random Struct. Algorithms **7**(3), 223–268 (1995). https://doi.org/10.1002/rsa.3240070304

23. Dvořák, P., Feldmann, A.E., Rai, A., Rzążewski, P.: Parameterized inapproximability of independent set in h-free graphs. Algorithmica **85**(4), 902–928 (2023). https://doi.org/10.1007/s00453-022-01052-5

24. Ge, Y., et al.: Can graph descriptive order affect solving graph problems with LLMs? arXiv (2024). https://doi.org/10.48550/arxiv.2402.07140

25. Goerttler, S., Wu, M., He, F.: Machine learning applications in medicine and biology. arXiv, pp. 1–41 (2024). https://doi.org/10.1007/978-3-031-51893-5_1

26. Grytczuk, J., Pawlik, B., Ruciński, A.: Shuffle squares and nest-free graphs. arXiv (2025). https://doi.org/10.48550/arxiv.2503.22043

27. Guerra, L., Klee, S.: Betti numbers of order-preserving graph homomorphisms. Involve, J. Math. **5**(1), 67–80 (2012)

28. Guo, H., Song, X., Lindner, A.B.: Seq2dfunc: 2-dimensional convolutional neural network on graph representation of synthetic sequences from massive-throughput assay. bioRxiv p. 2019.12.22.886085 (2019). https://doi.org/10.1101/2019.12.22.886085

29. Hedrlín, Z., Pultr, A., Trnková, V.: Concerning a categorial approach to topological and algebraic theories. In: General Topology and its Relations to Modern Analysis and Algebra, pp. 176–181. Academia Publishing House of the Czechoslovak Academy of Sciences (1967). http://eudml.org/doc/220068

30. Hell, P., Nešetřil, J.: On the complexity of h-coloring. J. Comb. Theory **48**, 92–110 (1986)

31. Hell, P., Nešetřil, J.: Graphs and Homomorphisms. Oxford University Press (2004)

32. Kourani, H., Schuster, D., Aalst, W.V.D.: Scalable discovery of partially ordered workflow models with formal guarantees. In: 2023 5th International Conference on Process Mining (ICPM), vol. 00, pp. 89–96 (2023). https://doi.org/10.1109/icpm60904.2023.10271941

33. Kourani, H., Zelst, S.J.V.: Business process management. Lecture Notes in Computer Science, pp. 92–108 (2023). https://doi.org/10.1007/978-3-031-41620-0_6

34. Kun, G., Nešetřil, J.: Dichotomy for orderings? arXiv (2025). https://doi.org/10.48550/arxiv.2504.13268

35. Kuzmina, E.A., Nizamova, G.F.: Curriculum development based on the graph model. Inf. Educ. (5), 33–43 (2020). https://doi.org/10.32517/0234-0453-2020-35-5-33-43

36. Li, X., Yao, P., Pan, Y.: Understanding complex systems. Underst. Complex Syst. 341–371 (2015). https://doi.org/10.1007/978-3-662-47824-0_13

37. Lim, K., Shin, N.H., Young-Yoon, Kim, C.S.: Order learning and its application to age estimation. In: International Conference on Learning Representations (2020). https://openreview.net/forum?id=HygsuaNFwr

38. Malkhi, D., Stathakopoulou, C., Yin, M.: BBCA-CHAIN: low latency, high throughput BFT consensus on a DAG. arXiv (2023). https://doi.org/10.48550/arxiv.2310.06335

39. Nešetřil, J.: Ramsey Theory, pp. 1331–1403. MIT Press, Cambridge, MA, USA (1996)

40. Nešetřil, J., de Mendez, P.O.: Towards a characterization of universal categories (2016)

41. Nie, J., Surya, E., Zeng, J.: On asymptotic packing of convex geometric and ordered graphs. J. Graph Theory **104**(4), 836–850 (2023). https://doi.org/10.1002/jgt.23002

42. Pach, J., Tardos, G.: Forbidden paths and cycles in ordered graphs and matrices. Israel J. Math. (2006). https://doi.org/10.1007/BF02773960, https://doi.org/10.1007/BF02773960

43. Romansky, R.: An approach for mathematical modeling and investigation of computer processes at a macro level. Mathematics **8**(10), 1838 (2020). https://doi.org/10.3390/math8101838

44. Thomas, T., Surendran, R., John, T.S., Alazab, M.: Intelligent mobile malware detection, pp. 69–78 (2023). https://doi.org/10.1201/9781003121510-5

45. Verbytskyi, A., et al.: HepMC3 event record library for monte carlo event generators. J. Phys: Conf. Ser. **1525**(1), 012017 (2020). https://doi.org/10.1088/1742-6596/1525/1/012017
46. Wang, D., Tian, F., Wei, D.: An improved ordered visibility graph aggregation operator for MADM. IEEE Access **10**, 93464–93474 (2022). https://doi.org/10.1109/access.2022.3172684
47. Xing, Z., Zhao, W.: Block-diagonal structure learning for subspace clustering. Expert Syst. Appl. **285**, 127767 (2025). https://doi.org/10.1016/j.eswa.2025.127767
48. Zhang, Q., et al.: Building dynamic system call sandbox with partial order analysis. Proc. ACM Program. Lang. **7**(OOPSLA2), 1253–1280 (2023). https://doi.org/10.1145/3622842

Solution Discovery for Vertex Cover, Independent Set, Dominating Set, and Feedback Vertex Set

Rin Saito[1], Anouk Sommer[2], Tatsuhiro Suga[1], Takahiro Suzuki[1(✉)], and Yuma Tamura[1]

[1] Graduate School of Information Sciences, Tohoku University, Sendai, Japan
{rin.saito,suga.tatsuhiro.p5,takahiro.suzuki.q4}@dc.tohoku.ac.jp,
tamura@tohoku.ac.jp
[2] Karlsruhe Institute of Technology, Karlsruhe, Germany
anouk.sommer@student.kit.edu

Abstract. In the solution discovery problem for a search problem on graphs, we are given an initial placement of k tokens on the vertices of a graph and asked whether this placement can be transformed into a feasible solution by applying a small number of modifications. In this paper, we study the computational complexity of solution discovery for several fundamental vertex-subset problems on graphs, namely VERTEX COVER DISCOVERY, INDEPENDENT SET DISCOVERY, DOMINATING SET DISCOVERY, and FEEDBACK VERTEX SET DISCOVERY. We first present XP algorithms for all four problems parameterized by clique-width. We then prove that VERTEX COVER DISCOVERY, INDEPENDENT SET DISCOVERY, and FEEDBACK VERTEX SET DISCOVERY are NP-complete for chordal graphs and graphs of diameter 2, which have unbounded clique-width. In contrast to these hardness results, we show that all three problems can be solved in polynomial time on split graphs. Furthermore, we design an FPT algorithm for FEEDBACK VERTEX SET DISCOVERY parameterized by the number of tokens.

Keywords: solution discovery · graph algorithm · parameterized complexity

1 Introduction

Many conventional problems in algorithmic theory ask for a feasible solution from a given graph. The feasible solution corresponds to a configuration of a real-world system, such as the placement of surveillance robots in a museum or factory. However, in practical scenarios, a current configuration of a system often already exists, and the task is to obtain a feasible one starting from it. In particular, the current configuration may become infeasible due to system failures and must be restored to a feasible configuration as soon as possible.

This work was partially supported by JST SPRING Grant Number JPMJSP2114, the *Baden-Württemberg-STIPENDIUM* from the Baden-Württemberg-Stiftung, and JSPS KAKENHI Grant Number JP25K21148.

© The Author(s), under exclusive license to Springer Nature Switzerland AG 2026
J. Kozik and A. Wolff (Eds.): SOFSEM 2026, LNCS 16448, pp. 432–446, 2026.
https://doi.org/10.1007/978-3-032-17801-5_32

To capture this type of situation, Fellows et al. proposed the framework of *solution discovery* [9], and recently, several researchers have studied it in various problems [12,13]. In this framework, the problem Π-DISCOVERY is defined with respect to a combinatorial problem Π together with the upper bound of *modification steps*, which specifies how many steps are allowed to modify a given current configuration into a feasible one. Formally, given an instance of Π, an initial configuration S, and a budget b, Π-DISCOVERY asks whether S can be transformed into a feasible solution to Π within b modification steps. Note that neither the initial nor the intermediate configurations are required to be feasible. Most studies of solution discovery focus on vertex-subset problems on graphs, such as VERTEX COVER and INDEPENDENT SET, and an atomic modification step follows the *token sliding model* [20].[1] In the token sliding model of Π-DISCOVERY, a vertex subset is regarded as a set of tokens placed on vertices, where each vertex is occupied by at most one token. Then, at each step, one token may slide along an edge. Recalling the example of placing surveillance robots, this model naturally represents the movement of robots between adjacent locations.

1.1   Our Contribution

This paper studies the computational complexity of VERTEX COVER DISCOVERY, INDEPENDENT SET DISCOVERY, DOMINATING SET DISCOVERY, and FEEDBACK VERTEX SET DISCOVERY (VC-D, IS-D, DS-D, and FVS-D, respectively) under the token sliding model. An overview of our results is provided in Table 1.[2]

The first three problems have been widely studied in previous work from the viewpoint of parameterized complexity [9,13]. We first show that VC-D, IS-D, and DS-D all admit XP algorithms when parameterized by clique-width. Consequently, these problems can be solved in polynomial time on graphs of bounded clique-width, including graphs of bounded treewidth [6], cographs, and distance-hereditary graphs. In particular, this result extends the earlier result of Fellows et al. that provided XP algorithms for VC-D, IS-D, and DS-D parameterized by treewidth [9]. We emphasize that this result is the best possible for clique-width: Grobler et al. proved the XNLP-hardness for these problems parameterized by pathwidth [13], which implies that no FPT algorithm exists for clique-width unless FPT = W[1], since graphs with bounded pathwidth have bounded treewidth, and then bounded clique-width [6].

We next turn to graph classes of unbounded clique-width, in particular, chordal graphs and their subclass, split graphs. We first observe that DS-D remains NP-complete even on split graphs. This follows immediately from the fact that DOMINATING SET is NP-complete on split graphs [2]. Remarkably, we prove that VC-D and IS-D are also NP-complete on chordal graphs, even

---

[1]   We use the term "token sliding" following the terminology in the paper by Fellows et al. [9] that introduced the solution discovery framework.

[2]   The XNLP-hardness of FVS-D parameterized by pathwidth follows by a simple reduction from VC-D; see Sect. 3.

Table 1. The parameterized complexity and complexity on graph classes for VC-D, IS-D, DS-D, and FVS-D. The parameter k denotes the number of tokens. The characters "h" and "c" represent "hard" and "complete", respectively.

		Problem			
		VC-D	FVS-D	IS-D	DS-D
Parameter	k	FPT [9]	FPT [Theorem 2]	W[1]-h [9]	W[1]-h [9]
	clique-width	XP [Theorem 1]	XP [Theorem 1]	XP [Theorem 1]	XP [Theorem 1]
	pathwidth	XNLP-h [13]	XNLP-h	XNLP-h [13]	XNLP-h [13]
Graph Class	chordal	NP-c [Theorem 4]	NP-c [Theorem 1]	NP-c [Corollary 1]	NP-c [18]
	diameter 2	NP-c [Corollary 2]	NP-c [Corollary 2]	NP-c [Corollary 2]	NP-c [18]
	split	P [Theorem 5]	P [Corollary 3]	P [Corollary 3]	NP-c [18]

though their underlying problems VERTEX COVER and INDEPENDENT SET are solvable in linear time on chordal graphs [11,23]. To complement this hardness result, we further show that VC-D and IS-D can be solved in polynomial time on split graphs. From the positive results on split graphs (excluding DS-D) and cographs, which have small diameter, a natural research direction is to design polynomial-time algorithms for graphs of small diameter. However, we observe that VC-D, IS-D, and DS-D are NP-complete even for graphs of diameter 2.

In addition, this paper initiates the study of FVS-D. FVS is one of the most fundamental vertex-subset problems, appearing in the list of Karp's 21 NP-complete problems [17], and has been extensively studied in parameterized complexity (see, e.g., [7]). Following the approach for VC-D, we provide an XP algorithm for FVS-D parameterized by clique-width and a polynomial-time algorithm on split graphs, whereas FVS-D is shown to be NP-complete on chordal graphs or graphs of diameter 2. An interesting question is whether FVS-D admits an FPT algorithm when parameterized by the number of tokens k, since VC-D, IS-D, and DS-D are W[1]-hard when parameterized by the feedback vertex set number [13], whereas VC-D is known to be fixed-parameter tractable when parameterized by k [9]. Note that the FPT algorithm for VC-D in [9] crucially relies on enumerating all minimal vertex covers of size at most k. This strategy does not carry over to FVS-D, since the number of minimal feedback vertex sets of size at most k can be as large as $O(n^k)$, and hence such enumeration cannot be done within FPT time. To overcome this obstacle, we exploit the concept of *k-compact representations* of minimal feedback vertex sets [14], which yields a single-exponential FPT algorithm parameterized by k.

Related Work. Fellows et al. initiated the study of a solution discovery framework for several classical NP-complete graph problems, including VC, IS, and DS, under the token sliding model [9]. They analyzed the parameterized complexity with respect to various parameters, such as the configuration size k and the budget b, and also considered it using notions of graph sparsity. In subsequent work, Grobler et al. investigated the kernelization complexity of VC-D, IS-D, DS-D, and other problems under the token sliding model, and provided a refined classification with respect not only to k and b, but also to structural parameters (e.g., pathwidth and feedback vertex set number). In another line of research, Grobler et al. investigated solution discovery for polynomial-time solvable problems [12].

Very recently, Bousquet et al. [4] provided a meta-algorithm for solution discovery based on model checking of monadic second-order logic (MSO). Note that their results do not imply an XP algorithm for clique-width, which is one of our contributions.

The token sliding model originates from research on *combinatorial reconfiguration* problems (see the surveys by van den Heuvel [15] and Nishimura [20]), which ask whether, given two feasible solutions of a combinatorial problem, there is a step-by-step transformation from one into another without losing the feasibility. In contrast, solution discovery does not specify a target solution, and intermediate configurations may be infeasible.

Organization. In Sect. 2, we give preliminaries and define our problems. In Sect. 3, we present XP algorithms for VC-D, IS-D, DS-D, and FVS-D parameterized by clique-width, and an FPT algorithm for FVS-D parameterized by the number of tokens. In Sect. 4, we analyze the complexity of VC-D, IS-D, and FVS-D on graph classes of unbounded clique-width, including chordal graphs, split graphs, and graphs of small diameter. Due to the space limitation, the proofs of claims marked $(*)$ are omitted in this paper, which can be found in the full version [24].

2   Preliminaries

For a positive integer p, we write $[p] = \{1, 2, \ldots, p\}$. Let $G = (V, E)$ be an undirected graph, and let $V(G)$ and $E(G)$ denote its vertex set and edge set, respectively. For a vertex $v \in V(G)$, we define the *open neighborhood* of v as $N(v) = \{u \in V(G) \colon uv \in E(G)\}$, and the *closed neighborhood* as $N[v] = N(v) \cup \{v\}$. For vertices $u, v \in V(G)$, we write $\mathsf{dist}(u, v)$ for the number of edges in a shortest path between u and v. The *diameter* of a connected graph G, denoted by $\mathsf{diam}(G)$, is $\max_{u,v \in V} \mathsf{dist}(u, v)$.

A *vertex cover* of G is a set $C \subseteq V(G)$ such that every edge has an endpoint in C. An *independent set* of G is a set $I \subseteq V(G)$ such that no two vertices in I are adjacent. A *dominating set* of G is a set $D \subseteq V(G)$ such that every vertex of G belongs to D or has a neighbor in D. A *feedback vertex set* of G is a set $F \subseteq V(G)$ such that every cycle of G contains at least one vertex from F.

Given a graph G and an integer k, the problems Vertex Cover (VC), Independent Set (IS), Dominating Set (DS), and Feedback Vertex Set (FVS) ask whether G contains such a set of size at most k (for VC, DS, FVS) or at least k (for IS).

A *split graph* is a graph whose vertex set can be partitioned into a clique and an independent set. A *chordal graph* is a graph in which every cycle of length at least 4 has a chord, i.e., an edge between two non-consecutive vertices of the cycle.

Our problems. Let G be a graph. In this work, a *configuration* of G is simply a vertex subset of G. Two configurations S and S' of G are *adjacent* if there exist $x \in S$ and $y \in V(G) \setminus S$ with $xy \in E(G)$ such that $S' = S \cup \{y\} \setminus \{x\}$. (In particular, $|S| = |S'|$ holds.) A configuration S_t is *reachable in ℓ steps* from S_s if there exists a sequence of configurations $\langle S_s = S_0, S_1, S_2, \ldots S_\ell = S_t \rangle$ such that S_{i-1} and S_i are adjacent for all $i \in [\ell]$. We now define four discovery problems: VERTEX COVER DISCOVERY (VC-D), INDEPENDENT SET DISCOVERY (IS-D), DOMINATING SET DISCOVERY (DS-D), and FEEDBACK VERTEX SET DISCOVERY (FVS-D). Given a graph G, an initial configuration S of size k, and a non-negative integer b (called the *budget*), the problem VC-D (IS-D, DS-D, and FVS-D, respectively) asks whether there exists a configuration S_t that forms a vertex cover (independent set, dominating set, and feedback vertex set, respectively) of G and is reachable from S in at most b steps.

Throughout this paper, we assume that the input graph is connected; otherwise, the problems can be solved independently on each connected component. We also assume $b \leq k \cdot \mathsf{diam}(G)$ because each token can traverse a path of length at most $\mathsf{diam}(G)$. Moreover, when $b = k \cdot \mathsf{diam}(G)$, the problem Π-DISCOVERY is equivalent to the corresponding underlying problem Π: given an instance (G, k) of Π, we construct an instance (G, S, b) of Π-DISCOVERY, where S is an arbitrary vertex-subset of size k and $b = k \cdot \mathsf{diam}(G)$. Since every configuration of size k is reachable from S within b steps, (G, k) is a yes-instance of Π if and only if (G, S, b) is a yes-instance of Π-DISCOVERY.

Finally, note that all four problems are in NP. Therefore, to prove their NP-completeness, it suffices to show their NP-hardness.

3   Parameterized Algorithms

In this section, we investigate the parameterized complexity of VC-D, IS-D, DS-D, and FVS-D.

It is known that VC-D, IS-D, and DS-D are XNLP-hard when parameterized by pathwidth [13], which implies that no FPT algorithm exists for pathwidth unless FPT = W[1]. We further observe that FVS-D is also XNLP-hard when parameterized by pathwidth: given an instance of VC-D, we can construct an equivalent instance of FVS-D by replacing each edge uv of the input graph with a triangle on vertices u, v, e_{uv}. This transformation increases the pathwidth by exactly one (see also the definition of pathwidth in [7]) and yields a parameterized logspace reduction (see the definition in [3]) from VC-D to FVS-D.

In contrast to these hardness results, we show that all four problems admit XP algorithms when parameterized by clique-width. Recall that the clique-width of a graph is always bounded from above by its treewidth and pathwidth. Therefore, our result strictly strengthens the XP algorithm of Fellows et al. [9] for VC-D, IS-D, and DS-D parameterized by treewidth.

Finally, we show that FVS-D admits a single-exponential FPT algorithm when parameterized by the number of tokens k. Note that the one for VC-D is already known, and IS-D and DS-D are W[1]-hard when parameterized by $k + b$ [9].

3.1   XP Algorithm Parameterized by Clique-Width

In this subsection, we first present an XP algorithm for VC-D parameterized by clique-width, based on dynamic programming on a tree structure known as a *w-expression*. Then, by applying the same framework, we also show XP algorithms for FVS-D, IS-D, and DS-D parameterized by clique-width.

We now give the definition of the clique-width of graphs. For a positive integer w, a *w-graph* is a graph in which each vertex is assigned a label from $\{1, 2, \ldots, w\}$. We define the following four operations on w-graphs.

Introduce $i(v)$**:** Create a graph consisting of a single vertex v labeled with $i \in [w]$.

Union $\oplus$**:** Take the disjoint union of two w-graphs.

Relabel $\rho_{i \to j}$**:** For distinct $i, j \in [w]$, replace every label i in the graph with j.

Join $\eta_{i,j}$**:** For distinct $i, j \in [w]$, add all possible edges between vertices labeled i and vertices labeled j.

The *clique-width* of a graph G is the smallest integer w such that G can be constructed from w-graphs by a sequence of these operations. Such a construction can be represented by a rooted tree, where each node corresponds to one of the above operations. This tree is called a *w-expression tree* of G, and its nodes are referred to as *introduce, union, relabel,* or *join* nodes, respectively. For a node t of the w-expression tree T, let G_t denote the labeled graph associated with the subtree rooted at t. For node t of T and $i \in [w]$, we call a vertex of G_t with label i an *i-vertex*, and U_i^t denotes the set of i-vertices in G_t. It is known that, given a graph G of clique-width cw, one can compute a $(2^{\mathrm{cw}+1} - 1)$-expression of G in $O(|V(G)|^3)$ time [16, 21, 22]. A w-expression tree is called *irredundant* if, for each edge uv of G, there exists exactly one join node $\eta_{i,j}$ that introduces the edge uv. Moreover, any w-expression tree can be transformed in linear time into an irredundant w-expression tree with $O(n)$ nodes [6]. Hence, in the following, we may always assume without loss of generality that the w-expression tree is irredundant.

Dynamic Programming. Let (G, S, b) be an instance of VC-D, where G has n vertices and T is a w-expression tree of G. We perform a bottom-up dynamic programming over T, where the update rules depend on the type of each node in T.

To simplify the description of our dynamic programming, we introduce a hypothetical vertex v^* that can hold an arbitrary number of tokens and is adjacent to every vertex of G_t for each node t of T. Intuitively, the vertex v^* virtually represents all vertices of G "outside" G_t. We then place $k - |V(G_t) \cap S|$ tokens on v^* in the initial configuration. Let G_t^* denote the graph obtained from G_t by adding v^* and connecting it to all vertices of G_t.

For each node t of T, we maintain a Boolean DP table indexed by tuples $s = (\ell, \mathtt{K}, \mathtt{A}, \mathtt{P})$, where $\mathtt{K}, \mathtt{A}, \mathtt{P}$ are functions such that $\mathtt{K} \colon [w] \to [k]$ and $\mathtt{A}, \mathtt{P} \colon [w] \to [b]$. Here, ℓ denotes the number of token moves used in G_t ($0 \le \ell \le b$). For each label $i \in [w]$, $\mathtt{K}(i)$ represents the number of tokens placed on vertices with label i as the final configuration in G_t^*. The values $\mathtt{A}(i)$ and $\mathtt{P}(i)$ indicate, respectively, the numbers of incoming and outgoing additional moves (*absorption* and *projection*) between i-vertices and v^* in G_t^*. Note that, for each node t, the number of possible tuples is at most $b \cdot k^w \cdot b^{2w} \leq n^{5w+2}$ since $k \leq n$ and $b \leq k \cdot \mathrm{diam}(G) \leq n^2$.

For each tuple s, the table entry $c_t[s]$ is set to **true** if and only if there exists a vertex cover C of G_t and a sequence $\mathcal{S}$ of configurations in G_t^* such that (i) $|C \cap U_z^t| = \mathsf{K}(z)$ for every $z \in [w]$; (ii) $\mathcal{S}$ performs exactly ℓ token moves within G_t, with $\ell \leq b$; (iii) exactly $\mathsf{A}(z)$ moves are performed from v^* to the vertices in U_z^t; and (iv) exactly $\mathsf{P}(z)$ moves are performed from the vertices in U_z^t to v^*.

A tuple $s = (\ell, \mathsf{K}, \mathsf{A}, \mathsf{P})$ is said to be *valid* if it satisfies the following conditions: (a) $\sum_{z \in [w]} \mathsf{K}(z) \leq k$; (b) if $|U_z^t| = 0$ for $z \in [w]$, then $\mathsf{K}(z) = \mathsf{A}(z) = \mathsf{P}(z) = 0$; (c) $\ell \leq b$; and (d) $|S \cap U_z^t| + \mathsf{A}(z) - \mathsf{P}(z) = \mathsf{K}(z) \leq |U_z^t|$ for all $z \in [w]$. Intuitively, condition (a) ensures that the number of tokens in G_t does not exceed $k \ (= |S|)$. Condition (b) states that if U_z^t is empty, then there is no token on z-vertices as a final configuration, and no token moves occur between z-vertices and v^*. Condition (c) guarantees that the total number of token moves occurring within G_t does not exceed the budget b. Condition (d) enforces the consistency of the number of tokens between the initial and final configurations. Note that, since $|S \cap U_z^t|$ and $|U_z^t|$ are already known for all $z \in [w]$, one can check whether a tuple s is valid in $O(w)$ time. At the root node r of T, our algorithm outputs **YES** if and only if there exists a valid tuple s with $c_r[s] = $ **true** such that $\mathsf{A}(z) = \mathsf{P}(z) = 0$ for all $z \in [w]$. As we will see later, the DP entry corresponding to any invalid tuple is always assigned the value **false**, since such moves cannot be realized.

Introduce node $i(v)$: Let t be an introduce node of T, and let $s = (\ell, \mathsf{K}, \mathsf{A}, \mathsf{P})$ be a tuple for t. In this case, the subgraph G_t consists only of a single vertex v. Since both the empty set $\emptyset$ and singleton $\{v\}$ form vertex covers of G_t, we set $c_t[s] = $ **true** if and only if s is a valid tuple and $\ell = 0$.

Union node $\oplus$: Let t be a union node of T with children t_1 and t_2, and let $s = (\ell, \mathsf{K}, \mathsf{A}, \mathsf{P})$ be a tuple for t. As G_t is the disjoint union of G_{t_1} and G_{t_2}, no tokens move between them. Recall that the union of a vertex cover of G_{t_1} and a vertex cover of G_{t_2} forms a vertex cover of G_t. Thus, each value of $\ell, \mathsf{K}(i), \mathsf{A}(i), \mathsf{P}(i)$ ($i \in [w]$) must be the sum of the corresponding values for t_1 and t_2, respectively. Consequently, we set $c_t[s] = $ **true** if and only if s is valid and there exist tuples $s_1 = (\ell_1, \mathsf{K}_1, \mathsf{A}_1, \mathsf{P}_1)$ for t_1 and $s_2 = (\ell_2, \mathsf{K}_2, \mathsf{A}_2, \mathsf{P}_2)$ for t_2 such that

1. $c_{t_1}[s_1] = $ **true** and $c_{t_2}[s_2] = $ **true**,
2. $\ell = \ell_1 + \ell_2$, and
3. $\mathsf{K}(z) = \mathsf{K}_1(z) + \mathsf{K}_2(z)$, $\mathsf{A}(z) = \mathsf{A}_1(z) + \mathsf{A}_2(z)$, and $\mathsf{P}(z) = \mathsf{P}_1(z) + \mathsf{P}_2(z)$ for all $z \in [w]$.

Relabel node $\rho_{i \to j}$: Let t be a relabel node of T with child t', and let $s = (\ell, \mathsf{K}, \mathsf{A}, \mathsf{P})$ be a tuple for t. Since all i-vertices of $G_{t'}$ are relabeled as j-vertices in G_t (i.e., $U_i^t = \emptyset$), we must have $\mathsf{K}(i) = \mathsf{A}(i) = \mathsf{P}(i) = 0$, and the values of i-vertices for t' are merged into those of j-vertices. Consequently, we set $c_t[s] = $ **true** if and only if s is valid, $\mathsf{K}(i) = \mathsf{A}(i) = \mathsf{P}(i) = 0$, and there exists a valid tuple $s' = (\ell', \mathsf{K}', \mathsf{A}', \mathsf{P}')$ for t' such that

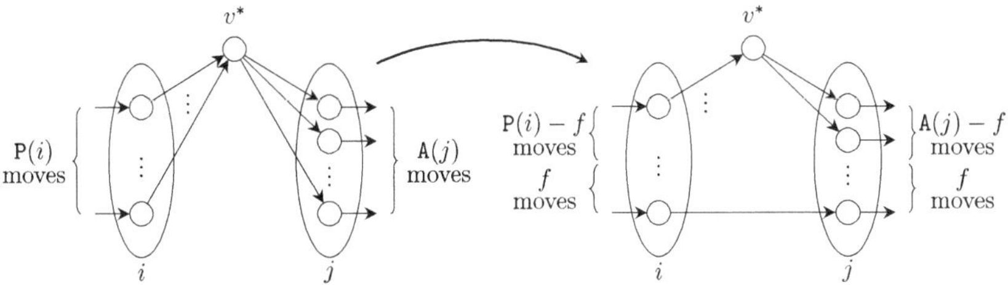

Fig. 1. A rough illustration for a join node. If f tokens directly move from the i-vertices to the j-vertices, we decrease both $\mathsf{P}(i)$ and $\mathsf{A}(j)$ by f, and increase the total number of moves ℓ by f.

1. $c_{t'}[s'] = \texttt{true}$,
2. $\ell = \ell'$,
3. $\mathsf{K}(j) = \mathsf{K}'(i) + \mathsf{K}'(j)$, $\mathsf{A}(j) = \mathsf{A}'(i) + \mathsf{A}'(j)$, $\mathsf{P}(j) = \mathsf{P}'(i) + \mathsf{P}'(j)$, and
4. $\mathsf{K}(z) = \mathsf{K}'(z)$, $\mathsf{A}(z) = \mathsf{A}'(z)$, $\mathsf{P}(z) = \mathsf{P}'(z)$ for all $z \in [w] \setminus \{i, j\}$.

Join node $\eta_{i,j}$: Let t be a join node of T with child t', and let $s = (\ell, \mathsf{K}, \mathsf{A}, \mathsf{P})$ be a tuple for t. Since T is irredundant, all edges between i-vertices and j-vertices are added in G_t. Thus, in any vertex cover C of G_t, at least one of the equalities $\mathsf{K}(i) = |U_i^t \cap C|$ or $\mathsf{K}(j) = |U_j^t \cap C|$ must hold; otherwise, some edge between U_i^t and U_j^t is uncovered. Therefore, if both $\mathsf{K}(i) < |U_i^t|$ and $\mathsf{K}(j) < |U_j^t|$, then we set $c_t[s] = \texttt{false}$.

Otherwise (i.e., $\mathsf{K}(i) = |U_i^t|$ or $\mathsf{K}(j) = |U_j^t|$), we consider the situation that some tokens move between i-vertices and j-vertices. Let $f \geq 0$ denote the number of additional moves from i-vertices to j-vertices, and let $g \geq 0$ denote the number of additional moves in the opposite direction. In other words, we allow tokens to move directly between i-vertices and j-vertices, which were previously treated as moves from i-vertices to j-vertices (or opposite direction), passing through v^*. Therefore, we then set $c_t[s] = \texttt{true}$ if and only if s is valid and there exist integers $f, g \geq 0$ together with a valid tuple $s' = (\ell', \mathsf{K}', \mathsf{A}', \mathsf{P}')$ for t' such that

1. $c_{t'}[s'] = \texttt{true}$,
2. $\ell = \ell' + f + g$,
3. $\mathsf{K}(z) = \mathsf{K}'(z)$ for all $z \in [w]$,
4. $\mathsf{A}(z) = \mathsf{A}'(z)$ and $\mathsf{P}(z) = \mathsf{P}'(z)$ for all $z \in [w] \setminus \{i, j\}$, and
5. $\mathsf{A}(i) = \mathsf{A}'(i) - g \geq 0$, $\mathsf{A}(j) = \mathsf{A}'(j) - f \geq 0$, $\mathsf{P}(i) = \mathsf{P}'(i) - f \geq 0$, $\mathsf{P}(j) = \mathsf{P}'(j) - g \geq 0$.

Regarding Conditions 2 and 5, note that there are f tokens that were moving from the i-vertices to the j-vertices via v^*, and g tokens that were moving from the j-vertices to the i-vertices via v^*. Since they originally passed through v^* without incurring any moves, condition 2 makes this correction. In condition 5, we distribute the tokens that were passing through v^* at node t'. A brief illustration is given in Fig. 1.

Running Time. Observe that the validity of a given tuple s can be verified in $O(w)$ time, and for each type of node t, the value $c_t[s]$ can also be computed within $n^{O(w)}$ time. Since the DP table of a node t contains at most n^{5w+2} entries and the expression tree T has $O(n)$ nodes, VC-D can be solved in $n^{O(w)}$ time.

Remark. The central idea of our algorithm is to perform dynamic programming over a w-expression by guessing the two pieces of information: the feasibility of solutions and the number of token movements. More precisely, we guess the number of tokens on each i-vertex ($i \in [w]$) in the final configuration for VC-D, and furthermore, we maintain auxiliary information that records the number of token movements between i-vertices and v^* in G_t^*, specifically ℓ, A, and P.

In other solution discovery problems, the first part of our approach can be inherited from the well-established dynamic programming approach for the underlying search problem parameterized by clique-width, which has already been known for FVS [1,5], IS, and DS. For instance, to solve FVS-D, one additionally keeps track of information for the connectivity constraints. Consequently, the same dynamic programming technique as VC-D can also be applied to FVS-D, IS-D, and DS-D. This yields the following result.

Theorem 1. *Given a w-expression tree of the input graph, the problems* VC-D, *FVS-D, IS-D, and DS-D can be solved in $n^{O(w)}$ time.*

3.2   FPT Algorithm Parameterized by the Number of Tokens

This section presents an FPT algorithm for FVS-Dparameterized by the number of tokens k.

Theorem 2. FVS-D *parameterized by the number of tokens k can be solved in $O(23.1^k n^3)$ time, where n denotes the number of vertices in the input graph.*

Our algorithm uses an approach for VC-D provided in [9]. We first generate all candidate solutions to FVS. Then, for each candidate solution, construct an edge-weighted complete bipartite graph H and find a minimum weight matching of H saturating one side of H. The key idea is that if one could enumerate all inclusion-wise minimal feedback vertex sets of size at most k, then the above approach yields an FPT algorithm for k. Unfortunately, in general, enumerating minimal feedback vertex sets cannot be done in FPT time with respect to k. For example, consider a graph consisting of k vertex-disjoint cycles of length n/k. The graph contains $\Omega((n/k)^k)$ distinct minimal feedback vertex sets.

We overcome this difficulty by utilizing the concept of *compact representations* of minimal feedback vertex sets [14]. The family $\mathcal{Y} = \{Y_1, Y_2, \dots\}$ of pairwise disjoint vertex subsets of a graph G is a *compact representation* of minimal feedback vertex sets if every set F with $|F \cap Y| = 1$ for all $Y \in \mathcal{Y}$ forms a minimal feedback vertex set of G. In particular, if $|\mathcal{Y}| \leq k$, then $\mathcal{Y}$ is called a *k-compact representation*. We say that $\mathcal{Y}$ *represents* a minimal feedback vertex set F of G if $|F \cap Y| = 1$ for every $Y \in \mathcal{Y}$. A list $\mathfrak{L}$ of k-compact representations is said to be *complete* if, for every minimal feedback vertex set F of size at most

k, there is $\mathcal{Y} \in \mathcal{L}$ that represents F. In other words, every feedback vertex set of size at most k is represented by at least one compact representation in $\mathcal{L}$.

We use the following result as a subroutine for our algorithm.

Theorem 3 (Theorem 3 of [19]). *Given a graph G with m edges and a positive integer k, there exists an algorithm that computes a complete list of k-compact representations of G in $O(23.1^k m)$ time. Moreover, the number of k-compact representations produced by the algorithm is at most $O(23.1^k)$.*

Let (G, S, b) be an instance of FVS-D. In our algorithm, we first construct a complete list $\mathcal{L}$ of k-compact representations of minimal feedback vertex sets in G by Theorem 3. Second, for each k-compact representation $\mathcal{Y} \in \mathcal{L}$, we examine whether there exists at least one feedback vertex set represented by $\mathcal{Y}$ that is reachable from S within b steps.

Now, we explain the second step of our algorithm. Fix a k-compact representation $\mathcal{Y} \in \mathcal{L}$. We construct an edge-weighted complete bipartite graph $H_{\mathcal{Y}}$ with bipartitions S and $\mathcal{Y}$. To distinguish the vertices of $H_{\mathcal{Y}}$ and the vertices of G, we refer to the vertices of $H_{\mathcal{Y}}$ as *nodes*. For $u \in S$ and $Y \in \mathcal{Y}$, we assign the weight of the edge (u, Y) to $\min_{y \in Y} \mathsf{dist}(u, y)$, which corresponds to the minimum number of steps required to move the token initially placed on u to some vertex in Y. We then consider a minimum-weight matching M^* of $H_{\mathcal{Y}}$ such that each node in $\mathcal{Y}$ of $H_{\mathcal{Y}}$ is incident to some edge in M^*.

The following Lemma 1 establishes the correspondence between the total weight of M and the minimum number of steps to reach some minimal feedback vertex sets represented by $\mathcal{Y}$ from S. The proof can be found in the full version.

Lemma 1 (∗). *The two statements are equivalent:*

(1) *There exists a matching $M = \{(u_1, Y_1), (u_2, Y_2), \ldots, (u_{|\mathcal{Y}|}, Y_{|\mathcal{Y}|})\}$ of $H_{\mathcal{Y}}$ with total weight at most b, where $u_i \in S$ and $Y_i \in \mathcal{Y}$ for $i \in [|\mathcal{Y}|]$; and*
(2) *There exists a feedback vertex set F that is reachable in b steps from S and a minimal feedback vertex set $F' \subseteq F$ represented by $\mathcal{Y}$.*

Our algorithm outputs YES if there exists a k-compact representation $\mathcal{Y} \in \mathcal{L}$ such that the corresponding bipartite graph $H_{\mathcal{Y}}$ admits a minimum-weight matching M^* of total weight at most b, and outputs NO otherwise.

Finally, we analyze the running time of our algorithm. A complete list $\mathcal{L}$ of k-compact representations such that $|\mathcal{L}| = O(23.1^k)$ can be obtained in $O(23.1^k m)$ time [19]. To construct the bipartite graph $H_{\mathcal{Y}}$ for $\mathcal{Y} \in \mathcal{L}$, we precompute in $O(k(n + m))$ time the distances from every pair of vertices $u \in S$ and $v \in V$ by applying a breadth-first search algorithm k times. For each $\mathcal{Y} \in \mathcal{L}$, the graph $H_{\mathcal{Y}}$ is constructed in $O(kn)$ time; for each $u \in S$, the weights of the edges incident to u can be decided in $O(n)$ time because $\sum_{\mathcal{Y} \in \mathcal{L}} |\mathcal{Y}| = O(n)$. Since $H_{\mathcal{Y}}$ has at most $2k$ vertices, the desired minimum-weight matching M^* of $H_{\mathcal{Y}}$ can be obtained in $O(|V(H_{\mathcal{Y}})|^3) = O(k^3)$ time using the Hungarian algorithm [8]. Therefore, the overall running time of our algorithm is $O(23.1^k m + k(n + m) + 23.1^k(kn + k^3)) = O(23.1^k(n^2 + k^3)) = O(23.1^k n^3)$.

4   Graphs of Unbounded Clique-Width

In this section, we analyze the computational complexity of VC-D, IS-D, and FVS-D on chordal graphs, split graphs, and graphs of bounded diameter, which have unbounded clique-width. In Sect. 4.1, we establish the NP-completeness of these problems on chordal graphs and graphs of diameter 2. In Sect. 4.2, we show the polynomial-time solvability of these problems on split graphs. We remark that DS-D is NP-complete (more precisely, W[2]-hard for the number of tokens k) for split graphs of diameter 2, which follows directly from the known result that DS is NP-complete on those graphs [18].

4.1   NP-Completeness

We show the NP-completeness of VC-D, IS-D, and FVS-D for chordal graphs and graphs of diameter 2. We first show the following theorem.

Theorem 4. VC-D *is NP-complete for chordal graphs.*

For the proof of Theorem 4, we reduce the EXACT COVER BY 3-SET problem to VC-D. In EXACT COVER BY 3-SET, we are given a ground set U of $3n$ elements together with a family $\mathcal{X}$ of three-element subsets of U. The task is to decide whether there exists a subfamily of $\mathcal{X}$ consisting of exactly n sets whose union forms U. It is known that EXACT COVER BY 3-SET is NP-complete [10].

Let $(U, \mathcal{X})$ be an instance of EXACT COVER BY 3-SET, where the ground set $U = \{u_1, \ldots, u_{3n}\}$ and the family $\mathcal{X} = \{X_1, \ldots, X_m\}$. We construct a corresponding instance (G, S, b) of VC-D as follows (see Fig. 2 for an illustration).

Let $Q = \{q_0, \ldots, q_{3n}\}$ be a set of $3n + 1$ vertices. For each $i \in [3n]$, the vertex q_i corresponds to the element $u_i \in U$. Next, let $I = \{v_1, \ldots, v_m\}$ be a set of m vertices, where each v_i corresponds to the set $X_i \in \mathcal{X}$. We then define a split graph G' with $V(G') = Q \cup I$ and $E(G') = \{qq' : q, q' \in Q\} \cup \{v_i q_j : v_i \in I, q_j \in Q, u_j \in S_i\}$. Since I forms an independent set and Q forms a clique in G', it follows that G' is indeed a split graph.

We construct the graph G from G' as follows: for each $v_i \in I$, add the four vertices $v_i^1, v_i^2, v_i^3, v_i^4$ so that $N(v_i^j) = \{v_i\} \cup (N(v_i) \cap Q)$ for each $j \in \{1, 2, 3, 4\}$ in G. Note that each v_i^j is a true twin[3] of v_i in G. That is, the induced subgraph $G[\{v_i, v_i^1, v_i^2, v_i^3, v_i^4\}]$ forms a star rooted at v_i with leaves $v_i^1, v_i^2, v_i^3, v_i^4$, and all these vertices share the same neighbors in Q. We denote this star by A_i. Finally, we set $S = \{v_i^1, v_i^2, v_i^3, v_i^4 : i \in [m]\}$ and $b = 3n + n = 4n$.

We observe that the constructed graph G is chordal, since G is obtained from a split graph G' by adding true twins, and split graphs preserve chordality under the operation of adding true twins. Clearly, the instance (G, S, b) for VC-D is constructed in polynomial time. To complete the polynomial-time reduction, we prove the following lemma. The proof can be found in the full version.

---

[3] Two vertices v, v' are *true twins* if $N[v] = N[v']$.

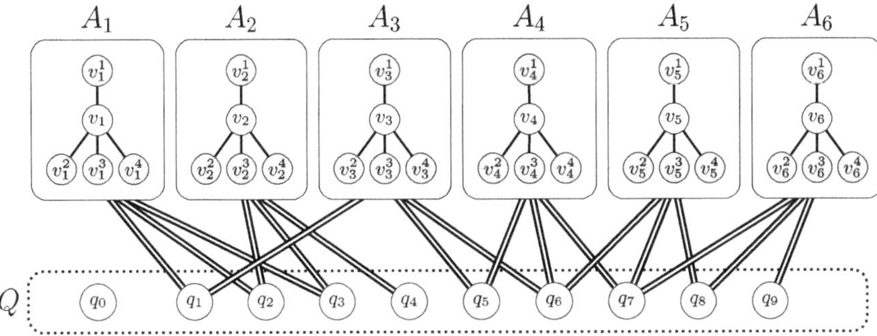

Fig. 2. Construction of an instance (G, S, b) of VC-D from an instance $(U, \mathcal{X})$ of EXACT COVER BY 3-SET, where $U = \{1, \ldots, 9\}$ and $\mathcal{X} = \{X_1 = \{1, 2, 3\}, X_2 = \{2, 3, 4\}, X_3 = \{1, 5, 6\}, X_4 = \{5, 6, 7\}, X_5 = \{6, 7, 8\}, X_6 = \{7, 8, 9\}\}$. The vertices $q_0, \ldots, q_9$ inside the dotted box form a clique of G. Each double line between a vertex q_i and a solid box labeled A_j represents all edges between q_i and the vertices in A_j. The vertices marked in gray constitute the initial configuration S, and $b = 9 + 3 = 12$.

Lemma 2 (∗). *An instance $(U, \mathcal{X})$ of* EXACT COVER BY 3-SET *is a yes-instance if and only if an instance (G, S, b) of* VC-D *is a yes-instance.*

This completes the proof of Theorem 4.

By the equivalence between VC-D and IS-D, we immediately obtain the NP-hardness of IS-D on chordal graphs. For FVS-D, given an instance of VC-D on a chordal graph G, we construct an equivalent instance of FVS-D by replacing each edge uv of G with a triangle on vertices u, v, and e_{uv}. The resulting graph remains chordal. Thus, we have the following corollary.

Corollary 1. IS-D *and* FVS-D *are* NP-*complete for chordal graphs.*

Next, we observe the NP-completeness of VC-D, IS-D, and FVS-D on graphs of diameter 2. Recall that if the budget b equals $k \cdot \mathsf{diam}(G)$, then the problem Π-DISCOVERY is equivalent to the underlying problem Π. Therefore, to prove NP-completeness for VC-D, IS-D, and FVS-D, it suffices to show the NP-completeness of the corresponding underlying problems VC, IS, and FVS on graphs of diameter 2. The details can be found in the full version.

Lemma 3 (∗). VC, IS, *and* FVS *are* NP-*complete on graphs of diameter* 2.

The following Corollary 2 is immediately obtained from Lemma 3.

Corollary 2. VC-D, IS-D, *and* FVS-D *are* NP-*complete on graphs of diameter* 2.

4.2 Polynomial-Time Algorithms

In this subsection, we show that VC-D, IS-D, and FVS-D can be solved in polynomial time on split graphs. We first focus on VC-D and then observe that similar results hold for IS-D and FVS-D. Note that split graphs are a subclass of both chordal graphs and graphs of diameter at most 3. The details can be found in the full version.

Theorem 5 (∗). VC-D *can be solved in* $O(n^4)$ *time on split graphs, where n is the number of vertices in an input graph.*

To show Theorem 5, we enumerate all minimal vertex covers in a given split graph in polynomial time, and then reduce our problem to the problem of finding a minimum-weight matching, in the same manner as Sect. 3.2.

We remark that this approach is universal: for any vertex subset problem Π, if one can enumerate candidate solutions of Π or construct a data structure that encodes such information (e.g., compact representations), Π-DISCOVERY can be solved and its running time mainly depends on the enumeration of solutions. In particular, we can observe that maximal independent sets and minimal feedback vertex sets in a split graph can be enumerated in polynomial time. Therefore, we have the following corollary.

Corollary 3. IS-D *and* FVS-D *can be solved in polynomial time on split graphs.*

We also emphasize that Theorem 5 and Corollary 3 can be extended to any graph class in which minimal (maximal for IS-D) solutions can be enumerated in polynomial time.

5   Conclusion

In this paper, we investigated the complexity of VC-D, IS-D, DS-D, and FVS-D from the perspective of parameterization and graph classes. We gave a general scheme for solving Π-DISCOVERY for various vertex-subset problems Π in Sect. 3.1 and Sect. 4.2. A promising direction for future research is to identify graph classes for which Π-DISCOVERY is NP-complete, while the underlying problem Π is polynomial-time solvable, as demonstrated for chordal graphs. For interval graphs, which are a subclass of chordal graphs, it is still open whether VC-D, IS-D, DS-D, and FVS-D are solvable in polynomial time. Furthermore, it remains open whether there exists an FPT algorithm for FVS-D parameterized by feedback vertex set number.

Acknowledgements. We thank Akira Suzuki and Xiao Zhou for fruitful discussions. We are also grateful to the anonymous reviewers for their constructive comments, which improved the presentation.

References

1. Bergougnoux, B., Kanté, M.M.: Fast exact algorithms for some connectivity problems parameterized by clique-width. Theoretical Computer Science **782**, 30–53 (2019), https://doi.org/10.1016/j.tcs.2019.02.030

2. Bertossi, A.A.: Dominating sets for split and bipartite graphs. Inf. Process. Lett. **19**(1), 37–40 (1984). https://doi.org/10.1016/0020-0190(84)90126-1

3. Bodlaender, H.L., Groenland, C., Nederlof, J., Swennenhuis, C.: Parameterized problems complete for nondeterministic FPT time and logarithmic space. Inf. Comput. **300**, 105195 (2024). https://doi.org/10.1016/j.ic.2024.105195

4. Bousquet, N., Mouawad, A.E., Maaz, S., Nishimura, N., Siebertz, S.: On algorithmic meta-theorems for solution discovery: Tractability and barriers (2025). https://arxiv.org/abs/2510.17344

5. Bui-Xuan, B., Suchý, O., Telle, J.A., Vatshelle, M.: Feedback vertex set on graphs of low clique-width. Euro. J. Comb. **34**(3), 666–679 (2013). https://doi.org/10.1016/j.ejc.2012.07.023

6. Courcelle, B., Olariu, S.: Upper bounds to the clique width of graphs. Discrete Appl. Math. **101**(1-3), 77–114 (2000). https://doi.org/10.1016/S0166-218X(99)00184-5

7. Cygan, M., et al.: Parameterized Algorithms. Springer, Cham (2015). https://doi.org/10.1007/978-3-319-21275-3

8. Edmonds, J., Karp, R.M.: Theoretical improvements in algorithmic efficiency for network flow problems. J. ACM **19**(2), 248–264 (1972). https://doi.org/10.1145/321694.321699

9. Fellows, M.R., et al.: On solution discovery via reconfiguration. In: ECAI 2023 - 26th European Conference on Artificial Intelligence, September 30–October 4, 2023, Kraków, Poland - Including 12th Conference on Prestigious Applications of Intelligent Systems (PAIS 2023). pp. 700–707 (2023). https://doi.org/10.3233/FAIA230334

10. Garey, M.R., Johnson, D.S.: Computers and Intractability: A Guide to the Theory of NP-Completeness. Freeman, W. H (1979)

11. Gavril, F.: Algorithms for minimum coloring, maximum clique, minimum covering by cliques, and maximum independent set of a chordal graph. SIAM J. Comput. **1**(2), 180–187 (1972). https://doi.org/10.1137/0201013

12. Grobler, M., et al..: Solution discovery via reconfiguration for problems in P. In: 51st International Colloquium on Automata, Languages, and Programming, ICALP 2024, July 8–12, 2024, Tallinn, Estonia. pp. 76:1–76:20 (2024). https://doi.org/10.4230/LIPIcs.ICALP.2024.76

13. Grobler, M., Maaz, S., Mouawad, A.E., Nishimura, N., Ramamoorthi, V., Siebertz, S.: Kernelization complexity of solution discovery problems. In: 35th International Symposium on Algorithms and Computation, ISAAC 2024, December 8–11, 2024, Sydney, Australia. pp. 36:1–36:17 (2024). https://doi.org/10.4230/LIPIcs.ISAAC.2024.36

14. Guo, J., Gramm, J., Hüffner, F., Niedermeier, R., Wernicke, S.: Compression-based fixed-parameter algorithms for feedback vertex set and edge bipartization. J. Comput. Sys. Sci. **72**(8), 1386–1396 (2006). https://doi.org/10.1016/j.jcss.2006.02.001

15. van den Heuvel, J.: The complexity of change. In: Surveys in Combinatorics 2013, London Mathematical Society Lecture Note Series, vol. 409, pp. 127–160. Cambridge University Press (2013). https://doi.org/10.1017/CBO9781139506748.005

16. Hliněný, P., Oum, S.: Finding branch-decompositions and rank-decompositions. SIAM J. Comput. **38**(3), 1012–1032 (2008). https://doi.org/10.1137/070685920

17. Karp, R.M.: Reducibility Among Combinatorial Problems. In: ünger, M., et al. (eds.) 50 Years of Integer Programming 1958-2008, pp. 219–241. Springer, Heidelberg (2010). https://doi.org/10.1007/978-3-540-68279-0_8

18. Lokshtanov, D., Misra, N., Philip, G., Ramanujan, M.S., Saurabh, S.: Hardness of r-dominating set on graphs of diameter $(r + 1)$. In: Gutin, G.Z., Szeider, S. (eds.) Parameterized and Exact Computation - 8th International Symposium, IPEC 2013, Sophia Antipolis, France, September 4–6, 2013, Revised Selected Papers. Lecture Notes in Computer Science, vol. 8246, pp. 255–267. Springer (2013). https://doi.org/10.1007/978-3-319-03898-8_22
19. Misra, N., Philip, G., Raman, V., Saurabh, S., Sikdar, S.: FPT algorithms for connected feedback vertex set. J. Combin. Optim. **24**(2), 131–146 (2012). https://doi.org/10.1007/s10878-011-9394-2
20. Nishimura, N.: Introduction to reconfiguration. Algorithms **11**(4), 52 (2018). https://doi.org/10.3390/a11040052
21. Oum, S.: Approximating rank-width and clique-width quickly. ACM Trans. Algorithms **5**(1), 10:1–10:20 (2008). https://doi.org/10.1145/1435375.1435385
22. Oum, S., Seymour, P.D.: Approximating clique-width and branch-width. J. Combin. Theory Series B. **96**(4), 514–528 (2006). https://doi.org/10.1016/j.jctb.2005.10.006
23. Rose, D.J., Tarjan, R.E., Lueker, G.S.: Algorithmic aspects of vertex elimination on graphs. SIAM J. Comput.**5**(2), 266–283 (1976). https://doi.org/10.1137/0205021
24. Saito, R., Sommer, A., Suga, T., Suzuki, T., Tamura, Y.: Solution discovery for vertex cover, independent set, dominating set, and feedback vertex set (2025). https://arxiv.org/abs/2511.23012

Quantified Colouring and H-Free Algorithmics

Kristina Asimi[1]($\boxtimes$), Tala Eagling-Vose[1], Santiago Guzmán-Pro[2],
Barnaby Martin[1], and Yiming Qiu[1]

[1] Durham University, Durham, UK
{kristina.asimi,tala.j.eagling-vose,barnaby.d.martin,
yiming.qiu}@durham.ac.uk
[2] Institut für Algebra, TU Dresden, Dresden, Germany
santiago.guzman_pro@tu-dresden.de

Abstract. A well-known consequence of Brooks' Theorem is that 3-Colouring, a.k.a. $\mathrm{CSP}(K_3)$, is solvable in polynomial time for subcubic graphs. We begin our investigation by observing that $\mathrm{QCSP}(K_3)$, a.k.a. Quantified 3-Colouring, is Pspace-complete on subcubic planar graphs. This allows us to give a complexity classification for bounded alternation Σ_i-$\mathrm{QCSP}(K_3)$ ($i \geq 3$, odd) for $\mathcal{H}$-topological-minor-free graphs. For all such (maybe infinite) sets $\mathcal{H}$, we delineate between those for which Σ_i-$\mathrm{QCSP}(K_3)$ is solvable in polynomial time, and those for which it is $\Sigma_{i-2}^{\mathrm{P}}$-hard.

We continue our investigation of $\mathrm{QCSP}(K_3)$ on $\mathcal{H}$-subgraph-free graphs, contrasting with both $\mathrm{CSP}(K_3)$, as well as the recently introduced C123-framework of [10].

Next, we turn our attention to P_5- and P_4-free graphs. We prove that $\mathrm{QCSP}(H)$ is in NP, for all finite graphs H, when the input is restricted to P_5-free graphs. For $\mathrm{QCSP}(K_3)$, on P_4-free graphs, we are able to improve this to a polynomial time algorithm.

Finally, we propose a certain template, the line graph of a subdivided star $L(K_{1,4}^r)$, so that $\mathrm{QCSP}(L(K_{1,4}^r))$ behaves curiously on P_m-free graphs. When m is small, this problem is solvable in polynomial time. For some medium m, it is NP-complete, while for large enough m, it is Pspace-complete.

Keywords: Quantified constraints · Forbidden subgraph · Computational complexity

1 Introduction

A fundamental result from graph theory is Brooks' Theorem which implies that all connected graphs with maximum degree 3 (we term these *subcubic*) can be properly 3-coloured, except K_4. It thus follows that 3-Colouring is solvable in polynomial time when restricted to subcubic instances. 3-Colouring may be seen as a *homomorphism* or *constraint satisfaction problem* (CSP), where the question is whether there is a homomorphism from an input graph G to K_3

© The Author(s), under exclusive license to Springer Nature Switzerland AG 2026
J. Kozik and A. Wolff (Eds.): SOFSEM 2026, LNCS 16448, pp. 447–460, 2026.
https://doi.org/10.1007/978-3-032-17801-5_33

or whether the corresponding canonical query of G is true on K_3, respectively. Throughout this article, we will use the notation of $\mathrm{CSP}(K_3)$ for this problem.

A popular generalisation of the CSP is the *quantified CSP* (QCSP) [2]. This is born out of augmenting primitive positive logic (which can express the canonical query of a structure) with universal quantification. The Boolean QCSP is better known as QUANTIFIED BOOLEAN FORMULAS (QBF) including such problems as QUANTIFIED 3-SATISFIABILITY. This latter is a protypical Pspace-complete problem (see, e.g., [14]). Indeed, if H is a finite structure, $\mathrm{QCSP}(H)$ is always in Pspace. For many years, only the complexities Pspace-complete and NP-complete were known for $\mathrm{QCSP}(H)$ above polynomial time. However, a breakthrough result uncovered various other complexities: including co-NP-complete, DP-complete and Θ_2^P-complete [18].[1]

A typical way to mitigate intractability of a problem is to introduce restrictions on the input instances, for example by omitting some kind of structural trait. Various definitions for this omitted structural trait exist in the literature, for example planar graphs can be defined as those which omit K_5 and $K_{3,3}$ as *minors*. In this article we will be most interested in structural traits involving the omission of graphs from some set $\mathcal{H}$ as minors, topological minors, induced subgraphs and (not necessarily induced) subgraphs. We term these $\mathcal{H}$-minor-free, $\mathcal{H}$-topological-minor-free, $\mathcal{H}$-free and $\mathcal{H}$-subgraph-free, respectively. Since the work of Robertson and Seymour [16], the minor-free case is the best understood. For example, if Π is a problem that is in polynomial time on graphs of bounded treewidth, yet NP-hard on planar graphs, then the complexity of Π on $\mathcal{H}$-minor-free graphs is completely classified (even when $\mathcal{H}$ is infinite). If $\mathcal{H}$ contains a planar graph, then Π on $\mathcal{H}$-minor-free graphs is solvable in polynomial time; otherwise it is NP-hard [16]. While minors derive from edge contraction, edge deletion and vertex deletion, *topological minors* derive from edge dissolution (degree 2 only), edge deletion and vertex deletion. In [10], the authors make a similar meta-classification explicit that has likely been obscured in the literature.

Theorem 1 (Theorem 2 in [10]). *Let Π be a problem that is computationally hard on planar subcubic graphs, but efficiently solvable for every graph class of bounded treewidth. For any set of graphs $\mathcal{H}$, the problem Π on $\mathcal{H}$-topological-minor-free graphs is efficiently solvable if $\mathcal{H}$ contains a planar subcubic graph (or equivalently, if the class of $\mathcal{H}$-topological-minor-free graphs has bounded treewidth) and is computationally hard otherwise.*

The frameworks reference the gap between "efficiently solvable" and "hard". Usually, but not always, this is polynomial time versus NP-complete. In fact, we will apply the previous theorem with easy meaning polynomial time but hardness rising in the polynomial hierarchy.

The principal object of study of [10] is the $\mathcal{H}$-subgraph-free case, hitherto more neglected in the literature than the $\mathcal{H}$-free case. The authors elaborate a

---

[1] It is not known if such exotic complexity classes can be achieved with graph templates.

further meta-classification that had lain implicit in the literature for years. It contains every graph problem Π satisfying the following three conditions (the operation of *edge subdivision* replaces an edge with a path of some fixed length):

C1. Π is efficiently solvable for every graph class of bounded treewidth;
C2. Π is computationally hard for the class of subcubic graphs; and
C3. Π is computationally hard under edge subdivision of subcubic graphs.

A problem Π that satisfies conditions C1–C3 is called a *C123-problem*. Let $\mathcal{S}$ be the class of disjoint unions of subdivided paths or claws.

Theorem 2 (Theorem 3 in [10]). *Let Π be a C123-problem. For any finite set of graphs $\mathcal{H}$, the problem Π on $\mathcal{H}$-subgraph-free graphs is efficiently solvable if $\mathcal{H}$ contains a graph from $\mathcal{S}$ (or equivalently, if the class of $\mathcal{H}$-subgraph-free graphs has bounded treewidth) and computationally hard otherwise.*

In [6], the authors noted the following.

Theorem 3 (Theorem 5 in [6]). *There exists some r such that $\mathrm{QCSP}(K_3)$ is Pspace-complete on the class of graphs of treewidth r.*

This means that there exists some subdivided claw H so that $\mathrm{QCSP}(K_3)$ is Pspace-complete on H-subgraph-free graphs. In this article, we continue the study on these and related problems restricted to graph classes omitting some structural traits. Owing to Theorem 3, we sometimes move to the bounded alternation QCSP in order to get stronger results.

A dominant theme in H-free algorithmics of colouring problems concerns $\mathrm{CSP}(K_n)$ on P_m-free graphs [8]. Let us note that there is not known to be an m so that $\mathrm{CSP}(K_3)$ is NP-complete on P_m-free graphs. Indeed, this problem is always quasipolynomial [15], so it is doubtful that such can exist. The situation for $\mathrm{CSP}(K_4)$ is not only different, but for this problem the dichotomy is fully classified. It was known some time ago that $\mathrm{CSP}(K_4)$ is NP-complete on P_9-free graphs [11]. This was later improved in two papers for P_8-free [3] and P_7-free [9], respectively. A similar sequence of papers tightening the polynomial time cases climaxed in [17], which showed that $\mathrm{CSP}(K_4)$ is in P on P_6-free graphs. In this paper, we are not interested in perfect boundaries. Therefore, we will use in the sequel the proof for P_9-free graphs [11], because it is relatively straightforward.

Our Results

We begin by observing the following two results.

Proposition 1. $\mathrm{QCSP}(K_3)$ *is* Pspace-*complete on subcubic planar graphs.*

Proposition 2. *For $r \geq 3$ odd, Σ_r-$\mathrm{QCSP}(K_3)$ is $\Sigma_{r-2}^{\mathrm{P}}$-hard on subcubic planar graphs.*

This allows us to derive the following through Theorem 1 together with the knowledge that bounded alternation QCSP can be solved in polynomial time on graphs of bounded treewidth [5].

Theorem 4. *For $r \geq 3$ odd, the complexity of Σ_r-QCSP(K_3) on $\mathcal{H}$-topological-minor-free graphs is completely classified (even when $\mathcal{H}$ is infinite). If $\mathcal{H}$ contains some planar subcubic graph, then Σ_r-QCSP(K_3) is solvable in polynomial time; otherwise it is $\Sigma_{r-2}^{\mathrm{P}}$-hard.*

The graphs $\mathbb{H}_i$, for $i \geq 1$, are built from two P_3s with a path of length i joining the middle vertices. The graphs H_i and C_i play a key role in the C123-framework. In particular, problems in this framework remain hard on $\{K_{1,4}, \mathbb{H}_1, \ldots, \mathbb{H}_i, C_3, \ldots, C_i\}$-subgraph-free graphs, for all $i \geq 3$. QCSP(K_3) manifests behaviours both similar and dissimilar to problems in the C123-framework, as the following results, respectively, attest.

Proposition 3. QCSP(K_3) *is* Pspace-*complete on $\mathbb{H}_1$-subgraph-free graphs.*

Let us note that CSP(K_3) is in polynomial time for $\mathbb{H}_1$-subgraph-free graphs.

Proposition 4. QCSP(K_3) *is in polynomial time on subcubic $\{\mathbb{H}_1, \mathbb{H}_2, \mathbb{H}_3, \mathbb{H}_4, \mathbb{H}_5, C_3, C_4\}$-subgraph-free graphs.*

It follows that QCSP(K_3) is not a C123-problem. The next part of the paper considers P_5- and P_4-free graphs (the latter coincides with *cographs*). We prove the following two results.

Theorem 5. *If H is a finite graph, then* QCSP(H) *restricted to inputs that are P_5-free is in* NP.

Proposition 5. QCSP(K_3) *is in polynomial time for inputs that are restricted to be P_4-free.*

In the final part of the paper, we consider a curious graph template, namely the line graph of a star on five vertices that has been subdivided $r - 1$ times $L(K_{1,4}^r)$. This graph has a K_4 in the middle with long paths of length r emanating from each of the 4 vertices (see Fig. 1). Recalling what we said earlier about the complexity of CSP(K_4) on P_m-free graphs, together with what is known about QCSPs (since at least [12]), this template has been engineered to have the following curious behaviour on P_m-free graphs, as m grows (assume in all cases $m \geq 8$).

Proposition 6. QCSP($L(K_{1,4}^r)$) *is solvable in polynomial time when restricted to P_3-free instances.*

Proposition 7. QCSP($L(K_{1,4}^r)$) *is* NP-*complete when restricted to P_{r-1}-free instances (for all $r \geq 8$).*

Proposition 8. QCSP($L(K_{1,4}^r)$) *is* Pspace-*complete when restricted to P_{2r+10}-free instances (for $r \geq 1$).*

Note that this sort of 3-way behaviour has not been observed in H-free algorithmics hitherto. We summarise the previous three results in the three regimes of Table 1.

Table 1. The three regimes of $\mathrm{QCSP}(L(K^r_{1,4}))$ for $r \geq 8$. Note that hardness propagates upwards in m while easiness propagates downwards.

m	The complexity of $\mathrm{QCSP}(L(K^r_{1,4}))$ on P_m-free graphs
≤ 3	P
$r - 1$	NP-complete
$\geq 2r + 10$	Pspace-complete

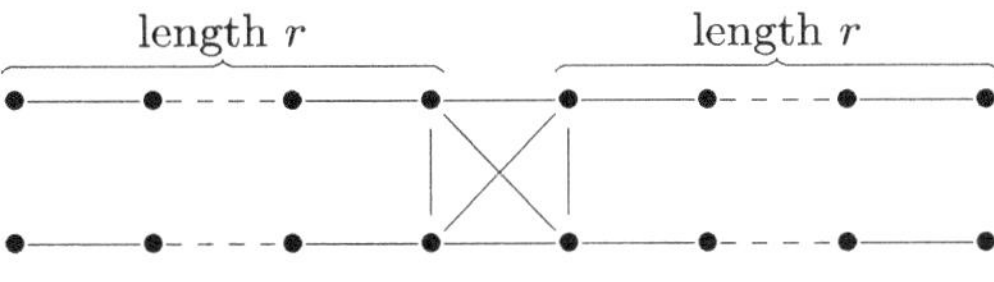

Fig. 1. The graph $L\left(K^r_{1,4}\right)$

2   Preliminaries

We denote by P_m a path with m vertices (the length of P_m is $m - 1$).

A graph is *subcubic* if its maximum degree is ≤ 3. This is true precisely if it omits $K_{1,4}$ as a subgraph. Let $K^r_{1,s}$ be built from $K_{1,s}$ by subdividing each edge $r - 1$ times. Let $\mathbb{H}_1$ be the graph on six vertices in the form of an "H". Let $\mathbb{H}_r$ be the graph $\mathbb{H}_1$ with the central, horizontal edge subdivided $r - 1$ times.

The chromatic number of a graph H is denoted by $\chi(H)$. A vertex set $X \subseteq V(H)$ is *dominating* in H if, for all $y \in V(H)$, either $y \in X$ or there exists $x \in X$ with $(x, y) \in E(H)$.

$\mathrm{QCSP}(H)$ is the problem to evaluate on H a sentence ϕ of the form

$$\forall x_1 \exists y_1 \forall x_2 \exists y_2 \ldots \forall x_n \exists y_n \Phi,$$

where Φ is a conjunction of positive atoms. H is referred to as the *template*. If the input is restricted to at most r alternations of quantifier with existential leading, we speak of Σ_r-$\mathrm{QCSP}(H)$. Note that each quantifier block may contain any number of variables (though in the example just given it is always one). In general, for example at $H = K_3$, Σ_r-$\mathrm{QCSP}(H)$ (for r odd) is complete for the complexity class Σ^{P}_r in the polynomial hierarchy [2].

A solution to an instance ϕ of $\mathrm{QCSP}(H)$ is therefore a sequence of (Skolem) functions $f_1, \ldots, f_n$ such that

$$(x_1, f(x_1), x_2, f_2(x_1, x_2), \ldots, x_n, f_n(x_1, \ldots, x_n))$$

is a solution of Φ for all $x_1, \ldots, x_n$ (i.e. $y_i = f_i(x_1, \ldots, x_i)$). This belies a (Hintikka) game semantics for the truth of a QCSP instance in which a player called Universal (male) plays the universal variables and a player called Existential (female) plays the existential variables, one after another, from the outside in. The Skolem functions above give a strategy for Existential. In our proofs we may revert to a game-theoretical parlance. We may refer to this game as the (ϕ, H) game, when we want to make instance and template explicit. Whilst the input for, e.g., QCSP(K_3) is a sentence ϕ, we often refer to the variables of ϕ as vertices in the underlying graph $G(\phi)$. We may also refer to elements of the domain $\{0, 1, 2\}$ of K_3 as colours, in deference to CSP(K_3), a.k.a. 3-COLOURABILITY.

A function $f : V(G) \to V(H)$ is a *homomorphism* if, for each $(x, y) \in E(G)$, $(f(x), f(y)) \in E(H)$.

3   QCSP(K_3)

It is known that CSP(K_3) can be solved in polynomial time for subcubic graphs [4] but is NP-complete in general for planar graphs [7]. To the best of our knowledge, it has not been noted that QCSP(K_3) is Pspace-complete on planar graphs. Moreover, the best-known proof that QCSP(K_3) is Pspace-complete, in [2], uses a reduction from QUANTIFIED NOT-ALL-EQUAL 3-SATISFIABILITY. This is problematic as NOT-ALL-EQUAL 3-SATISFIABILITY is solvable in polynomial time on planar instances [13]. However, one can prove QCSP(K_3) is Pspace-complete on planar graphs by self-reduction, by replacing crossing over edges in a non-planar representation with the original crossover widget (gadget) from Fig. 2 in [7].

Proposition 1. QCSP(K_3) is Pspace-complete on subcubic planar graphs.

Proof. We argue by self-reduction from QCSP(K_3) on planar graphs, that is we reduce a planar instance ϕ of QCSP(K_3) of arbitrary degree to a subcubic instance ϕ' of QCSP(K_3). For each vertex x of degree > 3, we replace x by the vertex gadget drawn in Fig. 2. We follow the quantifier order of the vertices of ϕ. Existential vertices x of ϕ become existential vertices $\exists x_1 \ldots \exists x_{d(x)}$ of ϕ' (where $d(x)$ is the degree of x). Universal vertices x of ϕ become vertices $\forall x_1 \exists x_2 \ldots \exists x_{d(x)}$ in ϕ'. The remaining vertices are quantified innermost: firstly the universal vertices labelled that way in the gadget, and then the (existential) bullet vertices of the gadget.

The reduction works in the following fashion. Suppose some two adjacent (via a $\bullet$) x_i and x_{i+1} in a vertex gadget are not played to the same colour i, but are rather played to i and j. Let k be the unique colour in $\{0, 1, 2\} \setminus \{i, j\}$. If universal plays the vertex labelled $\forall$ in between and above x_i and x_{i+1} (see Fig. 2) as k, then existential has no move for the vertex labelled $\bullet$ adjacent to these three vertices. On the other hand, if x_i and x_{i+1} are both played to colour i, then no matter what universal plays to the vertex labelled $\forall$, there will be a response for existential on the vertex labelled $\bullet$. Thus, the vertex gadget is both sound and complete for the reduction it is used in.

The same method will allow us to derive the following refinement of the previous proposition.

Proposition 2. For $r \geq 3$ odd, Σ_r-QCSP(K_3) is Σ^P_{r-2}-hard on subcubic planar graphs.

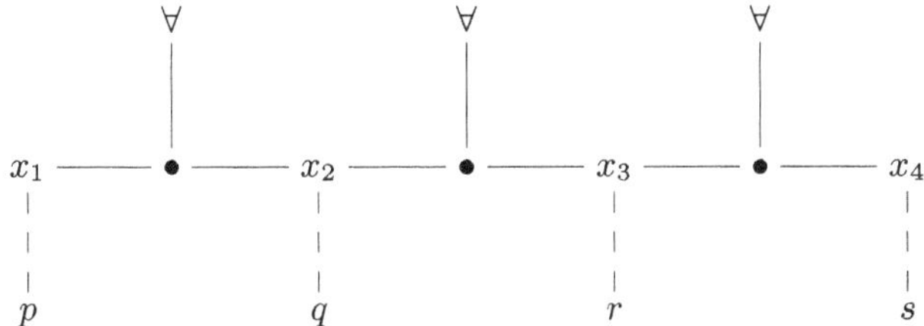

Fig. 2. Gadget for the subcubic self-reduction of Propositions 1 and 2. Dashed edges represent edges in the original graph. One could attach two such edges at both x_1 and x_4 but we have drawn only one (i.e. simulating degree 4). Universal vertices (labelled with $\forall$) are quantified innermost, except for vertices labelled with the bullet ($\bullet$) which are existentially quantified after. This innermost Π_2 quantification ensures that existential played x_1, x_2, x_3, x_4 all to the same colour.

Proof. We use the construction of the previous proof, noting that it maps Σ_{r-2} instances to Σ_r instances. The proof concludes by using the Σ^P_{r-2}-hardness of Σ_{r-2} instances of QCSP(K_3) arising from the preservation of the quantifier structure in [2].

Proposition 3. QCSP(K_3) is Pspace-complete on $\mathbb{H}_1$-subgraph-free graphs.

Proof. We argue by self-reduction, that is we reduce an instance ϕ of QCSP(K_3) to another instance ϕ' of QCSP(K_3) that omits $\mathbb{H}_1$ as a subgraph. Let x be a vertex in ϕ. We replace x by the gadget drawn in Fig. 3. We follow the quantifier order of the vertices of ϕ. Existential vertices x of ϕ become existential vertices $\exists x_1 \ldots \exists x_{d(x)}$ of ϕ' (where $d(x)$ is the degree of x). Universal vertices x of ϕ become vertices $\forall x_1 \exists x_2 \ldots \exists x_{d(x)}$ in ϕ'. The remaining vertices are quantified as follows. The diamond vertices are quantified existentially after their associated x_i. Then all other vertices are quantified innermost: firstly the universal vertices labelled that way in the gadget, and then the (existential) bullet vertices of the gadget. The proof concludes as that for Proposition 1.

3.1   A Polynomial Case

In this subsection we will give some propositions whose proofs are deferred to the full version of the paper.

Proposition 9. *Let ϕ be an input for subcubic* QCSP(K_3) *whose graph ϕ_G is a tree in which at most one vertex has degree 3. Then determining if $K_3 \models \phi$ is in* P.

Proof. Check if ϕ contains any of the finite set of minimal no-instances from Σ (given in the full version of the paper). Reject if it does; else accept.

Proposition 10. *Let ϕ be an input for subcubic* QCSP(K_3) *whose graph ϕ_G is such that*

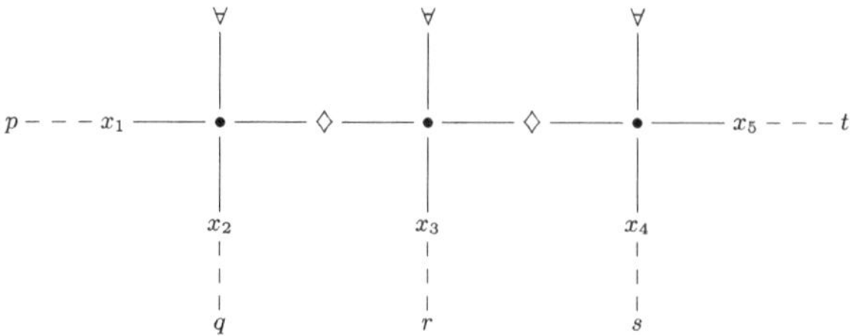

Fig. 3. Gadget for the H_1-subgraph-free reduction of Proposition 3. Dashed edges represent edges in the original graph. Universal vertices (labelled with $\forall$) are quantified innermost, except for vertices labelled with the bullet ($\bullet$) which are existentially quantified after. This quantification ensures that Existential played x_1, x_2, x_3, x_4, x_5 and the two vertices labelled diamond ($\Diamond$) all to the same colour.

(i). all vertices of degree 3 are at distance > 5; and
(ii). it contains no induced subgraphs C_3 or C_4.

Then determining if $K_3 \models \phi$ is in P.

The main result of this section, which we now recall, follows directly.

Proposition 4. QCSP(K_3) is in polynomial time on subcubic $\{\mathbb{H}_1, \mathbb{H}_2, \mathbb{H}_3, \mathbb{H}_4, \mathbb{H}_5, C_3, C_4\}$-subgraph-free graphs.

Proof. Since the graph underlying the input ϕ is subcubic $\{\mathbb{H}_1, \mathbb{H}_2, \mathbb{H}_3, \mathbb{H}_4, \mathbb{H}_5, C_3, C_4\}$-subgraph-free we can deduce that all vertices of degree 3 are at distance > 5. We now apply Proposition 10.

4 P_5- and P_4-Free Graphs

Lemma 1. *Let H be a loopless finite graph and ϕ a connected instance of* QCSP(H). *If ϕ is a P_5-free yes-instance, then the number of universal quantified variables in ϕ is at most $\chi(H)(\chi(H) - 1)$.*

Proof. Let G be the primal graph of ϕ, and k the chromatic number of H. Since G is P_5-free, it follows from Theorem 8 in [1], that G contains either a dominating clique, or a dominating P_3. Also, ϕ is a yes-instance of QCSP(K_k), and so, G contains a dominating set on at most k vertices. Finally, every vertex in G has at most $k - 1$ neighbours representing a universally quantified variable (otherwise, ϕ can easily be seen to be a no-instance of QCSP(H)). Putting both observations together, we conclude that ϕ has at most $k(k-1)$ universally quantified variables.

Applying the previous lemma, we arrive at the following, which we reproduce from the introduction.

Theorem 5. *If H is a finite graph, then* QCSP(H) *restricted to inputs that are P_5-free is in* NP.

Proof. According to the previous lemma we may assume that the number of universal variables in each connected component of $G(\phi)$ is bounded above by $\chi(H)(\chi(H)-1)$. We may now check all possibilities for these variables by brute force, while using an NP oracle for the existential variables.

Consider a quantified conjunctive positive formula ϕ with primal graph G. We say that a vertex v of G is a *universal dominating* vertex, if for every universally quantifier variable u of ϕ, there is an edge (u,v) in G.

Lemma 2. *Let ϕ be a quantified conjunctive positive formula of graphs. If the primal graph of G is a connected cograph, and the set of universal vertices is an independent set, then G contains a universal dominating vertex.*

Proof. Since G is a connected cograph, then $V(G)$ admits a partition (X,Y) such that every pair of vertices $x \in X$ and $y \in Y$ are adjacent. Using the fact that there are no pair of adjacent universal vertices, we assume without loss of generality that the set U of vertices representing a universally quantified is contained in X. Since every $y \in Y$ is adjacent to all vertices in U, we conclude that any vertex $y \in Y$ is a universal dominating vertex.

Given a positive integer N, we say that a graph G is N-*dominating* if for every subset of vertices U with $|U| \leq N$, there is a vertex v such that $U \subseteq N(v)$. For instance, a graph G is not 1-dominating if and only if G contains an isolated vertex, and clearly, if G is N-dominating and $N \geq 2$, then G is $(N-1)$-dominating. For convenience, we say that every graph is 0-dominating.

Lemma 3. *Let H be a finite graph, and let N be the maximum integer such that H is N-dominating. If ϕ is a P_4-free connected yes-instance, then the number of universal quantified variables in ϕ is at most N.*

Proof. Let G be the primal graph of ϕ, and let V' be any subset of vertices of H with as many vertices as quantified variables in ϕ. By evaluating the universally quantified variables of G with distinct elements of V', we conclude via Lemma 2 that H must contain a vertex v that dominates V'.

The (general) diamond contraction rule on a graph H with diamond D on $\{a,b,c,d\}$ with central edge (b,c) identifies vertices a and d as a new vertex ad whose neighbours are all of the vertices that were neighbours of a or d. The remainder of the proofs in this section are deferred to the full version of the paper.

Lemma 4. *Let H be a graph (containing a diamond but not a K_4) and let H' be the graph after contracting this diamond. If H is P_4-free then H' is P_4-free.*

Call a graph *star-cross'd* if it is built from a collection of K_2s and K_3s by identifying them all on some chosen vertex. See Fig. 4 for an example.

Lemma 5. *Let H be a graph without K_4 that is not complete bipartite or star-cross'd and is diamond-free. H is P_4-free iff H is P_4-subgraph-free.*

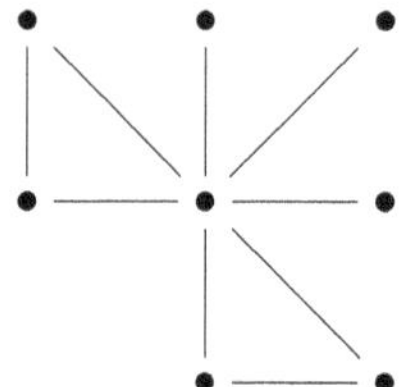

Fig. 4. An example of a star cross'd graph.

Lemma 6. *ϕ is a yes-instance of* QCSP(K_3) *iff ϕ_D is a yes-instance of* QCSP(K_3).

We can now prove (see the full version of the paper) the main result of this section, which we now recall.

Proposition 5. QCSP(K_3) is in polynomial time for inputs that are restricted to be cographs.

5   The Line Graph of $K_{1,4}^r$

Let us recall the following propositions from the introduction.

Proposition 6. QCSP($L(K_{1,4}^r)$) is solvable in polynomial time when restricted to P_3-free instances.

Proof. Since the graph underlying the input ϕ is P_3-free, every connected component is a clique. Let us deal with the connected components individually. If there are two universal variables, or a universal variable is following an existential variable, then this will be a no-instance. If the clique is bigger than 4 it will also be a no-instance. Suppose the clique is of size 1 or 2, then this is a yes-instance. If it is of size 3 or 4, then it is a yes-instance, if $r = 0$, and otherwise it is a no-instance (as one vertex may be evaluated to a leaf in $(K_{1,4}^r)$. If there are no universal variables, then it is a yes-instance iff the clique is of size at most 4. Discerning these cases can clearly be done in polynomial time, and the result follows.

Proposition 7. QCSP($L(K_{1,4}^r)$) is NP-complete when restricted to P_{r-1}-free instances (for all $r \geq 8$).

Proof. It suffices to prove that the problem is in NP, since on purely existential instances CSP($L(K_{1,4}^r)$) = CSP(K_4) which is NP-complete by [9]. Let ϕ be a connected instance involving a universal variable x. If ϕ involves more than one universal variable then, since ϕ is P_{r-1}-free, it is a no-instance (imagine evaluating two different universal variables to two different leaves in $L(K_{1,4}^r)$).

We argue that $L(K_{1,4}^r) \models \phi$ iff $P_{r-1} \models \phi$. For the forward direction, imagine that the universal variable is evaluated to a leaf node in $L(K_{1,4}^r)$. Then the whole of the component must be evaluated to the path of that leaf, because it is of bounded diameter (P_{r-1}-free). It follows that this component must be bipartite. It therefore follows that $P_{r-1} \models \phi$. For the backward direction, if $P_{r-1} \models \phi$, and ϕ is connected, then ϕ may have only one universal variable and this must be first in the quantifier order. For every $y \in V(L(K^r_{1,4}))$ and $x \in V(P_{r-1})$ there exists a homomorphism f from P_{k-1} to $L(K^r_{1,4})$ so that $f(x) = y$, therefore we find $L(K^r_{1,4}) \models \phi$.

It follows that we may reduce $\mathrm{QCSP}(L(K^r_{1,4}))$ to $\mathrm{CSP}(L(K^r_{1,4}))$ on P_{r-1}-free instances. W.l.o.g. we assume the input ϕ to be connected. If ϕ has more than one universal variable, we reduce to a fixed no-instance (e.g. K_5). If ϕ has one universal variable, then we evaluate ϕ as an instance of the tractable $\mathrm{QCSP}(P_{k-1}) = \mathrm{QCSP}(K_2)$ (see [12]). If it is a no-instance, we reduce to K_5, else we reduce to (e.g.) K_2. If ϕ has no universal variables, then we apply the identity reduction.

Let us name the vertices of K_4 as $\{0, 1, 2, 3\}$. Then let us name the vertices in $L(K^r_{1,4})$ on the paths emanating from i in the K_4 as $i = i(0), i(1), \ldots, i(r)$. The following is easy to verify.

Lemma 7. *In $L(K^r_{1,4})$, there is a path of length $r+1$ from every vertex to $\{0, 1\}$. Further, there is a path of length $r + 1$ from $0(r)$ to 1 but not to 0, and a path of length $r + 1$ from $1(r)$ to 0 but not to 0.*

We now set about describing a reduction from *Quantified 3-Satisfiability* (Q3SAT) to $\mathrm{QCSP}(L(K^r_{1,4}))$. Consider an instance $\phi := \exists x_1 \forall y_1 \ldots \exists x_n \forall y_n \Phi(\boldsymbol{x}, \boldsymbol{y})$ of Q3SAT (where we assume w.l.o.g. that the first variable is existential) where $\Phi(\boldsymbol{x}, \boldsymbol{y}) = (C_1 \wedge C_2 \wedge \cdots \wedge C_m)$, $C_i = (l_{i1} \vee l_{i2} \vee l_{i3})$, $l_{ij} \in (X \cup \overline{X}) \cup (Y \cup \overline{Y}), 1 \leq j \leq 3$). We will produce an instance

$$\psi := \exists x_1, \overline{x_1} \forall y_1^{r+1} \exists y_1^r \ldots y_1^1 y_1 \overline{y}_1 \ldots \ldots \exists x_n, \overline{x_n} \forall y_n^{r+1} \exists y_n^r \ldots y_n^1 y_n \overline{y}_n \exists z \Psi(\boldsymbol{x}, \boldsymbol{y}, \boldsymbol{z})$$

of $\mathrm{QCSP}(L(K^r_{1,4}))$. We borrow heavily from Sect. 2 in [11].[2] The reduction goes as follows.

For each existential variable $x \in X$, let $G_1(r)$ denote the P_2 with vertices $x, \bar{x}$. These are the $\boldsymbol{x}$ variables in ϕ.

For each universal variable $y \in Y$, let $G_2(y)$ denote the P_{r+3} with vertices $y^{r+1}, y^r, \ldots, y^1, y, \bar{y}$. These are the $\boldsymbol{y}$ variables in ϕ.

The graph $G(C_i)$ is the ten-vertex clause graph depicted in Fig. 5 (from [11]).

The graph G is obtained from all $G(x)$, all $G(y)$, all $G(C_i)$, and the dummy vertex d by adding the following additional edges:

- $\{c_i, l\}, 1 \leq i \leq m, l \in (X \cup \overline{X}) \cup (Y \cup \overline{Y})$
- $\{b_{ij}, l\}, 1 \leq i \leq m, 1 \leq j \leq 3$, and $l \in (X \cup \overline{X}) \cup (Y \cup \overline{Y})$
- $\{d, v\}, v \in G(C_i), 1 \leq i \leq m$
- $\{c_{ij}, l\}$ for all $1 \leq i \leq m, 1 \leq j \leq 3$, and $l \in (X \cup \overline{X}) \cup (Y \cup \overline{Y})$ with $l_{ij} = l$ (where l_{ij} are the literals of Q3SAT).

---

[2] The notation is also from [11]. For example, $\bar{x}$ is a variable distinct from x, so vector notation $\boldsymbol{x}$ indicates tuples rather than overline.

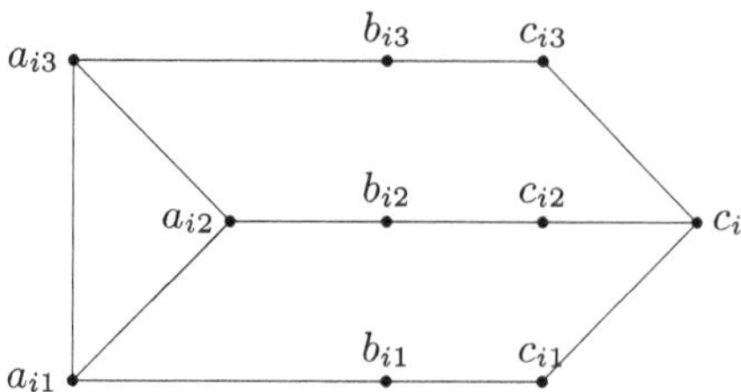

Fig. 5. The ten-vertex clause gadget $G(C_i)$

Note that z contains the dummy variable d and also all the new vertices associated with clause gadgets. We now give another supporting lemma, whose proof we defer to the full version of the paper.

Lemma 8. *Let ψ be a sentence built according to the reduction just specified, and Ψ be its quantifier-free part. Let $T = (X \cup \overline{X}) \cup (Y \cup \overline{Y})$ be the set of variables in ψ of the form $x_i, \overline{x_i}, y_i, \overline{y_i}$ (there are $2n$ such variables). In any homomorphism h from $G(\Phi)$ to $L(K_{1,4}^r)$, $h(T) = \{i, j\}$ for some $i \neq j \in \{0, 1, 2, 3\}$.*

Let us recall the following result from the introduction, whose proof we defer to the full version of the paper.

Proposition 8. QCSP($L(K_{1,4}^r)$) is Pspace-complete when restricted to P_{2r+10}-free instances (for $r \geq 1$).

6   Final Remarks

The principal aim of this paper was to compare CSP(K_3) and QCSP(K_3) through the lens of H-free and H-subgraph-free algorithmics. Since QCSP(K_3) is Pspace-hard on some class of bounded treewidth, we restricted to Σ_r-QCSP(K_3) to obtain stronger results. Unlike CSP(K_3), Σ_r-QCSP(K_3) already achieves hardness on subcubic instances, as well as those that are $\mathbb{H}_1$-subgraph-free. The outstanding open line of research from this paper is to classify fully QCSP(K_3) on H-subgraph-free graphs. For the reasons already mentioned, it is likely to be easier to do this for Σ_r-QCSP(K_3). Though Σ_r-QCSP(K_3) has much in common with C123-problems, we proved (in Proposition 4) that it is not a C123-problem.

A second line of research is to study QCSP(H) on P_4-free graphs (cographs). Though we proved this problem to be in NP, it may be possible to improve this to polynomial time.

A final line of enquiry asks for tight bounds on the transition between P, NP-complete and Pspace-complete for the problem QCSP($L(K_{1,4}^r)$).

Acknowledgments. Kristina Asimi and Barnaby Martin were funded by EPSRC Grant EP/X03190X/1.

Santiago Guzmán-Pro was funded by the European Research Council (Project POCOCOP, ERC Synergy Grant 101071674). Views and opinions expressed are however those of the authors only and do not necessarily reflect those of the European

Union or the European Research Council Executive Agency. Neither the European Union nor the granting authority can be held responsible for them.

We are grateful for corrections from Jan Bok and Nikola Jedličková as well as feedback from several anonymous reviewers.

Disclosure of Interests. The authors have no competing interests to declare that are relevant to the content of this article.

References

1. Bacsó, G., Tuza, Z.S.: Dominating Cliques in P_5-free Graphs. Period. Math. Hung. **21**, 303–308 (1990)
2. Börner, F., Bulatov, A.A., Chen, H., Jeavons, P., Krokhin, A.A.: The complexity of constraint satisfaction games and QCSP. Inf. Comput. **207**(9), 923–944 (2009)
3. Broersma, H., Golovach, P.A., Paulusma, D., Song, J.: Updating the complexity status of coloring graphs without a fixed induced linear forest. Theoret. Comput. Sci. **414**(1), 9–19 (2012)
4. Brooks, R.L.: On colouring the nodes of a network. Math. Proc. Cambridge Philos. Soc. **37**(2), 194–197 (1941)
5. Chen, H.: Quantified constraint satisfaction and bounded treewidth. In: de Mántaras, R.L., Saitta, L. (eds.) Proceedings of the 16th Eureopean Conference on Artificial Intelligence, ECAI 2004, including Prestigious Applicants of Intelligent Systems, PAIS 2004, Valencia, Spain, 22–27 August 2004, pp. 161–165. IOS Press (2004)
6. Eagling-Vose, T., Martin, B., Paulusma, D., Smith, S.: Graph homomorphism, monotone classes and bounded pathwidth. In: 20th Conference on Computability in Europe, CiE 2016, Amsterdam, Netherlands, 8–12 July 2024, Proceedings (2024)
7. Garey, M.R., Johnson, D.S., Stockmeyer, L.: Some simplified NP-complete graph problems. Theoret. Comput. Sci. **1**(3), 237–267 (1976)
8. Golovach, P.A., Johnson, M., Paulusma, D., Song, J.: A survey on the computational complexity of coloring graphs with forbidden subgraphs. J. Graph Theory **84**(4), 331–363 (2017)
9. Huang, S.: Improved complexity results on k-coloring P_t-free graphs. Eur. J. Comb. **51**, 336 346 (2016)
10. Johnson, M., et al.: Complexity framework for forbidden subgraphs I: the framework. Algorithmica **87**(3), 429–464 (2025)
11. Randerath, B., Schiermeyer, I.: On the complexity of 4-coloring graphs without long induced paths. Theor. Comput. Sci. **389**(1-2), 330–335 (2007)
12. Martin, B., Madelaine, F.R.: Towards a trichotomy for quantified H-coloring. In: Beckmann, A., Berger, U., Löwe, B., Tucker, J.V. (eds.) Logical Approaches to Computational Barriers, Second Conference on Computability in Europe, CiE 2006, Swansea, UK, June 30-July 5, 2006, Proceedings, volume 3988 of *Lecture Notes in Computer Science*, pp. 342–352. Springer (2006)
13. Moret, B.M.E.: Planar NAE3SAT is in P. SIGACT News **19**(2), 51–54 (1988)
14. Christos, H.: Papadimitriou. Computational Complexity. Addison-Wesley (1994)
15. Pilipczuk, M., Pilipczuk, M., Rzążewski, P.: Quasi-polynomial-time algorithm for independent set in P_t-free graphs via shrinking the space of induced paths. In: Le, H.V., King, V. (eds.) 4th Symposium on Simplicity in Algorithms, SOSA 2021, Virtual Conference, 11–12 January 2021, pp. 204–209. SIAM (2021)

16. Robertson, N., Seymour, P.D.: Graph minors. V. excluding a planar graph. J. Comb. Theory B **41**(1), 92–114 (1986)
17. Spirkl, S., Chudnovsky, M., Zhong, M.: Four-coloring P_6-free graphs. In: Chan, T.M. (ed.) Proceedings of the Thirtieth Annual ACM-SIAM Symposium on Discrete Algorithms, SODA 2019, San Diego, California, USA, 6–9 January 2019, pp. 1239–1256. SIAM (2019)
18. Zhuk, D., Martin, B.: QCSP monsters and the demise of the Chen conjecture. J. ACM **69**(5), 35:1–35:44 (2022)

Bounds and Hardness Results for Conflict-Free Choosability

Shiwali Gupta$^{(\boxtimes)}$ and Rogers Mathew$^{(\boxtimes)}$ iD

Department of Computer Science and Engineering, Indian Institute of Technology, Hyderabad, Hyderabad, India
`{cs21resch11002,rogers}@iith.ac.in`

Abstract. A *(partial) conflict-free coloring* of a hypergraph $\mathcal{H}$ is an assignment of colors to (a subset of) the vertex set of $\mathcal{H}$ such that every hyperedge in $\mathcal{H}$ has a vertex whose color is distinct from every other vertex in that hyperedge. The minimum number of colors required for such a coloring is known as the *(partial) conflict-free chromatic number* of $\mathcal{H}$. It is easy to see that the conflict-free chromatic number of a hypergraph is at most its partial conflict-free chromatic number plus one. Conflict-free coloring has also been studied on the open/closed neighborhood hypergraphs of a given graph under the name open/closed neighborhood conflict-free coloring. In this paper, we study partial and full list variants of conflict-free coloring where, for every vertex v, we are given a list of admissible colors L_v such that v is allowed to be colored only from L_v.

It was shown by Pach and Tardos [Combinatorics, Probability and Computing, 2009] that for any constant $\epsilon > 0$, the closed-neighborhood conflict-free chromatic number of a graph G is at most $O(\ln^{2+\epsilon} \Delta)$, where Δ represents the maximum degree of G. Later, Glebov, Szabó, and Tardos [Combinatorics, Probability and Computing, 2014] showed that there exist graphs G that require $\Omega(\ln^2 \Delta)$ colors for a closed neighborhood conflict-free coloring. Bhyravarapu, Kalyanasundaram, and Mathew [Journal of Graph Theory, 2021] bridged the gap between the upper and the lower bound. They showed that the closed-neighborhood conflict-free chromatic number of any graph G is at most $O(\ln^2 \Delta)$.

In this paper, we extend the $O(\ln^2 \Delta)$ upper bound to the partial list variant of the closed-neighborhood conflict-free chromatic number. Further, we establish computational complexity results concerning the list open/closed-neighborhood conflict-free chromatic numbers.

Keywords: Conflict-free coloring · list conflict-free coloring · conflict-free choosability · choice number · maximum degree · minimum degree · hardness

1 Introduction

1.1 Conflict-Free Coloring

A *partial conflict-free coloring* (or *CF* coloring*) of a hypergraph $\mathcal{H} = (V, \mathcal{E})$ using k colors is an assignment $f : V' \rightarrow \{1, \ldots, k\}$, where $V' \subseteq V$, such that

© The Author(s), under exclusive license to Springer Nature Switzerland AG 2026
J. Kozik and A. Wolff (Eds.): SOFSEM 2026, LNCS 16448, pp. 461–474, 2026.
https://doi.org/10.1007/978-3-032-17801-5_34

every $E \in \mathcal{E}$ contains a point whose color is distinct from that of every other point in E. If $V' = V$, then we call f a *conflict-free coloring* (or *CF coloring*) of $\mathcal{H}$. The minimum k for which there is a (partial) CF coloring of $\mathcal{H}$ using k colors is called the (resp., *partial*) *conflict-free chromatic number* of $\mathcal{H}$. We shall use $\chi_{\mathrm{ON}}(\mathcal{H})$ (resp., $\chi^*_{ON}(\mathcal{H})$) to denote (resp., partial) conflict-free chromatic number of $\mathcal{H}$. Giving a new unused color to all the uncolored vertices in a partial CF coloring of $\mathcal{H}$ gives a CF coloring of $\mathcal{H}$. The observation below captures this.

Observation 1. $\chi_{\mathrm{ON}}(\mathcal{H}) \leq \chi^*_{ON}(\mathcal{H}) + 1$.

The notion of conflict-free coloring is well studied with respect to *open/closed neighborhood hypergraphs of graphs* (the edge set of such a hypergraph is the set of open/closed neighborhoods of each vertex of the graph under consideration). Let $V(G)$ and $E(G)$ denote the vertex set and edge set of a graph G, respectively. For a vertex $v \in V(G)$, the set of neighbors of v in G is called the *open neighborhood* of v. It is denoted by $N_G(v)$. The *closed neighborhood* of v is defined as $\{v\} \cup N_G(v)$. It is denoted by $N_G[v]$.

A *partial conflict-free open neighborhood coloring* (or *CFON* coloring*) of G is an assignment of colors to $V' \subseteq V(G)$ such that every vertex in G sees a uniquely colored vertex in its open neighborhood. If $V' = V(G)$, then we call it a *CFON coloring*. The minimum number of colors required for a CFON* coloring (resp., CFON coloring) of G is called the *CFON* chromatic number* (resp., *CFON chromatic number*) of G, denoted by $\chi^*_{\mathrm{ON}}(G)$ (resp., $\chi_{\mathrm{ON}}(G)$). Analogously, by replacing 'open neighborhood' with 'closed neighborhood' in the above definitions, we define CFCN* coloring, CFCN coloring, CFCN* chromatic number (denoted $\chi^*_{\mathrm{CN}}(G)$), and CFCN chromatic number (denoted $\chi_{\mathrm{CN}}(G)$).

Observation 1 implies that for any graph G, $\chi_{\mathrm{ON}}(G) \leq \chi^*_{\mathrm{ON}}(G) + 1$ and $\chi_{\mathrm{CN}}(G) \leq \chi^*_{\mathrm{CN}}(G) + 1$.

The following inequality connects the CF open and closed chromatic numbers of a graph with each other.

Proposition 1 (Inequality 1.3 in [20]). *(i)* $\chi_{\mathrm{CN}}(G) \leq 2\chi_{\mathrm{ON}}(G)$, *and (ii)* $\chi^*_{\mathrm{CN}}(G) \leq 2\chi^*_{\mathrm{ON}}(G)$.

The example given below shows that there are graphs whose CFON chromatic number is arbitrarily large compared to its CFCN chromatic number.

Example 1 ([20]). Let K_n denote the complete graph on n vertices. Let $K_n^{1/2}$ denote the graph obtained by subdividing every edge of K_n exactly once. Then, it is known that (i) $\chi_{\mathrm{ON}}(K_n^{1/2}) = n$, and (ii) $\chi_{\mathrm{CN}}(K_n^{1/2}) = 2$.

Partial conflict-free coloring on open/closed neighborhood is NP-complete for planar [1] and chordal graphs [3,6], efficiently solvable for outerplanar [1], and interval graphs [3,5]. It was shown in [12] that deciding whether a graph has a conflict-free coloring with two colors is NP-complete for both closed and open neighborhoods. Moreover, CFCN variant is hard to approximate within a factor less than $3/2$, unless $\mathsf{P} = \mathsf{NP}$. The CFON variant cannot be approximated within a factor $n^{1/2-\varepsilon}$ for any $\varepsilon > 0$, unless $\mathsf{P} = \mathsf{NP}$.

Conflict-free coloring was first studied by Even et al. [11] in 2004 and was further studied in many contexts (see [2,4,5,7,9,10,13,17,19–21]). See Smorodinsky [22] for a survey on conflict-free coloring and its applications. Conflict-free coloring has found applications in frequency assignment problems for cellular networks, in battery consumption-related optimization problems in sensor networks, in vertex ranking problems that find applications in VLSI design and operations research, in pliable index coding problem in coding theory [18] etc.

The study on conflict-free coloring was originally motivated by its application in the frequency assignment problem for cellular networks. There are two types of nodes in a cellular network: base stations and mobile clients. Base stations serve as the network's backbone and have a fixed position. Mobile clients are connected to the base station. The client and base station are connected by a radio link. Each base station has a fixed frequency to transmit or to enable a radio link for the clients. Clients are continuously scanning the frequencies in search of base stations with a good range. When a client is in the range of more than one base station, then mutual interference occurs if the same frequency is assigned to these base stations which will make the links noisy. The primary challenge in the frequency assignment problem in cellular networks is allocating frequencies to base stations in a way that ensures each client is served by some base station whose frequency is distinct from that of the other base stations in its range. The goal is to minimize the number of assigned frequencies because frequencies are expensive and limited; it is preferable to have a scheme that reuses frequencies wherever possible.

We can formulate the frequency assignment problem as a hypergraph coloring problem. Let $\mathcal{H} = (V, \mathcal{E})$ be a hypergraph, where vertices are the given set of base stations and each hyperedge denotes the set of base stations in the range of each client. Thus, we try to find the minimum number of colors required to color the vertices such that every hyperedge has a vertex with a color distinct from the color of all the other vertices in it. Research on conflict-free coloring was performed under the assumption that any color from a global range of colors can be used and the goal was to minimize the number of colors used. Assume that the wireless network's base stations are further limited to using a subset of the available frequencies. As a result, different base stations may have access to different subsets of frequencies. Studying the list variant of conflict-free coloring is relevant in this context. In this variant, each vertex has a list of colors attached to it. Each vertex receives a color from its list. In the next section, we formally define this notion.

1.2   List Conflict-Free Coloring

Let $\mathcal{H} = (V, \mathcal{E})$ be a hypergraph, and let k be a positive integer. Let $\mathcal{L} = \{L_v : v \in V\}$, where each L_v is a list of colors.

Definition 1 (k-assignment). *Let $\mathcal{L} = \{L_v : v \in V\}$ denote an assignment of a list of admissible colors to each vertex of $\mathcal{H}$. We say $\mathcal{L}$ is a k-assignment for $\mathcal{H}$ if $|L_v| = k$, for every $v \in V$.*

Definition 2 (k-CF*-choosable, k-CF-choosable). *Given a list assignment* $\mathcal{L} = \{L_v : v \in V\}$, *we say that* $\mathcal{H}$ *admits an* $\mathcal{L}$-CF*-*coloring if there exists a coloring* $f : V' \to \bigcup_{v \in V'} L_v$, *where* $V' \subseteq V$, *such that* $f(v) \in L_v$, $\forall v \in V'$, *and every hyperedge* E *in* $\mathcal{H}$ *contains a point whose color is distinct from that of every other point in* E. *When* $V' = V$, *we say* $\mathcal{H}$ *admits an* $\mathcal{L}$-CF-*coloring. We say that* $\mathcal{H}$ *is* k-CF*-*choosable (resp., k-CF-choosable) if for every k-assignment* $\mathcal{L}$, $\mathcal{H}$ *admits an* $\mathcal{L}$-CF*-*coloring (resp., $\mathcal{L}$-CF-coloring).*

Definition 3 (CF* choice number, CF choice number). *The minimum k for which* $\mathcal{H}$ *is* k-CF*-*choosable (resp., k-CF-choosable) is called the* CF* *choice number (resp.,* CF *choice number) of* $\mathcal{H}$. *This is denoted by* $ch^*_{\mathrm{ON}}(\mathcal{H})$ *(resp.,* $ch_{\mathrm{ON}}(\mathcal{H})$).

The following theorem due to Cheilaris, Smorodinsky, and Sulovský [8] gives us a relation between the CF chromatic number and CF choice number of a hypergraph $\mathcal{H}$.

Theorem 1. *[8] For any hypergraph* $\mathcal{H}$ *with n vertices,*

(i) $\chi_{\mathrm{CF}}(\mathcal{H}) \leq ch_{\mathrm{ON}}(\mathcal{H}) \leq \chi_{\mathrm{CF}}(\mathcal{H}) \cdot \ln n + 1$, *and*

(ii) $\chi^*_{\mathrm{ON}}(\mathcal{H}) \quad \leq \quad ch^*_{\mathrm{ON}}(\mathcal{H}) \quad \leq \quad \chi^*_{\mathrm{ON}}(\mathcal{H}) \cdot \ln n_1 + 1$, *where* $n_1 = \min\{$*no. of colored vertices under* $f : f\,is\,a\,CF^*\,coloring\,of\,\mathcal{H}\,using\,only$ $\chi^*_{\mathrm{ON}}(\mathcal{H})$ *no. of colors*$\}$.

Similar to CFON and CFCN coloring, we define their list analogues below. Before we go into their definitions, let us look at the definition of the choice number of a graph G, denoted by $ch(G)$. Given $\mathcal{L} = \{L_v : v \in V(G)\}$, we say G admits $\mathcal{L}$-*coloring*, if there exists a proper coloring $f : V(G) \to \bigcup_{v \in V(G)} L_v$ such that $f(v) \in L_v$, $\forall v \in V(G)$. We say G is k-*choosable* if G admits an $\mathcal{L}$-coloring, for all k-assignments $\mathcal{L}$. The minimum k for which G is k-choosable is called the *choice number* of G.

Definition 4 (CFON* choice number, CFON choice number). *Given a list assignment* $\mathcal{L} = \{L_v : v \in V(G)\}$ *for a graph* G, *we say* G *is* $\mathcal{L}$-CFON*-*colorable if there exists a CFON* coloring* $f : V' \to \bigcup_{v \in V'} L_v$, *where* $V' \subseteq V(G)$, *such that* $f(v) \in L_v$, $\forall v \in V'$. *When* $V' = V$, *we say* G *is* $\mathcal{L}$-CFON-*colorable. We say that* G *is* k-CFON* *(resp., k-CFON)-choosable if for every k-assignment* $\mathcal{L}$, G *is* $\mathcal{L}$-CFON* *(resp., $\mathcal{L}$-CFON)-colorable. The minimum k for which G is* k-CFON* *(resp., k-CFON)-choosable is called the* CFON**(resp., CFON) choice number of* G. *This is denoted by* $ch^*_{\mathrm{CN}}(G)$ *(resp.,* $ch_{CN}(G)$).

Analogously, one can define the CFCN* choice number and CFCN choice number of a graph G which will be denoted by $ch^*_{\mathrm{ON}}(G)$ and $ch_{\mathrm{ON}}(G)$, respectively.

To the best of our knowledge, the only paper that studies conflict-free choosability is due to Cheilaris, Smorodinsky, and Sulovský in [8]. Using the potential method, they show that (see Theorem 2.5 [8]) the CF choice number of a hereditarily k-colorable hypergraph on n vertices is $O(k \ln n)$. Further, they show that the geometric hypergraphs (i) induced by a set of n planar discs, and (ii) induced by n points with respect to planar regions like discs, halfplanes, or intervals can be hereditarily colored using a constant number of colors. Thus, all these geometric hypergraphs have their CF choice number equal to $O(\ln n)$. As for a general hypergraph $\mathcal{H}$ with m edges having a maximum degree of Δ, they show:

$$ch_{ON}(\mathcal{H}) \le 1/2 + \sqrt{2\,m + 1/4} \tag{1}$$

$$ch_{ON}(\mathcal{H}) \le \Delta + 1 \tag{2}$$

Further in [8], the authors show that $ch_{ON}(\mathcal{H}_n) = \lfloor \ln_2 n \rfloor + 1$, where $\mathcal{H}_n$ is a 'discrete interval hypergraph' on n vertices. For a graph G on n vertices, let $H_G^{path} := (V(G), \{S \mid S \text{ is the vertex set of a path in } G\})$. Theorem 4.1 in [8] says that $ch_{CF}(H_G^{path}) = O(\sqrt{n})$.

1.3 Some Quick Observations

In the Table 1 below, we present a collection of known observations, along with some new ones, in connection with various conflict-free coloring and conflict-free choosability parameters. We also provide tight examples that illustrate the relationships between these parameters. Here, $\mathcal{H}$ denotes a hypergraph on n vertices, and G denotes a graph on n vertices. For detailed proofs of each of these observations, see [14].

Table 1. A summary of the quick observations

Relation	Asymptotic Tightness
$\chi_{CF}(\mathcal{H}) \le ch_{ON}(\mathcal{H}) \le \chi_{CF}(\mathcal{H}) \cdot \ln n + 1$ [Known]	Up to multiplicative factor of $\ln\ln n$ [Expl. 2]
$\chi^*_{CF}(\mathcal{H}) \le ch^*_{ON}(\mathcal{H}) \le \chi^*_{CF}(\mathcal{H}) \cdot \ln n + 1$ [Known]	Up to multiplicative factor of $\ln\ln n$ [Expl. 3]
$\chi_{ON}(G) \le 2\chi^*_{ON}(G)$ [Known]	Tight [Known]
$ch_{ON}(G) \le 2ch^*_{ON}(G) \cdot \ln n + 1$ [Prop. 2]	Not known
$\chi_{CF}(\mathcal{H}) \le \chi^*_{ON}(\mathcal{H}) + 1$ [Known]	Tight [Known]
$ch_{ON}(\mathcal{H}) = O(ch^*_{ON}(\mathcal{H}) + \ln n)$ [Thm. 3]	Up to additive factor of $(\ln n - \frac{\ln n}{\ln\ln n})$ [Expl. 2]

Unlike in classical CF coloring (where $\chi_{ON}(\mathcal{H}) \le \chi^*_{ON}(\mathcal{H}) + 1$), there are hypergraphs $\mathcal{H}$ for which $ch_{ON}(\mathcal{H})$ is arbitrarily larger than $ch^*_{ON}(\mathcal{H})$. See Example 2. Example 3 gives graphs whose CFON* chromatic number and CFON* choice number are far apart.

Example 2 ($\star$).[1] Let $d > 2$. Let K_{d,d^d} denote the complete bipartite graph with bipartition $\{A, B\}$ having d vertices in Part A and d^d vertices in Part B. Let

---

[1] The proofs of the statements marked with ($\star$) appear in the full version [14].

$K_{d,d^d}^{1/2}$ denote the graph obtained by subdividing every edge of K_{d,d^d} exactly once. Then, (i) $ch_{\mathrm{ON}}^*(K_{d,d^d}^{1/2}) \leq 2$, (ii) $\chi_{\mathrm{ON}}(K_{d,d^d}^{1/2}) \leq 3$, and (iii) $ch_{\mathrm{ON}}(K_{d,d^d}^{1/2}) \geq d+1$.

Example 3 ($\star$). Let $d > 2$. Let $P_{d,d^d}^{1/2}$ denote the graph obtained by adding exactly one pendant vertex to every vertex in $K_{d,d^d}^{1/2}$ whose degree is greater than 2. Then, (i) $\chi_{\mathrm{ON}}^*(P_{d,d^d}^{1/2}) \leq 3$, and (ii) $ch_{\mathrm{ON}}^*(P_{d,d^d}^{1/2}) \geq d+1 = \Omega(\frac{\ln n}{\ln \ln n})$, where n denotes the number of vertices in $P_{d,d^d}^{1/2}$.

The following proposition connects the CFON and CFCN choice numbers of a graph.

Proposition 2. *For any graph G on n vertices,*
(i) $ch_{\mathrm{CN}}(G) \leq 2ch_{\mathrm{ON}}(G) \cdot \ln n + 1$,
(ii) $ch_{\mathrm{CN}}^*(G) \leq 2ch_{\mathrm{ON}}^*(G) \cdot \ln n_1 + 1$, where n_1 is the minimum number of colored vertices over all possible CFCN colorings of G that use only $\chi_{\mathrm{CN}}^*(G)$ colors.*

Proof. We shall prove only (i) as the proof of (ii) is similar. We have, $ch_{\mathrm{CN}}(G) \leq \chi_{\mathrm{CN}}(G) \cdot \ln n + 1 \leq 2\chi_{\mathrm{ON}}(G) \cdot \ln n + 1 \leq 2ch_{\mathrm{ON}}(G) \cdot \ln n + 1$, where the first and last inequalities follow from Proposition 3 in [14] and the second inequality follows from Proposition 1. $\qquad\square$

A path on two vertices is a graph G with $2 = ch_{\mathrm{CN}}(G) > ch_{\mathrm{ON}}(G) = 1$. We do not know of any graph G for which $ch_{\mathrm{CN}}(G)/ch_{\mathrm{ON}}(G)$ is arbitrarily large. Thus, it is possible that the bound given in Proposition 2 is far from tight.

1.4   Preliminaries

We study simple, finite, and undirected graphs throughout this paper. When discussing open neighborhood coloring, we assume the graph under consideration has no isolated vertices. Let *palette* of $\mathcal{L}$, denoted $\mathcal{P}_{\mathcal{L}}$, be defined as $\mathcal{P}_{\mathcal{L}} := \bigcup_{v \in V(G)} L_v$. Given an $S \subseteq V(G)$, we shall use (i) $G[S]$ to denote the subgraph of G induced by the vertices in S, and (ii) $G - S$ to denote the subgraph induced by the vertices in $V(G) \backslash S$. The *degree* of an element $v \in V$ in a given hypergraph $\mathcal{H} = (V, \mathcal{E})$, denoted by $d_{\mathcal{H}}(v)$, is the number of hyperedges that v is present in. The *maximum degree* of $\mathcal{H}$ is defined as $\max\{d_{\mathcal{H}}(v) : v \in V\}$. A *planar graph* is a graph that can be drawn on a plane in such a way that no two distinct edges intersect, except at a shared vertex.

1.5   Our Results

In this paper, we prove upper bounds for CF* choice numbers of general hypergraphs and CFON*/CFCN* choice numbers of general graphs. One can use Theorem 3 to extend these results to CF choice numbers and CFON/CFCN choice numbers.

Our first result comes from observing that the proof of the general upper bound for the conflict-free chromatic number of a hypergraph due to Pach and Tardos [20] can be easily extended to partial list CF coloring. We thus show in Sect. 3 that for a hypergraph $\mathcal{H}$, $ch^*_{\mathrm{ON}}(\mathcal{H}) = (t\Gamma^{1/t}\ln\Gamma)$, where every hyperedge in $\mathcal{H}$ is of size at least $2t-1$ and every hyperedge overlaps with at most Γ other hyperedges. Let G be a graph with maximum degree Δ and minimum degree $\Omega(\ln\Delta)$. Applying the above result on the open/closed neighborhood hypergraphs of G with $\Gamma = \Delta^2$ and $t = \frac{\ln\Delta+1}{2}$, we show in Sect. 3 that (i) $ch^*_{\mathrm{ON}}(G) = O(\ln^2\Delta)$, and (ii) $ch^*_{\mathrm{CN}}(G) = O(\ln^2\Delta)$.

For any graph G with maximum degree Δ, Pach and Tardos [20] in 2009 showed that for any constant $\epsilon > 0$, $\chi_{\mathrm{CN}}(G) = O(\ln^{2+\epsilon}\Delta)$. Later, Glebov, Szabó, and Tardos [13] in 2014 showed that $\chi_{\mathrm{CN}}(G) = \Omega(\ln^2\Delta)$. Bhyravarapu, Kalyanasundaram, and Mathew [4] in 2021 bridged the gap between the upper and the lower bound. They showed that $\chi_{\mathrm{CN}}(G) = O(\ln^2\Delta)$. We generalize the upper bound of [4] to list CFCN* coloring by proving $ch^*_{\mathrm{CN}}(G) = O(\ln^2\Delta)$ in Sect. 3. The proof of $\chi_{\mathrm{CN}}(G) = O(\ln^2\Delta)$ given in [4] uses the idea of maximal distance-3 sets and cannot be adapted to prove the list version. Our proof crucially uses the extension (mentioned in the paragraph above) of the theorem due to Pach and Tardos.

We study the computational complexity of the CFON/CFON*/CFCN choosability problems in Sect. 4. Let k be a constant integer. We show that the k-CFON-choosability problem, the k-CFON*-choosability problem, and the $g^G_{\{2,k\}}$-CFCN-choosability problem (defined in Sect. 4.2) are Π^P_2-complete on (i) bipartite graphs, for any $k \geq 3$, (ii) planar triangle-free graphs, when $k = 3$, and (iii) planar graphs, when $k = 4$.

2   Auxiliary Results

In this section, we state some known results (and a new result, Theorem 3) that we use later.

Theorem 2. *[8] For any hypergraph $\mathcal{H}$ with maximum degree Δ, $ch_{\mathrm{ON}}(\mathcal{H}) \leq \Delta + 1$.*

Below, we state the theorem that gives an upper bound on the CF choice number of a hypergraph in terms of its CF* choice number. Thus, all the bounds that we prove in this paper for CF*/CFON*/CFON* choice numbers can be extended to CF/CFON/CFON choice numbers.

Theorem 3 ($\star$). *For any hypergraph $\mathcal{H} = (V, \mathcal{E})$ with $|V| = n$, $ch_{\mathrm{ON}}(\mathcal{H}) = O(ch^*_{\mathrm{ON}}(\mathcal{H}) + \ln n)$.*

Additional results in this section are provided in [14].

3   Upper Bounds

The following theorem is an extension of Theorem 1.2 in [20] from CF coloring to list CF coloring of hypergraphs. The proof below is also a straightforward extension of the proof in [20].

Theorem 4 ($\star$). *For any positive integers t and Γ, the partial list conflict-free chromatic number of any hypergraph $\mathcal{H} = (V, \mathcal{E})$ in which each edge is of size at least $2t - 1$ and intersects at most Γ other hyperedges is $O(t\Gamma^{1/t} \ln \Gamma)$.*

Proof Here is a brief outline of the proof. Let $\mathcal{L} = \{L_v : v \in V\}$, where each list L_v ($\subseteq \mathbb{N}$) is sorted in increasing order. We first take a subset V' of V such that for every hyperedge $f \in \mathcal{E}$, $2t - 1 \leq |f \cap V'| \leq (2t - 1)(\Gamma + 1)$. We color the points in V' using the following iterative process until all the points are colored. In Round i, each uncolored point then is independently assigned the i-th color in its list with a probability $q := \frac{1}{30t\Gamma^{1/t}}$. We will then show that this procedure yields a partial list CF coloring of $\mathcal{H}$ with a non-zero probability. See [14] for the full proof. $\square$

Corollary 1. *Let G be a graph with maximum degree Δ and minimum degree $\Omega(\ln \Delta)$. Then, (i) $ch^*_{\mathrm{ON}}(G) = O(\ln^2 \Delta)$. (ii) $ch^*_{\mathrm{CN}}(G) = O(\ln^2 \Delta)$.*

Proof. Consider the open/closed neighborhood hypergraph of G. Every hyperedge in this hypergraph is of size at least $2t - 1$ and overlaps with Γ other hyperedges, where $t = \Omega(\ln \Delta)$ and $\Gamma = \Delta^2$. Apply Theorem 4 to obtain the desired result. $\square$

Below we prove the main result of this section.

Theorem 5. *For any graph G with maximum degree Δ, $ch^*_{\mathrm{CN}}(G) = O(\ln^2 \Delta)$.*

Proof. We first partition $V(G)$ into two parts. Let $A := \{v \in V(G) : d_G(v) \geq \ln \Delta\}$ and $B := V(G) \backslash A$. We further partition A into two components, namely A_1 and A_2. Let $A_1 := \{v \in A : d_A(v) \geq \ln \Delta\}$ and $A_2 := A \backslash A_1$, where $d_A(v)$ denotes the number of neighbors of v in A. Thus every vertex in A_2 has some neighbor in B, whereas a vertex in A_1 may or may not have a neighbor in B.

Let $\mathcal{L} = \{L_v : v \in V(G)\}$ be a r-assignment for G given by the adversary, where $r = K \cdot \lceil \ln^2 \Delta \rceil$ and K is a sufficiently large constant. We obtain the desired $\mathcal{L}$-CFCN*-coloring of G by list CF*-coloring hypergraphs $\mathcal{H}_1, \mathcal{H}_2$ that are defined below. A list CF*-coloring of the hypergraph $\mathcal{H}_1$ will ensure that every vertex in A_1 sees a unique color in its closed neighborhood. In a similar way, a list CF*-coloring of the hypergraph $\mathcal{H}_2$ will ensure that every vertex in $A_2 \cup B$ sees a unique color in its closed neighborhood. Before coloring $\mathcal{H}_2$, we update the lists of its vertices to avoid 'conflicts'. We explain this in detail below.

– Let $\mathcal{H}_1 = (V_1, \mathcal{E}_1)$ be a hypergraph, where $V_1 = A$ and $\mathcal{E}_1 = \{N_G[v] \cap A : v \in A_1\}$. From the definition of A_1, the minimum size of a hyperedge in $\mathcal{H}_1$ is at least $\ln \Delta$. Each hyperedge in $\mathcal{H}_1$ overlaps with at most Δ^2 other hyperedges. Hence, by Theorem 4, we have $ch^*_{\mathrm{ON}}(\mathcal{H}_1) = O(\ln^2 \Delta)$. Let f_1 denote this coloring. Under f_1, every vertex in A_1 sees a unique color in its closed neighborhood.

Next, by list CF* coloring a hypergraph $\mathcal{H}_2$ (defined below) having vertex set B, we intend to take care of every vertex in $A_2 \cup B$. This however can lead to problems of two types which are described below.

(i) Let c_v be the unique color seen by a vertex $v \in A_1$ under the coloring f_1. We should ensure that none of the neighbors of v in B receive the color c_v. For this, we update the lists of vertices $b \in B$ as $L_b^1 = L_b \backslash S_b$, where $S_b = \{c_v : v \in N_G(b) \cap A_1\}$. Note that $|S_b| < \ln \Delta$.

(ii) For every $w \in A_2 \cup B$, let c_w be the unique color to be seen by w under the list CF* coloring of $\mathcal{H}_2$ (to be defined) having vertex set B. We need to ensure that such a vertex w does not have a neighbor in A which received the color c_w under the coloring f_1. For this purpose, for each vertex $b \in B$, we define $T_b = \{f_1(a) : a \in N_G[w] \cap A, w \in N_G[b] \cap (A_2 \cup B)\}$. Since $|N_G[b]| \leq \ln \Delta$ and $|N_G[w] \cap A| \leq \ln \Delta$, for all $w \in (A_2 \cup B)$, we have $|T_b| \leq \ln^2 \Delta$. Let $L_b^2 = L_b^1 \backslash T_b$, for every $b \in B$. By taking a sufficiently large constant K (recall that, every list L_v is of size $K \cdot \lceil \ln^2 \Delta \rceil$), we can ensure that $|L_b^2| \geq \ln \Delta + 1$.

- Let $\mathcal{L}^2 = \{L_b^2 : b \in B\}$. Let $\mathcal{H}_2 = (V_2, \mathcal{E}_2)$, where $V_2 = B$ and $\mathcal{E}_2 = \{N_G[v] \cap B : v \in A_2 \cup B\}$. From the definition of B, the maximum degree of $\mathcal{H}_2$ is at most $\ln \Delta$. Hence, by Theorem 2, $\mathcal{H}_2$ admits an $\mathcal{L}^2$-CF*-coloring.

This completes the proof. $\hfill\square$

4   Hardness Results

4.1   Conflict-Free Open Neighborhood Choosability

The only connected graph with a CFON-choice-number equal to 1 is a path on 2 vertices. Therefore, deciding whether a graph is 1-CFON choosable can be done in polynomial time.

Let $k \geq 3$ be an integer. We now show the Π_2^P-hardness of the k-CFON-CHOOSABILITY PROBLEM and the k-CFON*-CHOOSABILITY PROBLEM by reducing from the k-CHOOSABILITY PROBLEM (which is defined below). See Sect. 1.2 for the definition of k-choosability.

---

k-CHOOSABILITY PROBLEM (k-CH PROBLEM)
Input: A graph G and an integer $k \geq 3$.
Question: Is G k-choosable?

---

k-CFON*-CHOOSABILITY PROBLEM (k-CFON*-CH PROBLEM)
Input: A graph G and a positive integer k.
Question: Is G k-CFON*-choosable?

---

k-CFON-CHOOSABILITY PROBLEM (k-CFON-CH PROBLEM)
Input: A graph G and a positive integer k.
Question: Is G k-CFON-choosable?

Theorem 6. *The k-CHOOSABILITY PROBLEM is Π_2^P-complete on:*

1. *bipartite graphs, for any constant $k \geq 3$ (see [16]).*
2. *planar triangle-free graphs, when $k = 3$ (see [15]).*
3. *planar graphs, when $k = 4$ (see [15]).*

It is easy to see that the k-CFON-CH PROBLEM is in Π_2^P. Given a graph G, a k-assignment $\mathcal{L} = \{L_v : v \in V(G)\}$, and a coloring function $f : V(G) \longrightarrow \bigcup_{v \in V(G)} L_v$, a polynomial time verifier verifies the following (i) every vertex $v \in V(G)$ is assigned a color from its list, i.e., $f(v) \in L(v)$, (ii) every vertex has some unique color in its open neighborhood. In a similar way, we can show that k-CFON*-CH PROBLEM is in Π_2^P.

Below, we present two constructions, one for a graph H_G and another for a graph H'_G, both derived from a given graph G.

Construction of H_G: We construct H_G by subdividing each edge of G exactly once. In other words, for every edge $uv \in E(G)$ with the endpoints u and v, we introduce a new vertex named x_{uv}, such that $N_{H_G}(x_{uv}) = \{u, v\}$. Note that the vertex x_{uv} is also known as x_{vu}. The vertices of G are called *original vertices* in H_G, and the remaining vertices in H_G are called *subdivided vertices*. It is easy to see that H_G can be constructed from G in $O(|V(G)| + |E(G)|)$ time.

Construction of H'_G: We first construct the graph H_G as described above. Then, we construct the graph H'_G from the graph H_G by attaching a pendant vertex to each original vertex in H_G. We leave the subdivided vertices untouched. Thus, H'_G has three types of vertices: original vertices (which are vertices of G), subdivided vertices (which are the newly introduced vertices in H_G), and pendant vertices (which are the newly introduced vertices in H'_G). It is easy to see that H'_G can be constructed from H_G in $O(|V(H_G)|)$ time.

We make the following observations concerning graphs H_G and H'_G.

Observation 2. *For each subdivided vertex x_{uv} to see a unique color in its open neighborhood, the vertices u and v should receive distinct colors in H_G.*

Observation 3. *Let p_i be the pendant vertex attached to v_i in H'_G. For each pendant vertex p_i to see a unique color in its open neighborhood, the vertex v_i must be colored in H'_G.*

Observation 4. *Both H_G and H'_G preserve graph properties such as bipartiteness, triangle freeness, and planarity.*

We now prove the following lemmas.

Lemma 1. *A graph G is k-choosable if and only if the graph H_G is k-CFON-choosable, where $k \geq 3$ is an integer.*

Proof. Let n denote $|V(G)|$. Suppose G is k-choosable. We will show that H_G is k-CFON-choosable. Let $\mathcal{L} = \{L_v : v \in V(H_G)\}$ be a k-assignment for H_G. Let $\mathcal{L}_G$ denote $\mathcal{L}$ restricted to the original vertices in H_G. Let f be a proper

$\mathcal{L}_G$-coloring of G. We now obtain an $\mathcal{L}$-CFON coloring f' for H_G as follows. For each original vertex $v \in V(H_G)$, let $f'(v) = f(v)$. By Observation 2, each subdivided vertex sees a unique color in its open neighborhood. Now, we have to handle the needs of the original vertices in H_G. With each original vertex $v \in V(H_G)$, we associate a subdivided vertex x_{uv}, where $u \neq v$. We assign some color to x_{uv} from its list. Finally, we color all the so far uncolored subdivided vertices $x_{pq} \in V(H)$ in such a way that x_{pq} does not receive the unique color of the vertex p or the vertex q. It is left to the reader to verify that this is indeed a valid CFON coloring of H_G.

Now we prove that if H_G is k-CFON-choosable, then G is k-choosable. By Observation 2, in any valid CFON coloring of H_G, the two neighbors of each subdivided vertex must have different colors. Specifically, every pair of original vertices in H_G that have an edge between them in G must have different colors. This ensures that a CFON coloring induced on original vertices in H_G is a proper coloring of G. $\qquad\square$

Lemma 2. *A graph G is k-choosable if and only if the graph H'_G is k-CFON*-choosable, where $k \geq 3$ is an integer.*

Proof. Since a pendant vertex is attached to every original vertex in H'_G, Observation 3 forces us to color every original vertex in any valid CFON* coloring of H'_G. Once we keep this observation in mind, the rest of the proof is similar to the proof of Lemma 1 and hence we omit the proof. $\qquad\square$

Combining Theorem 6 with Lemmas 1 and 2, we have the following theorem.

Theorem 7. *Both the k-CFON-CHOOSABILITY PROBLEM and the k-CFON*-CHOOSABILITY PROBLEM are Π_2^P-complete on:*

1. *bipartite graphs, for any constant $k \geq 3$.*
2. *planar triangle-free graphs, when $k = 3$.*
3. *planar graphs, when $k = 4$.*

4.2 Conflict-Free Closed Neighborhood Choosability

The only graph with a CFCN-choice-number equal to 1 is the edgeless graph, i.e., a graph consisting solely of isolated vertices. Therefore, deciding whether a graph is 1-CFCN-choosable can be done in polynomial time.

Let S be a finite subset of positive integers. Given a graph G, let g_S^G be a function whose domain is $V(G)$ and co-domain is S. That is, $g_S^G : V(G) \to S$. A list assignment $\mathcal{L} = \{L_v : v \in V(G)\}$ for G is a g_S^G-*assignment for G* if $\forall v \in V(G)$, $|L_v| = g_S^G(v)$. For a given function g_S^G, we say that G is g_S^G-*CFCN-choosable* if for every g_S^G-assignment $\mathcal{L}$, G is $\mathcal{L}$-CFCN-colorable. Thus, a k-assignment for G is a $g_{\{k\}}^G$-assignment, and the statement 'G is k-CFCN-choosable' is equivalent to the statement 'G is $g_{\{k\}}^G$-CFCN-choosable'. For a given function $g_{\{2,k\}}^G$, we define the $g_{\{2,k\}}^G$-CFCN-CHOOSABILITY PROBLEM below.

$g^G_{\{2,k\}}$-CFCN-CHOOSABILITY PROBLEM ($g^G_{\{2,k\}}$-CFCN-CH PROBLEM)

Input: A graph G, a function $g^G_{\{2,k\}} : V(G) \to \{2,k\}$.

Question: Is G $g^G_{\{2,k\}}$-CFCN-choosable?

It is easy to see that the $g^G_{\{2,k\}}$-CFCN-CH PROBLEM is in Π^P_2. The input is a graph G and a function $g^G_{\{2,k\}} : V(G) \to \{2,k\}$. Given a $g^G_{\{2,k\}}$-assignment $\mathcal{L} = \{L_v : v \in V(G)\}$, and a coloring function $f : V(G) \longrightarrow \bigcup_{v \in V(G)} L_v$, a polynomial time verifier verifies the following (i) every vertex $v \in V(G)$ is assigned a color from its list, i.e., $f(v) \in L(v)$, (ii) every vertex has some unique color in its closed neighborhood.

Let $k \geq 3$ be an integer constant. Let G be a graph on n vertices. Below, we present the construction of a graph H^k_G from G.

Construction of H^k_G: Let $N = 2\binom{nk}{2}$. We construct H^k_G by attaching N distinct pendants to each vertex in G. Thus, H^k_G has $n+Nn$ vertices in total. The vertices of G are called *original vertices* in H^k_G, and the remaining vertices in H^k_G are called *pendant vertices*. It is easy to see that H^k_G can be constructed from G in $O(n^3)$ time as k is a constant.

Observation 5. *H^k_G preserves graph properties like bipartiteness, triangle-freeness, and planarity.*

We now prove the following lemma.

Lemma 3. *G is k-choosable iff H^k_G is $g^G_{\{2,k\}}$-CFCN choosable, where*

$$g^G_{\{2,k\}}(v) = \begin{cases} 2, & \text{if } v \text{ is a pendant vertex in } H^k_G \\ k, & \text{otherwise} \end{cases}$$

Proof. Suppose G is k-choosable. We are given a function $g^G_{\{2,k\}} : V(H^k_G) \to \{2,k\}$ that satisfies $g^G_{\{2,k\}}(v) = 2$, if v is a pendant vertex in H^k_G and $g^G_{\{2,k\}}(v) = k$, otherwise. Let $\mathcal{L} = \{L_v : v \in V(H^k_G)\}$ be a $g^G_{\{2,k\}}$-assignment for H^k_G. Let $\mathcal{L}_G$ be $\mathcal{L}$ restricted to the original vertices in H^k_G. Let f be a proper $\mathcal{L}_G$-coloring of G. We now obtain an $\mathcal{L}$-CFCN coloring f' for H^k_G as follows. For each original vertex $v \in V(H^k_G)$, let $f'(v) = f(v)$. For each pendant vertex $p \in V(H^k_G)$, $f'(p)$ can be any color (from the list L_p) other than the color received, under f', by p's only neighbor. It can be verified that f' is a valid $\mathcal{L}$-CFCN coloring of H^k_G.

Suppose H^k_G is $g^G_{\{2,k\}}$-CFCN choosable, where $g^G_{\{2,k\}}(v) = 2$ if and only if v is a pendant vertex in H^k_G. Let $\mathcal{L} = \{L_v : v \in V(G)\}$ be a k-assignment for G. Let palette of $\mathcal{L}$ be defined as $\mathcal{P}_\mathcal{L} := \bigcup_{v \in V(G)} L_v$. Clearly, $P := |\mathcal{P}_\mathcal{L}| \leq nk$. Below, we define a $g^G_{\{2,k\}}$-assignment $\mathcal{L}' = \{L'_v : v \in V(H^k_G)\}$ for H^k_G. For each original vertex $v \in H^k_G$, we have $L'_v = L_v$. Recall that, each original vertex has N (which is at least $2\binom{P}{2}$) pendants attached to it. For the pendant vertices, we assign 2-sized lists that are subsets of $\mathcal{P}_\mathcal{L}$ in such a way that for every 2-sized subset S of $\mathcal{P}_\mathcal{L}$ and for every original vertex $v \in V(H^k_G)$, at least two pendant vertices attached to v have S as its list under $\mathcal{L}'$. Thus, $\mathcal{P}_{\mathcal{L}'} = \mathcal{P}_{\mathcal{L}}$. Let f' be an $\mathcal{L}'$-CFCN coloring of H_G^k. Recall that $L'_v = L_v$, for all original vertices $v \in V(H_G^k)$. We claim that f' restricted to the original vertices of H_G^k is an $\mathcal{L}$-coloring of G. In other words, under f', every original vertex in H_G^k sees its own color as the unique color in its closed neighborhood. This is because it sees every other color in the palette $\mathcal{P}_{\mathcal{L}}$ at least twice among the pendants attached it. $\qquad\square$

Combining Theorem 6, Lemma 3, and Observation 5, we have the following theorem.

Theorem 8. *The $g_{\{2,k\}}^G$-CFCN-CHOOSABILITY PROBLEM is Π_2^P-complete on:*

1. *bipartite graphs, for any constant $k \geq 3$.*
2. *planar triangle-free graphs, when $k = 3$.*
3. *planar graphs, when $k = 4$.*

5   Concluding Remarks

We have established hardness results for k-CFON/CFON*-choosability and for $(2, k)$-CFCN-choosability, for all $k \geq 3$. However, the complexity of the following problems remain open : (i) k-CFON*-CHOOSABILITY when $k = 1, 2$, (ii) k-CFON-CHOOSABILITY when $k = 2$, (iii) k-CFCN*-CHOOSABILITY when $k \geq 1$, and (iv) k-CFCN-CHOOSABILITY when $k \geq 2$.

References

1. Abel, Z., et al.: Conflict-free coloring of graphs. SIAM J. Discrete Math. **32**(4), 2675–2702 (2018)
2. Alon, N., Smorodinsky, S.: Conflict-free colorings of shallow discs. In: Proceedings of the Twenty-Second Annual Symposium on Computational geometry, pp. 41–43 (2006)
3. Bhyravarapu, S., Hartmann, T.A., Kalyanasundaram, S., Vinod Reddy, I.: Conflict-free coloring. graphs of bounded clique width and intersection graphs. In: Flocchini, P., Moura, L. (eds.) IWOCA 2021. LNCS, vol. 12757, pp. 92–106. Springer, Cham (2021). https://doi.org/10.1007/978-3-030-79987-8_7
4. Bhyravarapu, S., Kalyanasundaram, S., Mathew, R.: A short note on conflict-free coloring on closed neighborhoods of bounded degree graphs. J. Graph Theory **97**(4), 553–556 (2021)
5. Bhyravarapu, S., Kalyanasundaram, S., Mathew, R.: Conflict-free coloring on claw-free graphs and interval graphs. In: 47th International Symposium on Mathematical Foundations of Computer Science (MFCS 2022). Schloss Dagstuhl-Leibniz-Zentrum für Informatik (2022)
6. Bhyravarapu, S., Kalyanasundaram, S., Mathew, R.: Conflict-free coloring on subclasses of perfect graphs and bipartite graphs. Theor. Comput. Sci. 115080 (2025)
7. Cheilaris, P.: Conflict-free coloring. City University of New York (2009)
8. Cheilaris, P., Smorodinsky, S., Sulovský, M.: The potential to improve the choice: list conflict-free coloring for geometric hypergraphs. In: Proceedings of the Twenty-Seventh Annual Symposium on Computational Geometry, pp. 424–432 (2011)

9. Dębski, M., Przybyło, J.: Conflict-free chromatic number versus conflict-free chromatic index. J. Graph Theory **99**(3), 349–358 (2022)
10. Elbassioni, K., Mustafa, N.H.: Conflict-free colorings of rectangles ranges. In: Durand, B., Thomas, W. (eds.) STACS 2006. LNCS, vol. 3884, pp. 254–263. Springer, Heidelberg (2006). https://doi.org/10.1007/11672142_20
11. Even, G., Lotker, Z., Ron, D., Smorodinsky, S.: Conflict-free colorings of simple geometric regions with applications to frequency assignment in cellular networks. SIAM J. Comput. **33**(1), 94–136 (2004)
12. Gargano, L., Rescigno, A.A.: Complexity of conflict-free colorings of graphs. Theoret. Comput. Sci. **566**, 39–49 (2015)
13. Glebov, R., Szabó, T., Tardos, G.: Conflict-free colouring of graphs. Comb. Probab. Comput. **23**(3), 434–448 (2014)
14. Gupta, S., Mathew, R.: Bounds and hardness results for conflict-free choosability. https://arxiv.org/abs/2409.12672 (2025)
15. Gutner, S.: The complexity of planar graph choosability. Discret. Math. **159**(1–3), 119–130 (1996)
16. Gutner, S., Tarsi, M.: Some results on (a: b)-choosability. Discret. Math. **309**(8), 2260–2270 (2009)
17. Har-Peled, S., Smorodinsky, S.: On conflict-free coloring of points and simple regions in the plane. In: Proceedings of the Nineteenth Annual Symposium on Computational Geometry, pp. 114–123 (2003)
18. Krishnan, P., Mathew, R., Kalyanasundaram, S.: Pliable index coding via conflict-free colorings of hypergraphs. IEEE Trans. Inf. Theory **70**(6), 3903–3921 (2024)
19. Lev-Tov, N., Peleg, D.: Conflict-free coloring of unit disks. Discret. Appl. Math. **157**(7), 1521–1532 (2009)
20. Pach, J., Tardos, G.: Conflict-free colourings of graphs and hypergraphs. Comb. Probab. Comput. **18**(5), 819–834 (2009)
21. Pach, J., Tóth, G.: Conflict-free colorings. In: Discrete and Computational Geometry: The Goodman-Pollack Festschrift, pp. 665–671. Springer, Cham (2003)
22. Smorodinsky, S.: Conflict-Free Coloring and its Applications, p. 331–389. Springer Heidelberg (2013)

The Rectilinear Steiner Forest Arborescence

Łukasz Mielewczyk[1] , Leonidas Palios[2] , and Paweł Żyliński[1(✉)]

^{1} University of Gdańsk, 80-308 Gdańsk, Poland
`{lukasz.mielewczyk,pawel.zylinski}@ug.edu.pl`
^{2} University of Ioannina, 45110 Ioannina, Greece
`palios@cs.uoi.gr`

Abstract. Given two point sets P and R lying in the first quadrant Q_1 and such that $(0,0) \in R$, the Rectilinear Steiner Forest Arborescence (RSFA) problem is to find the minimum-length spanning forest F such that for each point $p \in P$, there exists a root $r \in R$, with $x(r) \leq x(p)$ and $y(r) \leq y(p)$, and a path in F connecting p to r whose length is equal to $(x(p) - x(r)) + (y(p) - y(r))$. The RSFA problem is a natural generalization of the Rectilinear Steiner Arborescence (RSA) problem, where $R = \{(0,0)\}$, and thus it is NP-hard. Herein, we briefly discuss a polynomial time approximation scheme for the RSFA problem, present a fast 2-approximation algorithm and provide a fixed-parameter algorithm.

Keywords: Rectilinear Steiner arborescence · PTAS · Approximation algorithm · FPT · Dynamic programming · Manhattan network

1 Introduction

A *rectilinear* graph $G = (V(G), E(G))$ is a plane graph with (weighted) edges corresponding to horizontal or vertical line segments in the plane $\mathbb{R}^2$, intersecting only at their endpoints (vertices). The weight $l(e)$ of an edge $e \in E(G)$ is equal to the length of the segment it corresponds to, and the *weight* of G, denoted by $l(G)$, is defined as the sum of all edge weights in G. With a slight abuse of notation, we also use G to denote the union of all the line segments (edges) in G, regarded as a point set, and refer to a vertex in $V(G)$ as a *point*.

A *rectilinear Steiner tree* for a set P of points in the plane is a rectilinear acyclic connected graph such that each point in P is an endpoint of some edge in the tree. Given a point r in the first quadrant Q_1 of the plane $\mathbb{R}^2$ and a subset $P \subset Q_1$ of points such that for any $p \in P$, its x- and y-coordinate is at least as that of r, a *rectilinear Steiner arborescence* (RSA) for P with root r is a rectilinear Steiner tree T for $P \cup \{r\}$ such that for each point $p \in P$, the length of the (unique) path in T from p to the root r equals $(x(p) - x(r)) + (y(p)) - (y(r))$, where $x(q)$ and $y(q)$ denotes the x- and y-coordinate, respectively, of point $q \in P \cup \{r\}$. A *minimum rectilinear Steiner arborescence* for P with root r is an RSA for P with root r that has the minimum weight $l(\cdot)$ over all RSAs for P (with root r).

© The Author(s), under exclusive license to Springer Nature Switzerland AG 2026
J. Kozik and A. Wolff (Eds.): SOFSEM 2026, LNCS 16448, pp. 475–490, 2026.
https://doi.org/10.1007/978-3-032-17801-5_35

The problem of determining the minimum rectilinear Steiner arborescence – having applications in the field of performance-driven VLSI design [12,13] – was first studied by Nastansky et al. [38] who proposed an integer programming formulation with exponential time complexity. In 1979, Laderia de Matos [30] proposed an exponential time dynamic programming algorithm. In 1985, Trubin [44] claimed that the problem is polynomially solvable, however, in 1992, Rao et al. [40] showed Trubin's algorithm to be incorrect; Rao et al. also provided a simple greedy 2-approximation for this problem, with $O(n \log n)$ running time, and their result was extended by Cordova and Lee [14] to the more general case where points can be located in all four quadrants, with the root located at the origin of $\mathbb{R}^2$. Another fast 2-approximation, even in the presence of obstacles, was presented by Ramanth [39]. Some other optimal exponential time algorithms can be found in [25,31] and some heuristics in [1,12,43]. More recently, [32,46] presented polynomial time approximation schemes for the problem, and the NP-hardness of the problem was shown by Shi and Su [41]. In addition, Fomin et al. [19] proposed the first subexponential algorithm for the problem, inspired by the work of Klein and Marx [28], who obtained a subexponential algorithm for the Subset Traveling Salesman Problem on planar graphs. The depth-restricted variant of the rectilinear Steiner arborescence problem was studied by Maßberg in [33], and the delay-restricted variant by Held and Rockel in [24], while the angle-restricted – in [6,45]. Finally, a more general problem, the Generalized Minimum Manhattan Network problem, was defined by Chepoi et al. [10] and then studied in [16,36].

The Rectilinear Steiner Arborescence with a Prespecified Sub-RSA. From a practical point of view, in real life, when considering optimization problems, we sometimes have to deal with scenarios in which we want to temporarily add extra points without however modifying the solution that we already have. Therefore, it is natural to consider such a scenario for the RSA problem as well.

Suppose that we are given a (not necessarily minimum) rectilinear Steiner arborescence T for a point set P_T with root r, which we will call *prespecified sub-RSA*, and let P be a set of additional points; w.l.o.g. we assume that none of the points in P lies on an edge of T. Now, we would like to connect the points in P to the arborescence T, thus getting a rectilinear arborescence for all points in $P_T \cup P$ (see Fig. 1). Formally, we are interested in computing the minimum rectilinear Steiner arborescence $\widehat{T}$ for $P_T \cup P$ with root r such that $T \subset \widehat{T}$.

The above problem formulation is natural; so, let us examine some basic properties of an (optimal) solution $\widehat{T}$. For a rectilinear Steiner arborescence T and a point set P, the bounding box $\mathrm{BBox}(T \cup P)$ is the smallest axis-aligned rectangle within which all the points in $V(T) \cup P$ lie. Now, for a point $p \in V(T) \cup P$, define its *left extension* as the maximal horizontal line segment l with right endpoint at p such that l has no points in common, except its endpoints, with T and the boundary of the bounding box $\mathrm{BBox}(T \cup P)$ of the arborescence T and the points in P; the *right, upward* and *downward extensions*, respectively, of p are defined analogously, see Fig. 2(a) for an illustration. Note that an extension may degenerate to a single point. Let $S(T, P)$ be the subgrid consisting of all

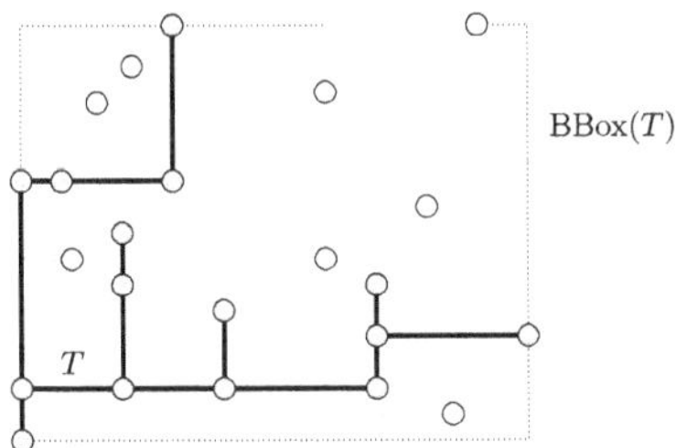

Fig. 1. The RSA problem with a prespecified sub-RSA: an optimal solution is shown in gray.

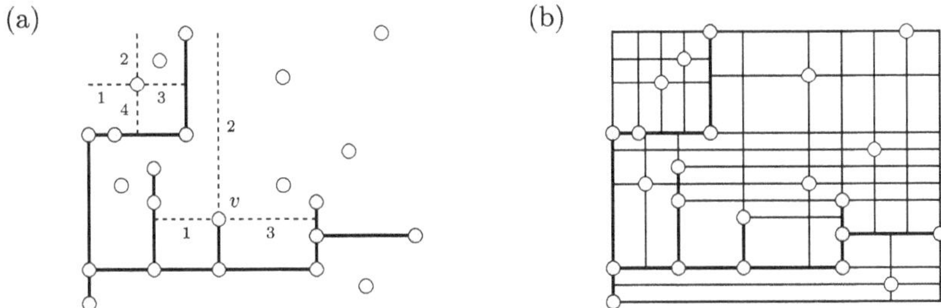

Fig. 2. (a) An illustration of the left (1), upward (2), right (3) and downward (4) extensions. Notice that the downward extension of v degenerates to a single point. (b) The subgrid $S(T, P)$.

points of (i) the boundary of the bounding box $\mathrm{BBox}(T \cup P)$, (ii) the edges of T, (iii) all extensions of all points in P, and (iv) only right and upward extensions of all vertices of T; see Fig. 2(b) for an illustration. The following result can be proven using standard methods for establishing Hanan grid-type results (see, for example, [15, 21, 23, 32]).

Lemma 1. *There exists a minimum rectilinear Steiner arborescence for $P_T \cup P$ with root r and with a prespecified sub-RSA T for P_T with root r such that it uses only subsegments of the grid $S(T, P)$.*

Let R be the set of degree-1 or degree-2 vertices of T and all left/bottom endpoints of the left/downward extensions from each point of P provided that they belong to T (i.e., the endpoints lie on an edge of T, see Fig. 3(a)); notice that $|R| = O(|V(T)| + |P|)$. Taking into account the above lemma, one can observe that there exists an optimal solution $\widehat{T}$, that is, a minimum rectilinear Steiner arborescence $\widehat{T}$ for $P_T \cup P$ with root r and with the pre-specified sub-RSA T for P_T (with root r), such that each point in P is connected to T with a rectilinear path ending at a point in R. And so, rather than considering the pre-specified rectilinear Steiner arborescence T, we may consider a rectlinear Steiner "forest" arborescence for the set P of points with possible roots among the elements of the set R. This observation motivates us to introduce the following general problem (see Fig. 3(b)).

The Rectilinear Steiner Forest Arborescence (RSFA) problem

Given two disjoint sets P and R of n points and m roots, respectively, lying in the first quadrant of $\mathbb{R}^2$ and such that $(0,0) \in R$, find a minimum rectilinear spanning forest F, called a *rectilinear Steiner forest arborescence for P with root set R*, such that for each point $p \in P$, there exists a root $r \in R$, with $x(r) \leq \mathrm{x}(p)$ and $\mathrm{y}(r) \leq \mathrm{y}(p)$, and a path π in F connecting p to r whose length is equal to $(\mathrm{x}(p) - \mathrm{x}(r)) + (\mathrm{y}(p) - \mathrm{y}(r))$.

As alluded above, our restatement of the problem definition preserves the optimum solution and asymptotically preserves the size of the input, and thus, any efficient algorithm for the RSFA problem can be applied to the RSA problem with a pre-specified sub-RSA, the problem defined formerly. Next, since the RSA problem is a special case of the RSFA problem where $R = \{(0,0)\}$, the NP-hardness of the RSA problem [41] directly implies the NP-hardness of the RSFA problem.

Corollary 1. *The RSFA problem is NP-hard.*

It is worth pointing out that a restricted variant of the RSFA problem, called the $(1,k)$-MRDPT problem (the Minimum Cost Rectilinear Distance Preserving Tree problem), where k is the number of non-root input points, has been studied in [43], for the purpose of a new efficient heuristic for the four-quadrant RSA problem. In this variant, we have one layer of *independent* roots (i.e., none of them covers another one; for the definition of 'covering' see the Notation section below) and several independent points in a second layer. And, based upon dynamic programming, Téllez and Sarrafzadeh provided a quadratic time algorithm that optimally solves the $(1,k)$-MRDPT problem; see [43] for more details.

Our Contribution. Taking into account Corollary 1 and following [32], we first discuss a PTAS (Sect. 2) for the RSFA problem. Next, we propose a fast 2-approximation algorithm (Sect. 3) and provide an FPT algorithm (Sect. 4) by extending the algorithmic results for the RSA problem [7,31,39]. Finally, some open problems related to Manhattan networks are discussed.

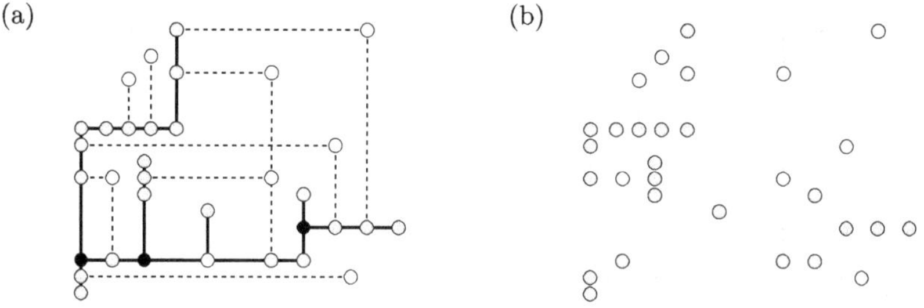

Fig. 3. (a) Gray points constitute the set R of roots. (b) The RSFA problem corresponding to the RSA problem with the pre-specified RSA in Fig. 1.

Notation. Unless otherwise stated, all points lie in the first quadrant of $\mathbb{R}^2$. For a point p, let $||p||$ denote the sum of coordinates $\mathrm{x}(p) + \mathrm{y}(p)$; $||p||$ is called the *z-coordinate* of p [40]. The function $|| \cdot ||$ imposes an ordering $\prec$ of points: for any two distinct points p and q, we have $p \prec q$ if and only if either $||p|| < ||q||$ or $||p|| = ||q||$ and $\mathrm{x}(p) < \mathrm{x}(q)$. Next, for any points p and q:

- the L_1-distance between p and q is denoted with $\mathrm{dist}(p, q)$;
- the point with coordinates $(\min\{\mathrm{x}(p), \mathrm{x}(q)\}, \min\{\mathrm{y}(p), \mathrm{y}(q)\})$ is denoted with $\langle p, q \rangle$
- a point p is said to *cover* q if and only if $\mathrm{x}(p) \leq \mathrm{x}(q)$ and $\mathrm{y}(p) \leq \mathrm{y}(q)$.

Finally, we shall use the abbreviation (P, R)-RSFA for a rectilinear Steiner forest arborescence for a given point set P with the root set R. In addition, we shall skip "for a given point set P with the root set R" when no confusion arises.

2 Polynomial Time Approximation Scheme

A standard approach for approximation algorithms for geometric optimization problems builds on the techniques developed for polynomial-time approximation schemes (PTAS) for geometric optimization problems due to Arora [2] and Mitchell [35]. In the former approach, in order to obtain a $(1 + \epsilon)$-approximation in $n^{O(1/\epsilon)}$ time, the instance is first scaled and perturbed to vertices of a polynomial-size integer grid. Next, the grid is recursively partitioned into dissection squares using a quadtree of logarithmic depth, and the so-called Structure Theorem is proven, which guarantees the existence of an almost-optimal solution that crosses the boundary of each dissection square only a few times and only in a number (depending on $\frac{1}{\epsilon}$) of pre-specified portals. Finally, to find a solution satisfying the Structure Theorem, dynamic programming is employed over the recursive decomposition, and by introducing the concept of portals, the running time becomes $n^c(\log n)^{O(\frac{1}{\epsilon})}$, where n is the size of an input instance, ϵ is any given constant $0 < \epsilon < 1$, and c is a constant.

In particular, Arora's method has been successfully applied to solve the RSA problem [32, 46], in fact without the portal technique[1], leading to a $(1 + \epsilon)$-approximation in $n^{O(1/\epsilon)}$ time. It is natural then to try to apply this technique for the RSFA problem as well. However, rounding in Arora's technique may result to optimal solutions that cannot be transformed to a sub-optimal one (by a small relative cost increase) for the original input (see Fig. 4 for an illustration). Consequently, rather then taking an effort in adapting Arora's technique, one can use the latter aforementioned method, that is, the idea of c-guillotine subdivisions proposed by Mitchell [35], thus getting a PTAS for the RSFA problem (see [34] for more details); we note that the same method has also been successfully applied to the *symmetric* rectilinear Steiner arborescence problem [8].

---

[1] As observed in [9], the portal technique seems to be non-applicable to a class of optimization problems with the "monotone path" requirement.

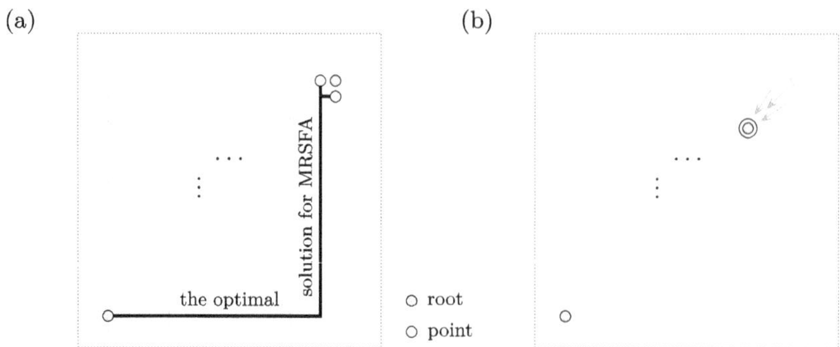

Fig. 4. Rounding in Arora's technique may result in the perturbed instance whose optimal solution for the RSFA problem cannot be transformed to a sub-optimal one for the original input, by a small relative cost increase. (b) In the perturbed instance of (a), the non-root (perturbed) points coincides with the (perturbed) root, so the cost of the optimal solution for the perturbed instance is 0.

Theorem 1. *Given any fixed positive integer k, any set P of n points, and any set R of m roots, all lying in the first quadrant of the plane, there is an $O((n+m)^{10k+5})$-time algorithm to compute an approximate (P, R)-RSFA whose length is within a factor of $(1 + \frac{1}{k})$ from the optimal.*

3   Efficient 2-Approximation Algorithm

From a practical point of view, the running time of the above PTAS for the RSFA problem is unsatisfactory which motivates us to design a fast constant factor approximation algorithm. For the RSA problem, there are two greedy 2-approximation algorithms [39,40]. It is natural then to try to apply similar approaches for the RSFA problem.

The idea of the 2-approximation algorithm of Rao et al. [40] for the RSA problem is as follows. Given a set P of points and the root r, the arborescence is generated by iteratively replacing the pair of points p and q in P by $\langle p, q \rangle$ until a single point (which has to be r) remains. The points p and q are chosen to maximize the z-coordinate $||\langle p, q \rangle||$ over all points in the current set P, and the resulting arborescence consists of all relevant edges from $\langle p, q \rangle$ to p and q (a degenerate case is possible if $\langle p, q \rangle$ is either p or q). Now, for the RSFA problem, when considering a set of roots R, a natural idea is to apply the above heuristic with the only difference that we must also handle each element of R as a possible root of a tree in the current partial solution. Unfortunately, observe that by replacing points p and q as suggested does not have a reasonable relative performance guarantee (for an example, see Fig. 5; clearly, one needs to avoid this replacement, since the sum of distances from p to the nearest root r_1 and from q to the nearest root r_2 is smaller than the sum of distances from p and q to $\langle p, q \rangle$ and from $\langle p, q \rangle$ to r).

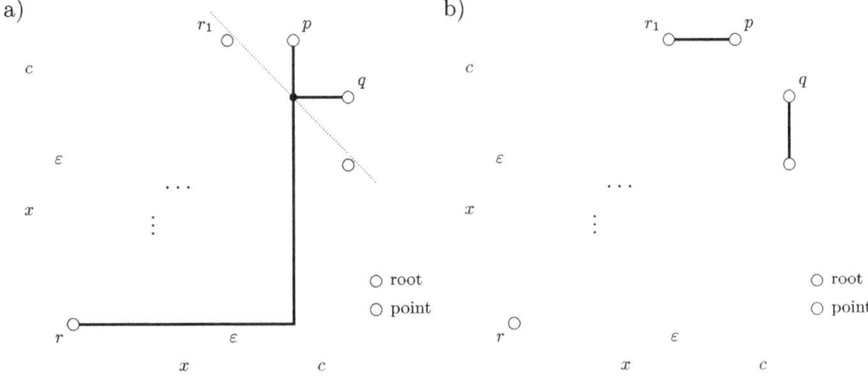

Fig. 5. (a) Replacing points p and q with $u = \langle p, q \rangle$ results in a rectilinear Steiner arborescence of the cost $2(c + x)$, while (b) the cost of the optimal solution is $2(c + \varepsilon)$.

The above observation introduces an issue to consider in the (possible) correctness proof and leads us to rely directly on the 2-approximation method in [39], using that algorithm as a subroutine. And indeed, a simple modification of this method results in an efficient 2-approximation algorithm for the RSFA problem as well. In the next subsection, we provide some intuition and an outline of the modified algorithm, while its correctness proof is presented in Subsect. 3.2.

3.1 The Outline and the Algorithm

A set C of points constitutes a *cover* for another set S if for every $s \in S$, there exists a $c \in C$ such that c covers s, and C is a *minimal* cover for S if C covers S and for every $c \in C$, the set $C \setminus \{c\}$ does not cover S. Note that if C is a minimal cover for S, then no point in C covers another point in C. Next, for a real $t \geq 0$, let L_t denote the line $y = -x + t$. Now, a point p is said to lie *below* the line L_t if $||p|| < t$. Likewise, if $||p|| = t$ then p is on L_t, and if $||p|| > t$ then p lies *above* L_t. Finally, given a (P, R)-RSFA F for a set P with a root set R, with $(0, 0) \in R$, let $P_{\geq t}$ denote the subset of points in P that lie on or above L_t and are not covered by any root in R lying above L_t, and let F_t denote the set of points in $F \cap L_t$.

So let P and R be two disjoint non-empty sets of n points and m roots, respectively, all lying in the first quadrant of $\mathbb{R}^2$ (with $(0, 0) \in R$). Let $H = H(P \cup R)$ be the Hanan grid of $P \cup R$, that is, the grid formed by the vertical and horizontal straight lines passing through all points in $P \cup R$ [23]; notice that each point in $P \cup R$ corresponds to some intersection grid point of H. Let Z be the set of distinct z-coordinates of all intersection points of $H(P \cup R)$ and let $z_0, z_1, z_2, \ldots, z_l$ be the ordering of elements of Z by increasing value; notice that $z_0 = 0$ since $(0, 0) \in R$ and recall that all points are lying in the first quadrant of $\mathbb{R}^2$. Finally, for $i = 1, \ldots, l$, define $\delta_i = z_i - z_{i-1}$ (notice $l \geq 1$ since the set P is non-empty). Now, following [39], one can observe that any (P, R)-

RSFA F with $(0,0) \in R$, can be constructed by appropriately connecting points in (an already specified) F_{z_i} to non-root points in $L_{z_{i+1}}$ by segments of length δ_{i+1}, for every z-coordinate $z_i \in Z$, $i = 0, 1, \ldots, l-1$. Consequently, the idea of the algorithm is to maintain an *expanding triangle* (the right triangle with vertices at $(0,0), (0, z_i)$ and $(z_i, 0)$), which initially consists of a single point, the root $(0,0)$. The (P, R)-RSFA is then constructed as the triangle expands. At any intermediate stage, the forest arborescence may contain several incomplete edges, all of which terminate at the boundary of the triangle. These edges are then extended as the triangle expands. There are four kinds of events that may take place as the triangle expands: (1) an incomplete edge is extended, (2) a Steiner point is introduced, (3) a point of P is encountered, and (4) a root is encountered. (Encountering a root is the only new event that is required in adapting the algorithm of Ramnath, but our modification affects also the original invariant in [39].) The idea is to deal with all four events in such a manner that the following invariant is maintained:

Invariant FPC: F_z *forms a minimal cover for* $P_{\geq z}$ *for each* $z \in Z$.

Let us now give a formal description of the algorithm. Assume that we have already computed the set F_{z_i} and let the points in $F_{z_i} = \{a_1^i, a_2^i, \ldots, a_{k_i}^i\}$ be ordered by increasing x-coordinate. For a point $a_j^i \in F_{z_i}$ ($1 \leq j \leq k_i = |F_{z_i}|$), define its *vertical* (resp. *horizontal*) *arm* as the vertical (resp. horizontal) half-line with endpoint at a_j^i, infinitely pointing up (resp. right). If there is a point $p \in P$ between or on the arms of a_j^i such that no root (in R) above the line $L_{z_{i+1}}$ covers p, we then define the set A_j^i as the set of points on the line $L_{z_{i+1}}$ that are at distance δ_{i+1} from a_j^i, that is, $A_j^i = \{(\mathrm{x}(a_j^i), \mathrm{y}(a_j^i) + \delta_{i+1}), (\mathrm{x}(a_j^i) + \delta_{i+1}, \mathrm{y}(a_j^i))\}$; otherwise, we set $A_j^i = \emptyset$. Finally, we set $C_{z_{i+1}} = \bigcup_{j=1}^{k_i} A_j^i$. Assuming now the aforementioned Invariant FPC for F_{z_i}, one can observe that $C_{z_{i+1}} \cup (R \cap L_{z_{i+1}})$ forms a cover for $P_{\geq z_{i+1}}$, which directly results in the following greedy algorithm for determining an RSFA; the algorithm can be simply implemented to have $O(n+m)^4$ running time.

Algorithm RSFA

1. $F_{z_0} := (0,0)$;

2.   **for** $i = 1$ **to** $|Z|$ **do**

2.1   choose any $F_{z_i} \subseteq C_{z_i} \cup (R \cap L_{z_i})$ forming a minimal cover for $P_{\geq z_i}$;

2.2   add only edges between points in $F_{z_{i-1}}$ and points in $F_{z_i} \setminus (R \cap L_{z_i})$;

3.2   Approximation Ratio

Let $t \geq 0$ be a real. By the definition of a rectilinear Steiner forest arborescence, given any RSFA F for a point set P with a root set R, where $(0,0) \in R$, we first observe that each point in $P_{\geq t}$ must be covered by a point in F_t; recall that $P_{\geq t}$ denotes the subset of points in P that lie on or above the line L_t and are not covered by any root in R lying above L_t. Next, following [39], given a set $U = \{u_1, \ldots, u_k\}$ of points satisfying $||u_i|| = t$ for each $u_i \in U$, and a set $W = \{w_1, \ldots, w_k\} \subseteq P_{\geq t}$, the set W is said to be *critical* for U if for each point $u_i \in U$, there exists a point in $w_i \in W$, referred to as the *critical point* for u_i, such that u_i covers w_i, and for any $u_j \in U$, $j \neq i$, u_j does not cover w_i. Clearly, if W is critical for U and U covers a set S with $W \subseteq S$, then U is a minimal cover for S. On the other hand, if U is a minimal cover for $P_{\geq t}$, then there exists a subset $V \subseteq P_{\geq t}$ such that V is critical for U. Next, for a point $u_i \in U$, the set of points (in the plane) not covered by any other point in U is called the *critical region for* u_i and is denoted by R_i; note that the region R_i is a half-open rectangular region (excluding the top-left corner (provided it exists), but including the bottom right (provided it exists)) with the apex at u_i. One can observe that if W is a critical set for U, then the critical point w_i for u_i must lie in the critical region R_i. Furthermore, we have the following lemma (see Lemma 2 in [39]).

Lemma 2. *If W is a critical set for U, then any point c with $||c|| = t$ (i.e., c is lying on the line L_t) covers at most two points in W.*

Proof. [39] Assume that points in $U = \{u_1, \ldots, u_l\}$ and $W = \{w_1, \ldots, w_l\}$ are ordered increasingly with respect to their x-coordinates; notice that w_i is then the critical point for u_i for each $i = 1, \ldots, l$. Suppose now to the contrary that there is a point c with $||c|| = t$ and such that it covers three critical points w_i, w_j and w_k, $1 \leq i < j < k \leq l$. Then we must have $\mathrm{x}(c) \leq \mathrm{x}(w_i)$ and $\mathrm{y}(c) \leq \mathrm{y}(w_k)$. Now, since W is critical for U, we obtain $||c|| = \mathrm{x}(c) + \mathrm{y}(c) \leq \mathrm{x}(w_i) + \mathrm{y}(w_k) < \mathrm{x}(u_j) + \mathrm{y}(u_j) = t$, a contradiction. $\qquad\square$

Let F be the final (P, R)-RSFA constructed by Algorithm RSFA with respect to the relevant sets F_{z_i}, $z_i \in Z = \{z_0, z_1, \ldots, z_l\}$. It follows from the construction that if (a real) $t \in (z_i, z_{i+1})$, then $|F_t| = |F_{z_{i+1}} \setminus R|$ (recall $F_t = F \cap L_t$), and moreover, F_t is a minimal cover for $P_{\geq t}$, and thus a critical set $W_t \subseteq P_{\geq t}$ for F_t exists. Now, let F^* be an optimal solution for the RSFA problem for the pair (P, R). Then any point in F_t^* covers at most two points in W_t by Lemma 2, and thus $|F_t^*| \geq |W_l|/2 = |F_t|/2$ which immediately implies

$$\int_{z_i}^{z_{i+1}} |F_t^*|\, \mathrm{d}t \geq \frac{1}{2} \int_{z_i}^{z_{i+1}} |F_t|\, \mathrm{d}t$$

($|F_{z_i}^*|$ and $|F_{z_i}|$ are finite). Consequently, since any (P, R)-RSFA $\widehat{F}$ satisfies $l(\widehat{F}) = \int_0^{z_l} |\widehat{F}_t|\, \mathrm{d}t$ [39, 40], we obtain

$$l(F^*) = \int_0^{z_l} |F_t^*|\, \mathrm{d}t \geq \frac{1}{2} \int_0^{z_l} |F_t|\, \mathrm{d}t = l(F)$$

which results in the following theorem.

Theorem 2. *The (P, R)-RSFA returned by Algorithm RSFA has a total length that is at most twice the length of the minimum RSFA.*

As regards an efficient implementation of Algorithm RSFA, we follow the sweep line method in [39] with some modifications and we have

Theorem 3. *There is a 2-approximation algorithms that solves the RSFA problem in $O((n + m)\log(n + m))$ time and $O(n + m)$ space.*

4   Fixed-Parameter Algorithm

Since there has been recent interest in the development of parameterized complexity results, in this section, following [3,7,29], we show that the RSFA problem is fixed-parameter tractable with respect to the parameter ℓ, being the minimum between the number h of horizontal lines and the number v of vertical lines on which all points are lying. In particular, we show that the RSFA problem can be solved in $O(2^\ell(n + m)\ell)$ time and $O(2^\ell)$ space. Recall that the idea behind fixed-parameter tractability is to separate out the complexity into two parts, – one part that depends purely and polynomially (if possible) on the size of the input, and another that depends on some "parameter" to the problem.

We note that in 1992, Ho et al. presented an $O(3^n n^2)$ time algorithm for the problem of determining a minimum rectilinear Steiner arborescence of n points in the plane [25]. Their algorithm is based upon a dynamic programming approach, it originates from a classical algorithm for the minimum Steiner tree problem in [18], and its time complexity can be slightly improved to $O(3^n)$ – see [34] for more details – which then leads to an exponential time algorithm that optimally solves the RSFA problem in $O(3^n + mn^2)$ time, where n and m is the number of input points and roots, respectively. However, more recently, inspired by the work of Klein and Marx [28] on a subexponential-time algorithm for the Subset Traveling Salesman problem on planar graphs, Fomin et al. [19] proposed the first subexponential-time algorithm for the RSA problem with $O(2^{\sqrt{n}\log n})$ running time, and we believe that their result can be extended to an exact algorithm for the RSFA problem running in $O(2^{\sqrt{n+m}\log(n+m)})$ time.

4.1   The Algorithm

Let P and R be two disjoint sets of n points and m roots, respectively, all lying in the first quadrant of $\mathbb{R}^2$ (with $(0,0) \in R$), and let $H = H(P \cup R)$ be the Hanan grid of $P \cup R$. We have the following lemma (again, it can be easily proven using standard methods for establishing Hanan grid-type result as we already did for the purpose of the proof of Lemma 1; details omitted).

Lemma 3. *There exists an optimal (P, R)-RSFA such that it uses only subsegments of the Hanan grid $H(P \cup R)$.*

Assume now that all intersection points of $H(P \cup R)$ are located on v vertical lines and h horizontal lines, and let $c_1, c_2, \ldots, c_{vh}$ be their lexicographic ordering; without loss of generality, assume that $2 \leq h \leq v$.

For each $k \in \{1, 2, \ldots, vh\}$, set $j := \max\{1, k - h + 1\}$; we consider functions

$$f_k \colon \{c_j, c_{j+1}, \ldots, c_k\} \to \{0, 1\}^{k-j+1}$$

where the intersection points c_i $(j \leq i \leq k)$ with $f_k(c_i) = 1$ play a role of gates, that is, these are the only points through which a partial solution may expand towards a global solution. Note that for $k \geq h$, we have $j = k - h + 1$, $f_k \colon \{c_{k-h+1}, c_{k-h+2}, \ldots, c_k\} \to \{0, 1\}^h$, and each horizontal line of $H(P \cup R)$ contains an intersection point among $c_{k-h+1}, c_{k-h+2} \cdots, c_k$.

Among all such functions f_k, we focus on those for which $f_k(c_i) = 1$ for each $c_i \in \{c_j, \ldots, c_k\} \cap (P \cup R)$; we call these functions k-*eligible*. Note that for each $k \in \{1, 2, \ldots, vh\}$, there may exist several k-eligible functions, which we will denote by $f_{k,1}, f_{k,2}, \ldots$. For each k-eligible function $f_{k,t}$, let $F_{f_{k,t}}$ denote an optimal rectilinear Steiner forest arborescence for $(P \cap \{c_1, c_2, \ldots, c_k\}) \cup f_{k,t}^{-1}(1)$ with the root set R (note that $f_{k,t}^{-1}(1) = \{\, c_i \mid c_i \in \{c_{k-h+1}, c_{k-h+2}, \ldots, c_k\}$ and $f_{k,t}(c_i) = 1 \,\}$). Clearly, in order to solve the RFSA problem for P with root set R, we need to compute a minimum-weight $F_{f_{k,t}}$ for a k-eligible function $f_{k,t}$ where $k = vh$.

All this can be done by dynamic programming, and the rationale for such an approach is the following: for each k-eligible function $f_{k,t}$, we consider all $O(1)$ "consistent" $(k-1)$-eligible functions $f_{k-1,s}$ (i.e., the functions $f_{k,t}$ and $f_{k-1,s}$ have the same values for $\{c_{\max\{1,k-h+1\}}, \ldots, c_{k-1}\}$) and an optimal RSFA $F_{f_{k,t}}$ is either a minimum-weight RSFA among all these optimal RSFAs $F_{f_{k-1,s}}$ if $f_{k,t}(c_k) = 0$ or a minimum-weight extension of the RSFAs $F_{f_{k-1,s}}$ to reach c_k if $f_{k,t}(c_k) = 1$.

To be more specific, the dynamic programming recursion (Bellman equation) is defined as follows. The base of the recursion is $k = 1$, for which the only 1-eligible function $f_{1,1} \colon \{c_1\} \to \{0, 1\}$ is the one with $f_{1,1}(c_1) = 1$ (since $c_1 = (0,0) \in R$), and so $F_{f_{1,1}}$ constitutes of the singleton c_1. Next, consider $2 \leq k \leq h$. For each k-eligible function $f_{k,t}$, let $f_{k-1,s}$ be the restriction of $f_{k,t}$ to $\{c_1, \ldots, c_{k-1}\}$, that is, $f_{k-1,s}(c_i) = f_{k,t}(c_i)$ for each $i \in \{1, \ldots, k-1\}$. (Notice that $f_{k-1,s}$ is a $(k-1)$-eligible function due to the k-eligibility of $f_{k,t}$ and thus it has been considered and the optimal RSFA $F_{f_{k-1,s}}$ has already been computed and tabulated.) Then if $c_k \in R$ or $f_{k,t}(c_k) = 0$, the optimal RSFA $F_{f_{k,t}}$ is identical to the RSFA $F_{f_{k-1,s}}$; on the other hand, if $c_k \notin R$ and $f_{k,t}(c_k) = 1$, then we must have $f_{k,t}(c_{k-1}) = 1$ and $F_{f_{k,t}}$ is the RSFA constructed from the RSFA $F_{f_{k-1,s}}$ by adding the edge $c_{k-1}c_k$.

Assume now that $k > h$. For each k-eligible function $f_{k,t}$, let Φ_{k-1} be the set of $(k-1)$-eligible functions such that for each $f_{k-1,s} \in \Phi_{k-1}$, it holds that $f_{k-1,s}(c_i) = f_{k,t}(c_i)$ for each $i \in \{k-h+1, k-h+2, \ldots, k-1\}$; since the functions in Φ_{k-1} differ only in their values at c_{k-h}, $|\Phi_{k-1}| \leq 2$, whereas the k-eligibility of $f_{k,t}$ implies that $|\Phi_{k-1}| \geq 1$. If $c_k \in R$ or $f_{k,t}(c_k) = 0$, then the optimal RSFA $F_{f_{k,t}}$ coincides with the already computed and tabulated RSFA $F_{f_{k-1,s}}$, where $f_{k-1,s}$ is a $(k-1)$-eligible function in Φ_{k-1} whose optimal RSFA $F_{f_{k-1,s}}$ is of minimum weight. If $c_k \notin R$ and $f_{k,t}(c_k) = 1$, then the RSFA $F_{f_{k,t}}$ is constructed from the RSFAs of the functions $f_{k-1,s}$ in Φ_{k-1} by adding the edge $c_j c_k$, where:

a) $j = k - h$ if $f_{k-1,s}(c_{k-h}) = 1$ (i.e., leftward segment)

   or

b) $j = k - 1$ if $k \neq 1 \bmod h$ and $f_{k-1,s}(c_{k-1}) = 1$ (i.e., downward segment).

Clearly, among cases (a-b) we choose the one that minimizes the sum

$$l(F_{f_{k-1,s}}) + \mathrm{dist}(c_j, c_k),$$

taken over all such $(k-1)$-eligible functions $f_{k-1,s}$.

The correctness of the above approach, in particular, the fact that at each single step, the constructed object is an optimal RSFA, directly follows from the construction (by induction on k). And, since we tabulate the values/RSFAs F_{f_k} for each $k \in \{1, \ldots, vh\}$ and each k-eligible function f_k, which are at most $O(2^h)$ of them for each k, and the RSFA F_{f_k} can be computed in $O(1)$ time from all the already tabulated data, the overall time complexity of our approach is $O(2^h vh) = O(2^h (n + m) h)$ since $v \leq n + m$. Consequently, the RSFA problem is fixed-parameter tractable with respect to the parameter ℓ, being the minimum between the number h of horizontal lines and the minimum number v of vertical lines that all points are lying on.

Theorem 4. *Given a set P of n points and a set R of m roots, with $(0,0) \in R$, lying on h horizontal lines and v vertical lines in the first quadrant of the plane, the optimal (P, R)-RSFA can be computed in $O(2^\ell (n + m)\ell)$ time and $O(2^\ell)$ space, where $\ell = \min(h, v)$.*

Our dynamic programming approach may produce a forest with many superfluous degree-2-vertices, forming a long path or a staircase, which results in a solution of size $\omega(|R| + |P|)$ (but still polynomial in $|R| + |P|$), whereas a solution of size $O(|P| + |R|)$ is sought. However, we can do a simple post-processing in order to delete these vertices, without affecting the (optimal) cost of the solution.

5   Conclusions

The above approaches are expected to extend for the case when points and roots can be located in any of the four quadrants, leading to the variant in which the following constraint is imposed on the output forest: for each point $p \in P$, there exists a path π to some root $r \in R$ in the same quadrant as p, with $|x(r)| \leq |x(p)|$ and $|y(x)| \leq |y(p)|$, such that the length of π is equal to $|x(p) - x(r)| + |y(p) - y(r)|$. In particular, following for example [39], the idea is to replace the expanding triangle with an *expanding diamond* (the square with vertices at $(0, z), (-z, 0), (0, -z)$ and $(z, 0)$), which initially consists of a single root $(0,0)$. The (P, R)-RSFA is then constructed by extending partial edges as the diamond expands. Again, the same four kinds of events are to be handled – (1) an incomplete edge is extended, (2) a Steiner point is introduced, (3) a point from P is encountered, and (4) a root is encountered – and the same invariant must be maintained: *the (partial) Steiner forest arborescence F_z forms a minimal cover for $P_{\geq z}$ for each $z \in Z$.*

Manhattan Networks. Given a set of points P in $\mathbb{R}^2$, a network (G, P) is said to be a *Manhattan network* for P if for all $p, q \in P$, there exists a Manhattan path between p and q with all its segments in G; the length of (G, P), denoted by $L(G)$, is the total length of all segments in G. For the given set S, the *Minimum Manhattan Network* (MNN) problem is to find a Manhattan network (G, S) of minimum length $L(G)$. The MNN problem was first introduced by Gudmundson et al. [21], who proposed an $O(n^3)$-time 4-approximation algorithm and an $O(n \log n)$-time 8-approximation algorithm. After [21], much research has been devoted to finding better approximation algorithms – see for example [4,5, 10,22,27,42] – and only in 2009, Chin et al. [11] proved the MMN problem to be NP-hard. In addition, a fixed-parameter algorithm running in $O(2^{14h} p(n))$ time where $p(n)$ is a polynomial in n and the parameter h is the minimum number of axis-parallel straight-lines (either all horizontal or all vertical) containing all the points in P was proposed by Knauer and in [29], the MNN problem in higher dimensions has been considered in [17,37], while [20,26] consider a variant of the problem where the goal is to minimize the number of vertices (Steiner points) and edges. Finally, when discussing the Minimum Manhattan Network problem, it is worth pointing out that the existence of a PTAS for it is still open.

Keeping in mind the motivation of RSFA, it is natural to define a relevant variant of MNN. Namely, given a minimum Manhattan network (G_1, P_1) and the input set P_2 of points within the bounding box of P_1, the *Minimum Manhattan Network with Backbone G_1* (MMNB) problem is to compute the minimum Manhattan network $(G_2, P_1 \cup P_2)$ such that $G_1 \subseteq G_2$. In particular, it is interesting to design a fast 2-approximation algorithm for the MMNB problem.

Disclosure of Interests. The authors declare that they have no relevant or material financial interests that relate to the research described in this paper.

References

1. Alexander, M.J., Robins, G.: New performance-driven FPGA routing algorithms. IEEE Trans. Comput. Aided Des. Integr. Circuits Syst. **15**(12), 1505–1517 (1996). https://doi.org/10.1109/43.552083
2. Arora, S.: Polynomial time approximation schemes for Euclidean traveling salesman and other geometric problems. J. ACM **45**(5), 753–782 (1998). https://doi.org/10.1145/290179.290180
3. Brazil, M., Thomas, D.A., Weng, J.F.: Rectilinear Steiner minimal trees on parallel lines. In: Du, D.-Z., Smith, J.M., Rubinstein, J.H. (eds.) Advances in Steiner Trees. Combinatorial Optimization, vol. 6, pp. 27–37. Springer, Boston (2000). https://doi.org/10.1007/978-1-4757-3171-2_3
4. Benkert, M., Wolff, A., Widmann, F.: The minimum Manhattan network problem: a fast factor-3 approximation. In: Akiyama, J., Kano, M., Tan, X. (eds.) DCG 2004. LNCS, vol. 3742, pp. 16–28. Springer, Heidelberg (2005). https://doi.org/10.1007/11589440_2
5. Benkert, M., Wolff, A., Widmann, F., Shirabe, T.: The minimum Manhattan network problem: approximations and exact solutions. Comput. Geom. Theory Appl. **35**(3), 188–208 (2006). https://doi.org/10.1016/j.comgeo.2005.09.004

6. Buchin, K., Speckmann, B., Verbeek, K.: Angle-restricted Steiner arborescences for flow map layout. Algorithmica **72**(2), 656–685 (2015). https://doi.org/10.1007/s00453-013-9867-z
7. Cambazard, H., Catusse, N.: Fixed-parameter algorithms for rectilinear Steiner tree and rectilinear traveling salesman problem in the plane. Eur. J. Oper. Res. **270**(2), 419–429 (2018). https://doi.org/10.1016/j.ejor.2018.03.042
8. Cheng, X., DasGupta, B., Lu, B.: Polynomial time approximation scheme for the symmetric rectilinear Steiner arborescence problem. J. Global Optim. **21**(4), 385–396 (2001). https://doi.org/10.1023/A:1012730702524
9. Cheng, X., Du, D.-Z., Kim, J.-M., Ngo, H.Q.: Guillotine cut in approximation algorithms. In: Murphey, R., Pardalos, P.M. (eds.) Cooperative Control and Optimization. Applied Optimization, vol. 66, pp. 21–34. Springer, Boston (2002). https://doi.org/10.1007/0-306-47536-7_2
10. Chepoi, C., Nouioua, K., Vaxes, Y.: A rounding algorithm for approximating minimum Manhattan networks. Theoret. Comput. Sci. **390**(1), 56–69 (2008). https://doi.org/10.1016/j.tcs.2007.10.013
11. Chin, F.Y.L., Guo, Z., Sun, H.: Minimum Manhattan network problem is NP-complete. Discret. Comput. Geom. **45**(4), 701–722 (2011). https://doi.org/10.1007/s00454-011-9342-z
12. Cong, J., Kahng, A.B., Leung, K.-S.: Efficient algorithms for the minimum shortest path Steiner arborescence problem with applications to VLSI physical design. IEEE Trans. Comput. Aided Des. Integr. Circuits Syst. **17**(1), 24–39 (1999). https://doi.org/10.1109/43.673630
13. Cong, J., Leung, K.-S., Zhou, D.: Performance-driven interconnect design based on distributed RC delay model. In: Proceedings of the 30th international Design Automation Conference, pp. 606–611 (1993). https://doi.org/10.1145/157485.165065
14. Córdova, J., Lee, Y.-H.: Fast optimal algorithms for the minimum rectilinear Steiner arborescence problem. Technical report TR-94-25, University of Florida (1994)
15. Czumaj, A., Czyzowicz, J., Gąsieniec, L., Jansson, J., Lingas, A., Żyliński, P.: Approximation algorithms for buy-at-bulk geometric network design. Int. J. Found. Comput. Sci. **22**(8), 1949–1969 (2011). https://doi.org/10.1007/978-3-642-03367-4_15
16. Das, A., Fleszar, K., Kobourov, S.G., Spoerhase, J., Veeramoni, S., Wolff, A.: Approximating the generalized minimum Manhattan network problem. Algorithmica **80**, 1170–1190 (2018). https://doi.org/10.1007/s00453-017-0298-0
17. Das, A., Gansner, E.R., Kaufmann, M., Kobourov, S., Spoerhase, J., Wolff, A.: Approximating minimum Manhattan networks in higher dimensions. Algorithmica **71**, 36–52 (2015). https://doi.org/10.1007/s00453-013-9778-z
18. Dreyfus, S.E., Wagner, R.A.: The Steiner problem in graphs. Networks **1**(3), 195–207 (1971). https://doi.org/10.1002/net.3230010302
19. Fomin, F.V., Kolay, S., Lokshtanov, D., Panolan, F., Saurabh, S.: Subexponential algorithms for rectilinear Steiner tree and arborescence problems. ACM Trans. Algorithms **16**(2), 1–37, Article no. 21 (2020). https://doi.org/10.1145/338142
20. Gudmundsson, J., Klein, O., Knauer, C., Smid, M.: Small Manhattan networks and algorithmic applications for the Earth mover's distance. In: Proceedings of the 23rd European Workshop on Computational Geometry, Graz, Austria, pp. 174–177 (2007)
21. Gudmundsson, J., Levcopoulos, Ch., Narasimhan, G.: Approximating a minimum Manhattan network. Nordic J. Comput. **8**(2), 219–232 (2001)

22. Guo, Z., Sun, H., Zhu, H.: Greedy construction of 2-approximation minimum Manhattan network. Int. J. Comput. Geom. Appl. **21**(3), 331–350 (2011). https://doi.org/10.1142/S0218195911003688
23. Hanan, M.: On Steiner's problem with rectilinear distance. SIAM J. Appl. Math. **14**, 255–265 (1966). https://doi.org/10.1137/011402
24. Held, S., Rockel, B.: Exact algorithms for delay-bounded Steiner arborescences. In: Proceedings of the 55th Annual Design Automation Conference, San Francisco, USA, Article no. 44, pp. 1–6 (2018). https://doi.org/10.1109/DAC.2018.8465902
25. Ho, J.-M., Ko, M.T., Ma, T.H., Sung, T.Y.: Algorithms for rectilinear optimal multicast trees. In: Ibaraki, T., Inagaki, Y., Iwama, K., Nishizeki, T., Yamashita, M. (eds.) ISAAC 1992. LNCS, vol. 650, pp. 106–115. Springer, Heidelberg (1992). https://doi.org/10.1007/3-540-56279-6_63
26. Jana, S., Maheshwari, A., Roy, S.: Linear-size planar Manhattan network for convex point sets. Comput. Geom. **100**, 101819 (2022). https://doi.org/10.1016/j.comgeo.2021.101819
27. Kato, R., Imai, K., Asano, T.: An improved algorithm for the minimum Manhattan network problem. In: Bose, P., Morin, P. (eds.) ISAAC 2002. LNCS, vol. 2518, pp. 344–356, Springer, Heidelberg (2002). https://doi.org/10.1007/3-540-36136-7_31
28. Klein, P.N., Marx, D.: A subexponential parameterized algorithm for Subset TSP on planar graphs. In: Proceedings of the 25th Annual ACM-SIAM Symposium on Discrete Algorithms, Portland, USA, pp. 1812–1830 (2014). https://doi.org/10.1137/1.9781611973402.13
29. Knauer, C., Spillner, A.: A fixed-parameter algorithm for the minimum Manhattan network problem. J. Comput. Geom. **2**(1), 189–204 (2011). https://doi.org/10.20382/jocg.v2i1a10
30. Laderia de Matos, R.R.: A rectilinear arborescence problem. Ph.D. thesis, University of Alabama, USA (1979)
31. Leung, K.-S., Cong J.: Fast optimal algorithms for the minimum rectilinear Steiner arborescence problem. In: Proceedings of the IEEE International Symposium on Circuits and Systems, Hong Kong, China, pp. 1568–1571 (1997). https://doi.org/10.1109/ISCAS.1997.621429
32. Lu, B., Ruan, L.: Polynomial time approximation scheme for the rectilinear Steiner arborescence problem. J. Comb. Optim. **4**(3), 357–363 (2000). https://doi.org/10.1023/A:1009826311973
33. Maßberg, J.: The depth-restricted rectilinear Steiner arborescence problem is NP-complete. arXiv:1508.06792 (2015)
34. Mielewczyk, Ł., Palios, L., Żyliński, P.: The rectilinear Steiner forest arborescence problem. Technical report, arXiv:2210.04576 (2022)
35. Mitchell, J.S.B.: Guillotine subdivisions approximate polygonal subdivisions: a simple polynomial-time approximation scheme for geometric TSP, k-MST, and related problems. SIAM J. Comput. **28**(4), 1298–1309 (1999). https://doi.org/10.1137/S009753979630976
36. Masumura, Y., Oki, T., Yamaguchi, Y.: Dynamic programming approach to the generalized minimum Manhattan network problem. Algorithmica **83**(12), 3681–3714 (2021). https://doi.org/10.1007/s00453-021-00868-x
37. Muñoz, X., Seibert, S., Unger, W.: The minimal Manhattan network problem in three dimensions. In: Das, S., Uehara, R. (eds.) WALCOM 2009. LNCS, vol. 5431, pp. 369–380. Springer, Heidelberg (2019). https://doi.org/10.1007/978-3-642-00202-1_32

38. Nastansky, L., Selkow, S.M., Stewart, N.F.: Cost minimum trees in directed acyclic graphs. Math. Methods Oper. Res. **18**, 59–67 (1974). https://doi.org/10.1007/BF01949715

39. Ramnath, S.: New approximations for the rectilinear Steiner arborescence problem [VLSI layout]. IEEE Trans. Comput.-Aided Des. Integr. Circuits Syst. **22**(7), 859–869 (2003). https://doi.org/10.1109/TCAD.2003.814249

40. Rao, S.K., Sadayappan, P., Hwang, F.K., Shor, P.W.: The rectilinear Steiner arborescence problem. Algorithmica **7**(1–6), 277–288 (1992). https://doi.org/10.1007/BF01758762

41. Shi, W., Su, C.: The rectilinear Steiner arborescence problem is NP-complete. SIAM J. Comput. **35**(3), 729–740 (2006). https://doi.org/10.1137/S0097539704371353

42. Seibert, S., Unger, W.: A 1.5-approximation of the minimal Manhattan network problem. In: Deng, X., Du, D.-Z. (eds.) ISAAC 2005. LNCS, vol. 3827, pp. 246–255. Springer, Heidelberg (2005). https://doi.org/10.1007/11602613_26

43. Téllez, G.E., Sarrafzadeh, M.: On rectilinear distance-preserving trees. VLSI Des. **7**(1), 15–30 (1998). https://doi.org/10.1109/ISCAS.1995.521476

44. Trubin, V.A.: Subclass of the Steiner problems on a plane with rectilinear metric. Cybernetics **21**, 320–322 (1995). https://doi.org/10.1007/BF01078826

45. Verbeek, K., Buchin, K., Speckmann, B.: Flow map layout via spiral trees. IEEE Trans. Visual Comput. Graph. **17**(12), 2536–2544 (2011). https://doi.org/10.1109/TVCG.2011.202

46. Zachariasen, M.: On the approximation of the rectilinear Steiner arborescence problem in the plane. Manuscript (2000)

Face-Hitting Dominating Sets in Plane Graphs: Alternative Proof and Linear-Time Algorithm

Therese Biedl[(✉)]

David R. Cheriton School of Computer Science, University of Waterloo, Waterloo, ON N2L 3G1, Canada
biedl@uwaterloo.ca

Abstract. In a recent paper, Francis, Illickan, Jose and Rajendraprasad showed that every n-vertex plane graph G has (under some natural restrictions) a vertex-partition into two sets V_1 and V_2 such that each V_i is *dominating* (every vertex of G contains a vertex of V_i in its closed neighbourhood) and *face-hitting* (every face of G is incident to a vertex of V_i). Their proof works by considering a super-graph G' of G that has certain properties, and among all such graphs, taking one that has the fewest edges. As such, their proof is not algorithmic. Their proof also relies on the 4-color theorem, for which a quadratic-time algorithm exists, but it would not be easy to implement.

In this paper, we give a new proof that every n-vertex plane graph G has (under the same restrictions) a vertex-partition into two dominating face-hitting sets. Our proof is constructive, and requires nothing more complicated than splitting a graph into 2-connected components, finding an ear decomposition, and computing a perfect matching in a 3-regular plane graph. For all these problems, linear-time algorithms are known and so we can find the vertex-partition in linear time.

1 Introduction

One of the oldest problems in graph theory is how to find a small *dominating set* in a graph $G = (V, E)$, i.e., a set S of vertices such that for every vertex v, either v is in S, or some neighbour of v is in S. Finding the minimum dominating set is NP-complete [11]. Related to finding the smallest dominating set is the question how big it may be required to be. Since a graph consisting of n isolated vertices clearly requires size n for any dominating set, one typically imposes some restriction on the class of graphs studied. One interesting result here was given by Matheson and Tarjan [14], who showed that every n-vertex simple triangulated planar graph G has a dominating set of size at most $n/3$. (See Sect. 2 for definitions.) In fact, they show the stronger result that the vertices

Supported by NSERC. The author would like to thank Daniel Rutschmann for helpful discussions.

© The Author(s), under exclusive license to Springer Nature Switzerland AG 2026
J. Kozik and A. Wolff (Eds.): SOFSEM 2026, LNCS 16448, pp. 491–504, 2026.
https://doi.org/10.1007/978-3-032-17801-5_36

of G can be partitioned into three disjoint sets that each are dominating. It is easy to see that some planar triangulated graphs require at least $n/4$ vertices in any dominating set, and the quest to bring the upper bound closer to $n/4$ has given rise to a number of interesting research insights. (See the paper by Christiansen, Rotenberg and Rutschmann [6] for the currently best upper bound of $\frac{22}{7}n$, and the references therein for more results for other related classes of planar graphs.)

In a recent paper, Francis, Illickan, Jose and Rajendraprasad [10] studied an interesting variant of this problem, where they are looking for a vertex set that is not only a dominating set, but that also is *face-hitting*, i.e., they presume that the planar graph G comes with a fixed crossing-free drawing (it is *plane*), and they want a dominating set S where additionally every face of the drawing is incident to at least one vertex of S. It is known from an earlier construction by Bose, Shermer, Toussaint and Zhu [5] that there are simple triangulated planar graphs where every face-hitting vertex set has size at least $\lfloor n/2 \rfloor$. (In a triangulated planar graph, such a set is automatically also dominating.) Francis et al. match this bound, and in particular, show that any plane graph (under some natural restrictions, such as "no isolated vertices") has a vertex-partition into two face-hitting dominating sets; the smaller of these sets has size at most $\lfloor n/2 \rfloor$. This turns out to improve the bound for dominating sets in triangulated planar graphs that have large independent sets, or small maximal independent sets.

From a mathematical point of view the result of [10] is very nice since it matches the lower bound. But from an algorithmic point of view it is somewhat disappointing, since the authors do not discuss how to find the vertex-partition efficiently, and it is not obvious how to do so. In particular, a crucial ingredient of their proof is to use a plane super-graph G' of G that has the maximum number of so-called "happy vertices" and "happy faces" (we will define a similar concept in Sect. 3.2), and among all such graphs, take the one with the fewest added edges. Clearly G' always exists, but it is not clear how to find it. From later properties that they show about G', it seems likely that one could simply start with $G' := G$, search for a violation of those properties, and then update G' according to the steps in their proofs. It is not clear how fast such a procedure would be, but it would surely take at least quadratic time.

A second bottleneck is that Francis et al. use the 4-color theorem to color the graph G' (and extract the vertex-partition by combining two color-classes). While there is an algorithm to find the 4-coloring of a planar graph [16], the run-time is quadratic (with a very large constant factor).

In this paper, we therefore re-prove the result by Francis et al. in a different way that immediately leads to a linear-time algorithm to find the vertex-partition. Rather than relying on the 4-color theorem, we find a bipartite subgraph H of G (whose 2-coloring then gives the vertex-partition). We ensure that for every vertex and every face of G at least one incident edge is retained in H, so either color-class will be a face-hitting dominating set. We sketch here an outline of how to find subgraph H. This can be done easily if G is triangulated, by virtue of removing the edges of a perfect matching in the dual of G (this always exists by Petersen's theorem). If G is not triangulated but 2-connected, then we first create a triangulated planar super-graph G^+ of G where all vertices but one are "happy". This is done by computing an ear decomposition of G, and triangulating each face suitably when its corresponding ear is added. Since vertices of G^+ are happy, removing the edges of a perfect matching in the dual of G^+ then again gives subgraph H. Finally, for a graph G that is not 2-connected, we simply combine the subgraphs of the 2-connected components. All steps can easily be implemented in linear time, leading us to our main result:

Theorem 1. *Let $G = (V, E)$ be a plane graph with n vertices that has no isolated vertices or faces incident to at most two vertices. Then in linear time we can partition V into two vertex sets V_1, V_2 that both are dominating as well as face-hitting.*

2 Preliminaries

This section clarifies some notation for a graph $G = (V, E)$ (for most standard notation, see for example Diestel [7]). Throughout, n is the number of vertices. A *loop* is an edge where the two endpoints coincide, while two *parallel edges* are two edges with the same set of endpoints. A graph is *simple* if it has no loops and no parallel edges. We do not assume that G is simple (except where explicitly stated below). Recall that a vertex set S is *dominating* if for every $v \in V$ either $v \in S$ or some neighbour of v is in S. An *edge cover* is a set S of edges such that for every $v \in V$ at least one incident edge is in S. A *perfect matching* is a set M of edges where every vertex is incident to exactly one edge in M.

Throughout the paper, we will assume that G is connected, since otherwise we can obtain a partition into face-hitting dominating sets in each component and combine them to get one for G. Graph G is called *2-connected* if it does not contain a *cut-vertex*, i.e., a vertex v such that $G \setminus v$ has more connected components than G. A *2-connected component* C is a maximal subgraph of G that is 2-connected. For the borderline case of a vertex v with an incident loop e, we consider v to be a cut-vertex and the graph formed by v and e to be a *trivial* 2-connected component.

We also assume that G is *plane*, i.e., comes with a fixed crossing-free drawing Γ on the sphere $\mathbb{S}^2$. The maximal regions of $\mathbb{S}^2 \setminus \Gamma$ are called the *faces* of G. Since G is connected, drawing Γ can be described by specifying the clockwise order of edges at each vertex, and every face can be described via the walk that defines its *facial boundary*. An *angle* $\angle uvw$ *at vertex* v consists of two edges (u, v) and (v, w) that are consecutive in the order of edges at v. Equivalently, the two edges are consecutive on the facial boundary of some face F; we hence also call this an *angle of* F.

The *degree* of a face F is the number of angles on it, while the *size* of F is the number of vertices incident to F. For a 2-connected plane graph the two concepts coincide since the boundary of F is a simple cycle. A *bigon* is a face whose boundary consists of two parallel edges and so has degree 2. A *triangle* is a closed walk with three edges. A *triangulated planar graph* is a plane graph where every facial boundary is a triangle. Note that we specifically *allow* a triangulated planar graphs to have parallel edges, as long as they do not form bigons. (The above definition principally also allows loops, but our later application will forbid these.) A 3^+-*face* is a face with size 3 or more; this excludes bigons, but it also excludes triangular faces that are bounded by a loop and two parallel edges.

Let S be either a vertex-set or an edge-set. A face F is *hit* by S if it is incident to at least one element of S. Set S is called *face-hitting* if it hits every face, and 3^+-*face hitting* if it hits every 3^+-face.

It will often be convenient to project the spherical drawing Γ of G into the plane (for our choice of projection point); this singles out the face containing the projection point to become the unbounded *outer face* while all other faces are *interior*. An *interior vertex* is a vertex not incident to the outer face.

3   Bipartite Edge Covers

To prove Theorem 1, it suffices to prove the following result.

Theorem 2. *BipEdgeCover Let $G = (V, E)$ be a connected plane graph with $n \geq 2$ vertices. Then in linear time we can find a bipartite subgraph H of G whose edges form a 3^+-face-hitting edge cover of G. Furthermore, we can pre-specify a reference-edge $\hat{e}$ of G and find such a graph H that includes $\hat{e}$.*

To see why this implies Theorem 1, consider a 2-coloring of H, say into vertex-sets V_1 and V_2, and let $i \in \{1, 2\}$. Since $E(H)$ is an edge cover, for every vertex v there exists an edge (v, w) in H; either $v \in V_i$ or $w \in V_i$, and so V_i is dominating. Since $E(H)$ is 3^+-face-hitting, for every 3^+-face F of G there exists an edge (x, y) incident to F in H; either $x \in V_i$ or $y \in V_i$, and so V_i is 3^+-face-hitting. Since by assumption of Theorem 1 all faces of G are 3^+-faces, the result holds. (We note here that Theorem 1, which we intentionally kept very similar to the statement in [10], could be strengthened ever so slightly to permit faces of size 1 or 2, as long as they need not be hit.)

We first prove Theorem 2 for a triangulated planar graph, where this is very easy. For later use as a subroutine, we actually prove a stronger claim on the edges $E(H)$ of H (which immediately implies that $E(H)$ is a face-hitting edge cover of G since every vertex and every face is incident to an angle).

Lemma 1. *Let G be a triangulated planar graph that has no loops. Fix a reference-edge $\hat{e}$. We can find in linear time a bipartite subgraph H of G that contains $\hat{e}$, and where for every angle $\angle uvw$ of G, at least one of the edges (u, v) and (v, w) is in H.*

Proof. The technique to find H is well-known (see e.g. [2,4]), but we review it here to prove the additional claims and verify that it works even if G has parallel edges. Since G is triangulated, its dual graph G^* is 3-regular. (Recall that the *dual graph* G^* has a vertex v_F for every face F of G, and an edge $e^* = (v_F, v_{F'})$ whenever faces F, F' are both incident to an edge e.) Graph G^* is also 2-connected, for otherwise it (as a 3-regular graph) would have an edge e^* whose removal disconnects G^*, and this edge is dual to a loop of G. By Petersen's theorem [15], G^* therefore has a perfect matching M^*. In fact (see e.g. the proof in [2]) we can require one edge of our choice to be in the perfect matching M^*; we choose it here to be an edge that shares an endpoint with $\hat{e}^*$ so that $\hat{e}^*$ is *not* in M^*. This matching can be computed in linear time [2].

Define M to be those edges of G whose dual edges are in M^*, and let $H :=$ $(V, E \setminus M)$. Since M^* is a perfect matching, all faces of H have degree 4, and since G has no loops this makes H a bipartite subgraph of G. Since the dual of $\hat{e}$ is not in M^*, reference-edge $\hat{e}$ is in H. Consider any angle $\angle uvw$ of G, say at face F, and observe that the two edges are distinct since G has no loops. Since M^* contains exactly one edge incident to v_F, at most one of (u, v) and (v, w) can be in M, and the other is retained in H. $\qquad\qquad\square$

3.1 2-Connected Plane Graphs

In this section, we prove Theorem 2 for a 2-connected plane graph G. To do so, we first consider the case when there are no bigons. We can then find a super-graph G^+ that is triangulated planar and has special properties, and apply Lemma 1 Formally, we say that a vertex v is *happy* (in some super-graph G^+ that will be clear from context) if v is incident to a face F of G^+ that is a triangle, and the angle of F at v consists of two edges of G. (In our figures, happy vertices are indicated by shading the corresponding angle.) Our goal is to find a triangulation of the input graph where every vertex (or nearly every vertex) is happy. We note here that this is vaguely related to other papers that dealt with triangulating a planar graph under some constraint (see e.g. [3,13]) or with papers that assigned angles to vertices under some constraints (see e.g. [8]), but to our knowledge this particular problem has not been studied except by Francis et al. [10].

We triangulate the graph with many happy vertices by adding the vertices and faces incrementally while maintaining a suitable invariant. We need a definition.

Definition 1. *An* ear decomposition *of a graph G is a collection $\langle P_1, \ldots, P_f \rangle$ of paths ('ears') in G with the following properties:*

- *Every edge of G belongs to exactly one path.*
- *The first path P_1 is a single edge.*
- *For $i > 1$, P_i is a path with two distinct endpoints. The endpoints of P_i are in $V_{i-1} := V(P_1) \cup \ldots \cup V(P_{i-1})$, while all other vertices of P_i are not in V_{i-1}.*

It is well-known that a graph G is 2-connected if and only if it has an ear decomposition [18]. Given an ear decomposition, we define (for $i \geq 1$) G_i to be the graph with vertices V_i and all edges of $P_1, \ldots, P_i$; note that G_i is 2-connected since it has an ear decomposition. It is known that if G is additionally a plane

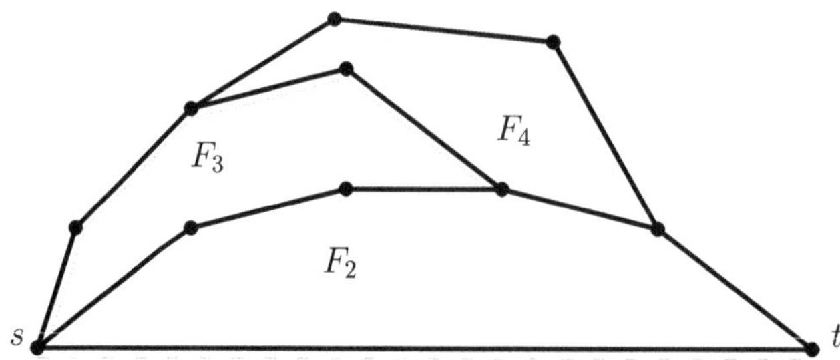

Fig. 1. An ear decomposition of a plane 2-connected graph.

graph, then we can choose an ear decomposition and outer face such that all interior faces of G_i are faces of G; in particular therefore ear P_i must lie entirely on the outer face of G_i and (for $i > 1$) adding P_i to G_{i-1} must have added an interior face F_i of G to G_i. See also Fig. 1. Furthermore, for any face F and any edge (s, t) on F, we can find (after a possible change of outer face of G) an ear decomposition such that $P_1 = (s, t)$ and $F_2 = F$.[1]

Fix such an ear decomposition $\langle P_1, \ldots, P_f \rangle$ of the given plane graph G; the goal is to build a super-graph G_i^+ of G_i with many happy vertices by induction on i. Specifically, call a vertex of G_i^+ v *interior-happy* if it is happy due to an interior face. We maintain the following **invariant** of G_i^+ for $i = 1, \ldots, n$:

(a) The outer face of G_i^+ is the same as the outer face of G_i.
(b) Every interior face of G_i^+ is a triangle.
(c) Every interior vertex of G_i^+ is interior-happy.
(d) For any edge e on the outer face, at least one endpoint of e is interior-happy, with the possible exception of the edge (s, t) that is the first ear P_1.
(e) G_i^+ may have parallel edges, but it has no loops.

Note that $G_1^+ := G_1$ satisfies the invariant since it has no interior faces or vertices, and the only edge on the outer face is (s, t). Now fix some $i > 1$, and let $P_i = \langle y_0, y_1, \ldots, y_k, y_{k+1} \rangle$ where $y_0 \neq y_{k+1}$ are vertices of G_{i-1} while all other vertices are new. Let F_i be the face added to G_{i-1} by P_i, and enumerate it as $\langle y_0, y_1, \ldots, y_k, y_{k+1} = z_0, z_1, \ldots, z_\ell, z_{\ell+1} = y_0 \rangle$, so $\langle z_0, z_1, \ldots, z_\ell, z_{\ell+1} \rangle$ was a path along the outer face of G_{i-1} between the ends of P_i. We add edges inside F_i in three phases (see also Fig. 2):

1. First ensure (c), i.e., all interior vertices of G_i^+ are interior-happy. This holds for all such vertices except $z_1, \ldots, z_\ell$ by induction. For $1 \leq j \leq \ell$, if z_j is

---

[1] We are not aware of an explicit reference for this result, but it can be extracted using standard graph ordering tools. In particular, compute an st-order with first vertex s and last vertex t in linear time [9], convert it to a bipolar orientation [17], choose the outer face to be the face different from F incident to (s, t), compute the dual bipolar orientation G^*, enumerate the interior faces of G via a topological order of G^*, and verify that each interior face then adds an ear. We refer the reader to [17] for definitions of these terms, and leave the details of correctness as an exercise.

unhappy, then add an edge (z_{j-1}, z_{j+1}) inside F_i. See for example vertex z_1 in Fig. 2. We call this operation *cutting z_j off F_i*, and note that it makes z_j interior-happy due to the triangle at angle $\angle z_{j-1}z_jz_{j+1}$. Adding the edge can be done without violating planarity since by (d) vertices z_{j-1} and z_{j+1} were interior-happy in G_{i-1}^+ and so we did not cut them off F_i.

2. Next ensure (d), i.e., for any edge on the outer face at least one endpoint is interior-happy. This holds for all edges except the ones of P_i by induction. If k is odd, then add a path y_0-y_2-y_4-$\ldots$-y_{k+1} inside F_i, and note that these edges do not interfere with the edges added in Phase 1. See the green (dashed) edges added inside F_3 in Fig. 2. This makes $y_1, y_3, \ldots, y_k$ interior-happy, and so (d) holds.

If k is even, then we first ensure that $y_{k+1}=z_0$ is interior-happy. Assume it was not interior-happy in G_{i-1}^+. Then by (d) either z_1 was interior-happy in G_{i-1}^+, or $(z_0, z_1) = (s, t)$. Either way we did *not* cut z_1 off face F_i in Phase 1 and can add an edge (z_1, y_k), which makes y_{k+1} interior-happy. See for example face F_2 in Fig. 2 where we make t interior-happy by adding an edge incident to s. Now that y_{k+1} is interior-happy (or was interior-happy in G_{i-1}^+ already), we add a path y_0-y_2-y_4-$\ldots$-y_k, which makes $y_1, y_3, \ldots, y_{k-1}$ interior-happy. With this (d) holds.

3. Finally we ensure (b), by triangulating what remains of F_i arbitrarily.

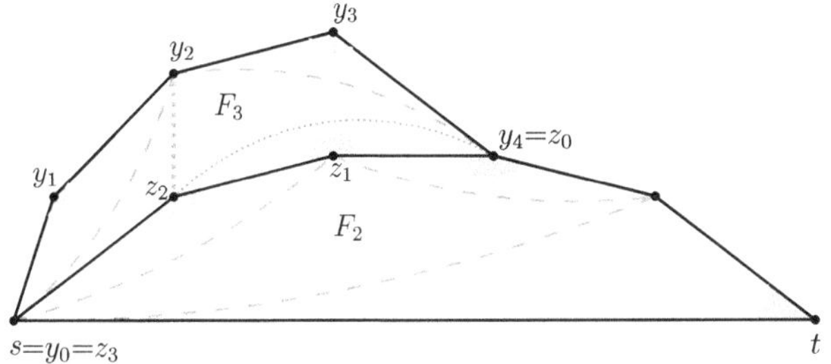

Fig. 2. The edges added for G_2^+ and G_3^+. Phase-1 edges are blue (dotted), Phase-2 edges are green (dashed), Phase-3 edges are brown (small squares). The angle that makes a vertex happy is shaded. (Color figure online)

Since all added edges were inside F_i, we never change the outer face of G_i. Also the boundary of F_i was a simple cycle of length 3 or more by 2-connectivity and since we have no bigons. Since we add edges only between vertices of distance 2 along the boundary of F_i, we never add loops.

This shows how to build G_n^+, which clearly can be done in linear time since the step of adding P_i takes time proportional to the degree of F_i. By triangulating its outer face, we can now obtain our desired super-graph G^+ as described below.

Lemma 2. *Let G be a 2-connected plane graph with $n \geq 3$ and no bigons. Fix a vertex s. Then in linear time we can find a triangulated super-graph G^+ that has no loops, and where all vertices except s are happy.*

Proof. Compute an ear decomposition where the first ear is an edge (s, t) for some t, and compute the super-graph G_n^+ of G as explained above. Enumerate the outer face as $\langle s{=}z_0, z_1, \ldots, z_\ell, z_{\ell+1}{=}t \rangle$. As in Phase 1, if z_j is unhappy for some $1 \leq j \leq \ell$, then add edge (z_{j-1}, z_{j+1}) through the outer face to make z_j interior-happy. See Fig. 3. Finally we can ensure that t is happy (we note that symmetrically we could have made s happy, but we cannot guarantee that both s and t are happy). Namely, if t is unhappy, then z_ℓ was interior-happy in G_n^+. So we did not add an edge $(z_{\ell-1}, z_{\ell+1})$, and we can add edge (z_ℓ, s) through the outer face to make t happy. Triangulate the remaining outer face arbitrarily to obtain G^+. Clearly all steps can be implemented in linear time.   □

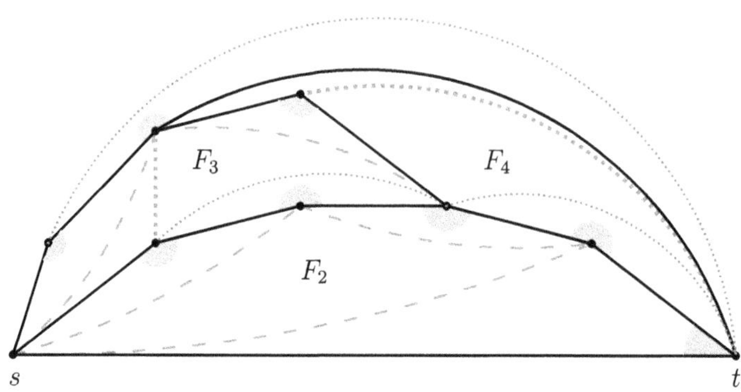

Fig. 3. Graph G_n^+ and triangulating the outer face. Vertex s is happy (though not interior-happy) due to the added edge.

The potentially unhappy vertex s will not bother us (because Lemma 1 allows us to specify one edge that must be in H), but it is a natural question whether it could be avoided. An example of an odd-length cycle C shows that in general this is impossible: No matter how we add edges, we can make at most $(|C| - 1)/2$ vertices on each side happy while staying planar, and so one vertex must remain unhappy. But we can show that for all other 2-connected plane graphs, *all* vertices can be made happy. Since this result is irrelevant for Theorem 2, we defer its proof to Sect. 4.

Before proving Theorem 2, we need an easy observation about angles of faces. (This slightly simplifies the approach of [10]; there is no need to explicitly ensure 'happy' faces.)

Observation 1. *Let G be a 2-connected plane graph without bigons, and let G^+ be any planar super-graph of G. Then for every face F of G, there exists in G^+ an angle that was also an angle of G at F.*

Proof. The boundary B of F is a simple cycle with at least three vertices since G is 2-connected without bigons. Consider the subgraph G_F of G^+ formed by B

and all edges that were inserted inside F. This is an *outerplanar* graph (it has a planar drawing where all vertices are on one face), and since it is additionally 2-connected with at least three vertices, it has a vertex w of degree 2. Vertex w lies on B, hence its two incident edges in G_F belong to B, and form an angle in G^+ since G_F includes all edges that we added inside F. $\qquad\square$

Corollary 1. *Theorem 2 holds for any 2-connected plane graph G without bigons.*

Proof. Clearly the theorem holds if $n = 2$ since G itself is then bipartite, so we may assume that $n \geq 3$. Let s be an endpoint of the reference-edge $\hat{e}$. Using Lemma 2, find a triangulated super-graph G^+ of G where all vertices except s are happy. Apply Lemma 1 to G^+ with respect to edge $\hat{e}$. This gives us, in linear time, a bipartite subgraph H^+ of G^+ that contains $\hat{e}$. Also, for every vertex $v \neq s$ there exists an incident angle in G^+ that makes v happy, i.e., both incident edges of v at the angle belong to G. At least one of these edges is retained in H^+, so every vertex $v \neq s$ has at least one incident edge in $E(H^+) \cap E(G)$. Define H to be the bipartite subgraph of H^+ with edge set $E(H^+) \cap E(G)$; by the above $E(H)$ is an edge-cover. By Observation 1 edge set $E(H)$ is also face-hitting: For any face F of G, super-graph G^+ has an angle where both edges were on F (in particular they were both in $E(G)$). At least one of them is retained in H^+, and therefore belongs to $E(H)$. $\qquad\square$

As it turns out, bigons do not pose a problem either.

Corollary 2. *Theorem 2 holds for any 2-connected plane graph G.*

Proof. Let G' be the subgraph of G obtained by deleting, for every bigon of G, one of the parallel edges that bound the bigon (choosing one that is not reference-edge $\hat{e}$, if $\hat{e}$ is in a bigon). Graph G' is 2-connected (adding parallel edges cannot increase vertex-connectivity) and has no bigons, and so by Corollary 1 we can find a bipartite subgraph H' such that $E(H')$ is a face-hitting edge cover of G'. Let H be the bipartite subgraph of G obtained by adding to H' all bigon-edges where one edge of the bigon was in H'. Clearly H is bipartite and all steps to find it can be implemented in linear time.

We claim that $E(H)$ is 3^+-face-hitting (and of course it is an edge cover since already $E(H')$ was one). Let F be any 3^+-face of G. If F was also a face of G', then it was hit by $E(H')$ and so also by $E(H)$. If F was not a face of G', then it was replaced in G' by a face F' that is the union of F and a number of bigons. Some edge e' of F' was in $E(H')$, hence e' is either on F or it belongs to a bigon, in case of which the other edge of that bigon was added to H. Either way therefore some edge of F is in $E(H)$. $\qquad\square$

3.2 Putting It All Together

We now put all ingredients together for the proof of Theorem 2.

Proof. We already know the result for 2-connected graphs (Corollary 2). The way to generalize it to arbitrary connected graphs is very standard: Compute the bipartite subgraph for each 2-connected component of G, and combine them to obtain a bipartite subgraph of G that (as one argues) satisfies all conditions. However, if G is not simple then we have to be a bit more careful. Therefore we give below the details of how to define the subgraphs and how to combine them.

For ease of description choose the outer face of G such that it contains the reference-edge $\hat{e}$. Compute the 2-connected components of G; this can be done in linear time [12]. Consider a non-trivial 2-connected component C (i.e., C has at least two vertices). If C does not contain $\hat{e}$, then choose as reference-edge $\hat{e}_C$ of C an arbitrary edge on the outer face of C; otherwise set $\hat{e}_C = \hat{e}$. Apply Corollary 2 to C (with $\hat{e}_C$ as reference-edge) to obtain a bipartite subgraph H_C of C for which $E(H_C)$ is a 3^+-face hitting edge cover. Let H be the union of all these subgraphs; since 2-connected components intersect in at most one vertex and H has no loops, this is bipartite. Since (in a connected graph with $n \geq 2$ vertices) every vertex belongs to at least one non-trivial 2-connected component C, and is incident to an edge of H_C, the set $E(H)$ is an edge cover.

Now we must show that $E(H)$ hits any 3^+-face F of G. If F is a face of one 2-connected component C, then $E(H_C)$ (and hence $E(H)$) hits F. If F is not a face of one 2-connected component, then let B be the boundary walk of F and let $C_1, \ldots, C_\ell$ for $\ell \geq 2$ be the 2-connected components intersected by B. Note that $B \cap E(C_i)$ is a face of C_i (in the induced planar embedding of C_i) for all i. If this is a 3^+-face of C_i then $E(H_{C_i})$ will include an edge of it and so $E(H)$ hits F. If $B \cap E(C_i)$ is the outer face of C_i, and C_i is not a loop, then reference edge $\hat{e}_{C_i}$ will be chosen on B and included in $E(H)$ and hit F. So the only non-trivial case is when for all i either $B \cap E(C_i)$ is an inner face of C_i of size at most 2, or C_i is a loop. But then F has size at most 2 and is not a 3^+-face. $\qquad\square$

4   Making Everyone Happy

In this section, we prove a strengthening of Lemma 2 which shows that we can make all vertices happy except in the case of an odd-length cycle.

Lemma 3. *Let G be a 2-connected plane graph with $n \geq 3$ and no bigons. Assume that G is not a cycle of odd length $d \geq 5$. Then G has a triangulated super-graph G^+ that has no loops, and where all vertices are happy.*

Proof. We continue in the notation of Sect. 3.1. The main idea is to choose the ear decomposition $P_1, \ldots, P_f$ (and perhaps also the outer face) carefully so that in the computed augmentation G_n^+, at least one of the two endpoints s, t of P_1 is interior-happy. Then when triangulating the outer face (as done in the proof of Lemma 2) we can ensure that the other vertex of $\{s, t\}$ is happy and the result holds. To show how to compute this augmentation G_n^+, we distinguish cases.

Case 1: G has a face F that either has even degree or is a triangle. (In particular this covers the case where G is an even-length cycle.) Choose the outer face of G

and an ear decomposition such that face F becomes F_2 and (s, t) is an arbitrary edge on it. If F_2 is a triangle then without adding further edges both s and t are interior-happy in $G_2 = G_2^+$. If F_2 is not a triangle, then ear P_2 has $\{s, t\}$ as its endpoints, and hence adds an even number of vertices. Inspecting Phase 2 of the computation of G_2^+ (see also the added edges in face F_2 of Fig. 2), we see that we make one of $\{s, t\}$ interior-happy.

Case 2: All faces of G have odd degree, but there exists a vertex x of degree 2. Since G is not a cycle, we can choose x such that it has a neighbour t with $\deg(t) \geq 3$. Choose an angle $\angle xts$, which defines vertex s. Choose an ear decomposition such that (s, t) is the first ear and F_2 is incident to $\angle xts$. See Fig. 4.

Face F_2 has odd degree, hence P_2 adds an odd number of vertices to the graph, and applying Phase 2 to face F_2 hence makes x interior-happy in G_2^+. Let F' be the other face incident to (x, t), and observe that this is not the outer face since $\deg(t) \geq 3$. Hence $F' = F_j$ for some $j \geq 3$, and it is added when handling ear $P_j = \langle y_0{=}s', \ldots, y_k, y_{k+1}{=}t \rangle$, where s' is some vertex of G_{j-1}. If k is even then adding P_j will make t interior-happy and we are done, so assume that k is odd. Enumerate the path from t to y_0 on the outer face of G_{j-1}^+ as $z_0{=}t, z_1, \ldots, z_{\ell+1}{=}s$ as before, and note that $z_1 = x$ had two incident edges in G_{j-1} already, so $x \neq s'$ and $\ell \geq 1$. We perform Phase 1 as before; this does not cut x off F_j since x was already interior-happy. We next add (x, y_k) to make t interior-happy. We next ensure that s' is interior-happy, by adding (z_ℓ, y_1) if it is not. To see that this is feasible in our special case here, observe first that if s' was unhappy in G_{j-1}^+ then z_ℓ was interior-happy and therefore is still on F_j after Phase 1. Also, by $\ell \geq 1$, adding (z_1, y_k) does not remove z_ℓ from F_j, and so we can add the edge inside F_j. Finally we add path y_1-y_3-$\ldots$-y_k to make $y_2, y_4, \ldots, y_{k-1}$ interior-happy, and triangulate what remains of F_j arbitrarily. With this the invariant holds for G_j^+ and t is interior-happy.

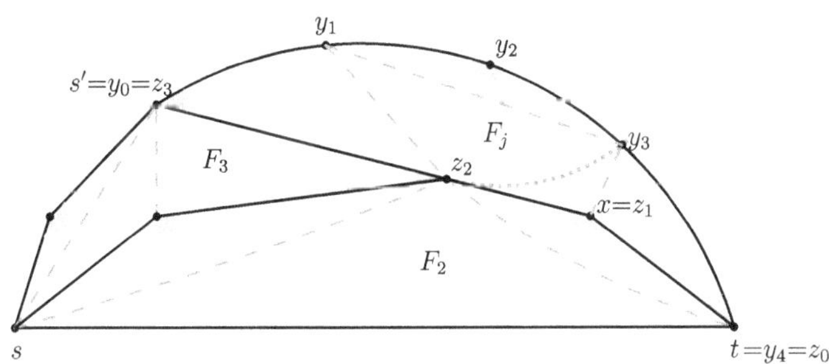

Fig. 4. Using a vertex x of degree 2 to make t happy. We have $k = 3$ and $\ell = 2$.

Case 3: None of the previous cases applies. Compute an arbitrary ear decomposition. Since G has no vertices of degree 2, we must have more than one interior face, and therefore have an ear P_3. We perform Phase 1 and 2 on faces F_2 and

F_3 together as follows (see also Fig. 5). The endpoints $\{s', t'\}$ of P_3 must be two *consecutive* vertices of F_2, for otherwise the vertices of P_2 between s' and t' would not be on the outer face of G_3, hence receive no further incident edges, and would have degree 2. Also both F_2 and F_3 have odd degree. Temporarily delete edge $e = (s', t')$ to turn the union of faces F_2 and F_3 into one face $\hat{F}$ of even degree, and enumerate it as $\langle y_0{=}s, y_1, \ldots, y_k, y_{k+1}{=}t \rangle$ with $(s', t') = (y_i, y_j)$; both k and $j - i$ are even. Consider the two cycles $C_0 := y_0\text{-}y_2\text{-}\ldots\text{-}y_k\text{-}y_0$ and $C_1 := y_1\text{-}y_3\text{-}\ldots\text{-}y_{k+1}\text{-}y_1$. Each cycle could be added inside $\hat{F}$ and make every other vertex on $\hat{F}$ interior-happy; in particular each cycle makes one of $\{s, t\}$ interior-happy. One of these two cycles would not cut y_i and y_j off $\hat{F}$ since $j - i$ is even. The edges of this cycle can hence be inserted inside F_2 and F_3 in G_3 and give the desired graph G_3^+ (after triangulating the rest of F_2 and F_3 arbitrarily) where one of s, t is interior-happy. $\qquad\qquad\square$

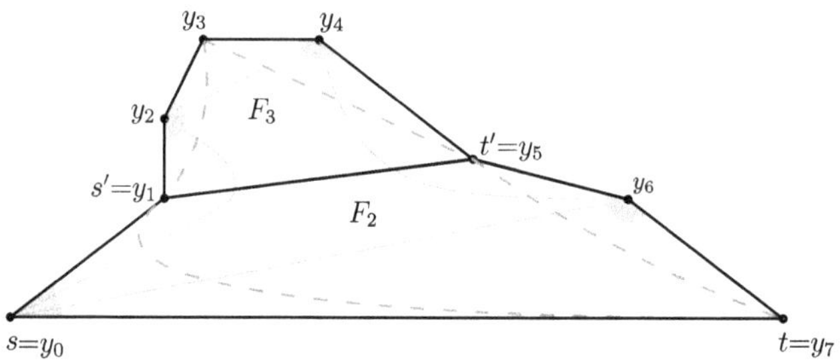

Fig. 5. If there are no vertices of degree 2, and all faces have odd degree, then applying Phase 2 to the union of F_2 and F_3 suitably makes one of $\{s, t\}$ interior-happy. The alternate cycle to consider is thin dotted.

5   Further Thoughts

In this paper, we gave a new proof that every simple planar connected graph with $n \geq 2$ vertices has a vertex partition into two face-hitting dominating sets, with the goal of developing a linear-time algorithm to find this vertex partition. The main tool was to find a bipartite subgraph whose edges are both an edge cover and hit every face. The proof extends to graphs that are not simple, as long as we are content with hitting only those faces that are incident to at least three distinct vertices.

We briefly discuss here what changes if we want to hit *all* faces in a non-simple graph using a dominating set S. (We continue to assume $n \geq 2$ and connectivity, since otherwise already being dominating may require $|S| = n$.) If we permit faces whose boundary is a loop, then we must include the endpoint of the loop in S, and so again $|S| = n$ may be required. If we permit bigons, then $|S| \geq \frac{3}{4}n$ may be required: Take $n/4$ copies of K_4, replace every edge by a bigon, and then connect the copies with arbitrary further edges to make the graph connected and planar. To hit the bigons at each K_4, set S must include a vertex cover of K_4, which requires 3 out of 4 vertices at each K_4 and hence at least $\frac{3}{4}n$ in total. On the positive side, a face-hitting dominating set of size at most $\frac{3}{4}n$ exists: take any vertex cover. This can be found by 4-coloring the graph and taking the three smallest color classes, or for a faster algorithm one can use Baker's approximation scheme [1] to find a vertex cover of size at most $(1 + \varepsilon)\frac{3}{4}n$, for $\varepsilon > 0$ arbitrarily small, in linear time.

As for open problems, a first question concerns whether adding parallel edges is ever needed in Lemma 2 and 3. Does every simple 2-connected graph G that is not an odd cycle have a *simple* triangulated supergraph G^+ where all vertices are happy? (This clearly holds if G is 3-connected, since a parallel edge can only exist in a planar graph if its ends are a cutting pair.) Another open problem concerns bounds for 4-connected planar graphs. The lower-bound example of Bose et al. [5] has triangles that are not faces, i.e., it is not 4-connected. Can we improve the size of face-hitting dominating sets for 4-connected planar graphs, at least for triangulations?

References

1. Baker, B.: Approximation algorithms for NP-complete problems on planar graphs. J. ACM **41**(1), 153–180 (1994)
2. Biedl, T., Bose, P., Demaine, E., Lubiw, A.: Efficient algorithms for Petersen's theorem. J. Algorithms **38**(1), 110–134 (2001)
3. Biedl, T., Kant, G., Kaufmann, M.: On triangulating planar graphs under the four-connectivity constraint. Algorithmica **19**(4), 427–446 (1997)
4. Bose, P., Kirkpatrick, D., Li, Z.: Efficient algorithms for guarding and illuminating the surface of a polyhedral terrain. In: CCCG 1996. International Informatics Series, vol. 5, pp. 217–222. Carleton University Press (1996)
5. Bose, P., Shermer, T., Toussaint, G., Zhu, B.: Guarding polyhedral terrains. Comput. Geom. Theory Appl. **6**(3), 173–195 (1997)
6. Christiansen, A., Rotenberg, E., Rutschmann, D.: Triangulations admit dominating sets of size $2n/7$. In: Proceedings of the 2024 ACM-SIAM Symposium on Discrete Algorithms, SODA 2024, pp. 1194–1240. SIAM (2024)
7. Diestel, R.: Graph Theory. Graduate Texts in Mathematics, vol. 173, 4th edn. Springer, Cham (2012)
8. Evans, W., Gethner, E., Spalding-Jamieson, J., Wolff, A.: Angle covers: algorithms and complexity. J. Graph Algorithms Appl. **25**(2), 643–661 (2021)
9. Even, S., Tarjan, R.E.: Computing an st-numbering. Theoret. Comput. Sci. **2**, 436–441 (1976)
10. Francis, P., Illickan, A., Jose, L., Rajendraprasad, D.: Face-hitting dominating sets in planar graphs. In: Graph-Theoretic Concepts in Computer Science - 50th International Workshop, WG 2024. LNCS, vol. 14760, pp. 211–219. Springer, Cham (2024)
11. Garey, M.R., Johnson, D.S.: Computers and Intractability: A Guide to the Theory of NP-Completeness. Freeman (1979)
12. Hopcroft, J.E., Tarjan, R.E.: Efficient algorithms for graph manipulation. Commun. ACM **16**(6) (1973)

13. Kant, G., Bodlaender, H.: Triangulating planar graphs while minimizing the maximum degree. Inf. Comput. **135**(1), 1–14 (1997)
14. Matheson, L., Tarjan, R.: Dominating sets in planar graphs. Eur. J. Comb. **17**(6), 565–568 (1996)
15. Petersen, J.: Die Theorie der regulären graphs (The theory of regular graphs). Acta Math. **15**, 193–220 (1891)
16. Robertson, N., Sanders, D., Seymour, P., Thomas, R.: The four-colour theorem. J. Combin. Theory Ser. B **70**(1), 2–44 (1997)
17. Rosenstiehl, P., Tarjan, R.E.: Rectilinear planar layouts and bipolar orientation of planar graphs. Discret. Comput. Geom. **1**, 343–353 (1986)
18. Whitney, H.: Non-separable and planar graphs. Proc. Natl. Acad. Sci. **17**(2), 125–127 (1931)

k-Planar and Fan-Crossing Drawings and Transductions of Planar Graphs

Petr Hliněný[ID] and Jan Jedelský[(✉)][ID]

Faculty of Informatics, Masaryk University, Brno, Czech Republic
{hlineny,xjedelsk}@fi.muni.cz

Abstract. We introduce a two-way connection between FO transductions (first-order logical transformations) of planar graphs and a certain variant of fan-crossing (fan-planar) drawings of graphs which for bounded-degree graphs essentially reduces to being k-planar for fixed k. For graph classes, this connection allows us to derive non-transducibility results from nonexistence of the said drawings and, conversely, from nonexistence of a transduction to derive nonexistence of the said drawings. For example, the class of 3D-grids is not k-planar for any fixed k. We hope that this connection will help to draw a path to a possible proof that not all toroidal graphs are transducible from planar graphs.

Our characterization can be extended to any fixed surface instead of the plane. The result is based on a very recent characterization of weakly sparse FO transductions of classes of bounded expansion by [Gajarský, Gładkowski, Jedelský, Pilipczuk and Toruńczyk, https://arxiv.org/abs/2505.15655].

1   Logic, Interpretations and Graph Drawings

Briefly introducing (cf. Sect. 2 for more details), a simple *first-order (FO) interpretation* of graphs is given by a binary FO formula ξ over graphs, and for a graph G the result of the interpretation is the simple graph $\xi(G)$ on the same vertex set and the edge relation determined by $uv \in E(\xi(G)) \iff G \models \xi(u,v)$ (assumed symmetric). For example, $\xi(u,v) \equiv \neg edge(u,v)$ interprets the complement of a graph, and $\xi(u,v) \equiv edge(u,v) \vee \exists x\big[u \neq v \wedge edge(u,x) \wedge edge(x,v)\big]$ interprets the square (second power) of it.

In an *FO transduction* τ, one can additionally multiplicate the vertices of G and assign arbitrary vertex colors before applying ξ, and then take an induced subgraph of the result (in particular, $\tau(G)$ is actually a hereditary set of graphs – unlike $\xi(G)$). These notions are naturally extended to graph classes; here $\tau(\mathcal{C})$ denotes the class of all graphs obtained by applying τ to all (colored) graphs from a class $\mathcal{C}$. All transductions in this paper are first-order.

We say that a graph class $\mathcal{D}$ is *transducible* from a class $\mathcal{C}$ if there is a transduction τ such that $\mathcal{D} \subseteq \tau(\mathcal{C})$. We interpret this relation as that $\mathcal{D}$ is not "logically richer" than $\mathcal{C}$ (although $\mathcal{D}$ may look combinatorially a lot more complicated class than $\mathcal{C}$). Alternatively, the aim is that if we understand and

© The Author(s), under exclusive license to Springer Nature Switzerland AG 2026
J. Kozik and A. Wolff (Eds.): SOFSEM 2026, LNCS 16448, pp. 505–516, 2026.
https://doi.org/10.1007/978-3-032-17801-5_37

can handle the class $\mathcal{C}$, we would be able to do so with $\mathcal{D}$. We have recently seen a surge of interest in FO transductions of graph classes, e.g., [3,12]. However, while transducibility is relatively easy to establish in particular cases of interest (essentially, guess a coloring and the formula), the opposite (not transducible) is usually much harder to prove and existing results are scarce.

Typically, to prove that a class $\mathcal{D}$ is not transducible from a class $\mathcal{C}$, one finds a suitable property of $\mathcal{C}$ which is preserved under transductions, but $\mathcal{D}$ does not possess this property. The rather few published examples of properties preserved under transductions include, e.g.; bounded clique-width [4], each value of shrub-depth [10], near-uniformness [8], bounded twin-width [2] and bounded flip-width [13]. However, neither of those is of much help in approaching such a basic and intriguing question as what characterizes transductions of the *class of planar graphs*. So far, to our knowledge, no useful direct connection (of transducibility) to graph topological properties and graph drawings has been published.

Very recently, there has been a notable progress in two directions. First, two groups [9,11] independently exploited different properties related to the product structure (of planar graphs) to prove, among other results, that 3D-grids are not transducible from planar graphs. Second, and most importantly for our research, Gajarský, Gładkowski, Jedelský, Pilipczuk and Toruńczyk [7] have given a new characterization of transducibility in sparse classes, stated in Theorem 1.1 below, which resolved several further open questions, such as that the graphs of tree-width $t + 1$ are not transducible from the graphs of tree-width t for any t.

We refer to Sect. 2 for the definitions of the used technical terms, and only briefly remark that planar and surface-embeddable graphs always are weakly sparse and of bounded expansion.

Theorem 1.1 (Gajarský et al.[7]). *Let $\mathcal{C}$ be a graph class of bounded expansion and $\mathcal{D}$ be a graph class transducible from $\mathcal{C}$ such that $\mathcal{D}$ is weakly sparse. Then there exists a $k \in \mathbb{N}$ such that $\mathcal{D}$ is contained in the class of congestion-k depth-k minors of the class $\mathcal{C}^\bullet$, where $\mathcal{C}^\bullet$ is the class obtained by adding a universal vertex to every graph of $\mathcal{C}$.*

Whereas Gajarský et al. [7] use the Theorem 1.1 primarily in connection with properties of weak colorings in the concerned graph classes, we give a different, topological, perspective of it. The aim is to provide tools from the graph drawing area for proving non-transducibility results.

We consider only (the traditional) drawings of graphs in which no edge passes through another vertex, no three edges meet in the same point except their common end, and there are finitely many intersection points (*crossings* or common end vertices) of pairs of the edges. We do *not* assume our drawings to be simple. For a drawing D of a graph, the *crossing graph C of D* has the edges of D as its vertices, and two edges are adjacent in C iff they cross in D.

A drawing of a graph G in the plane is *k-planar* if every edge carries at most k crossings. A *fan* in G is any subset of edges incident to the same vertex. A drawing is *fan-crossing* if every edge is only crossed by edges of a fan.

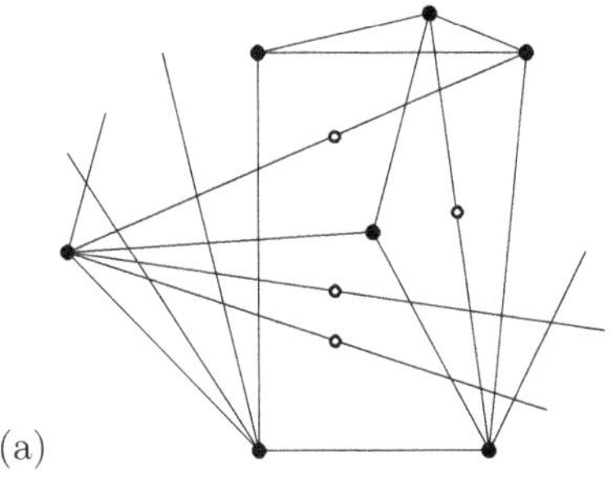
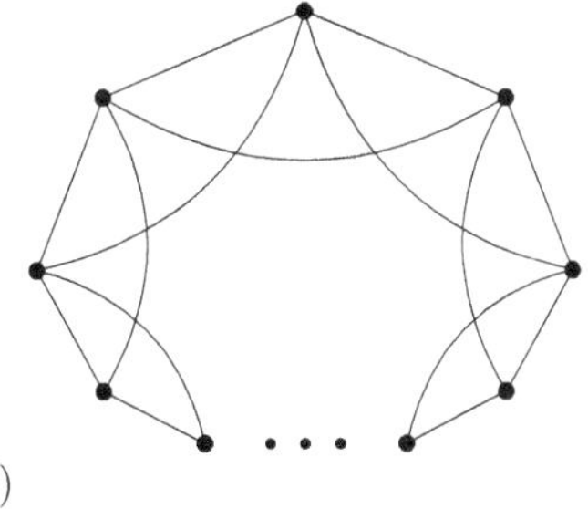

Fig. 1. (a) An example of a 2-fold 2-clustered fan-crossing drawing D. The four subdivision vertices (i.e., the set $V(D') \setminus V(D)$ from Definition 1.2) are hollow, and the three components of the crossing graph of the subdivided drawing D' are emphasized by shade colors. (b) An ordinary fan-crossing drawing of a large m-edge graph (actually, strongly fan-planar) which is not 1-fold ℓ-clustered for any $\ell < \frac{m}{4}$, because no subdivisions are allowed and each fan has at most 4 edges.

Our main result, Theorem 1.3, is based on a variant of fan-crossing drawings and fan-planarity[1] which is analogous to the generalization from 1-planar to k-planar graphs.

Definition 1.2. A drawing D of a graph G in the plane is called *k-fold ℓ-clustered fan-crossing* if there is a drawing D' obtained from D by subdividing each edge with at most $k-1$ new vertices, such that in every connected component $M \subseteq E(D')$ of the crossing graph of D', the edges forming M are subsets (as point sets) of the edges of at most ℓ fans of D.

See Fig. 1. We remark that 1-fold ℓ-clustered fan-crossing drawings are not directly comparable to ordinary fan-crossing drawings which, in general, do no fulfil the ℓ-clustered condition for any ℓ. However, similar 'clustering' property is naturally satisfied in 1-planar drawings and in natural $(k-1)$-subdivisions of k-planar drawings.

Theorem 1.3. *Let C be the class of planar graphs and D be a weakly sparse graph class. Then D is transducible from C, if and only if there exists $k \in \mathbb{N}$ such that every graph of D has a k-fold k-clustered fan-crossing drawing in the plane after deleting at most k of its vertices.*

Since any k-fold k-clustered fan-crossing drawing of a graph G is trivially $(\Delta(G) \cdot k^2)$-planar, we also have the following corollary interesting by itself:

Corollary 1.4. *Let C be the class of planar graphs and D be a graph class of bounded maximum degree. Then D is transducible from C, if and only if there exists $k \in \mathbb{N}$ such that every graph of D is k-planar after deleting at most k of its edges.*

---

[1] This generalization somehow resembles k-fan-bundle-planar drawings of [1], but the core difference is that our "bundles" are allowed to branch and cross in multiple sections. See also Sect. 3.

Remark 1.5. If the 'target' class $\mathcal{D}$ has the *duplication property*, that is with each $H \in \mathcal{D}$ there is $H' \in \mathcal{D}$ such that H' contains two disjoint copies of H as an induced subgraph, then the part 'after deleting at most k of its vertices/edges' can be safely removed from both Theorem 1.3 and Corollary 1.4. (The duplication property is weaker than being closed on disjoint unions.)

2   Transductions; Congested Shallow Minors

We consider only finite simple graphs (simplicity of graphs is natural and necessary when dealing with FO transductions as we do here).

For an FO formula ξ of (vertex) colored graphs with two free variables, and a colored graph G, let $\xi(G) = H$ be the graph with $V(H) = V(G)$ and $E(H) = \{xy : G \models \xi(x, y) \wedge \xi(y, x)\}$. A *transduction* τ (determined by ξ) maps a graph G to a set of graphs $\tau(G)$ such that $H \in \tau(G)$ if and only if, for some coloring of G, the graph H is an induced subgraph of $\xi(G)$. This definition actually describes non-copying transductions, and we stick with this simplified view here since copying transductions (allowing to duplicate the vertices of G before the rest) coincide with non-copying ones on planar graphs and graphs in surfaces (via adding leaves).

Consider a graph G and a collection $\mathcal{A} \subseteq 2^{V(G)}$ of vertex subsets such that, for each $A \in \mathcal{A}$, the induced subgraph $G[A]$ is connected. Then $\mathcal{A}$ is called a *congestion-c depth-d minor model* of a graph H on $V(H) = \mathcal{A}$ (and the members of $\mathcal{A}$ are called *model sets*) if: (i) every induced subgraph $G[A]$, $A \in \mathcal{A}$ is of radius $\leq d$, (ii) for every vertex $v \in V(G)$, at most c sets of $\mathcal{A}$ contain v, and (iii) for every edge $AB \in E(H)$ (where $A, B \in \mathcal{A}$), the sets A and B *touch* in G – meaning that $A \cap B \neq \emptyset$ or some two vertices $a \in A$ and $b \in B$ are adjacent in G. Note that if we set $c = 1$, we get the usual (depth-d or unlimited d) minor model with disjoint model sets.

A graph G_1 is a *congestion-c depth-d minor* of G if there exists a congestion-c depth-d minor model in G of a graph isomorphic to G_1. If $c = 1$, we speak simply about a *depth-d minor* of G.

A graph class $\mathcal{D}$ is *weakly sparse* if there is an integer t such that no member of $\mathcal{D}$ contains a $K_{t,t}$ subgraph.

A graph class $\mathcal{D}$ is of *bounded expansion* if there exists a function f such that every depth-d minor of a graph from $\mathcal{D}$ has average degree at most $f(d)$.

3   Proofs of Theorem 1.3 and Corollary 1.4

We start with the forward direction '$\Rightarrow$' of the proofs.

We actually prove a stronger conclusion formulated towards fan-planarity, cf. Fig. 2. For a drawing D of a graph G, a set F of edges in D crossing an arc α has the *fan property* if F is a fan with a root x, all edges of F intersect α once, and they come from x to α on the same side[2] of α. The fan property

---

[2] Recall that we are drawing on a plane, so both "enclosing" and "coming from the same side" are well-defined.

of F is *strong* if, additionally, the union of α and the edges of F (as a point set) does not enclose (see footnote 2) the two ends of α in the plane. With respect to Definition 1.2, a drawing D is *k-fold ℓ-clustered strongly fan-planar* if, additionally, each of the fans covering the crossing-graph component M has the strong fan property towards the other arcs of M.

 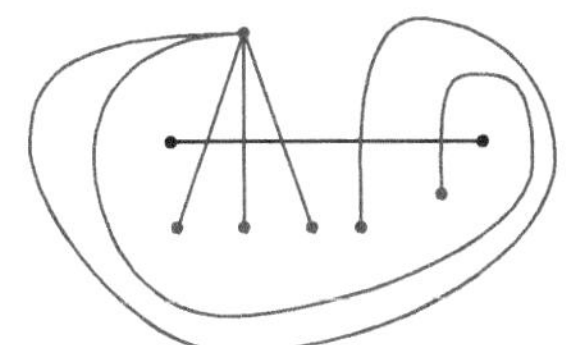 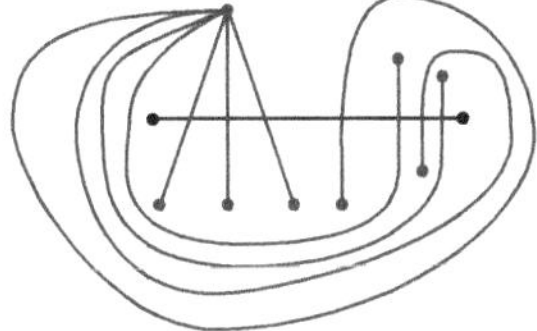

Fig. 2. (left) A strongly fan-planar drawing. (middle) A weakly fan-planar drawing that is not strongly fan-planar. (right) A fan-crossing drawing that is not even weakly fan-planar.

Lemma 3.1. *Let G be a planar graph and H a congestion-k depth-k minor of G. Then H has a k'-fold k'-clustered strongly fan-planar drawing in the plane for some $k' \in \mathcal{O}(k)$.*

Proof. Let $\mathcal{A}$ be a congestion-k depth-k minor model of H in G, and for every $v \in V(H)$, let $T_v \subseteq G$ be a rooted BFS tree of $G[A_v]$ (where A_v represents v) of depth $\leq k$. Let the root of T_v be $r_v \in V(G)$ (note that $r_v = r_w$ is possible for distinct $v, w \in V(H)$).

Fix a plane embedding D of G, and pick a collection of small pairwise disjoint disks δ_u around each $u \in V(G)$, and δ_e around an internal ('middle') point of each $e \in E(G)$, such that δ_u intersects D only in edges incident to u and δ_e intersects D only in e.

By a *branching arc* with a root x in the plane we mean a plane embedding of a tree with x as its root. The first step is to represent H as a subgraph of the intersection graph of a collection of branching arcs $\mathcal{R} = \{\varrho_v : v \in V(H)\}$ such that the root of ϱ_v lies in δ_{r_v} and the root is disjoint from other arcs, and that no three branching arcs meet in the same point. Moreover, for any $v, w \in V(H)$, the branching arcs ϱ_v and ϱ_w are allowed to intersect only in the disks δ_u for $u \in A_v \cap A_w$, or in the disk δ_e if disjoint A_v, A_w touch via an edge e in G.

Getting the sought drawing is nearly straightforward. We draw each ϱ_v of $\mathcal{R}$ closely along the plane drawing of T_v in D, but avoiding δ_e for $e \in E(T_v)$, such that for every $vw \in E(H)$;

- if A_v and A_w share a vertex $u \in A_v \cap A_w$ in the minor model in G, then the arcs ϱ_v and ϱ_w intersect (not necessarily cross) inside δ_u, and
- if disjoint A_v, A_w only touch via an edge $e \in E(G)$, then new branches of arcs are added to ϱ_v and ϱ_w along the drawing of e such that they intersect inside δ_e.

Moreover, this drawing can clearly avoid any crossings of arcs outside of the disks δ_u and δ_e. Note that we only require H to be a subgraph of the intersection graph of $\mathcal{R}$, and so superfluous intersections between the arcs do not pose a problem. Finally, since multiple intersections between pairs of arcs are allowed, this drawing satisfies all desired properties.

Secondly, as H is a subgraph of the intersection graph of $\mathcal{R}$, for every edge $f = vw \in E(H)$ we have a simple arc $\alpha_f \subseteq \varrho_v \cup \varrho_w$ between the roots of ϱ_v and ϱ_w. We arbitrarily pick a point in common $b_f \in \alpha_f \cap \varrho_v \cap \varrho_w$. Then, in a tiny neighborhood of every branching arc $\varrho_v \in \mathcal{R}$ we draw (with no self-intersections) a 'half-fan' with the center vertex v in the root of ϱ_v and rays closely following each α_f for $f \ni v$ from v to the point b_f. This is possible by our choices of α_f and b_f (in particular, since the chosen 'meeting points' b_f are pairwise distinct by the property of no three branching arcs meeting in the same point), and the union is a drawing D_1 of the graph H. (We note in passing that the obtained drawing easily avoids tangential intersections between edges, and so has only proper crossings.)

So, in D_1, every edge $f = vw \in E(H)$ follows a path $P \subseteq G$ such that $P \subseteq T_v \cup T_w (+e)$, where e is a possible edge connecting A_v and A_w if they are disjoint. Hence the length of P is at most $2k + 1$ and we subdivide f once along each edge of $P - e$ and twice (before and after δ_e) at e if it is present. After doing so for all $f \in E(H)$, we get a drawing D_2 subdividing D_1.

By the construction, each connected component $M \subseteq E(D_2)$ of the crossing graph of D_2 consists of edges intersecting inside one of the disks δ_u or δ_e defined above. Each edge of M comes with a branching arc $\varrho_v \in \mathcal{R}$ where $u \in A_v$ (in case of δ_e this u is either end of e), and since $\mathcal{A}$ is a congestion-k model, there are $\leq k$ arcs of $\mathcal{R}$ involved in the component M altogether. In other words, D_2 is an $\mathcal{O}(k)$-fold $\mathcal{O}(k)$-clustered fan-crossing drawing. Furthermore, by our construction, every arc $\varrho_v \in \mathcal{R}$ involved in M enters (meaning in direction from the root of ϱ_v) the respective disk δ_u or δ_e in one point (possibly the root of ϱ_v itself). We denote by $P_{u,v}$ the set of paths following the same branching arc ϱ_v and intersecting δ_u. By local modifications just inside the disks δ_u, it is easy to ensure that the paths $P_{u,v}$, viewed as edges of H, have the strong fan property towards every other arc drawn by M. Since u is the only endpoint in δ_u, it suffices to ensure that u is arc-connected to the boundary of δ_u in $\delta_u - (\bigcup P_{u,v} - \epsilon_u)$, where $\epsilon_u \subseteq \delta_u$ is a small disk around u intersected only by the paths having u as endpoint. $\qquad\square$

The '$\Rightarrow$' direction of Theorem 1.3 is finished easily from Theorem 1.1 and Lemma 3.1. Let $H' \in \mathcal{D}$ be a graph. By Theorem 1.1, there is $G \in \mathcal{C}^\bullet$, and a congestion-$k$ depth-k minor model of H' in G. Let $u \in V(G)$ be a universal vertex, i.e., $G - u$ is planar. Let $U \subseteq V(H')$ be the set of at most k vertices of H' represented in this model by sets containing u. Then $H := H' - U$ is a congestion-k depth-k minor of planar $G - u$, and we conclude with the drawing and bound k' by Lemma 3.1. The '$\Rightarrow$' direction of Corollary 1.4 then follows immediately as mentioned in Sect. 1.

The converse '⇐' directions of Theorem 1.3 and Corollary 1.4 are together stated next.

Lemma 3.2. *Let $k \in \mathbb{N}$, H be a graph and $X \subseteq V(H)$, $|X| \leq k$ a set such that $H - X$ has (a) a k-planar drawing, or (b) k-fold k-clustered fan-crossing drawing in the plane. Then there is an FO formula $\xi(x,y)$ depending only on k, and a colored planar graph G on vertex set $V(G) \supseteq V(H)$ such that $\xi(G)[V(H)] = H$.*

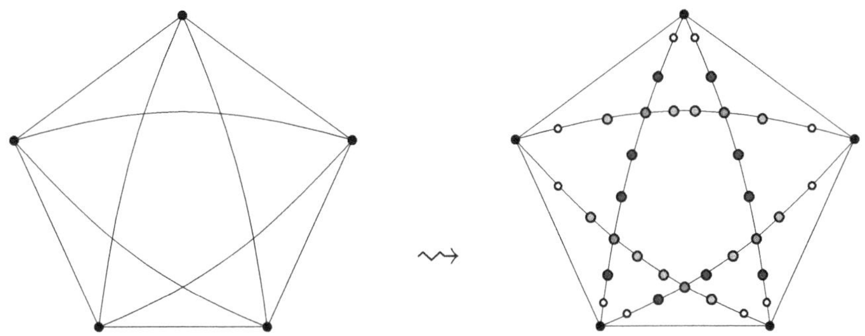

Fig. 3. An illustration of the proof of Lemma 3.2(a) (here with $X = \emptyset$); turning a 2-planar drawing D into a planar colored drawing D''. Color b_0 is red, and b_1 and b_2 are green and blue.(Color figure online)

Proof. (a) We let D be a k-planar drawing of $H - X$, and make a planar drawing D' by turning each crossing of D into a new vertex. Let D'' subdivide every edge of D' incident to a former crossing twice and G'' be the graph of D''. Let $X = \{x_1, \ldots, x_m\}$, $m \leq k$, and G be formed from G'' by adding $x_1, \ldots, x_m$ as isolated vertices. Note that $V(H) \subseteq V(G)$ and every edge of $H - X$ has a corresponding (i.e., formed by the said subdivisions) path in G of length at most $3(k+1)$. See Fig. 3.

We introduce colors c_i and c_i' for $1 \leq i \leq m$, and three more colors b_0, b_1 and b_2. Color c_i is given to x_i and c_i' to every neighbor of x_i in H (the colors c_i' are combined together into one resulting color at each vertex of H). Color b_0 is given to every vertex v of G coming from a crossing in D. If such v comes from a crossing of two edges in D, we arbitrarily order these edges as e_1, e_2, and for each $i \in \{1, 2\}$, we color by b_i the two neighbors of v in G which belong to the former edge e_i of D.

The formula $\xi(x,y)$ depending on k is constructed as a disjunction of the following possibilities:

- For some $1 \leq i \leq m$, x has color c_i and y color c_i' or vice versa, or
- xy is an edge of G, or
- there exists a path $P = (x, z_1, \ldots, z_p, y)$ in G of length at most $3(k+1)$, such that for every $1 < i < p$ the vertex z_i has color b_1 or b_2, or z_i has color b_0 and both its neighbors z_{i-1} and z_{i+1} have the same color among b_1 or b_2.

The said properties are routinely expressible in FO logic (in particular, we use a separate existential quantifier for every internal vertex of the path P). It is also immediate that, when restricted to the ground set $V(H)$, $\xi(x,y)$ captures precisely the edges xy of H.

(b) Let D be a k-fold k-clustered fan-crossing drawing of $H - X$, and let D' be the subdivision of D as assumed by Definition 1.2. First of all, by local perturbations, we may assume that there are no tangential intersections between edges in D', only proper crossings. Let D_1 further subdivide every crossed edge of D' incident to a vertex $v \in V(H)$ right next to v (for technical reasons, we do not want original vertices of H to be incident to crossed edges). Let H_1 be the graph drawn by D_1 and $M_1, \ldots, M_a \subseteq E(H_1)$ be the connected components of the crossing graph of D_1.

For each (now fixed) $i \in \{1, \ldots, a\}$ we do the following. By Definition 1.2, there are at most k fans $F_i^1, \ldots, F_i^k \subseteq E(H - X)$ "covering" together all arcs of M_i. Let w_i^j be the center vertex of the fan F_i^j. Viewed in H_1, for each $1 \leq j \leq k$ we have a set $\mathcal{F}_i^j$ of paths in H_1 starting in w_i^j, where every edge $f \in F_i^j$ of $H - X$ is subdivided into the path $P \in \mathcal{F}_i^j$ in H_1. Hence, denoting by $E(\mathcal{F}_i^j) := \bigcup\{E(P) : P \in \mathcal{F}_i^j\}$, we have $M_i \subseteq E(\mathcal{F}_i^1) \cup \cdots \cup E(\mathcal{F}_i^k)$. We denote by $R_i^j \subseteq V(H_1)$ the subset of the vertices incident to $M_i \cap E(\mathcal{F}_i^j)$ which are reachable from w_i^j via a subpath $P - M_i$ of some $P \in \mathcal{F}_i^j$, and by $T_i^j \subseteq V(H_1)$ the subset of the vertices incident to $M_i \cap E(\mathcal{F}_i^j)$ which are not in R_i^j.

Next, we turn every crossing of M_i into a vertex and contract all these new vertices into one vertex m_i (and remove loops). Hence, we get an induced star S_i centered in m_i and all leaves being from $V(H_1) \setminus V(H)$, and denote by S_i' the graph obtained by subdividing each ray of S_i with one new vertex. Doing this for all $i = 1, \ldots, a$ (the steps are clearly independent of each other), we altogether get a drawing D_2, which is actually a planar embedding since we have destroyed all crossings of D_1 in the process. See Fig. 4.

Similarly as in (a), we let G be the planar graph obtained from the drawing D_2 by adding the isolated vertices of $X = \{x_1, \ldots, x_m\}$, $m \leq k$. In G, we introduce colors c_i and c_i' for $1 \leq i \leq m \leq k$, and colors b_j and b_j' for $0 \leq j \leq k$:

- Color c_i is given to $x_i \in X$ and c_i' to every neighbor of x_i in H.
- Color b_0' is given to every vertex of $V(G) \cap \big(V(H_1) \setminus V(H)\big)$ (these are the degree-2 vertices created by subdivisions in D_1), and color b_0 is given to all vertices $m_1, \ldots, m_a$ (these were created by contractions of the components of the crossing graph of D_1).
- For every $i \in \{1, \ldots, a\}$ and $j \in \{1, \ldots, k\}$, color b_j is given to the vertices of S_i' which subdivide the edges of S_i from m_i to R_i^j, and color b_j' is given to the vertices of S_i' which subdivide the edges of S_i from m_i to T_i^j.

The formula $\xi(x,y)$ depending on k is constructed as a disjunction of the following possibilities:

- For some $1 \leq i \leq m$, x has color c_i and y color c_i' or vice versa, or

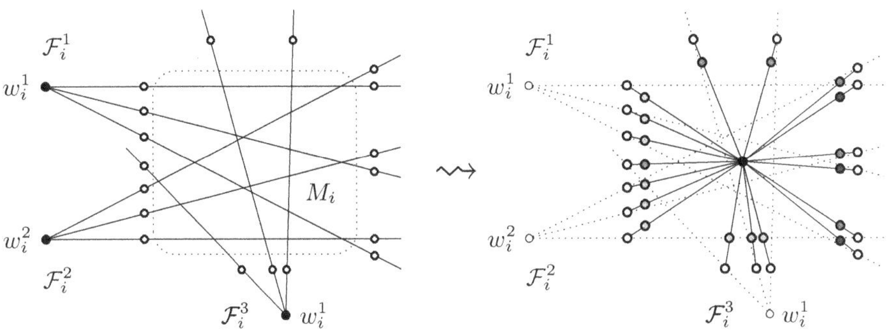

Fig. 4. An illustration of the proof of Lemma 3.2(b) (here with $X = \emptyset$). (left) Component M_i of the crossing graph of the drawing D_1 from the proof, with sets $\mathcal{F}_i^1, \mathcal{F}_i^2, \mathcal{F}_i^3$ of paths subdividing original fans centered at w_i^1, w_i^2, w_i^3. (right) The corresponding fragment of the planar drawing D_2, displaying the subdivided star S_i' and the coloring assigned by the proof. Color b_0 is black, b_0' is white, and $b_j, b_j', j = 1, 2, 3$, are light and dark (resp.) shades of colors in order blue, red and green.(Color figure online)

- xy is an edge of G, or
- there exists a path $P = (x, z_1, \ldots, z_p, y)$ in G of length at most $4k+1$, such that for every $1 \leq i \leq p$ the vertex z_i has color in the set $\{b_0', b_1, b_1', \ldots, b_k, b_k'\}$, or (for $1 < i < p$) z_i has color b_0 and for some $j \in \{1, \ldots, k\}$, its neighbors z_{i-1} and z_{i+1} have (in some order) colors b_j and b_j'.

Again, the said properties are routinely expressible in FO logic. If $xy \in E(H)$ and $\{x, y\} \cap X \neq \emptyset$, then $G \models \xi(x, y)$ by the first point of the definition of ξ. Otherwise, xy is an uncrossed edge of D which exists also in G, or xy is crossed in D and by our construction and coloring of G, there is a corresponding subdivided path $P' \subseteq E(H_1)$ from x to y, and so a path $P \subseteq E(G)$ satisfying the third point of the definition of ξ, that is, $G \models \xi(x, y)$ again.

If, on the other hand, $G \models \xi(x, y)$ for some $x, y \in V(H)$, then xy is an edge of H incident to the set X, or $xy \in E(G)$ and then $xy \in E(H)$, or the following holds. The path P claimed by the definition of $\xi(x, y)$ has internal vertices only from $V(G) \setminus V(H)$ thanks to the coloring, only the internal vertices of P colored b_0 may have degree greater than 2, and the coloring ensures "proper traversal" of the path P through every former component of the crossing graph of D_1 (up to possible exchange of sections of the same fan, which is insignificant). Hence, then $xy \in E(H)$. Altogether, $\xi(G)[V(H)] = H$.

$\square$

4   Conclusions

Inspired by the recent interest in studying (FO) transducibility between graph classes, and namely from the class of planar graphs, we have proved (based on [7]) an asymptotic characterization of transducibility from planar graphs in the weakly sparse world – Theorem 1.3. There is nothing special about the plane in this result, and since all our arguments use planarity "locally", Theorem 1.3 and Corollary 1.4 can be easily extended to any fixed surface.

For example, from [9,11], Corollary 1.4 and Remark 1.5 we immediately get:

Corollary 4.1. *The class of 3D-grids (maximum degree 6) is not k-planar for any fixed k.*

Likewise, there is no surface Σ and integer k such that the class of 3D-grids is drawable with $\leq k$ crossings per edge in Σ (a natural generalization of k-planarity).

We are not aware of any published elementary proof of Corollary 4.1, though, this claim can be easily derived from published results: Dujmović et al. [5] proved that every n-vertex k-planar graph has tree-width $\mathcal{O}\big(\sqrt{(k+1)n}\big)$ and, on the other hand, Dvořák and Wood showed [6] that any balanced bipartition of an n-vertex 3D grid is crossed by $\Omega(n^{2/3})$ edges, and so the tree-width is at least $\Omega(n^{2/3})$. Knowing validity of Corollary 4.1 via other means, we can also use Corollary 1.4 to prove that 3D-grids are not transducible from planar graphs (this is, in addition to [9,11], a third way of proving the statement which was still open a year ago).

Nevertheless, our main desire is to draw a path to a possible solution of the following problem, in which *toroidal graphs* are graphs embeddable (without crossings) in the torus.

Problem 4.2. (cf. [12]). Is the class of toroidal graphs transducible from that of planar graphs?

By Theorem 1.3, answer 'Yes' to Problem 4.2 is equivalent to having a k-fold k-clustered fan-crossing drawing (after deleting $\leq k$ of its vertices) for fixed k and every toroidal graph, which does not seem likely to us. An even more presentable connection exists in the case of bounded degrees, via Corollary 1.4 (note that Remark 1.5 does not apply here):

Problem 4.3. For which (or all) $d \geq 3$ there is k such that every toroidal graph of maximum degree $\leq d$ is k-planar after deleting $\leq k$ of its edges?

Independently of Problem 4.2, Problem 4.3 makes good sense also without allowing to delete $\leq k$ edges.

Answer 'No' to Problem 4.3 (for any d) would readily give answer 'No' to Problem 4.2. On the other hand, answer 'Yes' to Problem 4.3, even for any single $d \geq 3$, confirms an affirmative answer to Problem 4.2 for all bounded-degree classes of toroidal graphs; see next Proposition 4.4. If, at the same time, Problem 4.2 had answer 'No' in general, this would uncover an interesting structural difference between bounded-degree and all toroidal graphs.

Proposition 4.4. *Assume there exist integers $d \geq 3$ and ℓ such that every toroidal graph of maximum degree $\leq d$ is ℓ-planar after deleting $\leq \ell$ of its edges. Then every class of toroidal graphs of bounded maximum degree is transducible from the class of planar graphs.*

Proof. We build on the following two standard observations:

(i) If H is an ℓ-planar graph, then H is a congestion-2 depth-ℓ minor of a planar graph H_1.

(ii) If H is a depth-ℓ_1 minor of a congestion-c depth-ℓ_2 minor of a graph H_1, then H is a congestion-c depth-$(2\ell_2 + 1)\ell_1$ minor of H_1.

As for (i), we construct the planar graph H_1 by introducing one new vertex for every crossing of the ℓ-planar drawing of H and replacing every edge e of H with a path Q_e whose internal vertices are the vertices of the crossings on e in order. A minor model of H then assigns the internal vertices of every such path Q_e to the model set of an arbitrary one of the ends of e. As for (ii), we simply "stack" one minor model on top of the other, which does not increase the congestion.

Let now $\mathcal{D}$ be a class of toroidal graphs of maximum degree Δ, and let $G \in \mathcal{D}$ be embedded in the torus. For every vertex $v \in V(G)$, we replace v with a path P_v on $deg_G(v)$ vertices, and make edges formerly incident to v now incident each to a different vertex of P_v in order given by the rotation of these edges in embedded G. Denoting the resulting graph by G_1, we easily get that G_1 is toroidal of maximum degree $\Delta_1 = 3 \leq d$, and G is a depth-Δ minor of G_1.

By the assumption (of Proposition 4.4), there is an induced subgraph $G_1' \subseteq G_1$ obtained by deleting $\leq \ell$ vertices, such that G_1' has a ℓ-planar drawing, and so G_1' is a congestion-2 depth-ℓ minor of a planar graph G_2 by (i). Moreover, by the definition of a minor model, there is an induced subgraph $G' \subseteq G$ obtained by deleting $\leq \ell$ vertices, such that G' is a depth-Δ minor of G_1'. Consequently, by (ii), G' is a congestion-2 depth-$(2\ell + 1)\Delta$ minor of G_2, and by Lemma 3.1, G' has a k'-fold k'-clustered strongly fan-planar drawing for some $k' \in \mathcal{O}(\ell\Delta)$. Therefore, in particular, every $G \in \mathcal{D}$ has a k-fold k-clustered fan-crossing drawing after deleting at most k of its vertices for $k = \max(\ell, k')$. By Theorem 1.3, we conclude that $\mathcal{D}$ is transducible from the class of planar graphs. $\qquad\square$

Lastly, we comment on some graph-drawing related aspects and questions of this research.

– Our results generally allow non-simple drawings (that is, two edges of the same fan may cross, and two independent edges may cross multiple times). We tend to believe that Theorem 1.3 fails to be true when restricted to only simple drawings, but finding a concrete counterexample seems to be a challenge.

– Definition 1.2 does not include all fan-crossing drawings for $k = 1$ (cf. Fig. 1), but we conjecture that there exist small integers k, ℓ such that every fan-crossing drawing is a k-fold ℓ-clustered fan-crossing drawing (at least in the subcase of simple drawings).

– Let a drawing D be called *strictly k-fold fan-crossing* if there is a drawing D' obtained from D by subdividing each edge at most $k - 1$ times, such that in every connected component $M \subseteq E(D')$ of the crossing graph of D', every edge $e \in M$ is crossed by edges coming from *one* fan of D. Can Theorem 1.3 (and mainly Lemma 3.1) be proved for strictly k-fold fan-crossing drawings?

References

1. Angelini, P., Bekos, M.A., Kaufmann, M., Kindermann, P., Schneck, T.: 1-fan-bundle-planar drawings of graphs. Theor. Comput. Sci. **723**, 23–50 (2018). https://doi.org/10.1016/J.TCS.2018.03.005
2. Bonnet, É., Kim, E.J., Thomassé, S., Watrigant, R.: Twin-width I: tractable FO model checking. J. ACM, **69**(1), 3:1–3:46 (2022). https://doi.org/10.1145/3486655
3. Braunfeld, S., Nešetřil, J., de Mendez, P.O., Siebertz, v S.: On first-order transductions of classes of graphs. CoRR, abs/2208.14412, 2022. http://arxiv.org/abs/2208.14412   arXiv:2208.14412,   https://doi.org/10.48550/ARXIV.2208.14412
4. Courcelle, B., Makowsky, J.A., Rotics, U.: Linear time solvable optimization problems on graphs of bounded clique-width. Theory Comput. Syst. **33**(2), 125–150 (2000). https://doi.org/10.1007/S002249910009
5. Dujmovic, V., Eppstein, D., Wood, D.R.: Structure of graphs with locally restricted crossings. SIAM J. Discret. Math. **31**(2), 805–824 (2017). https://doi.org/10.1137/16M1062879
6. Dvořák, Z., Wood, D.R.: Product structure of graph classes with strongly sublinear separators. CoRR, abs/2208.10074, 2022. http://arxiv.org/abs/2208.10074 arXiv:2208.10074, https://doi.org/10.48550/ARXIV.2208.10074
7. Gajarský, J., Gładkowski, J., Jedelský, J., Pilipczuk, M., Toruńczyk, S.: First-order transducibility among classes of sparse graphs. CoRR, abs/2505.15655, 2025.   http://arxiv.org/abs/2505.15655   arXiv:2505.15655,   https://doi.org/10.48550/ARXIV.2505.15655
8. Gajarský, J., Hliněný, P., Obdržálek, J., Lokshtanov, D., Ramanujan, M.S.: A new perspective on FO model checking of dense graph classes. ACM Trans. Comput. Log. **21**(4), 28:1–28:23 (2020). https://doi.org/10.1145/3383206
9. Gajarský, J., Pilipczuk, M., Pokrývka, F.: 3D-grids are not transducible from planar graphs. In: 40th Annual ACM/IEEE Symposium on Logic in Computer Science, LICS 2025, Singapore, June 23-26, 2025, pp. 831–842. IEEE (2025). https://doi.org/10.1109/LICS65433.2025.00068
10. Ganian, R., Hliněný, P., Nešetřil, J., Obdržálek, J., de Mendez, P.O.: Shrub-depth: Capturing height of dense graphs. Log. Methods Comput. Sci. **15**(1) (2019). https://doi.org/10.23638/LMCS-15(1:7)2019
11. Hlinený, P., Jedelský, J.: Transductions of graph classes admitting product structure. In: 40th Annual ACM/IEEE Symposium on Logic in Computer Science, LICS 2025, Singapore, June 23-26, 2025, pp. 843–855. IEEE (2025). https://doi.org/10.1109/LICS65433.2025.00069
12. Pilipczuk, M.: Graph classes through the lens of logic. CoRR, abs/2501.04166 (2025). http://arxiv.org/abs/2501.04166 arXiv:2501.04166, https://doi.org/10.48550/ARXIV.2501.04166
13. Toruńczyk, S.: Flip-width: cops and robber on dense graphs. In 64th IEEE Annual Symposium on Foundations of Computer Science, FOCS 2023, Santa Cruz, CA, USA, November 6-9, 2023, pp. 663–700. IEEE (2023). https://doi.org/10.1109/FOCS57990.2023.00045

Hypergraphs as Metro Maps: Drawing Paths with Few Bends in Trees, Cacti, and Plane 4-Graphs

Sabine Cornelsen[1] , Henry Förster[2] , Siddharth Gupta[3] ,
Stephen Kobourov[2] , and Johannes Zink[2(✉)]

[1] University of Konstanz, Konstanz, Germany
[2] TU Munich, Heilbronn, Germany
`zink@algo.cit.tum.de`
[3] BITS Pilani, K K Birla Goa Campus, Goa, India

Abstract. A *hypergraph* consists of a set of vertices and a set of subsets of vertices, called *hyperedges*. In the metro map metaphor, each hyperedge is represented by a path (the metro line) and the union of all these paths is the support graph (metro network) of the hypergraph. Formally speaking, a *path-based support* is a graph together with a set of paths. We consider the problem of constructing drawings of path-based supports that (i) minimize the sum of the number of bends on all paths, (ii) minimize the maximum number of bends on any path, or (iii) maximize the number of 0-bend paths, then the number of 1-bend paths, etc. We concentrate on straight-line drawings of path-based tree and cactus supports as well as orthogonal drawings of path-based plane supports with maximum degree 4.

Keywords: hypergraphs · metro map metaphor · bend minimization

1 Introduction

A *hypergraph* $H = (V, A)$ is a set V of vertices and a set A of subsets of V, called hyperedges. Among others, hypergraphs can be visualized as bipartite graphs with vertices $V \cup A$, as *Hasse diagrams* (upward drawing of the transitive reduction of the inclusion relation between hyperedges), as Euler diagrams (enclosing the vertices of each hyperedge by a simple closed curve), or as incidence matrices. For an overview on more visualization styles, also see [1,14]. We consider the metro map metaphor, where each hyperedge h is visualized with a metro line along which the vertices in the hyperedge are the stations; see Fig. 1 for an examples of a hypergraph made by Simpson characters and an example of a hypergraph constructed from the authors of this paper and its references. That is, each metro line is a simple path p_h with the same vertices as h, the union of the paths is a *path-based support* of the hypergraph $H = (V, A)$: a *support graph* $G = (V, E)$ and a set $\mathcal{P} = \{p_h; h \in A\}$ of paths of G such that each edge of G is

The full version of this paper is available on arXiv [10].

© The Author(s), under exclusive license to Springer Nature Switzerland AG 2026

J. Kozik and A. Wolff (Eds.): SOFSEM 2026, LNCS 16448, pp. 517–531, 2026.
https://doi.org/10.1007/978-3-032-17801-5_38

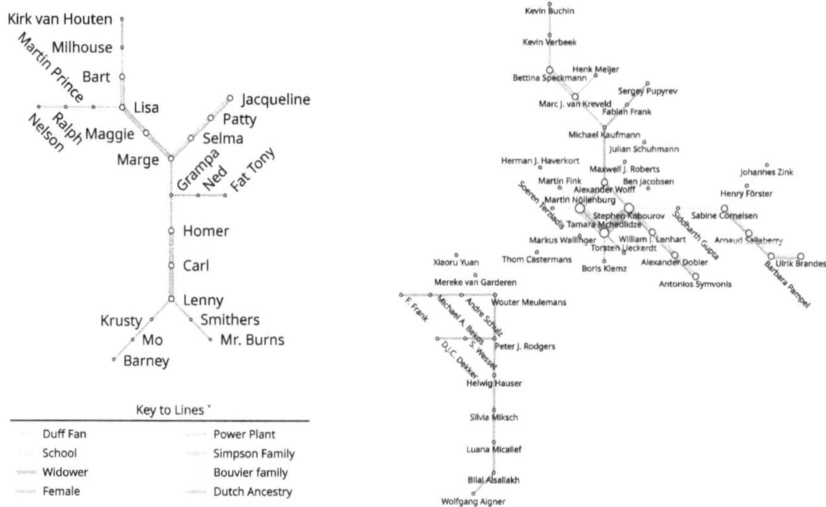

(a) Hypergraph of Simpson characters with a tree support.

(b) Largest component of the co-authorship network of this paper and its references.

Fig. 1. Metro map representations from metrosets.ac.tuwien.ac.at [18].

contained in at least one path of $\mathcal{P}$. A *cactus* is a connected graph where each 2-connected component is an edge (bridge) or a simple cycle. A *plane 4-graph* is a planar graph of maximum degree 4 with a fixed planar embedding.[1] We say that G is a *path-based planar support*, a *path-based tree support*, a *path-based cactus support*, or a *path-based plane 4-graph support*, if G is a planar graph, a tree, a cactus, or a plane 4-graph, respectively.

Planarity and bends are the most obvious measures of visual complexity for paths. Let n_i with $i = 0, 1, 2, \dots$ be the number of paths with i bends. We study how to draw a support graph subject to the following objectives (see also Fig. 2):

(i)   minimizing the *total number of bends*, i.e., $\sum_{i=0}^{\infty} i \cdot n_i$.
(ii) minimizing the *curve complexity*, i.e., $\max_i i$ such that $n_i > 0$.
(iii) lexicographically maximizing the *bend vector* $(n_0, n_1, n_2, \dots)$, i.e., first maximizing the number n_0 of paths without bends, then maximizing the number n_1 of paths with one bend, then two bends, etc.

We focus on straight-line drawings where paths can only bend at vertices and *orthogonal drawings* (see Fig. 3c) where edges are drawn as polylines of horizontal and vertical line segments and a path can bend at vertices and bends of edges. The *(planar) curve complexity* of a path-based support is the minimum curve complexity over all its (planar) straight-line drawings.

A special case are so-called *linear hypergraphs/supports* where any two hyperedges/paths share at most one vertex. If in a planar embedding of a linear path-

---

[1] We use *embedding* in the combinatorial sense: it prescribes at each vertex the order of incident edges (rotation system) and defines the outer face. We do not assume a given rotation system for trees and cacti.

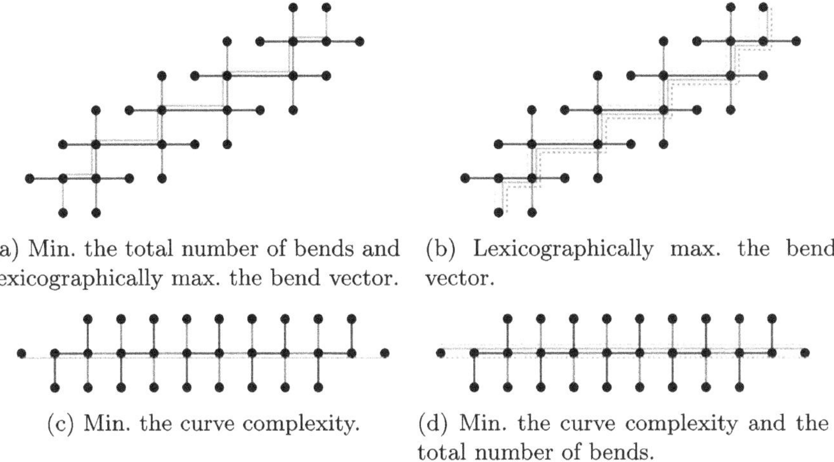

(a) Min. the total number of bends and lexicographically max. the bend vector.

(b) Lexicographically max. the bend vector.

(c) Min. the curve complexity.

(d) Min. the curve complexity and the total number of bends.

Fig. 2. The caterpillars in (a) and (c) have the same set of paths. In (b) and (d) there are two more paths on the spine, affecting the optimality of the drawings.

based support $(G, \mathcal{P})$ no two paths touch, i.e., they are either disjoint or properly intersect (in a single vertex), the question if G is drawable such that no path has a bend corresponds to the $\exists\mathbb{R}$-hard problem of pseudo-segment stretchability [24]. So, we consider bend-minimization of path-based supports for restricted classes of graphs. Path-based tree supports (if they exist) have the advantage that (i) they can be computed in polynomial time [25], (ii) have the minimum number of edges among all supports, and (iii) display also intersections of any two hyperedges as a path. Cactus supports are a natural generalization of tree supports.

Related Work. Visualizing hypergraphs with the metro map metaphor has practical applications [18] and gives rise to theoretical considerations [15]. Even though the choice of the paths for the hyperedges can impact the number of bends, computing a visualization of a hypergraph in the metro map metaphor has often been considered as a combination of two formal problems of independent interest.

The first problem is the computation of a suitable support graph. More precisely, given a hypergraph the goal is to decide if it admits an – ideally path-based – $\mathcal{G}$ support where $\mathcal{G}$ is a family of graphs. In this context, it is NP-complete to decide if there is a path-based planar support and NP-hard to find a path-based support with the minimum number of edges [7]. However, a path-based tree support [25] and a (not necessarily path-based)[2] cactus support [6] can be constructed in polynomial time, if they exist. Cactus supports can be used to store the set of all minimum cuts of an undirected graph [11]. More work on computing planar supports can be found in [8,9,19–21].

---

[2] More generally, a graph is a *support* of a hypergraph if every hyperedge induces a connected subgraph.

The second problem is to compute a high-quality geometric embedding for a given path-based support. The problem of embedding the support graph with few bends has been studied intensely from a practical point of view: Bast, Brosi, and Storandt [2] show how to automatically generate metro maps with few bends on an octilinear grid using both, an exact ILP and a shortest-path based approximation algorithm. They also consider metro maps on triangular, octilinear, hexalinear, and ortho-radial grids [3], while Nickel and Nöllenburg [22] generalized the mixed-integer programming approach of [23] from octilinear grids to k slopes through any vertex. Hong et al. [17] apply spring embedders to compute metro maps that mostly follow the octilinear grid. Fink et al. [13] guarantee bend-free metro-lines by using Bézier curves. Bekos et al. [4] study drawing supports for spatial hypergraphs such that the vertices are displaced on a rectangular or an ortho-radial grid. Among others they provide a simulated-annealing algorithm for tree-based supports and show that bend minimization is NP-hard. From a theoretical point of view, Dobler et al. [12] consider the general problem of drawing hyperedges as lines or line-segments where the order of the vertices of the hyperedges is not given. They show that it is $\exists\mathbb{R}$-hard to decide whether a hypergraph admits such representations. We remark that one of their proofs extends to the scenario where the order of vertices is fixed.

Our Contribution. We investigate the second problem, i.e., minimizing the bends in a given path-based support graph – in contrast to most earlier attempts, from a more theoretic point of view.

In Sect. 3, we consider straight-line drawings of path-based tree and cactus supports. For path-based tree supports one among our three objective functions can be optimized in polynomial time while the other two are hard: Minimizing the curve complexity or lexicographically maximizing the bend vector is NP-hard (Thms. 1 and 2) even for very restricted classes of path-based tree supports. However, minimizing the total number of bends is polynomial-time solvable using maximum weighted matching (Thm. 3). Deciding whether the curve complexity is 0 or 1 can also be done in polynomial time via a 2-SAT formulation (Thm. 4). We then study the parameterized complexity. Deciding whether a path-based tree support has curve complexity b is fixed-parameter tractable if parameterized by the number of paths (Thm. 5), the vertex cover number (Thm. 6), or the number of paths per vertex plus the curve complexity (Thm. 7). The latter is a dynamic-programming approach that also yields FPT algorithms for lexicographically maximizing the bend vector (Cor. 1).

For a path-based cactus support we can test in near-linear time whether its planar curve complexity is zero (Thm. 8). Combining this with the dynamic program of Thm. 7, we get FPT algorithms for determining the curve complexity and for lexicographically maximizing the bend vector of cactus supports.

Finally, in Sect. 4, we show how to construct an orthogonal drawing of a path-based plane 4-graph support, minimizing the total number of bends on all paths. Proof details can be found in the full version [10] if marked with a ($\star$).

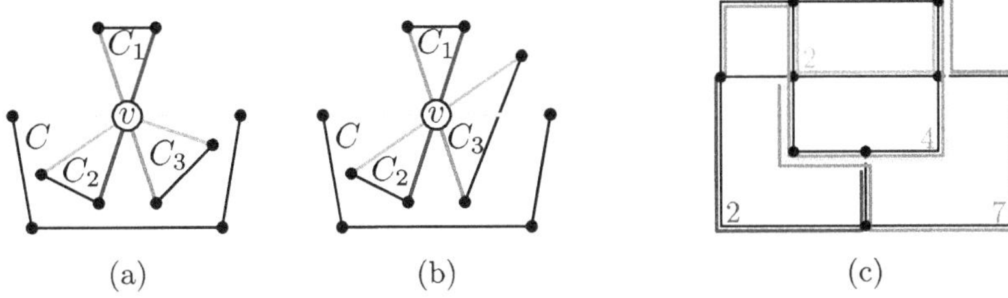

Fig. 3. (a) Planar curve complexity 1 and (b) curve complexity 0 for the same cactus support. (c) orthogonal drawing of a plane 4-graph with 18 bends.

2   Preliminaries

We use standard graph terminology. A *binary tree* (of height h) is a rooted tree where every vertex has at most two children (and there is no vertex with distance $h + 1$ to the root). A binary tree of height h is *complete* if for $i \in \{0, \dots, h - 1\}$, there are 2^i vertices with distance i to the root and *perfect* if in addition it has also 2^h leaves with distance h to the root. A tree is a *caterpillar* if removing its leaves makes it a path, its *spine*. All considered paths and cycles are *simple*.

A problem Π is *fixed-parameter tractable* (FPT) or *slice-wise polynomial* (XP) parameterized by a parameter k if there is an algorithm that solves an instance I of Π in time $\mathcal{O}(f(k) \cdot |I|^c)$ or $\mathcal{O}(|I|^{f(k)})$, respectively, where f is a computable function and c is a constant.

For a graph G, a vertex v and a subgraph C of G, we write $G - v$, $G + v$, and $G - C$ as a shorthand for the graph arising from G after removing v, adding v, and removing all vertices of C, respectively. The size of a path-based support $(G, \mathcal{P})$ is $\|\mathcal{P}\| := \sum_{P \in \mathcal{P}} |P|$, where $|P|$ is the number of edges on the path P. In a drawing of a path-based support, we say that two incident edges are *aligned* if they are contained in one line and disjoint except for their common endpoint. An *alignment (requirement)* of the set E of edges of a support is a set of pairs of incident edges from E such that for each vertex v and for each edge e incident to v there is at most one other edge e' incident to v such that e and e' are a pair. A *realization* of an alignment is a drawing of the support in which the two edges of any pair are aligned. Since we are not restricted to a specific grid, any alignment requirement for a path-based tree support can be realized in a planar way. Hence, the problem of drawing a path-based tree support is purely combinatorial as it suffices to compute a suitable alignment. In particular, the planar curve complexity of a tree equals its curve complexity. However, this is not true for path-based cactus supports (Fig. 3). Moreover, each cycle needs at least three bends in any (non-degenerate) straight-line drawing.

3   Straight-Line Drawings of Tree and Cactus Supports

We first start with straight-line drawings of the support graph. The curve complexity in trees and cacti is in general unbounded: In a complete binary tree with paths from each leaf to the root, some path must bend at each inner

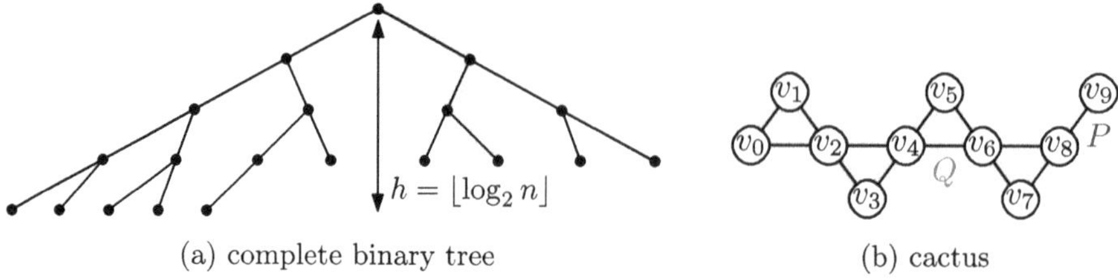

(a) complete binary tree   (b) cactus

Fig. 4. The curve complexity of complete binary trees and cacti is unbounded.

vertex. Thus, in a complete binary tree with n vertices there is a path with $\lfloor \log_2 n \rfloor - 2 \in \Omega(\log n)$ bends; see Fig. 4a. In the cactus induced by the two paths $P : \langle v_0, v_1, v_2, \ldots, v_{n-1} \rangle$ and $Q : \langle v_0, v_2, v_4, \ldots v_{2 \cdot \lfloor \frac{n-1}{2} \rfloor} \rangle$, the path P bends at least $\lfloor \frac{n-1}{2} \rfloor \in \Omega(n)$ times; see Fig. 4b.

3.1   NP-Completeness

We first show that minimizing the curve complexity is NP-hard.

Theorem 1 ($\star$). *It is NP-complete to decide whether the curve complexity of a path-based tree support is at most b even if (i) $b = 3$, every path has length at most 5, and the diameter of the support graph is at most 6; or (ii) the maximum degree of the support graph is at most 3.*

Proof (Overview & Case (i).). Containment in NP is clear. To show NP-hardness, we reduce from 3-SAT. Given a 3-SAT formula Φ with n variables and m clauses, we construct a tree support with (i) diameter 6 or (ii) maximum degree 3, respectively, that admits a drawing with at most $b = 3$ or $b = 2 \cdot (\lceil \log_2 n \rceil + \lceil \log_2 m \rceil + 1)$ bends per path if and only if Φ has a satisfying truth assignment. See Fig. 5 for Case (i) and see Fig. 6 for Case (ii).

For each variable v_i, $i = 1, \ldots, n$, the variable gadget is a star with central vertex x_i and two leaves labeled v_i and $\neg v_i$. A clause gadget of a clause c_j, $j = 1, \ldots, m$, is a perfect binary tree of height two rooted at a vertex labeled c_j whose leaves are labeled with the literals of c_j. One literal appears twice, once in each subtree rooted at the children of c_j. Labels referring to a variable appear multiple times, once in a variable gadget and at most twice in each clause gadget.

Case (i). For each clause c_j, $j = 1, \ldots, m$, there are four copies of the clause gadget. A central vertex x connects to all $4m$ central vertices of the clause gadgets and to all n central vertices of the variable gadgets. The set of paths contains a path from any leaf labeled v_i or $\neg v_i$ of a clause gadget of a clause c_j to the leaf labeled v_i or $\neg v_i$ in the variable gadget of v_i, depending on whether v_i or $\neg v_i$ appears in c_j. Clearly, the diameter of this graph is 6.

In an optimal drawing, we may assume that, at every vertex of degree 3, exactly two edges are aligned. Hence, for each clause gadget c, exactly one sub-path from x to one leaf of c has two bends, while the remaining sub-paths ending

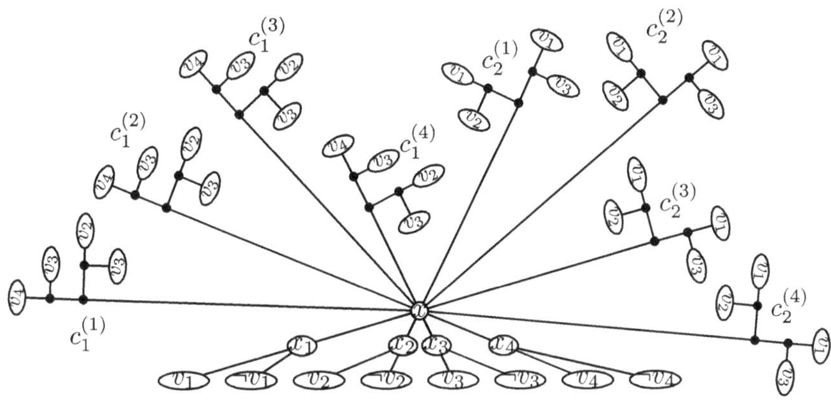

Fig. 5. Graph with diameter 6 corresponding to clauses $c_1 = v_2 \vee v_3 \vee \neg v_4$ and $c_2 = v_1 \vee \neg v_2 \vee v_3$, drawn corresponding to $v_1 = v_3 = v_4 = \top$, $v_2 = \bot$. Among the $2 \cdot 4 \cdot 4 = 32$ paths only the four to one copy of a clause gadget for c_1 are drawn.

in the other three leaves of c have zero or one bend each. There is at least one copy of each clause gadget such that each path ending there bends also in x. Now consider the variable gadgets. We may assume that, for each $i = 1, \ldots, n$, the edge xx_i is aligned with either $x_i v_i$ or $x_i \neg v_i$ since x_i is a degree-3 vertex. Interpreting the former as v_i being true, there is a satisfying truth assignment if and only if there is a drawing of curve complexity 3:

Assume first that there is a drawing with at most three bends per path. For a clause c_j, consider the clause gadget H of c_j such that each path ending in H bends in x. Consider the leaf u of H such that the path P ending in u bends twice between x and u. Then P has already three bends and cannot have a further bend in the variable gadget. Thus, the respective literal is true. Assume now that Φ has a satisfying truth assignment. We align the variable gadgets according to the truth assignment. For a clause c_j let v_i be a variable that makes c_j true. Draw the four clause gadgets associated with c_j such that the path from x with two bends ends at a leaf labeled v_i. Then each path has at most 3 bends.

Case (ii). We consider this case in the full version [10]. The main idea is to realize the connections at the high-degree vertex x by binary trees. See Fig. 6 for an illustration. $\qquad\square$

Similarly, lexicographically maximizing the bend vector is also NP-hard.

Theorem 2 ($\star$). *It is NP-complete to decide if a path-based tree support admits a drawing in which n_0 paths have zero bends each and n_1 paths have one bend each even if the diameter of the support graph is at most four.*

3.2 Tree Supports – Exact Algorithms

The total number of bends can be minimized in polynomial time in trees.

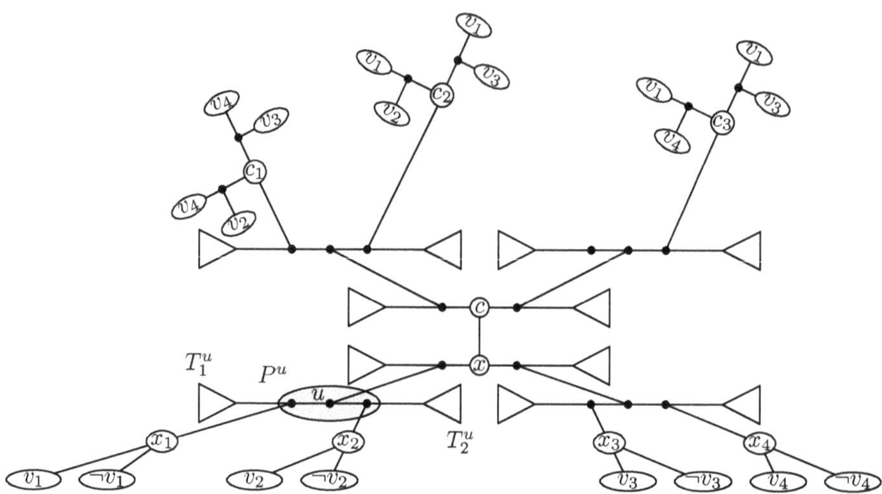

Fig. 6. NP-hardness construction with maximum degree 3. Triangles indicate perfect binary trees of height $\lceil \log_2 n \rceil + \lceil \log_2 m \rceil + 1$ for n variables and m clauses; here $n = 4$ and $m = 3$. These perfect binary trees come in pairs T_1^u, T_2^u with paths between all leaves in T_1^u and all leaves in T_2^u.

Theorem 3 ($\star$). *Given a path-based tree support $(G = (V, E), \mathcal{P})$, an alignment of E that yields a realization where the sum of bends in all paths is minimum, can be computed in $\mathcal{O}(|V| \cdot |\mathcal{P}| + |V|^{2.5} \log^2(\max\{|V|, |\mathcal{P}|\}))$ time.*

Proof. For each vertex v, let $K(v)$ be the complete graph on neighborhood $N(v)$ of v. The weight of an edge uw of $K(v)$ is the number of paths in $\mathcal{P}$ containing uv and vw. Let $M(v)$ be a maximum-weight matching on $K(v)$ and align uv and vw if and only if $uw \in M(v)$. We compute a maximum-weight matching [16] for each vertex. It is easy to see that such matchings yield an optimal solution for aligning edges at vertices. See the full version [10] for the running time. $\square$

The curve-complexity of a path-based tree support can be determined in polynomial time if it is very small.

Theorem 4. *Given a path-based tree support $(G = (V, E), \mathcal{P})$, it can be decided in $\mathcal{O}((\|\mathcal{P}\| + |V|^2) \cdot |V|)$ time whether its planar curve complexity is 0 or 1.*

Proof. We use a 2-SAT formulation with a variable $x_{e_1 e_2}$ for each pair e_1, e_2 of adjacent edges. We interpret $x_{e_1 e_2}$ as true if and only if e_1 and e_2 are not aligned. For any three distinct edges e_1, e_2, e_3 sharing a common end vertex, we need the consistency condition that whenever e_1 is aligned to e_2 then e_1 cannot be aligned to e_3, i.e., the clause $x_{e_1 e_2} \vee x_{e_1 e_3}$. By the same argument, we also have the clauses $x_{e_1 e_2} \vee x_{e_2 e_3}$ and $x_{e_1 e_3} \vee x_{e_2 e_3}$. If we insist on having no bends, then we need for any two adjacent edges e_1, e_2 on the same path the condition $\neg x_{e_1 e_2}$. If we aim for at most one bend per path, we add for any four edges e_1, e_2, e_3, e_4 on a path where e_1, e_2 are adjacent and e_3, e_4 are adjacent, the condition that at least one among the two pairs must be aligned, i.e., the clause $\neg x_{e_1 e_2} \vee \neg x_{e_3 e_4}$. This yields a 2-SAT formulation with $\sum_{v \in V} \deg(v)(\deg(v) - 1) \in \mathcal{O}(|V|^2)$ variables and $\mathcal{O}(\sum_{P \in \mathcal{P}} |P|^2 + \sum_{v \in V} \deg^3(v)) \subseteq \mathcal{O}(\|\mathcal{P}\| \cdot |V| + |V|^3)$ clauses. The number of constraints determines the running time since constructing the instance can be done in the same time and solving 2-SAT can be done in linear time. $\square$

The curve complexity is at most two if the support graph is a caterpillar: draw the spine on a straight line as in Fig. 2d. So, the curve complexity of a path-based caterpillar support can be determined in polynomial time.

3.3   Tree Supports – Parameterized Algorithms

We now consider the parameterized complexity of the problems.

Theorem 5. *Determining the (planar) curve complexity for a path-based tree support $(G, \mathcal{P})$ is in FPT parameterized by $|\mathcal{P}|$.*

Proof. For each leaf v of G, there is a path in $\mathcal{P}$ that ends in v, thus, the number of leaves is at most $2|\mathcal{P}|$. Iteratively replace each vertex v of degree 2 and its incident edges by a single edge connecting the neighbors of v, extending paths ending at v to one of the neighbors. Now the number of inner vertices is less than the number of leaves, i.e. the size of the resulting tree is less than $4|\mathcal{P}|$. This kernel can be solved brute-force testing all edge alignments. $\square$

Theorem 6. *Determining the (planar) curve complexity for a path-based tree support $(G, \mathcal{P})$ is in FPT parameterized by the vertex cover number k.*

Proof. Let S be a minimum vertex cover of G. If G has at most two vertices, then the curve complexity is 0. Otherwise, w.l.o.g. S contains no leaves of G. Let G' be the graph obtained from G by removing all its leaves. All leaves of G' must be in S and S is a vertex cover of G'. In the full version [10], we show inductively:

Claim 1 ($\star$). *The number n' of vertices of G' is at most $2k - 1$.*

Thus, the number of distinct paths in the tree G' is at most $\binom{2k-1}{2}$ – one for each pair of vertices. We show that it suffices to maintain at most six paths of $\mathcal{P}$ for each path of G' in order to obtain a support with the same curve complexity as $(G, \mathcal{P})$. We call this set of retained paths $\widetilde{\mathcal{P}}$. Let $\widetilde{G}$ be the graph induced by $\widetilde{\mathcal{P}}$. In order to make sure that the curve complexity of $(G, \mathcal{P})$ does not exceed the curve complexity of $(\widetilde{G}, \widetilde{\mathcal{P}})$, we construct $\widetilde{\mathcal{P}}$ with the following property: Let $\widetilde{\Gamma}$ be a drawing of $(\widetilde{G}, \widetilde{\mathcal{P}})$ with optimal curve complexity, let Γ be a drawing of $(G, \mathcal{P})$ with optimal curve complexity among all drawings extending $\widetilde{\Gamma}$, let $\pi' \in \mathcal{P} \setminus \widetilde{\mathcal{P}}$, and let π be the subpath of π' in G'. Then $\widetilde{\mathcal{P}}$ contains a path $\widetilde{\pi}$ with the same subpath π in G' that has at least as many bends as π'. We ensure this using the following observations. (i) Any path in $\mathcal{P}$ has at most two bends more than its subpath in G'. (ii) Let $e = vx$ be an edge of G'. If there are two leaves u and w adjacent to v in G such that $\mathcal{P}$ contains a path through e and vu and a path through e and vw, then at least one of the two paths must bend at v. (iii) If at least three paths share their sub-path in G' and extend on both ends to distinct leaves, at least one of these paths has two bends next to its leaves.

We start with $\widetilde{\mathcal{P}} = \emptyset$. Let π be any path of G' and let v, v' be its end vertices. Let $\mathcal{P}(\pi)$ be the set of paths of $\mathcal{P}$ whose intersection with G' is precisely π. Moreover, for $i = 0, 1, 2$, let $\mathcal{P}_i(\pi)$ be the set of paths in $\mathcal{P}(\pi)$ containing i leaves of G. Let $L_v(\pi), L_{v'}(\pi)$ be the set of leaves adjacent to v and v', respectively, such that some path of $\mathcal{P}_2(\pi)$ ends there and assume w.l.o.g. that $|L_v(\pi)| \leq |L_{v'}(\pi)|$. We distinguish several cases. (a) If $\mathcal{P}_1(\pi) \cup \mathcal{P}_2(\pi) = \emptyset$ and $\mathcal{P}_0(\pi) \neq \emptyset$ add π to $\widetilde{\mathcal{P}}$. (b) If $\mathcal{P}_2(\pi)$ contains at most one path π, we add π and up to four paths from $\mathcal{P}_1(\pi)$ to $\widetilde{\mathcal{P}}$, up to two containing a leaf incident to v and up to two containing a leaf incident to v'. (c) If $|L_v(\pi)| \leq 3$ then add for each vertex $u \in L_v(\pi)$ up to two paths of $\mathcal{P}_2(\pi)$ ending in u to $\widetilde{\mathcal{P}}$. (d) If there are two vertices $u_1, u_2 \in L_v(\pi)$ such that at least two paths of $\mathcal{P}_2(\pi)$ end in u_1 and u_2 each, add two for both of them to $\widetilde{\mathcal{P}}$; proceed symmetrically for $L_{v'}(\pi)$. (e) Otherwise, if there are six vertices $u_1, u_2, u_3 \in L_v(\pi)$ and $u'_1, u'_2, u'_3 \in L_{v'}(\pi)$ such that there are u_i-u'_i-paths for $i = 1, 2, 3$ in $\mathcal{P}_2(\pi)$ add these three paths to $\widetilde{\mathcal{P}}$. (f) In the remaining cases, $|L_v(\pi)| \geq 4$, $|L_{v'}(\pi)| \geq 4$, and there is at most one vertex $u \in L_v(\pi)$ and at most one vertex $u' \in L_{v'}(\pi)$ where more than one path ends. (If in all vertices of $L_v(\pi)$ $(L_{v'}(\pi))$ only one path ends, let u (u') be any of them.) In this case, we add up to two paths of $\mathcal{P}_2(\pi)$ ending in u or u', respectively, to $\widetilde{\mathcal{P}}$, and, if it exists, additionally one path of $\mathcal{P}_2(\pi)$ with no end vertex in $\{u, u'\}$. Observe that there cannot be two such paths since otherwise Case (e) would hold.

Finally, remove all leaves of G where none of the paths of $\widetilde{\mathcal{P}}$ ends. Let the resulting graph be $\widetilde{G}$. This yields a kernel $(\widetilde{G}, \widetilde{\mathcal{P}})$ with $\mathcal{O}(k^2)$ vertices and $\mathcal{O}(k^2)$ paths having the same curve complexity as $(G, \mathcal{P})$. $\qquad\square$

Path p *passes through* vertex v, if v is contained in p, but not as an end vertex.

Theorem 7. *Determining the (planar) curve complexity for a path-based tree support $(G, \mathcal{P})$ is in XP parameterized by the maximum number k of paths in $\mathcal{P}$ passing through a vertex. Deciding if the (planar) curve complexity is b is in FPT parameterized by $k + b$.*

Proof. We use bottom-up dynamic programming. Root G at any vertex. For each vertex v, we consider feasible drawings of the subtree G_v rooted at v including the edge e_v between v and its parent p_v. For each such drawing, we store a record that contains for each path π of $\mathcal{P}(v) := \{\pi \in \mathcal{P} \mid e_v \in \pi\}$ the number b_π of bends in G_v. Since $b_\pi \in \{0, 1, \ldots, b\}$, this yields $\mathcal{O}(b^k)$ different records (b is upper bounded by n). Let $B(v)$ be the set of records that we store for v. For each record $\langle b_\pi \mid \pi \in \mathcal{P}(v) \rangle$ in $B(v)$, we also store the minimum curve complexity $C_v(\langle b_\pi \mid \pi \in \mathcal{P}(v) \rangle)$ that can be achieved in G_v with b_π bends on $\pi \in \mathcal{P}(v)$.

If v is a leaf, $B(v)$ contains a single record where all paths have zero bends. Let v be a non-leaf vertex. Let c_i, $i = 1, \ldots, s$ denote the *active* children of v, i.e., the children of v that occur on a path in $\mathcal{P}$ with internal vertex v. Since there are at most k paths through v and each such path contains at most two edges incident to v, it follows that $s \leq 2k$. We have already recursively computed the set $B(c_i)$ of all records for c_i. We now try each combination of records in $B(c_1), \ldots, B(c_s)$ combined with each possible alignment of edges in $\{vc_1, \ldots, vc_s, vp_v\}$. More precisely, for $i = 1, \ldots, s$ let $\langle b_\pi^i \mid \pi \in \mathcal{P}(c_i) \rangle$ be a record in $B(c_i)$ and consider a possible alignment of the edges in $\{vc_1, \ldots, vc_s, vp_v\}$. We compute the record in $B(v)$ and the respective curve complexity C_v as follows. Set $b_\pi = 0$ if π starts at v and contains e_v. For each path π through v do the following:

1. If π contains c_i, v, p_v, we set $b_\pi = b_\pi^i$ if we aligned the edges vc_i and vp_v, and $b_\pi = b_\pi^i + 1$ otherwise. If $b_\pi > b$, we reject the resulting record.
2. If π contains c_i, v, c_j, for two children c_i and c_j of v, we set $b_\pi = b_\pi^i + b_\pi^j$ if we aligned the edges vc_i and vc_j, and $b_\pi = b_\pi^i + b_\pi^j + 1$ otherwise. If $b_\pi > b$, we reject the resulting record.

If we did not reject the record when handling a path π through v, add the record $\langle b_\pi \mid \pi \in \mathcal{P}(v) \rangle$ to $B(v)$. Obtain the curve complexity from the curve complexities of the newly combined paths, the respective records from $c_1, \ldots, c_s$, and the maximum curve complexity $C_{\max}$ used for inactive child trees. Namely, we set

$$C_v(\langle b_\pi \mid \pi \in \mathcal{P}(v) \rangle) = \max(\max_{i=1,\ldots,s} C_{c_i}(\langle b_\pi^i \mid \pi \in \mathcal{P}(c_i) \rangle), \max_{\pi \text{ path through } v} b_\pi, C_{\max}).$$

In the process, we may create a record with the same number of bends per path through e_v several times. We update the curve complexity if the newly computed record achieves better curve complexity, otherwise, we keep the old solution. Amongst the valid solutions for the root, we select the one optimizing the curve complexity. The correctness follows from the fact that we exhaustively considered all possibilities, i.e., it is obvious that each subtree of a vertex v can be aligned with at most one sibling subtree or with the parent p_v of v. Otherwise, new bends are created as computed in the algorithm. Finally, a drawing always exists for each record that does not exceed the bend limit b along a single path.

The runtime is dominated by the number of alignments of subtrees and by the number of records. Each of the at most k subtrees may be aligned to another subtree or do nothing. This equals the number of matchings in the complete graph K_k, for which a (non-tight) upper bound is $k! \in 2^{\mathcal{O}(k \log k)}$. The number of records is in $\mathcal{O}((b^k)^{2k})$ in each of the n steps as seen before. In particular, note that the optimal solution for each inactive child is always the same since we do not need to align it anymore. The computations for each combination can be performed in polynomial time. Hence, we obtain an XP algorithm if b is unbounded (i.e., $b \in \mathcal{O}(n)$) and an FPT algorithm if b is a parameter. We want to point out that the proof is constructive and that an alignement yielding the computed curve complexity can be stored throughout the construction. $\square$

Keeping track of the number of paths with a certain number of bends yields:

Corollary 1. *Given a path-based tree support $(G, \mathcal{P})$, finding a drawing that lexicographically maximizes the bend vector is in XP parameterized by k where k is the maximum number of paths in $\mathcal{P}$ passing through a vertex. Moreover, there is an FPT algorithm parameterized by $k + b$ where b denotes the maximum number of bends allowed on each path.*

Proof. We slightly modify the dynamic program from Thm. 7. Instead of evaluating the quality of the record $\langle b_p \mid p \in \mathcal{P}(v) \rangle$ on the curve complexity $\mathcal{C}_v(\langle b_p \mid p \in \mathcal{P}(v) \rangle)$, we evaluate it in terms of the number of paths with exactly β bends that are entirely located within the already finished subtree. Thus, the best record first maximizes the number n_0 of 0-bend paths, then amongst records with n_0 0-bend paths, it maximizes the number n_1 of 1-bend paths, and so on, up to b (in the FPT case) or n (in the XP case). Observe that we still keep a record for each combination of $\langle b_p \mid p \in \mathcal{P}(v) \rangle$ for each allowed number β of bends, i.e., we still explore the full solution space. This shows the XP algorithm in terms of k and the FPT algorithm in terms of $k + b$. $\qquad\square$

3.4 Cactus Supports

We show in the full version [10] that every linear path-based cactus support has curve-complexity 0 and how to determine in linear time whether its planar curve-complexity is 0. We use this result for the following theorem.

Theorem 8 ($\star$). *Given a path-based cactus support $(G = (V, E), \mathcal{P})$, it can be decided in near-linear time whether its (planar) curve complexity is 0.*

Proof. If there are three edges e, e_1, and e_2 incident to a vertex v and two paths in $\mathcal{P}$, one containing e and e_1 and the other containing e and e_2 then there is no drawing without bends. Otherwise we merge paths that share an edge into one path. Thus, in the obtained cactus support $(G = (V, E), \mathcal{P}')$ no two paths share an edge and it has (planar) curve complexity zero if and only if $(G = (V, E), \mathcal{P})$ does. If there are two paths in $\mathcal{P}'$ that intersect twice, then the two paths would contain a cycle and, thus, cannot both be drawn on a straight line. Otherwise, the support is linear and we can solve the problem in linear time. A naïve implementation of this algorithm yields cubic running time. Using union-find data structures, we prove the near-linear running time in the full version [10]. $\square$

Using Thm. 8 and also keeping track of the number of bends per cycle, we generalize the dynamic program from trees to cactus supports.

Theorem 9 ($\star$). *Thm. 7 and Cor. 1 generalize to path-based cactus supports.*

4 Orthogonal Drawings for Plane 4-Graph Supports

Metro maps are often layed out on an octilinear grid. Here, we consider the more restrictive orthogonal grid. See Fig. 3c.

Theorem 10 ($\star$). *Given a path-based plane 4-graph support $(G, \mathcal{P})$, an orthogonal drawing of G with the property that the sum of the bends in all paths is minimimum among all orthogonal drawings of G with the given embedding can be computed in near-linear time.*

Proof (Sketch). We adapt methods for bend minimization for orthogonal drawings of plane 4-graphs. The key observation is that we now count bends on paths of $\mathcal{P}$ instead of bends on edges of the support graph G. This leads to two key differences in the evaluation of an orthogonal drawing of G: First, bends on an edge e actually correspond to $k(e)$ bends where $k(e)$ denotes the number of paths in $\mathcal{P}$ containing e. Second, while the alignment of edges at a vertex v is not directly contributing to the bends in the drawing of the support graph, vertex v might be an interior vertex of some paths in $\mathcal{P}$ and we have to account for the number of paths that bend at each vertex v. To this end, we observe that the evaluation depends on the degree of v: 1. If v has degree 1, it is not an interior vertex of any path. 2. If v has degree 4, there is no choice to be made. 3. If v has degree 2, we can choose whether to align its two incident edges (in this case, no path bends at v) or not to align them (in this case, all paths with interior vertex v bend at v). 4. If v has degree 3, we can align exactly one pair of its three incident edges. All paths traversing the other two pairs of incident edges bend at v. In the full version [10], we describe how to adjust Tamassia's [26] approach for bend-minimum orthogonal drawings to correctly evaluate the number of bends along the paths of $\mathcal{P}$ which results in a near-linear time [5] algorithm for minimizing the total number of bends along the paths in $\mathcal{P}$. $\qquad\square$

5 Open Problems

Our results motivate open problems for drawing path-based tree or cactus supports with few bends. What is the complexity of (a) minimizing the total number of bends for a path-based cactus support, (b) minimizing the curve complexity and lexicographically maximizing the bend vector for a plane 4-graph support, (c) deciding whether the curve complexity is two for a tree support, or one or two for a cactus support, (d) maximizing the number of paths with no bends, (e) lexicographically maximizing the bend vector for trees of maximum degree 3, and (f) computing a path-based cactus support. Moreover, (g) is the curve complexity W[1]-hard if parameterized by the maximum number of paths through a vertex?

Acknowledgments. This work was initiated at the Annual Workshop on Graph and Network Visualization (GNV 2024), Heiligkreuztal, Germany, June 2024. S. Gupta is supported, in part, by the Anusandhan National Research Foundation (ANRF) grant ANRF/ECRG/2024/006849/PMS.

Disclosure of Interests. The authors have no competing interests to declare that are relevant to the content of this article.

References

1. Alsallakh, B., Micallef, L., Aigner, W., Hauser, H., Miksch, S., Rodgers, P.J.: The state-of-the-art of set visualization. Comput. Graph. Forum **35**(1), 234–260 (2016). https://doi.org/10.1111/cgf.12722
2. Bast, H., Brosi, P., Storandt, S.: Metro maps on octilinear grid graphs. Comput. Graph. Forum **39**(3), 357–367 (2020). https://doi.org/10.1111/CGF.13986
3. Bast, H., Brosi, P., Storandt, S.: Metro maps on flexible base grids. In: Hoel, E., Oliver, D., Wong, R.C., Eldawy, A. (eds.) Proceedings of the 17th International Symposium on Spatial and Temporal Databases, SSTD 2021, pp. 12–22. ACM (2021). https://doi.org/10.1145/3469830.3470899
4. Bekos, M.A., et al.: Computing schematic layouts for spatial hypergraphs on concentric circles and grids. Comput. Graph. Forum **41**(6), 316–335 (2022). https://doi.org/10.1111/CGF.14497
5. van den Brand, J., et al.: A deterministic almost-linear time algorithm for minimum-cost flow. In: Proceedings of the 64th Annual Symposium on Foundations of Computer Science, FOCS 2023, pp. 503–514. IEEE (2023). https://doi.org/10.1109/FOCS57990.2023.00037
6. Brandes, U., Cornelsen, S., Pampel, B., Sallaberry, A.: Blocks of hypergraphs - applied to hypergraphs and outerplanarity. In: Iliopoulos, C.S., Smyth, W.F. (eds.) Proceedings of the 21st International Workshop on Combinatorial Algorithms, IWOCA 2010. Lecture Notes in Computer Science, vol. 6460, pp. 201–211. Springer (2010). https://doi.org/10.1007/978-3-642-19222-7_21
7. Brandes, U., Cornelsen, S., Pampel, B., Sallaberry, A.: Path-based supports for hypergraphs. J. Discrete Algorithms **14**, 248–261 (2012). https://doi.org/10.1016/J.JDA.2011.12.009
8. Buchin, K., van Kreveld, M.J., Meijer, H., Speckmann, B., Verbeek, K.: On planar supports for hypergraphs. J. Graph Algorithms Appl. **15**(4), 533–549 (2011). https://doi.org/10.7155/JGAA.00237
9. Castermans, T., van Garderen, M., Meulemans, W., Nöllenburg, M., Yuan, X.: Short plane supports for spatial hypergraphs. J. Graph Algorithms Appl. **23**(3), 463–498 (2019). https://doi.org/10.7155/JGAA.00499
10. Cornelsen, S., Förster, H., Gupta, S., Kobourov, S., Zink, J.: Hypergraphs as metro maps: drawing paths with few bends in trees, cacti, and plane 4-graphs. arXiv report (2025). http://arxiv.org/abs/2511.22508
11. Dinitz, E.A., Karzanov, A.V., Lomonosov, M.V.: On the structure of a family of minimal weighted cuts in a graph. In: Fridman, A.A. (ed.) Studies in Discrete Optimization, Nauka, pp. 290–306 (1976). in Russian
12. Dobler, A., Kobourov, S.G., Mondal, D., Nöllenburg, M.: Representing hypergraphs by point-line incidences. In: Královic, R., Kurková, V. (eds.) Proceedings of the 50th International Conference on Current Trends in Theory and Practice of Computer Science, SOFSEM 2025. Lecture Notes in Computer Science, vol. 15538, pp. 241–254. Springer (2025). https://doi.org/10.1007/978-3-031-82670-2_18
13. Fink, M., et al.: Drawing metro maps using Bézier curves. In: Didimo, W., Patrignani, M. (eds.) Proceedings of the 20th International Symposium on Graph Drawing, GD 2012. Lecture Notes in Computer Science, vol. 7704, pp. 463–474. Springer (2012). https://doi.org/10.1007/978-3-642-36763-2_41
14. Fischer, M.T., Frings, A., Keim, D.A., Seebacher, D.: Towards a survey on static and dynamic hypergraph visualizations. In: Proceedings of 2021 IEEE Visualization Conference, VIS, pp. 81–85. IEEE (2021). https://doi.org/10.1109/VIS49827.2021.9623305

15. Frank, F., et al.: Using the metro-map metaphor for drawing hypergraphs. In: Bureš, T., et al., (eds.) Proceedings of the 47th International Conference on Current Trends in Theory and Practice of Computer Science, SOFSEM 2021. LNCS, vol. 12607, pp. 361–372. Springer, Cham (2021). https://doi.org/10.1007/978-3-030-67731-2_26

16. Gabow, H.N., Tarjan, R.E.: Faster scaling algorithms for general graph-matching problems. J. ACM **38**(4), 815–853 (1991). https://doi.org/10.1145/115234.115366

17. Hong, S., Merrick, D., do Nascimento, H.A.D.: The metro map layout problem. In: Churcher, N., Churcher, C. (eds.) Proceedings of the Australasian Symposium on Information Visualisation, InVis.au 2004. CRPIT, vol. 35, pp. 91–100. Australian Computer Society (2004). http://crpit.scem.westernsydney.edu.au/abstracts/CRPITV35Hong.html

18. Jacobsen, B., Wallinger, M., Kobourov, S., Nöllenburg, M.: Metrosets: Visualizing sets as metro maps. IEEE Trans. Visual Comput. Graphics **27**(2), 1257–1267 (2021). https://doi.org/10.1109/TVCG.2020.3030475

19. Johnson, D.S., Pollak, H.O.: Hypergraph planarity and the complexity of drawing Venn diagrams. J. Graph Theory **11**(3), 309–325 (1987). https://doi.org/10.1002/JGT.3190110306

20. Kaufmann, M., van Kreveld, M.J., Speckmann, B.: Subdivision drawings of hypergraphs. In: Tollis, I.G., Patrignani, M. (eds.) Proceedings of the 16th International Symposium on Graph Drawing, GD 2008. Lecture Notes in Computer Science, vol. 5417, pp. 396–407. Springer (2008). https://doi.org/10.1007/978-3-642-00219-9_39

21. Klemz, B., Mchedlidze, T., Nöllenburg, M.: Minimum tree supports for hypergraphs and low-concurrency Euler diagrams. In: Ravi, R., Gørtz, I.L. (eds.) Proceedings of the 14th Scandinavian Symposium and Workshops on Algorithm Theory, SWAT 2014. LNCS, vol. 8503, pp. 265–276. Springer, Cham (2014). https://doi.org/10.1007/978-3-319-08404-6_23

22. Nickel, S., Nöllenburg, M.: Towards data-driven multilinear metro maps. In: Pietarinen, A., et al., (eds.) Proceedings of the 11th International Conference on Diagrammatic Representation and Inference, Diagrams 2020. Lecture Notes in Computer Science, vol. 12169, pp. 153–161. Springer (2020). https://doi.org/10.1007/978-3-030-54249-8_12

23. Nöllenburg, M., Wolff, A.: Drawing and labeling high-quality metro maps by mixed-integer programming. IEEE Trans. Visual Comput. Graphics **17**(5), 626–641 (2011). https://doi.org/10.1109/TVCG.2010.81

24. Schaefer, M.: Complexity of some geometric and topological problems. In: Eppstein, D., Gansner, E.R. (eds.) Proceedings of the 17th International Symposium of Graph Drawing, GD 2009. Lecture Notes in Computer Science, vol. 5849, pp. 334–344. Springer (2009). https://doi.org/10.1007/978-3-642-11805-0_32

25. Swaminathan, R., Wagner, D.K.: On the consecutive-retrieval problem. SIAM J. Comput. **23**(2), 398–414 (1994). https://doi.org/10.1137/S0097539792235487

26. Tamassia, R.: On embedding a graph in the grid with the minimum number of bends. SIAM J. Comput. **16**(3) (1987). https://doi.org/10.1137/0216030

Edge-Constrained Hamiltonian Paths on a Point Set

Todor Antić[1] , Aleksa Džuklevski[1] , Jiří Fiala[1] , Jan Kratochvíl[1] ,
Giuseppe Liotta[2] , Morteza Saghafian[3] , Maria Saumell[4] ,
and Johannes Zink[5]([✉])

[1] Department of Applied Mathematics, Faculty of Mathematics and Physics,
Charles University, Prague, Czech Republic
{todor,fiala,honza}@kam.mff.cuni.cz, aleksa@matfyz.cz
[2] Università degli Studi di Perugia, Perugia, Italy
giuseppe.liotta@unipg.it
[3] Institute of Science and Technology Austria, Klosterneuburg, Austria
morteza.saghafian@ist.ac.at
[4] Department of Theoretical Computer Science, Faculty of Information Technology,
Czech Technical University in Prague, Prague, Czech Republic
maria.saumell@fit.cvut.cz
[5] Technische Universität München, Heilbronn, Germany
zink@algo.cit.tum.de

Abstract. Let S be a set of distinct points in general position in the
Euclidean plane. A plane Hamiltonian path on S is a crossing-free geometric path such that every point of S is a vertex of the path. It is
known that, if S is sufficiently large, there exist three edge-disjoint plane
Hamiltonian paths on S. In this paper we study an edge-constrained
version of the problem of finding Hamiltonian paths on a point set. We
first consider the problem of finding a single plane Hamiltonian path π
with endpoints $s, t \in S$ and constraints given by a segment $\overline{ab}$, where
$a, b \in S$. We consider the following scenarios: (i) $\overline{ab} \in \pi$; (ii) $\overline{ab} \notin \pi$. We
characterize those quintuples $\langle S, a, b, s, t \rangle$ for which π exists. Secondly,
we consider the problem of finding two plane Hamiltonian paths π_1, π_2
on a set S with constraints given by a segment $\overline{ab}$, where $a, b \in S$. We
consider the following scenarios: (i) π_1 and π_2 share no edges and $\overline{ab}$ is
an edge of π_1; (ii) π_1 and π_2 share no edges and none of them includes
$\overline{ab}$ as an edge; (iii) both π_1 and π_2 include $\overline{ab}$ as an edge and share no
other edges. In all cases, we characterize those triples $\langle S, a, b \rangle$ for which
π_1 and π_2 exist.

Keywords: plane Hamiltonian paths · point sets · geometric graph
theory

The Full Version of This Paper Is Available on ArXiv [7].

© The Author(s), under exclusive license to Springer Nature Switzerland AG 2026
J. Kozik and A. Wolff (Eds.): SOFSEM 2026, LNCS 16448, pp. 532–546, 2026.
https://doi.org/10.1007/978-3-032-17801-5_39

1   Introduction

Let S be a set of points in the plane in general position, that is, with no three collinear points. A *geometric graph* on S is a graph with vertex set S whose edges are segments between points in S. A geometric graph on S is *plane* if no two edges cross each other. A *Hamiltonian path* on S is a geometric graph on S that is a path and contains all points from S. Our goal is to compute one or two *plane Hamiltonian paths* on S that satisfy certain *geometric constraints*. In the case where the constraint is that the two Hamiltonian paths are *edge-disjoint*, the existence of two such paths was shown by Aichholzer, Hackl, Korman, van Kreveld, Löffler, Pilz, Speckmann and Welzl [3]. This result was recently extended by Kindermann, Kratochvíl, Liotta and Valtr [19], who showed that in fact three edge-disjoint plane Hamiltonian paths on S always exist if $|S|$ is large enough.

When working on problems involving point sets in the plane, we might need some different or additional properties that the paths need to satisfy besides being plane, Hamiltonian, and edge-disjoint. We might want to find a path with prescribed endpoints or a prescribed (first) edge. In some cases, finding paths with such properties is straightforward and serves more as an exercise, while in others it can pose a significant challenge and requires more complex proofs. Such problems often arise when studying reconfigurations of plane paths, see [5, Lemmas 1 and 2] or [20, Lemma 9]. In a different direction, determining if it is possible to find a single plane Hamiltonian path that does not contain a prescribed set of segments has been studied by several authors [17,18,26,27], but the exact solution is known only when the point set is in convex position.

In this paper, we investigate the problem with the natural constraints that segments or endpoints are given and must be included or avoided. We restrict to the case where just one segment is given, which already turns out to be non-trivial. We remark that the existence of edge-restricted plane Hamiltonian paths has already been studied in the more general setting of *simple drawings* of complete graphs (see, for example, [6, Theorem 3.13] or [8, Theorem 2.4]), but these results are not useful for the problems that we study.

Our problems are also related to a well-studied topic of geometric graph theory, called *geometric graph packing problem*. Given a geometric *host* graph G and a subgraph H of G, we want to find as many edge-disjoint copies of H inside G as possible. In our case, the host graph is the complete geometric graph with vertex set S and H is a plane Hamiltonian path, that is, a plane spanning path of G. Bose, Hurtado, Rivera-Campo, and Wood [13,14] characterize those plane trees that can be packed into a complete geometric graph whose vertex set is in convex position. The authors of [3,4] prove that $\Omega(\sqrt{n})$ plane trees can be packed into a complete geometric graph with n vertices. This was later improved to $\lfloor n/3 \rfloor$ by Biniaz and García [11,12]. However, it remains an open question whether this lower bound also applies to plane Hamiltonian paths. Biniaz, Bose, Maheshwari, and Smid [10] show that any set of n points in general position admits at least $\lceil \log_2 n \rceil - 1$ plane perfect matchings, and that, for some point sets, the maximum number of such matchings is at most $\lceil n/3 \rceil$.

In the non-geometric setting, our problem is related to a line of research investigating a generalization of Hamiltonian paths on planar graphs that are called *Tutte paths* [9, 22–25]. Tutte paths are paths in planar graphs that decompose the graph into components with at most three attachments to the path. In some scenarios, it is asked for a Tutte path that starts and ends at given vertices and contains a given edge.

Our contribution. We study the existence of one or two plane Hamiltonian paths under various geometric constraints. We first consider the problem of finding a single plane Hamiltonian path π with endpoints $s, t \in S$ and constraints given by a segment $\overline{ab}$, where $a, b \in S$; see Sect. 3. We examine the following scenarios: (i) $\overline{ab} \notin \pi$ (Theorem 1) and (ii) $\overline{ab} \in \pi$ (Theorem 2). We characterize those quintuples $\langle S, a, b, s, t \rangle$ for which π exists.

Secondly, as our main results, we consider the problem of finding two plane Hamiltonian paths π_1, π_2 with constraints given by a segment $\overline{ab}$, where $a, b \in S$; see Sect. 4. We study the following scenarios: (i) π_1 and π_2 share no edges and none of them includes $\overline{ab}$ as an edge (Proposition 1), (ii) π_1 and π_2 share no edges and $\overline{ab}$ is an edge of π_1 (Theorem 3), and (iii) both π_1 and π_2 include $\overline{ab}$ as an edge and share no other edges (Theorem 4). In all cases, we characterize those triples $\langle S, a, b \rangle$ for which π_1 and π_2 exist.

Statements whose complete proofs are available in the full version on arXiv are marked with a clickable ($\star$).

2   Preliminaries

Throughout the paper we always assume that S is a set of n points in the plane in general position. By general position, we mean that there are no three collinear points. We write $\mathrm{CH}(S)$ for the convex hull of S and $\partial\,\mathrm{CH}(S)$ for its boundary. For two points a, b, we write $\ell(ab)$ to denote the line through a and b, and $\overline{ab}$ to denote the line segment between a and b. Given four distinct points a, b, x, y, we call a line segment $\overline{xy}$ a *bridge* over $\ell(ab)$ if it crosses $\ell(ab)$ but not $\overline{ab}$. If π_1 and π_2 are two (Hamiltonian) paths on S, we write $\pi_1 \cap \pi_2 = \emptyset$ if they are edge-disjoint. This is somewhat unusual, but convenient for our application. Usually, we call $v \in S$ a point, but we call a point $v \in S \cap \partial\,\mathrm{CH}(S)$ a *vertex* of $\partial\,\mathrm{CH}(S)$ and a segment between two vertices of $\partial\,\mathrm{CH}(S)$ an *edge* of $\mathrm{CH}(S)$. We often refer to two results from [19] stated next. Inspecting their constructive proofs from an algorithmic point of view, it is easy to see that the described procedures can be realized as $O(n \log n)$-time algorithms. The running time comes from sorting a linear number of points by their radial coordinates.

Lemma 1 ([19], Lemma 5).   *Let S be a set of points in general position in the plane, and let s and t be two distinct points in S. Then, there is a plane Hamiltonian path π in S such that π starts in s and ends in t. Such a path can be found in $O(n \log n)$ time.*

Lemma 2 ([19], Theorem 5). *Let S be a set of at least 5 points in general position in the plane and let s and t be two (not necessarily distinct) points of $\partial\,\mathrm{CH}(S)$. Then, S contains two edge-disjoint plane Hamiltonian paths, one starting at s and the other one at t. Moreover, if the points s and t are distinct, then the paths can be chosen so that none of them contains the edge st. Such paths can be found in $O(n\log n)$ time.*

For our purposes, Lemma 2 is often too strong and cannot be applied to four or fewer points. We prove a weaker statement that holds already for four points:

Lemma 3. *Let S be a set of at least 4 points in general position in the plane and let s and t be two consccutive points of $\partial\,\mathrm{CH}(S)$. Then, S contains two edge-disjoint plane Hamiltonian paths, one starting at s and the other one at t. Such paths can be found in $O(n\log n)$ time.*

Proof. If $|S| \geq 5$, the result follows from Lemma 2. If $|S| = 4$, there are only two sets to consider (see Fig. 1), and for both we can find the desired paths.

Fig. 1. Two Hamiltonian paths on 4 points with neighboring starting points s, t.

Often, we use the following lemma to construct a path:

Lemma 4 ($\star$). *Let S be a set of at least three points in general position in the plane, and let $a, b, z \in S$ be distinct with $z \in \partial\,\mathrm{CH}(S)$. Then, there is a plane Hamiltonian path π in S such that $\overline{ab} \in \pi$ and π starts in z. Such a path can be found in $O(n\log n)$ time.*

3   Finding One Plane Hamiltonian Path

In this section, we consider the following problem: We are given a point set S and two distinct points $s, t \in S$. Our goal is to find a plane Hamiltonian s–t path π in S. Additionally, we are given two distinct points $a, b \in S$ and we require that π contains or avoids the segment $\overline{ab}$. In each of these two situations, we fully characterize the quintuples (S, a, b, s, t) for which such a path exists.

3.1   One Path with Prescribed Endpoints and Avoiding a Segment

We consider the problem of finding an s–t path π that does not contain $\overline{ab}$. To this end, we first observe that, in some situations, such a path does not exist. The proofs of the following two lemmas describing such situations can be found in the full version [7]. See Fig. 2 for examples of sets satisfying the conditions of Lemma 5 and 6.

Lemma 5 ($\star$). *Let S be a set of $n \geq 3$ points in convex position and $a, b, s, t \in S$ points such that $\overline{ab}$ and $\overline{st}$ are disjoint edges of $\partial\,\mathrm{CH}(S)$. Then, π does not exist.*

Lemma 6 ($\star$). *Let S be a set of $n \geq 5$ points with $n - 1$ vertices on $\partial\,\mathrm{CH}(S)$. Let s be the point in the interior of $\mathrm{CH}(S)$, a, b, t be three distinct vertices of $\partial\,\mathrm{CH}(S)$ and let x, y be the two vertices of $\partial\,\mathrm{CH}(S)$ adjacent to t such that a, b, x, t, y appear in this order on $\partial\,\mathrm{CH}(S)$. If $\overline{ab}$ is an edge of $\partial\,\mathrm{CH}(S)$ and s lies on the same side of $\ell(ax)$ and $\ell(by)$ as t, then π does not exist.*

Fig. 2. Sets satisfying the conditions of Lemma 5 (left) and Lemma 6 (right).

The main result of this section is that, except in the cases described above, the desired path π always exists.

Theorem 1 ($\star$). *Let S be a set of at least 5 points in general position in the plane and $a, b, s, t \in S$ such that $a \neq b$, $s \neq t$. Then, there is a plane Hamiltonian path π in S with endpoints s and t such that $\overline{ab} \not\in \pi$, unless the quintuple $\langle S, a, b, s, t \rangle$ satisfies the conditions of Lemma 5 or Lemma 6. We can decide if π exists, and construct it if it does, in $O(n \log n)$ time, where $n = |S|$.*

To prove Theorem 1, we will need the following less restricted statement.

Lemma 7 ($\star$). *Let S be a set of $n \geq 4$ points in general position in the plane, $a \neq b \in S$, and let $s \in S$ be any point. Then, there is a plane Hamiltonian path π in S such that one endpoint of π is s, the other endpoint is a or b, and $\overline{ab} \not\in \pi$. Such a path can be computed in $O(n \log n)$ time.*

Proof [of Theorem 1] Assume that $n \geq 5$ and the quintuple $\langle S, a, b, s, t \rangle$ does not satisfy the conditions of Lemma 5 or Lemma 6. Further, assume that $\overline{ab}$ is horizontal and that a lies to the left of b. Also, assume that $\{a, b\} \neq \{s, t\}$, as otherwise the result follows from Lemma 1. We now split into cases.

Case A: s and t lie in different closed halfplanes determined by $\ell(ab)$.
Note that the assumption of this case guarantees that $s, t \notin \{a, b\}$. Without loss of generality we assume that s lies above and t lies below $\ell(ab)$, and that s is not the only point of S lying above $\ell(ab)$. We first apply Lemma 7 and construct a plane path starting in s and collecting all the points of S that are in the same closed halfplane determined by $\ell(ab)$ as s, but not using the segment $\overline{ab}$. Such path is either an s–a or an s–b path, so assume that the former is true. Then, we construct an a–t path for the remaining points using Lemma 1. Finally, concatenating these two paths gives π. See Fig. 3a.

Case B: s and t lie in the same closed halfplane determined by $\ell(ab)$ and $\overline{ab}$ is not an edge of $\partial\,\mathrm{CH}(S)$. See the full version [7].
For the remaining cases, we assume that $\overline{ab}$ is an edge of $\partial\,\mathrm{CH}(S)$.

Case C: S is in convex position and $\overline{st}$ is not an edge of $\partial\,\mathrm{CH}(S)$.
See Fig. 3b and in the full version [7].

Case D: S contains only one point in the interior of $\mathrm{CH}(S)$.
See Fig. 3c and in the full version [7].

Case E: S contains at least two points in the interior of $\mathrm{CH}(S)$.
If the only two points in the interior of $\mathrm{CH}(S)$ are s and t, π can be obtained from the a–b path in $\partial\,\mathrm{CH}(S)$ not containing $\overline{ab}$, by appending one of s, t to the front and the other to the end. At least one of the two possible cases yields a non-crossing path (if the segments $\overline{sa}$ and $\overline{bt}$ cross, $\overline{ta}$ and $\overline{bs}$ are disjoint). From now on, we may assume that there is at least one point distinct from s, t in the interior. Let x be a point in the interior of $\mathrm{CH}(S)$ such that the triangle $\triangle(a, b, x)$ contains no other points of S. We now need to consider two subcases.

Case E1: We can choose $x \neq s$. Then, the lines $\ell(ax)$ and $\ell(bx)$ split $\mathrm{CH}(S)$ into 4 convex sets. If one of them contains both s and t, we split it into two using a suitable ray from x in such a way that s and t are in separate convex sets. We then apply Lemma 1 (at most) three times to obtain suitable s–a and t–b paths[1] and concatenating them with the segments $\overline{ax}$ and $\overline{xb}$ finally gives us the desired path π. Observe that, in the construction of the t–b path (or, depending on the situation, one of the other paths), we can always find a suitable bridge over $\ell(bx)$ (or $\ell(ax)$); see Fig. 3d.

Case E2: s is the only point such that the triangle $\triangle(a, s, b)$ is empty.
See Fig. 3e and 3f and in the full version [7].

Note that we only apply Lemma 1 and make local modifications or simple tests for the cases, which allows for a running time in $O(n \log n)$. $\qquad\qquad\square$

3.2   One Path with Prescribed Endpoints and Including a Segment

Next, we consider the problem of finding an s–t path π that contains $\overline{ab}$. As in the previous problem, in some situations such a path does not exist. The first claim is immediate.

---

[1] Or suitable t–a and s–b paths.

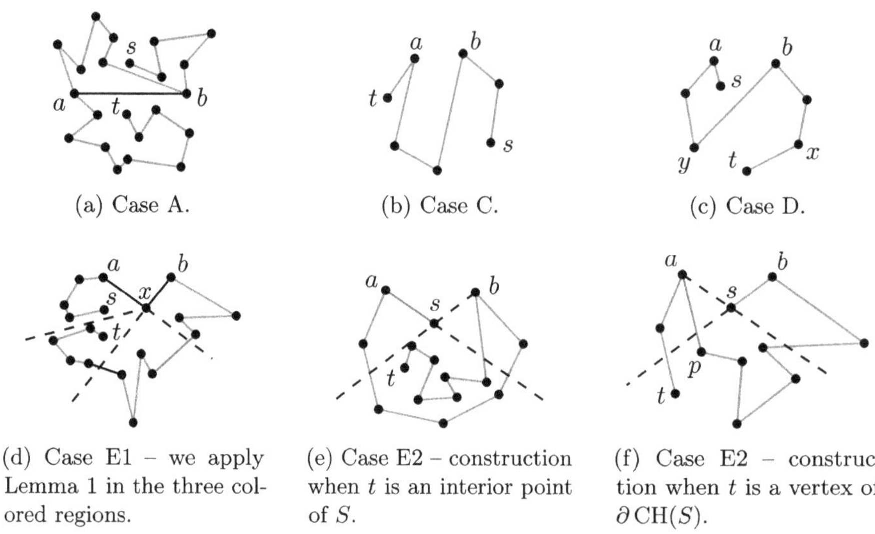

(a) Case A.

(b) Case C.

(c) Case D.

(d) Case E1 – we apply Lemma 1 in the three colored regions.

(e) Case E2 – construction when t is an interior point of S.

(f) Case E2 – construction when t is a vertex of $\partial\,\mathrm{CH}(S)$.

Fig. 3. Construction of path π in Theorem 1.

Lemma 8. *If S is a set of more than two points in general position and $\{s,t\} = \{a,b\}$, then π does not exist.*

The proofs of the next three lemmas can be found in the full version [7].

Lemma 9 ($\star$). *If S is a set of n points in general position, a,b are non-consecutive vertices of $\partial\,\mathrm{CH}(S)$, and s and t lie in the same closed halfplane determined by $\ell(ab)$, then π does not exist.*

Before describing the remaining situations in which π does not exist, recall that a bridge over $\ell(ab)$ is a segment that crosses $\ell(ab)$ but not $\overline{ab}$.

Lemma 10 ($\star$). *Let S be a set of n points in general position. Suppose that $\{a,b\} \not\subset \partial\,\mathrm{CH}(S)$, one of s,t is in $\{a,b\}$, in the open halfplane determined by $\ell(ab)$ containing the other of s,t there is at least one more point of S, and every bridge $\overline{xy}$ over $\ell(ab)$ is incident to the other of s,t. Then, π does not exist.*

Lemma 11 ($\star$). *Let S be a set of n points in general position. Suppose that $\overline{ab}$ is not an edge of $\partial\,\mathrm{CH}(S)$, $s \in \{a,b\}$, and for every bridge $\overline{xy}$ over $\ell(ab)$, the following holds:*

- *both x and y are vertices of $\partial\,\mathrm{CH}(S)$,*
- *if $\overline{xy}$ is an edge of $\partial\,\mathrm{CH}(S)$, then $t \in \{x,y\}$ and*
- *the open halfplane determined by $\ell(xy)$ that contains a,b also contains at least one point on each side of $\ell(ab)$.*

Then, π does not exist.

Fig. 4. Sets satisfying the conditions of Lemma 10 (left) and Lemma 11 (right), with possible bridges drawn in dashed blue. (Color figure online)

See Fig. 4 for examples of sets satisfying the conditions of Lemma 10 and 11.

The main result of this section is that, except in the cases described above, π always exist. We give the details in the full version [7].

Theorem 2 ($\star$). *Let S be a set of at least 2 points in general position in the plane and $a, b, s, t \in S$ such that $a \neq b$, $s \neq t$. Then there is a plane Hamiltonian path π in S with endpoints s and t such that $\overline{ab} \in \pi$, unless the quintuple $\langle S, a, b, s, t \rangle$ satisfies the conditions of Lemma 8, 9, 10 or 11. We can decide if π exists, and construct it if it does, in $O(n \log n)$ time, where $n = |S|$.*

4   Finding Two Plane Hamiltonian Paths

In this section, we consider the following problem: We are given a point set S and two distinct points $a, b \in S$. Our goal is to find two plane Hamiltonian paths π_1 and π_2 in S. Additionally, we require that neither, one or both of π_1, π_2 contain $\overline{ab}$ and are otherwise edge-disjoint.

We first observe that, if we want both π_1 and π_2 to avoid $\overline{ab}$, we can always find suitable paths.

Proposition 1 ($\star$). *Let S be a set of at least 5 points in general position in the plane and $a, b \in S$ such that $a \neq b$. Then there exist two plane Hamiltonian paths π_1, π_2 in S such that $\pi_1 \cap \pi_2 = \emptyset$ and $\overline{ab} \notin \pi_1 \cup \pi_2$. Such paths can be computed in polynomial time.*

We remark that, for $n \geq 7$, the result of Proposition 1 follows from the main result of [19]. For $n \in \{5, 6\}$, the number of cases that need to be checked can be reduced using the framework of Pilz and Welzl [21]; see the full version [7] for details.

The remaining two situations require a more involved treatment.

4.1   Two Edge-Disjoint Paths with One of Them Including
a Segment

We consider the problem of finding two edge-disjoint plane Hamiltonian paths on a point set S, with the constraint that one of them needs to include a segment $\overline{ab}$, where $a, b \in S$ are distinct. We show that this is always possible.

To prove Theorem 3, we build on Lemma 1 and 4. In addition to these, we need the following well-known result by Abellanas, García-López, Hernández-Peñalver, Noy, and Ramos [1,2], which can also be deduced from the earlier results by Hershberger and Suri [15,16]. It was used to show the existence of two and three plane Hamiltonian paths in [3] and [19].

Lemma 12 ([2]⋆). *Let S_{down} and S_{up} be two sets of points in general position such that $|S_{\mathrm{down}}| \leq |S_{\mathrm{up}}| \leq |S_{\mathrm{down}}| + 1$ and there is a line that separates S_{down} from S_{up}. Further, let $z \in S_{\mathrm{up}}$ be any vertex (of two possible candidates) on $\partial\,\mathrm{CH}(S_{\mathrm{down}} \cup S_{\mathrm{up}})$ that is neighboring a vertex from S_{down} on $\partial\,\mathrm{CH}(S_{\mathrm{down}} \cup S_{\mathrm{up}})$. Then, there exists a plane Hamiltonian path π on $S_{\mathrm{down}} \cup S_{\mathrm{up}}$ such that π starts in z and every edge in π has one endpoint in S_{down} and one endpoint in S_{up}. Such a path can be found in $O(n \log n)$ time, where $n = |S_{\mathrm{down}}| + |S_{\mathrm{up}}|$.*

We can now prove Theorem 3.

Theorem 3 (⋆). *Let S be a set of at least 4 points in general position in the plane and $a, b \in S$ such that $a \neq b$. Then, there exist two plane Hamiltonian paths π_1, π_2 in S such that $\pi_1 \cap \pi_2 = \emptyset$ and $\overline{ab} \in \pi_1$. Such paths can be computed in $O(n \log n)$ time, where $n = |S|$.*

Proof. Let us first discuss the small values of n. To be able to have two edge-disjoint Hamiltonian paths, we need at least four points (otherwise, there are not enough edges between the points to host the edges of the two paths). For $n = 4$, all six possible edges between the points are needed to host two edge-disjoint Hamiltonian paths. The four points can have three or four points in the convex hull. In both cases, we can find two edge-disjoint plane Hamiltonian paths, as illustrated in Fig. 1. From now on, we assume that $n \geq 5$.

We first rotate the point set S so that $\overline{ab}$ is horizontal. We then sort in $O(n \log n)$ time the points in S in increasing order by y-coordinate (pairs of points with the same y-coordinate are sorted arbitrarily). Let S_{down} be the set of the first $\lfloor n/2 \rfloor$ points (the lower points) and let $S_{\mathrm{up}} = S \setminus S_{\mathrm{down}}$ (the upper points). Clearly, S_{down} and S_{up} can be separated by a line. We refer to S_{down} and S_{up} as *partitions* (of S). Since $n \geq 5$, we have that $|S_{\mathrm{down}}| \geq 2$, and $|S_{\mathrm{up}}| \geq 3$.

By Lemma 12, there is a plane Hamiltonian path π on S such that every edge of π has an endpoint in S_{down} and an endpoint in S_{up}, and π starts in a point $z \in S_{\mathrm{up}}$ contained in $\partial\,\mathrm{CH}(S_{\mathrm{down}} \cup S_{\mathrm{up}})$. Such a path can be found in $O(n \log n)$ time.

We say that a point $p \in S_{\mathrm{down}}$ *sees* a point $q \in S_{\mathrm{up}}$ if and only if $\overline{pq} \cap \mathrm{CH}(S_{\mathrm{down}}) = p$ and $\overline{pq} \cap \mathrm{CH}(S_{\mathrm{up}}) = q$. If p sees q, then p lies on $\partial\,\mathrm{CH}(S_{\mathrm{down}})$, q lies on $\partial\,\mathrm{CH}(S_{\mathrm{up}})$, and $\overline{pq}$ intersects $\mathrm{CH}(S_{\mathrm{down}})$ and $\mathrm{CH}(S_{\mathrm{up}})$ at a single point. We use the symmetric definition if $p \in S_{\mathrm{up}}$ and $q \in S_{\mathrm{down}}$. We consider two cases:

Case A: a and b are in distinct partitions. We distinguish whether $\overline{ab} \in \pi$ or not.

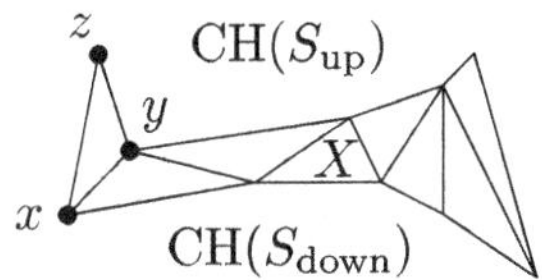

Fig. 5. In Case A2.2, x sees at least two points of S_{up}.

Case A1: $\overline{ab} \notin \pi$. Let $\pi_2 = \pi$. Notice that, since $\overline{ab}$ is horizontal, one of a, b is the topmost point of S_{down} and the other is the bottommost point of S_{up}. Let π_1^{down} and π_1^{up} be the paths traversing the points of S_{down} and S_{up}, respectively, such that vertices are traversed by increasing y-coordinate. Let π_1 be the concatenation of π_1^{down}, $\overline{ab}$, and π_1^{up}. Clearly, π_1 and π_2 are two plane Hamiltonian paths such that $\overline{ab} \in \pi_1$ and $\pi_1 \cap \pi_2 = \emptyset$. Additionally, constructing π_1 takes $O(n)$ time after S is sorted.

Case A2: $\overline{ab} \in \pi$. Let $\pi_1 = \pi$. For the second path, we again aim to find two plane Hamiltonian paths on S_{down} and S_{up}. The challenging part is to find a suitable "bridge" that connects the two paths and is not contained in π_1. To this end, we analyze the set of points that z (the starting point of π_1) sees. The conditions of the following cases can be checked in $O(n \log n)$ time.

Case A2.1: z sees two points x, y of S_{down}. At most one of $\{\overline{xz}, \overline{yz}\}$ can lie in π_1; without loss of generality, let this be $\overline{xz}$. Let π_2^{down} and π_2^{up} be two plane Hamiltonian paths on S_{down} and S_{up} starting in y and z, respectively. Such paths exist thanks to Lemma 1, and can be found in $O(n \log n)$ time. Let π_2 be the concatenation of π_2^{down}, $\overline{yz}$, and π_2^{up}. Clearly, π_1 and π_2 are two plane Hamiltonian paths such that $\overline{ab} \in \pi_1$ and $\pi_1 \cap \pi_2 = \emptyset$.

Case A2.2: z sees one point x of S_{down} and $|S_{\mathrm{down}}| = |S_{\mathrm{up}}|$. We show that, if z sees only one point x of S_{down}, then x sees at least two points of S_{up} (see Fig. 5): Let X be the closure of $\mathrm{CH}(S) \setminus (\mathrm{CH}(S_{\mathrm{down}}) \cup \mathrm{CH}(S_{\mathrm{up}}))$. Clearly, X is a polygonal region with vertices x and z, and z is visible to only two vertices of X: x and one of z's neighbors on $\partial\,\mathrm{CH}(S_{\mathrm{up}})$, which we call y. Since X can be triangulated, x sees y. We now perform a reflection of the point set S through the line $\ell(ab)$. Taking the reverse order of the points sorted by y-coordinate and splitting them into two halves exactly switches the sets S_{down} and S_{up}. Since $|S_{\mathrm{down}}| = |S_{\mathrm{up}}|$, the new point set fulfills the hypothesis of our case distinction. Thus, x takes the role of z and we are in Case A2.1.

Case A2.3: z sees one point of S_{down} and $|S_{\mathrm{down}}| + 1 = |S_{\mathrm{up}}|$. We perform a reflection of S through the line $\ell(ab)$. Afterwards, we partition the new point set into S_{up} and S_{down} as described above. Since $|S_{\mathrm{up}}| > |S_{\mathrm{down}}|$, a and b end up in S_{up} and we are in Case B.

Case B: a and b are in the same partition. In this case, the main idea is similar as before: to use π as π_2 (since π does not contain $\overline{ab}$), to find spanning paths of S_{up} and S_{down} including $\overline{ab}$, and to connect them as π_1. If the connection edge is easy to find, we are done immediately. This is the case, e.g., if the topmost point x of S_{down} sees a point s of S_{up} such that $\overline{xs} \notin \pi$ and neither x nor s are identical to a or b. However, there are some more subtle cases, e.g., if the candidates for the connection edges are all contained in π or if points like x and s are identical to a or b. We exhaustively investigate all these special cases and their treatment in the full version [7]. There, we show that we can always find in $O(n \log n)$ the desired paths π_1 and π_2.   □

4.2 Two Paths Sharing a Single Segment

Lastly, we consider the problem of finding two plane Hamiltonian paths on a point set S that overlap on a prescribed segment $\overline{ab}$ and are otherwise disjoint. Unlike the previous case, such paths do not always exist. We fully characterize the triples $\langle S, a, b \rangle$ for which such paths exist. We distinguish three cases, based on whether both, one or none of a, b are vertices of $\partial \mathrm{CH}(S)$.

Lemma 13. *Let S be a set of n points in the plane in general position, and let $a, b \in S$ such that $a, b \in \partial \mathrm{CH}(S)$. There exist two plane Hamiltonian paths π_1, π_2 on S such that $\pi_1 \cap \pi_2 = \overline{ab}$ if and only if one of the following holds:*

- *one of the two open halfplanes bounded by $\ell(ab)$ contains 2 or 3 points of S, and the other open halfplane contains no points of S; or*
- *both open halfplanes bounded by $\ell(ab)$ contain at most 1 or at least 4 points of S.*

Proof. As a, b are vertices of $\partial \mathrm{CH}(S)$, they split $S \setminus \{a, b\}$ into sets S_1, S_2; see Fig. 6f. Assume that $\overline{ab}$ is horizontal and S_1 lies above $\overline{ab}$. We get these cases:

Case A: $|S_2| = 0$, $|S_1| = 1$. Let x be the point in S_1. We define the two paths as $\pi_1 = x, a, b$ and $\pi_2 = x, b, a$.

Case B: $|S_2| = 0$, $|S_1| > 1$. If $|S_1| = 2$ or 3, we can construct the paths as in Fig. 6a and 6b. If $|S_1| \geq 4$, note that both a and b see at least 2 vertices of $\partial \mathrm{CH}(S_1)$. Let x be a vertex visible by a, and y a vertex visible by b, with $y \neq x$. Note that, if $|S_1| = 4$, we can choose x, y to be two consecutive vertices of the convex hull. Then let π_1', π_2' be two plane Hamiltonian paths on S_1 starting at x and y, respectively, as guaranteed by Lemma 2 and 3. We can obtain our desired paths π_1 and π_2 by extending π_1', π_2' as in Figure 6c.

Case C: $|S_2| = 1$, $|S_1| = 1$. We can construct the paths as in Fig. 6d.

Case D: $|S_2| = 1$, $|S_1| \geq 4$. We find π_1', π_2' as in Case C, and extend them as in Fig. 6e.

Case E: $|S_2| \geq 4$, $|S_1| \geq 4$. Let x_1 and y_1 (with $x_1 \neq y_1$) be two vertices of $\partial \mathrm{CH}(S_1)$ visible from a and b, respectively, and x_2 and y_2 be two vertices of $\partial \mathrm{CH}(S_2)$ defined analogously. Let π_1^1, π_2^1 be two plane Hamiltonian paths on S_1 starting at x_1 and y_1, respectively, as guaranteed by Lemma 2 and 3. Analogously, let π_1^2, π_2^2 to be two plane Hamiltonian paths on S_2 starting at x_2 and y_2. Finally, we connect π_1^1 to π_2^2 and π_2^1 to π_1^2 as in Fig. 6f, and obtain our desired paths π_1 and π_2.

In the remaining cases it is impossible to find π_1 and π_2.

Case F: $|S_2| = 2$, $|S_1| \geq 1$. Let $S_2 = \{x, y\}$. Assume that there are two paths π_1, π_2 with the desired properties. Then at most one of them can contain the segment $\overline{xy}$. Assume that π_2 does not contain this segment. Then π_2 must contain edges $\overline{ax}$ and $\overline{by}$. However, it is then clear that we cannot continue π_2 as either a or b would have degree 3 or the edge $\overline{ab}$ would be crossed in π_2. Therefore, it is impossible to find such two paths.

Case G: $|S_2| = 3$, $|S_1| \geq 1$. Let $S_2 = \{x, y, z\}$. Assume that there are two paths π_1, π_2 with the desired properties. We can assume that π_2 contains at most one of the segments $\overline{xy}, \overline{xz}, \overline{yz}$. Then, by a similar argument as in the previous case, we can see that it is impossible to complete π_2, contradicting the assumption that two such paths exist. $\qquad\square$

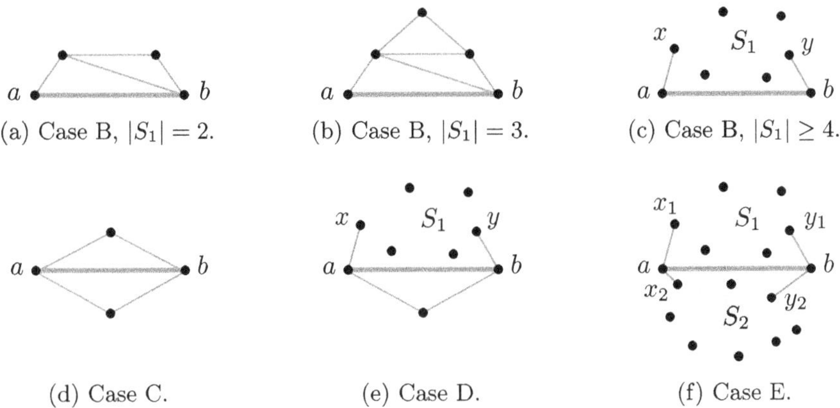

(a) Case B, $|S_1| = 2$.   (b) Case B, $|S_1| = 3$.   (c) Case B, $|S_1| \geq 4$.

(d) Case C.   (e) Case D.   (f) Case E.

Fig. 6. Construction of π_1 and π_2 when both a and b are vertices of $\partial\,\mathrm{CH}(S)$.

The proofs of Lemma 14 and 15 are available in the full version [7].

Lemma 14 ($\star$). *Let S be a set of n points in the plane in general position, and let $a, b \in S$ such that $a \notin \partial\,\mathrm{CH}(S)$ and $b \in \partial\,\mathrm{CH}(S)$. There exist two plane Hamiltonian paths π_1, π_2 on S such that $\pi_1 \cap \pi_2 = \overline{ab}$ if and only if one of the following holds:*

- *there are at least two bridges over $\ell(ab)$; or*
- *there is only one bridge over $\ell(ab)$ and at least three additional points of S on both sides of $\ell(ab)$.*

Lemma 15 ($\star$). *Let S be a set of n points in the plane in general position, and let $a, b \in S$ be two points not lying on the convex hull of S. There exist two plane Hamiltonian paths π_1, π_2 on S such that $\pi_1 \cap \pi_2 = \overline{ab}$.*

Now, we can state the main result of this section.

Theorem 4. *Let S be a set of at least 4 points in the plane in general position, and let $a, b \in S$ such that $a \neq b$. There exist two plane Hamiltonian paths π_1, π_2 on S such that $\pi_1 \cap \pi_2 = \overline{ab}$ if and only if the triple $\langle S, a, b \rangle$ satisfies one of the Lemmas 13,14,15. Moreover, we can decide if π_1, π_2 exist, and construct them if they do, in $O(n \log n)$ time, where $n = |S|$.*

Proof. The existence part of the statement follows from Lemmas 13,14,15. Note that the construction of the paths π_1, π_2 in the proofs of Lemmas 13,14 and 15 is based on multiple applications of Lemmas 2,3, and ad-hoc constructions for small point sets. Thus, the running time is $O(n \log n)$. $\square$

5   Open Problems

We conclude by listing some open problems. Firstly, a natural way to unify and extend Theorems 2 and 3 would be to determine under what conditions two paths can be found when one or both endpoints are prescribed to one or both of the paths. Doing this would probably require significant refinements of the arguments presented in the proof of Theorem 3. A somewhat simpler problem might be to extend Proposition 1 and Theorem 4 in a similar way by prescribing one or both endpoints to one or both paths. Another way to generalize our results is by prescribing more than one edge to be taken or avoided by one or two paths.

Lastly, the most challenging open problem is to reduce the gap between the upper and lower bounds for the geometric graph packing problem for plane Hamiltonian paths. The best known lower bound is 3 and the best upper bound is $\lceil \frac{n}{3} \rceil$, both proven in [19].

Acknowledgments. We thank the organizers of the HOMONOLO 2024 workshop in Nová Louka, Czech Republic, for the fruitful atmosphere where the research on this project was initiated.

T. Antić, A. Džuklevski, J. Kratochvíl and M. Saumell received funding from GAČR grant 23–04949X, T.A and A.Dž were additionally supported by GAUK grant SVV–2025–260822. G. Liotta was supported in part by MUR of Italy, PRIN Project no. 2022TS4Y3N – EXPAND and PON Project ARS01_00540. J. Fiala was in part supported by GAČR grant 25-16847S.

Disclosure of Interests. The authors have no competing interests to declare that are relevant to the content of this article.

References

1. Abellanas, M., Garcia-Lopez, J., Hernández-Peñalver, G., Noy, M., Ramos, P.A.: Bipartite embeddings of trees in the plane. In: North, S.C. (ed.) Proceedings of the 4th International Symposium on Graph Drawing (GD '96). Lecture Notes in Computer Science, vol. 1190, pp. 1–10. Springer (1996). https://doi.org/10.1007/3-540-62495-3_33

2. Abellanas, M., Garcia-Lopez, J., Hernández-Peñalver, G., Noy, M., Ramos, P.A.: Bipartite embeddings of trees in the plane. Discret. Appl. Math. **93**(2–3), 141–148 (1999). https://doi.org/10.1016/S0166-218X(99)00042-6
3. Aichholzer, O., et al.: Packing plane spanning trees and paths in complete geometric graphs. Inf. Process. Lett. **124**, 35–41 (2017). https://doi.org/10.1016/J.IPL.2017.04.006
4. Aichholzer, O., et al.: Packing plane spanning trees and paths in complete geometric graphs. In: Proceedings of the 26th Canadian Conference on Computational Geometry (CCCG '14), pp. 233–238 (2014). https://cccg.ca/proceedings/2014/papers/paper34.pdf
5. Aichholzer, O., et al.: Flipping plane spanning paths. In: Lin, C.C., Lin, B.M.T., Liotta, G. (eds.) Proceedings of the 17th International Conference and Workshops on Algorithms and Computation (WALCOM'23), pp. 49–60. Springer Nature Switzerland, Cham (2023). https://doi.org/10.1007/978-3-031-27051-2_5
6. Aichholzer, O., Orthaber, J., Vogtenhuber, B.: Towards crossing-free Hamiltonian cycles in simple drawings of complete graphs. Comput. Geom. Topoloy **3**(2), 5:1–5:30 (2024). https://doi.org/10.57717/CGT.V3I2.47
7. Antić, T., et al.: Edge-constrained Hamiltonian paths on a point set. arXiv report (2025). https://arxiv.org/abs/2511.22526
8. Bergold, H., Felsner, S., Reddy, M., Orthaber, J., Scheucher, M.: Plane Hamiltonian cycles in convex drawings. Discrete and Computational Geometry (2025). https://doi.org/10.1007/s00454-025-00752-3
9. Biedl, T., Kindermann, P.: Finding Tutte paths in linear time. In: Baier, C., Chatzigiannakis, I., Flocchini, P., Leonardi, S. (eds.) Proceedings of the 46th International Colloquium on Automata, Languages, and Programming (ICALP '19). LIPIcs, vol. 132, pp. 23:1–23:14. Schloss Dagstuhl - Leibniz-Zentrum für Informatik (2019). https://doi.org/10.4230/LIPICS.ICALP.2019.23
10. Biniaz, A., Bose, P., Maheshwari, A., Smid, M.H.M.: Packing plane perfect matchings into a point set. Discrete Math. Theor. Comput. Sci. **17**(2), 119–142 (2015). https://doi.org/10.46298/DMTCS.2132
11. Biniaz, A., García, A.: Packing plane spanning trees into a point set. In: Durocher, S., Kamali, S. (eds.) Proceedings of the 30th Canadian Conference on Computational Geometry (CCCG'18), pp. 49–53 (2018). http://www.cs.umanitoba.ca/%7Ecccg2018/papers/session2A-p1.pdf
12. Biniaz, A., García, A.: Packing plane spanning trees into a point set. Comput. Geom. **90**, 101653 (2020). https://doi.org/10.1016/j.comgeo.2020.101653
13. Bose, P., Hurtado, F., Rivera-Campo, E., Wood, D.R.: Partitions of complete geometric graphs into plane trees. In: Pach, J. (ed.) Proceedings of the 12th International Symposium on Graph Drawing (GD '04). Lecture Notes in Computer Science, vol. 3383, pp. 71 – 81. Springer (2004). https://doi.org/10.1007/978-3-540-31843-9_9
14. Bose, P., Hurtado, F., Rivera-Campo, E., Wood, D.R.: Partitions of complete geometric graphs into plane trees. Comput. Geom. **34**(2), 116–125 (2006). https://doi.org/10.1016/j.comgeo.2005.08.006
15. Hershberger, J., Suri, S.: Applications of a semi-dynamic convex hull algorithm. In: Gilbert, J.R., Karlsson, R.G. (eds.) Proceedings of the 2nd Scandinavian Workshop on Algorithm Theory (SWAT '90). Lecture Notes in Computer Science, vol. 447, pp. 380–392. Springer (1990). https://doi.org/10.1007/3-540-52846-6_106
16. Hershberger, J., Suri, S.: Applications of a semi-dynamic convex hull algorithm. BIT **32**(2), 249–267 (1992). https://doi.org/10.1007/BF01994880

17. Keller, C., Perles, M.A.: Blockers for simple Hamiltonian paths in convex geometric graphs of even order. Discrete Comput. Geom. **60**(1), 1–8 (2017). https://doi.org/10.1007/s00454-017-9921-8

18. Keller, C., Perles, M.A.: Blockers for simple Hamiltonian paths in convex geometric graphs of odd order. Discrete Comput. Geom. **65**(2), 425–449 (2019). https://doi.org/10.1007/s00454-019-00155-1

19. Kindermann, P., Kratochvíl, J., Liotta, G., Valtr, P.: Three edge-disjoint plane spanning paths in a point set. In: Bekos, M.A., Chimani, M. (eds.) Proceedings of the 31st International Symposium on Graph Drawing and Network Visualization (GD '23). Lecture Notes in Computer Science, vol. 14465, pp. 323–338. Springer (2023). https://doi.org/10.1007/978-3-031-49272-3_22

20. Kleist, L., Kramer, P., Rieck, C.: On the connectivity of the flip graph of plane spanning paths. In: Král, D., Milanič, M. (eds.) Proceedings of the 50th International Workshop on Graph-Theoretic Concepts in Computer Science (WG'24), pp. 327–342. Springer-Verlag, Berlin, Heidelberg (2024). https://doi.org/10.1007/978-3-031-75409-8_23

21. Pilz, A., Welzl, E.: Order on order types. Discrete Comput. Geom. **59**(4), 886–922 (2017). https://doi.org/10.1007/s00454-017-9912-9

22. Sanders, D.P.: On paths in planar graphs. Journal of Graph Theory **24**(4), 341–345 (1997). https://doi.org/10.1002/(sici)1097-0118(199704)24:4<341::aid-jgt6>3.0.co;2-o

23. Schmid, A., Schmidt, J.M.: Computing Tutte paths. In: Chatzigiannakis, I., Kaklamanis, C., Marx, D., Sannella, D. (eds.) Proceedings of the 45th International Colloquium on Automata, Languages, and Programming (ICALP '18). LIPIcs, vol. 107, pp. 98:1–98:14. Schloss Dagstuhl - Leibniz-Zentrum für Informatik (2018). https://doi.org/10.4230/LIPICS.ICALP.2018.98

24. Thomassen, C.: A theorem on paths in planar graphs. J. Graph Theory **7**(2), 169–176 (1983). https://doi.org/10.1002/JGT.3190070205

25. Tutte, W.T.: Bridges and Hamiltonian circuits in planar graphs. Aequationes Math. **15**(1), 1–33 (1977). https://doi.org/10.1007/bf01837870

26. Černý, J., Dvořák, Z., Jelínek, V., Kára, J.: Noncrossing Hamiltonian Paths in Geometric Graphs. In: Liotta, G. (ed.) GD 2003. LNCS, vol. 2912, pp. 86–97. Springer, Heidelberg (2004). https://doi.org/10.1007/978-3-540-24595-7_8

27. Černý, J., Dvořák, Z., Jelínek, V., Kára, J.: Noncrossing Hamiltonian paths in geometric graphs. Discret. Appl. Math. **155**(9), 1096–1105 (2007). https://doi.org/10.1016/j.dam.2005.12.010

Optimal Approximations
for the Requirement Cut Problem
on Sparse Graph Classes

Nadym Mallek[1(✉)] and Kirill Simonov[2]

[1] Hasso Plattner Institute, University of Potsdam, Potsdam, Germany
`nadym.mallek@hpi.de`
[2] Department of Informatics, University of Bergen, Bergen, Norway
`k.simonov@uib.no`

Abstract. We study the Requirement Cut problem, a generalization of numerous classical graph partitioning problems including Multicut, Multiway Cut, k-Cut, and Steiner Multicut among others. Given a graph with edge costs, terminal groups $S_1, \ldots, S_g$ and integer requirements $r_1, \ldots, r_g$; the goal is to compute a minimum-cost edge cut that separates each group S_i into at least r_i connected components. Despite many efforts, the best known approximation for Requirement Cut yields a double-logarithmic $\mathcal{O}(\log(g) \cdot \log(n))$ approximation ratio as it relies on embedding general graphs into trees and solving the tree instance.

In this paper, we explore two largely unstudied structural parameters in order to obtain single-logarithmic approximation ratios: (1) the number of minimal Steiner trees in the instance, which in particular is upper-bounded by the number of spanning trees of the graphs multiplied by g, and (2) the *depth* of series-parallel graphs. Specifically, we show that if the number of minimal Steiner trees is polynomial in n, then a simple LP-rounding algorithm yields an $\mathcal{O}(\log n)$-approximation, and if the graph is series-parallel with a constant depth then a refined analysis of a known probabilistic embedding yields a $\mathcal{O}(depth \cdot \log g)$-approximation on series-parallel graphs of bounded depth. Both results extend the known class of graphs that have a single-logarithmic approximation ratio.

Keywords: Requirement Cut · LP rounding · Randomized rounding · Spanning trees · Series-parallel graphs · Approximation algorithms · Graph partitioning

1 Introduction

Graph partitioning problems are a central topic in combinatorial optimization, with applications such as from network design [15], VLSI Layout [11], image segmentation [4,24], parallel computing [17] and even bioinformatics [31]. In this paper, we study the Requirement Cut problem, a generalization of several well-known cut problems. Given an undirected graph G with vertex set $V(G)$,

© The Author(s), under exclusive license to Springer Nature Switzerland AG 2026
J. Kozik and A. Wolff (Eds.): SOFSEM 2026, LNCS 16448, pp. 547–562, 2026.
https://doi.org/10.1007/978-3-032-17801-5_40

edge set $E(G)$, non-negative (?) edge costs, and a collection $S = (S_1, \ldots, S_g)$ of g terminal groups with $S_i \subseteq V(G)$ and with a requirement $r_i > 1$ for each group, the goal is to compute a minimum-cost edge cut $C \subseteq E(G)$ that separates each group S_i into at least r_i components in $G \setminus C$.

The Requirement Cut problem is a natural extension of multiple classical problems that have been studied for decades. In the following, we present those problems, reference the paper that introduced it and summarize the current best general and parameterized approximation ratios:

- Multicut [29]: Given a set of terminal pairs $(s_1, t_1), \ldots, (s_k, t_k)$, the goal is to find a minimum-cost edge cut that separates each pair. This corresponds to the Requirement Cut problem when each terminal group consists of exactly two vertices and has a requirement of $r_i = 2$. Garg et al. [22] gives the best known approximation for general graphs with a $\mathcal{O}(\log(k))$-approximation ratio via LP rounding (region growing). Friedrich et al. give a constant approximation for series-parallel graphs [20] and a more general [21] $\mathcal{O}(\log(tw))$ for graphs of treewidth tw.
- Multiway Cut [13]: Given a set of terminals $S = \{s_1, \ldots, s_k\}$, the objective is to find a minimum-cost edge cut that separates all terminals into distinct components. This is a special case of Requirement Cut where there is only one group and $r = k$. Chopra et al. [10] studied the linear program of this problem in depth and got a $2 - \frac{2}{k}$ approximation ratio. After several improvements and new techniques used [6,12,32], this chain culmianted with a 1.2965 approximation by Sharma and Vondràk [39]. This problem also has seen a relatively new variant in the past few years called the Norm Multiway Cut [7,8] which is a variant where the goal is to minimize the norm of the edges in the boundaries of k-parts of the graph.
- Steiner-Multicut [34]: Given sets of terminals $\mathcal{S} = (S_1, \ldots, S_k)$ each of size $(t_1, \ldots, t_k)$, the goal is to find a minimum-cost edge cut that separates each set in at least 2 components. This corresponds to the Requirement Cut problem when each requirement is 2. This problem has particularly many applications has even been studied in a parametrized complexity setting [5]. Its best current approximation comes directly from the Requirement Cut and comes with a $\mathcal{O}(\log(k)\log(n))$ ratio. Moreover, it is a direct generalization of a less studied variant when $k = 1$, the Steiner-Cut problem [30].
- Multi-Multiway Cut [2]: A generalization of Multiway Cut where multiple terminal groups must be separated into connected components. This corresponds to the Requirement Cut problem when each terminal group has a requirement equal to its size. The best known approximation comes from the paper that introduced the problem [2] and is $\log(2k)$. Slightly later two more papers have studied the problem under different angles, [40] by adding label edges and [14] studying the problem on bounded branch-width graphs.
- k-Cut [23]: The problem of partitioning a graph into at least k connected components using a minimum-cost edge cut. This can be viewed as a Requirement Cut instance where there is only one terminal set covering all the vertices and its requirement $r = k$ (k parameter given by the problem). First, the 3-cut

was introduced and solved in polynomial time by Hochbaum and Schmoys [28]. It was then extended to k-cut problem by Goldsmith and Hochbaum [23] who gave a polynomial time algorithm for any fixed k. Later, by better understanding its parametrized complexity, Saran and Vazirani [38] found the current best approximation ratio $2 - \frac{2}{k}$. More recently, Gupta et al. [25], gave tight bounds for the run-time of computing an optimal solution.

- Steiner k-Cut [9]: This is a generalization of the k-Cut problem, where $S_1 \subseteq V, |S_1| \geq k$ and the objective is to find a minimum cost set of edges whose removal results in at least $r_1 = k$ disconnected components, each containing a terminal. When $k = n$ this is exactly the k-Cut problem. The problem has been established and approximated best by Chekuri et al. [9] to a $2 - \frac{2}{k}$ ratio. It is worth noting that their work has established the Linear Program formulation used to solve the Requirement Cut Problem introduced later (Fig. 1).

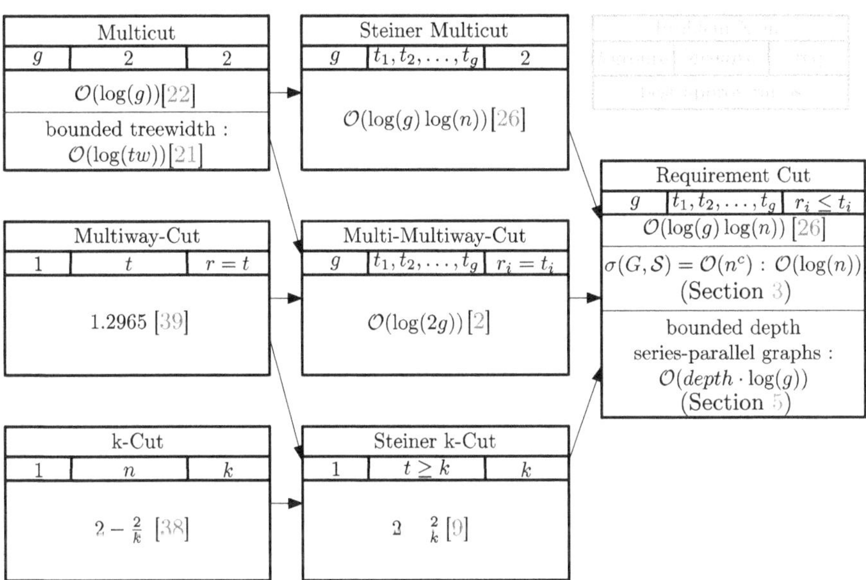

Fig. 1. Hierarchy of graph cut problems captured by the Requirement Cut framework putting our results in perspective. Boxes contain parameters, and best known approximation results for general graphs (and for more particular cases when specified). Arrows denote special cases of the problems (i.e., the source is a particular case of the target).

These problems have been studied mostly through the lens of approximation algorithms. Some of them have received particular attention on specific classes of graphs with parameters such as treewidth, branchwidth or pathwidth.

The hardness of the Requirement Cut can be shown through any of these problems since they are all hard and are particular cases of it. A particular reduction that helps us with bounding the approximability on trees is a reduction from Set Cover taken from [36]. Consider a reduction from Set Cover to the Requirement Cut problem on trees. We construct a star graph where each leaf node corresponds to a set in the Set Cover instance, and the center node of the star connects to each of these leaves via an edge of cost 1. For each element j in the universe, we define a group consisting of the root and all leaf nodes whose corresponding sets contain element j. We assign a requirement of 2 to each group. In this construction, selecting an edge in the tree (i.e., choosing a set in the Set Cover instance) ensures that the corresponding group is cut (i.e., its requirement is satisfied) if and only if that set covers one of the group's elements. Hence, there is a one-to-one correspondence between feasible solutions of the Requirement Cut instance and valid set covers, which cost the same. Feige et al. [19] showed that Set Cover cannot be approximated in polynomial time to a $\Omega(\log(n))$ ratio where n is the size of the universe to cover. It follows that Requirement Cut on trees is hard to $\Omega(\log(g))$- approximate where g is the number of groups.

Now that we have seen that the best approximation possible on trees for our problem is $\log(g)$, let us put the results of our paper in context. Note that, the reduction from Set Cover to Requirement Cut on trees works even when all the requirements are 2 (i.e. the Requirement Cut instance is a Steiner Multicut instance). In 1997, Klein, Plotkin, Rao and Tardos [34] presented an algorithm that yields $\mathcal{O}(\log^3(gt))$-approximation with $t = \max(t_i)$ (the sizes of the sets) for the Steiner Multicut problem. This "triple log" approximation ratio remained the best for this problem until 2010. In 2010, the Requirement Cut problem was introduced by Nagarajan and Ravi [36] who proved a $\mathcal{O}(\log(n) \cdot \log(gR))$-approximation with $R = \max(r_i)$. This approximation became, at the same time, the best approximation ratio for the Steiner Multicut problem. Their bound was later improved by the same set of authors along with Gupta [26] that got the approximation ratio down to $\mathcal{O}(\log(n) \cdot \log(g))$, effectively removing all dependency on the requirements. Both involve very similar algorithmic steps:

1. Finding a good approximation algorithm for trees, $\log(gR)$ for [36] and $\log(g)$ for [26].
2. Using Fakcharoenphol, Rao and Talwar's algorithm (FRT) [18], they embed the graph into a tree and pay a $\log(n)$ factor for it.
3. They use the the first approximation on the resulting tree and get their $\mathcal{O}(\log(gR) \cdot \log(n))$ or $\mathcal{O}(\log(g) \cdot \log(n))$-approximation.

It is also worth noting that Gupta et al. [26] technically has a slightly better $\mathcal{O}(\log(k) \cdot \log(g))$-approximation with k being the total number of terminals. They obtain this improvement through something that they call a slight generalization of FRT [18] in which they embed into a tree only the terminal vertices and hence only pay $\log(k)$ and not $\log(n)$. But this remains secondary in their paper since their main contribution was closing the gap between the $\log(gR)$ approximation on trees and the known lower bound of $\log(g)$ making it tight.

For the past few years, one goal of the community has been to beat those "double log" approximations with "single log" ones. Since we know that embedding into trees has a $\log(n)$ cost, (surprisingly) even for treewidth 2 graphs [27], and that a Requirement Cut problem instance on trees cannot be approximated better than with a $\log(g)$ factor [19], we know that continuing this path of using tree embeddings will not result into a "single log" approximation. In this paper, we try something different over two different routes that we present in the next section.

1.1 Our Contributions

We address the question: *For which graph classes does the* Requirement Cut *problem admit optimal (single-logarithmic) approximations?* We identify two structural parameters—(1) the **number of spanning or minimal Steiner trees** and (2) the **depth** of series-parallel graphs—and show that both yield improved guarantees for Requirement Cut.

First, we show that graphs with a polynomial number of spanning trees (or, more generally, **polynomially many minimal Steiner trees**) admit an $\mathcal{O}(\log n)$-approximation via a randomized LP rounding in the style of Nagarajan and Ravi [36].

Second, we prove that series-parallel graphs of constant depth admit an $\mathcal{O}(\log g)$-approximation by refining the embedding analysis of Emek and Peleg [16], extending the well-known tree-based bounds to a significantly larger class.

Together, these results broaden the set of instances that allow "single-log" approximation ratios and highlight two parameters—series-parallel *depth* and the order of the spanning-tree count—that to the best of our knowledge have not been exploited previously.

- **Instances with Polynomially Many Steiner Trees.** For a given instance, let $\sigma(G, \mathcal{S})$ denote the number of distinct minimal Steiner trees. Although computing $\sigma(G, \mathcal{S})$ is #P-hard [41], we upper-bound it by $g \cdot \tau(G)$, where $\tau(G)$ is the number of spanning trees (computable in polynomial time via Kirchhoff's Matrix Tree Theorem [33]). We show that if $\tau(G) = \mathcal{O}(n^{c_1})$, then necessarily $\sigma(G, \mathcal{S})$ is also polynomial, enabling an $\mathcal{O}(\log n)$-approximation via our rounding algorithm (Sect. 4).
- **Bounded-Depth Series-Parallel Graphs.** Series-parallel graphs form a rich minor-closed class (treewidth 2, K_4-minor-free). We introduce their **depth**—a natural parameter reflecting the height of their series/parallel decomposition—and show that graphs of depth m admit an $\mathcal{O}(m \log g)$-approximation. In particular, for constant depth we obtain an $\mathcal{O}(\log g)$-approximation, extending the tree case to a significantly broader family. This perspective also connects to the recursive *melonic* graphs studied in physics.

Section 3 presents several classes of graphs and instances where our guarantees apply. Section 4 states and analyzes our rounding algorithm, including a structural LP lemma of independent interest (Lemma 1). Section 5 gives the depth-sensitive refinement of Emek and Peleg's embedding. Due to space constraints, some proofs are omitted or sketched and the following sections are shortened; full details appear in the extended version [35].

2   Preliminaries

Definition 1 (Cut). *Let G be an undirected graph. A cut in G is a set of edges $C \subseteq E(G)$ such that removing the edges in C disconnects some specified structure in the graph. We say that a cut separates a set of vertices $S \subseteq V(G)$ if the vertices in S lie in at least two distinct connected components of $G \setminus C$.*

Definition 2 (Requirement Cut Instance). *A Requirement Cut instance is defined by an undirected graph G, a cost function $c : E(G) \to \mathbb{R}_{\geq 0}$, a collection of terminal sets $\mathcal{S} = \{S_1, \ldots, S_g\} \subseteq V(G)$ with $S_i \subseteq V$, and integer requirements $r_1, \ldots, r_g \geq 2$. The goal is to find a minimum-cost edge set $C \subseteq E(G)$ such that in the graph $G \setminus C$, each terminal set S_i is partitioned into at least r_i connected components.*

Definition 3 (Steiner Tree). *Let G be an undirected graph and a set of terminal sets $\mathcal{S} = \{S_1, \ldots, S_g\} \subseteq V(G)$. A Steiner tree for a terminal set S is a minimally connected subgraph $T \subseteq G$ that spans all vertices in S and is called an S-Steiner tree. T may include additional non-terminal vertices, called Steiner vertices. An edge connecting two terminals (resp. Steiner) vertices may be called* terminal edge *(resp.* Steiner edge*).*

Additionally, we define the length of a Steiner tree T as $\sum_{e \in T} d_e$ for a given length function d and d_e the length of edge e.

Definition 4 (Minimal Steiner Tree). *A Steiner tree is said to be minimal if all its leaves are terminal vertices. The number of distinct minimal Steiner trees in a graph G with terminal sets $\mathcal{S} = \{S_1, \ldots, S_g\} \subseteq V(G)$, will be noted as $\sigma(G, \mathcal{S})$.*

With this definition of Minimal Steiner tree, let us prove the following claim:

Claim 1. In a graph G with $\mathcal{S} = \{S_1, \ldots, S_g\}$ terminal sets, we have $\sigma(G, \mathcal{S}) \leq g \cdot \tau(G)$. With $\tau(G)$ the number of distinct spanning trees in G.

Proof. Let T be a minimal Steiner tree of set S_k for some $1 \leq k \leq g$. By adding edges and vertices to T until it covers $V(G)$, one can create at least one spanning tree per minimal Steiner tree. Also, two distinct minimal Steiner trees for the same terminal set cannot extend to the same spanning tree in G, as they differ in at least one edge. Therefore, there exist at most $\tau(G)$ minimal Steiner trees for each terminal set $S_i \in \mathcal{S}$. $\qquad\square$

Definition 5 (Series-Parallel Graphs). *A series–parallel graph (SP graph) is any graph that can be obtained from a single edge K_2 by a finite sequence of the following operations:*

- ***Parallel composition:** Replace an edge (u, v) into a set of two parallel edges that connect u and v.*
- ***Series composition:** Replace an (u, v) into a degree 2 vertex connected to u and v.*

Definition 6 (Composition Trace). *Let G be a series-parallel graph constructed via a sequence of series and parallel compositions. The **composition trace** of G is a directed tree T whose:*

- *leaves correspond to the individual edges of G,*
- *internal nodes represent subgraphs formed by (series or parallel) composition of their children*
- *root corresponds to G itself.*

We group consecutive series or parallel compositions into single nodes, so that every path from a leaf to the root alternates between series and parallel composition steps. A node in T may have arbitrarily many children.

Definition 7 (Depth of a Series-Parallel Graph). *Let G be a series-parallel graph and let T be its composition trace. The depth of G, is defined as the depth of the tree T, that is, if l is the length of the longest path from a leaf (corresponding to a single edge) to the root (corresponding to G itself): depth $= l - 1$.*

For example, a single edge would be a series-parallel graph of depth 0, a path or a many parallel edges would have depth 1 and a composition of parallel paths would have depth 2. This means that this definition makes each class of bounded depth series-parallel graphs an infinite class of graphs.

3 Graphs with Polynomial Number of Steiner Trees

We now identify classes of instances for which our rounding algorithm from Sect. 4 yields a single-logarithmic approximation. The key quantity is the number $\sigma(G, \mathcal{S})$ of minimal Steiner trees (cf. Definition 4). By Sect. 1, it is always upper-bounded by $g \cdot \tau(G)$, where $\tau(G)$ is the number of spanning trees of G.

3.1 Graphs with Polynomially Many Spanning Trees

We first consider graphs with $\tau(G) = \mathcal{O}(n^c)$ for some constant c. By Kirchhoff's Matrix Tree Theorem [33], $\tau(G)$ can be computed in polynomial time, and Sect. 1 then implies $\sigma(G, \mathcal{S}) = \mathcal{O}(n^c g)$, so our algorithm yields an $\mathcal{O}(\log n)$-approximation.

Two simple examples of such graph classes are:

- **Graphs with small feedback edge set.** If removing at most k edges makes G acyclic, then any spanning tree can be obtained by choosing, for each of the at most k cycles, which edge to delete. This gives $\tau(G) = \mathcal{O}(n^k)$, hence polynomial.
- **Graphs with few short cycles.** Suppose G has at most $\mathcal{O}(\log n)$ cycles, each of length at most k. Again, each cycle contributes a factor of at most k to the number of spanning trees, so $\tau(G) = \mathcal{O}(k^{\log n}) = \mathcal{O}(n^{\log k})$, which is polynomial for fixed k (Fig. 2).

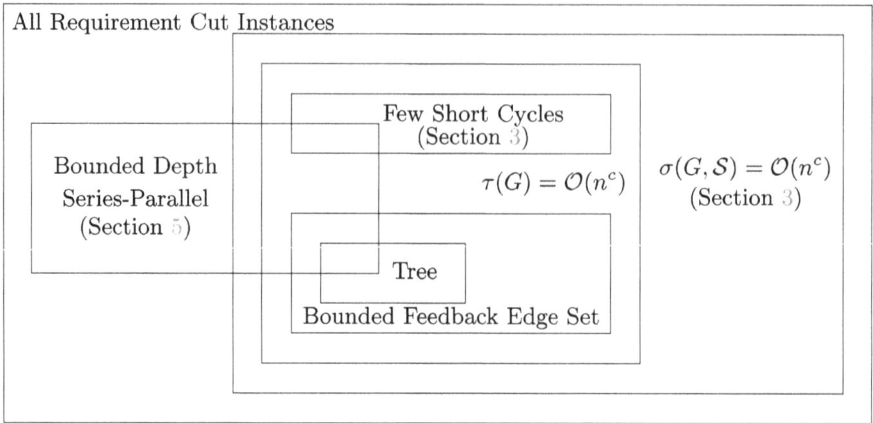

Fig. 2. Overview of graph classes and instances for which the Requirement Cut problem admits a single-logarithmic approximation in this paper.

3.2   Instances with Polynomially Many Minimal Steiner Trees

The parameter $\sigma(G, \mathcal{S})$ can be much smaller than $g \cdot \tau(G)$, because it depends on the placement of the terminal sets $\mathcal{S}$ rather than on the whole graph. In particular, our analysis only needs $\sigma(G, \mathcal{S}) = \text{poly}(n)$, regardless of how large $\tau(G)$ is.

One natural situation where this happens is when all terminal sets lie in "almost pendant" subgraphs that each have few spanning trees.

In such cases, even if the global number of spanning trees $\tau(G)$ is super-polynomial, the number of relevant minimal Steiner trees remains polynomial, and our rounding algorithm still guarantees a single-logarithmic approximation ratio.

4   Requirement Cut Algorithm

4.1   Linear Programs

We use essentially the same LPs as Nagarajan and Ravi [36] and Gupta et al. [26], but Lemma 1 provides an additional structural property crucial for our rounding analysis. Both LPs are defined on slight modifications of G. For the first LP, IP-Steiner, we complete G by adding 0-cost edges, obtaining a clique G_V. For the second, Relaxed LP, we only add 0-cost edges inside each terminal set S_i, making S_i a clique; we denote this by $G_{S_i} = (V, E_{S_i})$.

$$\text{(IP-Steiner)} \quad \min \quad \sum_{e \in E} c_e d_e$$

$$\text{s.t.} \quad \sum_{e \in \mathcal{T}_i} d_e \geq r_i - 1 \quad \forall \mathcal{T}_i : S_i\text{-Steiner tree in } G_V, \quad \forall i \in [g]$$

$$d_e \in \{0, 1\} \quad \forall e \in E$$

$$\text{(Relaxed LP)} \quad \min \quad \sum_{e \in E} c_e d_e$$

$$\text{s.t.} \quad \sum_{e \in \mathcal{T}_i} d_e \geq r_i - 1 \quad \forall \mathcal{T}_i : \text{spanning tree in } G_{S_i}, \ \forall i \in [g]$$

$$d_{\{u,w\}} \leq d_{\{u,v\}} + d_{\{v,w\}} \quad \forall u, v, w \in V$$

$$0 \leq d_{\{u,v\}} \leq 1 \quad \forall u, v \in V$$

IP-Steiner is an exact formulation of **Requirement Cut**. The nontrivial point is why constraints only on Steiner trees already enforce that each S_i is split into r_i components. This is precisely where completeness of G_V (or of the cliques G_{S_i} in Relaxed LP) matters: without the added 0-cost edges, one could cut $r_i - 1$ edges along a long Steiner path and still leave all terminals in only two components. Lemma 1 formalizes that, in the completed graphs, enforcing $d(T) \geq r_i - 1$ for every S_i-Steiner tree T is equivalent to satisfying the requirement for S_i.

Our rounding algorithm is a randomized rounding of the solution of Relaxed LP, which can be solved via the ellipsoid method with a minimum spanning tree oracle. It ensures that every Steiner tree is cut at least $r_i - 1$ times, but unlike in the tree setting of [26, 36], we cannot directly relate the number of cuts to the number of components. Instead, we use the following lemma. As in previous work, let d^* be an optimal solution to Relaxed LP and define $d_{\{u,v\}} = \min(2d^*_{\{u,v\}}, 1)$, which is still a metric.

Claim 2 [36]. For any group S_i, the minimum Steiner tree on S_i under the metric $d = \min(2 \cdot d^*, 1)$ has length at least $r_i - 1$.

Lemma 1. *Let G an undirected graph, $\mathcal{S} = \{S_1, \ldots, S_g\}$ with $S_i \subseteq V(G)$, and integer requirements $r_1, \ldots, r_g \geq 2$. For each i, let G_{S_i} be obtained from G by adding 0-cost edges between nonadjacent pairs in S_i, making S_i a clique. Let $d : E(G) \rightarrow \{0,1\}$ be a metric such that for every S_i-Steiner tree T, $d(T) \geq r_i - 1$. Then the following are equivalent:*

- *every spanning tree of G_{S_i} is cut at least $r_i - 1$ times, and*
- *the vertices of S_i lie in at least r_i connected components.*

Proof (Short proof of Lemma 1; see the full version on Arxiv [35]). Starting from any S-Steiner tree T_0 whose cuts do *not* split S into r terminal components, we iteratively "shortcut" non-terminal components using 0-cost edges inside S. Each step produces a new S-Steiner tree, also with total cut-length at least $r - 1$, and strictly reduces (or locally prepares to reduce) the number of non-terminal components. After finitely many steps we obtain a family of S-Steiner trees whose cut edges partition S into at least r components, as claimed. $\qquad \square$

Lemma 1 shows that it suffices for our rounding algorithm to cut every Steiner tree enough times (with high probability) to obtain a feasible solution to **Requirement Cut**. Although the LPs are solved on graphs augmented with 0-cost edges, the final cut is taken in the original graph G; we may always add all 0-cost edges to the cut without affecting its cost. Thus cutting all Steiner trees in the modified graphs is equivalent to satisfying all requirements in G.

4.2 Algorithm

In this section we will see how our algorithm can give us a $\mathcal{O}(\log(\sigma(G,\mathcal{S})))$-approximation for the Requirement Cut problem where $\sigma(G,\mathcal{S})$ is the number of distinct minimal Steiner trees in the graph with respect to the given terminal sets. Computing $\sigma(G,\mathcal{S})$ is #P-Complete since it is a more general case of counting paths which is #P-Complete [41]. However, we can upper-bound it by $\tau(G) \cdot g$ (Sect. 1) which is easy to compute through Kirchhoff's Theorem [33]. When clear from the context, we shall denote the number $\sigma(G,\mathcal{S})$ simply by σ.

Algorithm 1. Randomized Rounding for the Requirement Cut problem

---

1: **Input:** Instance of the Requirement Cut problem on graph G
2: Solve the LP relaxation to obtain fractional values $\{d(e) \in [0,1] \mid e \in E(G)\}$
3: Initialize cut set $C \leftarrow \emptyset$
4: Choose $\alpha = \frac{1}{c \cdot \log(\sigma)}$ with c large enough (see Choosing alpha)
5: **for** each edge $e \in E$ **do**
6:     Draw $x_e \sim \text{Uniform}(0, \alpha)$
7:     **if** $x_e \leq d(e)$ **then**
8:         Add e to C
9:     **end if**
10: **end for**
11: **Output:** Cut set $C \subseteq E(G)$

The rest of this section of the paper will be dedicated to proving the following theorem:

Theorem 1. *Let I be a Requirement Cut Instance (Definition 2) with σ being the number of minimal Steiner trees in I. Then, there is an LP rounding algorithm for the Requirement Cut problem that returns a solution of cost $\mathcal{O}(\log(\sigma)) \cdot LP_{\text{opt}}$ with probability $1 - \frac{1}{poly(\sigma)}$, where LP_{opt} is the value returned by Relaxed LP.*

The algorithm provides an $\mathcal{O}(\log(n))$-factor approximation for all graphs with $\tau(G) = \mathcal{O}(n^c)$ (see Sect. 3 for examples). It also matches the $\mathcal{O}(\log(g))$-approximation on trees given by [26], while having a simpler process, as it only involves one of their two phases.

Lemma 2. *The expected cost of C, the set of edges returned by Algorithm 1 is $\mathbb{E}(c(C)) \leq \frac{1}{\alpha} \sum_{e \in E} c_e d_e = \mathcal{O}(\log(\sigma)) \cdot LP_{\text{opt}}$.*

Proof. The probability of a given edge e being in the cut C is $\mathbb{P}(e \in C) = \min\left\{1, \frac{d(e)}{\alpha}\right\}$. Thus, the expected cost of C is $\mathbb{E}(c(C)) \leq \frac{1}{\alpha} \sum_{e \in E} c_e d_e = \mathcal{O}(\log(\sigma)) \cdot LP_{\text{opt}}$ by linearity of expectations. $\square$

Lemma 3 ($\star$). *After the execution of Algorithm 1 the probability that a group S_i lies in less than r_i is $\frac{1}{\sigma^{\frac{c}{2}-1}}$ with $c \geq 4$ a constant chosen arbitrarily.*

Proof (Short Proof of Lemma 3 see version on Arxiv [35] for complete proof).
Let $\alpha = \frac{1}{c \log \sigma}$ with c a sufficiently large constant and sample $x_e \sim \mathrm{Unif}(0, \alpha)$
independently for $e \in E$; include e in the cut iff $x_e \leq d(e)$. For every S_i-Steiner
tree t, LP feasibility gives $\sum_{e \in t} d(e) \geq r_i - 1$. Split the cut into $\mathcal{C}_1 = \{e :$
$d(e) \geq \alpha\}$ and $\mathcal{C}_2 = \{e : d(e) < \alpha\}$, remove $\mathcal{C}_1$, and let $r_i' \leq r_i$ be the residual
requirement still unsucceded after the removal of $\mathcal{C}_1$.

For $e \in E \setminus \mathcal{C}_1$ we have $\Pr[e \in \mathcal{C}_2] = \mathbb{E}[Z_e] = d(e)/\alpha$. Writing $Y_t := \sum_{e \in t} Z_e$,

$$\mathbb{E}[Y_t] = \sum_{e \in t} \mathbb{E}[Z_e] \geq \frac{r_i' - 1}{\alpha}.$$

By Chernoff (for $0 < \delta < 1$),

$$\Pr(Y_t < (1 - \delta)\,\mathbb{E}[Y_t]) \leq \exp\left(-\frac{\delta^2}{2}\,\mathbb{E}[Y_t]\right).$$

Taking $\delta = 1 - \frac{r_i' - 2}{\mathbb{E}[Y_t]}$ yields $\Pr(Y_t \leq r_i' - 2) \leq \exp\left(-\frac{(\mathbb{E}[Y_t] - r_i' - 2)^2}{2\,\mathbb{E}[Y_t]}\right)$. Using $\mathbb{E}[Y_t] \geq$
$\frac{r_i' - 1}{\alpha}$, $r_i' \geq 2$, and $\alpha \leq \frac{1}{4}$ (achieved by increasing c), we get the convenient bound

$$\Pr(Y_t \leq r_i' - 2) \leq \exp\left(-\frac{(1 - 4\alpha)^2}{2\alpha}\right) \leq \exp\left(4 - \frac{c \log \sigma}{2}\right) = \frac{e^4}{\sigma^{c/2}}.$$

A union bound over at most σ relevant trees across all groups gives a failure
probability at most $\frac{e^4}{\sigma^{\frac{c}{2} - 1}}$. Since $\sigma = \mathrm{poly}(n)$, this is $n^{-\Omega(1)}$. Hence the cut
satisfies every group with high probability, proving the claim. $\qquad\square$

This also completes the proof of Theorem 1.

Choosing Alpha. Let us now discuss the details of fixing $\alpha = \frac{1}{c \cdot \log(\sigma)}$. For some
part of our analysis (see long version of proof of Lemma 3) we need $c \geq 4$ assum-
ing $\log(n) \geq 1$. Technically speaking, one could have a $\log(\sigma)$-approximation
even for instances where σ is not $poly(n)$. But this would not lead to a "single-
log" approximation, as the factor is only $\mathcal{O}(\log(n))$ when $\sigma = poly(n)$ or smaller.

5   Depth of Series Parallel Graphs

In this section we refine the approximation bounds for the Requirement Cut
problem on series-parallel graphs by reanalyzing the embedding algorithm of
Emek and Peleg [16]. Combined with the $\mathcal{O}(\log g)$-approximation for trees [26],
this yields an $\mathcal{O}(m \log g)$-approximation on series-parallel graphs of depth m (as
in Definition 7).

Series-parallel graphs in general cannot be embedded into trees with expected
distortion $o(\log n)$ [27]. Hence, for unrestricted series-parallel graphs, the best
known ratio for Requirement Cut matches the general case, namely $\mathcal{O}(\log g \cdot \log n)$
via tree embeddings [26,36]. The lower-bound construction of Gupta et al. [27] in

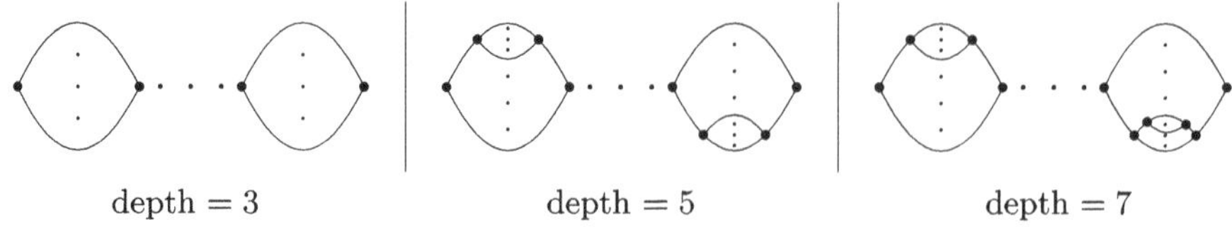

depth = 3   depth = 5   depth = 7

Fig. 3. Series-parallel graphs of depth 3, 5, and 7, obtained by nested series/parallel compositions. Each black segment denotes an arbitrary path, so the number of distinct x–y paths can be exponential in n.

fact has depth $\Theta(\log n)$, so our result gives a strictly better guarantee whenever the depth m is $o(\log n)$ (and in particular when m is constant) (Fig. 3).

The depth parameter coincides with the height of a standard decomposition (composition trace) of the series-parallel graph and thus naturally relates to tree-depth [37]. Since series-parallel graphs have treewidth 2, bounded-depth series-parallel graphs form a subclass of graphs of bounded tree-depth. They also strictly generalize several recursively defined series-parallel families studied in physics (e.g. melonic graphs [1,3]).

Emek–Peleg Embedding Revisited. Emek and Peleg [16] give a randomized algorithm CONSTRUCT_TREE that, given a two-terminal series-parallel graph G with terminals x, y, outputs a spanning tree T so that the expected stretch of each edge (u, v) is at most $\mathcal{O}(\log n)$. The algorithm proceeds bottom-up along the composition trace, treating series and parallel compositions separately: series compositions glue recursive trees in series, while parallel compositions keep one shortest x–y branch intact and delete one random edge on the x–y path of each remaining branch.

Using the notation of [16], for any fixed pair (u, v) one can associate a sequence of at most m parallel-composition steps and nonnegative integers $a_1, \ldots, a_m, b_1, \ldots, b_m, d$ satisfying $b_1 = 1$, $b_{i+1} \le a_i + b_i$, $d \le a_m + b_m$, and $\sum_{i=1}^{m} a_i \le n$. Their Lemma 4.1 expresses the expected distance in the random tree as

$$\mathbb{E}[dist_T(u, v)] = \sum_{i=1}^{m} \frac{a_i}{a_i + b_i} \left(\prod_{j=1}^{i-1} \frac{b_j}{a_j + b_j} \right) (b_i + c_i) + \left(\prod_{i=1}^{m} \frac{b_i}{a_i + b_i} \right) (a_m + c_m + d),$$

where $c_i = \sum_{j=1}^{i-1} a_j$. They bound this by $\mathcal{O}(\log n)$ using $\sum_i a_i \le n$.

We instead bound it in terms of the depth m.

Theorem 2. *Let* T *be the random spanning tree returned by* CONSTRUCT_TREE *on a series-parallel graph* G *of depth* m. *Then*

$$\mathbb{E}[dist_T(u, v)] \le \mathcal{O}(m) \quad \text{for all } (u, v) \in E(G).$$

Proof. Using $b_{i+1} \le a_i + b_i$ and $b_1 = 1$, we obtain $b_i \le 1 + \sum_{j=1}^{i-1} a_j$ for all i. Substituting $c_i = \sum_{j=1}^{i-1} a_j$ into the expression above and applying the bounds

$b_i \leq 1 + \sum_{j<i} a_j$ and $d \leq a_m + b_m \leq 1 + 2\sum_{j=1}^m a_j$, one checks that each summand in the first term is at most 2, and the final term is also at most 2. Hence

$$\mathbb{E}[dist_{\mathcal{T}}(u, v)] \leq 2\,m + 2 = \mathcal{O}(m).$$

$\square$

Since any $\mathcal{O}(\log g)$-approximation for Requirement Cut on trees extends to distributions over trees with distortion $\mathcal{O}(m)$, we obtain an $\mathcal{O}(m \log g)$-approximation for Requirement Cut on series-parallel graphs of depth m. In particular, bounded-depth series-parallel graphs admit an $\mathcal{O}(\log g)$-approximation.

6   Conclusion

We introduced two structural parameters—the number of (minimal) Steiner trees and the depth of series-parallel graphs—that allow single-logarithmic approximations for the Requirement Cut problem on natural classes of graphs. Instances with polynomially many minimal Steiner trees admit $\mathcal{O}(\log n)$-approximations via LP rounding, while bounded-depth series-parallel graphs admit $\mathcal{O}(\log g)$-approximations through a depth-aware embedding analysis.

These results raise several directions for further work. One is to identify additional "tree-like" parameters beyond spanning-tree counts and depth that still permit single-log approximations, possibly for broader cut or connectivity problems. Another is to understand whether similar techniques extend to other sparse graph classes (e.g. bounded tree-depth or minor-closed families) or to vertex-cut variants. Finally, closing the gap between these structured settings and general graphs may require fundamentally new tools, either improving embeddings beyond known lower bounds or bypassing them altogether with non-embedding-based rounding paradigms.

References

1. Aluffi, P., Marcolli, M., Qaisar, W.: Motives of melonic graphs. Annales de l'Institut Henri Poincaré D Comb. Phys. Interact. **10**(3), 503–554 (2022). https://doi.org/10.4171/aihpd/156
2. Avidor, A., Langberg, M.: The multi-multiway cut problem. Theoret. Comput. Sci. **377**(1–3), 35–42 (2007)
3. Baratin, A., Carrozza, S., Oriti, D., Ryan, J., Smerlak, M.: Melonic phase transition in group field theory. Lett. Math. Phys. **104**(8), 1003–1017 (2014). https://doi.org/10.1007/s11005-014-0699-9
4. Boykov, Y., Veksler, O., Zabih, R.: Fast approximate energy minimization via graph cuts. IEEE Trans. Pattern Anal. Mach. Intell. **23**(11), 1222–1239 (2001). https://doi.org/10.1109/34.969114
5. Bringmann, K., Hermelin, D., Mnich, M., van Leeuwen, E.J.: Parameterized complexity dichotomy for Steiner Multicut. J. Comput. Syst. Sci. **82**(6), 1020–1043 (2016)

6. Buchbinder, N., Seffi, J., Schwartz, R.: Simplex partitioning via exponential clocks and the multiway-cut problem. SIAM J. Comput. **47**(4), 1463–1482 (2018). https://doi.org/10.1137/15M1045521

7. Carlson, C., Jafarov, J., Makarychev, K., Makarychev, Y., Shan, L.: Approximation Algorithm for Norm Multiway Cut. LIPIcs, vol. 274, ESA 2023 **274**, 32:1–32:14 (2023). https://doi.org/10.4230/LIPICS.ESA.2023.32, artwork Size: 14 pages, 676151 bytes ISBN: 9783959772952 Medium: application/pdf Publisher: Schloss Dagstuhl – Leibniz-Zentrum für Informatik

8. Chandrasekaran, K., Wang, W.: l_p-norm multiway cut. LIPIcs, vol. 204, ESA 2021 **204**, 29:1–29:15 (2021). https://doi.org/10.4230/LIPICS.ESA.2021.29, artwork Size: 15 pages, 879891 bytes ISBN: 9783959772044 Medium: application/pdf Publisher: Schloss Dagstuhl – Leibniz-Zentrum für Informatik

9. Chekuri, C., Guha, S., Naor, J.S.: The Steiner k-cut problem. SIAM J. Discret. Math. **20**(1), 261–271 (2006). https://doi.org/10.1137/S0895480104445095

10. Chopra, S., Rao, M.R.: On the multiway cut polyhedron. Networks **21**(1), 51–89 (1991). https://doi.org/10.1002/net.3230210106

11. Cong, J., Shinnerl, J.R., Du, D.Z., Pardalos, P.M. (eds.): Multilevel Optimization in VLSICAD, Combinatorial Optimization, vol. 14. Springer, Boston (2003). https://doi.org/10.1007/978-1-4757-3748-6

12. Călinescu, G., Karloff, H., Rabani, Y.: An improved approximation algorithm for multiway cut. J. Comput. Syst. Sci. **60**(3), 564–574 (2000)

13. Dahlhaus, E., Johnson, D.S., Papadimitriou, C.H., Seymour, P.D., Yannakakis, M.: The complexity of multiterminal cuts. SIAM J. Comput. **23**(4), 864–894 (1994). https://doi.org/10.1137/S0097539792225297

14. Deng, X., Lin, B., Zhang, C.: Multi-multiway cut problem on graphs of bounded branch width. In: FAW-AAIM (2013). https://api.semanticscholar.org/CorpusID: 943999

15. Donde, V., et al.: Identification of severe multiple contingencies in electric power networks. In: Proceedings of the 37th Annual North American Power Symposium, pp. 59–66. IEEE, Ames, IA, USA (2005). https://doi.org/10.1109/NAPS. 2005.1560502

16. Emek, Y., Peleg, D.: A tight upper bound on the probabilistic embedding of series-parallel graphs. SIAM J. Discret. Math. **23**(4), 1827–1841 (2010). https://doi.org/10.1137/090745441

17. Fagginger Auer, B., Bisseling, R.: Graph coarsening and clustering on the GPU. In: Bader, D., Meyerhenke, H., Sanders, P., Wagner, D. (eds.) Contemporary Mathematics, vol. 588, pp. 223–240. American Mathematical Society, Providence, Rhode Island (2013). https://doi.org/10.1090/conm/588/11706. http://www.ams. org/conm/588

18. Fakcharoenphol, J., Rao, S., Talwar, K.: A tight bound on approximating arbitrary metrics by tree metrics. J. Comput. Syst. Sci. **69**(3), 485–497 (2004)

19. Feige, U.: A threshold of ln n for approximating set cover. J. ACM **45**(4), 634–652 (1998). https://doi.org/10.1145/285055.285059

20. Friedrich, T., Issac, D., Kumar, N., Mallek, N., Zeif, Z.: A primal-dual algorithm for multicommodity flows and multicuts in treewidth-2 graphs. In: Chakrabarti, A., Swamy, C. (eds.) Approximation, Randomization, and Combinatorial Optimization. Algorithms and Techniques (APPROX/RANDOM 2022). Leibniz International Proceedings in Informatics (LIPIcs), vol. 245, pp. 55:1–55:18. Schloss Dagstuhl – Leibniz-Zentrum für Informatik, Dagstuhl, Germany (2022). https://doi.org/10.4230/LIPIcs.APPROX/RANDOM.2022.55. https://drops.dagstuhl.de/entities/document/10.4230/LIPIcs.APPROX/RANDOM.2022.55, iSSN: 1868-8969
21. Friedrich, T., Issac, D., Kumar, N., Mallek, N., Zeif, Z.: Approximate max-flow min-multicut theorem for graphs of bounded treewidth. In: Proceedings of the 55th Annual ACM Symposium on Theory of Computing, pp. 1325–1334. ACM, Orlando, FL, USA (2023). https://doi.org/10.1145/3564246.3585150
22. Garg, N., Vazirani, V.V., Yannakakis, M.: Approximate max-flow min-(multi)cut theorems and their applications. In: Proceedings of the Twenty-Fifth Annual ACM Symposium on Theory of Computing, STOC 1993, pp. 698–707. ACM Press, San Diego, California, USA (1993). https://doi.org/10.1145/167088.167266
23. Goldschmidt, O., Hochbaum, D.S.: A polynomial algorithm for the k-cut problem for fixed k. Math. Oper. Res. **19**(1), 24–37 (1994). https://www.jstor.org/stable/3690374
24. Grady, L., Schwartz, E.: Isoperimetric graph partitioning for image segmentation. IEEE Trans. Pattern Anal. Mach. Intell. **28**(3), 469–475 (2006)
25. Gupta, A., Harris, D.G., Lee, E., Li, J.: Optimal bounds for the k-cut problem. J. ACM **69**(1) (2021). https://doi.org/10.1145/3478018
26. Gupta, A., Nagarajan, V., Ravi, R.: An improved approximation algorithm for requirement cut. Oper. Res. Lett. **38**(4), 322–325 (2010)
27. Gupta, A., Newman, I., Rabinovich, Y., Sinclair, A.: Cuts, trees and L1-embeddings of graphs*. Combinatorica **24**(2), 233–269 (2004). https://doi.org/10.1007/s00493-004-0015-x
28. Hochbaum, D.S., Shmoys, D.B.: An $O(|V|^2)$ algorithm for the planar 3-cut problem. SIAM J. Algebraic Discret. Methods **6**(4), 707–712 (1985). https://doi.org/10.1137/0606068
29. Hu, T.C.: Multi-commodity network flows. Oper. Res. **11**(3), 344–360 (1963). https://doi.org/10.1287/opre.11.3.344
30. Jue, S., Klein, P.N.: A near linear time minimum Steiner cut algorithm for planar graphs (2019). https://doi.org/10.48550/arXiv.1912.11103, [cs]
31. Junker, B.H., Schreiber, F. (eds.): Analysis of Biological Networks. Wiley Series on Bioinformatics. Wiley-Interscience, Hoboken (2008), oCLC: ocn166454234
32. Karger, D.R., Klein, P., Stein, C., Thorup, M., Young, N.E.: Rounding algorithms for a geometric embedding of minimum multiway cut. Math. Oper. Res. **29**(3), 436–461 (2004). https://www.jstor.org/stable/30035660
33. Kirchhoff, G.: Ueber die Auflösung der Gleichungen, auf welche man bei der Untersuchung der linearen Vertheilung galvanischer Ströme geführt wird. Ann. Phys. **148**(12), 497–508 (1847). https://doi.org/10.1002/andp.18471481202
34. Klein, P.N., Plotkin, S.A., Rao, S., Tardos, E.: Approximation algorithms for steiner and directed multicuts. J. Algorithms **22**(2), 241–269 (1997)
35. Mallek, N., Simonov, K.: Optimal approximations for the requirement cut problem on sparse graph classes (2025). https://arxiv.org/abs/2505.21433, [cs.DS]
36. Nagarajan, V., Ravi, R.: Approximation algorithms for requirement cut on graphs. Algorithmica **56**(2), 198–213 (2010). https://doi.org/10.1007/s00453-008-9171-5

37. Nešetřil, J., de Mendez, P.O.: Bounded height trees and tree-depth. In: Sparsity. AC, vol. 28, pp. 115–144. Springer, Heidelberg (2012). https://doi.org/10.1007/978-3-642-27875-4_6
38. Saran, H., Vazirani, V.V.: Finding k cuts within twice the optimal. SIAM J. Comput. **24**(1), 101–108 (1995). https://doi.org/10.1137/S0097539792251730
39. Sharma, A., Vondrák, J.: Multiway cut, pairwise realizable distributions, and descending thresholds. In: Proceedings of the Forty-Sixth Annual ACM Symposium on Theory of Computing, STOC 2014, pp. 724–733. Association for Computing Machinery, New York (2014). https://doi.org/10.1145/2591796.2591866
40. Shuguang, L., Xiao, X.: Multi-multiway cuts with edge labels. In: 2009 4th International Conference on Computer Science & Education, pp. 1860–1863 (2009). https://doi.org/10.1109/ICCSE.2009.5228230
41. Valiant, L.G.: The complexity of enumeration and reliability problems. SIAM J. Comput. **8**(3), 410–421 (1979). https://doi.org/10.1137/0208032

Clique-Free t-Matchings
in Degree-Bounded Graphs

Katarzyna Paluch[1(✉)] and Mateusz Wasylkiewicz[2]

[1] Institute of Computer Science, University of Wrocław, Wrocław, Poland
`abraka@cs.uni.wroc.pl`
[2] Institute of Computer Science, Department of Fundamentals of Computer Science,
Wrocław University of Science and Technology, Wrocław, Poland
`mateusz.wasylkiewicz@pwr.edu.pl`

Abstract. We consider problems of finding a maximum size/weight t-matching without forbidden subgraphs in an undirected graph G with the maximum degree bounded by $t + 1$, where t is an integer greater than 2. A t-matching is a subset of edges such that each vertex is incident to at most t edges of the subset. Depending on the variant, forbidden subgraphs denote certain subsets of t-regular complete partite subgraphs of G. A graph is complete partite if there exists a partition of its vertex set such that every pair of vertices from different sets is connected by an edge and vertices from the same set form an independent set. A clique K_{t+1} and a bipartite clique $K_{t,t}$ are examples of complete partite graphs. These problems are natural generalizations of the triangle-free and square-free 2-matching problems in subcubic graphs. In the weighted setting we assume that the weights of edges of G are *vertex-induced* on every forbidden subgraph. We present simple and fast combinatorial algorithms for these problems. The presented algorithms are the first ones for the weighted versions, and for the unweighted ones, are faster than those known previously. Additionally, the running times of our algorithms are independent of t. Our approach relies on the use of gadgets with so-called *half-edges*. A *half-edge* of edge e is, informally speaking, a half of e containing exactly one of its endpoints.

1 Introduction

We consider several variants of the problem of finding a maximum size or weight t-matching without forbidden subgraphs. Given a positive integer t, a subset M of edges of an undirected simple graph is a t-*matching* if every vertex is incident to at most t edges of M. The set of t-matchings is contained in a wider class of b-matchings, where for every vertex v in the set of vertices V of the graph, we are given a natural number $b(v)$ (note that $b(v)$s may be different for different vertices) and a subset of edges is a b-*matching* if every vertex is incident to at most $b(v)$ of its edges. A b-matching of maximum size/weight can be found in

---

Partly supported by Polish National Science Center grant 2018/29/B/ST6/02633.

© The Author(s), under exclusive license to Springer Nature Switzerland AG 2026
J. Kozik and A. Wolff (Eds.): SOFSEM 2026, LNCS 16448, pp. 563–577, 2026.
https://doi.org/10.1007/978-3-032-17801-5_41

polynomial time by a reduction to a classical matching. We refer to Lovász and Plummer [23] for further background on b-matchings.

In the problems that we study in this paper we are given an undirected graph $G = (V, E)$ in which each vertex has degree at most $t + 1$ and the goal is to find a maximum size/weight t-matching that does not contain certain complete t-regular partite graphs. A graph $H = (V_H, E_H)$ is **complete partite** if there exists a partition $V_1, V_2, \ldots, V_p$ of V_H of size $p \geq 2$ such that $E_H = \{(v, u) : \exists i \neq j \; v \in V_i \wedge u \in V_j\}$. Each such V_i is called a **color class** of H. We denote the i-th color class of H by $V_i(H)$. If all color classes of H have the same size, we denote H by K_q^p where $q = |V_1|$. Notice that K_q^p is t-regular for $t = (p - 1)q$. Observe that a clique K_a consisting of a vertices and a bipartite clique $K_{a,b}$ with color classes consisting of a and b vertices are special cases of complete partite graphs. We say that a t-matching of G is **restricted** if it does not contain an edge set of any K_{t+1} and $K_{t,t}$ of G and that it is K_q^p - **free** for $t = (p - 1)q$ if it does not contain an edge set of any subgraph of G isomorphic to t-regular K_q^p. The restricted t-matching problem and the K_q^p-free t-matching problem consist in finding a maximum size restricted/K_q^p-free t-matching, respectively. Depending on the variant of the problem, we refer to all respective subgraphs K_q^p, K_{t+1} and $K_{t,t}$ as **forbidden** subgraphs. We say that the restricted/K_q^p-free t-matching problem is **bounded** if every vertex of the input graph is assumed to have degree at most $t + 1$. Polynomial time algorithms for the bounded restricted t-matching problem and the bounded K_q^p-free t-matching problem were presented by Bérczi and Végh [3] and Kobayashi and Yin [22], respectively.

In the weighted versions of these problems, each edge e is associated with a nonnegative weight $w(e)$ and we are interested in computing a restricted/K_q^p-free t-matching of maximum weight. For a subgraph H of G, we say that the weight function is **vertex-induced on** H if real (possibly negative) values called **potentials** can be assigned to the vertices of H in such a way that the weight of every edge of H is equal to the sum of the potentials of its endpoints. Vertex-induced edge weight functions are a (slight) generalization of vertex-weighted edge functions, which are quite common and which were considered among others in [1, 6, 32]. Notice that in our setting weights of edges not belonging to any forbidden subgraphs may be arbitrary. Vornberger [35] showed that the weighted restricted 2-matching problem is $\mathcal{NP}$-hard. However, the weighted **bounded** restricted 2-matching problem is solvable in polynomial time when the weights of the edges are vertex-induced on every subgraph isomorphic to a square, which was shown by Paluch and Wasylkiewicz [29]. The $\mathcal{NP}$-hardness result from [35] has been strengthened - the weighted version of the restricted 2-matching problem is $\mathcal{NP}$-hard for general weights, even in subcubic and bipartite graphs (see [2, Theorem 5.1.]). For this reason an assumption is made in [2, 12, 24, 28, 29, 31, 33] that the weight function is vertex-induced on every forbidden square. In fact, it follows from the proof of Theorem 5.2 in [2] that finding a maximum weight restricted 2-factor (i.e. a 2-matching in which every vertex has degree exactly two) is $\mathcal{NP}$-hard even if the input graph is cubic, bipartite and planar and whose edge weights belong to $\{0, 1, 3n + 1\}$. This does not leave much space for the choice of a weight function that would admit polynomial solution. On the other hand, the weighted restricted 2-matching problem in cubic graphs for 0–1 weights can be solved in polynomial time by a reduction to its unweighted variant in subcubic graphs. Király [17] proved that the restricted 2-factor and 2-matching problems are $\mathcal{NP}$-hard in bipartite graphs, even for 0–1 and 1–2 weights, respectively.

1.1 Our Results

We present simple combinatorial algorithms for the weighted and unweighted versions of both the bounded restricted t-matching problem and the bounded K_q^p-free t-matching problem, both for $t \geq 3$. In the weighted setting we assume that the weights of the edges are nonnegative and vertex-induced on every forbidden subgraph. In these algorithms, instead of computing a restricted or K_q^p-free t-matching directly, we find its complement. Such a complement has fewer edges than the sought-after restricted or K_q^p-free t-matching which results in faster running time of the algorithm. To accomplish this, we augment some forbidden subgraphs of the input graph with gadgets containing so-called *half-edges* and define a function b on the set of vertices in such a way that, any b-matching in the thus constructed graph G' yields a complement of restricted or K_q^p-free t-matching. Half-edges have already been introduced in [27] and used in several subsequent papers. The presented algorithms are the first ones for the weighted versions of these problems. The running time of our algorithms for the weighted versions is $\mathcal{O}(\min\{nm \log n, n^3\})$. Moreover, for the unweighted versions, our algorithms are faster than those known previously. The algorithms in [3, 22] were stated to have running time, respectively, $\mathcal{O}(t^3 n + nm \log m)$ and $\mathcal{O}(pq^3 n + nm \log m)$ but by combining these algorithms with our fast method of finding cliques and by using a better estimation of Gabow's algorithm, they actually can run in $\mathcal{O}(m^{3/2})$ time. For the unweighted versions of both problems, the running times of our algorithms are $\mathcal{O}(\sqrt{n}m)$. Previous approaches to these problems relied on the shrinking of parts of or subsets of forbidden subgraphs and because of this do not lend themselves to the weighted setting. Moreover, we show that all forbidden subgraphs can be found in optimal $\mathcal{O}(m)$ time which allows us to remove the dependence on t, p and q in the time complexity of our algorithms.

Recently, Iwamasa, Kobayashi and Takazawa [16] presented an algorithm which solves the restricted/K_q^p-free t-matching problem in graphs of degree at most $2t - 1$ in polynomial time for any fixed t (but exponential in t). Although our algorithm works only in graphs of degree at most $t + 1$, it runs in polynomial time even when t is part of the input.

The algorithms presented in this paper extend and generalize those from [29]. The generalizations require some new ideas as well as making observations regarding the structure. Also, instead of computing a restricted/K_q^p-free t-matching directly, we deal with finding its complement. This change of focus and an efficient method of identifying forbidden subgraphs allow us to considerably improve the running time and make it independent of t. It turns out that the assumption $t \geq 3$ allows us to simplify some arguments. The most involved is the case of $q = 2$ in the bounded K_q^p-free t-matching problem. Notice that this case was considered separately also in the previous algorithm for this problem given by Kobayashi and Yin [22].

1.2   Related Work

Restricted 2-matchings and restricted t-matchings are classical problems of combinatorial optimization. Notice that a 2-matching is a set of vertex-disjoint cycles and paths whereas a restricted 2-matching is a 2-matching whose every cycle has length at least five since K_3 is a triangle, i.e. a cycle of length three, and $K_{2,2}$ is a square, i.e. a cycle of length four. Therefore, the restricted 2-matching problem can be used to approximate some variants of the travelling salesman problem (see [9] for details). Hartvigsen [11,13] gave the first polynomial time (but very complicated) algorithm for the problem of finding a maximum size triangle-free 2-matching. Recently, Paluch [26] presented a simpler and faster algorithm for this problem which employs half-edges. For the same problem, Bosch-Calvo, Grandoni and Ameli [4] presented a PTAS and a simpler proof of correctness of this algorithm was given by Kobayashi and Noguchi [20]. Papadimitriou [7] showed that it is $\mathcal{NP}$-hard to find a maximum size 2-matching without any cycle of length at most five. Kobayashi [19] gave a polynomial algorithm for finding a maximum weight 2-matching that does not contain any triangle from a given set of forbidden edge-disjoint triangles. Polynomial algorithms for the square-free and/or triangle-free 2-matchings in subcubic graphs were presented in [2,3,14,15,18,22,29].

Regarding the square-free 2-matching problem in general graphs, Nam [25] constructed a complex algorithm for it for graphs, in which all squares are vertex-disjoint. Finding a polynomial algorithm for the square-free 2-matching problem in general graphs is an open problem.

A natural generalization of the restricted 2-matching problem is the restricted t-matching problem. Polynomial algorithms for restricted t-matchings in bipartite graphs were presented in [12,24,28,31,33]. In all previous algorithms for the weighted version of the restricted t-matching problem in bipartite graphs, an assumption is made that the weight function is vertex-induced on every forbidden $K_{t,t}$ since this version is $\mathcal{NP}$-hard for $t = 2$. It is an interesting question to generalize the known results for the square-free and/or triangle-free 2-matching problem in subcubic graphs to restricted t-matchings in graphs of maximum degree at most $t+1$. Since all of these algorithms used the fact that every vertex of a forbidden subgraph is incident to at most one edge which does not belong to this forbidden subgraph, it is natural to consider restricted t-matching problem in graphs of the maximum degree at most $t+1$, i.e. the bounded restricted t-matching problem. As stated before, the unweighted version of this problem was considered by Bérczi and Végh [3].

It is worth mentioning that the bounded restricted t-matching problem has some connections to the connectivity augmentation problem in which we want to make a given undirected graph G k-connected by adding a minimum number of different new edges. As pointed out by Bérczi and Végh [3], this problem is equivalent to finding a maximum size t-matching which contains no complete bipartite subgraphs $K_{a,b}$ consisting of $t + 2$ vertices in the complement of G for $t = n - k - 1$. A special case of the connectivity augmentation problem is increasing connectivity by one problem when given graph G is already $(k - 1)$-connected for which an algorithm which works in $\mathcal{O}(kn^7)$ time was given by Végh [34]. The maximum degree of the complement of a $(k - 1)$-connected graph is at most $t + 1$ for $t = n - k - 1$, so the increasing connectivity by one problem can be reduced to a problem of finding a maximum size t-matching which contains no complete bipartite subgraphs of size $t + 2$ in the graphs of the maximum degree at most $t + 1$. Weighted versions of these problems are also of interest since, in the same paper, Végh considered a generalization of the increasing connectivity by one problem when the edges of the complement of G are given weights and we want to minimize the total sum of added edges. This version is $\mathcal{NP}$-hard for general weights, however Végh presented a polynomial algorithm for the weighted version if the weights are vertex-induced on the whole complement of G.

It turns out that the polynomial solvability of the restricted t-matching problem is also related to jump systems. Jump systems were introduced in [5] as a common generalization of (among others) the sets of bases of a matroid and degree sequences of subgraphs of a graph. In connection to our problems, it was conjectured by Cunningham [8] that for every natural number k, the degree sequences of 2-matchings with no cycles of length at most k of any graph form a jump system if and only if there exists a polynomial algorithm for finding such 2-matchings of maximum size. The only open case of this conjecture is $k = 4$. It was shown by Kobayashi et al. [21] that the degree sequences of restricted 2-matchings of any graph form a jump system which suggests that there should exist a polynomial algorithm for the restricted 2-matching problem in general graphs. A generalization of Cunningham's conjecture to the t-matchings seems to be also true. In the same paper, Kobayashi et al. proved that the degree sequences of restricted t-matchings in any graph form a jump system. In fact, they proved an even stronger result. They showed that the degree sequences of t-matchings with no subgraphs from a given family $\mathcal{H}$ of t-regular connected graphs form a jump system in any graph if and only if every member of $\mathcal{H}$ is a complete partite graph. Moreover, it was shown by Kobayashi and Yin [22] that this problem is $\mathcal{NP}$-hard, even in the graphs of maximum degree at most $t + 1$, if $\mathcal{H}$ is a singleton of any t-regular connected graph which is not complete partite. Notice that a t-regular complete partite graph is exactly K_q^p for $t = (p - 1)q$, hence it is natural to consider K_q^p-free t-matchings for such values of t.

2   Preliminaries

Let $G = (V, E)$ be an undirected graph with vertex set V and edge set E. We denote the number of vertices of G by n and the number of edges of G by m. We assume that all graphs are **simple**, i.e., they contain neither loops nor parallel edges. We denote an edge connecting vertices v and u by (v, u). For a subgraph H of G, we denote the vertex set of H by $V(H)$ and the edge set by $E(H)$. Given a weight function $w : E \to \mathbb{R}$, the **weight of** H, denoted by $w(H)$, is defined as the sum of weights of edges of H. For an edge set $F \subseteq E$ and $v \in V$, we denote by $\deg_F(v)$ the number of edges of F incident to v. With a slight abuse of notation, for a graph H, we sometimes write $\deg_H(v)$ instead of $\deg_{E(H)}(v)$. For any natural number k, we say that G is k-regular if every vertex of G has degree exactly k. For a vertex set $A \subseteq V$, we denote by $G[A]$ a subgraph of G induced by A. Given two graphs $H_1 = (V_1, E_1)$ and $H_2 = (V_2, E_2)$, we denote by $H_1 \cap H_2$ a graph $(V_1 \cap V_2, E_1 \cap E_2)$. With a slight abuse of notation, we denote that H_1 and H_2 are isomorphic by $H_1 = H_2$.

An instance of each of the two problems that we consider in the paper consists of an undirected graph $G = (V, E)$ whose every vertex has degree at least one and at most $t + 1$, and a weight function $w : E \to \mathbb{R}_{\geq 0}$. In the bounded K_q^p-free t-matching problem we are also given natural numbers $p \geq 2$, $q \geq 1$ such that $t = (p - 1)q$. In the bounded restricted (resp. K_q^p-free) t-matching problem we assume that w is vertex-induced on every K_{t+1} and $K_{t,t}$ (resp. on every K_q^p) of G and the goal is to find a maximum weight restricted (resp. K_q^p-free) t-matching of G.

For a weight function $w : E \to \mathbb{R}$ which is vertex-induced on some subgraph H of G, we say that a function $r_H : V(H) \to \mathbb{R}$ which assigns a potential to every vertex of H is a **potential function** of H.

We say that an edge set $M \subseteq E$ is a **co-t -matching** of G if $E \setminus M$ is a t-matching of G. Additionally, we say that a co-t-matching $M \subseteq E$ is **covering** (resp. K_q^p-**covering**) if $E \setminus M$ is a restricted (resp. K_q^p-free) t-matching of G. We say that an edge set $M \subseteq E$ **covers** a forbidden subgraph H of G, or that H **is covered by** M, if M contains at least one edge of H. Notice that if the degree of every vertex of G is at most $t + 1$, then an edge set $M \subseteq E$ is a co-t-matching of G if every vertex of G of degree $t + 1$ is incident to at least one edge of M. Moreover, M is a K_q^p-covering (resp. covering) co-t-matching if, in addition, M covers every K_q^p (resp. every K_{t+1} and $K_{t,t}$) of G. A $(K_q^p$-)covering co-t-matching M of G is said to be **minimum weight** if there is no $(K_q^p$-)covering co-t-matching N of G of weight smaller than $w(M)$.

We will need to compute a b-matching of a graph G where we are given vectors $l, b \in \mathbb{N}^V$ and a weight function $w : E \to \mathbb{R}$. (We allow negative weights here.) For a vertex $v \in V$, $[l(v), b(v)]$ is said to be a **capacity interval** of v. An edge set $M \subseteq E$ is said to be an (l, b)-matching if $l(v) \leq \deg_M(v) \leq b(v)$ for every $v \in V$. Given an (l, b)-matching M and an edge $e = (u, v) \in M$, we say that u is **matched to** v in M. Moreover, we say that a vertex v of G is **unmatched** in M if v is not incident to any edge of M. A **matching** and a **perfect matching** of G is any (l, b)-matching of G where the capacity interval of every vertex v of

G is equal to, respectively, $[0, 1]$ and $[1, 1]$. An (l, b)-matching M is said to be a ***maximum weight*** (l, b)-matching if there is no (l, b)-matching M' of G of weight greater than $w(M)$. A maximum weight (l, b)-matching can be computed efficiently, as shown by the following theorem.

Theorem 1. ([10]). *There is an algorithm that, given a multigraph $G = (V, E)$ (i.e. G may contain parallel edges), a weight function $w : E \to \mathbb{R}$ and vectors $l, b \in \mathbb{N}^V$, finds a maximum weight (l, b)-matching of G in time $\mathcal{O}\left(\sum_{v \in V} b(v) \min\{|E| \log |V|, |V|^2\}\right)$, assuming that G admits any (l, b)-matching.*

We define a ***minimum weight*** (l, b)-matching of G analogously. Notice that calculating a minimum weight (l, b)-matching of G can be reduced to calculating a maximum weight (l, b)-matching of G by negating the weight of every edge of G.

3   Outline of the Algorithm

We give an outline of our algorithms for the weighted versions of the bounded restricted and the bounded K_q^p-free t-matching problem assuming that the weight function is vertex-induced on every forbidden subgraph. Recall that we assume that $t \geq 3$. In the K_q^p-free t-matching problem we additionally assume that $t = (p - 1)q$.

Algorithm 1. Computing a maximum weight restricted or K_q^p-free t-matching of graph G of maximum degree at most $t + 1$ given a weight function $w : E \to \mathbb{R}_{\geq 0}$.

1: Construct an auxiliary multigraph $G' = (V', E')$ consisting of $\mathcal{O}(n)$ vertices and $\mathcal{O}(m)$ edges by augmenting some forbidden subgraphs of G with gadgets containing half-edges. (Both gadgets and half-edges are defined later.)

2: Define a weight function $w' : E' \to \mathbb{R}$ and vectors $l, b \in \mathbb{N}^{V'}$.

3: Compute a minimum weight (l, b)-matching M' of G'.

4: Construct a co-t-matching $\bar{M}$ of G which covers all augmented forbidden subgraphs by replacing all half-edges of M' with some edges of G in such a way that $w(\bar{M}) \leq w'(M')$.

5: Add some edges of the remaining forbidden subgraphs to $\bar{M}$ by replacing some of its edges with other ones without increasing the weight of $\bar{M}$ in order to obtain $(K_q^p\text{-})$covering co-t-matching of G.

6: Return $E \setminus \bar{M}$.

The precise construction of G', w', l and b for the bounded restricted (resp. K_q^p-free) t-matching problem is given in Sect. 4 (resp. Section 5).

For the bounded restricted t-matching problem, an implementation of Step 4 is given in the proof of Theorem 3 whereas an implementation of Step 5 consists of subsequent applications of the proof of Lemma 5. An implementation of these steps for the bounded K_q^p-free t-matching problem is given in the proof of Theorem 4.

Claim. Algorithm 1 runs in time $\mathcal{O}(\min\{nm \log n, n^3\})$ in the weighted version and $\mathcal{O}(\sqrt{n}m)$ in the unweighted.

The proof of this claim can be found in the full version of the paper [30]. The main ingredients are the following. We prove that Step 3 of Algorithm 1 can be implemented by a slight modification of the algorithms presented by Gabow in [10]. We use the fact that the size of the sought-after co-t-matching $\bar{M}$ of G is $\mathcal{O}(n)$ to achieve the desired running times. We show that all forbidden subgraphs of G can be found in $\mathcal{O}(m)$ time.

4   The Bounded Restricted t-Matching Problem

In this section we address the weighted bounded restricted t-matching problem. At the beginning we define the gadgets for forbidden subgraphs. In Subsect. 4.1 we consider the case when all forbidden subgraphs of G are pairwise vertex-disjoint. We deal with overlapping forbidden subgraphs in Subsect. 4.2.

We build a graph $G' = (V', E')$ together with a weight function $w' : E' \to \mathbb{R}$ and capacity intervals $[l(v), b(v)]$ for every $v \in V'$, in which subgraphs, called **gadgets**, are added to some forbidden subgraphs of G. The precise construction of gadgets is the following.

Let H be any K_{t+1} of G. We define a **gadget for** H as follows. Let v_1, v_2, ..., v_{t+1} be vertices of H. We introduce one new vertex u_H called a **subdivision vertex**. For each vertex v_i of H we add an edge (v_i, u_H) called a **half-edge** of H (see Fig. 1). We set the capacity interval of u_H to $[2, 2]$.

Let H be any $K_{t,t}$ of G. We define a **gadget for** H as follows. Let a_1, a_2, ..., $a_t \in V_1(H)$ and b_1, b_2, ..., $b_t \in V_2(H)$. We introduce two new subdivision vertices u_H^1 and u_H^2. We connect u_H^1 with all vertices of $V_1(H)$ by half-edges. Symmetrically, we connect u_H^2 with all vertices of $V_2(H)$ by half-edges (see Fig. 2). We set the capacity intervals of u_H^1 and u_H^2 to $[1, 1]$.

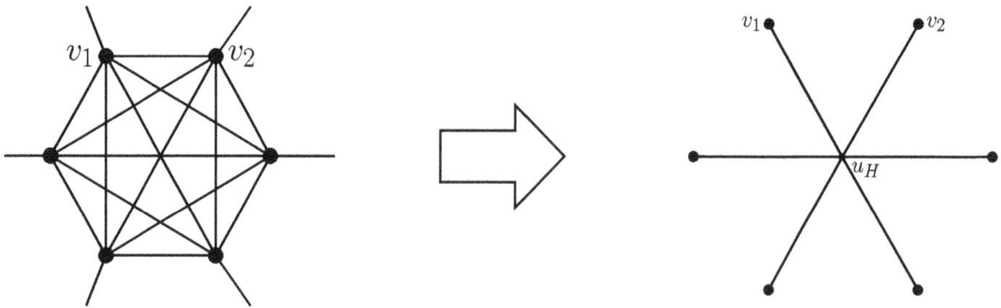

Fig. 1. A gadget for K_{t+1} for $t = 5$.

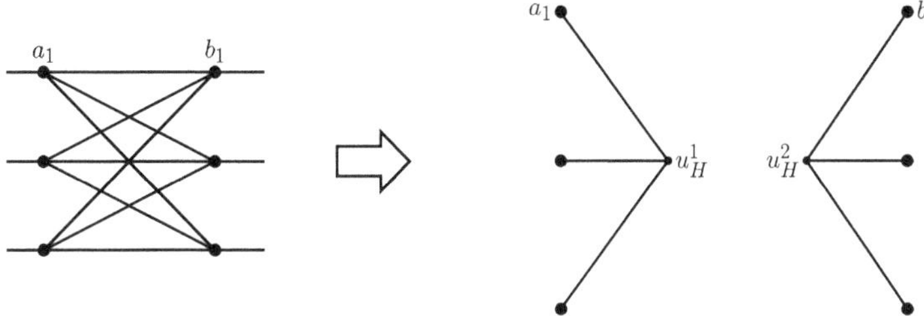

Fig. 2. A gadget for $K_{t,t}$ for $t = 3$.

The main ideas behind these gadgets are the following. An (l, b)-matching M' of G' is to represent roughly a covering co-t-matching $\bar{M}$ of G. We want to ensure that at least one edge of H belongs to $\bar{M}$. Notice that M' contains exactly two half-edges of H incident to some vertices z_1 and z_2 of H. We add an edge (z_1, z_2) to $\bar{M}$. Moreover, we add to $\bar{M}$ all edges of H which belong to M'. In this way, H is guaranteed to be covered by $\bar{M}$.

Regarding weights w' of the edges in the gadgets, we assign them as follows. Let r_H be any potential function of H. Every half-edge of H incident to vertex v_i gets weight $r_H(v_i)$.

4.1   Disjoint Forbidden Subgraphs

In this subsection we consider the case when all forbidden subgraphs of G are pairwise vertex-disjoint. We show how to drop this assumption later, in Subsect. 4.2.

The precise construction of G', w', l and b is the following. We start off with $G' = G$. We add a gadget to every forbidden subgraph of G. This operation is well-defined because these forbidden subgraphs are pairwise vertex-disjoint. We set the capacity interval of every vertex v of degree $t + 1$ of G to $[1, t + 1]$. We set the capacity interval of every remaining vertex v of G to $[0, \deg_G(v)]$. We set the weight of every edge e of G' which belongs to G to $w(e)$. In theorems below we show the correspondence between covering co-t-matchings of G and (l, b)-matchings of G'.

Theorem 2. *Let $\bar{M}$ be any covering co-t-matching of G. Then we can find an (l, b)-matching M' of G' such that $w'(M') = w(\bar{M})$.*

Proof. We initialize M' as the empty set. We add every edge of $\bar{M}$ that does not belong to any forbidden subgraph of G to M'.

Consider any $H = K_{t+1}$ of G. Since $\bar{M}$ is covering, there exists an edge of H which belongs to $\bar{M}$. If more than one edge of H belongs to $\bar{M}$, we choose one of them. Suppose that we chose $(z_1, z_2) \in \bar{M}$. Then we add edges (z_1, u_H) and (z_2, u_H) to M'. We add every remaining edge of H which belongs to $\bar{M}$ to M'.

Consider any $H = K_{t,t}$ of G. Since $\bar{M}$ is covering, there exists an edge (a, b) of H which belongs to $\bar{M}$. Suppose that $a \in V_1(H)$ and $b \in V_2(H)$. We add edges

(a, u_H^1) and (b, u_H^2) to M'. We add every remaining edge of H which belongs to $\bar{M}$ to M'.

Since, from the definition of w', the weight of any replaced edge in G is equal to the sum of the weights of its corresponding half-edges in G', we get that $w'(M') = w(\bar{M})$.

Theorem 3. *Let M' be any (l, b)-matching of G'. Then we can find a covering co-t-matching $\bar{M}$ of G such that $w(\bar{M}) \leq w'(M')$.*

Proof. We initialize $\bar{M}$ as the empty set. We add every edge of M' which belongs to G to $\bar{M}$. For every forbidden subgraph of G we add some of its edges to $\bar{M}$ as follows.

Let H be any K_{t+1} of G. Notice that exactly two vertices of H, say v_1 and v_2, are matched to u_H in M'. We add (v_1, v_2) to $\bar{M}$ if it does not belong to $\bar{M}$ already.

For any $H = K_{t,t}$ of G we proceed analogously, because in the gadget for H the subdivision vertices u_H^1 and u_H^2 are matched to two vertices $a \in V_1(H)$ and $b \in V_2(H)$.

Notice that, from construction, every vertex of degree exactly $t + 1$ in G is incident to an edge of $\bar{M}$. Moreover, $w(\bar{M}) \leq w'(M')$ because we assume that $w(e)$ is nonnegative for every edge e of G.

4.2   Non-Disjoint Forbidden Subgraphs

In the following lemmas we examine the ways in which forbidden subgraphs of G may overlap.

Lemma 1. *Let H_1 and H_2 be any two different K_{t+1}'s of G with a common vertex. Then $H_1 \cap H_2 = K_t$.*

Lemma 2. *Let H_1 and H_2 be any two different $K_{t,t}$'s of G with a common vertex. Then $H_1 \cap H_2$ contains $K_{t-1,t-1}$.*

Lemma 3. *Let H_1 and H_2 be any, respectively, K_{t+1} and $K_{t,t}$ of G with a common vertex. Then $t = 3$ and $H_1 \cap H_2 = K_{2,2}$.*

We define the following family of **problematic** subgraphs of G.

Definition 1. Let H be any forbidden K_{t+1} of G. We say that H is **unproblematic** if H has a common vertex with another forbidden subgraph H' of G such that:

1. $H' = K_{t+1}$ and $w(H) \leq w(H')$ (if $w(H) = w(H')$, H' is unproblematic as well), or
2. $H' = K_{t,t}$.

Otherwise, we say that H is **problematic**.

Definition 2. Let H be any forbidden $K_{t,t}$ of G. We say that H is ***unproblematic*** if H has a common vertex with another forbidden subgraph $H' = K_{t,t}$ of G such that $w(H) \leq w(H')$. Otherwise, we say that H is ***problematic***.

Lemma 4. *Any two different problematic subgraphs of G are vertex-disjoint.*

To find a minimum weight covering co-t-matching of G, we use the graph G' from Subsect. 4.1. However, we only augment every problematic subgraph with its gadget. In the such constructed G' we compute an (l, b)-matching of G', which corresponds to a co-t-matching $\bar{M}$ of G. The co-t-matching $\bar{M}$ may not cover some unproblematic forbidden subgraphs. Below we show that it is possible to modify $\bar{M}$ so that it covers every forbidden subgraph.

Lemma 5. *Let $\bar{M}$ be any co-t-matching of G. If $\bar{M}$ does not cover some unproblematic subgraph H, then we can find a co-t-matching $\bar{N}$ of G such that $w(\bar{N}) \leq w(\bar{M})$ which covers H and every forbidden subgraph covered by $\bar{M}$.*

Proof. If H is a K_{t+1} of G sharing a vertex with some $H' = K_{t,t}$ of G, then by Lemma 3, $t = 3$, $H = K_4$, $H' = K_{3,3}$ of G and $H \cap H' = K_{2,2}$. Let $v_1 \in V(H) \cap V_1(H')$, $v_2 \in V(H) \cap V_2(H')$, $u_1 \in V_1(H') \setminus V(H)$ and $u_2 \in V_2(H') \setminus V(H)$. We set $\bar{N} = \bar{M} \setminus \{(v_1, u_2), (v_2, u_1)\} \cup \{(v_1, v_2), (u_1, u_2)\}$. Notice that the only edge of H' incident to u_1 or u_2 which can possibly not belong to $\bar{M}$ is (u_1, u_2) since every vertex of H is incident to $t = 3$ edges of H which do not belong to $\bar{M}$. Thus, both u_1 and u_2 are incident to some edge of $\bar{N}$. Moreover, $w(\bar{N}) \leq w(\bar{M})$ because w is vertex-induced on H'.

If H is a K_{t+1} of G which does not share a vertex with any $K_{t,t}$ of G, then H shares a vertex with another $H' = K_{t+1}$ of G such that $w(H) \leq w(H')$. Let u (resp. u') be the only vertex of H (resp. H') which does not belong to H' (resp. H). Notice that every edge of H' incident to u' belongs to $\bar{M}$, so $\deg_{\bar{M}}(u') \geq t$. Since $w(H) \leq w(H')$, the sum of the weights of edges of H incident to u is not greater than the sum of weights of edges of H' incident to u'. Therefore, there exists a common vertex of H and H', say z, such that $w(u, z) \leq w(u', z)$. We define $\bar{N}$ as $\bar{M}$ with the edge (u', z) replaced by (u, z). As a result, $\bar{N}$ covers H. It still covers H' since every vertex of $V(H) \cap V(H')$ different than z is matched to u' in $\bar{N}$. Notice that $\bar{N}$ is still a co-t-matching since $\deg_{\bar{N}}(u') = \deg_{\bar{M}}(u') - 1 \geq t - 1 \geq 1$ since $t \geq 3$.

Let H be a $K_{t,t}$ of G sharing a vertex with another $H' = K_{t,t}$ of G such that $w(H) \leq w(H')$. If $H \cap H' = K_{t-1,t-1}$, we proceed analogously as in the case when $H = K_{t+1}$ shares a vertex with $H' = K_{t,t}$. Otherwise, $H \cap H' = K_{t,t-1}$ and we proceed analogously as in the case when $H = K_{t+1}$ shares a vertex with another $H' = K_{t+1}$.

5   The Bounded K_q^p-free t-Matching problem

In this section we consider the weighted bounded K_q^p-free t-matching problem. Recall that $t = (p - 1)q$ for this problem. Let us observe that for $q = 1$, $K_q^p =$

K_{t+1} and for $p = 2$, $K_q^p = K_{t,t}$. Since the bounded K_q^p-free t-matching problem for these cases can be solved by a slight modification of the algorithm for the bounded restricted t-matching problem, we assume that $q \geq 2$ and $p \geq 3$.

We denote the empty graph on n vertices by I_n. For graphs $G_1 = (V_1, E_1)$ and $G_2 = (V_2, E_2)$ such that V_1 and V_2 are disjoint, we say that a graph $G_1 \times G_2 = (V_1 \cup V_2, E_1 \cup E_2 \cup (V_1 \times V_2))$ is a **product** of G_1 and G_2.

Similarly as in Sect. 4, we build a graph $G' = (V', E')$ together with a weight function $w' : E' \to \mathbb{R}$, in which some K_q^p's of G are augmented with gadgets. Below we consider the case when case $q \geq 3$ and the case of $q = 2$ is considered in the full version of the paper [30].

Let $q \geq 3$ and let H be any K_q^p of G. We define a **gadget for** H as follows. We introduce p subdivision vertices u_H^1, u_H^2, ..., u_H^p and a **global vertex** z_H. For every i, we connect u_H^i with every vertex of $V_i(H)$ by half-edges and with z_H (see Fig. 3). Let r_H be any potential function of H. A half-edge of H incident to vertex v_i gets weight $r_H(v_i)$. The remaining edges of the gadget get weight 0. We set the capacity interval of vertex z_H to $[p-2, p-2]$ and of every subdivision vertex of the gadget for H to $[1, 1]$. This gadget ensures that at least one edge of H belongs to the co-t-matching $\bar{M}$.

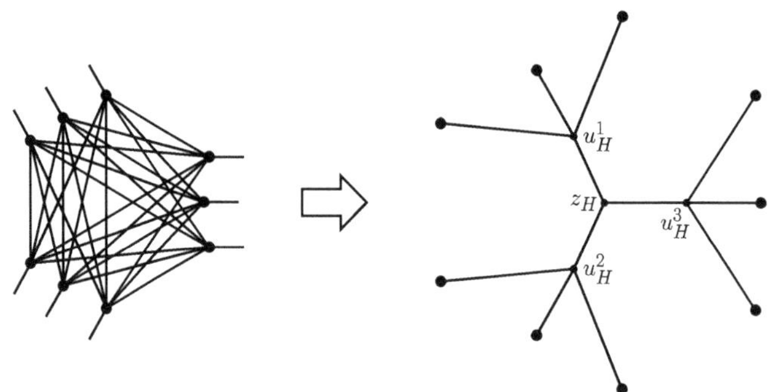

Fig. 3. A gadget for K_q^p for $p = 3$ and $q = 3$.

Lemma 6. *Let H_1 and H_2 be any two different K_q^p's of G with a common vertex. Then $H_1 \cap H_2$ contains $I_{q-1} \times K_q^{p-1}$.*

Definition 3. Let H be any K_q^p of G. We say that H is **unproblematic** if there exists another $H' = K_q^p$ of G that shares a vertex with H and such that $w(H) \leq w(H')$. Otherwise, we say that H is **problematic**.

Lemma 7. *Any two different problematic K_q^p's of G are vertex-disjoint.*

When constructing G', we only augment every problematic K_q^p of G with its gadget, analogously as in Subsect. 4.2.

Theorem 4. *For any K_q^p-covering co-t-matching $\bar{M}$ of G there exists an (l,b)-matching M' of G' such that $w'(M') = w(\bar{M})$.*
For any (l,b)-matching M' of G' there exists a K_q^p-covering co-t-matching $\bar{M}$ of G such that $w(\bar{M}) \leq w'(M')$.

The proof of Theorem 4 is similar to the ones of its analogues from Sect. 4.

6   Conclusion

We presented the first polynomial algorithms for the weighted versions of the bounded restricted and K_q^p-free t-matching problems in which the weight function is vertex-induced on every forbidden subgraph. Moreover, these algorithms for the unweighted versions are faster than those known previously.

In fact, previous algorithms by Bérczi and Végh [3] and Kobayashi and Yin [22] were designed for a bit wider setting when vertices not belonging to forbidden subgraphs may have arbitrary degrees, degree constraints and parallel edges. Our algorithm can be easily adapted to this more general case to achieve $\mathcal{O}(m^{3/2})$ running time in the unweighted setting, as opposed by $\mathcal{O}(t^3 n + m^{3/2})$ or $\mathcal{O}(t^3 n/p^2 + m^{3/2})$ time with previous algorithms, improving running time for large values of t. In the weighted setting the running time of our algorithm in this more general setting becomes $\mathcal{O}(\min\{m^2 \log n, n^2 m\})$. Notice that we can still find all forbidden subgraphs in this case since any vertex of degree more than $t + 1$ cannot belong to any forbidden subgraph and thus can be ignored.

Our algorithms extend to the slightly more general setting (considered in [3,22]) when only some complete t-regular partite subgraphs of G have to be excluded from the sought-after t-matching. We refer to such subgraph as ***forbidden***. In this case, we modify Algorithm 1 for the bounded restricted t-matching problem and the bounded K_q^p-free t-matching problem for $q \geq 3$ as follows. In Step 1, we augment only forbidden problematic subgraphs. We proceed similarly as in [22] for the K_2^p-free t-matching problem. The time complexity of Algorithm 1for this setting depends on the representation of forbidden subgraphs in the input.

It is imaginable that our algorithms can be adapted to find maximum size or weight t matchings which contain no complete bipartite subgraphs consisting of $t + 2$ vertices in the graphs of maximum degree at most $t + 1$, possibly leading to a faster algorithm for the problem of increasing connectivity by one. Of course, an assumption that the weight function is vertex-induced on every forbidden complete bipartite subgraph has to be made in the weighted version of this problem.

Regarding the unbounded case, we suppose that it might be possible to generalize the algorithm in [26] for maximum triangle-free 2-matchings to solve the problem of maximum K_{t+1}-free t-matching.

Based on our result, one can suspect that the degree sequences of maximum weight restricted t-matchings form an M-concave function on the constant-parity jump system in any graph of maximum degree at most $t+1$ if the weight function is vertex-induced on every forbidden subgraph.

References

1. Aggarwal, G., Goel, G., Karande, C., Mehta, A.: Online vertex-weighted bipartite matching and single-bid budgeted allocations. In: Proceedings of the Twenty-Second Annual ACM-SIAM Symposium on Discrete Algorithms, pp. 1253–1264. Society for Industrial and Applied Mathematics (2011)
2. Bérczi, K., Kobayashi, Y.: An algorithm for $(n-3)$-connectivity augmentation problem: jump system approach. J. Combin. Theo. Series B **102**(3), 565–587 (2012)
3. K. Bérczi and L. Végh. Restricted b-matchings in degree-bounded graphs. In Integer Programming and Combinatorial Optimization, pp. 43–56 (2010)
4. Bosch-Calvo, M., Grandoni, F., Ameli, A.J.: A PTAS for triangle-free 2-matching. arXiv:2311.11869 (2023)
5. Bouchet, A., Cunningham, W.H.: Delta-matroids, jump systems, and bisubmodular polyhedra. SIAM J. Discret. Math. **8**(1), 17–32 (1995)
6. Chekuri, C., Ene, A., Vakilian, A.: Node-weighted network design in planar and minor-closed families of graphs. In: Czumaj, A., Mehlhorn, K., Pitts, A., Wattenhofer, R. (eds.) Automata. Languages, and Programming, pp. 206–217. Springer, Berlin Heidelberg (2012). https://doi.org/10.1007/978-3-642-31594-7_18
7. Cornuéjols, G., Pulleyblank, W.: A matching problem with side conditions. Discret. Math. **29**(2), 135–159 (1980)
8. Cunningham, W.: Matching, matroids, and extensions. Math. Program. **91**, 515–542 (2002)
9. Fisher, M., Nemhauser, G., Wolsey, L.: An analysis of approximations for finding a maximum weight hamiltonian circuit. Oper. Res. **27**(4), 799–809 (1979)
10. Gabow, H.: An efficient reduction technique for degree-constrained subgraph and bidirected network flow problems. In: Proceedings of the Fifteenth Annual ACM Symposium on Theory of Computing, pp. 448–456 (1983)
11. Hartvigsen, D.: Extensions of Matching Theory. PhD thesis, Carnegie-Mellon University (1984)
12. Hartvigsen, D.: Finding maximum square-free 2-matchings in bipartite graphs. J. Combin. Theo. Series B **96**(5), 693–705 (2006)
13. Hartvigsen, D.: Finding triangle-free 2-factors in general graphs. J. Graph Theory **106**(3), 581–662 (2024)
14. Hartvigsen, D., Li, Y.: Maximum cardinality simple 2-matchings in subcubic graphs. SIAM J. Optim. **21**(3), 1027–1045 (2011)
15. Hartvigsen, D., Li, Y.: Polyhedron of triangle-free simple 2-matchings in subcubic graphs. Math. Program. **138**, 43–82 (2013)
16. Iwamasa, Y., Kobayashi, Y., Takazawa, K.: Finding a maximum restricted t-matching via boolean edge-CSP. In: 32nd Annual European Symposium on Algorithms, pp. 75:1–75:15 (2024)
17. Király, Z.: Restricted t-Matchings Bipartite Graphs. Technical report, Egerváry Research Group (2009)
18. Kobayashi, Y.: A simple algorithm for finding a maximum triangle-free 2-matching in subcubic graphs. Discret. Optim. **7**, 197–202 (2010)
19. Kobayashi, Y.: Weighted triangle-free 2-matching problem with edge-disjoint forbidden triangles. In: Integer Programming and Combinatorial Optimization, pp. 280–293 (2020)
20. Kobayashi, Y., Noguchi, T.: Validating a PTAS for triangle-free 2-matching via a simple decomposition theorem. In 2025 Symposium on Simplicity in Algorithms, pp. 281–289 (2025)

21. Kobayashi, Y., Szabo, J., Takazawa, K.: A proof of Cunningham's conjecture on restricted subgraphs and jump systems. J. Combin. Theo. Series B **102**, 948–966 (2012)
22. Kobayashi, Y., Yin, X.: An algorithm for finding a maximum t-matching excluding complete partite subgraphs. Discret. Optim. **9**(2), 98–108 (2012)
23. Lovász, L., Plummer, M.: Matching theory. AMS Chelsea Publishing, corrected reprint of the 1986 original edition (2009)
24. Makai, M.: On maximum cost $K_{t,t}$-free t-matchings of bipartite graphs. SIAM J. Discret. Math. **21**, 349–360 (2007)
25. Nam, Y.: Matching Theory: Subgraphs with Degree Constraints and other Properties. PhD thesis, University of British Columbia (1994)
26. Paluch, K.: Triangle-free 2-matchings. arXiv:2311.13590 (2023)
27. Paluch, K., Elbassioni, K., van Zuylen, A.: Simpler approximation of the maximum asymmetric traveling salesman problem. In 29th International Symposium on Theoretical Aspects of Computer Science, pp. 501–506 (2012)
28. Paluch, K., Wasylkiewicz, M.: Restricted t-matchings via half-edges. In: 29th Annual European Symposium on Algorithms, pp. 73:1–73:17 (2021)
29. Paluch,K., Wasylkiewicz, M.: A simple combinatorial algorithm for restricted 2-matchings in subcubic graphs - via half-edges. Inform. Process. Lett. 171 (2021)
30. Paluch, K., Wasylkiewicz, M.: Clique-free t-matchings in degree-bounded graphs. arXiv:2405.00429 (2024). https://doi.org/10.48550/arXiv.2405.00429
31. Pap, G.: Combinatorial algorithms for matchings, even factors and square-free 2-factors. Math. Program. **110**, 57–69 (2007)
32. Svensson, O.: Approximating $ATSP$ by relaxing connectivity. In: 56th Annual Symposium on Foundations of Computer Science, pp. 1–19 (2015)
33. Takazawa, K.: A weighted $K_{t,t}$-free t-factor algorithm for bipartite graphs. Math. Oper. Res. **34**(2), 351–362 (2009)
34. Végh, L.: Augmenting undirected node-connectivity by one. SIAM J. Discret. Math. **25**(2), 695–718 (2011)
35. Vornberger, O.: Easy and hard cycle covers. Technical report, Universität Paderborn (1980)

Spanning Trees with a Small Vertex Cover: The Complexity on Specific Graph Classes

Toranosuke Kokai[1]([⊠]), Akira Suzuki[2] [ID], Takahiro Suzuki[1] [ID], Yuma Tamura[1] [ID], and Xiao Zhou[1]

[1] Graduate School of Information Sciences, Tohoku University, Sendai, Japan
{toranosuke.kokai.s6,takahiro.suzuki.q4}@dc.tohoku.ac.jp,
{tamura,zhou}@tohoku.ac.jp
[2] Center for Data-driven Science and Artificial Intelligence, Tohoku University,
Sendai, Japan
akira@tohoku.ac.jp

Abstract. In the context of algorithm theory, various studies have been conducted on spanning trees with desirable properties. In this paper, we consider the MINIMUM COVER SPANNING TREE problem (MCST for short). Given a graph G and a positive integer k, the problem determines whether G has a spanning tree with a vertex cover of size at most k. We reveal the equivalence between MCST and the DOMINATING SET problem when G is of diameter at most 2 or P_5-free. This provides the intractability for these graphs and the tractability for several subclasses of P_5-free graphs. We also show that MCST is NP-complete for bipartite planar graphs of maximum degree 4 and unit disk graphs. These hardness results resolve open questions posed in prior research. Finally, we present an FPT algorithm for MCST parameterized by clique-width and a linear-time algorithm for interval graphs.

Keywords: Spanning tree · Graph algorithms · NP-completeness · Fixed-parameter tractability

1 Introduction

A *spanning tree* of a graph is a fundamental concept in graph theory. For a connected graph G, a spanning tree T is an acyclic connected subgraph of G that contains all vertices in G. It is well known that a spanning tree of G can be found in polynomial time. Spanning trees play a crucial role in both theoretical science and practical applications. Theoretically, spanning trees serve as a powerful tool for developing algorithms and are deeply connected to more advanced mathematical theories such as matroids. In terms of applications, spanning trees are essential for achieving efficient communication in networks.

This work was partially supported by JSPS KAKENHI Grant Numbers JP25K14980 and JP25K21148.

© The Author(s), under exclusive license to Springer Nature Switzerland AG 2026
J. Kozik and A. Wolff (Eds.): SOFSEM 2026, LNCS 16448, pp. 578–592, 2026.
https://doi.org/10.1007/978-3-032-17801-5_42

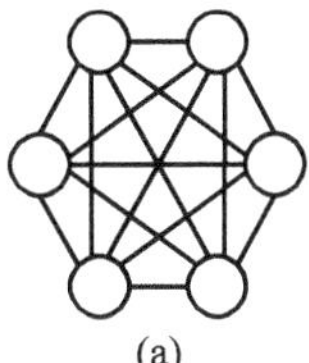 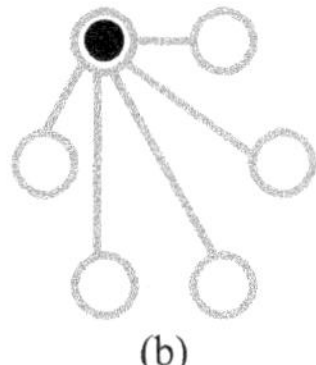 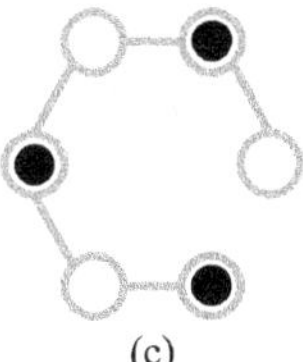

(a)   (b)   (c)

Fig. 1. (a) A complete graph G, (b) a spanning tree of G with a vertex cover of size 1, and (c) a spanning tree of G with a vertex cover of size 3. The spanning trees are shown in bold green lines, and the corresponding vertex covers are indicated by black circles. (Color figure online)

Spanning trees with specific restrictions have also attracted attention, such as those with few leaves [23], many leaves [22], bounded diameter [12], or small poise (the maximum degree plus the diameter) [21]. In particular, a spanning tree with two leaves is also known as a Hamiltonian path, a topic of long-standing research in graph theory. The maximum number of leaves among spanning trees of G is called the *max-leaf number*. The max-leaf number is closely related to the connected domination number (for example, see [3]). Moreover, the max-leaf number is also known as a structural parameter of graphs, and efficient FPT algorithms have been developed for various problems [7].

In this paper, we deal with spanning trees with a small vertex cover. A vertex subset S of a graph G is a *vertex cover* if at least one endpoint of every edge of G belongs to S. Given a positive integer k, the MINIMUM COVER SPANNING TREE problem (MCST for short) asks for a spanning tree T with a vertex cover of size at most k. Intuitively, spanning trees with a small vertex cover number tend to have a star-like structure and a small diameter. Spanning trees of such type are considered easy to handle in practical applications. For example, a complete graph G with n vertices admits a spanning star T with a vertex cover of size 1, whereas it also admits a spanning path T' with a vertex cover of size $\lfloor n/2 \rfloor$ (see Fig. 1). If the vertices and edges of G represent the communication nodes and links of a network, then communication along T is expected to have lower latency and to be easier to monitor.

1.1   Known Results

Angel et al. showed that any graph G satisfies $\gamma(G) \leq \tau_{ST}(G) \leq 2\gamma(G) - 1$ [1], where $\tau_{ST}(G)$ denotes the minimum size of vertex covers among spanning trees of G and $\gamma(G)$ denotes the minimum size of dominating sets in G. From the (in)approximability of the DOMINATING SET problem, it follows that a polynomial-time $O(\log n)$-approximation algorithm exists for MCST, whereas there exists a constant c such that MCST is NP-hard to approximate within a factor of $c \log n$.

Fukunaga and Maehara gave an alternative proof of the above inapproximability of MCST [9]. They also provided a polynomial-time constant-factor

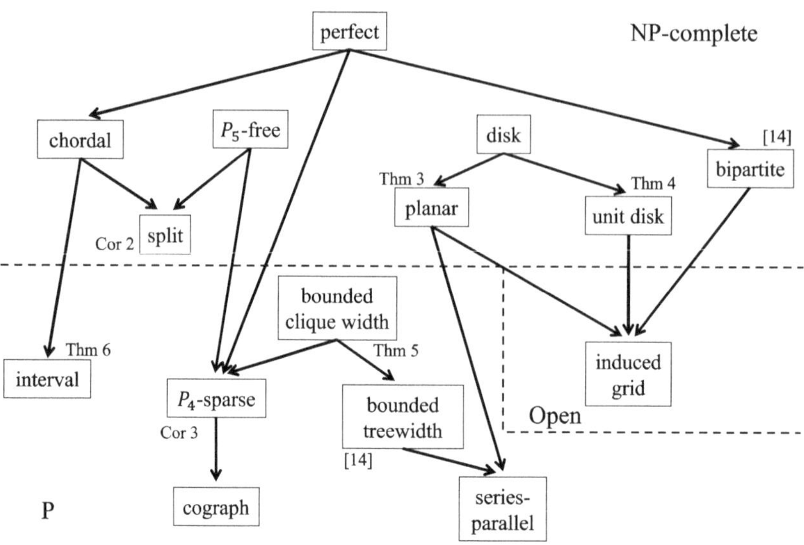

Fig. 2. Known and our results with respect to graph classes. Each arrow $A \to B$ represents that the graph class B is a subclass of the graph class A.

approximation algorithm for MCST on unit disk graphs and graphs excluding a fixed minor.

Kaur and Misra investigated MCST from the perspective of parameterized complexity [14]. MCST was shown to be W[2]-hard when parameterized by the vertex cover size k in spanning trees, even for bipartite graphs. On the other hand, an FPT algorithm parameterized by treewidth was provided. They also studied a kernelization and showed that the problem parameterized by k does not admit a polynomial kernel unless coNP $\subseteq$ NP/poly.

1.2   Our Contribution

This paper provides a detailed study of MCST from the viewpoint of graph classes. (See Fig. 2.) Denote P_n by the path with n vertices. We first investigate the relationship between $\tau_{ST}(G)$ and $\gamma(G)$. We reveal that $\tau_{ST}(G) = \gamma(G)$ holds when a graph G has diameter at most 2 or is P_5-free. These findings have various algorithmic consequences. We show that MCST on split graphs with n vertices cannot be approximated to within a factor of $(1-\varepsilon)\log n$ in polynomial time for any constant $0 < \varepsilon < 1$ unless NP $\subseteq$ DTIME$(n^{O(\log \log n)})$. Moreover, it can be shown that MCST remains W[2]-hard even for split graphs of diameter 2 when parameterized by $\tau_{ST}(G)$. On the other hand, if DOMINATING SET is solvable in polynomial time for a subclass of P_5-free graphs, then MCST can also be solved in polynomial time for the same graph class. Such graph classes include cographs, P_4-sparse graphs, chain graphs, co-bipartite graphs, and co-interval graphs. We emphasize that our result is tight in the sense that there is a P_6-free graph G of diameter 3 such that $\tau_{ST}(G) \neq \gamma(G)$.

The above result motivates us to analyze the computational complexity of MCST on graph classes that admit long paths: planar graphs, bipartite graphs, unit disk graphs, bounded clique-width graphs, and interval graphs. We show that MCST is NP-complete for planar bipartite graphs with maximum degree at most 4 and unit disk graphs. These results resolve the open question posed by Fukunaga and Maehara [9]. On the other hand, we reveal that MCST is fixed-parameter tractable when parameterized by clique-width, which extends the known result that MCST is fixed-parameter tractable for treewidth [14]. This can be proved by defining a problem equivalent to MCST that is expressible in monadic second-order logic (MSO_1). A result of Courcelle et al. implies that the equivalent problem is fixed-parameter tractable for clique-width [5]. We also design an FPT algorithm parameterized by clique-width to improve the running time. Finally, we present a linear-time algorithm for interval graphs.

Due to the space limitation, proofs marked ($*$) are omitted. The complete proofs can be found in the full version [16].

Future Work. One interesting open question is whether MCST is solvable in polynomial time for induced grid graphs (i.e. induced subgraphs of the Cartesian product of two paths). We note that DOMINATING SET is NP-complete for induced grid graphs, whereas VERTEX COVER can be solved in polynomial time for bipartite graphs.

2   Preliminaries

For a positive integer n, we write $[n] = \{1, \ldots, n\}$. Let $G = (V, E)$ be a graph: we also denote by $V(G)$ and $E(G)$ the vertex set and the edge set of G, respectively. All the graphs considered in this paper are finite, simple, and undirected. For two vertices u, v in a graph G, denote by $\text{dist}_G(u, v)$ the length of a shortest path between u and v in G. The *diameter* of G is defined as $\max_{u,v \in V(G)} \text{dist}_G(u, v)$. A graph is said to be *connected* if there exists a path between every pair of vertices $u, v \in V(G)$. A *component* of G is a maximal connected subgraph of G. For a vertex v of G, we denote by $N_G(v)$ and $N_G[v]$ the open and closed neighborhood of v in G, respectively, that is, $N_G(v) = \{u \in V \mid uv \in E\}$ and $N_G[v] = N_G(v) \cup \{v\}$. We sometimes drop the subscript G if it is clear from the context. The *degree* $\deg_G(v)$ of v is the size of $N_G(v)$, that is, $\deg_G(v) = |N_G(v)|$. The *maximum degree* of G is defined as $\max_{v \in V} \deg_G(v)$. A vertex subset $I \subseteq V$ is called an *independent set* if $uv \notin E$ for any two vertices $u, v \in I$. A *dominating set* of G is a subset $D \subseteq V$ such that every $v \in V \setminus D$ is adjacent to some vertex in D. A graph is *bipartite* if there exists a partition (I_1, I_2) of V such that both I_1 and I_2 are independent sets. A graph is *planar* if it can be embedded in the plane without crossing edges. The graph G is *acyclic*, or equivalently a *forest*, if G contains no cycles. A *spanning forest* F of G is an acyclic subgraph with $V(F) = V(G)$. A *spanning tree* T is a connected spanning forest of G. A vertex subset $C \subseteq V(G)$ is called a *vertex cover* if every edge $uv \in E(G)$ satisfies $u \in C$ or $v \in C$. We also say that the graph G is *covered by* C.

Given a connected graph $G = (V, E)$ and a positive integer k, the Minimum Cover Spanning Tree (MCST) problem determines whether G admits a spanning tree T with a vertex cover of size at most k.

Here, we introduce a useful notion that will be used throughout the paper. For a graph $G = (V, E)$ and a subset $S \subseteq V$, the *boundary subgraph* $H = (V, E_H)$ for S is defined as a subgraph of G such that E_H is the set of edges incident to some vertex in S. Note that S is a vertex cover of every subgraph of H.

Observation 1. Let S be a vertex subset of a graph G. There exists a spanning tree of G covered by S if and only if the boundary subgraph for S is connected.

3   Relationship to the Domination Number

The *domination number* of G, denoted by $\gamma(G)$, is the size of a minimum dominating set in G. Denote by $\tau_{ST}(G)$ the minimum size of a vertex cover among spanning trees in G. In this section, we show the following theorems.

Theorem 1. *For any graph G with diameter at most 2, $\gamma(G) = \tau_{ST}(G)$.*

Proof. Since $\gamma(G) \leq \tau_{ST}(G)$ holds for general graphs [1], the task here is to show that $\tau_{ST}(G) \leq \gamma(G)$. Let S be a minimum dominating set of $G = (V, E)$, where $|S| = \gamma(G)$. Let H be the boundary subgraph for S. From Observation 1, it suffices to prove that H is connected, which leads to $\tau_{ST}(G) \leq |S| = \gamma(G)$.

For the sake of contradiction, assume that H has two connected components H_1 and H_2. Let $S_1 = S \cap V(H_1)$ and $S_2 = S \cap V(H_2)$. Note that $S_1 \neq \emptyset$ and $S_2 \neq \emptyset$ from the construction of H. Consider a vertex $v_1 \in S_1$ and a vertex $v_2 \in S_2$. Since G has diameter at most 2, the vertices v_1 and v_2 are adjacent or there exists a vertex u in G such that $\langle v_1, u, v_2 \rangle$ forms a path of G. In both cases, from the definition of H, the vertices v_1 and v_2 are contained in the same connected component in H, a contradiction.   □

For a graph H, a graph G is said to be *H-free* if G contains no induced subgraph isomorphic to H.

Theorem 2. *For any connected P_5-free graph G, $\gamma(G) = \tau_{ST}(G)$.*

Proof. Let H be the boundary subgraph for a minimum dominating set S of G, where $|S| = \gamma(G)$. Suppose that H has $c \geq 2$ connected components. Let $H_1, \ldots, H_c$ be connected components of H and denote $S_i = S \cap V(H_i)$ and $R_i = V(H_i) \setminus S$ for $i \in [c]$. If $S_i = \emptyset$, then the vertices in R_i are not dominated by the vertices in S, contradicting that S is a dominating set of G. If $R_i = \emptyset$, then it contradicts the minimality of S. Thus, we have $S_i \neq \emptyset$ and $R_i \neq \emptyset$.

Consider H_1. Then, there exists another connected component, say H_2, such that there exist two vertices $r_1 \in R_1$ and $r_2 \in R_2$ with $r_1 r_2 \in E(H)$; otherwise, the graph G would not be connected. Moreover, from the construction of H, there exist two vertices $s_1 \in S_1$ and $s_2 \in S_2$ such that $s_1 r_1 \in E(H)$ and $r_2 s_2 \in E(H)$. In other words, $P = \langle s_1, r_1, r_2, s_2 \rangle$ forms a path of G. Observe that

$s_1 r_2, s_1 s_2, r_1 s_2 \notin E(G)$, since H_1 and H_2 are distinct connected components. Consequently, P is in fact an induced path of G.

Let $S' = S \cup \{r_1, r_2\} \setminus \{s_1, s_2\}$, and let H' be the boundary subgraph for S'. We show the following two claims: (1) S' is a minimum dominating set of G; and (2) H' has fewer connected components than the graph H. Iterating the above arguments, eventually the boundary subgraph H^* for a minimum dominating set S^* of G can be obtained. It follows from Observation 1 that G has a spanning tree T covered by S^*. In conclusion, we have $\tau_{ST}(G) \le |S^*| = \gamma(G)$, and Theorem 2 follows from the discussion in Sect. 1.1.

We give the proof of claim (1). For the sake of contradiction, assume that there exists a vertex $w \notin S'$ that is not adjacent to any vertex in S'. Since S is a dominating set of G, the vertex w is adjacent to s_1 or s_2. Without loss of generality, suppose that w is adjacent to s_1. From the assumption of S', w is adjacent to neither r_1 nor r_2. Furthermore, w is not adjacent to s_2; otherwise, s_1 and s_2 belong to the same connected component of H, a contradiction. Therefore, $\langle w, s_1, r_1, r_2, s_2 \rangle$ forms an induced path of G. However, this contradicts that G is P_5-free. This completes the proof of claim (1).

We now prove the claim (2). First, we show that any distinct vertices in $V(H_1) \cup V(H_2)$ are in the same connected component of H'. To this end, it suffices to claim that for each vertex $v \in V(H_1)$ (resp. $v \in V(H_2)$), there is a (v, r_1)-path (resp. a (v, r_2)-path) in H'. This is because $r_1 r_2 \in E(H')$ and the existence of a path is transitive; that is, if there is an (x, y)-path and a (y, z)-path, then an (x, z)-path exists.

Suppose the case $v \in V(H_1)$. If v is adjacent to r_1 in H, the claim is trivial because v is also adjacent to r_1 in H'. Suppose otherwise. Since H_1 is connected, there is a shortest (v, s_1)-path $P' = \langle v, p_1, \ldots, p_\ell, s_1 \rangle$ of H_1, where ℓ is a positive integer and $p_i \in V(H_1)$ for $i \in [\ell]$. Note that $p_i \ne s_1$ for $i \in [\ell]$. It follows from $(S \setminus S') \cap V(H_1) = \{s_1\}$ that we have $v p_1 \in E(H')$ and $p_i p_{i+1} \in E(H')$ for $i \in [\ell - 1]$. However, it is possible that $p_\ell s_1 \notin E(H')$. To obtain a (v, r_1)-path in H', let us turn our attention to the fact that $p_\ell s_2 \notin E(G)$ from the definition of H_1 and H_2. This implies that the vertex p_ℓ is adjacent to r_1 or r_2 in G (and more specifically in H' due to S'); otherwise, $\langle p_\ell, s_1, r_1, r_2, s_2 \rangle$ forms an induced P_5 of G, since $P = \langle s_1, r_1, r_2, s_2 \rangle$ is an induced path of G. Thus, $\langle v, p_1, \ldots, p_\ell, r_1 \rangle$ or $\langle v, p_1, \ldots, p_\ell, r_2, r_1 \rangle$ forms a (v, r_1)-path of H'. The same can be applied for $V(H_2)$ and r_2, and we can see that there is a (v, r_2)-path for every $v \in V(H_2)$.

Furthermore, it should be noted that every vertex $v \in V(H_i) \cap S$ also belongs to $V(H_i) \cap S'$ for $i \in [c] \setminus \{1, 2\}$. This implies that H_i is also a subgraph of H'. Thus, any two vertices in $V(H_i)$ for $i \in [c] \setminus \{1, 2\}$ are contained in the same connected component of H'. (It is possible that every vertex in $V(H_i)$ is in the same connected components as the vertices $V(H_1) \cup V(H_2)$.) Therefore, H' has at most $c - 1$ connected components, as claimed. $\qquad\square$

Remark 1. Denote by C_6 the cycle of length 6. Observe that $\gamma(C_6) = 2$ and $\tau_{ST}(C_6) = 3$, and hence $\gamma(C_6) \ne \tau_{ST}(C_6)$. Furthermore, C_6 has diameter 3 and is P_6-free. In that sense, Theorems 1 and 2 are tight.

Algorithmic Consequences. The DOMINATING SET problem asks for a minimum dominating set D of a graph G. It is known that DOMINATING SET on split graphs with n vertices cannot be approximated to within a factor of $(1 - \varepsilon) \log n$ in polynomial time for any constant $0 < \varepsilon < 1$ unless $\mathrm{NP} \subseteq \mathrm{DTIME}(n^{O(\log \log n)})$ [4]. Moreover, when parameterized by $\gamma(G)$, DOMINATING SET remains W[2]-hard even for split graphs of diameter 2 [18], which implies that FPT algorithms are unlikely to exist for these graphs. Note that split graphs are P_5-free. Consequently, we obtain the following two corollaries from Theorems 1 and 2.

Corollary 1. *MCST on split graphs with n vertices cannot be approximated to within a factor of $(1 - \varepsilon) \log n$ in polynomial time for any constant $0 < \varepsilon < 1$ unless* $\mathrm{NP} \subseteq \mathrm{DTIME}(n^{O(\log \log n)})$.

Corollary 2. *MCST remains W[2]-hard even for split graphs G of diameter 2 when parameterized by* $\tau_{ST}(G)$.

On the other hand, DOMINATING SET is solvable in polynomial time for cographs, P_4-sparse graphs, chain graphs [5],[1] co-bipartite graphs [17] and co-interval graphs [15]. These graph classes are subclasses of P_5-free graphs. The following corollary is immediately obtained from Theorem 2.

Corollary 3. *MCST is solvable in polynomial time for cographs, P_4-sparse graphs, chain graphs, co-bipartite graphs, and co-interval graphs.*

4   NP-Completeness

In this section, we prove that MCST is NP-complete for several graph classes. We begin by describing the source problem for our reduction.

4.1   SIMPLE PLANAR MONOTONE (SPM) 3-BOUNDED 3-SAT

Let φ be a conjunctive normal form (CNF) formula over Boolean variables $\mathcal{X} = \{x_1, \dots, x_n\}$, given by a conjunction of clauses $\mathcal{C} = \{C_1, \dots, C_m\}$, that is, $\varphi = C_1 \wedge \dots \wedge C_m$. A CNF formula φ is *satisfiable* if there exists a truth assignment to $\mathcal{X}$ that makes φ evaluate to true. A 3-CNF formula is a CNF formula in which each clause $C_j \in \mathcal{C}$ is the disjunction of at most three literals. Given a 3-CNF formula φ, the 3-SAT problem decides whether φ is satisfiable. A clause containing only positive literals is called a *positive clause*, and one containing only negative literals is called a *negative clause*. For a CNF formula φ, the *incidence graph* $\mathcal{G}(\varphi) = (\mathcal{V}, \mathcal{E})$ is a bipartite graph with $\mathcal{V} = \mathcal{X} \cup \mathcal{C}$ and $\mathcal{E} = \{(x_i, C_j) \mid x_i \in C_j \text{ or } \overline{x_i} \in C_j\}$. A *rectilinear representation* of $\mathcal{G}(\varphi)$ is a planar drawing in which each variable and clause is represented by a rectangle, all variable vertices lie along a horizontal line, and each edge connecting a variable

---

[1] This follows from the fact that cographs, P_4-sparse graphs, and chain graphs have bounded clique-width [5,10].

to a clause is drawn as a vertical line segment with no edge crossings. (See Fig. 3(a).) SIMPLE PLANAR MONOTONE (SPM) 3-BOUNDED 3-SAT is a variant of 3-SAT with the following additional restrictions:

- Each clause in $\mathcal{C}$ contains at most one occurrence of each variable (simple);
- The incidence graph $\mathcal{G}(\varphi)$ of φ has a rectilinear representation (planar);
- Each clause is either a positive or a negative clause (monotone); and
- Each variable occurs in at most three clauses (3-bounded).

Given a simple planar monotone 3-bounded 3-CNF formula φ, the SIMPLE PLANAR MONOTONE 3-BOUNDED 3-SAT problem determines whether φ is satisfiable. This problem is known to be NP-complete [6].

When φ is planar and monotone, it may be supposed that a rectilinear representation of $\mathcal{G}(\varphi)$ is drawn with all positive clauses placed above the variables and all negative clauses placed below them [2]. (See Fig. 3(a) again.) Note that any variable occurring only in positive clauses can be assigned to True. Similarly, any variable occurring only in negative clauses can be assigned to False. Thus, we may assume that each variable appears in exactly one or two positive clauses and in exactly one or two negative clauses.

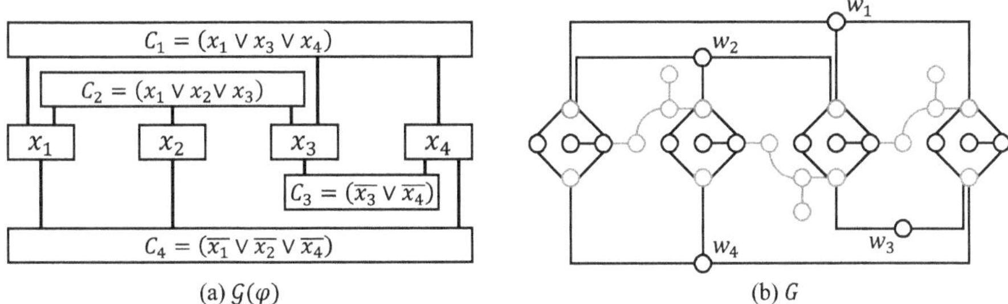

(a) $\mathcal{G}(\varphi)$ (b) G

Fig. 3. (a) An example of illustrating a rectilinear representation of a formula $\varphi = (x_1 \vee x_3 \vee x_4) \wedge (x_1 \vee x_2 \vee x_3) \vee (\overline{x_3} \vee \overline{x_4}) \wedge (\overline{x_1} \vee \overline{x_2} \vee \overline{x_4})$ and (b) the graph G constructed from φ.

4.2 Planar Bipartite Graphs with Maximum Degree 4

We now turn to prove the following theorem.

Theorem 3. *MCST is NP-complete even for planar bipartite graphs with maximum degree at most 4.*

Clearly, MCST belongs to the class NP. To prove Theorem 3, we provide a polynomial-time reduction from SPM 3-Bounded 3-SAT to MCST. Let φ be an instance of SPM 3-Bounded 3-SAT and let $\mathcal{G}(\varphi)$ be its rectilinear representation. We first construct an instance (G, k) of MCST from $\mathcal{G}(\varphi)$ with the following gadgets.

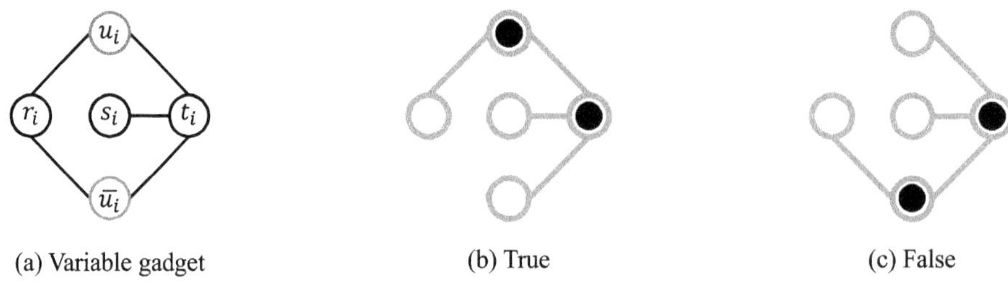

(a) Variable gadget   (b) True   (c) False

Fig. 4. (a) The variable gadget U_i. If the variable x_i is assigned True (resp. False), then the corresponding tree is depicted in green and its vertex cover is indicated by black circles in (b) (resp. (c)). (Color figure online)

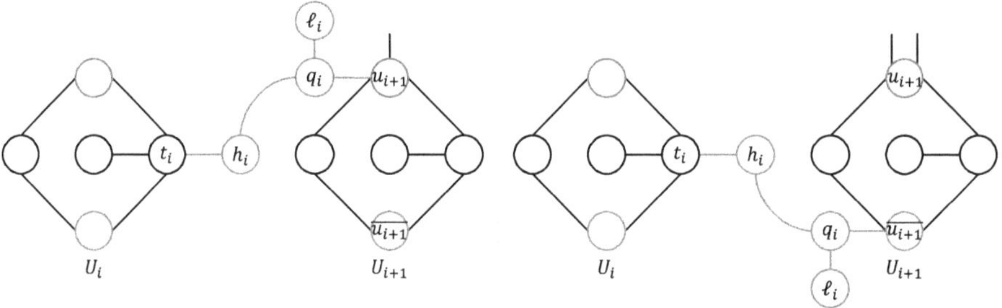

Fig. 5. A connector gadget between variable gadgets. There are two ways to connect the connector gadget depending on the degree of u_{i+1}.

Variable Gadget. The variable gadget U_i for a Boolean variable $x_i \in \mathcal{X}$ consists of the cycle $\langle u_i, r_i, \overline{u_i}, t_i \rangle$ and the vertex s_i adjacent to t_i, as illustrated in Fig. 4(a). Assigning the value True (resp. False) to the variable x_i corresponds to including the vertex u_i (resp. $\overline{u_i}$) in a vertex cover of a spanning tree of U_i.

Connector Gadget. The connector gadget is used to connect spanning trees in U_i and U_{i+1}. The connector gadget H_i with $i \in [n-1]$ consists of the path $\langle h_i, q_i, \ell_i \rangle$ together with the edge $h_i t_i$ and the edge either $q_i u_{i+1}$ or $q_i \overline{u_{i+1}}$, as shown in Fig. 5. The choice of which edge to include depends on the occurrences of the variable x_{i+1}. Recall that each variable appears in exactly one or two positive clauses and in exactly one or two negative clauses. If x_{i+1} occurs exactly once in a positive clause, then we connect q_i to u_{i+1}. Conversely, if x_{i+1} occurs twice in positive clauses, then it occurs exactly once in a negative clause by the 3-bounded restriction. In this case, we connect q_i to $\overline{u_{i+1}}$.

Overall Construction. Using the gadgets introduced above, we construct the graph G for the instance of MCST from $\mathcal{G}(\varphi)$. We first replace each vertex of $\mathcal{G}(\varphi)$ corresponding to $x_i \in \mathcal{X}$ with the variable gadget U_i. Furthermore, for each $i \in [n-1]$, connect variable gadgets U_i and U_{i+1} using the connector gadget H_i.

Then, we introduce a vertex set $W = \{w_j \mid j \in [m]\}$ corresponding to clauses $\mathcal{C} = \{C_j \mid j \in [m]\}$, and add the edges $\{u_i w_j \mid x_i \in C_j\} \cup \{\overline{u_i} w_k \mid \overline{x_i} \in C_k\}$ for each $i \in [n]$. Figure 3 shows the resulting graph G constructed from $\mathcal{G}(\varphi)$ by the above procedure. Finally, we set $k = 3n - 1$, completing the construction of the instance (G, k).

It is clear that the instance (G, k) can be obtained in polynomial time. The following two lemmas complete the proof of Theorem 3.

Lemma 1 (∗). *The constructed graph G is planar, bipartite, and of maximum degree 4.*

Lemma 2 (∗). *The Boolean formula φ is satisfiable if and only if the instance (G, k) of MCST has a solution.*

4.3   Unit Disk Graphs

A graph G is a *unit disk graph* if each vertex of G is associated with a (closed) disk with diameter 1 on the plane, and two vertices $u, v \in V(G)$ are adjacent if and only if the two disks corresponding to u and v have non-empty intersection. Such a set of unit disks on the plane is called the *geometric representation* of G.

Theorem 4. *MCST is NP-complete for unit disk graphs.*

To prove Theorem 4, we provide a polynomial-time reduction from MCST on planar graphs with maximum degree 4, which was shown to be NP-complete in Theorem 3, to MCST on unit disk graphs. Let (G, k) be an instance of MCST on planar graphs with maximum degree 4. We construct an instance (G', k') of MCST on unit disk graphs from (G, k).

First, we embed G in the plane using the following lemma.

Lemma 3 ([24]). *A planar graph G with maximum degree 4 can be embedded in the plane using $O(|V(G)|)$ area in such a way that its vertices are at integer coordinates and its edges drawn so that they are made up of line segments of the form $x = i$ or $y = j$, for integers i and j.*

Figure 6(a) illustrates an example of Lemma 3. This embedding allows us to represent each vertex $v_i \in V(G)$ with $i \in [n]$ by a coordinate (x_i, y_i), where x_i and y_i are integers. For each edge $v_i v_j \in E(G)$, we define the length function as $M(v_i, v_j) = |x_i - x_j| + |y_i - y_j|$, which corresponds to the Manhattan distance. Using this length function, we construct the graph G'. For each edge uv, let $\alpha = M(u, v)$, and introduce a set $\{\langle \ell_i, s_i, r_i \rangle \mid i \in [\alpha]\}$ of α copies of a 3-cycle. Then, add the edges $u\ell_1$, $r_\alpha v$, and $r_i \ell_{i+1}$ for each $i \in [\alpha - 1]$, which replaces the original edge uv with a graph formed by connecting the 3-cycles. Figure 6(b) depicts the resulting graph G' constructed from G by replacing each edge according to the above procedure. Finally, we set $k' = k + \sum_{uv \in E(G)} M(u, v)$, completing the construction of the instance (G', k').

It is clear that the instance (G', k') can be obtained in polynomial time. The following two lemmas complete the proof of Theorem 4.

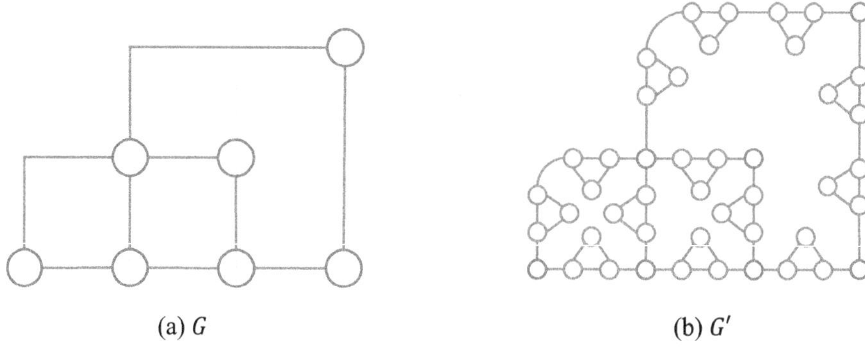

(a) G (b) G'

Fig. 6. (a) A planar graph G with maximum degree 4 embedded according to Lemma 3, and (b) the unit disk graph G' constructed from G.

Lemma 4 (∗)**.** *The constructed graph G' is a unit disk graph.*

Lemma 5 (∗)**.** *(G, k) has a solution to MCST if and only if (G', k') has a solution to MCST.*

5   Algorithms

5.1   Clique-Width

In this section, we present an FPT algorithm parameterized by clique-width. For a positive integer w, a *w-labeled graph* is a graph whose vertices are labeled with integers from $[w] = \{1, 2, \ldots, w\}$. We consider the following four operations on w-labeled graphs.

> **Introduce** $i(v)$: Create a graph consisting of a single vertex v labeled with $i \in [w]$.
>
> **Union** $\oplus$: Take the disjoint union of two w-labeled graphs.
>
> **Relabel** $\rho_{i \to j}$: Replace every label i with j.
>
> **Join** $\eta_{i,j}$: Add all possible edges between vertices labeled i and j.

The *clique-width* of a graph G is the smallest integer w such that a w-labeled graph isomorphic to G can be obtained from a sequence of these operations. The sequence can be represented by a rooted tree, called a *w-expression tree* $\mathcal{T}$ of G, whose nodes are classified according to the above operations, namely, *introduce nodes, union nodes, relabel nodes,* and *join nodes,* respectively. It is known that, given a graph G of clique-width at most w, there exists an algorithm that computes a $(2^{w+1} - 1)$-expression tree for G in time $O(|V(G)|^3)$ [11,19,20].

First, we define a problem equivalent to MCST and prove that a solution to the problem can be expressed in monadic second-order logic, called MSO_1. To show this, we demonstrate the following equivalence.

Lemma 6 (∗)**.** *Let $G = (V, E)$ be a connected graph with at least two vertices, and let S be a vertex subset of G. Then the following claims (a) and (b) are equivalent:*

(a) there exists a spanning tree T of G covered by S; and

(b) for any non-empty vertex subset $C \subset V$, there exist two vertices u and v that satisfy the following four conditions: (1) $uv \in E$; (2) $u \in C$; (3) $v \in V \setminus C$; and (4) $u \in S$ or $v \in S$.

A vertex set S that satisfies the property (b) in Lemma 6 can be expressed in MSO_1. The result of Courcelle et al. implies that the minimization problem of S, as well as MCST, admits an FPT algorithm parameterized by clique-width [5]. However, it is known that its running time can take the form of an exponential tower of $|\varphi|$, where $|\varphi|$ is the length of an MSO_1 formula φ [8]. Therefore, we will design a faster FPT algorithm.

Our main theorem of this section is as follows.

Theorem 5. *Given a w-expression tree of an n-vertex graph, MCST can be solved in time $2^{2^{O(w)}} n$.*

To show Theorem 5, we present an algorithm using dynamic programming over a given w-expression tree $\mathcal{T}$. Let t be a node in $\mathcal{T}$. We denote by G_t the labeled graph constructed by operations in the subtree rooted at t. We compute candidate solutions on G_t for each node t of $\mathcal{T}$ in a bottom-up manner. Recall that, by Observation 1, there exists a spanning tree of G covered by a vertex set S if and only if the boundary subgraph for S is connected. Thus, we will store candidate solutions as boundary subgraphs instead of spanning trees.

Consider the boundary subgraph H^* for a vertex subset S^* of G. The restriction of S^* to the subgraph G_t yields the boundary subgraph H for $S = V(G_t) \cap S^*$. Hence, we guess S at each node t. Moreover, since H is required to be connected at the root of $\mathcal{T}$, we also need to record the connected components of H. To obtain an FPT algorithm for clique-width, these pieces of information are managed using labels from $[w]$ instead of vertex sets. Let $[w]$ be the set of labels, and let $\mathcal{P} = 2^{[w]}$ denote its power set. For the boundary subgraph H for a vertex set S of G_t and a pair $(C, X) \in \mathcal{P} \times \mathcal{P}$, we say that a component H' of H *matches* (C, X) if

- the label set from $V(H')$ are exactly C; and
- the label set from $V(H') \cap S$ are exactly X.

A function $f : \mathcal{P} \times \mathcal{P} \to \{0, 1, 2\}$ is *valid* if there exists the boundary subgraph H for a vertex set S of G_t such that every pair $(C, X) \in \mathcal{P} \times \mathcal{P}$ satisfies $f(C, X) = \min\{2, h'\}$, where h' is the number of components H' of H. Otherwise, f is called *invalid*. Note that we do not distinguish cases where there are at least two such components. We also note that whenever a component H' matches (C, X), we have $X \subseteq C$. Consequently, it is not necessary to consider a pair (C, X) with $X \setminus C \neq \emptyset$.

For each node t of $\mathcal{T}$, we define dp_t as a DP table that, for each valid function $f : \mathcal{P} \times \mathcal{P} \to \{0, 1, 2\}$, stores the minimum size of vertex sets S such that the number of components in the boundary subgraph for S corresponds to f. For an invalid function f, we set $\mathrm{dp}_t(f) = +\infty$. The number of possible pairs $(C, X) \in \mathcal{P} \times \mathcal{P}$ is bounded by 4^w. The function f independently assigns one

---

Algorithm 1. Solving MCST on interval graphs

1: **Input:** An interval ordering $I = (v_1, v_2, \ldots, v_n)$ of a connected interval graph G
2: **Output:** A minimum vertex cover S among spanning trees of G
3: $S \leftarrow \emptyset$, $V_T \leftarrow \{v_1\}$
4: **while** $V_T \neq V(G)$ **do**
5:     $t_1 \leftarrow \min\{i \mid v_i \notin V_T\}$
6:     $t_2 \leftarrow \max\{i \mid v_i \in V_T\}$
7:     $t \leftarrow \min\{t_1, t_2\}$
8:     $s \leftarrow \max\{i \mid v_i \in N[v_t]\}$
9:     $S \leftarrow S \cup \{v_s\}$
10:     $V_T \leftarrow V_T \cup N[v_s]$
11: **end while**
12: **return** S

---

of three values $\{0, 1, 2\}$ to each pair (C, X), and hence the number of possible functions is 3^{4^w}. Therefore, the size of the DP table is $2^{2^{O(w)}}$ for each node t.

At the root node r of $\mathcal{T}$, the desired boundary subgraph H must be connected. Therefore, we look up $\mathrm{dp}_r(f)$ for every function $f : \mathcal{P} \times \mathcal{P} \to \{0, 1, 2\}$ such that $f(C, X) = 1$ for exactly one pair $(C, X) \in \mathcal{P} \times \mathcal{P}$ and $f(C', X') = 0$ for every other pair (C', X'). Let F be the set of functions f satisfying the above conditions. An instance (G, k) of MCST is a yes-instance if and only if $\min\{\mathrm{dp}_r(f) \mid f \in F\} \leq k$.

5.2 Interval Graphs

This section shows the following theorem.

Theorem 6. *MCST can be solved in linear time for interval graphs.*

A sequence $(v_1, v_2, \ldots, v_n)$ of the vertices in an n-vertex graph G is an *interval ordering* I if it satisfies the following condition: for any three integers $a, b, c \in [n]$ with $a < b < c$, if $v_a v_c \in E(G)$, then $v_b v_c \in E(G)$. A graph is an *interval graph* if it admits an interval ordering. An interval ordering of a given interval graph can be computed in linear time [13]. We give Algorithm 1 to solve MCST for interval graphs as the proof of Theorem 6. Observe that Algorithm 1 can be implemented to run in linear time. The following lemma completes the proof of Theorem 6.

Lemma 7 (∗). *Algorithm 1 outputs a minimum vertex cover among spanning trees of a connected interval graph.*

Acknowledgments. We thank anonymous reviewers for their valuable comments and suggestions which greatly helped to improve the presentation of this paper.

References

1. Angel, E., Bampis, E., Chau, V., Kononov, A.V.: Min-power covering problems. In: Elbassioni, K.M., Makino, K., (eds.) Algorithms and Computation - 26th International Symposium, ISAAC 2015, volume 9472 of Lecture Notes in Computer Science, pp. 367–377. Springer (2015). https://doi.org/10.1007/978-3-662-48971-0_32
2. Mark de Berg and Amirali Khosravi. Optimal binary space partitions in the plane. In: Thai, M.T., Sahni, S., (eds.) Computing and Combinatorics, 16th Annual International Conference, COCOON 2010, Nha Trang, Vietnam, July 19–21, 2010. Proceedings, volume 6196 of Lecture Notes in Computer Science, pp. 216–225. Springer (2010). https://doi.org/10.1007/978-3-642-14031-0_25
3. Caro, Y., West, D.B., Yuster, R.: Connected domination and spanning trees with many leaves. SIAM J. Discret. Math. **13**(2), 202–211 (2000). https://doi.org/10.1137/S0895480199353780
4. Chlebík, M., Chlebíková, J.: Approximation hardness of dominating set problems in bounded degree graphs. Inf. Comput. **206**(11), 1264–1275 (2008). https://doi.org/10.1016/J.IC.2008.07.003
5. Courcelle, B., Makowsky, J.A., Rotics, U.: Linear time solvable optimization problems on graphs of bounded clique-width. Theory Comput. Syst. **33**(2), 125–150 (2000). https://doi.org/10.1007/S002249910009
6. Darmann, A., Döcker, J., Dorn, B.: The monotone satisfiability problem with bounded variable appearances. Int. J. Found. Comput. Sci. **29**(06), 979–993 (2018). https://doi.org/10.1142/S0129054118500168
7. Fellows, M.R., et al.: The complexity ecology of parameters: an illustration using bounded max leaf number. Theory Comput. Syst. **45**(4), 822–848 (2009). https://doi.org/10.1007/S00224-009-9167-9
8. Frick, M., Grohe, M.: The complexity of first-order and monadic second-order logic revisited. Ann. Pure Appl. Log. **130**(1–3), 3–31 (2004). https://doi.org/10.1016/J.APAL.2004.01.007
9. Fukunaga, T., Maehara, T.: Computing a tree having a small vertex cover. Theor. Comput. Sci. **791**, 48–61 (2019). https://doi.org/10.1016/J.TCS.2019.05.018
10. Golumbic, M.C., Rotics, U.: On the clique-width of some perfect graph classes. Int. J. Found. Comput. Sci. **11**(3), 423–443 (2000). https://doi.org/10.1142/S0129054100000260
11. Hliněný, P., Oum, S.: Finding branch-decompositions and rank-decompositions. In: Arge, L., Hoffmann, M., Welzl, E. (eds.) ESA 2007. LNCS, vol. 4698, pp. 163–174. Springer, Heidelberg (2007). https://doi.org/10.1007/978-3-540-75520-3_16
12. Ho, J., Lee, D.T., Chang, C., Wong, C.K.: Minimum diameter spanning trees and related problems. SIAM J. Comput. **20**(5), 987–997 (1991). https://doi.org/10.1137/0220060
13. Hsu, W.L., Ma, T.H.: Fast and simple algorithms for recognizing chordal comparability graphs and interval graphs. SIAM J. Comput. **28**(3), 1004–1020 (1999). https://doi.org/10.1137/S0097539792224814
14. Kaur, C., Misra, N.: On the parameterized complexity of spanning trees with small vertex covers. In Changat, M., Das, S., (eds.) Algorithms and Discrete Applied Mathematics - 6th International Conference, CALDAM 2020, volume 12016 of Lecture Notes in Computer Science, pp. 427–438. Springer (2020). https://doi.org/10.1007/978-3-030-39219-2_34

15. Keil, J.M., Belleville, P.: Dominating the complements of bounded tolerance graphs and the complements of trapezoid graphs. Discrete Applied Math. **140**(1), 73–89 (2004). https://doi.org/10.1016/j.dam.2003.04.004
16. Kokai, T., Suzuki, A., Suzuki, T., Tamura, Y., Zhou, X.: Spanning trees with a small vertex cover: the complexity on specific graph classes (2025). https://arxiv.org/abs/2511.22912
17. Kratsch, D., Stewart, L.: Domination on cocomparability graphs. SIAM J. Discret. Math. **6**(3), 400–417 (1993). https://doi.org/10.1137/0406032
18. Lokshtanov, D., Misra, N., Philip, G., Ramanujan, M.S., Saurabh, S.: Hardness of r-dominating set on graphs of diameter $(r + 1)$. In: Gutin, G.Z., Szeider, S., (eds.) Parameterized and Exact Computation - 8th International Symposium, IPEC 2013, volume 8246 of Lecture Notes in Computer Science, pp. 255–267. Springer (2013). https://doi.org/10.1007/978-3-319-03898-8_22
19. Oum, S.: Approximating rank-width and clique-width quickly. In: Kratsch, D., (ed.) Graph-Theoretic Concepts in Computer Science, pp. 49–58, Berlin, Heidelberg, 2005. Springer Berlin Heidelberg
20. Oum, S., Seymour, P.: Approximating clique-width and branch-width. J. Combinatorial Theor. Ser. B **96**(4), 514–528 (2006). https://doi.org/10.1016/j.jctb.2005.10.006
21. Ravi, R.: Rapid rumor ramification: approximating the minimum broadcast time (extended abstract). In: 35th Annual Symposium on Foundations of Computer Science, pp. 202–213. IEEE Computer Society, 1994. https://doi.org/10.1109/SFCS.1994.365693
22. Rosamond, F.A.: Max leaf spanning tree. In: Kao, M., (ed.) Encyclopedia of Algorithms, pp. 1211–1215. Springer (2016). https://doi.org/10.1007/978-1-4939-2864-4_228
23. Salamon, G., Wiener, G.: On finding spanning trees with few leaves. Inf. Process. Lett. **105**(5), 164–169 (2008). https://doi.org/10.1016/J.IPL.2007.08.030
24. Valiant, L.G.: Universality considerations in VLSI circuits. IEEE Trans. Comput. **30**(2), 135–140 (1981)

Efficient Trace Frequency Queries in Sparse Graphs

Christine Awofeso[1], Pål Grønås Drange[2], Patrick Greaves[1],
Oded Lachish[1], and Felix Reidl[1(✉)]

[1] Birkbeck, University of London, London, UK
`{cawofe01,pgreav01}@student.bbk.ac.uk`, `{o.lachish,f.reidl}@bbk.ac.uk`
[2] University of Bergen, Bergen, Norway
`pal.drange@uib.no`

Abstract. Understanding how a vertex relates to a set of vertices is a fundamental task in graph analysis. Given a graph G and a vertex set $X \subseteq V(G)$, consider the collection of subsets of the form $N(u) \cap X$ where u ranges over all vertices outside X. These intersections, which we call the *traces* of X, capture all ways vertices in G connect to X, and in this paper we consider the problem of listing these traces efficiently, and the related problem of recording the multiplicity (*frequency*) of each trace.For a given query set X, both problems have obvious algorithms with running time $O(|N(X)| \cdot |X|)$ and conditional lower bounds suggest that, on general graphs, one cannot expect better. However, in certain sparse graph classes, more efficient algorithms are possible: Drange *et al.*(IPEC 2023) used a data structure that answers trace queries in d-degenerate graphs with linear initialisation time and query time that only depends on the query set X and d. However, the query time is exponential in $|X|$, which makes this approach impractical.

By using a stronger parameter than degeneracy, namely the strong 2-colouring number s_2, we construct a data structure in $O(d \cdot \|G\|)$ time, which answers subsequent trace frequency queries in time $O\big((d^2 + s_2^{d+2})|X|\big)$, where $\|G\|$ is the number of edges of G, s_2 is the strong 2-colouring number and d the degeneracy of a suitable ordering of G. We demonstrate that this data structure is indeed practical and that it beats the simple, obvious alternative in almost all tested settings, using a collection of 217 real-world networks with up to 1.1M edges. As part of this effort, we demonstrate that computing an ordering with a small strong 2-colouring number is feasible with a simple heuristic.

1 Introduction

Imagine a database with products and customers, modelled as a bipartite graph where an edge indicates that a customer bought a product. We would like to understand the buying behaviour of a target segment X of customers better.A natural operation here is to group the members of X according to their purchases, specifically, each subset $X' \subseteq X$ is of interest if there exists a product y such

© The Author(s), under exclusive license to Springer Nature Switzerland AG 2026
J. Kozik and A. Wolff (Eds.): SOFSEM 2026, LNCS 16448, pp. 593–606, 2026.
https://doi.org/10.1007/978-3-032-17801-5_43

that y was bought exactly by X' among customers in X. This allows us to investigate what differentiates X' from $X \setminus X'$ and, given *how many* products were bought exactly by X', determine the relevance of this subgroup. In general, we call the subsets $X' \subseteq X$ for which there exists a vertex $y \notin X$ with $N(y) \cap X = X'$ the *traces* of X.

Example

In this example, the *traces* of X are the sets

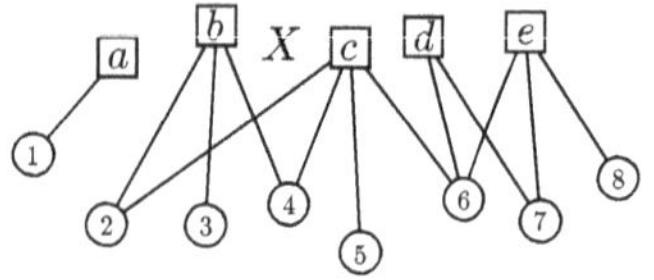

- $\{b, c\}$ (neighbours of 2 and 4),
- $\{b\}$ (neighbour of 3),
- $\{c\}$ (neighbour of 5),
- $\{c, d\}$ (neighbours of 6),
- $\{d\}$ (neighbour of 7), and
- $\emptyset$ (vertices 1 and 8 have no neighbours in X).

The *frequencies* of each trace are the number of vertices that generate it, for example, the frequency of $\{b, c\}$ is $|\{2, 4\}| = 2$.

The problem we investigate here is as follows: given a graph G (not necessarily bipartite) and a query set $X \subseteq V(G)$, list the traces of X and count their frequencies. Additionally, we will look at the related problems of only listing the traces, as well as the simpler problem of counting the number of neighbours of X, i.e. the size of $|N(X)|$. Note that any algorithm or data structure that counts trace frequencies implicitly solves this latter problem as well.

A natural way to count trace frequencies is to compute for each $y \in N(X)$ the trace $N(y) \cap X$ and store it in a suitable data structure. This results in a running time of at least $\Omega(|X| \cdot |N(X)|)$, where $|N(X)|$ is potentially very large. Since the output of this query can have size $2^{|X|}$, it is clear that we need to restrict ourselves to graphs without such adverse structures to arrive at a reasonable running time. Maybe more surprisingly, we show via a simple reduction that already the task of counting the number of neighbours of X cannot be solved in time $O(2^{o(|X|)})$, unless the Hitting Set Conjecture fails (see Sect. 5).

The Sparse Setting. We therefore focus on sparse graph classes that are relevant in practice. One important property of many such classes is precisely that the number of traces of any given set X is bounded by a function of $|X|$. In d-degenerate graphs[1], it is well-known that the number of traces is bounded by $O(|X|^d)$ (we include a proof in Sect. 3 for completeness). Since degeneracy of real-world networks is typically small—the average degeneracy in our dataset is around 21—this class is a good starting point. In a previous paper [6] with a very different problem setting, a subset of the authors together with Muzi designed a data structure that, after an initialisation time of $O(d2^d|G|)$ in a d-degenerate graph G, answers trace frequency queries in time $O(|X|2^{|X|}+d|X|^2)$. The factor $2^{|X|}$ in this specific data structure is unavoidable, as it uses inclusion-exclusion over $|X|$ to answer the query. There were two key problems with this approach we look to overcome. First, while in the context of [6] the query sets

---

[1] Recall that a graph is d-degenerate if every subgraph contains a vertex of degree at most d.

were known to be small, we here are interested in scenarios where X is potentially very large. Second, the data structure in [6] uses $\Omega(2^d|G|)$ space, which is impractical for larger networks with even moderate degeneracy.

Motivated by this, we focus on sparseness parameters that provide tighter bounds on the number of traces. Many graph classes—like planar graphs, graphs of bounded degree, or graphs excluding a (topological) minor—admit a linear bound, i.e. every set $|X|$ has at most $O(|X|)$ traces. A unified 'explanation' of this fact can be found in the theory of *bounded expansion classes* (which include the above classes) as described by Nešetřil and Ossonda de Mendez [10]. Specifically, they note that the *strong 2-colouring number* $\mathrm{scol}_2(G)$ governs the number of traces in a graph G, and that $\mathrm{scol}_2(G)$ is a constant for all graphs from bounded expansion classes [12]. We define scol_2 in Sect. 2 and show how it bounds the number of traces in Sect. 3.

Our Contribution. We design a data structure that, given a suitable ordering of the input graph with degeneracy d and strong 2-colouring number s_2, is initialised with time and space complexity $O(d\|G\|)$ and then answers trace frequency queries in time $O((d^2 + s_2^{d+2})|X|)$. Further, the properties of the strong 2-colouring number allow us to store the necessary information using bit vectors of length s_2, greatly reducing the memory consumption in practice.

To demonstrate that the data structure is indeed practical, we conducted several computational experiments on a set of 217 real-world networks from various domains, with up to 1.1M edges. The results are summarized in Sect. 6 and a more detailed breakdown can be found in the full version.

In the first set of experiments, we show that finding suitable orderings with low strong 2-colouring numbers is achievable with a very simple heuristic, even though the problem is NP-hard [4]. We use a related measure that provides upper and lower bounds for the strong 2-colouring number and which can be computed exactly in practice [2] to argue that the heuristic works well on the 217 instances.

In the second set, we compare the performance of our data structure in answering trace, trace frequency, and neighbourhood counting queries against the obvious baseline algorithms for different query sizes. We deemed that a comparison against the data structure by Drange *et al.* [6] was not necessary as the results in that paper already show that the exponential dependency on the query size is prohibitive in practice.

The results show that for trace listing and trace frequency queries, our data structure outperform the naive algorithms in almost all scenarios. For neighbourhood counting, the naive algorithm is superior. A more comprehensive breakdown of the results can be found in the full version of the paper[2]

2   Preliminaries

For an integer k, we use $[k]$ as a short-hand for the set $\{0, 1, 2, \ldots, k-1\}$. We use blackboard bold letters like $\mathbb{X}$ to denote totally ordered sets, that is, some

---

[2] The full version of the paper can be found here: http://arxiv.org/abs/2511.22289..

underlying set X associated with a total order $<_{\mathbb{X}}$. We further, for $Y \subseteq X$, use the notations $\max_{\mathbb{X}} Y$ to mean the maximum vertex in Y under $<_{\mathbb{X}}$ and the similarly defined $\min_{\mathbb{X}} Y$. If the context allows it, we will sometimes drop this subscript. All graphs in this paper will be undirected and simple. For a graph G we use $V(G)$ and $E(G)$ to refer to its vertex- and edge-set. We use the shorthands $|G| := |V(G)|$ and $\|G\| := |E(G)|$.

The *trace* $\mathrm{tr}_G(X)$ of vertex set X in graph G is the set of all subsets $X' \subseteq X$ for which there exists a vertex $y \in G \setminus X$ such that $N(y) \cap X = X'$. That is,

$$\mathrm{tr}_G(X) := \{X' \subseteq X \mid \exists y \in V(G) \setminus X \text{ with } N(y) \cap X = X'\}.$$

In the following it will be useful to constrain the set $Y \subseteq V(G)$ from which we choose the vertices y. To that end, we use the notation

$$\mathrm{tr}_G(X \mid Y) := \{X' \subseteq X \mid \exists y \in Y \setminus X \text{ with } N(y) \cap X = X'\}.$$

A useful observation is that $\mathrm{tr}_G(X \mid Y)$ and $\mathrm{tr}_G(X \mid Y \cap N(X))$ can only differ by at most one set and that this is exactly the case when $\emptyset \in \mathrm{tr}_G(X \mid Y)$:

Observation 1. *For all $X, Y \subseteq V(G)$ it holds that $\mathrm{tr}_G(X \mid Y)$ contains the same traces as $\mathrm{tr}_G(X \mid Y \cap N(X))$, with the exception of the empty set which might be contained in the former but never appears in the latter. Therefore,*

$$|\mathrm{tr}_G(X \mid Y \cap N(X))| \le |\mathrm{tr}_G(X)| \le 1 + |\mathrm{tr}_G(X \mid Y \cap N(X))|.$$

An important related quantity is the *trace frequency* $\mathrm{tr}_G^{\#}(X)$, which not only records which traces in X appear in G but also how often. We understand $\mathrm{tr}_G^{\#}(X)$ as a multiset of traces which implicitly records the frequency of each trace. $\mathrm{tr}_G^{\#}(X \mid Y)$ similarly counts the frequencies of traces induced in X by vertices from Y.

An *ordered graph* is a pair $\mathbb{G} = (G, <)$ where G is a graph and $<$ a total ordering of $V(G)$. We write $<_{\mathbb{G}}$ to denote the ordering for a given ordered graph and extend this notation to the derived relations $\le_{\mathbb{G}}, >_{\mathbb{G}}, \ge_{\mathbb{G}}$. For simplicity we will call $\mathbb{G}$ an *ordering of* G and we write $\pi(G)$ to denote the set of all possible orderings.

We use the same notations for graphs and ordered graphs, additionally we write $N_{\mathbb{G}}^-(u) := \{v \in N(u) \mid v <_{\mathbb{G}} u\}$ for the *left neighbourhood* of a vertex $u \in \mathbb{G}$. We write $N_{\mathbb{G}}^-[u] := N_{\mathbb{G}}^-(u) \cup \{u\}$ for the closed left neighbourhood which we extend to vertex sets X via $N_{\mathbb{G}}^-[X] := \bigcup_{u \in X} N_{\mathbb{G}}^-[u] \cup X$. We write $\Delta^-(\mathbb{G}) := \max\{|N_{\mathbb{G}}^-(u)| \mid u \in \mathbb{G}\}$ to denote the maximum left-degree of an ordered graph.

A graph G is *d-degenerate* if there exists an ordering $\mathbb{G}$ such that $\Delta^-(\mathbb{G}) \le d$. We say an ordering $\mathbb{G}$ of a graph G is *d*-degenerate if $\Delta^-(\mathbb{G}) \le d$, and just *degenerate* if $\Delta^-(\mathbb{G})$ is minimum. The degeneracy ordering of a graph can be computed in time $O(n + m)$ and $O(dn)$ for *d*-degenerate graphs [9].

The *strongly 2-reachable set* for a vertex u in an ordered graph $\mathbb{G}$ is the set $S_{\mathbb{G}}^2(u) := N^-(N^+(u) \cup u) \cap V_{\le u}$. That is, $S_{\mathbb{G}}^2(u)$ contains all vertices that are smaller than u and are neighbours of u, or can be reached via a path of length 2 where the mid-point is larger than u.

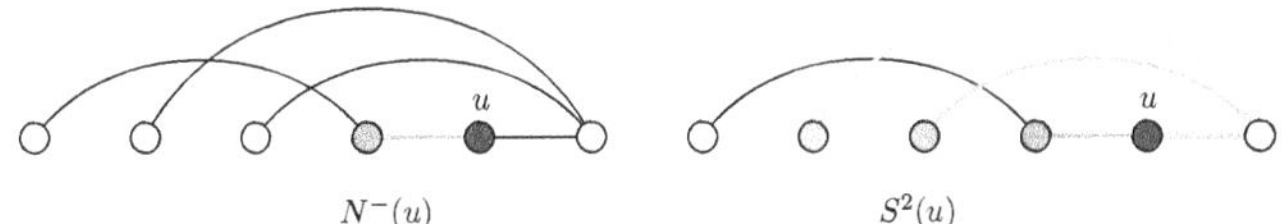

We will often need the set $S^2_{\mathbb{G}}(u) \cup \{u\}$, therefore we write $S^2_{\mathbb{G}}[u]$ for this 'closed' version of the set. We usually drop the subscript $\mathbb{G}$ if the ordering is clear from the context. The *strong 2-colouring number* $\mathrm{scol}_2(\mathbb{G})$ is then defined as the maximum size of S^2, i.e. $\mathrm{scol}_2(\mathbb{G}) := \max_{u \in \mathbb{G}} |S^2_{\mathbb{G}}(u)|$. For an unordered graph G, the strong 2-colouring number is defined as the minimum over all possible ordering of G, i.e. $\mathrm{scol}_2(G) := \min_{\mathbb{G} \in \pi(G)} \mathrm{scol}_2(\mathbb{G})$. The following observation describes the important property of S^2 that lies at the heart of our trace data structure:

Observation 2. *Let* $\mathbb{G}$ *be an ordered graph,* $u \in \mathbb{G}$ *and* $x \in N^-(u)$. *Then* $N^-(u) \cap V_{\leq x} \subseteq S^2[x]$.

Proof. Every vertex $y \in V_{\leq x}$ is no larger than x in $\mathbb{G}$ and hence every $y \in N^-(u) \cap V_{\leq x}$ is no larger than x. Since $xu \in E(\mathbb{G})$, the path xuy exists. Thus $y \in S^2[x]$.

One important special case is when $x = \max_{\mathbb{G}} N^-(u)$ is the largest among the left neighbours of u, in that case we have that $N^-(u) \subseteq S^2[x]$.

3   Bounding the Number of Traces by s_2 and d

The first important observation is that in a degenerate ordering $\mathbb{G}$, the number of traces induced by vertices that are smaller than the query set's maximum cannot be large—simply because the number of such vertices is bounded:

Lemma 1. *Let* $\mathbb{G}$ *be a d-degenerate ordering of a graph* G *and let* $X \subseteq V(G)$. *Let further* $z := \max_{\mathbb{G}} X$ *be the largest vertex in* X *under* $<_{\mathbb{G}}$. *Then* $|\mathrm{tr}_G(X \mid V_{\leq z})| \leq d \cdot |X| + 1$.

Proof. Simply note that $|N(X) \cap V_{\leq z}| \leq \Delta^-(\mathbb{G}) \cdot |X| = d \cdot |X|$. By Observation 1,

$$|\mathrm{tr}_G(X \mid V_{\leq z})| \leq |\mathrm{tr}_G(X \mid V_{\leq z} \cap N(X))| + 1 \leq d \cdot |X| + 1. \qquad \square$$

The interesting bounds therefore concern the traces induced by vertices that are *larger* than the maximum of the query set. The bounds based on the degeneracy and the strong 2-colouring number illustrate already why the latter might be more beneficial for our purposes. Note that the degeneracy-bound is well-known, we include it here for completeness. Bounds using s_2 are known indirectly ([5,8,11] prove bounds for measures related to s_2), but as far as we are aware the below bound is more precise:

Lemma 2. *Let $\mathbb{G}$ be an ordering of a graph G and let $X \subseteq V(G)$. Let d be the degeneracy and s_2 the strong 2-colouring number of $\mathbb{G}$. Then the following bounds hold: 1) $|\operatorname{tr}_G(X)| \leq |X|^d + d \cdot |X| + 1$ and 2) $|\operatorname{tr}_G(X)| \leq |X| s_2^d + d \cdot |X| + 1$.*

Proof. Let $z := \max_{\mathbb{G}} X$ be the largest vertex in the query set X. We prove the bounds by bounding $|\operatorname{tr}(X \mid V_{>z})|$ by $|X|^d$ and $s_2^d |X|$, respectively. The claimed bounds then follow by seeing that $\operatorname{tr}(X) = \operatorname{tr}(X \mid V_{\leq z}) \cup \operatorname{tr}(X \mid V_{>z})$ and applying Lemma 1 to the first set.

For the first bound, simply note that for every vertex $y \in V_{>z}$ it holds that $N(y) \cap X = N^-(y) \cap X$. Since $|N^-(y)| \leq d$, it follows that these vertices can induce at most $\sum_{i=0}^{d} \binom{|X|}{i} \leq |X|^d + 1$ traces and hence $|\operatorname{tr}(X \mid V_{>z})| \leq |X|^d + 1$. Note that this bound counts the empty set which is already counted by the bound from Lemma 1. We can therefore combine the two bounds and subtract one to arrive at the claimed bound.

For the second bound we will set up a recursive inequality over the size of the trace set and the largest vertex in the trace. To that end, let $t_i^u := |\{S \in \operatorname{tr}(G \mid V_{>z}) \mid |S| = i \text{ and } \max_{\mathbb{G}} S = u\}|$ denote the number of traces in $\operatorname{tr}(G \mid V_{>z})$ of size exactly i and with u as their largest vertex. As observed above, the maximum size is d and thus $t_i^u = 0$ for $i > d$. As the induction basis we use $t_1^u = 1$ for all $u \in X$. Note that if vertices v and u with $v \leq_{\mathbb{G}} u$ appear in a trace together, then $v \in S^2(u)$. Therefore we can bound t_i^u by

$$t_i^u \leq \sum_{v \in S^2(u)} t_{i-1}^v \leq \sum_{v \in S^2(u)} \sum_{w \in S^2(v)} t_{i-2}^w \leq \cdots \leq s_2^{i-1}.$$

Therefore, $|\operatorname{tr}_G(X \mid V_{>z})| \leq 1 + \sum_{u \in X} \sum_{i=1}^{d} t_i^u \leq 1 + |X| \sum_{i=1}^{d} s_2^{i-1} \leq 1 + |X| s_2^d$. This bound again counts the empty set and we can subtract one when combining it with the bound from Lemma 1.

Note that it is trivial to construct a d-degenerate graph G where, for some set X, $|\operatorname{tr}_G(X)| \geq |X|^d$. As such, for the task of *outputting* traces a data structure based on d-degeneracy cannot have a query complexity lower than $O(|X|^d)$. Whether it is possible to *count* the number of traces faster than that is an interesting open question, in Sect. 5 we show some conditional lower bounds for data structure that only count the number of neighbours.

4   The Trace Data Structure

In this section we fix an ordering $\mathbb{G}$ of our input graph G and let d be the degeneracy[3] and s_2 the strong 2-colouring number of $\mathbb{G}$. While computing an optimal ordering $\mathbb{G}$ with $s_2 = \operatorname{scol}_2(\mathbb{G})$ is NP-hard [3], as we demonstrate below in Sect. 6, in practice it is easy to find orderings with low scol_2 values.

---

[3] Note that d is the degeneracy of the ordering and might be larger than the graph's actual degeneracy. A theoretical worst-case here is that $d = s_2$. In practice, we find that d is much smaller than s_2.

Input: An ordered graph $\mathbb{G}$
Initialize R as an empty associative array;
for $u \in \mathbb{G}$ **do**
    for $x \in N^-(u)$ **do**
        $K \leftarrow N^-(u) \cap V_{\leq x}$;   // takes time $O(d)$
        $\mathsf{R}[x][K] \leftarrow \mathsf{R}[x][K] + 1$;   // Non-existing keys are treated as zero

return R;

Algorithm 1: Initialisation of the data structure R

In the following, we will assume that $d, s_2 \geq 1$ as $d = 0$ and $s_2 = 0$ are only possible in edgeless graphs which we will exclude from this discussion. We first describe the underlying data structure R in theory and discuss its implementation further below. On a high level, R is a two-level associative array with vertices of G as the first key, vertex subsets as the second key, and an integer as the stored value. We initialise R as described in Algorithm 1.

Observation 3. *After initialisation, for every vertex u the data structure $\mathsf{R}[u]$ contains exactly the sets* $\mathrm{tr}_G(S^2[u] \mid V_{>u})$ *as keys and their multiplicities in* $\mathrm{tr}_G^{\#}(S^2[u] \mid V_{>u})$ *as values,* $\mathsf{R}[u][X]$ *tells us how many vertices in $V_{>u}$ have X as their trace in $S^2[u]$.*

Proof. By Observation 2 for every vertex $y \in V_{>u}$ we have that $N^-(y) \cap V_{\leq u}$ is a subset of $S^2[u]$, thus the keys of $\mathsf{R}[u]$ are exactly $\mathrm{tr}_G(S^2[u] \mid V_{>u})$. The claim about the multiplicity is easy to verify from the code. $\qquad\square$

Lemma 3. *The initialisation of R takes time $O(d\|G\|) = O(nd^2)$ and the complete data structure uses the same space; $O(d\|G\|) = O(nd^2)$. Moreover, for every vertex $u \in V(G)$, $\mathsf{R}[u]$ stores at most $3s_2^{d+1}$ keys.*

Proof. The running time is straightforward. For every vertex u in the outer loop, we store $|N^-(u)| \leq d$ prefix sets as keys in R. The number stored in the dataset are all bounded above by $|G|$ and therefore take unit space in the RAM model. Accordingly, we use a total of $O(d|G|)$ space.

We are left to bound the number of keys in $\mathsf{R}[u]$. By Observation 3, for every vertex u the data structure $\mathsf{R}[u]$ has $\mathrm{tr}_G(S^2[u] \mid V_{>u})$ as keys. In the proof of Lemma 2, we showed that for any set X and vertex z the quantity $\mathrm{tr}_G(X \mid V_{>z})$ is bounded by $|X|s_2^d + 1$. Accordingly, the number of keys in $\mathsf{R}[u]$ is at most

$$\mathrm{tr}_G(S^2[u] \mid V_{>u}) \leq |S^2[u]|s_2^d + 1 \leq (s_2 + 1)s_2^d + 1 \leq 3s_2^{d+1},$$

where we used that $d, s_2 \geq 1$ in the last step. $\qquad\square$

Input: A vertex set $X \subseteq G$ with ordering $x_1, \ldots, x_\ell$ in $\mathbb{G}$
Output: An associative array Tr containing $\mathrm{tr}^{\#}(X)$
Initialize Tr as an empty associative array storing integers
$\mathsf{Tr}[\emptyset] \leftarrow |V(G) \setminus X|$

① Collect 'right' traces

for $i \in [\ell]$ **do**
    for $A \in \mathsf{R}[x_i]$ **do**
        $S \leftarrow A \cap X$
        $\mathsf{Tr}[S] \leftarrow \mathsf{Tr}[S] + \mathsf{R}[x_i][A]$
        $S' \leftarrow S \setminus \max_{\mathbb{G}} S$
        $\mathsf{Tr}[S'] \leftarrow \mathsf{Tr}[S'] - \mathsf{R}[x_i][A]$

② Collect and correct 'left' traces

Initialize L as an empty associative array storing vertex lists
for $i \in [l]$ **do**
    for $u \in N^-[x_i]$ **do**
        if $u \notin \mathsf{L}$ **then**
            $S \leftarrow N^-(u) \cap X$
            $\mathsf{Tr}[S] \leftarrow \mathsf{Tr}[S] - 1$ ②a Remove incorrect trace
            $\mathsf{L}[u] \leftarrow S$ ②b Insert left neighbours of u
        $\mathsf{L}[u] \leftarrow \mathsf{L}[u] \cup \{x_i\}$

for $u \in \mathsf{L}$ **do**
    if $u \notin X$ **then**
        $S \leftarrow \mathsf{L}[u]$
        $\mathsf{Tr}[S] \leftarrow \mathsf{Tr}[S] + 1$ ②c Count correct trace

return Tr

Algorithm 2: Answering trace queries using the data structure R.

Lemma 4. *Algorithm 2 answers the trace frequency query for a vertex subset $X \subseteq V(G)$ in time $O\big((d^2 + s_2^{d+2})|X|\big)$.*

Proof. We first prove the correctness and then the running time of the algorithm.

Correctness. Let in the following $z := \max_{\mathbb{G}} X$ be the largest vertex in X. Fix $X' \subseteq X$. We now show that at the end of Algorithm 2, the data structure contains the correct count for X', that is, $\mathsf{Tr}[X']$ exactly the multiplicity of X' in $\mathrm{tr}^{\#}(X)$. Let Y' contain all vertices whose trace in X is X', including vertices in X.

Claim. After completion of part ①, $\mathsf{Tr}[X']$ contains the number of vertices $y \in Y'$ for which $N^-(y) \cap X = X'$.

Proof. Let $x_i := \max_{\mathbb{G}} X'$, then when the outer loop of part ① reaches i, by Observation 3, $\mathsf{R}[x_i]$ contains exactly $\mathrm{tr}(S^2[x_i] \mid V_{>x_i})$ as keys. Note that in all previous iterations, neither the set S nor the set S' can be equal to X' since in those iterations, $x_i \notin S$.

Consider first the case where no vertex in $V_{>x_i}$ has X' as their trace in $x_1, \ldots, x_i$. In that case, neither the set S nor the set S' will equal X' and therefore $\mathsf{Tr}[X']$ remains unchanged.

If there exists any vertex $y \in V_{>x_i}$ with $N(y) \cap \{x_1, \ldots, x_i\} = X'$, then note that $X' \subseteq S^2[x_i]$. Let Y_t contain all vertices of $V_{>x_i}$ where $N^-(y) \cap X = X'$ and Y_f all vertices of $V_{>x_i}$ where $N^-(y) \cap X \supset X'$. We now show that $\mathsf{Tr}[X']$ equals $|Y_t|$ at the end of part ①.

By Observation 3, $\mathsf{R}[x_i][A]$ contains the number of vertices in $V_{>x_i}$ whose trace in $S^2[x]$ is A. So every vertex $y \in Y_y \cup Y_f$ is counted in $\mathsf{Tr}[X']$ at exactly the iteration where $N^-(y) \cap S^2[x_i] = X'$. For a vertex $y \in Y_f$, let x_j be the next vertex in $N^-(y) \cap X$ in the ordering $x_1, \ldots, x_\ell$, by construction of Y_f such an index must exist. Consider the iteration of the inner loop when $A = N^-(y) \cap \{x_1, \ldots, x_j\}$. Note that in this iteration $S' = X'$ since x_i is the immediate predecessor of x_j in $A \cap X$ under $\mathbb{G}$. In this iteration, the contribution of y to $\mathsf{Tr}[X']$ is removed. Therefore, after iterating through all $x_1, \ldots, x_\ell$, we conclude that $\mathsf{Tr}[X']$ indeed contains the number of vertices vertices $y \in Y'$ for which $N^-(y) \cap X = X'$.

Note that at this point, the traces of all vertices in $V_{>z}$ are correctly recorded in Tr. However, the data structure also contains incorrect traces for some vertex in $V_{\leq z}$. Namely, those vertices y who lie somewhere between x_1 and x_ℓ in $\mathbb{G}$. These vertices currently contribute to $\mathsf{Tr}[N^-(y) \cap X]$ but their actual trace in X might be different to $N^-(y) \cap X \neq N(y)$. This is corrected in part ②: note that for any such vertex y with $N^-(y) \cap X \neq N(y) \cap X$ it follows that $y \in N^-[X]$. Their contribution to $\mathsf{Tr}[N^-(y) \cap X]$ is removed in line ②ₐ, note that this line is executed exactly once for each vertex in $N^-[X]$ as the vertex is inserted into L immediately afterwards. To see that the correct trace is counted for each $u \in N^-[X]$, we first show the following:

Claim. After the first loop in part ② has finished, L contains exactly $N^-[X]$ as keys and for all $u \in N^-[X]$, we have that $\mathsf{L}[u] = N(u) \cap X$.

Proof. It is clear that all keys of L are indeed from $N^-[X]$ as these are the only vertices used for insertions. To see that it is all of $N^-[X]$, simply note that for each $u \in N^-[X]$ there exists at least one $i \in [l]$ such that $u \in N^-[x_i]$. Once i takes this value in the loop, u is inserted into L.

Consider now $u \in N^-[X]$. When u is inserted into in line ②ₐ we have that $\mathsf{L}[u] = N^-(u) \cap X$. Then for every right neighbour $x_i \in N^+(u) \cap X$, at loop iteration i, x_i is added to $\mathsf{L}[u]$. We conclude that when the loop terminates, $\mathsf{L}[u] = (N^-(u) \cap X) \cup (N^+(u) \cap X) = N(u) \cap X$ and the claim holds. $\qquad\square$

Finally, in the second loop of part ②, the $\mathsf{L}[u]$ for $u \in N^-(X)$ (so excluding X) is inserted into Tr. We conclude that after part ②, Tr indeed contains the trace frequencies of X.

Running Time Analysis. As stated, we assume that that the associative arrays Tr, R are constructed using a hash map and therefore have constant expected query time. The inner data structures $R[x]$ for $x \in G$ have a query time proportionally to the size of the key set, which can be achieved using prefix tries.

Let us begin by analysing the running time of part ①. By Lemma 3, the number of elements in $R[x_i]$ as at most $3s_2^{d+1}$. Accordingly, both loops taken together perform at most $|X|3s_2^{d+1}$ iterations. In each iteration, note that we can bound the size of all three sets (S, S', A) by $|A|$ and since $A \subseteq S^2[x_i]$ this in turn is bounded by $O(s_2)$. Therefore all access operations $(Tr[S], Tr[S'], R[x_i][A])$ are bounded by $O(s_2)$ and the total running time of part ① is at most $O(s_2^{d+2}|X|)$.

Let us now analyse the running of part ②. The two nested loops run in time $\sum_{x_i \in X} |N^-[x_i]| \leq d|X|$. For every vertex in $N^-[X]$, the if-statement inside these loops is executed exaclty once. Note that S has size at most d, thus constructing S and accessing $Tr[S]$ is possible in time $O(|S|) = O(d)$. The update of $L[u]$ after the if-statement takes constant time, so in total the running time of the first loop is bounded by $O(d^2|X|)$.

For the second loop it costs $O(|S|)$ to update every trace S of vertices in $N^-(X)$. That is, the running time cost is proportional to $\sum_{u \in N^-(X)} N(u) \cap X$ which is equal to the total number of edges between X and $N^-(X)$. This in turn is bounded by $d(|X| + |N^-(X)|) = O(d^2|X|)$ and the claimed running time follows. $\square$

Implementation notes

While the theoretical analysis already suggests that the data structure is quite space-efficient, Observation 2 holds the key for great practical gains: Because the data structure $R[x]$, $x \in G$, only stores keys for subset of $S^2[x]$ and the size of S^2 is usually small (in the tens or hundreds for most networks, see Section 6), we can store all keys in $S^2[x]$ as bit vectors of length $|S^2[x]|$ over the ground set $S^2[x]$. This is orders of magnitudes smaller than storing vertex sets that contain vertex ids, where each vertex id realistically has a size of at least 32 bit, and has second-order effects on cache efficiency. This also means that the set intersection in loop ① of Algorithm 2 can be computed as a bitwise AND between two bit vectors, which again is an order of magnitude faster than doing the same with two hash sets.

5   A Simple Conditional Lower Bound

As discussed above, answering trace queries in d-degnerate graphs is lower bounded by $\Omega(|X|^d)$ simply by the potential output size, though there is no immediate reason why *counting* the number of traces should not be possible in less time. Here we provide some weaker, conditional lower bounds already for counting the number of neighbours of a query set. To that end, recall the following conjecture formulated by Abboud, V. Williams, and Wang [1]:

Definition 1 (Hitting Set Conjecture (HSC)). *There is no $\epsilon > 0$ such that for all $c \geq 1$ there exists an algorithm that given two lists $\mathcal{A}, \mathcal{B}$ of n subsets of a universe U of size $c \log n$, can decide in $O(n^{2-\epsilon})$ time whether there exists a set in $\mathcal{A}$ that intersects every set in $\mathcal{B}$.*

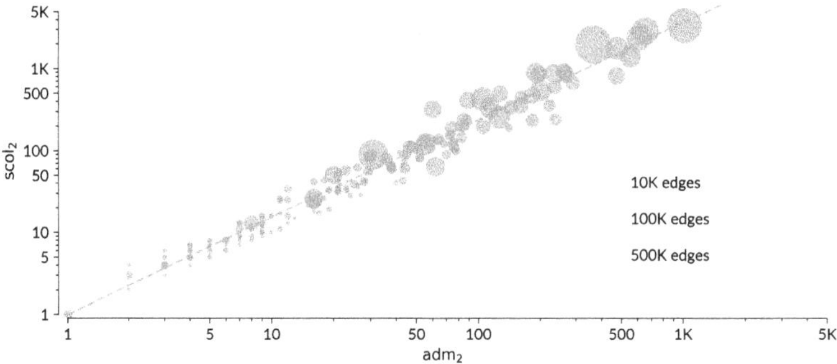

Fig. 1. Strong 2-colouring numbers of degeneracy/degree orderings for all 217 networks in the corpus, compared to their exact 2-admissibility. The shaded region indicates the lower and upper bounds based on 2-admissibility orderings, proving that degeneracy/degree orders have very low strong 2-colouring numbers in practice. The red dashed line is the fitted polynomial $\mathrm{adm}_2(G)^{1.188}$.

The problem described above is called the *Hitting Set Existence* (HSE) problem. The following conditional lower bound for neighbourhood-counting data structures follows from a straightforward adaption of a construction by Abboud *et al.* [1]:

Lemma 5. *For every data structure that can be initialised in time $f_1(d)|G|$ on a d-degenerate graph G and then count the size of $N[X]$ for a given query set $X \subseteq V(G)$ in time $f_2(d + |X|)$, it holds that one of $f_1(x), f_2(x)$ cannot be in $O(2^{o(x)})$ unless the HSC fails.*

A proof can be found in the full version of the paper. The same construction shows that for every data structure that can be initialised in time $f_1(|G|)|G|$ on a graph G and count the neighbourhood sizes for query sets $X \subseteq V(G)$ in time $f_2(|X|)$, it is not possible that $f_1(x) \in o(x)$ and $f_2(x) \in O(2^{o(x)})$ unless the HSC fails.

6 Experiments

We implemented[4] the data structure and comparison algorithms in Rust (v1.88.0) and ran the experiments on an Apple M2 @3.49 GHz with 8GB RAM on a single core. Our test dataset[5] contains 217 networks ranging from eight to 1.1M edges from various domains like biology, infrastructure, sociology, and communication.

---

[4] Code available at https://github.com/microgravitas/exp-trace-sampling.
[5] Networks available at https://github.com/microgravitas/network-corpus.

Computing strong 2-colouring orders in practice.

An immediate practical concern of using decompositions like the strong 2-colouring order is the decomposition has to be readily computable. As mentioned above, the problem of finding an optimal $scol_2$-ordering is NP-hard, however, two simple heuristics work very well in practice, namely by computing either a degeneracy ordering (which is possible in linear time [9] and very fast in practice) or by simply ordering the vertices by ascending degree. In our experiments, we computed both and chose the ordering with the better $scol_2$-value.

Since we do not have an optimal $scol_2$-values as a baseline to compare these heuristics, we instead use an algorithm recently published by a subset of the authors [2] that can compute the *2-admissibility* adm_2 exactly, even on very large networks.

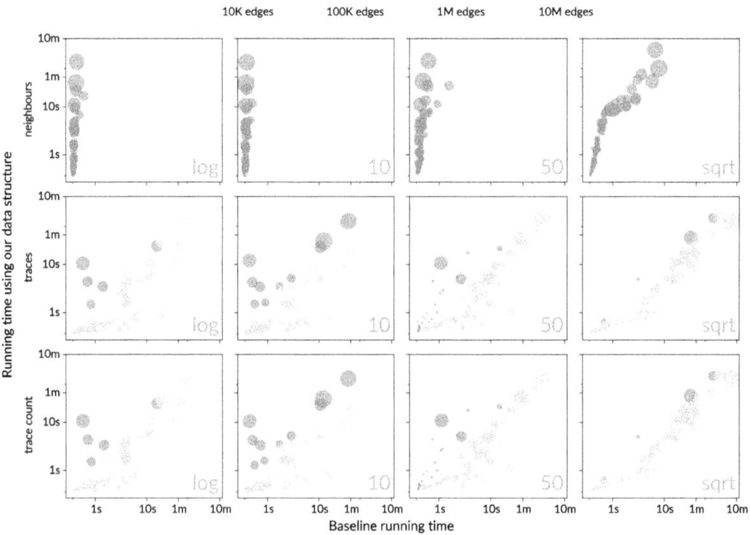

Fig. 2. Comparison of baseline algorithms to algorithms using our data structure for neighbourhood-counting (top), trace enumeration (middle), and trace frequency (bottom). The running time for the baseline algorithm is on the x-axis, the running time of our algorithm on the y-axis. The time was measured for 1000 query sets chosen uniformly at random of size $\log_2 |G|$, 10, 50, and $\sqrt{|G|}$. Red indicates that the baseline algorithm was faster, orange that it was faster only if including the setup time of our data structure, and green that our algorithm was faster including the setup time. Marker sizes correspond to $\|G\|$.

For any graph G it holds that $adm_2(G) \leq scol_2(G) \leq adm_2(G) \cdot (adm_2(G) - 1)$ [7], therefore the 2-admissibility provides a context to judge the strong 2-colouring number of heuristically computed orderings.

For the 217 networks in our corpus, we found that the degeneracy/degree heuristic computed orderings with $scol_2$-values much closer to the lower bound given by adm_2 (Fig. 1). Testing different fitting functions (linear, polynomial, exponential) between s_2 and the 2-admissibility a_2, we found the best fit ($R^2 = 0.917$) for $s_2 \sim a_2^{1.188}$ in our dataset. The average scol_2-value computed was 155.8, with a 75% of the networks having a value below 110 and 50% below 25. The largest value we found was 3379 in a network with 426K vertices and 8M edges (`dogster_friendships`). s_2 has almost no correlation with the number of vertices ($r \approx 0.24$) and correlates positively with the number of edges ($r \approx 0.73$). There is only a weak positive correlation with the average degree ($r \approx 0.39$), so density is not the only factor at play here, indicating that the networks in the dataset have various structural differences.

Comparing Queries Against Baseline Algorithms. We compare our data structure on three tasks to generic baseline algorithms: counting the number of neighbours for a query set X, computing $\mathrm{tr}(X)$, and computing $\mathrm{tr}^{\#}(X)$. We implemented the baseline algorithms using sensible defaults, like using state-of-the art hash maps, but otherwise stuck to the 'obvious' algorithm one would implement to solve the particular problem.

Since all three tasks are sensitive to the query size, we tried four different query set sizes: $\log_2 |G|$ vertices, ten vertices, fifty vertices, and $\sqrt{|G|}$ vertices. On about half of the networks we have that $\log_2 |G| \leq 10$, and the overall maximum is 21. Larger query sets favour our algorithm, therefore we did not test sets beyond $\sqrt{|G|}$.

For each network G, we generated 1000 query sets chosen uniformly at random from $V(G)$ and ran our algorithm and the baseline algorithm on the same set. The setup cost for our data structure is included in the running time, but it only has to be computed once to answer all 1000 queries.

As Fig. 2 illustrates, our algorithm is preferable in almost all cases for the trace counting and listing task, in particular for larger query sizes (50, $\sqrt{|G|}$). On smaller queries ($\log_2 |G|$, 10), it performs worse on very large graphs which either have a large s_2, `livemocha` with 2.1M edges and $s_2 = 2347$, or have such low edge density that most queries will be answered very quickly by the basic algorithm, `mag_geology_coauthor` with 4.4M edges and an average degree of 3.1. Regarding the neighbourhood counting, the basic algorithm seems preferable in all circumstances.

7   Conclusion

We designed and tested a data structure to answer trace listing and trace frequency queries based on orderings with small 2-colouring number. We experimentally confirmed that the data structure is indeed practicable, as already suggested by its favourable theoretical properties. In our experiments, the data structure performed better than the 'obvious' basic algorithms in almost all settings, and we presented a very simple heuristic to compute the necessary ordering that performs well on all tested networks.

We pose as an open question whether it is possible to improve these types of queries on d-degenerate graphs, specifically whether there exists a data structure that can be initialised in linear time and answers trace frequency queries in time $O(f(d)|H|^d)$.

References

1. Abboud, A., Williams, V., Wang, J.: Approximation and fixed parameter subquadratic algorithms for radius and diameter in sparse graphs. In: Proceedings of the Twenty-Seventh Annual ACM-SIAM Symposium on Discrete Algorithms, SODA '16, pp. 377–391, USA, January 2016. Society for Industrial and Applied Mathematics
2. Awofeso, C., Greaves, P., Lachish, O., Reidl, F.: A practical algorithm for 2-admissibility. In: 23rd International Symposium on Experimental Algorithms, SEA 2025, July 22-24, 2025, Venice, Italy, volume 338 of LIPIcs, pp. 3:1–3:19. Schloss Dagstuhl - Leibniz-Zentrum für Informatik (2025)
3. Breen-McKay, M., Lavallee, B., Sullivan, B.D.: Hardness of the generalized coloring numbers. Eur. J. Comb. p. 103709 (2023)
4. Chen, J., Huang, X., Kanj, I.A., Xia, G.: Strong computational lower bounds via parameterized complexity. J. Comput. Syst. Sci. **72**(8), 1346–1367 (2006)
5. Drange, P. G., et al.: Kernelization and sparseness: the case of dominating set. In: 33rd Symposium on Theoretical Aspects of Computer Science, STACS 2016, February 17-20, 2016, Orléans, France, volume 47 of LIPIcs, pp. 31:1–31:14. Schloss Dagstuhl - Leibniz-Zentrum für Informatik (2016)
6. Drange,P. G., Greaves, P., Muzi, I., Reidl, F.: Computing complexity measures of degenerate graphs. In: 18th International Symposium on Parameterized and Exact Computation (IPEC 2023), volume 285 of Leibniz International Proceedings in Informatics (LIPIcs), pp. 14:1–14:21, Dagstuhl, Germany, (2023). Schloss Dagstuhl – Leibniz-Zentrum für Informatik
7. Dvořák, Z.:. Constant-factor approximation of the domination number in sparse graphs. Eur. J. Comb., 34(5):833–840 (2013)
8. Gajarský, J., et al.: Kernelization using structural parameters on sparse graph classes. In: Bodlaender, H.L., Italiano, G.F. (eds.) ESA 2013. LNCS, vol. 8125, pp. 529–540. Springer, Heidelberg (2013). https://doi.org/10.1007/978-3-642-40450-4_45
9. Matula, D.W., Beck, L.L.: Smallest-last ordering and clustering and graph coloring algorithms. J. ACM **30**(3), 417–427 (1983)
10. Sparsity. AC, vol. 28. Springer, Heidelberg (2012). https://doi.org/10.1007/978-3-642-27875-4
11. Reidl, F., Villaamil, F., Stavropoulos, K, S.: Characterising bounded expansion by neighbourhood complexity. Eur. J. Comb., 75:152–168 (2019)
12. Zhu, X.: Colouring graphs with bounded generalized colouring number. Discret. Math. **309**(18), 5562–5568 (2009)

Counting Large Patterns in Degenerate Graphs

Christine Awofeso[iD], Patrick Greaves[iD], Oded Lachish[iD], and Felix Reidl[(✉)][iD]

Birkbeck, University of London, London, UK
{cawofe01,pgreav01}@student.bbk.ac.uk, {o.lachish,f.reidl}@bbk.ac.uk

Abstract. Subgraph counting is a fundamental algorithmic problem with many applications, including in the analysis of social and biological networks. The problem asks for the number of occurrences of a pattern graph H as a subgraph of a host graph G and is known to be computationally challenging: it is $\#W[1]$-hard even when H is restricted to simple structures such as cliques or paths. Curticapean and Marx (FOCS'14) show that if the graph H has vertex cover number τ, subgraph counting has time complexity $O(|H|^{2^{O(\tau)}}|G|^{\tau+O(1)})$.

This raises the question of whether this upper bound can be improved for input graphs G from a restricted family of graphs. Earlier work by Eppstein (IPL'94) shows that this is indeed possible, by proving that when G is a d-degenerate graph and H is a biclique of arbitrary size, subgraph counting has time complexity $O(d3^{d/3}|G|)$.

We show that if the input is restricted to d-degenerate graphs, the upper bound of Curticapean and Marx can be improved for a family of graphs H that includes all bicliques and satisfies a property we call (c,d)-*locatable*. Importantly, our algorithm's running time only has a polynomial dependence on the size of H.

A key feature of (c,d)-locatable graphs H is that they admit a vertex cover of size at most cd. We further characterize $(1,d)$-locatable graphs, for which our algorithms achieve a linear running time dependence on $|G|$, and we establish a lower bound showing that counting graphs which are barely *not* $(1,d)$-locatable is already $\#W[1]$-hard.

We note that the restriction to d-degenerate graphs has been a fruitful line of research leading to two very general results (FOCS'21, SODA'25) and this creates the impression that we largely understand the complexity of counting substructures in degenerate graphs. However, all aforementioned results have an exponential dependency on the size of the pattern graph H.

1   Introduction

Recall that a graph is d-degenerate iff its vertices can be ordered in such a way that every vertex has at most d neighbours preceding it in the order. Bressan and Roth [5] recently proved several lower and upper bounds related to counting substructures—subgraphs, induced subgraphs, homomorphisms—in degenerate graphs which, up to some minor factors, provide a characterisation of

© The Author(s), under exclusive license to Springer Nature Switzerland AG 2026
J. Kozik and A. Wolff (Eds.): SOFSEM 2026, LNCS 16448, pp. 607–620, 2026.
https://doi.org/10.1007/978-3-032-17801-5_44

these problems: Given a graph H and a d-degenerate graph G, they show that counting how often H appears as a subgraph in G is possible in expected time $f(|H|, d)|G|^{\text{im}(H)}$ and counting how often it appears as an induced subgraph in time $f(|H|, d)|G|^{\alpha(H)}$, where $\text{im}(H)$ is the induced matching number and $\alpha(H)$ the independence number of H. Assuming the exponential time hypothesis (ETH), Bressan and Roth further show almost matching lower bounds[1]. Previous work had focused on hardness classifications for counting substructures without any degeneracy restrictions on G [6–8], parameterising the treewidth of G [12], and the generalised subgraph problem [14].

Regarding the *linear-time* countability of subgraphs in degenerate graphs, Bera *et al.* [3] proved that all graphs H on at most five vertices can be counted in time $O(\|G\| d^{|H|-2})$. Moreover, they showed that counting cycles on six or more vertices is not possible in linear time unless the Triangle Detection Conjecture[2] (TDC) fails. Bressan [4] designed an algorithm that counts subgraphs in time $O(f(d, |H|)|G|^{\tau_1(H)})$, where $\tau_1(H)$ is a width-measure called the *DAG treewidth* of H. Regarding lower bounds, Bera *et al.* [2] provided a complete characterisation of the graphs H that can be counted in linear time in a degenerate host graph as a subgraph/induced subgraph under the TDC[3].

These results create the impression that we largely understand the complexity of counting substructures in degenerate graphs. The problem is, in all aforementioned upper bounds the running times have an exponential dependence on $|H|$. However, in the general case (*cf.*Theorem 2.8 in [7]), if H has vertex cover number τ, subgraph counting can be solved in time $O(|H|^{2^{O(\tau)}}|G|^{\tau+O(1)})$, and much earlier Eppstein [10] showed that when G is a d-degenerate graph and H is a biclique, subgraph counting can be solved in time $O(d3^{d/3}|G|)$. This leaves the question of whether the upper bound of Curticapean and Marx can be improved for a larger family of graphs than just bicliques.

Our Contribution: We show that if the input is restricted to d-degenerate graphs, the upper bound of Curticapean and Marx can be improved for a family of graphs H satisfying a property we call (c, d)-*locatable*, which includes all bicliques. Specifically, we show that for every such graph H subgraph counting can done in time $O\big((cd)^2 f(c, d)|H|^{f(c,d)}|G|^c\big)$ where $f(c, d) = 2^d (cd)^{d+1}$ (Theorem 4, Sect. 3.3) and as a semi-induced subgraph[4] in time $O\big((cd)^{2cd+d}|G|^c\big)$ (Theorem 3, Sect. 3.3). If the host graph G is bipartite, the time complexity holds for counting induced subgraphs. We note that our result is incompara-

---

[1] $f(|H|, d)|G|^{o(\text{im}(H)/\log \text{im}(H))}$ and $f(|H|, d)|G|^{o(\alpha(H)/\log \alpha(H))}$, respectively.

[2] The conjecture states that there exists a constant δ such that any algorithm under the word-RAM model takes times at least $|G|^{1+\delta-o(1)}$ in expectation to detect whether G contains a triangle [1].

[3] The exact characterisations are somewhat technical, but they revolve round the existence of C_k, $k \geq 6$, as certain types of substructures in H.

[4] A pattern H with vertex cover S and independent set T appears *semi-induced* in G if some supergraph $\hat{H}$ obtained by only adding edges between T-vertices in H appears as an induced subgraph of G.

ble to Bressan's algorithm, since our algorithm can count patterns with large independent sets which have unbounded DAG treewidth.

To compare these results to those of Curticapean and Marx, note that the cd in our result bounds the vertex cover number τ of H. Expressing our running time with this parameter, we have that $O\big((cd)^2 f(c,d)|H|^{f(c,d)}|G|^c\big)$ translates to $O\big((\tau)^2 f(\tau,d)|H|^{f(\tau,d)}|G|^{\tau/d}\big)$ where $f(\tau,d) = 2^d(\tau)^{d+1}$ and $O\big((cd)^{2cd+d}|G|^c\big)$ translates to $O\big((\tau)^{2\tau+d}|G|^{\tau/d}\big)$, so in both cases this improves the exponent of $|G|$ by a factor of $1/d$.

We are specifically interested in $(1,d)$-locatable patterns as these can be counted with only a linear dependence on $|G|$ in the running time. We characterise these patterns (Sect. 3.4) and supplement our positive algorithmic results with a lower bound that shows that even counting patterns very close to being $(1,d)$-locatable is already $\#W[1]$-hard (Sect. 4). Note that we explicitly link the degeneracy d to the pattern via the locatability property, our setting is different from the characterisation by Bera $et\ al.$ [2], who ask which patterns can be counted in linear time for $all\ d$. This means that $(1,d)$-locatable patterns can be counted in linear time in graphs of degeneracy $\leq d$, but might not be in graphs of degeneracy $> d$. In this short version of the paper, proofs marked with $\star$ can be found in the full version[5]

2   Preliminaries

We will often use the placeholder   for variables whose value is irrelevant in the given context.

For an integer k, we use $[k]$ as a short-hand for the set $\{0,1,2,\ldots,k-1\}$. We use blackboard bold letters like $\mathbb{X}$ to denote totally ordered sets, that is, some underlying set X associated with a total order $<_\mathbb{X}$. We further, for $Y \subseteq X$, use the notations $\max_\mathbb{X} Y$ to mean the maximum member in Y under $<_\mathbb{X}$ and the similarly defined $\min_\mathbb{X} Y$. If the context allows it, we will sometimes drop this subscript, $e.g.$we shorten $\max_\mathbb{X} X_1 <_\mathbb{X} \max_\mathbb{X} X_2$ to $\max X_1 <_\mathbb{X} \max X_2$ or $\max_\mathbb{X} \mathbb{X}$ to simply $\max \mathbb{X}$. Finally, for a real-valued function f we write $\mathrm{supp}(f)$ to denote its support, $i.e.$the set of all values for which f is non-zero.

The $index\ function$ $\iota_\mathbb{X}\colon X \to \mathbb{N}$ maps elements of X to their corresponding position in $\mathbb{X}$, where $\mathbb{X}$ is the set X imbued with some linear order. We extend this function to sets via $\iota_\mathbb{X}(S) = \{\iota_\mathbb{X}(s) \mid s \in S\}$. For any integer $i \in [\|X\|]$ we write $\mathbb{X}[i]$ to mean the ith element in the ordered set. An $index\ set$ I for $\mathbb{X}$ is simply a subset of $[\|X\|]$ and we extend the index notation to sets via $\mathbb{X}[I] := \{\mathbb{X}[i] \mid i \in I\}$.

For a graph G we use $V(G)$ and $E(G)$ to refer to its vertex- and edge-set, respectively and $\delta(G)$ to denote its minimum degree. We use the shorthands $|G| := |V(G)|$ and $\|G\| := |E(G)|$. We write $\mathrm{aut}(G)$ for the number of automorphisms of G.

An $ordered\ graph$ is a pair $\mathbb{G} = (G,<)$ where G is a graph and $<$ a total ordering of $V(G)$. We write $<_\mathbb{G}$ to denote the ordering for a given ordered graph

---

[5] Full version can be found here: http://arxiv.org/abs/2511.20385..

and extend this notation to the derived relations $\leq_\mathbb{G}, >_\mathbb{G}, \geq_\mathbb{G}$. For simplicity we will call $\mathbb{G}$ an *ordering of* G.

We use the same notations for graphs and ordered graphs, additionally we write $N_\mathbb{G}^-(u) := \{v \in N(u) \mid v <_\mathbb{G} u\}$ for the *left neighbourhood* of a vertex $u \in \mathbb{G}$. We write $N_\mathbb{G}^-[u] := N_\mathbb{G}^-(u) \cup \{u\}$ for the closed left neighbourhood which we extend to vertex sets X via $N_\mathbb{G}^-[X] := \bigcup_{u \in X} N_\mathbb{G}^-[u] \cup X$. We write $\Delta^-(\mathbb{G}) := \{|N_\mathbb{G}^-(u)| \mid u \in \mathbb{G}\}$ to denote the maximum left-degree of an ordered graph.

Two ordered graphs $\mathbb{G}_1, \mathbb{G}_2$ are *isomorphic* if the function that maps the ith vertex of $\mathbb{G}_1$ to the ith vertex of $\mathbb{G}_2$ is a graph isomorphism. We write $\mathbb{G}_1 \simeq \mathbb{G}_2$ to indicate that two ordered graphs are isomorphic.

A graph G is *d-degenerate* if there exists an ordering $\mathbb{G}$ such that $\Delta^-(\mathbb{G}) \leq d$. We say an ordering $\mathbb{G}$ of a graph G is d-degenerate if $\Delta^-(\mathbb{G}) \leq d$, and just *degenerate* if $\Delta^-(\mathbb{G})$ is minimum over all orders of G. The degeneracy ordering of a graph can be computed in time $O(|G| + \|G\|)$ in general and $O(d|G|)$ for d-degenerate graphs [13].

The Degeneracy Toolkit. We will make use of the following data structures for degenerate graphs by Drange *et al.*:

Lemma 1. ([9]). *Let $\mathbb{G}$ be an ordered graph with degeneracy d. Then in time $O(d2^d n)$ we can compute a subset dictionary R over $V(G)$ which for any $X \subseteq V(G)$ answers the query $R[X] := \big|\{v \in G \mid X \subseteq N^-(v)\}\big|$ in time $O(|X|)$.*

Theorem 1. ([9]). *Let $\mathbb{G}$ be an ordered graph on n vertices with degeneracy d. After a preprocessing time of $O(d2^d n)$, we can, for any given $S \subseteq V(G)$, compute a subset dictionary Q_S in time $O(|S|2^{|S|} + d|S|^2)$ which for any $X \subseteq S \subseteq V(G)$ answers the query[6] $Q_S[X] := \big|\{v \in V(G) \setminus X \mid S \cap N(v) = X\}\big|$ in time $O(|X|)$.*

In the following we will need the concept of a *neighbourhood class*. For a bipartite graph with sides S, T we partition T into sets such that $u, v \in T$ are in the same class iff $N(u) = N(v)$, that is, u and v have exactly the same neighbours in S.

In the following we will make use of this simple observation about degenerate graphs and neighbourhood classes:

Lemma 2. ($\star$). *Let $H = (S, T, E)$ be a bipartite d-degenerate graph. Then the number of vertices in T of degree at least d is at most $d|S|$ and the number of neighbourhood classes in T is at most $4|S|^d$.*

We note that the neighbourhood classes of a bipartite graph H can be computed in time $O(|H| + \|H\|)$ using the partition-refinement data structure [11].

---

[6] The data structure in the referenced paper counts all vertices $v \in V(G)$ with the given property but it is trivial to exclude vertices contained in X.

3   Counting Large Patterns

Let us review how bicliques of arbitrary size can be counted in linear time in degenerate graphs. There are three ideas to this algorithm: First, bicliques with two large sides cannot exist in degenerate graphs. Second, if a biclique has at least one small side, than this side can be located in the left-neighbourhood of a vertex and therefore can be found in linear time. Third, once we have found one side of a biclique we can use a suitable data structure to count how many joint neighbours this side has in the host graph and therefore count the number of bicliques 'rooted' at this vertex set.

Theorem 2. (Adapted from[10]). *There exists an algorithm that for any $s \leq t \in \mathbb{N}$ and any d-degenerate graph G counts the number of (non-induced) $K_{s,t}$ in G in time $O(d^2 2^d |G|)$.*

Proof. We begin with the simple case of $s \geq d + 1$, in which case $K_{s,t}$ is not d-degenerate and therefore cannot occur in G. In all remaining cases, our algorithm computes a d-degeneracy ordering $\mathbb{G}$ of G in time $O(d|G|)$ and preprocesses the graph in time $O(d2^d|G|)$ to compute the data structure R described in Lemma 1.

Next, consider the regime where $s \leq d$ and $t \geq d+1$. Then for any occurrence of $K_{s,t}$ as $\mathbb{K} \subseteq \mathbb{G}$ it holds that the last vertex $x = \max \mathbb{K}$ must be part of the t-side of the $K_{s,t}$ since otherwise we would find a vertex with left-degree $t > d$. We can therefore locate the s-side of the $K_{s,t}$ by guessing x and then guessing an s-sized subset $S \subseteq N^-[x]$ in total time $O(n\binom{d}{s}) = O(2^d n)$. Let c_L denote the number of joint neighbours of S who appear not to the right of $\max_{\mathbb{G}} S$, i.e.$c_l := |\{v \in N_G(S) \mid v < \max_{\mathbb{G}} S\}|$. Note that we can compute c_L by simply inspecting each of the $\leq d$ vertices in $N^-(\max_{\mathbb{G}} S)$ in time $O(ds) = O(d^2)$. Let c_r denote the number of joint neighbours of S who appear to the right of $\max_{\mathbb{G}} S$ and note that $c_r = R[S]$, meaning we can retrieve it in time $O(s) = O(d)$. Given c_l and c_r, the number of $K_{s,t}$ in G whose s-side is exactly S is then given by $\binom{c_l + c_r}{t}$.

Finally, consider the regime where $s \leq t \leq d$. Let $\mathbb{K} \subseteq \mathbb{G}$ be any occurrence of $K_{s,t}$ with sides S, T. Let $x := \max \mathbb{K}$, then either $S \subseteq N^-(x)$ or $T \subseteq N^-(t)$. In either case, we can proceed exactly as in the previous case to count the occurrence of *all* $K_{s,t}$ with this specific S- or T-set in G. $\square$

Based on these core ideas we now develop our algorithm to count *large* graphs within degenerate host graph. These graphs will have a specific structure, which we formalise here:

Definition 1. (Pattern). *A pattern is a connected graph H whose vertex set is partitioned into sets S, T, called its sides, where the set T is independent in H, $T \subseteq N(S)$ and $|S| < |T|$. We call S the small and T the large side of the pattern.*

We will look at two variants of how patterns might appear in a host graph. The first corresponds to a 'semi-induced' subgraph, meaning that all edges and non-edges inside S and between S and T must appear exactly as in the pattern. However, we allow for arbitrary edges to appear between T-vertices.

Definition 2. (Strong Pattern Containment, – Frequency). *We say that a pattern H with sides S, T is* strongly contained *in another graph G if there exist disjoint vertex sets $\tilde{S}, \tilde{T} \subseteq V(G)$ and a bijection $\phi\colon S \cup T \to \tilde{S} \cup \tilde{T}$ with the following properties:*

- *$\phi(S) = \tilde{S}$ and $\phi(T) = \tilde{T}$, and*
- *for every pair $u \in S$, $v \in S \cup T$ it holds that $uv \in H \iff \phi(u)\phi(v) \in G$.*

We call ϕ a strong embedding *of H into G and use the shorthand $H \xrightarrow{\phi} G$ to denote this fact. We further define $\#(H \hookrightarrow G)$ as the number of strong embeddings that exist of H into G.*

The second notion of pattern containment corresponds to subgraphs, for a unified presentation we use a different terminology and notation:

Definition 3. (Weak Pattern Containment, – Frequency). *We say that a pattern H with sides S, T is* weakly contained *in another graph G if there exist disjoint vertex sets $\tilde{S}, \tilde{T} \subseteq V(G)$, an edge set $\tilde{F} \subseteq E(G[S \cup T])$, and a bijection $\psi\colon V(H) \cup E(H) \to \tilde{S} \cup \tilde{T} \cup \tilde{F}$ with the following properties:*

- *$\phi(S) = \tilde{S}$ and $\phi(T) = \tilde{T}$, and*
- *for every edge $uv \in H$ it holds that $\phi(u)\phi(v) = \phi(uv)$ and $\phi(uv) \in \tilde{F}$.*

We call ψ a weak embedding *of H into G and use the shorthand $H \overset{\phi}{\dashrightarrow} G$ to denote this fact. We further define $\#(H \dashrightarrow G)$ as the number of weak embeddings that exist of H into G.*

Since weak embeddings not only map vertices but also edges, we will use the notation $\psi(H)$ to denote the subgraph of G onto which H is mapped by ψ.

In the following we will often count embeddings with certain additional properties. In these cases, we will use the notation
$$\#(H \xrightarrow{\phi} G \mid P(\phi)) := |\{H \xrightarrow{\phi} G \mid P(\phi)\}|, \quad \text{where } P \text{ is some predicate}$$
about ϕ, to mean the number of strong embeddings ϕ for which P holds. We use an analogous notation for weak embeddings.

3.1   Computing and Encoding Pattern Automorphisms

In contrast to the nicely symmetric bicliques, arbitrary patterns have much more complicated automorphisms that we have to compute explicitly. Since S is a vertex cover of the pattern, the automorphisms of H are pretty much fixed by the orbit of S under the automorphism action. In particular, they are very tame if the orbit of S only contains S itself. However, this is not true for all patterns and thus we have to explicitly compute the orbit in order to proceed. The S-orbit $\mathrm{orb}_S(H)$ contains all sets $X \subseteq V(G)$ onto which S is mapped by some automorphism of H, that is $\mathrm{orb}_S(H) := \{X \subseteq V(H) \mid \exists \text{ automorphism } \lambda \text{ of } H \text{ with } \lambda(S) = X\}$.

Lemma 3. $(\star)$ *For every pattern H the S-orbit $\mathrm{orb}_S(H)$ can be enumerated in time $O(|S|^{|S|} \cdot |H|^4)$. If H is d-degenerate, the running time improves to $O(d^2 |S|^{|S|+d} \cdot |H|^2)$.*

The second quantity we will need are automorphisms that stabilise S, which we will denote by $\mathrm{aut}_S(H)$. That is, $\mathrm{aut}_S(H)$ contains all automorphisms $\lambda \in \mathrm{aut}(H)$ with $\lambda(S) = S$.

Lemma 4. ($\star$). *For every pattern H the number of automorphisms $\mathrm{aut}_S(H)$ can be computed in time $O(|S|^{|S|} \cdot |H|)$. If H is d-degenerate, the running time improves to $O(\|H\| + |S|^{|S|+d})$.*

Corollary 1. *For every pattern H the number of automorphisms $\mathrm{aut}(H)$ can be computed in time $O(|S|^{|S|} \cdot |H|^4)$. If H is d-degenerate, the running time improves to $O(d^2|S|^{|S|+d} \cdot |H|^2)$.*

Proof. If we view $\mathrm{aut}(H)$ as acting on $2^{V(G)}$ instead of $V(G)$, then by the Orbit–Stabilizer Theorem: $\mathrm{aut}(H) = |\mathrm{orb}_S(H)| \cdot \mathrm{aut}_S(H)$. Therefore we simply combine Lemma 3 and Lemma 4. $\qquad\square$

We will need to keep track of embeddings, automorphisms, and orderings which poses a challenge for the presentation of our work and the patience of the reader. The following notion of an *index representation* encodes a specific automorphism of H under a specific ordering in which it might appear in a host graph.

Definition 4. (Index Representation). *Let H be a pattern with sides S, T and let $\mathbb{S}$ be an ordering of S. Let further $s := |S|$. The* index representation *of H under $\mathbb{S}$ is a tuple $\mathrm{rep}(H, \mathbb{S}) = (\mathbb{H}_S, \mathrm{J})$ where $\mathbb{H}_S$ is an ordered graph with vertices $[s]$ and $\mathrm{J}\colon 2^{[s]} \to \mathbb{N}$ is a function such that*

- $\mathbb{H}_S$ *is the ordered graph obtain from $H[S]$ by relabelling vertices according to their index in $\mathbb{S}$ and applying the natural order on $[s]$, and*
- *for all $I \subseteq [s]$ it holds that $\mathrm{J}(I) = \big|\{v \in T \mid N_H(v) = \mathbb{S}[I]\}\big|$.*

We write $\mathrm{rep}(H) := \{\mathrm{rep}(H, \mathbb{S}) \mid \mathbb{S} \in \pi(S)\}$ for the set of all index representations of H.

Put more simply, an index representation provides a uniform encoding of how the S-part of a pattern is ordered and accordingly how the neighbourhoods of T-vertices will be labelled. We will frequently need the index representation for the image of a given embedding of H into G, which is why we introduce the following shorthand:

Definition 5. (Index Representation of Embedding). *Let λ be a either a strong or weak embedding of H into G and let $\mathbb{G}$ be an ordering of G. Let further $\tilde{H}$ be the pattern in G defined by the image of λ, meaning $\tilde{H} = G[\lambda(V(H))] \setminus E(G[\lambda(T)]$ in the strong case and $\tilde{H} = \lambda(H)$ in the weak case. We define the notation $\mathrm{rep}(\lambda, \mathbb{G}) := \mathrm{rep}(\tilde{H}, \tilde{\mathbb{S}})$ where $\tilde{\mathbb{S}}$ is the set $\lambda(S)$ ordered by $<_{\mathbb{G}}$.*

Index representation are semi-canonical representations of patterns because we can use them to describe embeddings: if ϕ is a strong embedding of H into G and $\mathbb{G}$ is an ordering of G, then there exists an ordering $\mathbb{S}$ of S such that $\mathrm{rep}(H, \mathbb{S}) = \mathrm{rep}(\phi, \mathbb{G})$. More specifically, we can use representations to distinguish certain embeddings:

Lemma 5. ($\star$). *Let ϕ_1, ϕ_2 be embeddings (weak or strong) of H into G with $\phi_1(S) = \phi_2(S)$ and let $\mathbb{G}$ be an ordering of G. If $\mathrm{rep}(\phi_1, \mathbb{G}) \neq \mathrm{rep}(\phi_2, \mathbb{G})$ then $\phi_1 \neq \phi_2$.*

3.2 Counting Fixed Occurrences

We now describe the parts of our algorithm which count the number of strong/weak embeddings whose S-side coincides with some fixed set $\tilde{S}$ in the host graph G. More concretely, these subroutines (Algorithm 1 and Algorithm 2) are given a specific index representation and only count embeddings that conform to this embedding. We deal with finding the sets $\tilde{S}$ in the second part of the algorithm.

Input: An ordered graph $\mathbb{G}$, an index representations $(\mathbb{H}_S, \mathrm{J})$ of a pattern H, a subset $\tilde{S} \subseteq V(G)$ of size $|S|$, and the data structure $Q_{\tilde{S}}$ from Theorem 1.

Output: $\#(H \overset{\phi}{\hookrightarrow} G \mid \phi(S) = \tilde{S} \text{ and } \mathrm{rep}(\phi, \mathbb{G}) = (\mathbb{H}_S, \mathrm{J}))$.

Function $CountStrong(\mathbb{G},\ (\mathbb{H}_S, \mathrm{J}),\ \tilde{S}, Q_{\tilde{S}},\ \mathrm{aut}_S(H))$

> $t = 0$
>
> Let $\tilde{\mathbb{S}}$ be S ordered by $<_{\mathbb{G}}$
>
> ① Check S-side of pattern
>
> **if** $\mathbb{H}_S \neq \mathbb{G}[\tilde{S}]$ **then**
> > **return** 0
>
> ② Count number of choices for T-side of pattern
>
> $k = 1$
>
> **for** $I \subseteq [s]$ **with** $\mathrm{J}(I) \neq 0$ **do**
> > $k = k \cdot \binom{Q_{\tilde{S}}[\tilde{\mathbb{S}}[I]]}{\mathrm{J}(I)}$
>
> **return** $\mathrm{aut}_S(H) \cdot k$ // Normalised for ease of presentation

Algorithm 1: Subroutine to count strong embeddings for a specific S-set.

Lemma 6. *Let $\mathbb{G}$ be an ordering of a graph G and let $(\mathbb{H}_S, \mathrm{J})$ be an index representations of a pattern H. Let further $\tilde{S} \subseteq V(G)$ and $\tilde{\mathbb{S}}$ be the ordering of S under $<_{\mathbb{G}}$. Let $Q_{\tilde{S}}$ be the data structure described in Theorem 1. Given $\mathbb{G}$, $(\mathbb{H}_S, \mathrm{J})$, $\tilde{S}$, and $Q_{\tilde{S}}$, the subroutine* CountStrong *(Algorithm 1) returns* $\#(H \overset{\phi}{\hookrightarrow} G \mid \phi(S) = \tilde{S} \text{ and } \mathrm{rep}(\phi, \mathbb{G}) = (\mathbb{H}_S, \mathrm{J}))$, *that is, the number of strong embeddings $H \overset{\phi}{\hookrightarrow} G$ with $\phi(S) = \tilde{S}$ and whose representative under $\mathbb{G}$ is exactly $(\mathbb{H}_S, \mathrm{J})$. The running time of* CountStrong *is* $O(|\mathrm{supp}(\mathrm{J})| \cdot |S|)$

Proof. Define the set $\Phi := \{H \overset{\phi}{\hookrightarrow} G \mid \phi(S) = \tilde{S} \text{ and } \mathrm{rep}(\phi, \mathbb{G}) = (\mathbb{H}_S, \mathrm{J})\}$ to contain all embeddings that map S onto $\tilde{S}$ and have the given index representation. Let further $\mathcal{H} := \{V(\phi(S \cup T)) \mid \phi \in \Phi\}$ contain all vertex subsets of $V(G)$ that are images of Φ.

First, we claim that $|\Phi| = \mathrm{aut}_S(H) \cdot |\mathcal{H}|$. This is the usual relationship between embeddings and substructures: Simply note that if we take an automorphism λ of H with $\lambda(S) = S$ and a strong embedding $\phi \in \Phi$ with $X = \phi(S \cup T)$, then $\phi \circ \lambda$ is a different strong embedding which maps onto X. In the other direction, we can take two distinct embeddings $\phi, \phi' \in \Phi$ and the function λ which satisfies $\phi \circ \lambda = \phi'$ will be an automorphism of H with $\lambda(S) = S$.

We claim that at the end of calling *CountStrong* with the above arguments, the variable k is equal to $|\mathcal{H}|$ and therefore *CountStrong* returns the value $|\Phi|$. Note that this is equivalent to claiming that $|\mathcal{H}| = \prod_{I \subseteq [s]} \binom{Q_{\tilde{S}}[\mathbb{S}[\tilde{I}]]}{\mathrm{J}(I)}$, as the right-hand side is precisely what the variable k contains at the end of the function. Recall that the data structure $Q_{\tilde{S}}[X]$ counts the number of joint vertices in G whose neighbourhood in $\tilde{S}$ is precisely X. For $X \subseteq \tilde{S}$, let us write $V_X := \{v \in V(G) \setminus S \mid N_G(v) \cap S = X\}$. With this notation, we see that $Q_{\tilde{S}}[X]$ is exactly $|V_X|$ and therefore the right-hand side is equal to

$$\prod_{I \subseteq [s]} \binom{Q_{\tilde{S}}[\mathbb{S}[\tilde{I}]]}{\mathrm{J}(I)} = \prod_{X \subseteq \tilde{S}} \binom{Q_{\tilde{S}}[X]}{\mathrm{J}(\iota_{\tilde{\mathbb{S}}}(X))} = \prod_{X \subseteq \tilde{S}} \binom{|V_X|}{\mathrm{J}(\iota_{\tilde{\mathbb{S}}}(X))}.$$

To see that this right-hand side is exactly $|\mathcal{H}|$, consider a vertex set $\tilde{H} \subseteq V(G)$ that is the image of at least one $\phi \in \Phi$. Since $\phi(S) = \tilde{S}$, we know that $\tilde{S} \subseteq \tilde{H}$ and hence $\phi(T) = \tilde{H} \setminus \tilde{S}$. Moreover, since $\mathrm{rep}(\phi, \mathbb{G}) = (\mathbb{H}_S, \mathrm{J})$, for each $X \subseteq \tilde{S}$ we have that $\phi(T)$ contains exactly $\mathrm{J}(\iota_{\tilde{\mathbb{S}}}(X))$ vertices of V_X. Accordingly, the total number of sets in $|\mathcal{H}|$ is exactly the product of choices for taking $\mathrm{J}(\iota_{\tilde{\mathbb{S}}}(X))$ out of V_X, which is exactly the right-hand side.

The running time is straightforward assuming that the function J is stored using *e.g.* a prefix trie and therefore allows us to iterate over its support in time $O(|\operatorname{supp}(\mathrm{J})|)$. $\square$

The proof for counting weak patterns is similar, but more involved since we have more choices of which edges to include. We defer the proof to the full version.

Lemma 7. ($\star$). *Let $\mathbb{G}$ be an ordering of a graph G and let $(\mathbb{H}_S, \mathrm{J})$ be an index representations of a pattern H. Let further $\tilde{S} \subseteq V(G)$ and $\tilde{\mathbb{S}}$ the ordering of S under $<_{\mathbb{G}}$. Let $Q_{\tilde{S}}$ be the data structure described in Theorem 1. Given $\mathbb{G}$, $(\mathbb{H}_S, \mathrm{J})$, $\tilde{S}$, and $Q_{\tilde{S}}$, the subroutine CountWeak$_d$ (Algorithm 2) returns $\#(H \overset{\psi}{\hookrightarrow} G \mid \psi(S) = \tilde{S} \text{ and } \mathrm{rep}(\psi, \mathbb{G}) = (\mathbb{H}_S, \mathrm{J}))$, that is, the number of weak embeddings $H \overset{\psi}{\hookrightarrow} G$ with $\psi(S) = \tilde{S}$ and whose representative under $\mathbb{G}$ is exactly $(\mathbb{H}_S, \mathrm{J})$. The running time of CountWeak$_d$ is $O(d2^d |S|^{d+2} \cdot |H|^{2^d |S|^{d+1}})$.*

3.3   Counting All Occurrences

Now that we have algorithms in place to count occurrence of patterns for specific sets $\tilde{S}$ in the host graph, we want to find potential candidate set in time better than $O(|G|^{|S|})$. This is clearly not possible for all patterns, as shown by the various existing lower bounds, and the following definitions capture the types of patterns our algorithm can count.

Definition 6. (Left-cover Number). *The left-cover number of a vertex set S in an ordered graph $\mathbb{H}$ is the size of the minimum vertex set C (not necessarily disjoint from S) such that $S \subseteq N_{\mathbb{H}}^-[C]$. We denote this number by $\gamma_S^-(\mathbb{H})$.*

Input: An ordered graph $\mathbb{G}$, an index representations $(\mathbb{H}_S, \mathrm{J})$ of a pattern H, a subset $\tilde{S} \subseteq V(G)$ of size $|S|$, and the data structure $Q_{\tilde{S}}$ from Theorem 1.

Output: $\#(H \overset{\psi}{\hookrightarrow} G \mid \psi(S) = \tilde{S} \text{ and } \mathrm{rep}(\psi, \mathbb{G}) = (\mathbb{H}_S, \mathrm{J}))$

Function $CountWeak_d(\mathbb{G}, (\mathbb{H}_S, \mathrm{J}), \tilde{S}, Q_{\tilde{S}}, \mathrm{aut}_S(H))$

    Let $\tilde{\mathbb{S}}$ be S ordered by $<_\mathbb{G}$

    if $\mathbb{H}_S \not\subseteq \mathbb{G}[\tilde{S}]$ **then**

        return 0

    (1) Construct flow graph (s.t are implicit, see proof of Lemma 7)

    Let $A = (L, R, F)$ with $L, R, F = \emptyset$

    Initialise empty dictionaries $\mathsf{supply}, \mathsf{demand}$

    for $I \subseteq [s]$ *with* $Q_{\tilde{S}}[\tilde{\mathbb{S}}[I]] > 0$ **do**

        $L = L \cup \{l_I\}$

        $\mathsf{supply}[l_I] = Q_{\tilde{S}}[\tilde{\mathbb{S}}[I]]$

    for $I \subseteq [s]$ *with* $\mathrm{J}(I) > 0$ **do**

        $R = R \cup \{r_I\}$

        $\mathsf{demand}[r_I] = \mathrm{J}(I)$

    for $l_I, r_{I'} \in L \times R$ *with* $I' \subseteq I$ **do**   *// See Lemma 7 for details*

        $F = F \cup (l_I, r_{I'})$

    (2) Enumerate embeddings

    $p = \max_{r_I \in R} \mathsf{demand}[r_I], \quad M = \sum_{r_I \in R} \mathsf{demand}[r_I], \quad t = 0$

    for $f \in [p]^F$ **do**

        // We treat edges in F as having infinite capacity

        if f *is not a flow on A or has value* $< M$ **then**

            continue

        (3) Compute number of embeddings represented by f

        $k = 1$

        for $l_I \in L$ **do**

            $k = k \cdot \dfrac{\mathsf{supply}[l]!}{(\mathsf{supply}[l] - f(sl))!}$

        $t = t + k$

    return $\mathrm{aut}_S(H) \cdot t$ *// Normalised for ease of presentation*

Algorithm 2: Subroutine to count weak embeddings for a specific S-set.

Definition 7. $((c, d)$**-locatable).** *A set $S' \subseteq S$ in a pattern H is (c, d)-locatable if for every d-degenerate ordering $\mathbb{H}$ of H it holds that $\gamma_{S'}^-(\mathbb{H}) \leq c$. We say that a pattern is (c, d)-locatable if S is (c, d)-locatable.*

Note that by this definition, all patterns with degeneracy larger than d are $(\cdot, d)$-locatable. This is intentional: if we ask how often a pattern H is contained in some d-degenerate graph G and H's degeneracy is larger than d, we can immediately answer this question. As a consequence, if a pattern is (c, d)-locatable, then it is (c', d)-locatable for every $c' \geq c$ and (c, d')-locatable for every $d' \leq d$. We note that every pattern is $(|S|, \infty)$-locatable. Let us next show how the left-cover number can be computed:

Input: A d-degenerate graph G, a pattern H
Output: $\#(H \hookrightarrow G)$ or $\#(H \dashrightarrow G)$

Compute d-degenerate ordering $\mathbb{G}$ of G
Initialise and fill subset dictionary R // Lemma 1
Compute $\mathrm{aut}_S(H)$ // Needed in CountStrong/CountWeak
    // Compute index reps. of d-degenerate orderings and left-cover number
Compute minimum c for which H is (c,d)-locatable   // Lemma 8
Compute the set $\mathcal{H} = \{\mathrm{rep}(H, \mathbb{H}[S]) \mid \mathbb{H}$ is d-degenerate$\}$ // Corollary 2
    // Count frequencies of H by index representation
$k = 0$
for $L \subseteq V(G)$ *with* $|L| \leq c$ **do**
    for $\tilde{S} \subseteq N_{\mathbb{G}}^-[L]$ **do**
        Compute $Q_{\tilde{S}}$ from R // Theorem 1
        for *every* $(\mathbb{H}_S, \mathrm{J}) \in \mathcal{H}$ **do**
            $k = k + CountStrong/CountWeak_d(\mathbb{G}, (\mathbb{H}_S, \mathrm{J}), \tilde{S}, Q_{\tilde{S}}, \mathrm{aut}_S(H))$

return k
Algorithm 3: Algorithm for counting (c,d)-locatable patterns in d-degenerate graphs.

Lemma 8. ($\star$). *For every pattern H and $d \in \mathbb{N}$, we can compute the minimum value $c \in \mathbb{N}$ such that H is (c,d)-locatable in time $O\big(d|H| + (cd)^{O((cd)^d)}\big)$.*

The proof of Lemma 8 can further be used to compute all relevant representations of a pattern, meaning those that can occur in a d-degenerate host graph.

Corollary 2. *For a pattern H and an integer $d \in N$ we can compute the set $\{\mathrm{rep}(H, \mathbb{H}[S]) \mid \mathbb{H}$ is a d-degenerate ordering$\}$ in time $O(d|H| + (cd)^{O(cd^2)})$.*

Proof. We proceed as in the proof of Lemma 8. For every ordering $\mathbb{S}$ of S and any placement of T_s among $\mathbb{S}$ we can check whether this placement corresponds to a d-degenerate ordering of H. If so, we add the representative $\mathrm{rep}(H, \mathbb{S})$ to the set. $\qquad\square$

Theorem 3. *There exists an algorithm that, given a d-degenerate graph G and a (c,d)-locatable pattern H as input, computes $\#(H \hookrightarrow G)$ in time[7] $O\big((cd)^{O((cd)^d)} + (cd)^{2cd+d}|G|^c\big)$.*

Proof. The algorithm is listed as Algorithm 3. We first compute a d-degeneracy ordering $\mathbb{G}$ of G in time $O(\|G\|)$, then $\mathrm{aut}_S(H)$ in time $O(d|S|^{|S|+d} + \|H\|)$ (Lemma 4), the minimum c for which H is (c,d)-locatable as well as the set $\mathcal{H}$ in time $O(d|H| + (cd)^{O((cd)^d)})$ (Lemma 8 and Corollary 2).

The algorithm now uses that H is (c,d)-locatable: for each strong embedding of $H \overset{\phi}{\hookrightarrow} G$ we can locate the set $\phi(S)$ in the left-neighbourhood of some set $L \subseteq$

---

[7] For ease of presentation we assume for the running time that $d \geq 2$. This is justified by the fact that graphs of degeneracy one are trees.

$V(G)$ of size at most c. Therefore, we enumerate all such potential sets L in time $O(|G|^c)$ and for each L check all $O((cd)^{|S|})$ candidate sets $\tilde{S} \subseteq N^-[L]$. For each set $\tilde{S}$, we invoke *CountStrong* a total number of $|\mathcal{H}| \leq |S|!$ times, resulting in a running time of

$$O\big(|G|^c (cd)^{|S|} |S|! |S|^{d+1}\big) = O\big((cd)^{cd}(cd)!(cd)^{d+1}|G|^c\big) = O\big((cd)^{2cd+d}|G|^c\big)$$

where we used that $|S| \leq cd$ and that $\mathrm{supp}(J) \leq 4|S|^d$ by Lemma 2 for each representative $(\mathbb{H}_S, J) \in \mathcal{H}$. The total running time is

$$O(\|G\|) + O(d|S|^{|S|+d} + \|H\|) + O(d|H| + (cd)^{O((cd)^d)}) + O\big((cd)^{2cd+d}|G|^c\big)$$
$$= O\big((cd)^{O((cd)^d)} + (cd)^{2cd+d}|G|^c\big)$$

$\square$

Theorem 4. ($\star$). *There exists an algorithm that, given a d-degenerate graph G and a (c,d)-locatable pattern H as input, computes $\#(H \hookrightarrow G)$ in time $(cd)^{O((cd)^d)} + O\big(cd^2 f(c,d) \cdot |H|^{f(c,d)} \cdot |G|^c\big)$ where $f(c,d) = 2^d(cd)^{d+1}$.*

3.4   The Structure of $(1,d)$-Locatable Pattern

While the general structure of these patterns appears to be rich, the structure of $(1,d)$-locatable patterns—which are probably the only patterns interesting in practical applications—can be characterised neatly:

Lemma 9. ($\star$). *A pattern H is $(1,d)$-locatable and d-degenerate iff one of the following holds: 1. $H[S]$ is a clique, or 2. T contains $d + 1 - \delta(H[S])$ vertices whose neighbourhood is S.*

The above characterisation of $(1,d)$-locatable patterns immediately poses the question whether a similar characterisation holds for $(1,d)$-locatable *subsets* $S' \subseteq S$. This turns out not to be true, we defer this discussion to the full version which also include some further notes on (c,d)-locatable pattern as well as remarks on locatability beyond distance one.

4   Lower Bounds

While the general structure of (c,d)-locatable patterns is complicated, from a practical perspective we are mostly interested in $(1,d)$-locatable patterns, since these are the *only* large patterns that can be counted in linear time in degenerate graphs[8], assuming the TDC [3]. We complement this lower bound by showing that patterns which are *barely* not $(1,d)$-locatable—adding one more vertex with neighbourhood S makes them $(1,d)$-locatable—are $\#W[1]$-hard to count.

Theorem 5. *For every $d \geq 5$ there exists a pattern H with sides S, T such that the problem of counting the number of strong or weak containments of H in a d-degenerate graphs is $\#W[1]$-hard when parametrised by $|S| + d$. Moreover, the pattern has exactly d vertices of degree $|S|$ in T and S is an independent set.*

---

[8] Excluding graphs on five or less vertices for which this is still possible.

5   Conclusion

We have initiated the study of counting *large* subgraphs in d-degenerate graphs and provided some positive and negative findings related to (c, d)-locatable patterns. We further provided a characterisation of $(1, d)$-locatable patterns which our algorithms count with only a linear dependence on $|G|$. In the future we plan on implementing and testing these algorithms on real-world networks, as well as investigating what type of patterns might be of interest in practical applications.

On the theoretical side, our main open question is whether $(1, d)$-locatable patterns are the only patterns (modulo small patterns) that can be counted in linear time in d-degenerate graphs. Further, we ask whether $(2, d)$-locatable patterns still have a simple characterisation. Finally, we note that our algorithm (Lemma 8) to decide whether a pattern is (c, d)-locatable has a surprisingly bad dependence on c and d. While for practical applications this should hardly matter, as c needs to be small anyway, it would be interesting to know whether better dependence on these parameters is possible.

References

1. Abboud, A., Williams, V.V.: Popular conjectures imply strong lower bounds for dynamic problems. In: 2014 IEEE 55th Annual Symposium on Foundations of Computer Science, pp. 434–443. IEEE (2014)
2. Bera, S.K., Gishboliner, L., Levanzov, Y., Seshadhri, C., Shapira, A.: Counting subgraphs in degenerate graphs. J. ACM, **69**(3) (2022)
3. Bera, S. K., Pashanasangi, N., Seshadhri, C.: Linear time subgraph counting, graph degeneracy, and the chasm at size six. In: Vidick, T., (ed.), 11th Innovations in Theoretical Computer Science Conference (ITCS 2020), volume 151 of Leibniz International Proceedings in Informatics (LIPIcs), pp. 38:1–38:20, Dagstuhl (2020). Schloss Dagstuhl – Leibniz-Zentrum für Informatik
4. Bressan, M.: Faster algorithms for counting subgraphs in sparse graphs. Algorithmica **83**(8), 2578–2605 (2021). https://doi.org/10.1007/s00453-021-00811-0
5. Bressan, M., Roth, M.: Exact and approximate pattern counting in degenerate graphs: new algorithms, hardness results, and complexity dichotomies. In: 2021 IEEE 62nd Annual Symposium on Foundations of Computer Science (FOCS), pp. 276–285. IEEE (2022)
6. Chen, Y., Thurley, M., Weyer, M.: Understanding the complexity of induced subgraph isomorphisms. In: Aceto, L., Damgård, I., Goldberg, L.A., Halldórsson, M.M., Ingólfsdóttir, A., Walukiewicz, I. (eds.) ICALP 2008. LNCS, vol. 5125, pp. 587–596. Springer, Heidelberg (2008). https://doi.org/10.1007/978-3-540-70575-8_48
7. Curticapean, R., Marx, D.: Complexity of counting subgraphs: only the boundedness of the vertex-cover number counts. In: 2014 IEEE 55th Annual Symposium on Foundations of Computer Science, pp. 130–139 (2014)
8. Dalmau, V., Jonsson, P.: The complexity of counting homomorphisms seen from the other side. Theor. Comput. Sci. **329**(1–3), 315–323 (2004

9. Drange, P.G., Greaves, P., Muzi, I., Reidl, F.: Computing complexity measures of degenerate graphs. In: 18th International Symposium on Parameterized and Exact Computation (IPEC 2023), volume 285 of Leibniz International Proceedings in Informatics (LIPIcs), pp. 14:1–14:21, Dagstuhl, Germany (2023). Schloss Dagstuhl – Leibniz-Zentrum für Informatik
10. Eppstein, D.: Arboricity and bipartite subgraph listing algorithms. Inf. Process. Lett. **51**(4), 207–211 (1994)
11. Habib, M., Paul, C., Viennoti, L.: A synthesis on partition refinement: a useful routine for strings, graphs, boolean matrices and automata. In: Morvan, M., Meinel, C., Krob, D. (eds.) STACS 1998. LNCS, vol. 1373, pp. 25–38. Springer, Heidelberg (1998). https://doi.org/10.1007/BFb0028546
12. Marx, D.: Can you beat treewidth? Theory Comput. **6**(1), 85–112 (2010)
13. Matula, D.W., Beck, L.L.: Smallest-last ordering and clustering and graph coloring algorithms. J. ACM (JACM) **30**(3), 417–427 (1983)
14. Roth, M., Schmitt, J., Wellnitz, P.: Counting small induced subgraphs satisfying monotone properties. SIAM J. Comput. **53**(6), S20-139 (2024)

Pinwheel Scheduling with Real Periods

Hiroshi Fujiwara[1]($\boxtimes$), Kota Miyagi[2], and Katsuhisa Ouchi[1]

[1] Shinshu University, Nagano, Japan
{fujiwara,k_ouchi}@shinshu-u.ac.jp
[2] The University of Electro-Communications, Chofu, Japan
m2530134@gl.cc.uec.ac.jp

Abstract. For a sequence of tasks, each with a positive integer period, the pinwheel scheduling problem involves finding a valid schedule in the sense that the schedule performs one task per day and each task is performed at least once every consecutive days of its period. It had been conjectured by Chan and Chin in 1993 that there exists a valid schedule for any sequence of tasks with density, the sum of the reciprocals of each period, at most $\frac{5}{6}$. Recently, Kawamura settled this conjecture affirmatively. In this paper we consider an extended version with real periods proposed by Kawamura, in which a valid schedule must perform each task i having a real period a_i at least l times in any consecutive $\lceil la_i \rceil$ days for all positive integer l. We show that any sequence of tasks such that the periods take three distinct real values and the density is at most $\frac{5}{6}$ admits a valid schedule. We hereby conjecture that the conjecture of Chan and Chin is true also for real periods.

Keywords: pinwheel scheduling · perpetual scheduling · density conjecture · task packing · Beatty sequence

1 Introduction

In the pinwheel scheduling problem [6], we are given a sequence of tasks, each with a positive integer period, and want to find a schedule that performs one task per day such that each task is performed at least once every consecutive days of its period. Let $\mathbb{N}$ and $\mathbb{Z}$ be the set of all positive integers and the set of all integers, respectively, and let $[k] = \{i \mid i \in \mathbb{N}, 1 \le i \le k\}$. Formally, an integer-valued instance of the pinwheel schedule problem is a sequence $A = (a_i)_{i \in [k]} \in \mathbb{N}^k$, where a_i is the *period* of task i. The goal is to find a *schedule* $S : \mathbb{Z} \to [k]$, which indicates that task $S(t)$ is performed on day t, such that the following condition is satisfied for each task $i \in [k]$:

for each $m \in \mathbb{Z}$, there exists a day $t \in [m, m + a_i) \cap \mathbb{Z}$ such that $S(t) = i$.

Such a schedule is said to be *valid* for instance A. In particular, when a schedule S is periodic, that is, there exists $p \in \mathbb{N}$ such that $S(t) = S(t+p)$ holds for all $t \in \mathbb{Z}$, we represent the schedule by a sequence of $|S(0)S(1)\cdots S(p-1)|$.

© The Author(s), under exclusive license to Springer Nature Switzerland AG 2026
J. Kozik and A. Wolff (Eds.): SOFSEM 2026, LNCS 16448, pp. 621–633, 2026.
https://doi.org/10.1007/978-3-032-17801-5_45

We say that an instance is *schedulable* if there exists a valid schedule for the instance. For example, instance $(2, 4, 4)$ is schedulable, since schedule $|1213|$ is valid for it.

The *density* of instance $A = (a_i)_{i \in [k]}$ is defined as $D(A) = \sum_{i \in [k]} \frac{1}{a_i}$. Intuitively, the density of an instance means how many tasks a valid schedule for the instance has to perform on average per day. Obviously, $D(A) \leq 1$ is necessary for instance A to admit a valid schedule. (See that the density of instance $(2, 4, 4)$ above is 1.) However, this is not sufficient. Chan and Chin conjectured as follows:

Conjecture 1 ([1]). If an integer-valued instance A satisfies $D(A) \leq \frac{5}{6}$, then A is schedulable.

This conjecture is best possible, since for any a_3, instance $(2, 3, a_3)$, whose density is $\frac{5}{6} + \frac{1}{a_3} > \frac{5}{6}$, is not schedulable. After much work by many researchers such as [2,3,11], Kawamura finally settled this conjecture in the affirmative.

Theorem 1 ([7, Theorem 1]). *If an integer-valued instance A satisfies $D(A) \leq \frac{5}{6}$, then A is schedulable.*

In his paper [7], Kawamura proposed an extended version of the pinwheel scheduling problem that allows *real* values at least 1 for periods of tasks, replacing the validity condition for each task $i \in [k]$ by:

for each $l \in \mathbb{N}$ and $m \in \mathbb{Z}$, there exist at least l values of $t \in [m, m + \lceil la_i \rceil) \cap \mathbb{Z}$ such that $S(t) = i$.

In other words, this condition is that for each $l \in \mathbb{N}$, task i is performed at least l times in any consecutive $\lceil la_i \rceil$ days. Please see that when a_i is an integer, the condition is equivalent to that of the original integer-valued problem. For example, a task with period $\frac{7}{2}$ has to be performed at least once every 4 days and at least twice every 7 days. Schedule $|1112112|$ is valid for instance $(2, \frac{7}{2})$.

It should be remarked that this is not a simple relaxation of the problem. More specifically, even if one identifies n tasks each with period a in a valid schedule, the identified task may not be performed according to a period of $\frac{a}{n}$. For example, schedule $|1111213|$ is valid for instance $(2, 7, 7)$. Identify tasks 2 and 3. Then, the identified task fails to meet the validity condition for a period of $\frac{7}{2}$; the task is not performed once every 4 days, though it is performed twice every 7 days.

Nevertheless, we conjecture that Theorem 1 holds true also for the real-valued version:

Conjecture 2. If a real-valued instance A satisfies $D(A) \leq \frac{5}{6}$, then A is schedulable.

This is also best possible for the same reason as the integer-valued problem. Our main theorem, Theorem 2, is a partial solution to Conjecture 2, which strengthens the plausibility of the conjecture.

Theorem 2. *If a real-valued instance A consists of three distinct values and satisfies $D(A) \leq \frac{5}{6}$, then A is schedulable.*

1.1   Previous Results

The Integer-Valued Pinwheel Scheduling Problem. The next theorem is one of the results aimed at solving Conjecture 1, which can be regarded as an integer-valued version of our main theorem.

Theorem 3. ([11, Theorem 4]). *If an integer-valued instance A consists of three distinct values and satisfies $D(A) \leq \frac{5}{6}$, then A is schedulable.*

We will see below that it seems hopeless to extend Theorem 3 directly to the real-valued problem. (Conversely, our proof of Theorem 2 can be seen as a simpler proof of Theorem 3.) The proof of Theorem 3 in the paper [11] constructs a valid schedule for a given integer-valued instance

$$(\underbrace{c_1, \cdots, c_1}_{q_1}, \underbrace{c_2, \cdots, c_2}_{q_2}, \underbrace{c_3, \cdots, c_3}_{q_3}).$$

In the schedule, for each $i = 1, 2,$ and 3, identify all the tasks with period c_i and rename them as task i. If the resulting schedule were valid for instance $(\frac{c_1}{q_1}, \frac{c_2}{q_2}, \frac{c_3}{q_3})$, this would give a solution to the problem with rational periods. However, this attempt fails. We have a counterexample: instance

$$(\underbrace{13, \cdots, 13}_{5}, \underbrace{14, \cdots, 14}_{5}, \underbrace{78, \cdots, 78}_{5}).$$

Apply this to the construction method and identify all the 5 tasks with period 14 in the resulting schedule. It will be found that the identified task is not performed at a period of $\frac{14}{5}$.

Gąsieniec et al. partially solved Conjecture 1 for the case of 12 or fewer tasks [3]. The authors provided a *Pareto surface* of the whole set of such instances, where a Pareto surface C for a set of pinwheel scheduling instances I is an inclusion minimal set of pairs of an instance with minimum periods and a schedule that is valid for the instance such that, for every instance A in I, C includes a valid schedule for A. We will also consider this point of view in Sects. 2 and 3.

We introduce a stronger conjecture recently made by Kawamura:

Conjecture 3 ([8, **Conjecture 1**]). If an integer-valued instance $A = (a_i)_{i \in [k]}$ satisfies $D(A) \leq 1 - \frac{a_1 - 1}{a_1(a_1 + 1)}$, then A is schedulable.

The Real-Valued Pinwheel Scheduling Problem. Kawamura [7] extended the pinwheel scheduling problem to real periods as one of the techniques for proving Theorem 1. The following theorem is a partial solution to Conjecture 2.

Theorem 4 ([7, Theorem 2]). *If a real-valued instance A contains at most two periods and satisfies $D(A) \leq 1$, then A is schedulable.*

Theorem 5 below can be seen as a partial solution to Conjecture 2 for the case where $a_1 > 103$, since the density threshold in the statement exceeds $\frac{5}{6}$ in this case.

Theorem 5 ([9, Theorem 4]). *If a real-valued instance $A = (a_i)_{i \in [k]}$ satisfies $D(A) \leq 1 - \frac{1 + \ln 2}{1 + \sqrt{a_1}} - \frac{3}{2a_1}$, then A is schedulable.*

Other Related Problems. The pinwheel scheduling problem can be viewed as a task packing problem. Its dual problem, the pinwheel covering problem, has recently been studied [9,10]. The bamboo garden trimming problem [4,7] is an optimization problem closely related to the pinwheel scheduling problem. Refer to the survey [8] for further variants and related problems.

2   A Warm-Up: Case of Two Distinct Real Values

To demonstrate the idea of the proof of our main theorem, Theorem 2, for a simpler case, we first attempt to prove a weaker theorem for the case of two distinct real values, Theorem 6. We mention that this can be obtained as a corollary of Theorem 4 and Lemma 2 below.

Theorem 6. *If a real-valued instance A consists of two distinct values and satisfies $D(A) \leq \frac{5}{6}$, then A is schedulable.*

The following two lemmas play an important role. All schedulability throughout this paper, except for that of some individual instances, is derived from these lemmas. Please see in Lemma 1 that the same schedule remains valid for the new instance.

Lemma 1 ([7, Lemma 3 (1)]). *If a schedule S is valid for a real-valued instance $(a_i)_{i \in [k]}$, then S is valid also for the instance obtained by replacing some period a_j $(j \in [k])$ by a real $b \geq a_j$.*

Lemma 2 ([7, Lemma 3 (2)]). *If a real-valued instance $(a_i)_{i \in [k]}$ is schedulable, so is the instance obtained by replacing some period a_j $(j \in [k])$ by q copies of qa_j for any $q \in \mathbb{N}$.*

Lemma 3 below states that any real-valued instance with two periods and density exactly $\frac{5}{6}$ is schedulable. Instead of such an instance $(a_1, \frac{1}{\frac{5}{6} - \frac{1}{a_1}})$ itself, we focus on its projection a_1. Note that the condition $\frac{6}{5} < a_1 \leq \frac{12}{5}$ in the statement ensures, without loss of generality, that the periods are ordered in non-decreasing order. Lemmas 1, 2, and 3 immediately imply Theorem 6.

Lemma 3. *Every real-valued instance $A = (a_1, \frac{1}{\frac{5}{6} - \frac{1}{a_1}})$ with $\frac{6}{5} < a_1 < \frac{12}{5}$ satisfies at least one of the following conditions and the schedule S defined there is valid for instance A:*
(i) If $\frac{6}{5} < a_1 \leq \frac{3}{2}$, then,

$$2S(t) = \begin{cases} 1, & t \equiv 0, 1, 2, 3, 4; \\ 2, & t \equiv 5 \ (mod\ 6), \end{cases} \tag{1}$$

which is represented as $|111112|$.

(ii) If $\frac{3}{2} \leq a_1 \leq 2$, then,

$$2S(t) = \begin{cases} 1, & t \equiv 0, 1; \\ 2, & t \equiv 2 \ (mod\ 3), \end{cases} \tag{2}$$

which is represented as $|112|$.
   (iii) If $2 \leq a_1 \leq 3$, then,

$$2S(t) = \begin{cases} 1, & t \equiv 0; \\ 2, & t \equiv 1 \ (mod\ 2), \end{cases} \tag{3}$$

which is represented as $|12|$.

It does not matter that the ranges of a_1 in the cases (i), (ii), and (iii) are not exclusive and even include values out of $\frac{6}{5} < a_1 \leq \frac{12}{5}$. The former fact means that there may exist multiple valid schedules for a single instance.

Now we are going to provide some lemmas for the proof of Lemma 3. We have found by hand that the three instances each with density 1 in the following Lemma 4 are schedulable and each period is minimum possible.

Lemma 4. *The schedules defined by (1), (2), and (3) are valid for real-valued instances $(\frac{6}{5}, 6)$, $(\frac{3}{2}, 3)$, and $(2, 2)$, respectively.*

Proof. It is immediate to see that each of the schedules satisfies the condition for the tasks having an integer period. We will demonstrate that S in (1) satisfies the condition for task 1 with period $\frac{6}{5}$ in instance $(\frac{6}{5}, 6)$.

We denote by $P_i(y, l)$ the proposition that for each $m \in \mathbb{Z}$, there exist at least l values of $t \in [m, m+y) \cap \mathbb{Z}$ such that $S(t) = i$, which means that schedule S performs task i at least l times in any consecutive y days. The schedulability of task 1 in instance $(\frac{6}{5}, 6)$ is represented as $P_1(\lceil l \cdot \frac{6}{5} \rceil, l)$ for all $l \in \mathbb{N}$. We are going to prove this.

For each $l = 1, 2, \ldots, 5$, we have $\lceil l \cdot \frac{6}{5} \rceil = l + 1$. According to (1), $S(t) = 1$ if $t = 0, 1, 2, 3, 4 \ (mod\ 6)$. It is seen that there are l days when task 1 is performed every consecutive $l + 1$ days. That is, $P_1(\lceil l \cdot \frac{6}{5} \rceil, l)$ is true.

For $l \geq 6$, let q be the quotient and r be the remainder when l is divided by 5. We have $\lceil l \cdot \frac{6}{5} \rceil = \lceil (5q + r) \cdot \frac{6}{5} \rceil = 6q + \lceil r \cdot \frac{6}{5} \rceil$. Using what we have already shown, we obtain that: Task 1 is done at least r times in any consecutive $\lceil r \cdot \frac{6}{5} \rceil$ days by $P_1(\lceil z \cdot \frac{6}{5} \rceil, z)$ for $z = 1, 2, 3, 4$, while it is done at least $5q$ times in any consecutive $6q$ days by $P_1(\lceil 5 \cdot \frac{6}{5} \rceil, 5)$. Therefore, task 1 is done at least $5q + r = l$ times in any consecutive $6q + \lceil r \cdot \frac{6}{5} \rceil$ days, which implies $P_1(\lceil l \cdot \frac{6}{5} \rceil, l)$. $\qquad \square$

Using the following helper lemma, we can further derive a set of schedulable instances from those found to be schedulable by Lemma 4. Note that the resulting set includes instances whose periods are not in non-decreasing order. The lemma follows immediately from Lemma 1.

Lemma 5. *If a schedule S is valid for real-valued instance (b_1, b_2), then S is valid also for real-valued instance $(a_1, \frac{1}{\frac{5}{6} - \frac{1}{a_1}})$ with $a_1 \geq b_1$ and $\frac{1}{\frac{5}{6} - \frac{1}{a_1}} \geq b_2$.*

Proof (of Lemma 3). Applying Lemma 5, the schedulability of instance $(\frac{6}{5}, 6)$ by Lemma 4 leads us to the fact that the schedule defined by (1) is valid for instance $(a_1, \frac{1}{\frac{5}{6} - \frac{1}{a_1}})$ if $a_1 \geq \frac{6}{5}$ and $\frac{1}{\frac{5}{6} - \frac{1}{a_1}} \geq 6$, equivalently,

$$\frac{6}{5} < a_1 \leq \frac{3}{2}. \tag{4}$$

Similarly, we derive that the schedule defined by (2) is valid for instance $(a_1, \frac{1}{\frac{5}{6} - \frac{1}{a_1}})$ if

$$\frac{3}{2} \leq a_1 \leq 2, \tag{5}$$

and that the schedule defined by (3) is valid for instance $(a_1, \frac{1}{\frac{5}{6} - \frac{1}{a_1}})$ if

$$2 \leq a_1 \leq 3. \tag{6}$$

It is obvious that a_1 such that $\frac{6}{5} < a_1 < \frac{12}{5}$ satisfies at least one of (4), (5), or (6). $\qquad\square$

We focus on real-valued instances with two periods. The proof of Theorem 6 based on Lemma 3 tells us that for any real-valued instance with two periods and density less than or equal to $\frac{5}{6}$, the schedule defined by either (1), (2), or (3) is valid. It is seen that each schedule is instance-insensitive in the sense that the same schedule remains valid as long as either (4), (5), or (6) holds for the instance. In the context of the paper [3], the result can be summarized as follows:

Corollary 1. $\{((\frac{6}{5}, 6), |111112|), ((\frac{3}{2}, 3), |112|), ((2, 2), |12|)\}$ *is a Pareto surface of the set of all real-valued instances with two periods and density less than or equal to* $\frac{5}{6}$.

Compare with the fact that $\{((2, 2), |12|)\}$ is a Pareto surface of the set of all *integer-valued* instances with two periods and density less than or equal to $\frac{5}{6}$ [3]. That is, adding two instance-schedule pairs to this yields a Pareto surface for the real-valued instances.

3   Proof of Theorem 2

Theorem 2 is proved in essentially the same way as Theorem 6. The following Lemma 6 is the heart of the argument. It states that any real-valued instance with three periods and density exactly $\frac{5}{6}$ is schedulable. We deal with a pair (a_1, a_2) as a projection of such an instance. Note that the conditions $a_1 < a_2$ and $\frac{5}{6} \leq \frac{1}{a_1} + \frac{2}{a_2}$ in the statement guarantee, without loss of generality, that the periods are ordered in non-decreasing order. Lemmas 1, 2, and 6 immediately imply Theorem 2.

Lemma 6. *Every real-valued instance* $A = (a_1, a_2, \frac{1}{\frac{5}{6} - \frac{1}{a_1} - \frac{1}{a_2}})$ *with* $a_1 < a_2$, $\frac{5}{6} < \frac{1}{a_1} + \frac{2}{a_2}$, *and* $\frac{1}{a_1} + \frac{1}{a_2} < \frac{5}{6}$ *satisfies at least one of the following conditions and the schedule* S *defined there is valid for instance* A:

(I) If $\frac{1}{a_1} + \frac{2}{a_2} \leq 1$, *then, letting* $a_2' = \frac{2}{1 - \frac{1}{a_1}}$,

$$S(t) = \begin{cases} 1, & t \in \{\lceil ja_1 \rceil - 1 \mid j \in \mathbb{Z}\}; \\ 2, & t \in \{\lfloor ja_2' \rfloor \mid j \in \mathbb{Z}\}; \\ 3, & t \in \{\lfloor (j + \frac{1}{2})a_2' \rfloor \mid j \in \mathbb{Z}\}. \end{cases} \tag{7}$$

(II) If $a_1 \geq \frac{3}{2}$, $a_2 > 5$, *and* $\frac{5}{6} - \frac{1}{9} \leq \frac{1}{a_1} + \frac{1}{a_2} < \frac{5}{6}$ *then*

$$S(t) = \begin{cases} 1, & t \equiv 0, 1, 3, 4, 6, 7; \\ 2, & t \equiv 2, 5; \\ 3, & t \equiv 8 \ (mod \ 9), \end{cases} \tag{8}$$

which is represented as $|112112113|$.

(III) If $a_1 \geq \frac{11}{7}$, $a_2 > 4$, *and* $\frac{5}{6} - \frac{1}{11} \leq \frac{1}{a_1} + \frac{1}{a_2} < \frac{5}{6}$ *then*

$$S(t) = \begin{cases} 1, & t \equiv 0, 1, 3, 4, 6, 7, 9; \\ 2, & t \equiv 2, 5, 8; \\ 3, & t \equiv 10 \ (mod \ 11), \end{cases} \tag{9}$$

which is represented as $|11211211213|$.

(IV) If $a_1 \geq \frac{12}{7}$, $a_2 > 3$, *and* $\frac{5}{6} - \frac{1}{12} \leq \frac{1}{a_1} + \frac{1}{a_2} < \frac{5}{6}$ *then,*

$$S(t) = \begin{cases} 1, & t \equiv 0, 2, 3, 5, 7, 8, 10; \\ 2, & t \equiv 1, 4, 6, 9; \\ 3, & t \equiv 11 \ (mod \ 12), \end{cases} \tag{10}$$

which is represented as $|121121211213|$.

(V) If $a_1 > 2$, $a_2 \geq 3$, *and* $\frac{5}{6} - \frac{1}{6} \leq \frac{1}{a_1} + \frac{1}{a_2} < \frac{5}{6}$ *then*

$$S(t) = \begin{cases} 1, & t \equiv 0, 2, 3; \\ 2, & t \equiv 1, 4; \\ 3, & t \equiv 5 \ (mod \ 6), \end{cases} \tag{11}$$

which is represented as $|121123|$.

(VI) If $a_1 > 2$, $a_2 \geq \frac{12}{5}$, *and* $\frac{5}{6} - \frac{1}{11} \leq \frac{1}{a_1} + \frac{1}{a_2} < \frac{5}{6}$, *then*

$$S(t) = \begin{cases} 1, & t \equiv 0, 2, 4, 5, 7, 9; \\ 2, & t \equiv 1, 3, 6, 8, 10; \\ 3, & t \equiv 11 \ (mod \ 12), \end{cases} \tag{12}$$

which is represented as $|121211212123|$.

(VII) If $a_1 \geq \frac{12}{5}$, $a_2 \geq \frac{12}{5}$, and $\frac{5}{6} - \frac{1}{6} \leq \frac{1}{a_1} + \frac{1}{a_2} < \frac{5}{6}$ then

$$S(t) = \begin{cases} 1, & t \equiv 0, 2, 4, 7, 9; \\ 2, & t \equiv 1, 3, 6, 8, 10; \\ 3, & t \equiv 5, 11 \ (mod\ 12), \end{cases} \tag{13}$$

which is represented as $|121213212123|$.

Lemma 6 splits the whole set of real-valued instances with three periods and density exactly $\frac{5}{6}$ into seven cases (I) through (VII). In particular, for the analysis of cases (II) through (VII), we discovered six real-valued instances of Lemma 8 for which the covering of Lemma 10 succeeds by trial and error while checking schedulability by an integer programming solver.

We discuss some lemmas for the proof of Lemma 6. We begin with instances that satisfy the condition of case (I) of Lemma 6, that is $\frac{1}{a_1} + \frac{2}{a_2} \leq 1$. We here insist on a valid schedule in an explicit form of $S(t) = \dots$. Otherwise, the schedulability itself can easily be shown by Lemmas 1 and 2, and the schedulability of instance $(a_1, \frac{a_2}{2})$ implied by Theorem 4.

Lemma 7. *The schedule S defined by (7) is valid for real-valued instance $(a_1, a_2, \frac{1}{\frac{5}{6} - \frac{1}{a_1} - \frac{1}{a_2}})$ for any a_1 and a_2 such that $a_1 < a_2$, $\frac{5}{6} < \frac{1}{a_1} + \frac{2}{a_2} \leq 1$, and $\frac{1}{a_1} + \frac{1}{a_2} < \frac{5}{6}$.*

Proof. First, we confirm that S performs a single task for every day. Since $a_2' = \frac{2}{1 - \frac{1}{a_1}} > 2$, the set of days when task 2 is performed $\{\lfloor ka_2' \rfloor \mid k \in \mathbb{Z}\}$ and the set of days when task 3 is performed $\{\lfloor (k + \frac{1}{2})a_2' \rfloor \mid k \in \mathbb{Z}\}$ have no intersection. Their union is $\{\lfloor k \cdot \frac{a_2'}{2} \rfloor \mid k \in \mathbb{Z}\}$. (See that S for tasks 2 and 3 is constructed by treating them as a single task with period $\frac{a_2'}{2}$ and then assigning tasks 2 and 3 in turn.) Moreover, for a_1 and a_2' such that $\frac{1}{a_1} + \frac{2}{a_2'} = 1$, $\{\lfloor k \cdot \frac{a_2'}{2} \rfloor \mid k \in \mathbb{Z}\}$ and the set of days when task 1 is performed $\{\lceil ja_1 \rceil - 1 \mid j \in \mathbb{Z}\}$ are known to be a partition of $\mathbb{Z}$ [5]. (When a_1 and a_2' are irrational numbers, this partition is called Rayleigh's or Beatty's theorem.)

Next, we show that S is valid for instance (a_1, a_2', a_2'), which is sufficient to prove the lemma due to $a_2 \geq a_2'$, $\frac{1}{\frac{5}{6} - \frac{1}{a_1} - \frac{1}{a_2}} \geq a_2'$, and Lemma 1.

According to (7), for any $l \in \mathbb{N}$, the maximum time span in which task 1 is performed exactly $(l-1)$ times is from day $\lceil ma_1 \rceil$ to day $(\lceil (m+l)a_1 \rceil - 2)$ for some integer m, and its length is $(\lceil (m+l)a_1 \rceil - \lceil ma_1 \rceil - 1)$ days. (i) If la_1 is an integer, we evaluate $\lceil (m+l)a_1 \rceil - \lceil ma_1 \rceil - 1 = ma_1 + \lceil la_1 \rceil - ma_1 - 1 = la_1 - 1 = \lceil la_1 \rceil - 1$. (ii) Otherwise, we get $\lceil (m+l)a_1 \rceil - \lceil ma_1 \rceil - 1 \leq (\lceil ma_1 \rceil + \lceil la_1 \rceil) - \lceil ma_1 \rceil - 1 = \lceil la_1 \rceil - 1$. In any case, we observe that task 1 is done at least l times in any consecutive $\lceil la_1 \rceil$ days.

We finally check the validity of tasks 2 and 3 together. Set $d = 0$ for task 2 and $\frac{1}{2}$ for task 3. For any $l \in \mathbb{N}$, the maximum time span in which the task is performed exactly $(l-1)$ times is from day $(\lfloor (m+d)a_2' \rfloor + 1)$ to day $(\lfloor (m+d+l)a_2' \rfloor - 1)$ for some integer m, and its length is $(\lfloor (m+d+l)a_2' \rfloor - \lfloor (m+d)a_2' \rfloor - 1)$ days. (i) If la_2' is an integer, we obtain $\lfloor (m+d+l)a_2' \rfloor - \lfloor (m+d)a_2' \rfloor - 1 = \lfloor (m+d)a_2' \rfloor + la_2' - \lfloor (m+d)a_2' \rfloor - 1 = la_2' - 1 = \lceil la_2' \rceil - 1$. (ii) Otherwise, we get $\lfloor (m+d+l)a_2' \rfloor - \lfloor (m+d)a_2' \rfloor - 1 \leq (\lfloor (m+d)a_2' \rfloor + \lfloor la_2' \rfloor + 1) - \lfloor (m+d)a_2' \rfloor - 1 = \lfloor la_2' \rfloor = \lceil la_2' \rceil - 1$. In any case, we see that the task is done at least l times in any consecutive $\lceil la_2' \rceil$ days. $\qquad\square$

The following lemma ensures the schedulability of the six real-valued instances. Each of the periods, except those involving ε, is minimum possible. Substituting zero for ε makes the instance non-schedulable.

Lemma 8. *The schedules defined by (8), (9), (10), (11), and (12) are valid for real-valued instances $(\frac{3}{2}, 5+\varepsilon, 9)$, $(\frac{11}{7}, 4+\varepsilon, 11)$, $(\frac{12}{7}, 3+\varepsilon, 12)$, $(2+\varepsilon, 3, 6)$, and $(2+\varepsilon, \frac{12}{5}, 12)$ for any $\varepsilon > 0$, respectively. The schedule defined by (13) is valid for $(\frac{12}{5}, \frac{12}{5}, 6)$.*

Proof. It is easy to check for the tasks with integer periods. For the tasks with rational periods, we can verify the validity as the same as in the proof of Lemma 4. We therefore focus on the tasks with periods that involve ε. We will demonstrate that for any $\varepsilon > 0$, S in (8) satisfies the condition for task 2 with period $5 + \varepsilon$ in instance $(\frac{3}{2}, 5+\varepsilon, 9)$.

Fix $\varepsilon > 0$ arbitrarily. As in the proof of Lemma 4, we denote by $P_i(y, l)$ the proposition that for each $m \in \mathbb{Z}$, there exist at least l values of $t \in [m, m+y) \cap \mathbb{Z}$ such that $S(t) = i$, which means that schedule S performs task i at least once every consecutive y days. Clearly, for $y \leq y'$ and $l \geq l'$, $P_i(y, l) \Rightarrow P_i(y', l')$ holds true. In addition, if $P_i(y, l)$ and $P_i(y', l')$ hold true, then so does $P_i(y+y', l+l')$.

The schedulability of task 2 in instance $(\frac{3}{2}, 5+\varepsilon, 9)$ is equivalent to $P_2(\lceil l \cdot (5+\varepsilon) \rceil, l)$ for all $l \in \mathbb{N}$. We are going to prove this. We know from (8) that S is itself has a period of 9 and $S(t) = 2$ for $t \equiv 2, 5 \pmod 9$, which implies that $P_2(9, 2)$ is true. See that $P_2(9, 2)$ indicates that task 2 is performed approximately on average once every $\frac{9}{2}(< 5 + \varepsilon)$ days. Roughly speaking, $P_2(\lceil l \cdot (5+\varepsilon) \rceil, l)$ holds true for large l and therefore we only need to check this for small l. We will often use the following inequality, $\lceil l \cdot (5+\varepsilon) \rceil = 5l + \lceil l\varepsilon \rceil \geq 5l + 1$ for any integer l.

For $l = 1$, (8) says that there is at least one day t satisfying $t \equiv 2, 5 \pmod 9$ in any consecutive $5l + 1 = 6$ days, thus $P_2(6, 1) \Rightarrow P_2(\lceil 1 \cdot (5+\varepsilon) \rceil, 1)$. For $l = 2$, since $\lceil 2 \cdot (5+\varepsilon) \rceil \geq 5 \cdot 2 + 1 = 11 > 1 \cdot 9$, $P_2(9, 2) \Rightarrow P_2(\lceil 2 \cdot (5+\varepsilon) \rceil, 2)$ holds. For $l = 3$, it follows that $\lceil 3 \cdot (5+\varepsilon) \rceil \geq 5 \cdot 3 + 1 = 16 = 1 \cdot 9 + 7$. Since $P_2(6, 1) \Rightarrow P_2(7, 1)$ and $P_2(9, 2)$, we have $P_2(9 + 7, 2 + 1) \Rightarrow P_2(\lceil 3 \cdot (5+\varepsilon) \rceil, 3)$. For $l = 4$, $\lceil 4 \cdot (5+\varepsilon) \rceil \geq 5 \cdot 4 + 1 = 21 > 2 \cdot 9$ holds. Using $P_2(9, 2)$, we derive $P_2(2 \cdot 9, 2 \cdot 2) \Rightarrow P_2(\lceil 4 \cdot (5+\varepsilon) \rceil, 4)$. For $l = 5$, we have $\lceil 5 \cdot (5+\varepsilon) \rceil \geq 5 \cdot 5 + 1 = 26 = 2 \cdot 9 + 8$. From $P_2(6, 1) \Rightarrow P_2(8, 1)$ and $P_2(9, 2)$, we get $P_2(2 \cdot 9 + 8, 2 \cdot 2 + 1) \Rightarrow P_2(\lceil 5 \cdot (5+\varepsilon) \rceil, 5)$. For $l = 6$, $\lceil 6 \cdot (5+\varepsilon) \rceil \geq 5 \cdot 6 + 1 = 31 > 3 \cdot 9$ holds. We have $P_2(3 \cdot 9, 3 \cdot 2) \Rightarrow P_2(\lceil 6 \cdot (5+\varepsilon) \rceil, 6)$ by $P_2(9, 2)$. Finally, for $l \geq 7$, we have $\lceil l \cdot (5+\varepsilon) \rceil \geq 5l + 1 \geq 9 \cdot \lfloor \frac{5l+1}{9} \rfloor$. It also holds true that $2 \cdot \lfloor \frac{5l+1}{9} \rfloor \geq l$, since $2 \cdot \lfloor \frac{5l+1}{9} \rfloor - l > 2 \cdot (\frac{5l+1}{9} - 1) - l = \frac{l-7}{9} \geq 0$. Applying $P_2(9, 2)$, we conclude $P_2(9 \cdot \lfloor \frac{5l+1}{9} \rfloor, 2 \cdot \lfloor \frac{5l+1}{9} \rfloor) \Rightarrow P_2(\lceil l \cdot (5+\varepsilon) \rceil, l)$. $\qquad\square$

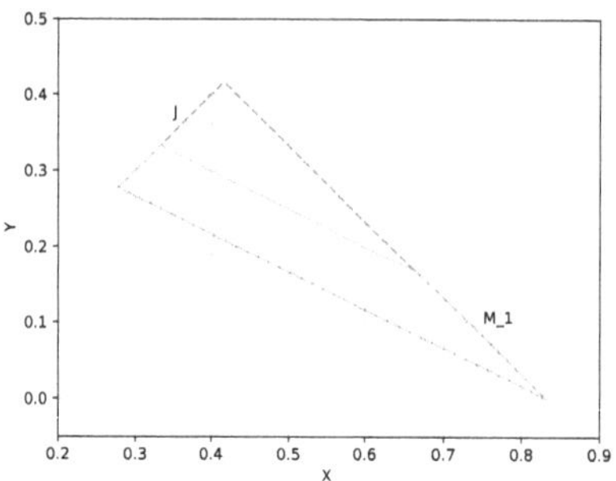

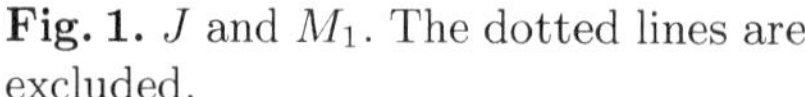

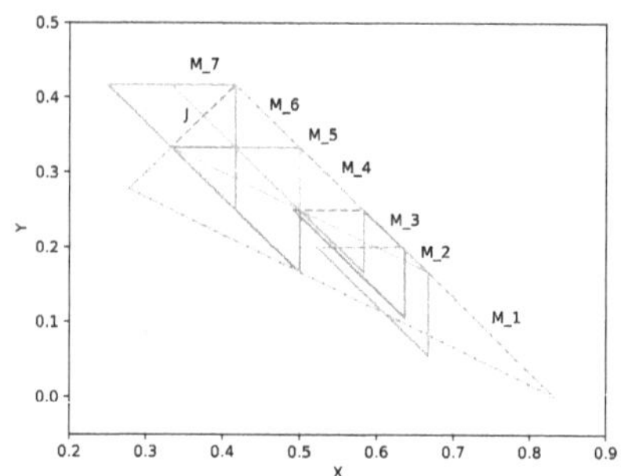

Fig. 1. J and M_1. The dotted lines are excluded.

Fig. 2. J and $M_1, M_2, \ldots, M_7$. The dotted lines are excluded.

Lemma 9 below is a helper lemma, which is the counterpart of Lemma 5 in Sect. 2, that derives a set of schedulable real-valued instances from those found to be schedulable by Lemma 8. Note that the resulting set includes instances whose periods are not in non-decreasing order. Lemma 9 follows immediately from Lemma 1.

Lemma 9. *If a schedule S is valid for real-valued instance (b_1, b_2, b_3), then S is valid also for real-valued instance $(a_1, a_2, \frac{1}{\frac{5}{6} - \frac{1}{a_1} - \frac{1}{a_2}})$ with $a_1 \geq b_1$, $a_2 \geq b_2$, and*

$$\frac{1}{\frac{5}{6} - \frac{1}{a_1} - \frac{1}{a_2}} \geq b_3.$$

The conditions of cases (II) through (VII) of Lemma 6 turn out to be the conditions that are fulfilled by schedulable real-valued instances obtained by applying Lemma 9 to the instances found to be schedulable by Lemma 8.

Lemma 6 holds if the union of such real-valued instances and real-valued instances that apply to case (I), which are shown to be schedulable by Lemma 7, contains any real-valued instance with three periods and density exactly $\frac{5}{6}$. To facilitate this discussion, we map a projection of a real-valued instance $(a_1, a_2) \in \{(x, y) \mid 1 \leq x, 1 \leq y\}$ by a bijection $(a_1, a_2) \mapsto (\frac{1}{a_1}, \frac{1}{a_2}) \in \{(X, Y) \mid 0 < X \leq 1, 0 < Y \leq 1\}$. In other words, we look at *frequencies* rather than periods.

The image of the whole set of real-valued instances with three periods and density exactly $\frac{5}{6}$ is

$$J := \left\{ (X, Y) \,\middle|\, X > Y, \frac{5}{6} < X + 2Y, X + Y < \frac{5}{6} \right\}. \tag{14}$$

(Refer to the beginning of the statement of Lemma 6 for the condition.) The images of the whole set of real-valued instances that apply to cases (I) through (VII) are, respectively,

$$M_1 := \left\{(X,Y) \,\middle|\, X > Y, \frac{5}{6} < X + 2Y \le 1, X + Y < \frac{5}{6}\right\},$$

$$M_2 := \left\{(X,Y) \,\middle|\, X \le \frac{2}{3}, Y < \frac{1}{5}, \frac{5}{6} - \frac{1}{9} \le X + Y < \frac{5}{6}\right\},$$

$$M_3 := \left\{(X,Y) \,\middle|\, X \le \frac{7}{11}, Y < \frac{1}{4}, \frac{5}{6} - \frac{1}{11} \le X + Y < \frac{5}{6}\right\},$$

$$M_4 := \left\{(X,Y) \,\middle|\, X \le \frac{7}{12}, Y < \frac{1}{3}, \frac{5}{6} - \frac{1}{12} \le X + Y < \frac{5}{6}\right\},$$

$$M_5 := \left\{(X,Y) \,\middle|\, X < \frac{1}{2}, Y \le \frac{1}{3}, \frac{5}{6} - \frac{1}{6} \le X + Y < \frac{5}{6}\right\},$$

$$M_6 := \left\{(X,Y) \,\middle|\, X < \frac{1}{2}, Y \le \frac{5}{12}, \frac{5}{6} - \frac{1}{12} \le X + Y < \frac{5}{6}\right\}, \text{ and}$$

$$M_7 := \left\{(X,Y) \,\middle|\, X \le \frac{5}{12}, Y \le \frac{5}{12}, \frac{5}{6} - \frac{1}{6} \le X + Y < \frac{5}{6}\right\}.$$

See Figs. 1 and 2.

Lemma 10 may be visually clear from the figures.

Lemma 10. *It holds that $J \subset \bigcup_{i=1}^{7} M_i$.*

Proof. First, we obtain $J \setminus M_1 = \{(X,Y) \,|\, X > Y, 1 < X + 2Y, X + Y < \frac{5}{6}\}$, which is the sides and interior of the triangle $(\frac{2}{3}, \frac{1}{6})$, $(\frac{5}{12}, \frac{5}{12})$, $(\frac{1}{3}, \frac{1}{3})$, excluding the two sides $(\frac{2}{3}, \frac{1}{6})$, $(\frac{5}{12}, \frac{5}{12})$ and $(\frac{2}{3}, \frac{1}{6})$, $(\frac{1}{3}, \frac{1}{3})$ (see Fig. 1). We will show that $J \setminus M_1 \subset \bigcup_{i=2}^{7} M_i$ (see Fig. 2) by dividing into regions $\frac{7}{11} < X \le 1$, $\frac{7}{12} < X \le \frac{7}{11}$, $\frac{1}{2} \le X \le \frac{7}{12}$, $\frac{5}{12} < X < \frac{1}{2}$, and $0 < X \le \frac{5}{12}$.

(i) $(J \setminus M_1) \cap \{(X,Y) \,|\, \frac{7}{11} < X \le 1, 0 < Y \le 1\}$ is the interior of the triangle $(\frac{2}{3}, \frac{1}{6})$, $(\frac{7}{11}, \frac{13}{66})$, $(\frac{7}{11}, \frac{2}{11})$. On the other hand, $M_2 \cap \{(X,Y) \,|\, \frac{7}{11} < X \le 1, 0 < Y \le 1\}$ is the sides and interior of the parallelogram $(\frac{2}{3}, \frac{1}{18})$, $(\frac{2}{3}, \frac{1}{6})$, $(\frac{7}{11}, \frac{13}{66})$, $(\frac{7}{11}, \frac{17}{198})$, excluding the side $(\frac{7}{11}, \frac{13}{66})$, $(\frac{7}{11}, \frac{17}{198})$. Noting that $\frac{17}{198} < \frac{2}{11} < \frac{13}{66}$ for the Y coordinates of the vertices, we obtain $(J \setminus M_1) \cap \{(X,Y) \,|\, \frac{7}{11} < X \le 1, 0 < Y \le 1\} \subset M_2 \cap \{(X,Y) \,|\, \frac{7}{11} < X \le 1, 0 < Y \le 1\}$.

(ii) $(J \setminus M_1) \cap \{(X,Y) \,|\, \frac{7}{12} < X \le \frac{7}{11}, 0 < Y \le 1\}$ is the sides and interior of the trapezoid $(\frac{7}{11}, \frac{2}{11})$, $(\frac{7}{11}, \frac{13}{66})$, $(\frac{7}{12}, \frac{1}{4})$, $(\frac{7}{12}, \frac{5}{24})$, excluding the three sides $(\frac{7}{11}, \frac{13}{66})$, $(\frac{7}{12}, \frac{1}{4})$, $(\frac{7}{12}, \frac{1}{4})$, $(\frac{7}{12}, \frac{5}{24})$, and $(\frac{7}{12}, \frac{5}{24})$, $(\frac{7}{11}, \frac{2}{11})$. $M_3 \cap \{(X,Y) \,|\, \frac{7}{12} < X \le \frac{7}{11}, 0 < Y \le 1\}$ is the sides and interior of the parallelogram $(\frac{7}{11}, \frac{7}{66})$, $(\frac{7}{11}, \frac{13}{66})$, $(\frac{7}{12}, \frac{1}{4})$, $(\frac{7}{12}, \frac{7}{44})$, excluding the two sides $(\frac{7}{11}, \frac{13}{66})$, $(\frac{7}{12}, \frac{1}{4})$ and $(\frac{7}{12}, \frac{1}{4})$, $(\frac{7}{12}, \frac{7}{44})$. Since $\frac{7}{66} < \frac{2}{11} < \frac{13}{66}$ and $\frac{7}{44} < \frac{5}{24} < \frac{1}{4}$, we have $(J \setminus M_1) \cap \{(X,Y) \,|\, \frac{7}{12} < X \le \frac{7}{11}, 0 < Y \le 1\} \subset M_3 \cap \{(X,Y) \,|\, \frac{7}{12} < X \le \frac{7}{11}, 0 < Y \le 1\}$.

(iii) $(J \setminus M_1) \cap \{(X,Y) \,|\, \frac{1}{2} \le X \le \frac{7}{12}, 0 < Y \le 1\}$ is the sides and interior of the trapezoid $(\frac{7}{12}, \frac{5}{24})$, $(\frac{7}{12}, \frac{1}{4})$, $(\frac{1}{2}, \frac{1}{3})$, $(\frac{1}{2}, \frac{1}{4})$, excluding the two sides $(\frac{7}{12}, \frac{1}{4})$, $(\frac{1}{2}, \frac{1}{3})$ and $(\frac{1}{2}, \frac{1}{4})$, $(\frac{7}{12}, \frac{5}{24})$. $M_4 \cap \{(X,Y) \,|\, \frac{1}{2} \le X \le \frac{7}{12}, 0 < Y \le 1\}$ is the sides and interior of the parallelogram $(\frac{7}{12}, \frac{1}{6})$, $(\frac{7}{12}, \frac{1}{4})$, $(\frac{1}{2}, \frac{1}{3})$, $(\frac{1}{2}, \frac{1}{4})$, excluding the side $(\frac{7}{12}, \frac{1}{4})$, $(\frac{1}{2}, \frac{1}{3})$. Due to $\frac{1}{6} < \frac{5}{24} < \frac{1}{4}$, it holds that $(J \setminus M_1) \cap \{(X,Y) \,|\, \frac{1}{2} \le X \le \frac{7}{12}, 0 < Y \le 1\} \subset M_4 \cap \{(X,Y) \,|\, \frac{1}{2} \le X \le \frac{7}{12}, 0 < Y \le 1\}$.

(iv) $(J \setminus M_1) \cap \{(X,Y) \,|\, \frac{5}{12} < X < \frac{1}{2}, 0 < Y \le 1\}$ is the interior of the trapezoid $(\frac{1}{2}, \frac{1}{4})$, $(\frac{1}{2}, \frac{1}{3})$, $(\frac{5}{12}, \frac{5}{12})$, $(\frac{5}{12}, \frac{7}{24})$. $M_5 \cap \{(X,Y) \,|\, \frac{5}{12} < X < \frac{1}{2}, 0 < Y < 1\}$ is the sides and interior of the trapezoid $(\frac{1}{2}, \frac{1}{6})$, $(\frac{1}{2}, \frac{1}{3})$, $(\frac{5}{12}, \frac{1}{3})$, $(\frac{5}{12}, \frac{1}{4})$, excluding the two sides $(\frac{1}{2}, \frac{1}{6})$, $(\frac{1}{2}, \frac{1}{3})$, and $(\frac{5}{12}, \frac{1}{3})$, $(\frac{5}{12}, \frac{1}{4})$. Besides, $M_6 \cap \{(X, Y) \mid \frac{5}{12} < X < \frac{1}{2}, 0 < Y \leq 1\}$ is the sides and interior of the parallelogram $(\frac{1}{2}, \frac{1}{4})$, $(\frac{1}{2}, \frac{1}{3})$, $(\frac{5}{12}, \frac{5}{12})$, $(\frac{5}{12}, \frac{1}{3})$, excluding the two sides $(\frac{1}{2}, \frac{1}{4})$, $(\frac{1}{2}, \frac{1}{3})$, and $(\frac{5}{12}, \frac{5}{12})$, $(\frac{5}{12}, \frac{1}{3})$. Note that $\frac{1}{6} < \frac{1}{4}$ and $\frac{1}{4} < \frac{7}{24}$ for the Y coordinates of the vertices. The part of region $(J \setminus M_1) \cap \{(X, Y) \mid \frac{5}{12} < X < \frac{1}{2}, 0 < Y \leq 1\}$ where $Y \leq \frac{1}{3}$ is a subset of $M_5 \cap \{(X, Y) \mid \frac{5}{12} < X < \frac{1}{2}, 0 < Y \leq 1\}$, while the rest of the region is a subset of $M_6 \cap \{(X, Y) \mid \frac{5}{12} < X < \frac{1}{2}, 0 < Y \leq 1\}$. Thus, $(J \setminus M_1) \cap \{(X, Y) \mid \frac{5}{12} < X < \frac{1}{2}, 0 < Y \leq 1\} \subset (M_5 \cup M_6) \cap \{(X, Y) \mid \frac{5}{12} < X < \frac{1}{2}, 0 < Y < 1\}$.

(v) $(J \setminus M_1) \cap \{(X, Y) \mid 0 < X \leq \frac{5}{12}, 0 < Y \leq 1\}$ is the sides and interior of the triangle $(\frac{5}{12}, \frac{7}{24})$, $(\frac{5}{12}, \frac{5}{12})$, $(\frac{1}{3}, \frac{1}{3})$, excluding the two sides $(\frac{5}{12}, \frac{5}{12})$, $(\frac{1}{3}, \frac{1}{3})$ and $(\frac{1}{3}, \frac{1}{3})$, $(\frac{5}{12}, \frac{7}{24})$. On the other hand, $M_7 \cap \{(X, Y) \mid 0 < X \leq \frac{5}{12}, 0 < Y \leq 1\}$ is the sides and interior of the triangle $(\frac{5}{12}, \frac{1}{4})$, $(\frac{5}{12}, \frac{5}{12})$, $(\frac{1}{4}, \frac{5}{12})$. Noting that $\frac{1}{4} < \frac{7}{24}$ for the Y coordinates of the vertices and the point $(\frac{1}{3}, \frac{1}{3})$ lies on the side $(\frac{1}{4}, \frac{5}{12})$, $(\frac{5}{12}, \frac{1}{4})$, we obtain $(J \setminus M_1) \cap \{(X, Y) \mid 0 < X \leq \frac{5}{12}, 0 < Y < 1\} \subset M_7 \cap \{(X, Y) \mid 0 < X \leq \frac{5}{12}, 0 < Y < 1\}$. $\square$

Proof (of Lemma 6). Lemmas 7, 8, and 9 imply that instances that apply to cases (I) through (VII) are schedulable and a valid schedule is given either of by (7), (8), (9), (10), (11), (12), or (13). Lemma 10 states that any real-valued instance with three periods and density less than or equal to $\frac{5}{6}$ satisfies one of the conditions of cases (I) through (VII). $\square$

We continue the analysis of real-valued instances with three periods. The proof of Theorem 2 based on Lemma 6 provides us with valid schedules for real-valued instances with three periods and density less than or equal to $\frac{5}{6}$. For an instance that applies to case (I) of Lemma 6, the schedule is determined by the individual period values of the instance. Namely, the schedule is instance-sensitive, which did not occur in the case of two periods in Sect. 2. (The same is true if the schedulability is derived from Lemmas 1 and 2, and Theorem 4.)

On the other hand, for an instance that applies to cases (II) through (VII), the schedule is independent of the individual period values of the instance. In particular, the last instance of Lemma 8, which corresponds to the condition of case (VII), has no ε, and the instance itself is the one with minimum periods among those schedulable. This fact can be expressed in the context of the paper [3] as follows:

Corollary 2. $\{((\frac{12}{5}, \frac{12}{5}, 6), |121213212123|)\}$ *is a Pareto surface of the set of all real-valued instances* (a_1, a_2, a_3) *such that* $a_1 \geq \frac{12}{5}$, $a_2 \geq \frac{12}{5}$, $a_3 \geq 6$, *and* $\frac{1}{a_1} + \frac{1}{a_2} + \frac{1}{a_3} \leq \frac{5}{6}$.

Unlike Corollary 1, we conjecture that there is no Pareto surface of the set of all real-valued instances with three periods and density less than or equal to $\frac{5}{6}$. In contrast, $\{((2, 4, 4), |1213|), (3, 3, 3), |123|)\}$ is known to be a Pareto surface of the set of all *integer-valued* instances with three periods and density less than or equal to $\frac{5}{6}$ [3].

Rather than a Pareto surface, let us focus on a common valid schedule for a given set of instances. As we have seen in Lemmas 6, 8, and 9, such a schedule is found for instances that apply to each of cases (II) through (VII). Then, what about the condition of case (I)? It is seen in Fig. 2 that almost half area of the trapezoid M_1 is covered by $M_2, M_3, \ldots, M_7$, which means that there is a finite set of valid schedules for instances that correspond to the covered region. However, we do not know anything about the uncovered region.

4   Future Work

Towards Conjecture 2, we would like to build a framework that comprehensively handles real-valued instance sets, analogous to the Pareto surface for integer-valued instance sets [3]. Although our current methodology could be extended to the case where the periods take four or five distinct values, solving the general case would be difficult.

The method of Kawamura [7] for deriving Theorem 4 seems hopeless to immediately apply. The method relies on the fact that: The subset of all integer-valued instances such that the periods and the density are less than a certain threshold is finite and therefore the schedulability can be investigated using a computer. However, a subset of such real-valued instances is in general infinite.

Acknowledgments. We are grateful to Akitoshi Kawamura for useful discussions. This work was supported by JSPS KAKENHI Grant Numbers 25K14990, 20K11689, and 20K11676.

References

1. Chan, M.Y., Chin, F.: Schedulers for larger classes of pinwheel instances. Algorithmica **9**, 425–462 (1993)
2. Fishburn, P.C., Lagarias, J.C.: Pinwheel scheduling: Achievable densities. Algorithmica **34**(1), 14–38 (2002)
3. Gąsieniec, L., Smith, B., Wild, S.: Towards the 5/6-density conjecture of pinwheel scheduling. In: Proc. ALENEX '22, pp. 91–103 (2022)
4. Gąsieniec, L., et al.: Perpetual maintenance of machines with different urgency requirements. J. Comput. Syst. Sci. **139**, 103476 (2024)
5. Golomb, S.W.: The "sales tax" theorem. Math. Magazine **49**(4), 187–189 (1976)
6. Holte, R., Mok, A., Rosier, L., Tulchinsky, I., Varvel, D.: The pinwheel: a real-time scheduling problem. In: Proc. 22nd Annual Hawaii International Conference on System Sciences, vol. 2, pp. 693–702 (1989)
7. Kawamura, A.: Proof of the density threshold conjecture for pinwheel scheduling. In: Proc. STOC '24, pp. 1816–1819 (2024)
8. Kawamura, A.: Perpetual Scheduling Under Frequency Constraints, pp. 353–361. Springer, Singapore (2025)
9. Kawamura, A., Kobayashi, Y., Kusano, Y.: Pinwheel covering. In: Proc. CIAC '25, Part II, pp. 185–199, Berlin, Heidelberg (2025)
10. Kawamura, A., Soejima, M.: Simple strategies versus optimal schedules in multi-agent patrolling. Theor. Comput. Sci. **839**, 195–206 (2020)
11. Lin, S.S., Lin, K.J.: A pinwheel scheduler for three distinct numbers with a tight schedulability bound. Algorithmica **19**(4), 411–426 (1997)

Limitations of Density-Based Heuristics and an Alternative Approach for Pinwheel Scheduling with Durations

Yosuke Kusano$^{(\boxtimes)}$

Kyoto University, Kyoto, Japan
`kusano.yosuke.48e@st.kyoto-u.ac.jp`

Abstract. In Duration-aware Pinwheel Scheduling (DAPS), each job J_i has a duration p_i (in days) and the objective is to find a schedule on a single machine such that each job J_i is completed at least once in every consecutive f_i days. This problem generalizes the well-known problem, (Ordinary) Pinwheel Scheduling, which assumes all jobs are unit-length (i.e., $p_i = 1$). Previous studies have shown that in order to schedule jobs $J_1, \ldots, J_n$ feasibly, the density of the instance, defined as $p_1/f_1 + \cdots + p_n/f_n$, must be ≤ 1. Additionally, in the case of Ordinary Pinwheel Scheduling, every instance with density $\leq 5/6$ is known to be always schedulable. However, it remains open whether a similar density threshold would lead to an efficient algorithm for deciding schedulability in DAPS. In this paper, we present a negative solution to this open problem under the assumption that $\mathsf{P} \neq \mathsf{NP}$, namely, we show that no constant density threshold leads to a polynomial-time algorithm for deciding schedulability in DAPS unless $\mathsf{P} = \mathsf{NP}$. Furthermore, we demonstrate that another heuristic from Ordinary Pinwheel Scheduling fails to extend to DAPS. After that, we present an online scheduler which always schedule jobs feasibly for some DAPS instance class, by inventing a new measurement of schedulability of instances. This scheduler can be regarded as a 3-approximation algorithm for the optimization version of DAPS – known as Continuous Bamboo Garden Trimming on star graphs – which improves upon the previous best approximation ratio $3 + 2\sqrt{2}$.

Keywords: Pinwheel Scheduling · Density · Localized Measure · Bamboo Garden Trimming · Scheduling Algorithms

1 Introduction

Consider a single machine that perpetually processes one of $n \in \mathbb{N} = \{1, 2, 3, \ldots\}$ jobs, $J_1, \ldots, J_n$, according to a schedule. In *Duration-aware Pinwheel Scheduling (DAPS)*, each job J_i has a duration $p_i \in \mathbb{N}$ (in days) and the objective is to find

This work was supported by JST ASPIRE Grant Number JPMJAP2302 and ISHIZUE 2025 of Kyoto University.

© The Author(s), under exclusive license to Springer Nature Switzerland AG 2026
J. Kozik and A. Wolff (Eds.): SOFSEM 2026, LNCS 16448, pp. 634–647, 2026.
https://doi.org/10.1007/978-3-032-17801-5_46

a schedule on the machine such that each job J_i is completed at least once in every consecutive $f_i \in \mathbb{N}$ days. This problem appeared in [6] and was motivated by applications such as the radar sensor management problem [7] and satellite communications. An input of DAPS consists of the number $n \in \mathbb{N}$ of jobs and pairs (f_i, p_i) for $i \in [n] = \{1, \ldots, n\}$, which we assume satisfy $f_1 \leq \cdots \leq f_n$ for convenience. We call such a tuple $(f_i, p_i)_{i \in [n]} \in \mathbb{N}^{2 \times n}$ a *DAPS instance* and each f_i the *frequency* of J_i. Given a DAPS instance $V = (f_i, p_i)_{i \in [n]}$, a *schedule* $S : \mathbb{Z}_{\geq 0} \to [n]$ is said to be *feasible* for V if for every $i \in [n]$, it satisfies

$$c_i(V, S) \leq f_i, \tag{1}$$

where $c_i(V, S)$ denotes the maximum length $\sum_{t=a+1}^{b} p_{S(t)}$ among the intervals between two consecutive completions $a, b \in S^{-1}(i)$ of J_i. Formally, the goal in DAPS is to find a feasible schedule for any given instance V. We say that a DAPS instance is *feasible* if it admits a feasible schedule. For example, a DAPS instance

$$V = \begin{pmatrix} f_1 & f_2 \\ p_1 & p_2 \end{pmatrix} = \begin{pmatrix} 4 & 4 \\ 1 & 3 \end{pmatrix}$$

is feasible because the schedule S assigning J_1, J_2 alternatively to the machine satisfies $c_1(V, S) = c_2(V, S) = p_1 + p_2 = 4$, and thus S is a feasible schedule for V. On the other hand, a DAPS instance

$$V' = \begin{pmatrix} f_1 & f_2 \\ p_1 & p_2 \end{pmatrix} = \begin{pmatrix} k & k^2 \\ 1 & k \end{pmatrix} \tag{2}$$

is *infeasible* for every $k \in \mathbb{N}$. In fact, once the machine starts processing J_2, it takes at least $p_2 + p_1 = k + 1$ days before the next completion of J_1. Thus, $c_1(V', S) \geq k + 1$ for every schedule S with $c_2(V', S) < \infty$.

A known necessary condition for the feasibility of $V = (f_i, p_i)_{i \in [n]}$ is that the *density* of V, defined as $\rho(V) = \sum_{i=1}^{n} p_i/f_i$, is ≤ 1. Also, previous researches have shown that DAPS is strongly NP-hard (i.e., NP-hard even if all numerical parameters are written in unary) even when restricted to instances only with density exactly equal to 1 [6], while DAPS is solvable in PSPACE by using a similar method in [11, Corollary 2.2]. However, few instance classes are known for which feasibility can be decided in polynomial time. On the other hand, in *(Ordinary) Pinwheel Scheduling*, where all jobs have unit length ($p_1 = \cdots = p_n = 1$), every instance with density $\leq 5/6$ is known to be feasible [14]. Furthermore, there is a simple greedy algorithm that constructs a feasible schedule for any instance $(f_i, 1)_{i \in [n]}$ with density ≤ 1 satisfying *frequency divisibility* $f_i | f_{i+1}$, i.e., f_i divides f_{i+1} for all $i \in [n-1]$ [11]. However, for DAPS, there exist infeasible instances even with arbitrarily small density; for example, the instance defined in (2).

Based on the above differences between DAPS and Ordinary Pinwheel Scheduling, the following two questions have been posed as open problems [6]:

1. Is DAPS NP-hard even when restricted to instances with density < 1?
2. Is it possible to develop a polynomial-time algorithm for deciding the feasibility of DAPS instances with density $\leq \alpha$ for some constant $0 < \alpha < 1$?

1.1   Our Contribution

In the next section, we address the above two questions simultaneously. Specifically, we demonstrate that for every $\varepsilon > 0$, DAPS is strongly NP-hard even when restricted to instances with density $< \varepsilon$, via a polynomial-time reduction from the known strongly NP-complete problem, *Restricted Numerical 3-Dimensional Matching (RN3DM)* [18]. Additionally, we show that DAPS remains a hard problem even when restricted to instances with frequency divisibility $f_i | f_{i+1}$ and density < 1. More precisely, we construct a polynomial-time reduction from the *Partition* problem, known to be (weakly) NP-hard, to DAPS instances with frequency divisibility and density $< 2/3 + \varepsilon$.

We also discuss several instance classes for which DAPS is efficiently solvable. At the beginning of Sect. 3, we show that the *localized measure* $\mu(V)$ of a DAPS instance $V = (f_i, p_i)_{i \in [n]}$, defined as

$$\mu(V) = \max_{i \in [n-1]} \frac{1}{f_i} \left(\sum_{j \le i} \left\lfloor \frac{f_i}{f_j} \right\rfloor p_j + \max_{j > i} p_j \right),$$

is required to be ≤ 1 for the feasibility of V. This measure $\mu(V)$ is derived from a local perspective, assuming that the density is a 'global measure'. As a corollary, we deduce that a DAPS instance in which every pair of frequencies differs by more than a factor of 2 is feasible if and only if its localized measure is ≤ 1. Second, we present an online scheduling algorithm that always constructs a feasible schedule for any DAPS instance with both density and localized measure $\le 1/3$. This algorithm is derived in a similar manner to a known approach in the optimization version of DAPS, known as *Continuous Bamboo Garden Trimming (Continuous BGT)* on star graphs [10]. In fact, our proof yields a 3-approximation algorithm for Continuous BGT on star graphs, improving upon the previous best known approximation ratio $3 + 2\sqrt{2} \approx 5.83$ [10].

1.2   Related Work

Since the introduction of Ordinary Pinwheel Scheduling in [11], two main research directions have emerged. The first concerns its computational complexity, which is conjectured to be PSPACE-complete [3]. While NP-hardness has shown under a compact instance representation [2], it remains open whether it is in NP, or whether it is NP-hard in the standard representation. The second direction focuses on identifying density thresholds that guarantee feasibility. It had been long conjectured [4], and was recently shown [14] that every Ordinary Pinwheel Scheduling instance with density $\le 5/6$ is feasible, (while there is an infeasible instance with density $5/6 + \varepsilon$ for any $\varepsilon > 0$). Additionally, there are also such density thresholds that depend on the smallest frequency f_1 [9,15].

Furthermore, there exist many variants of Ordinary Pinwheel Scheduling beyond DAPS. One such variant is the optimization version, known as (simply) *Bamboo Garden Trimming (BGT)* [8], in which the objective is to minimize the maximum height of bamboos growing with different rates (corresponding to a reciprocal $1/f_i$ of frequency f_i in Ordinary Pinwheel Scheduling). The Continuous BGT problem is a generalization where bamboos are located in a metric space; it corresponds to the optimization version of DAPS when metric spaces are restricted to star graphs. Other variants include cases with multiple machines [1], settings with explorable uncertainty [5], and formulations in which each job requires 'exactly once' rather than 'at least once' over some frequency [17]. More recently, the k-VISITS problem was introduced, where unit-length jobs must be completed exactly k times under frequency constraints. This problem has been shown to be strongly NP-hard via a reduction also from RN3DM [12].

2   Hardness Results

Let $\mathfrak{S}_n$ be the set of all permutations of $[n] = \{1, \ldots, n\}$ for each $n \in \mathbb{N}$. The following problem is the restricted version of Numerical 3-Dimensional Matching, which is strongly NP-complete [18].

Definition 1 (RN3DM [18]). *Given a non-increasing sequence of positive integers $(a_i)_{i=1}^n$ and a positive integer T such that $\sum_{i=1}^n (a_i + 2i) = nT$ and $a_1 < T$, the RN3DM problem asks whether there are two permutations $\sigma, \tau \in \mathfrak{S}_n$ such that*

$$a_i + \sigma(i) + \tau(i) = T$$

for every $i \in [n]$.

Our first main result is established via a polynomial-time reduction from RN3DM, as follows.

Theorem 1. *For every $\varepsilon > 0$, DAPS is strongly NP-hard even when restricted to instances with density $< \varepsilon$.*

Proof. For any fixed positive integer k, it is sufficient to present a polynomial-time reduction from any instance of RN3DM to a DAPS instance with density $< 1/k$.

Suppose that $A = (a_i)_{i=1}^n$ and $T > 0$ define an RN3DM instance. We construct a DAPS instance $V = V_{A,T}$ as

$$V = \begin{pmatrix} f_i \\ p_i \end{pmatrix}_{i \in [n+1]} = \begin{pmatrix} T - a_1 + 3nk - 1 & \cdots & T - a_n + 3nk - 1 & 6nk^2 \\ 1 & \cdots & 1 & 3nk \end{pmatrix}.$$

Since k is fixed, this construction is in polynomial-time with respect to the size of A, T. In addition, the density $\rho(V)$ satisfies that

$$\rho(V) = \sum_{i=1}^n \frac{1}{T - a_i + 3nk - 1} + \frac{3nk}{6nk^2} < \sum_{i=1}^n \frac{1}{2nk} + \frac{1}{2k} = \frac{1}{k}.$$

In what follows, we show that the instance (A, T) of RN3DM has a solution if and only if the constructed V is feasible.

First, we show the necessity ('only if' part). Suppose that permutations $\sigma, \tau \in \mathfrak{S}_n$ are a solution to (A, T), namely, they satisfy $a_i + \sigma(i) + \tau(i) = T$ for all $i \in [n]$. We define a new permutation $\omega \in \mathfrak{S}_n$ as $\omega(i) = n - \sigma(i) + 1$ for each $i \in [n]$. By definition, it holds that

$$f_i = T - a_i + 3nk - 1 = (a_i + \sigma(i) + \tau(i)) - a_i + 3nk - 1 = n - \omega(i) + 3nk + \tau(i)$$

for each $i \in [n]$. Using ω and τ, we construct a schedule $S : \mathbb{Z}_{\geq 0} \to [n+1]$ for V as follows (see Fig. 1):

$$S(t) = \begin{cases} i & \text{if } t \equiv \omega(i), \tau(i) + n + 1 \pmod{2n+1} \text{ for } i \leq n, \\ n+1 & \text{if } t \equiv n + 1 \pmod{2n+1}. \end{cases}$$

Taking into account the periodicity of S, for each $i \leq n$, the $c_i(V, S)$ is given by the maximum length of the following two time-intervals between

- the completion of J_i at $t \equiv \omega(i)$ and its next completion at $t \equiv \tau(i) + n + 1$,
- the completion of J_i at $t \equiv \tau(i) + n + 1$ and its next completion at $t \equiv \omega(i)$.

Hence, for each $i \leq n$, the $c_i(V, S)$ satisfies

$$\begin{aligned}
& c_i(V, S) \\
&= \max\left\{ \sum_{j=\omega(i)+1}^{n} p_{\omega^{-1}(j)} + p_{n+1} + \sum_{j=1}^{\tau(i)} p_{\tau^{-1}(j)}, \ \sum_{j=\tau(i)+1}^{n} p_{\tau^{-1}(j)} + \sum_{j=1}^{\omega(i)} p_{\omega^{-1}(j)} \right\} \\
&= \max\left\{ (n - \omega(i)) + 3nk + \tau(i), \ (n - \tau(i)) + \omega(i) \right\} \\
&\leq \max\left\{ (n - \omega(i)) + 3nk + \tau(i), \ 2n \right\} \\
&= (n - \omega(i)) + 3nk + \tau(i) = f_i,
\end{aligned}$$

where we assume $\sum_{i=n+1}^{n} p_{\omega^{-1}(i)} = 0$. Additionally, since S completes $J_1, \ldots, J_n$ exactly twice each, and J_{n+1} exactly once in every consecutive $2n+1$ completions,

$$c_{n+1}(V, S) = 2 \sum_{i=1}^{n} p_i + p_{n+1} = 2n + 3nk < 6nk^2 = f_{n+1}.$$

Therefore, S is a feasible schedule for V, and thus V is feasible.

Next, we show the sufficiency ('if' part). Suppose that $S_* : \mathbb{Z}_{\geq 0} \to [n+1]$ is a feasible schedule for V. The feasibility of S_* guarantees the existence of $u > f_n$ such that $S_*(u) = n + 1$. For each $i \in [n]$, noting that J_i is completed

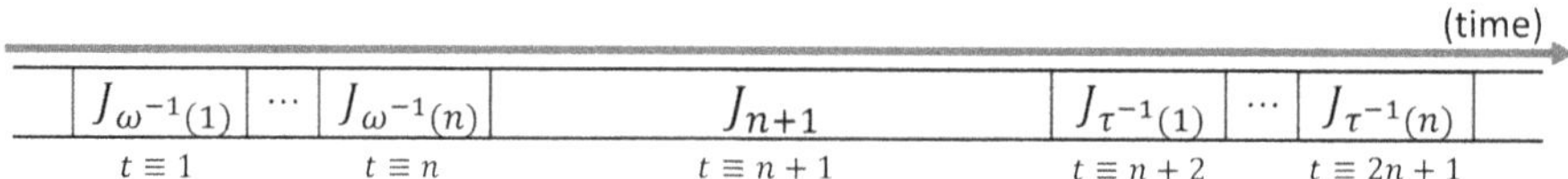

Fig. 1. The time schedule according to S.

at least once before position u, let $x_i = \max(S_*^{-1}(i) \cap [0, u-1])$ denote the last position before u at which J_i is processed, and let $y_i = \min(S_*^{-1}(i) \cap [u+1, \infty))$ denote the first position after u at which J_i is processed. With these notations, let $\omega_*, \tau_* \in \mathfrak{S}_n$ be the uniquely determined permutations such that

$$x_{\omega_*^{-1}(1)} < x_{\omega_*^{-1}(2)} < \cdots < x_{\omega_*^{-1}(n)}, \qquad y_{\tau_*^{-1}(1)} < y_{\tau_*^{-1}(2)} < \cdots < y_{\tau_*^{-1}(n)}.$$

Namely, we suppose that x_i and y_i are the $\omega_*(i)$-th and $\tau_*(i)$-th smallest values, respectively. Since the first position at which J_i is processed after position x_i is y_i,

$$c_i(V, S_*) \geq \sum_{t=x_i+1}^{y_i} p_{S_*(t)}$$

$$\geq \sum_{j=\omega_*(i)+1}^{n} p_{\omega_*^{-1}(j)} + p_{S_*(u)} + \sum_{j=1}^{\tau_*(i)} p_{\tau_*^{-1}(j)}$$

$$= n - \omega_*(i) + 3nk + \tau_*(i)$$

for each $i \in [n]$. The feasibility of S_* implies that $c_i(V, S_*) \leq f_i = T - a_i + 3nk - 1$, hence we obtain

$$T \geq a_i + (n - \omega_*(i) + 1) + \tau_*(i). \tag{3}$$

Summing both sides of (3) over $i = 1, \ldots, n$, we get

$$nT \geq \sum_{i=1}^{n} \big(a_i + (n - \omega_*(i) + 1) + \tau_*(i)\big) = \sum_{i=1}^{n} (a_i + 2i) = nT,$$

which shows that the inequality of (3) actually holds with equality. Thus, if we define a permutation $\sigma_* \in \mathfrak{S}_n$ as $\sigma_*(i) = n - \omega_*(i) + 1$, then it satisfies that $T = a_i + \sigma_*(i) + \tau_*(i)$ for all $i \in [n]$, and σ_*, τ_* are a solution for the instance (A, T) of RN3DM. $\qquad\square$

As stated in Sect. 1, one of the fundamental results in Ordinary Pinwheel Scheduling is the equivalence between the feasibility of $(f_i, 1)_{i \in [n]}$ and its density ≤ 1 when $f_i \mid f_{i+1}$ for all $i \in [n-1]$. We next demonstrate a hardness result along this line in the context of DAPS. Our reduction is constructed from the Partition problem, which is (weakly) NP-complete [13]:

Definition 2 (Partition). *Given positive integers $(a_i)_{i=1}^n$ such that the sum $\sum_{i=1}^n a_i$ is even, the Partition problem asks whether there is a partition of $[n]$ to two subsets U_1, U_2 such that*

$$\sum_{i \in U_1} a_i = \sum_{i \in U_2} a_i.$$

Theorem 2. *For every $\varepsilon > 0$, DAPS is (weakly) NP-hard even when restricted to instances $(f_i, p_i)_{i \in [n]}$ such that $f_i \mid f_{i+1}$ for all $i \in [n-1]$ and the density is $< 2/3 + \varepsilon$.*

Proof. For any fixed positive integer k, we present a polynomial-time reduction from any instance of Partition to a DAPS instance with frequency divisibility $f_i | f_{i+1}$ and density $< 2/3 + 2/k$.

For convenience, we assume in this proof that job indices start from 0, i.e., we consider job J_0. Given an instance $(a_i)_{i \in [n]}$ of Partition, we construct a DAPS instance V as

$$V = \begin{pmatrix} f_i \\ p_i \end{pmatrix}_{i=0,\ldots,n+1} = \begin{pmatrix} kT+1 & 3(kT+1) & \cdots & 3(kT+1) & 3kT(kT+1) \\ 1 & ka_1 & \cdots & ka_n & kT \end{pmatrix},$$

where $T = (\sum_{i=1}^{n} a_i)/2$. This construction is in polynomial-time with respect to the size of $(a_i)_{i \in [n]}$. In addition, the density $\rho(V)$ satisfies that

$$\begin{aligned} \rho(V) &= \frac{1}{kT+1} + \sum_{i=1}^{n} \frac{ka_i}{3(kT+1)} + \frac{kT}{3kT(kT+1)} \\ &= \frac{1}{kT+1} + \frac{2kT}{3(kT+1)} + \frac{1}{3(kT+1)} \\ &< \frac{2}{3} + \frac{2}{k}. \end{aligned}$$

In what follows, we show that $(a_i)_{i \in [n]}$ is a yes-instance of Partition if and only if V is feasible.

We first show the necessity ('only if' part). Suppose that disjoint subsets $\{i_1, \ldots, i_l\}$, $\{i_{l+1}, \ldots, i_n\} \subseteq [n]$ form a solution to $(a_i)_{i \in [n]}$, i.e., they satisfy $\sum_{j=1}^{l} a_{i_j} = \sum_{j=l+1}^{n} a_{i_j} = T$. We define a schedule $S : \mathbb{Z}_{\geq 0} \to \{0, 1, \ldots n+1\}$ as

$$S(t) = \begin{cases} 0 & \text{if } t \equiv 0, l+1, n+2, \\ i_j & \text{if } t \equiv j \text{ for } j \leq l, \\ i_j & \text{if } t \equiv j+1 \text{ for } j > l, \\ n+1 & \text{if } t \equiv n+3, \end{cases}$$

where all congruences $\equiv$ are taken modulo $n+4$. This schedule S completes J_0 exactly three times, and $J_1, \ldots J_{n+1}$ exactly once each, in every consecutive $n+4$ completions, hence

$$c_i(V, S) = 3p_0 + \sum_{i=1}^{n+1} p_i = 3 + \sum_{i=1}^{n} ka_i + kT = 3(kT+1) \leq f_i$$

holds for all $i \geq 1$. Additionally, by the periodicity of S, $c_0(V, S)$ is given by the maximum length among the following three time-intervals between:

- the completion of J_0 at $t \equiv 0$ and its next completion at $t \equiv l+1$,
- the completion of J_0 at $t \equiv l+1$ and its next completion at $t \equiv n+2$,
- the comnpletion of J_0 at $t \equiv n+2$ and its next completion at $t \equiv n+4 \equiv 0$.

Therefore, we obtain

$$c_0(V, S) = \max\left\{\sum_{t=1}^{l+1} p_{S(t)}, \ \sum_{t=l+2}^{n+2} p_{S(t)}, \ \sum_{t=n+3}^{n+4} p_{S(t)}\right\}$$

$$= \max\left\{\sum_{j=1}^{l} p_{i_j} + p_0, \ \sum_{j=l+1}^{n} p_{i_j} + p_0, \ p_{n+1} + p_0\right\}$$

$$= \max\left\{\sum_{j=1}^{l} ka_{i_j} + 1, \ \sum_{j=l+1}^{n} ka_{i_j} + 1, \ kT + 1\right\}$$

$$= kT + 1 = f_0.$$

Hence, S is a feasible schedule for V, and thus V is feasible.

We next show the sufficiency ('if' part). Suppose that S_* is a feasible schedule for V. The feasibility of S_* guarantees a position $u \geq 0$ such that $S_*(u+1) = n+1$. In addition, since $c_0(V, S_*) \leq f_0 = p_0 + p_{n+1}$ holds, job J_0 is processed immediately before and after each processing of J_{n+1} in the schedule S_*. Consequently, $S_*(u) = S_*(u+2) = 0$. Let x_1, x_2, x_3 denote the three consecutive positions at which J_0 is processed after u, with $x_1 = u + 2 < x_2 < x_3$. We define two subsets U_1, U_2 of $[n] = \{1, 2, \ldots, n\}$ as

$$U_m = \{i \in [n] \mid (x_m, x_{m+1}) \cap S_*^{-1}(i) \neq \emptyset\}$$

for $m = 1, 2$, where $(x_m, x_{m+1}) = \{x_m + 1, x_m + 2, \ldots, x_{m+1} - 1\}$. By definition, for every $i \in U_m$, J_i is completed during the interval between the completions of J_0 at x_m and at x_{m+1}, hence

$$\sum_{i \in U_m} p_i + p_0 \leq c_0(V, S_*) \leq f_0 = kT + 1,$$

for each $m = 1, 2$. Noting that $p_0 = 1$ and $p_i = ka_i$ for $i \in [n]$, we obtain that

$$\sum_{i \in U_m} a_i \leq T \tag{4}$$

for each $m = 1, 2$. Let J_{i_*} be the job among $J_1, \ldots, J_n$ whose first processing position y_{i_*} after position x_1 is the latest. If $U_1 \cup U_2$ were a proper subset of $[n]$, then $i_* \in [n] \setminus (U_1 \cup U_2)$ and $y_{i_*} > x_3$ would hold (see Fig. 2). Thus, we would have

$$c_{i_*}(V, S_*) \geq p_{S_*(u)} + p_{S_*(u+1)} + \sum_{t=x_1}^{y_{i_*}} p_{S_*(t)}$$

$$\geq p_0 + p_{n+1} + p_{S_*(x_1)} + p_{S_*(x_2)} + p_{S_*(x_3)} + \sum_{i \in [n]} p_i$$

$$= 4 + kT + \sum_{i \in [n]} ka_i = 3(kT + 1) + 1.$$

Since $f_{i_*} = 3(kT + 1)$, this contradicts the feasibility of S_* and we obtain $U_1 \cup U_2 = [n]$. Noting that $\sum_{i \in [n]} a_i = 2T$ and the inequality (4), we conclude that $\sum_{i \in U_1} a_i = \sum_{i \in U_2} a_i = T$ and $U_1 \cap U_2 = \emptyset$. Thus, the subsets $U_1, U_2 \subseteq [n]$ form a solution for the instance $(a_i)_{i \in [n]}$ of the Partition problem. $\qquad\square$

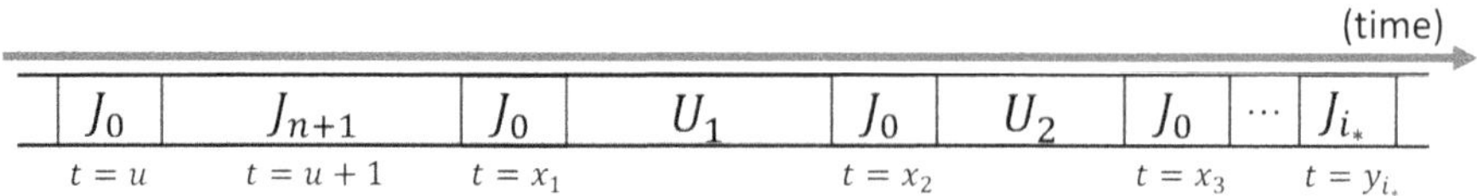

Fig. 2. The time schedule from position $t = u$ to $t = y_{i_*}$ according to S_*.

3   Efficiently Solvable Classes

As mentioned in Sect. 1, we first show that every feasible DAPS instance has its localized measure ≤ 1. We begin by introducing the optimization version of DAPS, in the context of which this statement is proven. By the definition (1), for any DAPS instance $V = (f_i, p_i)_{i \in [n]}$, a schedule $S : \mathbb{Z}_{\geq 0} \to [n]$ is feasible for V if and only if the value $\mathrm{val}(V, S)$ of V and S, defined as

$$\mathrm{val}(V, S) = \max_{i \in [n]} \frac{1}{f_i} c_i(V, S),$$

is ≤ 1. Therefore, V is feasible if and only if $OPT(V) \leq 1$, where

$$OPT(V) = \min_{S : \mathbb{Z}_{\geq 0} \to [n]} \mathrm{val}(V, S).$$

Given $n \in \mathbb{N}$ and a DAPS instance $V \in \mathbb{N}^{2 \times n}$, *Continuous BGT on star graphs* [10] is the problem to find a schedule S for V such that $\mathrm{val}(V, S) = OPT(V)$.

It is known that $\rho(V) \leq OPT(V)$ for every DAPS instance V [10, Lemma 9], which implies as a corollary that every feasible V satisfies $\rho(V) \leq 1$. We show a similar statement for the localized measure

$$\mu(V) = \max_{i \in [n-1]} \frac{1}{f_i} \left(\sum_{j \leq i} \left\lfloor \frac{f_i}{f_j} \right\rfloor p_j + \max_{j > i} p_j \right).$$

Theorem 3. *For every DAPS instance V, $\mu(V) \leq OPT(V)$. Especially, $\mu(V) \leq 1$ if V is feasible.*

Proof. Fix an arbitrary DAPS instance $V = (f_i, p_i)_{i \in [n]}$, and let $S : \mathbb{Z}_{\geq 0} \to [n]$ be a schedule such that $\mathrm{val}(V, S) < \infty$. In particular, S completes each job at least once.

We first show that

$$\sum_{j \leq i} \left\lfloor \frac{K}{c_j(V, S)} \right\rfloor p_j + \max_{j > i} p_j \leq K \tag{5}$$

for every $i \in [n - 1]$ and any positive number $K > \max_{j \in [n]} p_j$. Fix such i and K, and let $\alpha > i$ be an index such that $p_\alpha = \max_{j > i} p_j$. For each $j \in [n]$, since $c_j(V, S)$ bounds the gap between two consecutive completions of J_j, the schedule S completes J_j at least $\lfloor K/c_j(V, S) \rfloor$ times during any interval of length K days. Thus, within any interval of K days from the beginning of a processing J_α, the schedule S completes

- J_α at least once because $K > p_\alpha$, and
- each $J_j \in \{J_1, \ldots, J_i\}$ at least $\lfloor K/c_j(V, S) \rfloor$ times (see Fig. 3).

Therefore, we obtain the inequality (5) for every $i \in [n-1]$ and $K > \max_{j \in [n]} p_j$.

For each job J_i, once S starts processing another job $J_j \neq J_i$, it takes at least $p_j + p_i$ days before the next completion of J_i. Hence it holds that $c_i(V, S) \geq \max_{j \neq i} p_j + p_i$ for all $i \in [n]$, consequently

$$f_i \cdot \mathrm{val}(V, S) \geq f_i \cdot \frac{1}{f_i} c_i(V, S) \geq \max_{j \neq i} p_j + p_i > \max_{j \in [n]} p_j.$$

By substituting $K = f_i \cdot \mathrm{val}(V, S)$ into the inequality (5), we obtain

$$\frac{1}{f_i} \left(\sum_{j \leq i} \left\lfloor \frac{f_i}{f_j} \right\rfloor p_j + \max_{j > i} p_j \right) = \frac{1}{f_i} \left(\sum_{j \leq i} \left\lfloor \frac{f_i \cdot \mathrm{val}(V, S)}{f_j \cdot \mathrm{val}(V, S)} \right\rfloor p_j + \max_{j > i} p_j \right)$$

$$\leq \frac{1}{f_i} \left(\sum_{j \leq i} \left\lfloor \frac{f_i \cdot \mathrm{val}(V, S)}{c_j(V, S)} \right\rfloor p_j + \max_{j > i} p_j \right)$$

$$\leq \frac{1}{f_i} \left(f_i \cdot \mathrm{val}(V, S) \right)$$

$$= \mathrm{val}(V, S)$$

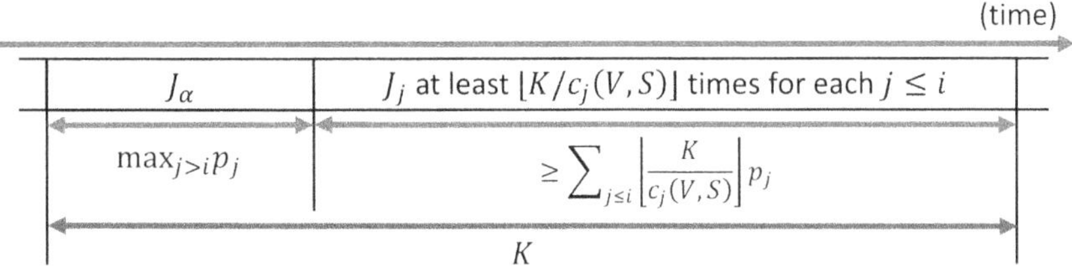

Fig. 3. An interval of K days from the beginning of a processing J_α.

for all $i \in [n-1]$, i.e.,

$$\mu(V) = \max_{i \in [n-1]} \frac{1}{f_i} \left(\sum_{j \leq i} \left\lfloor \frac{f_i}{f_j} \right\rfloor p_j + \max_{j > i} p_j \right) \leq \mathrm{val}(V, S).$$

Since S is an arbitrary schedule such that $\mathrm{val}(V, S) < \infty$, this implies $\mu(V) \leq OPT(V)$. $\qquad\square$

The following corollary presents an efficiently solvable class of DAPS instances.

Corollary 1. *A DAPS instance $V = (f_i, p_i)_{i \in [n]}$ satisfying $2f_i \leq f_{i+1}$ for all $i \in [n-1]$ is feasible if and only if $(p_1 + \max_{j>1} p_j)/f_1 \leq 1$.*

Proof. We show that the following schedule is optimal for any DAPS instance $V = (f_i, p_i)_{i \in [n]}$ satisfying $2f_i \leq f_{i+1}$ for all $i \in [n-1]$. Let S be the schedule that completes $J_1, J_2, \ldots, J_{n-1}, J_n$ exactly once every $2, 4, \ldots, 2^{n-1}, 2^{n-1}$ job completions, respectively. Concretely, S is defined as

$$S(t) = \begin{cases} i & \text{if } t \equiv 2^{i-1} - 1 \pmod{2^i} \text{ for } 1 \leq i \leq n-1, \\ n & \text{if } t \equiv 2^{n-1} - 1 \pmod{2^{n-1}}. \end{cases}$$

For every $i \in [n-1]$, in every consecutive 2^i job completions, this schedule completes J_j exactly 2^{i-j} times for each $j \leq i$, and exactly once for an arbitrary $j > i$, hence

$$\begin{aligned}
\frac{1}{f_i} c_i(V, S) &= \frac{1}{f_i} \left(\sum_{j \leq i} 2^{i-j} p_j + \max_{j > i} p_j \right) \\
&\leq \frac{1}{2^{i-1} f_1} \left(2^{i-1} p_1 + 2^{i-1} \max_{j > 1} p_j \right) \\
&= \frac{1}{f_1} \left(p_1 + \max_{j > 1} p_j \right).
\end{aligned}$$

Since it holds $c_n(V, S) = c_{n-1}(V, S)$ by definition, $\mathrm{val}(V, S) \leq (p_1 + \max_{j>1} p_j)/f_1$. Using Theorem 3, we have

$$\mu(V) \leq OPT(V) \leq \mathrm{val}(V, S) \leq \frac{1}{f_1} \left(p_1 + \max_{j > 1} p_j \right) \leq \mu(V).$$

Consequently, $(p_1 + \max_{j>1} p_j)/f_1 = OPT(V) \leq 1$ if and only if V is feasible. $\qquad\square$

Finally, we give an online scheduler which always constructs a feasible schedule for any DAPS instance V with $L(V) = \max\{\rho(V), \mu(V)\} \leq 1/3$. Given a real number $x > 0$, the scheduler, called *Reduce-Fastest(x)* [8,10] in BGT or Continuous BGT on star graphs, selects at each round the job with the minimum frequency among those J_i for which the time elapsed since the previous completion is $\geq x \cdot f_i$. Our proof of the following theorem is inspired by Kuszmaul's proof [16, Theorem 3.1] for BGT.

Theorem 4. *For any DAPS instance V, the schedule $\mathcal{A}(V)$ constructed by Reduce-Fastest$(2L(V))$ satisfies $\mathrm{val}(V, \mathcal{A}(V)) \leq 3L(V)$. Especially, every DAPS instance with $L(V) \leq 1/3$ is feasible.*

Proof. In this proof, we consider the time flow as continuous, and assume that the k-th day corresponds to the interval $(k, k+1] \subseteq \mathbb{R}_{>0}$; that is, the completion of a job on the k-th day corresponds to the time point $k + 1 \in \mathbb{R}_{>0}$.

Fix a DAPS instance $V = (f_i, p_i)_{i \in [n]}$ and an index $i \in [n]$. Let $N_1, N_2 \in \mathbb{R}_{>0}$ be an arbitrary pair of consecutive completions of J_i according to $\mathcal{A}(V)$, and assume $N_1 < N_2$. It is sufficient to show $(N_2 - N_1)/f_i \leq 3L$, where $L = L(V)$.

Let $T = N_1 + 2L \cdot f_i$ be the point at which J_i becomes a candidate for processing. At each round in the interval $(T, N_2 - p_i] \subseteq \mathbb{R}_{>0}$, the processing of J_i is "blocked" by processing of J_j for some $j < i$ (since $f_1 \leq \cdots \leq f_n$). For each $j < i$, we define m_j as the number of times when J_j can block the processing of J_i during $(T, T + Lf_i]$. We can deduce $m_j \leq f_i/f_j$ for each $j < i$ as follows:

- If $m_j \leq 1$, then obviously $m_j \leq f_i/f_j$.
- Assume $m_j \geq 2$. The interval $(T, T + Lf_i]$ includes at least $m_j - 1$ number of disjoint intervals with length $2L \cdot f_j$ during which J_j is not completed. This implies that $2Lf_j(m_j - 1) \leq Lf_i$, hence we obtain $m_j \leq 2(m_j - 1) \leq f_i/f_j$ by the assumption.

Especially, since each m_j is an integer, it satisfies $m_j \leq \lfloor f_i/f_j \rfloor$.

Let J_* denote the job being in process or completed at time T. In the interval $(T, T + Lf_i]$, the duration for which J_i is blocked is at most $p_* + \sum_{j<i} m_j p_j$. If $* < i$, then the interval $(T, T + Lf_i]$ includes m_* number of disjoint sub-intervals with length $2L \cdot f_*$ during which J_* is not completed. Consequently, $m_* + 1 \leq \lfloor f_i/f_* \rfloor$ as aforementioned. Thus we obtain

$$p_* + \sum_{j<i} m_j p_j \leq \sum_{j<i} \left\lfloor \frac{f_i}{f_j} \right\rfloor p_j \leq Lf_i - p_i \qquad\qquad \text{if } * < i,$$

$$p_* + \sum_{j<i} m_j p_j \leq \sum_{j<i} \left\lfloor \frac{f_i}{f_j} \right\rfloor p_j + \max_{j>i} p_j \leq Lf_i - p_i \qquad\qquad \text{if } * > i,$$

by using $\rho(V) \leq L$ and $\mu(V) \leq L$ for the first and the second bound, respectively. Therefore, the processing of J_i begins at some point in the interval $(T, T + Lf_i - p_i]$, ensuring $N_2 \leq T + Lf_i$ (see Fig. 4). Consequently, we have $(N_2 - N_1)/f_i \leq (N_1 + 3Lf_i - N_1)/f_i = 3L$. $\qquad\square$

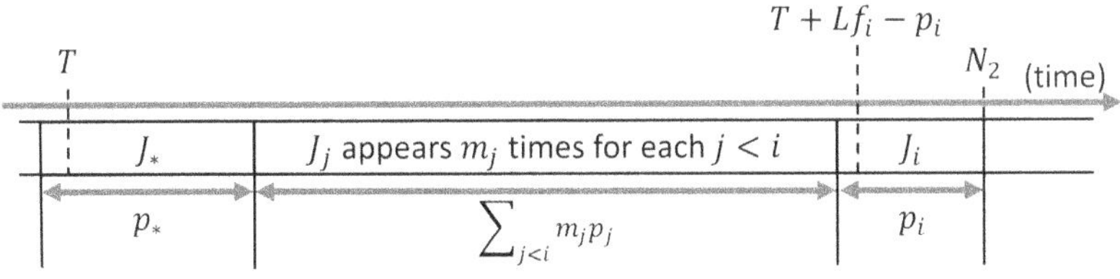

Fig. 4. The processing of J_i starts after being blocked m_j times by J_j for each $j < i$.

Noting that $L(V) \leq OPT(V)$ for every V, we conclude the following:

Corollary 2. *Reduce-Fastest*$(2L)$ *is a 3-approximation online algorithm for Continuous BGT on star graphs.*

4   Open Problems

We have shown that DAPS is strongly NP-hard even when restricted to instances with arbitrarily small density. On the other hand, the complexity analysis on instances with frequency divisibility remains incomplete. In particular, it is still open whether there exists a polynomial-time algorithm for deciding the feasibility of DAPS instances with frequency divisibility and density $\leq \alpha$ for some constant $0 < \alpha \leq 2/3$. Additionally, it is interesting to address whether DAPS is PSPACE-complete since Ordinary Pinwheel Scheduling is conjectured to be PSPACE-complete.

As stated in Sect. 1, it is known that the threshold $\rho(V) \leq 5/6$ guarantees the feasibility of any Ordinary Pinwheel Scheduling instance V, and this threshold is known to be best possible. Theorem 4 leads to a similar direction for future work; what is the maximum constant $\alpha < 1$ such that the threshold $L(V) \leq \alpha$ guarantees the feasibility of any DAPS instance V?

References

1. Bar-Noy, A., Ladner, R.E.: Windows scheduling problems for broadcast systems. SIAM J. Comput. **32**(4), 1091–1113 (2003). https://doi.org/10.1137/S009753970240447X
2. Bar-Noy, A., Ladner, R.E., Tamir, T.: Windows scheduling as a restricted version of bin packing. ACM Trans. Algorithms **3**(3), 28–es (2007). https://doi.org/10.1145/1273340.1273344
3. Bosman, T., van Ee, M., Jiao, Y., Marchetti-Spaccamela, A., Ravi, R., Stougie, L.: Approximation algorithms for replenishment problems with fixed turnover times. Algorithmica **84**(9), 2597–2621 (2022). https://doi.org/10.1007/s00453-022-00974-4
4. Chan, M., Chin, F.: Schedulers for larger classes of pinwheel instances. Algorithmica **9**, 425–462 (1993). https://doi.org/10.1007/BF01187034
5. Evans, W., Tabatabaee, S.A.: Perpetual scheduling with explorable uncertainty. In: Finocchi, I., Georgiadis, L. (eds.) Algorithms and Complexity, pp. 331–344. Springer Nature Switzerland, Cham (2025). https://doi.org/10.1007/978-3-031-92935-9_21
6. Feinberg, E., Curry, M.: Generalized pinwheel problem. Math. Methods Oper. Res. **62**, 99–122 (2005). https://doi.org/10.1007/s00186-005-0443-4
7. Feinberg, E.A., Bender, M.A., Curry, M.T., Huang, D., Koutsoudis, T., Bernstein, J.L.: Sensor resource management for an airborne early warning radar. In: Drummond, O.E. (ed.) Signal and Data Processing of Small Targets 2002. vol. 4728, pp. 145 – 156. International Society for Optics and Photonics, SPIE (2002). https://doi.org/10.1117/12.478500

8. Gąsieniec, L., Klasing, R., Levcopoulos, C., Lingas, A., Min, J., Radzik, T.: Bamboo garden trimming problem (perpetual maintenance of machines with different attendance urgency factors). In: Steffen, B., Baier, C., van den Brand, M., Eder, J., Hinchey, M., Margaria, T. (eds.) SOFSEM 2017. LNCS, vol. 10139, pp. 229–240. Springer, Cham (2017). https://doi.org/10.1007/978-3-319-51963-0_18

9. Gąsieniec, L., et al.: Perpetual maintenance of machines with different urgency requirements. J. Comput. Syst. Sci. **139**, 103476 (2024). https://doi.org/10.1016/j.jcss.2023.103476

10. Höhne, F., van Stee, R.: A 10/7-Approximation for discrete bamboo garden trimming and continuous trimming on star graphs. In: Megow, N., Smith, A. (eds.) Approximation, Randomization, and Combinatorial Optimization. Algorithms and Techniques (APPROX/RANDOM 2023). Leibniz International Proceedings in Informatics (LIPIcs), vol. 275, pp. 16:1–16:19. Schloss Dagstuhl – Leibniz-Zentrum für Informatik, Dagstuhl, Germany (2023). https://doi.org/10.4230/LIPIcs.APPROX/RANDOM.2023.16

11. Holte, R., Mok, A., Rosier, L., Tulchinsky, I., Varvel, D.: The pinwheel: a real-time scheduling problem. In: [1989] Proceedings of the Twenty-Second Annual Hawaii International Conference on System Sciences. Volume II: Software Track. vol. 2, pp. 693–702 (1989). https://doi.org/10.1109/HICSS.1989.48075

12. Kanellopoulos, S., Pergaminelis, C., Kokkou, M., Markou, E., Pagourtzis, A.: Finite pinwheel scheduling: the k-visits problem (2025). https://arxiv.org/abs/2507.11681

13. Karp, R.M.: Reducibility among combinatorial problems. In: Miller, R.E., Thatcher, J.W., Bohlinger, J.D. (eds.) Complexity of Computer Computations, pp. 85–103. Springer US, Boston (1972). https://doi.org/10.1007/978-1-4684-2001-2_9

14. Kawamura, A.: Proof of the density threshold conjecture for pinwheel scheduling. In: Proceedings of the 56th Annual ACM Symposium on Theory of Computing, pp. 1816–1819. STOC 2024, Association for Computing Machinery, New York (2024). https://doi.org/10.1145/3618260.3649757

15. Kawamura, A., Kobayashi, Y., Kusano, Y.: Pinwheel covering. In: Finocchi, I., Georgiadis, L. (eds.) Algorithms and Complexity, pp. 185–199. Springer Nature Switzerland, Cham (2025). https://doi.org/10.1007/978-3-031-92935-9_12

16. Kuszmaul, J.: Bamboo trimming revisited: simple algorithms can do well too. In: Proceedings of the 34th ACM Symposium on Parallelism in Algorithms and Architectures, pp. 411–417. SPAA '22, Association for Computing Machinery, New York (2022). https://doi.org/10.1145/3490148.3538580

17. Mok, A., Rosier, L., Tulchinsky, I., Varvel, D.: Algorithms and complexity of the periodic maintenance problem. Microprocess. Microprogram. **27**(1), 657–664 (1989). https://doi.org/10.1016/0165-6074(89)90128-2

18. Yu, W., Hoogeveen, H., Lenstra, J.K.: Minimizing makespan in a two-machine flow shop with delays and unit-time operations is np-hard. J. of Scheduling **7**(5), 333–348 (2004). https://doi.org/10.1023/B:JOSH.0000036858.59787.c2

Optimal-Length Labeling Schemes and Fast Algorithms for k-Gathering and k-Broadcasting

Adam Gańczorz$^{(\boxtimes)}$ and Tomasz Jurdziński

Institute of Computer Science, University of Wrocław, Wrocław, Poland
{adam.ganczorz,tju}@cs.uni.wroc.pl

Abstract. We consider basic communication tasks in arbitrary radio networks: k-broadcasting and k-gathering. In the case of k-broadcasting, messages from k sources have to get to all nodes in the network. The goal of k-gathering is to collect messages from k source nodes in a designated sink node. We consider these problems in the framework of distributed algorithms with advice.

Krisko and Miller showed in 2021 that the optimal size of advice for k-broadcasting is $\Theta(\min(\log \Delta,\ \log k))$, where Δ is equal to the maximum degree of a vertex of the input communication graph. We show that the same bound $\Theta(\min(\log \Delta, \log k))$ on the size of optimal labeling scheme holds also for the k-gathering problems. Moreover, we design fast algorithms for both problems with asymptotically optimal size of advice. For k-gathering, our algorithm works in at most $D + k$ rounds, where D is the diameter of the communication graph. This time bound is optimal even for centralized algorithms. We apply the k-gathering algorithm for k-broadcasting to achieve an algorithm working in time $O(D + \log^2 n + k)$ rounds. We also exhibit a logarithmic time complexity gap between distributed algorithms with advice of optimal size and distributed algorithms with distinct arbitrary labels.

Keywords: radio network · distributed algorithms · algorithms with advice · labeling scheme · broadcasting · gathering

1   Introduction

We consider two fundamental communication tasks in arbitrary radio networks called k-broadcasting and k-gathering. In the k-broadcasting problem messages from k sources has to be delivered to all nodes in the network. The goal of k-gathering is to collect messages from k source nodes in a fixed sink node.

We consider the distributed computing model of radio networks, defined as undirected connected graphs. In the absence of labels, most communication tasks including the problems considered in this paper are infeasible for deterministic algorithms in many networks, due to interferences. Hence we assume that nodes are assigned labels that are, not necessarily different, binary strings. At the

© The Author(s), under exclusive license to Springer Nature Switzerland AG 2026
J. Kozik and A. Wolff (Eds.): SOFSEM 2026, LNCS 16448, pp. 648–662, 2026.
https://doi.org/10.1007/978-3-032-17801-5_47

beginning of an execution of an algorithm each node knows only its own label and can use it as a parameter in a deterministic algorithm executed simultaneously by all nodes. The size of a labeling scheme is equal to the largest length of a label. Another complexity measure is time understood as the largest number of communication rounds of an execution of an algorithm for given network parameters.

1.1 The Model and the Problems

We consider radio networks (RN), where a network is modeled as a simple undirected connected graph $G = (V, E)$ with $n = |V|$ nodes. We also distinguish other parameters of the network, D which is equal to the diameter of G, and Δ which is its maximum degree. With a small abuse of notation, we sometimes use D to denote the height of a BFS spanning tree of a graph with a fixed root node. Observe however that the height of a BFS tree belongs to the range $[D/2, D]$, so it is of the same asymptotic size as D.

We use square brackets to denote sets of consecutive integers: $[i, j] = \{i, \ldots, j\}$ and $[i] = [1, i]$. All logarithms are to the base 2.

As usually assumed in the algorithmic literature on radio networks, nodes communicate in synchronous rounds also called steps. In each round, a node can either transmit a message to all its neighbors, or stay silent and listen. At the receiving end, a node v hears the message from a neighbor w in a given round, if v listens in this round, and if w is its only neighbor that transmits in the given round. Otherwise, v does not receive any message or information, and, in particular, cannot determine whether or not any of its neighbors transmitted. The *time* of an algorithm for a given problem is the largest number of rounds it takes to solve it, expressed as a function of the task and various network parameters.

If nodes are indistinguishable, i.e., in the absence of labels, the problems considered in the paper cannot be solved deterministically. Even 1-broadcasting and 1-gathering cannot be solved in the four-cycle due to the symmetry breaking problem. For example, for 1-broadcasting, the node antipodal to the source cannot get its message because, in any round, either both its neighbors transmit or both are silent. Therefore we consider labeled networks, i.e., we assign *labels* to nodes. A *labeling scheme* is a function $\mathcal{L}$ which, for a given network represented by a graph $G = (V, E)$, assigns binary strings to the nodes of the communication graph G. The string $\mathcal{L}(v)$ is called the *label* of the node v. Note that labels assigned by a labeling scheme are not necessarily distinct. The *length* or *size* of a labeling scheme $\mathcal{L}$ is equal to the maximum length of any label assigned by it. Every node knows initially only its label, and can use it as a parameter in the deterministic algorithm executed simultaneously and distributively by all nodes.

Solving distributed network problems with short labels can be seen in the framework of algorithms with *advice* (i.e., with *labeling schemes*). In this setting, nodes are anonymous and do not know anything about the network. An oracle that knows the network provides advice, i.e., a binary string, to each node before the beginning of the computation. A distributed algorithm executed at nodes uses these strings (each node uses its own advice string) to solve the problem efficiently. The required size of advice equal to the maximum length of the strings can be seen as a measure of the difficulty of the problem. In the radio network literature regarding algorithms with advice different strings may be given to different nodes [9–12] and we also make this assumption in the present paper. The scenario of the same advice given to all nodes would be useless in the case of deterministic algorithms in radio networks: no deterministic communication could occur. If advice strings may be different, they can be considered as labels assigned to nodes. Such labeling schemes permitting to solve a given network task efficiently are also called *informative labeling schemes*.

We now formally define communication problems considered in this paper. Let $G = (V, E)$ be the communication graph of a radio network. The *k-gathering* problem assumes that there are k nodes $s_1, \ldots, s_k \in V$, called *sources*, and one node $s \in V$, called the sink. Each source has a message, and all messages have to reach the *sink*. We distinguish the unit-message k-broadcasting and the multi-message k-broadcasting. In both cases there are k nodes $s_1, \ldots, s_k \in V$, called *sources*. In the case of *unit-message k-broadcasting* all the sources know the same broadcast message M while in the case of *multi-message k-broadcasting* the nodes $s_1, \ldots, s_k$ initially have possibly distinct messages $M_1, \ldots, M_k$. The goal is to deliver the message M (in the unit-message case) or all the messages $M_1, \ldots, M_k$ (for the multi-message k-broadcasting) to all nodes of the communication graph $G(V, E)$. The unit-message k-broadcasting problem with $k = 1$ is just called the *broadcasting problem*.

We assume the *non-spontaneous wake-up* model. In this model only source nodes start an execution of an algorithm at the fixed round 0 and each other node might join the execution only after receiving a message from other node. This model is in general more demanding than the *spontaneous wake-up* model, where all nodes start an execution of an algorithm at the same round (see e.g. [8]). However, as all problems considered in this paper assume that each node is either awaken already at the round 0 of the algorithm or it can only be awaken by a message from another node, we can think of the number of rounds elapsed from the round 0 of an execution of an algorithm as the current value of a *central clock*. Then, we can add the current value of the central clock to each transmitted message and guarantee that each node learns the state of the central clock.

As usual in algorithmic literature concerning radio networks we assume that many messages can be combined together into a single packet transmitted in a round, so a node can transmit several original messages m_i in one round. Thus, there is no restriction on the size of messages sent by nodes.

The ultimate goal for the framework with advice is to design algorithms which use optimal size of labels and simultaneously work in optimal time.

1.2   Our Results

We present an algorithm for k-gathering, using a labeling scheme of length $O(\min(\log k, \log \Delta))$, and running in at most $D + k$ rounds. We show that the length of our labeling scheme is optimal and the time is also asymptotically optimal, even for centralized algorithms in which the topology of the communication graph is *known* to nodes. For k such that $k = \omega(D\Delta)$,[1] we provide a k-gathering algorithm which works in time $O(D\Delta)$. Thus, we accomplish k-gathering in $O(\min(D + k, D\Delta))$ rounds. By combining k-gathering and broadcasting, we get an algorithm for the multi-message k-broadcasting problem. Thanks to the $O(D + \log^2 n)$-round algorithm for broadcasting with $O(1)$ labels [15] and $O(D + k)$-round k-gathering with advice of size $O(\min(\log k, \log \Delta))$ provided here, we get a $O(D + \log^2 n + k)$ time algorithm for multi-message k-broadcasting with labels of optimal size $\Theta(\min(\log k, \log \Delta))$, where the optimality of the size of labels follows from [24]. As for the unit-message broadcasting problem, our reduction to the standard broadcasting problem gives the algorithm with $O(1)$ length of labels and $O(D + \log^2 n)$ rounds.

Thus, our solutions for the considered communication tasks use labels of asymptotically optimal length. Moreover distributed algorithms for k-gathering using those labels work in optimal time, even for algorithms working in the scenario with known topology of the input graph (to each node).

1.3   Related Work

The most studied related tasks are broadcasting and gossiping. Here gossiping is the problem in which each node has its own message and all messages have to be distributed to all nodes. Optimal time centralized broadcasting algorithms were given in [17,23] and the best known gossiping time (without any extra assumptions on parameters) follows from [17]. They also consider gossiping for which they design an algorithm which works in $O(D + \Delta \log n)$ rounds. For large values of Δ, the gossiping from [17] was later improved in [6]. The best known deterministic distributed algorithm for broadcasting which time that depends only on n is $O(n \log n \log \log n)$ [25], later improved in [7] for some values of parameters D and Δ. For gossiping, the best known time in directed strongly connected graphs was given in [16,18] and for undirected graphs in [26]. The n-gathering problem in asymmetric directed graphs [4] and trees was considered in [3,5].

The paradigm of algorithms with advice has been applied to many different distributed network tasks: finding a minimum spanning tree [11], finding the topology of the network [12,14,21], determining the size of a network [14,20], single-message broadcasting [22], and leader election [19]. In [2,9,10,15], the broadcasting algorithm with advice of size $O(1)$ were provided working in time $O(n)$, $O(D \log n + \log^2 n)$ and optimal $O(D + \log^2 n)$, respectively.

Krisko and Miller [24] were the first who considered the k-broadcasting problem for radio networks with advice. They show that $\Theta(\min(\log k, \log \Delta))$ is the

---

[1] As an example of such a case one may consider a graph with a spanning tree of height $O(\log \Delta)$. Then D is also $O(\log \Delta)$, while k might be even equal to n if each node is a source of a message. And thus $O(D\Delta) = o(k)$ for wide range of values of Δ.

optimal size of labels and they design a k-broadcasting algorithm for general graphs and for trees without optimizing time of the algorithms. Bu et al. [1] introduced the study of the labeling for the the convergast problem which corresponds to the n-gathering problem. They however mainly consider radio networks with collision detection, a model which is stronger than the one considered in this paper. They also provide a solution which works in our model in $\Theta(n)$ time and $\Theta(\log n)$-bit labels, which matches the extreme case of k-gathering with $k = n$.

1.4   Organization of the Paper

In Sect. 2 we present k-gathering algorithms. Section 3 regards the k-broadcasting problem. We discuss further research directions in Sect. 4. Missing proofs are available in the full version of the paper [13].

2   The k-Gathering Problem

In this section we consider k-gathering. Section 2.1 gives a lower bound $k + D$ on the time needed for k-gathering and a lower bound $\min(\log k, \log \Delta)$ on the length of a labeling scheme. In Sect. 2.2, we introduce some technical tools used in a k-gathering algorithm that works in time $O(D + k)$, presented in Sect. 2.3. In Sect. 2.4 we present an algorithm which works in $O(D\Delta)$ rounds. Finally, in Sect. 2.5, we summarize the results and show a logarithmic gap in time complexity between algorithms with and without advice.

2.1   Lower Bounds for the k-Gathering Problem

In this section we show that $D + k$ is a lower bound on the number of rounds needed to accomplish k-gathering, and that $\min(\log \Delta, \log k)$ is a lower bound on the length of a labeling scheme sufficient to accomplish this task.

Consider the graph $G_{D,p}$ with nodes $V = \{v_1, v_2, \ldots, v_{D+p}\}$, and the edges
$$\{(v_i, v_{i+1}) \mid i \in [1, D]\} \cup \{(v_D, v_j) \mid j \in [D + 1, D + p]\}$$
Assume that $p = k$, k input messages are originally located in the nodes $v_{D+1}, \ldots, v_{D+k}$ and the messages should be gathered in v_1. As the unique path to v_1 of each message goes through v_D, we see that at least k rounds are needed in order to transfer all messages from their source nodes to v_D. Then, an additional $D-1$ rounds are necessary to pass any message from v_D to v_1. These observations imply the lower bound $D + k$ on the number of rounds for k-gathering.

For the lower bound on the size of labels, look again at the graph. Firstly assume that $p = k$, the source messages are originally located in the nodes $v_{D+1}, \ldots, v_{D+k}$ and they should be gathered at v_1. Then we need the labels of size $\log k$ in order to guarantee that all nodes $v_{D+1}, \ldots, v_{D+k}$ have distinct labels necessary for delivery of the source messages to v_1. Indeed, if two nodes v_{D+i} and v_{D+j} for $i \neq j$, $i, j \in [k]$ have equal labels and they are connected only to v_D, then both of them either transmit or listen in each round. Thus, none of them transmits its message successfully to its only neighbor v_D and therefore the given instance of k-gathering problem can not be accomplished. Secondly, if $p = \Delta < k$ and Δ of the k source messages are located originally in $v_{D+1}, \ldots, v_{D+\Delta}$, the labels of size $\log \Delta$ are necessary.

The above considerations lead to the following conclusion.

Theorem 1. *The k-gathering task requires at least $k + D$ rounds and a labeling scheme of size* $\min (\log k, \log \Delta)$.

We note that the above lower bound on the size of labels could be also obtained by a reduction of the k-broadcasting problem from [24] to a composition of a k-gathering and standard broadcasting and the fact that broadcasting can be done with $O(1)$-bit labels. The lower bound on k-broadcasting is given in [24].

2.2   Tools for k-Gathering Algorithms

Definition 1 (Children-parents assignment). *Let $G = (V, E)$ be a graph and let:*
– $X = \{x_1, \ldots, x_p\}$ such that $x_i \in V$ and $x_i \neq x_j$ for each $i \neq j$ be a set called children,
– $Y = \{y_1, \ldots, y_r\} \subseteq V \setminus \{x_1, \ldots, x_p\}$ such that $(x_i, y_j) \in E$ for each $i \in [p]$ and at least one $j \in [r]$ be a set called parents,
– $par : X \to Y$ such that for each $y \in Y$, there exists $x \in X$ such that $(x, y) \in E$ and $par(x) = y$ (i.e., par is a surjective function).
Then, we say that (X, Y, par) is a children-parents assignment, *x is a* child *of y and y is the* parent *of x iff $par(x) = y$. The function par is called a* parent *function.*

In the following part of the paper we use the notion of a *schedule* in order to describe some distributed algorithms for radio networks with advice. Formally, a *schedule S* might be seen as an assignment of all possible labels of nodes to $0/1$ sequences of a fixed size T, where S_λ denotes the $0/1$ sequence assigned to the label λ. Such the schedule S is understood as an algorithm in radio networks, where each node with the label λ transmits in the round i of the algorithm corresponding to S iff the value of the ith bit of S_λ is equal to 1. In general, the assumptions of the non-spontaneous wake-up model might prevent a possibility of an execution of a schedule as a distributed algorithm. However, as we explained earlier, all problems considered in this paper assume that each node is either awaken already at the first round of the algorithm (if it is a source of a message) or it can only be awaken by a message from another node. Thus, we can add the current value of the central clock to each message and guarantee in this way that each node learns the state of this central clock before it is supposed to send its first message.

Definition 2. *A schedule S is called an (X, Y, par) children-parents schedule in a graph $G(V, E)$ for a children-parents assignment (X, Y, par) iff x successfully sends a message to $par(x)$ for each $x \in X$ during an execution of S in G.*

Our ultimate goal is to use children-parents schedules as stages of k-gathering algorithms guaranteeing that each message gets closer to the sink node for the k-gathering task. Therefore, we would like to make progress towards finalizing k-gathering in a children-parents schedule either by minimizing the size of the schedule or by decreasing dispersion of messages measured by the number of nodes in which the messages to be gathered are located (here, decreasing dispersion corresponds to decreasing the number of parent nodes). To this aim, we will show that there is a way to rearrange a children-parents assignment such that the sum of the sizes of the optimal children-parents schedule and the number of target nodes $|Y|$ is bounded from above.

Lemma 1. *Let (X, Y, par) be a children-parents assignment in a graph $G(V, E)$. Then, there exists $Y_\star \subseteq Y$, $par_\star : X \to Y_\star$ and a schedule S such that*
– $(X, Y_\star, par_\star)$ is a children-parents assignment in G,
– S is a $(X, Y_\star, par_\star)$ children-parents schedule, such that each node from X transmits exactly once in S,
– $|Y_\star| \leq |X| - |S| + 1$, i.e., $|Y_\star| + |S| \leq |X| + 1$.

Proof. We will proceed by induction on the size of X.

Consider the base case that $|X| = 0$. Then S is an empty schedule. Thus $Y_\star = Y = \varnothing$, $|Y_\star| = |Y| = 0$, and therefore $|Y_\star| \leq |X|$ by Def. 1 and $|S| = 0$. These inequalities imply that $|Y_\star| \leq |X| - |S| + 1$ as required.

Now assume that the lemma holds for all children-parents assignments such that the number of children is smaller than $k > 0$. Let us take any children-parents assignment (X, Y, par) and let $|X| = k$. Let us look at the subgraph of G induced on $X \cup Y$. Let $y_k \in Y$ be a node with some degree $d > 0$, i.e., $\deg(y_k) = d$. Let $X' = \{x \in X | (x, y_k) \notin E\}$ and $Y' = \{y | \exists x \in X' \ par(x) = y\}$ (see Fig. 1). Let par' be the function par with the domain reduced to X'. Then (X', Y', par') is a correct children-parents assignment with $|X'| = |X| - d, |Y'| \leq |Y| - 1$. From the inductive hypothesis we get a children-parents assignment $(X', Y'_\star, par'_\star)$ for $Y'_\star \subseteq Y'$ and a $(X', Y'_\star, par'_\star)$ children-parents schedule S' such that $|Y'_\star| \leq |X'| - |S'| + 1$.

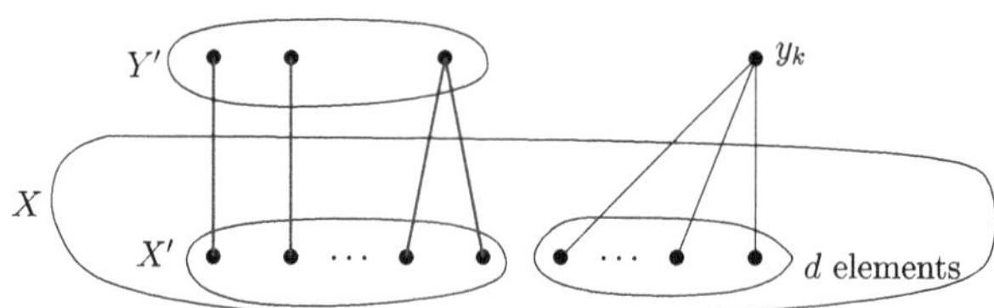

Fig. 1. Inductive step in the proof of Lemma 1, Case 1. Segments denote edges, fat blue segments correspond to edges connecting elements of X' with their parents. There are no edges connecting nodes from X' with y_k in $G(V, E)$.

Now let us look at two cases:

Case 1: There is a child x of y_k such that, for each node $y \in Y'_\star$, there is no edge (x, y) in G.

Then the node x does not collide with any node in X' at any element of $Y'_\star \cup \{y_k\}$ as all nodes of X' are not connected with y_k and x is not connected with nodes of $Y'_\star$ according to our assumption. Let us modify S' in the following way. Firstly, we send a message of x in the first round of the schedule S' in parallel of other transmission of S' of that round. According to our above assumption and reasoning, transmission of x is successfully delivered to y_k and it does not interfere (does not causes collisions) with any transmission of other nodes in the first round of S'. Secondly, after S', we assign a separate round for each of the remaining children of y_k in order to allow them to transmit without collisions. That gives us a schedule S of length $|S| \le |S'| + d - 1$ as y_k has $d > 0$ children. Moreover, we obtain the new set of parents $Y_\star = Y'_\star \cup \{y_k\}$ and therefore $|Y_\star| = |Y'_\star| + 1 \le |X'| - |S'| + 1 + 1 \le (|X| - d) - (|S| - d + 1) + 2 = |X| - |S| + 1$, where the first equality follows from the construction of $Y'_\star$, the first inequality follows from the inductive hypothesis, the second equality follows from the relationships $|X'| = |X| - d$ and $|S'| = |S| - d + 1$ which in turn follow from our construction of X' and S'.

Case 2: Each child x_i of y_k has an edge to some $y'_i \in Y'_\star$ for all nodes $i \in [d]$.

Then we will assign y'_i as the (possibly different than according to the function par) parent $\mathrm{par}'(x_i)$ of x_i for each $i \in [d]$. This enables us to avoid using y_k or any other node outside of $Y'_\star$ as a parent of any node. Our schedule S will be then just S' followed by d rounds such that the ith neighbor of y_k is the only transmitter in the round i. Then the size of S is $|S| = |S'| + d$ and $|Y_\star| = |Y'_\star| \le |X'| - |S'| + 1 = |X| - d - (|S| - d) + 1 = |X| - |S| + 1$, where the inequality follows from the inductive assumption and the equalities follow from the construction. In this way we finish the inductive proof of the lemma. $\square$

2.3   The k-Gathering in $D + k$ Rounds

In this section we design an algorithm which accomplishes k-gathering in $D + k$ rounds with labels of size $O(\min(\log \Delta, \log k))$. The idea of the algorithm is as follows. Let T be a BFS tree rooted at the sink node of a given instance of the k-gathering problem. The algorithm will work in phases. In the first phase we will build a children-parents assignment A, along with a children-parents schedule S on leaves of T (containing a message) as children, and their parents in T. Thus, after an execution of S, the parent nodes from A will get all messages from the children. More precisely, we will actually execute a schedule S' based on S that ensures that, for each leaf v in T, each message located in v is received by at least one parent of a leaf in T located on the same level as v (not necessarily by the parent of v as assigned by the par function of the children-parent assignment A).

After the first phase, we start the second phase which now is a children-parents schedule with children equal to the leaves of T', where T' is a subtree of T obtained by removing the leaves of T. The tricky parts are in the encodings of behaviors of nodes in the schedules using the labels of size $O(\min(\log \Delta, \log k))$. Other non-obvious observations are needed in the evaluation of time of the composition of the children-parents schedules, which requires amortized analysis.

First, we present a centralized implementation of the above strategy. Then, an optimal labeling scheme is presented. Finally, a distributed algorithm simulating the centralized one and using assigned labels is described and analyzed.

Centralized Transmission Scheme for k-gathering in $D + k$ Rounds.

In this section we describe more precisely the centralized algorithm, which will serve as a base for a distributed algorithm.

The algorithm works in phases. In the ith phase, we will create a children-parents schedule S_i. Let $d(v)$ for each node v be the distance from v to the sink node s. Let T_0 be a tree consisting of shortest paths from source nodes to the sink node s. Then, let X_0 be the set of leaves of T_0, let Y_0 be the set of parents of the nodes from X_0 and par_0 be the corresponding parent function. Let S_0, Y_0', par_0' be the schedule, the parent set, and the parent function obtained by an application of Lemma 1 for the children-parents assignment $(X_0, Y_0, \mathrm{par}_0)$. Moreover, let T_1 be a subtree of T_0 obtained by removing the set X_0 of leaves of T_0 – see Fig. 2 for an example illustration. If some nodes become leaves without any message after this removal, we remove them from the tree. For $i > 0$, let Z_i be the set of source nodes of T_i outside of Y_{i-1}'. Then, let $\mathrm{Leaves}(T_i)$ be the set of leaves in T_i, let $X_i = \mathrm{Leaves}(T_i) \cap \left(Y_{i-1}' \cup Z_i\right)$, let Y_i be the set of parents of X_i in T_i and par_i be the corresponding parent function. Then let S_i, Y_i', par_i' be the schedule, the parent set, and the parent function respectively obtained by an application of Lemma 1 for $(X_i, Y_i, \mathrm{par}_i)$. Then remove X_i from T_i to form T_{i+1}, and also remove nodes that become leaves without any message after this removal.

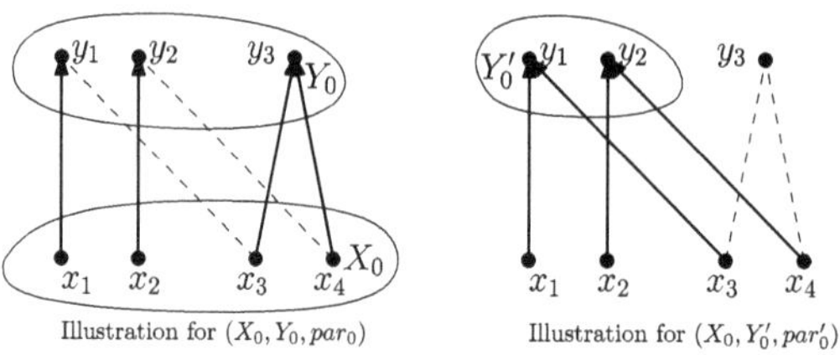

Illustration for (X_0, Y_0, par_0) Illustration for (X_0, Y_0', par_0')

Fig. 2. An example for $(X_0, Y_0, \mathrm{par}_0)$ and $(X_0, Y_0', \mathrm{par}_0')$. Fat edges with arrows connect children with their parents. Dashed edges do not connect parents and children. The schedule S_0 might be such that x_1 and x_2 transmit simultaneously in the first round, x_3 and x_4 transmit simultaneously in the second round. (Note that the parents of x_3 and x_4 in par_0' are y_1 and y_2, respectively.) If messages are originally located only in $x_1, \ldots, x_4$ then $x_1, \ldots, x_4$ as well as y_3 are removed from T_0 in order to obtain T_1.

The resulting centralized scheme for k-gathering S would be thus a concatenation of schedules $S_0, S_1, \ldots, S_{m-1}$. Below we prove that such S is actually a k-gathering schedule and that its time is at most $D + k$.

Lemma 2. *The broadcast schedule S is a correct schedule for k-gathering. The length of S is at most $D + k$.*

Proof. Let us analyze the path of some message a, located originally at a node u_0 of T_0, to the sink node s. The node u_0 eventually becomes a leaf in T_i for some i, so $u_0 \in X_i$ in phase i, and then the schedule S_i obtained from Lemma 1 guarantees that u_0 will successfully transmit its message to the parent u_1 with $d(u_1) = d(u_0) - 1$. We can continue this reasoning with each next u_l until the message of u_0 reaches the node $u_{d(u_0)} = s$. Thus, each message gets delivered to the source node.

Now, let us analyze the length of S. First observe that the number of phases is equal to the largest distance of a leaf of T_0 to s which is $m \leq D$. By Lemma 1, the length of S_i is at most $|X_i| - |Y_i'| + 1$ for $i \geq 0$. As $m \leq D$, $|X_0| + \sum_{i=1}^{m-1} |Z_i| \leq k$,[2] $|X_i| = |Y_{i-1}'| + |Z_i|$ for $i \geq 1$, and $Y_{m-1}' = \{s\}$, we get the following bound on the size of S: $|S| = \sum_{i=0}^{m-1} |S_i| \leq \sum_{i=0}^{m-1} (|X_i| - |Y_i'| + 1) = |X_0| + \sum_{i=1}^{m-1} |Z_i| - |Y_{m-1}'| + m \leq k + D - 1$. $\qquad \square$

Labeling Scheme for k-Gathering in $D + k$ Rounds. Now we will describe the labeling scheme for a distributed implementation of the above $O(D + k)$-round schedule for k-gathering. Then, in the next section, a k-gathering algorithm using this labeling scheme will be presented.

Firstly, let us assume that $k \leq \Delta$. In this case we will encode behaviors of nodes in the centralized schedule presented earlier, using $O(\log k)$-bit labels. Let v be a node which sends its message in the ith phase, that is during a schedule S_i. Note that each node transmits only once during the above described centralized schedule. (This property follows from two facts. Firstly, each node with a message to send is a leaf in T_i for exactly one value of i. Secondly, a node is only included in X_i when it is a leaf. So each node with a message to send transmits during exactly one schedule S_i. Moreover, by Lemma 1, each node in X_i transmits exactly once in S_i.) Let t_v be the round number during S_i when v transmits a message and T_i be the length of S_i. Let us assign unique identifiers from the set $[1, \ldots, k]$ to all original messages which are supposed to be gathered. Let ID_v be the identifier of the message, which v gets at the latest, during the schedule S_{i-1}, according to the centralized algorithm or ID_v is the identifier of the message originally located in v if v is is an element of Z_i. Then the label of v will be equal to $\mathcal{L}(v) = (0, \mathrm{ID}_v, t_v, T_i)$, where the value 0 at the first coordinate indicates that we are in the case $k \leq \Delta$. By Lemma 1 and the fact that there are only k messages, the length of each schedule S_i is $O(k)$ and thus the length of $\mathcal{L}(v)$ is $O(\log k)$.

Now, consider the case that $k > \Delta$. Firstly, let $c : V \rightarrow \{0, 1, \ldots, \Delta^2\}$ be a 2-hop $(\Delta^2 + 1)$-coloring of the input graph. That is, $c(v)$ is the color of v. Let v be a node which sends its message in the ith phase. As $k > \Delta$, the length T_i of S_i in our centralized schedule might be much larger than Δ as well as the round

---

[2] Recall the assumption that each pair of original messages are located in different nodes at the beginning.

t_v of the transmission of a node during S_i. Therefore, to encode T_i and t_v using $O(\min(\log k, \log \Delta)) = O(\log \Delta)$ bits, we will optimize the schedule S_i in such way that its length is always $O(\Delta^2)$, using the following observation.

Observation 1. *Each children-parents schedule can be optimized to be of length at most $\Delta^2 + 1$.*

Observation 1 is derived from 2-hop $(\Delta^2 + 1)$-coloring. If a children-parents schedule is longer than $\Delta^2 + 1$, we can exchange it with the round-robin protocol on the values $c(v)$ of the 2-hop $(\Delta^2 + 1)$-coloring. Given a 2-hop coloring, nodes with the same color can transmit simultaneously such that all their neighbors receive collision-free messages transmitted to them. This fact follows from the observation that, in the radio network model, messages of two distinct transmitters u, v might collide only at their common neighbor which implies that the distance between u and v is at most 2 in such a case.

Let S_i' be such an optimized schedule of length $\min(|S_i|, \Delta^2 + 1)$. Again, let t_v be the unique time step during S_i' when v transmits a message and T_i be the length of S_i'. Let c_v be the color of a node that sends its message to v at the latest during S_{i-1}'. Then the labeling for v is equal to $\mathcal{L}(v) = (1, c_v, t_v, T_i)$, where 1 at the first coordinate indicates that we are in the case $k > \Delta$. As the length of $\mathcal{L}(v)$ is $O(\log \Delta)$ in this case, we get the following lemma.

Lemma 3. *The length of the labeling scheme is $O(\min(\log \Delta, \log k))$.*

Distributed Algorithm for k-Gathering in $D + k$ Rounds. Now, we will describe the distributed algorithm for k-gathering using labels constructed according to the labeling scheme described above. We consider two cases determined by the first coordinate of the label $\mathcal{L}(v)$ of each node $v \in V$.

Firstly consider the case that $k \leq \Delta$. Each node from X_i will, apart from all received messages from the current instance of the k-gathering problem, add the value of $T_0 + \cdots + T_i$ to its messages. Thus each node v will be able to deduce the value of $T_0 + \cdots + T_{i-1} + t_v$ – the round number when v should transmit. Indeed, this value will be determined based on the content of received messages and the own label of the considered node v. However, although the nodes from X_i know T_i from their labels, they do not know the sum $\sum_{j=0}^{i-1} T_i$. This value might be too large to encode it in the label of size $O(\min(\log \Delta, \log k))$. Therefore, we do not store it in the label of a node from X_i. However, we show below that actually each node $v \in X_i$ will learn the value of $T_0 + \cdots + T_{i-1}$ before the round of its actual unique transmission according to the presented above centralized schedule for k-gathering and therefore it can determine its transmission round and send $\sum_{i=0}^{i} T_i$ with its message to its parent. According to the centralized schedule, $v \in X_i$ should broadcast in the phase i. If $i = 0$, v will just broadcast in the round t_v encoded in its label. If $i > 0$ then there is some node u relaying its messages to v in phase S_{i-1}, along with the value of $T_0 + \cdots + T_{i-1}$. The node v knows when it gathered all messages if it receives a message containing an ID equal to $\mathtt{ID}_v$ from the label $\mathcal{L}(v)$. Then v knows that it must transmit in the next phase and $\sum_{j=0}^{i-1} T_j + t_v$ determines the exact round number of its transmission. Moreover, as v knows T_i from its label, it can determine $\sum_{j=0}^{i} T_j$ to be transmitted by v to the next level.

We reason similarly in the case that $k > \Delta$. Each node v will, in addition to all transmitted messages, append to its broadcast messages the value of $T_0 + \cdots + T_i$. In this case however, it will also append $c(v)$. Thus the node v knows it gathered all messages if it receives a message from a node with its color equal to c_v from v's label $\mathcal{L}(v)$. (Note that there can only be one such message, because the distance-two coloring guarantees that all of v's neighbors have distinct colors.) The node v thus proceeds as in the case that $k \leq \Delta$.

The above reasoning lead to the following theorem.

Theorem 2. *There is a $O(\min(\log k, \log \Delta))$-bit labeling scheme and a distributed algorithm for k-gathering using this labeling running in at most $D + k$ rounds.*

Note that the upper bound on the number of rounds just $D + k$, not only an asymptotic $O(D + k)$. This bound matches the lower bound from Sect. 2.1.

2.4 k-Gathering for $k = \omega(D\Delta)$

Now, we give an algorithm for the specific case that k is large, $k = \omega(D\Delta)$.

For a given rooted tree T, let $p_T(v)$ denote the parent of the node v in T. If the considered tree is clear from the context, we write $p(v)$ instead of $p_T(v)$.

Lemma 6 from [14] states that there is an efficient way to assign the value $s(v) \in [0, \Delta - 1]$ to each node v of any BFS tree T so that
(a) $s(u) \neq s(w)$ for different children u, w of a node of v in T,
(b) there is no edge between u and $p(w)$ for any two nodes $u \neq w$ on the same level of T such that $s(u) = s(w)$.
Now, consider a fixed BFS tree with the sink node s as the root of T. Then, let T' be the smallest with respect to set inclusion subtree of T which contains the root ε of T and all the sources $s_1, \ldots, s_k$. Note that each leaf of T' is a source vertex. Let $s : V \rightarrow [0, \Delta - 1]$ be a function which satisfies the above properties (a)–(b). Let $c : V \rightarrow [0, \Delta^2]$ be a $(\Delta^2 + 1)$ distance-two coloring of the input graph. Let the label of a node v consist of the values of $s(v)$, $c(v)$, $c(p(v))$, Δ, the number of children of v in T', and $l(v) \in [0, 2]$ which is equal to the level of v in T (i.e., its distance to the root of T) mod 3.

Now, we describe a k-gathering algorithm A using the above labels. An execution is split into phases consisting of three rounds $\{0, 1, 2\}$. If v is supposed to send a message in a given phase, then it transmits in the round $l(v)$ of the phase. In this way messages from different levels do not collide with each other.

At the beginning of the algorithm, only the leaves of T' are active. A leaf v transmits its message in the phase $s(v)$. Each non-leaf node v (i.e., the number of its children stored in its label is larger than 0) does not send any message, until messages from all its children are received. If v receives its last message from its child in phase t, it transmits all received messages in the closest phase $t' > t$ such that $t' \bmod \Delta = s(v)$ and it remains inactive in other phases. Apart from the original messages to be gathered $s_1, \ldots, s_k$, each node v adds to a transmitted message the values of $c(v)$ and $c(p(v))$. Note that each node is able to distinguish messages received from its children from other messages thanks to the fact that $c(v) \neq c(w)$ if $w \neq v$ is the parent of a neighor u of v. Moreover, v is able to determine when it receives the last message from its children thanks to the fact that it stores the number of its children in the label and it can distinguish messages from different children, as $c(u) \neq c(u')$ for the children $u \neq u'$ of v. An inductive proof wrt to the levels of T' leads to Lemma 4.

Lemma 4. *The algorithm A is a k-gathering algorithm which works in time $O(D\Delta)$ with labels of size $O(\log \Delta)$.*

2.5   Summary and Separation from the Model Without Advice

Theorem 2 and Lemma 4 lead to the following result.

Corollary 1. *There exists a distributed algorithm for k-gathering in time $O\left(\min(D + k, D\Delta)\right)$ with a labeling scheme of optimal length $O(\min(\log \Delta, \log k))$.*

On the other hand, we prove the following result.

Corollary 2. *The distributed deterministic k-gathering problem with arbitrary distinct IDs of nodes requires $\Omega(D \log n + k \log(n/k))$ rounds.*

As our k-gathering algorithm with optimal advice works in $O(D+k)$ rounds, we obtain a $\Omega(\log n)$ gap between the algorithms with optimal advice and the ones with arbitrary distinct IDs for each $k = O(n^\varepsilon)$, where $0 < \varepsilon < 1$ is a constant.

3   The k-Broadcasting Problems

A simple composition of a k-gathering algorithm and a single-message broadcasting algorithm lead to the following corollaries.

Corollary 3. *There exists a unit-message k-broadcasting algorithm with advice size $O(1)$ which works in $O(D + \log^2 n)$ rounds.*

Corollary 4. *There exists a multi-message k-broadcasting algorithm with the size of the advice $O(\min(\log k, \log \Delta))$ which works in $O(D + \log^2 n + k)$ rounds.*

Let us also note here that the $\Omega(D + \log^2 n)$ lower bound holds even for centralized unit-message broadcasting.

4 Open Problems

An interesting further research direction is to determine optimal length of labeling schemes for k-gathering and multi-message k-broadcasting in the model without possibility of combining many source messages together.

Acknowledgments. Supported by the Polish National Science Centre grant 2020/39/B/ST6/03288.

We would like to thank anonymous reviewers for their useful comments and suggestions and to Andrzej Pelc for engaging us in this area of research.

References

1. Bu, G., Lotker, Z., Potop-Butucaru, M., Rabie, M.: Lower and upper bounds for deterministic convergecast with labeling schemes. Theor. Comput. Sci. **952**, 113775 (2023)
2. Bu, G., Potop-Butucaru, M., Rabie, M.: Wireless broadcast with short labels. In: Georgiou, C., Majumdar, R. (eds.) Networked Systems - 8th International Conference, NETYS 2020, Marrakech, Morocco, June 3-5, 2020, Proceedings. Lecture Notes in Computer Science, vol. 12129, pp. 146–169. Springer (2020). https://doi. org/10.1007/978-3-030-67087-0_10
3. Chrobak, M., Costello, K.P.: Faster information gathering in ad-hoc radio tree networks. Algorithmica **80**(3), 1013–1040 (2018)
4. Chrobak, M., Costello, K.P., Gasieniec, L.: Information gathering in ad-hoc radio networks. Inf. Comput. **281**, 104769 (2021)
5. Chrobak, M., Costello, K.P., Gasieniec, L., Kowalski, D.R.: Information gathering in ad-hoc radio networks with tree topology. Inf. Comput. **258**, 1–27 (2018). https://doi.org/10.1016/j.ic.2017.11.003
6. Cicalese, F., Manne, F., Xin, Q.: Faster deterministic communication in radio networks. Algorithmica **54**(2), 226–242 (2009)
7. Czumaj, A., Davies, P.: Deterministic communication in radio networks. SIAM J. Comput. **47**(1), 218–240 (2018)
8. Czumaj, A., Davies, P.: Exploiting spontaneous transmissions for broadcasting and leader election in radio networks. J. ACM **68**(2), 13:1–13:22 (2021). https://doi. org/10.1145/3446383
9. Ellen, F., Gilbert, S.: Constant-length labelling schemes for faster deterministic radio broadcast. In: Scheideler, C., Spear, M. (eds.) SPAA '20: 32nd ACM Symposium on Parallelism in Algorithms and Architectures, Virtual Event, USA, July 15-17, 2020, pp. 213–222. ACM (2020). https://doi.org/10.1145/3350755.3400238
10. Ellen, F., Gorain, B., Miller, A., Pelc, A.: Constant-length labeling schemes for deterministic radio broadcast. In: Scheideler, C., Berenbrink, P. (eds.) The 31st ACM on Symposium on Parallelism in Algorithms and Architectures, SPAA 2019, Phoenix, AZ, USA, June 22-24, 2019, pp. 171–178. ACM (2019).https://doi.org/ 10.1145/3323165.3323194
11. Fraigniaud, P., Korman, A., Lebhar, E.: Local MST computation with short advice. Theory Comput. Syst. **47**(4), 920–933 (2010)
12. Fusco, E.G., Pelc, A., Petreschi, R.: Topology recognition with advice. Inf. Comput. **247**, 254–265 (2016). https://doi.org/10.1016/j.ic.2016.01.005

13. Ganczorz, A., Jurdzinski, T.: Optimal-length labeling schemes and fast algorithms for k-gathering and k-broadcasting (2025). https://arxiv.org/abs/2512.02252
14. Ganczorz, A., Jurdzinski, T., Lewko, M., Pelc, A.: Deterministic size discovery and topology recognition in radio networks with short labels. Inf. Comput. **292**, 105010 (2023). https://doi.org/10.1016/J.IC.2023.105010
15. Ganczorz, A., Jurdzinski, T., Pelc, A.: Optimal-length labeling schemes for fast deterministic communication in radio networks. CoRR **abs/2410.07382, to appear in OPODIS 2025,** (2024). https://doi.org/10.48550/ARXIV.2410.07382
16. Gasieniec, L., Lingas, A.: On adaptive deterministic gossiping in ad hoc radio networks. Inf. Process. Lett. **83**(2), 89–93 (2002)
17. Gasieniec, L., Peleg, D., Xin, Q.: Faster communication in known topology radio networks. Distributed Comput. **19**(4), 289–300 (2007)
18. Gasieniec, L., Radzik, T., Xin, Q.: Faster deterministic gossiping in directed ad hoc radio networks. In: Hagerup, T., Katajainen, J. (eds.) Algorithm Theory - SWAT 2004, 9th Scandinavian Workshop on Algorithm Theory, Humlebaek, Denmark, July 8-10, 2004, Proceedings. Lecture Notes in Computer Science, vol. 3111, pp. 397–407. Springer (2004). https://doi.org/10.1007/978-3-540-27810-8_34
19. Glacet, C., Miller, A., Pelc, A.: Time vs. information tradeoffs for leader election in anonymous trees. ACM Trans. Algorithms **13**(3), 31:1–31:41 (2017). https://doi.org/10.1145/3039870
20. Gorain, B., Pelc, A.: Finding the size and the diameter of a radio network using short labels. Theor. Comput. Sci. **864**, 20–33 (2021)
21. Gorain, B., Pelc, A.: Short labeling schemes for topology recognition in wireless tree networks. Theor. Comput. Sci. **861**, 23–44 (2021)
22. Ilcinkas, D., Kowalski, D.R., Pelc, A.: Fast radio broadcasting with advice. Theor. Comput. Sci. **411**(14–15), 1544–1557 (2010)
23. Kowalski, D.R., Pelc, A.: Optimal deterministic broadcasting in known topology radio networks. Distrib. Comput. **19**(3), 185–195 (2007)
24. Krisko, C., Miller, A.: Labeling schemes for deterministic radio multi-broadcast. In: Kowalik, L., Pilipczuk, M., Rzazewski, P. (eds.) Graph-Theoretic Concepts in Computer Science - 47th International Workshop, WG 2021, Warsaw, Poland, June 23-25, 2021, Revised Selected Papers. Lecture Notes in Computer Science, vol. 12911, pp. 374–387. Springer (2021). https://doi.org/10.1007/978-3-030-86838-3_29
25. Marco, G.D.: Distributed broadcast in unknown radio networks. SIAM J. Comput. **39**(6), 2162–2175 (2010)
26. Vaya, S.: Round complexity of leader election and gossiping in bidirectional radio networks. Inf. Process. Lett. **113**(9), 307–312 (2013)

Distribution Testing Meets Sum Estimation

Sampriti Roy$^{(\boxtimes)}$

Department of Computer Science and Engineering, Indian Institute of Technology, Madras, India
`sampritiroy9@gmail.com`

Abstract. We study the problem of estimating the sum of n elements of a universe U, $|U| = n$, each associated with a weight $w(i) > 0$, within a structured universe. Specifically, our goal is to estimate the sum $W = \sum_{i=1}^{n} w(i)$ within a factor of $(1 \pm \epsilon)$ using a sublinear number of samples, under the assumption that the weights are non-increasing, i.e., $w(1) \geq w(2) \geq \ldots \geq w(n)$. The sum estimation problem is well-studied under different access models to the universe U. However, to the best of our knowledge, nothing is known about the sum estimation problem using non-adaptive conditional sampling. To address this gap, we explore the sum estimation problem using non-adaptive conditional weighted and non-adaptive conditional uniform samples, assuming that the underlying distribution $(D(i) = w(i)/W)$ is monotone. Furthermore, we extend our approach to the case where the underlying distribution of U is unimodal. We also address the problem of approximating the support size of U when $w(i) = 0$ or $w(i) \geq W/n$ using hybrid sampling, where both weighted and uniform sampling are available to access U.

We provide an algorithm for estimating the sum of n elements where the weights of the elements are non-increasing. Our algorithm requires $O(\frac{1}{\epsilon^3} \log n + \frac{1}{\epsilon^6})$ non-adaptive weighted conditional samples and $O(\frac{1}{\epsilon^3} \log n)$ non-adaptive uniform conditional samples. Our algorithm also follows the $\Omega(\log n)$ lower bound proposed by [2]. We also extend our algorithm when the underlying distribution D is unimodal. The sample complexity of the algorithm is the same as that of monotone with an additional $O(\log n)$ adaptive evaluation queries to find the minimum weighted point in the domain $[n]$. Additionally, we investigate the problem of estimating the support size of U, which consists of n elements such that the weight corresponding to them is either 0 or at least W/n. Our algorithm uses $O\big(\frac{\log^3 (n/\epsilon)}{\epsilon^8} \cdot \log^4 \frac{\log (n/\epsilon)}{\epsilon}\big)$ uniform samples and $O\big(\frac{\log (n/\epsilon)}{\epsilon^2} \cdot \log \frac{\log (n/\epsilon)}{\epsilon}\big)$ weighted samples from the universe and approximate the support size k such that $k - 2\epsilon n \leq \hat{k} \leq k + \epsilon n$.

1 Introduction

Distribution testing and sum estimation are two closely related problems with broad applications in statistics and data science. Distribution testing aims to

© The Author(s), under exclusive license to Springer Nature Switzerland AG 2026

J. Kozik and A. Wolff (Eds.): SOFSEM 2026, LNCS 16448, pp. 663–677, 2026.
https://doi.org/10.1007/978-3-032-17801-5_48

determine whether an unknown distribution satisfies a certain property $\mathcal{P}$ or if it is ϵ-far from $\mathcal{P}$, using only a limited number of samples. Sum estimation, on the other hand, aims to approximate the total weight $W = \sum_{i=1}^{n} w(i)$ of a universe $U = [n]$ from a small number of samples, where each sample returns a pair $(i, w(i))$. The goal is to output $\widehat{W}$ such that $(1 - \epsilon)W \leq \widehat{W} \leq (1 + \epsilon)W$ with high probability.

The sum estimation problem can be naturally viewed through the lens of distribution testing by interpreting the universe U with weights $w(i)$ as an unknown distribution D over $[n]$, where $D(i) = w(i)/W$. This perspective allows the application of tools and techniques from distribution testing to the study of sum estimation. Notably, recent advances in distribution testing have shown that classical properties can often be tested with optimal sample complexity under structural assumptions such as monotonicity, k-modality, or log-concavity.

In this paper, we build on this connection to design sample-efficient algorithms for sum estimation under monotone weights. Additionally, we study the problem of estimating the number of elements with non-zero weights, which corresponds to support size estimation in distribution testing. Our results show how structural assumptions can lead to stronger guarantees for both problems.

Motivation: A main goal of this work is to exploit monotonicity for efficient sum estimation. Exact sums over massive domains are often infeasible. Inspired by distribution testing where structural promises like monotonicity or unimodality can cut sample complexity from sublinear to $O(\log n)$, we ask whether similar savings are possible for sum estimation. This question matters in practice. Many real-world datasets are monotone or nearly so: pageview counts drop with page rank; network flows follow heavy-tailed patterns; and frequency counts in text or genomics often decrease predictably. Using such structures enables accurate estimates with far less memory and computation. We also address support size estimation, the task of counting elements with nonzero weight, which arises in tracking distinct users, identifying active sensors, or estimating sparsity in large datasets. As a special case of sum estimation with binary weights, it motivates designing algorithms that work under restricted sampling models.

Sampling Techniques: Sum estimation and distribution testing both aim to infer properties from restricted access to a dataset or distribution. This section reviews the sampling models used in our setting, many of which are inspired by analogous models in distribution testing.

Definition 1 (Standard Access Model (SAMP) [16]). *In the standard access model, the oracle returns an index $i \in [n]$ with probability $D(i)$.*

The direct analogue of **SAMP** model is the weighted sampling model in sum estimation, where $i \in [n]$ is returned with probability $w(i)/W$. This equivalence allows techniques developed in distribution testing to be translated naturally to the sum estimation setting. To capture more flexible access, both fields make use of *conditional access model.*

Definition 2 (Conditional Access Model *(COND)* [9,11]). *Given any subset $S \subseteq [n]$ with $D(S) > 0$, the oracle returns an element $i \in S$ with probability $D(i)/D(S)$. In the uniform conditional variant, the oracle returns $i \in S$ with probability $1/|S|$.*

Conditional access provides significantly more flexibility than standard access, enabling algorithms to focus on regions of interest. Sometimes, in addition to sampling, distribution testing includes access to an *evaluation model* (EVAL) [19], which returns $D(i)$ for any queried index $i \in [n]$.

Definition 3 (Evaluation Model *(EVAL)* [19]). *Given any index $i \in [n]$, the oracle returns its weight $D(i)$.*

This mirrors a similar model in sum estimation, which outputs $w(i)$ for a queried index. The evaluation model can be combined with the standard access model to form the *dual model*, where an index is sampled according to the distribution (or weights), and its value is then queried.

In the sum estimation setting, we work with a *dual* like model. Given a subset $S \subseteq [n]$, the oracle returns an element $i \in S$ with probability $w(i)/W(S)$, together with its weight $w(i)$, referred to as *weighted conditional* sampling. Under *uniform conditional sampling*, the oracle instead returns $i \in S$ uniformly at random, i.e., with probability $1/|S|$, again along with $w(i)$. When both types of access—uniform sampling over $[n]$ and weighted sampling over $[n]$ are available and return $(i, w(i))$ pairs, we refer to this combined access model as the *hybrid model*.

Overall, our techniques build on a unifying theme: translating the query models from distribution testing to sum estimation under structural constraints. By doing so, we present how conditional sampling, evaluation oracles, and hybrid model can be used for estimating sums and support size under different assumptions on the weight structure.

Background and Related Works: Distribution testing began with the work of Goldreich and Ron [16], who gave a sample-efficient algorithm for testing uniformity. Later work showed that structural promises reduce complexity. Daskalakis et al. [14] designed identity and closeness testers for monotone distributions, and Diakonikolas et al. [15] extended these ideas to log-concave and t-piecewise constant families. These results show how structure yields lower sample complexity—an idea central to our approach.

Conditional sampling has also been crucial in distribution testing. Canonne et al. [9] studied uniformity, identity, and closeness under conditional access. Canonne [8] designed a monotonicity tester in this model. Canonne and Rubinfeld [7] analyzed the evaluation model, giving efficient testers when direct access to weights is allowed.

Support size estimation, closely related to sum estimation, has been studied under many sampling access models. Raskhodnikova et al. [18] approximated support size assuming each element has mass at least $1/n$. Valiant and Valiant [20] proved lower bounds without such assumptions. Acharya et al. [1] gave non-adaptive lower bounds under conditional queries. Chakraborty et al. [10] addressed the gap between upper and lower bounds under conditional sampling. Canonne and Rubinfeld [7] studied support size in the EVAL model with zero-or-$1/n$ masses. These works highlight how access models and structure shape complexity.

Sum estimation of n variables with weights in $[0,1]$ has also been widely studied. Canetti et al. [6] gave uniform-sampling lower bounds. Motwani et al. [17] introduced proportional and hybrid sampling, achieving $O(\sqrt{n})$ and $O(n^{1/3})$ samples, respectively. Beretta and Tětek [4] tightened these to optimal $\Theta(\sqrt{n}/\varepsilon)$ for proportional sampling and nearly tight $O(n^{1/3}/\varepsilon^{4/3})$ in the hybrid model.

Closer to our work, Acharya et al. [2] studied sum estimation under conditional sampling and gave an adaptive algorithm with $O(\log^3 n/\varepsilon^2)$ weighted samples. Their method relies on adaptivity. We show that with monotonically non-increasing weights, non-adaptive conditional sampling suffices, leading to simpler and more efficient algorithms. Additionally, classical problems like estimating the number of distinct elements [3,13] can be viewed as special cases of sum estimation where weights are binary.

1.1 Our Contributions

- We provide an algorithm for estimating the sum of n elements where the weights of the elements are non-increasing. The algorithm requires $O(\frac{1}{\epsilon^3}\log n + \frac{1}{\epsilon^6})$ weighted conditional samples and $O(\frac{1}{\epsilon^3}\log n)$ uniform conditional samples. Our algorithm matches the $\Omega(\log n)$ non-adaptive lower bound for sum estimation proposed by Acharya et al. [2].
- We also extend our algorithm for the case when the distribution over the weights is unimodal. That is, the weights first decrease up to some point $j \in [n]$ and then increase from there to n. The number of samples required remains the same as in the monotone case, with an additional $O(\log n)$ adaptive EVAL queries to find the smallest weighted point in the domain $[n]$.
- We study the problem of estimating the size of the support of U, where U has n elements and each element's weight is either 0 or at least W/n. Our algorithm uses $O(\frac{\log^3(n/\epsilon)}{\epsilon^8} \cdot \log^4(\frac{\log(n/\epsilon)}{\epsilon}))$ uniform samples and $O(\frac{\log(n/\epsilon)}{\epsilon^2} \cdot \log(\frac{\log(n/\epsilon)}{\epsilon}))$ weighted samples over $[n]$. It outputs an estimate $\hat{k}$ of the support size k such that $k - 2\epsilon n \le \hat{k} \le k + \epsilon n$.

2 Notation and Preliminaries

Let U be a universe of n elements with non-negative weights $w(i)$ and total weight $W = \sum_{i \in [n]} w(i)$. Define the distribution D over $[n]$ by $D(i) = w(i)/W$. For a subset $S \subseteq [n]$, the conditional distribution D_S is $D_S(i) = w(i)/W(S)$, where $W(S) = \sum_{i \in S} w(i)$. The uniform distribution over S is denoted by $\mathcal{U}_S$. The support size of a distribution denoted by sup is the number of non-zero elements in the domain of D.

For two distributions D_1, D_2 over $[n]$, the total variation distance is $d_{TV}(D_1, D_2) = \frac{1}{2} \sum_{i \in [n]} |D_1(i) - D_2(i)|$. Let $\mathcal{D}$ be the set of distributions over $[n]$, and $\mathcal{P} \subseteq \mathcal{D}$ a property. A distribution D is ϵ-far from $\mathcal{P}$ if $d_{TV}(D, D') > \epsilon$ for all $D' \in \mathcal{P}$. A distribution D is monotone (non-increasing) if $D(1) \geq D(2) \geq \cdots \geq D(n)$. Given a partition $\mathcal{I} = \{I_j\}_{j=1}^{\ell}$ of $[n]$, the flattened distribution $(D^f)^{\mathcal{I}}$ is defined by $(D^f)^{\mathcal{I}}(i) = \sum_{t \in I_j} D(t)/|I_j|$ for $i \in I_j$. A key structural result for monotone distributions is the oblivious decomposition, first shown by Birgé [5] for non-increasing distributions over the continuous domain $[0, n]$. Daskalakis et al. [14] adapted Birgé's arguments to discrete distributions over $[n]$, showing that any monotone distribution is close to its flattened version with respect to an oblivious partition of exponentially growing intervals. Although often called Birgé's decomposition, we cite the discrete version from [14] formally below.

Theorem 1 (Oblivious partitioning [14]). *Let D be a non-increasing distribution over $[n]$ and $\mathcal{I} = \{I_1, ..., I_\ell\}$ be an interval partitioning of D from left to right such that $|I_j| = \lfloor (1+\epsilon)^j \rfloor$, for $0 < \epsilon < 1$. For $\ell = O(\frac{1}{\epsilon} \log (n\epsilon))$, the flattened distribution corresponding to $\mathcal{I}$, $(D^f)^{\mathcal{I}}$ is close to D, i.e., $d_{TV}(D, (D^f)^{\mathcal{I}}) \leq \epsilon$.*

The conclusion of Theorem 1 also holds for non-decreasing distributions:

Theorem 2 (Oblivious partitioning [14]). *Let D is non-decreasing over $[n]$ and $\mathcal{I} = \{I_1, \ldots, I_\ell\}$ is an interval partition with $|I_j| = \lfloor (1+\epsilon)^j \rfloor$ (for $0 < \epsilon < 1$) constructed in reverse order $(j = \ell, \ell - 1, \ldots, 1)$, then for $\ell = O(\frac{1}{\epsilon} \log(n\epsilon))$ the flattened distribution $(D^f)^{\mathcal{I}}$ satisfies $d_{TV}(D, (D^f)^{\mathcal{I}}) \leq \epsilon$.*

The oblivious property of monotone distributions extends to a broader class of flattened distributions where the flattened version $(D^f)^{\mathcal{I}}$ remains close to D under a given partition $\mathcal{I}$. We formalize this definition below.

Definition 4 ((ϵ, ℓ)-flattened distribution [12]). *A class of distributions $\mathcal{C}$, is said to be (ϵ, ℓ)-flattened if for every $D \in \mathcal{C}$, $d_{TV}(D, (D^f)^{\mathcal{I}}) \leq \epsilon$, where $(D^f)^{\mathcal{I}}$ is the flattened distribution of D according to a set of partitions $\mathcal{I} = \{I_1, ..., I_\ell\}$.*

The following inequality will be used in the analysis of the Algorithm 2.

Lemma 1 (Hoeffding's inequality). *Let $X_1, ..., X_m$ be independent random variables such that $X_i \in [a_i, b_i]$ for $1 \leq i \leq m$, and $X = (X_1 + X_2 + ... + X_m)/m$. Then $Pr[|X - \mathbb{E}[X]| \geq t] \leq 2exp(\frac{-2(mt)^2}{\sum_i (b_i - a_i)^2})$.*

3   Estimating Sum for Monotonically Non-Increasing Weights Using Conditional Sampling Model

In this section, we present an algorithm for estimating the sum of n elements with monotonically non-increasing weights. Let the universe be $[n]$ with weights $w(i)$, and let $W = \sum_{i=1}^{n} w(i)$. The goal is to compute an estimate $\widehat{W}$ such that $(1 - \epsilon)W \leq \widehat{W} \leq (1 + \epsilon)W$.

Our approach uses conditional sampling. Given a subset $S \subseteq [n]$, a weighted conditional sample returns $(i, w(i))$ with probability $w(i)/W(S)$, while a uniform conditional sample returns $(i, w(i))$ with probability $1/|S|$. The induced distribution $D(i) = w(i)/W$ inherits monotonicity, allowing us to apply the oblivious decomposition (Theorem 1) to partition $[n]$ into $\ell = O(\log n/\epsilon)$ intervals. We show that the intervals far from uniform contribute little weight, while approximately uniform intervals hold most of the total weight.

Our algorithm first identifies and discards intervals far from uniformity. Estimating the total weight over the remaining intervals provides a good approximation of W. To this end, we test whether a conditional distribution D_S is uniform or ϵ-far in total variation. Inspired by Canonne et al. [9], we compute the weight ratios using our query model, which returns $(i, w(i))$ directly. This lets us determine precisely whether D_S is uniform or far from uniformity.

Algorithm 1: Testing Uniformity

Input: Sample access to D_S, where $S \subseteq [n]$, error parameter $\epsilon \in (0, 1)$,
Output: Accept if D_S is uniform, Reject otherwise

1   Sample $T_1 = O\left(\frac{1}{\epsilon} \log\left(1/\epsilon\right)\right)$ pairs $(i, w(i))$ from D_S;
2   Sample $T_2 = O\left(\frac{1}{\epsilon} \log\left(1/\epsilon\right)\right)$ pairs $(j, w(j))$ uniformly at random from S;
3   **for** *each pair* (i, j), *where* $i \in T_1$ *and* $j \in T_2$ **do**
4   **if** $\frac{w(i)}{w(j)} > \left(1 + \frac{\epsilon}{2}\right)$ **then**
5   Reject; **EXIT**;

6   Accept;

Theorem 3. *The algorithm Testing Uniformity (Algorithm 1) samples $O(\frac{1}{\epsilon} \log\left(1/\epsilon\right))$ points from D_S and $O(\frac{1}{\epsilon} \log\left(1/\epsilon\right))$ points uniformly from S. It returns* **Accept** *with probability at least $2/3$ if D_S is uniform over S, and returns* **Reject** *with probability at least $2/3$ if $d_{TV}(D_S, \mathcal{U}_S) > \epsilon$.*

Proof. **Completeness:** Let D_S be uniform over $[S]$, then for every pair $(i, w(i))$ and $(j, w(j))$, $\frac{w(i)}{w(j)} = 1$. Hence the algorithm will not output Reject inside the for loop.

Soundness: Let $d_{TV}(D_S, \mathcal{U}_S) > \epsilon$. We start by defining the following sets, $H = \left\{h \in [S] | D_S(h) \geq \frac{1}{S}\right\}$ and $L = \left\{l \in [S] | D_S(l) < \frac{1}{S}\right\}$. It is easy to see that when D_S is ϵ-far from uniformity, $\sum_{h \in H}(D(h) - \frac{1}{S}) = \sum_{l \in L}(\frac{1}{S} - D(l)) > \epsilon/2$. Now, we have the following observations:

- Let $L' \subset L$ such that $L' = \left\{l' \in L | D_S(l') < \frac{1}{S} - \frac{\epsilon}{4S}\right\}$. Then $|L'| > (\epsilon/4)S$. While sampling $O(\frac{1}{\epsilon} \log\left(1/\epsilon\right))$ points uniformly from S, the probability that no point comes from L' is at most $(1 - \epsilon/4)^{O(1/\epsilon \log\left(1/\epsilon\right))} < \epsilon/100$. Applying union bound over all $O(1/\epsilon \log\left(1/\epsilon\right))$ points, with probability at least $9/10$, at least one point will come from the set L'.

– Let $H' \subset H$ such that $H' = \left\{ h' \in H | D_S(h') > \frac{1}{S} + \frac{\epsilon}{4S} \right\}$. Then $D_S(H') > \epsilon/4$. While sampling $O(\frac{1}{\epsilon} \log{(1/\epsilon)})$ points from D_S, using similar argument as that of the previous case, we argue that at least one point will occur from the set H'.

Consider a pair of points (i, j) where i is sampled from D_S and j is sampled uniformly from S, such that $i \in H'$ and $j \in L'$. Then $\frac{D_S(i)}{D_S(j)} > \frac{(1+\epsilon/4)}{(1-\epsilon/4)} > (1+\epsilon/2)$. Observe that $D_S(i) = \frac{w(i)}{W(S)}$ and $D_S(j) = \frac{w(j)}{W(S)}$. Therefore, when D_S is far from uniformity, there exists at least a pair (i, j) such that $\frac{w(i)}{w(j)} > (1 + \epsilon/2)$ and the algorithm returns Reject in this case.

Given that the weights $w(i)$ are monotonically non-increasing, the induced distribution $D(i) = \frac{w(i)}{W}$ is also non-increasing. Using this property, we apply the oblivious decomposition technique (Theorem 1) to partition the domain into intervals. We then use the Algorithm 1 to identify the intervals where the conditional distributions are far from uniformity. Prior work by Rubinfeld and Servedio [19] presented a one-query algorithm for testing the uniformity of monotone distributions when direct access to $D_S(i)$ is available. However, their procedure only distinguishes between the cases "D_S is uniform" and "D_S is not uniform" without quantifying the distance from uniformity. In contrast, our algorithm explicitly tests whether D_S is uniform or ϵ-far from uniformity, where the ϵ-farness criterion is essential for our analysis.

After removing the intervals that are ϵ-far from uniformity, we proceed to estimate the total weight over the remaining intervals as discussed previously. Before presenting the algorithm, we formalize the relevant parameters related to the oblivious decomposition when applied to a non-increasing distribution D.

Let D be a monotone distribution and let $\mathcal{I} = \{I_1, ..., I_\ell\}$ be the set of oblivious partitions (Theorem 1) with parameter $\epsilon_1 = \epsilon(1 - \delta)$, for $\delta = \sqrt{(1 - \epsilon)}$. Therefore, we have that $d_{TV}(D, (D^f)^{\mathcal{I}}) \leq \epsilon_1$. In other words, this implies that $\sum_{x \in [n]} |D(x) - (D^f)^{\mathcal{I}}(x)| \leq \epsilon_1$. Observe that by the definition of flattened distribution, $\forall x \in I_j$, $(D^f)^{\mathcal{I}}(x) = \frac{\sum_{x \in I_j} D(x)}{|I_j|} = \frac{D(I_j)}{|I_j|}$. Substituting the definition of flattened distribution, we get that

$$\sum_{j=1}^{\ell} \sum_{x \in I_j} |D(x) - \frac{D(I_j)}{|I_j|}| \leq \epsilon_1.$$

By factoring out $D(I_j)$ from the first summation, we get $\sum_{j=1}^{\ell} D(I_j) \sum_{x \in I_j} |\frac{D(x)}{D(I_j)} - \frac{1}{|I_j|}| \leq \epsilon_1$. Observe that $\forall x \in I_j$, $\frac{D(x)}{D(I_j)}$ defines the conditional distribution over I_j, which is denoted by D_{I_j}. Therefore, $\sum_{x \in I_j} |\frac{D(x)}{D(I_j)} - \frac{1}{|I_j|}| = d_{TV}(D_{I_j}, \mathcal{U}_{I_j})$, where $\mathcal{U}_{I_j}$ is the uniform distribution over I_j. Substituting back, we obtain that,

$$\sum_{j=1}^{\ell} D(I_j) d_{TV}(D_{I_j}, \mathcal{U}_{I_j}) \leq \epsilon_1 \tag{1}$$

Consider D be a distribution over $[n]$, where $D(x) = \frac{w(x)}{W}$, and D_{I_j} be a conditional distribution over I_j such that $D_{I_j}(x) = \frac{w(x)}{W(I_j)}$. Note that the total weight of an interval I_j is defined by $D(I_j) = \frac{W(I_j)}{W}$. Let $J \subseteq \mathcal{I}$ be the set of intervals where $\forall I_j \in J, d_{TV}(D_{I_j}, U_{I_j}) > \epsilon$. Substituting this to the Equation 1, we get that $\sum_{I_j \in J} D(I_j) \leq \epsilon_1/\epsilon = (1 - \delta)$. Where $I_j \in J$ are the set of intervals for which the conditional distributions are ϵ-far from uniformity.

Consider $J' \subset \mathcal{I}$ to be the set of remaining intervals where D_{I_j} is not ϵ-far from uniformity. Then $\sum_{I_j \in J'} D(I_j) > \delta$. In other words, we can consider J' to be a large bucket where each point $i \in \{[n] \setminus J\}$. We also argue that when D is monotone, $D(J') > \delta$. The core idea of our algorithm is to estimate the total weight of the elements that lie in the large bucket J'. The total weight of the points which lie in the intervals that are ϵ-far from uniformity is at most $(1 - \delta)$. Therefore, estimating the total weight of the set J' will suffice to estimate the total weight of all elements up to $(1 \pm \epsilon)$ multiplicative error. We present our algorithm with respect to a monotonically non-increasing universe over $[n]$ where for each $i \in [n]$, $D(i) = \frac{w(i)}{W}$ and $D_S(i) = \frac{w(i)}{W(S)}$.

Algorithm 2: Estimate Sum Monotone

Input: Weighted conditional sample access (over a set S) to a monotonically non-increasing universe, set of oblivious partitions $\{I_1, \ldots, I_\ell\}$, for $\ell = O\left(\frac{1}{\epsilon_1} \log n\right)$ where $\epsilon_1 = \epsilon(1 - \delta)$ for $\delta = \sqrt{(1 - \epsilon)}$

Output: Estimate $\widehat{W}$ for the sum

1   $J = \emptyset$;

2   **for** *each interval I_j* **do**

3   Run Algorithm 1;

4   **if** *Output is Reject* **then**

5   Add I_j to J;

6   **if** $J = \emptyset$ **then**

7   Check **if** $w(1) = w(n)$ **then**

8   Return $W = n \cdot w(1)$

9   **else**

10   Sample a pair $(i, w(i))$ conditioned on each interval I_j and return $W = \sum_{j \in [\ell]} |I_j| \cdot w(i)$

11   **else**

12   Sample $T = O\left(\frac{1}{\epsilon^6}\right)$ pairs of $(i, w(i))$ from n according to weighted sampling;

13   **for** *each $(i, w(i)) \in T$* **do**

14         $X_i = \begin{cases} w(i); \text{ when } i \notin J \\ 0; \text{ otherwise} \end{cases}$

15   Return $\widehat{W} = \frac{1}{T} \sum_{i=1}^{T} X_i$;

Theorem 4. *Given sample access to a set of elements over a universe $[n]$ whose weights are non-increasing, the algorithm Estimate Sum Monotone (Algorithm 2) uses total $O(\frac{1}{\epsilon^3}\log n + \frac{1}{\epsilon^6})$ conditional samples and returns an estimate $\widehat{W}$ such that with probability at least $2/3$, $(1-2\epsilon)W \leq \widehat{W} \leq (1+\epsilon)W$.*

Proof. The algorithm estimates the total weight $W = \sum_{i=1}^{n} w(i)$ when the weights induce a non-increasing distribution over the domain $[n]$. Depending on the structure of this distribution, we consider the following distinct cases: **Case 1: All weights are equal.** If $J = \emptyset$ and $w(1) = w(n)$, then by monotonicity all weights are equal and D is uniform. The algorithm returns $W = n \cdot w(1)$. **Case 2: Histogram Structure.** If $J = \emptyset$ but $w(1) \neq w(n)$, the distribution is glob-ally non-uniform but each interval is (nearly) uniform. The algorithm samples one point per interval and estimates $W = \sum_j |I_j|\, w(i)$.

Case 3: Strictly monotone. We now consider the case where the underlying distribution is monotone and the set J is non-empty. In this setting, oblivious decomposition ensures that the total probability mass of the intervals in J, those whose conditional distributions are ϵ-far from uniform, is bounded by $(1 - \delta)$, i.e., $\sum_{I_j \in J} D(I_j) \leq 1 - \delta$. Considering J' to be the set of intervals containing points from $[n] \setminus J$, $\sum_{I_j \in J'} D(I_j) > \delta$. Now, while sampling $O(\frac{1}{\epsilon^4})$ weighted samples from $[n]$, probability that no point comes from the set J' is at most $(1 - \delta)^{1/\epsilon^4}$. By applying a union bound over all samples, with high probability, at least one point will come from the set J'.

Observe that $\widehat{W} = \frac{1}{T}\sum_{i=1}^{T} X_i$, where $X_i = w(i)$; when $i \notin J$. Considering J' be the large bucket such that $J' = \{i : i \in [n] \setminus J\}$, the expectation of $\widehat{W}$ is

$$\mathbb{E}[\widehat{W}] = \sum_{i \in J'} \frac{W(J')}{W} \cdot w(i) = \frac{W^2(J')}{W}.$$

We apply Hoeffding inequality (Lemma 1) to bound the expectation. The random variables X_i takes the value in the range $[min_i\ w(i), max_i\ w(i)]$. Therefore, we have

$$Pr\left[|\widehat{W} - \frac{W^2(J')}{W}| > \epsilon \cdot \frac{W^2(J')}{W}\right] \leq \exp\left(-\frac{\epsilon^2 W^4(J')T}{W^2 \cdot [max_i\ w(i) - min_i\ w(i)]^2}\right).$$

We know $\sum_{I_j \in J'} D(I_j) = D(J') > \delta$. In other words, $D(J') = \frac{W(J')}{W} > \delta$. Also, $[max_i\ w(i) - min_i\ w(i)] < W$. Together with the facts, we get that for $T = O(\frac{1}{\epsilon^6})$, with high probability, $(1 - \epsilon)\frac{W^2(J')}{W} \leq \widehat{W} \leq (1 + \epsilon)\frac{W^2(J')}{W}$.

Now, $\delta^2 W^2 < W^2(J') < W^2$. For $\delta = \sqrt{(1 - \epsilon)}$, we conclude that $(1 - \epsilon)^2 W < \widehat{W} < (1+\epsilon)W$. Considering $(1-\epsilon)^2 > (1-2\epsilon)$, we prove that $(1-2\epsilon)W < \widehat{W} < (1 + \epsilon)W$.

Remark 1. We argue that monotonicity is a strong structural assumption. It enables oblivious decomposition such that the intervals far from unifor-mity contribute little mass, while the remaining intervals are approximately uniform and contain most of the total weight. Formally, for $\{I_1, \ldots, I_\ell\}$, $\sum_{j=1}^{\ell} D(I_j)\, d_{TV}\left(D_{I_j}, \mathcal{U}_{I_j}\right) \leq \epsilon_1$, which is the core ingredient of our analysis. Any distribution class whose flattened version is ϵ-close to the original with respect to a set of ℓ partitions, also satisfies this property. Monotone distributions are a special case with $\ell = O(\log n)$. Thus, although our algorithm is stated for monotone weights, it extends to broader classes of (ϵ, ℓ)-flattened distributions (Definition 4).

Discussion on the Lower Bound: We present an overview of the lower bound for the problem of sum estimation in the context of a monotonically non-increasing universe. To do so, we draw a connection between sum estimation and a special case of support size estimation studied by Acharya et al. [2], and first revisit their lower bound. When the universe consists solely of elements with weights in $\{0, 1\}$, the task of sum estimation reduces to estimating the number of non-zero entries in the domain—that is, estimating the support size. Acharya et al. [1] showed that, in the conditional sampling model with non-adaptive queries, any algorithm that estimates the support size up to a factor of $\log n$ requires at least $\Omega(\frac{\log n}{\log \log n})$ queries. In a subsequent work, Acharya et al. [2] extended this lower bound to sum estimation. Specifically, they considered estimating the sum $W_S = \sum_{i \in S} w(i)$, where S is a random subset of size 2^t, with t chosen uniformly at random from $\{1, 2, \ldots, \log n\}$, and where $w(i) = 1$ if $i \in S$, and $w(i) = 0$ otherwise. Under this construction, estimating the support size becomes equivalent to estimating the sum W_S. By adapting the similar argument for a smaller (constant) value of the approximation factor, the same techniques yield the following result presented by Acharya et al. [2].

Theorem 5 (Non-adaptive lower bound [2]). *Any non-adaptive algorithm for estimating W_S up to a factor 2 requires at least $\Omega(\log n)$ queries.*

Our non-adaptive algorithm for monotonically non-increasing weights uses $O(\frac{1}{\epsilon^3} \log n + \frac{1}{\epsilon^6})$ conditional samples, thus matching the $\Omega(\log n)$ dependence established by Acharya et al. and showing that the logarithmic term is inherent in the non-adaptive setting.

A Special Case of Unimodal Distribution: We now handle the case where the weight sequence over $[n]$ is unimodal, i.e., decreasing up to some index j and increasing thereafter. Such a distribution decomposes naturally into two monotone segments: a non-increasing part on $[1, j]$ and a non-decreasing part on $[j + 1, n]$. Our approach also applies symmetrically when the first half is increasing and the second half is decreasing. If the location of the minimum-weight element (the "mode" of the unimodal distribution) is known, we simply apply our monotone-sum estimator separately to each segment and return $\widehat{W} = \widehat{W}_1 + \widehat{W}_2$. To identify the point of minimum weight, we use a binary search strategy. At each step, we query the weight of the midpoint, $w(n/2)$, and compare it with its immediate neighbors, $w(n/2 - 1)$ and $w(n/2 + 1)$. If $w(n/2 - 1) < w(n/2) < w(n/2+1)$, the distribution is increasing in this region, so the minimum lies in the left half $[1, n/2 - 1]$. If $w(n/2 - 1) > w(n/2) > w(n/2 + 1)$, the distribution is still decreasing, and the minimum lies in the right half $[n/2+1, n]$. Repeat this process until there is a constant number of elements in an interval. Then query all the points and find the minimum point.

This process continues recursively until we narrow down the interval to a constant number of elements, at which point we query all remaining points to identify the exact minimum. This binary search requires only $O(\log n)$ additional EVAL queries. Once the minimum region is found, Algorithm 2 is applied to each segment using the decompositions of Theorems 1 and 2. By Theorem 4, both estimates satisfy $(1 - 2\epsilon)W_i \leq \widehat{W}_i \leq (1 - \epsilon)W_i$, and summing yields the following guarantee.

Theorem 6 *Let the weights over $[n]$ form a unimodal sequence (either decrease – increase or increase – decrease). There is an algorithm using $O(\frac{1}{\epsilon^3} \log n + \frac{1}{\epsilon^6})$ non-adaptive conditional samples and $O(\log n)$ adaptive evaluation queries that outputs $\widehat{W}$ with $(1 - -2\epsilon)W \leq \widehat{W} \leq (1 - \epsilon)W$, with probability at least $2/3$.*

4   Estimating Support Size Using Hybrid Model

In this section, we present an algorithm for estimating the support size of a universe U of n elements. Each element i has either zero weight or at least a minimum weight $w(i) \geq W/n$, where W is the total weight. The support size k is the number of non-zero elements, and our goal is to estimate k up to an additive error of $\pm \epsilon n$. We work in the hybrid sampling model, which combines uniform and weighted samples. At a high level, our approach estimates the support size by counting the size of the neighborhoods of the sampled points. Specifically, we sample a set of elements from the domain $[n]$, where each element x is selected with probability proportional to its weight $w(x)$. For each sampled element x, we define its neighborhood by the set of elements whose weights are within a multiplicative factor of $(1 \pm \epsilon)$ of $w(x)$. Since elements are sampled with probability proportional to their weights, non-zero weight elements are likely to be included in the samples. By aggregating the sizes of the neighborhoods of sampled points, we obtain an estimate of the total number of non-zero elements— that is, the support size. To formalize this, we introduce the procedure *estimate neighborhood fraction* (Algorithm 3), which, given a point $(x, w(x))$, uses uniform sampling to approximate the fraction $|U_\epsilon(x)|/n$.

Algorithm 3: Estimate Neighborhood Fraction

Input: A pair $(i, w(i))$, $\epsilon, \alpha \in (0, 1)$, where $\alpha = \frac{\epsilon^3}{\log(n/\epsilon)\log(\log n/\epsilon)}$

Output: Estimated neighborhood fraction $\hat{f}$

1 $X = 0$;

2 Sample $S = O\left(\frac{1}{\alpha^2}\right)$ set of $(j, w(j))$ pairs uniformly from $[n]$;

3 **for** *each $(j, w(j))$ pair* **do**

4   **if** $\frac{w(i)}{(1+\epsilon)} \le w(j) \le (1 + \epsilon)w(i)$ **then**

5   Set $X_j = 1$;

6     $X = X + X_j$;

7 Return $\hat{f} = \frac{1}{S} \cdot X$;

Theorem 7. *Let $U_\epsilon(i) = \{j \in [n] : \frac{1}{1+\epsilon}w(i) \le w(j) \le (1 + \epsilon)w(i)\}$ be the neighborhood of the point $i \in [n]$. The algorithm Estimate Neighborhood Fraction (Algorithm 3) returns $\hat{f}$, such that $\frac{|U_\epsilon(i)|}{n} - \frac{\epsilon^3}{\log(n/\epsilon)\log\log n/\epsilon} \le \hat{f} \le \frac{|U_\epsilon(i)|}{n} + \frac{\epsilon^3}{\log(n/\epsilon)\log\log n/\epsilon}$. The algorithm uses $O(\frac{\log^2(n/\epsilon)\log^2\log n/\epsilon}{\epsilon^6})$ uniform samples from $[n]$.*

Proof. The proof is straightforward with an application of additive Chernoff bound. Observe that $\hat{f} = \frac{\sum_{j \in S} X_j}{S}$. Therefore, we have, $\mathbb{E}[\hat{f}] = \mathbb{E}[X_j] = \sum_{j;j \in U_\epsilon(i)} \frac{1}{n} = \frac{|U_\epsilon(i)|}{n}$.

Now, we bound the expected value calculated above. $X_1, X_2, ..., X_S$ are all random variables taking values in $[0, 1]$. For $S = O(\frac{1}{\alpha^2}) = O(\frac{\log^2(n/\epsilon)\log^2\log n/\epsilon}{\epsilon^6})$, by applying Chernoff bound, we get $Pr[|\hat{f} - \frac{|U_\epsilon(i)|}{n}| > \alpha] \le e^{-\alpha^2 \cdot S} < 1/10$. Hence, with probability at least $9/10$, $|\hat{f} - \frac{|U_\epsilon(i)|}{n}| \le \frac{\epsilon^3}{\log(n/\epsilon)\log\log n/\epsilon}$.

The above theorem provides a way to approximate the fraction of points in the neighborhood of a given element. To implement this, we first identify a target set of points for which to estimate neighborhood fractions. Intuitively, sampling proportionally to weights favors elements with higher weights, making them a natural choice for this set. To formalize this, we use the concept of a *weight cover* from [9] to find the neighborhoods of a set R. Analyzing the sizes of these neighborhoods allows us to estimate the overall support size efficiently.

Definition 5. *[Weight Cover [9]] Let D be a distribution over $[n]$ and a parameter $\epsilon_1 > 0$, we say that a point $i \in [n]$ is ϵ_1-covered by a set $R = \{r_1, ..., r_t\} \subseteq [n]$ if there exists a point $r_j \in R$ such that $D(i) \in [1/(1 + \epsilon_1), (1 + \epsilon_1)]D(r_j)$. Let the set of points in $[n]$ that are ϵ_1-covered by R be denoted by $U_{\epsilon_1}(R)$. We say that R is an (ϵ_1, ϵ_2)-cover for D if $D([n] \setminus U_{\epsilon_1}(R)) \le \epsilon_2$.*

To find the set R, we state the following lemma analogous to Canonne et al. [9].

Lemma 2. *Let there be a set of size $|U| = n$, where each element i has an associated weight $w(i)$. Let $R = O\left(\frac{\log{(n/\epsilon)}}{\epsilon^2} \cdot \log\frac{\log{(n/\epsilon)}}{\epsilon}\right)$ samples are drawn according to the proportional sampling. If $U_{\epsilon/c}(R)$ be the set of points ϵ/c-covered by R, then $W([n] \setminus U_{\epsilon_1}(R)) \leq (\epsilon/c) \cdot W$.*

We now bring together the parts we described above and present the algorithm.

Algorithm 4: Estimate Support Size

Input: Hybrid sampling access to the universe, error parameter $\epsilon \in (0, 1)$
Output: Estimated support size $\hat{k}$

1   Sample $R = O\left(\frac{\log{(n/\epsilon)}}{\epsilon^2} \cdot \log\frac{\log{(n/\epsilon)}}{\epsilon}\right)$ pairs of $(i, w(i))$ according to weighted sampling;

2   Set $\hat{k} = 0$;

3   **for** *each* $(i, w(i)) \in R$ **do**

4   Run Algorithm 3 and let $\hat{f}_i$ be the output;

5   Calculate $\hat{k} = \hat{k} + n \cdot \hat{f}_i$;

6   Return $\hat{k}$;

Theorem 8. *Given sample access to a universe of n elements such that $\min_{x \in sup} w(i) \geq \frac{W}{n}$, the algorithm Estimate Support Size (Algorithm 4) returns an estimate of the support size such that $k - 2\epsilon n \leq \hat{k} \leq k + \epsilon n$. The algorithm uses $O\left(\frac{\log^3{(n/\epsilon)}}{\epsilon^8} \cdot \log^4\frac{\log{(n/\epsilon)}}{\epsilon}\right)$ uniform samples and $O\left(\frac{\log{(n/\epsilon)}}{\epsilon^2} \cdot \log\frac{\log{(n/\epsilon)}}{\epsilon}\right)$ weighted samples from the universe.*

Proof. Observe that $\hat{k} = \sum_{i \in R} n \cdot \hat{f}_i$. Also, we know from the Theorem 7,

$$\frac{|U_\epsilon(i)|}{n} - \frac{\epsilon^3}{\log{(n/\epsilon)}\log\log n/\epsilon} \leq \hat{f}_i \leq \frac{|U_\epsilon(i)|}{n} + \frac{\epsilon^3}{\log{(n/\epsilon)}\log\log n/\epsilon}.$$

Summing over $R = O\left(\frac{\log{(n/\epsilon)}}{\epsilon^2} \cdot \log\frac{\log{(n/\epsilon)}}{\epsilon}\right)$, we get $\sum_{i \in R}|U_\epsilon(i)| - \epsilon \cdot n \leq \hat{k} \leq \sum_{i \in R}|U_\epsilon(i)| + \epsilon \cdot n$. We know by Lemma 2, $W([n] \setminus U_\epsilon(R)) \leq (\epsilon/c) \cdot W$. As k denotes the set of elements with non-zero weights, we argue that $W([k] \setminus U_\epsilon(R)) \leq (\epsilon/c) \cdot W$. Let T be the set of non-zero elements that are not ϵ/c covered by R. Then by the Lemma 2, $W(T) \leq (\epsilon/c) \cdot W$. We are also given the promise that for $i \in T$, $\min_{x \in sup} w(i) \geq \frac{W}{n}$. Therefore, $|T| \cdot \frac{W}{n} \leq \frac{\epsilon W}{c}$, or, $|T| \leq n\epsilon/c \leq n\epsilon$ for $c \geq 1$. Now, observe that $\sum_{i \in R}|U_\epsilon(i)| = k - |T| \geq k - n\epsilon$. Additionally, $\sum_{i \in R}|U_\epsilon(i)| \leq k$. Substituting these values, finally we get $k - 2\epsilon \cdot n \leq \hat{k} \leq k + \epsilon \cdot n$.

Sample Complexity Analysis: The algorithm samples $O\left(\frac{\log(n/\epsilon)}{\epsilon^2}\cdot\log\frac{\log(n/\epsilon)}{\epsilon}\right)$ weighted points from the universe. For each point the algorithm runs"Estimate neighborhood fraction". Estimating neighborhood for a fixed point i requires $O\left(\frac{\log^2(n/\epsilon)\log^2\log n/\epsilon}{\epsilon^6}\right)$ uniform samples from $[n]$. To guarantee that the estimated neighborhood is a good estimator for every $i \in R$ points, we need to use a union bound over R. Hence, each run of "Estimate neighborhood fraction" uses $O\left(\frac{\log^2(n/\epsilon)\log^3\log n/\epsilon}{\epsilon^6}\right)$. Therefore, a total of $O\left(\frac{\log^3(n/\epsilon)}{\epsilon^8}\cdot\log^4\frac{\log(n/\epsilon)}{\epsilon}\right)$ uniform samples required for the algorithm.

5 Conclusion

We have presented algorithms for estimating the sum of weighted elements when the underlying distribution is monotone or unimodal, using a combination of weighted and uniform conditional queries. Our results achieve near-optimal sample complexity, though tightening the gap between upper and lower bounds for monotone distributions remains an open question. Future work includes extending these techniques to more complex structures, such as k-modal or histogram-like distributions, and exploring support size estimation using only weighted sampling. These directions could further improve efficiency in practical large-scale data applications.

References

1. Acharya, J., Canonne, C.L., Kamath, G.: A chasm between identity and equivalence testing with conditional queries. In: Approximation, Randomization, and Combinatorial Optimization. Algorithms and Techniques (APPROX/RANDOM 2015), volume 40 of Leibniz International Proceedings in Informatics (LIPIcs), pages 449–466 (2015). https://doi.org/10.4230/LIPIcs. APPROX-RANDOM.2015.449
2. Acharya, J., Canonne, C.L., Kamath, G.: Adaptive estimation in weighted group testing. In: 2015 IEEE International Symposium on Information Theory (ISIT), pages 2116–2120 (2015). https://doi.org/10.1109/ISIT.2015.7282829.
3. Bar-Yossef, Z., Kumar, R., Sivakumar, D.: Sampling algorithms: lower bounds and applications. In: Proceedings of the Thirty-Third Annual ACM Symposium on Theory of Computing, STOC '01, page 266–275, New York, NY, USA, 2001. Association for Computing Machinery. https://doi.org/10.1145/380752.380810
4. Beretta, L., Tětek, J.: Better sum estimation via weighted sampling. ACM Trans. Algorithms, **20**(3) (2024). https://doi.org/10.1145/3650030
5. Birge, L.: On the risk of histograms for estimating decreasing densities. Ann. Stat. **15**(3), 1013–1022 (1987)
6. Canetti, R., Even, G., Goldreich, O.: Lower bounds for sampling algorithms for estimating the average. Inf. Process. Lett. **53**(1), 17–25 (1995). https://doi.org/ 10.1016/0020-0190(94)00171-T
7. Canonne, C., Rubinfeld, R.: Testing probability distributions underlying aggregated data. In: Automata, Languages, and Programming, pages 283–295, Berlin, Heidelberg (2014). Springer Berlin Heidelberg

8. Canonne, C.L.: Big data on the rise: testing monotonicity of distributions. In: 42nd International Conference on Automata, Languages and Programming (ICALP) (2015)

9. Canonne, C.L., Ron, D., Servedio, R.A.: Testing probability distributions using conditional samples. SIAM J. Comput. **44**(3), 540–616 (2015)

10. Chakraborty, D., Kumar, G., Meel, K.S.: Support size estimation: the power of conditioning. In: 48th International Symposium on Mathematical Foundations of Computer Science (MFCS 2023), volume 272 of Leibniz International Proceedings in Informatics (LIPIcs), pages 33:1–33:13. Schloss Dagstuhl – Leibniz-Zentrum für Informatik (2023). https://doi.org/10.4230/LIPIcs.MFCS.2023.33.

11. Chakraborty, S., Fischer, E., Goldhirsh, Y., Matsliah, A.: On the power of conditional samples in distribution testing. SIAM J. Comput. **45**(4), 1261–1296 (2016)

12. Chan, S.-O., Diakonikolas, I., Servedio, R.A., Sun, X.: Learning mixtures of structured distributions over discrete domains. In: Proceedings of the Twenty-Fourth Annual ACM-SIAM Symposium on Discrete Algorithms, SODA '13, page 1380–1394, USA (2013). Society for Industrial and Applied Mathematics

13. Charikar, M., Chaudhuri, S., Motwani, R., Narasayya, V.: Towards estimation error guarantees for distinct values. In: Proceedings of the Nineteenth ACM SIGMOD-SIGACT-SIGART Symposium on Principles of Database Systems, PODS '00, pages 268–279, New York, NY, USA, 2000. Association for Computing Machinery. https://doi.org/10.1145/335168.335230

14. Daskalakis, C., Diakonikolas, I., Servedio, R.A., Valiant, G., Valiant, P.: Testing k-modal distributions: optimal algorithms via reductions. In: Conference, 24th Annual ACM-SIAM Symposium on Discrete Algorithms, SODA (2013)

15. Diakonikolas, I., Kane, D.M., Nikishkin, V.: Testing identity of structured distributions. In: Proceedings of the Twenty-Sixth Annual ACM-SIAM Symposium on Discrete Algorithms, SODA '15, pages 1841–1854, USA. Soc. Indus. Appl. Math. (2015)

16. Goldreich, O., Ron, D.: On testing expansion in bounded-degree graphs. Electron Colloq Comput Complexity **7**, 01 (2000)

17. Motwani, R., Panigrahy, R., Xu, Y.: Estimating sum by weighted sampling. In: Proceedings of the 34th International Conference on Automata, Languages and Programming, ICALP'07, pages 53–64, Berlin, Heidelberg (2007). Springer-Verlag

18. Raskhodnikova, S., Ron, D., Shpilka, A., Smith, A.: Strong lower bounds for approximating distribution support size and the distinct elements problem. SIAM J. Comput. **39**(3), 813–842 (2009). https://doi.org/10.1137/070701649

19. Rubinfeld, R., Servedio, R.A.: Testing monotone high-dimensional distributions. In: Proceedings of the Thirty-Seventh Annual ACM Symposium on Theory of Computing, STOC '05, pages 147–156, New York, NY, USA (2005). Association for Computing Machinery

20. Valiant, G., Valiant, P.: A CLT and tight lower bounds for estimating entropy. Electr. Colloquium Comput. Compl. (ECCC), **17**(183) (2010)

The Buffer Minimization Problem
for Scheduling Flow Jobs with Conflicts

Niklas Haas[ID], Sören Schmitt[ID], and Rob van Stee[(✉)][ID]

Department of Mathematics, University of Siegen, Siegen, Germany
niklas.haas@student.uni-siegen.de,
{soeren.schmitt,rob.vanstee}@uni-siegen.de

Abstract. We consider the online *buffer minimization in multiprocessor systems with conflicts problem* (in short, the *buffer minimization problem*) in the recently introduced flow model. In an online fashion, workloads arrive on some of the n processors and are stored in an input buffer. Processors can run and reduce these workloads, but conflicts between pairs of processors restrict simultaneous task execution. Conflicts are represented by a graph, where vertices correspond to processors and edges indicate conflicting pairs. An online algorithm must decide which processors are run at a time; so provide a valid schedule respecting the conflict constraints.

The objective is to minimize the maximal workload observed across all processors during the schedule. Unlike the original model, where workloads arrive as discrete blocks at specific time points, the flow model assumes workloads arrive continuously over intervals or not at all. We present tight bounds for all graphs with four vertices (except the path, which has been solved previously) and for the families of general complete graphs and complete bipartite graphs. We also recover almost tight bounds for complete k-partite graphs.

For the original model, we narrow the gap for the graph consisting of a triangle and an additional edge to a fourth vertex.

1 Introduction

In practical multiprocessor settings, it often happens that processors (also called machines) share resources. In such cases, only one of these machines can use the resource at any given time. We model such a situation using a *conflict graph* which specifies for the given set of machines, which are the vertices of the graph, which pairs of machines share resources. This is indicated by having an edge between two machines.

We consider a model with m parallel machines. Load arrives over time, separately to each of the machines. Time is continuous. Each machine has a buffer to store unprocessed load. If a machine is running, it processes load at a rate of 1. A machine can only run if none of the machines with which it has a conflict (that is, its neighbors in the conflict graph) is running. The goal is to minimize the maximum load that is ever stored in a single buffer.

© The Author(s), under exclusive license to Springer Nature Switzerland AG 2026
J. Kozik and A. Wolff (Eds.): SOFSEM 2026, LNCS 16448, pp. 678–692, 2026.
https://doi.org/10.1007/978-3-032-17801-5_49

We consider the online setting in which the future arriving load is unknown. An online algorithm is compared to the optimal offline solution that can only be achieved if the whole input is known in advance. Note that in the optimal solution, buffers are often also required: consider two neighboring machines that receive load at the same time.

The question now becomes: what size buffers does an online algorithm need, relative to the buffer size in an optimal solution, so that the online algorithm can process any input without overflowing any buffer? This number is known as the competitive ratio, and it depends on the conflict graph and on the number of machines. Chrobak et al. [7] who introduced this model considered two versions: one where the online algorithm knows the optimal buffer size in advance (the weak competitive ratio) and one where it does not (the strong competitive ratio). In this paper, we will focus on the first measure, which we will usually call simply *competitive ratio*.

In the original model, load arrives as discrete tasks of various sizes. Höhne and van Stee [11] later introduced the so-called *flow model* in which at all times, for any machine, load either arrives at a constant rate of 1 (not as an instantaneous block of load) or not at all. The model is a special case of the original model which maintains its overall complexity. An interesting feature of the flow model is that deterministic algorithms appear to be as powerful as randomized algorithms (against an oblivious adversary).

Previously only complete (k-partite) graphs and paths have been considered. We go the next natural step and focus on graphs with four vertices. Our goal is to investigate which factors contribute to the problem's difficulty. In this paper, we consider all graphs with at most four vertices and give tight results for the flow model in all cases. We also give tight or nearly tight results for larger graphs, as well as tight results for almost all graphs with four vertices in the original model. All of our algorithms are short to state, but are carefully constructed to handle various difficult inputs. The proofs involve intricate considerations of what combinations of delays are possible on the various machines. A main building block of the proof consists of invariants for the total delay on pairs or triples of adjacent machines. Our results are summarized in Table 1. Missing proofs can be found in the full version [10].

1.1   Related Work

The *buffer minimization in multiprocessor systems with conflicts* problem was introduced by Chrobak et al. in 2001 [7]. They present a universal algorithm showing that even the strong competitive ratio on any graph is bounded. Then they consider various families of graphs. For the complete graphs K_n the authors show that the weak and strong competitive ratios are H_n the n-th harmonic number, which is achieved by the natural greedy approach. They also show that this approach is almost optimal for complete k-partite graphs $K_{n_1,\ldots,n_k}$. If the graph is a tree with diameter at most Δ they provide an $(1 + \Delta/2)$-competitive algorithm. In the final part of their work, the authors give results for the remaining graphs with four vertices. For the path with four vertices

Table 1. Lower and upper bounds on the competitive ratio in the flow model and the original model. Here, μ is the number of the independent sets in the k-partition that consist of a single vertex in a complete k-partite graph.

Graph	Flow LB	Flow UB	Original LB	Original UB
$K_{1,2}$ (Path)	1 [11]	1 [11]	2 [6]	2 [7]
P_4 (Path)	4/3 [11]	4/3 [11]	9/4 [11]	9/4 [11]
K_n ($n \geq 2$)	$H_n - \frac{1}{2}$	$H_n - \frac{1}{2}$	H_n [7]	H_n [7]
$K_{k,n-k}$	$\mu \geq 1\colon 1$ $\mu = 0\colon 3/2$	$\mu \geq 1\colon 1$ $\mu = 0\colon 3/2$	$\mu = 2\colon 3/2$ [7] $\mu \leq 1\colon 2$ [6]	$\mu = 2\colon 3/2$ $\mu \leq 1\colon 2$ [7]
complete k-partite ($k \geq 3$)	$\mu \geq 1\colon$ $H_{k-1} - \frac{1}{2} + \frac{k-\mu}{k-1}$ $\mu = 0\colon$ $H_{k-1} + \frac{1}{2}$	$H_{k-1} + \frac{1}{2}$	$\mu \geq 1\colon$ $H_{k-1} + \frac{k-\mu}{k-1}$ [6] $\mu = 0\colon$ $H_{k-1} + 1$	$H_{k-1} + 1$ [7]
$K_3 + e$	4/3	4/3	13/6	9/4
$K_4 - e$	3/2	3/2	2 [6]	2 [6]

they claim that both the weak and strong competitive ratios are at least 2 and show an upper bound of $5/2$ on both. For the graph $K_3 + e$ which you get by connecting an additional vertex to one vertex of K_3, they claim a lower bound of $13/6$ for the weak competitive ratio and an upper bound of $5/2$ for the strong competitive ratio. Finally, for $K_4 - e$, the graph you get by deleting an edge from K_4, they claim that the weak competitive ratio is exactly 2 and show that the strong competitive ratio is below $5/2$.

Höhne and van Stee focus on graphs that are paths [11]. They close the gap for the path with four vertices by providing a $9/4$-competitive algorithm and a matching lower bound. For the path with 5 vertices they show that the competitive ratio is between $16/7$ and $5/2$. For the general paths with at least 6 vertices they provide a lower bound of $12/5$ and an upper bound linear in the number of vertices of $n/2 + 1/4$. They also introduce a new model called the flow model as described above. In this model the workload arrives at a constant rate over time in intervals instead of a block arriving at a specific point in time. So the workload function for the machines is continuous in time. This is a cleaner model which avoids the feature of the original model that the competitive ratio is usually achieved on the very last job that arrives, which then often increases the competitive ratio maintained so far by 1. Nevertheless, it is not the case that the competitive ratios for the two models simply differ by 1, and the algorithms for the two models are often quite different on the same graph.

For small paths with up to five vertices Höhne and van Stee present tight results. The paths with two or three vertices can be solved optimally (competitive ratio of 1). For the path with four vertices the competitive ratio is $4/3$. In contrast to the original model, they give a tight bound of $3/2$ for the path with five vertices. Finally, if the path consists of at least six vertices the competitive ratio is between $3/2$ and $n/2 - 2/3$.

The concept of conflicts in scheduling problems is well-established, with many variations having been extensively studied. These variations differ in several key aspects like objective function, job model, online/offline and conflict model. While we focus on conflicts between machines, another model that has been widely studied is one where conflicts arise between jobs. Several studies consider models where conflicting jobs cannot be executed concurrently [1,3,9,12,13], while others address scenarios in which conflicting jobs cannot be scheduled on the same machine [2,4,14–17]. In [5], conflicts occur between machines, but each job is associated with pre- and post-processing blocking times. The goal in this case is to find a schedule where the blocking times of jobs on conflicting machines do not overlap.

Dósa and Epstein consider a model in which, rather than having individual input buffers for each machine, a universal buffer with a fixed capacity is used to delay job execution [8].

2   Preliminaries

For simplicity we assume without loss of generality for the rest of the paper that the offline buffer size is 1. (Precisely, for any input I the online algorithm receives the scaled input I' such that $\text{OPT}(I') = 1$ along with the information that $\text{OPT}(I') = 1$.)

We refer to the processors as machines and to the workload as load. The vertices v_i of the conflict graph G represent the machines m_i and an edge indicates a conflict between the adjacent machines, restricting simultaneous task execution. The buffer size is the maximal load on a machine over time.

For the original model, an input can be described by a list of triples, where the triple (t, k, s) means that a job of size s arrives at time t on machine k. However, in many lower bound constructions load arrives only at integer times, and then we will just use a vector $(t; s_1, \ldots, s_n)$ to describe the loads arriving at some time t, and a list of such vectors to describe a complete input.

For the flow model, an input is a vector of load functions $\mathbb{R}_{\geq 0} \to \{0, 1\}$, one function for each machine which describes for each non-negative time t whether or not the machine is receiving load at time t. In lower bound constructions, similar to in the original model the incoming loads will usually be piecewise constant vectors, changing only at integer times. They will be described as lists of vectors, one vector for each time step.

Our timeline starts at time $t = 0$. Furthermore, as always we only consider finite inputs, meaning there exists a time T after which no machine receives any additional load. We also focus exclusively on connected graphs because the connected components of a graph can be analyzed independently without impacting the overall decision-making process.

Definition 1. *The (strong) competitive ratio of an algorithm* ALG *for graph G is defined as* $\sup_I \frac{\text{ALG}(I)}{\text{OPT}(I)}$, *where I is a feasible input. The (strong) competitive* *ratio of a graph G is defined as* $\inf_{\text{ALG}} \sup_I \frac{\text{ALG}(I)}{\text{OPT}(I)}$, *where* ALG *is a feasible algorithm and* I *a feasible input. The weak competitive ratio is defined analogously, where additionally* $\text{OPT}(I) = 1$.

In this paper, we use competitive ratio to refer to the weak competitive ratio and strong competitive to refer to the competitive ratio where the offline buffer size is unknown in advance.

Definition 2. *Let a_i^t be the load on machine m_i of an online algorithm* ALG *at time t and z_i^t be the load on machine m_i of an optimal offline algorithm* OPT *at time t. Define the delay of machine m_i at time t as $d_i^t := a_i^t - z_i^t$.[1]*

For notational convenience, we denote the load on machine m_i when ignoring the load that arrives exactly at time t by $a_i^{t^-}$.

Assumption 1. *We assume that a buffer size of 1 suffices if the input was known in advance. Precisely, we assume $z_i \leq 1$ at all times and there exists a time t and $i \in \{1, \ldots, n\}$ such that $z_i^t = 1$.*

2.1 Flow Model

In the *flow model* load on machines arrives as a continuous flow over time, in contrast to the original model where load arrives as a single block at a specific point in time. The input is revealed to the online algorithm as follows. At any time t, it is aware of the set of machines that are currently receiving load (at speed 1), without knowing for how long these machines will receive load. It can make its schedule starting from time t based on this knowledge; the schedule can change (possibly many times) without the set of machines that receive load changing. The set of machines that are receiving load can change arbitrarily often in any interval.

This restricts the set of allowed inputs; for instance, two connected machines cannot receive load at speed $1/2$ in this model. However, any feasible input can be approximated by an allowed input arbitrarily well by changing back and forth between different sets of machines receiving load. Hence, the power of the adversary is not restricted. We can approximate any feasible input for this model arbitrarily well by an input for the original model by letting very small blocks of load arrive over time. Comparing this to the flow model described above, it means that the algorithm has a small amount of lookahead (the size of these small blocks), as it is aware of some load that is still to arrive. As the size of the blocks tends to 0, the input tends to an input for our flow model. Conversely, for any input for the original model, a corresponding input for the flow model exists in which each arriving block of load arrives at a rate of 1 instead of instantaneously.

Hence, any upper bound that was proved for the original model still holds; however, we can usually achieve better results in this special case. Conversely,

---

[1] We may omit the exponent for time t if we do not need to specify a specific point in time.

any lower bound for the flow model also holds for the original model. Note that the lower bound for the original model in [11] crucially uses that some of the input arrives as a flow, which helps to keep various options open for the adversary.

An algorithm for the flow model must select at each time t an independent set of machines such that each machine either has nonzero load or is currently receiving load. We simplify the setting by allowing algorithms to instead set *speeds* for each machine which can change arbitrarily often. Such an algorithm can be implemented by rapidly changing back and forth between various independent sets of machines. The speeds must be such that there is no interval of nonzero length ℓ in which any clique of machines processes more than ℓ load. Furthermore, for simplicity we allow a machine with zero load to run at nonzero speed, but of course its load does not decrease in this case. Using speeds does not make it easier to design good algorithms (indeed, none of our algorithms set speeds explicitly), but it makes it easier to analyze them.

Initially, all machines are empty, so $a_i^0 = 0$ for $i \in \{1, \ldots, n\}$.

We define the natural GREEDY strategy that we will use as a (sub)routine in many of our presented algorithms. At any time t, GREEDY recursively selects a machine with highest load (prioritizing machines receiving load) for running at time t that is not in conflict with any previously selected machine. If there is a tie for the highest load (also in a later step of the recursion), GREEDY will run these machines at equal speeds as discussed above. In the flow model, GREEDY has the following nice property.

Lemma 1. *In the flow model, while* GREEDY *is running, there is never a single machine which has the unique highest load among all machines.*[2]

Proof. Assume that there exists at some time $t > 0$ a machine m, that has a strictly higher load than its neighboring machines. We show that this leads to a contradiction. Thus, for any machine, there always exists at least one neighbor with load at least as high as that of the machine itself.

There exists a point in time before t, for example at the beginning, at which there is no such machine m. Let $t_0 \geq 0$ be the supremum over all such times. For the load on m to differ from the loads on its neighbors, a measurable amount of time must have passed between t_0 and t. Let this time interval be $I := [t_0, t]$, with $t_0 < t$. There must exist a subset $J \subseteq I$ of nonzero measure during which machine m did not run. Else, if m had run throughout the entire interval I (except for a null set), its load would not have increased. Similarly, if the neighbors of m had not run during I (except for a null set), their loads would not have decreased. Hence, machine m would not have more load than its neighbors at time t.

By the intermediate value theorem, there exists a time $t' \in I$ at which m has a higher load than its neighbors, but the difference (to the machine with the closest load) is strictly smaller than at time t. Furthermore, we can choose

---

[2] This statement does not hold in the original model where jobs arrive as blocks of load.

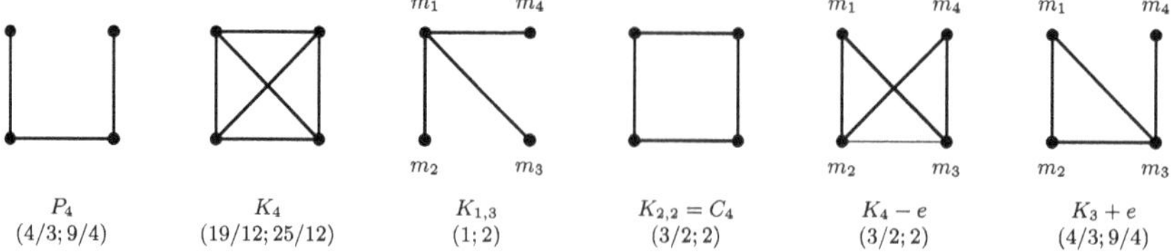

Fig. 1. Graphs with four vertices. The numbers $(R_f; R_o)$ represent the competitive ratios in the flow model (R_f) and the original model (R_o), respectively.

$t' \in J$, since the difference between the load on m and that of its neighbors increases only during J.

The fact that machine m at time t' has load no lower than that of all other machines, while its neighbors have strictly lower loads, but m was not running, contradicts the definition of GREEDY. Since GREEDY prefers machines with higher load, m would have been included in the independent set of machines selected to run.

3   Complete Graphs and Complete Partite Graphs

For families of complete graphs and complete k-partite graphs, the GREEDY algorithm performs remarkably well. For complete graphs, GREEDY achieves a competitive ratio of $H_n - \frac{1}{2}$, which cannot be improved even by randomized algorithms. For general complete k-partite graphs, we recover almost tight bounds, confirming that GREEDY remains highly effective. However, for complete bipartite graphs, adapting to their specific structure is necessary to obtain tight results.

If one independent set consists of a single vertex, a competitive ratio of 1 can be achieved. Else, GREEDY achieves a competitive ratio of $3/2$, which is optimal even for randomized algorithms.

We now consider general k-partite graphs and recover almost tight bounds, where the lower bound deteriorates as μ increases.

Theorem 2. GREEDY *is* $(H_{k-1} + \frac{1}{2})$-*competitive on complete k-partite graphs.*

Theorem 3. *For $k \geq 3$, each online algorithm has a competitive ratio of at least $H_{k-1} + 1/2$ (resp., $H_{k-1} - 1/2 + (k - \mu)/(k - 1)$) for complete k-partite graphs with $\mu = 0$ (resp., $\mu \geq 1$).*

4   Graphs with Four Vertices

We now focus on the graphs with four vertices. There exist six different graphs: $P_4, K_4, K_{1,3}, K_{2,2}, K_4 - e$ and $K_3 + e (Fig. 1)$.

We first consider some graphs that are already covered by previous results.

– The competitive ratio for the path P_4 is $4/3$ [11].

- K_4 is a complete graph, so GREEDY is $(H_4 - 1/2 = 19/12)$-competitive and no algorithm can perform better (Sect. 3).
- $K_{1,3}$ is a complete bipartite graph with $\mu = 1$, so PRIO_{V_ℓ} is optimal (Sect. 3).
- $K_{2,2} = C_4$ is a complete bipartite graph with $\mu = 0$, so GREEDY is $3/2$-competitive and no algorithm can perform better (Sect. 3).

4.1 $K_4 - e$

Since $K_4 - e$ is a complete tripartite graph, GREEDY is at worst 2-competitive on it. It is easy to verify that the lower bound of 2 does work for GREEDY (unlike on the complete tripartite graph K_3). However, a more sophisticated algorithm can achieve significantly improved competitiveness.

Algorithm 1: PRIOCENTER

> **Data:** $K_4 - e$ graph, where $\deg(m_1) = \deg(m_4) = 2$ and
> $\qquad \deg(m_2) = \deg(m_3) = 3$
> **if** $a_1 \geq \max\{1, a_2, a_3\}$ *or* $a_4 \geq \max\{1, a_2, a_3\}$ *or* $(a_2 = 0$ *and* $a_3 = 0)$
> **then**
> $\quad |$ run GREEDY on all machines;
> **else**
> $\quad \lfloor$ run GREEDY on m_2 and m_3

Theorem 4. *Algorithm* PRIOCENTER *is $3/2$-competitive on $K_4 - e$.*

Proof. We first show that PRIOCENTER maintains the invariant $d_1 + d_2 + d_3 \leq 0$. It is true at the start. Whenever $a_1 > 0$ and m_2 and m_3 are both not running, m_1 is running. So during times in which $a_1 > 0$, $d_1 + d_2 + d_3$ does not increase.

Suppose $d_1 + d_2 + d_3 \leq 0$ at the start of an interval in which $a_1 > 1$. Then since $a_1 > 0$, this remains true throughout the interval. Moreover, $a_1 > 1$ implies $d_1 > 0$. Hence, $d_2 + d_3$ remains negative during such an interval.

Finally, if $a_1 = 0$, then $d_1 \leq 0$ and $d_2 + d_3$ can only increase if m_4 is running. This can only happen if $a_2 = a_3 = 0$, in which case the invariants are maintained, or $a_4 > 1$, so $d_4 > 0$, which implies $d_2 + d_3 < 0$ as long as we have $d_2 + d_3 + d_4 \leq 0$ symmetrically to above.

We conclude that $d_1 + d_2 + d_3 \leq 0$ remains true as long as $d_2 + d_3 + d_4 \leq 0$ or $a_1 > 0$ and $d_2 + d_3 + d_4 \leq 0$ remains true as long as $d_1 + d_2 + d_3 \leq 0$ or $a_4 > 0$. Both are true at the start, so both remain true (also during times at which $a_1 = a_4 = 0$). We also see from this proof that $d_2 + d_3 \leq 0$ always holds.

By $d_1 + d_2 + d_3 \leq 0$ we have $d_1 + d_2 \leq -d_3 = z_i - a_i \leq 1$. By symmetry $d_i + d_j \leq 1$ follows for any combination $(i, j) \in \{1, 4\} \times \{2, 3\}$.

If $a_1 > 1$, then by the rules of the algorithm $a_1 = \max(a_2, a_3)$. Without loss of generality, let $a_2 = \max(a_2, a_3)$. Then $d_1 > 0$ and $d_2 > 0$. Hence $d_3 < 0$ and $a_3 < 1$. Therefore, as long as $a_1 > 1$, only m_1 (and m_4) and m_2 can be running, and $a_1 = a_2$. We also have that $a_2 > 1$ can only happen if $a_1 > 1$ (or $a_4 > 1$), as $d_2 + d_3 \leq 0$. In this case also $a_1 = a_2$ (or $a_4 = a_2$). Since $d_1 + d_2 \leq 1$, we conclude that $a_1 \leq 3/2$ at all times. By symmetry, the same holds for all machines.

Sending flow on m_1 and m_3 for one time unit followed by flow on m_1 and m_2 for two time units shows that this is tight. After one time unit we have $\text{ALG} = (1, 0, 0, 0)$ and $\text{OPT} = (0, 0, 1, 0)$. Then after two more time units the combined load on machines m_1 and m_2 increases by 2. Hence the algorithm balances the load and ends in state $\text{ALG} = (3/2, 3/2, 0, 0)$, while $\text{OPT} = (1, 1, 1, 0)$.

Theorem 5. *No randomized online algorithm can be better than 3/2-competitive for $K_4 - e$.*

4.2 $K_3 + e$

Original Model. We first observe that the path P_3 is an induced subgraph and hence a lower bound of 2 holds. The first natural approach is to focus on the machine with highest degree, which we call m_3, and run GREEDY on the remaining machines. The input $((0; 1, 1, 0, 0), (1; 0, 1, 1, 0), (2; 0, 1, 0, 0))$ shows that this is at best 5/2-competitive. The algorithm ends in state $(1/2, 5/2, 0, 0)$ and OPT in $(1, 1, 1, 0)$. This algorithm, however, puts too much emphasis on m_3.

Algorithm 2: PRIOGREEDY$_{\text{original}}$

Data: $K_3 + e$ graph, where $\deg(m_3) = 3$ and $\deg(m_4) = 1$
if $a_3 > 5/4$ **then**
 | run m_3
else
 | **if** $a_4 > 5/4$ *or* $a_1 + a_2 > 1$ *or* $a_3 = 0$ **then**
 | | run m_4 and run GREEDY on m_1 and m_2;
 | **else**
 | └ run m_3;

Theorem 6. *In the original model, algorithm* PRIOGREEDY$_{\text{original}}$ *is 9/4-competitive on $K_3 + e$.*

The input $I = ((0; 1, 1, 0, 0), (1; 0, 1, 1, 0), (2; 0, 0, 1, 1), (3; 0, 0, 0, 1), (5; 0, 0, 1, 1),$ $(6; 1, 1, 1, 0), (9; 1, 0, 1, 0), (10; 0, 1, 1, 0), (11; 0, 0, 1, 0))$ shows that the competitive result for PRIOGREEDY$_{\text{original}}$ is tight.

Theorem 7. *In the original model, no (deterministic) online algorithm can be better than 13/6-competitive for $K_3 + e$.*

Flow Model In contrast to the original model, for the path P_3 the offline optimum can be matched by an online algorithm and hence the path does not yield a lower bound. However, the complete graph K_3 is also an induced subgraph of $K_3 + e$ and yields a lower bound of 4/3.

The following algorithm manages to match this bound.

Algorithm 3: PRIOGREEDY

Data: $K_3 + e$ graph, where $\deg(m_3) = 3$ and $\deg(m_4) = 1$
if *(A)* $\max(a_1, a_2, a_4) \geq 4/3$ *and one of these machines with load* $\geq 4/3$
receives load or $\max(a_1, a_2, a_4) > 4/3$ **then**
| run m_4 and run GREEDY on m_1 and m_2;
else
   if *(B)* $a_3 > 1/3$ *or* $a_1 = a_2 = 0$ **then**
   | run m_3;
   else
      run GREEDY on m_1, m_2, m_3 and run m_4 if possible

We will show that our algorithm maintains the following two invariants.

$$d_1 + d_2 + d_3 \leq 0 \tag{1}$$
$$d_3 + d_4 \leq 1/3 \tag{2}$$

Lemma 2. *Invariant (1) can only become false if* $a_1 = a_2 = 0$ *and* $a_4 = 4/3$
and m_4 *receives load. As long as (2) holds, (1) remains true. Invariant (2) can
only become false if* $\max(a_1, a_2) \geq 4/3$, $\min(a_1, a_2) < 1/3$, $a_4 = 0$ *and* m_4 *does
not receive load.*

Proof. Both invariants hold at the start. If (A) is not satisfied, at least one
machine of m_1, m_2 or m_3 is run and hence $d_1 + d_2 + d_3$ does not increase,
or $a_1 = a_2 = a_3 = 0$ which immediately implies (1). We see that (1) can only
become false while (A) is satisfied and only if GREEDY runs none of the machines
m_1, m_2, m_3. This implies $a_1 = a_2 = 0$, $a_4 \geq 4/3$ and m_4 receives load. By (2),
$d_3 \leq 0$ at this time, so (1) continues to hold as long as (2) holds.

We now consider (2). If (A) is not satisfied, then if $a_4 > 0$, at least one
machine m_3 or m_4 is run and hence $d_3 + d_4$ does not increase, and while $a_4 = 0$
the value $d_3 + d_4$ can only increase while $a_3 \leq 1/3$, implying (2). So (2) can only
become false while (A) is satisfied. If additionally $a_4 > 0$ or m_4 receives load,
then m_4 is run and $d_3 + d_4$ does not increase. Hence for (2) to become false we
must have $a_4 = 0$ and m_4 does not receive load. To satisfy (A) with $a_4 = 0$ we
must have $\max\{a_1, a_2\} \geq 4/3$. Wlog let $a_2 = \max\{a_1, a_2\}$. Once invariant (2)
becomes false, we have $d_3 > 1/3, d_2 \geq 1/3$ and hence $d_1 < -2/3$ by invariant
(1) and it follows $a_1 < 1/3$. Because a_1 does not decrease as we run m_2 (which
is needed for (2) to become false), this was also true just before invariant (2)
became false. $\qquad\square$

Lemma 3. *During execution of the algorithm the following holds.*

1. *If* $a_3 \geq 1/3$ *and* $\max(a_1, a_2) > 1/3$, *then* $a_3 \geq 1/3$ *is maintained as long as*
   $\max(a_1, a_2) > 1/3$.
2. *If* $\min(a_1, a_2) \geq 1/3$, *then this is maintained as long as* $\max(a_1, a_2) \geq 1/3$.
3. *While* $a_3 > 1/3$ *and* $\max(a_1, a_2) < 4/3$, *if* $a_4 \geq 0$, *then* $a_4 \geq 0$ *continues to
   hold.*

4. *While $a_3 > 1/3$ and $a_4 < 4/3$, if $\max(a_1, a_2) \geq 4/3$, then $\max(a_1, a_2) \geq 4/3$ continues to hold.*

Proof. 1. Machine m_3 is not run if (A) holds. If (B) holds, we have $a_3 > 1/3$. Finally, if (A) and (B) do not hold a_3 cannot drop below $1/3$ as long as $\max(a_1, a_2) > 1/3$, since GREEDY would run one of those machines.

2. GREEDY is used to decide which machine m_1 or m_2 is run. As long as $\min(a_1, a_2) < \max(a_1, a_2)$ the machine with load $\max(a_1, a_2)$ is run. This changes once $\min(a_1, a_2) = \max(a_1, a_2)$, at which time both machines are run.

3. Condition (B) holds and hence if (A) does not hold m_4 is not run. If (A) holds, then by $\max(a_1, a_2) < 4/3$ we have $a_4 \geq 4/3$ and if $a_4 = 4/3$, m_4 must receive load. It follows that its load does not drop below $4/3$.

4. Condition (B) holds and hence if (A) does not hold machines m_1 and m_2 are not run. If (A) holds, then by $a_4 < 4/3$, there must be at least one machine m_1 or m_2 with load at least $4/3$. If $\max\{a_1, a_2\} = 4/3$, at least one of these machines receives load. It follows that its load does not drop below $4/3$. □

Definition 3. *A critical interval is a maximal open interval in which we have $\max(a_1, a_2) > 1/3$ and $\min(a_1, a_2) < 1/3$ throughout.*

A critical interval is always bounded because the input ends at some point, so the buffers eventually become empty.

Lemma 4. *Let (t_1, t_2) be a critical interval. We have $\max(a_1, a_2) \leq 4/3$ in the entire critical interval, and $a_2^{t_1} = 1/3$. If (2) holds at time t_1, then $d_3^{t_1} \leq 1/3$. If additionally $a_3^{t_1} > 1/3$, then $d_3^{t_1} \leq 0$ and $a_4^{t_1} \geq 4/3$.*

Proof. Without loss of generality, let $a_2 = \max(a_1, a_2)$ in this interval. Then $a_1 < 1/3 < a_2$ in the entire interval. We have $a_1^{t_1} < 1/3$ (by Lemma 3.2 and the previous line) and hence $a_2^{t_1} = 1/3$ by the maximality of a critical interval. By condition (A), $a_2 \leq 4/3$ is maintained during this interval: m_2 will always be running if $a_2 \geq 4/3$ and m_2 is receiving load, since $a_1 < 1/3$.

If $a_3^{t_1} \leq 1/3$ there is nothing more to show. Suppose $a_3^{t_1} > 1/3$. Let t_0 be the last time before t_1 such that $a_3^{t_0} = 1/3$. In this case condition (B) was satisfied in the interval (t_0, t_1), but still a_3 increased since then.

Suppose for a contradiction that $a_4 < 4/3$ in the entire interval (t_0, t_1). Then a_1 does not decrease below $1/3$ in (t_0, t_1) since $a_3 > 1/3$ and because (A) can only be true in this case if $a_2 = 4/3$, and then a_1 also does not decrease below $1/3$. Because $a_1^{t_1} < 1/3$ we have $a_1 < 1/3$ in the entire interval (t_0, t_2). We have $a_3 > 1/3$ during (t_0, t_1) and a_3 increased during some time in (t_0, t_1), meaning that $a_2 \geq 4/3$ for some time in (t_0, t_1), since $a_1 < 1/3$ and by assumption $a_4 < 4/3$ throughout. By Lemma 3.4 it follows that $a_2^{t_1} = 4/3$, contradicting the definition of t_1. Hence the assumption $a_4 < 4/3$ was wrong and there was a time $t' \in (t_0, t_1)$ at which $a_4 = 4/3$. From this point on a_4 remains equal to $4/3$ by Lemma 3.3 unless $\max(a_1, a_2) = 4/3$ at some point. In such a case, a_4 can only drop below $4/3$ if a_1 or a_2 remains $4/3$, and then that load cannot drop below $4/3$ anymore (unless a_4 becomes equal to $4/3$ again) by Lemma 3.4. However, since $a_1^{t_1} < 1/3$ and $a_2^{t_1} = 1/3$, we must have $a_4^{t_1} = 4/3$.

Therefore $d_4^{t_1} \geq 1/3$ and by the invariant $d_3 + d_4 \leq 1/3$, we have $d_3^{t_1} \leq 0$.

Observation 1. *In a critical interval, whenever a_4 increases, d_3 does not increase because* PRIOGREEDY *is running m_3. If we reach $a_4 = 4/3$ again, then $d_3 \leq 0$ as long as invariant 2 holds.*

Lemma 5. *Algorithm* PRIOGREEDY *maintains (1) and (2).*

Proof. By Lemma 2, we only need to show that (2) is maintained. Let (t_1, t_2) be a critical interval in which (2) holds at time t_1 and in which there is a time at which $\max(a_1, a_2) = 4/3$, $\min(a_1, a_2) < 1/3$, $a_4 = 0$ and $a_3 \geq 1/3$, and let T be the first such time in that interval. By Lemma 2 the invariants are maintained in the interval $(t_1, T]$. By Lemma 4, $a_2 \leq 4/3$ throughout this interval.

By lemma 4, we have $a_2^{t_1} = 1/3$, hence in the interval $[t_1, T]$, the value a_2 increases by 1. The value $d_2 + d_3$ does not increase because the algorithm either works on m_2 or m_3 as $a_2 > 1/3$; the algorithm never works on m_1 because $a_1 < 1/3 < a_2$. We have $a_2^T - a_2^{t_1} = 1$. There are two cases.

Case 1: $a_3^{t_1} \leq 1/3$ In this case $a_3^T \geq a_3^{t_1}$ and we get

$$d_2^T + d_3^T \leq d_2^{t_1} + d_3^{t_1} \Rightarrow z_2^T + z_3^T \geq 1 + a_3^T - a_3^{t_1} \geq 1 + a_3^T - 1/3.$$

Case 2: $a_3^{t_1} > 1/3$ After time t_1, a_3 decreases by exactly $a_3^{t_1} - a_3^T$ (this amount may be negative) whereas a_2 increases by exactly 1. The overall increase in the value $a_2 + a_3$ is therefore $1 + a_3^T - a_3^{t_1}$. Since $d_2 + d_3$ does not increase in the interval $[t_1, T]$, we have that $z_2 + z_3$ increases by at least the same amount. With Lemma 4 we also have $d_3^{t_1} \leq 0$, which leads to

$$z_2^T + z_3^T \geq a_3^{t_1} + 1 + a_3^T - a_3^{t_1} = 1 + a_3^T.$$

By Lemma 3.1, $a_3 \geq 1/3$ in $[T, t_2]$ because $a_2 > 1/3$ in $[T, t_2)$. Consider the machines m_2 and m_3 at a time $t \in [T, t_2)$. Suppose a_4 remains below $4/3$ in $[T, t]$. In any part of a critical interval in which $a_4 \leq 4/3$, machine m_2 only runs if it receives load and $a_2 = 4/3$, or if $a_3 \leq 1/3$. Let t_{23}^+ be the total length of time that both machines received load during $[T, t]$, and let t_{23}^- be the total length of time that both machines did not receive load during $[T, t]$. During the remaining time (if any), exactly one of these machines receives load and $z_2 + z_3$ does not decrease. If m_2 is the machine receiving load, a_3 does not increase; if m_3 is the machine receiving load, then ALG runs m_3 using condition (B) because $a_3 \geq 1/3$ in $[T, t_2]$ (and because it does not run m_2 in this case) and again a_3 does not increase.

Since $z_2^t \leq 1$, we have

$$z_3^t \geq t_{23}^+ - t_{23}^- + \begin{cases} a_3^T - 1/3 & \text{if } a_3^{t_1} \leq 1/3 \\ a_3^T & \text{if } a_3^{t_1} > 1/3 \end{cases} \tag{3}$$

whereas

$$a_3^t \leq a_3^T + t_{23}^+ - t_{23}^- \tag{4}$$

as long as the right hand side of (4) is more than $1/3$ because $\textsc{Alg}$ works on m_2 when both m_2 and m_3 receive load and $a_2 = 4/3$, as well as when $a_2 > 4/3$, and on m_3 otherwise by condition (B), since $a_4 < 4/3$.

By the definitions of t_{23}^+ and t_{23}^- and because $a_2^T = 4/3$, for $t \in [T, t_2)$ we also have

$$a_2^t + a_3^t = 4/3 + a_3^T + t_{23}^+ - t_{23}^-. \tag{5}$$

If $a_3 \leq 1/3$ and the right hand side of (4) is more than $1/3$, clearly (4) also holds. During intervals in which the right hand side of (4) is at most $1/3$, a_3 remains equal to $1/3$ because $a_2 > 1/3$, so $d_3 \leq 1/3$. The value a_3 can only become larger again once $a_2 = 4/3$ again, which happens when $a_3^t = a_3^T + t_{23}^+ - t_{23}^-$ by (5). We conclude that (4) holds throughout.

Since both (3) and (4) hold, we have $d_3^t = a_3^t - z_3^t \leq 0$ if $a_3^{t_1} > 1/3$, else $d_3^t \leq 1/3$.

If $a_4 \geq 4/3$ at any point in $[T, t]$, then after this $d_3 + d_4$ does not increase until possibly $a_4 = 0$ again, and this also requires that $\max(a_1, a_2) = 4/3$ again. So in this case both invariants hold at least until the next time that $\max(a_1, a_2) = 4/3, \min(a_1, a_2) < 1/3, a_4 = 0$ and $a_3 \geq 1/3$ by the above. If this happens before time t_2, then we define this to be our new time T and continue as above. In this case the value t_1 remains the same (so if we were previously in Case 2, we now still have $a_4^{t_1} \geq 4/3$ and $d_3^{t_1} \leq 0$). Otherwise we are done with the current interval.

This proof shows that (2) is maintained during any critical interval in which it holds at the start of that interval, so by Lemma 2 both invariants hold throughout.

Theorem 8. *Algorithm* $\textsc{PrioGreedy}$ *is $4/3$-competitive on $K_3 + e$.*

5   Conclusions

Our results on complete partite graphs show that the number of vertices n is certainly not the sole factor determining the competitive ratio. These results also show that the flow model is not significantly easier than the original model even though load appears more slowly; for complete k-partite graphs, the results are asymptotically the same. Generally, it seems that the density of a graph is an important factor in what competitive ratios can be achieved. Symmetry helps to analyze graphs, as we see for the graphs on four vertices. The special structure of each individual graph has a significant effect on the competitive ratio, and this complicates attempts to generalize these results further.

Acknowledgments. We gratefully acknowledge Marek Chrobak and Jiří Sgall for valuable discussions.

Disclosure of interests. The authors have no competing interests to declare that are relevant to the content of this article.

References

1. Baker, B.S., Coffman Jr, E.G.: Mutual exclusion scheduling. Theor. Comput. Sci., **162**(2), 225–243 (1996). https://doi.org/10.1016/0304-3975(96)00031-X

2. Bodlaender, H.L., Jansen, K.: On the complexity of scheduling incompatible jobs with unit-times. In: Borzyszkowski, A.M., Sokolowski, S., eds., Mathematical Foundations of Computer Science 1993, 18th International Symposium, MFCS'93, Gdansk, Poland, August 30 - September 3, 1993, Proceedings, volume 711 of Lecture Notes in Computer Science, pp. 291–300. Springer (1993). https://doi.org/10.1007/3-540-57182-5_21

3. Bodlaender, H.L., Jansen, K.: Restrictions of graph partition problems. part I. Theor. Comput. Sci., **148**(1), 93–109 (1995). https://doi.org/10.1016/0304-3975(95)00057-4

4. Bodlaender, H.L., Jansen, K., Woeginger, G.J.: Scheduling with incompatible jobs. Discret. Appl. Math., **55**(3), 219–232 (1994). https://doi.org/10.1016/0166-218X(94)90009-4

5. Buchem, M., Kleist, L., Waldschmidt, D.S.G.: Scheduling with machine conflicts. In: Chalermsook, P., Laekhanukit, B., eds., Approximation and Online Algorithms - 20th International Workshop, WAOA 2022, Potsdam, Germany, September 8-9, 2022, Proceedings, volume 13538 of Lecture Notes in Computer Science, pp. 36–60. Springer (2022). https://doi.org/10.1007/978-3-031-18367-6_3

6. Chrobak, M.: Personal communication.

7. Chrobak, M., et al.: The buffer minimization problem for multiprocessor scheduling with conflicts. In: Orejas, F., Spirakis, P.G., Leeuwen, J., eds., Automata, Languages and Programming, pp. 862–874, Berlin, Heidelberg (2001). Springer Berlin Heidelberg

8. D'osa, G., Epstein, L.: Online scheduling with a buffer on related machines. J. Comb. Optim., **20**(2), 161–179 (2010). https://doi.org/10.1007/S10878-008-9200-Y.

9. Even, G., Halld'orsson, M.M., Kaplan, L., Ron, D.: Scheduling with conflicts: online and offline algorithms. J. Sched., **12**(2), 199–224 (2009). https://doi.org/10.1007/S10951-008-0089-1

10. Haas, N., Schmitt, S., Stee, R.: The buffer minimization problem for scheduling flow jobs with conflicts (2025). https://arxiv.org/abs/2511.19690, arXiv:2511.19690

11. Höhne, F., Stee, R.: Buffer minimization with conflicts on a line. Theor. Comput. Sci., **876**, 25–33 (2021). https://doi.org/10.1016/J.TCS.2021.05.013

12. Hong, H.-C., Lin, B.M.T.: Parallel dedicated machine scheduling with conflict graphs. Comput. Ind. Eng., **124**, 316–321 (2018). https://doi.org/10.1016/J.CIE.2018.07.035

13. Irani, S., Leung, V.J.: Scheduling with conflicts, and applications to traffic signal control. In: Tardos, E., eds., Proceedings of the Seventh Annual ACMSIAM Symposium on Discrete Algorithms, 28-30 January 1996, Atlanta, Georgia, USA, pp. 85–94. ACM/SIAM 1(996). http://dl.acm.org/citation.cfm?id=313852.313892

14. Kowalczyk, D., Leus, R.: An exact algorithm for parallel machine scheduling with conflicts. J. Sched., **20**(4), 355–372 (2017). https://doi.org/10.1007/S10951-016-0482-0

15. Mallek, A., Bendraouche, M., Boudhar, M.: Scheduling identical jobs on uniform machines with a conflict graph. Comput. Oper. Res., **111**, 357–366 (2019). https://doi.org/10.1016/J.COR.2019.07.011

16. Mallek, A., Boudhar, M.: A branch-and-bound algorithm for the problem of scheduling with a conflict graph. In: 2020 International Conference on Decision Aid Sciences and Application (DASA), pp. 778–782 (2020). https://doi.org/10.1109/DASA51403.2020.9317026
17. Mallek, A., Boudhar, M.: Scheduling on uniform machines with a conflict graph: complexity and resolution. Int. Trans. Oper. Res., **31**(2), 863–888 (2024). https://doi.org/10.1111/ITOR.13170

Posters

Complete Receptive Fields for Circular Convolution Networks

Jan-Ole Perschewski[(✉)][iD] and Sebastian Stober[iD]

Otto-von-Guericke University Magdeburg, 39106 Magdeburg, Germany
`jan-ole.perschewski@ovgu.de`, `stober@ovgu.de`

Abstract. Convolution is one of the core building blocks of deep neural networks. An important aspect is the receptive field which describes the elements of the input that influence the output. We introduce the definition of complete circular convolutions schemes which use the maximum possible receptive field given their kernel sizes and depth. From this, we can infer that complete schemes can be permuted and include shifted patterns. Furthermore, initial searches show the existence of non-contiguous complete convolution schemes.

Keywords: Circular Convolution · Receptive Field · Completeness

Introduction. Convolution is a core idea in deep learning architectures. We consider the special case of circular convolution [1], $(x * w)_i = \sum_{j=1}^{k} w_j x_{[i+d \cdot j] \bmod n}$, where $x \in \mathbb{R}^n$ is the input sequence, $w \in \mathbb{R}^k$ is a kernel of size k and $d \in \mathbb{N}_0$ is the dilation factor. This is equivalent to a dotproduct between kernel and input sequence at each position of the input sequence. Usually, convolution is repeated sequentially to detect larger regions. Hence, an important question is how many values can be seen and if all values are seen from each output position.

Definitions. Applying l kernels one after another leads to at most $s = \prod_{i=1}^{l} k^{(i)}$ elements influencing the overall calculations, which is the full receptive field. In contrast, the effective receptive field refers to the subset of positions that contribute significantly [3]. However, trivially applying convolutions one after another with a step size of 1 and dilation of 1 leads to significantly smaller receptive fields. A common schedule for convolutional networks is a kernel size of 3 and a dilation scheduler $d^{(i)} = 2^{i-1}$ [2]. This schedule leads to a significant overlap decreasing the overal receptive field given the weights.

We model circular convolutions using circulant matrices. We define a convolution scheme for a l layer network as l tuples $(a_0^{(i)}, \ldots, a_{s-1}^{(i)}), \forall i = 1, \ldots, l$ with $a_i^{(l)} \in \{0, 1\}$ and kernel size $k^{(i)}$. Each pattern represents a sparse, non-contiguous kernel. The size of the kernels defines the maximum receptive field $s = \prod_{i=1}^{l} k^{(i)}$. The application of the pattern is equivalent to multiplication with a circulant matrix $(x * w) = xC^{(l)}$, where $C_{i,j}^{(l)} = a_{(i-j) \bmod s}^{(l)}$. Hence, a convolution that includes all positions would be equivalent to a circulant matrix

© The Author(s), under exclusive license to Springer Nature Switzerland AG 2026
J. Kozik and A. Wolff (Eds.): SOFSEM 2026, LNCS 16448, pp. 695–696, 2026.
https://doi.org/10.1007/978-3-032-17801-5

where every entry is 1. The reason is due to the length s which implies that at most s values can be distributed. If the resulting matrix contains zeros or values larger than one, some positions are ignored or used multiple times. With these explanations, we can define

Definition 1. *A convolution scheme with patterns* $(a_0^{(i)}, \ldots, a_{s-1}^{(i)})$ *and kernel sizes* $k^{(i)}$ *for* $i = 1, \ldots, l$ *is complete if* $\left(C^{(1)} C^{(2)} \ldots C^{(l)}\right)_{m,n} = 1, \forall m, n = 1, \ldots, s.$

Observations. Considering the reduction to circulant matrices, we can make three interesting observations. First, the circulant matrices build an abelian group given the matrix multiplication. Hence, a permutation of a complete scheme is complete as well. Interestingly, this reveals an unexplored convolution scheme inverting the order of dilation. Second, a shift matrix is a circulant matrix, which means that given a scheme, each pattern can be shifted leading to another complete scheme. This reduces the search space for valid patterns. Third, using an exhaustive search algorithm for $k^{(1)} = k^{(2)} = 3$, we found multiple complete schemes including non-contiguous kernels with different spacing between values seen in Fig. 1.

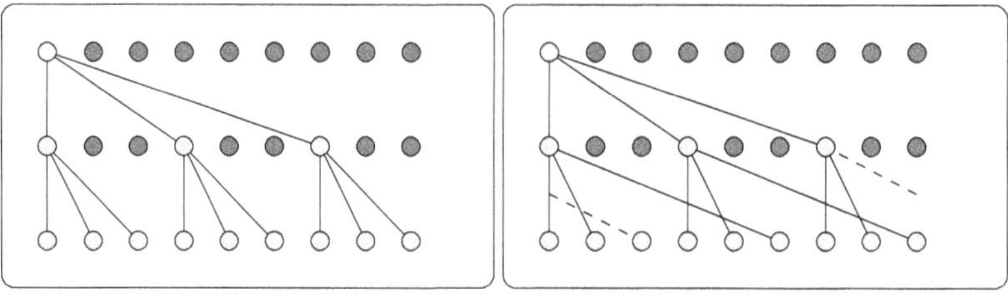

Fig. 1. Example for a complete circular convolution scheme only regular patterns (left) or one irregular pattern (right).

Conclusion. We define complete convolution schemes for circular convolution and use this definition to derive further experimental research directions and possible directions to reduce the search space for possible schemes.

References

1. Cheng, L., Khalitov, R., Yu, T., Zhang, J., Yang, Z.: Classification of long sequential data using circular dilated convolutional neural networks. Neurocomputing **518**, 50–59 (2023). https://doi.org/https://doi.org/10.1016/j.neucom.2022.10.054
2. Dieleman, S., et al.: WaveNet: a generative model for raw audio. arXiv preprint arXiv:1609.03499 (2016)
3. Luo, W., Li, Y., Urtasun, R., Zemel, R.: Understanding the effective receptive field in deep convolutional neural networks. In: Advances in Neural Information Processing Systems. vol. 29. Curran Associates, Inc. (2016)

A Proposal for Failover Routing with Few Bits: Combining Outerplanarity with Arborescences

Erik van den Akker[(✉)] and Klaus-Tycho Foerster

TU Dortmund University, Dortmund, Germany
erik.vandenakker@tu-dortmund.de

1 Motivation and Background

Computer networks have become mission critical in today's world, and hence in particular their resilience to e.g. link failures is of paramount importance. As long as the packets' destination is still reachable, (logical) connectivity will eventually be restored by (failover) routing protocols, but in the meantime, data will be lost, service level agreements violated, and especially the performance of overlaying protocols and applications will be impacted, up to the point of critical failures. While corresponding distributed algorithms and logically centralized control plane applications have been optimized over decades, they are still orders of magnitudes slower than the common first line of defense, namely preinstalled conditional failover routing rules [4]. These deterministic rules are preinstalled on routers and only match on locally available information, i.e., the packet's incoming port, the incident link failures, and the packet's header fields. For the latter fields, we assume that we follow the common assumption of reading the static destination information, i.e., following destination-based routing.

Previous work showed that for such a setting, where the destination must only be reached as long it is still connected in the graph, surviving more than 1 link failure is impossible in general [3, 6]. However, special graph classes allow for higher resilience: outerplanar graphs allow an arbitrary number of failures [7] (perfect resilience) and k-(edge)-connected graphs allow for $(k-1)$-resilience for $k \leq 5$. It is still an open question if $(k-1)$-resilience can be achieved for values of $k > 6$ [2]. However, recent work improved these bounds by modifying few bits in the packet header, allowing for 2-resilience in many practical cases [1] and on planar graphs [5].

2 Our Algorithm Proposal

A good amount of real-world wide-area networks exhibit outerplanar or near outerplanar graph characteristics [8], where the more complex details of the individual nodes might be hidden to scientific mapping efforts. We hence propose to combine the two corresponding algorithmic ideas, namely fast failover routing on outerplanar and on k-edge-connected graphs.

© The Author(s), under exclusive license to Springer Nature Switzerland AG 2026
J. Kozik and A. Wolff (Eds.): SOFSEM 2026, LNCS 16448, pp. 697–699, 2026.
https://doi.org/10.1007/978-3-032-17801-5

We thus assume an outerplanar graph $H = (V, E)$ embedded in the plane, where the individual nodes v_i correspond to graphs G_i. The routing on the graph H is performed clockwise-circular on each node v_i, where the next outgoing link is picked in clockwise direction based on the incoming port, possibly skipping failed links, resulting in perfect resilience on H [7]. The routing on the corresponding graphs G_i needs to match the above behavior and to this end, for each exit-link e in H incident to v_i, we can create k_i arc-disjoint spanning arborescences in the k_i-edge-connected graph G_i that route towards e. By circularly switching through these arborescences after each encountered link failure, we can reach $(k_i - 1)$-resilience, for $k_i \leq 5$ [2]. However, the desired link e in H might not be reachable on its arborescences or has failed itself. For the first case, we need to decide when to stop circling through its arborescences. Here Chiesa et al. [2] provided an upper bound of at most $2f$ switches for f failures for $k_i \leq 5$ arborescences, i.e., $\log(2k_i)$ bits suffice for k_i-resiliency towards e, such that we can switch to the next set of arborescences towards the link clockwise-circular to e incident to v_i in H. It remains to enumerate these sets of arborescences in G_i with $\log(\Delta(H))$ bits, where $\Delta(H)$ denotes the maximum node degree in H. Lastly, we need to cover the case on how to reach the destination d in the final node v_d in H. To this end, we proceed as before described for a node v_i and its corresponding graph G_i, but create an extra set of k_d arborescences routing towards d, increasing the number of needed bits from $\log(\Delta(H))$ to $\log(\Delta(H) + 1)$. Overall, we hence are able to combine the benefits of both routing schemes, by leveraging $\log(\Delta(H) + 1) + \log(2k)$ bits, where $k = \max\{5\} \cup \{k_1, \ldots, k_{|V|}\}$. Moreover, the arborescences to route to the exits of each v_i and routing schemes can be re-used for different destinations, as only the specific arborescences for each destination v_d differ.

Outlook We next would like to explore the practical benefits of our proposal on real-world and synthetic networks, for graph-based evaluations as well as with hardware experiments. Furthermore, we plan to explore how k can be efficiently increased as well as how to leverage host networks beyond outerplanar graphs.

References

1. van den Akker, E., Foerster, K.: Short paper: towards 2-resilient local failover in destination-based routing. In: Doka, K., Tsiropoulou, E.E. (eds.) ALGOCLOUD 2024. LNCS, vol. 15455, pp. 17–25. Springer, Cham (2024). https://doi.org/10.1007/978-3-031-94677-6_2
2. Chiesa, M., et al.: Exploring the limits of static failover routing (2016). arxiv.org/abs/1409.0034
3. Chiesa, M., et al.: On the resiliency of static forwarding tables. IEEE/ACM Trans. Netw. **25**(2), 1133–1146 (2017)
4. Chiesa, M., et al.: A survey of fast-recovery mechanisms in packet-switched networks. IEEE Commun. Surv. Tutorials **23**(2), 1253–1301 (2021)
5. Dai, W., et al.: Fast rerouting against dynamic failures: 2-resilience via eardecomposition and planarity. In: OPODIS (2025)
6. Feigenbaum, J., et al.: Brief announcement: on the resilience of routing tables. In: PODC, pp. 237–238. ACM (2012)

7. Foerster, K., et al.: On the feasibility of perfect resilience with local fast failover. In: APOCS, pp. 55–69. SIAM (2021)
8. Foerster, K., et al.: On the price of locality in static fast rerouting. In: DSN, pp. 215–226. IEEE (2022)

Exact Exponential Algorithms for Colorful 3-Rainbow Domination

Tetiana Lavynska$^{(\boxtimes)}$ (iD)

Otto von Guericke University Magdeburg, Magdeburg, Germany
`tetiana.lavynska@ovgu.de`

Abstract. Through a novel perspective that frames colorful rainbow domination as a cover with dominating sets, we develop exact exponential time algorithms, achieving runtime $O^*(4.4074^n)$.

Keywords: colorful rainbow domination · exact exponential algorithms

Given a graph $G = (V, E)$, and a set of k colors, we consider functions that assign a subset of colors to each vertex. For a function f to be a *colorful k-rainbow dominating function*, every vertex v must get at least one color and have all k colors present in its closed neighborhood. The weight of f is the total number of colors assigned: $w(f) = \sum_{v \in V} |f(v)|$. The goal is to find a minimum-weight function. The colorful k-rainbow domination problem is NP-complete for $k \geq 3$ [2]. For $k = 3$, it remains NP-complete in split and bipartite graphs [2], and in maximum-degree-3 graphs [3], but is solvable in linear time in trees and interval graphs [2], and in block and cactus graphs [3].

We look at the problem from a different perspective. Let f be some colorful k-rainbow dominating function. Let $i \in \{1, \ldots, k\}$ and let S_i be the subset of vertices that have color i in their color set. For each vertex v we have $\bigcup_{x \in N[v]} f(x) = \{1, \ldots, k\}$. Therefore, for each vertex v and for each color i, we have $i \in \bigcup_{x \in N[v]} f(x)$ and thus $N[v] \cap S_i \neq \emptyset$, i.e. each S_i is a dominating set. Each vertex v belongs to at least one S_i since $f(v) \neq \emptyset$. For the weight of f we get: $w(f) = \sum_{i \in \{1, \ldots, k\}} |S_i|$. Thus, the problem of finding an optimal colorful k-rainbow dominating function is equivalent to the problem of finding a cover of the vertex set with k dominating sets such that the sum of their sizes is minimal.

A straightforward algorithm for colorful 3-rainbow domination uses an exhaustive search. Each triple of subsets (S_1, S_2, S_3) of the vertex set V is a possible candidate for being an optimal solution if S_1, S_2, S_3 are dominating sets and $S_1 \cup S_2 \cup S_3 = V$. Choose a candidate with smallest weight. The runtime is $O^*((2^n)^3) = O^*(8^n)$. Next, we reduce the search space. For each fixed pair of dominating sets S_1 and S_2, we search for a dominating set S_3 with smallest cardinality such that $V - (S_1 \cup S_2) \subseteq S_3$. Thus, we need to solve the following problem. Given a graph and a subset of vertices U, the goal of the *Dominating Set Completion problem (DSC)* is to find a dominating set S with smallest cardinality

© The Author(s), under exclusive license to Springer Nature Switzerland AG 2026
J. Kozik and A. Wolff (Eds.): SOFSEM 2026, LNCS 16448, pp. 700–702, 2026.
https://doi.org/10.1007/978-3-032-17801-5

such that $U \subseteq S$. Let $\mathcal{A}$ be an algorithm that solves the Dominating Set problem for a given graph $G = (V, E)$ with n vertices in time $T(n)$. Then an algorithm $\mathcal{A}'$ exists that solves DSC in time $T(n - |U| + 2) + O(|V| + |E|)$. Let the input of DSC be a graph $G = (V, E)$ and a subset of vertices $U = \{u_1, \ldots, u_{|U|}\} \subseteq V$. We construct a new graph G_U by adding two new vertices w and w' and deleting the vertices of U. We connect w to each neighbour of $u_i, 1 \leq i \leq |U|$ that is not in U and to w'. Algorithm $\mathcal{A}$ finds solution S_U for the Dominating Set problem for G_U in time $T(n - |U| + 2)$. W.l.o.g. $w' \notin S_U$. Thus, set $(S_U - \{w\}) \cup U$ is an optimal solution of DSC. We use the $O(1.4969^n)$-time algorithm by van Rooij et al. [4] as our black-box algorithm $\mathcal{A}$ and get an algorithm $\mathcal{A}'$ with the same runtime. So, we get an improved algorithm for our problem. For each pair of subsets S_1 and S_2 of V, solve DSC for $U = V - (S_1 \cup S_2)$. Let S_3 be the corresponding solution. Triple (S_1, S_2, S_3) is a possible candidate. Choose a candidate with smallest weight. The runtime is $O^*(4^n \cdot 1.4969^n) = O^*(5.9876^n)$.

We consider another approach to reduce the search space. Let (S_1, S_2, S_3) be an optimal solution and let us fix some $i \in \{1, 2, 3\}$. If a vertex $v \in S_i$ exists such that $S_i - \{v\}$ is a dominating set, then $v \notin S_j$ for $j \neq i$, otherwise, there is a solution with smaller weight. So, for each $S_i' \subseteq S_i$ such that S_i' is a minimal dominating set, the vertices of $S_i - S_i'$ do not appear in any other set $S_j, j \neq i$. Thus, the color of such vertices is not important for their neighborhood. Hence, for each optimal solution triple (S_1, S_2, S_3), a triple of minimal dominating sets (S_1', S_2', S_3') exists such that $S_i' \subseteq S_i$ for each $i \in \{1, 2, 3\}$ and $\sum_{i \in \{1,2,3\}} |S_i| = (\sum_{i \in \{1,2,3\}} |S_i'|) + |V - (S_1' \cup S_2' \cup S_3')|$. Vertices of $V - (S_1' \cup S_2' \cup S_3')$ must get exactly one color, whereby it is not important which one. This leads to the following algorithm. For each triple of minimal dominating sets (S_1', S_2', S_3'), triple $(S_1', S_2', S_3' \cup (V - (S_1' \cup S_2' \cup S_3')))$ is a possible candidate. Choose among the candidates a triple with smallest weight. For iterating through all minimal dominating sets, we use an algorithm by Fomin et al. [1] that lists these sets in time $O(1.7159^n)$. Thus, runtime is $O^*((1.7159^n)^3) = O^*(5.0522^n)$.

Finally, we combine both ideas. For each pair of minimal dominating sets S_1 and S_2, solve DSC for $U = V - (S_1 \cup S_2)$ in time $O^*(1.4969^n)$. Let S_3 be the corresponding solution. Then triple (S_1, S_2, S_3) is a possible candidate. Choose among the candidates a triple with smallest weight. Again, using the algorithm by Fomin et al. [1], we iterate through all pairs of minimal dominating sets. Thus, the resulting runtime is $O^*((1.7159^n)^2 \cdot 1.4969^n) = O^*(4.4074^n)$.

References

1. Fomin, F.V., Grandoni, F., Pyatkin, A.V., Stepanov, A.A.: Combinatorial bounds via measure and conquer: bounding minimal dominating sets and applications. ACM Trans. Algorithms **5**(1), 1–17 (2008). https://doi.org/10.1145/1435375.1435384
2. Lavynska, T.: Colorful 3-rainbow domination. In: Královič, R., Kůrková, V. (eds.) SOFSEM 2025. LNCS, vol. 15539, pp. 99–111. Springer, Cham (2025). https://doi.org/10.1007/978-3-031-82697-9_8

3. Lavynska, T.: Linear time algorithms for colorful 3-rainbow domination in block and cactus graphs. In: Jeż, A., Otop, J. (eds.) FCT 2025. LNCS, vol. 16106, pp. 336–349. Springer, Cham (2025). https://doi.org/10.1007/978-3-032-04700-7_25
4. Van Rooij, J.M., Bodlaender, H.L.: Exact algorithms for dominating set. Discrete Appl. Math. **159**(17), 2147–2164 (2011). https://doi.org/10.1016/J.DAM.2011.07.001

A Learning Framework for Twin-Width and Related Problems

Ryan O'Connor[1]([⊠]), Johannes Meintrup[2], Maximilian Huber[2],
Alexander Leonhardt[3], Manuel Penschuck[3], Yosuke Mizutani[4], Oscar Yeoh[1],
and Deepak Ajwani[1]

[1] University College Dublin, Dublin, Ireland
ryan.o-connor@ucdconnect.ie
ryan.o-connor@ucdconnect.ie
[2] University of Applied Sciences Mittelhessen, Giessen, Germany
[3] Goethe University, Frankfurt am Main, Germany
[4] University of Utah, Salt Lake City, USA

1 Introduction

Practical parameterized algorithms require structural measures that restrict exponential costs to a single parameter, which must be quickly computable and well-characterized across instance distributions. This process has traditionally demanded years of manual effort and engineering We investigate whether machine learning (ML) can help to automate this process by studying twin-width [1], a novel graph parameter of considerable recent interest. Informally, twin-width measures how "close" a graph is to a cograph, i.e., a graph that reduces to a single vertex by repeatedly merging vertices with identical closed neighborhoods. We also apply our methodology to treewidth, whose properties and algorithms are well established. We show that classifiers can accurately predict both parameters, and that ML-guided solvers enable competitive computation of certificates. Feature-importance analyses provide insight into the structural factors underlying twin-width in particular, suggesting that these techniques may be valuable tools for researchers in other novel settings.

2 Methodology

Our methodology has two stages: parameter estimation and certificate computation. We train classifiers on simple graph features to predict twin-width and treewidth. For twin-width, these classifiers guide a priority-queuebased search over partially contracted graphs: at each step, a Jaccard-filtered set of vertex-pair contractions is scored and inserted into the queue. The treewidth pipeline is analogous, with classifiers steering an ML heuristic and serving as tie-breakers for the *MinDegree* and *MinFill* heuristics. Labels are computed using exact solvers on multiple graph classes. Of note, we introduce "Perturbed Rook Graphs", a type of grid graph that provides a novel hightwin-width benchmark. Apart from ErdősRényi graphs at $p = 0.5$, we know no other consistent hightwin-width generators. The treewidth classifier is trained on perturbed *k-trees* to produce a dataset spanning a wide range of parameter values.

© The Author(s), under exclusive license to Springer Nature Switzerland AG 2026

J. Kozik and A. Wolff (Eds.): SOFSEM 2026, LNCS 16448, pp. 703–705, 2026.
https://doi.org/10.1007/978-3-032-17801-5

3   Results – Classification

Trained classifiers for both parameters achieve strong metrics across all datasets. Predictions on all test sets are either exact or deviate by at most one. Feature selection across graph distributions identifies which properties most influence the parameters. Notably, selected features align with known closed-form twin-width approximations for several graph classes, suggesting feature selection as an effective exploratory tool in new domains.

4   Results – Computing Parameter Certificates

We compare our ML-guided twin-width solver with a greedy baseline, as well as the winning solver from the PACE 2023 heuristic track, *GUTHM* [2]. Twin-width graphs vary between $30 - 100$ nodes (for treewidth we include graphs up to 150 nodes). Results across multiple graph classes are presented in Table 1

Table 1. Comparison of Twin-width Solvers on Different Graph Classes

Solver	G(n,p)		Planar		RHG		Perturbed Rook Graphs		G(n,p=0.5)	
	Opt	Time	Opt	Time	Opt	Time	Opt	Time	Opt	Time
Greedy	1.65	300.0	1.72	300.0	1.78	300.0	1.27	300.0	1.24	300.0
ML	1.20	13.72	1.33	75.65	1.095	43.41	1.13	65.81	1.058	108.1
GUTHM	1.04	10.0	1.04	10.0	1.004	10.0	1.05	10.0	1.06	10.0

Table 2. Treewidth Solver Results

Dataset	Full		Sampled	
	Opt Ratio	Time (sec)	Opt Ratio	Time (sec)
MinDegree	1.031	0.0002	1.031	0.0002
MinDegree-ML	1.031	30.6	1.033	7.04
MinFill	1.038	0.0004	1.038	0.0004
MinFill-ML	1.017	30.1	1.017	6.9
ML-select	1.354	22.2	1.315	5.2

Relative solver performance is consistent across datasets: GUTHM achieves the best optimality ratios, though its advantage over our ML solver diminishes on highertwin-width instances. This is expected, reflecting GUTHM's extensive competitive tuning, whereas our ML framework uses minimal domain knowledge and is therefore more broadly applicable to other graph parameters.

Treewidth solver results are presented in Table 2. A pure ML method, *ML Select*, chooses contractions solely from classifier output, while two ML-augmented heuristics query the *ML Select* classifier to break ties in the *MinDegree* and *MinFill* heuristics. We compare these to the standard heuristics.

ML-select is slow and produces relatively poor solutions. The ML-extension for *MinDegree* does not consistently improve optimality, but applying it to *MinFill* yields marked improvement, at the cost of increased runtime.

References

1. Bonnet., Kim, E.J., Thomassé, S., Watrigant, R.: Twinwidth I: tractable FO model checking. ACM J. ACM (JACM), **69**(1), 1–46 (2021)
2. Leonhardt, A., et al.: Pace solver description: exact (GUTHMI) and heuristic (GUTHM). In 18th International Symposium on Parameterized and Exact Computation (IPEC 2023), pp. 37–1. Schloss Dagstuhl–Leibniz-Zentrum für Informatik (2023)

Three Colors Are Sometimes Necessary

Tetiana Lavynska$^{(\boxtimes)}$ [ORCID]

Otto von Guericke University Magdeburg, Magdeburg, Germany
tetiana.lavynska@ovgu.de

Abstract. We consider the colorful 3-rainbow domination problem and construct a family of connected planar graphs for which every optimal solution contains at least one vertex that gets all three colors.

Keywords: domination · colorful rainbow domination

We consider a variation of the rainbow domination problem [1] called colorful rainbow domination. Given a graph $G = (V, E)$, a function $f\colon V \to 2^{\{1,2,3\}}$ that assigns a subset of the colors $\{1, 2, 3\}$ to each vertex is a *colorful 3-rainbow dominating function* if every vertex v gets at least one color: $f(v) \neq \emptyset$ and has all 3 colors present in its closed neighborhood: $\bigcup_{x \in N[v]} f(x) = \{1, 2, 3\}$. The *weight* of f is the total number of colors assigned: $w(f) = \sum_{v \in V} |f(v)|$. The minimum weight achieving this condition for G is called the *colorful 3-rainbow domination number* of G and is denoted by $\gamma_{\mathrm{cr}3}(G)$. The colorful 3-rainbow domination problem is NP-complete in split graphs, in bipartite graphs [2], and in maximum-degree-3 graphs [3], but is solvable in linear time for trees and interval graphs [2]. Furthermore, linear time algorithms exist for generalizations of trees: for block and cactus graphs [3]. The optimal solutions for different graph classes computed by algorithms in [2, 3] all share a common feature. No vertex gets all three colors, except for several special cases. For such cases, solutions with the same optimal weight and desired property can be computed. Therefore, a hypothesis arises that, for graphs without isolated vertices, we can always find an optimal solution, where each vertex gets at most two colors. We disprove this hypothesis by the following theorem.

Theorem 1. *A family of connected planar graphs exists for which every optimal solution of the colorful 3-rainbow domination problem contains at least one vertex that gets all three colors.*

Proof. We consider the following gadget with 13 vertices given in Fig. 1. Vertices x, y, z are connected pairwise by a path of length 2 containing one vertex u_{xy}, u_{xz}, and u_{yz}, respectively. Vertex v is connected to vertices x, y, z by paths of length 3. Vertices w_{vx1} and w_{vx2} lie on the path between v and x. Analogously, we have vertices w_{vy1} and w_{vy2} and vertices w_{vz1} and w_{vz2}. In the following, we will use several copies of this gadget, all sharing the same vertex v. Thus, v will have more than three neighbors. Neighborhoods of all other vertices remain

© The Author(s), under exclusive license to Springer Nature Switzerland AG 2026
J. Kozik and A. Wolff (Eds.): SOFSEM 2026, LNCS 16448, pp. 706–708, 2026.
https://doi.org/10.1007/978-3-032-17801-5

the same. We analyze which colorings of this gadget exist such that all vertices except maybe v have all three colors in their neighborhood. If we assign only one color to each vertex except v, then the following conditions must hold. The colors of x and y must be different since they are the only two neighbors of vertex u_{xy}. As well as colors of x and z, and colors of y and z. Due to the path of length 3 connecting x and v, the color of x must also be a part of the color set of v. The same holds for the color of y and the color of z. To summarize, if each vertex except v gets one color, then v must get all three colors. Such a coloring f is a colorful 3-rainbow dominating function for this gadget with weight 15.

Now we take $m \geq 3$ copies of this gadget such that all copies share the same vertex v. This yields a graph G_m with $12m + 1$ vertices, as shown in Fig. 1. We use the described above colorful 3-rainbow dominating function f and color the vertices of each gadget according to it. We get a solution with weight $12m + 3$. First, we show that no solution with smaller weight exists. There is no solution with weight $12m + 1$, which is equal to the number of vertices in G_m. Let f' be a colorful 3-rainbow dominating function for G_m with weight $12m + 2$. Function f' assigns one color to each vertex except one. We know that this cannot be v. Thus, it is a vertex that is a part of only one gadget. Therefore, we have at least one gadget in which f' assigns one color to each vertex. This leads to a contradiction. Second, we show that in the solution with weight $12m + 3$, vertex v must always get three colors. Let f' be a colorful 3-rainbow dominating function for G_m with weight $12m + 3$, whereby vertex v gets at most two colors. At most two vertices may get two colors. Thus, we have at least one gadget in which f' assigns one color to each vertex except maybe v. This again leads to a contradiction. $\square$

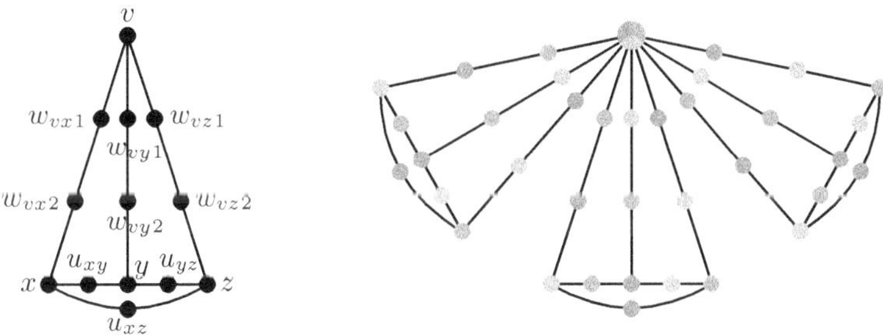

Fig. 1. A gadget and an example of graph G_m for $m = 3$ with its optimal coloring.

References

1. Brešar, B., Šumenjak, T.K.: On the 2-rainbow domination in graphs. Discrete Appl. Math. **155**(17), 2394–2400 (2007). https://doi.org/10.1016/J.DAM.2007.07.018

2. Lavynska, T.: Colorful 3-rainbow domination. In: Královič, R., Kůrková, V. (eds.) SOFSEM 2025. LNCS, vol. 15539, pp. 99–111. Springer, Cham (2025). https://doi.org/10.1007/978-3-031-82697-9_8
3. Lavynska, T.: Linear time algorithms for colorful 3-rainbow domination in block and cactus graphs. In: Jeż, A., Otop, J. (eds.) FCT 2025. LNCS, vol. 16106, pp. 336–349. Springer, Cham (2025). https://doi.org/10.1007/978-3-032-04700-7_25

Geometric Graph Modification: Modeling and Algorithms

Nicolás Honorato-Droguett[1(✉)], Kazuhiro Kurita[2], Tesshu Hanaka[3], and Hirotaka Ono[1]

[1] Nagoya University, Nagoya, Japan
honorato.droguett.nicolas.n7@s.mail.nagoya-u.ac.jp, ono@i.nagoya-u.ac.jp
[2] Okayama University, Okayama, Japan
k-kurita@okayama-u.ac.jp
[3] Kyushu University, Fukuoka, Japan
hanaka@inf.kyushu-u.ac.jp

Introduction. In *graph modification* problems, vertices and edges are treated as abstract entities and insertion or deletion are used as edit operations. When vertices represent geometric objects and edges their intersections, overlooking the underlying geometry may lead to unsuitable editings. In particular, *geometric intersection graphs* naturally capture spatial relationships and arise in *scheduling*, *map labelling* and *visibility*, where overlaps are modelled as edges. This motivates us to consider graph modification from a *geometric perspective*: instead of directly editing the graph structure, we modify the positions of the objects so that the resulting intersection graph is in a specific graph class, as illustrated in Fig. 1.

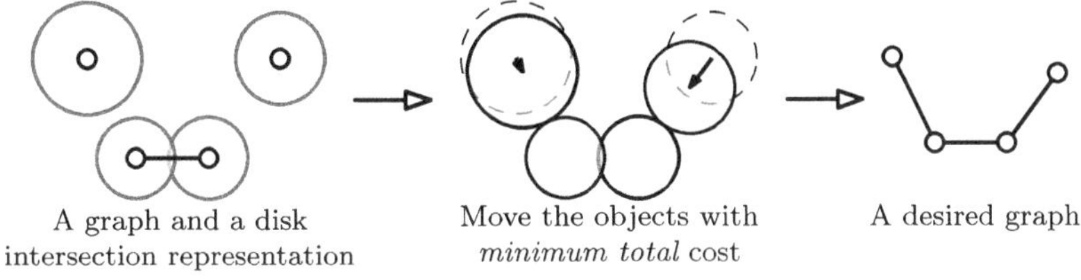

Fig. 1. Overview of GEOMETRIC GRAPH EDIT DISTANCE: The disks of a given collection are moved with minimum total moving distance to obtain a connected graph.

Depending on the choice of Π, GGED models diverse optimisation tasks: removing all overlaps yields edgeless or acyclic graphs, while increasing intersections yields complete graphs or graphs with a certain degree of connectivity. We have algorithms for the following classes (see Table 1): *complete graphs* (Π_{comp}), *graphs with a k-clique* ($\Pi_{k\text{-clique}}$), *k-connected graphs* ($\Pi_{k\text{-conn}}$), *edgeless graphs* (Π_{edgeless}), *acyclic graphs* (Π_{acyc}) and $\overline{\Pi_{k\text{-clique}}}$. For 1-dimensional objects, GGED admits efficient algorithms for several target classes. However, it becomes intractable once length or weight are unrestricted. For higher-dimensional objects, we obtain a similar situation depending on the metric and target class.

© The Author(s), under exclusive license to Springer Nature Switzerland AG 2026
J. Kozik and A. Wolff (Eds.): SOFSEM 2026, LNCS 16448, pp. 709–710, 2026.
https://doi.org/10.1007/978-3-032-17801-5

Table 1. Main results for dense graphs [1] and sparse graphs [2, 3]: In this table, L_1 and L_2 are the Manhattan and Euclidean distances, respectively. In the fifth row, k is the number of maximal cliques. Results with a $\star$-mark also hold for Π_{acyc} and $\overline{\Pi_{k\text{-clique}}}$.

Object type	Distance	Weight	Graph Class	Complexity
Interval	-	Yes	Π_{comp}	$O(n)$
Unit Interval	-	No	$\Pi_{k\text{-clique}}$, $\Pi_{k\text{-conn}}$	$O(n \log n)$
Unit Circular Arc	-	No	Π_{edgeless} $\star$	$O(n \log n)$
Unit Interval	-	Yes	Π_{edgeless}	$(1 + \frac{n}{k})^k \cdot \mathrm{poly}(n)$
d-cube, d-ball, $d \geq 2$	L_1 and L_2	No	Π_{edgeless} $\star$	strongly NP-hard

Open Problems. The main open question is whether GGED(Π_{edgeless}) is tractable for unweighted unit intervals (for $k = n$, the complexity in Table 1 becomes exponential). For dense graphs, can GGED($\Pi_{k\text{-clique}}$) and GGED($\Pi_{k\text{-conn}}$) be solved in $O(n \log n)$ time even with arbitrary lengths or weights? For strongly NP-hard cases, for which parameters does the problem become fixed-parameter tractable? Noteworthy parameters include the number of moved objects, the number of weights and lengths, the number of maximal cliques and the number of distance range. Figure 2 illustrates other special cases, restrictions and geometric transformations not covered in the present research. Finally, other fundamental graph classes and approximate algorithms also remain open directions.

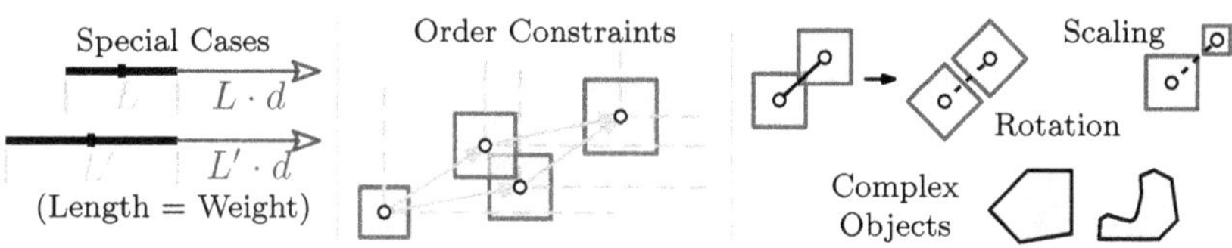

Fig. 2. Potential directions for further research.

References

1. Honorato-Droguett, N., Kurita, K., Hanaka, T., Ono, H.: Algorithms for optimally shifting intervals under intersection graph models. In: Li, B., Li, M., Sun, X. (eds.) IJTCS-FAW. LNCS, vol. 14752, pp. 66–78. Springer, Singapore (2024). https://doi.org/10.1007/978-981-97-7752-5_5
2. Honorato-Droguett, N., Kurita, K., Hanaka, T., Ono, H.: On the complexity of minimising the moving distance for dispersing objects. In: WADS. vol. 349, pp. 36:1–36:14. Schloss Dagstuhl – LZI (2025)
3. Honorato-Droguett, N., Kurita, K., Hanaka, T., Ono, H., Wolff, A.: Further results on rendering geometric intersection graphs sparse by dispersion (2025). arxiv.org/abs/2509.20903. to appear at WALCOM 2026

Rectilinear Drawings
of the *d*-dimensional cube

Todor Antić[1]([✉]) [iD], Niloufar Fuladi[2] [iD], Anna M. Limbach[2] [iD],
and Pavel Valtr[1] [iD]

[1] Department of Applied Mathematics, Faculty of Mathematics and Physics, Charles
University, Prague, Czech Republic
todor@kam.mff.cuni.cz, valtr@kam.mff.cuni.cz
[2] Computer Science Institute, Faculty of Mathematics and Physics, Charles
University, Prague, Czech Republic
nfuladi@iuuk.mff.cuni.cz, limbach@iuuk.mff.cuni.cz

1 Introduction

In this work, we investigate rectilinear drawings of the d-dimensional cube Q_d,
that is, a graph with 2^d vertices labeled by binary strings of length d in which
two vertices are connected by an edge if their labels differ exactly in one digit. A
rectilinear drawing is a drawing of a graph with vertices represented by points
in the plane and edges represented by line segments between the vertices. We
say that a drawing is *plane* if no two edges cross.

The *maximal rectilinear crossing number* of a graph G (denoted by
$\mathrm{CR_{max}}(G)$) is the maximal number of crossings, taken over all possible rectilinear
drawings of G. This parameter has been considered for many graph classes, such
as cycles [4], complete graphs [5] and k-regular graphs [2], among others. Alpert,
Feder, Harborth and Klein [3] investigated this question for Q_d. They proved
that $\mathrm{CR_{max}}(Q_d) \geq 2^{d-2}(2^{d-1}(d^2 - 2d + 3) - d^2 - 1)$ and conjectured that this is
tight. We give an alternative shorter proof of a more general result. Interestingly,
even though our construction is obtained by a different iterative process than
the one presented by Alpert et al., the resulting drawings are combinatorially
equivalent.

We also consider the problem of finding plane subgraphs in (rectilinear) drawings of Q_d. This problem has been previously considered for complete (bipartite)
graphs extensively (see [1, 6] for some recent progress). We provide constructions
of drawings of Q_d which do not contain any plane subgraphs with more than
$2d - 2$ edges and no plane path with more than $2d - 3$ edges. On the other hand,
we prove that every rectilinear drawing of Q_d with vertices in convex position
contains a plane path of length $d - 1$ (if d is odd) or d (if d is even). If the
vertices are not in convex position, we can prove that a path of length 4 exists
and proving any better lower bound seems to be a very challenging problem.

© The Author(s), under exclusive license to Springer Nature Switzerland AG 2026
J. Kozik and A. Wolff (Eds.): SOFSEM 2026, LNCS 16448, pp. 711–713, 2026.
https://doi.org/10.1007/978-3-032-17801-5

2 Our Constructions

We now define two rectilinear drawings of Q_d, denoted by $\mathcal{Q}_d$ and $\mathcal{R}_d$, which we build iteratively using drawings $\mathcal{Q}_{d-1}$ and $\mathcal{R}_{d-1}$. In our drawings, the vertices of Q_d will be placed on a circle C indexed by the elements of $\mathbb{Z}/2^d\mathbb{Z}$.

Let $\mathcal{Q}_1$ be the drawing of Q_1 given by mapping $0 \mapsto \overline{0}$ and $1 \mapsto \overline{1}$ and let $\mathcal{R}_2$ be the drawing of Q_2 given by mapping the vertices $00 \mapsto \overline{0}, 01 \mapsto \overline{1}, 11 \mapsto \overline{2}, 10 \mapsto \overline{3}$.

For $d \geq 2$, we construct the drawing $\mathcal{Q}_d$ by taking two copies $\mathcal{Q}', \mathcal{Q}''$, of $\mathcal{Q}_{d-1}$, multiplying the vertices in both of them by 2 and rotating $\mathcal{Q}''$ by $2^{d-1} - 1$ (which is slightly less than a $180°$ rotation). We then connect every vertex in $\mathcal{Q}'$ to its copy in $\mathcal{Q}''$. Now, we map vertices of Q_d to $\mathcal{Q}_d$ by mapping a vertex $0x$ (the string we get from concatenating a 0 before x) to the image of x in $\mathcal{Q}'$ and $1x$ to the image of x in $\mathcal{Q}''$. Note that the construction of $\mathcal{Q}_d$ is equivalent to the construction of Alpert et al. [3]. For $d \geq 3$, we analogously build $\mathcal{R}_d$ out of two copies of $\mathcal{R}_{d-1}$.

Theorem 1. *If every rectilinear drawing of Q_d contains a fixed plane graph G, then G is a forest of caterpillars.*

Theorem 2. *For $d \geq 2$, the drawing $\mathcal{R}_d$ does not contain a plane subgraph with more than $2d - 2$ edges. Similarly, it does not contain a plane path with more than $2d - 3$ edges.*

Given a drawing of a graph whose vertices lie on a circle C, the length of an edge uv is the minimum over the number of segments on C (the arcs between two consecutive vertices) that lie in the circular arcs (u, v) and (v, u). For each vertex v, its *length profile* is the multiset $(\ell_1^v \geq \ell_2^v \geq \ldots \geq \ell_d^v)$ of the lengths of its incident edges. If every vertex of a drawing has the same length profile, we call it the *length profile* of the drawing and the drawing is called *length regular*.

Theorem 3. *Every length regular drawing of Q_d with length profile $(\ell_1 \geq \ell_2 \geq \ldots \geq \ell_d)$ has exactly $2^{d-1} \sum_{i=1}^{d}(\ell_i - 1)(i - \frac{1}{2})$ crossings.*

Corollary 1. *As the drawing $\mathcal{Q}_d$ is length regular with lengths $\ell_1 = 2^{d-1}$ and $\ell_i = 2^{d-1} - 2^{i-2}$ for $2 \leq i \leq d$, the maximal rectilinear crossing number $CR_{\max}(Q_d)$ is at least $2^{d-2}(2^{d-1}(d^2 - 2d + 3) - d^2 - 1)$.*

References

1. Aichholzer, O., García, A., Tejel, J., Vogtenhuber, B., Weinberger, A.: Twisted ways to find plane structures in simple drawings of complete graphs. DCG **71**(1), 40–66 (2024). https://doi.org/10.1007/s00454-023-00610-0
2. Alpert, M., Feder, E., Harborth, H.: The maximum of the maximum rectilinear crossing numbers of d-regular graphs of order n. Electron. J. Comb. **16** (2008). https://api.semanticscholar.org/CorpusID:9548696

3. Alpert, M., Feder, E., Harborth, H., Klein, S.: The maximum rectilinear crossing number of the n dimensional cube graph. In: Proceedings of the Fortieth Southeastern International Conference on Combinatorics, Graph Theory and Computing. vol. 195, pp. 147–158 (2009)
4. Bode, J.P., Feder, E., Harborth, H., Horowitz, D., Lichter, T.: Extremal values of the maximum rectilinear crossing number of cycles with diagonals. Congr. Numer. **220**, 33–48 (2014)
5. Harborth, H., Thürmann, C.: Maximum rectilinear crossing number in drawings of the complete graph with a given convex hull. Congr. Numer. **221**, 121–127 (2014)
6. Soukup, J.: Bicolored point sets admitting non-crossing alternating Hamiltonian paths (2024). https://doi.org/10.48550/ARXIV.2404.06105

Author Index

© The Editor(s) (if applicable) and The Author(s), under exclusive license
to Springer Nature Switzerland AG 2026
J. Kozik and A. Wolff (Eds.): SOFSEM 2026, LNCS 16448, pp. 715–717, 2026.
https://doi.org/10.1007/978-3-032-17801-5

GPSR Compliance
The European Union's (EU) General Product Safety Regulation (GPSR) is a set
of rules that requires consumer products to be safe and our obligations to
ensure this.

If you have any concerns about our products, you can contact us on

ProductSafety@springernature.com

In case Publisher is established outside the EU, the EU authorized
representative is:

Springer Nature Customer Service Center GmbH
Europaplatz 3
69115 Heidelberg, Germany

www.ingramcontent.com/pod-product-compliance
Ingram Content Group UK Ltd.
Pitfield, Milton Keynes, MK11 3LW, UK
UKHW020813080726
473059UK00007B/2213